HUMAN MOLECULAR GENETICS

4TH EDITION

TOM STRACHAN AND ANDREW READ

Garland Science

Garland Science
Vice President: Denise Schanck
Senior Editor: Elizabeth Owen
Senior Editorial Assistant: David Borrowdale
Development Editor: Mary Purton
Deputy Production Manager: Simon Hill
Illustrator: Nigel Orme
Typesetting: EJ Publishing Services
Cover Design: Andrew Magee
Copyeditor: Bruce Goatly
Proofreader: Jo Clayton
Indexer: Merrall-Ross International Ltd

Front Cover Image is adapted from the image created for the 1000 Genomes Project of the National Human Genome Institute, National Institutes of Health, courtesy of Jane Ades.

Back Cover Images Massively parallel pyrosequencing relies on DNA capture beads. Single beads carrying a unique type of single-stranded DNA library fragment are placed in individual wells of a fiber-optic slide into which are deposited smaller beads containing immobilized enzymes needed for pyrophosphate sequencing. Images courtesy of 454 Sequencing © 2010 Roche Diagnotics.

Tom Strachan is Scientific Director of the Institute of Human Genetics and Professor of Human Molecular Genetics at Newcastle University, UK, and is a Fellow of the Academy of Medical Sciences and a Fellow of the Royal Society of Edinburgh. Tom's early research interests were in multigene family evolution and interlocus sequence exchange, notably in the HLA and 21-hydroxylase gene clusters. While pursuing the latter, he became interested in medical genetics and disorders of development. His most recent research has focused on developmental control of the cohesin regulators Nipbl and Mau-2.

Tom Strachan and **Andrew Read** were recipients of the European Society of Human Genetics Education Award in 2007.

Andrew Read is Emeritus Professor of Human Genetics at Manchester University, UK and a Fellow of the Academy of Medical Sciences. Andrew has been particularly concerned with making the benefits of DNA technology available to people with genetic problems. He established one of the first DNA diagnostic laboratories in the UK over 20 years ago (it is now one of two National Genetics Reference Laboratories), and was founder chairman of the British Society for Human Genetics, the main professional body in this area. His own research is on the molecular pathology of various hereditary syndromes, especially hereditary hearing loss.

ISBN 978- 0-815-34149-9

Library of Congress Cataloging-in-Publication Data
Strachan, T.
  Human molecular genetics / Tom Strachan and Andrew Read. --
4th ed.
    p. ; cm.
  Rev. ed. of: Human molecular genetics 3 / Tom Strachan and Andrew Read. 3rd ed. c2004.
  Includes bibliographical references and indexes.
  ISBN 978-0-8153-4149-9
  1. Human molecular genetics. I. Read, Andrew P., 1939- II. Strachan, T. Human molecular genetics 3. III. Title.
  [DNLM: 1. Genome, Human. 2. Molecular Biology. QU 470 S894h 2010]
  QH431.S787 2010
  611'.01816--dc22
                    2010004605

Published by Garland Science, Taylor & Francis Group, LLC, an informa business, 270 Madison Avenue, New York NY 10016, USA, and 2 Park Square, Milton Park, Abingdon, OX14 4RN, UK.

Printed in the United States of America
15 14 13 12 11 10 9 8 7 6 5 4 3 2 1

Taylor & Francis Group, an informa business

Visit our website at http://www.garlandscience.com

Mixed Sources
Product group from well-managed forests, controlled sources and recycled wood or fiber
www.fsc.org   Cert no. SW-COC-002985
©1996 Forest Stewardship Council

HUMAN MOLECULAR GENETICS

4TH EDITION

Dedication

This book is dedicated to Rodney Harris and to the memory of Margaret Weddle (1943–2009).

Preface

The pace of scientific and technical advance in human genetics has not slackened since our third edition appeared in 2004. This has mandated a thorough revision and reorganization of *Human Molecular Genetics* with much of the text being completely rewritten. While only a few of the basic introductory chapters retain their identity from the third edition, the aims of the text remain the same: to provide a framework of principles rather than a list of facts, to provide a bridge between basic textbooks and the research literature, and to communicate our continuing excitement about this very fast-moving area of science.

The 'finished' human genome reference sequence was published in 2004 and we are now entering an era where vast DNA sequence datasets will be produced annually. The game changer is the advent of massively parallel DNA sequencing which is already transforming how we approach genetics. Single molecule sequencing will lead to a dramatic reduction in DNA sequencing costs and promises the ability to sequence a human genome in hours. We can confidently expect that the genomes of huge numbers of organisms and individuals will have been completed before the next edition of this book.

Powerful bioinformatics programs are already being pressed into service to compare our genome with that of a burgeoning number of other organisms. Comparative genomics is helping us understand the forces that have shaped the evolution of our genome and that of many model organisms that are so important to research and various biomedical applications. These studies have already been extremely helpful in defining the most highly conserved and presumably important parts of our genome. They are also helping us to identify the fastest changing components of our genome and what it is that makes us unique.

Sequence-based transcriptomic analysis will become a major industry. It will be an important player in our quest to understand human gene function within the context of large projects, such as the ENCODE project that aims to create an encyclopedia of DNA elements of known function. Eventually, as vast datasets are accumulated on gene function, the stage will truly be set for systems biology to develop.

Other large scale projects such as HapMap have been exploring the range of genetic variation across the world's populations. In disease-related research, genome-wide screening for copy number variants has identified the problems affecting many individual patients and led to the delineation of new microdeletion and microduplication syndromes. Whole exome sequencing is now poised to explain the causes of many rare recessive conditions. In cancer, the first full genome sequences of tumors are starting to reveal the landscape of carcinogenesis in unprecedented detail.

For common complex diseases, however, the picture is less pleasing. A combination of new science (HapMap) and new technology (high-throughput SNP genotyping) has finally allowed researchers to identify genetic susceptibility factors for common diseases, but it has become apparent that the variants revealed by genome-wide association studies explain only a very small part of the overall genetic susceptibility to most complex diseases. We are left with a problem—where is the hidden heritability? Will it be found by large-scale resequencing, or perhaps might it lie in epigenetic effects?

All these developments have affected both the way genetics research is done, and the way we think about our genome. Genetics is more than ever about processing and collating vast amounts of private and public data to extract meaningful patterns. The data have also been forcing us to revise some of our basic ideas about human genetics. Humans are more variable than we thought, with copy number variants accounting for more variable nucleotides than SNPs. We transcribe almost all of our genome, and the old picture of discrete genes thinly scattered across a sea of junk DNA is starting to look untenable. Cells are now known to be awash with a startling variety of noncoding RNAs of unknown function. Perhaps our genome might be primarily an RNA, rather than a protein, machine.

The fourth edition of *Human Molecular Genetics* has therefore been heavily updated in order to maintain its hallmark currency and continue to provide a framework for understanding this exciting and rapidly advancing subject. Coverage of epigenetics, noncoding RNAs, and cell biology, including stem cells, have all been expanded. Greater detail has been provided on the major animal models used in genetic studies and how they are used as models for human disease. The most recent developments in next generation sequencing and comparative genomics have been included. The text closes by looking at the development of therapies to treat human disease. Genetic testing and screening, stem cells and cell therapy, and personalized medicine are all discussed together with a balanced view of the ethical issues surrounding these issues.

We would like to thank the staff at Garland Science who have undertaken the job of converting our drafts into the finished product, Elizabeth Owen, Mary Purton, David Borrowdale, and Simon Hill, and hope readers will appreciate all the work they have put into this. As ever we are grateful to our respective families for their forbearance and support.

TEACHING RESOURCES

The images from the book are downloadable from the web in JPEG and Powerpoint® formats via the Classwire™ course management system. The system also provides access to instructional resources for other Garland Science books. In addition to serving as an online archive of electronic teaching resources, Classwire allows instructors to build customized websites for their classes. Please visit www.classwire.com/garlandscience or email science@garland.com for further information. (Classwire is a trademark of Chalkfree, Inc.)

Acknowledgments

In writing this book we have benefited greatly from the advice of many geneticists and biologists. We are grateful to many colleagues at Newcastle and Manchester Universities who commented on some aspects of the text notably: C. Brooks, K. Bushby, A. Knight, M. Lako, H. Middleton-Price, S. Ramsden, G. Saretzki, N. Thakker, and A. Wallace. We are also appreciative of help and data provided by various staff members at the European Bioinformatics Institute notably: Xose Fernandez, Javier Herrero, Julio Fernandez Banet, Simon White, and Ewan Birney. In addition, we would like to thank the following for their suggestions in preparing this edition.

Alexandra I.F. Blakemore (Imperial College London, UK); Daniel A. Brazeau (University at Buffalo, USA); Carolyn J. Brown (University of British Columbia, Canada); Frederic Chedin (University of California Davis, USA); Edwin Cuppen (University Medical Center, Utrecht, The Netherlands); Ken Dewar (McGill University, Canada); Ian Dunham (European Bioinformatics Institute, UK); T. Mary Fujiwara (McGill University, Canada); Adrian J. Hall (Sheffield Hallam University, UK); Lise Lotte Hansen (Aarhus University, Denmark); Verle Headings (Howard University, USA); Graham Heap (Barts and The London School of Medicine and Dentistry, UK); Mary O. Huff (Bellarmine University, USA); David L. Hurley (East Tennessee State University, USA); David Iles (University of Leeds, UK); Colin A. Johnson (University of Leeds, UK); Bobby P.C. Koeleman (University Medical Center, Utrecht, The Netherlands); Michael R. Ladomery (University of the West of England, UK); Dick Lindhout (University Medical Center, Utrecht, The Netherlands); John Loughlin (Newcastle University, UK); Donald Macleod (University of Edinburgh, UK); Eleanor M. Maine (Syracuse University, USA); Rhayza Maingon (Keele University, UK); Gudrun Moore (Institute of Child Health, UK); Tom Moore (University College Cork, Ireland); Kenneth Morgan (McGill University, Canada); Karen E. Morrison (University of Birmingham, UK); Brenda Murphy (University of Western Ontario, Canada); Roberta Palmour (McGill University, Canada); Nollaig Parfrey (University College Cork, Ireland); Aimee K. Ryan (McGill University, Canada); Jennifer Sanders (Brown University, USA); Sharon Shriver (Pennsylvania State University, USA); John J. Taylor (Newcastle University, UK); Leo P. ten Kate (VU University Medical Center, The Netherlands); Jürgen Tomiuk (University of Tubingen, Germany); Patricia N. Tonin (McGill University, Canada); David A. van Heel (Barts and The London School of Medicine and Dentistry, UK); Malcolm von Schantz (University of Surrey, UK).

Contents

Detailed Contents

Chapter 5
Principles of Development   133

Chapter 8
Analyzing the Structure and Expression of Genes and Genomes 213

Chapter 11
Human Gene Expression   345

Chapter 12
Studying Gene Function in the
Post-Genome Era   381

Chapter 13
Human Genetic Variability and Its
Consequences   405

Chapter 16
Identifying Human Disease Genes and Susceptibility Factors — 497

Chapter 19
Pharmacogenetics, Personalized Medicine, and Population Screening   605

Chapter 20
Genetic Manipulation of Animals for Modeling Disease and Investigating Gene Function 639

Chapter 1

Nucleic Acid Structure and Gene Expression

1

KEY CONCEPTS

- The great bulk of eukaryotic genetic information is stored in the DNA found in the nucleus. A tiny amount is also stored in mitochondrial and chloroplast DNA.

- DNA molecules are polymers of nucleotide repeat units that consist of one of four types of nitrogenous base, plus a sugar, plus a phosphate.

- The backbone of any DNA molecule is a sugar–phosphate polymer, but it is the sequence of the bases attached to the sugars that determines the identity and genetic function of any DNA sequence.

- DNA normally occurs as a double helix, comprising two strands that are held together by hydrogen bonds between pairs of complementary nitrogenous bases.

- Transmission of genetic information from cell to cell is normally achieved by copying the complementary DNA molecules that are then shared equally between two daughter cells.

- Genes are discrete segments of DNA that are used as a template to synthesize a functional complementary RNA molecule.

- Most genes make an RNA that will serve as a template for making a polypeptide.

- Various genes make RNA molecules that do not encode polypeptide. Such noncoding RNA often helps regulate the expression of other genes.

- Like DNA, RNA molecules are polymers of nucleotide repeat units that consist of one of four types of nitrogenous base (three of these are the same as in DNA), plus a slightly different sugar, plus a phosphate.

- Unlike DNA, RNA molecules are usually single-stranded.

- To become functional, newly synthesized RNA must undergo a series of maturation steps such as excising unwanted intervening sequences and chemical modification of certain bases.

- Polypeptide synthesis occurs at ribosomes, either in the cytoplasm or inside mitochondria and chloroplasts.

- The sequential information encoded in the RNA is interpreted at the ribosome via a triplet genetic code, determining the basic structure of the polypeptide.

- Polypeptides often undergo a variety of chemical modifications.

- Proteins display extraordinary structural and functional diversity.

1.1   DNA, RNA, AND POLYPEPTIDES

Molecular genetics is primarily concerned with the inter-relationship between two nucleic acids, DNA and RNA, and how these are used to synthesize polypeptides, the basic component of all proteins. RNA may have been the hereditary material at a very early stage of evolution, but now, except in certain viruses, it no longer serves this role. Genetic information is instead stored in more chemically stable DNA molecules that can be copied faithfully and transmitted to daughter cells.

Nucleic acids were originally isolated from the nuclei of white blood cells, but are found in all cells and in viruses. In eukaryotes, DNA molecules are found mainly in the chromosomes of the nucleus, but each mitochondrion also has a small DNA molecule, as do the chloroplasts of plant cells.

A **gene** is a part of a DNA molecule that serves as a template for making a functionally important RNA molecule. In simple organisms such as bacteria, the DNA is packed with genes (typically at least several hundred up to a few thousand different genes). In eukaryotes, the small DNA molecules of the mitochondrion or chloroplast contain a few genes (tens up to hundreds) but the nucleus often contains thousands of genes, and complex eukaryotes typically have tens of thousands. In the latter case, however, much of the DNA consists of repetitive sequences whose functions are not easily identified. Some of the repetitive DNA sequences support essential chromosomal functions, but there are also many defective copies of functional genes.

There are many different types of RNA molecule but they can be divided into two broad classes. In one class, each RNA molecule contains a **coding RNA** sequence that can be decoded to generate a corresponding polypeptide sequence. Because this class of RNA carries genetic information from DNA to the protein synthesis machinery, it is described as **messenger RNA** (**mRNA**). Messenger RNA made in the nucleus needs to be exported to the cytoplasm to make proteins, but the messenger RNA synthesized in mitochondria and chloroplasts is used to make proteins within these organelles. Most gene expression is ultimately dedicated to making polypeptides, so proteins represent the major functional endpoint of the information stored in DNA.

The other RNA class is **noncoding RNA**. Such molecules do not serve as templates for making polypeptides. Instead they are often involved in assisting the expression of other genes, sometimes in a fairly general way and sometimes by regulating the expression of a small set of target genes. These regulatory processes may involve catalytic RNA molecules (ribozymes).

Most genetic information flows in the sequence DNA → RNA → polypeptide

Genetic information generally flows in a one-way direction: DNA is decoded to make RNA, and then RNA is used to make polypeptides that subsequently form proteins. Because of its universality, this flow of genetic information has been described as the central dogma of molecular biology. Two processes are essential in all cellular organisms:

- **transcription**, by which DNA is used by an RNA polymerase as a template for synthesizing one of many different types of RNA;
- **translation**, by which mRNA is decoded to make polypeptides at ribosomes, which are large RNA–protein complexes found in the cytoplasm, and also in mitochondria and chloroplasts.

Genetic information is encoded in the linear sequence of nucleotides in DNA and is decoded in groups of three nucleotides at a time (triplets) to give a linear sequence of nucleotides in RNA. This is in turn decoded in groups of three nucleotides (codons) to generate a linear sequence of amino acids in the polypeptide product.

Eukaryotic cells, including mammalian cells, contain nonviral chromosomal DNA sequences, such as members of the mammalian LINE-1 repetitive DNA family, that encode cellular reverse transcriptases, which can produce DNA sequences from an RNA template. The central dogma of unidirectional flow of genetic information in cells is therefore not strictly valid.

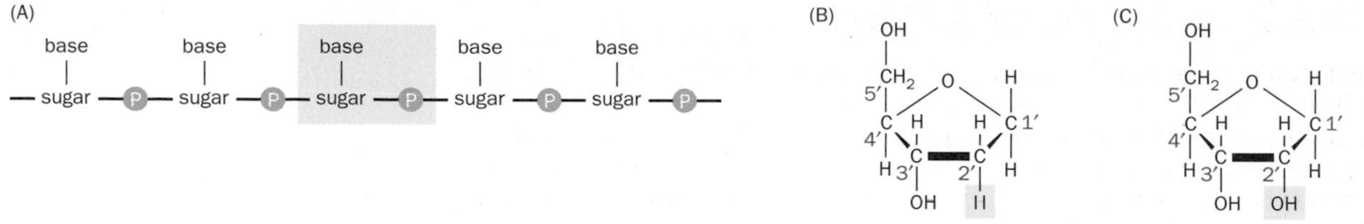

Figure 1.1 Repeat units in nucleic acids. (A) The linear backbone of nucleic acids consists of alternating phosphate and sugar residues. Attached to each sugar is a base. The basic repeat unit (pale peach shading) consists of a base + sugar + phosphate = a nucleotide. The sugar has five carbon atoms numbered 1′ to 5′. (B) In DNA, the sugar is deoxyribose. (C) In RNA, the sugar is ribose, which differs from deoxyribose in having a hydroxyl (OH) group attached to carbon 2′.

Nucleic acids and polypeptides are linear sequences of simple repeat units

Nucleic acids

DNA and RNA have very similar structures. Both are large polymers with long linear backbones of alternating residues of a phosphate and a five-carbon sugar. Attached to each sugar residue is a nitrogenous base (**Figure 1.1A**). The sugars in DNA and RNA differ, in either lacking or possessing, respectively, an –OH group at their 2′-carbon positions (Figure 1.1B, C). In deoxyribonucleic acid (DNA), the sugar is deoxyribose; in ribonucleic acid (RNA), the sugar is ribose.

Unlike the sugar and phosphate residues, the bases of a nucleic acid molecule vary. The sequence of bases identifies the nucleic acid and determines its function. Four types of base are commonly found in DNA: adenine (A), cytosine (C), guanine (G), and thymine (T). RNA also has four major types of base. Three of them (adenine, cytosine, and guanine) also occur in DNA, but in RNA uracil (U) replaces thymine (**Figure 1.2A**).

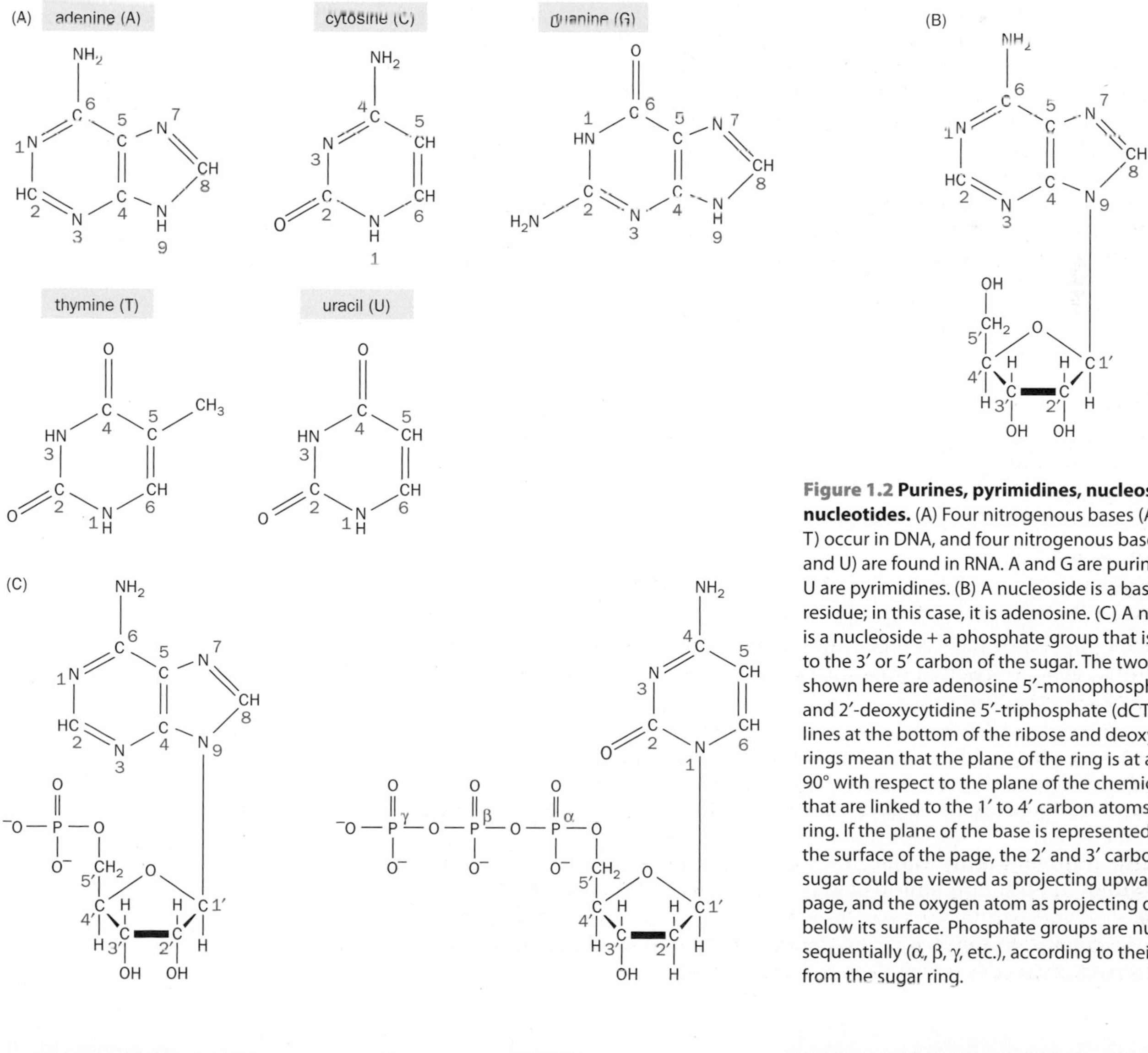

Figure 1.2 Purines, pyrimidines, nucleosides, and nucleotides. (A) Four nitrogenous bases (A, C, G, and T) occur in DNA, and four nitrogenous bases (A, C, G, and U) are found in RNA. A and G are purines; C, T, and U are pyrimidines. (B) A nucleoside is a base + sugar residue; in this case, it is adenosine. (C) A nucleotide is a nucleoside + a phosphate group that is attached to the 3′ or 5′ carbon of the sugar. The two examples shown here are adenosine 5′-monophosphate (AMP) and 2′-deoxycytidine 5′-triphosphate (dCTP). The bold lines at the bottom of the ribose and deoxyribose rings mean that the plane of the ring is at an angle of 90° with respect to the plane of the chemical groups that are linked to the 1′ to 4′ carbon atoms within the ring. If the plane of the base is represented as lying on the surface of the page, the 2′ and 3′ carbons of the sugar could be viewed as projecting upward out of the page, and the oxygen atom as projecting downward below its surface. Phosphate groups are numbered sequentially (α, β, γ, etc.), according to their distance from the sugar ring.

TABLE 1.1 NOMENCLATURE FOR BASES, NUCLEOSIDES, AND NUCLEOTIDES

Base	Nucleoside = base + sugar		Nucleotide = nucleoside + phosphate(s)		
	Ribose	Deoxyribose	Monophosphate	Diphosphate	Triphosphate
Purine					
Adenine	adenosine		adenosine monophosphate (AMP)[a]	adenosine diphosphate (ADP)	adenosine triphosphate (ATP)
		deoxyadenosine	deoxyadenosine monophosphate (dAMP)[b]	deoxyadenosine diphosphate (dADP)	deoxyadenosine triphosphate (dATP)
Guanine	guanosine		guanosine monophosphate (GMP)[a]	guanosine diphosphate (GDP)	guanosine triphosphate (GTP)
		deoxyguanosine	deoxyguanosine monophosphate (dGMP)	deoxyguanosine diphosphate (dGDP)	deoxyguanosine triphosphate (dGTP)
Pyrimidine					
Cytosine	cytidine		cytidine monophosphate (CMP)[a]	cytidine diphosphate (CDP)	cytidine triphosphate (CTP)
		deoxycytidine	deoxycytidine monophosphate (dCMP)	deoxycytidine diphosphate (dCDP)	deoxycytidine triphosphate (dCTP)
Thymine	thymidine		thymidine monophosphate (TMP)[a]	thymidine diphosphate (TDP)	thymidine triphosphate (TTP)
		deoxythymidine	deoxythymidine monophosphate (dTMP)	deoxythymidine diphosphate (dTDP)	deoxythymidine triphosphate (dTTP)
Uracil	uridine		uridine monophosphate (UMP)[a]	uridine diphosphate (UDP)	uridine triphosphate (UTP)
		deoxyuridine	deoxyuridine monophosphate (dUMP)	deoxyuridine diphosphate (dUDP)	deoxyuridine triphosphate (dUTP)

[a]Nucleoside monophosphates are alternatively named as follows: AMP, adenylate; GMP, guanylate; CMP, cytidylate; TMP, thymidylate; UMP, uridylate.
[b]Where the sugar is ribose, the nucleotide is AMP; where the sugar is deoxyribose, the nucleotide is dAMP. This pattern applies throughout the table. Note that TMP, TDP, and TTP are not normally found in cells.

Bases consist of heterocyclic rings of carbon and nitrogen atoms, and can be divided into two classes: **purines** (A and G), which have two interlocked rings, and **pyrimidines** (C, T, and U), which have a single ring. In nucleic acids, each base is attached to carbon 1′ (one prime) of the sugar; a sugar with an attached base is called a nucleoside (Figure 1.2B). A nucleoside with a phosphate group attached at the 5′ or 3′ carbon of the sugar is the basic repeat unit of a DNA strand and is called a **nucleotide** (Figure 1.2C and **Table 1.1**).

Polypeptides

Proteins are composed of one or more **polypeptide** molecules that may be modified by the addition of carbohydrate side chains or other chemical groups. Like DNA and RNA, polypeptide molecules are polymers that are a linear sequence of repeating units. The basic repeat unit is called an **amino acid**. An amino acid has a positively charged amino group ($-NH_2$) and a negatively charged carboxylic acid (carboxyl) group (–COOH). These are connected by a central α-carbon atom that also bears an identifying side chain that determines the chemical nature of the amino acid. Polypeptides are formed by a condensation reaction between the amino group of one amino acid and the carboxyl group of the next, to form a repeating backbone, where the side chain (called an R-group) can differ from one amino acid to another (**Figure 1.3**).

The 20 different common amino acids can be categorized according to their side chains:

- basic amino acids (**Figure 1.4A**) carry a side chain with a net positive charge at physiological pH;

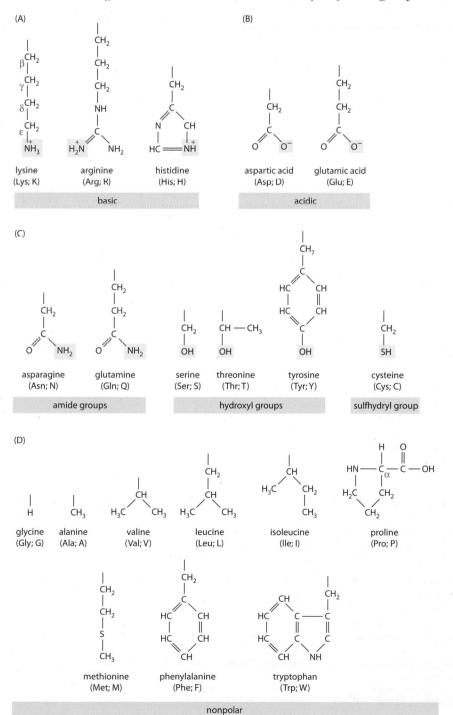

Figure 1.3 The basic repeat structure of polypeptides. A polypeptide is a polymer consisting of amino acid repeat units (pale peach shading). Amino acids have the general formula $H_2N–CH(R)–COOH$, where R is the side chain, $H_2N–$ is the amino group, and –COOH is the carboxyl group. The central α carbon carries all three groups, in each amino acid. The blue shading illustrates one of the peptide bonds that link adjacent amino acids.

- acidic amino acids (Figure 1.4B) carry a side chain with a net negative charge at physiological pH;

- uncharged polar amino acids (Figure 1.4C) are electrically neutral overall, although their side chains carry polar electrical groups with fractional electrical charges (denoted as δ+ or δ–);

- nonpolar neutral amino acids (Figure 1.4D) are hydrophobic (repelling water), often interacting with one another and with other hydrophobic groups.

(A)

lysine (Lys; K)　arginine (Arg; R)　histidine (His; H)

basic

(B)

aspartic acid (Asp; D)　glutamic acid (Glu; E)

acidic

(C)

asparagine (Asn; N)　glutamine (Gln; Q)

amide groups

serine (Ser; S)　threonine (Thr; T)　tyrosine (Tyr; Y)

hydroxyl groups

cysteine (Cys; C)

sulfhydryl group

(D)

glycine (Gly; G)　alanine (Ala; A)　valine (Val; V)　leucine (Leu; L)　isoleucine (Ile; I)　proline (Pro; P)

methionine (Met; M)　phenylalanine (Phe; F)　tryptophan (Trp; W)

nonpolar

Figure 1.4 R groups of the 20 common amino acids, grouped according to chemical class. There are 11 polar amino acids, divided into three classes: (A) basic amino acids (positively charged); (B) acidic amino acids (negatively charged); and (C) uncharged polar amino acids bearing three different types of chemical group. Polar chemical groups are highlighted. (D) In addition, a fourth class is composed of nine nonpolar neutral amino acids. Amino acids within each class are chemically very similar. Side-chain carbon atoms are numbered from the central α-carbon atom (see the lysine side chain). In proline, the R group's side chain connects to the amino acid's –NH₂ group as well as to its central α-carbon atom.

TABLE 1.2 WEAK NONCOVALENT BONDING BONDS AND FORCES	
Type of bond	Nature of bond
Hydrogen	Hydrogen bonds form when a hydrogen atom interacts with electron-attracting atoms, usually oxygen or nitrogen atoms
Ionic	Ionic interactions occur between charged groups. They can be very strong in crystals, but in an aqueous environment the charged groups are shielded by both water molecules and ions in solution and so are quite weak. Nevertheless, they can be very important in biological function, as in enzyme–substrate recognition
Van der Waals forces	Any two atoms in close proximity show a weak attractive bonding interaction (van der Waals attraction) as a result of their fluctuating electrical charges. When atoms become extremely close, they repel each other very strongly (van der Waals repulsion). Although individual van der Waals attractions are very weak, the cumulative effect of many such attractive forces can be important when there is a very good fit between the surfaces of two macromolecules
Hydrophobic forces	Water is a polar molecule. Hydrophobic molecules or chemical groups in an aqueous environment tend to cluster. This minimizes their disruptive effects on the complex network of hydrogen bonds between water molecules. Hydrophobic groups are said to be held together by hydrophobic bonds, although the basis of their attraction is their common repulsion by water molecules

In general, polar amino acids are hydrophilic, and nonpolar amino acids are hydrophobic. Glycine, with its very small side chain, and cysteine (whose –SH group is not as polar as an –OH group) occupy intermediate positions on the hydrophilic–hydrophobic scale.

As described below, the side chains can be modified by the addition of various chemical groups or sugar chains.

The type of chemical bonding determines stability and function

The stability of nucleic acid and protein polymers is primarily dependent on strong covalent bonds between the atoms of their linear backbones. In addition to covalent bonds, weak noncovalent bonds (Table 1.2) are important both between and within nucleic acids or protein molecules (Box 1.1). Individual noncovalent bonds are typically more than 10 times weaker than individual covalent bonds.

The structure of water is particularly complex, with a rapidly fluctuating network of noncovalent bonding occurring between water molecules. The predominant force in this structure is the **hydrogen bond**, a weak electrostatic bond between fractionally positive hydrogen atoms and fractionally negative atoms (oxygen atoms, in the case of water molecules).

BOX 1.1 THE IMPORTANCE OF HYDROGEN BONDING IN NUCLEIC ACIDS AND PROTEINS

Intermolecular hydrogen bonding in nucleic acids
This is important in permitting the formation of the following double-stranded nucleic acids:

- *Double-stranded DNA*. The stability of the double helix is maintained by hydrogen bonding between A–T and C–G base pairs. The individual hydrogen bonds are weak, but in eukaryotic cells the two strands of a DNA helix are held together by between tens of thousands and hundreds of millions of hydrogen bonds.
- *DNA–RNA duplexes*. Hydrogen bonds form naturally between DNA and RNA during transcription, but the base pairing is transient because the RNA migrates away from DNA as it matures.
- *Double-stranded RNA*. This occurs stably in the genomes of some viruses. It also arises transiently in cells during gene expression. For example, during RNA splicing, small nuclear RNA molecules bind to complementary sequences in pre-mRNA, and mRNA

codons bind to tRNA during translation. Many regulatory RNAs, such as microRNAs, control the expression of selected target genes by base pairing to complementary sequences at the RNA level.

Intramolecular hydrogen bonding in nucleic acids
This is particularly prevalent in RNA molecules. Intramolecular base pairing can form hairpins that may be crucially important to the structure of some RNAs such as rRNA and tRNA (see Figure 1.9), and as targets for gene regulation.

Intramolecular hydrogen bonding in proteins
Several characteristic elements of protein secondary structure, such as α-helices and β-pleated sheets, arise because of hydrogen bonding between side chains of different amino acids on the same polypeptide chain.

Charged molecules are highly soluble in water. Because of the phosphate groups in their component nucleotides, both DNA and RNA are negatively charged polyanions. Depending on their amino acid composition, proteins may be electrically neutral, or they may carry a net positive charge (basic protein) or a net negative charge (acidic protein). All of these molecules can form multiple interactions with the water during their solubilization. Even electrically neutral proteins are readily soluble, if they contain sufficient charged or neutral polar amino acids. In contrast, membrane-bound proteins with many hydrophobic amino acids are thermodynamically more stable in a hydrophobic environment.

Although individually weak, the numerous noncovalent bonds acting together make large contributions to the stability of the **conformation** (structure) of these molecules and are important for specifying the shape of a macromolecule. Covalent bonds are comparatively stable, so a high input of energy is needed to break them. Noncovalent bonds, however, are constantly being made and broken at physiological temperatures (see Box 1.1).

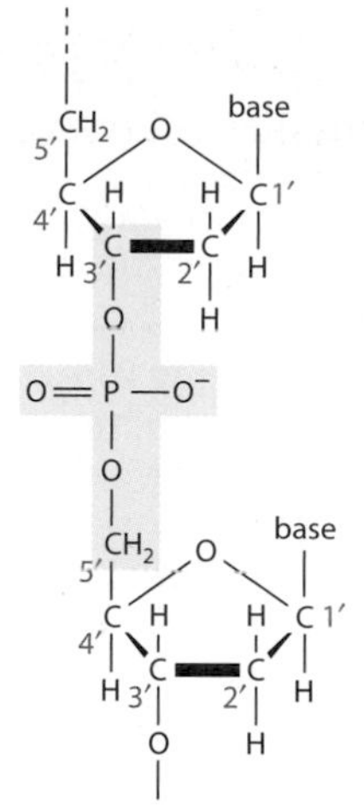

1.2 NUCLEIC ACID STRUCTURE AND DNA REPLICATION

DNA and RNA structure

DNA and RNA molecules have linear backbones consisting of alternating sugar residues and phosphate groups. The sugar residues are linked by **3′, 5′-phosphodiester bonds**, in which a phosphate group links the 3′ carbon atom of one sugar to the 5′ carbon atom of the next sugar in the sugar–phosphate backbone (**Figure 1.5**).

Although certain viral genomes are composed of single-stranded DNA, cellular DNA forms a **double helix**: two strands of DNA are held together by hydrogen bonds to form a duplex. Hydrogen bonding occurs between the laterally opposed **complementary base pairs** on the two strands of the DNA duplex. Such base pairs form according to Watson–Crick rules: A pairs with T, while G pairs with C (**Figure 1.6**).

Because of base pairing, the base composition of DNA is not random: the amount of A equals that of T, and the amount of G equals that of C. The base composition of DNA can therefore be specified by quoting the percentage of GC (= percentage of G + percentage of C) in its composition. For example, DNA with 42% GC has the following base composition: G, 21%; C, 21%; A, 29%; T, 29%.

The two strands of a DNA double helix curve around each other to produce a minor groove and a major groove in the double helix, where the distance occupied by a single complete turn of the helix (its pitch) is 3.6 nm (**Figure 1.7**). DNA can adopt different types of helical structure. Under physiological conditions, most DNA in bacterial or eukaryotic cells adopts the B form, which is a right-handed helix (it spirals in a clockwise direction away from the observer) and has 10 base pairs per turn. Rarer forms are A-DNA (right-handed helix with 11 base pairs per turn) and Z-DNA (a left-handed helix with 12 base pairs per turn).

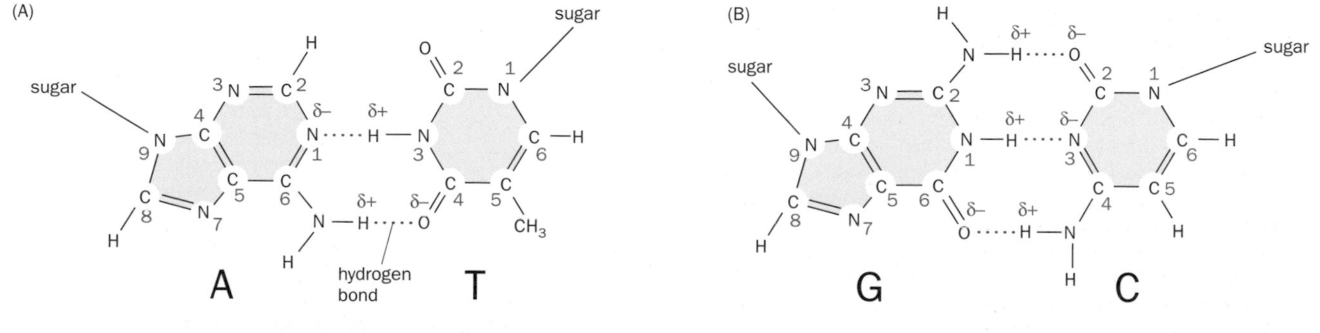

Figure 1.5 A 3′, 5′-phosphodiester bond. The phosphodiester bond (pale peach shading) joins the 3′ carbon atom of one sugar to the 5′ carbon atom of the next sugar in the sugar–phosphate backbone of a nucleic acid.

Figure 1.6 AT and GC base pairs. (A) AT base pairs have two connecting hydrogen bonds (dotted red lines); (B) GC base pairs have three. Fractional positive charges and fractional negative charges are shown by δ+ and δ−, respectively.

Because the phosphodiester bonds link carbon atoms number 3′ and number 5′ of successive sugar residues, the two ends of a linear DNA strand are different. The **5′ end** has a terminal sugar residue in which carbon atom number 5′ is not linked to another sugar residue. The **3′ end** has a terminal sugar residue whose 3′ carbon is not involved in phosphodiester bonding. The two strands of a DNA duplex are described as being anti-parallel to each other because the 5′→3′ direction of one DNA strand is the opposite to that of its partner, according to Watson–Crick base-pairing rules (**Figure 1.8**).

Genetic information is encoded by the linear sequence of bases in the DNA strands. The two strands of a DNA duplex have complementary sequences, so the sequence of bases of one DNA strand can therefore readily be inferred from that of the other strand. It is usual to describe DNA by writing the sequence of bases of one strand only, in the 5′→3′ direction, which is the direction of synthesis of new DNA or RNA from a DNA template. When describing the sequence of a DNA region encompassing two neighboring bases (a dinucleotide) on one DNA strand, it is usual to insert a 'p' to denote a connecting phosphodiester bond. So, a CG base pair means a C on one DNA strand is hydrogen-bonded to a G on the complementary strand, but CpG represents a deoxycytidine covalently linked to a neighboring deoxyguanosine on the same DNA strand (see Figure 1.8).

Unlike DNA, RNA is normally single-stranded except for certain viruses that have double-stranded RNA genomes. However, to perform certain cell functions two RNA molecules may need to associate transiently to form base pairs, and intermolecular hydrogen bonding also permits the formation of transient RNA–DNA duplexes (see Box 1.1).

In addition, hydrogen bonding can occur between bases within a single-stranded RNA (or DNA) molecule to produce structurally and functionally important stretches of double-stranded sequence. Hairpin structures may be formed with stems that are stabilized by hydrogen bonding between bases (**Figure 1.9A**). Intrachain base pairing causes certain RNA molecules to have complex structures (Figure 1.9B).

In double-stranded RNA, A pairs with U instead of T. Although G usually pairs with C, sometimes G–U base pairs are formed (see the example in Figure 1.9B). Although not particularly stable, G–U base pairing does not significantly distort the RNA–RNA helix.

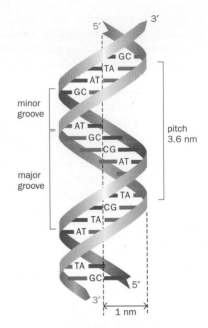

Figure 1.7 Features of the DNA double helix. The two DNA strands wind round each other, producing a minor groove and a major groove in the double helix. The double helix has a pitch of 3.6 nm and a radius of 1 nm per turn.

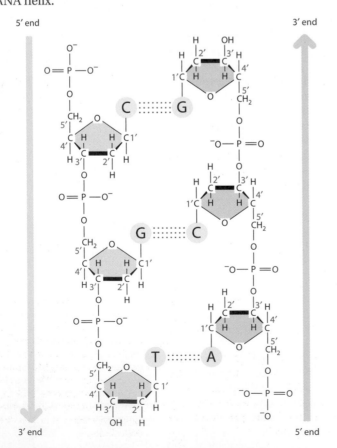

Figure 1.8 Anti-parallel nature of the DNA double helix. The two anti-parallel DNA strands run in opposite directions in linking 3′ to 5′ carbon atoms in the sugar residues. This double-stranded trinucleotide has the sequence 5′ pCpGpT–OH 3′/5′ pApCpG–OH 3′, where p stands for a phosphate group and –OH 3′ represents the 3′ terminal hydroxyl group. This is conventionally abbreviated to give the 5′→3′ sequence of nucleotides on only one strand, either as 5′-CGT-3′ (blue strand) or as 5′-ACG-3′ (purple strand).

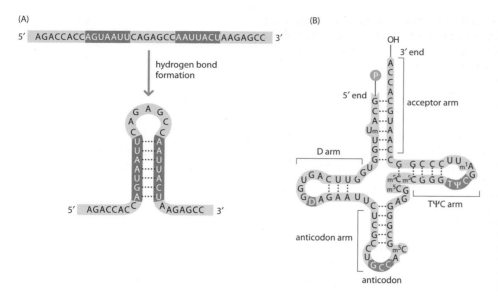

(A)

5′ AGACCACCAGUAAUUCAGAGCCAAUUACUAAGAGCC 3′

hydrogen bond formation

(B)

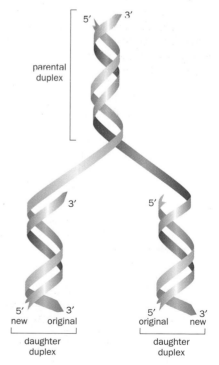

Figure 1.9 Base pairing within single-stranded nucleic acids. (A) Hairpin formation by intramolecular hydrogen bonding. Hydrogen bonds between the sequences highlighted by dark pink shading within this single-stranded nucleic acid (shown here as RNA) can stabilize folding back to form a hairpin with a double-stranded stem. (B) Extensive intramolecular base pairing in transfer RNA. The tRNA^{Gly} shown here as an example illustrates the classical cloverleaf tRNA structure. There are three hairpins (the D arm, anticodon arm, and the TψC arm) plus a stretch of base pairing between 5′ and 3′ terminal sequences (called the acceptor arm because the 3′ end is used to attach an amino acid). Note that tRNAs always have the same number of base pairs in the stems of the different arms of their cloverleaf structure and that the anticodon at the center of the middle loop identifies the tRNA according to the amino acid it will bear. The minor nucleotides depicted are: D, 5,6-dihydrouridine; ψ, pseudouridine (5-ribosyluracil); m^{5}C, 5-methylcytidine; m^{1}A, 1-methyladenosine; Um, 2′-O-methyluridine.

Replication is semi-conservative and semi-discontinuous

For new DNA synthesis (**replication**) to begin, the two DNA strands of a helix need to be unwound by the enzyme helicase. The two unwound DNA strands then each serve as a template for DNA polymerase to make complementary DNA strands, using the four deoxynucleoside triphosphates (dATP, dCTP, dGTP, and dTTP). Two daughter DNA duplexes are formed, each identical to the parent molecule (**Figure 1.10**). Each daughter DNA duplex contains one strand from the parent molecule and one newly synthesized DNA strand, so the replication process is **semi-conservative**.

DNA replication is initiated at specific points, called **origins of replication**, generating Y-shaped replication forks, where the parental DNA duplex is opened up. The anti-parallel parental DNA strands serve as templates for the synthesis of complementary daughter strands that run in opposite directions.

The overall direction of chain growth is 5′→3′ for one daughter strand, the **leading strand**, but 3′→5′ for the other daughter strand, the **lagging strand** (**Figure 1.11**). The reactions catalyzed by DNA polymerase involve the addition of a deoxynucleoside monophosphate (dNMP) residue to the free 3′ hydroxyl group of the growing DNA strand. However, only the leading strand always has a free 3′ hydroxyl group that allows continuous elongation in the same direction in which the replication fork moves.

The direction of synthesis of the lagging strand is opposite to that in which the replication fork moves. As a result, strand synthesis needs to be accomplished in a progressive series of steps, making DNA segments that are typically 100–1000 nucleotides long (Okazaki fragments). Successively synthesized fragments are eventually joined covalently by the enzyme DNA ligase to ensure the creation of two complete daughter DNA duplexes. Only the leading strand is synthesized continuously, so DNA synthesis is therefore **semi-discontinuous**.

DNA polymerases sometimes work in DNA repair and recombination

The machinery for DNA replication relies on a variety of proteins (**Box 1.2**) and RNA primers, and has been highly conserved during evolution. However, the complexity of the process is greater in mammalian cells, in terms of the numbers of different DNA polymerases (**Table 1.3**), and of their constituent proteins and subunits.

Most DNA polymerases in mammalian cells use an individual DNA strand as a template for synthesizing a complementary DNA strand and so are DNA-directed DNA polymerases. Unlike RNA polymerases, DNA polymerases normally require the 3′-hydroxyl end of a base-paired primer strand as a substrate. Therefore, an RNA primer, synthesized by a primase, is needed to provide a free 3′ OH group for the DNA polymerase to start synthesizing DNA.

Figure 1.10 Semi-conservative DNA replication. The parental DNA duplex consists of two complementary, anti-parallel DNA strands that unwind to serve as templates for the synthesis of new complementary DNA strands. Each completed daughter DNA duplex contains one of the two parental DNA strands plus one newly synthesized DNA strand, and is structurally identical to the original parental DNA duplex.

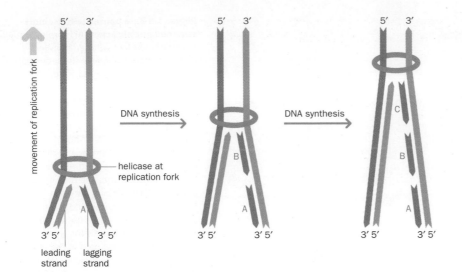

Figure 1.11 Semi-discontinuous DNA replication. The enzyme helicase opens up a replication fork, where synthesis of new daughter DNA strands can begin. The overall direction of movement of the replication fork matches that of the continuous 5′→3′ synthesis of the leading daughter DNA strand. Replication is semi-discontinuous because the lagging strand, which is synthesized in the opposite direction, is built up in pieces (Okazaki fragments, shown here as fragments A, B, and C), that will later be stitched together by a DNA ligase.

There are close to 20 different types of DNA polymerase in mammalian cells. Most use DNA as a template to synthesize DNA and they have been grouped into four families—A, B, X, and Y—on the basis of sequence comparisons (see Table 1.3).

Members of family B are classical (high-fidelity) DNA polymerases and include the enzymes devoted to replicating nuclear DNA. They mostly have an associated 3′-5′ exonuclease activity that is important in **proofreading**: if the wrong base is inserted at the 3′ OH group of the growing DNA chain the 3′-5′ exonuclease snips it out. This results in high-fidelity replication, because base misincorporation errors are extremely infrequent. DNA polymerase α is a complex of a polymerase and a primase and is devoted to initiating DNA synthesis and initiating Okazaki fragments. DNA polymerases δ and ε carry out most of the DNA synthesis and are strand-specific (see Table 1.3).

Many DNA polymerases work in DNA repair or recombination. They include classical high-fidelity DNA polymerases that are also involved in replication (DNA polymerases δ and ε) and others that are dedicated to DNA repair or recombination. Some of the latter are high-fidelity polymerases but many of them are comparatively prone to base misincorporation, notably members of family X and especially family Y members. For example, DNA polymerase ι (iota) can have an error rate 20,000 or more times that of DNA polymerase ε.

The high error rate in some DNA polymerases is tolerated because they work in DNA repair processes and so are used to synthesize only small stretches of DNA. In other cases, high error rates are advantageous. For example, low-fidelity

---

BOX 1.2 MAJOR CLASSES OF PROTEINS INVOLVED IN DNA REPLICATION

- Topoisomerases—start the process of DNA unwinding by breaking a single DNA strand, releasing the tension holding the helix in its coiled and supercoiled form.
- Helicases—unwind the double helix at the replication fork, once supercoiling has been eliminated by a topoisomerase.
- Single-stranded binding proteins—maintain the stability of the replication fork. Single-stranded DNA is very vulnerable to enzymatic attack; the bound proteins protect it from being degraded.
- Primases—enzymes that attach a small complementary RNA sequence (a **primer**) to single-stranded DNA at the replication fork. The RNA primer provides the 3′ OH needed by DNA polymerase to begin synthesis (unlike RNA polymerases, DNA polymerases cannot initiate new strand synthesis from a bare single-stranded template but require an initiating molecule with a free 3′ OH group onto which deoxynucleoside triphosphates can be attached to build a complementary strand).

- DNA polymerases—for synthesizing new DNA strands. New cellular DNA synthesis normally depends on an existing DNA strand template that is read by a DNA-directed DNA polymerase. This complex aggregate of protein subunits often also provides DNA proofreading and DNA repair functions (see Table 1.3). This means that any wrongly incorporated bases can be identified, removed, and repaired. DNA can also be synthesized from an RNA template, using an RNA-directed DNA polymerase (a reverse transcriptase). The ends of linear chromosomes are copied using a reverse transcriptase (telomerase).
- DNA ligases—needed to seal nicks that remain in newly synthesized DNA after the RNA primers have been removed and the small gaps filled by DNA polymerase. The DNA ligases catalyze the formation of a phosphodiester bond between unattached but adjacent 3′ hydroxyl and 5′ phosphate groups.

TABLE 1.3 MAMMALIAN DNA POLYMERASES

DNA-DIRECTED DNA POLYMERASES

Polymerase	Family	Standard DNA replication	Additional or alternative roles in DNA repair, recombination, etc.
α (alpha)	B	initiates synthesis at replication origins and initiates synthesis of Okazaki fragments on lagging strand	
β (beta)	X		base excision repair[b]
γ (gamma)	A	mitochondrial DNA synthesis	mitochondrial DNA repair
δ (delta)	B	main polymerase that synthesizes lagging strand	multiple roles in DNA repair
ε (epsilon)	B	synthesizes leading strand	multiple roles in DNA repair
ζ (zeta)	B		translesion synthesis[c]
η (eta)	Y		translesion synthesis[c]
θ (theta)	A		possible role in interstrand crosslink repair[d]; base excision repair[b]; translesion synthesis[c]; somatic hypermutation[g]
ι (iota)	Y		translesion synthesis[c]; possible roles in base excision repair[b] and mismatch repair[e]
κ (kappa)	Y		translesion synthesis[c]; nucleotide excision repair[f]
λ (lambda)	X		double-strand break repair; VDJ recombination[g]; base excision repair[b]
μ (mu)	X		
ν (nu)	A		possible role in interstrand crosslink repair[d]
Rev1	Y		translesion synthesis[c]
TdT[a]	X		VDJ recombination[g]

RNA-DIRECTED DNA POLYMERASES (REVERSE TRANSCRIPTASES)

Interspersed repeat reverse transcriptases (LINE-1 or endogenous retrovirus elements)	occasionally converts mRNA and other RNA into cDNA, which can integrate elsewhere into the genome
Telomerase reverse transcriptase (Tert)	replicates DNA at the ends of linear chromosomes

[a]Terminal deoxynucleotide transferase. [b]Base excision repair identifies and removes inappropriate bases or inappropriately modified bases. [c]Translesion synthesis involves the replication of DNA past damaged DNA (lesions) on the template strand. [d]Interstrand crosslink repair is the repair of highly cytotoxic lesions where covalent DNA bonds have been formed between the DNA strands. [e]Mismatch repair is a form of DNA repair that corrects mistakes arising when noncomplementary nucleotides form a base pair. [f]Nucleotide excision repair is used to fix helix-distorting lesions. [g]Somatic hypermutation and VDJ recombination are mechanisms used in B cells to diversify immunoglobulin sequences.

DNA polymerases can continue to synthesize new DNA strands opposite a lesion in the template DNA (translesion synthesis) and they can contribute to the sequence diversity of immunoglobulins (e.g. by introducing many base changes in coding sequences) and so assist in the recognition of numerous foreign antigens by the immune system.

Many viruses have RNA genomes

DNA is the hereditary material in all present-day cells and we generally think of the **genome** as the collective term for the different hereditary DNA molecules of an organism or cell. However, many viruses have an RNA genome. These RNA molecules can undergo self-replication, although the 2′ OH group on their ribose residues makes the sugar–phosphate bonds rather unstable chemically. By contrast, in DNA, the deoxyribose residues carry only hydrogen atoms at the 2′ position, making DNA a more stable carrier of genetic information.

TABLE 1.4 DIFFERENT CLASSES OF GENOME

	Single linear	Multiple linear	Single circular	Multiple circular	Mixed (linear + circular)
DNA GENOMES					
Single-stranded (ss)DNA	some viruses	segmented ssDNA viruses	some viruses	–	–
Double-stranded (ds)DNA	some viruses; a very few bacteria, e.g. *Borrelia*	segmented dsDNA viruses; eukaryotic nuclei	mitochondria; chloroplasts; many bacteria and Archaea	multipartite viruses; some bacteria	a very few bacteria, e.g. *Agrobacterium tumefaciens*
RNA GENOMES					
Single-stranded (ss)RNA	some viruses	segmented ssRNA viruses	a very few viruses	–	–
Double-stranded (ds)RNA	a few viruses	segmented dsRNA viruses	–	–	–

See Figure 1.12 for examples of viral genomes and for explanations of segmented and multipartite viruses.

Viruses have developed many different strategies to infect and subvert cells, and their genomes show extraordinary diversity when compared with cellular genomes (**Table 1.4** and **Figure 1.12**). Because RNA replication has a much higher error rate than DNA replication, viral RNA genomes have a higher mutational load than DNA genomes. Although viral RNA genomes are generally quite small, the elevated mutation rate permits more rapid adaptation to changing environmental conditions. RNA viruses usually replicate in the cytoplasm; DNA viruses generally replicate in the nucleus.

Retroviruses are unusual RNA viruses both because they replicate in the nucleus and also because their RNA replicates via a DNA intermediate. The single-stranded RNA genome is converted into a single-stranded cDNA using a viral reverse transcriptase. The single-stranded viral cDNA is then converted into double-stranded DNA, by using a DNA polymerase from the host cell. Other viral proteins then help insert this double-stranded DNA into the host cell's chromosomal DNA. It can remain there for long periods or be used to synthesize new viral RNA genomes that are packaged as new virus particles.

1.3   RNA TRANSCRIPTION AND GENE EXPRESSION

As well as having global roles in storing and transmitting genetic information and supporting chromosome function, DNA can have cell-specific functions because it contains sequences that can be used to make RNA and polypeptides in ways that differ from cell to cell. **Genes** are discrete DNA segments that are spaced at irregular intervals along the DNA sequence and serve as templates for making complementary RNA sequences (transcription). The initial primary RNA transcript must then undergo a series of maturation steps that ultimately result in a mature, functional noncoding RNA or a messenger RNA that will in turn serve as a template to make a polypeptide. Some of the gene products are needed by essentially all cells for a variety of vital cell processes (such as DNA replication or protein synthesis). However, other RNA and protein products are made in some cell types but not others and may even be specific for individual cells in some exceptional cases, as in individual B and T lymphocytes.

The DNA compositions of the different cell types in a multicellular organism are essentially identical. The variation between cells happens because of differences in gene expression, primarily at the level of transcription: different genes are transcribed in different cells according to the needs of the cells. Some genes, known as housekeeping genes, need to be expressed in essentially all cells, but other genes show tissue-specific gene expression or they may be expressed at specific times (e.g. at specific stages of development or of the cell cycle).

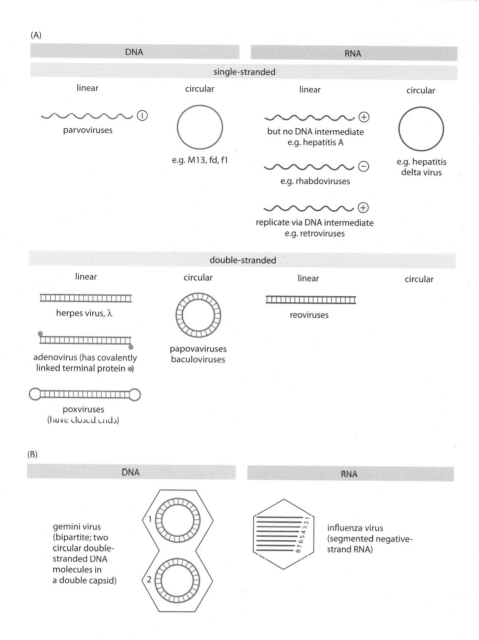

Figure 1.12 The extraordinary variety of viral genomes. (A) Strandedness and topology. In single-stranded viral genomes, the RNA used to make protein products may have the same sense as the genome, which is therefore a *positive-strand genome* (+), or be the opposite sense (antisense) of the genome, a *negative-strand genome* (–). Some single-stranded (+) RNA viruses (e.g. retroviruses) go through a DNA intermediate, and some double-stranded DNA viruses (e.g. hepatitis B) go through a replicative RNA form. (B) Segmented and multipartite genomes. *Segmented genomes* are ones in which the genome is divided into multiple different nucleic acid molecules, each specifying a messenger RNA devoted to making a single polypeptide. For example, the genome of an influenza virus is partitioned into eight different negative single-strand RNA molecules. In some segmented genomes, each of the different molecules is packaged in a separate virus particle (capsid). Such genomes are described as *multipartite genomes,* as illustrated here by the bipartite genome of the gemini virus.

Normally, only one of the two DNA strands in a duplex serves as a template for RNA synthesis. During transcription, double-stranded DNA is bound by RNA polymerase. The DNA is then unwound, enabling the DNA strand that will act as a template for RNA synthesis (the template strand) to form a transient double-stranded RNA–DNA hybrid with the growing RNA strand.

The RNA transcript is complementary to the template strand of the DNA and has the same 5′→3′ direction and base sequence (except that U replaces T) as the opposite, nontemplate DNA strand. The nontemplate strand is often called the **sense strand**, and the template strand is often called the **antisense strand** (**Figure 1.13**).

In documenting gene sequences, it is customary to show only the DNA sequence of the sense strand. The orientation of sequences relative to a gene normally refers to the sense strand. For example, the 5′ end of a gene refers to

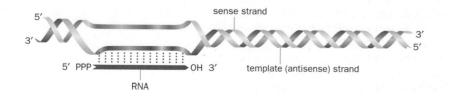

Figure 1.13 Transcribed RNA is complementary to one strand of DNA. The nucleotide sequence of transcribed RNA is normally identical to that of the sense strand, except that U replaces T, and is complementary to that of the template strand. The nucleotide at the extreme 5′ end of a primary RNA transcript carries a 5′ triphosphate group that may later undergo modification; the 3′ end has a free hydroxyl group.

sequences at the 5′ end of the sense strand, and upstream or downstream sequences flank the gene at its 5′ or 3′ ends, respectively, with reference to the *sense* strand. For transcription to proceed efficiently, various proteins (transcription factors) must bind to particular DNA sequence elements (collectively called a **promoter**) that are often located close to and upstream of a gene. The bound transcription factors serve to position and guide the RNA polymerase.

RNA polymerases synthesize RNA from four nucleotide precursors: ATP, CTP, GTP, and UTP. Elongation involves the addition of the appropriate ribonucleoside monophosphate residue (AMP, CMP, GMP, or UMP) to the free 3′ hydroxyl group at the 3′ end of the growing RNA strand. These nucleotides are derived by splitting a pyrophosphate residue (PP_i) from their appropriate ribonucleoside triphosphate (rNTP) precursors. Only the initiator nucleotide at the extreme 5′ end of a primary transcript carries a 5′ triphosphate group.

Most genes are expressed to make polypeptides

Most eukaryotic genes are expressed to produce polypeptides using RNA polymerase II, one of three RNA polymerases (Table 1.5). All three RNA polymerases cannot initiate transcription by themselves: they require regulatory factors. A crucial regulatory element is the **promoter**, a collection of closely spaced short DNA sequence elements in the immediate vicinity of a gene. Promoters are recognized and bound by transcription factors that then guide and activate the polymerase. Transcription factors are said to be ***trans*-acting**, because they are produced by remote genes and need to migrate to their sites of action. In contrast, promoter sequences are ***cis*-acting** because they are located on the same DNA molecule as the genes they regulate.

Promoters recognized by RNA polymerase II often include the following elements:

- The *TATA box*. Often TATAAA, or a variant sequence, this element is usually found about 25 base pairs (bp) upstream from the transcriptional start site (designated by –25; Figure 1.14A). It usually occurs in genes that are actively transcribed by RNA polymerase II only at a particular stage in the cell cycle (e.g. histone genes) or in specific cell types (e.g. the β-globin gene). A mutation in the TATA box does not prevent the initiation of transcription but causes transcription to begin at an incorrect location.

- The *GC box*. Usually a variant of the sequence GGGCGG, the GC box occurs in a variety of genes, many of which lack a TATA box. This is the case for the 'housekeeping' genes that perform the same function in all cells (such as those encoding DNA and RNA polymerases, histones, or ribosomal proteins). Although the GC box sequence is sequentially asymmetrical, it seems to function in either orientation (Figure 1.14B).

- The *CAAT box*. Often located at position –80 it is usually the strongest determinant of promoter efficiency. Like the GC box, it functions in either orientation.

TABLE 1.5 THE THREE CLASSES OF EUKARYOTIC RNA POLYMERASE		
RNA polymerase	**RNA synthesized**	**Notes**
I	28S rRNA[a], 18S rRNA[a], 5.8S rRNA[a]	localized in the nucleolus; RNA polymerase I produces a single primary transcript (45S rRNA) that is cleaved to give the three rRNA classes listed here
II	mRNA[b], miRNA[c], most snRNAs[d] and snoRNAs[e]	RNA polymerase II transcripts are unique in being subject to capping and polyadenylation
III	5S rRNA[a], tRNA[f], U6 snRNA[g], 7SL RNA[h], various other small noncoding RNAs	the promoter for some genes transcribed by RNA polymerase III (e.g. 5S rRNA, tRNA, 7SL RNA) is internal to the gene; for others, it is located upstream of the gene (see Figure 1.15)

[a]Ribosomal RNA. [b]Messenger RNA. [c]MicroRNA. [d]Small nuclear RNAs. [e]Small nucleolar RNAs. [f]Transfer RNA. [g]U6 snRNA is a component of the spliceosome, an RNA–protein complex that removes unwanted noncoding sequences from newly formed RNA transcripts. [h]7SL RNA forms part of the signal recognition particle, which has an important role in the transport of newly synthesized proteins.

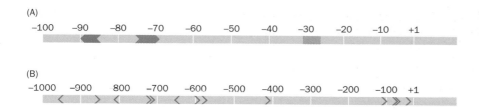

Figure 1.14 Promoters for two eukaryotic genes encoding polypeptides.
Polypeptide-encoding genes are transcribed by RNA polymerase II. The promoters are defined by short sequence elements located in regions just upstream of the transcription start site (+1). (A) The β-globin gene promoter includes a TATA box (orange), a CAAT box (purple), and a GC box (blue). (B) The glucocorticoid receptor gene is unusual in possessing 13 upstream GC boxes: 10 in the normal orientation, and 3 in the reverse orientation (alternative orientations for GC box elements are indicated by chevron directions).

For a gene to be transcribed by RNA polymerase II the DNA must first be bound by general transcription factors, to form a preinitiation complex. General transcription factors required by RNA polymerase II include TFIIA, TFIIB, TFIID, TFIIE, TFIIF, and TFIIH. These transcription factors may themselves comprise several components. For example, TFIID consists of the TATA-box-binding protein (TBP; also found in association with RNA polymerases I and III) plus various TBP-associated factors (TAF proteins). The complex that is required to initiate transcription by an RNA polymerase is known as the **basal transcription apparatus** and consists of the polymerase plus all of its associated general transcription factors.

In addition to the general transcription factors required by RNA polymerase II, specific recognition elements are recognized by tissue-restricted transcription factors. For example, an **enhancer** is a cluster of *cis*-acting short sequence elements that can enhance the transcriptional activity of a specific eukaryotic gene. Unlike a promoter, which has a relatively constant position with regard to the transcriptional initiation site, enhancers are located at variable (often considerable) distances from their transcriptional start sites. Furthermore, their function is independent of their orientation. Enhancers do, however, also bind gene regulatory proteins. The DNA between the promoter and enhancer sites loops out, which brings the two different DNA sequences together and allows the proteins bound to the enhancer to interact with the transcription factors bound to the promoter, or with the RNA polymerase.

A silencer has similar properties to an enhancer but it inhibits, rather than stimulates, the transcriptional activity of a specific gene.

Different sets of RNA genes are transcribed by the three eukaryotic RNA polymerases

Genes that encode polypeptides are always transcribed by RNA polymerase II. However, RNA genes (genes that make noncoding RNA) may be transcribed by polymerases I, II, or III, depending on the type of RNA (see Table 1.5). RNA polymerase I is unusual because it is dedicated to transcribing RNA from a single transcription unit, generating a large transcript that is then processed to yield three types of ribosomal RNA (see below).

RNA polymerase II synthesizes various types of small noncoding RNA in addition to mRNA. They include many types of small nuclear RNA (snRNA) and of small nucleolar RNA (snoRNA) that are involved in different RNA processing events. In addition, it synthesizes many microRNAs (miRNAs) that can show tissue-specific expression and typically regulate the expression of distinctive sets of target genes.

RNA polymerase III transcribes a variety of small noncoding RNAs that are typically expressed in almost all cells, including the different transfer RNA species, 5S ribosomal RNA (rRNA), and some snRNAs. The genes for transfer RNAs (tRNAs) and 5S rRNA are unusual in that the promoters lie within, rather than upstream of, the transcribed sequence (**Figure 1.15**).

Internal promoters are possible because the job of a promoter is simply to attract transcription factors that will guide the RNA polymerase to the correct transcriptional start site. By the time the polymerase is in place and ready to initiate transcription, any transcription factors previously bound to downstream promoter elements will have been removed from the template strand. As an example, transcription of a tRNA gene begins with the following sequence:

- TFIIIC (transcription factor for polymerase IIIC) binds to the A and B boxes of the internal promoter of a tRNA gene (see Figure 1.15).

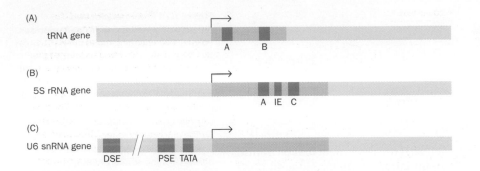

Figure 1.15 Promoter elements in three genes transcribed by RNA polymerase III.
(A) tRNA genes have an internal promoter consisting of an A box (located within the D arm of the tRNA; see Figure 1.9B) and a B box that is usually found in the TψC arm. (B) The promoter of the *Xenopus* 5S rRNA gene has three components: an A box (+50 to +60), an intermediate element (IE; +67 to +72), and the C box (+80 to +90). (C) The human U6 snRNA gene has an external promoter consisting of three components. A distal sequence element (DSE; –240 to –215) enhances transcription and works alongside a core promoter composed of a proximal sequence element (PSE; –65 to –48) and a TATA box (–32 to –25). Arrows mark the +1 position.

- Bound TFIIIC guides the binding of another transcription factor, TFIIIB, to a position upstream of the transcriptional start site; TFIIIC is no longer required and any bound TFIIIC is removed from the internal promoter.
- TFIIIB guides RNA polymerase III to bind to the transcriptional start site.

1.4   RNA PROCESSING

The RNA transcript of most eukaryotic genes undergoes a series of processing reactions to make a mature mRNA or noncoding RNA.

RNA splicing removes unwanted sequences from the primary transcript

For most vertebrate genes—almost all protein-coding genes and some RNA genes—only a small portion of the gene sequence is eventually decoded to give the final product. In these cases the genetic instructions for making an mRNA or mature noncoding RNA occur in **exon** segments that are separated by intervening **intron** sequences that do not contribute genetic information to the final product.

Transcription of a gene initially produces a primary transcript RNA that is complementary to the entire length of the gene, including both exons and introns. This primary transcript then undergoes **RNA splicing**, which is a series of reactions whereby the intronic RNA segments are removed and discarded while the remaining exonic RNA segments are joined end-to-end, to give a shorter RNA product (**Figure 1.16**).

RNA splicing requires recognition of the nucleotide sequences at the boundaries of transcribed exons and introns (splice junctions). The dinucleotides at the ends of introns are highly conserved: the vast majority of introns start with a GT (becoming GU in intronic RNA) and end with an AG (the **GT–AG rule**).

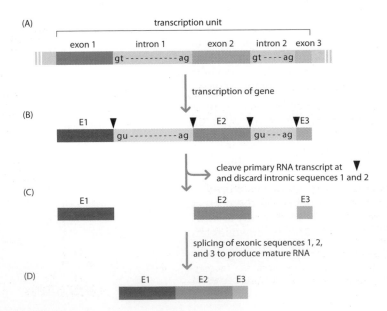

Figure 1.16 The process of RNA splicing.
(A) In this example, the gene contains three exons and two introns. (B) The primary RNA transcript is a continuous RNA copy of the gene and contains sequences transcribed from exons (E1, E2, and E3) and introns. (C) The primary transcript is cleaved at regions corresponding to exon–intron boundaries (splice junctions). The RNA copies of the introns are snipped out and discarded. (D) The RNA copies of the exons are retained and then fused together (spliced) in the same linear order as in the genomic DNA sequence.

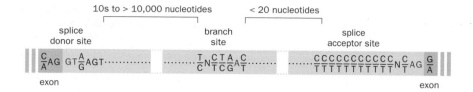

Although the conserved GT and AG dinucleotides are crucial for splicing, they are not sufficient to mark the limits of an intron. The nucleotide sequences that are immediately adjacent to them are also quite highly conserved, constituting splice junction consensus sequences (**Figure 1.17**). A third conserved intronic sequence that is also important in splicing is known as the branch site and is typically located no more than 40 nucleotides upstream of the intron's 5′ terminal AG (see Figure 1.17). Other exonic and intronic sequences can promote splicing (splice enhancer sequences) or inhibit it (splice silencer sequences), and mutations in these sequences can cause disease.

The essential steps in splicing are as follows:

- Nucleophilic attack of the intron's 5′ terminal G nucleotide by the invariant A of the branch site consensus sequence, to form a lariat-shaped structure.

- Cleavage of the exon/intron junction at the splice donor site.

- Nucleophilic attack by the 3′ end of the upstream exon of the splice acceptor site, leading to cleavage and release of the intronic RNA in the form of a lariat, and the splicing together of the two exonic RNA segments (**Figure 1.18**).

For genes residing in eukaryotic nuclei, RNA splicing is mediated by a large RNA protein complex, called the spliceosome. Spliceosomes have five types of snRNA (small nuclear RNA) and more than 50 proteins. The snRNA molecules associate with proteins to form small nuclear ribonucleoprotein (snRNP, or snurp) particles. The specificity of the splicing reaction is established by RNA–RNA base pairing between the RNA transcript to be spliced and snRNA molecules within the spliceosome. There are two types of spliceosome:

- The *major (GU-AG) spliceosome* processes transcripts corresponding to classical GT-AG introns. It contains five types of snRNA. U1 and U2 snRNAs recognize and bind the splice donor and branch sites, respectively. U4, U5, and U6 snRNAs subsequently bind to cause looping out of the intronic RNA (**Figure 1.19**).

- The *minor (AU-AC) spliceosome* processes transcripts corresponding to rare AU-AC introns. It also has five snRNAs but uses U11 and U12 snRNA instead of U1 and U2 and has variants of U4 and U6 snRNA.

Once a splice donor site is recognized by the spliceosome, it scans the RNA sequence until it meets the next splice acceptor site (signaled as a target by the upstream presence of the branch site consensus sequence).

Specialized nucleotides are added to the ends of most RNA polymerase II transcripts

In addition to RNA splicing, the ends of RNA polymerase II transcripts undergo modifications: the 5′ end is capped by adding a variant guanine by using an unusual phosphodiester bond, and a long sequence of adenines is added to the 3′ end. As well as protecting the ends from cellular exonucleases, these modifications may assist the correct functioning of the RNA transcripts.

Figure 1.17 Three consensus DNA sequences in introns of complex eukaryotes. Most introns in eukaryotic genes contain conserved sequences that correspond to three functionally important regions. Two of the regions, the splice donor site and the splice acceptor site, span the 5′ and 3′ boundaries of the intron. The branch site is an additional important region that typically occurs less than 20 nucleotides upstream of the splice acceptor site. The nucleotides shown in red in these three consensus sequences are almost invariant. The other nucleotides detailed in both the intron and the exons are those most commonly found at each position. In some instances, two nucleotides may be equally common, as in the case of C and T near the 3′ end of the intron. Where N appears, any of the four nucleotides may occur.

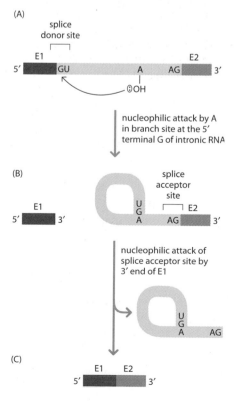

Figure 1.18 The mechanism of RNA splicing. (A) The unprocessed primary RNA transcript with intronic RNA separating sequences E1 and E2 that correspond to exons in DNA. The splicing mechanism involves a nucleophilic attack on the G of the 5′ GU dinucleotide. This is carried out by the 2′ OH group on the conserved A of the branch site and results in the formation of a lariat structure (B), and cleavage of the splice donor site. The 3′ OH at the 3′ end of the E1 sequence performs a nucleophilic attack on the splice acceptor site, causing release of the intronic RNA (as a lariat-shaped structure) and (C) fusion (splicing) of E1 and E2.

Figure 1.19 Role of small nuclear ribonucleoprotein (snRNPs) in RNA splicing. (A) The unprocessed primary RNA transcript as in Figure 1.18. (B) Within the spliceosome, part of the U1 snRNA is complementary in sequence to the splice donor site consensus sequence. As a result, the U1 snRNA-protein complex (U1 snRNP) binds to the splice junction by RNA–RNA base pairing. The U2 snRNP complex similarly binds to the branch site by RNA–RNA base pairing. Interaction between the splice donor and splice acceptor sites is stabilized by (C) the binding of a multi-snRNP particle that contains the U4, U5, and U6 snRNAs. The U5 snRNP binds simultaneously to both the splice donor and splice acceptor sites. Their cleavage releases the intronic sequence and allows (D) E1 and E2 to be spliced together.

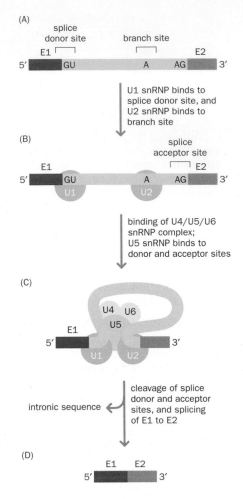

5′ capping

Shortly after the initiation of synthesis of primary RNA transcripts that will become mRNA, a methylated nucleoside (7-methylguanosine, m^7G) is linked by a 5′–5′ phosphodiester bond to the first 5′ nucleotide. This is described as the **capping** of the 5′ end of the transcript (**Figure 1.20**); the caps of snRNA gene transcripts may undergo additional modification. The 5′ cap may have several functions:

- to protect the transcript from 5′→3′ exonuclease attack (uncapped mRNA molecules are rapidly degraded);
- to facilitate transport of mRNAs from the nucleus to the cytoplasm;
- to facilitate RNA splicing; and
- to facilitate attachment of the 40S subunit of cytoplasmic ribosomes to mRNA during translation.

3′ polyadenylation

Transcription by both RNA polymerase I and III stops after the enzyme recognizes a specific transcription termination site. However, the 3′ ends of mRNA molecules are determined by a post-transcriptional cleavage reaction. The sequence AAUAAA (sometimes AUUAAA) signals the 3′ cleavage for most polymerase II transcripts.

Cleavage occurs at a specific site 15–30 nucleotides downstream of the AAUAAA sequence, although the primary transcript may continue for hundreds or even thousands of nucleotides past the cleavage point. After cleavage has occurred, the enzyme poly(A) polymerase sequentially adds adenylate (AMP) residues to the 3′ end (about 200 in the case of mammalian mRNA). This polyadenylation reaction (**Figure 1.21**) produces a **poly(A) tail** that is thought to:

- Help transport mRNA to the cytoplasm.
- Stabilize at least some mRNA molecules in the cytoplasm.
- Enhance recognition of mRNA by the ribosomal machinery.

Histone genes are unique in producing mRNA that does not become polyadenylated; termination of their transcription nevertheless also involves 3′ cleavage of the primary transcript.

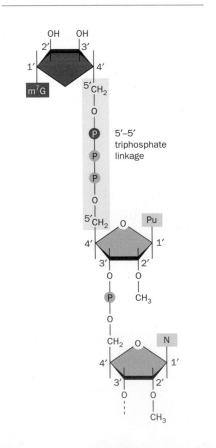

Figure 1.20 The 5′ cap of a eukaryotic mRNA. The nucleotide components in pink represent the residue of the original 5′ end of a eukaryotic pre-mRNA. The primary pre-mRNA transcript begins with a nucleotide that contains a purine (Pu) base and a 5′ triphosphate group. However, as the pre-mRNA undergoes processing, the end phosphate group at the 5′ end is excised with a phosphatase to leave a 5′ diphosphate group, and a specialized nucleotide is covalently joined to form a *cap* that will protect mRNA from exonuclease attack and assist in the initiation of translation. The cap nucleotide (with base shown in red) is first formed when a GTP residue is cleaved to generate a guanosine monophosphate that is then added through a *5′–5′ triphosphate linkage* (pale peach shading) to the diphosphate group of the original purine end nucleotide. Subsequently nitrogen atom 7 of the new 5′ terminal G is methylated. In mRNAs synthesized in vertebrate cells, the 2′ carbon atom of the ribose of each of the two adjacent nucleotides, the original purine end-nucleotide and its neighbor, are also methylated, as illustrated in this example. m^7G, 7-methylguanosine; N, any nucleotide.

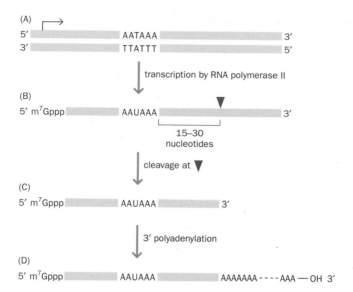

Figure 1.21 Polyadenylation of 3′ ends of eukaryotic mRNAs. (A, B) As RNA polymerase II advances to transcribe a gene it carries at its rear two multiprotein complexes required for polyadenylation: CPSF (cleavage and polyadenylation specificity factor) and CStF (cleavage and stimulation factor) that cooperate to identify a polyadenylation signal downstream of the termination codon in the RNA transcript and to cut the transcript. The polyadenylation signal comprises an AAUAAA sequence or close variant and some poorly understood downstream signals. (C) Cleavage occurs normally about 15–30 nucleotides downstream of the AAUAAA element, and (D) AMP residues are subsequently added by poly(A) polymerase to form a poly(A) tail.

rRNA and tRNA transcripts undergo extensive processing

Four major classes of eukaryotic rRNA have been identified: 28S, 18S, 5.8S, and 5S rRNA (S is the Svedberg coefficient, a measure of how fast large molecular structures sediment in an ultracentrifuge, corresponding directly to size and shape). 18S rRNA is found in the small subunits of ribosomes; the other three are components of the large subunit. Very large amounts of rRNA are required for cells to perform protein synthesis, so many genes are devoted to making rRNA in the nucleolus, a visibly distinct compartment of the nucleus.

In human cells a cluster of approximately 250 genes synthesizes 5S rRNA using RNA polymerase III, which also transcribes some other small RNA species. The 28S, 18S, and 5.8S rRNAs are encoded by consecutive genes on a common 13 kb transcription unit (**Figure 1.22**) that is transcribed by RNA polymerase I. A compound unit of the 13 kb transcription unit and an adjacent 27 kb non-transcribed spacer is tandemly repeated about 30–40 times at the nucleolar organizer regions on the short arms of each of the five human acrocentric chromosomes (13, 14, 15, 21, and 22). These five clusters of rRNA genes, each about 1.5 million bases (Mb) long, are sometimes referred to as **ribosomal DNA (rDNA)**.

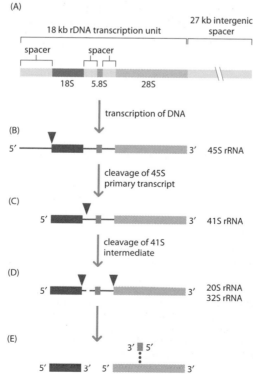

Figure 1.22 The major rRNA species are synthesized by cleavage of a shared primary transcript. (A) In human cells, the 18S, 5.8S, and 28S rRNAs are encoded by a single transcription unit that is 13 kb long. It occurs within tandem repeat units of about 40 kb that also includes a roughly 27 kb non-transcribed (intergenic) spacer. (B) Transcription by RNA polymerase I produces a 13 kb primary transcript (45S rRNA) that then undergoes a complex series of post-transcriptional cleavages. (C–E) Ultimately, individual 18S, 28S, and 5.8S rRNA molecules are released. The 18S rRNA will form part of the small ribosomal subunit. The 5.8S rRNA binds to a complementary segment of the 28S rRNA; the resulting complex will form part of the large ribosomal subunit. The latter also contains 5S rRNA, which is encoded separately by dedicated genes transcribed by RNA polymerase III.

In addition to the sequence of cleavage reactions (see Figure 1.22), the primary rRNA transcript also undergoes a variety of base-specific modifications. This extensive RNA processing is undertaken by many different small nucleolar RNAs that are encoded by about 200 different genes in the human genome.

Mature tRNA molecules also undergo extensive base modifications, and about 10% of the bases in any tRNA are modified versions of A, C, G, or U. Common examples of modified nucleosides include dihydrouridine, which has extra hydrogens at carbons 5 and 6; pseudouridine, an isomer of uridine; inosine (deaminated guanosine); and N,N'-dimethylguanosine.

1.5   TRANSLATION, POST-TRANSLATIONAL PROCESSING, AND PROTEIN STRUCTURE

The mRNA produced by genes in the nucleus migrates to the cytoplasm, where it engages with ribosomes and other components to initiate translation and polypeptide synthesis. Messenger RNA transcribed from genes in the mitochondria and chloroplasts is translated on dedicated ribosomes within these organelles.

Only a central segment of a eukaryotic mRNA molecule is translated to make a polypeptide. The flanking **untranslated regions** (the 5′ UTR and 3′ UTR) are transcribed from exon sequences present at the 5′ and 3′ ends of the gene. They assist in binding and stabilizing the mRNA on the ribosomes, and promote efficient translation (**Figure 1.23**).

Ribosomes are large RNA–protein complexes composed of two subunits. In eukaryotes, cytoplasmic ribosomes have a large 60S subunit and a smaller 40S subunit. The 60S subunit contains three types of rRNA molecule: 28S rRNA, 5.8S rRNA, and 5S rRNA, as well as about 50 ribosomal proteins. The 40S subunit contains a single 18S rRNA and more than 30 ribosomal proteins. Ribosomes provide the structural framework for polypeptide synthesis. The RNA components are predominantly responsible for the catalytic function of the ribosome; the protein components are thought to enhance the function of the rRNA molecules, although a surprising number of them do not seem to be essential for ribosome function.

mRNA is decoded to specify polypeptides

The assembly of a new polypeptide from its constituent amino acids is governed by a triplet genetic code. Within an mRNA the central nucleotide sequence that is used to make polypeptide is scanned from 5′ to 3′ on the ribosome in groups of three nucleotides (**codons**). Each codon specifies an amino acid, and the decoding process uses a collection of different tRNA molecules, each of which binds one type of amino acid. An amino acid–tRNA complex is known as an aminoacyl tRNA and is formed when a dedicated aminoacyl tRNA synthetase covalently links the required amino acid to the terminal adenosine in the conserved CCA trinucleotide at the 3′ end of the tRNA.

Figure 1.23 Transcription and translation of the human β-globin gene. (A) The β-globin gene comprises three exons (E1–E3) and two introns. The 5′ end sequence of E1 and the 3′ end sequence of E3 are noncoding sequences (unshaded sections). (B) These sequences are transcribed and so occur at the 5′ and 3′ ends (unshaded sections) of the β-globin mRNA that emerges from RNA processing. They are not, however, translated and so do not specify any part of the precursor polypeptide (C). This figure also illustrates that some codons can be specified by bases that are separated by an intron. The arginine at position 104 in the β-globin polypeptide is encoded by the last three nucleotides (AGG) of exon 2 but the arginine at position 30 is encoded by an AGG codon whose first two bases are encoded by the last two nucleotides of exon 1 and whose third base is encoded by the first nucleotide of exon 2. (D) During post-translational modification the 147-amino acid precursor polypeptide undergoes cleavage to remove its N-terminal methionine residue, to generate the mature 146-residue β-globin protein. The flanking N and C symbols to the left and right, respectively, in (C) and (D) depict the N-terminus (N) and C-terminus (C).

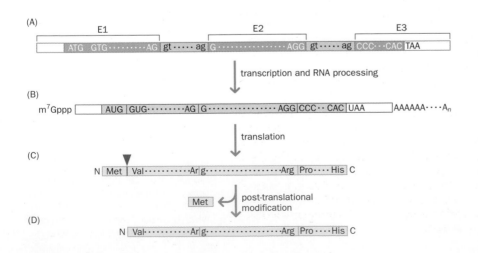

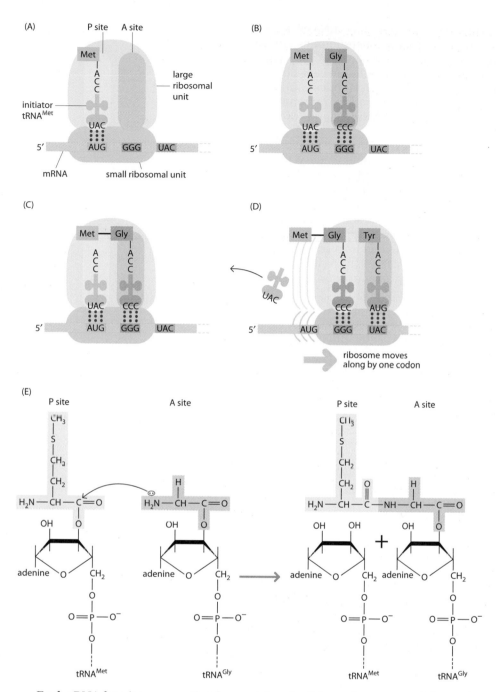

Figure 1.24 In translation, the genetic code is deciphered on ribosomes by codon–anticodon recognition. (A) The large ribosomal subunit (60S in eukaryotes) has two sites for binding an aminoacyl tRNA (a transfer RNA with its attached amino acid): the P (peptidyl) site and the A (aminoacyl) site. The small ribosomal subunit (40S in eukaryotes) binds mRNA, which is scanned along its 5′ UTR in a 5′→3′ direction until the start codon is identified, an AUG located within a larger consensus sequence (see the text). An initiator tRNA^Met carrying a methionine residue binds to the P site with its anticodon in register with the AUG start codon. (B) The appropriate aminoacyl tRNA is bound to the A site with its anticodon base-pairing with the next codon (GGG in this case, specifying glycine). (C) The rRNA in the large subunit catalyzes peptide bond formation, resulting in the methionine detaching from its tRNA and being bound instead to the glycine attached to the tRNA held at the A site. (D) The ribosome translocates along the mRNA so that the tRNA bearing the Met-Gly dipeptide is bound by the P site. The next aminoacyl tRNA (here, carrying Tyr) binds to the A site in preparation for new peptide bond formation. (E) Peptide bond formation. The N atom of the amino group of the amino acid bound to the tRNA in the A site makes a nucleophilic attack on the carboxyl C atom of the amino acid held by the tRNA bound to the P site.

Each tRNA has its own **anticodon**, a trinucleotide at the center of the anti-codon arm (see Figure 1.9B) that provides the necessary specificity to interpret the genetic code. For an amino acid to be added to a growing polypeptide, the relevant codon of the mRNA molecule must be recognized by base pairing with a complementary anticodon on the appropriate aminoacyl tRNA molecule. This happens on the ribosome. The small ribosomal subunit binds the mRNA, and the large subunit has two sites for binding aminoacyl tRNAs, namely a P (peptidyl) site and an A (aminoacyl) site (**Figure 1.24**).

The cap at the 5′ end of messenger RNA molecules is important in initiating translation. It is recognized by certain key proteins that bind the small ribosomal subunit, and these initiation factors hold the mRNA in place. In cap-dependent translation initiation, the ribosome scans the 5′ UTR of the mRNA in the 5′→3′ direction to find a suitable initiation codon, an AUG that is found within the Kozak consensus sequence 5′-GCC**Pu**CC**AUG**G-3′ (where Pu represents purine). The most important determinants are the G at position +4 (immediately follow-ing the AUG codon), and the purine (preferably A) at −3 (three nucleotides upstream of the AUG codon).

When a suitable initiation codon is identified, an initiating $tRNA^{Met}$ with its attached methionine binds to the P site on the large ribosomal subunit so that its anticodon base-pairs with the AUG initiator codon on the mRNA (see Figure 1.24). Once this has happened, the transcriptional reading frame is established and codons are interpreted as successive groups of three nucleotides continuing in the 5′→3′ direction downstream of the initiating AUG codon. An aminoacyl tRNA for the second codon (a $tRNA^{Gly}$ to recognize GGG in the example of Figure 1.24) binds to the neighboring A site in the large subunit.

Once the P and A sites are occupied by aminoacyl tRNAs, the largest rRNA component within the large subunit of the ribosome is thought to act as a peptidyl transferase. It catalyzes the formation of a peptide bond by a condensation reaction between the amino group of the amino acid held by the tRNA in the A site and the carboxyl group of the methionine held by the $tRNA^{Met}$. The net result is to detach the initiator methionine from its tRNA and attach it to the second amino acid, forming a dipeptide (see Figure 1.24). Now without any attached amino acid, the $tRNA^{Met}$ migrates away from the P site and its place is taken by the tRNA with the attached dipeptide that formerly occupied the A site. The liberated A site is now filled by an aminoacyl tRNA carrying an anticodon that is complementary to the third codon, and a new peptide bond is formed to make a tripeptide, and so on.

After a ribosome has initiated translation of an mRNA and has then moved along the mRNA, other ribosomes can engage with the same mRNA. The resulting polyribosome structures (polysomes) make multiple copies of a polypeptide from the one mRNA molecule. Polypeptide chain elongation occurs until a termination codon is met. For mRNA transcribed from nuclear genes, termination codons come in three varieties: UAA (ochre), UAG (amber), and UGA (opal), but there are some differences for mitochondrial mRNA as described in the next section.

In response to a termination codon a protein release factor enters the A site instead of an aminoacyl tRNA to signal that the polypeptide should disengage from the ribosome. The completed polypeptide will then undergo processing that can include cleavage and modification of the side chains. Its backbone will have a free amino group at one end (the N-terminal end) and a free carboxyl group at the other end (the C-terminal end).

The genetic code is degenerate and not quite universal

The genetic code is a three-letter code, and there are four possible bases to choose from at each of the three base positions in a codon. There are therefore $4^3 = 64$ possible codons, which is more than sufficient to encode the 20 major types of amino acid. The genetic code is degenerate because, on average, each amino acid is specified by about three different codons. Some amino acids (such as leucine, serine, and arginine) are specified by as many as six codons; others are much more poorly represented (**Figure 1.25**). The degeneracy of the genetic code most often involves the third base of the codon.

mtDNA variants									mtDNA variant
	AAA	Lys	CAA	Gln	GAA	Glu	UAA	STOP	
	AAG		CAG		GAG		UAG		
	AAC	Asn	CAC	His	GAC	Asp	UAC	Tyr	
	AAU		CAU		GAU		UAU		
	ACA	Thr	CCA	Pro	GCA	Ala	UCA	Ser	
	ACG		CCG		GCG		UCG		
	ACC		CCC		GCC		UCC		
	ACU		CCU		GCU		UCU		
STOP	AGA	Arg	CGA	Arg	GGA	Gly	UGA	STOP	Trp
	AGG		CGG		GGG		UGG	Trp	
	AGC	Ser	CGC		GGC		UGC	Cys	
	AGU		CGU		GGU		UGU		
Met	AUA	Ile	CUA	Leu	GUA	Val	UUA	Leu	
	AUG	Met	CUG		GUG		UUG		
	AUC	Ile	CUC		GUC		UUC	Phe	
	AUU		CUU		GUU		UUU		

Figure 1.25 The genetic code. All 64 possible codons of the genetic code and the amino acid specified by each, as read in the 5′→3′ direction from the mRNA sequence. The interpretations of the 64 codons in the 'universal' genetic code are shown in black immediately to the right of the codons. Sixty-one codons specify an amino acid. Three STOP codons (UAA, UAG, and UGA) do not encode any amino acid. The genetic code for mitochondrial mRNA (mtDNA) conforms to the universal code except for a few variants. For example, in the mitochondrial genetic code in humans and many other species four codons are used differently: UGA encodes tryptophan instead of being a STOP codon, AUA encodes methionine, and instead of encoding arginine, AGA and AGG are STOP codons.

Although more than 60 codons can specify an amino acid, the number of different cytoplasmic tRNA molecules is quite a bit smaller, and only 22 types of mitochondrial tRNA are made. The interpretation of more than 60 sense codons with a much smaller number of different tRNAs is possible because base pairing in RNA is more flexible than in DNA. Pairing of codon and anticodon follows the normal A–U and G–C rules for the first two base positions in a codon. However, at the third position there is some flexibility (base wobble), and GU base pairs are tolerated here (Table 1.6).

The genetic code is the same throughout nearly all life forms. However, mitochondria and chloroplasts have a limited capacity for protein synthesis, and during evolution their genetic codes have diverged slightly from that used at cytoplasmic ribosomes. Translation of nuclear-encoded mRNA continues until one of three stop codons is encountered (UAA, UAG, or UGA) but in mammalian mitochondria there are four possibilities (UAA, UAG, AGA, and AGG).

The meaning of a codon can also be dependent upon the sequence context; that is, the nature of the nucleotide sequence in which it is embedded. Depending on the surrounding sequence, some codons in a few types of nuclear-encoded mRNA can be interpreted differently from normal. For example, in a wide variety of cells the stop codon UGA is alternatively interpreted as encoding selenocysteine with some nuclear-encoded mRNAs, and UAG can sometimes be interpreted to encode glutamine.

Post-translational processing: chemical modification of amino acids and polypeptide cleavage

Primary translation products often undergo a variety of modifications during or after translation. Simple or complex chemical groups are often covalently attached to the side chains of certain amino acids (Table 1.7). In addition, polypeptides may occasionally be cleaved to yield one or more active polypeptide products.

Addition of carbohydrate groups

Glycoproteins have oligosaccharides covalently attached to the side chains of certain amino acids. Few proteins in the cytosol are glycosylated (carry an

TABLE 1.6 RULES FOR BASE PAIRING CAN BE RELAXED (WOBBLE) AT POSITION 3 OF A CODON

Base at 5′ end of tRNA anticodon	Base recognized at 3′ end of mRNA codon
A	U only
C	G only
G (or I)[a]	C or U
U	A or G

[a]Inosine (I) is a deaminated form of guanosine.

TABLE 1.7 MAJOR TYPES OF MODIFICATION OF POLYPEPTIDES

Type of modification (group added)	Target amino acid(s)	Notes
Phosphorylation (PO_4^-)	Tyr, Ser, Thr	achieved by specific kinases; may be reversed by phosphatases
Methylation (CH_3)	Lys	achieved by methylases; reversed by demethylases
Hydroxylation (OH)	Pro, Lys, Asp	hydroxyproline (Hyp) and hydroxylysine (Hyl) are particularly common in collagens
Acetylation (CH_3CO)	Lys	achieved by an acetylase; reversed by deacetylase
Carboxylation (COOH)	Glu	achieved by γ-carboxylase
N-glycosylation (complex carbohydrate)	Asn[a]	takes place initially in the endoplasmic reticulum, with later additional changes occurring in the Golgi apparatus
O-glycosylation (complex carbohydrate)	Ser, Thr, Hyl[b]	takes place in the Golgi apparatus; less common than *N*-glycosylation
Glycosylphosphatidylinositol (glycolipid)	Asp[c]	serves to anchor protein to outer layer of plasma membrane
Myristoylation (C_{14} fatty acyl group)	Gly[d]	serves as membrane anchor
Palmitoylation (C_{16} fatty acyl group)	Cys[e]	serves as membrane anchor
Farnesylation (C_{15} prenyl group)	Cys[c]	serves as membrane anchor
Geranylgeranylation (C_{20} prenyl group)	Cys[c]	serves as membrane anchor

[a]This is especially common when Asn is in the sequence: Asn-X-(Ser/Thr), where X is any amino acid other than Pro. [b]Hydroxylysine. [c]At C-terminus of polypeptide. [d]At N-terminus of polypeptide. [e]To form *S*-palmitoyl link.

attached carbohydrate); if they are, they have a single sugar residue, *N*-acetyl-glucosamine, attached to a serine or threonine residue. However, proteins that are secreted from cells or transported to lysosomes, the Golgi apparatus, or the plasma membrane are routinely glycosylated. In these cases, the sugars are assembled as oligosaccharides before being attached to the protein.

Two major types of glycosylation occur. Carbohydrate *N*-glycosylation involves attaching a carbohydrate group to the nitrogen atom of an asparagine side chain, and *O*-glycosylation entails adding a carbohydrate to the oxygen atom of an OH group carried by the side chains of certain amino acids (see Table 1.7).

Proteoglycans are proteins with attached glycosaminoglycans (polysaccharides) that usually include repeating disaccharide units containing glucosamine or galactosamine. The best-characterized proteoglycans are components of the extracellular matrix, a complex network of macromolecules secreted by, and surrounding, cells in tissues or in culture systems.

Addition of lipid groups

Some proteins, notably membrane proteins, are modified by the addition of fatty acyl or prenyl groups. These added groups typically serve as membrane anchors, hydrophobic amino acid sequences that secure a newly synthesized protein within either a plasma membrane or the endoplasmic reticulum (Table 1.8).

Anchoring a protein to the outer layer of the plasma membrane involves the attachment of a glycosylphosphatidylinositol (GPI) group. This glycolipid group contains a fatty acyl group that serves as the membrane anchor; it is linked successively to a glycerophosphate unit, an oligosaccharide unit, and finally—through a phosphoethanolamine unit—to the C-terminus of the protein. The entire protein, except the GPI anchor, is located in the extracellular space.

Post-translational cleavage

The primary translation product may also undergo internal cleavage to generate a smaller mature product. Occasionally the initiating methionine is cleaved from

TABLE 1.8 LEVELS OF PROTEIN STRUCTURE		
Level	Definition	Notes
Primary	the linear sequence of amino acids in a polypeptide	can vary enormously in length from a few to thousands of amino acids
Secondary	the path that a polypeptide backbone follows within local regions of the primary structure	varies along the length of the polypeptide; common elements of secondary structure include the α-helix and β-pleated sheet
Tertiary	the overall three-dimensional structure of a polypeptide, arising from the combination of all of the secondary structures	can take various forms (e.g. globular, rod-like, tube, coil, sheet)
Quaternary	the aggregate structure of a multimeric protein (comprising more than one subunit, which may be of more than one type)	can be stabilized by disulfide bridges between subunits or ligand binding, and other factors

the primary translation product, as during the synthesis of β-globin (see Figure 1.23C, D). More substantial polypeptide cleavage is observed during the maturation of many proteins, including plasma proteins, polypeptide hormones, neuropeptides, and growth factors. Cleavable signal sequences are often used to mark proteins either for export or for transport to a specific intracellular location. A single mRNA molecule can sometimes specify more than one functional polypeptide chain as a result of post-translational cleavage of a large precursor polypeptide (**Figure 1.26**).

The complex relationship between amino acid sequence and protein structure

Proteins can be composed of one or more polypeptides, each of which may be subject to post-translational modification. Interactions between a protein and either of the following may substantially alter the conformation of that protein:

- A **cofactor**, such as a divalent cation (such as Ca^{2+}, Fe^{2+}, Cu^{2+}, or Zn^{2+}), or a small molecule required for functional enzyme activity (such as NAD^+).

- A **ligand** (any molecule that a protein binds specifically).

Four different levels of structural organization in proteins have been distinguished and defined (see Table 1.8).

Even within a single polypeptide, there is ample scope for hydrogen bonding between different amino acid residues. This stabilizes the partial polar charges along the backbone of the polypeptide and has profound effects on that protein's overall shape. With regard to a protein's conformation, the most significant hydrogen bonds are those that occur between the oxygen of one peptide bond's carbonyl (C=O) group and the hydrogen of the amino (NH) group of another peptide bond. Several fundamental structural patterns (motifs) stabilized by hydrogen bonding within a single polypeptide have been identified, the most fundamental of which are described below.

The α-helix

This is a rigid cylinder that is stabilized by hydrogen bonding between the carbonyl oxygen of a peptide bond and the hydrogen atom of the amino nitrogen of a peptide bond located four amino acids away (**Figure 1.27**). α-Helices often occur

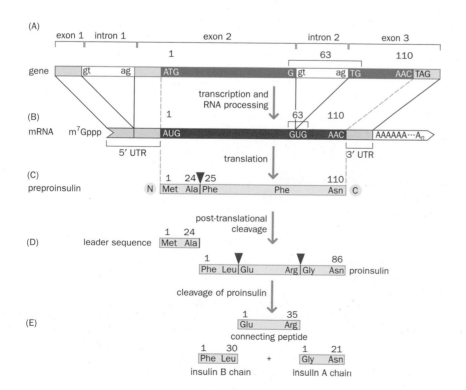

Figure 1.26 Insulin synthesis involves multiple post-translational cleavages of polypeptide precursors. (A) The human insulin gene comprises three exons and two introns. The coding sequence (the part that will be used to make polypeptide) is shown in deep blue. It is confined to the 3′ sequence of exon 2 and the 5′ sequence of exon 3. (B) Exon 1 and the 5′ part of exon 2 specify the 5′ untranslated region (5′ UTR), and the 3′ end of exon 3 specifies the 3′ UTR. The UTRs are transcribed and so are present at the ends of the mRNA. (C) A primary translation product, preproinsulin, has 110 residues and is cleaved to give (D) a 24-residue N-terminal *leader sequence* (that is required for the protein to cross the cell membrane but is thereafter discarded) plus an 86-residue proinsulin precursor. (E) Proinsulin is cleaved to give a central segment (the connecting peptide) that may maintain the conformation of the A and B chains of insulin before the formation of their interconnecting covalent disulfide bridges (see Figure 1.29).

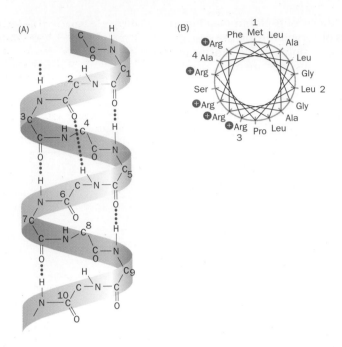

Figure 1.27 The structure of a standard α-helix and an amphipathic α-helix.
(A) The structure of an α-helix is stabilized by hydrogen bonding between the oxygen of the carbonyl group (C=O) of each peptide bond and the hydrogen on the peptide bond amide group (NH) of the fourth amino acid away, making the helix have 3.6 amino acid residues per turn. The side chains of each amino acid are located on the outside of the helix; there is almost no free space within the helix. Note that only the backbone of the polypeptide is shown, and some bonds have been omitted for clarity. (B) An amphipathic α-helix has tighter packing and has charged amino acids and hydrophobic amino acids located on different surfaces. Here we show an end view of such a helix: five positively charged arginine residues are clustered on one side of the helix, whereas the opposing side has a series of hydrophobic amino acids (mostly Ala, Leu, and Gly). The lines within the circle indicate neighboring residues—the initiator methionine (position 1) is connected to a leucine (2), which is connected to an arginine (3), which is adjacent to an alanine (4), and so on.

in proteins that perform key cellular functions (such as transcription factors, where they are usually represented in the DNA-binding domains). Identical α-helices with a repeating arrangement of nonpolar side chains can coil round each other to form a particularly stable coiled coil. Coiled coils occur in many fibrous proteins, such as collagen of the extracellular matrix, the muscle protein tropomyosin, α-keratin in hair, and fibrinogen in blood clots.

The β-pleated sheet

β-Pleated sheets are also stabilized by hydrogen bonding but, in this case, they occur between opposed peptide bonds in parallel or anti-parallel segments of the same polypeptide chain (**Figure 1.28**). β-Pleated sheets occur—often together with α-helices—at the core of most globular proteins.

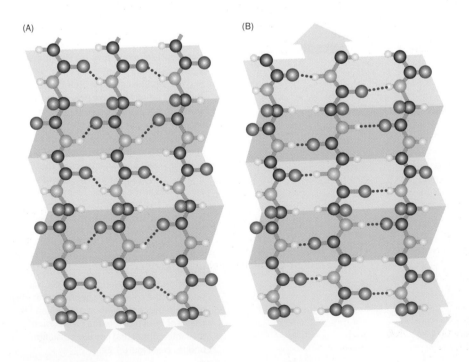

Figure 1.28 The structure of a β-pleated sheet. Hydrogen bonding occurs here between the carbonyl (C=O) oxygens and amide (NH) hydrogens on adjacent segments of (A) parallel and (B) anti-parallel β-pleated sheets. [Adapted from Lehninger AL, Nelson DL & Cox MM (1993) Principles of Biochemistry, 2nd ed. With permission from WH Freeman and Company.]

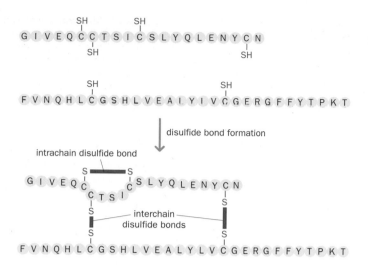

Figure 1.29 Intrachain and interchain disulfide bridges in human insulin. Disulfide bridges (–S–S–) form by a condensation reaction between the sulfhydryl (–SH) groups on the side chains of cysteine residues. They can form between cysteine side chains within the same polypeptide (such as between positions 6 and 11 within the insulin A chain) and also between cysteine side chains on different interacting polypeptides such as the insulin A and B chains.

The β-turn

Hydrogen bonding can occur between amino acids that are even nearer to each other within a polypeptide. When this arises between the peptide bond CO group of one amino acid residue and the peptide bond NH group of an amino acid residue three places farther along, this results in a hairpin β-turn. Abrupt changes in the direction of a polypeptide enable compact globular shapes to be achieved. These β-turns can connect parallel or anti-parallel strands in β-pleated sheets.

Higher-order structures

Many more complex structural motifs, consisting of combinations of the above structural modules, form **protein domains**. Such domains are often crucial to a protein's overall shape and stability and usually represent functional units involved in binding other molecules. Another important determinant of the structure (and function) of a protein is **disulfide bridges**. They can form between the sulfur atoms of sulfhydryl (–SH) groups on two amino acids that may reside on a single polypeptide chain or on two polypeptide chains (**Figure 1.29**).

In general, the primary structure of a protein determines the set of secondary structures that, together, generate the protein's tertiary structure. Secondary structural motifs can be predicted from an analysis of the primary structure, but the overall tertiary structure cannot easily be accurately predicted. Finally, some proteins form complex aggregates of polypeptide subunits, giving an arrangement known as the quaternary structure.

FURTHER READING

Agris PF, Vendeix FA & Graham WD (2007) tRNA's wobble decoding of the genome: 40 years of modification. *J. Mol. Biol.* 366, 1–13.

Calvo O & Manley JL (2003) Strange bedfellows: polyadenylation factors at the promoter. *Genes Dev.* 17, 1321–1327.

Carradine CR, Drew HR, Luisi B & Travers AA (2004) Understanding DNA. The Molecule And How It Works, 3rd ed. Academic Press.

Crain PF, Rozenski J & McCloskey JA. RNA Modification Database. http://library.med.utah.edu/RNAmods

Fedorova O & Zingler N (2007) Group II introns: structure, folding and splicing mechanisms. *Biol. Chem.* 388, 665–678.

Garcia-Diaz M & Bebenek K (2007) Multiple functions of DNA polymerases. *Crit. Rev. Plant Sci.* 26, 105–122.

Preiss T & Hentze MW (2003) Starting the protein synthesis machine: eukaryotic translation initiation. *BioEssays* 25, 1201–1211.

Sander DM. Big Picture Book of Viruses. http://www.virology.net/Big_Virology/BVHomePage.html

Wear MA & Cooper JA (2004) Capping protein: new insights into mechanism and regulation. *Trends Biochem. Sci.* 29, 418–428.

Whitford D (2005) Protein Structure And Function. John Wiley.

Chapter 2

Chromosome Structure and Function

2

KEY CONCEPTS

- Chromosomes have two fundamental roles: the faithful transmission and appropriate expression of genetic information.

- Prokaryotic chromosomes contain circular double-stranded DNA molecules that are relatively protein-free, but eukaryotic chromosomes consist of linear double-stranded DNA molecules complexed throughout their lengths with proteins.

- Chromatin is the DNA–protein matrix of eukaryotic chromosomes. The complexed proteins serve structural roles, including compacting the DNA in different ways, and also regulatory roles.

- Chromosomes undergo major changes in the cell cycle, notably at S phase when they replicate and at M phase when the replicated chromosomes become separated and allocated to two daughter cells.

- DNA replication at S phase produces two double-stranded daughter DNA molecules that are held together at a specialized region, the centromere. When the daughter DNA molecules remain held together like this they are known as sister chromatids, but once they separate at M phase they become individual chromosomes.

- At the metaphase stage of M phase the chromosomes are so highly condensed that gene expression is uniformly shut down. But this is the optimal time for viewing them under the microscope. Staining with dyes that bind preferentially to GC-rich or AT-rich regions can give reproducible chromosome banding patterns that allow different chromosomes to be differentiated.

- During interphase, the long period of the cell cycle that separates successive M phases, chromosomes have generally very long extended conformations and are invisible under optical microscopy. The extended structure means that genes can be expressed efficiently.

- Even during interphase some chromosomal regions always remain highly condensed and transcriptionally inactive (heterochromatin), whereas others are extended to allow gene expression (euchromatin).

- Sperm and egg cells have one copy of each chromosome (they are haploid), but most cells are diploid, having two sets of chromosomes.

- Fertilization of a haploid egg by a haploid sperm generates the diploid zygote from which all other body cells arise by cell division.

- In mitosis a cell divides to give two daughter cells, each with the same number and types of chromosomes as the original cell.

- Meiosis is a specialized form of cell division that occurs in certain cells of the testes and ovaries to produce haploid sperm and egg cells. During meiosis new genetic combinations are randomly created, partly by exchanging sequences between maternal and paternal chromosomes.

- Three types of functional element are needed for eukaryotic chromosomes to transmit DNA faithfully from mother cell to daughter cells: the centromere (ensures correct chromosome segregation at cell division); replication origins (initiate DNA replication); and telomeres (cap the chromosomes to stop the internal DNA from being degraded by nucleases).

- An abnormal number of chromosomes can sometimes occur but this is often lethal if present in most cells of the body.

- Structural chromosome abnormalities arising from breaks in chromosomes can cause genes to be deleted or incorrectly expressed.

- Having the correct number and structure of chromosomes is not enough. They must also have the correct parental origin because certain genes are preferentially expressed on either paternally or maternally inherited chromosomes.

The underlying structure and fundamental functions of DNA—replication and transcription—were introduced in the previous chapter. But DNA functions in a context. In eukaryotic cells, the very long DNA molecules in the nucleus are complexed with a variety of structural and regulatory proteins and structured into linear chromosomes. The DNA molecules in mitochondria are different: they are comparatively short, have little protein attached to them, and are circular.

This chapter introduces the life cycle of chromosomes in eukaryotic cells that are usually formed from other cells by cell division. The process of cell division is a small component of the cell cycle, the process in which chromosomes and their constituent DNA molecules need to make perfect copies of themselves and then segregate into daughter cells. There are important differences between how this occurs in routine cell division and in the specialized form of cell division that gives rise to sperm and egg cells.

A feature common to both types of cell division is the importance to the cell of chromosome condensation. This affects the expression of information encoded in the DNA and makes long and fragile DNA strands resilient to breakage during the dramatic rearrangements that occur in cell division. The use of dyes to stain condensed chromosomes has revealed patterns that, like fingerprints, can be used to distinguish between them. Careful examination of these and other patterns can reveal evidence of chromosomal abnormalities, such as breakages and rearrangements that have occurred and survived but may cause disease.

2.1   PLOIDY AND THE CELL CYCLE

The chromosome and DNA content of cells is defined by the number (n) of different chromosomes, the **chromosome set**, and its associated DNA content (C). For human cells, $n = 23$ and C is about 3.5 pg (3.5×10^{-12} g). Different cell types in an organism, however, may differ in **ploidy**—the number of copies they have of the chromosome set. Sperm and egg cells carry a single chromosome set and are said to be **haploid** (they have n chromosomes and a DNA content of C). Most human and mammalian cells carry two copies of the chromosome set and are **diploid** (having $2n$ chromosomes and a DNA content of $2C$). However, in several non-mammalian animal species most of the body cells are not diploid but instead are either haploid or polyploid. In the latter case, some species are tetraploid ($4n$) and others have a ploidy of more than $4n$, but triploidy ($3n$) is less common in animals because it can give rise to problems in producing sperm and egg cells.

The cells of our body are all derived ultimately from a single diploid cell, the **zygote**, that is formed when a sperm fertilizes an egg. Starting from the zygote, organisms grow by repeated rounds of cell division. Each round of cell division is a **cell cycle** and comprises a brief M phase, during which cell division occurs, and the much longer intervening **interphase**, which has three parts (**Figure 2.1**). They are: S phase (during which DNA synthesis occurs), G_1 phase (the gap between M phase and S phase), and G_2 phase (the gap between S phase and M phase).

We will describe the cell biology underlying the phases of the cell cycle in a later chapter. Here we are concerned with the life cycle of chromosomes. During each cell cycle, chromosomes undergo profound changes to their structure, number, and distribution within the cell. From the end of M phase right through until DNA duplication in S phase, a chromosome of a diploid cell contains a single DNA double helix and the total DNA content is $2C$ (see Figure 2.1). After DNA duplication, the total DNA content is $4C$, but the duplicated double helices are held together along their lengths so that each chromosome has double the DNA content of a chromosome in early S phase. During M phase the duplicated double helices separate, generating two daughter chromosomes, giving $4n$ chromosomes. After equal distribution of the chromosomes to the two daughter cells, both cells will have $2n$ chromosomes and a DNA content of $2C$ (see Figure 2.1).

G_1 is the normal state of a cell, and is the long-term end state of non-dividing cells. Cells enter S phase only if they are committed to mitosis; as will be described in more detail in Chapter 4, non-dividing cells remain in a modified G_1 stage, sometimes called the G_0 phase. The cell cycle diagram can give the impression that all the interesting action happens in S and M phases—but this is an illusion. A cell spends most of its life in G_0 or G_1 phase, and that is where the genome does most of its work.

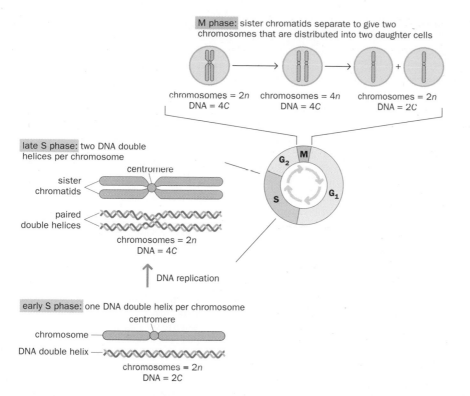

M phase: sister chromatids separate to give two chromosomes that are distributed into two daughter cells

chromosomes = 2n
DNA = 4C

chromosomes = 4n
DNA = 4C

chromosomes = 2n
DNA = 2C

late S phase: two DNA double helices per chromosome

centromere

sister chromatids

paired double helices

chromosomes = 2n
DNA = 4C

DNA replication

early S phase: one DNA double helix per chromosome

centromere

chromosome

DNA double helix

chromosomes = 2n
DNA = 2C

Figure 2.1 Changes in chromosomes and DNA content during the cell cycle. The cell cycle shown at the right includes a very short M phase, when the chromosomes become extremely highly condensed in preparation for nuclear and cell division. Afterwards, cells enter a long period of growth called interphase, during which chromosomes are enormously extended so that genes can be expressed. Interphase is divided into three phases: G_1, S (when the DNA replicates), and G_2. Chromosomes contain one DNA double helix from the end of M phase right through until just before the DNA is duplicated in S phase. After the DNA double helix has been duplicated, the two resulting double helices are held together tightly along their lengths (by specialized protein complexes called cohesins) until M phase. As the chromosomes condense at M phase they are now seen to consist of two sister chromatids, each containing a DNA duplex, that are bound together only at the centromeres. During M phase the two sister chromatids separate to form two independent chromosomes that are then equally distributed into the daughter cells.

A small subset of diploid body cells constitute the **germ line** that gives rise to **gametes** (sperm cells or egg cells). In humans, where $n = 23$, each gamete contains one sex chromosome plus 22 non-sex chromosomes (**autosomes**). In eggs, the sex chromosome is always an X; in sperm it may be either an X or a Y. After a haploid sperm fertilizes a haploid egg, the resulting diploid zygote and almost all of its descendant cells have the chromosome constitution 46,XX (female) or 46,XY (male) (**Figure 2.2**).

Cells outside the germ line are **somatic cells**. Human somatic cells are usually diploid but, as will be described later, there are notable exceptions. Some types of non-dividing cell lack a nucleus and any chromosomes, and so are nulliploid. Other cell types have multiple chromosome sets; they are naturally polyploid as a result of multiple rounds of DNA replication without cell division.

2.2   MITOSIS AND MEIOSIS

Mitosis and meiosis are both cell division processes that involve chromosome replication and cell division. However, the products of mitosis have the same ploidy as the initiating cell, whereas meiosis halves the cell's ploidy. Furthermore, whereas mitosis gives rise to genetically identical products, meiosis generates genetic diversity to ensure that offspring are genetically different from their parents.

Mitosis is the normal form of cell division

As an embryo develops through fetus, infant, and child to adult, many cell cycles are needed to generate the required number of cells. Because many cells have a limited life span, there is also a continuous requirement to generate new cells,

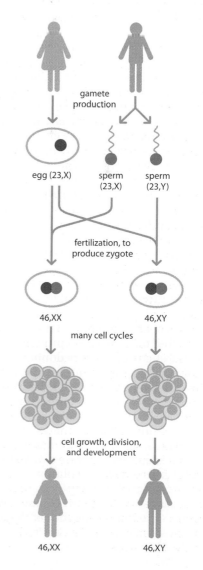

gamete production

egg (23,X)

sperm (23,X)

sperm (23,Y)

fertilization, to produce zygote

46,XX

46,XY

many cell cycles

cell growth, division, and development

46,XX

46,XY

Figure 2.2 The human life cycle, from a chromosomal viewpoint. Haploid egg and sperm cells originate from diploid precursors in the ovary and testis in women and men, respectively. All eggs have a 23,X chromosome constitution, representing 22 autosomes plus a single X sex chromosome. A sperm can carry either sex chromosome, so that the chromosome constitution is 23,X (50%) and 23,Y (50%). After fertilization and fusion of the egg and sperm nuclei, the diploid zygote will have a chromosome constitution of either 46,XX or 46,XY, depending on which sex chromosome the fertilizing sperm carried. After many cell cycles, this zygote gives rise to all cells of the adult body, almost all of which will have the same chromosome complement as the zygote from which they originated.

(A)

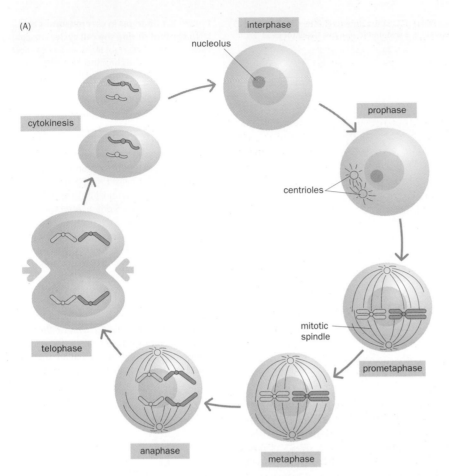

(B)

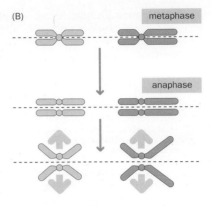

Figure 2.3 Mitosis and cytokinesis.
(A) Mitotic stages and cytokinesis. Late in interphase, the duplicated chromosomes are still dispersed in the nucleus, and the nucleolus is distinct. Early in prophase, the first stage of mitosis, the centrioles (which were previously duplicated in interphase) begin to separate and migrate to opposite poles of the cell, where they will form the spindle poles. In prometaphase, the nuclear envelope breaks down, and the now highly condensed chromosomes become attached at their centromeres to the array of microtubules extending from the mitotic spindle. At metaphase, the chromosomes all lie along the middle of the mitotic spindle. At anaphase, the sister chromatids separate and begin to migrate toward opposite poles of the cell, as a result of both shortening of the microtubules and further separation of the spindle poles. The nuclear envelope forms again around the daughter nuclei during telophase and the chromosomes decondense, completing mitosis. Constriction of the cell begins. During cytokinesis, filaments beneath the plasma membrane constrict the cytoplasm, ultimately producing two daughter cells. (B) Metaphase–anaphase transition. Metaphase chromosomes aligned along the equatorial plane (dashed line) have their sister chromatids held tightly together (by cohesin protein complexes) at the centromeres. The transition to anaphase is marked by disruption of the cohesin complexes, thereby releasing the sister chromatids to form independent chromosomes with their own centromeres, which are then pulled by the microtubules of the spindle in the direction of opposing poles (arrows).

even in an adult organism. All of these cell divisions occur by **mitosis**, which is the normal process of cell division throughout the human life cycle. Mitosis ensures that a single cell gives rise to two daughter cells that are each genetically identical to the parent cell, barring any errors that might have occurred during DNA replication. During a human lifetime, there may be something like 10^{17} mitotic divisions.

The M phase of the cell cycle includes various stages of nuclear division (prophase, prometaphase, metaphase, anaphase, and telophase), and also cell division (**cytokinesis**), which overlaps the final stages of mitosis (**Figure 2.3**). In preparation for cell division, the previously highly extended duplicated chromosomes contract and condense so that, by the metaphase stage of mitosis, they are readily visible when viewed under the microscope.

The chromosomes of early S phase have one DNA double helix, but after DNA replication two identical DNA double helices are produced (see Figure 2.1), and they are held together along their lengths by multisubunit protein complexes called cohesins. Recent data suggest that individual cohesin subunits are linked together to form a large protein ring. Some models suggest that multiple cohesin rings encircle the two double helices to entrap them along their lengths, or cohesin rings form round the individual double helices and then interact to ensure that the two double helices are held tightly together.

Later, when the chromosomes have undergone compaction in preparation for cell division, the cohesins are removed from all parts of the chromosomes apart from the centromeres. As a result, by prometaphase when the chromosomes can now be viewed under the light microscope, individual chromosomes can be seen to comprise two **sister chromatids** that are attached at the centromere by the residual cohesin complexes that continue to bind the two DNA helices at this position.

Later still, at the start of anaphase, the residual cohesin complexes holding the sister chromatids together at the centromere are removed. The two sister chromatids can now disengage to become independent chromosomes that will be pulled to opposite poles of the cell and then distributed equally to the daughter cells (see Figure 2.3). Interaction between the mitotic spindle and the centromere is crucial to this process and we will consider this in detail in Section 2.3.

Meiosis is a specialized reductive cell division that gives rise to sperm and egg cells

Diploid primordial germ cells migrate into the embryonic gonad and engage in repeated rounds of mitosis, to generate spermatogonia in males and oogonia in females. Further growth and differentiation produce primary spermatocytes in the testis, and primary oocytes in the ovary. This process requires many more mitotic divisions in males than in females, and probably contributes to differences in mutation rate between the sexes.

The diploid spermatocytes and oocytes can then undergo **meiosis**, the cell division process that produces haploid gametes. Meiosis is a *reductive* division because it involves two successive cell divisions (known as meiosis I and II) but only one round of DNA replication. As a result, it gives rise to four haploid cells. In males, the two meiotic cell divisions are each symmetrical, producing four functionally equivalent spermatozoa. Female meiosis is different because at each meiosis asymmetric cell division results in an unequal division of the cytoplasm. The products of female meiosis I (the first meiotic division) are a large secondary oocyte and a small cell (**polar body**) that is discarded. During meiosis II the secondary oocyte then gives rise to the large mature egg cell and a second polar body, which again is discarded (**Figure 2.4**).

In humans, primary oocytes enter meiosis I during fetal development but are then all arrested at prophase until after the onset of puberty. After puberty in females, one primary oocyte completes meiosis with each menstrual cycle. Because ovulation can continue up to the fifth and sometimes sixth decades of life, this means that meiosis can be arrested for many decades in primary oocytes that are used in ovulation in later life. While arrested in prophase, the primary oocytes continue to grow to a large size, acquiring an outer jelly coat, cortical granules, and reserves of ribosomes, mRNA, yolk, and other cytoplasmic resources that would sustain an early embryo. In males, huge numbers of sperm are produced continuously from puberty onward.

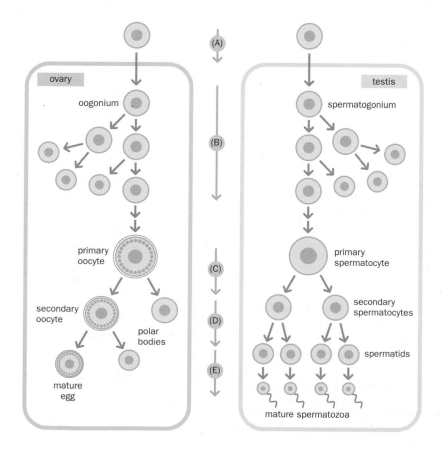

Figure 2.4 Male and female germ line development and gametogenesis.
(A) Diploid primordial germ cells migrate to the embryonic gonad [the female ovary (left) or the male testis (right)], and enter rounds of mitosis that establish spermatogonia (in males) and oogonia (in females). (B) These undergo further mitotic divisions, growth, and differentiation to produce diploid primary spermatocytes and diploid primary oocytes, which can enter meiosis. (C) Meiosis I. After DNA duplications, the cells become tetraploid but then divide to produce two diploid cells. In male gametogenesis, the cell division is symmetrical, generating identical diploid secondary spermatocytes. In female meiosis I, by contrast, the division is asymmetric; the secondary oocyte is much larger than the first polar body, which is discarded. (D) Meiosis II. The diploid secondary spermatocyte and secondary oocyte divide without prior DNA synthesis to give haploid cell products. In male gametogenesis, this division is again symmetrical, producing two haploid spermatids from each secondary spermatocyte. In female meiosis II, the egg produced is much larger than the second (also discarded) polar body. (E) Maturation of spermatids produces four spermatozoa.

diploid primary spermatocytes

| 1 | 2 | 3 | 4 | 5 | 6 | 7 | 8 | 9 | 10 | 11 | 12 | 13 | 14 | 15 | 16 | 17 | 18 | 19 | 20 | 21 | 22 | X | maternal |
| 1 | 2 | 3 | 4 | 5 | 6 | 7 | 8 | 9 | 10 | 11 | 12 | 13 | 14 | 15 | 16 | 17 | 18 | 19 | 20 | 21 | 22 | Y | paternal |

↓ meiosis

haploid sperm cells

1	2	3	4	5	6	7	8	9	10	11	12	13	14	15	16	17	18	19	20	21	22	Y	sperm 1
1	2	3	4	5	6	7	8	9	10	11	12	13	14	15	16	17	18	19	20	21	22	X	sperm 2
1	2	3	4	5	6	7	8	9	10	11	12	13	14	15	16	17	18	19	20	21	22	Y	sperm 3
1	2	3	4	5	6	7	8	9	10	11	12	13	14	15	16	17	18	19	20	21	22	X	sperm 4
1	2	3	4	5	6	7	8	9	10	11	12	13	14	15	16	17	18	19	20	21	22	X	sperm 5

Figure 2.5 Independent assortment of maternal and paternal homologs during meiosis. The figure shows a random selection of just 5 of the 8,388,608 (2^{23}) theoretically possible combinations of homologs that might occur in haploid human spermatozoa after meiosis in a diploid primary spermatocyte. Maternally derived homologs are represented by pink boxes, and paternally derived homologs by blue boxes. For simplicity, the diagram ignores recombination.

The second division of meiosis is identical to mitosis, but the first division has important differences. Its purpose is to generate *genetic diversity* between the daughter cells. This is done by two mechanisms: independent assortment of paternal and maternal homologs, and recombination.

Independent assortment

Every diploid cell contains two chromosome sets and therefore has two copies (**homologs**) of each chromosome (except in the special case of the X and Y chromosomes in males). One homolog is paternally inherited and the other is maternally inherited. During meiosis I the maternal and paternal homologs of each pair of replicated chromosomes undergo **synapsis** by pairing together to form a **bivalent**. After DNA replication, the homologous chromosomes each comprise two sister chromatids, so each bivalent is a four-stranded structure at the metaphase plate. Spindle fibers then pull one complete chromosome (two chromatids) to either pole. In humans, for each of the 23 homologous pairs, the choice of which daughter cell each homolog enters is independent. This allows 2^{23} or about 8.4×10^6 different possible combinations of parental chromosomes in the gametes that might arise from a single meiotic division (**Figure 2.5**).

Recombination

The five stages of prophase of meiosis I (**Figure 2.6**) begin during fetal life and, in human females, can last for decades. During this extended process, the homologs within each bivalent normally exchange segments of DNA at randomly positioned but matching locations. At the zygotene stage (Figure 2.6B), a proteinaceous **synaptonemal complex** forms between closely apposed homologous chromosomes. Completion of the synaptonemal complex marks the start of the pachytene stage (Figure 2.6C), during which recombination (**crossover**) occurs. Crossover involves physical breakage of the DNA in one paternal and one maternal chromatid, and the subsequent joining of maternal and paternal fragments.

The mechanism that allows alignment of the homologs is not known (see Figure 2.6A, B), although such close apposition is required for recombination. Located at intervals on the synaptonemal complex are very large multiprotein assemblies, called recombination nodules, that may mediate recombination

Figure 2.6 The five stages of prophase in meiosis I. (A) In leptotene, the duplicated homologous chromosomes begin to condense but remain unpaired. (B) In zygotene, duplicated maternal and paternal homologs pair to form bivalents, comprising four chromatids. (C) In pachytene, recombination (crossing over) occurs by means of the physical breakage and subsequent rejoining of maternal and paternal chromosome fragments. There are two crossovers in the bivalent on the left and one in the bivalent on the right. For simplicity, both crossovers on the left involve the same two chromatids. In reality, more crossovers may occur, involving three or even all four chromatids in a bivalent. (D) During diplotene, the homologous chromosomes may separate slightly, except at the chiasmata. (E) Diakinesis is marked by contraction of the bivalents and is the transition to metaphase I. In this figure, only 2 of 23 possible pairs of homologs are illustrated (with the maternal homolog colored light blue, and the paternal homolog dark blue).

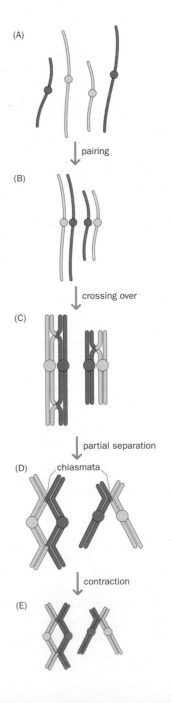

(A)

↓ pairing

(B)

↓ crossing over

(C)

↓ partial separation

(D)   chiasmata

↓ contraction

(E)

Figure 2.7 Metaphase I to production of gametes. (A) At metaphase I, the bivalents align on the metaphase plate, at the center of the spindle apparatus. Contraction of spindle fibers draws the chromosomes in the direction of the spindle poles (arrows). (B) The transition to anaphase I occurs at the consequent rupture of the chiasmata. (C) Cytokinesis segregates the two chromosome sets, each to a different primary spermatocyte. Note that, as shown in this panel, after recombination during prophase I the chromatids share a single centromere but are no longer identical. (D) Meiosis II in each primary spermatocyte, which does not include DNA replication, generates unique genetic combinations in the haploid secondary spermatocytes. Only 2 of the possible 23 different human chromosomes are depicted, for clarity, so only 2^2 (i.e. 4) of the possible 2^{23} (8,388,608) possible combinations are illustrated. Although oogenesis can produce only one functional haploid gamete per meiotic division (see Figure 2.4), the processes by which genetic diversity arises are the same as in spermatogenesis.

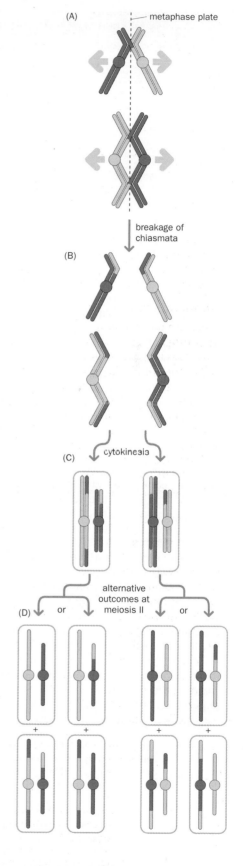

events. Recombined homologs seem to be physically connected at specific points. Each such connection marks the point of crossover and is known as a **chiasma** (plural chiasmata). There are an average of 55 chiasmata per cell in human male meiosis, and maybe 50% more in female meiosis.

In addition to their role in recombination, chiasmata are thought to be essential for correct chromosome segregation during meiosis I. By holding together maternal and paternal homologs of each chromosome pair on the spindle until anaphase I, they have a role analogous to that of the centromeres in mitosis and in meiosis II. Children with incorrect numbers of chromosomes have been shown genetically to be often the product of gametes in which a bivalent lacked chiasmata.

Meiosis II resembles mitosis, except that there are only 23 chromosomes instead of 46. Each chromosome already consists of two chromatids that become separated at anaphase II. However, whereas the sister chromatids of a mitotic chromosome are genetically identical, the two chromatids of a chromosome entering meiosis II (**Figure 2.7**) are usually genetically different from each other, as a result of recombination events that took place during meiosis I.

Together, the effects of recombination between homologs (during prophase I) as well as independent assortment of homologs (during anaphase I) ensure that a single individual can produce an almost unlimited number of genetically distinct gametes. The genetic consequences of recombination are considered more fully in a later chapter.

X–Y pairing

During meiosis I in a human primary oocyte, each chromosome has a fully homologous partner, and the two X chromosomes synapse and engage in crossover just like any other pair of homologs. In male meiosis there is a problem. The human X and Y sex chromosomes are very different from one another. Not only is the X very much larger than the Y but it has a rather different DNA content and very many more genes than the Y. Nevertheless, the X and Y do pair during prophase I, thus ensuring that at anaphase I each daughter cell receives one sex chromosome, either an X or a Y.

Human X and Y chromosomes pair end-to-end rather than along the whole length, thanks to short regions of homology between the X and Y chromosomes at the very ends of the two chromosomes. Pairing is sustained by an obligatory crossover in a 2.6 Mb homology region at the tips of the short arms, but crossover also sometimes occurs in a second homology region, 0.32 Mb long, at the tips of the long arms. Genes in the terminal X–Y homology regions have some interesting properties:

- They are present as homologous copies on the X and Y chromosomes.

- They are mostly not subject to the transcriptional inactivation that affects most X-linked genes as a result of the normal decondensation of one of the two X chromosomes in female mammalian somatic cells (**X-inactivation**).

- They display inheritance patterns like those of genes on autosomal chromosomes, rather than X-linked or Y-linked genes.

As a result of their autosomal-like inheritance, the terminal X–Y homology regions are known as **pseudoautosomal regions**. We will describe them in more detail in a later chapter when we consider how sex chromosomes evolved in mammals.

TABLE 2.1 COMPARISON OF MITOSIS AND MEIOSIS

Characteristic	Mitosis	Meiosis
Location	all tissues	specialized germ line cells in testis and ovary
Products	diploid somatic cells	haploid sperm and eggs
DNA replication and cell division	normally one round of replication per cell division	only one round of replication per two cell divisions
Duration of prophase	short (~30 min in human cells)	can take decades to complete
Pairing of maternal and paternal homologs	no	yes, during meiosis I
Recombination	rare and abnormal	during each meiosis; normally occurs at least once in each chromosome arm after pairing of maternal and paternal homologs
Relationship between daughter cells	genetically identical	genetically different as a result of independent assortment of homologs and recombination

Mitosis and meiosis have key similarities and differences

Mitosis involves a single turn of the cell cycle. After the DNA is replicated during S phase, the two sister chromatids of each chromosome are divided equally between the daughter cells during M phase. Meiotic cell division also involves one round of DNA synthesis, but this is followed by two cell divisions *without* an intervening second round of DNA synthesis, allowing diploid cells to generate haploid products. Although the second cell division of meiosis is identical to that of mitosis, the first meiotic division has distinct features that enable genetic diversity to arise. This relies on two mechanisms: independent assortment of paternal and maternal homologs, and recombination (**Table 2.1**).

2.3   STRUCTURE AND FUNCTION OF CHROMOSOMES

Chromosomes have two fundamental roles: faithful transmission and appropriate expression of genetic information. The processes of cell division are fascinating, and changes to the arrangement of chromosomes can have profound medical consequences. Knowledge of the detailed structure of chromosomes is crucial to understanding these vital processes.

Chromosome structure as generally illustrated in textbooks represents only the state that occurs during metaphase, while cells are preparing to undergo the last stages of cell division. At this time the chromatids are still connected to each other at their centromeres and they are so condensed that they can be seen with a light microscope. But metaphase chromosomes are so tightly packed that their genes cannot be expressed. Chromosomes have a quite different structure during most of the cell cycle. Throughout interphase, most chromosome regions are comparatively very highly extended, allowing genes to be expressed.

For a chromosome to be copied and transmitted accurately to daughter cells, it requires just three types of structural element, each of which is discussed in this section of the chapter:

- A centromere, which is most evident at metaphase, where it is the narrowest part of the chromosome and the region at which spindle fibers attach.
- Replication origins, certain DNA sequences along each chromosome at which DNA replication can be initiated.
- Telomeres, the ends of linear chromosomes that have a specialized structure to prevent internal DNA from being degraded by nucleases.

Artificial chromosomes that include large introduced DNA fragments function normally in both yeast and mammalian cells if, and only if, they contain all three of these elements.

Chromosomal DNA is coiled hierarchically

In the eukaryotic cell the structure of each chromosome is highly ordered and during mitosis each chromosome undergoes several levels of compaction. To achieve this, DNA is complexed with various proteins and subject to coiling and supercoiling to form **chromatin**. Even in the interphase nucleus, when the DNA is in a very highly extended form, the 2 nm thick DNA double helix undergoes at least two levels of coiling that are directed by binding of basic histone proteins. First, a 10 nm thick filament is formed that is then coiled into a 30 nm thick chromatin fiber. The chromatin fiber undergoes looping and is supported by a scaffold of nonhistone proteins (**Figure 2.8A**).

The **nucleosome** is the fundamental unit of DNA packaging: a stretch of 147 bp of double-stranded DNA is coiled in just less than two turns around a central core of eight histone proteins (two molecules each of the core histones H2A, H2B, H3, and H4, all highly conserved proteins) (Figure 2.8B). Adjacent nucleosomes are connected by a short length (8–114 bp) of linker DNA; the length of the linker DNA varies both between species and between regions of the genome. Electron micrographs of suitable preparations show nucleosomes to have a *string of beads* appearance (Figure 2.8C). This first level of DNA packaging is the only one that still allows transcriptional activity.

The N-terminal tails of the core histones protrude from the nucleosomes (see Figure 2.8B). Specific amino acids in the histone tails can undergo various types of post-translational modification, notably acetylation, phosphorylation, and methylation. As a result, different proteins can be bound to the chromatin in a way that affects chromatin condensation and the local level of transcriptional activity. Additional histone genes encode variant forms of the core histones that

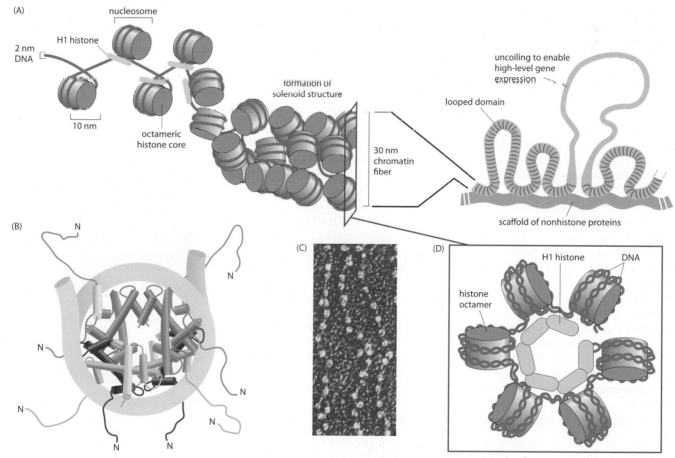

Figure 2.8 From DNA double helix to interphase chromatin. (A) Binding of basic histone proteins. The 2 nm thick DNA double helix binds basic histones to undergo coiling, forming a 10 nm thick nucleosome filament that is further coiled into the 30 nm chromatin fiber. In interphase the chromatin fiber is organized into looped domains with ~50–200 kb of DNA attached to a scaffold of nonhistone acidic proteins. High levels of gene expression require local uncoiling of the chromatin fiber to give the 10 nm nucleosomal filaments. (B) A nucleosome consists of almost two turns of DNA wrapped round an octamer of core histones (two each of H2, H3A, H3B, and H4). Note the extensive α-helical structure of histones and their protruding N-terminal tails. (C) Electron micrograph of 10 nm nucleosomal filaments. (D) Cross section view of the 30 nm chromatin fiber showing one turn of the solenoid [the outer red box corresponds to that shown in (A)]. The additional H1 histone, which binds to linker DNA, is important in organizing the structure of the 30 nm chromatin fiber. [(A) adapted from Grunstein M (1992) *Sci. Am.* 267, 68. With permission from Scientific American Inc., and Alberts B, Johnson A, Lewis J et al. (2008) Molecular Biology of the Cell, 5th ed. Garland Science/Taylor & Francis LLC. (B) adapted from Alberts B, Johnson A, Lewis J et al. (2008) Molecular Biology of the Cell, 5th ed. Garland Science/Taylor & Francis LLC from figures by Jakob Waterborg, University of Missouri—Kansas City. (C) courtesy of Jakob Waterborg, University of Missouri—Kansas City. (D) adapted from Klug A (1985) Proceedings, RW Welch Federation Conference, *Chem. Res.* 39, 133. With permission from The Welch Foundation.]

may be associated with particular chromosomal regions and specialized functions. For example, centromeres have a histone H3 variant known as CENP-A, and the H2AX variant of histone H2 is associated with DNA repair and recombination.

In addition to the four core histones and their variants, a fifth histone, H1, binds to the linker DNA and is thought to be important for chromosome condensation and, in particular, the formation of the 30 nm chromatin fiber. In this structure, nucleosomes are packed into a spiral or *solenoid* arrangement with six to eight nucleosomes per turn, and the H1 proteins are bound to the DNA on the inside of the solenoid with one H1 molecule associated with each nucleosome (Figure 2.8D).

Packaging DNA into first nucleosomes and then the solenoids of the 30 nm chromatin fiber results in a linear condensation of about fiftyfold. During M phase, the DNA in a human metaphase chromosome is compacted even further to about 1/10,000 of its stretched-out length. The scaffold of metaphase chromosomes contains high levels of topoisomerase II and notably of condensins, protein complexes that are structurally and evolutionarily related to cohesins. Condensins organize tight packaging of the chromatin, but the way in which the chromatin is so greatly compacted is not well understood.

Interphase chromatin varies in its degree of compaction

During interphase, most chromatin exists in an extended state that is diffusely staining and dispersed through the nucleus (**euchromatin**). Euchromatin is marked by relatively weak binding by histone H1 molecules and by extensive acetylation of the four types of nucleosomal histone. Euchromatin is not uniform. Some euchromatin regions are more condensed than others, and genes may or may not be expressed, depending on the cell type and its functional requirements.

Some chromatin, however, remains highly condensed throughout the cell cycle and forms dark-staining regions (**heterochromatin**). Genes that are naturally located within heterochromatin, or that become incorporated into heterochromatin through chromosomal rearrangement, are typically not expressed. Heterochromatin is associated with tight histone H1 binding, and two classes have been defined:

- Constitutive heterochromatin is condensed and generally genetically inactive, consisting largely of repetitive DNA. It is generally found in and around the centromeres and at telomeres, and also accounts for much of the Y chromosome in mammals. In human chromosomes, it is also notably found in the short arms of the acrocentric chromosomes and at secondary constrictions on the long arms of chromosomes 1, 9, and 16.

- Facultative heterochromatin is sometimes genetically inactive (condensed) and sometimes active (decondensed). In each somatic cell of female mammals, for example, one of the two X chromosomes is randomly inactivated and condensed (**X-inactivation**). In addition, both the X and the Y chromosomes become reversibly condensed for about 15 days during meiosis in spermatogenesis, forming the XY body which is segregated into a special nuclear compartment.

Each chromosome has its own territory in the interphase nucleus

The nucleus is highly organized, with many subnuclear compartments in addition to the nucleolus, where rRNA is transcribed and ribosomal subunits are assembled. The positioning of the chromosomes within the nucleus is also highly organized, as revealed by specialized techniques that analyze the movements of individual chromosomes during interphase within living cells.

The centromeres of different interphase chromosomes in human cells are less clearly aligned than in other organisms. They tend to cluster together at the periphery of the nucleus during G_1 phase before becoming much more dispersed during S phase. Although the chromosomes are all in a highly extended form, they are not extensively entwined. Instead, they seem to occupy relatively small non-overlapping territories (**Figure 2.9**).

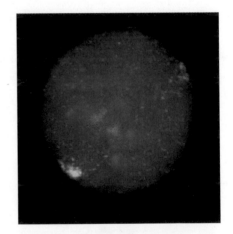

Figure 2.9 Individual chromosomes occupy distinct chromosome territories in the interphase nucleus. The nucleus of this human cell at interphase appears blue as a result of staining with DAPI, a fluorescent DNA-binding dye. DNA probes specific for human chromosome 18 or chromosome 19 were labeled with, respectively, green or red fluorescent dyes. Within the nucleus both copies of chromosome 18 (green signal) are seen to be located at the periphery of the nucleus, but the chromosome 19 copies (red signal) are shown to be within the interior of the nucleus. (Courtesy of Wendy Bickmore, MRC Human Genetics Unit, Edinburgh.)

Although interphase chromosomes do not seem to have favorite nuclear locations, chromosome positioning is nevertheless non-random. The human chromosomes that have the highest gene density tend to concentrate at the center of the nucleus; gene-poor chromosomes are located toward the nuclear envelope. Chromosome movements are probably restrained by telomere interaction with the nuclear envelope and also by internal nuclear structures (including the nucleolus in the case of chromosomes containing ribosomal RNA genes).

Centromeres have a pivotal role in chromosome movement but have evolved to be very different in different organisms

Chromosomes normally have a single **centromere**, the region where duplicated sister chromatids remain joined until anaphase. In metaphase chromosomes, the centromere is readily apparent as the primary constriction that separates the short and long arms. The centromere is essential for attaching chromosomes to the mitotic spindle and for chromosome segregation during cell division. Abnormal chromosome fragments that lack a centromere (**acentric** fragments) cannot attach to the spindle and fail to be correctly segregated to the nuclei of either daughter cell.

Late in prophase large multiprotein complexes, known as **kinetochores**, form at each centromere, one attached to each sister chromatid. Microtubules attach to each kinetochore, linking the centromeres to the spindle poles (**Box 2.1**). At

BOX 2.1 COMPONENTS OF THE MITOTIC SPINDLE

The mitotic spindle is formed from **microtubules** (polymers of a heterodimer of α-tubulin and β-tubulin) and microtubule-associated proteins. At each of the two spindle poles in a dividing cell is a **centrosome** that seeds the outward growth of microtubule fibers and is the major microtubule-organizing center of the cell. Because their constituent tubulins are synthesized in a particular direction, the microtubule fibers are polar, with a minus (–) end (the one next to the centromere) and a plus (+) end (the distal growing end).

Each centrosome is composed of a fibrous matrix containing a pair of **centrioles**, which are short cylindrical structures, composed of microtubules and associated proteins; the two centrioles are arranged at right angles (**Figure 1A**). During G_1, the two centrioles in a pair separate, and during S phase, a daughter centriole begins to grow at the base of each mother centriole until it is fully formed during G_2. The two centriole pairs remain close together in a single centrosomal

complex until the beginning of M phase. At that point the centrosome complex splits in two and the two halves begin to separate. Each daughter centrosome develops its own array of microtubules and begins to migrate to one end of the cell, where it will form a spindle pole (Figure 1C).

Three different forms of microtubule fiber occur in the fully formed mitotic spindle:

- Polar fibers, which develop at prophase, extend from the two poles of the spindle toward the equator.
- Kinetochore fibers, which develop at prometaphase, connect the large multiprotein structure at the centromere of each chromatid (the kinetochore; Figure 1B) and the spindle poles.
- Astral fibers form around each centrosome and extend to the periphery of the cell.

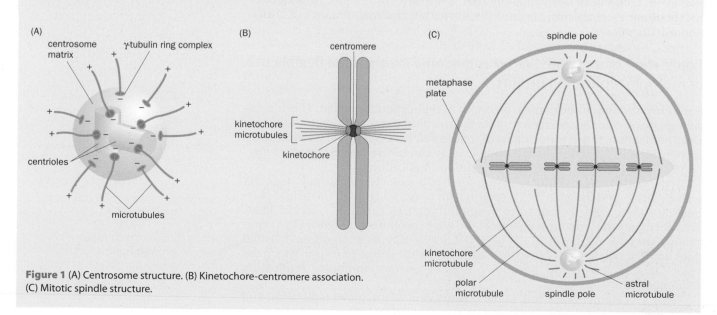

Figure 1 (A) Centrosome structure. (B) Kinetochore-centromere association. (C) Mitotic spindle structure.

anaphase, the kinetochore microtubules pull the previously paired sister chromatids toward opposite poles of the spindle. Kinetochores at the centromeres control assembly and disassembly of the attached microtubules, which drives chromosome movement.

In the budding yeast *Saccharomyces cerevisiae*, the sequences that specify centromere function are very short, as are other functional chromosomal elements (**Figure 2.10**). The centromere element (CEN) is about 120 bp long and contains three principal sequence elements, of which the central one, CDE II, is particularly important for attaching microtubules to the kinetochore. A centromeric CEN fragment derived from one *S. cerevisiae* chromosome can replace the centromere of another yeast chromosome with no apparent consequence.

S. cerevisiae centromeres are highly unusual because they are very small and the DNA sequences specify the sites of centromere assembly. They do not have counterparts in the centromeres of the fission yeast *Schizosaccharomyces pombe* or in those of multicellular animals. Centromere size has increased during eukaryote evolution and, in complex organisms, centromeric DNA is dominated by repeated sequences that evolve rapidly and are species-specific. The relatively rapid evolution of centromeric DNA and associated proteins might contribute to the reproductive isolation of emerging species.

Although centromeric DNA shows remarkable sequence heterogeneity across eukaryotes, centromeres are universally marked by the presence of a centromere-specific variant of histone H3, generically known as CenH3 (the human form of CenH3 is named CENP-A). At centromeres, CenH3/CENP-A replaces the normal histone H3 and is essential for attachment to spindle microtubules. Depending on centromere organization, different numbers of spindle microtubules can be attached (**Figure 2.11**).

Mammalian centromeres are particularly complex. They often extend over several megabases and contain some chromosome-specific as well as repetitive DNA. A major component of human centromeric DNA is α-satellite DNA, whose structure is based on tandem repeats of a 171 bp monomer. Adjacent repeat units can show minor variations in sequence, and occasional tandem amplification of a sequence of several slightly different neighboring repeats results in a higher-order repeat organization. This type of α-satellite DNA is characteristic of centromeres and is marked by 17 bp recognition sites for the centromere-binding protein CENP-B.

Unlike the very small discrete centromeres of *S. cerevisiae*, the much larger centromeres of other eukaryotes are not dependent on sequence organization alone. Neither specific DNA sequences (e.g. α-satellite DNA) nor the DNA-binding protein CENP-B are essential or sufficient to dictate the assembly of a functional mammalian centromere. Poorly understood DNA characteristics specify a chromatin conformation that somehow, by means of sequence-independent mechanisms, controls the formation and maintenance of a functional centromere.

Replication of a mammalian chromosome involves the flexible use of multiple replication origins

For a chromosome to be copied, an origin of replication is needed. This is a *cis*-acting DNA sequence to which protein factors bind in preparation for initiating DNA replication. Eukaryotic origins of replication have been studied most comprehensively in yeast, in which a genetic assay can be used to test whether fragments of yeast DNA can promote autonomous replication.

In the yeast assay, test fragments are stitched into bacterial plasmids, together with a particular yeast gene that is essential for yeast cell growth. The hybrid plasmids are then used to transform a mutant yeast lacking this essential gene. Transformants that can form colonies (and have therefore undergone DNA replication) are selected. Because the bacterial origin of replication on the plasmid does not function in yeast cells, the identified colonies must be cells in which the yeast DNA in the hybrid plasmid possesses an **autonomously replicating sequence (ARS)** element.

Yeast ARS elements are functionally equivalent to origins of replication and are thought to derive from authentic replication origins. They are only about

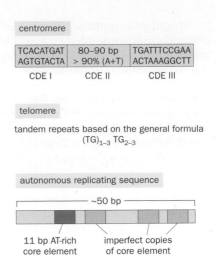

centromere

TCACATGAT	80–90 bp	TGATTTCCGAA
AGTGTACTA	> 90% (A+T)	ACTAAAGGCTT
CDE I	CDE II	CDE III

telomere

tandem repeats based on the general formula $(TG)_{1-3} \, TG_{2-3}$

autonomous replicating sequence

~50 bp

11 bp AT-rich core element   imperfect copies of core element

Figure 2.10 In *S. cerevisiae*, chromosome function is uniquely dependent on short defined DNA sequence elements. *S. cerevisiae* centromeres are very short (often 100–110 bp) and, unusually for eukaryotic centromeres, are composed of defined sequence elements. There are three contiguous centromere DNA elements (CDEs), of which CDE II and CDE III are the most important functionally. Telomeres are composed of tandem TG-rich repeats. Autonomous replicating sequences are defined by short AT-rich sequences. The three types of short sequence can be combined with foreign DNA to make an artificial chromosome in yeast cells.

50 bp long and consist of an AT-rich region with a conserved 11 bp sequence plus some imperfect copies of this sequence (see Figure 2.10). An ARS encodes binding sites for both a transcription factor and a multiprotein origin of replication complex (ORC).

In mammalian cells, the absence of a genetic assay has made it more difficult to define origins of DNA replication, but DNA is replicated at multiple initiation sites along each chromosome. Reported human replication origins are often several kilobases long, and their ORC-binding sites seem to be less specific than those in yeast. Unlike in yeast, mammalian artificial chromosomes seem not to require specific ARS sequences. Instead, the speed at which the replication fork moves in mammalian cells seems to determine the spacing between the regions to which chromatin loops are anchored, and this spacing in turn controls the choice of sites at which DNA replication is initiated.

Telomeres have specialized structures to preserve the ends of linear chromosomes

Telomeres are specialized heterochromatic DNA–protein complexes at the ends of linear eukaryotic chromosomes. As in centromeres, the nucleosomes around which telomeric DNA is coiled contain modified histones that promote the formation of constitutive heterochromatin.

Telomere structure, function, and evolution

Telomeric DNA sequences are almost always composed of moderately long arrays of short tandem repeats that, unlike centromeric DNA, have generally been well conserved during evolution. In all vertebrates that have been examined, the repeating sequence is the hexanucleotide TTAGGG (**Table 2.2**). The repeats are G-rich on one of the DNA strands (the G-strand) and C-rich on the complementary strand. On the centromeric side of the human telomeric TTAGGG repeats are a further 100–300 kb of telomere-associated repeat sequences (**Figure 2.12A**). These have not been conserved during evolution, and their function is not yet understood.

The (TTAGGG)$_n$ array of a human telomere often spans about 10–15 kb (see Figure 2.12A). A very large protein complex (called shelterin, or the **telosome**) contains several components that recognize and bind to telomeric DNA. Of these components, two telomere repeat binding factors (TRF1 and TRF2) bind to double-stranded TTAGGG sequences.

As a result of natural difficulty in replicating the lagging DNA strand at the extreme end of a telomere (discussed in the next section), the G-rich strand has a single-stranded overhang at its 3′ end that is typically 150–200 nucleotides long (see Figure 2.12A). This can fold back and form base pairs with the other, C-rich, strand to form a telomeric loop known as the T-loop (Figure 2.12B, C).

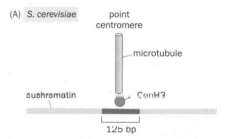

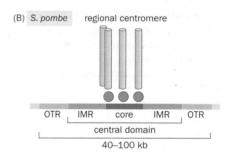

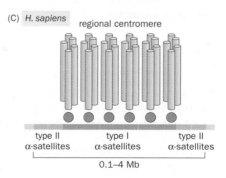

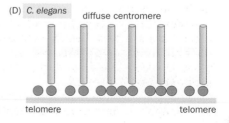

Figure 2.11 Differences in eukaryotic centromere organization. In all eukaryotic centromeres a centromere-specific variant of histone H3 (generically called CenH3) is implicated in microtubule binding. However, the extent to which the centromere extends across the DNA of a chromosome and the number of microtubules involved can vary enormously. (A) The budding yeast *S. cerevisiae* has the simplest form of centromere organization, a point centromere, with just 125 bp of DNA wrapped around a single nucleosome; each kinetochore makes only one stable microtubule attachment during metaphase. (B) In the fission yeast *S. pombe*, centromeres have multiple microtubule attachments clustered in a region occupying 35–110 kb of DNA (the average chromosome has more than 4 Mb of DNA). The microtubule attachment sites are clustered in a non-repetitive core sequence which is flanked by different types of repeat sequence (IMR, innermost; OTR, outermost). (C) Human centromeres have multiple microtubule attachment sites and are clustered over regions of up to 4 Mb of DNA. Higher-order α-satellite DNA repeats are prominent at the centromeres. In addition to binding to nucleosomes containing a CenH3 protein, CENP-A, they also have binding sites for the CENP-B protein. (D) In some eukaryote species, such as the nematode *Caenorhabditis elegans*, the centromere is very diffuse: kinetochores that bind spindle microtubules are distributed across the whole length of the chromosome, and the chromosomes are said to be *holocentric*. [Adapted from Vagnarelli P, Ribeiro SA & Earnshaw WC (2008) *FEBS Lett.* 582, 1950–1959. With permission from Elsevier.]

TABLE 2.2 EVOLUTIONARY CONSERVATION OF TELOMERIC REPEAT SEQUENCES

Occurrence	Consensus telomere repeat sequence[a]
Saccharomyces cerevisiae	TG_{1-3}
Schizosaccharomyces pombe	$TTACAG_{1-8}$
Neurospora crassa	TTAGGG
Paramecium	TTGGGG
Trypanosoma	TAGGGG
Chlamydomonas	TTTTAGGG
Arabidopsis	TTTAGGG
Nematodes	TTAGGC
Vertebrates	TTAGGG

[a]In the direction toward the end of the chromosome. Note: although telomere function is conserved throughout eukaryotes and telomeric repeat sequences are generally strongly conserved, the telomeres of arthropods, such as *Drosophila*, are radically different in structure, being composed of long DNA repeats that are unrelated to the TG-rich oligonucleotide repeats found in other eukaryotes.

The T-loop probably represents a conserved mechanism for protecting chromosome ends. If a telomere is lost after chromosome breakage, the resulting chromosome end is unstable; it tends to fuse with the ends of other broken chromosomes, or to be involved in recombination events, or to be degraded. Telomere-binding proteins, notably the telosome component POT1, bind to single-stranded TTAGGG repeats and can protect the terminal DNA *in vitro* and perhaps also *in vivo*.

Telomerase and the chromosome end-replication problem

During DNA synthesis the DNA polymerase extends the growing DNA chains in the 5′→3′ direction. One of the new DNA strands, the leading strand, grows in the 5′→3′ direction of DNA synthesis, but the other strand, the lagging strand, is synthesized in pieces (Okazaki fragments) because it must grow in a direction opposite to that of the 5′→3′ direction of DNA synthesis. A succession of 'backstitching' syntheses are required to produce a series of DNA fragments whose ends are then sealed by DNA ligase (see Figure 1.11).

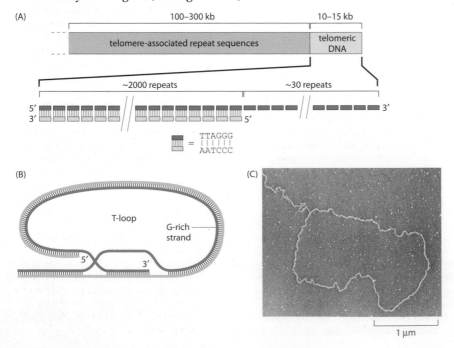

Figure 2.12 At telomeres, highly conserved oligonucleotide repeats are bound by specialized proteins to form a protective loop. (A) Telomere structure. The DNA at the very ends of human chromosomes is defined by a tandem array of roughly 1700–2500 copies of the hexanucleotide TTAGGG (which is conserved in vertebrates; see Table 2.2). The G-rich strand, however, protrudes at the terminus to form a single-stranded region composed of about 30 TTAGGG repeats. The array of distinctive conserved short repeats is bound by the shelterin (or **telosome**) complex (not shown, for simplicity; two of its subunits, the telomere repeat binding factors TRF1 and TRF2, bind directly to double-stranded regions, whereas POT1 can bind to the single-stranded repeats). Like centromeric DNA, telomeric DNA has modified histones that act as signals for forming constitutive heterochromatin. (B) T-loop formation. The single-stranded terminus of the G-rich strand can loop back and invade the double-stranded region by base-pairing with the complementary C-rich strand sequence. The resulting T-loop is thought to protect the telomere DNA from natural cellular mechanisms that repair double-stranded DNA breaks. (C) Electron micrograph showing the formation of a roughly 15 kb T-loop at the end of an interphase human chromosome (after fixing, deproteination, and artifical thickening to assist viewing). [From Griffith JD, Comeau L, Rosenfield S et al. (1999) *Cell* 97, 503–514. With permission from Elsevier.]

Unlike RNA polymerases, DNA polymerases *absolutely* require a free 3′ OH group from a double-stranded nucleic acid from which to extend synthesis. This is achieved by employing an RNA polymerase to synthesize a complementary RNA primer that primes the synthesis of each of the DNA fragments used to make the lagging strand. In these cases the RNA primer requires the presence of some DNA *ahead of* the sequence to be copied, to serve as its template. However, at the extreme end of a linear DNA molecule, there can never be such a template, and a different mechanism is required to solve the problem of completing replication at the ends of a linear DNA molecule.

A solution to the ***end-replication problem*** is provided by a specialized reverse transcriptase (RNA-dependent DNA polymerase) that completes leading-strand synthesis. Telomerase is a ribonucleoprotein enzyme whose polymerase function is critically dependent on an RNA subunit, TERC (telomerase RNA component), and a protein subunit, TERT (telomerase reverse transcriptase). At the 5′ end of vertebrate TERC RNA is a hexanucleotide sequence that is complementary to the telomere repeat sequence (**Figure 2.13**). It will act as a template to prime the extended DNA synthesis of telomeric DNA sequences on the leading strand. Further extension of the leading strand provides the necessary template for DNA polymerase to complete the synthesis of the lagging strand.

In humans, telomere length is known to be highly variable, and telomerase activity is largely absent from adult cells except for certain cells in highly proliferative tissues such as the germ line, blood, skin, and intestine. In cells that lack telomerase, the extreme ends of telomeric DNA do not get replicated at S phase and their telomeres shorten progressively. Telomere shortening is effectively a way of counting cell divisions and has been related to cell senescence and aging. Cancer cells find ways of activating telomerase, leading to uncontrolled replication.

2.4   STUDYING HUMAN CHROMOSOMES

Human chromosomes have been analyzed for research and diagnostic purposes for many decades. A succession of technological advances has permitted chromosome analyses with ever-increasing resolution and structural discrimination.

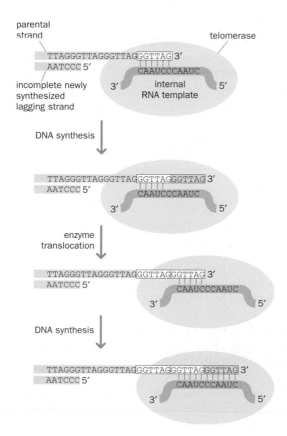

Figure 2.13 Telomere DNA replication. The example shows the situation for human telomeres; as in other vertebrates, telomeric DNA consists of TTAGGG repeats (Figure 2.12). The 3′ end of the parental DNA is extended by DNA synthesis using the RNA component of telomerase as a template. The human telomerase RNA component has an almost perfect tandem repeat (<u>CUAACCCUAAC</u>G) near its 5′ end that contains a complementary template sequence for guiding the synthesis of additional telomeric hexanucleotide sequences at the terminal end. Recently synthesized hexanucleotide repeats are shown in an open box, and the most recently synthesized repeats are shown in orange shading; here, the array is suggested to be extended by tandem copies of a GGTTAG hexanucleotide, but vertebrate telomere repeats can equally be represented by (GGTTAG)$_n$ as by (TTAGGG)$_n$. By extending the leading strand in this way, telomerase provides a template for the lagging strand to be extended (by a standard DNA-dependent DNA polymerase). Eventually, after the telomerase has finished adding hexanucleotide repeats, the lagging strand cannot be extended to the extreme 5′ end, leaving the G-rich strand with an overhanging 3′ end close to 200 nucleotides long (as shown in Figure 2.12).

Chromosome analysis is easier for mitosis than meiosis

Analysis of chromosomes (**cytogenetics**) normally requires the viewing of dividing cells, but obtaining these directly from the human body can be difficult. Bone marrow is a possible source, but it is more convenient to obtain non-dividing cells and then propagate them in cell culture under laboratory conditions. Circulating blood cells and skin fibroblasts are the most commonly used sources of human cells for cytogenetic analyses. People rarely mind giving a small blood sample, and the T lymphocytes in blood can be easily induced to divide by treatment with lectins (such as phytohemagglutinin). Alternatively, fibroblasts are cultured from skin biopsies. In addition, prenatal diagnosis routinely involves chromosome analyses on fetal cells shed into the amniotic fluid or removed from chorionic villi.

Although chromosomes were described accurately in some organisms as early as the 1880s, for many decades all attempts to prepare spreads of human chromosomes produced a tangle that defied analysis. To obtain analyzable chromosome spreads, the cells were grown in liquid suspension and then treated with hypotonic saline to make them swell. This allowed the first good-quality preparations to be made in 1956. White blood cells are put into a rich culture medium laced with phytohemagglutinin and allowed to grow for 48–72 hours, by which time they should be dividing freely. Nevertheless, because M phase occupies only a small part of the cell cycle, few cells will be actually dividing at any one time.

The proportion of cells in mitosis (the mitotic index) can be increased by treating the culture with a spindle-disrupting agent, such as colcemid. Cells enter metaphase but are then unable to progress through the rest of M phase, so they accumulate at the metaphase stage of mitosis.

Chromosomes from a slightly earlier stage of M phase (prometaphase) are less contracted and show more detail, making analysis easier. To obtain substantial cell numbers, the cells are synchronized by temporarily preventing them from progressing through the cell cycle. Often this is achieved by adding excess thymidine (to provoke a decrease in dCTP concentration, causing DNA synthesis to slow down so that cells remain in S phase). When the thymidine effect is removed, the cells progress through the cycle synchronously. After release from arrest in S phase, by trial and error an optimal time can be determined when a good proportion of cells are in the desired prometaphase stage.

Meiosis can be studied only in testicular or ovarian samples. Female meiosis is especially difficult to study, because it is active only in fetal ovaries, whereas male meiosis can be studied in a testicular biopsy from any post-pubertal male who is willing to give one. The results of meiosis can be studied by analyzing chromosomes from sperm, although the methodology for this is cumbersome. Meiotic analysis is used for some investigations of male infertility.

Chromosomes are identified by size and staining pattern

Until the 1970s, human chromosomes were identified on the basis of their size and the position of the centromeres. This allowed chromosomes to be classified into groups (**Table 2.3**) but not unambiguously identified.

Chromosome banding

The introduction of techniques that revealed chromosome banding patterns allowed each chromosome to be identified and permitted more accurate definition of chromosomal abnormalities. The banding techniques require the chromosomes to undergo denaturation or enzyme digestion, followed by exposure to a DNA-specific dye. The procedures produce alternating light and dark bands in mitotic chromosomes (**Figure 2.14** and **Figure 2.15**). Features of some banding techniques follow.

- *G-banding*—the chromosomes are subjected to controlled digestion with trypsin before being stained with Giemsa, a DNA-binding chemical dye. Positively staining dark bands are known as G bands. Pale bands are G negative.

- *Q-banding*—the chromosomes are stained with a fluorescent dye that binds preferentially to AT-rich DNA, notably quinacrine, DAPI (4′, 6-diamidino-2-phenylindole), or Hoechst 33258, and viewed by ultraviolet fluorescence.

TABLE 2.3 HUMAN CHROMOSOME GROUPS

Group	Chromosomes[a]	Description
A	1–3	largest; 1 and 3 are metacentric[b] but 2 is submetacentric[c]
B	4, 5	large; submetacentric with two arms very different in size
C	6–12, X	of medium size; submetacentric
D	13–15	of medium size; acrocentric[d] with satellites[e]
E	16–18	small; 16 is metacentric but 17 and 18 are submetacentric
F	19, 20	small; metacentric
G	21, 22, Y	small; acrocentric, with satellites on 21 and 22 but not on Y

[a]Autosomes are numbered from largest to smallest, except that chromosome 21 is slightly smaller than chromosome 22. [b]A *metacentric* chromosome has its centromere at or near the middle. [c]A *submetacentric* chromosome has its centromere placed so that the two arms are of clearly unequal length. [d]An *acrocentric* chromosome has its centromere at or near one end. [e]A satellite, in this context, is a small segment separated by a non-centromeric constriction from the rest of a chromosome; these occur on the short arms of most acrocentric human chromosomes.

Fluorescing bands are called Q bands and mark the same chromosomal segments as G bands.

- *R-banding*—this is essentially the reverse of the G-banding pattern. The chromosomes are heat-denatured in saline before being stained with Giemsa. The heat treatment denatures AT-rich DNA, and R bands are Q negative. The same pattern can be produced by binding GC-specific dyes such as chromomycin A$_3$, olivomycin, or mithramycin.

- *T-banding*—identifies a subset of the R bands that are especially concentrated close to the telomeres. The T bands are the most intensely staining of the R bands and are revealed by using either a particularly severe heat treatment of the chromosomes before they are stained with Giemsa, or a combination of standard dyes and fluorescent dyes.

- *C-banding*—this is thought to demonstrate constitutive heterochromatin, mainly at the centromeres. The chromosomes are typically exposed to denaturation with a saturated solution of barium hydroxide, before Giemsa staining.

Banding patterns correlate with functional elements of chromosome structure. The DNA of G bands replicates late in S phase and is relatively condensed, whereas the DNA of R bands (which are G negative) generally replicates early in S phase and is less condensed. G-band DNA contains relatively few genes and is less active transcriptionally. Although the AT content of human G-band DNA is only slightly higher than that of R-band DNA, individual G bands consistently have lower GC content than their immediate flanking sequences. This may correlate with how the DNA becomes organized as it condenses.

Banding resolution can be increased by using more elongated chromosomes, at or before prometaphase rather than metaphase. High-resolution procedures for human chromosomes can identify 400, 550, or 850 bands (see Figures 2.14 and 2.15; see the inside back cover of this book for an ideogram for all human chromosomes). Analysis of banding patterns can also be used to investigate chromosomal abnormalities involving missing and mispositioned genetic material.

Reporting of cytogenetic analyses

A minimal cytogenetic analysis report provides a text-only statement that always gives the total number of chromosomes and the sex chromosome constitution (a **karyotype**). The normal karyotypes for human females and males are 46,XX and 46,XY, respectively. When there is a chromosomal abnormality, the karyotype also describes the type of abnormality and the chromosome bands or sub-bands affected (**Box 2.2**).

BOX 2.2 HUMAN CHROMOSOME NOMENCLATURE

The International System for Human Cytogenetic Nomenclature (ISCN) is fixed by the Standing Committee on Human Cytogenetic Nomenclature that issues detailed nomenclature reports, most recently in 2005 (see Further Reading). The basic terminology for banded chromosomes was decided at a meeting in Paris in 1971, and is often referred to as the Paris nomenclature.

Short arm locations are labeled **p** (*petit*) and long arms **q** (*queue*). Each chromosome arm is divided into regions labeled p1, p2, p3, etc., and q1, q2, q3, etc., counting outward from the centromere. Regions are delimited by specific landmarks, which are consistent and distinct morphological features, such as the ends of the chromosome arms, the centromere, and certain bands. Regions are divided into bands labeled p11 (one-one, *not* eleven!), p12, p13, etc., sub-bands labeled p11.1, p11.2, etc., and sub-sub-bands, for example p11.21, p11.22, in each case counting outward from the centromere (see Figure 2.14 and also the inside back cover of this book, where ideograms are shown for all human chromosomes).

Relative distance from the centromere is described by the words **proximal** and **distal**. Thus, proximal Xq means the segment of the long arm of the X that is *closest to the centromere*, and distal 2p means the portion of the short arm of chromosome 2 that is most *distant from the centromere*, and therefore closest to the telomere.

When comparing human chromosomes with those of another species, the convention is to use the first letter of the genus name and the first two letters of the species name, for example:

- HSA18: human chromosome 18 (*Homo sapiens*)
- PTR10: chimp chromosome 10 (*Pan troglodytes*)
- MMU6: mouse chromosome 6 (*Mus musculus*)

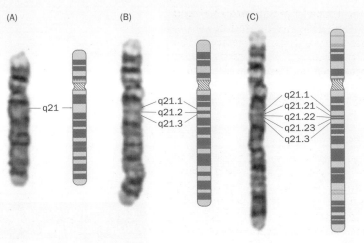

(A)   (B)   (C)

q21

q21.1
q21.2
q21.3

q21.1
q21.21
q21.22
q21.23
q21.3

Figure 2.14 Different chromosome banding resolutions can resolve bands, sub-bands, and sub-sub-bands. G-banding patterns for human chromosome 4 (with accompanying ideogram at the right) are shown at increasing levels of resolution. The levels correspond approximately to (A) 400, (B) 550, and (C) 850 bands per haploid set, allowing the visual subdivision of bands into sub-bands and sub-sub-bands as the resolution increases. [Adapted from Cross & Wolstenholme (2001). Human Cytogenetics: Constitutional Analysis, 3rd ed. (DE Rooney, ed.). With permission of Oxford University Press.]

More informative cytogenetic reports also show the banding structures of individual chromosomes. A **karyogram** is a graphical representation of an individual's full set of mitotic chromosomes displayed as homologous pairs (see Figure 2.15); confusingly, this is also often loosely called a karyotype. Karyograms used to be prepared by cutting up a photograph of metaphase chromosomes that had been spread out on a microscope slide (a *chromosome spread*) and matching up homologous chromosomes; now image analysis software is used instead.

As will be described in the next section, molecular cytogenetic analyses are now also widely used, particularly in analyzing the chromosomes of tumor cells.

Molecular cytogenetics locates specific DNA sequences on chromosomes

Chromosome banding techniques allow analysis of the gross structural organization of chromosomes, with a resolution of several megabases. For higher-resolution analyses, specific DNA sequences within chromosomes need to be detected.

To detect a desired DNA sequence (the **target** DNA sequence), a complementary oligonucleotide or short nucleic acid sequence is first labeled in some way to generate an oligonucleotide or nucleic acid **probe**. The longer the probe, the more specific is its binding. Oligonucleotide probes are often 15–50 nucleotides long and are chemically synthesized. Nucleic acid probes often range in size from several hundred nucleotides to a few kilobases long and are often purified fragments of genomic DNA or DNA copies of RNA transcripts.

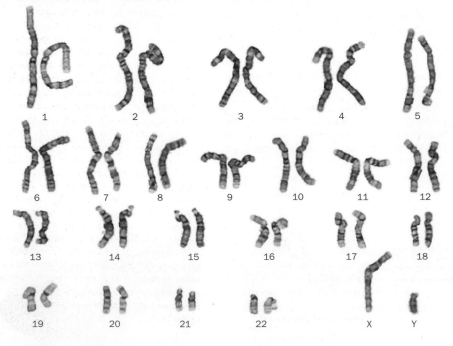

1   2   3   4   5

6   7   8   9   10   11   12

13   14   15   16   17   18

19   20   21   22   X   Y

Figure 2.15 Prometaphase mitotic chromosomes from lymphocytes of a normal human male. The karyogram shows high-resolution G-banding (between 550 and 850 bands per haploid set). Compare with the idealized ideograms in the inside back cover page of this book. [From Cross & Wolstenholme (2001). Human Cytogenetics: Constitutional Analysis, 3rd ed. (DE Rooney, ed.). With permission of Oxford University Press.]

If initially double-stranded, the labeled probe is made single-stranded and then added to a chromosome preparation or to cells that are treated in such a way so as to make the chromosomal DNA single-stranded (**denaturation**). The aim is to allow single-stranded probe sequences to bind specifically by hydrogen bonding to complementary single-stranded target DNA sequences within chromosomes (**molecular hybridization**). During this procedure the chromosomes are immobilized in some way, either within a cell-free chromosome preparation, or within a cell preparation.

Chromosome fluorescence *in situ* hybridization (FISH)

In standard chromosome *in situ* hybridization, a labeled probe is hybridized to denatured chromosomal DNA present in an air-dried microscope slide preparation of metaphase chromosomes. Modern methods use a fluorescence labeling system so that ultimately a **fluorophore** (fluorescent dye) becomes bound to the target DNA sequence, enabling it to be efficiently tracked by fluorescence microscopy. The probe can be labeled directly by incorporating a fluorescently labeled nucleotide precursor. Alternatively, a nucleotide containing a *reporter molecule* (such as biotin or digoxigenin) is incorporated into the DNA after which it can be specifically bound by a fluorescently labeled *affinity molecule* that binds strongly and specifically to the reporter molecule.

In conventional chromosome FISH, homogeneous DNA probes are hybridized to fixed metaphase or prometaphase chromosome spreads on a glass slide. Hybridization of the probe to the DNA being investigated is often marked as double spots, indicating that the labeled probe has bound to both sister chromatids (**Figure 2.16A** and **Figure 2.17A**). By using sophisticated image-processing equipment and reporter-binding molecules carrying different fluorescent dyes, several DNA probes can be hybridized simultaneously so that the location of specific sequences can be identified in relation to each other.

The maximum resolution of metaphase FISH is several megabases, but the use of the more extended prometaphase chromosomes can permit 1 Mb resolution. Variations of the standard method involve artificial stretching of DNA or chromatin fibers (**fiber FISH**) so that the resolution can be increased to the kilobase level.

Interphase FISH also offers high-resolution chromosome FISH because during interphase the chromosomes are naturally much less condensed than during metaphase. In this technique the probes hybridize to target chromosomal DNA within cellular or nuclear preparations that are treated with digestive enzymes to permit probe access—see Figure 2.17B for an application. Interphase FISH can be performed on fresh or frozen samples or on material that has been preserved in paraffin blocks.

Chromosome painting and molecular karyotyping

In **chromosome painting**, a special application of chromosome FISH, the probe is a cocktail of multiple different DNA fragments that derive from many different

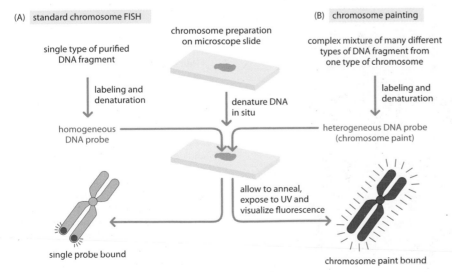

Figure 2.16 Basis of the chromosome FISH and chromosome painting methods. Both methods are used on interphase chromosomes as well as the metaphase chromosomes illustrated here, where chromosome spreads are prepared and fixed on a microscope slide and the DNA is denatured *in situ*. (A) In standard chromosome FISH, homogeneous DNA probes are usually used to hybridize to a single type of chromosomal DNA sequence. Individual probes are labeled with a specific fluorophore and each probe has the possibility of hybridizing to two sister chromatids to give a double signal. Practical applications of chromosome FISH used in metaphase and interphase chromosomes are shown in Figures 2.17A and 2.17B, respectively. (B) Chromosome painting uses a heterogeneous collection of labeled DNA probes that derive originally from multiple regions of a defined type of chromosome. The fragments in the probe cocktail are simultaneously labeled with a single type of fluorophore and will hybridize to multiple regions of a single chromosome, giving an aggregate fluorescent signal that spans an entire chromosome. Figures 2.18 and 2.9 show practical applications of painting metaphase and interphase chromosomes, respectively.

(A) standard chromosome FISH

single type of purified DNA fragment

chromosome preparation on microscope slide

labeling and denaturation

denature DNA in situ

homogeneous DNA probe

(B) chromosome painting

complex mixture of many different types of DNA fragment from one type of chromosome

labeling and denaturation

heterogeneous DNA probe (chromosome paint)

allow to anneal, expose to UV and visualize fluorescence

single probe bound

chromosome paint bound

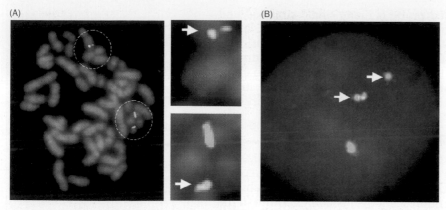

locations in a single type of chromosome (Figure 2.16B). The resulting aggregate hybridization signal typically spans a whole chromosome, causing it to fluoresce.

Figure 2.9 shows an example of chromosome painting in interphase nuclei, but most chromosome painting is performed on metaphase chromosomes. It is widely employed in investigating or defining abnormal chromosome rearrangements in clinical and cancer cytogenetics (**Figure 2.18**), and is particularly helpful with preparations from tumors, which are often of poor quality.

Chromosome painting was initially limited by the small number of differently colored fluorophores available. Now multiple targets can be detected simultaneously, by using different probes, each labeled with its own fluorophore, or by employing different *balances* of particular fluorophores. Analysis of the resulting color mixtures requires automated digital-image analysis, in which particular combinations of fluorophores are assigned artificial *pseudocolors*.

One application is genome-wide chromosome painting. Thus, 24 different chromosome paints can be used to distinguish between the 24 different types of human chromosome. This multiplex FISH (M-FISH) method constitutes a form of molecular karyotyping that is sometimes called spectral karyotyping (SKY) (**Figure 2.19**). M-FISH/SKY is valuable in analyzing tumor samples. Complex chromosome rearrangements frequently occur in tumors and are generally difficult to interpret by standard karyotyping with G-banding, but the use of 24 different chromosome paints facilitates interpretation.

Comparative genome hybridization (CGH)

Sometimes a form of *genome painting* is performed with cocktails of DNA fragments representing all chromosomes but labeled with a single fluorophore. The aim is to compare the genomic DNA of two closely related sources of cellular DNA that are suspected to differ, through gain or loss of either subchromosomal regions or whole chromosomes.

In standard **comparative genome hybridization** (**CGH**), total genomic DNA is isolated from two such cellular sources, independently labeled with two

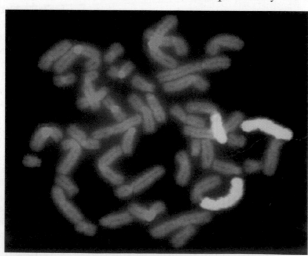

Figure 2.17 Two-color FISH to detect BCR–ABL1 rearrangement in chronic myeloid leukemia. Most cases of chronic myeloid leukemia (CML) result from a t(9;22) reciprocal translocation, with breakpoints that interrupt the *ABL1* oncogene at 9q34 and the *BCR* gene at 22q11. As a result of the CML translocation, there is one normal copy each of chromosomes 9 and 22, carrying normal *ABL1* and *BCR* alleles, respectively, and two derivative chromosomes carrying *ABL1–BCR* and *BCR–ABL1* fusion genes. The figure shows examples of chromosome FISH analysis of CML using (A) metaphase FISH and (B) interphase FISH with an *ABL1* probe (red signal) and a *BCR* probe (green signal). The normal *ABL1* and *BCR* alleles give standard red and green signals, respectively [with signals from both sister chromatids visible for at least *ABL1* in (A)]. The panels to the right represent blow-ups of the areas in the dashed circles in (A). The white arrows show characteristic signals for the fusion genes on the two translocation chromosomes [der(9) and der(22)]. These can be identified because of the very close positioning of the red and green signals, with overlapping red and green signals appearing orange–yellow. (Courtesy of Fiona Harding, Northern Genetics Service, Newcastle upon Tyne.)

Figure 2.18 Defining chromosome rearrangements by chromosome painting. By karyotyping a peripheral blood sample, an abnormal X chromosome was identified with extra chromosomal material present on the short arm. Follow-up chromosome painting investigations showed that the additional material present on the short arm of the abnormal X chromosome originated from chromosome 4, as revealed here with a chromosome X paint (red signal) and a chromosome 4 paint (green signal). The background stain for the chromosomes is the blue DAPI stain. (Courtesy of Gareth Breese, Northern Genetics Service, Newcastle upon Tyne.)

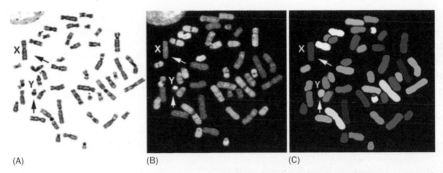

(A)   (B)   (C)

Figure 2.19 Molecular karyotyping with the SKY procedure. The analyzed chromosomes are from a normal human male. (A) Metaphase chromosome banding by the inverted DPI method, which is similar to G-banding but accentuates the heterochromatic region of some chromosomes (e.g. 1, 9, 16, and Y). (B) The same metaphase as in (A) after hybridization with chromosome paints. To obtain suitable probes, individual human chromosomes were first fractionated according to size and base composition by flow cytometry, and sorted into pools of the same chromosome type. DNA was purified and amplified from the individual chromosome pools and then fluorescently labeled for use as hybridization probes. Probes were labeled with at least one, and as many as five, fluorophore combinations and then pooled and hybridized to the metaphase chromosomes. Outputs are RGB display colors. (C) The same metaphase spread, but here digital imaging allows the fluorescence outputs from the different chromosome paint signals in (B) to be assigned artificial pseudocolors. The pseudocolors are assigned according to threshold wavelength values for the detected fluorescence and help to distinguish the different chromosomes. Arrows indicate the sex chromosomes. [From Padilla-Nash HM, Barenboim-Stapleton L, Difilippantonio MJ & Ried T (2007) *Nature Protocols* 1, 3129–3142. With permission from Macmillan Publishers Ltd.]

different fluorophores, and simultaneously hybridized to normal metaphase chromosomes. The ratio of the two color signals is then compared along the length of each chromosome to identify chromosomal or subchromosomal differences between the two sources. Often a test source, such as a tumor cell's DNA, is referenced against a normal control source, to identify regions of the genome that have been amplified or lost in the test sample (**Figure 2.20**).

As an alternative to standard CGH, variant methods have more recently been developed in which the two genomic DNAs to be compared are hybridized to large collections of purified DNA fragments representing uniformly spaced regions of the genome, instead of to metaphase chromosomes. As a spin-off from the Human Genome Project, comprehensive sets of large purified DNA fragments (*DNA clones*) have been sequenced and ordered into linear maps corresponding to each chromosome. In **array CGH** the target DNA is a collection of such clones, typically at a resolution of one clone per megabase, that are deposited in defined geometric arrays of microscopic spots on a solid support (a **microarray**).

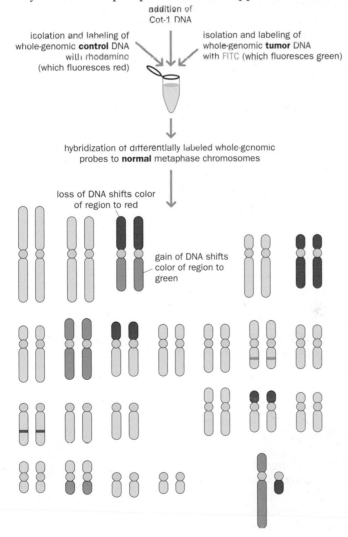

addition of
Cot-1 DNA

isolation and labeling of whole-genomic **control** DNA with rhodamine (which fluoresces red)

isolation and labeling of whole-genomic **tumor** DNA with FITC (which fluoresces green)

hybridization of differentially labeled whole-genomic probes to **normal** metaphase chromosomes

loss of DNA shifts color of region to red

gain of DNA shifts color of region to green

Figure 2.20 Comparative genome hybridization (CGH) using whole genomic DNA probes. CGH begins with the isolation of both genomic DNA from a tumor sample and genomic DNA from an individual who has a normal karyotype (the control DNA). Next, the two genomes are differentially labeled; for example, the control DNA with the red fluorochrome rhodamine, and the tumor DNA with the green fluorochrome fluorescein isothiocyanate (FITC). Cot-1 DNA is a fraction of purified DNA that is experimentally enriched for highly repetitive DNA. An excess of unlabeled human Cot-1 DNA is added to the labeled probe mixtures to suppress the repetitive DNA sequences that are present in both genomes (to obtain the desired hybridization signals with a minimum of background noise). The differentially labeled genomes are then combined and hybridized to normal metaphase chromosomes. The relative intensities of the green and red fluorochromes reflect the actual copy-number changes that have occurred in the tumor genome. DNA losses and gains are indicated by a shift to red and green fluorescence, respectively. [From McNeil & Ried (2000) *Expert Rev. Mol. Med.* 14, 1–14. With permission from Cambridge University Press.]

As in conventional CGH, the sample being investigated is labeled with one fluorescent dye, and the reference DNA is labeled with a different fluorescent dye. The two DNA samples are then simultaneously hybridized to the microarray. Any difference in the number of copies of a particular DNA sequence in the unknown sample versus the reference DNA will affect the ratio of the two bound fluorescent dyes. Although it is technically demanding both to undertake and to analyze the results, array CGH is beginning to be widely developed and used because of its considerable diagnostic potential.

2.5 CHROMOSOME ABNORMALITIES

Chromosome abnormalities might be defined as changes that produce a visible alteration of the chromosomes. How much can be seen depends on the technique used. The smallest loss or gain of material visible by traditional methods on standard cytogenetic preparations is about 4 Mb of DNA. However, FISH allows much smaller changes to be seen, and the use of molecular cytogenetics has removed any clear dividing line between changes described as chromosomal abnormalities and changes thought of as molecular or DNA defects.

An alternative definition of a chromosomal abnormality is an abnormality produced by specifically chromosomal mechanisms, such as misrepair of broken chromosomes or improper recombination events, or by incorrect segregation of chromosomes during mitosis or meiosis.

Chromosomal abnormalities can be classified into two types according to their distribution in cells of the body. A **constitutional abnormality** is present in all cells of the body. Where this occurs, the abnormality must have been present very early in development, most probably the result of an abnormal sperm or egg, or maybe abnormal fertilization or an abnormal event in the very early embryo.

A somatic (or acquired) abnormality is present in only certain cells or tissues of an individual. An individual with a somatic abnormality is said to be a **mosaic** as a result of possessing two populations of cells with different chromosome constitutions, each deriving from the same zygote.

Chromosomal abnormalities, whether constitutional or somatic, mostly fall into two categories, according to whether the copy number is altered (numerical abnormalities) or their structure is abnormal (structural abnormalities) (Table 2.4).

Numerical chromosomal abnormalities involve gain or loss of complete chromosomes

Three classes of numerical chromosomal abnormality can be distinguished: polyploidy, aneuploidy, and mixoploidy.

Polyploidy

Out of all recognized human pregnancies, 1–3% produce a triploid embryo (Figure 2.21A). The usual cause is two sperm fertilizing a single egg (**dispermy**), but triploidy is sometimes attributable to fertilization involving a diploid gamete. Triploids very seldom survive to term, and the condition is not compatible with life. Tetraploidy (Figure 2.21B) is much rarer and always lethal. It is usually due to failure to complete the first zygotic division: the DNA has replicated to give a content of $4C$ but cell division has not then taken place as normal. Although constitutional polyploidy is rare and lethal, all normal people have some polyploid cells.

Aneuploidy

Euploidy means having *complete* chromosome sets (n, $2n$, $3n$, and so on). **Aneuploidy** is the opposite: one or more individual chromosomes are present in an extra copy or are missing. In trisomy there are three copies of a particular chromosome in an otherwise diploid cell; an example is trisomy 21 (47,XX,+21 or 47,XY,+21) in Down syndrome. In monosomy a chromosome is lacking from an otherwise diploid state, as in monosomy X (45,X) in Turner syndrome. Cancer cells often show extreme aneuploidy, with multiple chromosomal abnormalities.

TABLE 2.4 NOMENCLATURE OF CHROMOSOME ABNORMALITIES

Type of abnormality	Examples	Explanation/notes
NUMERICAL		
Triploidy	69,XXX, 69,XXY, 69,XYY	a type of polyploidy
Trisomy	47,XX,+21	gain of a chromosome is indicated by +
Monosomy	45,X	a type of aneuploidy; loss of an autosome is indicated by –
Mosaicism	47,XXX/46,XX	a type of mixoploidy
STRUCTURAL		
Deletion	46,XY,del(4)(p16.3)	terminal deletion (breakpoint at 4p16.3)
	46,XX,del(5)(q13q33)	interstitial deletion (5q13–q33)
Inversion	46,XY,inv(11)(p11p15)	paracentric inversion (breakpoints on same arm)
Duplication	46,XX,dup(1)(q22q25)	duplication of region spanning 1q22 to 1q25
Insertion	46,XX,ins(2)(p13q21q31)	a rearrangement of one copy of chromosome 2 by insertion of segment 2q21–q31 into a breakpoint at 2p13
Ring chromosome	46,XY,r(7)(p22q36)	joining of broken ends at 7p22 and 7q36
Marker	47,XX,+mar	indicates a cell that contains a marker chromosome (an extra unidentified chromosome)
Reciprocal translocation	46,XX,t(2;6)(q35;p21.3)	a balanced reciprocal translocation with breakpoints at 2q35 and 6p21.3
Robertsonian translocation (gives rise to one derivative chromosome)	45,XY,der(14;21)(q10;q10)	a balanced carrier of a 14;21 Robertsonian translocation. q10 is not really a chromosome band, but indicates the centromere; der is used when one chromosome from a translocation is present
	46,XX,der(14;21)(q10;q10),+21	an individual with Down syndrome possessing one normal chromosome 14, a Robertsonian translocation 14;21 chromosome, and two normal copies of chromosome 21

This is a short nomenclature; a more complicated nomenclature is defined by the ISCN that allows complete description of any chromosome abnormality; see Shaffer LG, Tommerup N (eds) (2005) ISCN 2005: An International System for Human Cytogenetic Nomenclature. Karger.

Aneuploid cells arise through two main mechanisms. One mechanism is **nondisjunction**, in which paired chromosomes fail to separate (*disjoin*) during meiotic anaphase I, or sister chromatids fail to disjoin at either meiosis II or mitosis. Nondisjunction during meiosis produces gametes with either 22 or 24 chromosomes, which after fertilization with a normal gamete produce a trisomic or monosomic zygote. Nondisjunction during mitosis produces a mosaic individual.

Anaphase lag is another mechanism that results in aneuploidy. If a chromosome or chromatid is delayed in its movement during anaphase and lags behind the others, it may fail to be incorporated into one of the two daughter nuclei. Chromosomes that do not enter the nucleus of a daughter cell are eventually degraded.

Mixoploidy

Mixoploidy means having two or more genetically different cell lineages within one individual. The genetically different cell populations usually arise from the same zygote (**mosaicism**). More rarely, they originate from different zygotes (**chimerism**); spontaneous chimerism usually arises by the aggregation of

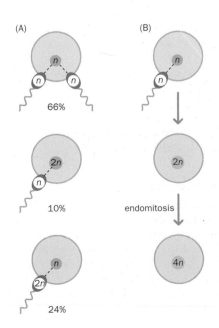

Figure 2.21 Origins of triploidy and tetraploidy. (A) Origins of human triploidy. Dispermy is the principal cause, accounting for 66% of cases. Triploidy is also caused by diploid gametes that arise by occasional faults in meiosis; fertilization of a diploid ovum and fertilization by a diploid sperm account for 10% and 24% of cases, respectively. (B) Tetraploidy involves normal fertilization and fusion of gametes to give a normal zygote. Subsequently, however, tetraploidy arises by endomitosis when DNA replicates without subsequent cell division.

fraternal twin zygotes, or immediate descendant cells therefrom, within the early embryo. Abnormalities that would otherwise be lethal (such as triploidy) may not be lethal in mixoploid individuals.

Aneuploidy mosaics are common. For example, mosaicism resulting in a proportion of normal cells and a proportion of aneuploid (e.g. trisomic) cells can be ascribed to nondisjunction or chromosome lag occurring in one of the mitotic divisions of the early embryo (any monosomic cells that are formed usually die out). Polyploidy mosaics (e.g. human diploid/triploid mosaics) are occasionally found. As gain or loss of a haploid set of chromosomes by mitotic nondisjunction is extremely unlikely, human diploid/triploid mosaics most probably arise by fusion of the second polar body with one of the cleavage nuclei of a normal diploid zygote.

Clinical consequences

Having the wrong number of chromosomes has serious, usually lethal, consequences (Table 2.5). Even though the extra chromosome 21 in a person with trisomy 21 (Down syndrome) is a perfectly normal chromosome, inherited from a normal parent, its presence causes multiple abnormalities that are present from birth (congenital). Embryos with trisomy 13 or trisomy 18 can also survive to term, but both result in severe developmental malformations, respectively Patau syndrome and Edwards syndrome. Other autosomal trisomies are not compatible with life. Autosomal monosomies have even more catastrophic consequences than trisomies, and are invariably lethal at the earliest stages of embryonic life.

The developmental abnormalities associated with monosomies and trisomies must be the consequence of an imbalance in the levels of gene products encoded on different chromosomes. Normal development and function depend on innumerable interactions between gene products that are often encoded on different chromosomes. For at least some of these interactions, balancing the levels of gene products encoded by genes on different chromosomes is critically important; each chromosome probably contains several genes for which the amount of

TABLE 2.5 CLINICAL CONSEQUENCES OF NUMERICAL CHROMOSOME ABNORMALITIES

Abnormality	Clinical consequences
POLYPLOIDY	
Triploidy (69,XXX or 69,XYY)	1–3% of all conceptions; almost never born live and do not survive long
ANEUPLOIDY (AUTOSOMES)	
Nullisomy (lacking a pair of homologs)	lethal at pre-implantation stage
Monosomy (one chromosome missing)	lethal during embryonic development
Trisomy (one extra chromosome)	usually lethal during embryonic or fetal[a] stages, but individuals with trisomy 13 (Patau syndrome) and trisomy 18 (Edwards syndrome) may survive to term; those with trisomy 21 (Down syndrome) may survive beyond age 40
ANEUPLOIDY (SEX CHROMOSOMES)	
Additional sex chromosomes	individuals with 47,XXX, 47,XXY, or 47,XYY all experience relatively minor problems and a normal lifespan
Lacking a sex chromosome	although 45,Y is never viable, in 45,X (Turner syndrome), about 99% of cases abort spontaneously; survivors are of normal intelligence but are infertile and show minor physical diagnostic characteristics

[a]In humans, the embryonic period spans fertilization through to the end of the eighth week of development. Fetal development then begins and lasts until birth.

gene product made is tightly regulated. Altering the relative numbers of chromosomes will affect these interactions, and reducing the gene copy number by 50% could be expected to have more severe consequences than providing three gene copies in place of two.

Having extra sex chromosomes can have far fewer ill effects than having an extra autosome. People with 47,XXX or 47,XYY karyotypes often function within the normal range, and 47,XXY men have relatively minor problems in comparison with people with any autosomal trisomy. Even monosomy (in 45,X women) can have remarkably few major consequences—although 45,Y is always lethal.

Because normal people can have either one or two X chromosomes, and either no or one Y, there must be special mechanisms that allow normal function with variable numbers of sex chromosomes. For the Y chromosome, this is because the very few genes it carries are very largely focused on determining maleness. The mammalian X chromosome is also a special case because X-chromosome inactivation in females controls the level of X-encoded gene products independently of the number of X chromosomes in the cell.

It is not obvious why triploidy is lethal in humans and other animals. With three copies of every autosome, the dosage of autosomal genes is balanced and should not cause problems. Triploids are always sterile because triplets of chromosomes cannot pair and segregate correctly in meiosis, but many triploid plants are in all other respects healthy and vigorous. The lethality in animals may be due to an imbalance between products encoded on the X chromosome and autosomes, for which X-chromosome inactivation would be unable to compensate.

A variety of structural chromosomal abnormalities result from misrepair or recombination errors

Chromosome breaks occur either as a result of damage to DNA (by radiation or chemicals, for example) or through faults in the recombination process. Chromosome breaks occurring during G_2 phase affect only one of the two sister chromatids and are called **chromatid breaks**. Breaks arising during G_1 phase, if not repaired before S phase, become *chromosome breaks* (affecting both chromatids). Cellular enzyme systems recognize and try to repair broken chromosomes. Repairs involve either joining two broken ends together or adding a telomere to a broken end.

During the cell cycle, special mechanisms normally prevent cells with unrepaired chromosome breaks from entering mitosis; however, as long as there are no free broken ends, cells can move toward cell division as required. When cellular repair mechanisms encounter unwanted free chromosome ends produced by chromosome breakage, three outcomes are possible. First, the damage may be irreparable, whereupon the cell is programmed to self-destruct. Alternatively, the damage may be correctly repaired, or it may be incorrectly repaired and can result in chromsomes with structural abnormalities.

In addition to faulty DNA repair, errors in recombination also produce structural chromosome abnormalities. After homologs pair during meiosis, they are subjected to recombination mechanisms that ensure breakage and re-joining of non-sister chromatids. If, however, recombination occurs between mispaired homologs, the resulting products may have structural abnormalities.

In addition to meiotic recombination, other natural cellular processes involve the breakage and rejoining of DNA. Notably, a form of recombination also occurs naturally in B and T cells in which the cellular DNA undergoes programmed rearrangements to make antibodies and T cell receptors. Abnormalities in these recombination processes can also cause structural chromosomal abnormalities that may be associated with cancer.

Structural chromosome abnormalities are often the result of incorrect joining together of two broken chromosome ends. If a single chromosome sustains two breaks, incorrect joining of fragments can result in chromosome material being lost (deletion), switched round in the reverse direction (inversion), or included in a circular chromosome (a ring chromosome) (**Figure 2.22**). Structurally abnormal chromosomes with a single centromere can be stably propagated through successive rounds of mitosis. However, any repaired chromosome that lacks a centromere (acentric) or possesses two centromeres (dicentric) will normally not segregate stably at mitosis and will eventually be lost.

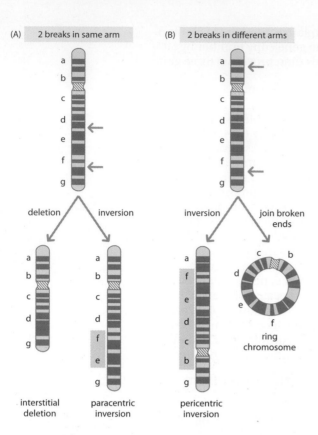

(A) 2 breaks in same arm   (B) 2 breaks in different arms

deletion   inversion   inversion   join broken ends

ring chromosome

interstitial deletion   paracentric inversion   pericentric inversion

Figure 2.22 Stable outcomes after incorrect repair of two breaks on a single chromosome. (A) Incorrect repair of two breaks (green arrows) occurring in the same chromosome arm can involve loss of the central fragment (here containing hypothetical regions e and f) and re-joining of the terminal fragments (deletion), or inversion of the central fragment through 180° and rejoining of the ends to the terminal fragments (called a paracentric inversion because it does not involve the centromere). (B) When two breaks occur on different arms of the same chromosome, the central fragment (encompassing hypothetical regions b to f in this example) may invert and re-join the terminal fragments (pericentric inversion). Alternatively, because the central fragment contains a centromere, the two ends can be joined to form a stable ring chromosome, while the acentric distal fragments are lost. Like other repaired chromosomes that retain a centromere, ring chromosomes can be stably propagated to daughter cells.

If two different chromosomes each sustain a single break, incorrect joining of the broken ends can result in a movement of chromosome material between chromosomes (**translocation**). A *reciprocal translocation* is the general term used to describe an exchange of fragments between two chromosomes (**Figure 2.23A**). If an acentric fragment from one chromosome is exchanged for an acentric fragment from another, the products are stable in mitosis. However, exchange of an acentric fragment for a centric fragment results in acentric and dicentric chromosomes that are unstable in mitosis.

A *Robertsonian translocation* is a specialized type of translocation between two of the five types of acrocentric chromosome in humans (chromosomes 13, 14, 15, 21, and 22). The short arm of each of these chromosomes is very small and very similar in DNA content: each contains 1–2 Mb of tandemly repeated rRNA genes sandwiched between two blocks of heterochromatic DNA (Figure 2.23B). Breaks in the short arms of two different acrocentric chromosomes followed by exchange of acentric and centric fragments results in acentric and dicentric products.

The acentric chromosome produced by a Robertsonian translocation contains only highly repetitive noncoding DNA and rRNA genes that are also present at high copy number on the other acrocentric human chromosomes. It is lost at mitosis without consequence. The other product is an unusual dicentric chromosome that is stable in mitosis. In this case the two centromeres are in close proximity (centric fusion) and often function as one large centromere so that the chromosome segregates regularly. Nevertheless, as will be described in the next section, such a chromosome may present problems during gametogenesis.

More complex translocations can involve multiple chromosome breakages. Insertions typically require at least three breaks—often a fragment liberated by two breaks in one chromosome arm inserts into another break that may be located in another region of the same chromosome or another chromosome.

An additional rare class of structural abnormality can arise from recombination after an aberrant chromosome division. The product is a symmetrical **isochromosome** consisting of either two long arms or two short arms of a particular chromosome. Human isochromosomes are rare except for i(Xq) and also i(21q), an occasional contributor to Down syndrome.

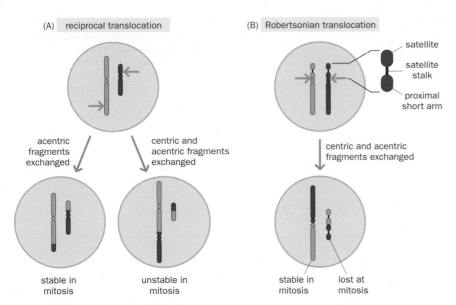

Figure 2.23 Reciprocal and Robertsonian translocations. (A) Reciprocal translocation. The derivative chromosomes are stable in mitosis when one acentric fragment is exchanged for another; when a centric fragment is exchanged for an acentric fragment, unstable acentric and dicentric chromosomes are produced. (B) Robertsonian translocation. This is a highly specialized reciprocal translocation in which exchange of centric and acentric fragments produces a dicentric chromosome that is nevertheless stable in mitosis, plus an acentric chromosome that is lost in mitosis without any effect on the phenotype. It occurs exclusively after breaks in the short arms of the human acrocentric chromosomes 13, 14, 15, 21, and 22. As illustrated on the right, the short arm of the acrocentric chromosomes consists of three regions: a proximal heterochromatic region (composed of highly repetitive noncoding DNA), a distal heterochromatic region (called a chromosome satellite), and a thin connecting region of euchromatin (the satellite stalk) composed of tandem rRNA genes. Breaks that occur close to the centromere can result in a dicentric chromosome in which the two centromeres are so close that they can function as a single centromere. The loss of the small acentric fragment has no phenotypic consequences because the only genes lost are rRNA genes that are also present in large copy number on the other acrocentric chromosomes.

Different factors contribute to the clinical consequences of structural chromosome abnormalities

Structural chromosome abnormalities are balanced if there is no net gain or loss of chromosomal material, and unbalanced if there is a net gain or loss. Standard cytogenetic analyses have routinely been used to define whether a chromosome abnormality is balanced or unbalanced, but higher-resolution analyses (notably array CGH) can detect submicroscopic losses or gains in what initially seemed to be balanced chromosome abnormalities. Truly balanced abnormalities (some inversions and balanced translocations) are less likely to affect the phenotype except in the following circumstances:

- A chromosome break disrupts an important gene.
- A chromosome break affects the expression of a gene without disrupting the coding sequence by, for example, separating a gene from a control element, or translocating an active gene into heterochromatin.
- Balanced X-autosome translocations cause non-random X-inactivation.

Although some chromosomal material is lost in Robertsonian translocations (see Figure 2.23B), they are regarded as balanced because there is no phenotypic effect; the loss of rRNA genes in the acentric recombination product is not significant because other acrocentric chromosomes retain sufficient rRNA genes.

Unbalanced chromosomal abnormalities can arise directly through deletion or, rarely, by duplication. They may also arise indirectly by malsegregation of chromosomes during meiosis, in a carrier of a balanced abnormality. In such carriers, meiosis is abnormal because the structures of homologous pairs of

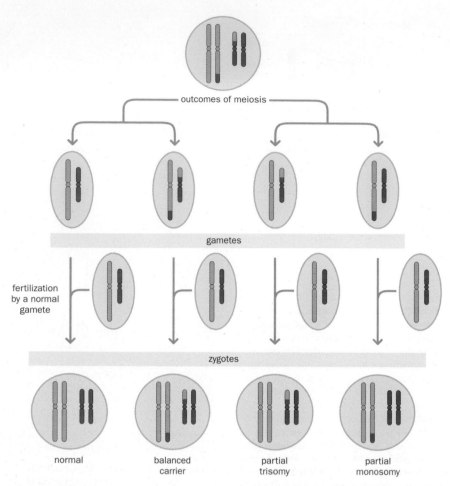

Figure 2.24 Possible outcomes of meiosis in a carrier of a balanced reciprocal translocation. Other modes of segregation are also possible, for example 3:1 segregation. The relative frequency of each possible gamete is not readily predicted. The risk of a carrier having a child with each of the possible outcomes depends on its frequency in the gametes and also on the likelihood of a conceptus with that abnormality developing to term. See the book by Gardner and Sutherland (in Further Reading) for discussion.

chromosomes do not correspond, with a variety of possible consequences, such as the following:

- A carrier of a balanced reciprocal translocation can produce gametes that give rise to an entirely normal child, a phenotypically normal balanced carrier, or to various unbalanced karyotypes that always combine monosomy for part of one of the chromosomes with trisomy for part of the other (**Figure 2.24**). It is not possible to make general statements about the relative frequencies of these outcomes. The size of the unbalanced segments depends on the position of the breakpoints. If the unbalanced segments are large, the fetus will probably abort spontaneously; a smaller imbalance may result in a live-born baby with abnormalities.

- A carrier of a balanced Robertsonian translocation can produce gametes that after fertilization give rise to an entirely normal child, a phenotypically normal balanced carrier, or a conceptus with full trisomy or full monosomy for one of the chromosomes involved (**Figure 2.25**).

- A carrier of a pericentric inversion may produce offspring with duplicated or deleted chromosomal segments. This is because, when the inverted and non-inverted homologs pair during meiosis, they form a loop to enable matching segments to pair along their whole length (**Figure 2.26A**). If recombination occurs within this loop, the consequence is an unbalanced deletion and duplication. Paracentric inversions form similar loops, but any recombination within this loop generates an acentric or dicentric chromosome (Figure 2.26B), which is unlikely to survive.

Incorrect parental origins of chromosomes can result in aberrant development and disease

Certain rare abnormalities demonstrate that it is not enough to have the correct number and structure of chromosomes: the chromosomes must also have the

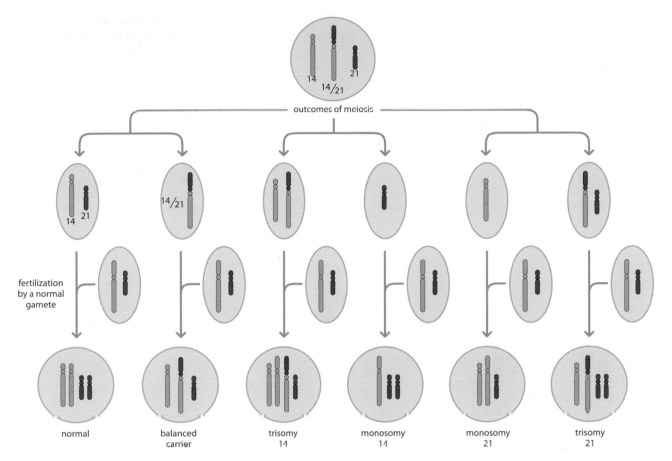

Figure 2.25 Possible outcomes of meiosis in a carrier of a Robertsonian translocation. Carriers are asymptomatic but often produce unbalanced gametes that can result in a monosomic or trisomic zygote. The two monosomic zygotes and the trisomy 14 zygote in this example would not be expected to develop to term.

correct parental origin. Rare 46,XX conceptuses in which both complete haploid genomes originate from the same parent (**uniparental diploidy**) never develop correctly, and experimentally induced uniparental diploidy in other mammals also results in abnormal development. Even for particular chromosomes, having both homologs derived from the same parent (**uniparental disomy**) can cause abnormality. Parental origin of chromosomes is important because some genes are expressed differently according to the parent of origin; such genes are subject to **imprinting**.

Uniparental diploidy in humans arises most often by fertilization of a faulty egg that lacks chromosomes by a diploid sperm (produced by chance chromosome doubling) or by the activation of an unovulated oocyte (parthenogenesis), resulting in each case in a 46,XX karyotype. The resulting conceptuses develop abnormally.

Normal human (and mammalian) development involves the production of both an embryo that develops into a fetus and also extra-embryonic membranes, including the trophoblast, that act to support development and give rise to the placenta. Uniparental diploidy changes the important balance between the embryo or fetus and its supporting membranes. Paternal uniparental diploidy produces hydatidiform moles, abnormal conceptuses that develop to show widespread hyperplasia (overgrowth) of the trophoblast but no fetal parts; they have a significant risk of transformation into choriocarcinoma. Maternal uniparental diploidy results in ovarian teratomas, rare benign tumors of the ovary that consist of disorganized embryonic tissues but are lacking in vital extra-embryonic membranes.

Uniparental disomy is thought to arise in most cases by trisomy rescue. In such cases a trisomic conceptus that would otherwise probably die before birth gives rise to a 46,XX or 46,XY embryo by chromosome loss. The chromosome loss occurs through faulty mitotic division very early in development in a cell that has

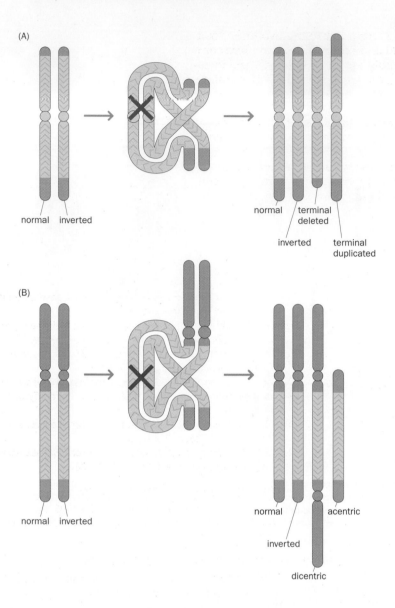

(A)

normal   inverted

normal   terminal
          deleted

inverted   terminal
            duplicated

(B)

normal   inverted

normal   acentric

inverted

dicentric

Figure 2.26 Possible outcomes of meiosis in inversion carriers. During meiosis in a carrier of an inversion chromosome, pairing of the homologs requires looping of the inverted region. The figure shows the four possible meiotic products after a single crossover (red cross) occurring in the inversion loop in (A) pericentric inversion and (B) paracentric inversion. In addition to non-recombinant normal and inversion chromosomes, the recombinant chromosomes undergo terminal duplication and deletion in pericentric inversions, or are dicentric and acentric in paracentric inversions.

the capacity to give rise to all the different cell types in the organism (a **totipotent** cell). The progeny of this cell are euploid and form the embryo, and all the aneuploid cells die. If each of the three copies of a trisomic chromosome has an equal chance of being lost, there is a two in three chance that a single chromosome loss will lead to the normal chromosome constitution and a one in three chance of uniparental disomy (either paternal or maternal).

Uniparental disomy may often go undiagnosed because there are no obvious phenotypic effects, and it is detected only when it results in characteristic syndromes. In such cases, the chromosome involved contains imprinted genes that are expressed incorrectly. Uniparental disomy can involve heterodisomy (one parent contributes non-identical homologs, most probably by trisomy rescue). Sometimes, however, the two chromosome copies inherited from one parent are identical (isodisomy). This can happen by monosomy rescue when a monosomic embryo that would otherwise die achieves euploidy by selective duplication of the monosomic chromosome.

CONCLUSION

In this chapter we have described fundamental aspects of chromosome structure and function, with an emphasis on human chromosomes and the generation of chromosome abnormalities in humans. There are close similarities in the ways that eukaryotic chromosomes are structured, notably in the way that DNA is complexed with histones and organized into higher-order structures.

But there are also important differences. During eukaryote evolution, centromeres have become progressively larger and more complex, and in more complex eukaryotes they are dominated by rapidly evolving species-specific repetitive DNA sequences. The mammalian sex chromosomes have unique characteristics and we will consider them in greater detail in Chapter 10 when we consider chromosome evolution. Specialized mammalian X-inactivation and imprinting systems will also be explored in greater detail in the context of gene expression in Chapter 11, as will the relationship between chromatin structure and gene expression.

We will also provide many illustrative examples on how chromosome abnormalities cause inherited disease and cancer in Chapters 16 and 17 that also contain fuller descriptions of some of the technologies used to analyze chromosomes, such as array CGH.

FURTHER READING

Chromosome structure and function

Belmont AS (2006) Mitotic chromosome structure and condensation. *Curr. Opin. Cell Biol.* 18, 632–638.

Blackburn EH (2001) Switching and signaling at the telomere. *Cell* 106, 661–673.

Courbet S, Gay S, Arnoult N et al. (2008) Replication fork movement sets chromatin loop size and origin choice in mammalian cells. *Nature* 455, 557–560.

Haering CH, Farcas AM, Arumugam P et al. (2008) The cohesin ring concatenates sister DNA molecules. *Nature* 454, 297–301.

Herrick J & Bensimon A (2008) Global regulation of genome duplication in eukaryotes: an overview from the epifluorescence microscope. *Chromosoma* 117, 243–260.

Louis EJ & Vershinin AV (2005) Chromosome ends: different sequences may provide conserved functions. *BioEssays* 27, 685–697.

Meaburn KJ & Misteli T (2007) Chromosome territories. *Nature* 445, 379–381.

Riethman H (2008) Human telomere structure and biology. *Annu. Rev. Genomics Hum. Genet.* 9, 1–19.

Vagnarelli P, Ribeiro SA & Earnshaw WC (2008) Centromeres: old tales and new tools. *FEBS Lett.* 582, 1950–1959.

Human chromosome analyses and chromosome abnormalities

Gardner RJM & Sutherland GR (2003) Chromosome Abnormalities and Genetic Counseling, 3rd ed. Oxford University Press.

McNeil N & Ried T (2000) New molecular cytogenetic techniques for identifying complex chromosomal rearrangements: technology and applications in molecular medicine. *Expert Rev. Mol. Med.* 14, 1–14.

Padilla-Nash H, Barenboim-Stapleton L, Difilippantonio MJ & Ried T (2006) Spectral karyotyping analysis of human and mouse chromosomes. *Nat. Protoc.* 1, 3129–3142.

Rooney DE (2001) (ed.) Human Cytogenetics: Constitutional Analysis, 3rd ed. Oxford University Press.

Rooney DE (2001) (ed.) Human Cytogenetics: Malignancy and Acquired Abnormalities, 3rd ed. Oxford University Press.

Schinzel A (2001) Catalogue of Unbalanced Chromosome Aberrations in Man. Walter de Gruyter.

Shaffer LG & Bejani BA (2006) Medical applications of array CGH and the transformation of clinical cytogenetics. *Cytogenet. Genome Res.* 115, 303–309.

Therman E & Susman M (1992) Human Chromosomes: Structure, Behavior and Effects, 3rd ed. Springer.

Trask BJ (2002) Human cytogenetics: 46 chromosomes, 46 years and counting. *Nat. Rev. Genet.* 3, 769–778.

Human cytogenetic nomenclature

Shaffer LG & Tommerup N (eds) (2005) ISCN 2005: An International System for Human Cytogenetic Nomenclature. Karger.

Web-based cytogenetic resources

Useful compilations are available on certain sites such as www.kumc.edu/gec/prof/cytogene.html

Chapter 3

Genes in Pedigrees and Populations

3

KEY CONCEPTS

- The pattern of inheritance of a character follows Mendel's rules if its presence, absence, or specific nature is normally determined by the genotype at a single locus.

- Mendelian characters can be dominant, recessive, or co-dominant. A character is dominant if it is evident in a heterozygous person, recessive if not.

- Human Mendelian characters give pedigree patterns that are often recognizable, although seldom as unambiguous as the results of laboratory breeding experiments because of the limited size and non-ideal structure of most human families.

- Genetically controlled characters can be dichotomous (present or absent) or continuous (quantitative). The genes controlling quantitative characters are called quantitative trait loci.

- Most characters depend on more than one genetic locus, and often also on environmental factors. Such characters are called complex or multifactorial.

- Polygenic theory explains why many quantitative characters show a normal (Gaussian) distribution in a population.

- Polygenic threshold theory provides a framework for understanding the genetics of dichotomous multifactorial characters, but it is not a tool for predicting the characteristics of individuals.

- Subject to certain conditions, gene frequencies and genotype frequencies are related by the Hardy–Weinberg formula.

- Consanguineous matings contribute disproportionately to the incidence of rare autosomal recessive conditions; ignoring this when using the Hardy–Weinberg formula leads to incorrect gene frequencies.

- Gene frequencies in populations are the result of a combination of random processes (genetic drift) and the effects of new mutation and natural selection.

- Heterozygote advantage often explains why an autosomal recessive condition is particularly common in a population.

Genetics as a science started with Gregor Mendel's experiments in the 1860s. Mendel bred and crossed pea plants, counted the numbers of plants in each generation that manifested certain traits, and deduced mathematical rules to explain his results. Patterns of inheritance in humans (and in all other diploid organisms that reproduce sexually) follow the same principles that Mendel identified in his peas, and are explained by the same basic concepts.

Genes can be defined in two ways:

- A gene is a determinant, or a co-determinant, of a character that is inherited in accordance with Mendel's rules.

- A gene is a functional unit of DNA.

In this chapter we will be exploring genes through the first of these definitions. Later chapters will consider genes as DNA. A major aim of human molecular genetics is to understand how genes as functional DNA sequences determine the observable characters of a person.

A **locus** (plural **loci**) is a unique chromosomal location defining the position of an individual gene or DNA sequence. Thus, we can speak of the ABO blood group locus, the Rhesus blood group locus, and so on.

Alleles are alternative versions of a gene. For example, A, B, and O are alternative alleles at the ABO locus. **Box 3.1** shows the formal rules for nomenclature of alleles. However, for pedigree interpretation, alleles are usually simply denoted by upper- and lower-case versions of the same letter, so that the genotypes at a locus would be written *AA*, *Aa*, or *aa*. The upper-case letter is conventionally used for the allele that determines the dominant character (see below).

The **genotype** is a list of the alleles present at one or a number of loci.

Phenotypes, **characters**, or **traits** are the observable properties of an organism. The means of observation may range from simple inspection to sophisticated laboratory investigations.

A person is **homozygous** at a locus if both alleles at that locus are the same, and **heterozygous** if they are different. For different purposes we may check more or less carefully whether the two alleles at a locus are really the same. For the pedigree interpretation in this chapter, we will be concerned only with the phenotypic consequences of a person's genotype. We will describe a person as homozygous if both alleles at the locus in question have the same phenotypic effect, even though inspection of the DNA sequence might reveal differences between them.

A person is **hemizygous** if they have only a single allele at a locus. This may be because the locus is on the X or Y chromosome in a male, or it may be because one copy of an autosomal locus is deleted.

A character is **dominant** if it is manifested in a heterozygous person, **recessive** if not.

3.1 MONOGENIC VERSUS MULTIFACTORIAL INHERITANCE

The simplest genetic characters are those whose presence or absence depends on the genotype at a single locus. That is not to say that the character itself is programmed by only one pair of genes: expression of any human character is likely to depend on the action of a large number of genes and environmental factors. However, sometimes a particular genotype at one locus is both necessary and sufficient for the character to be expressed, given the normal range of human genetic and environmental backgrounds. Such characters are called *Mendelian*. Mendelian characters can be recognized by the characteristic pedigree patterns they give, as described in the next section. The best starting point for acquiring information on any such character, whether pathological or non-pathological, is the Online Mendelian Inheritance in Man (OMIM) database (**Box 3.2**). Where appropriate, throughout this book the OMIM reference number is quoted for each human character when it is first described.

Most human genetic or partly genetic characters are not Mendelian. They are governed by genes at more than one locus. The more complex the pathway between a DNA sequence and an observable trait, the less likely it is that the trait will show a simple Mendelian pedigree pattern. Thus, DNA sequence variants are almost always inherited in a cleanly Mendelian manner—which explains why

BOX 3.2 DATABASES OF HUMAN GENETIC DISEASES AND MENDELIAN CHARACTERS

This is a short selective list of especially useful, reliable, and stable resources; many other useful databases may be found by searching.

OMIM (http://www.ncbi.nlm.nih.gov/omim). The Online Mendelian Inheritance in Man database is the most reliable single source of information on human Mendelian characters and the underlying genes. The index numbers quoted throughout this book (e.g. OMIM 193500) give direct access to the relevant entry. OMIM contains about 20,000 entries, which may be sequenced genes, characters or diseases associated with known sequenced genes, or characters that are inherited in a Mendelian way but for which no gene has yet been identified. Some entries describe characters that are not normally Mendelian. In those cases the OMIM entry will concentrate on any Mendelian or near-Mendelian subset and may therefore not give a balanced picture of the overall etiology. Each entry is a detailed historically ordered review of the genetics of the character, with subsidiary clinical and other information, and a very useful list of references. Entries have accumulated text over many years with only patchy rewrites, so that the early part of an entry may not reflect current understanding.

The **Genetic Association Database** (http://geneticassociationdb.nih.gov), maintained by the US National Institute on Aging, can be searched for a list of genes and publications reporting possible genetic susceptibility factors for multifactorial diseases. At the time of writing it is at an early stage of development, but it should become a valuable resource for accessing information that is otherwise dispersed over many individual publications.

Genecards (http://www.genecards.org), from the Weizmann Institute in Israel, contains about 50,000 automatically generated entries, mostly relating to specific human genes. It gives access to a large amount of biological information about each gene.

GeneTests (http://www.geneclinics.org) is a database of human genetic diseases, maintained by the US National Institutes of Health and aimed mainly at clinicians. It includes brief clinical and genetic reviews of about 500 of the most common Mendelian diseases. There is more clinical information than in OMIM.

---

they are the characters of choice for following the transmission of a chromosomal segment through a pedigree, as described in Chapter 13. Directly observed characteristics of proteins (such as electrophoretic mobility or enzyme activity) usually display Mendelian pedigree patterns, but these characters can be influenced by more than one locus because of post-translational modification of gene products. Disease states or other typical traits that reflect the biochemical action of a protein within a cellular context are less often entirely determined at a single genetic locus. The failure or malfunction of a developmental pathway that results in a birth defect is likely to involve a complex balance of factors. Thus, the common birth defects (for example, cleft palate, spina bifida, or congenital heart disease) are rarely Mendelian. Behavioral traits such as IQ test performance or schizophrenia are still less likely to be Mendelian—but they may still be genetically determined to a greater or smaller extent.

Non-Mendelian characters may depend on two, three, or many genetic loci. We use **multifactorial** here as a catch-all term covering all these possibilities. More specifically, the genetic determination may involve a small number of loci (**oligogenic**) or many loci each of individually small effect (**polygenic**); or there may be a single major locus with a polygenic background—that is, the genotype at one locus has a major effect on the phenotype, but this effect is modified by the cumulative minor effects of genes at many other loci. In reality, characters form a continuous spectrum, from perfectly Mendelian through to truly polygenic (**Figure 3.1**). Superimposed on this there may be a greater or smaller effect of environmental factors.

For **dichotomous characters** (characters that you either have or do not have, such as extra fingers) the underlying loci are envisaged as **susceptibility genes**, whereas for **quantitative** or **continuous characters** (such as height or weight) they are seen as **quantitative trait loci** (QTLs). Any of these characters may tend to run in families, but the pedigree patterns are not Mendelian and do not fit any standard pattern.

In a further layer of complication, a common human condition such as diabetes is likely to be very heterogeneous in its causation. Some cases may have a simple Mendelian cause, some might be entirely the result of environmental factors, while the majority of cases may be multifactorial. Such conditions are called **complex**.

Figure 3.1 Spectrum of genetic determination. ABO blood group depends (with rare exceptions) on the genotype at just one locus, the *ABO* locus at chromosome 9q34. Rhesus hemolytic disease of the newborn depends on the genotypes of mother and baby at the *RHD* locus at chromosome 1p36, but also on mother and baby's being ABO compatible. Hirschsprung disease depends on the interaction of several genetic loci. Adult stature is determined by the cumulative small effects of many loci. Environmental factors are also important in the etiology of Rhesus hemolytic disease, Hirschsprung disease, and adult height.

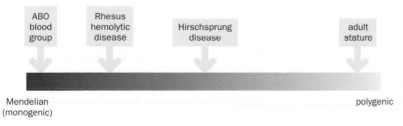

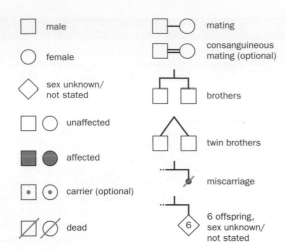

Figure 3.2 Main symbols used in pedigrees. The dot symbol for a carrier, and the double marriage lines for consanguineous matings, are used to draw attention to those features, but their absence does not necessarily mean that a person is not a carrier, or that a union is not consanguineous.

3.2   MENDELIAN PEDIGREE PATTERNS

There are five basic Mendelian pedigree patterns

Mendelian characters may be determined by loci on an autosome or on the X or Y sex chromosomes. Autosomal characters in both sexes, and X-linked characters in females, can be dominant or recessive. Males are hemizygous for loci on the X and Y chromosomes; that is, they have only a single copy of each gene. Thus, they are never heterozygous for any X-linked or Y-linked character (in the rare XYY males, the two Y chromosomes are duplicates), and it is not necessary to know whether such a character is dominant or recessive to predict a man's phenotype from his genotype. There are five archetypal Mendelian pedigree patterns:

- Autosomal dominant
- Autosomal recessive
- X-linked dominant
- X-linked recessive
- Y-linked

Figure 3.2 shows the symbols commonly used for pedigree drawing. Generations are usually labeled in Roman numerals, and individuals within each generation in Arabic numerals (see examples in **Figures 3.3–3.7**). Thus, III-7 or III₇ is the seventh person from the left (unless explicitly numbered otherwise) in generation III. An arrow can be used to indicate the **proband** or **propositus** (female **proposita**) through whom the family was ascertained. The five patterns are illustrated in Figures 3.3–3.7 and described in **Box 3.3**. These basic patterns are subject to various complications that are discussed below and illustrated in Figures 3.13–3.21.

Note that dominance and recessiveness are properties of characters, not genes. Thus, sickle-cell anemia is recessive because only Hb^S homozygotes manifest it, but sickling trait, which is the phenotype of Hb^S heterozygotes, is

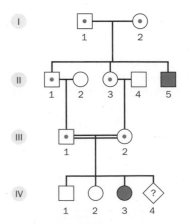

Figure 3.4 Pedigree pattern of an autosomal recessive condition. People who must be carriers are indicated with dots; IV₁ and/or IV₂ might also be carriers, but we do not know. The risk for the individual marked with a query is 1 in 4.

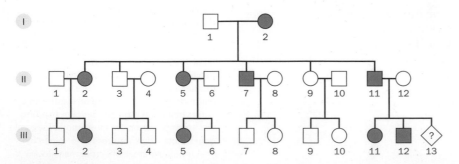

Figure 3.3 Pedigree pattern of an autosomal dominant condition. The risk for the individual marked with a query is 1 in 2.

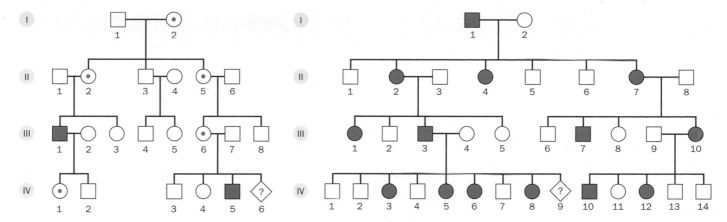

Figure 3.5 Pedigree pattern of an X-linked recessive condition. The females marked with dots are definitely carriers; individuals III₃ and/or IV₄ may also be carriers, but we do not know. The risk for the individual marked with a query is 1 in 2 males or 1 in 4 of all offspring.

Figure 3.6 Pedigree pattern of an X-linked dominant condition. The risk for the individual marked with a query is negligibly low if male, but 100% if female.

dominant. In experimental organisms, geneticists use the term **co-dominant** when the heterozygote has an intermediate phenotype and the homozygote is more severely affected. The term **dominant** is reserved for conditions in which the homozygote is indistinguishable from the heterozygote. However, most human dominant syndromes are known only in heterozygotes. Sometimes homozygotes have been described, born from matings of two heterozygous affected people, and often the homozygotes are much more severely affected. Examples are achondroplasia (short-limbed dwarfism; OMIM 100800) and type 1 Waardenburg syndrome (deafness with pigmentary abnormalities; OMIM 193500). Nevertheless we describe achondroplasia and Waardenburg syndrome as dominant because these terms describe the phenotypes that are seen in hetero-zygotes.

X-inactivation

Animals, including humans, do not readily tolerate having wrong numbers of chromosomes. Chromosomal aneuploidies (extra or missing chromosomes, as in Down syndrome—see Chapter 2) have severe, usually lethal, consequences. Nevertheless, in organisms with an XX/XY sex determination system, males and females must be able to develop normally despite having different numbers of sex chromosomes. This requires special arrangements. For the human Y chromo-some, the solution is to carry very few genes; those few are mainly related to male sexual function (see below). The human X chromosome, in contrast, carries many essential genes, as demonstrated by the many severe or lethal X-linked condi-tions. Different organisms solve the problem of coping with either XX or XY chro-mosome constitutions in different ways. In male *Drosophila* flies, genes on the single X chromosome are transcribed at double the rate of those on the X chro-mosomes of females. Mammals, including humans, take a different approach:

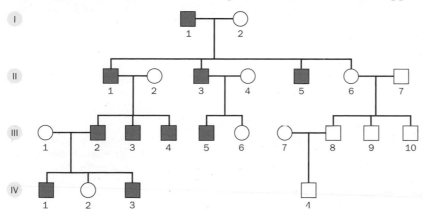

Figure 3.7 Hypothetical pedigree pattern of a Y-linked condition.

BOX 3.3 SUMMARY OF PATTERNS OF INHERITANCE

Autosomal dominant inheritance (Figure 3.3):
- An affected person usually has at least one affected parent (for exceptions see Figures 3.14 and 3.21).
- It affects either sex.
- It is transmitted by either sex.
- A child with one affected and one unaffected parent has a 50% chance of being affected (this assumes that the affected person is heterozygous, which is usually true for rare conditions).

Autosomal recessive inheritance (Figure 3.4):
- Affected people are usually born to unaffected parents.
- Parents of affected people are usually asymptomatic carriers.
- There is an increased incidence of parental consanguinity.
- It affects either sex.
- After the birth of an affected child, each subsequent child has a 25% chance of being affected (assuming that both parents are phenotypically normal carriers).

X-linked recessive inheritance (Figure 3.5):
- It affects mainly males.
- Affected males are usually born to unaffected parents; the mother is normally an asymptomatic carrier and may have affected male relatives.
- Females may be affected if the father is affected and the mother is a carrier, or occasionally as a result of non-random X-inactivation.
- There is no male-to-male transmission in the pedigree (but matings of an affected male and carrier female can give the *appearance* of male-to-male transmission; see Figure 3.20).

X-linked dominant inheritance (Figure 3.6):
- It affects either sex, but more females than males.
- Usually at least one parent is affected.
- Females are often more mildly and more variably affected than males.
- The child of an affected female, regardless of its sex, has a 50% chance of being affected.
- For an affected male, all his daughters but none of his sons are affected.

Y-linked inheritance (Figure 3.7):
- It affects only males.
- Affected males always have an affected father (unless there is a new mutation).
- All sons of an affected man are affected.

Mitochondrial inheritance (Figure 3.10):
- It affects both sexes.
- It is usually inherited from an affected mother (but is often caused by *de novo* mutations with the mother unaffected).
- It is not transmitted by a father to his children.
- There are highly variable clinical manifestations.

they use **X-inactivation** (sometimes called **lyonization** after its discoverer, Dr Mary Lyon).

Early in embryogenesis each cell somehow counts its number of X chromosomes, and then permanently inactivates all X chromosomes except one in each somatic cell. Inactivated X chromosomes are still physically present, and on a standard karyotype they look entirely normal, but the inactive X remains condensed throughout the cell cycle. Most genes on the chromosome are permanently silenced in somatic cells. The mechanism is discussed in Chapter 11. In interphase cells the inactive X may be seen under the microscope as a **Barr body** or **sex chromatin** (Figure 3.8). Regardless of the karyotype, each somatic cell retains a single active X:

- XY males keep their single X active (no Barr body).
- XX females inactivate one X in each cell (one Barr body).
- Females with Turner syndrome (45,X) do not inactivate their X (no Barr body).
- Males with Klinefelter syndrome (47,XXY) inactivate one X (one Barr body).
- 47,XXX females inactivate two X chromosomes (two Barr bodies).

Mosaicism due to X-inactivation

Which X (the paternal or maternal copy) is chosen for inactivation in an XX embryo is determined randomly and independently by each cell of the embryo, probably at the 10–20-cell stage (some exceptions to this are known, but they do not affect discussion of human pedigrees). However, once a cell has chosen which X to inactivate, that X remains inactive in all its daughter cells. Within the clonal descendants of one embryonic cell, each cell expresses the same X chromosome. Thus, a heterozygous female is a mosaic of clones in which alternative alleles are expressed. Each cell expresses either the normal or the abnormal allele, but not both.

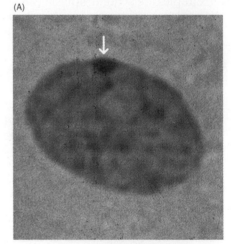

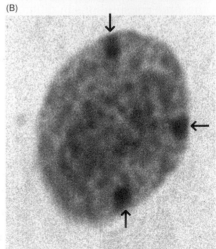

Figure 3.8 A Barr body. (A) A cell from an XX female has one inactivated X chromosome and shows a single Barr body. (B) A cell from a 49,XXXXY male has three inactivated X chromosomes and shows three Barr bodies. (Courtesy of Malcolm Ferguson-Smith, University of Cambridge.)

When the phenotype depends on a circulating product, as in hemophilia (failure of blood to clot; OMIM 306700, 306900), there is an averaging effect between the normal and abnormal clones. Female carriers may be biochemically abnormal but are usually clinically unaffected. When the phenotype is a localized property of individual cells, as in hypohidrotic ectodermal dysplasia (missing sweat glands, and abnormal teeth and hair; OMIM 305100), female carriers show patches of normal and abnormal tissue. If an X-linked condition is pathogenic in males because of the absence of some category of cell, those cells will show highly skewed X-inactivation in carrier females. For example, males with X-linked Bruton agammaglobulinemia (OMIM 300300) lack all mature B lymphocytes. Carrier females have B lymphocytes, but in every such cell the X chromosome that carries the mutation is the one that has been inactivated. During embryogenesis some cells will have inactivated the normal X chromosome, but the descendants of those cells will have been unable to develop into mature B lymphocytes. Therefore lymphocytes, but not other tissues, show completely skewed X-inactivation in these women.

Female carriers of recessive X-linked conditions often show some minor signs of the condition. Occasional heterozygous women may be quite severely affected because by bad luck most cells in some critical tissue have inactivated the normal X. They are known as **manifesting heterozygotes**. Equally, women who are heterozygous for a dominant X-linked condition are usually more mildly and variably affected than men, because many of their cells express only the normal X.

Few genes on the Y chromosome

No known human character, apart from maleness itself, is known to give the stereotypical Y-linked pedigree of Figure 3.7. Claims of Y-linkage for 'porcupine men' and hairy ears are questionable (see OMIM 146600 and 425500, respectively). Because normal females lack all Y-linked genes, any such genes must code either for non-essential characters or for male-specific functions. Some genes exist as functional copies on both the Y and the X; they might prove an exception to this argument, but they would not give a classical Y-linked pedigree pattern. Interstitial deletions of the Y-chromosome long arm are an important cause of male infertility, but of course infertile males will not produce pedigrees like that in Figure 3.7. Jobling and Tyler-Smith provide a useful review of the gene content of the Y chromosome and its possible involvement in disease (see Further Reading). As described in Chapter 10, loss of genes from the Y chromosome has been a steady evolutionary process, which may eventually culminate in loss of the entire chromosome.

Genes in the pseudoautosomal region

As described in Chapter 2 (Section 2.2), the distal 2.6 Mb of Xp and Yp contain homologous DNA sequences, and in males they pair in prophase I of meiosis. This ensures that at anaphase I each daughter cell receives one sex chromosome, either the X or the Y. Pairing is sustained by an obligatory crossover in this region. Thus, the few genes in these regions segregate in a **pseudoautosomal** and not a sex-linked pattern (**Figure 3.9**). Leri–Weill dyschondrosteosis (OMIM 127300) is one of the few pseudoautosomal conditions described. Note that the OMIM

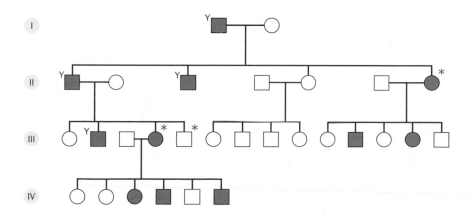

Figure 3.9 Pedigree of a pseudoautosomal condition. The allele causing this dominant condition is on the Y chromosome in the affected males marked Y, but on the X chromosome in all other affected individuals. The three individuals marked with a red asterisk are the result of an X–Y crossover in their father. Note how the pedigree is indistinguishable from an autosomal dominant pedigree, although the causative gene is located on Xp or Yp.

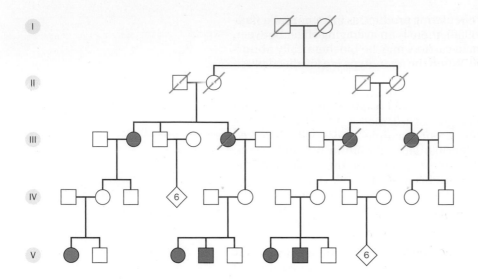

Figure 3.10 Pedigree of a mitochondrially determined disease. Note the very incomplete penetrance: although every child of an affected mother inherits mutant mitochondria through the ovum, only some are clinically affected. One cause of this can be variable heteroplasmy. [Family reported in Prezant TR, Agapian JV, Bohlman MC et al. (1993) *Nat. Genet.* 4, 289–294. With permission from Macmillan Publishers Ltd.]

number starts with 1, indicating autosomal dominant inheritance, although the causative genes are on the X and Y chromosomes. There is a second, much smaller, region of X–Y homology at the distal ends of the long arms, but there is not normally any crossing over between those short sequences, and so there is no special pedigree pattern associated with genes in that second pseudoautosomal region.

Conditions caused by mutations in the mitochondrial DNA

In addition to the mutations in genes carried on the nuclear chromosomes, mitochondrial mutations are a significant cause of human genetic disease. The mitochondrial genome is small (only 16.5 kb) but it is highly mutable in comparison with the nuclear genome, probably because mitochondrial DNA replication is more error-prone and the number of replications is much higher. Mitochondrially encoded diseases have two unusual features, **matrilineal inheritance** and frequent **heteroplasmy**.

Inheritance is matrilineal, because sperm do not contribute mitochondria to the zygote (some exceptions to this have been described, but paternally derived mitochondrial variants are not normally detected in children). Thus, a mitochondrially inherited condition can affect both sexes but is passed on only by affected mothers. At first sight the pedigree pattern (**Figure 3.10**) can seem to be autosomal dominant, because an affected parent can have affected children of either sex—but affected fathers never transmit the condition to their offspring.

Cells contain many mitochondrial genomes, and there may be multiple mitochondrial species in one individual. In some patients with a mitochondrial disease, every mitochondrial genome carries the causative mutation (**homoplasmy**), but in other cases a mixed population of normal and mutant genomes is seen within each cell (heteroplasmy). A single-celled zygote cannot be mosaic for nuclear genes, but it contains many mitochondria; a heteroplasmic mother can therefore have a heteroplasmic child. In such cases the proportion of abnormal mitochondrial genomes can vary widely between the mother and child. This happens because during early development the germ line passes through a stage in which cells contain very few mitochondria (the 'mitochondrial bottleneck'). Additionally, mitochondrial variants, especially those involving large deletions or duplications of the mitochondrial genome, often evolve rapidly within an individual, so that different tissues or the same tissue at different times may contain different spectra of variants. The overall result of all these factors is that clinical phenotypes are often highly variable, even within a family. The MITOMAP database (www.mitomap.org) is the best first source of information on all these matters, and gives many useful examples.

The mitochondrion is not a self-sufficient organelle. Many essential mitochondrial functions are provided by nuclear-encoded genes. It follows that many diseases that are the result of mitochondrial dysfunction are nevertheless caused by mutations in nuclear genes and follow standard Mendelian patterns.

The mode of inheritance can rarely be defined unambiguously in a single pedigree

Given the limited size of human families, it is rarely possible to be completely certain of the mode of inheritance of a character from an inspection of a single pedigree. For many of the rarer conditions, the stated mode is no more than an informed guess. Assigning modes of inheritance is important, because that is the basis of the risk estimates used in genetic counseling. However, it is important to recognize that until the underlying gene has been convincingly identified, the mode of inheritance deduced from an examination of pedigrees is often a working hypothesis rather than established fact. OMIM uses specific symbols before the number of an entry to denote the status of each entry:

* denotes a gene of known sequence.

\# denotes a phenotype with a known molecular basis.

\+ denotes a gene of known sequence and a phenotype.

% denotes a confirmed Mendelian phenotype for which the gene has not yet been definitely identified.

In experimental animals one could quickly check the mode of inheritance by setting up a suitable cross and checking for a 1 in 2 or 1 in 4 ratio of phenotypes in the offspring. In human pedigrees, the proportion of affected children (the **segregation ratio**) is not a very reliable indicator of the mode of inheritance. Mostly this is because the numbers are too small, but the way in which affected families are collected can also bias the observed ratio of affected to unaffected children.

Getting the right ratios: the problem of bias of ascertainment

Suppose we wish to show that a condition is autosomal recessive. We could collect a set of families and check that the segregation ratio is 1 in 4. At first sight this would seem a trivial task, provided that the condition is not too rare. But in fact the expected proportion of affected children in our sample is not 1 in 4. The problem is **bias of ascertainment** (ascertainment in this context means finding the people who form the study sample).

Assuming there is no independent way of recognizing carriers, the families will be identified through an affected child. Thus, the families shown in the unshaded area in **Figure 3.11** will not be ascertained, and the observed

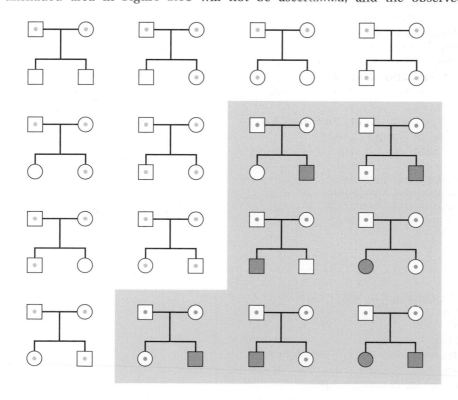

Figure 3.11 Biased ascertainment. In each of these 16 pedigrees, both parents are carriers of an autosomal recessive condition. Overall, their 32 children show the expected 1:2:1 distribution of genotypes (*AA*, 8/32; *Aa*, 16/32; *aa*, 8/32). However, if families can be recognized only through affected children, only the families shown in the shaded area will be picked up, and the proportion of the children in that sample who are affected will be 8/14, not 1/4. Statistical methods are available for correcting such biased ascertainment and recovering the true ratio.

segregation ratio in the two-child families collected is not 1 in 4 but 8 in 14. Families with three children, ascertained in the same way, would give a different segregation ratio, 48 in 111.

The examples above presuppose **complete truncate ascertainment**: we collect all the families in some defined population who have at least one affected child. But this is not the only possible way of collecting families. We might have ascertained affected children by taking the first 100 to be seen in a busy clinic (so that many more could have been ascertained from the same population by carrying on for longer). Under these conditions, a family with two affected children is twice as likely to be picked up as one with only a single affected child, and one with four affected is four times as likely. **Single selection**, where the probability of being ascertained is proportional to the number of affected children in the family, introduces a different bias of ascertainment, and requires a different statistical correction. Box 3.4 shows how some of these effects may be predicted and corrected.

The examples above are for recessive conditions, in which bias is most obvious, but more subtle biases can distort the estimated ratios in any condition. In general it is best not to rely too much on ratios of affected children when deciding the likely mode of inheritance of a condition.

The relation between Mendelian characters and gene sequences

Pedigree patterns provide an essential entry point into human genetics, but there is no simple one-to-one correspondence between the characters defined as Mendelian by pedigree analysis and genes defined as functional DNA units. It would be a serious error to imagine that the 4000 or so OMIM entries marked with a # or % symbol necessarily define 4000 DNA coding sequences. This would be an unjustified extension of the **one gene–one enzyme hypothesis** of Beadle and Tatum. Back in the 1940s this hypothesis allowed a major leap forward in understanding how genes determine phenotypes. Since then it has been extended: some genes encode nontranslated RNAs, some proteins are not enzymes, and many proteins contain several separately encoded polypeptide

---

BOX 3.4 CALCULATING AND CORRECTING THE SEGREGATION RATIO FOR A RECESSIVE CONDITION

The segregation ratio for a family with n children, under complete truncate ascertainment, can be estimated from the following formulae, which give the proportions of sibships with different numbers of affected children:

$(1/4)^n$	all children affected
$(n,1)(1/4)^{n-1}(3/4)$	all except one child affected
$(n,2)(1/4)^{n-2}(3/4)^2$	all except two children affected
etc.	

(n,x) means $n!/[x!(n-x)!]$, where $n!$ (n factorial) means

$n \times (n-1) \times (n-2) \times (n-3) \times \ldots \times 2 \times 1$

For the mathematically inclined, this is an example of a truncated binomial distribution, a binomial expansion of $(1/4 + 3/4)^n$ in which the last term (no affected children) is omitted.

The overall proportion of affected children can now be calculated. For example, for every 64 three-child families we have:

1 family with 3 affected	total 3 affected, 0 unaffected
9 families with 2 affected	total 18 affected, 9 unaffected
27 families with 1 affected	total 27 affected, 54 unaffected
[27 families with no affected—but these families will not be ascertained]	
Overall (among ascertained families)	48 affected, 63 unaffected
Apparent segregation ratio 48/(48 + 63) = 48/111 = 0.432	

Correcting the segregation ratio

If: p = the true (unbiased) segregation ratio
  R = the number of affected children
  S = the number of affected singletons (children who are the only affected child in the family)
  T = the total number of children
  N = the number of sibships

For complete truncate ascertainment, $p = (R - S)/(T - S)$.

For example, for the three-child families illustrated above, $p = (48 - 27)/(111 - 27) = 21/84 = 0.25$.

For single selection, $p = (R - N)/(T - N)$.

chains. But even with these extensions, Beadle and Tatum's hypothesis cannot be used to imply a one-to-one correspondence between entries in the OMIM catalogue and DNA transcription units.

The genes of classical genetics are abstract entities. Any character that is determined at a single chromosomal location will segregate in a Mendelian pattern—but the determinant may not be a gene in the molecular geneticist's sense of the word. Fascio-scapulo-humeral muscular dystrophy (severe but non-lethal weakness of certain muscle groups; OMIM 158900) is caused by small deletions of sequences at 4q35—but, at the time of writing, nobody has yet found a relevant protein-coding sequence at that location, despite intensive searching and sequencing. This example is unusual—most OMIM entries probably do describe the consequences of mutations affecting a single transcription unit. However, there is still no one-to-one correspondence between phenotypes and transcription units because of three types of heterogeneity:

- **Locus heterogeneity** is where the same clinical phenotype can result from mutations at any one of several different loci. This may be due to **epistasis**, where gene A controls the action of gene B, or lies upstream of gene B in a pathway; alternatively genes A and B may function in separate pathways that affect the same phenotype.

- **Allelic heterogeneity** is where many different mutations within a given gene can be seen in different patients with a certain genetic condition (explored more fully in Chapter 16). Many diseases show both locus and allelic heterogeneity.

- **Clinical heterogeneity** is used here to describe the situation in which mutations in the same gene produce two or more different diseases in different people. For example, mutations in the *HPRT* gene can produce either a form of gout or Lesch–Nyhan syndrome (severe mental retardation with behavioral problems; OMIM 300322). Note that this is not the same as **pleiotropy**. That term means that one mutation has multiple effects in the same organism. Most mutations are pleiotropic if you look carefully enough at the phenotype.

Locus heterogeneity

Hearing loss provides good examples of locus heterogeneity. When two people with autosomal recessive profound congenital hearing loss marry, as they often do, the children most often have normal hearing. It is easy to see that many different genes would be needed to construct so exquisite a machine as the cochlear hair cell, and a defect in any of those genes could lead to deafness. If the mutations causing deafness are in different genes in the two parents, all their children will be double heterozygotes—but assuming that both parental conditions are recessive, the children will have normal hearing (**Figure 3.12**).

When two recessive characters may or may not be caused by mutations at the same locus, a test cross between homozygotes for the two characters (in experimental organisms) can provide the answer. This is called a **complementation test**. If the mutations in the two parental stocks are at the same locus, the progeny will not have a wild-type allele, and so will be phenotypically abnormal. If there are two different loci, the progeny are heterozygous for each of the two recessive characters, and therefore phenotypically normal, as illustrated in **Table 3.1**.

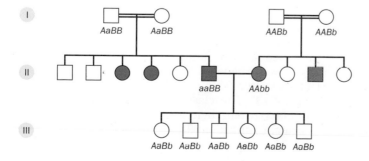

Figure 3.12 Complementation. Affected individuals all have profound congenital hearing loss. II$_6$ and II$_7$ are offspring of unaffected but consanguineous parents, and each has affected sibs, making it likely that all the affected individuals in generation II are homozygous for autosomal recessive hearing loss. However, all the children of II$_6$ and II$_7$ are unaffected. This shows that the mutations in II$_6$ and II$_7$ must be non-allelic, as indicated by the genotypes.

TABLE 3.1 RESULTS OF MATING BETWEEN TWO HOMOZYGOTES FOR A RECESSIVE CHARACTER		
	One locus (*A*)	Two loci (*A, B*)
Parental genotypes	*aa* × *aa*	*aaBB* × *AAbb*
Offspring genotype	*aa*	*AaBb*
Offspring phenotype	abnormal	normal

Very occasionally alleles at the same locus can complement each other (**interallelic complementation**). This might happen if the gene product is a protein that dimerizes, and the mutant alleles in the two parents affect different parts of the protein, such that a heterodimer may retain some function. In general, however, if two mutations complement each other, it is reasonable to assume that they involve different loci. Cell-based complementation tests (fusing cell lines in tissue culture) have been important in sorting out the genetics of human phenotypes such as DNA repair defects, in which the abnormality can be observed in cultured cells. Hearing loss provides a rare opportunity to see complementation in action in human pedigrees.

Locus heterogeneity is only to be expected in conditions such as deafness, blindness, or learning disability, in which a rather general pathway has failed; but even with more specific pathologies, multiple loci are very frequent. A striking example is Usher syndrome, an autosomal recessive combination of hearing loss and progressive blindness (retinitis pigmentosa), which can be caused by mutations at any of 10 or more unlinked loci. OMIM has separate entries for known examples of locus heterogeneity (defined by linkage or mutation analysis), but there must be many undetected examples still contained within single entries.

Clinical heterogeneity

Sometimes several apparently distinct human phenotypes turn out all to be caused by different allelic mutations at the same locus. The difference may be one of degree—mutations that partly inactivate the dystrophin gene produce Becker muscular dystrophy (OMIM 300376), whereas mutations that completely inactivate the same gene produce the similar but more severe Duchenne muscular dystrophy (lethal muscle wasting; OMIM 310200). At other times the difference is qualitative—inactivation of the androgen receptor gene causes androgen insensitivity (46,XY embryos develop as females; OMIM 313700), but expansion of a run of glutamine codons within the same gene causes a very different disease, spinobulbar muscular atrophy or Kennedy disease (OMIM 313200). These and other genotype–phenotype correlations are discussed in more depth in Chapter 13.

3.3   COMPLICATIONS TO THE BASIC MENDELIAN PEDIGREE PATTERNS

In real life various complications often disguise a basic Mendelian pattern. Figures 3.13–3.21 illustrate several common complications.

A common recessive condition can mimic a dominant pedigree pattern

If a recessive trait is common in a population, there is a good chance that it may be brought into the pedigree independently by two or more people. A common recessive character such as blood group O may be seen in successive generations because of repeated matings of group O people with heterozygotes. This produces a pattern resembling dominant inheritance (**Figure 3.13**). The classic Mendelian pedigree patterns are best seen with rare conditions, where there is little chance that somebody who marries into the family might coincidentally also carry the disease mutation that is segregating in the family.

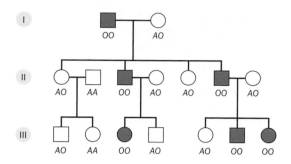

Figure 3.13 Complications to the basic Mendelian patterns (1): a common recessive condition giving an apparently dominant pedigree. If a recessive trait is sufficiently common that unrelated people marrying in to the family often carry it, the pedigree may misleadingly resemble that of a dominant trait. The condition in the figure is blood group O.

A dominant condition may fail to manifest itself

The **penetrance** of a character, for a given genotype, is the probability that a person who has the genotype will manifest the character. By definition, a dominant character is manifested in a heterozygous person, and so should show 100% penetrance. Nevertheless, many human characters, although generally showing dominant inheritance, occasionally skip a generation. In **Figure 3.14**, II_2 has an affected parent and an affected child, and almost certainly carries the mutant gene, but is phenotypically normal. This would be described as a case of **non-penetrance**.

There is no mystery about non-penetrance—indeed, 100% penetrance is the more surprising phenomenon. Very often the presence or absence of a character depends, in the main and in normal circumstances, on the genotype at one locus, but an unusual genetic background, a particular lifestyle, or maybe just chance means that the occasional person may fail to manifest the character. Non-penetrance is a major pitfall in genetic counseling. It would be an unwise counselor who, knowing that the condition in Figure 3.14 was dominant and seeing that III_7 was free of signs, told her that she had no risk of having affected children. One of the jobs of genetic counselors is to know the usual degree of penetrance of each dominant condition.

Frequently, of course, a character depends on many factors and does not show a Mendelian pedigree pattern even if entirely genetic. There is a continuum of characters from fully penetrant Mendelian to multifactorial (**Figure 3.15**), with increasing influence of other genetic loci and/or the environment. No logical break separates imperfectly penetrant Mendelian from multifactorial characters; it is a question of which is the most useful description to apply.

Age-related penetrance in late-onset diseases

A particularly important case of reduced penetrance is seen with late-onset diseases. Genetic conditions are not necessarily **congenital** (present at birth). The genotype is fixed at conception, but the phenotype may not manifest until adult life. In such cases the penetrance is age-related. Huntington disease (progressive

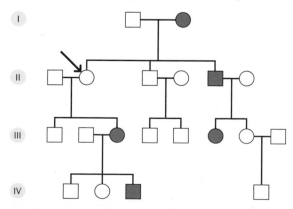

Figure 3.14 Complications to the basic Mendelian patterns (2): non-penetrance. Individual II_2 carries the disease gene but does not show symptoms. Other unaffected family members, such as II_3, III_1, IV_1, or IV_2, might also be non-penetrant gene carriers.

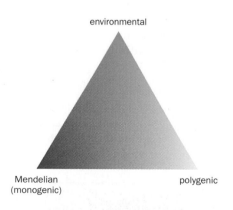

Figure 3.15 A two-dimensional spectrum of human characters. The variable influence of environmental factors is added to the spectrum of genetic determination shown in Figure 3.1. The mix of factors determining any particular trait could be represented by a point located somewhere within the triangle.

neurodegeneration; OMIM 143100) is a well-known example (**Figure 3.16**). Delayed onset might be caused by the slow accumulation of a noxious substance, by incremental tissue death, or by an inability to repair some form of environmental damage. Hereditary cancers are caused by a chance second mutation affecting a cell of a person who already carries one mutation in a tumor suppressor gene in every cell (see Chapter 17). That second mutation could occur at any time, and so the risk of having acquired it is cumulative and increases through life. Depending on the disease, the penetrance may become 100% if the person lives long enough, or there may be people who carry the gene but who will never develop symptoms no matter how long they live. Age-of-onset curves such as those in Figure 3.16 are important tools in genetic counseling, because they enable the geneticist to estimate the chance that an at-risk but asymptomatic person will subsequently develop the disease.

Many conditions show variable expression

Related to non-penetrance is the **variable expression** frequently seen in dominant conditions. **Figure 3.17** shows an example from a family with Waardenburg syndrome. Different family members show different features of the syndrome. The cause is the same as with non-penetrance: other genes, environmental factors, or pure chance have some influence on the development of the symptoms. Non-penetrance and variable expression are typically problems with dominant, rather than recessive, characters. In part this reflects the difficulty of spotting non-penetrant cases in a typical recessive pedigree. However, as a general rule, recessive conditions are less variable than dominant ones, probably because the phenotype of a heterozygote involves a balance between the effects of the two alleles, so that the outcome is likely to be more sensitive to outside influence than the phenotype of a homozygote. However, both non-penetrance and variable expression are occasionally seen in recessive conditions.

These complications are much more conspicuous in humans than in experimental organisms. Laboratory animals and crop plants are far more genetically uniform than humans, and live in much more constant environments. What we see in human genetics is typical of a natural mammalian population. Nevertheless, mouse geneticists are familiar with the way in which the expression of a mutant gene can change when it is bred onto a different genetic background—an important consideration when studying mouse models of human diseases.

Anticipation

Anticipation describes the tendency of some conditions to become more severe (or have earlier onset) in successive generations. Anticipation is a hallmark of conditions caused by a very special genetic mechanism (**dynamic mutation**) whereby a run of tandemly repeated nucleotides, such as CAGCAGCAGCAGCAG, is meiotically unstable. A $(CAG)_{40}$ sequence in a parent may appear as a $(CAG)_{55}$ sequence in a child. Certain of these unstable repeat sequences become pathogenic above some threshold size. Examples include fragile X syndrome (mental retardation with various physical signs; OMIM 309550), myotonic dystrophy (a

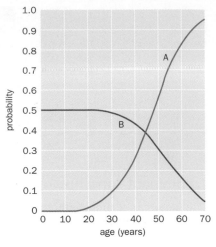

Figure 3.16 Age-of-onset curves for Huntington disease. Curve A shows the probability that an individual carrying the disease gene will have developed symptoms by a given age. Curve B shows the risk at a given age that an asymptomatic person who has an affected parent nevertheless carries the disease gene. [From Harper PS (1998) Practical Genetic Counselling, 5th ed. With permission from Edward Arnold (Publishers) Ltd.]

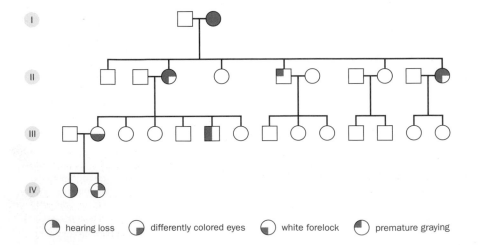

hearing loss   differently colored eyes   white forelock   premature graying

Figure 3.17 Complications to the basic Mendelian patterns (3): variable expression. Different affected family members show different features of type 1 Waardenburg syndrome, an autosomal dominant trait.

very variable multisystem disease with characteristic muscular dysfunction; OMIM 160900), and Huntington disease. Severity or age of onset of these diseases correlates with the repeat length, and the repeat length tends to grow as the gene is transmitted down the generations. These unusual diseases are discussed in more detail in Chapter 16.

Dominant pedigrees often give the appearance of anticipation, but this is usually a false impression. True anticipation is very easily mimicked by random variations in severity. A family comes to clinical attention when a severely affected child is born. Investigating the history, the geneticist notes that one of the parents is affected, but only mildly. This looks like anticipation but may actually be just a bias of ascertainment. Had the parent been severely affected, he or she would most probably never have become a parent; had the child been mildly affected, the family might not have come to notice. Claims of anticipation without evidence of a dynamic mutation should be treated with caution. To be credible, a claim of anticipation requires careful statistical backing or direct molecular evidence, not just clinical impression.

Imprinting

Certain human characters are autosomal dominant, affect both sexes, and are transmitted by parents of either sex—but manifest only when inherited from a parent of one particular sex. For example, there are families with autosomal dominant inheritance of glomus body tumors or paragangliomas (OMIM 168000). In these families the tumors occur only in men or women who inherit the gene from their father (**Figure 3.18A**). Beckwith–Wiedemann syndrome (BWS; OMIM 130650) shows the opposite effect. This combination of congenital abdominal wall defects (exomphalos), an oversized tongue (macroglossia), and excessive birth weight is sometimes inherited as a dominant condition, but it is expressed only in babies who inherit it from their mother (Figure 3.18B).

(A)

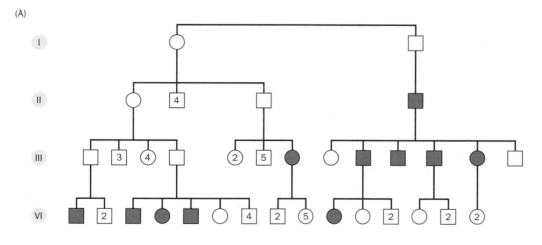

(B)

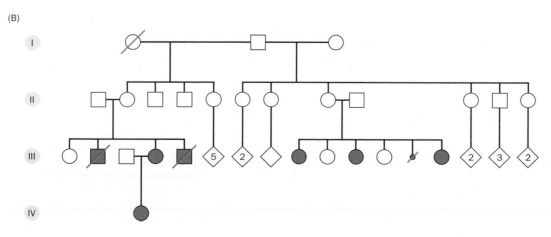

Figure 3.18 Complications to the basic Mendelian patterns (4): imprinted gene expression. (A) In this family, autosomal dominant glomus tumors manifest only when the gene is inherited from the father. (B) In this family, autosomal dominant Beckwith–Wiedemann syndrome manifests only when the gene is inherited from the mother. [(A) family reported in Heutink P, van der Mey AG, Sandkuijl LA et al. (1992) *Hum. Molec. Genet.* 1, 7–10. With permission from Oxford University Press. (B) family reported in Viljoen & Ramesar (1992) *J. Med. Genet.* 29, 221–225. With permission from BMJ Publishing Group Ltd.]

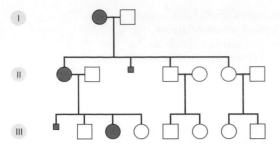

Figure 3.19 Complications to the basic Mendelian patterns (5): a male-lethal X-linked condition. In this family with X-linked dominant incontinentia pigmenti (OMIM 308300), affected males abort spontaneously (small squares).

These parental sex effects are evidence of **imprinting**, a poorly understood phenomenon whereby certain genes are somehow marked (imprinted) with their parental origin. Many questions surround the mechanism and evolutionary purpose of imprinting. Imprinting is an example of an **epigenetic** mechanism—a heritable change in gene expression that does not depend on a change in DNA sequence. Epigenetic mechanisms and effects are discussed in more detail in Chapter 11.

Male lethality may complicate X-linked pedigrees

For some X-linked dominant conditions, absence of the normal allele is lethal before birth. Thus affected males are not born, and we see a condition that affects only females, who pass it on to half their daughters but none of their sons (**Figure 3.19**). If the family were large enough, one might notice that there are only half as many boys as girls, and a history of miscarriages (because the 50% of males who inherited the mutant allele miscarry before birth). An example is incontinentia pigmenti (linear skin defects following defined patterns known as Blaschko lines, often accompanied by neurological or skeletal problems; OMIM 308300). Rett syndrome (OMIM 312750) shows a characteristic developmental regression in females: these girls are normal at birth, develop normally for the first year or two, but then stop developing, and eventually regress, losing speech and other abilities that they acquired in early life. In males, Rett syndrome is usually lethal before birth, but rare survivors have a severe neonatal encephalopathy. Until the *RTT* gene was cloned, it was not recognized that these males had the same gene defect as females with classical Rett syndrome.

Inbreeding can complicate pedigree interpretation

The absence of male-to-male transmission is a hallmark of X-linked pedigree patterns—but if an affected man marries a carrier woman, he may have an affected son. Naturally this is most likely to happen as a result of inbreeding in a family in which the condition is segregating. Such matings can also produce homozygous affected females. **Figure 3.20** shows an example.

New mutations and mosaicism complicate pedigree interpretation

Many cases of severe dominant or X-linked genetic disease are the result of fresh mutations, striking without warning in a family with no previous history of the condition. A fully penetrant lethal dominant condition would necessarily always occur by fresh mutation, because the parents could never be affected—an example is thanatophoric dysplasia (severe shortening of long bones and abnormal fusion of cranial sutures; OMIM 187600). For a non-lethal but deleterious dominant condition a similar argument applies, but to a smaller degree. If the condition prevents many affected people from reproducing, but if nevertheless fresh cases keep occurring, many or most of these must be caused by new mutations. Serious X-linked recessive diseases also show a significant proportion of fresh mutations, because the disease allele is exposed to natural selection whenever it is in a male. Autosomal recessive pedigrees, in contrast, are not significantly affected. Ultimately there must have been a mutational event, but the mutant allele can propagate for many generations in asymptomatic carriers, and we can reasonably assume that the parents of an affected child are both carriers.

When a normal couple with no relevant family history have a child with severe abnormalities (**Figure 3.21A**), deciding the mode of inheritance and recurrence

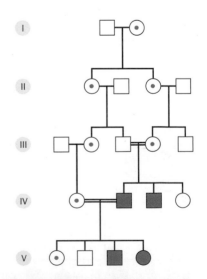

Figure 3.20 Complications to the basic Mendelian patterns (6): an X-linked recessive pedigree with inbreeding. There is an affected female and apparent male-to-male transmission.

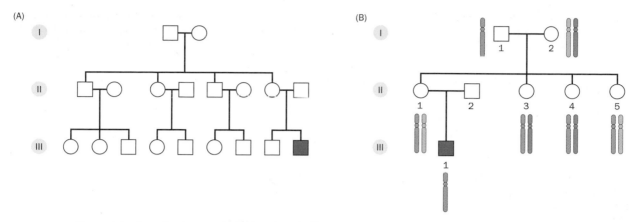

Figure 3.21 Complications to the basic Mendelian patterns (7): new mutations. (A) A new autosomal dominant mutation. The pedigree pattern mimics an autosomal or X-linked recessive pattern. (B) A new mutation in X-linked recessive Duchenne muscular dystrophy. The three grandparental X chromosomes were distinguished by using genetic markers; here we distinguish them with three different colors (ignoring recombination). III_1 has the grandpaternal X, which has acquired a mutation at one of four possible points in the pedigree—each with very different implications for genetic counseling:

- If III_1 carries a new mutation, the recurrence risk for all family members is very low.
- If II_1 is a germinal mosaic, there is a significant (but hard to quantify) risk for her future children, but not for those of her three sisters.
- If II_1 was the result of a single mutant sperm, her own future offspring have the standard recurrence risk for an X-linked recessive trait, but those of her sisters are free of risk.
- If I_1 was a germinal mosaic, all four sisters in generation II have a significant (but hard to quantify) risk of being carriers of the condition.

risk can be very difficult—the problem might be autosomal recessive, autosomal dominant with a new mutation, X-linked recessive (if the child is male), or nongenetic. A further complication is introduced by germinal mosaicism (see below).

Mosaics

We have seen that in serious autosomal dominant and X-linked diseases, in which affected people have few or no children, the disease alleles are maintained in the population by recurrent mutation. A common assumption is that an entirely normal person produces a single mutant gamete. However, this is not necessarily what happens. Unless there is something special about the mutational process, such that it can happen only during gametogenesis, mutations could arise at any time during post-zygotic life. Post-zygotic mutations produce *mosaics* with two (or more) genetically distinct cell lines. Mosaicism has already been described in Chapter 2 in connection with chromosomal aberrations; it can equally be the result of gene mutations.

Mosaicism can affect somatic and/or germ-line tissues. Post-zygotic mutations are not merely frequent, they are inevitable. Human mutation rates are typically 10^{-6} per gene per generation. That is to say, a person who carried a certain gene in its wild-type form at conception has a chance of the order of one in a million of passing it to a child in mutant form. The chain of cell divisions linking the two events would typically be a few hundred divisions long (longer in males than in females, and longer with age in males—see Chapter 2). But overall, something of the order of 10^{14} divisions would be involved in getting from a single-cell zygote to a person's complete adult body. Considering the likely mutation rate per cell division, it follows that every one of us must be a mosaic for innumerable mutations. This should cause no anxiety. If the DNA of a cell in your finger mutates to the Huntington disease genotype, or a cell in your ear picks up a cystic fibrosis mutation, there are absolutely no consequences for you or your family. Only if a somatic mutation results in the emergence of a substantial clone of mutant cells is there a risk to the whole organism. This can happen in two ways:

- The mutation occurs in an early embryo, affecting a cell that is the progenitor of a significant fraction of the whole organism. In that case the mosaic individual may show clinical signs of disease.

- The mutation causes abnormal proliferation of a cell that would normally replicate slowly or not at all, thus generating a clone of mutant cells. This is how cancer happens, and this whole topic is discussed in detail in Chapter 17.

A mutation in a germ-line cell early in development can produce a person who harbors a large clone of mutant germ-line cells [**germinal** (or **gonadal**) **mosaicism**]. As a result, a normal couple with no previous family history may produce more than one child with the same serious dominant disease. The pedigree mimics recessive inheritance. Even if the correct mode of inheritance is realized, it is very difficult to calculate a recurrence risk to use in counseling the parents. The problem is discussed by van der Meulen et al. (see Further Reading). Usually an empiric risk (see below) is quoted. Figure 3.21B shows an example of the uncertainty that germinal mosaicism introduces into counseling, in this case in an X-linked disease.

Molecular studies can be a great help in these cases. Sometimes it is possible to demonstrate directly that a normal father is producing a proportion of mutant sperm. Direct testing of the germ line is not feasible in women, but other accessible tissues such as fibroblasts or hair roots can be examined for evidence of mosaicism. A negative result on somatic tissues does not rule out germ-line mosaicism, but a positive result, in conjunction with an affected child, proves it. Individual II$_2$ in **Figure 3.22** is an example.

Chimeras

Mosaics start life as a single fertilized egg. **Chimeras**, in contrast, are the result of fusion of two zygotes to form a single embryo (the reverse of twinning), or alternatively of limited colonization of one twin by cells from a non-identical co-twin (**Figure 3.23**). Chimerism is proved by the presence of too many parental alleles at several loci (in a sample that is prepared from a large number of cells). If just one locus were involved, one would suspect mosaicism for a single mutation, rather than the much rarer phenomenon of chimerism. Blood-grouping centers occasionally discover chimeras among normal donors, and some intersex patients turn out to be XX/XY chimeras. A fascinating example was described by Strain et al. (see Further Reading). They showed that a 46,XY/46,XX boy was the result of two embryos amalgamating after an *in vitro* fertilization in which three embryos had been transferred into the mother's uterus.

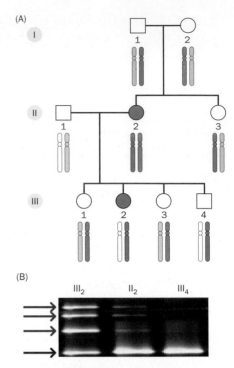

Figure 3.22 Germ-line and somatic mosaicism in a dominant disease.
(A) Although the grandparents in this pedigree are unaffected, individuals II$_2$ and III$_2$ both suffer from familial adenomatous polyposis, a dominant inherited form of colorectal cancer (OMIM 175100; see Chapter 17). The four grandparental copies of chromosome 5 (where the pathogenic gene is known to be located) were distinguished by using genetic markers, and are color-coded here (ignoring recombination). (B) The pathogenic mutation could be detected in III$_2$ by gel electrophoresis of blood DNA (red arrows; the black arrow shows the band due to the normal, wild-type allele). For II$_2$ the gel trace of her blood DNA shows the mutant bands, but only very weakly, showing that she is a somatic mosaic for the mutation. The mutation is absent in III$_4$, even though marker studies showed that he inherited the high-risk (blue in this figure) chromosome from his mother. Individual II$_2$ must therefore be both a germ-line and a somatic mosaic for the mutation. (Courtesy of Bert Bakker, Leiden University Medical Center.)

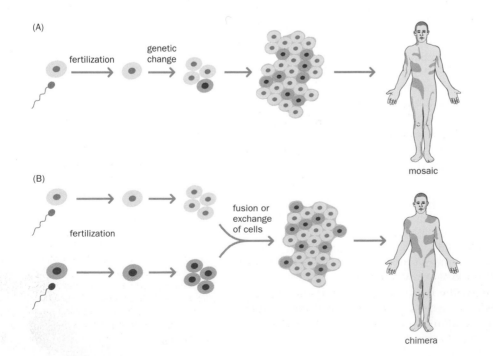

Figure 3.23 Mosaics and chimeras.
(A) Mosaics have two or more genetically different cell lines derived from a single zygote. The genetic change indicated may be a gene mutation, a numerical or structural chromosomal change, or in the special case of lyonization, X-inactivation. (B) A chimera is derived from two zygotes, which are usually both normal but genetically distinct.

3.4 GENETICS OF MULTIFACTORIAL CHARACTERS: THE POLYGENIC THRESHOLD THEORY

In the early twentieth century there was controversy between proponents of Mendelian and quantitative models of inheritance

By the time that Mendel's work was rediscovered in 1900, a rival school of genetics was well established in the UK and elsewhere. Francis Galton, the remarkable and eccentric cousin of Charles Darwin, devoted much of his vast talent to systematizing the study of human variation. Starting with an article, Hereditary talent and character, published in the same year, 1865, as Mendel's paper (and expanded in 1869 to a book, *Hereditary Genius*), he spent many years investigating family resemblances. Galton was devoted to quantifying observations and applying statistical analysis. His Anthropometric Laboratory, established in London in 1884, recorded from his subjects (who paid him threepence for the privilege) their weight, sitting and standing height, arm span, breathing capacity, strength of pull and of squeeze, force of blow, reaction time, keenness of sight and hearing, color discrimination, and judgments of length. In one of the first applications of statistics, he compared physical attributes of parents and children, and established the degree of correlation between relatives. By 1900 he had built up a large body of knowledge about the inheritance of such attributes, and a tradition (**biometrics**) of their investigation.

When Mendel's work was rediscovered, a controversy arose. Biometricians accepted that a few rare abnormalities or curious quirks might be inherited as Mendel described, but they pointed out that most of the characters likely to be important in evolution (fertility, body size, strength, and skill in catching prey or finding food) were continuous or quantitative characters and not amenable to Mendelian analysis. We all have these characters, only to different degrees, so you cannot define their inheritance by drawing pedigrees and marking in the people who have them. Mendelian analysis requires dichotomous characters that you either have or do not have. A controversy, heated at times, ran on between Mendelians and biometricians until 1918. That year saw a seminal paper by RA Fisher in which he demonstrated that characters governed by a large number of independent Mendelian factors (polygenic characters) would display precisely the continuous nature, quantitative variation, and family correlations described by the biometricians. Later, DS Falconer extended this model to cover dichotomous characters. Fisher's and Falconer's analyses created a unified theoretical basis for human genetics. The next sections set out their ideas, in a nonmathematical form. A more rigorous treatment can be found in textbooks of quantitative or population genetics.

Polygenic theory explains how quantitative traits can be genetically determined

Any variable quantitative character that depends on the additive action of a large number of individually small independent causes (whether genetic or not) will show a normal (Gaussian) distribution in the population. **Figure 3.24** gives a highly simplified illustration of this for a genetic character. We suppose the character to depend on alleles at a single locus, then at two loci, then at three. As more loci are included, we see two consequences:

- The simple one-to-one relationship between genotype and phenotype disappears. Except for the extreme phenotypes, it is not possible to infer the genotype from the phenotype.

- As the number of loci increases, the distribution looks increasingly like a Gaussian curve. The addition of a little environmental variation would smooth out the three-locus distribution into a good Gaussian curve.

A more sophisticated treatment, allowing dominance and varying gene frequencies, leads to the same conclusions. Because relatives share genes, their phenotypes are correlated, and Fisher's 1918 paper predicted the size of the correlation for different relationships.

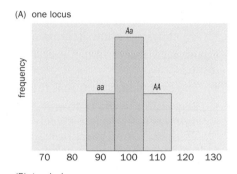

(A) one locus

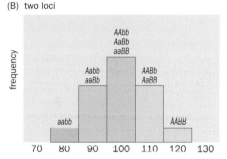

(B) two loci

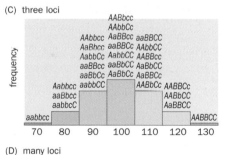

(C) three loci

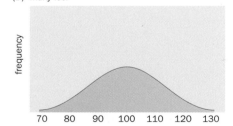

(D) many loci

Figure 3.24 Successive approximations to a Gaussian distribution. The charts show the distribution in the population of a hypothetical character that has a mean value of 100 units. The character is determined by the additive (co-dominant) effects of alleles. Each upper-case allele adds 5 units to the value, and each lower-case allele subtracts 5 units. All allele frequencies are 0.5. (A) The character is determined by a single locus. (B) Two loci. (C) Three loci. (D) The addition of a minor amount of 'random' (environmental or polygenic) variation produces the Gaussian curve.

Regression to the mean

A much misunderstood feature, both of biometric data and of polygenic theory, is **regression to the mean**. Imagine, for the sake of example only, that variation in IQ were entirely genetically determined. **Figure 3.25** shows that in our simplified two-locus model, for each class of mothers, the average IQ of their children is halfway between the mother's value and the population mean. This is regression to the mean—but its implications are often misinterpreted. Two common misconceptions are:

- After a few generations everybody will be exactly the same.
- If a character shows regression to the mean, it must be genetic.

Figure 3.25 shows that the first of these beliefs is wrong. In a simple genetic model:

- The overall distribution is the same in each generation.
- Regression works both ways: for each class of *children*, the average for their *mothers* is halfway between the children's value and the population mean. This may sound paradoxical, but it can be confirmed by inspecting, for example, the right-hand column of the bottom histogram in the figure (children of IQ 120). One-quarter of their mothers have IQ 120, half 110, and one-quarter 100, making an average of 110.

Regarding the second of these beliefs, regression to the mean is not a genetic mechanism but a purely statistical phenomenon. Whether the determinants of IQ are genetic, environmental, or any mix of the two, if we take an exceptional group of mothers (for example, those with an IQ of 120), then these mothers must have had an exceptional set of determinants. If we take a second group who share half those determinants (their children, their sibs, or either of their parents), the average phenotype in this second group will deviate from the population mean by half as much. Genetics provides the figure of one-half—it is because each child inherits one-half of his or her genes from his mother that the average IQ of the children (in this model) is half of the way between the mother's IQ and the population mean—but genetics does not supply the principle of regression.

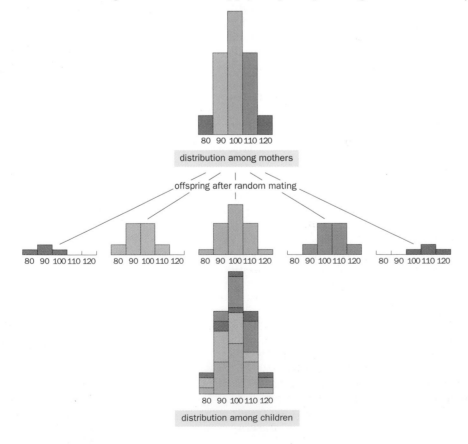

distribution among mothers

offspring after random mating

distribution among children

Figure 3.25 Regression to the mean. The same character as in Figure 3.24B: mean 100, determined by co-dominant alleles *A*, *a*, *B*, and *b* at two loci, all gene frequencies = 0.5. Top: distribution in a series of mothers. Middle: distributions in children of each class of mothers, assuming random mating. Bottom: summed distribution in the children. Note that: (a) the distribution in the children is the same as the distribution in the mothers; (b) for each class of mothers, the mean for their children is halfway between the mothers' value and the population mean (100); and (c) for each class of children (bottom), the mean for their mothers is halfway between the children's value and the population mean.

Hidden assumptions

In the simple model of Figure 3.25 there is a hidden assumption: that there is random mating. For each class of mothers, the average IQ of their husbands is assumed to be 100. Thus, the average IQ of the children is actually the mid-parental IQ, as common sense would suggest. In the real world, highly intelligent women tend to marry men of above average intelligence (**assortative mating**). The regression would therefore be less than halfway to the population mean, even if IQ were a purely genetic character.

A second assumption of our simplified model is that there is no dominance. Each person's phenotype is assumed to be the sum of the contribution of each allele at each relevant locus. If we allow dominance, the effect of some of a parent's genes will be masked by dominant alleles and invisible in their phenotype, but they can still be passed on and can affect the child's phenotype. Given dominance, the expectation for the child is no longer the mid-parental value. Our best guess about the likely phenotypic effect of the masked recessive alleles is obtained by looking at the rest of the population. Therefore, the child's expected phenotype will be displaced from the mid-parental value toward the population mean. How far it will be displaced depends on how important dominance is in determining the phenotype.

Heritability measures the proportion of the overall variance of a character that is due to genetic differences

Gaussian curves are specified by just two parameters, the mean and the variance (or the standard deviation, which is the square root of the variance). Variances have the useful property of being additive when they are due to independent causes. Thus, the overall variance of the phenotype V_P is the sum of the variances due to the individual causes of variation—the environmental variance V_E and the genetic variance V_G:

$$V_P = V_G + V_E$$

V_G can in turn be broken down to a variance V_A, which is due to simply additive genetic effects, and two extra terms. V_D accounts for dominance effects: because of dominance, the effect of a certain combination of alleles at a locus may not be simply the sum of their individual effects. V_I is an interaction variance: the overall effect of genes at several loci may not be simply the sum of the effects that each would have if present alone:

$$V_G = V_A + V_D + V_I$$

Therefore:

$$V_P = V_A + V_D + V_I + V_E$$

The **heritability** (h^2) of a trait is the proportion of the total variance that is genetic:

$$h^2 = V_G / V_P$$

For animal breeders interested in breeding cows with higher milk yields, the heritability is an important measure of how far a breeding program can create a herd in which the average animal resembles today's best. For humans, heritability figures require very careful interpretation. Heritabilities of human traits are often estimated as part of segregation analysis, which is described in more detail in Chapter 15. However, it should be borne in mind that for many human traits, especially behavioral traits, the simple partitioning of variance into environmental and genetic components is not applicable. Different genotypes are likely to respond differently to different environments. Moreover, we give our children both their genes and their environment. Genetic disadvantage and social disadvantage can often go together, so genetic and environmental factors may not be independent. If genetic and environmental factors are not independent, V_P does not equal $V_G + V_E$; we need to introduce additional variances to account for the correlations or interactions between specific genotypes and specific environments. A proliferation of variances can rapidly reduce the explanatory power of the models, and in general this has been a difficult area in which to work.

Misunderstanding heritability

The term *heritability* is often misunderstood. Heritability is quite different from the mode of inheritance. The mode of inheritance (autosomal dominant, polygenic, etc.) is a fixed property of a trait, but heritability is not. Heritability of IQ is shorthand for heritability of variations in IQ. Contrast the following two questions:

- To what extent is IQ genetic? This is a meaningless question.

- How much of the differences in IQ between people in a particular country at a particular time is caused by their genetic differences, and how much by their different environments and life histories? This is a meaningful question, even if difficult to answer.

In different social circumstances, the heritability of IQ will differ. The more equal a society is, the higher the heritability of IQ should be. If everybody has equal opportunities, several of the environmental differences between people have been removed. Therefore more of the remaining differences in IQ will be due to the genetic differences between people.

The threshold model extended polygenic theory to cover dichotomous characters

Some continuously variable characters such as blood pressure or body mass index are of great importance in public health, but for medical geneticists the innumerable diseases and malformations that tend to run in families but do not show Mendelian pedigree patterns are of greater concern. A major conceptual tool in non-Mendelian genetics was provided by the extension of polygenic theory to dichotomous or discontinuous characters (those that you either have or do not have).

The key concept is that even for a dichotomous character, there is an underlying continuously variable **susceptibility**. You may or may not have a cleft palate, but every embryo has a certain susceptibility to cleft palate. The susceptibility may be low or high; it is polygenic and follows a Gaussian distribution in the population. Together with the polygenic susceptibility, we postulate the existence of a *threshold*. Embryos whose susceptibility exceeds a critical threshold value develop cleft palate; those whose susceptibility is below the threshold, even if only just below, develop a normal palate. Stripped of mathematical subtlety, the model can be represented as in **Figure 3.26**. The threshold can be imagined as the neutral point of the balance. Changing the balance of factors tips the phenotype one way or the other.

For cleft palate, a polygenic threshold model seems intuitively reasonable. All embryos start with a cleft palate. During early development the palatal shelves must become horizontal and fuse together. They must do this within a specific developmental window of time. Many different genetic and environmental factors influence embryonic development, so it seems reasonable that the genetic part of the susceptibility should be polygenic. Whether the palatal shelves meet and fuse with time to spare, or whether they only just manage to fuse in time, is unimportant—if they fuse, a normal palate forms; if they do not fuse, a cleft palate results. There is therefore a natural threshold superimposed on a continuously variable process.

Using threshold theory to understand recurrence risks

Threshold theory helps explain how recurrence risks vary in families. Affected people must have an unfortunate combination of high-susceptibility alleles. Their relatives who share genes with them will also, on average, have an increased susceptibility, with the divergence from the population mean depending on the proportion of shared genes. Thus, polygenic threshold characters tend to run in families (**Figure 3.27**). Parents who have had several affected children may have just been unlucky, but on average they will have more high-risk alleles than parents with only one affected child. The threshold is fixed, but the average susceptibility, and hence the recurrence risk, increases with an increasing number of previous affected children.

Many supposed threshold conditions have different incidences in the two sexes. Threshold theory accommodates this by postulating sex-specific thresholds.

Figure 3.26 Multifactorial determination of a disease or malformation. The angels and devils can represent any combination of genetic and environmental factors. Adding an extra devil or removing an angel can tip the balance, without that particular factor being the cause of the disease in any general sense. (Courtesy of the late Professor RSW Smithells.)

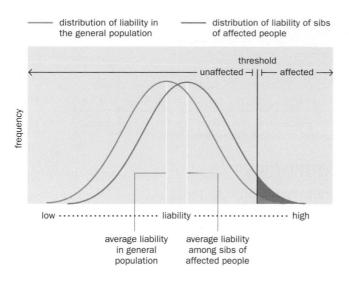

Figure 3.27 A polygenic threshold model for dichotomous non-Mendelian characters. Liability to the condition is polygenic and normally distributed (green curve). People whose liability is above a certain threshold value (the balance point in Figure 3.26) are affected. The distribution of liability among sibs of an affected person (purple curve) is shifted toward higher liability because they share genes with their affected sib. A greater proportion of them have liability exceeding the (fixed) threshold. As a result, the condition tends to run in families.

Congenital pyloric stenosis, for example, is five times more common in boys than girls. The threshold must be higher for girls than boys. Therefore, to be affected, a girl has on average to have a higher liability than a boy. Relatives of an affected girl therefore have a higher average susceptibility than relatives of an affected boy (**Figure 3.28**). The recurrence risk is correspondingly higher, although in each case a baby's risk of being affected is five times higher if it is a boy because a less extreme liability is sufficient to cause a boy to be affected (**Table 3.2**).

Counseling in non-Mendelian conditions is based on empiric risks

In genetic counseling for non-Mendelian conditions, risks are not derived from polygenic theory; they are **empiric risks** obtained through population surveys such as those in Table 3.2. This is fundamentally different from the situation with Mendelian conditions, where the risks (1 in 2, 1 in 4, and so on) come from theory. The effect of family history is also quite different. Consider two examples:

- If a couple have had a baby with cystic fibrosis, an autosomal recessive Mendelian condition, we can safely assume that they are both carriers. The risk of their next child being affected is 1 in 4. This remains true regardless of how many affected or normal children they have already produced (summed up in the phrase 'chance has no memory').

- If a couple have had a baby with neural tube defect, a complex non-Mendelian character, survey data suggest that the recurrence risk is about 2–4% in most populations. But if they have already had two affected babies, the survey data suggest the recurrence risk is substantially higher, often about 10%. It is not that having a second affected baby has caused their recurrence risk to increase, but it has enabled us to recognize them as a couple who always had been at particularly high risk. For multifactorial conditions, bad luck in the past is a predictor of bad luck in the future. A cynic would say it is simply the counselor being wise after the event—but the practice accords with our understanding based on threshold theory, as well as with epidemiological data, and it represents the best we can offer in an imperfect state of knowledge.

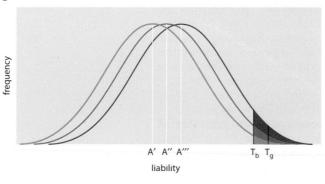

Figure 3.28 A polygenic dichotomous character with sex-specific thresholds. The figure shows a model that explains data such as those in Table 3.2. For all three curves, girls with a liability above the threshold value T_g and boys above the threshold value T_b manifest the condition, and are represented by shaded areas under the curves. As in Figure 3.27, the general population (green curve) displays a liability to this polygenic disease condition that is normally distributed, with an average liability of A'. The liability among siblings of affected boys (blue curve) is higher, with average A'', and a greater proportion of these brothers and sisters have a liability that exceeds the respective threshold levels. Among siblings of affected girls the liability is still higher (red curve, average liability A'''), and an even greater proportion of these brothers and sisters will be affected because they have a liability that exceeds their sex-specific threshold levels.

TABLE 3.2 RECURRENCE RISKS FOR PYLORIC STENOSIS				
Relatives of	Sons	Daughters	Brothers	Sisters
Male proband	19/296 (6.42%)	7/274 (2.55%)	5/230 (2.17%)	5/242 (2.07%)
Female proband	14/61 (22.95%)	7/62 (11.48%)	11/101 (10.89%)	9/101 (8.91%)

More boys than girls are affected, but the recurrence risk is higher for relatives of an affected girl. The data fit a polygenic threshold model with sex-specific thresholds (Figure 3.28). [Data from Fuhrmann and Vogel (1976) Genetic Counselling. Springer.]

3.5 FACTORS AFFECTING GENE FREQUENCIES

Over a whole population there may be many different alleles at a particular locus, although each individual person has just two alleles, which may be identical or different. The **gene pool** for the A locus consists of all alleles for that locus in the population. The **gene frequency** of allele A_1 is the proportion of all A alleles in the gene pool that are A_1. Strictly this should be called the *allele frequency*, but the term *gene frequency* is widely used, and we use it here. The following sections consider the relationship between gene frequencies and genotype frequencies. After that we will discuss the factors that can cause gene frequencies to change.

A thought experiment: picking genes from the gene pool

Consider two alleles, A_1 and A_2 at the A locus. Let their gene frequencies be p and q, respectively (p and q are each between 0 and 1). Let us perform a thought experiment:

- Pick an allele at random from the gene pool. There is a chance p that it is A_1 and a chance q that it is A_2.
- Pick a second allele at random. Again, the chance of picking A_1 is p and the chance of picking A_2 is q (we assume that the gene pool is sufficiently large that removing the first allele has not significantly changed the gene frequencies in the remaining pool). It follows that:
  - The chance that both alleles were A_1 is p^2.
  - The chance that both alleles were A_2 is q^2.
  - The chance that the first allele was A_1 and the second A_2 is pq. The chance that the first was A_2 and the second A_1 is qp. Overall, the chance of picking one A_1 and one A_2 allele is $2pq$.

The Hardy–Weinberg distribution relates genotype frequencies to gene frequencies

If we pick a person at random from the population, this is equivalent to picking two genes at random from the gene pool. Staying with our alleles A_1 and A_2, the chance that the person is A_1A_1 is p^2, the chance that they are A_1A_2 is $2pq$, and the chance that they are A_2A_2 is q^2. This simple relationship between gene frequencies and genotype frequencies is called the **Hardy–Weinberg distribution**. It holds whenever a person's two genes are drawn independently and at random from the gene pool. A_1 and A_2 may be the only alleles at the locus (in which case $p + q = 1$) or there may be other alleles and other genotypes, in which case $p + q < 1$. For X-linked loci, males, being hemizygous (having only one allele), are A_1 or A_2 with frequencies p and q, respectively, whereas females can be A_1A_1, A_1A_2, or A_2A_2 (**Box 3.5**).

The Hardy–Weinberg relationship is very useful for predicting risks in genetic counseling. Genotype frequencies in a population are expected to follow the Hardy–Weinberg distribution. Some special explanation is required if they do not (see below).

Using the Hardy–Weinberg relationship in genetic counseling

Gene frequencies or genotype frequencies are essential inputs into many forms of genetic analysis, such as linkage and segregation analysis (described in Chapters 14 and 15, respectively), and they have a particular importance in calculating genetic risks. Three typical examples follow.

BOX 3.5 THE HARDY–WEINBERG RELATIONSHIP BETWEEN GENE FREQUENCIES AND GENOTYPE FREQUENCIES

Consider two alleles at a locus. Allele A_1 has frequency p, and allele A_2 has frequency q.

				X-linked locus				
	Autosomal locus			Males		Females		
Genotype	A_1A_1	A_1A_2	A_2A_2	A_1	A_2	A_1A_1	A_1A_2	A_2A_2
Frequency	p^2	$2pq$	q^2	p	q	p^2	$2pq$	q^2

Note that these genotype frequencies will be seen whether or not A_1 and A_2 are the only alleles at the locus.

Example 1

An autosomal recessive condition affects 1 newborn in 10,000. What is the expected frequency of carriers?

Hardy–Weinberg relationships give the following relationships:

Phenotypes:	Unaffected		Affected
Genotypes:	*AA*	*Aa*	*aa*
Frequencies:	p^2	$2pq$	$q^2 = 1/10,000$

q^2 is 1 in 10,000 or 10^{-4}, and therefore $q = 10^{-2}$ or 1 in 100. One in 100 genes at the *A* locus are *a*, 99 in 100 are *A*. The carrier frequency, $2pq$, is therefore $2 \times 99/100 \times 1/100$, which is very nearly 1 in 50.

This calculation assumes that the frequency of the condition has not been increased by inbreeding. This is an important proviso, which is considered in more detail below.

Example 2

If a parent of a child affected by the above condition remarries, what is the risk of producing an affected child in the new marriage?

To produce an affected child, both parents must be carriers, and the risk is then 1 in 4. We know that the parent of the affected child is a carrier. If the new spouse is, from the genetic point of view, a random member of the population, his or her carrier risk is $2pq$, which we have calculated as 1 in 50. The overall risk is then $1/50 \times 1/4 = 1/200$.

This assumes that there is no family history of the same disease in the new spouse's family.

Example 3

X-linked red–green color blindness affects 1 in 12 Swiss males. What proportion of Swiss females will be carriers and what proportion will be affected?

Assuming that we are dealing with a single X-linked recessive phenotype, the Hardy–Weinberg relationships are as follows:

	Males		Females		
Genotypes:	A_1	A_2	A_1A_1	A_1A_2	A_2A_2
Frequencies:	p	$q = 1/12$	p^2	$2pq$	q^2

$q = 1/12$; therefore $p = 1 - q = 11/12$. The frequency of carriers, $2pq$, is then $2 \times 1/12 \times 11/12 = 22/144$, and the frequencies of homozygous affected females is $q^2 = 1$ in 144. Thus, this single-locus model predicts that 15% of females will be carriers and 0.7% will be affected.

In fact, the frequency of affected females is rather lower than this calculation predicts. This is part of the evidence that there are two forms of red–green blindness, protan and deutan. These are caused by variants in different but adjacent genes at Xq28. The mathematics and genetics are discussed in OMIM entry 303900.

Inbreeding

The simple Hardy–Weinberg calculations are invalid if the underlying assumption, that a person's two alleles are picked independently from the gene pool, is violated. In particular, there is a problem if there has not been random mating.

Assortative mating can take several forms, but the most generally important is **inbreeding**. A marriage in which there is a previous genetic relationship between the partners is called a **consanguineous** marriage.

If you marry a relative you are marrying somebody whose genes resemble your own:

- **First-degree relatives** (parents, children, full sibs) share half their genes (always for parents and children; on average for sibs).
- **Second-degree relatives** (grandparents, grandchildren, uncles, aunts, nephews, nieces, half-sibs) share one-quarter of their genes, on average.
- **Third-degree relatives** (first cousins etc.) share one-eighth of their genes on average.

In each case the sharing refers to the proportion of genes that are identical by descent, as a result of that relationship. We all share many genes simply by virtue of all being human. The **coefficient of relationship** between two people is defined as the proportion of their genes that they share by virtue of their relationship. The **coefficient of inbreeding** of a person can be defined as the probability that they receive at a given locus two alleles that are identical by descent. It is equal to half the coefficient of relationship of the parents. Popular myth would have it that many rural or isolated populations are heavily inbred and backward because of that. In fact, average inbreeding coefficients in most populations are well below 0.01, and almost never exceed 0.03 even in the most isolated groups.

Although inbreeding is seldom significant at the population level, it can have major consequences for individuals. The calculations in **Box 3.6** show that the rarer a recessive condition is, the greater will be the proportion of all cases that are the product of cousin or other consanguineous marriages. For most Western societies only 1% or less of all marriages would be between first cousins. Nevertheless, for a recessive condition with gene frequency $q = 0.01$, the formula shows that 6% of all affected children would be born in that 1% of marriages. The converse also holds. For example, among white northern Europeans there is little increased consanguinity among parents of children with cystic fibrosis, because the condition is so common. Tay–Sachs disease (OMIM 272800) is strongly associated with parental consanguinity among non-Jews, in whom it is rare, but much less so among Ashkenazi Jews, in whom it is rather common.

For rare autosomal recessive conditions, a basic Hardy–Weinberg calculation that ignores inbreeding will badly overestimate the carrier frequency in the population at large. For a couple who have already had an affected child, the recurrence risk is 1 in 4, regardless of whether they are cousins or unrelated. But for situations in which the risk of an affected child depends on estimating the frequency of carriers in the population, a correct calculation requires a knowledge of population genetic theory, and of the genetics of the particular population in question, that goes beyond the scope of this book.

Other causes of departures from the Hardy–Weinberg relationship

Inbreeding is by far the most important cause of deviations from the Hardy–Weinberg relationship between gene frequencies and genotype frequencies. Other possibilities include selective migration and mortality. If people with a certain genotype are more likely to be removed from the population, by migration or death, then those remaining will be depleted of that genotype, in comparison with Hardy–Weinberg predictions. Large-scale immigration of people from a population with a different gene frequency would equally upset the relationship. One generation of random mating would suffice to reestablish a Hardy–Weinberg relationship, albeit with a new gene frequency. However, if the population remained stratified into groups that did not interbreed freely, the Hardy–Weinberg relationship might apply within each group, but not to the overall population.

For genetic marker studies, systematic laboratory error is an important possible cause of an apparent deviation from the expected Hardy–Weinberg ratios. For example, the assay used might score heterozygotes unreliably. Observed proportions can be compared with the Hardy–Weinberg prediction by a simple χ^2 test with $k(k-1)/2$ degrees of freedom, where there are k alleles at the locus being tested. More sophisticated statistical treatments are needed if the numbers are small.

BOX 3.6 EFFECTS OF INBREEDING

Suppose a man marries his first cousin. Consider the risk of their having a baby affected with an autosomal recessive condition that has gene frequency q. If we know nothing about him, there is a chance $2pq$ that he is a carrier of the condition. If he is a carrier, there is a 1 in 8 chance that his cousin also is, by virtue of their common ancestry (**Figure 1**). The overall risk of an affected child is $2pq \times 1/8 \times 1/4 = pq/16$. Had he married an unrelated woman the risk would have been q^2, the same as for any other couple. The ratio of risk is

$$pq/16 : q^2 = p : 16q$$

For a rare recessive condition, q will be small; p will therefore be almost 1 and the ratio simplifies to

$$1 : 16q$$

Suppose a proportion c of all marriages in a population are between first cousins. Let each marriage produce one child. Suppose the risk that child would be affected by a certain rare autosomal recessive condition, when the parents are first cousins, is x. The risk for a child from an outbred marriage is $16qx$. The c cousin marriages produce xc affected children, and the $(1 - c)$ outbred marriages produce $16qx(1 - c)$ affected children. A proportion $c/[c + 16q(1 - c)]$ of all affected children will be born to first cousins. The table shows some examples.

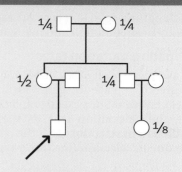

Figure 1 Relatives share genes. Figures show the proportion of their genes that relatives share with a proband (arrow) by virtue of their relationship.

Proportion of all marriages that are between first cousins	Frequency of disease allele	Percentage of all affected children whose parents are first cousins
0.01	0.01	6
0.01	0.005	11
0.01	0.001	39
0.05	0.01	25
0.05	0.005	40
0.05	0.001	77

Note that we have used several simplifying assumptions:
- In deriving the 1:16q ratio of risks we used the approximation $p = 1$, which would be reasonable only for rare recessive conditions.
- We ignored the possibility that the cousin might carry the mutant allele inherited independently, and not as a result of her relationship to the carrier man. Again, that is reasonable for a rare condition, but not for a very common one.
- Finally, we assumed that all unions were either between first cousins or fully outbred.

Despite these approximations, it remains generally true that consanguineous unions contribute disproportionately to the incidence of rare recessive diseases.

A final point to note is that unrecognized locus heterogeneity can upset the use of the Hardy–Weinberg relation to calculate genetic risks. Suppose a recessive disease with frequency F is actually caused by homozygosity at any one of 10 loci. The example of Usher syndrome, quoted earlier, shows that this is not an unrealistic scenario. To keep the mathematics simple, we will suppose that each of the 10 loci contributes equally to the incidence of the disease, so each separate locus contributes $F/10$ to the incidence. We wish to advise a known carrier of his risk of having an affected child if his partner is unrelated and has no family history of the disease. As we calculated previously, the risk is $2pq \times 1/4$. For each locus, the disease allele has frequency $\sqrt{(F/10)}$, and the carrier frequency is approximately $2\sqrt{(F/10)}$. If we were unaware of the locus heterogeneity, we would assume the frequency of the disease allele to be $\sqrt{F}$, and the carrier frequency to be approximately $2\sqrt{F}$. For equally frequent disease alleles at n loci, the single-locus calculation overstates the risk by a factor $\sqrt{n}$.

Gene frequencies can vary with time

One generation of random mating is sufficient to establish a Hardy–Weinberg relationship between gene frequencies and genotype frequencies. In the absence of any disturbing factors, the frequencies will remain unchanged over the generations (Hardy–Weinberg equilibrium). However, many factors can cause gene frequencies to change with time. The general mechanisms that affect the population frequency of alleles are discussed in Box 10.5. Random changes (genetic drift) are often important in small populations. However, for a strongly disadvantageous disease allele, its frequency will primarily depend on a balance between the rate at which mutation is creating fresh examples and the rate at which natural selection is removing them. New alleles are constantly being created by fresh mutation and being removed (if deleterious) by natural selection.

Estimating mutation rates

For any given level of selection, we can calculate the mutation rate that would be required to replace the genes lost by selection. If we assume that, averaged over time, new mutations exactly replace the disease alleles lost through natural selection, the calculation tells us the present mutation rate. We can define the **coefficient of selection** (s) as the relative chance of reproductive failure of a genotype due to selection (the fittest type in the population has $s = 0$, a genetic lethal has $s = 1$).

- For an autosomal recessive condition, a proportion q^2 of the population is affected. The loss of disease alleles each generation is sq^2. This could be balanced by mutation at the rate of $\mu(1 - q^2)$, where μ is the mutation rate per gene per generation. For equilibrium, $sq^2 = \mu(1 - q^2)$. If q is small, $1 - q^2$ is very close to 1, so to a close approximation $\mu = sq^2$.

- For a rare autosomal dominant condition, homozygotes are excessively rare. Heterozygotes occur with frequency $2pq$ (frequency of disease gene = p). Only half the genes lost through the reproductive failure of heterozygotes are the disease allele, so the rate of gene loss is very nearly sp. This could be balanced by a rate of new mutation of μq^2. If q is almost 1, μq^2 is very nearly equal to μ. Thus, if there is mutation-selection equilibrium, $\mu = sp$.

- For an X-linked recessive disease, the rate of gene loss through affected males is sq. This could be balanced by a mutation rate 3μ, because all X chromosomes in the population are available for mutation, but only the one-third of X chromosomes that are in males are exposed to selection. Therefore $\mu = sq/3$.

An alternative, perhaps more intuitive, formulation of these relationships expresses μ in terms of F, the frequency of the condition in the population, and f, the biological fitness of affected people (defined simply as $1 - s$):

- For an autosomal recessive condition, $F = q^2$; therefore $\mu = sq^2 = F(1 - f)$.

- For an autosomal dominant condition, $F = 2pq$, which is very nearly $2p$ for a rare condition; $\mu = sp = F(1 - f)/2$.

- For an X-linked recessive condition, $F = q$ (in males), so $\mu = sq/3 = F(1 - f)/3$.

For many dominant and X-linked conditions mutation rates estimated in this way are in line with the general expectation, from studies in many organisms, that mutation rates are typically 10^{-5} to 10^{-6} per gene per generation. However, for many autosomal recessive conditions the calculated mutation rate is remarkably high. Consider cystic fibrosis (CF), for example.

Until very recently, virtually nobody with CF lived long enough to reproduce; therefore $s = 1$. Cystic fibrosis affects about one birth in 2000 in many northern European populations. Thus, $q^2 = 1/2000$, and the formula gives $\mu = 5 \times 10^{-4}$. This would be a strikingly high mutation rate for any gene—but in fact there is good evidence that new CF mutations are very rare. The ethnic distribution of CF is very uneven: it is primarily a disease of northern Europeans, but even among those people, certain groups such as the Finns have a very low incidence. It is hard to understand how this could happen if new mutations were very frequent—mutation rates are unlikely to differ greatly between populations. Moreover, there is evidence that the commonest CF mutation in northern Europeans has been in those populations for many centuries (detailed in Chapter 14). This points to the fact that the mutation rate calculation is invalid. The next section explains why.

The importance of heterozygote advantage

We saw that the formula $\mu = sq^2$ gives an unrealistically high mutation rate for CF (and for many other common autosomal recessive conditions). The calculation is invalid because it ignores heterozygote advantage. Cystic fibrosis carriers have, or had in the past, some selective advantage over normal homozygotes. There has been debate over what this advantage might be. The *CFTR* gene encodes a membrane chloride channel, which is required by *Salmonella typhi* to enable it

---

BOX 3.7 SELECTION IN FAVOR OF HETEROZYGOTES FOR CYSTIC FIBROSIS

For CF, the disease frequency in Denmark is about one in 2000 births.

Phenotypes:	Unaffected		Affected
Genotypes:	AA	Aa	aa
Frequencies:	p^2	$2pq$	$q^2 = 1/2000$

q^2 is 5×10^{-4}; therefore $q = 0.022$ and $p = 1 - q = 0.978$.
$p/q = 0.978/0.022 = 43.72 = s_2/s_1$.
If $s_2 = 1$ (affected homozygotes never reproduce), $s_1 = 0.023$.
The present CF gene frequency will be maintained, even without fresh mutations, if Aa heterozygotes have on average 2.3% more surviving children than AA homozygotes.

---

to enter epithelial cells, so maybe Aa heterozygotes are relatively resistant to typhoid fever, compared with AA or aa homozygotes. Whatever the cause of the heterozygote advantage, if s_1 and s_2 are the coefficients of selection against the AA and aa genotypes respectively, then an equilibrium is established (without recurrent mutation) when the ratio of the gene frequencies of A and a, p/q, is s_2/s_1. Box 3.7 illustrates the calculation for CF, and shows that a heterozygote advantage too small to observe in population surveys can have a major effect on gene frequencies. Autosomal recessive conditions frequently show a very uneven ethnic distribution, being common in some populations and rare in others. The explanation is often a combination of a **founder effect** (the population being derived from a small number of founders, one of whom happened to be a carrier of the gene) and heterozygote advantage.

It is worth remembering that the medically important Mendelian diseases are those that are both common and serious. They must all have some special feature to allow them to remain common in the face of the selection pressure against them. This may be an exceptionally high mutation rate (Duchenne muscular dystrophy), propagation of non-pathological premutations (non-pathogenic sequence changes that are unstable and have a high rate of conversion to pathogenic mutations, best exemplified in fragile X mental retardation syndrome; OMIM 300624), or onset of symptoms after reproductive age (Huntington disease)—but, for common serious recessive conditions, it is most often heterozygote advantage.

CONCLUSION

In this chapter we have considered genes in a way that Gregor Mendel would have recognized. Genes, as treated here, are recognized through the pedigree patterns they give rise to as they segregate through multigeneration families. Characters that do not give Mendelian pedigree patterns are explained with mathematical tools that stem from the work of Mendel's contemporary, Francis Galton. Further mathematical tools developed by Hardy and Weinberg 100 years ago can be used to infer aspects of population genetics. All this can be done without ever asking what the physical nature of genes is. Even when the relationship of genes to chromosomes was worked out early in the twentieth century, this was done without any knowledge of their physical nature, or any need to understand it, as we will see in Chapter 14. But of course we are interested in understanding what genes do and what they are, and for this we need to get physical—first with cells, then with DNA. This is the subject of the following chapters.

FURTHER READING

Introduction: some basic definitions

Druery CT & Bateson W (translators, revised by Blumberg R) (1901) Mendel's paper in English. http://www.mendelweb.org/ Mendel.html [An English translation of Mendel's original 1865 paper, with notes.]

Read AP & Donnai D (2007) The New Clinical Genetics. Scion. [A basic undergraduate textbook illustrating clinical aspects of the topics covered in this chapter.]
University of Kansas Genetics Education Center. http://www.kumc.edu/gec/ [A portal to a large range of Web resources covering basic genetics.]

Mendelian pedigree patterns

Epstein MP, Lin X & Boehnke M (2002) Ascertainment-adjusted parameter estimates revisited. *Am. J. Hum. Genet.* 70, 886–895. [A difficult mathematical paper that nevertheless gives useful references and discussion of the general problem of correcting bias of ascertainment; the full text is available on PubMed Central.]

Jobling MA & Tyler-Smith C (2000) New uses for new haplotypes: the human Y chromosome, disease and selection. *Trends Genet.* 16, 356–362.

Wilkie AOM (1994) The molecular basis of dominance. *J. Med. Genet.* 31, 89–98. [An excellent review of why some characters are dominant and others recessive, with the emphasis on human clinical conditions.]

Zschocke J (2008) Dominant versus recessive: Molecular mechanisms in metabolic disease. *J. Inherit. Metab. Dis.* 31, 599–618. [A detailed discussion, with many examples, of the limitations of the simple division of Mendelian characters into dominant or recessive.]

Complications to the basic Mendelian pedigree patterns

Strain L, Dean JC, Hamilton MP & Bonthron DT (1998) A true hermaphrodite chimera resulting from embryo amalgamation after in vitro fertilization. *N. Engl. J. Med.* 338, 166–169.

Van der Meulen MA, van der Meulen MJP & te Meerman GJ (1995) Recurrence risk for germinal mosaics revisited. *J. Med. Genet.* 32, 102–104. [Mathematical modeling of the risks.]

Genetics of multifactorial characters

Falconer DS & Mackay TFC (1996) Introduction To Quantitative Genetics, 4th ed. Longmans Green. [An approachable text covering all aspects of quantitative genetics; not specifically focused on human genetics.]

Biographical detail, commentaries, and a large range of facsimile documents on the life and achievements of Francis Galton are available from http://galton.org/

Factors affecting gene frequencies

Harper PS (2001) Genetic Counselling, 5th ed. Hodder Arnold. [The standard text; includes many examples of pedigrees.]

The book by Falconer & Mackay referenced above has a very useful introductory section on gene frequencies and the factors determining and changing them.

Chapter 4

Cells and Cell–Cell Communication

4

KEY CONCEPTS

- Cells show extraordinary diversity in size, form, and function. Histology recognizes more than 200 different cell types in adult humans, but this is a massive underestimate—the number of different neurons alone is likely to be in the thousands.

- Cells make connections with each other (cell adhesion); transient connections allow cells to perform various functions, while stable connections allow functionally similar cells to form tissues.

- Tissues are composed of cells of one or more types, plus an extracellular matrix (ECM), a complex network of secreted macromolecules that supports cells and interacts with them to regulate many aspects of their behavior.

- Cell adhesion is regulated by a limited number of different types of cell adhesion molecule that link cells to neighboring cells or to ECM. Neighboring cells also make contact through different types of cell junction.

- Many cell functions require that cells signal to each other over both short and long distances. Signals transmitted by one cell change the behavior of responding cells by altering the activity of proteins, notably transcription factors, that bring about a change in gene expression.

- Transmitting cells often secrete a ligand molecule that binds to a receptor on the surface of responding cells to initiate a downstream signal transduction pathway.

- Other small signaling molecules pass through the plasma membranes of responding cells to bind to intracellular receptors, or are anchored in the membrane of the transmitting cell and interact with receptors on the surfaces of adjacent cells.

- Cell proliferation is regulated at different stages in the cell cycle. During development, cell proliferation causes rapid growth of an organism. At maturity, cell proliferation is limited to certain cell types that need to be renewed, and there is an equilibrium between cell birth and cell death.

- Programmed cell death is functionally important in development and throughout life. There are different pathways for getting rid of cells that are unwanted, unnecessary, or potentially dangerous.

- During development, cells become progressively more restricted in the range of cell types that they give rise to. At maturity, most cells are terminally differentiated.

- Stem cells are comparatively unspecialized cells that can give rise to differentiated cells and yet are able to renew themselves, often by asymmetric cell division.

- Stem cells derived from the early embryo can give rise to all the cells of the body. Those derived from cells later in development have less differentiation potential.

- Immune system cells are diverse and highly specialized. Those of the innate immune system work in the nonspecific recognition of foreign or altered host molecules known as antigens.

- B and T lymphocytes are the core of the adaptive immune system, which mounts strong, highly specific immune responses.

- Unique DNA rearrangements in B and T cells means that different immunoglobulins are made in different B cells and that different T-cell receptors are made in different T cells. As a result, a single individual can generate huge ranges of immunoglobulins and T-cell receptors.

All living organisms consist of **cells**—aqueous, membrane-enclosed compartments that interact with each other and the environment. Every cell arises by either the division or the fusion of existing cells. Ultimately, there must be an unbroken chain of cells leading back to the first successful primordial cell that lived maybe 3.5 billion years ago.

Some cells are independent unicellular organisms. Such organisms must perform all the activities that are necessary to sustain life, and they must also be capable of reproduction. They are therefore extremely sensitive to changes in their environment, and they typically have very short life cycles and are suited to rapid proliferation. As a result, they can adapt quickly to changes around them—mutants that are more able to survive in a particular environment can flourish quickly. This has led to an enormous range of single-celled organisms that have evolved to fit different, sometimes extreme, environmental niches. Although very successful, inevitably they are of limited complexity.

Multicellular organisms have comparatively greater longevity, and changes in phenotype are correspondingly slow. Their success has been based on the partitioning of different functions into different cell types. The numerous forms of interaction between cells provide huge potential for functional complexity. During evolution, the transition between unicellular and multicellular forms must have been a multi-step process but the initial cell aggregation is modeled by some present-day organisms. The slime mold, *Dictyostelium discoideum*, for example, exists predominantly as single cells but has a multicellular stage in its life cycle. Because eukaryotes require single cells for meiosis, multicellular organisms have specialized *germ cells* set aside to facilitate reproduction.

In this chapter we look at the structure, number, and diversity of human cells and how these come together to form the tissues of the human body. Cell specialization depends on signaling between cells, and so we review the principles underlying cell signaling. The proliferation of cells is controlled by various checkpoints in the cell cycle, and programmed cell death allows the culling of unwanted or damaged cells during tissue formation. Next, we describe the role of stem cells in both the development and the renewal of mature tissues. Finally, we consider the cells of the immune system and the basis of the unique diversity of B and T cells that permits effective immune responses to harmful cells and viruses.

4.1   CELL STRUCTURE AND DIVERSITY

All cells contain an aqueous cytoplasm surrounded by a membrane of phospholipids and proteins, but cells can vary greatly in size and complexity.

Prokaryotes and eukaryotes represent a fundamental division of cellular life forms

Cells can be classified into broad taxonomic groups according to differences in their internal organization and functions (see **Figure 4.1**). The major division of organisms into prokaryotes and eukaryotes is founded on fundamental differences in cell architecture.

Prokaryotes have a simple internal organization, with a single, membrane-enclosed compartment that is not subdivided by any internal membranes, as occurs in eukaryotes (see below). Under the electron microscope, prokaryotic cells appear relatively featureless. However, prokaryotes are far from primitive: they have been through many more generations of evolution than humans. All prokaryotes are unicellular, and they comprise two kingdoms of life:

- **Bacteria** (formerly called *eubacteria* to distinguish them from archaebacteria) are found in many environments. Some cause disease; others perform tasks that are useful or essential to human survival. Huge numbers of bacteria inhabit our bodies. Comprising about 500–1000 different species, the vast majority of such *commensal* bacteria live in the gut, and many are beneficial: they ferment complex indigestible carbohydrates and synthesize various vitamins, including folic acid, vitamin K, and biotin.

- **Archaea** are a poorly understood group of organisms that superficially resemble bacteria and so were formerly termed *archaebacteria*. They are often found in extreme environments (such as hot acid springs), but some species are found in more convivial locations such as in the guts of cows.

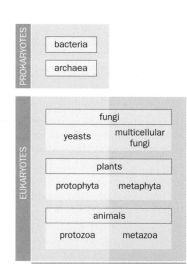

Figure 4.1 Classification of unicellular and multicellular organisms. The first two domains of life—bacteria and archaea—are also known as prokaryotes because they lack internal membranes. Eukaryotes, which form the third domain, have membrane-enclosed organelles. This domain is further subdivided into unicellular and multicellular fungi, plants, and animals. The name **protist** has been used as a generic term for any unicellular eukaryote but has also been used to describe unicellular eukaryotes that can not easily be classified as animals, plants, or fungi.

Prokaryotes have no defined nucleus and their chromosomal DNA does not seem to be highly organized; instead it exists as a nucleoprotein complex known as the *nucleoid* (Box 4.1). The typical prokaryote genome is represented as a single, circular chromosome containing less than 10 Mb of DNA. However, prokaryotes have recently been identified that possess multiple circular chromosomes, multiple linear chromosomes, mixtures of circular and linear chromosomes, and genomes up to 30 Mb in size in the case of *Bacillus megaterium*.

Eukaryotes are thought to have first appeared about 1.5 billion years ago. They have a much more complex organization than their prokaryote counterparts, with internal membranes and membrane-enclosed organelles including a nucleus (see Box 4.1). There is only one kingdom of eukaryotic organisms—the Eukarya—but it includes both unicellular organisms (yeasts, protozoans, and so on) and multicellular fungi, plants, and animals.

In the vast majority of eukaryotes, the genome comprises two or more linear chromosomes contained within the nucleus. Each chromosome is a single, very long DNA molecule packaged with histones and other proteins in an elaborate and highly organized manner. The number and DNA content of the chromosomes vary greatly between species (see also below).

The membranes surrounding and dividing the cell are selectively permeable, regulating the transport of a variety of ions and small molecules into and out of the cell and between compartments.

The soluble portions of both the cytoplasm and the nucleoplasm are highly organized. In the nucleus, a variety of subnuclear structures have been identified and are thought to be arranged in the context of a *nuclear matrix* (see Box 4.1). In the cytoplasm, the cytoskeleton is an internal scaffold of protein filaments that provides stability, generates the forces needed for movement and changes in cell shape, facilitates the intracellular transport of organelles, and allows communication between the cell and its environment.

The extraordinary diversity of cells in the body

Cell diversity is reflected in differences in cell size and shape. A typical bacterial cell is 1 μm in diameter, and whereas the average diameter of eukaryotic cells is about 10–30 μm, some specialized cells can grow much larger. Mammalian egg cells are about 100 μm in diameter, but other eggs that store nutrients required for development can be much larger (the ostrich egg can be up to 20 cm long). Some cells are very long: human muscle fiber cells can extend as long as 30 cm, and human neurons can reach up to 1 meter in length.

Complex animals have many highly specialized cells. Histology textbooks recognize more than 200 different cell types in adult humans (Table 4.1). However, histology is a comparatively crude way of classifying cells, relying heavily on differences in cell size, morphology, and ability to take up certain stains. Some cells have been difficult to access and study, and the true extent of cell diversity is very much higher than that recognized by histology.

Neurons are now known to be extremely diverse, and although the number of different neuron types remains unknown, it is likely to be very large (some recent estimates suggest more than 10,000). They are linked to each other by astonishingly complex connections—some individual neurons can be connected to 100,000 other neurons. B and T **lymphocytes** also show huge diversity. They are extremely unusual cells because, as they mature, each B cell and each T cell undergoes cell-specific DNA rearrangements to assemble functional genes to make immunoglobulins and T-cell receptors, respectively. As a result, individual B cells from a single individual produce different immunoglobulins, and individual T cells exhibit different T-cell receptors.

Germ cells are specialized for reproductive functions

In multicellular organisms, development and growth are separated from reproductive functions. A specialized population of *germ cells* is set aside to perform reproductive functions; in evolutionary terms, the remaining *somatic cells* provide a vessel to carry these reproductive cells for the purpose of achieving reproduction. In plants and primitive animals, ordinary somatic cells can give rise to germ cells throughout the life of the organism. However, in most of the animals

BOX 4.1 INTRACELLULAR ORGANIZATION WITHIN ANIMAL CELLS

Eukaryotic cells have many different types of membrane-enclosed structures within their cytoplasm. Not all the structures listed below are present in every cell type; some human cells are so specialized to perform a single function that the nucleus and other organelles are discarded and the cells rely on pre-synthesized gene products.

The **plasma membrane** provides a protective barrier around the cell. It is based on a double layer of phospholipids (**Figure 1**). Hydrophobic lipid 'tails' are sandwiched between hydrophilic phosphate groups that are in contact with polar aqueous environments, the cytoplasm and extracellular environment. The plasma membrane is *selectively permeable*, regulating the transport of a variety of ions and small molecules into and out of the cell.

The **cytosol**, the aqueous component of the cytoplasm, makes up about half the volume of the cell and is the site of major metabolic activity, including most protein synthesis. The cytosol is very highly organized by a series of protein filaments, collectively called the **cytoskeleton**, that have a major role in cell movement, cell shape, and intracellular transport. There are three types of cytoskeletal filament:

- *Microfilaments* are polymers of the protein actin (and so are also known as actin filaments). They provide mechanical support to the cell, allow controlled changes to cell shape, and facilitate cell movement by forming structures such as *filopodia* and *lamellipodia* (extensions to the cell that allow it to crawl along surfaces).
- *Microtubules* are much more rigid than actin filaments and are polymers of tubulin proteins. They are important constituents of the centrosome and mitotic spindle (see Box 2.1) and also form the core of cilia and flagella.
- *Intermediate filaments* have predominantly structural roles. Examples include the neurofilaments of nervous system cells and keratins in epithelial cells.

The **endoplasmic reticulum** (ER) consists of flattened, single-membrane vesicles whose inner compartments (*cisternae*) are interconnected to form channels throughout the cytoplasm. The ER has two key functions: the intracellular storage of Ca^{2+}, which is widely used in cell signaling; and the synthesis, folding, and modification of proteins and lipids destined for the cell membrane or for secretion. The *rough endoplasmic reticulum* is studded with ribosomes that synthesize proteins that will cross the membrane into the intracisternal space before being transported to the periphery of the cell. Here, they can be incorporated into the plasma membrane, retrieved to the ER, or secreted from the cell. The attachment of the sugar residues (glycosylation) that adorn many human proteins begins in the ER.

The **Golgi complex** consists of flattened single-membrane vesicles, which are often stacked. Its primary function is to secrete cell products, such as proteins, to the exterior and to help form the plasma membrane and the membranes of lysosomes. Some of the small vesicles that arise peripherally by a pinching-off process contain secretory products (*secretory vacuoles*). Glycoproteins arriving from the ER are further modified in the Golgi complex.

The **nucleus** contains the chromosomes and the vast majority of the DNA of an animal cell. It is surrounded by a *nuclear envelope*, composed of two membranes separated by a narrow space and continuous with the ER. Openings in the nuclear envelope (*nuclear pores*) are lined with specialized protein complexes that act as specific transporters of macromolecules between the nucleus and cytoplasm. Within the nucleus, the chromosomes are arranged in a highly ordered way. Some evidence supports the existence of a *nuclear matrix*, or scaffold, a protein network to which chromosomes are attached. Additional nuclear substructure includes these structures:

- The *nucleolus* is a discrete region where ribosomal RNA (rRNA) is synthesized and processed and where ribosome subunits are first assembled. Human rRNA genes are clustered on the short arms of chromosomes 13, 14, 15, 21, and 22, and are brought close together within the nucleolus.
- *Cajal bodies* (or coiled bodies) are thought to be sites where small nuclear/nucleolar ribonucleoprotein (snRNP/snoRNP) particles are assembled, and may also be involved in gene regulation.
- *Speckles* (or interchromatin granules) are believed to be regions where fully mature snRNPs assemble in readiness for splicing pre-mRNA.
- *PML bodies* appear as rings and are composed predominantly of the premyelocytic leukemia (PML) protein, but their functions are unknown.
- *Perichromatin fibrils* are sites where nascent RNA accumulates.
- *Cleavage bodies* are sites of polyadenylation and cleavage.

Mitochondria are sites of oxidative phosphorylation, by which organic nutrients are oxidized to generate ATP that is then used to power the different functions of a cell. They have two membranes: a comparatively smooth outer membrane, and a complex, highly folded inner mitochondrial membrane. The inner compartment, the *mitochondrial matrix*, contains enzymes and chemical intermediates involved in energy metabolism. Mitochondria, and also the chloroplasts of plants cells, contain DNA. They also have their own ribosomes that are dedicated to translating mRNA transcribed from mitochondrial DNA. However, most mitochondrial proteins are encoded by nuclear genes, synthesized on cytoplasmic ribosomes, and then imported into mitochondria.

Peroxisomes (**microbodies**) are small single-membrane vesicles containing enzymes that use molecular oxygen to oxidize their substrates and generate hydrogen peroxide.

Lysosomes are small membrane-enclosed vesicles containing hydrolytic enzymes that digest materials brought into the cell by *phagocytosis* or *pinocytosis*. Lysosomes also help in the degradation of cell components after cell death.

Cilia are small structures containing microtubule filaments that extend from the plasma membrane and beat backward and forward, or rotate. In vertebrates, a very few specialized cell types have multiple cilia that are used to generate movement. They include epithelial cells lining the lungs and oviduct, where the cilia beat together to move mucus away from the lungs or the egg toward the uterus. Most vertebrate cells, however, have a single cilium, known as the *primary cilium*, whose function is not involved in generating movement—instead, it is packed with many different kinds of receptor molecule and acts as a sensor of the cell's environment. Sperm cells have a single, rather large and much longer version of a cilium, known as a *flagellum*. The flagellum moves in a whip-like fashion to propel the sperm cell forward.

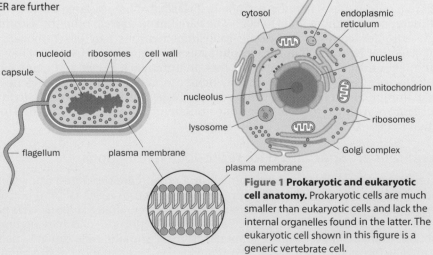

Figure 1 Prokaryotic and eukaryotic cell anatomy. Prokaryotic cells are much smaller than eukaryotic cells and lack the internal organelles found in the latter. The eukaryotic cell shown in this figure is a generic vertebrate cell.

TABLE 4.1 A CLASSIFICATION OF CELLS IN ADULT HUMANS

Group and subgroup (number of major classes)	Examples
BLOOD/IMMUNE SYSTEM CELLS (>30)	
Megakaryocytes (2)	megakaryocyte, platelet (Figures 4.2A, 4.17)
Red blood cell (1)	erythrocyte (Figure 4.17)
Stem cells/committed progenitors (various)	hematopoietic stem cell (Figure 4.17)
Lymphocytes (>10)	B cells, T cells, natural killer cell (Figure 4.17)
Monocytes and macrophages (>6)	
Dendritic cells (>4)	
Blood granulocytes (3)	basophil, neutrophil, eosinophil (Figure 4.17)
Mast cells	
CILIATED CELLS, PROPULSIVE FUNCTION (4)	
In respiratory tract, oviduct/endometrium, testis, CNS	oviduct ciliated cell
CONTRACTILE CELLS (MANY)	
Heart muscle cells (3)	myoblast, syncytial muscle fiber cell (Figure 4.2B)
Skeletal muscle cells (6)	
Smooth muscle cells (various)	see Figure 4.4
Myoepithelial cells (2)	
EPITHELIAL CELLS (>80)	
Exocrine secretion specialists (>27)	goblet (mucus-secreting) and Paneth (lysozyme-secreting) cells, of intestine (Figure 4.18B)
Keratinizing (12)	keratinocyte, basal cell of epidermis (Figure 4.18A)
Primary barrier function or involved in wet stratified barrier (11)	collecting duct cell of kidney
Absorptive function in gut, exocrine glands, and urogenital tract (8)	intestinal brush border cell (with microvilli) (Figures 4.3, 4.4)
Lining closed internal body cavities (>20)	vascular endothelial cell
EXTRACELLULAR MATRIX SECRETION SPECIALISTS (MANY)	
Connective tissue (many)	fibroblasts, including chondrocyte (cartilage) and osteoblast/osteocyte (bone) (Figures 4.4, 4.5)
Epithelial (3)	ameloblast (secretes tooth enamel)
GERM CELLS (>7)	
Female-specific (3); male-specific (3)	oocyte, spermatocyte (Figure 5.8)
NEURONS AND SENSORY TRANSDUCERS (VERY MANY)	
Photoreceptors and cells involved in perception of acceleration and gravity, hearing, taste, touch, temperature, blood pH, pain, etc. (many)	rod cell
Autonomic neurons (multiple)	cholinergic neuron
CNS neurons (large variety)	neuron (Figure 4.6)
Supporting cells of sense organs and peripheral neurons (12)	Schwann cell (Figure 4.6)
CNS glial cells (many)	astrocyte, oligodendrocyte
OTHER CELLS (>40)	
Hormone-secreting specialists (>30)	Leydig cell of testis, secreting testosterone
Lens cells (2)	lens fiber (crystallin-containing)
Metabolism and storage specialists (4)	liver hepatocyte and lipocyte, brown and white fat cells
Pigment cells (2)	melanocyte, retinal pigmented epithelial cell

CNS, central nervous system. Note that some cells in early development are not represented in adults. A full list of the human adult cell types recognized by histology is available on the Garland Scientific Web site at: http://www.garlandscience.com/textbooks/0815341059 .asp?type=supplements that we understand in detail—insects, nematodes, and vertebrates—the germ cells are set aside very early in development as a dedicated **germ line** and represent the sole source of gametes.

The germ cells are the only cells in the body capable of meiosis. To produce haploid sperm and egg cells, the precursor germ cells must undergo two rounds of cell division but only one round of DNA synthesis. In mammals, germ-line cells derive from *primordial germ cells* that are induced in the early embryo.

Cells in an individual multicellular organism can differ in DNA content

Cells differ in DNA content between organisms, within a species, and within an individual. For any species, the reference DNA content of cells, the *C value*, is the amount of DNA in the haploid chromosome set of a sperm or egg cell. *C* values vary widely for different organisms, but there is no direct relationship between the *C* value and biological complexity (the **C value paradox**). While most mammals have a *C* value of about 2500–3500 Mb of DNA, the human *C* value is only 19% of that of an onion, 4% of that of some lily plants, and—remarkably—only 0.5% of that of the single-celled *Amoeba dubia* (Table 4.2)!

The DNA content of cells within a single individual can also show variation as a result of differences in *ploidy* (the number of chromosome sets). Some cells, for example erythrocytes, platelets, and mature keratinocytes, lose their nucleus and so are nulliploid. Sperm and egg cells are haploid ($1C$). The majority of cells are diploid ($2C$), but some undergo several rounds of DNA replication without cell division (**endomitosis**) and so become polyploid. Examples are hepatocytes (less than $8C$) in the liver, cardiomyocytes ($4C$–$8C$) in heart muscle, and megakaryocytes ($16C$–$64C$) (Figure 4.2A). Skeletal muscle fiber cells are a striking example of *syncytial cells*, cells that are formed by multiple rounds of cell fusion. The individual cells can become very long and contain very many diploid nuclei (Figure 4.2B).

The DNA sequence also varies from cell to cell, between species, between individuals of one species, and even between cells within a single multicellular organism. As a result of mutation, differences in DNA sequence between cells from different species can be very significant, depending on the evolutionary distance separating the species under comparison. DNA from cells of different individuals of the same species also show mutational differences. The DNA from two unrelated humans contains approximately one change in every 1000 nucleotides.

TABLE 4.2 GENOME SIZE IS NOT SIMPLY RELATED TO THE COMPLEXITY OF AN ORGANISM

Organism	Genome size (Mb)	Gene number
UNICELLULAR		
Escherichia coli	4.6	~5000
Saccharomyces cerevisiae	13	6200
Amoeba dubia	670,000	?
MULTICELLULAR		
Caenorhabditis elegans	95	~21,190
Drosophila melanogaster	180	~14,400
Allium cepa (onion)	15,000	?
Mus musculus	2900	>25,000
Homo sapiens	3200	>23,000

Note that gene numbers are best current estimates, partly because of the difficulty in identifying genes encoding functional RNA products. For genome sizes on a wide range of organisms, see the database of genome sizes at http://www.cbs.dtu.dk/databases/DOGS/index.php

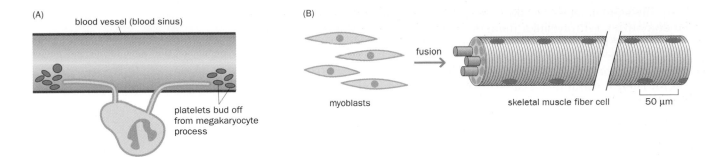

There are also very small differences in the DNA sequence of cells from a single individual. Such differences can arise in three ways:

- *Programmed differences in specialized cells.* Sperm cells have either an X or a Y chromosome. Other examples are mature B and T lymphocytes, in which *cell-specific* DNA rearrangements occur so that different B cells and different T cells have differences in the arrangement of the DNA segments that will encode immunoglobulins or T-cell receptors, respectively. This is described in more detail in Section 4.6.

- *Random mutation and instability of DNA.* The DNA of all cells is constantly mutating because of environmental damage, chemical degradation, genome instability, and small but significant errors in DNA replication and DNA repair. During development, each cell builds up a unique profile of mutations.

- *Chimerism and colonization.* Very occasionally, an individual may naturally have two or more clones of cells with very different DNA sequences. Fraternal (non-identical twin) embryos can spontaneously fuse in early development, or one such embryo can be colonized by cells derived from its twin.

Figure 4.2 Examples of polyploid cells arising from endomitosis or cell fusion. (A) The megakaryocyte is a giant polyploid (16C–64C) bone marrow cell that is responsible for producing the thrombocytes (platelets) needed for blood clotting. It has a large multi-lobed nucleus as a result of undergoing multiple rounds of DNA replication without cell division (*endomitosis*). Multiple platelets are formed by budding from cytoplasmic processes of the megakaryocyte and so have no nucleus. (B) Skeletal muscle fiber cells are polyploid because they are formed by the fusion of large numbers of myoblast cells to produce extremely long multinucleated cells. A multinucleated cell is known as a *syncytium*.

4.2 CELL ADHESION AND TISSUE FORMATION

The cells of a multicellular organism need to be held together. In vertebrates and other complex organisms, cells are assembled to make **tissues**—collections of interconnected cells that perform a similar function—and organs. Various levels of interaction contribute to this process:

- As they move and assemble into tissues and organs, cells must be able to recognize and bind to each other, a process known as **cell adhesion**.

- Cells in animal tissues frequently form **cell junctions** with their neighbors, and these can have different functions.

- The cells of tissues are also bound by the **extracellular matrix** (**ECM**), the complex network of secreted macromolecules that occupies the space between cells. Most human and adult tissues contain ECM, but the proportion can vary widely.

Even where cells do not form tissues—as in blood cells—cell adhesion is vitally important, permitting transient cell–cell interactions that are required for various cell functions.

During embryonic development, groups of similar cells are formed into tissues. For even simple tissues such as epithelium, the descendants of the progenitor cells must not be allowed to simply wander off. The requirement becomes more critical when the tissue is formed after some of the progenitor cells arrive from long and complicated cell migration routes in the developing embryo. Cells are kept in place by cell adhesion, and the architecture of the tissue is developed and maintained by the specificity of cell adhesion interactions.

Cell adhesion molecules work by having a receptor and a complementary ligand attached to the surfaces of adjacent cells. There may be hundreds of thousands of such molecules per cell, and so binding is very strong. Cells may stick together directly and/or they may form associations with the ECM. During development, changes in the expression of adhesion molecules allow cells to make and break connections with each other, facilitating cell migration. In the mature organism, adhesion interactions between cells are generally strengthened by the formation of cell junctions.

Cell adhesion molecules (CAMs) are typically transmembrane receptors with three domains: an intracellular domain that interacts with the cytoskeleton, a transmembrane domain that spans the width of the phospholipid bilayer, and an extracellular domain that interacts either with identical CAMs on the surface of other cells (*homophilic binding*) or with different CAMs (*heterophilic binding*) or the ECM. There are four major classes of cell adhesion molecule:

- *Cadherins* are the only class to participate in homophilic binding. Binding typically requires the presence of calcium ions.

- *Integrins* are adhesion heterodimers that usually mediate cell–matrix interactions, but certain leukocyte integrins are also involved in cell–cell adhesion. They are also calcium-dependent.

- *Selectins* mediate transient cell–cell interactions in the bloodstream. They are important in binding leukocytes (white blood cells) to the endothelial cells that line blood vessels so that blood cells can migrate out of the bloodstream into a tissue (*extravasation*).

- *Ig-CAMs* (immunoglobulin superfamily cell adhesion molecules) are calcium-independent and possess immunoglobulin-like domains (see Section 4.6).

Cell junctions regulate the contact between cells

Vertebrate cell junctions act as barriers, help to anchor cells, or permit the direct intercellular passage of small molecules.

Tight junctions

Tight junctions are designed to act as barriers; they are prevalent in the epithelial cell sheets that line the free surfaces and all cavities of the body, and serve as selective permeability barriers, separating fluids on either side that have different chemical compositions. Central to their barrier role is the ability of tight junctions to effect such tight seals between the cells that they can prevent even small molecules from leaking from one side of the epithelial sheet to the other (**Figure 4.3**). Tight junctions seem to be formed by *sealing strands* made up of transmembrane proteins embedded in both plasma membranes, with extracellular domains joining one another directly. The sealing strands completely encircle the apical ends of each epithelial cell (the *apical end* is the end facing outward, toward the surface).

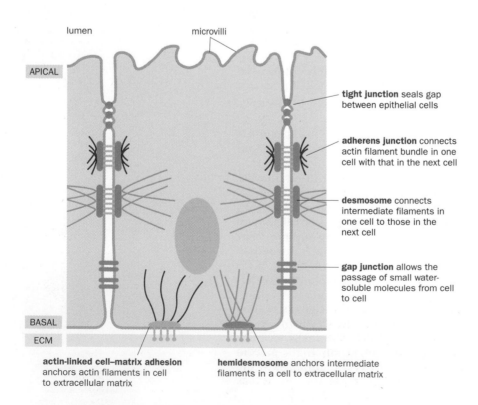

lumen   microvilli

APICAL

tight junction seals gap between epithelial cells

adherens junction connects actin filament bundle in one cell with that in the next cell

desmosome connects intermediate filaments in one cell to those in the next cell

gap junction allows the passage of small water-soluble molecules from cell to cell

BASAL

ECM

actin-linked cell–matrix adhesion anchors actin filaments in cell to extracellular matrix

hemidesmosome anchors intermediate filaments in a cell to extracellular matrix

Figure 4.3 The principal classes of cell junctions found in vertebrate epithelial cells. This example shows intestinal epithelial cells that are arranged in a sheet overlying a thin layer of extracellular matrix (ECM), known as the *basal lamina*. Depending on whether actin filaments or intermediate filaments are involved, cells are anchored to the ECM using two types of junction, and also to their cell neighbors using another two types of cell junction, as shown. Individual cells are symmetric along the axes that are parallel to the ECM layer, but they show polarity along the axis from the top (apical) end of the cell that faces the lumen to the bottom (basal) part of the cell. The tight junctions occupy the most apical position and divide the cell surface into an apical region (rich in intestinal microvilli) and the remaining basolateral cell surface. Immediately below the tight junctions are adherens junctions and then a special parallel row of desmosomes. Gap junctions and additional desmosomes are less regularly organized. [From Alberts B, Johnson A, Lewis J et al. (2008) Molecular Biology of the Cell, 5th ed. Garland Science/ Taylor & Francis LLC.]

Anchoring cell junctions

Other cell junctions mechanically attach cells (and their cytoskeletons) to their neighbors by using cadherins, or they attach cells to the ECM by using integrins. In each case, the components of the cytoskeleton that are linked from cell to cell can be actin filaments or intermediate filaments. There are four types (see Figure 4.3):

- *Adherens junctions.* Cadherins on one cell bind to cadherins on another. The cadherins are linked to actin filaments using anchor proteins such as catenins, vinculin, and α-actinin.

- *Desmosomes.* Desmocollins and desmogleins on one cell bind to the same on another. They are linked to intermediate filaments using anchor proteins such as desmoplakins and plakoglobin.

- *Focal adhesions.* Integrins on a cell surface bind to ECM proteins. The integrins are connected internally to actin filaments using anchor proteins such as talin, vinculin, α-actinin, and filamin.

- *Hemidesmosomes.* Integrins on epithelial cell surfaces bind to a protein component, laminin, of the basal lamina. The integrins are connected internally to intermediate filaments by using anchor proteins such as plectin.

Communicating cell junctions

Gap junctions permit inorganic ions and other small hydrophilic molecules (less than 1 kD) to pass directly from a cell to its neighbors (see Figure 4.3). The plasma membranes of participating cells come into close contact, establishing a uniform gap of about 2–4 nm. The gap is bridged by contact between a radial assembly of six connexin molecules on each plasma membrane; when orientated in the correct register, they form an intercellular channel. Gap junctions allow electrical coupling of nerve cells (see electrical synapses in Section 4.3) and coordinate cell functions in a variety of other tissues.

The extracellular matrix regulates cell behavior as well as acting as a scaffold to support tissues

The ECM comprises a three-dimensional array of protein fibers embedded in a gel of complex carbohydrates called *glycosaminoglycans*. The ECM can account for a substantial amount of tissue volume, especially in connective tissues, which are the major component of cartilage and bone and provide the framework of the body. The molecular composition of the ECM dictates the physical properties of connective tissue. It can be calcified to form very hard structures (such as bones and teeth), it can be transparent (cornea), and it can form strong rope-like structures (tendons).

The ECM is not just a scaffold for supporting the physical structure of tissues. It also regulates the behavior of cells that come into contact with it. It can influence their shape and function, and can affect their development and their capacity for proliferation, migration, and survival. Cells can, in turn, modify the structure of ECM by secreting enzymes such as proteases.

In accordance with its diverse functions, the ECM contains a complex mixture of macromolecules that are mostly made locally by some of the cells within the ECM. In connective tissue, the matrix macromolecules are secreted largely by fibroblast-type cells. In addition to proteins, the ECM macromolecules include glycosaminoglycans and proteoglycans. *Glycosaminoglycans* are very long polysaccharide chains assembled from tandem repeats of particular disaccharides. Hyaluronic acid is the only protein-free glycosaminoglycan in the ECM. *Proteoglycans* have a protein core with covalently attached glycosaminoglycans and exist in various different forms in the ECM.

Being extremely large and highly hydrophilic, glycosaminoglycans readily form hydrated gels that generally act as cushions to protect tissues against compression. Tissues such as cartilage, in which the proteoglycan content of the ECM is particularly high, are highly resistant to compression. Proteoglycans can form complex superstructures in which individual proteoglycan molecules are arranged around a hyaluronic acid backbone. Such complexes can act as

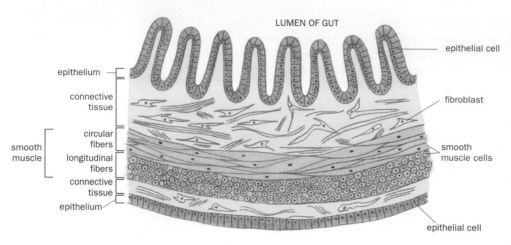

Figure 4.4 The gut as an example of the relationships between cells, tissues, and organs. The gut is a long tube-shaped organ largely constructed from three tissues. Epithelial tissues form the inner and outer surfaces of the tube and are separated from internal layers of muscle tissue by connective tissue. The latter is mostly composed of extracellular matrix (extracellular fluid containing a complex network of secreted macromolecules; see Figure 4.5). The inner epithelial layer (top) is a semi-permeable barrier, keeping the gut contents within the gut cavity (the lumen) while transporting selected nutrients from the lumen through into the extracellular fluid of the adjacent layer of connective tissue. [From Alberts B, Johnson A, Lewis J et al. (2002) Molecular Biology of the Cell, 4th ed. Garland Science/Taylor & Francis LLC.]

biological reservoirs by storing active molecules such as growth factors, and proteoglycans may be essential for the diffusion of certain signaling molecules.

The ECM macromolecules have different functional roles. There are structural proteins, such as collagens, and also elastin, which allows tissues to regain their shape after being deformed. Various proteins are involved in adhesion; for example, fibronectin and vitrinectin facilitate cell–matrix adhesion, whereas laminins facilitate the adhesion of cells to the basal lamina of epithelial tissue (see below). Proteoglycans also mediate cell adhesion and can bind growth factors and other bioactive molecules. Hyaluronic acid facilitates cell migration, particularly during development and tissue repair, and the tenascin protein also controls cell migration.

Specialized cell types are organized into tissues

There are many different types of cell in adult humans (see Table 4.1), but they are mostly organized into just a few major types of tissue. Organs are typically composed of a few different tissue types; for example, the gut comprises layers of epithelium, connective tissue, and smooth muscle (**Figure 4.4**). The common tissues—epithelium, muscle, nerve, and connective tissues—are described below, and lymphoid tissue is described in Section 4.6.

Epithelium

Epithelial tissue has little ECM and is characterized by tight cell binding between adjacent cells, forming cell sheets on the surface of the tissue. The cells are bound to their neighbors by strong adhesive forces that permit the cells to bear most of the mechanical stress that the tissue is subjected to. Here, the ECM mostly consists of a thin layer, the **basal lamina**, that is secreted by the cells in the overlying layer of epithelium (**Figure 4.5**). The epithelial cells show consistent internal cell

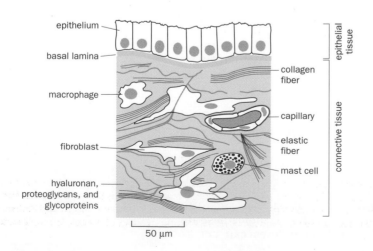

Figure 4.5 Connective tissue: cells and structure. The figure shows an example of connective tissue underlying epithelium. Connective tissue is dominated by an extracellular matrix (ECM) consisting of a three-dimensional array of protein fibers embedded in a gel of complex carbohydrates (glycosaminoglycans). Cells are sparsely distributed within the ECM and comprise indigenous cells and various immigrant blood/immune system cells (such as monocytes, macrophages, T cells, plasma cells, and leukocytes). The indigenous cells include *fibroblasts* (cells that synthesize and secrete most of the ECM macromolecules), fat cells, and *mast cells* (which secrete histamine-containing granules in response to insect bites or exposure to allergens). [From Alberts B, Johnson A, Lewis J et al. (2002) Molecular Biology of the Cell, 4th ed. Garland Science/Taylor & Francis LLC.]

asymmetry (polarity) in a plane that is at right angles to the cell sheet, with a *basolateral* part of the cell adjacent to and interacting with the basal lamina, and an *apical* part at the opposing end that faces the exterior or the lumen of a cylindrical tube.

Connective tissue

Connective tissue is largely composed of ECM that is rich in fibrous polymers, notably collagen. Sparsely distributed within connective tissue is a remarkable variety of specialized cells, including both indigenous cells and also some immigrant cells, notably immune-system and blood cells. The indigenous cells comprise *primitive mesenchymal cells* (undifferentiated multipotent stem cells) and various differentiated cells that they give rise to, notably fibroblasts that synthesize and secrete most of the ECM macromolecules (see Figure 4.5).

Cells are sparsely distributed in the supporting ECM, and it is the ECM rather than the cells it contains that bears most of the mechanical stress on connective tissue (see Figure 4.5). *Loose connective tissue* has fibroblasts surrounded by a flexible collagen fiber matrix; it is found beneath the epithelium in skin and many internal organs and also forms a protective layer over muscle, nerves, and blood vessels. In *fibrous connective tissue* the collagen fibers are densely packed, providing strength to tendons and ligaments. Cartilage and bone are rigid forms of connective tissue.

Muscle tissue

Muscle tissue is composed of contractile cells that have the special ability to shorten or contract so as to produce movement of the body parts. Skeletal muscle fibers are cylindrical, striated, under voluntary control, and multinucleated because they arise by the fusion of precursor cells called myoblasts (see Figure 4.2B). Smooth muscle cells are spindle-shaped, have a single, centrally located nucleus, lack striations, and are under involuntary control (see Figure 4.4). Cardiac muscle has branching fibers, striations, and intercalated disks; the component cells, cardiomyocytes, each have a single nucleus, and contraction is not under voluntary control.

Nervous tissue

Nervous tissue is limited to the brain, spinal cord, and nerves. Neurons are electrically excitable cells that process and transmit information via electrical signals (impulses) and secreted neurotransmitters. They have three principal parts: the cell body (the main part of the cell, performing general functions); a network of dendrites (extensions of the cytoplasm that carry incoming impulses to the cell body); and a single long axon that carries impulses away from the cell body to the end of the axon (**Figure 4.6**).

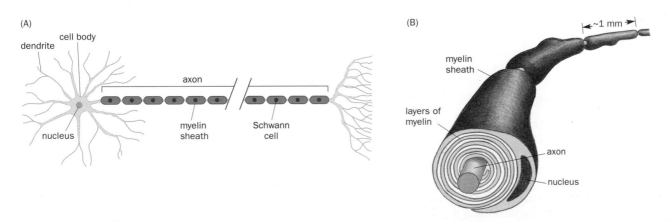

Figure 4.6 Neurons and myelination. (A) Neuron structure. Each neuron has a single long axon with multiple axon termini (dendrites) that are connected to other neurons or to an effector cell such as a muscle cell. Neurons are insulated by certain glial cells such as Schwann cells that form a myelin sheath. (B) Myelination of an axon from a peripheral nerve. Each Schwann cell wraps its plasma membrane concentrically around the axon, forming a myelin sheath covering 1 mm of the axon. [From Alberts B, Johnson A, Lewis J et al. (2008) Molecular Biology of the Cell, 5th ed. Garland Science/Taylor & Francis LLC.]

Neurons account for less than 10% of cells in the nervous system; the other 90% are *glial cells*. Glial cells do not transmit impulses, but instead support the activities of the neurons in a variety of ways. The axons of neurons have an insulating sheath of a phospholipid, myelin, that is produced by certain glial cells: *oligodendrocytes* (in the central nervous system) and *Schwann cells* (in the peripheral nervous system). *Astrocytes* are small star-shaped glial cells that ensheath synapses and regulate neuronal function. *Microglial cells* are phagocytic and protect against bacterial invasion. Other glial cells provide nutrients by binding blood vessels to the neurons.

Neurons communicate with each other at two types of synapse, as detailed in the next section.

4.3 PRINCIPLES OF CELL SIGNALING

All cells receive and respond to signals from their environment, such as changes in the extracellular concentrations of certain ions and nutrient molecules, or temperature shocks. Whereas single-celled organisms largely function independently, the cells of multicellular organisms must cooperate with each other for the benefit of the organism. To coordinate and regulate physiological and biochemical functions they must communicate effectively by sending and receiving signals. The role of cell signaling during early development will be explained in Chapter 5. In this section, we give some of the principles underlying cell signaling.

Signaling molecules bind to specific receptors in responding cells to trigger altered cell behavior

In a multicellular organism virtually all aspects of cell behavior—metabolism, movement, proliferation, differentiation—are regulated by cell signaling. Transmitting cells produce signaling molecules that are recognized by responding cells, causing them to change their behavior. Intercellular signaling can take place over long distances, as in the endocrine signaling used to transmit hormones, or over short distances by several different mechanisms:

- *Paracrine signaling* involves signaling between neighboring cells. A cell sends a secreted signaling molecule that diffuses over a short distance to responding cells in the local neighborhood.

- *Synaptic signaling* is a specialized form of signaling that involves signaling across an extremely narrow gap, the synaptic cleft, between the terminus of an axon and the cell body of a communicating neuron (see below).

- *Juxtacrine signaling* occurs when the transmitting cell is in direct contact with the responding cell; the signaling molecule is tethered to the surface of the transmitting cell and is bound by a receptor on the surface of the responding cell.

Cell signaling is initiated when transmitting cells produce signaling molecules that are recognized and bound by specific receptors in the responding cells. The transmitting cells and the responding cells are usually different cell types, but in *autocrine signaling* a cell produces a signaling molecule that can bind to a receptor on its own cell surface or on identical neighbor cells. Autocrine signaling can be used to reinforce a signaling decision or to coordinate decisions by groups of cells.

Some small hydrophobic signaling molecules can pass directly through the plasma membrane of the responding cell and bind to intracellular receptors (**Figure 4.7B**). In many examples of cell signaling, however, the signaling molecule cannot cross the cell membrane—it may be a soluble molecule (Figure 4.7A) or be anchored in the plasma membrane of the transmitting cell (Figure 4.7C)—and so binds a receptor spanning the plasma membrane of the responding cell. Most vertebrate cell signaling pathways involve the migration of signaling molecules that then bind to receptors on the surface of responding cells. **Table 4.3** provides examples of the different cell signaling systems in vertebrates, and we consider details of some mechanisms in the sections below.

The endpoint of most cell signaling is altered gene expression, producing behavioral changes in the responding cells. The altered gene expression is usually

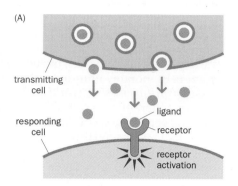

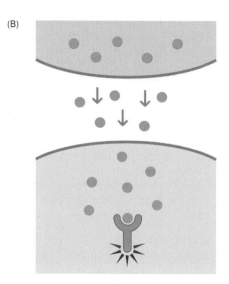

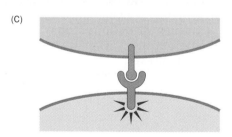

Figure 4.7 Three types of relationship between ligand and receptor in cell signaling. (A) Soluble ligand and cell surface receptor. The ligand is often a protein that binds to a transmembrane protein receptor, activating its cytoplasmic tail; see Figures 4.10 and 4.15 for examples. (B) Small soluble ligand and intracellular receptor. The ligand may be a gas (such as nitric oxide), or a small protein or steroid that can freely pass through membranes. The intracellular receptor is often in the nucleus but may be in the cytoplasm, as in the example shown in Figure 4.9. (C) Ligand and cell surface receptor anchored in the plasma membrane of adjacent cells; see the interaction between the Fas receptor and its ligand in Figure 4.16 for an example.

TABLE 4.3 IMPORTANT CLASSES OF VERTEBRATE CELL SIGNALING MOLECULES AND THEIR RECEPTORS

Signaling molecule	Receptor	Examples/comments
Small signaling molecules that cross the cell membrane	**Internal receptors**	**See Figure 4.7A**
Steroid hormones, retinoids	can reside in cytoplasm or nucleus; are converted to transcription factors when bound by their ligand	signaling using nuclear hormone receptors (Figures 4.8 and 4.9)
Signaling molecules that migrate to reach cell surface	**Cell surface receptors**	**See Figure 4.7B**
Some growth factors (FGFs, EGFs), ephrins, and some hormones (insulin)	receptors with intrinsic kinase activity	many; FGF, EGF, and ephrin signaling are widely used during embryonic development
Cytokines (interleukins, interferons, etc.); some hormones and growth factors (growth hormone, prolactin, erythropoietin, etc.)	receptors with associated tyrosine kinase activity; internal domains have associated JAK protein (Figure 4.10)	many
Various hormones (epinephrine, serotonin, glucagon, FSH, etc.), histamine, opioids, neurokinins, etc.	G-protein-coupled receptors (Figure 4.11); they span the cell membrane seven times and their internal domains have associated GTP-binding proteins	signaling involving olfactory receptors, taste receptors, rhodopsin receptors, etc.
Hedgehog family	Patched	many examples in signaling during embryonic development
TGF-β family	receptors with associated serine/threonine kinase activity, internal domain associated with SMAD proteins	many examples in signaling during embryonic development
Wnt family	Frizzled	many examples in signaling during embryonic development
Neurotransmitters	ion-channel-coupled receptors	many
Signaling molecules immobilized on cell surface	**Cell surface receptors**	**See Figure 4.7C**
Death signals	death receptors	Fas signaling in apoptosis (Figure 4.16)
Delta/Serrate family	Notch	important in neural development

EGF, epidermal growth factor; FGF, fibroblast growth factor; FSH, follicle-stimulating hormone; TGF-β, transforming growth factor-β.

induced by the activation of a (previously inactive) specific *transcription factor*, a protein that selectively binds to the DNA of certain target genes to modulate gene expression. One part of a transcription factor protein is used to recognize and bind the target DNA sequence; another part is used to activate gene expression (**Box 4.2**).

Binding of a signaling molecule to a transmembrane receptor induces a change in the receptor's cytoplasmic domain. The alteration in the receptor activates a *signal transduction pathway* that typically culminates in the activation (or sometimes inhibition) of a transcription factor. A signaling molecule that passes through the cell membrane and binds directly to an intracellular receptor induces that receptor to become an active transcription factor.

Some signaling molecules bind intracellular receptors that activate target genes directly

Small hydrophobic signaling molecules such as steroid hormones are able to diffuse through the plasma membrane of the target cell and to bind intracellular receptors in the nucleus or cytoplasm. Thus, these receptors, which are often called *hormone nuclear receptors*, are inducible transcription factors. After ligand binding, the receptor protein is activated and associates with a specific DNA response element located in the promoter regions of perhaps 50–100 target genes and, with the help of suitable *co-activator* proteins, activates their transcription.

BOX 4.2 TRANSCRIPTION FACTOR STRUCTURE

Genes are regulated by a variety of protein transcription factors that recognize and bind a short nucleotide sequence in DNA. Eukaryotic transcription factors generally have two distinct functions located in different parts of the protein:

- a **DNA-binding domain** that allows the transcription factor to bind to a specific sequence element in a target gene;
- an **activation domain** that stimulates transcription of the target gene, probably by interacting with basal transcription factors in the transcription complex on the promoter.

Some proteins, such as steroid hormone receptors, have only a DNA-binding domain but, after binding a ligand, they can cooperate with proteins called **co-activators** to perform the activities of a transcription factor. See the example of the glucocorticoid receptor in Figure 4.9.

Common DNA-binding motifs in transcription factors

Several common protein structural motifs have been identified, most of which use α-helices (or occasionally β-sheets) to bind to the major groove of DNA. Although such structural motifs provide the basis for DNA binding, the precise sequence of the DNA-binding domain determines the DNA sequence-specific recognition. Most transcription factors bind to DNA as dimers (often *homodimers*), and the DNA-binding region is often distinct from the region specifying dimer formation.

The **helix–turn–helix** (**HTH**) **motif** (**Figure 1**) is a common motif found in transcription factors. It consists of two short α-helices separated by a short amino acid sequence that induces a turn, so that the two α-helices are orientated in different planes. Structural studies have suggested that the C-terminal helix (shown to the right) acts as a specific *recognition helix* because it fits into the major groove of the DNA controlling the precise DNA sequence that is recognized.

The **helix–loop–helix** (**HLH**) **motif** also consists of two α-helices, but this time connected by a flexible loop that, unlike the short turn in the HTH motif, is flexible enough to permit folding back so that the two helices can pack against each other (that is, the two helices lie in planes that are parallel to each other). The HLH motif mediates both DNA binding and protein dimer formation. Heterodimers comprising a full-length HLH protein and a truncated HLH protein that lacks the full length of the α-helix necessary to bind to the DNA are unable to bind DNA tightly. As a result, HLH heterodimers are thought to act as a control mechanism, by enabling the *inactivation* of specific gene regulatory proteins.

The **leucine zipper** is a helical stretch of amino acids rich in hydrophobic leucine residues, aligned on one side of the helix. These hydrophobic patches allow two individual α-helical monomers to join together over a short distance to form a *coiled coil*. Beyond this region, the two α-helices separate, so that the overall dimer is a Y-shaped structure. The dimer is thought to grip the double helix much like a clothes peg grips a clothes line. Leucine zipper proteins normally form homodimers but can occasionally form heterodimers. The latter provides an important combinatorial control mechanism in gene regulation.

The **zinc finger** motif involves the binding of a Zn^{2+} ion by four conserved amino acids (normally either histidine or cysteine) so as to form a loop (finger), which is often tandemly repeated. The so-called C2H2 (Cys_2/His_2) zinc finger typically comprises about 23 amino acids, with neighboring fingers separated by a stretch of about seven or eight amino acids. The structure of a zinc finger may consist of an α-helix and a β-sheet held together by coordination with the Zn^{2+} ion, or of two α-helices, as shown in Figure 1. In either case, the primary contact with the DNA is made by an α-helix binding to the major groove.

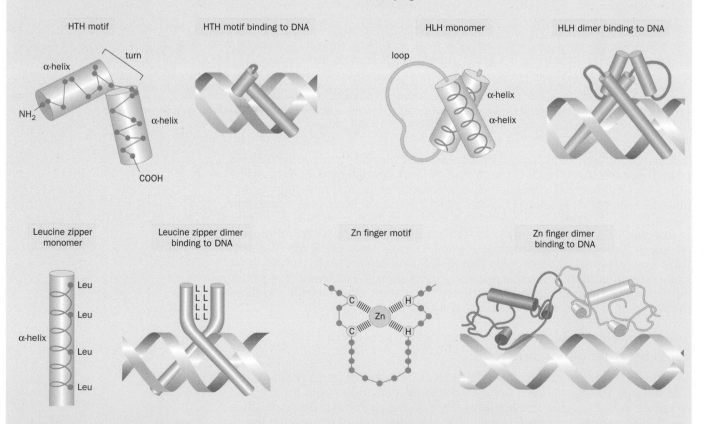

Figure 1 Structural motifs commonly found in transcription factors and DNA-binding proteins. HTH, helix–turn–helix motif; HLH, helix–loop–helix motif.

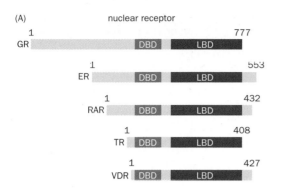

(A) nuclear receptor

(B) response element

Figure 4.8 The nuclear receptor superfamily. (A) Members of the nuclear receptor superfamily all have a similar structure, with a central DNA-binding domain (DBD) and a C-terminal ligand-binding domain (LBD). Numbers refer to the protein size in amino acid residues. GR, glucocorticoid receptor; ER, estrogen receptor; RAR, retinoic acid receptor; TR, thyroxine receptor; VDR, vitamin D receptor. (B) The response elements recognized by the nuclear receptors also have a conserved structure, with two hexanucleotide recognition sequences typically separated by either three or five nucleotides. The hexanucleotide sequences have the general consensus of AGNNCA with the two central nucleotides (pale shading) conferring specificity and belonging to one of three classes: AA, AC, or GT.

The receptors for steroid hormones and also those for the signaling molecules thyroxine and retinoic acid belong to a common nuclear receptor superfamily. Each receptor in this superfamily contains a centrally located DNA-binding domain of about 68 amino acids, and a ligand-binding domain of about 240 amino acids located close to the C terminus (**Figure 4.8A**). The DNA-binding domain contains structural motifs known as zinc fingers (see Box 4.2) and binds as a dimer, with each monomer recognizing one of two hexanucleotides in the response element. The two hexanucleotides are either inverted repeats or direct repeats that are usually separated by three or five nucleotides (Figure 4.8B).

The nuclear hormone receptors are normally found in the cytoplasm in an inactive state. Either the ligand-binding domain directly represses the DNA-binding domain or the receptor is bound to an inhibitory protein, as in the glucocorticoid receptor (**Figure 4.9**). Upon ligand binding, the inhibition is relieved and the activated ligand–receptor complex migrates to the nucleus.

Signaling through cell surface receptors often involves kinase cascades

The plasma membranes of animal cells positively bristle with transmembrane receptors for signaling molecules, and the *primary cilium*, which protrudes into the external environment, is studded with such receptors. This previously mysterious organelle is now thought to have a major role as a sensor, using the multiple different receptors to sense, and respond to, alterations in the extracellular environment.

When a receptor binds its signaling molecule on the external surface of the cell, a conformational change is induced in its internal (cytoplasmic) domain. This change alters the properties of the receptor, perhaps affecting its contact with another protein, or stimulating some latent enzyme activity. For many signaling receptors, either the receptor or an associated protein has integral kinase activity. The activated kinase often causes the receptor to phosphorylate itself, which then allows it to phosphorylate other proteins inside the cell, thereby activating them. These target proteins are themselves often kinases. They, in turn, can phosphorylate and thereby activate proteins further down a signal transduction pathway, resulting in a *kinase cascade*.

Eventually, a transcription factor is activated (or inhibited as appropriate), resulting in a change in gene expression. The length of the signaling cascade can be short (as in the cytokine-regulated JAK–STAT pathway; see **Figure 4.10**) or have many steps (such as the MAP kinase pathway). Pathways in which receptors do not have kinase activity often have several downstream components with either kinase or phosphorylase activity.

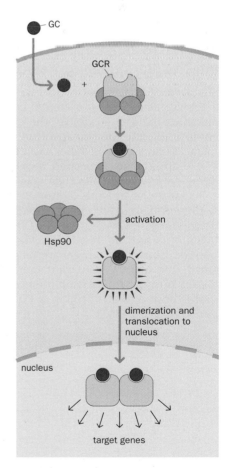

Figure 4.9 Cell signaling by ligand activation of an intracellular receptor. A glucocorticoid (GC), like other hydrophobic hormones, can pass through the plasma membrane and bind to a specific intracellular receptor. The glucocorticoid receptor (GCR) is normally bound to an Hsp90 inhibitory protein complex and is found within the cytoplasm. After binding to glucocorticoid, however, the inhibitor complex is released and the now activated receptor forms dimers and translocates to the nucleus. Here, it works as a transcription factor by specifically binding to a particular *response element* sequence (lower panel; see Figure 4.8B) in target genes at, and with the cooperation of, specific *co-activator* proteins, activating the target genes.

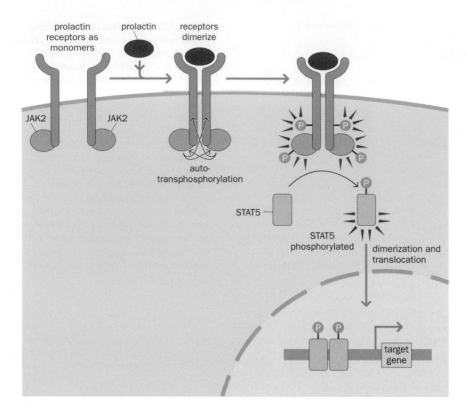

Figure 4.10 Ligand activation of a plasma membrane receptor. JAK–STAT signaling involves JAK kinases that are bound to the cytoplasmic domain of certain transmembrane receptors (notably cytokine receptors), and members of the STAT transcription factor family. For example, as shown here, JAK2 is bound to the prolactin receptor. Prolactin induces monomeric receptors to dimerize, bringing two JAK2 kinase molecules into close proximity. Each JAK2 molecule then cross-phosphorylates the other and the cytoplasmic domain of its attached receptor. Phosphorylation of the receptors activates a dormant receptor kinase activity that then targets a specific STAT protein, in this case STAT5. Once activated by phosphorylation, STAT5 dimerizes, translocates to the nucleus, and activates the transcription of various target genes.

Signal transduction pathways often use small intermediate intracellular signaling molecules

Signal transduction pathways initiated by the binding of a signaling molecule to a cell surface receptor are often complex. In addition to the kinases and other enzymes mentioned above, the cascade of interacting molecules involved in communicating the signal within the cell can include various small, diffusible intracellular signaling molecules that act as intermediates in signal transduction. These are known as **second messengers** (Table 4.4), with the first messenger being the extracellular signaling molecule.

G proteins are one common type of second messenger, which bind to *G-protein-coupled receptors* (*GPCRs*). Mammalian genomes typically have more than 1000 genes that encode GPCRs, many of them serving as receptors for prostaglandins and related lipids, various neurotransmitters, neuropeptides, and peptide hormones; others are responsible for relaying the sensations of sight, smell, and taste. GPCRs characteristically have a transmembrane domain that passes through the plasma membrane seven times and a cytoplasmic domain that is bound to a G protein, a GTP-binding protein with three subunits—α, β, and γ (Figure 4.11).

TABLE 4.4 EXAMPLES OF SECOND MESSENGERS IN CELL SIGNALING

Class	Examples	Origin	Role
Hydrophilic (cytosolic)	Cyclic AMP (cAMP)	produced from ATP by adenylate cyclase	effects are usually mediated through protein kinase
	Cyclic GMP (cGMP)	produced from GTP by guanylate cyclase	best-characterized role is in visual reception in the vertebrate eye
	Ca^{2+}	external sources or released from intracellular stores in the endoplasmic reticulum	Ca^{2+} channels can be voltage-gated and receptor-operated; probably the most widely used intracellular messenger (see Figure 4.12)
Hydrophobic (membrane-associated)	phosphatidylinositol 4,5-bisphosphate (PIP_2) and its cleavage products; diacylglycerol (DAG); inositol trisphosphate (IP_3)	PIP_2 is a phospholipid that is enriched at the plasma membrane and can be cleaved to generate DAG on the inner layer of the plasma membrane, releasing IP_3 into the cytoplasm	IP_3 binds to receptors in the endoplasmic reticulum, thereby causing the release of Ca^{2+} into the cytosol; Ca^{2+}-dependent protein kinase C is thereby recruited to the plasma membrane (see Figure 4.12)

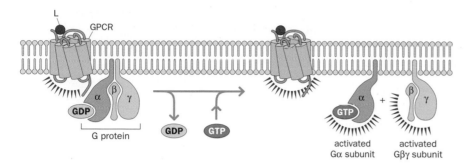

Figure 4.11 Cyclic nucleotide and lipid second messengers in G-protein signaling. G-protein-coupled receptors (GPCR) have seven transmembrane helices and a short cytoplasmic region bound by an associated G protein with three subunits: α, β, and γ. Binding of ligand (L) to the extracellular domain of a GPCR activates its cytoplasmic domain and causes exchange of GTP for GDP on the G protein. As a result, the Gα subunit is activated and dissociates from the Gβγ dimer, which in turn becomes activated. Different subtypes of G-protein subunits can perform different functions (see Figure 4.12).

In its inactive form, the G protein has GDP bound to its α subunit. On binding of a ligand to the GPCR, the G protein is stimulated and the α subunit releases GDP and binds GTP instead. Binding of GTP causes the activated α subunit to dissociate from the βγ dimer, which is then activated. Each of the activated α and βγ units can then interact with proteins downstream in the signal transduction pathway.

There are several different types of G protein, each of which associates with different second messengers. Some G proteins stimulate or inhibit the membrane-bound enzyme adenylate cyclase, causing a change in intracellular cAMP levels. Other G proteins stimulate the production of lipids, such as inositol 1,4,5-trisphosphate and diacylglycerol, and the release of calcium ions (**Figure 4.12**). In turn, the second messengers activate downstream protein kinases such as cAMP-dependent protein kinase A and calcium-dependent protein kinase C, which go on to phosphorylate (and change the activity of) particular transcription factors.

There is extensive crosstalk between different signaling pathways. At any moment, the response given by a particular cell depends on the sum of all signals that it receives and the nature of the receptors available to it.

Synaptic signaling is a specialized form of cell signaling that does not require the activation of transcription factors

Signaling between neurons needs to occur extremely rapidly and is achieved by synaptic signaling.

- In *chemical synapses* the axon termini of a transmitting neuron are closely apposed to dendrites of receiving neurons but are separated by a short gap, the synaptic cleft (**Figure 4.13**). A dendrite receives an incoming neurotransmitter signal, often glutamate or γ-aminobutyric acid (GABA) which are used extensively throughout the nervous system. In response, the local plasma membrane is depolarized to generate an action potential, an electrical impulse

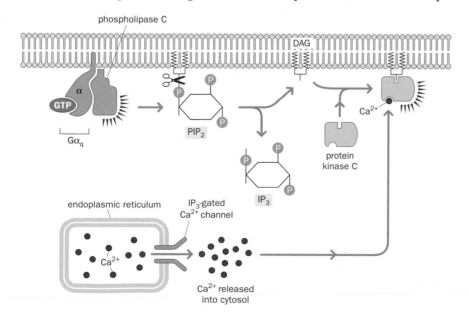

Figure 4.12 The G-protein subunit Gα_q uses various lipids and Ca^{2+} as second messengers. Activation of a type of G-protein α subunit known as Gα_q causes it to bind and activate phospholipase C. Activated phospholipase C then migrates along the plasma membrane to bind and cleave membrane-bound phosphatidylinositol 4,5-bisphosphate (PIP$_2$). The reaction leaves a diacylglycerol (DAG) residue embedded in the membrane and liberates inositol 1,4,5-trisphosphate (IP$_3$). The released IP$_3$ diffuses to the endoplasmic reticulum, where it promotes the opening of an IP$_3$-gated calcium ion channel, causing an efflux of Ca^{2+} from stores in the endoplasmic reticulum. With the help of Ca^{2+}, the membrane-bound DAG causes the activation of protein kinase C, which is then recruited to the plasma membrane, where it phosphorylates target proteins that differ according to cell type.

that spreads as a traveling wave along the plasma membrane of the axon. The change in electrical potential at the axon terminus causes a neurotransmitter to be released that diffuses across the synaptic cleft to bind to receptors on dendrites of interconnected neurons, causing local depolarization of the plasma membrane (see Figure 4.13). In a similar fashion, neurons also transmit signals to muscles (at *neuromuscular junctions*) and glands.

- *Electrical synapses* are less common. Here, the gap between the two connecting neurons is very small, only about 3.5 nm, and is known as a *gap junction*. Neurotransmitters are not involved; instead, gap junction channels allow ions to cross from the cytoplasm of one neuron into another, causing rapid depolarization of the membrane.

4.4 CELL PROLIFERATION, SENESCENCE, AND PROGRAMMED CELL DEATH

Cells are formed usually by cell division. Cell division may be common in certain tissues but especially so in early development, when rapid cell proliferation underlies the growth of multicellular organisms and the progression toward maturity. However, throughout development, cell death is common, and by the time of maturity an equilibrium is reached between cell proliferation and cell death.

Although some cell loss is accidental and is the result of injury or disease, planned or programmed cell death is very common and is functionally important. Like the organisms that contain them, cells age, and the aging process—*cell senescence*—is related to various factors, including the frequency of cell division.

Most of the cells in mature animals are non-dividing cells, but some tissues and cells turn over rapidly

The number of cells in a multicellular organism is determined by the balance between the rates of *cell proliferation* (which depends in turn on continued cell division) and cell death. Tracking the birth and death of mammalian cells *in vivo* is problematic. Most of our knowledge about rates of mammalian cell proliferation has therefore come from cultured cells, where, under normal circumstances, one turn of the mammalian **cell cycle** lasts approximately 20–30 hours.

M phase (mitosis and cytokinesis) lasts only about 1 hour, and cells spend the great majority of their time (and do most of their work) in interphase. Interphase comprises three cell cycle phases: *S phase* (DNA synthesis) and two intervening, or gap, phases that separate it from M phase: G_1 *phase* (cell growth, centrosome duplication, and so on) and G_2 *phase* (preparation of factors needed for mitosis). Of the two gap phases, G_1 is particularly important and its length can vary greatly, under the control of several regulatory factors. Furthermore, when the supply of nutrients is poor, progress through the G_1 phase of the cycle may be delayed.

If the cells receive an antiproliferative stimulus they may exit from the cell cycle altogether to enter a modified G_1 phase called G_0 *phase* (**Figure 4.14A**). Cells in G_0 phase are in a prolonged non-dividing state. But they are not dormant: they can become **terminally differentiated**; that is, irreversibly committed to serve a specialized function. Most cells in the body are in this state, but they often actively synthesize and secrete proteins and may be highly motile.

G_0 cells can also continue to grow. For example, after withdrawing from the cell cycle, neurons become progressively larger as they project long axons that continue to lengthen until growth stops at maturity. For some neurons the cytoplasm–nucleus ratio increases more than 100,000-fold during this period. Some G_0 cells do not become terminally differentiated but are *quiescent*. In response to certain external stimuli they can rejoin the cell cycle and start dividing again, to replace cells lost through accidental cell death or tissue injury.

Mature multicellular animals do contain some dividing cells that are needed to replace cells that naturally undergo a high turnover. Sperm cells are continuously being manufactured in mammals. There is also a high turnover of blood cells, and gut and skin epithelial cells are highly proliferative to compensate for the continuous shedding of cells from these organs. Even the adult mammalian

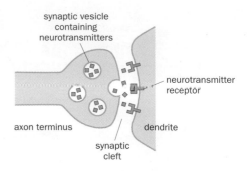

Figure 4.13 Synaptic signaling. At chemical synapses, a depolarized axon terminus of a transmitting (presynaptic) cell releases a neurotransmitter that is normally stored in vesicles. The release occurs by exocytosis: the vesicles containing the neurotransmitters fuse with the plasma membrane, releasing their contents into the narrow (about 20–40 nm) synaptic cleft. The neurotransmitters then bind to transmitter-gated ion channel receptors on the surface of the dendrites of a communicating neuron, causing local depolarization of the plasma membrane.

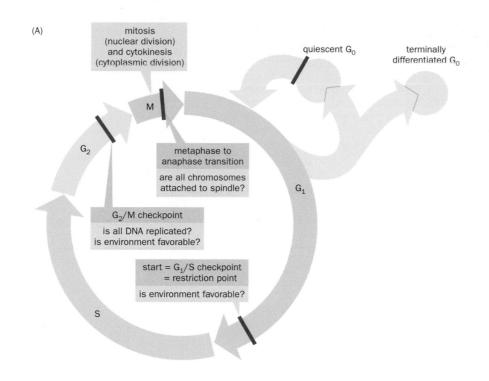

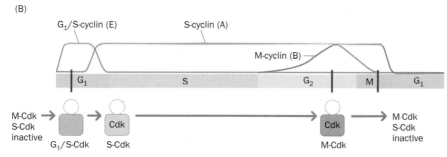

Figure 4.14 The cell cycle showing the three major checkpoints. (A) Passage through the cell cycle is controlled by checkpoints preceding transitions between phases. For example, cells cannot leave G_2 phase to proceed with mitosis (M phase) until the DNA has been replicated and the conditions are conducive to cell division. Cells in G_1 phase may enter either a quiescent G_0 phase, from which they can re-enter the cell cycle, or a terminally differentiated state. (B) Each checkpoint is regulated by a specific cyclin-dependent kinase (Cdk) bound to a cyclin protein. Cyclins are synthesized and degraded at specific times within the cell cycle, limiting the availability of each Cdk–cyclin complex.

brain—long believed to be unable to make new neurons—is now known to have three regions where new neurons are born.

Mitogens promote cell proliferation by overcoming braking mechanisms that restrain cell cycle progression in G_1

Cell proliferation is regulated by intrinsic (intracellular) factors and by extracellular signals. The intrinsic signals that regulate the cell cycle generally hold back (restrain) the cell cycle in response to sensors that indicate some fault, or unfavorable circumstance, at certain *cell cycle checkpoints* (see Figure 4.14A).

The transition from one phase of the cell cycle to the next is regulated by different cyclin-dependent kinases (Cdks): for example, Cdk1 and Cdk2 regulate entry into mitosis and S phase, respectively. Cdk concentrations are generally constant throughout the cell cycle, but the Cdks are active only when they are bound by a cyclin protein. Different cyclins are synthesized and degraded at specific points in the cell cycle (Figure 4.14B). Thus, the amounts of individual cyclin–Cdk complexes, and ultimately of activated Cdks, parallel the amounts of cyclins that they bind.

Unicellular organisms tend to grow and divide as rapidly as they can, but in multicellular organisms cells divide only when the organism needs more cells. The start checkpoint at the G_1/S boundary is a major target for regulators that prevent cell division. For example, in mammalian cells the Rb retinoblastoma protein and the p53 protein cause cells to arrest in G_1 if they contain damaged DNA.

In multicellular organisms, the cells that do divide must receive extracellular signals called **mitogens** that stimulate them to divide. Mitogens typically regulate cell division by overcoming intracellular braking mechanisms operating in G_1 that restrain progress through the cell cycle.

The braking mechanisms in G_1 are naturally overcome by factors promoting S phase, such as E2F, a regulator that controls the synthesis of many proteins needed for S phase. During G_1, E2F is initially inhibited by being bound by the negative regulator Rb. As G_1 progresses, regulatory protein complexes (cyclin D–Cdk4 and cyclin E–Cdk2) accumulate, resulting in phosphorylation of Rb. Phosphorylated Rb has a much lower affinity for E2F, freeing it to promote the synthesis of factors needed for S phase.

Mitogens increase the rate of cell division by easing the normal restraints on passage through G_1. They bind to transmembrane receptor tyrosine kinases, stimulating a signal transduction pathway that includes a small GTPase known as Ras and a MAP kinase (mitogen-activated protein kinase) cascade that ultimately activates transcription factors promoting the transition to S phase (**Figure 4.15**). Downstream targets include proteins such as Myc, which stimulates the production of both E2F and cyclin–Cdk complexes that phosphorylate Rb and so liberate E2F.

Cancer cells find ways of avoiding restrictions on the cell cycle, sometimes by mutating genes that code for checkpoint control proteins. This topic is covered in detail in Chapter 17.

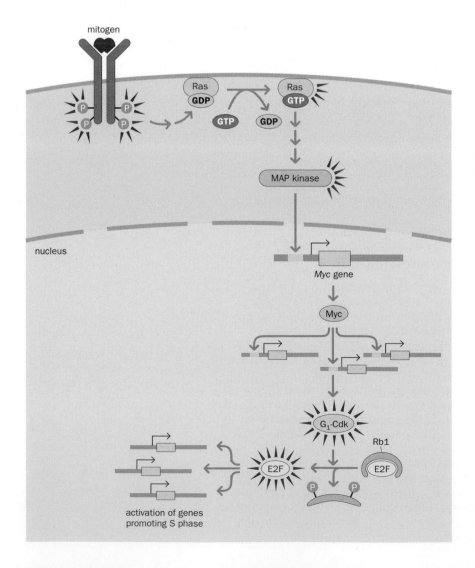

Figure 4.15 Mitogens promote cell proliferation through MAP kinase pathways. Mitogens bind to tyrosine receptor kinases, causing the monomers to dimerize and transphosphorylate each other. The activated receptor relays the signal through an accessory protein, leading ultimately to activation of a MAP kinase and subsequently activation of transcription of target genes, such as the gene encoding the Myc transcription factor. Myc in turn activates various genes, including some that lead to increased G_1-Cdk activity, which in turn causes phosphorylation of the retinoblastoma protein Rb1. Unphosphorylated Rb1 normally binds the transcription factor E2F and keeps it in an inactive state, but phosphorylation of Rb1 causes a conformational change so that it releases E2F. The activated E2F transcription factor then activates the transcription of genes that promote S phase, notably the gene encoding cyclin A. Black arrows signify transcriptional activation.

Cell proliferation limits and the concept of cell senescence

Intracellular mechanisms limit cell proliferation when not required. As organisms age (senesce), physiological deficits accumulate that undoubtedly have a cellular basis, including progressive oxidative damage to macromolecules. Cell senescence cannot easily be studied *in vivo*; instead, cell culture models have been used. Fibroblasts grown in culture from surgically removed human tissue typically achieve about only 30–50 population doublings in standard medium, the *Hayflick limit*. Proliferation rates are initially normal but gradually decline and then halt, as the cells become arrested in G_1. They enter a terminal non-dividing state in which they remain metabolically active for a while but eventually die.

How this *replicative cell senescence* relates to cells *in vivo* and to organismal aging is not fully understood. Links between cellular and organismal senescence were postulated, and for many years the Hayflick limit was viewed as being age-dependent. The modern consensus, however, is that there is no compelling evidence to support a relationship between replicative capacity *in vitro* and the age of the donors that provided the fibroblasts.

The phenomenon of cell senescence described above suggested the existence of a cellular biological clock or, more accurately, a replication counter. A characteristic of cells that undergo senescence is that their telomeres (chromosome ends) progressively shorten at each cell division (because of the problem of replicating chromosome ends). Eventually, the erosion of the telomeres destabilizes the telomeric T-loops (see Figure 2.12) and leads to removal of the protective telomere cap. When this happens, the uncapped DNA at the end of the chromosomes is no different from the double-stranded DNA breaks that can occur when DNA is damaged. As a result, cell surveillance systems that monitor DNA integrity are thought to induce the cells to enter a state of senescence. Senescence seems to be effectively a type of DNA damage response.

Not all cells are subject to cell senescence. Certain cells from embryos can be propagated for very long periods in culture and are effectively immortal, as are tumor cells, which subvert normal cell cycle controls. The idea that lack of telomere integrity is important in cell senescence is supported by studies of telomerase, an enzyme that counteracts telomere shortening by re-elongating telomeres. Whereas most normal human somatic cells have negligible or tightly regulated telomerase activity, immortal cells have constitutively high telomerase levels. Fibroblasts and other somatic cells can be artificially immortalized by transfecting them with a gene encoding the catalytic subunit of telomerase.

Large numbers of our cells are naturally programmed to die

Throughout the existence of a multicellular organism, individual cells are born and die. Cell death occasionally occurs because of irreversible accidental damage to cells (**necrosis**). Causes include physical trauma, exposure to extreme temperatures, and oxygen starvation. Typically, large groups of neighboring cells are simultaneously affected. The process leads to leakiness of the plasma membrane; water rushes in and causes the cells to swell up until the cellular membranes burst. Thereafter, the cells undergo autodigestion, producing local inflammation and attracting macrophages that ingest the cell debris.

In addition to accidental cell death, very large numbers of cells are also deliberately and naturally selected to die throughout the existence of a multicellular organism. Such **programmed cell death** (**PCD**) can occur in response to signals sent, or withheld, by other cells (during development or immune surveillance, for example), or they can arise after a cell's internal monitoring systems sense major damage to vital cell components such as its DNA or mitochondria.

A variety of different types of programmed cell death are known. Of these, **apoptosis**, or type I PCD, has been extensively studied and is characterized by very specific changes in cell structure. Typically, individual cells (rather than groups of cells) undergo apoptosis, and as they die the cells shrink rather than swell. A characteristic feature is that the chromatin condenses into compact patches that accumulate around the periphery of the nucleus. The nuclear DNA becomes fragmented and the nucleus breaks into discrete *chromatin bodies*. There is violent cellular movement (the cell appears to boil), and eventually the cell breaks apart into several membrane-lined vesicles called *apoptotic bodies* that are phagocytosed.

Another form of PCD is **autophagy**, a catabolic process in which the cell's components are degraded as a way of coping with adverse conditions such as nutrient starvation or infection by certain intracellular pathogens. Intracellular double-membrane structures engulf large sections of the cytoplasm and fuse with lysosomes, thereby targeting the enclosed proteins and organelles for degradation. Taken to an extreme, autophagy can lead to cell death. Other forms of PCD are known but are poorly characterized.

The importance of programmed cell death

Programmed cell death is crucially important in many aspects of the development of a multicellular organism and is also vital for the normal functioning of mature organisms. A quantitative illustration of its importance in development is provided by the nematode *Caenorhabditis elegans*, the only multicellular organism for which the lineage of all body cells is known. The adult worm has 959 somatic cells but is formed from a total of 1090 cells, 131 of which undergo apoptosis during embryonic development.

The selection of cells destined to die is highly specific: the same 131 cells die in different individuals. Although 959 out of the 1090 cells survive, apoptosis is the default cell fate: all 1090 cells are programmed to die. The cells that survive do so by signaling to each other: they secrete proteins called *survival factors* that bind to cell surface receptors and override default apoptosis pathways.

Programmed cell death is used to remove defective and excess/unwanted cells during embryonic development and is responsible for quality control of T and B lymphocytes and neurons (**Table 4.5**). In mature multicellular organisms, cell death and cell proliferation are carefully balanced. There is a high cell turnover in some systems, such as for blood cells and epithelial cells in gut and skin. About 100,000 cells are programmed to die each second in adult humans, but they are replaced by mitosis.

Programmed cell death is also important in defense mechanisms that rid the body of virus-infected cells, tumor cells, and other cells that are perceived to be abnormal in some way.

Programmed cell death has increasingly been recognized to be important in human disease. Aberrations in apoptosis are important in the etiology of autoimmune diseases, virally induced diseases, and cancer. For example, helper T lymphocytes are key cells in immunosurveillance systems that the human body uses to recognize and kill virally infected cells. HIV proteins cause apoptosis of these

TABLE 4.5 SOME IMPORTANT FUNCTIONS OF PROGRAMMED CELL DEATH		
Function	**Examples**	**Remarks**
Killing of defective cells during development	removal of defective immature lymphocytes	natural mistakes occur in the DNA rearrangements that give T and B lymphocytes their specific receptors; up to 95% of immature T cells have defective receptors and die by apoptosis
Killing of harmful cells	removal of harmful T lymphocytes	some immature T cells contain receptors that recognize self antigens instead of foreign antigens and need to be eliminated by apoptosis before they cause normal host cells to die
	removal of virally infected and tumor cells	a class of T lymphocytes known as killer T cells is responsible for inducing apoptosis in host cells that express viral antigens or altered antigens/neoantigens produced by tumor cells
	removal of cells with damaged DNA	cells with damaged DNA tend to accumulate harmful mutations and are potentially harmful; DNA damage often induces cell suicide pathways
Killing of excess, obsolete, or unnecessary cells	removal of surplus neurons during development	in the embryo, many more neurons are produced than are needed; those neurons that fail to make the right connections, for example to muscle cells, undergo apoptosis
	removal of interdigital cells	during development, fingers and toes are sculpted from spade-like handplates and footplates by causing apoptosis of the unnecessary interdigital cells (see Figure 5.7)

key immune system cells, allowing disease progression toward AIDS. Many cancer cells have devised strategies to oppose the immunosurveillance systems that normally induce apoptosis of cancer cells. Successful chemotherapy often relies on using chemicals that help induce cancer cells to apoptose. Other forms of PCD are important in neurodegenerative disease, such as in Huntington disease and Alzheimer disease, as well as in myocardial infarction and in stroke, in which secondary PCD caused by oxygen deprivation in the area surrounding the initially affected cells greatly increases the size of the affected area.

Apoptosis is performed by caspases in response to death signals or sustained cell stress

The key molecules that execute apoptosis are the caspase family of proteases. These enzymes have cysteine at their active site and cleave their substrates on the C-terminal side of aspartate residues. Inactive caspase precursors (procaspases) are synthesized naturally by all cells and have an N-terminal prodomain that needs to be cleaved off to activate the caspase. There are two classes of procaspase:

- *Initiator procaspases*, such as caspase 8 and caspase 9, have long prodomains and can undergo autoactivation. Their job is to initiate cell death pathways after they have been activated by signals transmitted through cell surface receptors or from internal sensors.

- *Effector procaspases*, such as caspases 3, 6, and 7, have short prodomains that are cleaved by initiator caspases. Effector caspases cleave about 100 different target proteins, including nuclear lamins (causing breakdown of the nuclear envelope) and cytoskeletal proteins (destroying cell architecture). In addition, cells naturally possess a cytoplasmic caspase-activated DNase. The DNase is normally kept inactive by being bound by an inhibitor protein. Effector caspases can cleave the inhibitor protein, however, thereby releasing the DNase, which migrates to the nucleus and fragments cellular DNA.

Protein modifications such as phosphorylation and ubiquitylation are reversible. By contrast, proteolysis is irreversible: once a peptide bond has been cleaved, cells cannot re-ligate the cleavage products. Once apoptosis has been initiated, therefore, there is no going back. There are two major types of apoptosis pathway, which use different initiator caspases. However, they then converge, using the same types of effector caspase to cleave target proteins.

Extrinsic pathways: signaling through cell surface death receptors

During development, neighboring cells exchange death signals and *survival factors* and have specific receptors for both types of signal. Cells that receive enough of a survival factor will live, because they initiate cell survival pathways that suppress programmed cell death pathways; those that do not receive enough survival factor will die. Competition between cells to receive enough survival factor is thought to control cell numbers both during development and in adulthood.

Many death receptors are members of the TNF (tumor necrosis factor) superfamily. Fas is a well-studied example. Its ligand (FasL) forms trimers and induces the Fas receptor to trimerize, causing clustering of death domains on the receptor's cytoplasmic tail. The clustered death domains attract the binding of an adaptor protein, FADD, that then recruits procaspase 8 to initiate a series of caspase cleavages (**Figure 4.16**).

Intrinsic pathways: intracellular responses to cell stress

Cells are induced to undergo apoptosis when they are sufficiently stressed that sensors indicate significant damage to certain key components. The integrity of genomic DNA needs to be protected, and mitochondria are vitally important energy producers. The endoplasmic reticulum is also crucial. As well as being required for protein and lipid synthesis, it is essential for correct protein folding and is the major intracellular depot for storing Ca^{2+}, the most widely used second messenger in cell signaling. Prolonged changes in Ca^{2+} concentration in the ER, or the accumulation of unfolded or misfolded proteins, can lead to apoptosis.

The mitochondrial pathway of apoptosis is initiated when pro-apoptosis cytoplasmic proteins such as Bax are activated. Bax then binds to the

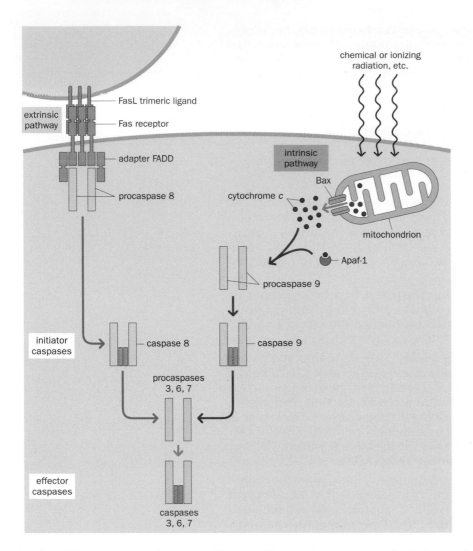

Figure 4.16 Apoptosis pathways. *Extrinsic apoptosis pathways* involve the activation of a cell surface death receptor by a ligand on a neighboring cell. The Fas death receptor is normally found as a monomer but is induced to form a trimer by its trimeric ligand, FasL. The resulting clustering of Fas receptors recruits the FADD adapter protein. FADD acts as a scaffold to recruit procaspase 8, which undergoes autoactivation to initiate a caspase cascade. *Intrinsic apoptosis pathways* are activated when vital cell components are damaged or stressed, for example in response to harmful radiation, chemicals, or hypoxia. These pathways are activated from within cells by using mitochondrial and endoplasmic reticulum components. In the mitochondrial pathways, pro-apoptosis factors, such as Bax shown here, form oligomers in the mitochondrial outer membrane, forming pores that allow the release of cytochrome *c*. In the cytosol, cytochrome *c* binds and activates the Apaf1 protein and induces the formation of an apoptosome that activates procaspase 9 and ultimately the same effector caspases (caspase 3, caspase 6, and caspase 7) as the death receptor pathways.

mitochondrial outer membrane and forms oligomers, permitting the release of cytochrome *c*, which in turn activates the cytoplasmic Apaf1 protein, causing activation of procaspase 9 (see Figure 4.16). Bax belongs to a large family of apoptosis regulators that include anti-apoptosis factors (such as Bcl-2) and pro-apoptosis factors.

4.5   STEM CELLS AND DIFFERENTIATION

Here we discuss principles underlying the basic biology of stem cells and cell differentiation. Applications toward disease treatment and ethical and societal issues are treated in Chapter 21. The regulation of cell fate in early development is considered in Chapter 5.

Cell specialization involves a directed series of hierarchical decisions

All our cells originate from the fertilized egg. The first few cell divisions that a mammalian zygote undergoes are symmetric, giving daughter cells with the same **potency**; that is, the same potential for developing into different cell types. The zygote and its immediate descendants are unspecialized and are said to be **totipotent**, because each cell retains the capacity to differentiate into all possible cells of the organism, including the extra-embryonic membranes (which will give rise to components of the placenta).

Thereafter, during development, cells become committed to alternative pathways that ultimately generate different cell types. The cells become progressively more specialized and at the same time *restricted* in their capacity to generate different types of descendent cell (**cell differentiation**).

TABLE 4.6 THE DIFFERENTIATION POTENTIAL OF PROGENITOR CELLS		
Progenitor cell type	Differentiation potential	Examples
Totipotent	the capacity to differentiate into all possible cells of the organism, including the extra-embryonic membranes	the zygote and its immediate descendants
Pluripotent	can give rise to all of the cells of the embryo, and therefore of a whole animal, but are no longer capable of giving rise to the extra-embryonic structures	cells within the inner cell mass of a blastocyst (Figure 4.19A)
Multipotent	can give rise to multiple lineages but has lost the ability to give rise to all body cells	hematopoietic stem cell (Figure 4.17)
Oligopotent	can give rise to a few types of differentiated cell	neural stem cell that can create a subset of neurons in brain
Unipotent	can give rise to one type of differentiated cell only	spermatogonial stem cell

Progenitor cells produced early in development have wide differentiation potential; those produced later have more limited potency (**Table 4.6**). The end-point of differentiation is a wide variety of different cell types (see Table 4.1), some of which are adapted to a specific function (such as hepatocyte or cardio-myocyte), whereas others, such as fibroblasts, have more general functions.

Many mature, *terminally differentiated*, cells do not divide. Other mature cells, often with names ending in *-blast*, are capable of actively dividing and acting as precursors of terminally differentiated cells, such as osteoblasts (which give rise to bone cells) and myoblasts (which are precursors of muscle cells).

Until the birth of an artificially cloned sheep called Dolly in 1996, the progressive specialization of cells during development was long thought to be irreversible in mammals. However, new methods allow terminally differentiated mammalian cells to be induced to *dedifferentiate* to give less specialized cell types, and even to produce pluripotent cells. We will consider these in Chapter 21 in the context of applications of stem cells.

Stem cells are rare self-renewing progenitor cells

Some progenitor cells that give rise to cells more specialized than themselves are also capable of undergoing self-renewal and have the capacity to be immortal. Such cells are known as **stem cells**. Stem cells differ in potency. Until recently, most pluripotent stem cells were derived from cells of the early embryo, but, as we will see in Chapter 21, differentiated cells from adults can be reprogrammed to give pluripotent stem cells. Other stem cells have been derived from naturally occurring stem cells in adult and fetal tissues and are typically more restricted in their differentiation potential and are often rather difficult to grow in culture. They are often loosely described as *adult stem cells*, but because they include fetal stem cells they have also been described as *tissue stem cells*.

Stem cells have attracted a great deal of attention because of their potential for treating diseases. There has been growing recognition, too, that cancers may often be the result of aberrant stem cells that have subverted normal constraints on cell proliferation. Certain types of pluripotent stem cell that are comparatively easy to culture have also been important in developing technologies for modeling disease and exploring gene function.

Tissue stem cells allow specific adult tissues to be replenished

In adults, dedicated tissue stem cells are regularly active in replenishing tissues that undergo rapid turnover: the epidermis of the skin (renewed roughly once every fortnight), the lining of the small intestine (villi are renewed about once every 3–5 days), and the bone marrow. However, some types of adult tissue, such as liver tissue, show minimal cell division or undergo cell division only when injured. Not all of these rarely regenerating tissues have dedicated stem cell populations; instead, it may well be that any of the tissue cells can participate as required in tissue regeneration. Although very little cell division occurs in the adult brain, dedicated stem cells continuously generate neurons in a few regions of the brain, notably the olfactory bulbs and hippocampus.

In adults, tissue stem cells are typically rare components of the tissue that they renew, and the lack of convincing stem cell-specific markers makes them generally difficult to purify. The ones that we know most about in mammals come from easily accessible tissues with rapidly proliferating cells. Two of the best-characterized systems are blood and epidermis, both of which are known to be maintained by a single type of progenitor cell. Blood cells are initially made in certain embryonic structures before the fetal liver takes over production. From fetal week 20 onward, blood cells originate from the bone marrow where two types of stem cell have long been known:

- **Hematopoietic stem cells** (**HSCs**). HSCs are anchored to fibroblast-like osteoblasts of the spongy inner marrow of the long bones. Grafting studies in mice provide especially persuasive evidence that a single type of HSC is responsible for making all the different blood cells. When a mouse that has been irradiated, to ensure destruction of the bone marrow, receives a graft of purified HSCs from another mouse strain, the incoming cells are able to differentiate and re-populate the blood. Only about 1 in 10,000 to 1 in 15,000 bone marrow cells is an HSC; the frequency in blood is about 1 in 100,000. In comparison with other stem cells, however, HSCs have been reasonably highly purified. They are multipotent, but the cell lineages to which they give rise become increasingly specialized, ultimately producing all of the terminally differentiated blood cells (**Figure 4.17**).

- **Mesenchymal stem cells** (**MSCs**). MSCs are primitive connective tissue stem cells. Bone marrow MSCs (also known as marrow *stromal cells*) are poorly defined, and so are heterogeneous. Unlike HSCs, they are not constantly self-renewing but are relatively long-lived. They are multipotent and give rise to a variety of cell types, including bone, cartilage, fat, and fibrous connective tissue.

MSCs are also found in umbilical cord blood, where they seem to have a wider differentiation potential than in the bone marrow. In addition, MSCs are known

Figure 4.17 A single type of self-renewing multipotent hematopoietic stem cell gives rise to all blood cell lineages.

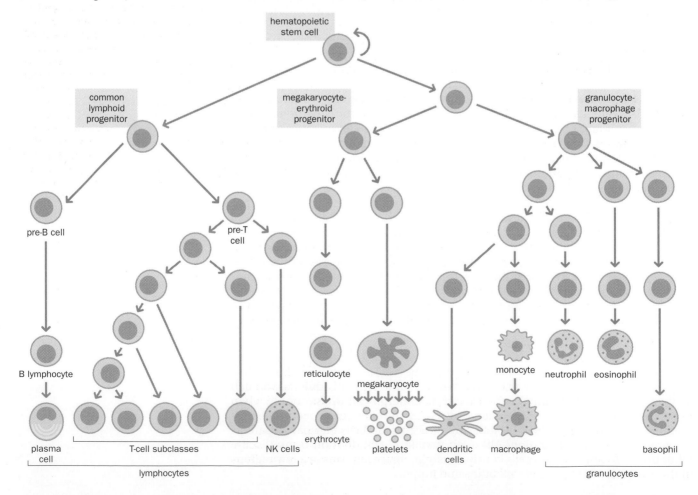

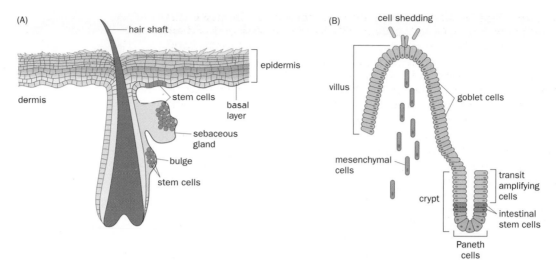

Figure 4.18 Locations of skin, hair follicle, and intestinal stem cells. (A) Location of epidermal stem cells. Epidermal stem cells can be found in three types of niche located in the bulge region of the hair follicle, the basal layer of the epidermis, and the sebaceous gland. Stem cells in the epidermal basal layer give rise to proliferating progenitor cells known as *transit amplifying cells*, which in turn give rise to increasingly differentiated upper cell layers. (B) Location of intestinal stem cells. Intestinal stem cells are located within narrow bands in *intestinal crypts*. They move downward to differentiate into lysozyme-secreting Paneth cells, or upward to generate transit amplifying (TA) cells. The nearby *intestinal villi* are regenerated every 3–5 days from the crypt TA cells and consist of three types of differentiated cell: mucus-secreting goblet cells, absorptive enterocytes (by far the most prevalent villus cell), and occasional hormone-secreting enteroendocrine cells. [(A) courtesy of The Graduate School of Biomedical Sciences, University of Medicine and Dentistry of New Jersey. (B) adapted from Blanpain C, Horsley V, & Fuchs E (2007) *Cell*, 128, 445–458. With permission from Elsevier.]

to occur in a wide variety of non-hematopoietic tissues. Recent evidence suggests that the MSCs in diverse tissue types may originate from progenitor cells found in the walls of blood vessels.

Stem cell niches

A variety of other adult tissues are also known to contain stem cells, including brain, skin, intestine, and skeletal muscle. Germ cells are sustained by dedicated male and female germ-line stem cells. Such stem cells exist in special supportive microenvironments, known as **stem cell niches**, where they are kept active and alive by substances secreted by neighboring cells.

The location of some stem cells has been studied, notably in model organisms. Epidermal stem cells are known to occur in three locations (**Figure 4.18A**), of which those in the bulge region of the hair follicles give rise to both the hair follicle and the epidermis. The stem cells that will form epidermis give rise to keratinocytes that progressively differentiate as they move toward upper layers. In the epithelium of the small intestine, stem cells are located near the base of *crypts*, which are small pits interspersed between projections called *villi* (Figure 4.18B).

One characteristic of stem cells that helps in identifying where they are located is that they divide only rarely. They do not undergo cell division very frequently, so as to conserve their proliferative potential and to minimize DNA replication errors. As a result, differentiation pathways generally begin with stem cells giving rise to proliferating progenitor cells known as **transit amplifying (TA) cells**. Unlike stem cells, TA cells are committed to differentiation and divide a finite number of times until they become differentiated.

Stem cell renewal versus differentiation

Some progress has been made in identifying factors that are important in stem cell renewal and differentiation potential, but much of the detail has yet to be worked out. Stem cells can in principle renew themselves by normal (symmetric) cell divisions. To produce cells that are more specialized than they are, and so launch differentiation pathways, stem cells need to undergo asymmetric cell divisions (**Box 4.3**). Initially, the stem cell gives rise to a cell that is committed to a different fate and that typically gives rise in turn to a series of progressively more specialized cells.

Embryonic stem cells and embryonic germ cells are pluripotent

Generally, tissue stem cells are difficult to grow and maintain in culture without differentiating, and their differentiation potential is normally quite limited. Cells from the early embryo, however, have very high differentiation potentials. In addition to inducing pluripotency by reprogramming differentiated cells, two types of embryonic cell have been used to establish pluripotent cell lines: embryonic stem cells and embryonic germ cells. Such pluripotent stem cells are theoretically capable of making all cells of the body.

BOX 4.3 ASYMMETRIC CELL DIVISION

Cell division is described as asymmetric when two daughter cells differ in their behaviors and/or fates. The cell division may be intrinsically *symmetric* to produce two identical daughter cells that nevertheless are exposed to different microenvironments, causing them to become different (**Figure 1A**). Alternatively, the cell division may be intrinsically *asymmetric* to produce two different daughters. The asymmetry may be introduced by positioning the spindle toward one end of the cell so that cell division gives rise to one large and one small daughter cell (Figure 1B). Alternatively, asymmetric localization of cell components or upstream cell polarity regulators dictate how cells develop. The spindle is deliberately oriented so that the daughter cells will differ in their content of cell fate determinants (Figure 1C).

Figure 1 Origins of cell asymmetry. (A) Exposure to different microenvironments can cause originally identical daughter cells to become different. (B) Eccentric positioning of the mitotic spindle results in one large and one small daughter cell. (C) Asymmetric localization of cell products such as of regulatory proteins (shown in small green filled circles) can result in asymmetric daughter cells. As shown on the right, this can happen after asymmetric localization of upstream polarity regulators (pink semicircles).

Meiotic divisions can be intrinsically asymmetric. The two successive meiotic divisions in female mammals each result in one small cell (a *polar body*) and one large cell (ultimately to become the egg). The large size difference ensures that most of the maternal stores are retained within the egg, and is due to positioning of the spindle away from the center of the cell.

Various mitotic divisions are intrinsically asymmetric. In mammals, the zygote undergoes successive symmetric divisions, but later in development and adulthood various asymmetric cell divisions are important in a variety of circumstances, for example in the adaptive immune system to ensure the balanced production of effector and memory T cells.

Asymmetric and symmetric stem cell division
Asymmetric cell division is often undertaken by stem cells, which need to both self-renew and give rise to differentiated cells. Each stem cell could achieve this economically by an asymmetric division to produce one daughter cell identical to the parent and one that is committed to a differentiation path (**Figure 2A**). However, this mechanism alone would be inefficient at generating the large number of stem cells that

are needed in early development or in response to depletion through injury, disease, or interventions such as chemotherapy, and there is evidence that stem cells proliferate rapidly *in vivo* by symmetric division.

Some mammalian stem cells seem to be programmed to switch between symmetric and asymmetric cell divisions during development. For example, both neural and epidermal progenitors undergo symmetric division to expand stem cell numbers during embryonic development. The daughter cells are in the same microenvironment as the original parental cell (Figure 2B). In mid to late gestation, however, asymmetric division is predominant. In both cases, this seems to be achieved by reorienting the mitotic spindle through 90° so that the two identical daughters now receive different signals. One of them retains the position of the parental stem cell in the stem cell niche and so receives signals to maintain stem cell identity by direct contact with signaling cells. The other daughter cell is geographically remote, does not receive the appropriate signals, and differentiates (orange cells in Figure 2C).

Cultured embryonic stem cells rely on symmetric divisions so that the daughter cells have the same potential to differentiate, depending on the experimental circumstances.

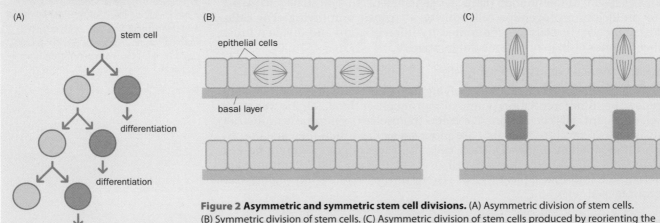

Figure 2 Asymmetric and symmetric stem cell divisions. (A) Asymmetric division of stem cells. (B) Symmetric division of stem cells. (C) Asymmetric division of stem cells produced by reorienting the spindle so that one of the two daughter cells is in a different microenvironment.

Origins of cultured embryonic stem cells

Embryonic stem (ES) cells were first obtained in the early 1980s after successful culturing of cells from the *inner cell mass* (ICM) of mouse *blastocysts*, hollow microscopic balls containing about 100 cells, of which about 30 or so internal cells constitute the ICM (**Figure 4.19A**). ES cells are pluripotent and were soon shown to be able to give rise to healthy adult mice after being injected into blastocysts from a different mouse strain that were then transferred into the oviduct of a foster mother (the technology and some applications are described in Chapter 20).

Human ES cells were first cultured in the late 1990s. They have traditionally been derived from embryos that develop from eggs fertilized *in vitro* in an IVF (*in vitro* fertilization) clinic. The IVF procedure is intended to help couples who have difficulty in conceiving children; it normally results in an excess of fertilized eggs that may then be donated for research purposes.

ES cells can be established in culture by transferring ICM cells into a dish containing culture medium. Culture dishes have traditionally been coated on their inner surfaces with a layer of *feeder cells* (such as embryonic fibroblasts). The overlying ICM cells attach to the sticky feeder cells and are stimulated to grow by nutrients released into the medium from the feeder cells (Figure 4.19B). After the ICM cells have proliferated, they are gently removed and subcultured by replating into several fresh culture dishes. After 6 months or so of repeated subculturing, the original small number of ICM cells can yield large numbers of ES cells. ES cells that have proliferated in cell culture for 6 months or more without differentiating, and that are demonstrably pluripotent while appearing genetically normal, are referred to as an *embryonic stem (ES) cell line*.

Pluripotency tests

To test whether ES cells are likely to be pluripotent, their differentiation potential is initially analyzed in culture. When cultured so as to favor differentiation, they first adhere to each other to form cell clumps known as *embryoid bodies*, then differentiate spontaneously to form cells of different types, but in a rather unpredictable fashion. By manipulating the growth media, they can also be tested for their ability to differentiate along particular pathways.

A more rigorous pluripotency test involves assaying the ability to differentiate *in vivo*. ES cells are injected into an immunosuppressed mouse to test for their ability to form a benign tumor known as a *teratoma*. Naturally occurring and experimentally induced teratomas typically contain a mixture of many differentiated or partly differentiated cell types from all three germ layers. If ES cells can form a teratoma they are considered to be pluripotent.

Embryonic germ cells

An alternative strategy for establishing pluripotent embryonic cell lines has involved isolating *primordial germ cells*, the cells of the gonadal ridge that will normally develop into mature gametes. When cultured *in vitro* primordial germ cells give rise to pluripotent **embryonic germ (EG) cells**. Human EG cells, derived from primordial germ cells or embryos and fetuses from 5 to 10 weeks old, were first cultured in the late 1990s.

EG cells behave in a very similar fashion to ES cells and share many marker proteins that are believed to be important in ES cell pluripotency and self-renewal. Currently, however, the molecular basis of pluripotency and of the capacity for self-renewal is imperfectly understood.

The exact relationship between cultured ES cells and pluripotent ICM cells is also not understood. ICM cells are not stem cells: they persist for only a limited number of cell divisions in the intact embryo before differentiating into other cell types with more restricted developmental potential. Even at the earliest stages ICM cells are heterogeneous, and some evidence suggests that cultured ES cells more closely resemble primitive ectoderm cells, whereas other evidence suggests that ES cells are of germ cell origin.

4.6 IMMUNE SYSTEM CELLS: FUNCTION THROUGH DIVERSITY

This section focuses on our immune system cells, both because they are medically important and because some immune system cells undergo extraordinary mechanisms that are designed to generate diversity. The vertebrate immune system is an intricate network of cells, proteins, and lymphoid organs that protects the body against infection by bacteria, viruses, fungi, and parasites, and can

(A)

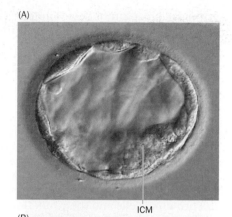

ICM

(B)

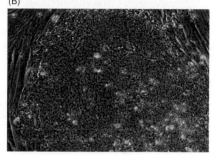

Figure 4.19 Origins and potential of embryonic stem cells. (A) A 6-day-old human blastocyst. The blastocyst is a fluid-filled hollow ball of cells containing about 100 cells. Most cells are located on the outside and will form trophoblast. The cells in the interior, the *inner cell mass (ICM)* will form all the cells of the embryo. (B) Cultured human ES (embryonic stem) cells. Cells surgically removed from the ICM can be used to give rise to a cultured human ES cell line. Human ES cell lines were traditionally established by growing the human ICM cells on a supporting layer of mouse embryonic fibroblast *feeder cells*. Human ES cells fill most of the image shown here, and the feeder cells are located at the extremities. (Courtesy of M Herbert (A) & M Lako (B), Newcastle University.)

protect against tumor development. Although we may think primarily of B and T lymphocytes, immune system cells include many other types of blood cell and a range of different tissue cells.

Not all actions of the immune system are beneficial. While fighting infection, substantial collateral damage can be inflicted on healthy tissues, and exaggerated or inappropriate immune responses against foreign substances such as pollen can cause allergic diseases. One of the main challenges that the immune system faces is distinguishing *self* (components of the host organism) from *nonself* (the foreign intruders), and although this complex task is generally accomplished with remarkable accuracy, mistakes occur, leading to autoimmune disease. The immune system is also responsible for the rejection of transplanted tissues. The remainder of this section will explain the workings of the immune system and introduce the principal cells involved. **Table 4.7** and **Table 4.8** give an overview of the principal cells and their functions.

Lymphocytes are key cells in our immune system, underlying its distinctive properties of diversity, specificity, and memory. They mature in the *primary lymphoid organs*, which in most mammals are the bone marrow and thymus. The *secondary lymphoid organs* are where lymphocytes encounter **antigens**, molecules that induce an immune response. The most organized of these are lymph nodes (which filter antigens from lymph) and the spleen (which is adapted to

TABLE 4.7 LYMPHOCYTE CLASSES AND THEIR FUNCTIONS

Lymphocyte class	Features	Location/production	Key function and roles
B cells	express immuno-globulins (Igs) and class II major histo-compatibility complex (MHC) proteins	originate in bone marrow and when mature migrate to blood and lymphoid organs	humoral (antibody) immunity; response to foreign antigens; important antigen-presenting cells
Naive B cell	express IgM/IgD on cell surface		
Plasma cell	secrete IgM, IgG, IgA, or IgE antibodies	produced by antigen-stimulated B cells in lymph nodes	secrete soluble antibodies
Memory B cells	clonally expanded B-cell populations	produced by antigen-stimulated B cells in lymph nodes	respond in secondary immune responses
T cells	express T-cell receptors (TCRs)	originate in bone marrow, carried by the blood to the thymus where they mature	cell-mediated immunity; important in antiviral and anti-tumor responses
Helper Th cell	$CD4^+$ receptors	derived from $\alpha\beta$ T cells	recognize exogenous antigens bound to class II MHC protein; signal to other immune system cells, stimulating them to proliferate and be activated
Cytotoxic Tc cell	$CD8^+$ receptors	derived from $\alpha\beta$ T cells	recognize endogenous proteins bound to class I MHC; kill virally infected host cells and tumor cells
Regulatory (Treg) cells (suppressor T cells)	$CD4^+$ and CD25	derived from $\alpha\beta$ T cells	suppress autoimmune responses
Memory T cells	clonally expanded populations of $\alpha\beta$ T cells	derived by clonal expansion of Tc, Th, or Treg cells	stimulated to expand rapidly in secondary immune responses
NKT cells	TCRs interact with CD1 and not MHC proteins		recognize glycolipid antigens; important regulatory cells
$\gamma\delta$ T cells	distinctive TCR has γ and δ chains		recognize glycolipid antigens; important regulatory cells
Natural killer (NK) cells	$CD16^+$ and $CD56^+$ Fc receptors for IgG; receptors that detect class I MHC	originate from bone marrow and are present in blood and spleen	innate immunity; first line of defense against many different viral infections; produce interferon-γ and TNF-α, which can stimulate dendritic cell maturation, activate macrophages, and regulate Th cells; recognize and kill stressed/diseased cells expressing reduced class I MHC proteins or stress proteins

TABLE 4.8 NON-LYMPHOCYTE IMMUNE SYSTEM CELLS AND THEIR FUNCTIONS		
Cell class	**Features/location**	**Functions**
MONONUCLEAR PHAGOCYTES		**key phagocytes; antigen-presenting cells**
Monocytes	in blood	
Motile macrophages	roam free; arise by division of monocytes	potent phagocytes
Tissue-specific macrophages	alveolar macrophages (lung), histocytes (connective tissue), intestinal macrophages, Kupffer cells (liver), mesangial cells (kidney), microglial cells (brain), osteoclasts (bone)	potent phagocytes
DENDRITIC CELLS (DCs)	**in tissues in contact with external environment (mainly in skin and inner linings of nose, lungs, stomach, intestines)**	**dedicated antigen-presenting cells; limited phagocytic activity; key coordinators of innate and adaptive immunity**
Langerhans cells	in skin but migrate to lymph nodes	
Interstitial DCs	move from interstitial spaces in most organs to lymph nodes/spleen	
BLOOD GRANULOCYTES	**in blood; migrate to sites of infection/inflammation during immune response**	**dedicated antigen-presenting cells**
Neutrophil	lysosome-like granules	active motile phagocytes; often the first cells to arrive at a site of inflammation after infection
Eosinophil	Fc receptors for IgE	important in killing antibody-coated large parasites
Basophil	Fc receptors for IgE; histamine-containing granules; are equivalents of the mast cells in tissues	important in mounting allergic responses and in responses to large parasites
MAST CELLS	**Fc receptors for IgE; histamine-containing granules; tissue equivalents of the basophils in blood**	**dedicated antigen-presenting cells; important in mounting allergic responses and in responses to large parasites**

filter blood). They are each packed with mature immune cells, predominantly lymphocytes but also including macrophages, dendritic cells, and other cells whose functions will be detailed below.

Rather less organized mucosa-associated lymphoid tissue is found in various sites in the body. Gut-associated lymphoid tissue collectively constitutes the largest lymphoid organ in the body, which is consistent with its need to interact with a huge load of antigens from food and commensal bacteria. It comprises the tonsils, adenoids, Peyer's patches in the small intestine, and lymphoid aggregates in the appendix and large intestine. Epithelium-associated lymphoid tissue is also found in the skin and in the mucous membranes lining the upper airways, bronchi, and genitourinary tract.

The first line of defense in vertebrate immune systems is physical. The epithelial linings of the skin, gut, and lungs act both as barriers (epithelial cells are firmly attached to each other by *tight junctions*) and as surfaces for trapping and killing pathogens. The epithelia can produce thick mucous secretions to trap pathogens, and can then kill them with lysozyme (in sweat) and antimicrobial peptides known as defensins (in mucus).

Vertebrates mount an immune response based on cooperation between the innate and adaptive immune systems. Cell signaling is a key element in this cooperation. Various cell surface molecules and secreted molecules called **cytokines** orchestrate the interactions between immune system cells that are needed for the proper functioning and regulation of the overall immune response. The most prevalent cytokines are members of the interleukin and interferon protein families.

The innate immune system provides a rapid response based on general pattern recognition of pathogens

The **innate immune system** is a primitive one, occurring in both invertebrates and vertebrates. It provides relatively nonspecific immune responses that are based on recognizing generic microbial structures, are identical in all individuals of a species, and are rapid, developing within minutes. The principal effector cells perform two major functions:

- Various cells engulf and kill bacteria and fungi by phagocytosis, a form of endocytosis. The major phagocytic cells are *neutrophils* (a white blood cell with many lysosomal-like cytoplasmic granules), tissue *macrophages* (ubiquitous, but concentrated in sites vulnerable to infection such as lungs and gut), and blood *monocytes* (precursors to macrophages).

- *NK* (*natural killer*) *cells* induce virus-infected cells and tumor cells to undergo apoptosis. Two principal pathways are used. First, NK cells bind to diseased cells, and small cytoplasmic granules are released by exocytosis into the intercellular space. Some of the released proteins (perforins) are inserted into the membranes of the target cell to create pores, whereas others (granzymes) enter the target cells, where they cleave critical substrates such as procaspase 3 that initiate apoptosis. In the second pathway, Fas ligands on NK cells activate Fas receptors on target cells to initiate apoptosis (see Figure 4.16).

Other cells help in various ways. In response to injury, *platelets* release coagulation factors and adhere to each other and to endothelial cells lining the walls of blood vessels to form clots and lessen bleeding. *Mast cells* release chemical messengers that cause blood vessels near a wound to constrict, reducing blood loss at the wound. They also secrete histamines and other factors that induce blood vessels slightly farther from the wound to dilate, increasing the delivery of blood components to the general region. Cells in tissue then release chemokines such as interleukin-8 to create a chemical gradient that attracts neutrophils and other white blood cells to the site.

The innate immune system must distinguish self from nonself. To do this, it recognizes conserved features of pathogens by using various cell surface receptors, and secreted recognition molecules in the blood complement system.

Toll-like receptors (TLRs) are a family of transmembrane proteins that are involved in the recognition of specific features of pathogens as being foreign. Examples of these features are the following: peptidoglycans (from bacterial cell walls); flagella; lipopolysaccharides (on Gram-negative bacteria); and double-stranded RNA, methylated DNA, and zymosan (in fungal cell walls). TLRs are abundant on macrophages and neutrophils, and on epithelial cells lining the lung and gut. Once stimulated, TLRs induce the surface expression of *co-stimulatory molecules* that are essential for initiating adaptive immune responses (see below) and stimulate the secretion of pharmacologically active molecules (prostaglandins and cytokines) that both initiate an inflammatory response and also help induce an adaptive immune response.

The **complement system** comprises 20 or so proteins that circulate in the blood and extracellular fluid. Complement proteins are made mostly by the liver, but blood monocytes, tissue macrophages, and epithelial cells of the gastrointestinal and genitourinary tracts can also make significant amounts. The proteins are proteases that interact in a proteolytic cascade on the surface of invading microbes. Eventually a membrane attack complex is formed that can punch holes through the outer membranes of the microbes, causing them to swell up and burst (**Figure 4.20**). Activation of complement is possible by three different pathways:

- The *classical pathway* involves interaction with the adaptive immune system, particularly invariant domains of antibodies, which bind to antigens on the surface of a microbe.

- The *alternative pathway* is directly triggered when the C3 complement protein binds to certain bacterial cell wall components.

- The *lectin pathway* uses nonspecific pattern recognition alone. The trigger is a mannan-binding lectin that binds to characteristic arrangements of mannose and fucose residues on bacterial cell walls.

The adaptive immune system mounts highly specific immune responses that are enhanced by memory cells

The **adaptive immune system** arose in early vertebrates, providing a powerful additional defense against pathogens. It is mobilized by components of the

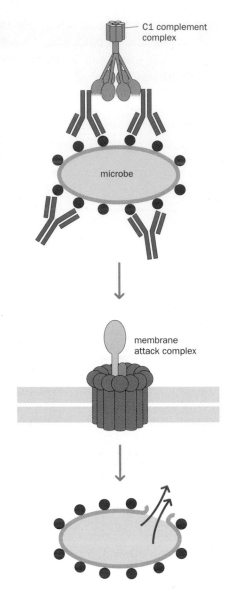

Figure 4.20 Activating the complement system. Antibodies bound to the surface of a microbe can be bound by complement via a component (C1q) of the C1 complement complex, activating the *classical complement pathway*. Eventually a membrane attack complex is formed: a ring of complement C9 proteins surrounding a core of complement proteins C5 to C8 integrates into the cell membrane of the microbe. The membrane attack complex then punches a hole in the microbe's cell membrane, and the cell contents are released, killing the microbe.

innate immune system when the latter fails to provide adequate protection against an invading pathogen. It has three key characteristics:

- Exquisite specificity: it can discriminate between tiny differences in molecular structure.
- Extraordinary adaptability: it can respond to an unlimited number of molecules.
- Memory: it can remember a previous encounter with a foreign molecule and respond more rapidly and more effectively on a second occasion.

Whereas innate immune responses are the same in all members of a species, adaptive immune responses vary between individuals: one individual may mount a strong reaction to a particular antigen that another individual may never respond to. There are two major arms to the adaptive immune response:

- *Humoral* (*antibody*) *immunity* is mediated by B lymphocytes (also called **B cells**).
- *Cell-mediated immunity* is effected by T lymphocytes (**T cells**) and is an important antiviral defense system.

Both B and T cells have dedicated cell surface receptors that can specifically recognize individual antigens.

What makes the adaptive immune system so proficient at recognizing antigens is that during its development in the primary lymphoid organs each naive B or T cell acquires a cell surface antigen receptor of a unique specificity. Binding of this receptor to its specific antigen activates the cell and causes it to proliferate into a clone of cells with the same immunological specificity as the parent cell (**clonal selection**). As a result, the number of lymphocytes that can recognize the specific antigen can be rapidly expanded.

Immunological memory is another important feature of the adaptive immune system. During the massive clonal expansion of antigen-specific lymphocytes that occurs after an initial encounter with antigen, some of the expanding daughter cells differentiate into *memory cells* that are able to respond to the antigen more rapidly or more effectively. Memory cells are endowed with higher-affinity antigen receptors, and also combinations of adhesion molecules, homing receptors, and cytokine receptors that direct the lymphocytes to migrate efficiently into specific tissues (*extravasation*).

Humoral immunity depends on the activities of soluble antibodies

B cells produced in the bone marrow have cell surface immunoglobulins (Igs) as their antigen receptors (*B-cell receptors*). When B cells are activated, however, they can differentiate into *plasma cells* that secrete their Ig receptors in a soluble form, known as **antibodies**. Igs are composed of two identical heavy chains and two identical light chains that are held together by disulfide bonding (**Figure 4.21**). The light chains can be one of two varieties (κ or γ) that are functionally equivalent. The heavy chains can be one of five functionally distinct types that define five immunoglobulin classes (**Table 4.9**).

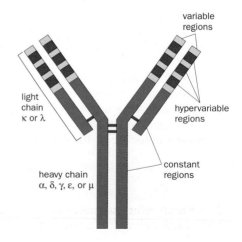

Figure 4.21 Antibody (immunoglobulin) structure. Antibodies are soluble immunoglobulins (Igs). Igs consist of two identical heavy chains, which can be one of five classes, and two identical light chains, which can be one of two classes, as shown. The two heavy chains are held together by disulfide bonds, and each light chain is linked to one heavy chain by disulfide bridges. As shown in Figure 4.22, each chain is composed of globular domains that are maintained by intrachain disulfide bridges. The N-terminal regions are known as *variable regions* because their sequence varies significantly from one antibody to another. Most of the sequence variation is concentrated in *hypervariable regions* (also called *complementarity-determining regions*), which are the sequences involved in antigen binding. The constant region of each heavy chain determines the class of the heavy chain and the response to bound antigen (see Table 4.9).

variable regions

light chain κ or λ

hypervariable regions

constant regions

heavy chain α, δ, γ, ε, or μ

TABLE 4.9 IMMUNOGLOBULIN ISOTYPES				
Isotype	H chain	Structure	Working form	Location and functions
IgM	μ	pentamer/monomer	B-cell receptor (monomer) and secreted antibody (pentamer)	first Ig to be produced (see Figure 4.26A); activates complement
IgD	δ	monomer	B-cell receptor only	second Ig to be produced (see Figure 4.26A); function unknown
IgA	α	often dimer or tetramer	B-cell receptor, but predominantly as antibodies secreted by plasma cells	predominant Ig in external secretions such as breast milk, saliva, tears, and mucus of the bronchial, genitourinary, and digestive tracts
IgG	γ	monomer		principal serum Ig; binds to Fc receptors on phagocytic cells; activates complement
IgE	ε	monomer		binds to Fc receptors of mast cells and blood basophils; mediates immediate hypersensitivity reactions responsible for symptoms of allergic conditions such as hay fever and asthma

Each Ig chain contains two distinct regions:

- An N-terminal *variable region* that is involved in antigen binding and has a variable sequence that accounts for the unique specificity of each B-cell receptor and antibody. The variability is not distributed evenly throughout the domain, but is concentrated in *hypervariable regions* (also called *complementarity-determining regions*), which are the regions that directly interact with antigen.

- A C-terminal *constant region*, a domain that is invariant within each class but differs significantly between the different heavy chain classes, accounting for the different functionality of each isotype.

Although heavy and light Ig chains are encoded at different genetic loci, they are structurally closely related. The light chains have two, and heavy chains four or five, *immunoglobulin domains*, domains of about 100 amino acids that are held together in a barrel-like structure by an internal disulfide bond. Many other cell surface proteins, including T-cell receptors and several other proteins involved in immune system functions, possess one or more of these domains and so define the *immunoglobulin superfamily* (**Figure 4.22**).

As described previously, when IgM or IgG antibodies coat microbes they can initiate the classical pathway of complement activation, causing lysis of the microbes (see Figure 4.20). Complement can also promote phagocytosis via complement receptors on macrophages and neutrophils. Antibodies also have several other important functions:

- *Blocking pathogen entry into cells*: viruses and certain microorganisms that can exist within animal cells enter into the cells by binding to certain preferred receptors on the cell surface. Antibodies can physically block this process (**Figure 4.23A**).

- *Neutralizing toxins*: antibodies can directly bind and neutralize toxins released by bacteria, inhibiting their enzymatic activity or their ability to bind to cell surface receptors (Figure 4.23B).

- *Activating effector cells*: many immune system cells express receptors that recognize the invariant domains of immunoglobulin molecules. Antibodies bound to the surface of microbes can therefore not only activate complement but also directly activate those immune system cells that carry appropriate Fc receptors. IgG antibodies, for example, promote both phagocytic uptake by neutrophils and macrophages and antibody-dependent cell-mediated cytotoxicity by NK cells (Figure 4.23C). IgE bound to IgE-specific Fc receptors on eosinophils, basophils, and mast cells can trigger the release of powerful pharmacological mediators and activate these cells to kill antibody-coated parasites.

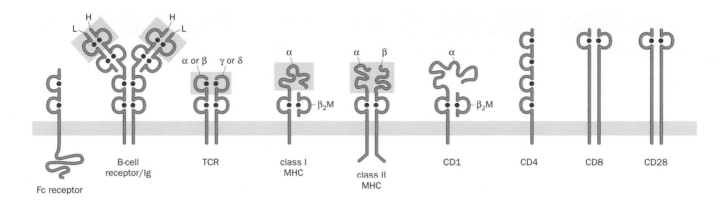

Figure 4.22 **Some representatives of the immunoglobulin superfamily.** The immunoglobulin domain is a barrel-like structure held together by a disulfide bridge (pink dot). Multiple copies of this structural domain are found in many proteins with important immune system functions, but other members of the immunoglobulin superfamily can have different functions, such as Ig-CAM cell adhesion molecules (Section 4.2). Shaded boxes show the variable domains of immune system proteins. $\beta_2 M$, β_2-microglobulin.

In cell-mediated immunity, T cells recognize cells containing fragments of foreign proteins

Microorganisms such as viruses that penetrate cells and multiply within them are out of reach of antibodies. T cells evolved to deal with this need; they carry on their surfaces dedicated T-cell receptors (TCRs) that recognize sequences from foreign antigens. TCRs are heterodimers that are evolutionarily related to both immunoglobulins and other cell receptors of the immunoglobulin superfamily (see Figure 4.22).

There are two types of TCR, composed either of α and β chains or of γ and δ chains. Unlike the immunoglobulins on B cells, most αβ TCRs and the cells that bear them are limited to recognizing foreign proteins, and only after they have been degraded intracellularly. Peptides of an appropriate size and sequence derived from this degradation bind to **major histocompatibility complex (MHC) proteins (Box 4.4)** that are then transported to the surface of the cell. With the peptide held in a cleft on its outer surface, the MHC protein presents the peptide to a T cell carrying an appropriate TCR$_{\alpha\beta}$ receptor (*antigen presentation*), thereby activating that T cell.

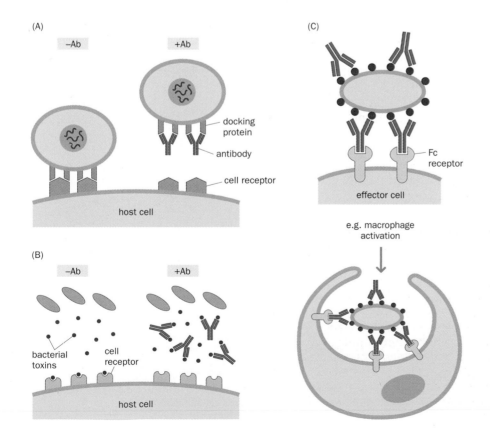

Figure 4.23 **Aspects of antibody function.** (A) Inhibiting viral infection. Viruses infect cells by first using docking proteins to bind to certain receptors on the plasma membrane of host cells. Antibodies (Ab) can bind to the viral docking proteins to prevent them from binding to host cell receptors. (B) Neutralizing toxins. Antibodies bind to toxins released from invading microbes and so stop them from binding to cell receptors. (C) Activating effector cells. Antibodies can bind and coat the surfaces of microbes and large target cells. Various immune system effector cells (notably macrophages, natural killer cells, neutrophils, eosinophils, and mast cells) carry Fc receptors that enable them to bind to the Fc region on IgA, IgG, or IgE antibodies. Antibody binding to an Fc receptor activates the effector cell and can lead to cell killing by phagocytosis, or release of lytic enzymes, death signals, and so on.

BOX 4.4 THE MAJOR HISTOCOMPATIBILITY COMPLEX, MHC PROTEINS, AND ANTIGEN PRESENTATION

The **major histocompatibility complex (MHC)** is a large gene cluster that contains genes mostly associated with cell-mediated immune responses and, in particular, with antigen processing and presentation. Among the protein products of the MHC are the class I and II MHC antigens, which are members of the *immunoglobulin superfamily* of cell surface proteins (see Figure 4.22). These transmembrane proteins are heterodimers. The MHC contains genes for the highly polymorphic α (heavy) chain of class I and both α and β chains of class II MHC proteins. The lighter chain of class I MHC protein, β$_2$-microglobulin, is not polymorphic and is encoded by a gene elsewhere in the genome.

The human MHC (known as the **HLA complex**; HLA stands for human leukocyte antigen) spans several megabases at 6p21.3. The class I MHC α chain genes (*HLA-A, HLA-B, HLA-C, HLA-E, HLA-F*, and *HLA-G*) are located at the telomeric side; class II MHC genes are tightly clustered at the centromeric end. An intervening region, sometimes known as the class III MHC gene region, contains various complement and other genes (**Figure 1**).

The MHC class I and class II proteins are among the most polymorphic vertebrate proteins known. Most individuals are heterozygous at the corresponding loci, and both allelic forms are functional and co-expressed at the cell surface. This extreme polymorphism is presumably driven by intense natural selection pressure from viruses and other parasites, which favors individuals with a greater capacity for recognizing infected host cells.

Because of their very high degree of polymorphism, class I and II MHC proteins are important in graft rejection after organ or tissue transplantation from one individual to another. If the donor and recipient (host) are unrelated, major differences could be expected in the MHC proteins expressed by transplanted cells and host cells. The graft will be rejected because it will be recognized as foreign and will be attacked by the immune system. To increase transplant success rates, **tissue typing** is performed to identify alleles at important MHC loci (in easily accessible blood cells) so that potential donors can be sought with a closely matching MHC profile, and drugs are administered that can suppress immune responses. However, MHC polymorphism remains a challenge for successful cell therapy.

Antigen presentation
The function of classical MHC proteins is to bind and transport peptides to the cell surface for **antigen presentation** to T cells with TCR$_{αβ}$ receptors.

- **Class I MHC proteins** bind peptides predominantly derived from *endogenous antigens* and present them to cytotoxic T (Tc) lymphocytes.
- **Class II MHC proteins** bind peptides predominantly derived from *exogenous antigens* and present them to helper T (Th) lymphocytes.

All cells with MHC proteins on their surface can present peptides to T cells and in this context can be referred to as **antigen-presenting cells**. However, when such cells are recognized and potentially killed by Tc cells they are often known as *target cells*. Almost all nucleated cells express class I MHC proteins and so can function as a *target cell* presenting endogenous antigens to Tc cells. Target cells are often cells infected with a virus or some other intracellular microorganism. In addition, altered self cells such as cancer cells and aging body cells can serve as target cells, as can *allogeneic* cells introduced via skin grafts or organ transplantation (between genetically non-identical individuals).

By contrast, relatively few cells express class II MHC molecules. Among the most important are dendritic cells, macrophages, and B cells. These cells are also among the few that can express the *co-stimulatory molecules* that are essential for initiating immune responses in both Tc and Th cells. For both these reasons they are sometimes referred to as *professional antigen-presenting cells*. Several

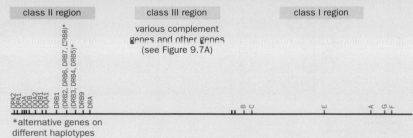

Figure 1 Organization of HLA genes in the human MHC.

other cell types (including fibroblasts in skin, vascular endothelial cells, glial cells, and pancreatic β cells) can be induced to express class II MHC proteins.

From antigen presentation to T-cell activation
Co-stimulatory molecules are surface molecules that interact with specific receptors on the T cell (**Figure 2**). The co-stimulatory signal delivered by this interaction is necessary to induce the Th cell to synthesize the T-cell growth factor interleukin-2 (IL-2) and to express high-affinity receptors for IL-2. The secreted IL-2 stimulates the proliferation of T cells expressing the CD4 or CD8 cell surface receptors. This complex way of inducing T-cell responses is presumably necessary to avoid the inadvertent activation of T cells, which would be potentially damaging, and to allow the detailed regulation of their proliferation and functional differentiation.

Co-stimulatory molecules are of such critical importance to the initiation and regulation of immune responses that they are not constitutively expressed, even by professional antigen-presenting cells. Their expression can be induced via the TCR-triggered transient expression of CD40 ligand on T cells and the subsequent interaction with and crosslinking of this ligand to CD40 on professional antigen-presenting cells. The expression of co-stimulators is also triggered by signaling arising from the recognition of the conserved features of pathogens that is part of the innate immune system, most importantly the signaling through Toll-like receptors expressed on dendritic cells, macrophages, and certain other cells.

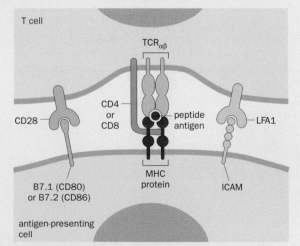

Figure 2 Presentation of peptide antigens to T cells. Cell adhesion (mediated by molecules such as ICAM1 and LAF1) brings the cells together. The peptides (about eight or nine amino acids long for class I MHC, and somewhat larger for class II MHC) are held in an outward-facing cleft in the MHC molecule. CD8 or CD4 proteins bind to non-polymorphic components of class I and class II MHC proteins, respectively. In addition to TCR–antigen recognition, a co-stimulatory signal is needed, such as a member of the B7 family that is recognized by CD28 receptors on the T-cell surface.

This process, in which most αβ T cells recognize antigens only after they have been degraded and become associated with MHC molecules, is often referred to as **MHC restriction**. It allows T cells to survey a peptide library derived from the entire set of proteins that is contained within the cell but is presented on the surface of the cell by MHC molecules, thereby providing a remarkable evolutionary solution to the problem of how to detect intracellular pathogens. At the same time, it restricts T cells to recognition of only those antigens that are associated with cell-surface MHC molecules and are derived from intracellular spaces. T cells therefore complement B cells and antibodies that can recognize only extracellular antigens and pathogens.

Cell-associated antigens that are synthesized within the cell are referred to as *endogenous* antigens. They derive from proteins produced within the cells (including normal cellular proteins, tumor proteins, or viral and bacterial proteins produced within infected cells). Such antigens are degraded within the cytosol by *proteasomes*, large complexes containing enzymes that cleave peptide bonds, converting proteins to peptides. These antigens are bound by class I MHC proteins. Antigens that are captured by endocytosis or phagocytosis are described as *exogenous* antigens. The internalized antigens move through several increasingly acidic compartments, where they encounter hydrolytic enzymes including proteases. These antigens are presented by class II MHC proteins.

An important aspect of this process is that MHC molecules cannot distinguish self from nonself antigens, and therefore, except on the rare occasions when a cell does indeed become infected or captures a microbial protein, the MHC molecules on its surface are presenting peptides derived from the degradation of self proteins. A key event during the development of αβ T cells in the thymus is that only those T cells that have receptors that potentially recognize foreign peptides in association with self MHC molecules are allowed to mature and be released into the periphery (positive selection). Those that recognize self peptides are induced to commit suicide by apoptosis (negative selection).

T-cell activation

Of the major classes of T cells (see Table 4.7), there are three main ones that participate in this process:

- *Helper T (Th) cells* (which usually carry a CD4 cell surface receptor) recognize peptides that are predominantly produced from exogenous protein antigens and bound to a class II MHC protein (see Box 4.4). Only a few types of cell—dendritic cells, macrophages, and B cells—express class II MHC proteins and act to present antigens to Th cells. Once activated, Th cells secrete cytokines to mobilize or regulate other immune system cells so as to generate an effective immune response. There are three types: Th1 cells activate macrophages and Tc cells (see below); Th2 cells stimulate activation of eosinophils and promote antibody production by B cells; Th17 cells promote autoimmunity and inflammation.

- *Cytotoxic T (Tc) cells* (which usually carry a CD8 cell surface receptor) recognize peptides that are predominantly produced from endogenous protein antigens and are bound to a class I MHC protein. Almost all nucleated cells express class I MHC proteins and so can present antigen to Tc cells. Once activated, and in association with cytokines secreted by Th cells, they differentiate into effector cells called *cytotoxic T lymphocytes (CTLs)*.

    CTLs are particularly important in recognizing and killing virally infected cells. Like NK cells, they induce apoptosis, either by the Fas pathway (see Figure 4.16) or often by delivering granules containing perforin and granzymes. Some viruses seek to avoid being killed by CTLs by downregulating the expression of class I MHC molecules on host cells. However, NK cells have receptors that screen host cells for the presence of class I MHC antigens and are activated when they sense host cells lacking class I MHC proteins.

- *Regulatory T (Treg) cells*. These cells generally express CD4 but are distinguished from helper T cells by their constitutive surface expression of CD25 (one of the chains of the interleukin-2 receptor) and intracellular expression of the transcription factor Foxp3. They suppress **autoimmunity** (immune responses to self antigens; **Box 4.5**)

The central characteristic of immunity is the capacity to distinguish between self and nonself. Ig and TCR gene loci are activated by DNA rearrangements that produce a random array of different B-cell and T-cell receptors. Inevitably, *self-reactive lymphocytes* are produced, cells that carry high-affinity receptors against antigens present on host cells. These potentially harmful self-reactive lymphocytes need to be eliminated or inactivated during the maturation process. For B cells this process is less critical, because their activation is often dependent on T cells and they lack cytotoxic activity. For T cells it is of central importance, and multiple mechanisms for preventing autoimmune damage by self-reactive T cells have been identified.

During their development in the thymus, immature T cells with high-affinity receptors for self-peptide:MHC complexes are eliminated. This is a central form of **self tolerance**. However, it is clearly not 100%

efficient because autoimmune diseases do occur. For example, in rheumatoid arthritis, self-reactive T cells attack the tissue in joints, causing an inflammatory response that results in swelling and tissue destruction. *Peripheral tolerance* is a back-up form of tolerance that is designed to inactivate, or otherwise regulate, autoreactive T or B cells in secondary lymphoid tissues.

One mechanism for achieving peripheral tolerance is **anergy**, in which self-reactive T cells that encounter antigen will generally do so without receiving co-stimulatory signals. In these circumstances, a permanent state of unresponsiveness is induced in which the cells are unable to respond even if they subsequently encounter antigen in the presence of co-stimulatory signals. Another mechanism is via regulatory T cells (Treg cells) that act in secondary lymphoid tissues and at sites of inflammation to down regulate autoimmune processes.

Not all $\alpha\beta$ T cells recognize peptides presented by MHC class I and class II molecules. In particular, a subclass called NKT cells can react with glycolipid antigens, which are presented by members of the CD1 protein family rather than by MHC proteins. CD1 proteins are non-polymorphic and structurally closely resemble class I MHC proteins; however, like class II MHC proteins, their expression is restricted to professional antigen-presenting cells.

Other T-cell populations express a $\gamma\delta$ T-cell receptor and are found in a wide variety of lymphoid tissues. Despite the abundance and presumed importance of these cells, very little is known about their antigen recognition and function except that in many cases it seems to be MHC-unrestricted. Possibly, $\gamma\delta$ T cells recognize antigens presented by non-MHC-encoded molecules, or perhaps like B cells they can recognize soluble antigens. Some human $\gamma\delta$ T cells resemble NKT cells in being able to recognize bacterial glycolipids bound to CD1 molecules. Other populations of $\gamma\delta$ T cells, such as those found in the skin, have essentially invariant TCRs and presumably recognize a common microbial antigen or a self molecule, but the nature of this is unknown.

The unique organization and expression of Ig and TCR genes

The diversity of B-cell receptors/Igs and TCRs is extraordinary. Each person needs to produce huge numbers of different Igs and TCRs to optimize the chances of recognizing any of a myriad of foreign antigens. Yet at the genome level we have only three immunoglobulin gene loci: *IGH* specifies the heavy chain, and *IGK* and *IGL* the alternative κ and λ light chains, respectively. Similarly, there are only four human TCR gene loci (*TRA*, *TRB*, *TRG*, and *TRD*), one each for making the four types of TCR chain: α, β, γ, and δ.

Despite the huge diversity of Igs and TCRs, an individual B or T cell is *monospecific*. Each such cell produces just one type of Ig or TCR heterodimer with a unique antigen-binding site, and so is specific for one type of antigen only. Different B (or T) cells make different Igs (or TCRs); it is the population of billions of different B and T cells each bearing a different antigen receptor that enables the immune system to respond to virtually any foreign material.

The diversity of Igs and TCRs follows on from the unique organization and expression of the respective genes. The variable region of each antigen receptor chain is specified by at least two different types of gene segment: *V gene segments* encode most of the variable region; *J gene segments* encode the joining region that forms a small part at the C-terminal end of the variable region. Additionally, *D gene segments* encode a small diversity region near the C-terminal ends of the variable region of Ig heavy chains, TCRβ chains, and TCRδ chains only. The organization of the gene segments is extraordinary: each one is usually present in numerous different copies in germ-line DNA (**Figure 4.24**).

Ig and TCR gene loci are inactive in all cells other than B and T lymphocytes, respectively, and only become active in B and T cells as they mature. As part of this process, random combinations of V + D (where present) + J gene segments are brought together by somatic recombination at the DNA level to create a novel

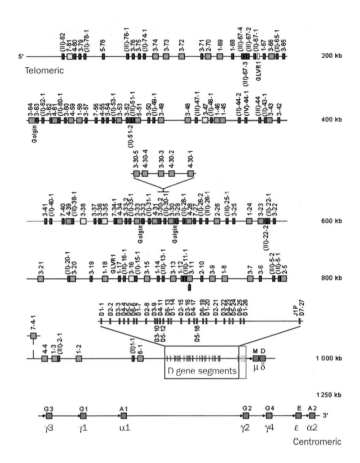

Figure 4.24 Multiple gene segments in the *IGH* Ig heavy-chain gene locus at 14q32. The *IGH* gene locus spans 1250 kb of 14q32.3 and, for clarification, is shown as six successive segments of about 200 kb from the telomeric end (top) to the centromeric end (bottom). V$_H$ gene segments span the first 900 kb. In addition to functional segments (green), there are many nonfunctional pseudogene segments (red) and some others whose functional status is uncertain (yellow). Next are 27 D$_H$ gene segments (dark blue within pink box) followed immediately by 9 J$_H$ gene segments (within pale blue box), and finally various transcription units (C$_\mu$, C$_\delta$, C$_{\gamma3}$, and so on), each of which contains several exons. The number of gene segments shows some variation, as represented by the variable presence of additional gene segments such as the five V$_H$ gene segments at about the 500 kb position. [Reproduced with permission from The International Immunogenetics Information System at http://www.imgt.org]

exon that potentially encodes the variable domain of a receptor chain (**Figure 4.25**). A typical receptor gene locus may contain 100 V, 5 D, and 10 J elements. Random recombination of these gives 5000 permutations, and because each immune receptor contains two chains, each formed from randomly rearranged receptor genes, the rearrangement process by itself creates a potential basic germ-line repertoire of (5000)^2 or about 25 million specificities.

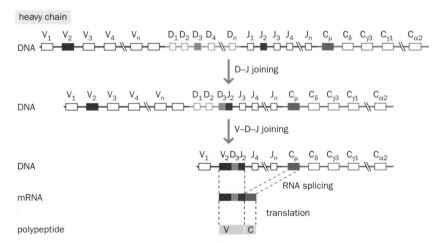

Figure 4.25 Cell-specific VDJ recombination as a prelude to making an Ig heavy chain. Two sequential somatic recombinations produce first D–J joining, then a mature VDJ exon. In this particular example, the second out of 129 different V segments (V$_2$) is fused to the third D region segment (D$_3$) and the second J region segment (J$_2$) to produce a functional V$_2$D$_3$J$_2$ exon, but the choice is cell-specific so that a neighboring B lymphocyte may have a functional V$_{129}$D$_{17}$J$_1$ exon, for example. Once the VDJ exon has been assembled, the gene can be transcribed using the VDJ exon as the first exon, with the subsequent exons provided by the nearest constant region (C) transcription unit.

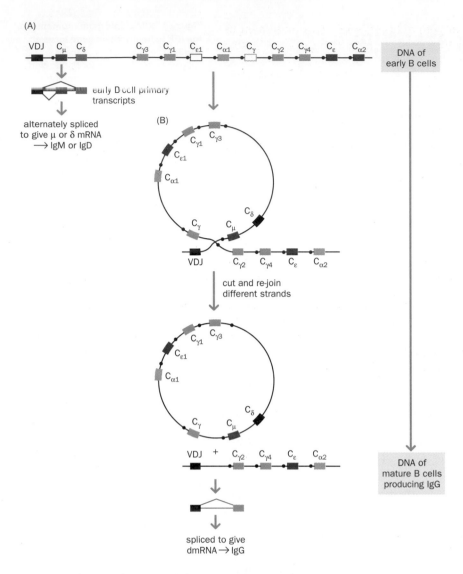

Figure 4.26 Ig heavy chain class switching is mediated by intrachromatid recombination. (A) Early (partial) switch to IgD. The constant region of a human Ig heavy chain is encoded by one of several different C transcription units: one each for the μ (IgM), δ (IgD) and ε (IgE) chains, four for γ (IgG), and two for α (IgA). The initial heavy chain class is IgM because RNA splicing brings together sequences transcribed from the VDJ exon and the exons of the neighboring C_μ transcription unit. As naïve B cells mature, however, alternative RNA splicing brings together sequences from the VDJ exon and exons of the C_δ transcription unit, leading to the additional production of IgD. (B) Late switch to IgA, IgG, and IgE. Later in the maturation process, class switching occurs by intrachromatid recombination, in which the same VDJ exon is brought close to the initially more distal constant region transcription units (C_α, C_γ, or C_ϵ) by a looping out of the intervening segments. The new VDJ-C combination is expressed to give IgA, IgG, or IgE. The combination for IgG is illustrated here.

Large as the combinatorial diversity above may seem, it represents only the tip of the iceberg of receptor diversity. A much bigger contribution to V domain variability is provided by mutational events that take place before the joining of V, D, and J elements and which create *junctional diversity*. First, exonucleases nibble away at the ends of each element, removing a variable number of nucleotides. Second, a template-independent DNA polymerase adds a variable number of random nucleotides to the nibbled ends, creating N regions. Combinatorial and junctional diversity together create an almost limitless repertoire of variable domains in immune receptors.

The genetic mechanisms leading to the production of functional VJ and VDJ exons often involve large-scale deletions of the sequences separating the selected gene segments (most probably by intrachromatid recombination; **Figure 4.26B**). Conserved *recombination signal sequences* flank the 3′ ends of each V and J segment and both the 5′ and 3′ ends of each D gene segment. They are recognized by RAG proteins that initiate the recombination process, and they are the only lymphocyte-specific proteins involved in the entire recombination process. The recombination signal sequences also specify the rules for recombination, enabling joining of V to J, or D to J followed by V to DJ, but never V to V or D to D.

Additional recombination and mutation mechanisms contribute to receptor diversity in B cells, but not T cells

The Ig heavy chain locus is unique in having a series of downstream constant region loci, each of which contains the various exons required to specify the constant domains of a separate class (isotype) of Ig. After VDJ rearrangement in immature B cells, transcription initially terminates downstream of the C_μ constant region so that the expressed heavy chain is of the IgM isotype. On maturation of the B cell, transcription extends to the end of the next downstream constant region, C_δ, resulting in a primary transcript containing at its 5′ end the VDJ exon followed by the various C_μ exons and then the various C_δ exons. This transcript undergoes alternate splicing and polyadenylation at the RNA level to generate mRNA species that have the same VDJ exon joined to either the C_μ or the C_δ exons. Thus, each individual mature B cell expresses IgM and IgD receptors of identical specificity. Exposure to foreign antigen during a *primary immune response* triggers a further change in the splicing pattern that deletes the exon specifying the transmembrane region of the IgM receptor, allowing the secretion of IgM (but not IgD).

After interaction with antigen, and under the direction of signals received from helper T cells, the descendants of a B cell can alter the structure of their antigen receptor, and ultimately of the antibody that they secrete, in two distinct ways. First, the B cell can switch to produce an Ig with the same antigen-binding site as before but using a different class of heavy chain (*class switching* or *isotype switching*). Such switching is mediated by switch regions upstream of each constant region locus (except C_δ) that allow a second and less well understood recombination process to take place at the DNA level in which the C_μ–C_δ module and a variable number of other constant regions are deleted, placing a new constant region immediately downstream of the VDJ exon (see Figure 4.26).

Second, again under the influence of signals from T cells, B cells can refine the affinity of Ig binding (affinity maturation) so that they can respond more effectively to foreign antigen on a future occasion (when they will be able to secrete large amounts of soluble antibody with a very high affinity for the foreign antigen in a powerful *secondary immune response*). This is brought about by *somatic hypermutation* in which DNA repair enzymes are targeted to the VDJ exon of the H chain and the VJ exon of the L chain where, in a poorly understood process, they create point mutations at individual bases in the sequence.

The monospecificity of Igs and TCRs is due to allelic exclusion and light chain exclusion

There are three Ig gene loci per haploid chromosome set, and so a total of six gene loci are potentially available to B cells for making Ig chains. However, an *individual* B cell is *monospecific*: it produces only one type of Ig molecule with a single type of heavy chain and a single type of light chain. There are two reasons for this:

- *Allelic exclusion*: a light chain or a heavy chain can be synthesized from a maternal chromosome or a paternal chromosome in any one B cell, but not from *both* parental homologs. As a result, there is *monoallelic expression* at the heavy chain gene locus in B cells. Most T cells that form αβ receptors also restrict expression to one of the two alleles.

- *Light chain exclusion*: a light chain synthesized in a single B cell may be a κ chain or a λ chain, but never both. As a result of this requirement plus that of allelic exclusion, there is monoallelic expression at one of the two functional light chain gene clusters and *no* expression at the other.

The decision to activate only one out of two possible heavy-chain alleles and only one out of four possible light-chain genes is not quite random. In each B-cell precursor, productive DNA rearrangements are attempted at *IGK* in preference to *IGL*, so that most B cells carry κ light chains. At least three mechanisms account for the unusual monoallelic expression of antigen receptors in lymphocytes. First, it seems that the rearrangement of receptor genes frequently goes wrong, destroying the first locus. It may also go wrong at the second attempt using the second locus, in which case the immature B or T cell dies. Second, even when it occurs correctly, as noted above only one-third of rearrangements will generate a functional protein. Third, once a chain has been successfully expressed on the cell surface, a feedback signal is delivered that inactivates the rearrangement process.

FURTHER READING

General

Alberts B, Johnson A, Lewis J et al. (2007) Molecular Biology Of The Cell, 5th ed. Garland Publishing.

Cells Alive! Collection of various videos and animations in cell biology, immunology, and related areas at http://www.cellsalive.com/toc.htm

Cooper GM & Hausman RE (2006) The Cell, A Molecular Approach, 4th ed. Sinauer Associates Inc.

Pollard TD & Earnshaw WC (2007) Cell Biology, 2nd ed. Saunders.

Cell structure and diversity

Handwerger KE & Gall JG (2006) Subnuclear organelles: new insights into form and function. *Trends Cell Biol.* 16, 19–26.

Liang B (2001) Construction of the cell membrane [animation]. http://www.wisc-online.com/objects/index_tj.asp?objID=AP1101

Singla V & Reiter JF (2006) The primary cilium as the cell's antennae: signaling at a sensory organelle. *Science* 313, 629–633.

Cell adhesion and tissue formation

Beckerle M (ed.) (2002) Cell Adhesion (Frontiers In Molecular Biology). Oxford University Press.

Extravasation Animation, available at: http://multimedia.mcb.harvard.edu/media.html [This video shows how regulating cell adhesion permits white blood cells to migrate into tissues.]

Gumbiner BM (1996) Cell adhesion: the molecular basis of tissue architecture and morphogenesis. *Cell* 84, 345–377.

Hynes RO (2002) Integrins: bidirectional allosteric signalling machines. *Cell* 110, 673–687.

Cell signaling

Brivanlou AH & Darnell JE (2002) Signal transduction and the control of gene expression. *Science* 295, 813–818.

Christensen ST & Ott CM (2007) Cell signaling: a ciliary signaling switch. *Science* 317, 330–331.

Database of Cell Signaling. *Science STKE (Signal Transduction Knowledge Environment)* at http://stke.sciencemag.org/cm/ [Permits browsing of details of signal transduction pathways.]

Gavi S, Shumay E, Wang HY & Malbon CC (2006) G-protein-coupled receptors and tyrosine kinases: crossroads in cell signaling and regulation. *Trends Endocr. Metab.* 17, 48–54.

Cell proliferation, senescence, and programmed cell death

Adams JM (2003) Ways of dying: multiple pathways to apoptosis. *Genes Dev.* 17, 2481–2495.

Berry D (2007) Molecular animation of cell death mediated by the Fas pathway. *Science STKE 3 April*, doi:10.1126/stke.3802007tr1. http://stke.sciencemag.org/cgi/content/full/sigtrans;2007/380/tr1/DC1

Elmore S (2007) Apoptosis: a review of programmed cell death. *Toxicol. Pathol.* 35, 495–516.

Morgan DO (2007) The Cell Cycle: Principles Of Control. New Science Press Ltd.

Raff M (1998) Cell suicide for beginners. *Nature* 396, 119–122.

Von Zglinicki T, Saretzki G, Ladhoff J et al. (2005) Human cell senescence as a DNA damage response. *Mech. Ageing Dev.* 126, 111–117.

Stem cells and differentiation

Moore KA & Lemischka IR (2006) Stem cells and their niches. *Science* 311, 1880–1885.

Nature Insight on Stem Cells (2001) *Nature* 414, 88–131.

Nature Insight on Stem Cells (2006) *Nature* 441, 1059–1102.

Weissman IL, Anderson DJ & Gage F (2001) Stem and progenitor cells: origins, phenotypes, lineage commitments and transdifferentiation. *Annu. Rev. Cell Dev. Biol.* 17, 387–403.

Immune system cells

Hayday AC & Pennington DJ (2007) Key factors in the organized chaos of early T cell development. *Nat. Immunol.* 8, 137–144.

Immune system cell lineages (2007) *Immunity* 26, 669–750.

Janeway C, Travers P, Walport M & Shlomchik M (2004) Immunobiology: The Immune System In Health And Disease, 6th ed. Garland Scientific.

Kindt TJ, Goldsby RA & Osborne BA (2007) Kuby Immunology, 6th ed. WH Freeman & Co.

Roitt IM, Martin SJ, Delves PJ & Burton D (2006) Roitt's Essential Immunology, 11th ed. Blackwell Scientific.

Chapter 5

Principles of Development

5

KEY CONCEPTS

- The zygote and each cell in very early stage vertebrate embryos (up to the 16-cell stage in mouse) can give rise to every type of adult cell; as development proceeds, the choice of cell fate narrows.

- In the early development of vertebrate embryos, the choice between alternative cell fates primarily depends on the position of a cell and its interactions with other cells rather than cell lineage.

- Many proteins that act as master regulators of early development are transcription factors; others are components of pathways that mediate signaling between cells.

- The vertebrate body plan is dependent on the specification of three orthogonal axes and the polarization of individual cells; there are very significant differences in the way that these axes are specified in different vertebrates.

- The fates of cells at different positions along an axis are dictated by master regulatory genes such as Hox genes.

- Cells become aware of their positions along an axis (and respond accordingly) as a result of differential exposure to signaling molecules known as morphogens.

- Major changes in the form of the embryo (morphogenesis) are driven by changes in cell shape, selective cell proliferation, and differences in cell affinity.

- Gastrulation is a key developmental stage in early animal embryos. Rapid cell migrations cause drastic restructuring of the embryo to form three germ layers—ectoderm, mesoderm, and endoderm—that will be precursors of defined tissues of the body.

- Mammals are unusual in that only a small fraction of the cells of the early embryo give rise to the mature animal; the rest of the cells are involved in establishing extraembryonic tissues that act as a life support.

- Developmental control genes are often highly conserved but there are important species differences in gastrulation and in many earlier developmental processes.

Many of the properties of cells described in Chapter 4 are particularly relevant to early development. Cell differentiation is especially relevant during embryonic development, when tissues and organs are being formed. The embryonic stages are characterized by dramatic changes in form that involve active cell proliferation, major cell migration events, extensive cell signaling, and programmed cell death. In this chapter we consider the details of some key aspects of development, with particular emphasis on vertebrate and especially mammalian development.

5.1   AN OVERVIEW OF DEVELOPMENT

Multicellular organisms can vary enormously in size, form, and cell number but, in each case, life begins with a single cell. The process of *development*, from single cell to mature organism, involves many rounds of cell division during which the cells must become increasingly specialized and organized in precise patterns. Their behavior and interactions with each other during development mold the overall morphology of the organism.

Traditionally, animal development has been divided into an *embryonic stage* (during which all the major organ systems are established), and a *postembryonic stage* (which in mammals consists predominantly of growth and refinement). Developmental biologists tend to concentrate on the embryonic stage because this is where the most exciting and dramatic events occur, but postembryonic development is important too.

Once the basic body plan has been established, it is not clear when development stops. In humans, postembryonic growth begins at the start of the fetal period in the fetus at about 9 weeks after fertilization but continues for up to two decades after birth, with some organs reaching maturity before others. It can be argued that human development ceases when the individual becomes sexually mature, but many tissues—skin, blood, and intestinal epithelium, for example—need to be replenished throughout life. In such cases, development never really stops at all; instead it reaches an equilibrium. Even aging, a natural part of the life cycle, can be regarded as a part of development.

Development is a gradual process. The fertilized egg initially gives rise to a simple embryo with relatively crude features. As development proceeds, the number of cell types increases and the organization of these cell types becomes more intricate. Complexity is achieved progressively. At the molecular level, development incorporates several different processes that affect cell behavior. The processes listed below are inter-related and can occur separately or in combination in different parts of the embryo.

- *Cell proliferation*—repeated cell division leads to an increase in cell number; in the mature organism, this is balanced by cell loss.
- *Growth*—this leads to an increase in overall organismal size and biomass.
- *Differentiation*—the process by which cells become structurally and functionally specialized.
- *Pattern formation*—the process by which cells become organized, initially to form the fundamental body plan of the organism and subsequently the detailed structures of different organs and tissues.
- *Morphogenesis*—changes in the overall shape or form of an organism. Underlying mechanisms include differential cell proliferation, selective cell–cell adhesion or cell–matrix adhesion, changes in cell shape and size, the selective use of programmed cell death, and control over the symmetry and plane of cell division.

Each of the processes above is controlled by genes that specify when and where in the embryo the particular gene products that will direct the behavior of individual cells are synthesized. The nucleated cells in a multicellular organism show only very minor differences in DNA sequence (see Chapter 4); essentially they contain the same DNA and the same genes. To allow cells to diversify into a large number of different types, *differential gene expression* is required.

Because gene expression is controlled by transcription factors, development ultimately depends on which transcription factors are active in each cell. To control transcription factors, cells communicate using complicated cell signaling pathways. Until quite recently, the gene products that were known to be involved in modulating transcription and in cell signaling were exclusively proteins, but now it is clear that many genes make noncoding RNA products that have important developmental functions. In this chapter we will focus on vertebrate, and particularly mammalian, development. Our knowledge of early human development is fragmentary, because access to samples for study is often restricted for ethical or practical reasons. As a result, much of the available information is derived from animal models of development.

Animal models of development

Because many of the key molecules, and even whole developmental pathways, are highly conserved, invertebrate models have been valuable in aiding the identification of genes that are important in vertebrate development (**Table 5.1**). They have also provided model systems that illuminate our understanding of vertebrate development. The fruit fly *Drosophila melanogaster* provides particularly useful models of neurogenesis and eye development. The nematode *Caenorhabditis elegans* has an almost invariant **cell lineage** and is the only organism for which the fate of all cells is known and for which there is a complete wiring diagram of the nervous system. Lineage mutants of *C. elegans* provide a useful means for studying cell memory in development, and the nematode vulva is a well established model of organogenesis.

At the levels of anatomy and physiology, vertebrate organisms are superior models of human development. All vertebrate embryos pass through similar stages of development, and although there are significant species differences in the detail of some of the earliest developmental processes, the embryos reach a *phylotypic stage* at which the body plan of all vertebrates is much the same.

Although mammalian models could be expected to provide the best models, mammalian embryos are not easily accessible for study, because development occurs inside the mother. Mammalian eggs, and very early stage embryos, are also tiny and difficult to study and manipulate. Nevertheless, as we shall see in Chapter 20, sophisticated genetic manipulations of mice have been possible, making the mouse a favorite model for human development. Although human

TABLE 5.1 PRINCIPAL ANIMAL MODELS OF HUMAN DEVELOPMENT

Group	Organism	Advantages	Disadvantages
Invertebrates	*Caenorhabditis elegans* (roundworm)	easy to breed and maintain (GT = 3–5 days); fate of every single cell is known	body plan different from that of vertebrates; difficult to do targeted mutagenesis
	Drosophila melanogaster (fruit fly)	easy to breed (GT = 12–14 days); sophisticated genetics; large numbers of mutants available	body plan different from that of vertebrates; cannot be stored frozen
Fish	*Danio rerio* (zebrafish)	relatively easy to breed (GT = 3 months) and maintain large populations; good genetics	small embryo size makes manipulation difficult
Frogs	*Xenopus laevis* (large-clawed frog)	transparent large embryo that is easy to manipulate	genetics is difficult because of tetraploid genome; not so easy to breed (GT = 12 months)
	Xenopus tropicalis (small-clawed frog)	diploid genome makes it genetically more amenable than *X. laevis*; easier to breed than *X. laevis* (GT < 5 months)	smaller embryo than *X. laevis*
Birds	*Gallus gallus* (chick)	accessible embryo that is easy to observe and manipulate	genetics is difficult; moderately long generation time (5 months)
Mammals	*Mus musculus* (mouse)	relatively easy to breed (GT = 2 months); sophisticated genetics; many different strains and mutants	small embryo size makes manipulation difficult; implantation of embryo makes it difficult to access
	Papio hamadryas (baboon)	extremely similar physiology to humans	very expensive to maintain even small colonies; difficult to breed (GT = 60 months); major ethical concerns about using primates for research investigations

GT, generation time (time from birth to sexual maturity).

development would be best inferred from studies of primates, primate models are disadvantaged by cost. Primates have also not been so amenable to genetic analyses, and their use as models is particularly contentious.

Other vertebrates offer some advantages. Amphibians, such as the frog *Xenopus laevis*, have comparatively large eggs (typically 1000–8000 times the size of a human egg) and develop from egg to tadpole outside the mother. Although the very earliest stages of development in birds also occur within the mother, much of avian development is very accessible. For example, in chick development, the embryo in a freshly laid hen's egg consists of a disk of cells that is only 2 mm in diameter, and development can be followed at all subsequent stages up to hatching. As a result, delicate surgical transplantation procedures can be conducted in amphibian and avian embryos that have been extremely important in our understanding of development. However, genetic approaches are often difficult in amphibian and avian models. The zebrafish *Danio rerio* combines genetic amenability with accessible and transparent embryos, and is arguably the most versatile of the vertebrate models of development.

5.2 CELL SPECIALIZATION DURING DEVELOPMENT

As animal embryos develop, their cells become progressively more specialized (*differentiated*), and their *potency*—the ability to give rise to different types of cell—gradually becomes more restricted.

Cells become specialized through an irreversible series of hierarchical decisions

The mammalian zygote and its immediate cleavage descendants (usually up to the 8–16-cell stage) are *totipotent*. Each such cell (or *blastomere*, as they are known) can give rise to all possible cells of the organism. Thereafter, however, cells become more restricted in their ability to give rise to different cell types. We provide below an overview of cell specialization in mammalian embryos. Only a small minority of the cells in the very early mammalian embryo give rise to the organism proper; most are devoted to making four kinds of extraembryonic membrane. As well as protecting the embryo (and later the fetus), the extraembryonic membranes provide it with nutrition, respiration, and excretion. Further details about their origins will be given in Section 5.5, when we consider the stages of early human development.

At about the 16-cell stage in mammals, the *morula* stage, the embryo appears as a solid ball of cells, but it is possible to discriminate between cells on the outside of the cluster and those in the interior. Fluid begins to be secreted by cells so that in the subsequent *blastocyst* stages, the ball of cells is hollow with fluid occupying much of the interior. A clear distinction can now be seen between two separate cell layers: an outer layer of cells (**trophoblast**) and a small group of internal cells (the **inner cell mass**). The outer cells will ultimately give rise to one of the extraembryonic membranes, the *chorion*, which later combines with maternal tissue to form the *placenta*. The inner cell mass will give rise to the embryo proper plus the other extraembryonic membranes (**Figure 5.1**).

Even before mammalian embryos implant in the uterine wall, the inner cell mass begins to differentiate into two layers, the *epiblast* and the *hypoblast*. The epiblast gives rise to some extraembryonic tissue as well as all the cells of the later stage embryo and fetus, but the hypoblast is exclusively devoted to making extraembryonic tissues (see Figure 5.1). The cells of the inner cell mass have traditionally been considered to be *pluripotent*: they can give rise to all of the cells of the embryo, but unlike totipotent cells they cannot give rise to extraembryonic structures derived from the trophoblast. At any time up until the late blastocyst, the potency of the embryonic cells is demonstrated by the ability of the embryo to form twins.

The embryo proper is formed from the embryonic epiblast. At an early stage, germ-line cells are set aside. Some of the embryonic epiblast cells are induced by signals from neighboring extraembryonic cells to become primordial germ cells. At a later stage, **gastrulation** occurs. Here, the embryo undergoes radical changes, and the non-germ-line cells are organized into three fundamental layers of cells.

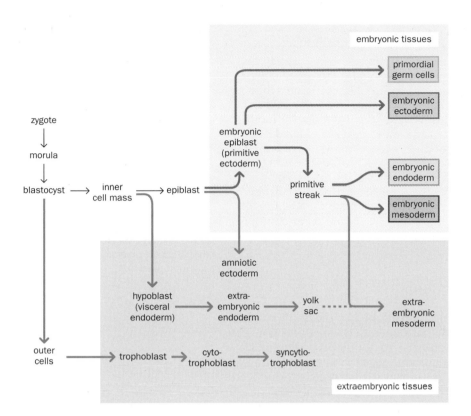

Figure 5.1 A road map for differentiation events in early human development. One of the first overt signs of tissue differentiation in mammalian embryos is apparent at the blastocyst stage, when there are two cell layers. The outer trophoblast cells give rise to extraembryonic tissue— cytotrophoblast, that will form chorionic villi, and syncytiotrophoblast, which will ingress into uterine tissue. The inner cells of blastocysts also give rise to different extraembryonic tissues plus the embryonic epiblast that will give rise to the embryo proper. Some embryonic epiblast cells are induced early on to form primordial germ cells, the precursors of germ cells, that will later migrate to the gonads (as detailed in Section 5.7). The other cells of the embryonic epiblast will later give rise to the three germ layers—ectoderm, mesoderm, and endoderm—that will be precursors of our somatic tissues (as detailed in Figure 5.13). The dashed line indicates a possible dual origin of the extraembryonic mesoderm. [Adapted from Gilbert (2006) Developmental Biology, 8th ed. With permission from Sinauer Associates, Inc.]

The three *germ layers*, as they are known, are ectoderm, mesoderm, and endoderm, and they will give rise to all the somatic tissues.

The constituent cells of the three germ layers are *multipotent*, and their differentiation potential is restricted. The ectoderm cells of the embryo, for example, give rise to epidermis, neural tissue, and neural crest, but they cannot normally give rise to kidney cells (mesoderm-derived) or liver cells (endoderm-derived). Cells from each of the three germ layers undergo a series of sequential differentiation steps. Eventually, *unipotent* progenitor cells give rise to *terminally differentiated* cells with specialized functions.

The choice between alternative cell fates often depends on cell position

As we will see in Section 5.3, a vertebrate zygote that seems symmetric will give rise to an organism that is clearly asymmetric. At a superficial level, vertebrates may seem symmetric around a longitudinal line, the *midline*, that divides the body into a left half and a right half, but clear asymmetry in two out of the three axes means that back (dorsal) can be distinguished from front (ventral), and top (anterior) from bottom (posterior).

In the very early stage mammalian embryo, cells are generally inherently flexible and the **fate** of a cell—the range of cell types that the cell can produce— seems to be determined largely by its position. Cell fates are often specified by signals from nearby cells, a process termed **induction**. Typically, cells in one tissue, the *inducer*, send signals to cells in another immediately adjacent tissue, the *responder*. As a result, the responder tissue is induced to change its behavior and is directed toward a new developmental pathway.

Solid evidence for induction has come from cell transplantation conducted by microsurgery on the comparatively large embryo of the frog *Xenopus laevis*. A good example is the formation of the *neural plate* that, as we will see in Section 5.6, gives rise to the neural tube and then to the central nervous system (brain plus spinal cord). The neural plate arises from ectoderm cells positioned along a central line (the *midline*) that runs along the back (*dorsal*) surface of the embryo; ectoderm cells on either side of the midline give rise to epidermis. Initially, however, all of the surface ectoderm is uncommitted (or *naive*): it is *competent* to give rise to either epidermis or neural plate.

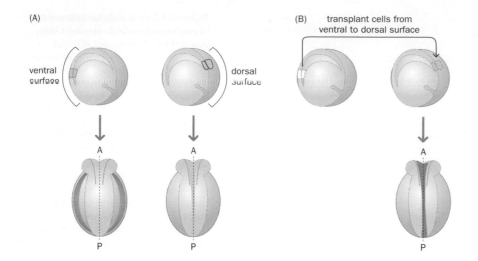

Figure 5.2 Ectoderm cells in the early _Xenopus_ embryo become committed to epidermal or neural cell fates according to their position. (A) Cells in the dorsal midline ectoderm (red) give rise to the neural plate that is formed along the anteroposterior (A–P) axis; other ectodermal cells such as ventral ectoderm (green) give rise to epidermis. (B) If the ventral ectoderm is grafted onto the dorsal side of the embryo, it is re-specified and now forms the neural plate instead of epidermis. This shows that the fate of early ectoderm cells is flexible and does not depend on their lineage but on their position. The fate of dorsal midline ectoderm cells is specified by signals from cells of an underlying mesoderm structure, the _notochord_, that forms along the anteroposterior axis.

The ectodermal cells are therefore flexible. If positioned on the front (_ventral_) surface of the embryo they will normally give rise to epidermis, but if surgically grafted to the dorsal midline surface they give rise to neural plate (**Figure 5.2**). Similarly, if dorsal midline ectoderm cells are grafted to the ventral or lateral regions of the embryo they form epidermis. The dorsal ectoderm cells are induced to form the neural plate along the midline because they receive specialized signals from underlying mesoderm cells. These signals are described in Section 5.3.

The fate of the ectoderm—epidermal or neural—depends on the _position_ of the cells, not their lineage, and is initially reversible. At this point cell fate is said to be _specified_, which means it can still be altered by changing the environment of the cell. Later on, the fate of the ectoderm becomes fixed and can no longer be altered by grafting. At this stage, the cells are said to be _determined_, irreversibly committed to their fate because they have initiated some molecular process that inevitably leads to differentiation. A new transcription factor may be synthesized that cannot be inactivated, or a particular gene expression pattern is locked in place through chromatin modifications. There may also be a loss of competence for induction. For example, ectoderm cells that are committed to becoming epidermis may stop synthesizing the receptor that responds to signals transmitted from the underlying mesoderm.

Sometimes cell fate can be specified by lineage rather than position

There are fewer examples of cell fate specified by lineage in vertebrate embryos. One explicit case is when stem cells divide by a form of asymmetric cell division that produces inherently different daughter cells. One daughter cell will have the same type of properties as the parent stem cell, ensuring stem cell renewal. The other daughter cell has altered properties, making it different from both the parent cell and its sister, and it becomes committed to producing a lineage of differentiated cells. In such cases, the fate of the committed daughter cell is not influenced by its position or by signals from other cells. The decision is _intrinsic_ to the stem cell lineage, and the specification of cell fate is said to be _autonomous_ (non-conditional).

The autonomous specification of cell fates in the above example results from the asymmetric distribution of regulatory molecules at cell division (see Box 4.3). Such asymmetry of individual cells is known as **cell polarity**. In neural stem cells, for example, there is asymmetric distribution of the receptor Notch-1 (concentrated at the apical pole) and also its intracellular antagonist Numb (concentrated at the basal pole). Division in the plane of the epithelial surface causes these determinants to be distributed equally in daughter cells, but a division at right angles to the epithelial surface causes them to be unequally segregated, and the daughter cells therefore develop differently (**Figure 5.3**). The polarization of cells is a crucial part of embryonic development.

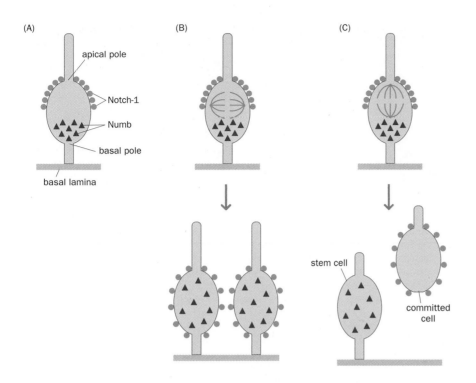

(A)

apical pole

Notch-1

Numb

basal pole

basal lamina

(B)

(C)

stem cell

committed cell

Figure 5.3 Cell fate in the descendants of neural stem cells depends on the plane of cell division. (A) Neural stem cells, occurring in neuroepithelium, have cytoplasmic processes at both the apical pole and the basal pole, with the latter providing an anchor that connects them to the underlying basal lamina. The Notch-1 cell surface receptor is concentrated at the apical pole, and its intracellular antagonist Numb is concentrated at the basal pole. (B) Symmetric divisions of neural stem cells occurring in the plane of the neuroepithelium result in an even distribution of Notch-1 and Numb in the daughter cells. (C) However, asymmetric divisions perpendicular to the neuroepithelium result in the formation of a replacement stem cell that remains anchored in the basal lamina and a committed neuronal progenitor that contains the Notch-1 receptor but not Numb. The latter cell is not anchored to the basal lamina and so can migrate and follow a pathway of neural differentiation.

5.3 PATTERN FORMATION IN DEVELOPMENT

Differentiation gives rise to cells with specialized structures and functions, but for an organism to function the cells also need to be organized in a useful way. Random organization of cells would give rise to amorphous heterogeneous tissues rather than highly ordered tissues and organs. A process is required that directs how cells should be organized during development, conforming to a *body plan*.

Although there are many minor differences between individuals, all members of the same animal species tend to conform to the same basic body plan. First, three orthogonal axes need to be specified, so that the cells of the organism can know their positions in relation to the three dimensions. The organs and tissues of the body are distributed in essentially the same way in relation to these axes in every individual, and this pattern emerges very early in development.

Later, defined patterns emerge within particular organs. A good example is the formation of five fingers on each hand and five toes on each foot. More detailed patterns are generated by the arrangement of cells within tissues. During development, such patterns emerge gradually, with an initially crude embryo being progressively refined like a picture coming into sharp focus. In this section, we discuss how pattern formation in the developing embryo is initiated, and the molecular mechanisms that are involved.

Emergence of the body plan is dependent on axis specification and polarization

Three axes need to be specified to produce an embryo and eventually a mature organism with a head and a tail, a back and a front, and left and right sides. In vertebrates the two major body axes, the *anteroposterior axis* (also known as the *craniocaudal axis*, from head to feet or tail) and the *dorsoventral axis* (from back to belly), are clearly asymmetric (**Figure 5.4A**).

The vertebrate left–right axis is different from the other two axes because it shows superficial bilateral symmetry—our left arm looks much the same as our right arm, for example. Inside the body, however, many organs are placed asymmetrically with respect to the longitudinal midline. Although major disturbances to the anteroposterior and dorsoventral axes are not compatible with life, various major abnormalities of the left–right axis are seen in some individuals (Figure 5.4B).

(A)

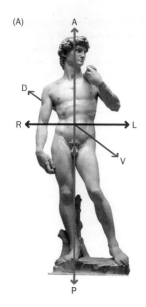

(B)

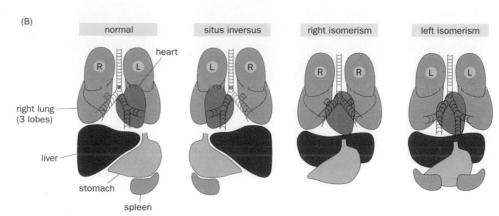

Figure 5.4 The three axes of bilaterally symmetric animals. (A) The three axes. The anteroposterior (A–P) axis is sometimes known as the craniocaudal or rostrocaudal axis (from the Latin *cauda* = tail and *rostrum* = beak). In vertebrates the A–P and dorsoventral (D–V) axes are clearly lines of asymmetry. By contrast, the left–right (L–R) axis is superficially symmetric (a plane through the A–P axis and at right angles to the L–R axis seems to divide the body into two equal halves). (B) Internal left–right asymmetry in humans. Vertebrate embryos are initially symmetric with respect to the L–R axis. The breaking of this axis of symmetry is evolutionarily conserved, so that the organization and placement of internal organs shows L–R asymmetry. In humans, the left lung has two lobes, the right lung has three. The heart, stomach, and spleen are placed to the left; the liver is to the right, as shown in the left panel. In about 1 in 10,000 individuals the pattern is reversed (situs inversus) without harmful consequences. Failure to break symmetry leads to isomerism: an individual may have two right halves (the liver and stomach become centralized but no spleen develops) or two left halves (resulting in two spleens). In some cases, assignment of left and right is internally inconsistent (*heterotaxia*), leading to heart defects and other problems. [(B) adapted from McGrath and Brueckner (2003) *Curr. Opin. Genet. Dev.* 13, 385–392. With permission from Elsevier.]

The establishment of polarity in the early embryo and the mechanisms of axis specification can vary significantly between animals. In various types of animal, asymmetry is present in the egg. During differentiation to produce the egg, certain molecules known collectively as *determinants* are deposited asymmetrically within the egg, endowing it with polarity. When the fertilized egg divides, the determinants are segregated unequally into different daughter cells, causing the embryo to become *polarized*.

In other animals, the egg is symmetric but symmetry is broken by an external cue from the environment. In chickens, for example, the anteroposterior axis of the embryo is defined by gravity as the egg rotates on its way down the oviduct. In frogs, both mechanisms are used: there is pre-existing asymmetry in the egg defined by the distribution of maternal gene products, while the site of sperm entry provides another positional coordinate.

In mammals the symmetry-breaking mechanism is unclear. There is no evidence for determinants in the zygote, and the cells of early mammalian embryos show considerable developmental flexibility when compared with those of other vertebrates (**Box 5.1**).

BOX 5.1 AXIS SPECIFICATION AND POLARITY IN THE EARLY MAMMALIAN EMBRYO

In many vertebrates, there is clear evidence of axes being set up in the egg or in the very early embryo. Mammalian embryos seem to be different. There is no clear sign of polarity in the mouse egg, and initial claims that the first cleavage of the mouse zygote is related to an axis of the embryo have not been substantiated.

The sperm entry point defines one early opportunity for establishing asymmetry. Fertilization induces the second meiotic division of the egg, and a second polar body (**Figure 1**) is generally extruded opposite the sperm entry site. This defines the *animal–vegetal axis* of the zygote, with the polar body at the animal pole. Subsequent cleavage divisions result in a blastocyst that shows bilateral symmetry aligned with the former animal–vegetal axis of the zygote (see Figure 1A).

The first clear sign of polarity in the early mouse embryo is at the blastocyst stage, when the cells are organized into two layers—an outer cell layer, known as *trophectoderm*, and an inner group of

cells, the *inner cell mass* (ICM), which are located at one end of the embryo, *the embryonic pole*. The opposite pole of the embryo is the abembryonic pole. The embryonic face of the ICM is in contact with the trophectoderm, whereas the abembryonic face is open to the fluid-filled cavity, the blastocele. This difference in environment is sufficient to specify the first two distinct cell layers in the ICM: primitive ectoderm (*epiblast*) at the embryonic pole, and primitive endoderm (*hypoblast*) at the abembryonic pole (Figure 1B, left panel). This, in turn, defines the dorsoventral (D–V) axis of the embryo. It is not clear how the ICM becomes positioned asymmetrically in the blastocyst in the first place, but it is interesting to note that when the site of sperm entry is tracked with fluorescent beads, it is consistently localized to trophectoderm cells at the embryonic–abembryonic border.

It is still unclear how the anteroposterior (A–P) axis is specified, but the position of sperm entry may have a defining role. The first overt indication of the A–P axis is the *primitive streak*, a linear structure that develops in mouse embryos at about 6.5 days after fertilization, and in human embryos at about 14 days after fertilization. The decision as to which end should form the head and which should form the tail rests with one of two major signaling centers in the early embryo, a region of extraembryonic tissue called the *anterior visceral endoderm* (*AVE*). In mice, this is initially located at the tip of the egg cylinder but it rotates toward the future cranial pole of the A–P axis just before gastrulation (Figure 1C). The second major signaling center, the *node*, is established at the opposite extreme of the epiblast.

The left–right (L–R) axis is the last of the three axes to form. The major step in determining L–R asymmetry occurs during gastrulation, when rotation of cilia at the embryonic node results in a unidirectional flow of perinodal fluid that is required to specify the L–R axis. Somehow genes are activated specifically on the embryo's left-hand side to produce Nodal and Lefty-2, initiating signaling pathways that activate genes encoding left-hand-specific transcription factors (such as *Pitx2*) and inhibiting those that produce right-hand-specific transcription factors.

Figure 1 Axis specification in the early mammalian embryo. (A) The animal–vegetal axis. The animal pole of the animal–vegetal axis in the mouse embryo is defined as the point at which the second polar body is extruded just after fertilization. (B) The embryonic–abembryonic axis and development of the anteroposterior axis in the mouse embryo. At the blastocyst stage, at about embryonic day 4 (E4; 4 days after fertilization) in mice, the inner cell mass (ICM) is oval with bilateral symmetry and consists of two layers, the outer epiblast layer and the more centrally located primitive (or visceral) endoderm (called hypoblast in human embryos). The ICM is confined to one pole of the embryo, the *embryonic pole*, and the resulting embryonic–abembryonic axis relates geometrically to the dorsoventral axis of the future epiblast. By 6.5 days after fertilization, the mouse epiblast is now shaped like a cup and is located at a position distal from what had been the embryonic pole. At this stage the primitive streak forms as a linear structure that is aligned along the posterior end of the anteroposterior axis. (C) The specification of the anterior visceral endoderm is a major symmetry-breaking event in the early mouse embryo. At about E5.5, before the primitive streak forms, signals from the epiblast induce a distally located region of the visceral endoderm, shown by green coloring, to proliferate and extend to one side, the anterior side, of the epiblast. Accordingly, this population of cells is known as the anterior visceral endoderm (AVE). At 6 days the extended AVE signals to the adjacent epiblast to specify the anterior ectoderm. As the anterior ectoderm becomes determined, proximal epiblast cells expressing genes characteristic of prospective mesoderm migrate to the posterior end and converge at a point to initiate the primitive streak, which forms at day 6.5. [(B) and (C) adapted from Wolpert L, Jessell T, Lawrence P et al. (2007) Principles of Development, 3rd ed. With permission from Oxford University Press.]

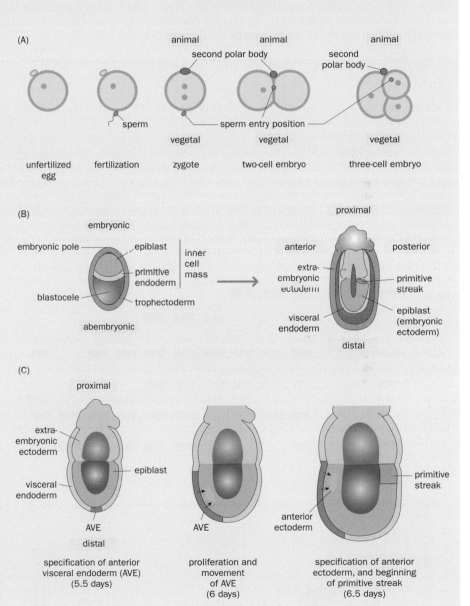

Pattern formation often depends on gradients of signaling molecules

Axis specification and polarization are important early events in development. If they are to generate the appropriate body plan, cells in different parts of the embryo must behave differently, ultimately by making different gene products. However, a cell can behave appropriately only if it knows its precise position in the organism. The major axes of the embryo provide the coordinates that allow the position of any cell to be absolutely and unambiguously defined.

How do cells become aware of their position along an axis and therefore behave accordingly? This question is pertinent because functionally equivalent cells at different positions are often required in the production of different structures. Examples include the formation of different fingers from the same cell types in the developing hand, and the formation of different vertebrae (some with ribs and some without) from the same cell types in the mesodermal structures known as somites.

In many developmental systems, the regionally specific behavior of cells has been shown to depend on a signal gradient that has different effects on equivalent target cells at different concentrations. Signaling molecules that work in this way are known as **morphogens**. In vertebrate embryos, this mechanism is used to pattern the main anteroposterior axis of the body, and both the anteroposterior and proximodistal axes of the limbs (the proximodistal axis of the limbs runs from adjacent to the trunk to the tips of the fingers or toes).

Homeotic mutations reveal the molecular basis of positional identity

In some rare *Drosophila* mutants, one body part develops with the likeness of another (a *homeotic transformation*). A mutant for the gene *Antennapedia*, for example, has legs growing out of its head in the place of antennae. It is therefore clear that some genes control the *positional identity* of a cell (the information that tells each cell where it is in the embryo and therefore how to behave to generate a regionally appropriate structure). Such genes are known as *homeotic genes*.

Drosophila has two clusters of similar homeotic genes, collectively called the homeotic gene complex (*HOM-C*; **Figure 5.5A**). Each of the genes encodes a

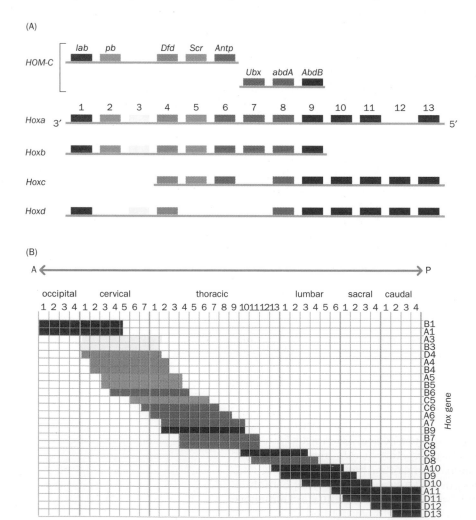

Figure 5.5 Hox genes are expressed at different positions along the anteroposterior axis according to their position within Hox clusters.
(A) Conservation of gene function within Hox clusters. In *Drosophila*, an evolutionary rearrangement of what was a single Hox cluster resulted in two subclusters collectively called *HOM-C*. Mammals have four Hox gene clusters, such as the mouse *Hoxa*, *Hoxb*, *Hoxc*, and *Hoxd* clusters shown here, that arose by the duplication of a single ancestral Hox cluster. Colors indicate sets of paralogous genes that have similar functions/expression patterns, such as *Drosophila labial* (*lab*) and *Hoxa1*, *Hoxb1*, and *Hoxd1*. (B) Mouse Hox genes show graded expression along the anteroposterior axis. Genes at the 3′ end, such as *Hoxa1* (A1) and *Hoxb1* (B1), show expression at anterior (A) parts of the embryo (in the head and neck regions); those at the 5′ end are expressed at posterior (P) regions toward the tail. [Adapted from Twyman (2000) Instant Notes In Developmental Biology. Taylor & Francis.]

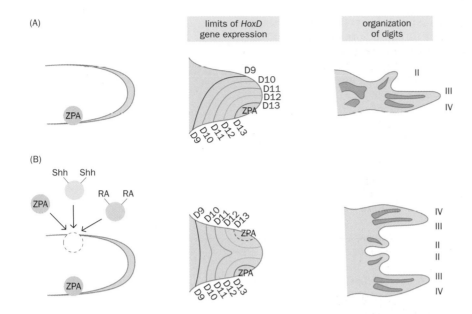

(A)

limits of *HoxD* gene expression

organization of digits

(B)

Shh   Shh

RA   RA

Figure 5.6 Digit formation in the chick limb bud is specified by the zone of polarizing activity. (A) Normal specification of digits. The zone of polarizing activity (ZPA) is located in the posterior margin of the developing limb bud (left panel). ZPA cells produce the morphogen Sonic hedgehog (Shh), and the ensuing gradient of Shh levels generates a nested overlapping pattern of expression patterns for the five different *HoxD* genes (middle panel), which is important for the specification of digits (right panel). (B) Duplication of ZPA signals causes mirror-image duplication of digits. Grafting of a second ZPA to the anterior margin of the limb bud (or placement of a bead coated with Shh or retinoic acid, RA) establishes an opposing morphogen gradient and results in a mirror-image reversal of digit fates. [From Twyman (2000) Instant Notes In Developmental Biology. Taylor & Francis.]

transcription factor containing a *homeodomain*, a short conserved DNA-binding domain. The DNA sequence specifying the homeodomain is known as a **homeobox**. The homeotic genes of *HOM-C* are expressed in overlapping patterns along the anteroposterior axis of the fly, dividing the body into discrete zones. The particular combination of genes expressed in each zone seems to establish a code that gives each cell along the axis a specific positional identity. When the codes are artificially manipulated (by disrupting the genes or deliberately overexpressing them), it is possible to generate flies with transformations of specific body parts.

Clustered homeobox genes that regulate position along the anteroposterior axis in this way are known as Hox genes and are functionally and structurally well conserved. Humans, and mice, have four unlinked clusters of Hox genes that are expressed in overlapping patterns along the anteroposterior axis in a strikingly similar manner to that of their counterparts in flies (Figure 5.5B). Studies in which the mouse genes have been knocked out by mutation or overexpressed have achieved body part transformations involving vertebrae. For example, mouse mutants with a disrupted *Hoxc8* gene have an extra pair of ribs resulting from the transformation of the first lumbar vertebra into the 13th thoracic vertebra.

Two of the mammalian *Hox* gene clusters, *HoxA* and *HoxD*, are also expressed in overlapping patterns along the limbs. Knocking out members of these gene clusters in mice, or overexpressing them, produces mutants with specific rearrangements of the limb segments. For example, mice with targeted disruptions of the *Hoxa11* and *Hoxd11* genes lack a radius and an ulna.

The differential expression of *Hox* genes is controlled by the action of morphogens. During vertebrate limb development, a particular subset of cells at the posterior margin of each limb bud—the **zone of polarizing activity** (**ZPA**)—is the source of a morphogen gradient (**Figure 5.6**). Cells nearest the ZPA form the smallest, most posterior digit of the hand or foot; those farthest away form the thumb or great toe. When a donor ZPA is grafted onto the anterior margin of a limb bud that already has its own ZPA, the limb becomes symmetric, with posterior digits at both extremities. Sonic hedgehog (Shh) seems to act as the morphogen. Because it cannot diffuse more than a few cell widths away from its source, its action seems to be indirect.

A bead soaked in Shh protein will substitute functionally for a ZPA, as will a bead soaked in retinoic acid, which is known to induce *Shh* gene expression. All five distal *HoxD* genes (*Hoxd9–Hoxd13*) are expressed at the heart of the ZPA. However, as the strength of the signal diminishes, the *HoxD* genes are switched off one by one, until at the thumb-forming anterior margin of the limb bud only *Hoxd9* remains switched on. In this way, signal gradients specifying the major embryonic axes are linked to the homeotic genes that control regional cell behavior.

The source of the morphogen gradient that guides *Hox* gene expression along the major anteroposterior axis of the embryo is thought to be a transient embryonic structure known as the *node*, which is one of two major signaling centers in the early embryo (see Box 5.1), and the morphogen itself is thought to be retinoic acid. The node secretes increasing amounts of retinoic acid as it regresses, such that posterior cells are exposed to larger amounts of the chemical than anterior cells, resulting in the progressive activation of more Hox genes in the posterior regions of the embryo.

5.4 MORPHOGENESIS

Cell division, with progressive pattern formation and cell differentiation, would eventually yield an embryo with organized cell types, but that embryo would be a static ball of cells. Real embryos are dynamic structures, with cells and tissues undergoing constant interactions and rearrangements to generate structures and shapes. Cells form sheets, tubes, loose reticular masses, and dense clumps. Cells migrate either individually or *en masse*. In some cases, such behavior is in response to the developmental program. In other cases, these processes drive development, bringing groups of cells together that would otherwise never come into contact. Several different mechanisms underlying **morphogenesis** are summarized in Table 5.2 and discussed in more detail below.

Morphogenesis can be driven by changes in cell shape and size

Orchestrated changes in cell shape can be brought about by reorganization of the cytoskeleton, and this can have a major impact on the structure of whole tissues. One of the landmark events in vertebrate development is the formation of the neural tube, which will ultimately give rise to the brain and spinal cord. As detailed in Section 5.6, a flat sheet of cells, the neural plate, is induced by signals from underlying cells to roll up into a tube, the neural tube. To achieve this, local contraction of microfilaments causes some columnar cells at the middle of the neural plate to become constricted at their apical ends, so that the top ends of the cells (the ends facing the external environment) become narrower. As a result, they become wedge-shaped and can now act as hinges. In combination with increased proliferation at the margins of the neural plate, this provides sufficient force for the entire neural plate to roll up into a tube. Similar behavior within any flat sheet of cells will tend to cause that sheet to fold inward (*invaginate*).

TABLE 5.2 MORPHOGENETIC PROCESSES IN DEVELOPMENT	
Process	**Example**
Change in cell shape	change from columnar to wedge-shaped cells during neural tube closure in birds and mammals
Change in cell size	expansion of adipocytes (fat cells) as they accumulate lipid droplets
Gain of cell–cell adhesion	condensation of cells of the cartilage mesenchyme in vertebrate limb bud
Loss of cell–cell adhesion	delamination of cells from epiblast during gastrulation in mammals
Cell–matrix interaction	migration of neural crest cells and germ cells
Loss of cell–matrix adhesion	delamination of cells from basal layer of the epidermis
Differential rates of cell proliferation	selective outgrowth of vertebrate limb buds by proliferation of cells in the progress zone, the undifferentiated population of mesenchyme cells from which successive parts of the limb are laid down
Alternative positioning and/or orientation of mitotic spindle	different embryonic cleavage patterns in animals; stereotyped cell divisions in nematodes
Apoptosis	separation of digits in vertebrate limb bud (Figure 5.7); selection of functional synapses in the mammalian nervous system
Cell fusion	formation of trophoblast and myotubes in mammals

TABLE 5.3 PROCESSES RESULTING FROM ALTERED CELL ADHESION

Process	Example
Migration	The movement of an individual cell with respect to other cells in the embryo. Some cells, notably the neural crest cells (Box 5.4) and germ cells (Section 5.7), migrate far from their original locations during development
Ingression	The movement of a cell from the surface of an embryo into its interior (Figure 5.13)
Egression	The movement of a cell from the interior of an embryo to the external surface
Delamination	The movement of cells out of an epithelial sheet, often to convert a single layer of cells into multiple layers. This is one of the major processes that underlie gastrulation (Figure 5.13) in mammalian embryos. Cells can also delaminate from a basement membrane, as occurs in the development of the skin
Intercalation	The opposite of delamination: cells from multiple cell layers merge into a single epithelial sheet
Condensation	The conversion of loosely packed mesenchyme cells into an epithelial structure; sometimes called a *mesenchymal-to-epithelial transition*
Dispersal	The opposite of condensation: conversion of an epithelial structure into loosely packed mesenchyme cells; an *epithelial-to-mesenchymal transition*
Epiboly	The spreading of a sheet of cells

Major morphogenetic changes in the embryo result from changes in cell affinity

Selective cell–cell adhesion and cell–matrix adhesion were described in Section 4.2 as mechanisms used to organize cells into tissues and maintain tissue boundaries. In development, regulating the synthesis of particular cell adhesion molecules allows cells to make and break contacts with each other and undergo very dynamic reorganization. Gastrulation is perhaps the most dramatic example of a morphogenetic process. The single sheet of epiblast turns in on itself and is converted into the three fundamental germ layers of the embryo, a process driven by a combination of changes in cell shape, selective cell proliferation, and differences in cell affinity. Various processes can result from altering the adhesive properties of cells (Table 5.3). For example, when a cell loses contact with those surrounding it (delamination), it is free to move to another location (migration, ingression, or egression). Conversely, an increase in cell–cell or cell–matrix adhesion allows new contacts to be made (intercalation or condensation).

Cell proliferation and apoptosis are important morphogenetic mechanisms

After an initial period of cleavage during which all cells divide at much the same rate, cells in different parts of the embryo begin to divide at different rates. This can be used to generate new structures. For example, rapid cell division in selected regions of the mesoderm gives rise to limb buds, whereas adjacent regions, dividing more slowly, do not form such structures.

The plane of cell division, which is dependent on the orientation of the mitotic spindle, is also important. For example, divisions perpendicular to the plane of an epithelial sheet will cause that sheet to expand by the incorporation of new cells. Divisions in the same plane as the sheet will generate additional layers. If the cells are asymmetric, as is true of some stem cells, then the plane of cell division can influence the types of daughter cell that are produced. Furthermore, asymmetric positioning of the mitotic spindle will result in a cleavage plane that is not in the center of the cell. The resulting asymmetric cell division will generate two daughter cells of different sizes. Asymmetric cell division in female gametogenesis produces a massive egg, containing most of the cytoplasm, and vestigial *polar bodies* that are essentially waste vessels for the unwanted haploid chromosome set (see Box 4.3). Contrast this with male gametogenesis, in which meiosis produces four equivalent spermatids.

Apoptosis is another important morphogenetic mechanism, because it allows gaps to be introduced into the body plan. The gaps between our fingers and toes

(A)

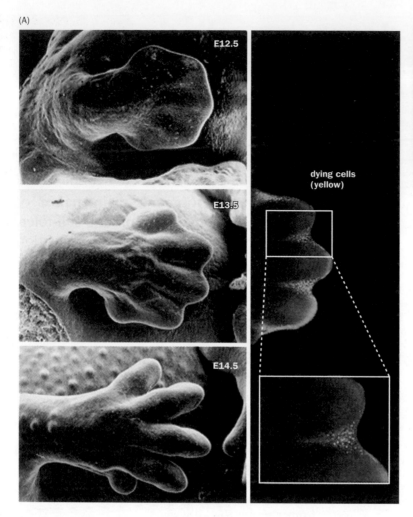

dying cells
(yellow)

(B)

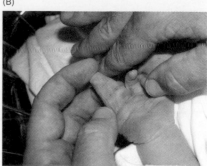

Figure 5.7 Programmed cell death sculpts fingers and toes during embryonic development. (A) The digits of a mouse paw are sculpted from the plate-like structure seen at embryonic day 12.5 (E12.5); the digits are fully connected by webbing. The cells within the webbing are programmed to die and have disappeared by E14.5. The dying cells are identified by acridine orange staining on the right panel. (B) Incomplete programmed cell death during human hand development results in webbed fingers. [(A) from Pollard TD & Earnshaw WC (2002) Cell Biology, 2nd ed. With permission from Elsevier.]

are created by the death of interdigital cells in the hand and foot plates beginning at about 45 days of gestation (**Figure 5.7**). In the mammalian nervous system, apoptosis is used to prune out the neurons with nonproductive connections, allowing the neuronal circuitry to be progressively refined. Remarkably, up to 50% of neurons are disposed of in this manner, and in the retina this can approach 80%.

5.5 EARLY HUMAN DEVELOPMENT: FERTILIZATION TO GASTRULATION

During fertilization the egg is activated to form a unique individual

Fertilization is the process by which two sex cells (*gametes*) fuse to create a new individual. The female gamete, the egg cell (or *oocyte*), is a very large cell that contains material necessary for the beginning of growth and development (**Figure 5.8A**). The cytoplasm is extremely well endowed with very large numbers of mitochondria and ribosomes, and large amounts of protein, including DNA and RNA polymerases. There are also considerable quantities of RNAs, protective chemicals, and morphogenetic factors. In many species, including birds, reptiles, fish, amphibians, and insects, the egg contains a significant amount of *yolk*, a collection of nutrients that is required to nourish the developing embryo before it can feed independently. Yolk is not required in mammalian eggs because the embryo will be nourished by the placental blood supply.

Outside the egg's plasma membrane is the *vitelline envelope*, which in mammals is a separate and thick extracellular matrix known as the **zona pellucida**. In mammals, too, the egg is surrounded by a layer of cells known as *cumulus cells* that nurture the egg before, and just after, ovulation.

(A)

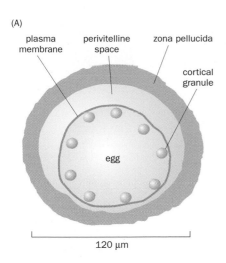

plasma membrane | perivitelline space | zona pellucida

cortical granule

egg

120 µm

(B)

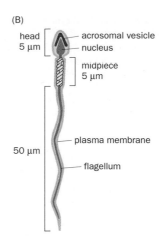

head 5 µm — acrosomal vesicle — nucleus

midpiece 5 µm

50 µm — plasma membrane — flagellum

Figure 5.8 The specialized sex cells. (A) The egg. The mammalian egg (oocyte) is a large cell, 120 µm in diameter, surrounded by an extracellular envelope, the *zona pellucida*, which contains three glycoproteins, ZP-1, ZP-2, and ZP-3, that polymerize to form a gel. The first polar body (the product of meiosis I; not shown here) lies under the zona within the *perivitelline space*. At ovulation, oocytes are in metaphase II. Meiosis II is not completed until after fertilization. (B) The sperm. This cell is much smaller than the egg, with a 5 µm head containing highly compacted DNA, a 5 µm cylindrical body (the midpiece, containing many mitochondria), and a 50 µm tail. At the front, the acrosomal vesicle contains enzymes that help the sperm to make a hole in the zona pellucida, allowing it to access and fertilize the egg. Fertilization triggers secretion of cortical granules by the egg that effectively inhibit further sperm from passing through the zona pellucida. [From Alberts B, Johnson A, Lewis J et al. (2002) *Molecular Biology of the Cell*, 4th ed. Garland Science/Taylor & Francis LLC.]

The male gamete, the sperm cell, is a small cell with a greatly reduced cytoplasm and a haploid nucleus. The nucleus is highly condensed and transcriptionally inactive because the normal histones are replaced by a special class of packaging proteins known as protamines. A long flagellum at the posterior end provides propulsion. The *acrosomal vesicle* (or *acrosome*) at the anterior end (Figure 5.8B) contains digestive enzymes. Human sperm cells have to migrate very considerable distances, and out of the 280 million or so ejaculated into the vagina only about 200 reach the required part of the oviduct where fertilization takes place.

Fertilization begins with attachment of a sperm to the zona pellucida followed by the release of enzymes from the acrosomal vesicle, causing local digestion of the zona pellucida. The head of the sperm then fuses with the plasma membrane of the oocyte and the sperm nucleus passes into the cytoplasm. Within the oocyte, the haploid sets of sperm and egg chromosomes are initially separated from each other and constitute, respectively, the male and female *pronuclei*. They subsequently fuse to form a diploid nucleus. The fertilized oocyte is known as the *zygote*.

Cleavage partitions the zygote into many smaller cells

Cleavage is the developmental stage during which the zygote divides repeatedly to form a number of smaller cells called **blastomeres**. The nature of the early cleavage divisions varies widely between different animal species. For example, in *Drosophila* and many other insects, the process does not even involve cell division; instead, the zygote nucleus undergoes a series of divisions in a common cytoplasm to generate a large, flattened multinucleated cell, the *syncytial blastoderm*. However, in many animals the result of cleavage is usually a ball of cells, often surrounding a fluid-filled cavity called the *blastocele*.

In most invertebrates, the ball of cells resulting from cleavage is called a *blastula*. In vertebrates the terminology varies. In amphibians and mammals, the term **morula** is used to describe the initial, loosely packed, ball of cells that results from early cleavage, and thereafter when the fluid-filled blastocele forms, the ball of cells is known as a blastula in amphibians but a **blastocyst** in mammals (**Figure 5.9** and **Figure 5.10**). The situation is different in birds, fish, and reptiles, in which the egg contains a lot of yolk, which inhibits cell division. Here the cleavage is restricted to a flattened *blastodisc* at the periphery of the cell.

For many species (but *not* mammals—see below), cleavage divisions are rapid because there are no intervening G_1 and G_2 gap phases in the cell cycle between DNA replication (S phase) and mitosis (M phase). In such cases there is no net growth of the embryo and so, as the cell number increases, the cell size decreases. Where this happens, the genome inherited from the zygote (the zygotic genome) is transcriptionally inactive during cleavage. Instead, there is heavy reliance on maternally inherited gene products distributed in the egg cytoplasm. The maternal gene products regulate the cell cycle and determine the rate of cleavage, and the cleavage divisions are synchronous. This type of regulation is often referred to

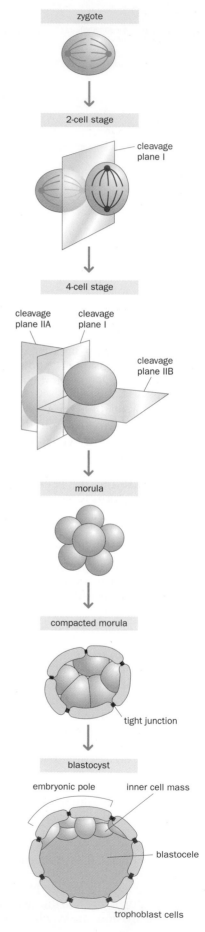

Figure 5.9 Early development of the mammalian embryo, from zygote to blastocyst. In mammals, the first cleavage is a normal meridional division (along the vertical axis), but in the second cleavage one of the two cells (blastomeres) divides meridionally while the other divides at right angles, equatorially (*rotational cleavage*). At the eight-cell stage, the mouse morula undergoes *compaction*, when the blastomeres huddle together to form a compact ball of cells. The tight packaging, stabilized by tight junctions formed between the outer cells of the ball, seals off the inside of the sphere. The inner cells form gap junctions, enabling small molecules and ions to pass between them. The morula does not have an internal cavity, but the outer cells secrete fluid so that the subsequent blastocyst becomes a hollow ball of cells with a fluid-filled internal cavity, the blastocele. The blastocyst has an outer layer of cells, the trophoblast, that will give rise to an extraembryonic membrane, the chorion, plus an inner cell mass (ICM) located at one end of the embryo, the *embryonic pole*. Cells from the ICM will give rise to the other extracellular membranes as well as the embryo proper and the subsequent fetus. For convenience, the zona pellucida that surrounds the early embryo (see Figure 5.10) is not shown. [Adapted from Twyman (2000) Instant Notes In Developmental Biology. Taylor & Francis.]

as maternal genome regulation, and the maternal gene products are often referred to as *maternal determinants*.

Mammalian eggs are among the smallest in the animal kingdom, and cleavage in mammals is exceptional in several ways

Cleavage in mammals is distinguished by several features, as follows:

- The zygotic genome is activated early, as early as the two-cell stage in some species. As a result, the cleavage divisions are controlled by the zygotic genome rather than by maternally inherited gene products. The divisions are slow because cell cycles include G_1 and G_2 phases and are asynchronous.

- The cleavage mechanism is unique. The first cleavage plane is vertical, but in the second round of cell division one of the cells cleaves vertically and the other horizontally (*rotational cleavage*—see Figure 5.9). Additionally, cells do not always divide at the same time to produce two-cell then four-cell then eight-cell stages, but can sometimes divide at different times to produce embryos with odd numbers of cells, such as three-cell or five-cell embryos.

- The embryo undergoes **compaction**. The loosely associated blastomeres of the eight-cell embryo flatten against each other to maximize their contacts and form a tightly packed morula. Compaction does occur in many non-mammalian embryos, but it is much more obvious in mammals.

Compaction has the effect of introducing a degree of cell polarity. Before compaction the blastomeres are rounded cells with uniformly distributed microvilli, and the cell adhesion molecule E-cadherin is found wherever the cells are in contact with each other. After compaction, the microvilli become restricted to the apical surface, while E-cadherin becomes distributed over the basolateral surfaces. Now the cells form tight junctions with their neighbors and the cytoskeletal elements are reorganized to form an apical band.

At about the 16-cell morula stage in mammals, it becomes possible to discriminate between two types of cell: external polarized cells and internal nonpolar cells. As the population of nonpolar cells increases, the cells begin to communicate with each other through gap junctions. The distinction between the two types of cells is a fundamental one and underlies the more overt distinction between the outer and inner cell layers of the subsequent blastocyst (see Figure 5.9).

Only a small percentage of the cells in the early mammalian embryo give rise to the mature organism

In many animal models of development, the organism is formed from cells that have descended from all the cells of the early embryo. Mammals are rather different: only a small minority of the cells of the early embryo give rise to the organism proper. This is so because much of early mammalian development is concerned with establishing the extraembryonic membranes—tissues that act as a life support but mostly do not contribute to the final organism (**Box 5.2**).

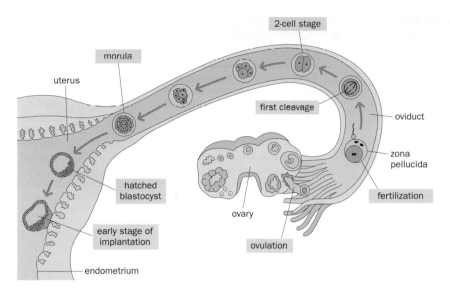

Figure 5.10 The physiological context of early embryonic development. Sperm deposited in the seminal fluid swim up into the uterus and then into the oviducts (= fallopian tubes). During ovulation, an egg is released from an ovary into the oviduct, where it may be fertilized by a sperm. The fertilized egg is slowly propelled along the oviduct by cilia on the inner lining of the oviduct. During its journey, the zygote goes through various cleavage divisions, but the zona pellucida usually prevents it from adhering to the oviduct walls (although this occasionally happens in humans, causing a dangerous condition, an *ectopic pregnancy*). Once in the uterus, the zona pellucida is partly degraded and the blastocyst is released to allow implantation in the wall of the uterus (endometrium). [From Gilbert (2006) Developmental Biology, 8th ed. With permission from Sinauer Associates, Inc.]

As the blastocele forms (at about the 32-cell stage in humans), the inner non-polar cells congregate at one end of the blastocele, the embryonic pole, to form an off-center inner cell mass (ICM). The outer trophoblast cells give rise to the *chorion*, the outermost extraembryonic membrane, while the cells of the ICM will give rise to all the cells of the organism plus the other three extra-embryonic membranes. At any time until the late blastocyst stage, splitting can lead to the production of identical (*monozygotic*) twins (**Box 5.3**).

BOX 5.2 EXTRAEMBRYONIC MEMBRANES AND PLACENTA

Early mammalian development is unusual in that it is concerned primarily with the formation of tissues that mostly do not contribute to the final organism. These tissues are the four extraembryonic membranes: yolk sac, amnion, chorion, and allantois. The chorion combines with maternal tissue to form the placenta. As well as protecting the embryo (and later the fetus), these life support systems are required to provide for its nutrition, respiration, and excretion.

Yolk sac
The most primitive of the four extraembryonic membranes, the yolk sac is found in all amniotes (mammals, birds, and reptiles) and also in sharks, bony fishes, and some amphibians. In bird embryos, the yolk sac surrounds a nutritive *yolk mass* (the yellow part of the egg, consisting mostly of phospholipids). In many mammals, including humans and mice, the yolk sac does not contain any yolk. The yolk sac is generally important because:

- the *primordial germ cells* pass through the yolk sac on their migration from the epiblast to the genital ridge (for more details, see Section 5.7);
- it is the source of the first blood cells of the conceptus and most of the first blood vessels, some of which extend themselves into the developing embryo.

The yolk sac originates from splanchnic (= visceral) lateral plate mesoderm and endoderm.

Amnion
The amnion is the innermost of the extraembryonic membranes, remaining attached to and immediately surrounding the embryo. It contains amniotic fluid that bathes the embryo, thereby preventing drying out during development, helping the embryo to float (and so reducing the effects of gravity on the body), and acting as a hydraulic cushion to protect the embryo from mechanical jolting. The amnion derives from ectoderm and somatic lateral plate mesoderm.

Chorion
The chorion is also derived from ectoderm and somatic lateral plate mesoderm. In the embryos of birds, the chorion is pressed against the shell membrane, but in mammalian embryos it is composed of trophoblast cells, which produce the enzymes that erode the lining of the uterus, helping the embryo to implant into the uterine wall. The chorion is also a source of hormones (chorionic gonadotropin) that influence the uterus as well as other systems. In all these cases, the chorion serves as a surface for respiratory exchange. In placental mammals, the chorion provides the fetal component of the placenta (see below).

Allantois
The most evolutionarily recent of the extraembryonic membranes, the allantois develops from the posterior part of the alimentary canal in the embryos of reptiles, birds, and mammals. It arises from an outward bulging of the floor of the hindgut and so is composed of endoderm and splanchnic lateral plate mesoderm. In most amniotes it acts as a waste (urine) storage system, but not in placental mammals (including humans). Although the allantois of placental mammals is vestigial and may regress, its blood vessels give rise to the umbilical cord vessels.

Placenta
The placenta is found only in some mammals and is derived partly from the conceptus and partly from the uterine wall. It develops after implantation, when the embryo induces a response in the neighboring maternal endometrium, changing it to become a nutrient-packed highly vascular tissue called the *decidua*. During the second and third weeks of development, the trophoblast tissue becomes vacuolated, and these vacuoles connect to nearby maternal capillaries, rapidly filling with blood.

As the chorion forms, it projects outgrowths known as *chorionic villi* into the vacuoles, bringing the maternal and embryonic blood supplies into close contact. At the end of 3 weeks, the chorion has differentiated fully and contains a vascular system that is connected to the embryo. Exchange of nutrients and waste products occurs over the chorionic villi. Initially, the embryo is completely surrounded by the decidua, but as it grows and expands into the uterus, the overlying decidual tissue (decidua capsularis) thins out and then disintegrates. The mature placenta is derived completely from the underlying decidua basalis.

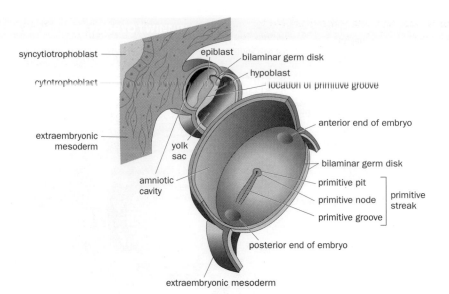

Figure 5.12 The primitive streak is an early marker of the anteroposterior axis. At about 15 days after fertilization, a transient linear structure called the primitive streak emerges. It appears on the dorsal surface of the human embryo, which is an oval bilaminar germ disk. It forms at the posterior end of the longitudinal midline, which will define the anteroposterior axis. The expanded view at the bottom right shows the dorsal surface of the embryo through the sectioned amnion and yolk sac (the inset at upper left shows the relationship of the embryo to the chorionic cavity). In this view the primitive streak is about 1 day old. [From Larsen (2001) Human Embryology, 3rd ed. With permission from Elsevier.]

the hypoblast (**Figure 5.13B**). Some of the ingressing epiblast cells invade the hypoblast and displace its cells, leading eventually to complete replacement of the hypoblast by a new layer of epiblast-derived cells, the definitive endoderm. Starting on day 16, some of the migrating epiblast cells diverge into the space between the epiblast and the newly formed definitive endoderm to form a third layer, the intraembryonic mesoderm (Figure 5.13C). When the intraembryonic mesoderm and definitive endoderm have formed, the residual epiblast is now described as the ectoderm, and the new three-layered structure is referred to as the *trilaminar germ disk* (see Figure 5.13C).

The ingressing mesoderm cells migrate in different directions, some laterally and others toward the anterior, and others are deposited on the midline (**Figure 5.14**).

The cells that migrate through the primitive pit in the center of the primitive node and come to rest on the midline form two structures:

- The *prechordal* (also called *prochordal*) *plate* is a compact mass of mesoderm to the anterior of the primitive pit. The prechordal plate will induce important cranial midline structures such as the brain.

- The *notochordal process* is a hollow tube that sprouts from the primitive pit and grows in length as cells proliferating in the region of the primitive node add on to its proximal end. The notochordal process and adjacent mesoderm

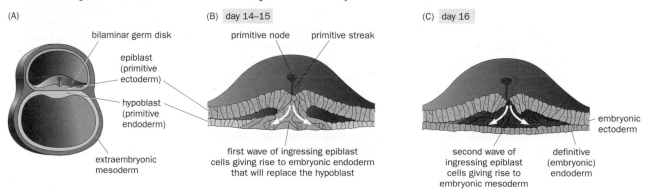

Figure 5.13 During human gastrulation, a flat bilaminar germ disk transforms into a trilaminar embryo. The outcome of gastrulation is similar in all mammals, but major differences may occur in the details of morphogenesis, particularly in how extraembryonic structures are formed and used. (A) In humans, the epiblast and hypoblast come into contact to form a flat *bilaminar germ disk*. The epiblast cells within this disk are described as the primitive ectoderm but will give rise to all three germ layers—ectoderm, endoderm, and mesoderm—as described in panels (B) and (C). The epiblast cells that are not in contact with the hypoblast will give rise to the ectoderm of the amnion. The hypoblast will give rise to the extraembryonic endoderm that lines the yolk sac. (B) The bilaminar germ disk at 14–15 days of human development. Along the length of the primitive streak, epiblast cells migrate downward to invade the hypoblast, and in so doing they become converted to embryonic endoderm and displace the cells of the hypoblast. (C) The bilaminar germ disk at 16 days of human development. A second wave of ingressing epiblast cells diverge into the space between the epiblast and the newly formed embryonic endoderm to form embryonic mesoderm. The remaining epiblast cells are now known as the embryonic ectoderm. [From Larsen (2001) Human Embryology, 3rd ed. With permission from Elsevier.]

Figure 5.14 Migration paths of mesoderm cells that ingress during gastrulation. Epiblast cells ingressing through the primitive pit migrate toward the anterior end (A) to form the notochordal process and the prechordal plate (oval shape at anterior end—see Figure 5.12) and laterally through the primitive groove to form the lateral mesoderm flanking the midline. The oval shape at the posterior end (P) marks the site of what will become the primordial anus. [From Larsen (2001) Human Embryology, 3rd ed. With permission from Elsevier.]

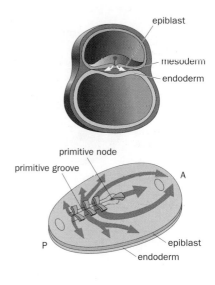

induce the overlying embryonic ectoderm to form the **neural plate**, the precursor of the central nervous system (**Figure 5.15**). At the same time as the notochordal process extends, the primitive streak regresses. By day 20 of human development, the notochordal process is completely formed, but it then transforms from a hollow tube to a solid rod, the **notochord**. The notochord will later induce the formation of components of the nervous system (as described in Section 5.6).

The ingressing mesoderm cells that migrate laterally condense into rodlike and sheetlike structures on either side of the notochord. There are three main structures (**Figure 5.16A**):

- The **paraxial mesoderm**, a pair of cylindrical condensations lying immediately adjacent to and flanking the notochord, first develops into a series of whorl-like structures known as *somitomeres*, which form along the antero-posterior axis through the third and fourth weeks of human development. The first seven cranial somitomeres will eventually go on to form the striated muscles of the face, jaw, and throat, but the other somitomeres develop further into discrete *blocks* of segmental mesoderm known as **somites** (see Figure 5.16C). Cervical, thoracic, lumbar, and sacral somites will establish the segmental organization of the body by giving rise to most of the axial skeleton (including the vertebral column), the voluntary muscles, and part of the dermis of the skin.

- The **intermediate mesoderm**, a pair of less pronounced cylindrical condensations, just lateral to the paraxial mesoderm, later develops into the urinary system, parts of the genital system, and kidneys.

- The remainder of the lateral mesoderm forms a flattened sheet, known as the **lateral plate mesoderm**. Starting on day 17 of human development, each of the lateral plates splits horizontally into two layers separated by a space that will become the body cavity, the *coelom*. The dorsal layer is known as the *somatic* (or *parietal*) *mesoderm* or **somatopleure**. It becomes applied to the inner surface of the ectoderm and will give rise to the inner lining of the body wall (see Figure 5.16B). The ventral layer, adjacent to the endoderm, is the *splanchnic* (or *visceral*) *mesoderm* or **splanchnopleure**, and it will give rise to the linings of the visceral organs.

Figure 5.17 summarizes how the three germ layers of the early embryo give rise to the many different tissues of the mature organism.

Figure 5.15 Progression of the human embryonic disk during week 3, showing development of the neural plate and notochord. The sketches are dorsal views of the embryonic disk, showing how it progresses from 15 to 21 days after fertilization. The primitive streak emerges at about day 15 along the posterior dorsal surface, advancing toward the center of the embryonic disk by the addition of cells to its posterior (caudal) end. Mesenchymal cells migrate from the anterior end of the primitive streak to form a midline cellular cord known as the notochordal process. The notochordal process grows cranially until it reaches the prechordal plate, the future site of the mouth. By day 18, the developing notochordal process can be seen to be accompanied by a thickening of the overlying epiblast to form the neural plate, which will eventually give rise to the brain and spinal cord. As the notochordal process develops, it changes from being a tube to become a solid rod, the notochord. The cloacal membrane is the primordial anus. A, anterior; P, posterior. [From Moore (1984) The Developing Human, 3rd ed. With permission from Elsevier.]

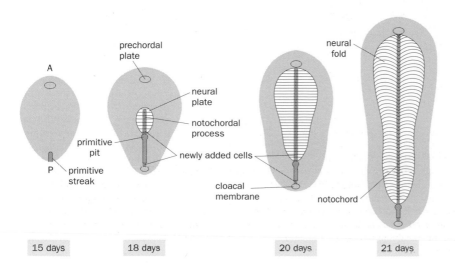

15 days   18 days   20 days   21 days

(A)

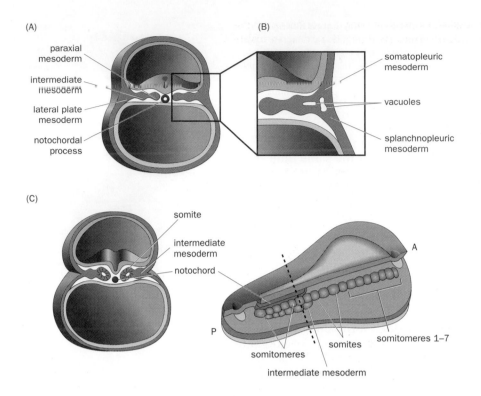

paraxial
mesoderm

intermediate
mesoderm

lateral plate
mesoderm

notochordal
process

(B)

somatopleuric
mesoderm

vacuoles

splanchnopleuric
mesoderm

(C)

somite

intermediate
mesoderm

notochord

somite

somitomeres 1–7

somites

somitomeres

intermediate mesoderm

A

P

Figure 5.16 Differentiation of lateral mesoderm. (A) Early differentiation of lateral mesoderm. Early on day 17, the mesoderm immediately flanking the notochordal process begins to differentiate and forms cylindrical condensations, the *paraxial mesoderm*. The neighboring, less pronounced, cylindrical condensations are the *intermediate mesoderm*. The rest of the lateral mesoderm forms a flattened sheet, the *lateral plate mesoderm*. (B) Differentiation of lateral plate mesoderm (LPM). Later on day 17, vacuoles form in the LPM, which begins to split into two layers. The dorsal layer, the *somatopleuric mesoderm*, will give rise to the inner lining of the body walls and to most of the dermis. The ventral layer, the *splanchnopleuric mesoderm*, will give rise to the *mesothelium*, the lining of embryonic mesoderm epithelium that covers the visceral organs. (C) Somites originate from paraxial mesoderm. The paraxial mesoderm goes on to form a series of rounded whorl-like structures, *somitomeres*. Somitomeres 1–7 at the anterior end will develop into structures of the head, and the rest will give rise to *somites*, blocks of segmental mesoderm that in turn give rise to the vertebral column and segmented muscles. In this diagram of a 21-day human embryo, six central somitomeres have already differentiated into somites; these will be followed later by the more posterior somitomeres. The dashed line indicates the axis where the section was made to give the cross-section shown on the left. A, anterior; P, posterior. [Adapted from Larsen (2001) Human Embryology, 3rd ed. With permission from Elsevier.]

5.6   NEURAL DEVELOPMENT

As described above, gastrulation results in a remarkable set of changes in the embryo, converting it from a two-layered structure to a three-layered one. The development of the embryo is now programmed toward organizing tissues into the precursors of the many organs and systems contained in the adult. The early

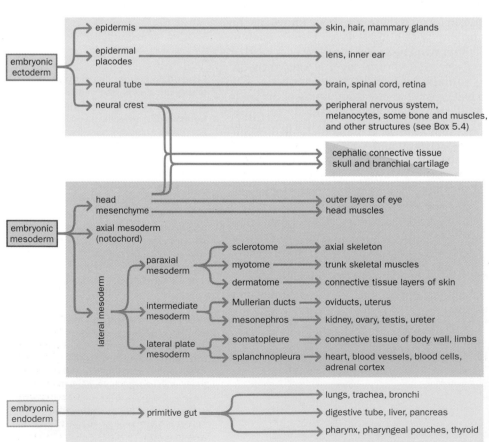

Figure 5.17 Principal derivatives of the three germ layers. The three germ layers formed during gastrulation will eventually form all tissues of the embryo. The connective tissue of the head and the cartilage of the skull and of structures derived from branchial arches are a mixture of ectodermal and mesodermal tissue, as shown. Although the notochord persists in adults in some primitive vertebrates, in mammals and other higher vertebrates it becomes ossified in regions of forming vertebrae and contributes to the center of the intervertebral disks. Note that some of the embryonic mesoderm cells go on to form extraembryonic mesoderm.

development of the nervous system is a good example of organogenesis because it shows how the processes of differentiation, pattern formation, and morphogenesis are exquisitely coordinated.

The axial mesoderm induces the overlying ectoderm to develop into the nervous system

The development of the nervous system marks the onset of organogenesis; it begins at the end of the third week of human development. The initiating event is the induction of the overlying ectoderm by *axial mesoderm* (mesoderm running along the anteroposterior axis). Within the axial mesoderm, cells of the prechordal plate and the anterior portion of the notochordal plate transmit signals to overlying ectoderm cells, causing them to differentiate into a thick plate of neuroepithelial cells (*neurectoderm*). The resulting *neural plate* appears at day 18 of human development but grows rapidly and changes proportions over the next 2 days (see Figure 5.15).

At about days 20–22 after fertilization, a process known as *neurulation* results in conversion of the neural plate into a **neural tube**, the precursor of the brain and spinal cord. The neural plate begins to crease ventrally along its midline to form a depression called the neural groove. This is thought to develop in response to inductive signals from the closely apposed notochord. Thick neural folds rotate around the neural groove and meet dorsally, initially at a mid-point along the anteroposterior axis (see Figure 5.15 and **Figure 5.18**). Closure of the neural tube proceeds in a zipperlike fashion and bidirectionally. Initially, the neural tube is open at both ends (the openings are called the anterior and posterior neuropores), but subsequently the openings need to be closed. Occasionally, there is a partial failure of neural tube closure, resulting in conditions such as spina bifida.

During neurulation, the **neural crest**, a specific population of cells arising along the lateral margins of the neural folds, detaches from the neural plate and migrates to many specific locations within the body. This highly versatile group of cells gives rise to part of the peripheral nervous system, melanocytes, some bone and muscle, and other structures (**Box 5.4**).

Pattern formation in the neural tube involves the coordinated expression of genes along two axes

As soon as the neural plate is formed, three large cranial vesicles (the future brain) become visible, as well as a narrower caudal section that will form the spinal cord. This anteroposterior polarity reflects the regional specificity of neural induction: the signals coming from the axial mesoderm contain **positional information** that causes the overlying ectoderm to form neural tissue specific for different parts of the axis. The precise nature of the signal in birds and mammals is not understood. In *Xenopus*, a model has been developed in which neural development is prevented by members of the bone morphogenetic protein (BMP) family of signaling proteins, and BMP antagonists are required to initiate neural induction. The mechanism in birds and mammals seems to be more complex and is an area of active research. It seems likely that a general neuralizing signal is released from the mesoderm that induces the formation of neural plate that is anterior in character. Another signal that originates in the caudal region of the embryo will 'posteriorize' that part of the neural plate. Molecules secreted by the anterior mesoderm are required for formation of the head.

Whatever the underlying mechanism, the signals activate different sets of transcription factors along the axis, and these confer positional identities on the cells and regulate their behavior. In the forebrain and midbrain, transcription

Figure 5.18 Morphogenesis of the nervous system. The change from a flat neural plate to a closed neural tube is caused by the formation of hinge points where cells become apically constricted, and by the proliferation of ectoderm at the margins of the neural plate pushing the sides together. As the neural folds come together, neural crest cells delaminate and begin to migrate away to diverse locations, as described later.

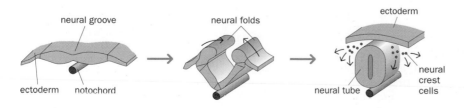

BOX 5.4 THE EXTRAORDINARY VERSATILITY OF THE VERTEBRATE NEURAL CREST

The vertebrate neural crest is quite extraordinary and, although derived from ectoderm, its importance has led to suggestions that it be recognized as a fourth germ layer. Neural crest cells originate during neurulation from dorsal ectoderm cells at the edges of the neural folds (Figure 5.18). The neural crest is a transient structure—the cells disperse soon after the neural tube closes. They undergo an epithelial-to-mesenchymal transition and they migrate away from the midline (**Figure 1**).

Neural crest cells migrate to peripheral locations, where they assume a variety of different ectodermal and mesodermal fates, giving rise to a quite prodigious number of differentiated cell types (**Table 1**). Most of our information comes from fate-mapping studies in the chick, which have been aided by the ability to transplant corresponding domains of the dorsal neural tube between chick and quail embryos and to identify the cellular derivatives of such domains.

Neural crest (NC) cells form at all levels along the anteroposterior axis of the neural tube; however, four main, but overlapping, regions are recognized with characteristic derivatives and functions as listed below.

Cranial neural crest
Some NC cells migrate dorsolaterally and give rise to the *craniofacial mesenchyme*, which in turn produces cartilage, bone, cranial neurons, glia, and connective tissues of the face. NC cells also give rise to structures within the pharyngeal (branchial) arches and pouches, including cartilaginous rudiments of several bones of the nose, face, middle ear, jaw, and neck and also cells of the thymus and odontoblasts of tooth primordia.

Trunk neural crest
An early wave of NC cells migrates ventrolaterally through the anterior half of the sclerotomes (the blocks of mesodermal tissue derived from somites that will differentiate into the vertebral cartilage). Some of the NC cells stay in the sclerotomes to form the *dorsal root ganglia* containing the sensory neurons. Other NC cells continue ventrally to form the sympathetic ganglia, the adrenal medulla, and nerve clusters surrounding the aorta. A later wave of NC cells migrates dorsolaterally

Table 1 Major classes of neural crest derivatives

Tissue/region	Cell types/structure
Peripheral nervous system (PNS) neurons	sensory ganglia; sympathetic ganglia; parasympathetic ganglia
PNS glial cells	Schwann cells; non-myelinating glial cells
Endocrine/ paraendocrine derivatives	adrenal medulla; calcitonin-secreting cells; carotid body type I cells
Head and neck	corneal endothelium and stroma; dermis, smooth muscle, and adipose tissue; facial and anterior ventral skull cartilage and bones; lachrymal gland, connective tissue; pituitary gland, connective tissue; salivary gland, connective tissue; thyroid, connective tissue; tooth papillae
Skin	dermis, smooth muscle, and adipose tissue; melanocytes
Others	arteries originating from aortic arch— connective tissue and smooth muscle; thymus, connective tissue; truncoconal septum

into the ectoderm, becoming melanocytes and going on to move through the skin toward the ventral midline of the body.

Vagal and sacral neural crest
The vagal (neck) NC lies opposite chick somites 1–7; the sacral NC lies posterior to somite 28. The neck and sacral NC cells generate the parasympathetic (enteric) ganglia of the gut.

Cardiac neural crest
The cardiac neural crest is a subregion of the vagal neural crest, extending from chick somites 1 to 3. Cardiac NC cells develop into melanocytes, neurons, cartilage, or connective tissue (of certain pharyngeal arches). In addition, this type of NC gives rise to the entire (muscular/connective tissue) walls of the large arteries as they arise from the heart (outflow tracts), and contribute to the septum separating the pulmonary circulation from the aorta.

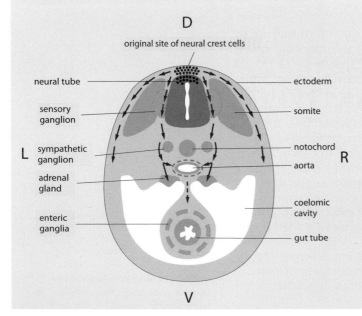

Figure 1 Main migratory pathways during neural crest migration. Schematic cross section through the middle part of the trunk of a chick embryo. Cells taking the pathway just beneath the ectoderm (outer red arrows) will form melanocytes; those that take the deep pathway via the somites (inner red arrows) will form the neurons and glial cells of sensory and sympathetic ganglia, and parts of the adrenal gland. The neurons and glia of the enteric ganglia, in the wall of the gut, are formed from neural crest cells that migrate along the length of the body from either the neck or sacral regions. [From Alberts B, Johnson A, Lewis J et al. (2002) Molecular Biology of the Cell, 4th ed. Garland Science/Taylor & Francis LLC.]

factors of the Emx and Otx families are expressed. In the hindbrain and spinal cord, positional identities are controlled by the Hox genes. In the hindbrain, it seems that the positional values of cells are fixed at the neural plate stage and that migrating neural crest cells carry this information with them and impose positional identities on the tissues surrounding their eventual resting place.

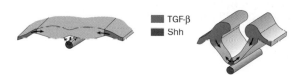

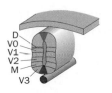

Conversely, positional identity in the spinal cord is imposed by signals from the surrounding paraxial mesoderm, which can be shown by transplanting cells to different parts of the axis. Cranial neural crest cells behave according to their lineage when moved to a new position; that is, they do the same things as they would in their original position. Trunk crest cells, in contrast, behave according to their new position—they do the same things as their neighbors.

The developing nervous system is a good model of pattern formation and differentiation, because various cell types (different classes of neurons and glia) arise along the dorsoventral axis. This process is controlled by a hierarchy of genetic regulators (**Figure 5.19**). Dorsoventral polarity is generated by opposing sets of signals originating from the notochord and the adjacent ectoderm. The notochord secretes the signaling molecule Sonic hedgehog (Shh), which has ventralizing activity, whereas the ectoderm secretes several members of the TGF-β (transforming growth factor) superfamily, including BMP4, BMP7, and a protein appropriately named Dorsalin.

As the neural plate begins to fold, these same signals begin to be expressed in the extreme ventral and dorsal regions of the neural tube itself—the floor plate and roof plate, respectively. The opposing signals have opposite effects on the activation of various homeodomain class transcription factors (such as Dbx1, Irx3, and NKx6.1). Expression of specific transcription factors divides the neural tube into discrete dorsoventral zones, which later become the regional centers of different classes of neuron (see Figure 5.19). For example, the zone defined by the expression of *Nkx6.1* alone becomes the region populated by motor neurons, which are residents of the ventral third of the neural tube.

Neuronal differentiation involves the combinatorial activity of transcription factors

Neurons do not arise uniformly in the neural ectoderm but are restricted to specific regions demarcated by the expression of *proneural genes* that encode transcription factors of the basic helix-loop-helix (bHLH) family, such as MASH1, MATH1, and neurogenin (**Figure 5.20**). Proneural gene expression confers on cells the ability to form neuroblasts, but not all proneural cells can adopt this fate. Instead, there is competition between the cells involving the expression of *neurogenic genes* such as *Notch* and *Delta* that are often involved in juxtacrine cell signaling. The successful cells form neuroblasts and inhibit the surrounding cells from doing the same. Therefore, neuroblasts arise in a precise spacing pattern. This pattern-forming process is termed **lateral inhibition**. The neurons then begin to differentiate according to their position with respect to the dorsoventral

Figure 5.19 Pattern formation of the nervous system. Dorsoventral pattern formation involves a competition between secreted signals from the notochord (Sonic hedgehog, Shh) and from the ectoderm (initially BMP4, but later other members of the TGF-β superfamily). As the neural plate folds up, the signals are expressed in the neural tube itself. The result is the activation of different transcription factors in different zones resulting in the regional specification of different neuronal dorsal (D) and ventral (V) subtypes and motor (M) neurons.

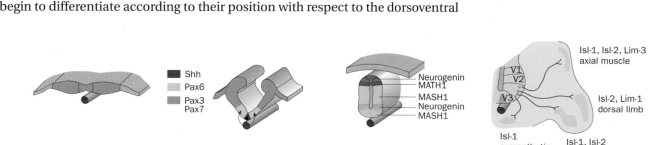

Figure 5.20 Differentiation of the nervous system. The genes that control neurogenesis in the developing nervous system are also expressed in discrete zones along the dorsoventral axis. They may be regulated by *Pax3*, *Pax6*, and *Pax7* genes, whose expression domains are in turn partly defined by the Sonic hedgehog signal emanating from the notochord. The expression domains of the proneural gene products, neurogenin, MATH1, and MASH1 are shown. Neural development is characterized by the expression of the transcription factor NeuroD. Then different classes of neuron begin to express distinct groups of transcription factors. Those of the LIM homeodomain family, which includes Isl-1, Isl-2, and Lim-3, are especially important in determining subneuronal fates by regulating the behavior of projected axons.

and anteroposterior axes of the nervous system, which is defined by the combination of transcription factors discussed in the previous section.

Further diversification is controlled by refinements in the expression patterns of the above transcription factors. For example, all motor neurons initially express two transcription factors of the LIM homoodomain family: Islet-1 and Islet-2. Later, only those motor neurons projecting their axons to the ventral limb muscles express just these two LIM transcription factors. Neurons expressing Isl-1, Isl-2, and a third transcription factor, Lim-3, project their axons to the axial muscles of the body wall. Neurons expressing Isl-2 and Lim-1 project their axons to dorsal limb muscles, whereas those expressing Isl-1 alone project their axons to sympathetic ganglia (see Figure 5.20). The transcription factors determine the particular combinations of receptors expressed on the axon growth cones and therefore determine the response of growing axons to different physical and chemical cues (such as chemoattractants). Once the first axons have reached their targets, further axons can find their way by growing along existing axon paths—a process termed *fasciculation*.

5.7 GERM-CELL AND SEX DETERMINATION IN MAMMALS

All mammals have male and female sexes. The decision between male and female development is made at conception, when the sperm delivers either an X chromosome or a Y chromosome to the egg, which always contains an X chromosome. The only exceptions occur when errors in meiosis produce gametes with missing or extra sex chromosomes, resulting in individuals with *sex-chromosome aneuploidies.*

Primordial germ cells are induced in the early mammalian embryo and migrate to the developing gonads

In insects, nematodes, and vertebrates there is a clear and early separation of germ cells from somatic cells, giving rise to a *germ line* that is distinct from somatic cells. Germ cells do not arise within the gonads; their precursors—the **primordial germ cells** (**PGCs**)—arise elsewhere and migrate into the developing gonads. In frogs, flies, and nematodes, the germ cells are determined by material (*germ plasm*) within the cytoplasm of the egg. In mammals, germ cells are induced in the early embryo. In the mouse, germ cells form in postimplantation embryos at the posterior region of the epiblast. BMP4/BMP8 signals transmitted from neighboring extraembryonic ectoderm cells induce the expression of *fragilis* in posterior epiblast cells and also of *blimp1* in a small subset of such cells lying immediately proximal to the ectoderm cells. The latter is a key regulator of PGC specification, and about 15 or so Blimp1⁺ cells will go on to give rise to about 40 PGCs at the posterior end of the primitive streak by 7 days of gestation (**Figure 5.21**).

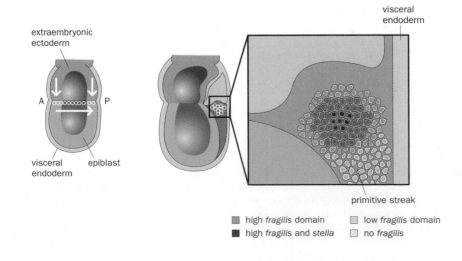

Figure 5.21 Mammalian primordial germ cells are induced in the early embryo. In the mouse embryo, signals from the extraembryonic ectoderm (vertical arrows) induce neighboring proximal epiblast cells (white circles) to become precursors of primordial germ cells (PGCs) and extraembryonic mesoderm. During gastrulation, these cells migrate to the posterior end of the embryo above the primitive streak. By embryonic day 7 in mouse development, the PGC precursor cells emerge from the posterior primitive streak and are identifiable because they express the *fragilis* gene. In the center of the cluster are PGCs that also express *stella* (and *blimp1*). A (anterior) and P (posterior) indicate the A–P axis. [Adapted from Wolpert L, Jessell T, Lawrence P et al. (2007) Principles of Development, 3rd ed. With permission from Oxford University Press.]

extraembryonic ectoderm

A   P

visceral endoderm   epiblast

visceral endoderm

primitive streak

■ high *fragilis* domain   ☐ low *fragilis* domain
■ high *fragilis* and *stella*   ☐ no *fragilis*

The PGCs then migrate away from the posterior region of the primitive streak into the endoderm and enter the developing hindgut for a short period. At the end of gastrulation the PGCs are determined and migrate into the *genital ridge*, a mesoderm structure that is part of the developing gonad. At this stage, the gonad is bipotential, being capable of developing into either testis or ovary.

Sex determination involves both intrinsic and positional information

Although the sex of the human embryo is established at conception, sexual differentiation does not begin to occur until the embryo is about 5 weeks old. *Primary sexual characteristics* (the development of the gonad and the choice between sperm and egg development) are dependent on the genotype of the embryo; *secondary sex characteristics* (the sex-specific structures of the genitourinary system and the external genitalia) are dependent on signals from the environment, mediated by hormonal signaling.

Male development depends on the presence or absence of the Y chromosome. A critical male-determining gene called *SRY* (sex-determining region of the Y chromosome) encodes a transcription factor that activates downstream genes required for testis development. The testis then produces sex hormones required for the development of male secondary sex characteristics.

Female gonad development was previously considered a default state. The *SRY* gene was thought sufficient to switch the bipotential embryonic gonad from female to male differentiation. Consistent with this is the fact that rare XX males often have a small fragment of the Y chromosome, including *SRY*, translocated onto the tip of one of their X chromosomes, and genetically female mice transgenic for the mouse *Sry* gene develop as males. More recent studies suggest that genes on the X chromosome and autosomes are also involved in the positive regulation of ovarian development. The overexpression of genes such as *DAX* and *WNT4A* can feminize XY individuals even if they possess a functional *SRY* gene.

Early gamete development seems to be controlled more by the environment than by the genotype of the germ cells. Female PGCs introduced into the testis will begin to differentiate into sperm, and male PGCs introduced into the ovary will begin to differentiate into oocytes. This may reflect the regulation of the cell cycle, because PGCs entering the testis arrest before meiosis; those entering the ovary commence meiosis immediately. Therefore, PGCs of either sex that colonize somatic tissue outside the gonad begin to differentiate into oocytes because there is no signal to arrest the cell cycle. In all of these unusual situations, however, functional gametes are not produced. Differentiation aborts at a relatively late stage, presumably because the genotype of the germ cells themselves also has a crucial role in gamete development.

Unlike primary sex characteristics, the default state is female for secondary sex characteristics. One of the genes regulated by SRY makes the SF1 transcription factor, which activates genes required for the production of male sex hormones, including *HSD17B3* (encoding hydroxysteroid-17-β-dehydrogenase 3—required for testosterone synthesis) and *AMH* (encoding anti-Mullerian hormone). Both hormones have important roles in the differentiation of the male genitourinary system. AMH, for example, causes the *Mullerian ducts* (which become the fallopian tubes and uterus in females) to break down.

Mutations inhibiting the production, distribution, elimination, or perception of such hormones produce feminized XY individuals. For example, androgen insensitivity syndrome results from defects in the testosterone receptor that prevent the body from responding to the hormone even if it is produced at normal levels. XY individuals with this disease appear outwardly as normal females but, owing to the effects of SRY and AMH, they possess undescended testes instead of ovaries, and they lack a uterus and fallopian tubes. Mutations that lead to the overproduction of male sex hormones in females have the opposite effect: XX individuals are virilized. Occasionally, this occurs in developing male/female fraternal twins, when the female twin is exposed to male hormones from her brother. The CYP19 enzyme converts androgens to estrogens, so mutations that increase its activity can result in the feminization of males; those decreasing or abolishing its activity can lead to the virilization of females.

5.8   CONSERVATION OF DEVELOPMENTAL PATHWAYS

The various mechanisms described above are critically important for the organism to develop, and they have generally been highly conserved during evolution. Nevertheless, developmental abnormalities of differing severity arise in some individuals as a result of mutations and/or environmental factors, and some developmental mechanisms show significant species differences.

Many human diseases are caused by the failure of normal developmental processes

The most extreme human diseases involve striking morphological abnormalities that are due to disruptions of the general processes of differentiation, pattern formation, and morphogenesis. An example is holoprosencephaly, a failure in the normal process of forebrain development, which in its severest form gives rise to individuals with a single eye (cyclopia) and no nose. As is the case for other human diseases, the phenotype can be influenced by both genetic and environmental factors. In some cases, it is clear that specific mutations have caused this defect. Examples include mutations in the *SHH* gene, which specifies the signaling protein Sonic hedgehog that acts as a morphogen in establishing the body plan in the developing embryo. Otherwise, the abnormality can be traced to an environmental cause, such as limited cholesterol intake in the maternal diet. In sex determination, as described above, both genes and the environment have a role.

Although some developmental abnormalities can be traced to the use of drugs (chemicals that are known to have *teratogenic* effects include alcohol, certain antibiotics, thalidomide, retinoic acid, and illegal drugs such as cocaine), many are the result of mutations in specific genes. As stated at the beginning of this chapter, some of the most important developmental genes are regulatory in nature, encoding either transcription factors or components of signaling pathways. Genes encoding structural components of the cell or extracellular matrix, and even metabolic enzymes, are also important in development and reveal important disease phenotypes.

Developmental processes are often highly conserved but some show considerable species differences

Where genes have been shown to cause developmental diseases, there is often a remarkable degree of evolutionary conservation among animals. This conservation applies not only to the genes but often also to entire pathways in which they are involved (**Figure 5.22**).

Conserved molecular pathways are often used for similar processes in very distantly related species (see Figure 5.22A). For example, as mentioned above, neural induction in *Xenopus* embryos involves a battle between the opposing effects of BMP4, which favors ventral and lateral fates, and dorsalizing (neuralizing) factors such as Chordin, Noggin, and Follistatin. The *Drosophila* orthologs of BMP4 and Chordin are proteins called Decapentaplegic (Dpp) and Short gastrulation (Sog), respectively. Remarkably, these proteins have equivalent roles in the formation of the *Drosophila* nervous system. Indeed, the relationship goes further. The activity of Dpp in *Drosophila* is enhanced by the protein Tolloid (Tol), which degrades Sog. In *Xenopus*, the role of Tol is taken by its ortholog Xolloid (Xol), and in zebrafish the equivalent molecule is BMP1. Both Xolloid and BMP1 degrade Chordin. There is also a cross-phylum pairing of *Xenopus* BMP7 and the *Drosophila* protein Screw, accessory proteins that are necessary for BMP4/Dpp activity.

In other cases, the same developmental pathway is used for very different purposes in different species (see Figure 5.22B). In mammals, epidermal growth factor (EGF) binds to its receptor, EGFR, to initiate a Ras–Raf–MAP kinase cascade and promote the proliferation of epidermal cells. Equivalent pathways exist in *D. melanogaster* and *C. elegans* (although the counterparts of the human pathway components may have different names), but the pathways perform different functions. In *Drosophila*, the equivalent pathway is used to promote differentiation of one of the eight photoreceptor cell types during eye development, and in *C. elegans* it is used to stimulate the division and differentiation of vulval cells.

(A)

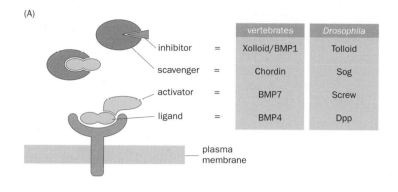

vertebrates	*Drosophila*
Xolloid/BMP1	Tolloid
Chordin	Sog
BMP7	Screw
BMP4	Dpp

inhibitor =
scavenger =
activator =
ligand =

plasma membrane

(B)

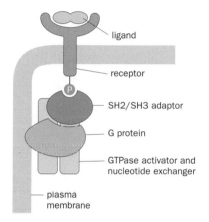

ligand
receptor
SH2/SH3 adaptor
G protein
GTPase activator and nucleotide exchanger
plasma membrane

human	fly	worm
EGF	BOSS	LIN-3
EGFR	Sevenless	LET-23
GRB2	Drk	SEM-5
Ras	Ras1	LET-60
GAP/GNRP	Gap1/Sos	GAP-1
epidermal cell proliferation	eye development	vulval cell differentiation

Figure 5.22 Evolutionary conservation of developmental pathways. (A) Components of the BMP4/Chordin pathway are conserved in *Drosophila* and vertebrates, where they have comparable roles in neural induction. (B) Although components of the epidermal growth factor signaling pathway are also conserved in several species, the pathway has diverse roles in vertebrates, flies (*D. melanogaster*), and worms (*C. elegans*).

One of the best examples of evolutionary conservation involves the homeotic genes, which seem to be present in all animals and to have very similar functions. The fundamental role of these genes in pattern formation is demonstrated by the ability of orthologous genes from very different species to substitute for each other. Complete rescue of the mutant phenotype has been achieved for some *Drosophila* mutants by introducing the orthologous human gene. For example, the *apterous* mutant has no wings, but addition of the normal allele of this *Drosophila* mutation to the mutant embryo, or the human counterpart *LHX2*, results in a normal phenotype.

However, there are also significant differences between species, even those that are closely related. There are very great differences between the vertebrate embryos before gastrulation, reflecting different strategies for nutrient acquisition, and the process of gastrulation is rather different between human and mouse embryos, with a flat bilaminar germ disk in humans and a cup-shaped one in mouse embryos. Sex determination mechanisms are also very diverse. Not all mammals use the human model of XY sex determination, and many reptiles dispense with the use of heteromorphic sex chromosomes altogether by relying on the temperature of the environment to specify the sex of the embryo.

FURTHER READING

General

Gilbert SF (2006) Developmental Biology, 8th ed. Sinauer Associates.

Hill M (2008) UNSW Embryology. http://embryology.med.unsw.edu.au/ [An educational resource for learning concepts in embryological development hosted by the University of New South Wales, Sydney.]

Wolpert L, Smith J, Jessell T et al. (2006) Principles Of Development, 3rd ed. Oxford University Press.

Noncoding RNA in development

Amaral PP & Mattick JS (2008) Noncoding RNA in development. *Mamm. Genome* 19, 454–492.

Stefani G & Slack FJ (2008) Small non-coding RNAs in animal development. *Nat. Rev. Mol. Cell Biol.* 9, 219–230.

Animal models of development

Bard J (ed.) (1994) Embryos: Color Atlas Of Development. Mosby.

Cell specialization during development

Stern CD (2006) Neural induction: 10 years on since the 'default model'. *Curr. Opin. Cell Biol.* 18, 692–697.

Pattern formation in development

Raya A & Belmonte JC (2006) Left–right asymmetry in the vertebrate embryo: from early information to higher-level integration. *Nat. Rev. Genet.* 7, 283–293.

Tabin C & Wolpert L (2007) Rethinking the proximodistal axis of the vertebrate limb in the molecular era. *Genes Dev.* 21, 1433–1442.

Takaoka K, Yamamoto M & Hamada H (2007) Origin of body axes in the mouse embryo. *Curr. Opin. Genet. Dev.* 17, 344–350.

Tam PP, Loebel DA & Tanaka SS (2006) Building the mouse gastrula: signals, symmetry and lineages. *Curr. Opin. Genet Dev.* 16, 419–425.

Morphogenesis

Aman A & Piotrowski T (2009) Cell migration during morphogenesis. *Dev. Biol.* (in press).

Kornberg TB & Guha A (2007) Understanding morphogen gradients: a problem of dispersion and containment. *Curr. Opin. Genet. Dev.* 17, 264–271.

Steinberg MS (2007) Differential adhesion in morphogenesis: a modern view. *Curr. Opin. Genet. Dev.* 17, 281–286.

Early human development: fertilization to gastrulation

Carlson BM (2004) Human Embryology And Developmental Biology, 3rd ed. Mosby.

Larsen WJ (2001) Human Embryology, 3rd ed. Elsevier.

Moore KL & Persaud TVN (2007) The Developing Human: Clinically Oriented Embryology, 8th ed. Saunders.

Neural development and neural crest

Copp AJ, Greene ND & Murdoch JN (2003) The genetic basis of mammalian neurulation. *Nat. Rev. Genet.* 4, 784–793.

Hall BK (2008) The neural crest and neural crest cells: discovery and significance for theories of embryonic organization. *J. Biosci.* 33, 781–793.

Sauka-Spengler T & Bronner-Fraser M (2006) Development and evolution of the migratory neural crest: a gene regulatory perspective. *Curr. Opin. Genet. Dev.* 16, 360–366.

Germ cell and sex determination in mammals

Hayashi K, Chuva de Sousa Lopes SM & Surani MA (2007) Germ cell specification in mice. *Science* 316, 394–396.

Wilhelm D, Palmer S & Koopman P (2007) Sex determination and gonadal development in mammals. *Physiol. Rev.* 87, 1–28.

Conservation of developmental pathways

Epstein CJ, Erickson RP & Wynshaw-Boris A (eds) (2004) Inborn Errors Of Development. Oxford University Press.

Rincon-Limas DE, Lu C-H, Canal I et al. (1999) Conservation of the expression and function of *apterous* orthologs in *Drosophila* and mammals. *Proc. Natl Acad. Sci. USA* 96, 2165–2170.

Chapter 6

Amplifying DNA: Cell-based DNA Cloning and PCR

6

KEY CONCEPTS

- DNA molecules are very long and complex and are prone to shearing when isolated from other cell constituents, making their purification difficult.

- DNA cloning means making multiple identical copies (clones) of a DNA sequence of interest, the target DNA; the increase in copy number is called amplification. The process requires a DNA polymerase to replicate a target DNA sequence repeatedly, either within cells or *in vitro*.

- In cell-based DNA cloning, target DNA is fractionated by transfer into bacterial or yeast cells (transformation). Each transformed cell typically takes up just one target DNA molecule and replicates it using the host cell's DNA polymerase.

- Before transfer into cells, a target DNA is joined to vector DNA molecules, a plasmid, or a phage DNA capable of self-replication inside cells. The target-vector DNA complexes are known as recombinant DNA.

- Restriction endonucleases cut DNA molecules at defined short recognition sequences and are needed to prepare target and vector DNA for joining together.

- After having been transferred into cells, recombinant DNA typically replicates independently of the host cell's chromosome(s).

- Large target DNAs are more stable in cells if carried by vectors whose copy number is constrained to one or two per cell.

- Megabases of DNA can be cloned in yeast by adding very short sequence elements necessary for chromosome function to form a linear artificial chromosome.

- In expression cloning the target DNA is designed to be expressed. It is converted to RNA and in many cases translated into a recombinant protein.

- Cell-based DNA cloning is slow and laborious; however, it can produce very large amounts of cloned DNA and is the first choice for cloning large DNA sequences, expressing genes, and making comprehensive sets of DNA clones (DNA libraries).

- DNA cloning can also be performed rapidly *in vitro* with the polymerase chain reaction (PCR). A purified DNA polymerase is usually used to replicate a specific DNA sequence selectively within a complex sample DNA. Oligonucleotides are designed to bind specifically to sites in the sample DNA that flank the target sequence, allowing the polymerase to initiate DNA replication at these sites.

- PCR can also be used to amplify many different target DNA fragments simultaneously, for example by using complex mixtures of numerous different primer sequences. This indiscriminate amplification can replenish many of the sequences in precious samples where there is little starting DNA.

- As well as being a DNA cloning procedure, PCR is also widely used as an assay to quantitate DNA and RNA after it has been converted by reverse transcriptase to complementary DNA (cDNA). Real-time PCR is the most efficient method, and quantitates PCR products while the PCR reaction is occurring.

The fundamentals of current DNA technology are very largely based on two quite different approaches to studying specific DNA sequences within a complex DNA population, as listed below (see also **Figure 6.1**):

- *DNA cloning.* DNA sequences of interest are *selectively replicated* in some way to produce very large numbers of identical copies (**clones**). The resulting huge increase in copy number (**amplification**) means that DNA cloning is effectively a way of purifying a desired DNA sequence so that it can be comprehensively studied.

- *Molecular hybridization.* The fragment of interest is not amplified or purified in any way; instead, it is *specifically detected* within a complex mixture of many different sequences. This will be covered in Chapter 7.

Before DNA cloning, our knowledge of DNA was extremely limited. DNA cloning technology changed all that and revolutionized the study of genetics. In comparison with protein sequences, DNA molecules are extremely large and complex—individual nuclear DNA molecules often contain hundreds of millions of nucleotides. When DNA is isolated from cells by standard methods, these huge molecules are fragmented by shear forces, generating complex mixtures of still very large DNA fragments (about 50–100 kb in length with standard DNA isolation methods). Given the complexity of the DNA isolated from the cells of a typical eukaryote, or even prokaryote, the challenge was how to analyze it.

For many eukaryotes, one early approach had been to separate different populations of DNA sequence by centrifugation. Ultracentrifugation in equilibrium density gradients (e.g. CsCl density gradients) typically fractionates the DNA from a eukaryotic cell into a major band representing the bulk of the DNA plus several minor satellite bands that are composed of classes of repetitive DNA. The DNA molecules of the satellite bands (*satellite DNA*) have different buoyant densities from that of the bulk DNA (and from each other) because they consist of very long arrays of short tandem repeats whose base composition is significantly different from that of the bulk DNA. The satellite DNA sequences were found to be involved in specific aspects of chromosome structure and function. Although valuable and interesting, the purified satellite DNAs were a minor component of the genome and did not contain genes.

What was needed were more general methods that allowed any DNA sequence to be purified, not just DNA sequences whose base compositions deviated significantly from that of the total DNA. Two major **DNA cloning** methodologies were developed that made this possible. Both use DNA polymerases to make multiple replica copies of DNA sequences (DNA clones), permitting in principle the purification of any DNA sequence within a complex starting DNA population.

Figure 6.1 General approaches for studying specific DNA sequences in complex DNA populations. DNA molecules are very long and complex, and any gene or exon or other DNA sequence of interest typically represents a tiny fraction of the starting DNA that we can isolate from cells. Two quite different approaches can be used to study a particular DNA sequence of interest. One way is to purify a specific DNA sequence by selectively replicating its sequence in some way (DNA cloning), thereby increasing its copy number (amplification). The specific DNA can either be cloned within bacterial or yeast cells by using cellular DNA polymerases, or by using purified DNA polymerases *in vitro*, as in the polymerase chain reaction (PCR). An alternative approach makes no attempt to purify the DNA sequence, but instead seeks to detect it specifically. It involves labeling a previously isolated nucleic acid sequence that is related in sequence to the desired DNA sequence so that it can specifically recognize it in a molecular hybridization reaction.

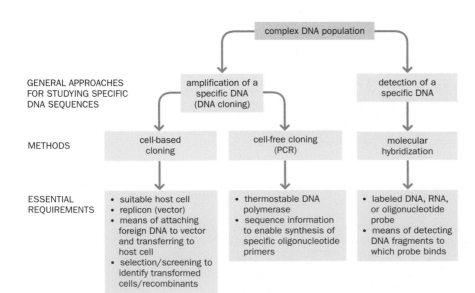

Cell-based DNA cloning

Cell-based DNA cloning was developed in the early 1970s and uses cells to fractionate a complex sample DNA. It relies on cutting the sample DNA into small pieces and transporting these fragments into suitable bacterial or yeast cells. Each host cell usually takes up just one fragment, and an individual fragment can be replicated many times within the cell by the host cell's DNA polymerase. Thereafter the cell is allowed to undergo many rounds of cell division to produce large numbers of cells with identical copies of just one small fragment of the sample DNA.

Cell-based DNA cloning revolutionized genetics and made it possible to purify and study any DNA sequence. Eventually it allowed vast numbers of DNA clones to be produced so that whole genomes and their RNA transcripts could be sequenced. It also allowed the production of purified proteins and specific functional RNA so that they could be studied at the basic research level while enabling various medical and biotechnological applications. We consider the principles of cell-based DNA cloning and basic cloning systems in Section 6.1 and more advanced cloning systems in Section 6.2 and Section 6.3.

The polymerase chain reaction (PCR)

PCR is a rapid cell-free DNA cloning method that was developed in the mid-1980s. It uses purified DNA polymerases to replicate defined DNA sequences selectively within a complex starting DNA population. Because it is a very rapid way of amplifying a specific DNA sequence, it allows quick screening of thousands of samples at a time. As a result, it has numerous applications in both basic and applied research; medical applications are based on its use as a diagnostic and rapid DNA-screening tool.

PCR is a very sensitive and robust DNA cloning method, and it has been widely used in forensic science and in amplifying DNA from tissue retrieved from historical or archaeological sites. As well as being a DNA cloning method, PCR is also widely used to quantitate both DNA and RNA in various assays in basic and applied research. We consider PCR in Section 6.4.

6.1 PRINCIPLES OF CELL-BASED DNA CLONING

Cell-based DNA cloning is mostly conducted in bacterial cells and was made possible by the discovery of type II **restriction endonucleases**. Restriction endonucleases serve to protect bacteria from invading pathogens, notably viruses. After recognizing *specific* short sequence elements in the foreign DNA, the restriction endonucleases cleave the foreign DNA in the vicinity of each such element, but the bacterial cell DNA is left uncut because it is protected by DNA methylation; **Box 6.1** gives more details and describes the different classes of restriction enzyme.

Type II restriction endonucleases offer two big advantages to molecular geneticists. First, they allow complex DNA molecules to be cut into defined fragments of manageable size. Second, they help the resulting DNA fragments to be joined to similarly cut **vector** molecules carrying an origin of replication, which enables them to self-replicate in cells. The resulting hybrid DNA molecules are known as **recombinant DNA**. This completes the first of four essential steps in cell-based DNA cloning (see **Figure 6.2A**).

The second essential step in cell-based DNA cloning involves fractionating a complex mixture of different recombinant DNA molecules by transferring them into suitable host cells, usually bacterial or yeast cells. This process is known as **transformation** and each host cell usually takes up just *one* type of recombinant DNA (Figure 6.2B). Cells that have taken up a recombinant DNA are known as **recombinants**.

The transformed cells are allowed to grow and multiply by repeated cell division. During this time the recombinant DNA typically replicates independently of the host cell's chromosome(s). Because an individual transformed cell normally contains just one type of recombinant DNA, repeated division of that cell generates a series of identical cells (**cell clones**) that each contain the same type of recombinant DNA. Populations of cell clones that arise from different transformed cells can be physically separated by a procedure known as *plating out*.

BOX 6.1 RESTRICTION ENDONUCLEASES AND RESTRICTION–MODIFICATION SYSTEMS

Restriction endonucleases (also called restriction enzymes) are a class of bacterial endonucleases that cleave double-stranded DNA on both strands, after first recognizing specific short sequence elements in the DNA. Wherever it encounters its specific recognition sequence, a restriction enzyme can cleave the DNA, often within the recognition sequence or close nearby.

Different bacterial strains produce different restriction enzymes, and more than 250 different sequence specificities have been identified (see REBASE, the Restriction Enzyme Database, at http://rebase.neb.com/rebase/rebase.html). The convention for naming restriction enzymes is to begin with three letters, a capital for the first letter of the bacterial genus followed by the first two letters of the species. Thereafter, one letter is used to denote the bacterial strain and finally a roman numeral is used to distinguish different restriction enzymes from the same bacterial strain. For example, *Eco*RI, which recognizes the sequence GAATTC, denotes the first restriction enzyme to be reported for *Escherichia coli* strain RY13.

The natural function of restriction enzymes is to protect bacteria from pathogens, notably bacteriophages (viruses that kill bacteria—known as phages for short). They do this by selectively cleaving the DNA of invading pathogens; the bacterial DNA is protected from cleavage because it has undergone site-specific methylation by sequence-specific host cell DNA methyltransferases (*modification*).

A particular strain of bacteria will produce a DNA methyltransferase with the *same* sequence specificity as a restriction enzyme produced by that strain. For example, *E. coli* strain RY13 produces the *Eco*RI methyltransferase that specifically recognizes the sequence GAATTC and methylates the central adenosine on both strands. As a result, the host cell's DNA is protected from cleavage by the *Eco*RI restriction enzyme (the methyl groups protrude into the major groove of the DNA at the binding site and so prevent the restriction enzyme from acting on it). The *Eco*RI endonuclease will, however, cut at unmethylated GAATTC sequences such as in the DNA of invading phage that has not previously been modified by the *Eco*RI methylase.

Together, a restriction enzyme and its cognate modification enzyme(s) form a restriction–modification (R-M) system. In some R-M systems, the two enzyme activities occur as separate subunits or as separate domains of a large combined restriction-and-modification enzyme. Restriction enzymes have been classified into various broad groupings, according to the nature of their R-M system and other characteristics. The most defined groups are listed below.

- Type I restriction enzymes are multisubunit, combined restriction-and-modification enzymes that cut far away from their recognition sequences at variable locations.
- Type II restriction enzymes are widely used in manipulating and analyzing DNA. They are typically single-subunit restriction endonucleases whose recognition sequences are often 4–8 bp long, and they cut at defined positions either within the recognition sequence or very close to it. Type IIR enzymes cut within their recognition sequences, which are often *palindromes* (the sequence in the 5′→3′ direction is the same on both strands). Type IIS enzymes cut outside asymmetric recognition sequences. Type IIB enzymes have bipartite recognition sequences. See Table 6.1 for examples.
- Type III restriction enzymes are multisubunit combined restriction-and-modification enzymes that recognize two separate nonpalindromic sequences that are inversely orientated. They cut the DNA about 20–30 bp outside their recognition sequence.

To do this, the mixture of transformed cells is spread out on an agar surface and allowed to grow to form well-separated cell colonies, each arising from a single transformed cell. Individual cell colonies can then be picked and allowed to undergo a second growth phase in liquid culture, resulting in very large numbers of cell clones (Figure 6.2C).

The final step is to harvest the expanded cell cultures and purify the recombinant DNA (Figure 6.2D). The sections below consider these steps in some detail.

Manageable pieces of target DNA are joined to vector molecules by using restriction endonucleases and DNA ligase

Of the different classes of restriction endonuclease (see Box 6.1), type I enzymes do not produce discrete restriction fragments and type III enzymes do not cut at all possible recognition sequences and rarely give complete digests. The most valuable for DNA manipulations and DNA analyses are type II restriction endonucleases. They cut at defined positions within their recognition sequences or very close to them, and so produce defined restriction fragments.

Many type II restriction endonucleases cut within sequences that are *palindromes* (the sequence of bases is the same on both strands when read in the 5′→3′ direction, as a result of a twofold axis of symmetry). The resulting restriction fragments may have blunt ends because the cleavage points occur exactly on the axis of symmetry. Often, however, cleavage is asymmetric, generating either overhanging 5′ ends or overhanging 3′ ends (see **Table 6.1**). As shown for the examples of *Bam*HI and *Pst*I in Table 6.1, asymmetric cutting of a palindromic sequence generates restriction fragments with two overhanging ends that are identical in sequence *and at the same time complementary in base sequence*. They will have a tendency to associate with each other (or with any other similarly complementary overhang) by forming base pairs. Such overhanging ends are thus often described as *sticky ends*.

Type II restriction endonucleases were crucially important for the development of cell-based DNA cloning because they could cut DNA to produce defined

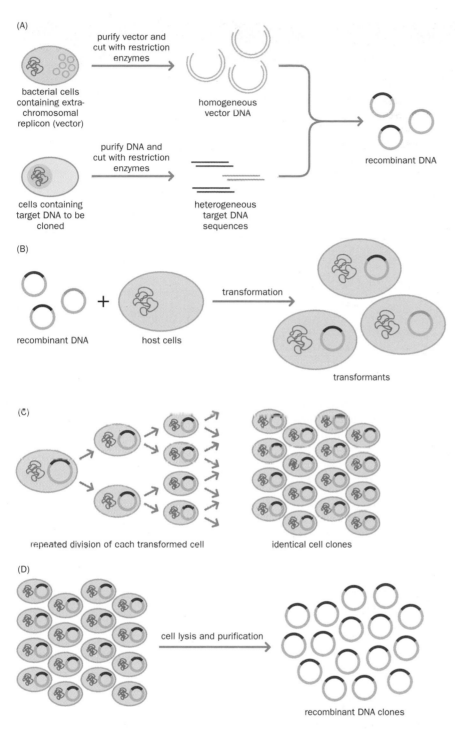

Figure 6.2 The four essential steps of cell-based DNA cloning. (A) Formation of recombinant DNA. Fragments of DNA cut with a restriction endonuclease are mixed with a homogeneous population of vector molecules that have been cut with a similar restriction endonuclease. The target DNA and vector are joined by DNA ligase to form recombinant DNA. Each vector contains an origin of replication that will allow it to be copied when inside a host cell. (B) Transformation. The recombinant DNA is mixed with host cells, which normally take up only one foreign DNA molecule. Thus, each cell normally contains a unique recombinant DNA. (C) Amplification. Individual transformed cells are allowed to undergo repeated cell division to give a colony of identical cell clones containing one type of recombinant DNA that are kept physically separate from other colonies containing cell clones with different recombinant DNA molecules. For convenience, we show the example of a vector whose copy number per cell is highly restrained but many plasmid vectors can reach quite high copy numbers per cell, constituting an additional type of amplification. (D) Isolation of recombinant DNA clones, after separation of the recombinant DNA from the host cell DNA.

DNA fragments of a manageable size. When DNA is isolated from tissues and cultured cells, the unpackaging of the huge DNA molecules and inevitable physical shearing results in heterogeneous populations of DNA fragments. The fragments are not easily studied because their average size is still very large and they are of randomly different lengths. With the use of restriction endonucleases it became possible to convert this hugely heterogeneous population of broken DNA fragments into sets of smaller restriction fragments of defined lengths.

Type II restriction endonucleases also paved the way for artificially recombining DNA molecules. By cutting a vector molecule and the target DNA with restriction endonuclease(s) producing the same type of sticky end, vector–target association is promoted by base pairing between their sticky ends (**Figure 6.3**). The hydrogen bonding between their termini facilitates subsequent covalent joining by enzymes known as DNA ligases. Intramolecular base pairing can also happen, as when the two overhanging ends of a vector molecule associate to form a circle

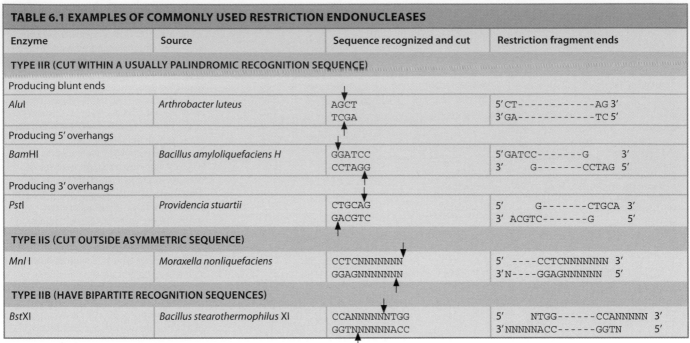

TABLE 6.1 EXAMPLES OF COMMONLY USED RESTRICTION ENDONUCLEASES

Enzyme	Source	Sequence recognized and cut	Restriction fragment ends
TYPE IIR (CUT WITHIN A USUALLY PALINDROMIC RECOGNITION SEQUENCE)			
Producing blunt ends			
*Alu*I	*Arthrobacter luteus*	AGCT TCGA	5′ CT-----------AG 3′ 3′ GA-----------TC 5′
Producing 5′ overhangs			
*Bam*HI	*Bacillus amyloliquefaciens* H	GGATCC CCTAGG	5′ GATCC-------G 3′ 3′ G-------CCTAG 5′
Producing 3′ overhangs			
*Pst*I	*Providencia stuartii*	CTGCAG GACGTC	5′ G-------CTGCA 3′ 3′ ACGTC-------G 5′
TYPE IIS (CUT OUTSIDE ASYMMETRIC SEQUENCE)			
Mnl I	*Moraxella nonliquefaciens*	CCTCNNNNNNN GGAGNNNNNNN	5′ ----CCTCNNNNNNN 3′ 3′ N----GGAGNNNNNN 5′
TYPE IIB (HAVE BIPARTITE RECOGNITION SEQUENCES)			
*Bst*XI	*Bacillus stearothermophilus* XI	CCANNNNNNTGG GGTNNNNNNACC	5′ NTGG------CCANNNNN 3′ 3′ NNNNNACC------GGTN 5′

For other enzymes see the REBASE database of restriction nucleases at http://rebase.neb.com/rebase/rebase.html. N = any nucleotide.

(*cyclization*), and ligation reactions are designed to promote the joining of target DNA to vector DNA and to minimize vector cyclization and other unwanted products such as linear **concatemers** produced by sequential end-to-end joining (see Figure 6.3).

Basic DNA cloning in bacterial cells uses vectors based on naturally occurring extrachromosomal replicons

Most cell-based DNA cloning uses modified bacteria as the host cells. Bacterial cell cycle times are short and so they multiply rapidly. They typically have a single circular double-stranded chromosome with a single origin of replication. Shortly after the host chromosome has replicated, the cell divides to give daughter cells that each have a single chromosome, like the parent cell.

Target DNA fragments typically lack a functional origin of replication, and so cannot replicate by themselves within bacterial cells. They need to be attached to a DNA **replicon**, a sequence that has a replication origin allowing propagation within cells. A replicon that carries target DNA as a passenger into cells and allows the target DNA to be replicated is known as a **vector** molecule. The most efficient way to propagate target DNA within cells is to use a vector with a replication origin that derives from a natural extrachromosomal replicon, one that replicates independently of the host cell's chromosome. Extrachromosomal replicons often go through several cycles of replication during a single cell cycle, and they can therefore reach high copy numbers (unlike the chromosomal DNA copy number, which is restricted to typically one per cell). As a result, large amounts of target DNA can be produced by being co-replicated with an extrachromosomal replicon.

The extrachromosomal replicons that are found in bacterial cells usually belong to one of two classes: plasmids and bacteriophages (phages). **Plasmids** are non-essential DNA molecules that replicate independently of the host cell's chromosome. They are vertically distributed to daughter cells after division of the host cell, but they can be transferred horizontally to neighboring cells during bacterial conjugation events. They often consist of small circular double-stranded DNA molecules, but some plasmids have linear DNA. They range in size from 2 kb to more than 200 kb and individually contain very few genes.

Plasmid copy number varies significantly: high-copy-number plasmids may reach more than 100 copies per cell, but other plasmids may be restricted to just one or two copies per cell. Different plasmids can coexist in a cell. A typical

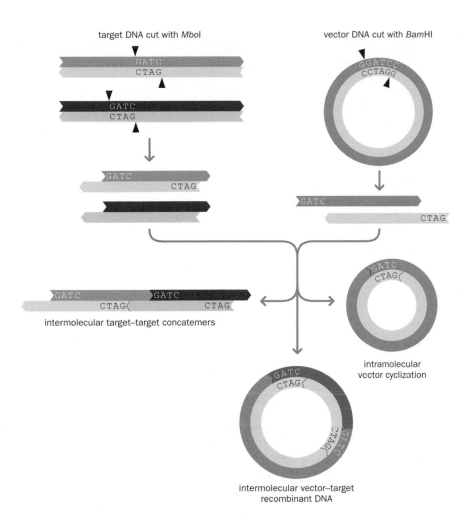

target DNA cut with *Mbo*I

vector DNA cut with *Bam*HI

intermolecular target–target concatemers

intramolecular
vector cyclization

intermolecular vector–target
recombinant DNA

Figure 6.3 Formation of recombinant DNA. Heterogeneous target DNA sequences cut with a type II restriction enzyme (*Mbo*I in this example) are mixed with vector DNA cut with another restriction enzyme such as *Bam*HI that generates the same type of sticky ends (here, a 5′ overhanging sequence of GATC). Ligation conditions are chosen that will allow target and vector molecules to combine, to give recombinant DNA. However, other ligation products are possible. Intermolecular target–target concatemers are shown on the left, and intramolecular vector cyclization on the right. Other possible ligation products not shown here include intermolecular vector concatemers, cyclization of multimers, and co-ligation events that involve two different target sequences being included with a vector molecule in the same recombinant DNA molecule.

Escherichia coli isolate, for example, might have three different small plasmids present in multiple copies and one large single-copy plasmid. Natural examples of bacterial plasmids include plasmids that carry the sex factor (F) and those that carry drug-resistance genes. Some plasmids sometimes insert their DNA into the bacterial chromosome (*integration*). Such plasmids, which can exist in two forms—extrachromosomal replicons or integrated plasmids—are known as **episomes**.

A **bacteriophage** (or **phage**, for short) is a virus that infects bacterial cells. Unlike plasmids, phages can exist extracellularly but, like other viruses, they must invade a host cell to reproduce, and they can only infect and reproduce within certain types of host cell. Phages may have DNA or RNA genomes, and in DNA phages the genome may consist of double-stranded DNA or single-stranded DNA and may be linear or circular. There is considerable variation in genome sizes, from a few kilobases to a few hundred kilobases.

In the mature phage particle, the genome is encased in a protein coat that can aid in binding to specific receptor proteins on the surface of a new host cell and entry into the cell. To escape from one cell to infect a new one, bacteriophages produce enzymes that cause the cell to burst (*cell lysis*). The ability to infect new cells depends on the number of phage particles produced, the *burst size*, and so high copy numbers are usually attained. Like episomal plasmids, some bacteriophages, such as phage λ, can also integrate into the host chromosome, whereupon the integrated phage sequence is known as a *prophage*.

Cloning in bacterial cells uses genetically modified plasmid or bacteriophage vectors and modified host cells

For natural plasmids and phages to be used as vector molecules, they need to be genetically modified in different ways. First, it is important to design the vector so that the target restriction fragments are inserted into a unique location (the *cloning site*) in the vector molecule. In addition, to accommodate different types of restriction fragments it is highly desirable to have a variety of different unique restriction sites that provide a variety of possible cloning sites. This is normally achieved by inserting into a vector a sequence that has been artificially synthesized to have the desired restriction sites. To do this, two long complementary oligonucleotides are synthesized so that when allowed to hybridize they form a 20–40 bp **polylinker** sequence with multiple desired restriction sites that is then inserted into the vector. To ensure that the desired cloning sites in the polylinker are unique, other examples of the same restriction sites that occur naturally elsewhere in the vector are eliminated by site-specific mutagenesis. **Figure 6.4** shows an example of a plasmid vector with a multiple cloning site region that originated by the insertion of a polylinker.

Another modification is to insert a **marker gene** within the vector. Plasmid vectors, for example, are widely modified to include an antibiotic-resistance gene, often a gene that confers resistance to ampicillin, chloramphenicol, kanamycin, or tetracycline (see Figure 6.4 for an example). In such cases the bacterial cells used for cloning need to be sensitive to the relevant antibiotic. To screen for cells transformed by the vector, the transformed cells are plated out on an agar surface that contains the relevant antibiotic; the resulting colonies should be descendants of originally antibiotic-sensitive cells that have been transformed by the vector to become resistant to antibiotics.

An additional widely used marker system derives from the *E. coli lacZ* gene and involves a blue–white color assay for β-galactosidase, the *lacZ* product. This enzyme can cleave the disaccharide lactose to give glucose plus galactose, but transcription of *lacZ* is normally repressed by a lac repressor protein encoded by the *lacI* sequence. Transcription of *lacZ* can, however, be induced by lactose, or a lactose analog such as isopropyl β-D-thiogalactoside (IPTG), which binds to the repressor protein and inactivates it. IPTG is generally used as the inducer because it cannot be metabolized, and therefore its concentration does not change as the cells grow. The marker assay is also designed to depend on complementation between different inactive β-galactosidase fragments produced by the host cell and the vector. Thus, the *E. coli* host is a mutant in which the 5′ end of the *lacZ* gene has been deleted, resulting in production of a large, but inactive, C-terminal fragment of β-galactosidase. Vectors such as pUC19 (see Figure 6.4) have been modified to contain *lacZ′*, a 5′ component of the *lacZ* gene that when transcribed produces a short inactive N-terminal fragment of β-galactosidase. This fragment, the α-fragment, complements the host cell's C-terminal fragment to produce a functional enzyme (α-*complementation*). The active enzyme is conveniently assayed by a chemical reaction as described in Figure 6.4.

A further refinement of the *lacZ* complementation assay permits the screening for transformed cells that have taken up recombinant DNA (vector plus insert DNA). In this case, the polylinker is inserted within the coding sequence that specifies the N-terminal β-galactosidase fragment, as in the example of the plasmid pUC19 (see Figure 6.4). In such cases the number of nucleotides in the polylinker is an exact multiple of three so that the vector's N-terminal β-galactosidase fragment can continue to complement the host's β-galactosidase fragment

Figure 6.4 An example of a high-copy-number plasmid vector, pUC19. The origin of replication (ori) enables more than 100 copies of pUC19 in a suitable *E. coli* host cell. The ampicillin resistance gene (*Amp*^R) permits selection for cells containing the vector molecule. *lacZ′* is a 5′ fragment of the *lacZ* gene that can be expressed to give a short N-terminal fragment of β-galactosidase, the α-fragment. *lacI* encodes a lac repressor protein that represses transcription of *lacZ′* until the inducer IPTG is added to the culture medium to bind to the repressor protein and inactivate it. The host cell has a mutant *lacZ* gene that makes an inactive C-terminal fragment of β-galactosidase, which can, however, be complemented by the α-fragment produced by the vector to produce a functional β-galactosidase. The assay uses a colorless substrate [5-bromo-4-chloro-3-indolyl-β-D-galactopyranoside (X-Gal)] that is converted to a non-toxic reaction product with an intense blue color. A polylinker inserted near the start of the vector's *lacZ′* coding sequence provides a *multiple cloning site* (MCS) region, a series of different unique restriction sites that allows alternative possibilities for cloning restriction fragments. The polylinker is small and inserted so that the *lacZ′* reading frame is preserved and functional. However, cloning of a comparatively large insert into the MCS typically causes insertional inactivation and an absence of β-galactosidase activity, and so cells with recombinant DNA will be colorless.

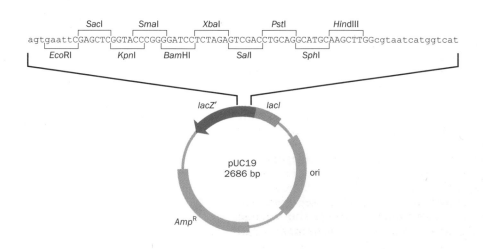

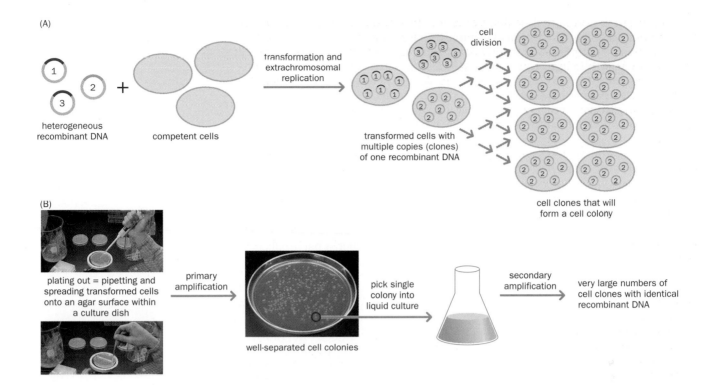

to produce blue colonies. However, recombinants with an insert in the polylinker can cause insertional inactivation so that no functional β-galactosidase is produced.

Transformation is the key DNA fractionation step in cell-based DNA cloning

The plasma membrane of cells is selectively permeable so that only certain small molecules are allowed to pass into and out of cells; large molecules such as long DNA fragments are normally unable to cross the membrane. However, cells can be treated in different ways so that the selective permeability of the plasma membranes is temporarily distorted. One way, for example, is to use **electroporation**, in which cells are exposed to a short high-voltage pulse of electricity. As a result of the temporary alteration to the cell membrane, some of the treated cells become *competent*, meaning that they are capable of taking up foreign DNA from the extracellular environment.

Only a small percentage of competent cells will take up recombinant DNA to become transformed. However, those that do will often take up just a single recombinant DNA molecule. This is the basis of the critical fractionation step in cell-based DNA cloning. The population of transformed cells can be thought of as a sorting office in which the complex mixture of DNA fragments is sorted by depositing *individual DNA molecules into individual recipient cells*. Extrachromosomal replication of the recombinant DNA often allows many identical copies of a single type of recombinant DNA in a cell (**Figure 6.5**).

Because circular DNA transforms much more efficiently than linear DNA, most of the cell transformants will contain cyclized products rather than linear recombinant DNA concatemers. Usually vector molecules are treated to reduce the chance of cyclizing and so most of the transformants will contain recombinant DNA. Note, however, that **co-transformation**, the occurrence of more than one type of introduced DNA molecule within a cell clone, can sometimes occur.

The transformed cells are allowed to multiply. In cloning using plasmid vectors and a bacterial cell host, a solution containing the transformed cells is simply spread over the surface of nutrient agar in a Petri dish and the cells are allowed to grow (a process called *plating out*—Figure 6.5). This usually results in the formation of bacterial colonies that consist of **cell clones** (all the cells in a single colony are identical because they are all descended from a single founder cell).

Figure 6.5 Transformation, the critical DNA fractionation step, and large-scale amplification of recombinant DNA.
(A) Cell-based DNA cloning typically starts with a complex population of target DNA fragments that are joined to a single type of vector molecule, producing a heterogeneous collection of recombinant DNA molecules. During transformation, the recombinant DNA molecules are transferred into competent cells, so that a cell typically takes up just one type of recombinant DNA molecule. As a result, the cell population acts as a fractionation system, allocating individual recombinant DNAs to individual cells. Within each cell, a recombinant DNA can replicate extrachromosomally, often producing many identical DNA clones. Thereafter each transformed cell undergoes repeated divisions to generate populations of identical cell clones that will form cell colonies. (B) The cell colonies can be physically separated by *plating out*, a procedure that involves growth of transformed cells on a nutrient agar surface so as to allow separation of cell colonies. Individual colonies are picked and added to liquid culture in flasks to allow secondary amplification. (Courtesy of James Stock, King's College. Used under the Creative Commons Attribution ShareAlike 3.0 license.)

Plating out is normally designed to produce well-separated colonies on the agar surface, allowing the physical separation of colonies containing different types of recombinant DNA. Picking an individual colony into a culture flask for subsequent growth in liquid culture permits a secondary expansion of cell numbers that can result in very large yields of cell clones, all identical to an ancestral single cell (see Figure 6.5). If the original cell contained a single type of foreign DNA fragment attached to a replicon, then so will the descendants, resulting in a huge amplification in the amount of the specific foreign fragment.

Recombinant DNA can be selectively purified after screening and amplifying cell clones with desired target DNA fragments

Some marker gene systems provide a general system for screening for recombinants, such as insertional inactivation in the case of the *lacZ* α-complementation system described above. To screen for an individual recombinant that has a specific target DNA of interest, it is usual to prepare a labeled probe that can specifically detect the target DNA in a hybridization reaction against bacterial colonies that have been separated on a membrane (this is known as *colony hybridization* and will be described in Chapter 7).

Desired bacterial colonies are picked and allowed to grow further in liquid culture, resulting in large numbers of cells and recombinant DNA. To recover the recombinant DNA, the cells are broken open (lysed). The host chromosome of bacterial cells is circular, like any plasmids containing introduced foreign DNA, but is very much larger than a plasmid. Although the smaller plasmid recombinant DNA remains circular, the large chromosomal DNA often undergoes breakage during cell lysis and subsequent DNA extraction, resulting in linear DNA fragments with free ends. The purification methods typically exploit the physical difference in lysed cells between chromosomal DNA (containing free DNA ends) and recombinant DNA (no free DNA ends). To do this, the DNA is first subjected to a denaturation step, for example by treatment with alkali, followed by a renaturation step. The linearized chromosomal DNA readily denatures and precipitates out of solution. However, the plasmid recombinant DNA is in the form of covalently closed circular DNA, and after the renaturation step it spontaneously forms supercoiled DNA and remains in solution.

As required, further purification is usually performed with anion-exchange chromatography. The widely used Qiagen anion-exchange resins consist of silica beads of defined size that have a hydrophilic surface coating, allowing dense coupling of positively charged diethylaminoethyl (DEAE) groups. Because of the very large number of negatively charged phosphate groups of the DNA backbone that are attracted to bind to the DEAE groups on the silica resin, DNA binds tightly to the resin, whereas impurities bind weakly or not at all. The tightly bound DNA can subsequently be eluted with high-salt buffers to disrupt the phosphate-DEAE bonding. The insert target DNA can be excised from a plasmid recombinant DNA and subcloned into other vectors as required that may offer the possibility to sequence the DNA or express any genes within the target, for example. We will consider some of these applications in the next sections.

6.2 LARGE INSERT CLONING AND CLONING SYSTEMS FOR PRODUCING SINGLE-STRANDED DNA

The first successful attempts at cloning human DNA fragments in bacterial cells took advantage of gene transcripts that were naturally highly enriched in some tissues. For example, much of the mRNA made in erythrocytes consists of α- and β-globin mRNA. When reverse transcriptase is used to copy erythrocyte mRNA, the resulting erythrocyte complementary DNA (cDNA) is greatly enriched in globin cDNA, facilitating its isolation.

To enable a more general method of cloning genes, and indeed all DNA sequences, technologies were subsequently developed with the aim of cloning all of the constituent DNA sequences in a starting population. The resulting comprehensive collections of DNA clones became known as **DNA libraries**. Genomic DNA libraries are prepared from total genomic DNA of any nucleated cells of an organism; cDNA libraries are prepared from cDNA copies of total RNA or mRNA from specific tissues or cells of an organism. DNA libraries have been vitally

TABLE 6.2 PROPERTIES OF CELL-BASED DNA CLONING SYSTEMS ARRANGED ACCORDING TO INSERT CAPACITY

Cloning system based on		Insert size	Copy number (per cell)	Applications
PLASMID VECTORS				
Plasmid vectors allowing replication to high-copy-numbers	Standard plasmid and phagemid vectors	Usually up to 5–10 kb	often high	Very versatile. Routinely used in subcloning of fragments amplified by PCR. Also widely used in making cDNA libraries and for expressing genes to give RNA or protein products. Phagemids can produce single-stranded recombinant DNA
	Cosmid vectors	30–44 kb	high	Used for making genomic DNA libraries with modest-sized inserts. However, high copy number makes inserts comparatively unstable
Plasmid vectors whose copy number is constrained	Fosmid vectors	30–44 kb	1–2	An alternative to cosmid cloning, providing much greater insert stability at the expense of lower yields of recombinant DNA
	PAC (P1 artificial chromosome) vectors	130–150 kb	1–2	Used for making genomic DNA libraries, often as a supplementary aid to BAC cloning
	BAC (bacterial artificial chromosome) vectors	up to 300 kb	1–2	Used extensively to make genomic libraries for use in genome projects
	YAC (yeast artificial chromosome) vectors	0.2–2.0 Mb	low	Used as an intermediate cloning system for some genome projects (e.g. Human Genome Project) but inserts are often unstable. Also useful in studying function of large genes or gene clusters
PHAGE VECTORS				
M13		small	low to medium	Useful in the past for making single-stranded recombinant DNA for sequencing templates and site-directed mutagenesis. Superseded by phagemids
λ insertion vectors e.g. λgt11		up to 10 kb	low	Useful for making cDNA libraries because they can accept quite large inserts and because large numbers of recombinants can be made
λ replacement vectors		9–23 kb[a]	low	Used for making genomic DNA libraries in the past; rarely used now
P1		70–100 kb	low	Used for making genomic libraries in the past; rarely used now

[a]Insert size is increased compared to insertion vectors by removing a central non-essential part of the λ genome and replacing it by the DNA to be cloned. Note that plasmid vectors are generally preferred; phage vectors are comparatively difficult to work with.

important tools in the Genome Projects and in working out gene structure and the relationship between genes and their transcripts. We will consider them in detail in Chapter 8.

Vertebrate genes are often very large with a high content of repetitive DNA, making them less amenable to cloning. Repetitive DNA is extremely rare in bacterial genomes, and large amounts of introduced recombinant DNA that is rich in repetitive DNA sequences are not well tolerated. As a result, standard plasmids can rarely be used to clone large (> 10 kb) fragments of animal DNA; instead, the inserts are usually less than 5 kb. Plasmids continue to be the workhorses of DNA cloning for many research purposes, however. Vectors such as pUC19 (see Figure 6.4) permit high-copy-number replication and large amounts of recombinant DNA, and, as we will see in Section 6.3, plasmid vectors are widely used to express genes of interest. To clone larger DNA fragments, different DNA cloning systems were needed (see **Table 6.2** for an overview). In addition, some applications for studying recombinant DNA, notably DNA sequencing and *in vitro* site-specific mutagenesis, required single-stranded DNA templates. As a result, cloning vectors were also developed that could be used to generate single-stranded recombinant DNA.

Early large insert cloning vectors exploited properties of bacteriophage λ

Bacteriophage λ (lambda) has a linear 48.5 kb double-stranded DNA genome. Within an *E. coli* host, λ DNA can replicate extrachromosomally or it can integrate into the host chromosome and be replicated as a chromosome component.

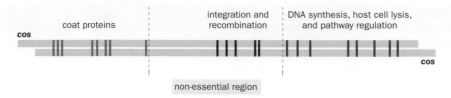

The central portion of the λ genome is a non-essential region: the genes it contains are required only for integrating λ DNA into the host chromosome. DNA cloning is therefore possible by removing a large central section of the λ genome DNA and replacing it by up to 23 kb of target DNA (**Figure 6.6**).

A further refinement that improved the upper size of fragments that could be cloned exploited the properties of the terminal *cos* sequences of λ. The *cos* sequences terminate in 5′ overhanging ends that are 12 nucleotides long and complementary in sequence so that they are cohesive ends; hence the name *cos* (see Figure 6.6). The λ genome needs to be linear to fit into its protein coat, but after the phage has attached to the membrane of a new host cell and inserted its linear DNA, the inserted λ DNA immediately cyclizes by base pairing between the *cos* sequences. Extrachromosomal replication of λ DNA occurs by a rolling-circle model that generates linear multimers of the unit length. The λ multimers are cleaved within the *cos* sequence to generate unit length linear genomes for packaging into viral protein coats.

For proper packaging of λ DNA within the phage protein coat the spacing between the *cos* sequences is constrained to be 35–53 kb. By inserting the *cos* sequence into a plasmid, a new type of cloning vector, the cosmid, was produced that, when used with an *in vitro* λ packaging system, could accept foreign fragments up to 40 kb in length. However, cosmid recombinants that contain vertebrate DNA 40 kb long are unstable. The inserts typically have repetitive sequences and when the recombinant DNA is present in a bacterial cell at high copy number, recombination makes the inserts prone to rearrangement and deletion.

To achieve cloning of larger inserts, cloning systems needed to use low-copy-number replicons. One successful approach uses bacterial hosts but with extrachromosomal replicons that are constrained in copy number. Another solution involves using yeast cells and seeks to construct artificial chromosomes to replicate target DNA. In this case, the vectors include sequences essential for chromosomal function, including an origin of replication that originates from a chromosome, resulting in low copy numbers. To generate very large target DNA fragments for cloning, DNA needs to be isolated from cells under gentle conditions and then subjected to *partial digestion* with a restriction endonuclease, so that only a small minority of the available restriction sites are actually cut.

Large DNA fragments can be cloned in bacterial cells by using extrachromosomal low-copy-number replicons

Many vectors used for DNA cloning in bacterial cells are based on extrachromosomal replicons that permit replication to high or medium copy numbers in host cells. The DNA of complex eukaryotes, notably mammals, contains numerous repetitive sequences that can lead to instability when propagated at high copy number in bacterial cells. Clones containing large eukaryotic DNA inserts with multiple repetitive sequences are particularly unstable, and propagating them at high copy numbers results in deletion of the insert or rearrangement of the cloned DNA. To overcome such a size limitation, plasmid and phage vectors were developed that used low-copy-number replicons.

Bacterial artificial chromosome (BAC) and fosmid vectors

The F factor, an *E. coli* fertility plasmid, contains two genes, *parA* and *parB*, that maintain the copy number at one or two per *E. coli* cell. Plasmid vectors based on the F-factor system are able to accept large foreign DNA fragments (up to 300 kb), and the resulting recombinants can be transferred with considerable efficiency into bacterial cells by using **electroporation**. The resulting plasmid recombinants are very large, and although not quite so large as a bacterial chromosome (which is also circular, double-stranded, and mostly protein-free) they became known as **bacterial artificial chromosomes** (**BACs**). BACs contain a low-copy-number

Figure 6.6 A map of the phage λ genome. Genes are shown as vertical bars. The central region of the 48.5 kb λ genome is non-essential for extrachromosomal replication, viral coat production, and lysis of the host cell—it contains genes that are required only when λ integrates into the chromosome of its host cell. When using λ replacement vectors, cloning occurs by replacing this central non-essential region by target restriction fragments that can be 9–23 kb long. The terminal *cos* sequences have 5′ overhanging ends that are 12 nucleotides long and complementary in sequence, enabling cyclization. The λ genome needs to be linear to fit into its protein coat, but after the phage has attached to the membrane of a new host cell and inserted its linear DNA, the inserted λ DNA immediately cyclizes by base pairing between the overhanging 5′ *cos* sequences, then replicates eventually by a rolling-circle model that generates linear multimers of the unit 48.5 kb length. Coat proteins are synthesized and the λ multimers are then snipped asymmetrically within the *cos* sequence to generate multiple copies of the unit genome that are then individually packed within protein coats.

replicon, and so only low yields of recombinant DNA can be recovered from the host cells. However, because of their great insert stability they were the most widely used clones that were selected for sequencing in the Human Genome Project.

Cosmid-like vectors that have been adapted for low-copy-number replication by using F-factor genes are known as fosmid vectors. The inserts have the same size range as in cosmid cloning but are much more stable.

Bacteriophage P1 vectors and P1 artificial chromosomes

An alternative approach to cloning large inserts in *E. coli* was to develop vectors based on bacteriophages that have naturally large genomes. For example, bacteriophage P1 has a 110–115 kb linear double-stranded DNA genome that, like phage λ, is packaged into a protein coat. Bacteriophage P1 cloning vectors have therefore been designed in which components of P1 are included in a circular plasmid. Subsequently, features of the P1 and F-factor systems were combined in another plasmid-based cloning system in which recombinants are known as a **P1 artificial chromosomes** (**PACs**). As in BACs, the inserts of PACs are very stable and they have also been used as direct templates for sequencing in some genome projects.

Yeast artificial chromosomes (YACs) enable cloning of megabase fragments of DNA

The most popular method for cloning extremely large DNA fragments is based on making artificial chromosomes that can be propagated in the budding yeast. The chromosomes of *Saccharomyces cerevisiae* are linear and vary in size from 200 kb to 1.5 Mb. As well as carrying large numbers of genes they also contain short sequence elements that are essential for their basic chromosomal functions, to maintain the DNA molecule, replicate it, and ensure faithful segregation into descendant cells after cell division. Only three types of short sequence, each at most only a few hundred base pairs, are needed for a DNA molecule to behave as a chromosome in *S. cerevisiae* (see Figure 2.10). *Centromeres* are required for the disjunction of sister chromatids in mitosis and also the disjunction of homologous chromosomes at the first meiotic division. *Telomeres* are needed for the complete replication of linear molecules and for protection of the ends of the chromosome from attack by nucleases. *Autonomous replicating sequence* (*ARS*) elements are required for autonomous replication of the chromosomal DNA and are thought to act as specific replication origins.

To make a **yeast artificial chromosome** (**YAC**), all that is needed is to combine a suitably sized foreign DNA fragment with four short sequences that can function in yeast cells (two telomeres, one centromere, and one ARS element). The resulting linear DNA molecule should have the telomere sequences correctly positioned at the termini (**Figure 6.7**). YACs cannot be transfected directly into

Figure 6.7 Making yeast artificial chromosomes (YACs). A gene conferring ampicillin resistance (*Amp*^R) and a plasmid-derived origin of replication (ori) allow the YAC vector to be replicated to a high copy number in an *E. coli* host. In addition, the vector contains the three types of element required for DNA to behave as a chromosome in yeast cells: a centromere (*CEN4*), an autonomous replicating sequence (*ARS1*), and two telomeres (TEL). The dominant alleles *TRP1* and *URA3* are included as marker genes: they complement recessive alleles (*trp1* and *ura3*) in the yeast host cell, providing a selection system for identifying transformed cells containing the YAC vector. The cloning site for foreign DNA is located within the *SUP4* gene. This gene compensates for a mutation in the yeast host cell that causes the accumulation of red pigment. The host cells are normally red, and those transformed with just the vector will form colorless colonies. Cloning of a foreign DNA fragment into the *SUP4* gene causes insertional inactivation, restoring the mutant (red) phenotype. Target DNA is partly digested with the restriction endonuclease *Eco*RI to give very large fragments (hundreds of kilobases) and mixed with linearized vector molecules cut with the restriction endonucleases *Bam*HI and *Eco*RI. Ligation occurs at the *Eco*RI ends, with vector sequences flanking the target DNA fragments. The recombinant DNA can then be used to transform specialized yeast host cells, and the selection systems described above can be used to identify transformed cells and those containing recombinant DNA.

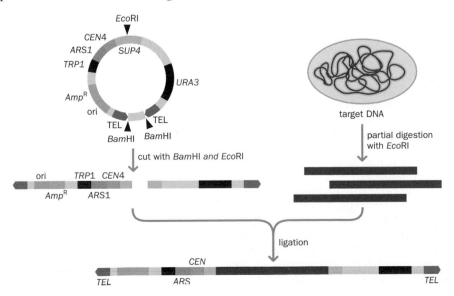

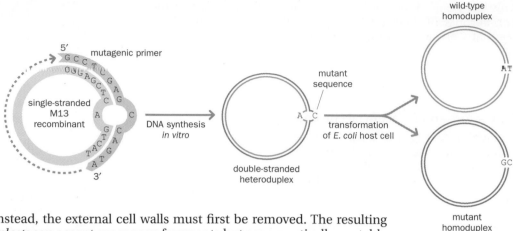

yeast cells; instead, the external cell walls must first be removed. The resulting yeast *spheroplasts* can accept exogenous fragments but are osmotically unstable and need to be embedded in agar. The overall transformation efficiency is very low, and the yield of cloned DNA is low (about one copy per cell). Nevertheless, the capacity to clone large DNA fragments (up to 2 Mb) made YACs a vital tool in the physical mapping of some complex genomes, notably the human genome.

Producing single-stranded DNA for use in DNA sequencing and *in vitro* site-specific mutagenesis

Single-stranded DNA clones have been preferred templates for DNA sequencing: the sequences obtained are clearer and easier to read than when double-stranded DNA is used. We will consider aspects of DNA sequencing in Chapter 8.

Single-stranded DNA templates are also ideal substrates for site-specific **in vitro** mutagenesis, an invaluable experimental tool in dissecting the functional contributions made by short sequences of nucleotides or amino acids. For example, the suspected contribution of a specific amino acid to the function of its protein can be tested by deleting the amino acid or mutating it to give a different amino acid with a rather different side chain. To do this, a single-stranded cloned DNA with the wild-type sequence is allowed to base-pair with a synthetic oligonucleotide that is designed to be perfectly complementary in base sequence except for some changes that will correspond to the desired mutation. The oligonucleotide is allowed to prime new DNA synthesis to create a complementary full-length sequence containing the desired mutation. The newly formed wild-type–mutant *heteroduplex* is used to transform cells, and the desired mutant sequence can be identified by screening for the mutation (**Figure 6.8**).

Single-stranded recombinant DNA is often obtained by using cloning systems based on filamentous bacteriophages such as phages M13, fd, and f1, which naturally adopt a single-stranded DNA form at some stage in their life cycle. Alternatively, plasmid vectors are used that contain filamentous phage DNA sequences that regulate the production of single-stranded DNA.

M13 vectors

Bacteriophage M13 has a 6.4 kb circular single-stranded DNA genome enclosed in a protein coat, forming a long filamentous structure. It infects certain strains of *E. coli*. After the M13 phage binds to the wall of a bacterial cell, it injects its DNA genome into the cell. Shortly after the single-stranded M13 DNA enters the cell it is converted to a double-stranded DNA, the replicative form that serves as a template for making numerous copies of the genome. After a short time, a phage-encoded product switches DNA synthesis toward the production of single strands that migrate to the cell membrane. Here they are enclosed in a protein coat, and hundreds of mature phage particles are extruded from the infected cell without cell lysis.

M13 vectors are based on the double-stranded replicative form, which has been modified to contain a multiple cloning site and a *lacZ* selection system for screening for recombinants. Double-stranded M13 recombinants can be transfected into a suitable *E. coli* strain; after a certain period, phage particles are harvested and used to retrieve single-stranded recombinant DNA (**Figure 6.9**).

Figure 6.8 Oligonucleotide mismatch mutagenesis. Many different methods of oligonucleotide mismatch mutagenesis can be used to create a desired point mutation at a unique predetermined site within a cloned DNA molecule. The example illustrates the use of a mutagenic oligonucleotide to direct a single nucleotide substitution in a gene. The gene is cloned into a suitable vector to generate a single-stranded recombinant DNA, such as an M13 vector (as shown here) or a phagemid vector. An oligonucleotide primer is designed to be complementary in sequence to a portion of the gene sequence encompassing the nucleotide to be mutated (in this case an adenine, A) and containing the desired non-complementary base at that position (C, not T). Despite the internal mismatch, annealing of the mutagenic primer is possible, and second-strand synthesis can be extended by DNA polymerase and the gap sealed by DNA ligase. The resulting heteroduplex can be transformed into *E. coli*, whereupon two populations of recombinants can be recovered: wild-type and mutant homoduplexes. The latter can be identified by PCR-based allele-specific amplification methods (see Figure 6.17).

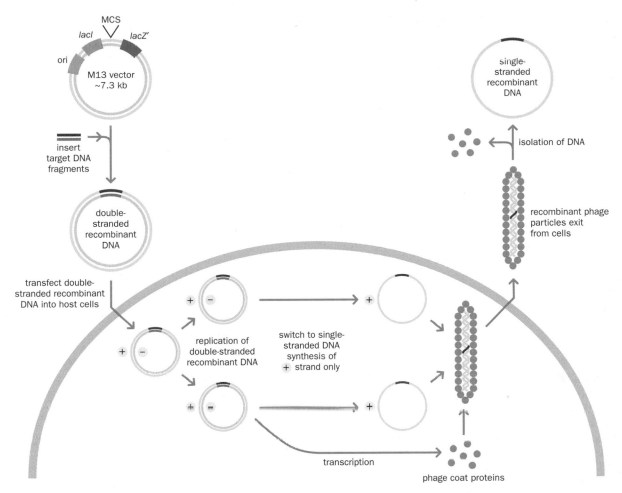

Figure 6.9 Producing single-stranded recombinant DNA. M13 vectors use the same *lacZ* assay for recombinant screening as pUC19 (see Figure 6.4). Target DNA is inserted in the multiple cloning site (MCS). The double-stranded M13 recombinant DNA enters the normal cycle of phage replication to generate numerous copies of the genome, before switching to production of single-stranded DNA (+ strand only). After packaging with M13 phage coat protein, mature recombinant phage particles exit from the cell without lysis. Single-stranded recombinant DNA is recovered by precipitating the extruded phage particles and chemically removing the protein.

Phagemid vectors

A small segment of a single-stranded filamentous bacteriophage genome can be inserted into a plasmid to form a hybrid vector known as a **phagemid**; **Figure 6.10** shows the pBluescript vector as an example. The chosen segment of the phage genome contains all the *cis*-acting elements required for DNA replication and assembly into phage particles. They permit successful cloning of inserts several kilobases long, unlike M13 vectors, in which inserts of this size tend to be unstable. After transformation of a suitable *E. coli* strain with a recombinant phagemid, the bacterial cells are *superinfected* with a filamentous *helper phage*, such as phage f1, which is required for providing the coat protein. Phage particles secreted from the superinfected cells will be a mixture of helper phage and recombinant phagemids. The mixed single-stranded DNA population can be used directly for DNA sequencing because the primer for initiating DNA strand synthesis is designed to bind specifically to a sequence of the phagemid vector adjacent to the cloning site.

Figure 6.10 pBluescript, a phagemid vector. The pBluescript series of phagemid vectors each contain two origins of replication, a standard plasmid one (ori) and a second one from the filamentous phage f1. The f1 origin of replication allows the production of single-stranded DNA, but the vector lacks the genes for phage coat proteins. Target DNA is inserted into the phagemid vector, which is then used to transform host *E. coli* cells. Superinfection of transformed cells with a helper M13 phage allows the recombinant phagemid DNA to be packaged within phage protein coats, and the recombinant single-stranded DNA can be recovered as in Figure 6.9.

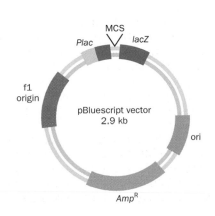

6.3 CLONING SYSTEMS DESIGNED FOR GENE EXPRESSION

The cloning systems described so far in this chapter are designed simply to amplify the target DNA to obtain sufficient quantities for structural and functional studies. However, in many circumstances it is also useful to be able to express the introduced gene in some way. In **expression cloning**, appropriate signals need to be provided alongside the introduced gene to enable the gene to be expressed in the transformed host cell.

Expression cloning is usually performed with cell-based cloning systems. Depending on the type of expression product required, and the purpose of the expression, many different cloning systems can be used. Sometimes an RNA product is sufficient, but in many cases the object is to produce a protein. In some cases, all that is required is to analyze gene expression, for which low expression levels are acceptable; in others, high expression is needed to retrieve, for example, large quantities of a specific protein. It is common to express the product in well-studied cells, but sometimes it may be sufficient to express the product *in vitro*.

Large amounts of protein can be produced by expression cloning in bacterial cells

Cloning of eukaryotic cDNA in an expression vector is often required for the production of proteins in large quantities for research purposes such as structural studies, or for biotechnology purposes, or as medically relevant compounds such as therapeutic proteins. Usually, a cDNA providing the genetic information specifying the protein sequences is inserted into a vector along with separate expression signals such as suitably strong promoters and other regulatory elements. Because the expression system is based on recombinant DNA, the resulting proteins are sometimes described, rather inaccurately, as **recombinant proteins**.

Bacterial cells have the advantage that they grow rapidly and can be expanded easily in culture to very large culture volumes. *Escherichia coli* has been a favorite host cell for the expression of introduced proteins.

The production of very large amounts of a non-endogenous protein can, however, be detrimental to the growth of a host cell, and can sometimes be toxic. Thus, it is often advantageous to use an *inducible promoter* so that expression can be delayed until the transformed cells have been identified and grown in bulk. For example, the pET series of vectors contain a bacteriophage T7 promoter that is not recognized by the endogenous *E. coli* RNA polymerase promoter (**Figure 6.11**). Such vectors are used to transform a genetically modified *E. coli* strain that contains a T7 RNA polymerase regulated by a *lac* promoter. The transformed cells can be selected and grown up in large quantities without expression of the foreign gene. Addition of the β-galactosidase inducer IPTG activates the *lac* promoter and expression of the adjacent foreign gene, and the cells can be harvested shortly afterward.

Bacteria have many advantages for expressing heterologous (foreign) proteins, but they also have limitations. Many eukaryotic proteins are modified by the addition of phosphate, lipid, or sugar groups after translation. These modifications are often essential for the biological function of the protein. Bacterial cells lack the enzymes needed for this post-translational processing and so certain eukaryotic proteins produced in bacterial cells become unstable, or show limited or no biological activity. Many eukaryotic proteins, notably some mammalian proteins, are also very much larger than bacterial proteins and cannot

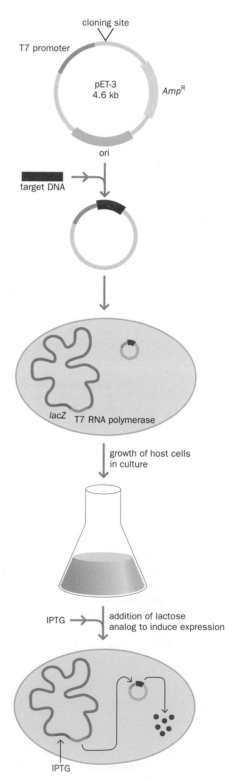

Figure 6.11 Inducible bacterial expression vectors. The pET-3 series of plasmid vectors contain a bacteriophage T7 promoter. They are used with a strain of *E. coli* that has been genetically modified to contain a gene for the phage T7 RNA polymerase under control of an inducible *lac* promoter. After transformation by a pET-3 recombinant DNA, the *E. coli* cells are allowed to grow to give large numbers of cells in culture. At a desired stage, the lactose analog isopropyl β-D-thiogalactoside (IPTG) is added to induce expression of the host's T7 RNA polymerase gene. The induced T7 RNA polymerase binds specifically to the T7 promoter on the recombinant DNA to give high-level expression of the insert. *Amp*^R, ampicillin resistance gene; ori, origin of replication.

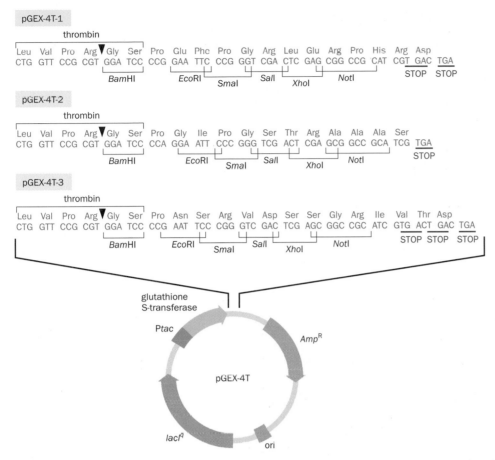

Figure 6.12 Fusion protein vectors.
The pGEX-4T series of vectors have a *tac* promoter (or P*tac*), a hybrid promoter with elements of the *trp* and *lac* promoters. Downstream gene expression is normally repressed by the repressor protein encoded by the *lacI*q gene but is inducible by the lactose analog IPTG. Immediately downstream of the *tac* promoter is a gene for the affinity tag glutathione S-transferase (GST) followed by a multiple cloning site (MCS). The object is to clone a target coding sequence for a protein into the MCS so that a fusion protein is produced, with an N-terminal GST sequence fused to the protein encoded by the target cDNA. Three alternative vectors, pGEX-4T-1, pGEX-4T-2, and pGEX-4T-3, have slightly different MCS sequences that differ in the translational reading frame. By using all three alternative vectors, cloned inserts can be expressed in each of the three amino acid reading frames so that in at least one of the three cases, the target DNA should be in frame with the GST sequence. The expressed fusion protein can be purified easily on a glutathione affinity purification column such as glutathione Sepharose 4B. Because the MCS is engineered to contain a thrombin cleavage site, the desired protein can be purified after cleavage at the thrombin cleavage site. *Amp*R, ampicillin resistance gene; ori, origin of replication; STOP, termination codon.

easily be synthesized in *E. coli*. Overexpression leads to the production of insoluble aggregates of misfolded protein (*inclusion bodies*). The inclusion bodies can easily be purified, but it is then difficult to solubilize protein and achieve efficient refolding *in vitro*.

Efforts to increase yield and solubility have often involved the production of **fusion proteins**. For example, the vector can be modified so that immediately adjacent to the cloning site it contains a cDNA sequence for all or part of an endogenous protein. Recombinants will therefore express the desired protein fused to an endogenous protein sequence. Many modern protein expression vectors are modified to contain a coding DNA sequence for a specific peptide or protein that is easily purified by *affinity chromatography*. In such cases the recombinants are expressed to give the desired protein but with a short peptide or protein tag attached that is known as an **affinity tag** because it is attached to assist purification of the recombinant protein by affinity chromatography.

Two favorite systems that allow affinity purification of expressed proteins are based on GST–glutathione affinity and polyhistidine–nickel ion affinity. Glutathione S-transferase (GST) is a small protein with a very high affinity for its substrate glutathione. The expression cloning vector positions the target DNA just after a gene encoding GST, so that a GST-fusion protein is produced in the transformed host cell (**Figure 6.12**). This fusion protein can be purified by selective binding to a column containing glutathione. Alternatively, an affinity tag of six consecutive histidine residues can be attached to a protein. The side chains of the His$_6$ tag bind selectively and strongly to nickel ions, assisting purification by affinity chromatography with a nickel–nitrilotriacetic acid matrix.

In phage display, heterologous proteins are expressed on the surface of phage particles

Phage display is a form of expression cloning in which the cloned gene is expressed to give a protein that is displayed on the surface of the phage particle. To do this, cloning sites in the phage vector are located within a phage gene that encodes a suitable coat protein. The modified gene can then be expressed as a

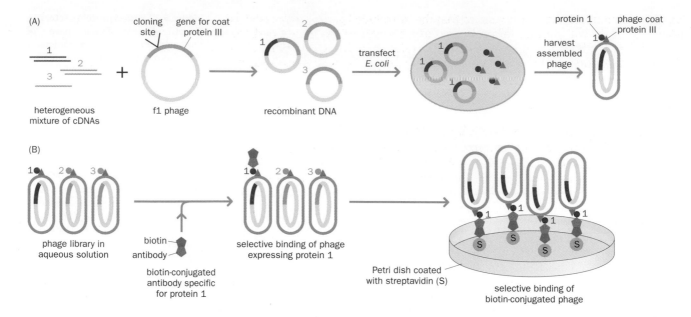

fusion protein that is incorporated into the phage's protein coat, so that it is displayed on the surface of the phage without affecting the phage's ability to infect cells. If an antibody is available for the expressed protein, phage displaying that protein can be selected by preferential binding to the antibody: affinity purification of virus particles bearing such a protein can be achieved from a 10^8-fold excess of phage not displaying the protein, using even minute quantities of the relevant antibody (**Figure 6.13**).

In protein engineering, phage display is a powerful adjunct to random mutagenesis programs as a way of selecting for desired variants from a library of mutants. It has also proved a powerful alternative source of constructing antibodies, bypassing normal immunization techniques and even hybridoma technology. Phage libraries can also be used to identify proteins that interact with a specific protein. In the same way in which antibodies can be used in affinity screening, a known protein (or any other molecule to which a protein can bind) can be used as a bait to select phages that display any other proteins that bind to the bait protein.

Eukaryotic gene expression is performed with greater fidelity in eukaryotic cell lines

Bacterial expression systems offer the great advantage of protein expression at very high levels. However, the biological properties of many eukaryotic proteins synthesized in bacteria may not be representative of the native molecules. When produced in bacterial cells, they do not undergo their normal post-translational processing, and the folding of the eukaryotic protein may be incorrect or inefficient. As an alternative, eukaryotic cells, including insect and mammalian cells, are widely used for expressing recombinant proteins. The expression vectors are typically designed to be able to be replicated both in the desired eukaryotic cells and in *E. coli*, where the intention is simply to produce large amounts of recombinant DNA that can subsequently be used for expression studies in eukaryotic cells.

Once transfected into animal cells, the recombinant DNA can be expressed for short or long periods, according to the expression vector used. For some purposes, all that is required is *transient expression*. Here, the recombinant DNA remains as an independent extrachromosomal genetic element (an *episome*) within the transfected cells. Expression of the introduced gene reaches a maximal level about two or three days after transfection of the expression vector into the mammalian cell line; thereafter, expression can diminish rapidly as a result of cell death or loss of the expression construct.

The alternative is to use *stable expression* systems in which the recombinant DNA can integrate into a host cell's chromosome. The advantage here is that if

Figure 6.13 Phage display. (A) Library construction. A heterogeneous mixture of target cDNAs is cloned into a phage vector, such as one based on phage M13 or f1, to express foreign proteins on the phage surface. In this example, DNA is inserted into a cloning site located within the initial part of the coding sequence for the gene from phage f1 that makes one of the phage coat proteins (protein III). The idea is that recombinants should express a fusion protein with the target protein fused to this phage coat protein, thereby enabling the phage particles to display the foreign protein on their surfaces. After transformation of host *E. coli* cells, the phage DNA replicates, and phage particles are assembled, extruded from the host cell, and harvested. The mixture of recombinant phages is known as a phage expression library. (B) Library screening with a specific antibody. An antibody specific for just one type of protein will bind only to phage particles that display that protein. If the antibody is complexed with biotin, phage particles carrying the desired protein can be purified by using streptavidin-coated Petri dishes: the desired recombinants will have an antibody complexed to a biotin group and the biotin group will bind strongly to streptavidin.

the recombinant DNA has integrated so that the introduced gene is expressed, all descendant cells will contain this gene and expression can be maintained over many cell generations. Some viruses are highly efficient at integrating into chromosomal DNA, and we will consider them in the context of gene therapy in a later chapter. But plasmid expression vectors can also randomly integrate at low frequencies, and stable cell lines can be developed with an integrated gene of interest.

Transient expression in insect cells by using baculovirus

Baculovirus gene expression is a popular method for producing large quantities of recombinant proteins in insect host cells: the protein yields are higher than in mammalian expression systems, and the costs are lower. In most cases, post-translational processing of eukaryotic proteins expressed in insect cells is similar to the protein processing that occurs in mammalian cells, and the expression of very large proteins is possible. The proteins produced in insect cells have comparable biological activities and immunological reactivities to those of proteins expressed in mammalian cells.

Autographa californica nuclear polyhedrosis virus (AcMNPV) is a baculovirus that can be propagated in certain insect cell lines and is often used as a cloning vector for protein expression systems. The viral polyhedrin protein is transcribed at high rates and, although essential for viral propagation in its normal habitat, it is not needed in culture. As a result, its coding sequence can be replaced by a coding DNA for a eukaryotic protein. The cloning vector is designed to express the inserted protein from the powerful polyhedrin promoter, resulting in expression levels accounting for more than 30% of the total cell protein.

Transient expression in mammalian cells

Expressing mammalian proteins in mammalian cells has the obvious advantage for expressing human or other mammalian proteins that correct protein folding and post-translational modification are generally not an issue, and it is possible to analyze downstream signals and cellular effects. Stable expression systems in mammalian cells, typically based on plasmid sequences that integrate within a chromosome, have delivered kilograms of complex proteins in industrial-scale bioreactors, but they have required large investments in time, resources, and equipment. As an alternative, large-scale transient expression systems have been developed for producing recombinant proteins in mammalian cells.

In addition to delivering protein expression, some mammalian cell lines have found major applications in screening the effects of *in vitro* manipulations on transcriptional and post-transcriptional control sequences. A good example is provided by COS cells, stable cell lines originally derived from an African green monkey kidney fibroblast cell line, CV-1. When CV-1 cells are infected with the monkey virus SV40, the normal SV40 lytic cycle ensues. However, when CV-1 cells were transformed with a strain of SV40 with a defective replication origin, a segment of the SV40 genome became integrated into the chromosomes of CV-1. The resulting COS cells (**C**V-1 with defective **o**rigin of **S**V40) stably express the SV40-encoded large T antigen, the only viral protein that is required for activation of the SV40 origin of replication.

By expressing the large SV40 T antigen in a stable way, COS cells permit any introduced circular DNA with a functional SV40 origin of replication to replicate independently of the host cell's chromosomes, with no clear size limitation. When transient expression vectors are transfected into COS cells, permanent cell lines do not result because massive vector replication makes the cells nonviable. Even though only a small proportion of cells are successfully transformed, the amplification of the introduced DNA to high copy numbers in those cells compensates for the low take-up rate.

Stable expression in mammalian cells

Genes introduced into mammalian cells can stably integrate into host chromosomal DNA, but the process is very inefficient. The rare stably transformed cells must be isolated from the background of nontransformed cells by selection for some marker. Two broad approaches have been used: functional complementation of mutant host cells and dominant selectable markers.

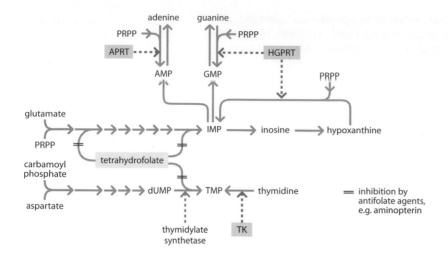

Figure 6.14 Purine and pyrimidine salvage and the basis of HAT selection in thymidine kinase marker assays. Cultured animal cells can normally synthesize purine nucleotides and thymidylate by *de novo* pathways that involve certain amino acid substrates. *De novo* synthesis of purine nucleotides involves the eventual synthesis of IMP (inosine monophosphate), which can be converted separately to AMP and GMP. *De novo* synthesis of pyrimidine nucleotides is dependent on synthesizing UMP and dUMP, which is converted to TMP (thymidylate) by thymidylate synthetase; UTP is deaminated to give CTP (not shown). An activated form of tetrahydrofolate is crucially required to provide a methyl or formyl group at three key steps that can be blocked by antifolate agents such as aminopterin, shutting down the *de novo* pathways. However, purine nucleotides and thymidylate can also be synthesized from preformed bases by salvage pathways that are not blocked by antifolate drugs such as aminopterin. The three salvage pathway enzymes are shown in pale pink boxes: thymidine kinase (TK), hypoxanthine–guanine phosphoribosyl transferase (HGPRT), and adenine phosphoribosyltransferase (APRT). The use of HAT medium (containing hypoxanthine, aminopterin, and thymidine) therefore selects for cells with active salvage pathway enzymes. Mutant Tk⁻ cells are genetically deficient in thymidine kinase and will not survive if grown in HAT medium unless they contain vector molecules containing a functional thymidine kinase gene. PRPP, 5-phosphoribosyl-1-pyrophosphate.

In functional complementation, host cells have a genetic deficiency that prevents a particular function; the original function can be restored if a functional copy of the defective gene is supplied by transformation by a vector. The transgene and the marker can be transferred as separate molecules by a process known as **co-transformation**. An example is the use of a *Tk* gene marker with cells that are genetically deficient in thymidine kinase (Tk⁻). Thymidine kinase is required to convert thymidine into thymidine monophosphate (TMP), but TMP can also be synthesized by enzymatic conversion from dUMP. The drug aminopterin blocks the dUMP→TMP reaction, so cells cannot grow in the presence of this drug unless they have a source of thymidine and a functional *Tk* gene. Selection for Tk⁺ cells is usually achieved in HAT (**h**ypoxanthine, **a**minopterin, **t**hymidine) medium (**Figure 6.14**).

The major disadvantage of complementation markers is that they are specific for a particular mutant host cell line in which the corresponding gene is nonfunctional. As a result, they have largely been superseded by **dominant selectable markers**, which confer a phenotype that is entirely novel to the cell and hence can be used in any cell type. Markers of this type are usually drug-resistance genes of bacterial origin that can confer resistance to drugs known to affect eukaryotic and bacterial cells. For example, the aminoglycoside antibiotics (which include neomycin and G418) are inhibitors of protein synthesis in both bacterial and eukaryotic cells. The neomycin phosphotransferase (*neo*) gene confers resistance to aminoglycoside antibiotics, so cells that have been transformed by the *neo* gene can be selected by growth in G418. **Figure 6.15** shows a typical mammalian expression vector that contains a *neo* marker, along with several other features that facilitate selection and purification.

6.4 CLONING DNA *IN VITRO*: THE POLYMERASE CHAIN REACTION

Cell-based DNA cloning opened many new avenues in the study of genetics, but the procedures involved are relatively laborious and time-consuming. To compensate for these deficiencies and provide a rapid way of studying DNA, alternative methods of cloning DNA were needed. The polymerase chain reaction (PCR), a way of cloning DNA *in vitro*, was first reported in the mid-1980s and had found numerous applications in both basic and clinical research. As well as being a DNA cloning method it also allows numerous assays that depend on quantitation of DNA or RNA.

Because of its simplicity, PCR is a popular technique with a wide range of applications that depend on essentially three major advantages: it is rapid, extremely sensitive, and robust. Its sensitivity allows the amplification of minute amounts of target DNA—even the DNA from a single cell—and various applications have been found in diagnosis, research, and forensic science. For example, an individual can be identified from PCR analyses of trace tissues such as a single hair or discarded skin cells. Its robustness means that it is often possible to amplify DNA from tissues or cells that are badly degraded, or embedded in some medium that makes it difficult to isolate DNA by the standard methods. Successful PCR amplification is possible from small amounts of degraded DNA extracted from decomposed tissues at archaeological or historical sites and from formalin-fixed tissue samples.

PCR can be used to amplify a rare target DNA selectively from within a complex DNA population

Usually, PCR is designed to permit the selective replication of one or more specific target DNA sequences within a heterogeneous collection of DNA sequences, enabling it to increase vastly in copy number (amplification). Often the starting DNA is total genomic DNA from a particular tissue or cultured cells, and the target DNA is typically a tiny fraction of the starting DNA. For example, if we want to amplify the 1.6 kb β-globin gene from human genomic DNA (with a haploid genome size of 3200 Mb), the ratio of target DNA to starting DNA is 1.6 kb:3200 Mb, or 1:2,000,000. Many PCR reactions involve the amplification of even smaller targets. For example, single human exons are on average about 290 bp long, and thus represent less than 0.000001% of a starting population of genomic DNA.

As an alternative to genomic DNA, the starting DNA may be total cDNA prepared by isolating RNA from a suitable tissue or cell line and then converting it into DNA with the enzyme reverse transcriptase. This is known as reverse transcriptase-PCR or RT-PCR for short. If the original RNA population has many copies of the transcript of the target DNA, there may be significant enrichment of the target sequence in comparison with its representation in genomic DNA. Under optimal conditions, RT-PCR can also sometimes be used to amplify target sequences from tissues in which they are expressed at only a basal transcription level.

The cyclical nature of PCR leads to exponential amplification of the target DNA

PCR depends on synthesizing oligonucleotide sequences to act as primers for new DNA synthesis at defined target sequences within the starting DNA; the idea is to selectively replicate only the target sequence(s). It consists of a series of cycles of three successive reactions conducted at different temperatures, so the process is sometimes described as *thermocycling*. First, the starting DNA is heated to a temperature that is high enough to break the hydrogen bonds holding the two complementary DNA strands together. As a result, the double strands separate to give single-stranded DNA (**denaturation**). For human genomic DNA, the reaction mixture is generally heated to about 93–95°C for the denaturation step. After cooling, the synthetic oligonucleotide **primers** are allowed to bind by base pairing to a complementary sequence on the single-stranded DNA (**annealing**). Next, in the presence of the four deoxynucleoside triphosphates dATP, dCTP, dGTP, and dTTP, a purified DNA polymerase initiates the synthesis of new DNA strands that are complementary to the individual DNA strands of the target DNA segment.

The orientation of the primers is deliberately chosen so that the direction of synthesis of new DNA strands is toward the other primer-binding site. As a result, the newly synthesized strands can, in turn, serve as templates for new DNA synthesis, causing a chain reaction with an exponential increase in product (**Figure 6.16**). The three successive reactions of denaturation, primer annealing, and DNA synthesis are repeated up to about 30–40 times in a standard PCR reaction.

The DNA synthesis step of PCR typically occurs at about 70–75°C, but the DNA polymerase used also needs to withstand the much higher temperature of the denaturation step. The requirement for a heat-stable enzyme drove researchers to isolate DNA polymerases from microorganisms whose natural habitat is hot springs. An early and widely used example is Taq polymerase from *Thermus aquaticus*. However, Taq polymerase lacks an associated 3′→5′ exonuclease activity to provide a proofreading activity. All polymerases make occasional errors by inserting the wrong base, but such copying errors can be rectified by a proofreading 3′→5′ exonuclease activity. As a result, other thermophilic bacteria were sought as sources of heat-stable polymerases with a 3′→5′ proofreading activity, such as Pfu polymerase from *Pyrococcus furiosus*.

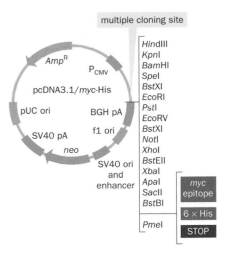

Figure 6.15 A mammalian expression vector, pcDNA3.1/*myc*-His. The Invitrogen pcDNA series of plasmid expression vectors offer high-level constitutive protein expression in mammalian cells. Shown here is the pcDNA3.1/*myc*-His expression vector. Cloned cDNA inserts can be transcribed from the strong cytomegalovirus (P_{CMV}) promoter, which ensures high-level expression in mammalian cells, and transcripts are polyadenylated with the help of a polyadenylation sequence element from the bovine growth hormone gene (BGH pA). The multiple cloning site region is followed by two short coding sequences that can be expressed to give peptide tags. The *myc* epitope tag allows recombinant screening with an antibody specific for the Myc peptide, and an affinity tag of six consecutive histidine residues (6 × His) facilitates purification of the recombinant protein by affinity chromatography with a nickel–nitrilotriacetic acid matrix. Translation ends at a defined termination codon (STOP). A *neo* gene marker (regulated by an SV40 promoter/enhancer and an SV40 poly(A) sequence) permits selection by the growth of transformed host cells on the antibiotic G418. Components for propagation in *E. coli* include a permissive origin of replication from plasmid ColE1 (pUC ori) and an ampicillin resistance gene (*Amp*^{R}) plus an origin of replication from phage f1 (f1 ori), which provides an option for producing single-stranded recombinant DNA.

(A)

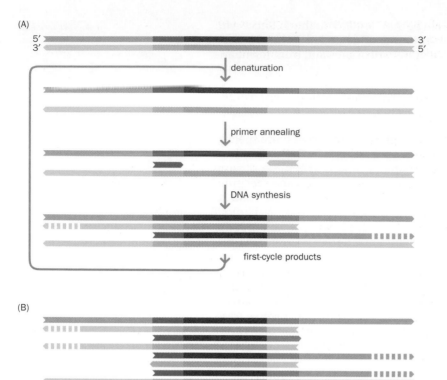

(B)

second-cycle products

third-cycle products

30th cycle

Figure 6.16 The polymerase chain reaction (PCR). (A) DNA containing the sequence to be amplified (the target sequence, shown here in red) is first denatured by heating. Oligonucleotide primers, designed to be complementary to DNA sequences flanking the target sequence, at the 3′ end on opposite strands (blue and green arrows), are then allowed to bind (primer annealing). Next, DNA synthesis occurs and the primers are incorporated into the newly synthesized DNA strands. The first cycle results in two new DNA strands whose 5′ ends are fixed by the position of the oligonucleotide primer but whose 3′ ends extend past the other primer (indicated by dotted lines). (B) After the second cycle, the four new strands consist of two more products with variable 3′ ends but now two strands of fixed length with both 5′ and 3′ ends defined by the primer sequences. After the third cycle, six out of the eight new strands consist of just the target and flanking primer sequences; after 30 cycles or so, with an exponential increase, this will be the predominant product.

Selective amplification of target sequences depends on highly specific binding of primer sequences

To permit selective amplification, some information about the sequence of the target DNA is required. The information is used to design two oligonucleotide primers, often about 18–25 nucleotides long, that are specific for two sequences immediately flanking the target sequence. To bind effectively, the primers must have sequences that have a very high degree of base complementarity to the sequences flanking the target DNA.

For most PCR reactions, the goal is to amplify a single DNA sequence, so it is important to reduce the chance that the primers will bind to other locations in the DNA than the desired ones. It is therefore important to avoid repetitive DNA sequences. To avoid the binding of the two primers to each other, the 3′ sequence of one primer should not be complementary in sequence to any region of the other primer in the same reaction. To ensure that a primer does not undergo internal base pairing, the primer sequence must not include inverted repeats or any self-complementary sequences more than 3 bp in length.

To ensure that the primer binds with a high degree of specificity to its intended complementary sequence, the temperature for the primer annealing step should be set as high as possible while still allowing base pairing between the primer and its binding site. A useful measure of the stability of any nucleic acid duplex is the **melting temperature** (T_m), the temperature corresponding to the mid-point in observed transitions from the double-stranded form to the single-stranded form. To ensure a high degree of specificity in primer binding, the primer annealing temperature is typically set to be about 5°C below the calculated melting temperature; if the annealing temperature were to be much lower, primers would tend to bind to other regions in the DNA that had partly complementary sequences.

The melting temperature, and so the temperature for primer annealing, depends on base composition. This is because GC base pairs have three hydrogen bonds and AT base pairs only two. Strands with a high percentage of GC base pairs are therefore more difficult to separate than those with a low composition of GC base pairs. Optimal priming uses primers with a GC content between 40% and 60% and with an even distribution of all four nucleotides. Additionally, the calculated T_m values for two primers used together should not differ by greater than 5°C, and the T_m of the amplified target DNA should not differ from those of the primers by greater than 10°C.

Primer design is aided by various commercial software programs and certain freeware programs that are accessible through the web at compilation sites such as http://www.humgen.nl/primer_ design.html. As we will see below, it is particularly important that the 3′ ends of primers are perfectly matched to their intended sequences. Various modifications are often used to reduce the chances of nonspecific primer binding, including hot-start PCR, touch-down PCR, and the use of nested primers (**Box 6.2**).

PCR is disadvantaged as a DNA cloning method by short lengths and comparatively low yields of product

PCR is the method of choice for selectively amplifying minute amounts of target DNA, quickly and accurately. It is ideally suited to conducting rapid assays on genomic or cDNA sequences. However, it has disadvantages as a DNA cloning method. In particular, the size range of the amplification products in a standard PCR reaction are rarely more than 5 kb, whereas cell-based DNA cloning makes it possible to clone fragments up to 2 Mb long. As the desired product length increases, it becomes increasingly difficult to obtain efficient amplification with PCR. To overcome this problem, long-range PCR protocols have been developed that use a mixture of two types of heat-stable polymerase in an effort to provide optimal levels of DNA polymerase and a proofreading 3′→5′ exonuclease activity. Using the modified protocols, PCR products can sometimes be obtained that are tens of kilobases in length; larger cloned DNA can be obtained only by cell-based DNA cloning.

PCR typically involves just 30–40 cycles, by which time the reaction reaches a plateau phase as reagents become depleted and inhibitors accumulate. Microgram quantities may be obtained of the desired DNA product, but it is time-consuming and expensive to scale the reaction up to achieve much larger DNA quantities. In addition, the PCR product may not be in a suitable form that will permit some subsequent studies. Thus, when particularly large quantities of DNA are required, it is often convenient to clone the PCR product in a cell-based cloning system. Plasmid cloning systems are used to propagate PCR-cloned DNA in bacterial cells. Once cloned, the insert can be cut out with suitable restriction endonucleases and transferred into other specialized plasmids that, for example, permit expression to give an RNA or protein product.

Several thermostable polymerases routinely used for PCR have a terminal deoxynucleotidyl transferase activity that selectively modifies PCR-generated fragments by adding a single nucleotide, generally deoxyadenosine, to the 3′ ends of amplified DNA fragments. The resulting 3′ dA overhangs can sometimes make it difficult to clone PCR products in cells, and various methods are used to increase cloning efficiency. Specialized vectors such as pGEM-T-easy can be treated so as to have complementary 3′ T overhangs in their cloning site that will encourage base pairing with the 3′ dA overhangs of the PCR product to be cloned (TA cloning). Alternatively, the overhanging nucleotides on the PCR products are removed with polishing enzymes such as T4 polymerase or Pfu polymerase. PCR primers can also be modified by designing a roughly 10-nucleotide extension containing

BOX 6.2 SOME COMMON PCR METHODS

Allele-specific PCR. Designed to amplify a DNA sequence while excluding the possibility of amplifying other alleles. Based on the requirement for precise base matching between the 3′ end of a PCR primer and the target DNA. See Figure 6.17.

Anchored PCR. Uses a sequence-specific primer and a universal primer for amplifying sequences adjacent to a known sequence. The universal primer recognizes and binds to a common sequence that is artificially attached to all of the different DNA molecules.

DOP-PCR (degenerate oligonucleotide-primed PCR). Uses partly *degenerate oligonucleotide primers* (sets of oligonucleotide sequences that have been synthesized in parallel to have the same base at certain nucleotide positions, while differing at other positions) to amplify a variety of related target DNAs.

Hot-start PCR. A way of increasing the specificity of a PCR reaction. Mixing all PCR reagents before an initial heat denaturation step allows more opportunity for nonspecific binding of primer sequences. To reduce this possibility, one or more components of the PCR are physically separated until the first denaturation step.

Inverse PCR. A way of accessing DNA that is immediately adjacent to a known sequence. In this case, the starting DNA population is digested with a restriction endonuclease, diluted to low DNA concentration, and then treated with DNA ligase to encourage the formation of circular DNA molecules by intramolecular ligation. The PCR primers are positioned so as to bind to a known DNA sequence and then initiate new DNA synthesis in a direction leading *away* from the known sequence and toward the unknown adjacent sequence, leading to amplification of the unknown sequence. See **Figure 1** (X and Y are uncharacterized sequences flanking a known sequence).

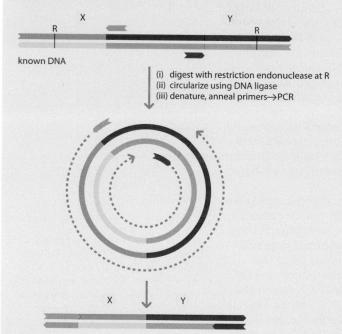

(i) digest with restriction endonuclease at R
(ii) circularize using DNA ligase
(iii) denature, anneal primers→PCR

known DNA

Figure 1 Inverse PCR.

Linker-primed PCR. A form of indiscriminate amplification that involves attaching oligonucleotide linkers to both ends of all DNA fragments in a starting DNA and amplifying all fragments by using a linker-specific primer (see Figure 6.18).

Nested primer PCR. A way of increasing the specificity of a PCR reaction. The products of an initial amplification reaction are diluted and used as the starting DNA source for a second reaction in which a different set of primers is used, corresponding to sequences located close, but *internal*, to those used in the first reaction.

RACE-PCR. A form of *anchor-primed PCR* (see above) for **r**apid **a**mplification of **c**DNA **e**nds. This will be described in a later chapter.

Real-time PCR (also called quantitative PCR or qPCR). Uses a fluorescence-detecting thermocycler machine to amplify specific nucleic acid sequences and simultaneously measure their concentrations. There are two major research applications: (1) to quantitate gene expression (and to confirm differential expression of genes detected by microarray hybridization analyses) and (2) to screen for mutations and single nucleotide polymorphisms. In analytical labs it is also used to measure the abundance of DNA or RNA sequences in clinical and industrial samples.

RT-PCR (reverse transcriptase PCR). PCR in which the starting population is total RNA or purified poly(A)$^+$ mRNA and an initial reverse transcriptase step to produce cDNA is required.

Touch-down PCR. A way of increasing the specificity of a PCR reaction. Most thermal cyclers can be programmed to perform runs in which the annealing temperature is lowered incrementally during the PCR cycling from an initial value above the expected T_m to a value below the T_m. By keeping the stringency of hybridization initially very high, the formation of spurious products is discouraged, allowing the expected sequence to predominate.

Whole genome PCR. Indiscriminate PCR. Formerly, this was performed by using comprehensively degenerate primers (see *degenerate PCR*) or by attaching oligonucleotide linkers to a complex DNA population and then using linker-specific oligonucleotide primers to amplify all sequences (*linker-primed PCR*). However, these methods do not amplify all sequences because secondary DNA structures pose a problem for the standard polymerases used in PCR, causing enzyme slippage or dissociation of the enzyme from the template, resulting in nonspecific amplification artifacts and incomplete coverage of loci. To offset these difficulties, a non-PCR isothermal amplification procedure is now commonly used, known as *multiple displacement amplification* (MDA). In the MDA method a strand-displacing DNA polymerase from phage phi29 is used in a rolling-circle form of DNA amplification at a constant temperature of about 30°C (see Further Reading).

a suitable restriction site at its 5′ end. The nucleotide extension does not base-pair to target DNA during amplification, but afterward the amplified product can be digested with the appropriate restriction enzyme to generate overhanging ends for cloning into a suitable vector.

A wide variety of PCR approaches have been developed for specific applications

The many applications of PCR and the need to optimize efficiency and specificity have prompted a wide variety of PCR approaches. We consider below and in Box 6.2 a few selected applications.

Allele-specific PCR

Many applications require the ability to distinguish between two alleles that may differ by only a single nucleotide. Diagnostic applications include the ability to distinguish between a disease-causing point mutation and the normal allele. Allele-specific PCR takes advantage of the crucial dependence of correct base pairing at the extreme 3′ end of bound primers. In the popular ARMS (amplification refractory mutation system) method, allele-specific primers are designed with their 3′ end nucleotides designed to base-pair with the variable nucleotide that distinguishes alleles. Under suitable experimental conditions amplification will not take place when the 3′ end nucleotide is not perfectly base paired, thereby distinguishing the two alleles (**Figure 6.17**).

Multiple target amplification and whole genome PCR methods

Standard PCR is based on the need for very specific primer binding, to allow the selective amplification of a desired known target sequence. However, PCR can also be deliberately designed to simultaneously amplify multiple different target DNA sequences. Sometimes there may be a need to amplify multiple members of a family of sequences that have some common sequence characteristic. Some highly repetitive elements such as the human Alu sequence occur at such a high frequency that it is possible to design primers that will bind to multiple Alu sequences. If two neighboring Alu sequences are in opposite orientations, a single type of Alu-specific primer can bind to both sequences, enabling amplification of the sequence between the two Alu repeats (Alu-PCR).

For some purposes, all sequences in a starting DNA population need to be amplified. This approach can be used to replenish a precious source of DNA that is present in very limiting quantities (*whole genome amplification*). As we will see in a later chapter, it has also been used for indiscriminate amplification to produce templates for DNA sequencing. One way of performing multiple target amplification is to attach a common double-stranded oligonucleotide **linker** covalently to the extremities of all of the DNA sequences in the starting

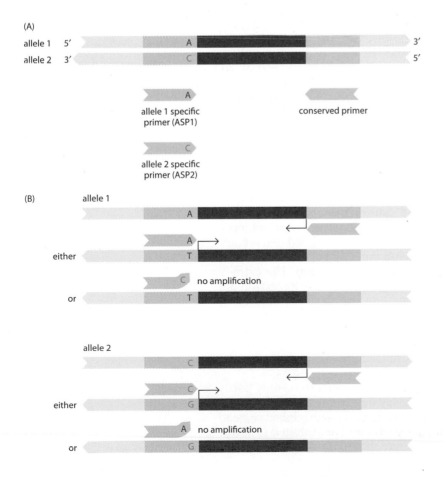

Figure 6.17 Allele-specific PCR is dependent on perfect base-pairing of the 3′ end nucleotide of primers. (A) Alleles 1 and 2 differ by a single base (A→C). Two allele-specific oligonucleotide primers (ASP1 and ASP2) are designed that are identical to the sequence of the two alleles over a region preceding the position of the variant nucleotide, but which differ and terminate in the variant nucleotide. ASP1 and ASP2 are used as alternative primers in PCR reactions with another primer that is designed to bind to a conserved region on the opposite DNA strand for both alleles. (B) ASP1 will bind perfectly to the complementary strand of the allele 1 sequence, permitting amplification with the conserved primer. However, the 3′-terminal C of the ASP2 primer mismatches with the T of the allele 1 sequence, making amplification impossible. Similarly, ASP2, but not ASP1, can bind perfectly to allele 2 and initiate amplification.

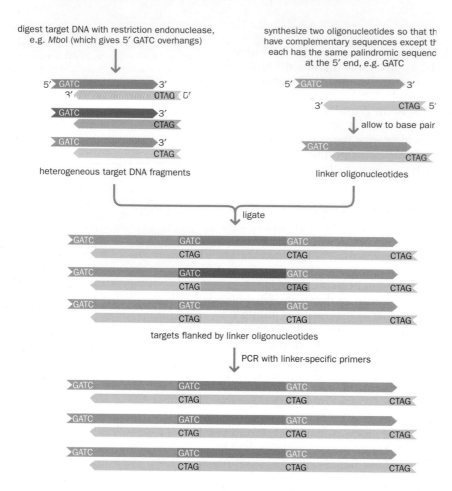

digest target DNA with restriction endonuclease, e.g. MboI (which gives 5′ GATC overhangs)

synthesize two oligonucleotides so that th have complementary sequences except th each has the same palindromic sequenc at the 5′ end, e.g. GATC

heterogeneous target DNA fragments

linker oligonucleotides

allow to base pair

ligate

targets flanked by linker oligonucleotides

PCR with linker-specific primers

Figure 6.18 Using linker oligonucleotides to amplify multiple target DNAs simultaneously. The DNA to be amplified is digested with a restriction endonuclease that produces overhanging ends. The linker is prepared by individually synthesizing two oligodeoxyribonucleotides that are complementary in sequence except that when allowed to combine they form a double-stranded DNA sequence with the same type of overhanging ends as the restriction fragments. Ligation of the linker oligonucleotides to the target DNA restriction fragments can result in fragments being flanked on each side by a linker. Primers that are specific for the linkers can then permit amplification of target DNA molecules with linkers at both ends (shown in the case of only one of the different fragments in this example). As a result, numerous DNA sequences in a starting population can be amplified simultaneously.

population and then use linker-specific primers to amplify all the DNA sequences (**Figure 6.18**). An alternative to using linker oligonucleotides is to use comprehensively degenerate oligonucleotides as primers. That is, when oligonucleotides are synthesized one nucleotide at a time, all four nucleotides can be inserted at specific positions instead of a single one, to give a parallel set of many different oligonucleotides. As a result, very large numbers of different oligonucleotides are synthesized and can bind to numerous different sequences in the genome, enabling PCR amplification of much of the genomic DNA. Note, however, that whole genome amplification is often now not performed by PCR but by an alternative form of *in vitro* DNA cloning (see Box 6.2).

PCR mutagenesis

PCR can be used to engineer various types of predetermined base substitutions, deletions, and insertions into a target DNA. In **5′ add-on mutagenesis** a desired sequence or chemical group is added in much the same way as can be achieved by ligating an oligonucleotide linker. A mutagenic primer is designed that is complementary to the target sequence at its 3′ end, and the 5′ end contains a desired novel sequence or a sequence with an attached chemical group. The extra 5′ sequence does not participate in the first annealing step of the PCR reaction but subsequently becomes incorporated into the amplified product, thereby generating a recombinant product (**Figure 6.19A**). The additional 5′ sequence may contain one or more of the following: a suitable restriction site that may facilitate subsequent cell-based DNA cloning; a modified nucleotide containing a reporter group or labeled group such as a biotinylated nucleotide or fluorophore; or a phage promoter to drive gene expression.

In *mismatched primer mutagenesis*, the primer is designed to be only partly complementary to the target site, but in such a way that it will still bind specifically to the target. Inevitably this means that the mutation is introduced close to the extreme end of the PCR product. This approach may be exploited to introduce an artificial diagnostic restriction enzyme site that permits screening for a

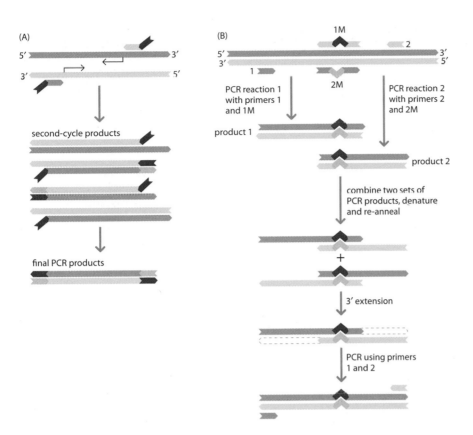

Figure 6.19 PCR mutagenesis. (A) 5′ add-on mutagenesis. Primers can be modified at the 5′ end to introduce a desired novel sequence or chemical group (red bars) that does not take part in the initial annealing to the target DNA but will be copied in subsequent cycles. The products of the second and final PCR cycles are shown in this example. (B) Mismatched primer mutagenesis. A specific predetermined mutation located in a central segment can be introduced by two separate PCR reactions, 1 and 2, each amplifying overlapping segments of DNA. Complementary mutant primers, 1M and 2M, introduce deliberate base mismatching at the site of the mutation. After the two PCR products are combined, denatured, and allowed to re-anneal, each of the two product 1 strands can base-pair with complementary strands from product 2 to form heteroduplexes with recessed 3′ ends. DNA polymerase can extend the 3′ ends to form full-length products with the introduced mutation in a central segment, and they can be amplified by using the outer primers 1 and 2 only.

known mutation. Mutations can also be introduced at any point within a chosen sequence by using mismatched primers. Two mutagenic reactions are designed in which the two separate PCR products have partly overlapping sequences containing the mutation. The denatured products are combined to generate a larger product with the mutation in a more central location (Figure 6.19B).

Real-time PCR (qPCR)

Real-time PCR (or qPCR, an abbreviation of quantitative PCR) is a widely used method to quantitate DNA or RNA. In the latter case, RNA is copied by using reverse transcriptase to make complementary DNA strands. Standard PCR reactions, in which the amplification products are analyzed at the end of the procedure, are not well suited to quantitation. Samples are removed from the reaction tubes and are usually size-fractionated on agarose gels and detected by binding ethidium bromide, which fluoresces under ultraviolet radiation. This procedure is both time-consuming and not very sensitive. In real-time PCR (qPCR), however, quantitation is done automatically, while the reaction is actually occurring, in specially designed PCR equipment. The reaction products are analyzed at an early stage in the amplification process when the reaction is still exponential, allowing more precise quantitation than at the end of the reaction. There are many applications, both in absolute quantitation of a nucleic acid and in comparative quantitation. One of the most important is in tracking gene expression, and we will describe in detail in Chapter 8 how qPCR works and is used in gene expression profiling.

FURTHER READING

General cell-based DNA cloning

Primrose SB, Twyman RM, Old RW & Bertola G (2006) Principles Of Gene Manipulation And Genomics, 7th ed. Blackwell Publishing Professional.

Roberts RJ (2009) Official REBASE Homepage. http://rebase .neb.com/rebase/rebase.html [database of restriction endonucleases.]

Sambrook J & Russell DW (2001) Molecular Cloning. A Laboratory Manual. Cold Spring Harbor Laboratory Press.

Vector databases: for compilation of websites see http://mybio .wikia.com/wiki/Vector_databases

Watson JD, Caudy AA, Myers RM & Witkowski JA (2007) Recombinant DNA: Genes And Genomes, 3rd ed. Freeman.

Large insert DNA cloning and production of single-stranded recombinant DNA

Iouannou PA, Amemiya CT, Garnes J et al. (1994) A new bacteriophage P1-derived vector for the propagation of large human DNA fragments. *Nat. Genet.* 6, 84–89.

Mead DA & Kemper B (1988) Chimeric single-stranded DNA phage-plasmid cloning vectors. *Biotechnology* 10, 85–102.

Schlessinger D (1990) Yeast artificial chromosomes: tools for mapping and analysis of complex genomes. *Trends Genet.* 6, 248–258.

Shizuya H, Birren B, Kim U-J et al. (1992) Cloning and stable maintenance of 300 kilobase-pair fragments of human DNA in *Escherichia coli. Proc. Natl Acad. Sci. USA* 89, 8794–8797.

Expression cloning and selectable markers in mammalian cells

Colosimo A, Goncz KK, Holmes AR et al. (2000) Transfer and expression of foreign genes in mammalian cells. *Biotechniques* 29, 314–331.

Hames BD (2004) Protein Expression: A Practical Approach. Oxford University Press.

Sidhu SS (2001) Engineering M13 for phage display. *Biomol. Eng.* 18, 57–63.

Szybalski W (1992) Use of the HPRT gene and the HAT selection technique in DNA-mediated transformation of mammalian cells. *BioEssays* 14, 495–500.

Wurm F & Bernard A (1999) Large-scale transient expression in mammalian cells for recombinant protein production. *Curr. Opin. Biotechnol.* 10, 156–159.

PCR

Kubista M, Andrade JM, Bengtsson M et al. (2006) The real-time polymerase chain reaction. *Mol. Aspects Med.* 27, 95–125.

McPherson MJ & Moller SG (2006) PCR: The Basics. Garland Scientific Press.

National Center for Biotechnology Information (2008) Primer-BLAST: primer designing tool. http://www.ncbi.nlm.nih .gov/tools/primer-blast/ [to find primers specific for a given sequence.]

In vitro isothermal DNA amplification using phage phi29 DNA polymerase

Dean FB, Hosono S, Fang L et al. (2002) Comprehensive human genome amplification using multiple displacement amplification. *Proc. Natl Acad. Sci. USA* 99, 5261–5266.

DNA mutagenesis

Ling MM & Robinson BH (1997) Approaches to DNA mutagenesis: an overview. *Anal. Biochem.* 254, 157–178.

Chapter 7

Nucleic Acid Hybridization: Principles and Applications

7

KEY CONCEPTS

- In nucleic acid hybridization, well-characterized nucleic acid or oligonucleotide populations (*probes*) are used to identify related sequences in test samples containing complex, often poorly understood nucleic acid populations.

- Nucleic acid hybridization relies on specificity of base pairing. Nucleic acids in both the probe and test sample populations are made single-stranded and mixed so that heteroduplexes can form between probe sequences and any complementary or partly complementary sequences (*targets*) in the test sample population.

- Heteroduplexes are detected by labeling one nucleic acid population in aqueous solution and allowing it to hybridize to an unlabeled nucleic acid population fixed to a solid support. After washing to remove unhybridized labeled nucleic acid, any label remaining on the solid support should come from a probe–target heteroduplex.

- The stability of a heteroduplex depends on the extent of base matching and is affected by parameters such as the length of the base-paired segment, the temperature, and the ionic environment.

- Commonly used DNA and RNA probes are hundreds of nucleotides long and are used to identify target sequences that show a high degree of sequence similarity to the probe.

- Short oligonucleotide probes (less than 20 nucleotides long) can be used to distinguish between targets that differ at single nucleotide positions.

- Nucleic acids are labeled by incorporating nucleotides containing radioisotopes or chemically modified groups that can be detected by a suitable assay.

- Many hybridization assays involve binding the nucleic acid population containing the target to a solid surface and then exposing it to a solution of labeled probe. The immobilized DNA may have been purified or may be present within immobilized cells or chromosomes.

- Microarray hybridization allows numerous hybridization assays to be performed simultaneously. Thousands of unlabeled DNA or oligonucleotide probes are fixed on a solid surface in a high-density grid format and are used to screen complex sample populations of labeled DNA or RNA in solution.

- The principal applications of microarray hybridization are in gene expression profiling and analysis of DNA variation.

DNA molecules are very large and break easily when isolated from cells, making them difficult to study. Chapter 6 looked at the way in which DNA cloning, either in cells or *in vitro*, allows individual DNA molecules to be selectively replicated to very high copy numbers. When amplified in this way, the DNA is effectively purified, and DNA cloning enables individual DNA sequences in genomic DNA to be studied, and also individual RNA molecules after they have been copied to give a cDNA. In this chapter, we consider an altogether different approach. Instead of trying to purify individual nucleic acid sequences, the object is to track them specifically within a complex population that represents a sample of biological or medical interest.

At the basic research level, nucleic acid hybridization is often used to track RNA transcripts and so obtain information on how genes are expressed. It is also used to identify relationships between DNA sequences from different sources. For example, a starting DNA sequence can be used to identify other closely related sequences from different organisms or from different individuals within the same species, or even related sequences from the same genome as the starting sequence. Nucleic acid hybridization is also often used as a way of identifying disease alleles and aberrant transcripts associated with disease.

7.1   PRINCIPLES OF NUCLEIC ACID HYBRIDIZATION

In nucleic acid hybridization a known nucleic acid population interrogates an imperfectly understood nucleic acid population

Nucleic acid hybridization is a fundamental tool in molecular genetics. It exploits the ability of single-stranded nucleic acids that are partly or fully complementary in sequence to form double-stranded molecules by base pairing with each other (**hybridization**). A glossary of relevant terms is provided in **Box 7.1** to assist readers who may be unfamiliar with aspects of terminology.

BOX 7.1 A GLOSSARY FOR NUCLEIC ACID HYBRIDIZATION

Anneal. To allow hydrogen bonds to form between two single strands. If two single-stranded nucleic acids share sufficient *base complementarity*, they will form a double-stranded nucleic acid duplex. Anneal has the opposite meaning to that of *denature*.

Antisense RNA. An RNA sequence that is complementary in sequence to a transcribed RNA, enabling base pairing between the two sequences.

Base complementarity. The degree to which the sequences of two single-stranded nucleic acids can form a double-stranded duplex by Watson–Crick base pairing (A binds to T or U; C binds to G).

Denature. To separate the individual strands of a double-stranded DNA duplex by breaking the hydrogen bonds between them. This can be achieved by heating, or by exposing the DNA to alkali or to a highly polar solvent such as urea or formamide. Denature has the opposite meaning to that of *anneal*.

DNA chip. Any very-high-density gridded array (*microarray*) of DNA clones or oligonucleotides that is used in a hybridization assay.

Feature (of a DNA or oligonucleotide microarray). The large number of identical DNA or oligonucleotide molecules at any one position in the microarray.

Heteroduplex. A double-stranded nucleic acid formed by base pairing between two single-stranded nucleic acids that do not originate from the same allele. There may be 100% base matching between the two sequences of a heteroduplex, notably when one of the hybridizing sequences is an oligonucleotide probe or when the sequences are from different alleles of the same gene or are evolutionarily closely related.

Homoduplex. A double-stranded DNA formed when two single-stranded sequences originating from the same allele are allowed to re-anneal. A homoduplex has 100% base matching and so is normally more stable than a heteroduplex.

Hybridization assay. An assay in which a population of well-characterized nucleic acids or oligonucleotides (the *probe*) is made single-stranded and used to search for complementary *target* sequences within an imperfectly or poorly understood population of nucleic acid sequences by annealing to form heteroduplexes.

Hybridization stringency. The degree to which hybridization conditions tolerate base mismatches in heteroduplexes. At high hybridization stringency, only the most perfectly matched sequences can base pair, but heteroduplexes with significant mismatches can be stable when the stringency is lowered by reducing the annealing temperature or by increasing the salt concentration.

In situ hybridization. A hybridization reaction in which a labeled probe is hybridized to nucleic acids so that their morphological location can be detected within fixed cells or chromosomes.

Melting temperature (T_m). The temperature corresponding to the mid-point in the observed transition from double-stranded to single-stranded forms of nucleic acids.

Microarray. A solid surface to which molecules of interest can be fixed at specific coordinates in a high-density grid format for use in some assay. An oligonucleotide or DNA microarray has numerous unlabeled DNA or oligonucleotide molecules affixed at precise positions on the array, to act as probes in a hybridization assay. Each specific position has many thousands of identical copies of a particular type of oligonucleotide or DNA molecule, constituting a *feature*.

Probe. A known nucleic acid or oligonucleotide population that is used in a hybridization assay to query an often complex nucleic acid population so as to identify related *target* sequences by forming heteroduplexes.

Riboprobe. An RNA probe.

Sequence similarity. The degree to which two sequences are identical in sequence or in base complementarity.

Target. Nucleic acid sequence that shows sufficient sequence similarity to a probe that it can base-pair with it in a hybridization assay to form a stable heteroduplex.

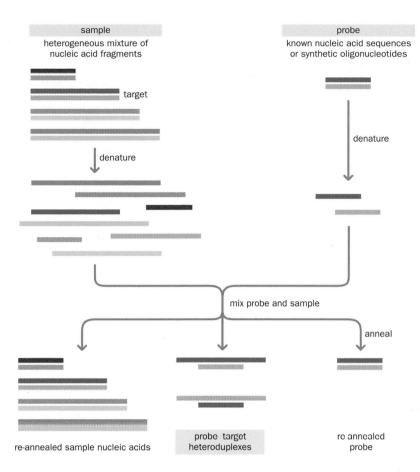

Figure 7.1 Formation of probe–target heteroduplexes in a nucleic acid hybridization assay. A test sample consisting of a complex mixture of nucleic acids and a defined probe population of known nucleic acid or oligonucleotide sequences are both made single-stranded, then mixed and allowed to anneal. Sequences that had previously been base-paired in the test sample and probe will re-anneal to form homoduplexes (bottom left and bottom right). In addition, new heteroduplexes will be formed between probe and target sequences that have complementary or partially complementary sequences (bottom center). The conditions of hybridization can be adjusted to favor the formation of heteroduplexes. In this way, probes selectively bind to and identify related nucleic acids within a complex nucleic acid population.

Nucleic acid hybridization can be performed in many different ways, but there is a common underlying principle: a *known*, well-characterized population of nucleic acid molecules or synthetic oligonucleotides is used to interrogate a complex population of nucleic acids in an imperfectly understood test sample of biological or medical interest.

The test sample to be studied may contain DNA molecules, for example total DNA from white blood cells from a single individual or from a particular type of tumor cell, or it may contain RNA, such as total RNA or mRNA expressed by a specific cell line or tissue. In either case, if they are to participate in nucleic acid hybridization, the nucleic acid molecules in the sample need to be made single-stranded (**denaturation**). Cellular DNA is naturally double-stranded but RNA too has significant amounts of internal double-stranded regions caused by intra-chain hydrogen bonding. The interchain and intrachain hydrogen bonding can be disrupted by various methods. Initial denaturation often involves heating or treatment with alkali; as required, nucleic acids may also be exposed to strongly polar molecules such as formamide or urea to keep them in a single-stranded state for long periods.

The known interrogating population consists of precisely defined nucleic acid sequences, which also need to be denatured, or synthetic single-stranded oligo-nucleotides. Each type of molecule in this population will act as a **probe** to locate complementary or partly complementary nucleic acids (**targets**) within the test sample. To do this, the single-stranded test sample and probe populations are mixed to allow base pairing between probe single strands and complementary target strands (**annealing**). The object is to form probe–target **heteroduplexes** (**Figure 7.1**); the specificity of the interaction between probe and target sequences depends on the degree of base matching between the two interacting strands.

Probe–target heteroduplexes are easier to identify after capture on a solid support

The efficiency of identifying probe–target heteroduplexes in solution is low. To assist in their identification, either the test sample nucleic acids or the probe

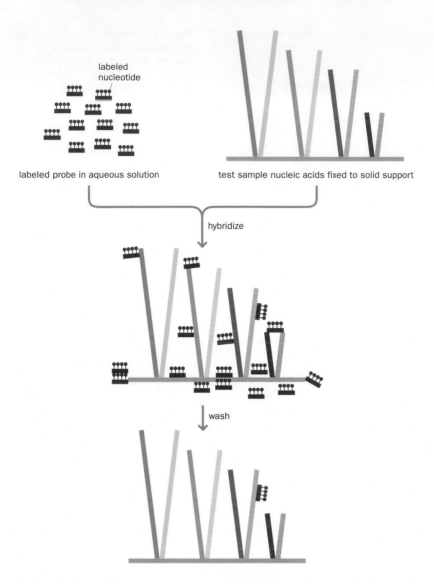

labeled
nucleotide

labeled probe in aqueous solution

test sample nucleic acids fixed to solid support

hybridize

wash

Figure 7.2 Probe–target heteroduplexes are most efficiently identified after capture on a solid support. Hybridization assays typically involve binding either the test sample population to the surface of a solid support (as shown here) or fixing the probe population to the solid support. The immobilized population is bound in such a way that different sequences are fixed to defined positions on the solid surface, but not necessarily tethered at one end as shown here. The other population is labeled then added to the immobilized population. Here, we use the example shown in Figure 7.1 in which the probe is imagined to be homogeneous and the test sample is imagined to be a heterogeneous mixture of different DNA molecules. For simplicity, we show just one of the labeled probe strands. After labeled single-stranded probe molecules pass over the solid support, they can be captured by hybridization to complementary target molecules bound to the support. After hybridization, washing will remove any excess probe that remains in solution or that is bound nonspecifically to the surface. Any label remaining on the solid support should now represent the probe–target heteroduplexes, and can lead to identification of the target sequence.

population is bound in some way to a solid support, often a plastic membrane, glass slide, or quartz wafer. The other population—respectively, probe or test sample—is provided in aqueous solution and is labeled by attaching a molecule that carries a distinctive radioisotope, or a chemical group that can be detected in some way. The labeled population is passed over the solid support so that the two populations can interact and form heteroduplexes.

We illustrate in **Figure 7.2** one of these possibilities, in which the probe population is labeled and in solution while the test sample is immobilized. Target DNA molecules in the test sample capture labeled probe molecules that have a complementary or partly complementary sequence, to form genuine probe–target heteroduplexes. In addition, some labeled probe molecules can bind nonspecifically to the support or to nontarget molecules on the support. However, after hybridization, the solid support is washed extensively so that the only label that can be detected on the support should come from genuine probe–target heteroduplexes. To maximize the chances of forming probe–target heteroduplexes, the target is normally present in great excess over the probe.

The alternative possibility—using immobilized probes and labeling nucleic acids of the test sample—will be considered later in this chapter, when we consider microarray hybridization.

Denaturation and annealing are affected by temperature, chemical environment, and the extent of hydrogen bonding

When complementary or partly complementary nucleic acid sequences associate to form duplexes, the number of new hydrogen bonds formed depends on the

7.2 LA
OLIGON

In nucleic a
tides is labe
other popul
and method
probe popu
oligonucleo
nucleic acid
usually com
Section 7.4.

Different o
DNA, RNA

When hybri
DNA, RNA,
ods. As requ
double stran

Convent
by amplifyir
stranded to
hundreds of
in length. Tl
nucleoside
they usually
are a mixtui
specific RN/
both sense a

RNA pro
cally a few l
from DNA c
contains a p
phage RNA
RNA synthe
labeled, spe
DNA. Single
complemen

Oligonuc
nucleotides
DNA. Synth
mononucle
typically lat
probes can
Figure 7.3).

Long nucl
labeled nt

For some pt
ends of the
using a kina
ments can a
fied primers
with its 5' ei

End-labe
inserted pe
percentage
As a result, t
thesis labeli
thesis labeli

TABLE 7.1 EQUATIONS FOR CALCULATING T_m

Hybrid	T_m (°C)
DNA–DNA	$81.5 + 16.6(\log_{10}[Na^+]^a) + 0.41(\%GC^b) - 500/L^c$
DNA–RNA or RNA–RNA	$79.8 + 18.5(\log_{10}[Na^+]^a) + 0.58(\%GC^b) + 11.8(\%GC^b)^2 - 820/L^c$
Oligo–DNA or oligo–RNA^d	
For < 20 nucleotides	$2(l_n)$
For 20–35 nucleotides	$22 + 1.46(l_n)$

^{a}Or for other monovalent cation, but only accurate in the range 0.01–0.4 M. ^{b}Only accurate for %GC from 30% to 75% GC. ^{c}L, length of duplex in base pairs. ^{d}Oligo, oligonucleotide; l_n, effective length of primer = 2 × (no. of G + C) + (no. of A + T). For each 1% formamide, T_m is decreased by about 0.6°C, and the presence of 6 M urea decreases T_m by about 30°C.

length of the duplex. There are more hydrogen bonds in longer nucleic acid molecules and so more energy is required to break them. The increase in duplex stability is not linearly proportional to length, and the effect of changing the length is particularly noticeable at shorter length ranges. As will be described in the next section, base mismatching reduces duplex stability.

Base composition and the chemical environment are also important factors in duplex stability. A high percentage of GC base pairs means greater difficulty in separating the strands of a duplex because GC base pairs have three hydrogen bonds and AT base pairs have only two. The presence of monovalent cations (e.g. Na$^+$) stabilizes the hydrogen bonds in double-stranded molecules, whereas strongly polar molecules (such as formamide and urea) disrupt hydrogen bonds and thus act as chemical denaturants.

A progressive increase in temperature also makes hydrogen bonds unstable, eventually disrupting them. The temperature corresponding to the mid-point in observed transitions from double-stranded form to single-stranded form is known as the **melting temperature** (T_m). For mammalian genomes, with a base composition of about 40% GC, the DNA denatures with a T_m of about 87°C in buffers whose pH and salt concentration approximate physiological conditions. The T_m of perfect hybrids formed by DNA, RNA, or oligonucleotide probes can be determined by using standard formulae (**Table 7.1**).

Stringent hybridization conditions increase the specificity of duplex formation

When nucleic acid hybridization is performed, the hybridization conditions are usually deliberately designed to maximize heteroduplex formation, even if it means that some nonspecific hybridization occurs. For example, the hybridization temperature may often be as much as 25°C below the T_m, and so probe molecules can base-pair with nucleic acid molecules that are distantly related in sequence in addition to the expected closely related target molecules. In such cases, the **hybridization stringency** is said to be low.

After encouraging strong probe–target binding, successive washes are often conducted under conditions that are less and less tolerant of base mismatching in heteroduplexes. This can be achieved by progressively increasing the temperature or incrementally reducing the concentration of NaCl in the buffer. The progressive increase in hybridization stringency can reveal different target sequences that are increasingly related to probe molecules. The last wash corresponds to a high hybridization stringency to ensure that heteroduplex formation is specific.

Probe–target heteroduplexes are most stable thermodynamically when the region of duplex formation contains perfect base matching. Mismatches between the two strands of a heteroduplex decrease the T_m, and for normal DNA probes each 1% of mismatching decreases the T_m by about 1°C. However, this effect diminishes with the length of the paired region. Thus, a considerable degree of mismatching can be tolerated if the overall region of base complementarity is

BOX 7.2 AUTORADIOGRAPHY

Autoradiography records the position of a radioactively labeled compound within a solid sample by producing an image in a photographic emulsion. In molecular genetic applications, the radiolabeled compounds are often DNA molecules or proteins, and the solid sample may be fixed chromatin or tissue samples mounted on a glass slide. Alternatively, DNA or protein samples can be embedded within a dried electrophoresis gel or fixed to the surface of a dried nylon membrane or nitrocellulose filter.

The solid sample is placed in intimate contact with an X-ray film, a plastic sheet with a coating of photographic emulsion. The photographic emulsion consists of silver halide crystals in suspension in a clear gelatinous phase. Radioactive emissions from the sample travel through the emulsion, converting Ag^+ ions to Ag atoms. The position of altered silver halide crystals can be revealed by development, an amplification process in which the rest of the Ag^+ ions in the altered crystals are reduced to give metallic silver. Any

unaltered silver halide crystals are then removed by the fixing process. The dark areas on the photographic film provide a two-dimensional representation of the distribution of the radiolabel in the original sample.

Direct autoradiography is best suited to detection of weak to medium-strength β-emitting radionuclides (such as ^3H or ^{35}S). However, high-energy β-particles (such as those from ^{32}P) will pass through the film, wasting most of the energy (**Table 1**). For samples emitting high-energy radiation, a modification is needed in which the emitted energy is converted to light by a suitable chemical (scintillator or fluor). Indirect autoradiography uses sheets of a solid inorganic scintillator as intensifying screens, which are placed behind the photographic film. Those emissions that pass through the photographic emulsion are absorbed by the screen and converted to light, which then also reduces the Ag^+ and intensifies the direct autoradiographic image.

TABLE 1 CHARACTERISTICS OF RADIOISOTOPES COMMONLY USED FOR LABELING DNA AND RNA PROBES

Isotope	Half-life	Decay type	Emission energy (MeV)	Exposure time	Suitability for high-resolution studies
^3H	12.4 years	β⁻	0.019	very long	excellent
^{32}P	14.3 days	β⁻	1.710	short	poor
^{33}P	25.5 days	β⁻	0.248	intermediate	intermediate
^{35}S	87.4 days	β⁻	0.167	intermediate	intermediate

in the original sample. The intensity of the autoradiographic signal is dependent on the energy of the radiation emitted and the duration of exposure.

The radioisotope ^{32}P has been used widely in nucleic acid hybridization assays because it emits readily detected high-energy β-particles. However, the high-energy particles also travel farther, spreading the signal, and so are disadvantageous when fine physical resolution is required. Alternative radioisotopes such as ^{35}S are more useful for techniques such as studying the expression of genes in cells and tissues in which morphological resolution is required.

Radioisotopes are readily detected, but they constitute health hazards and the radioactivity decays with time, making it necessary to synthesize fresh probes before each experiment. Thus, nonisotopic labels containing distinctive chemical groups that are both stable and efficiently detected are now routinely employed.

Fluorophores are commonly used in nonisotopic labeling of nucleic acids

Nonisotopic labeling of nucleic acids involves the incorporation of nucleotides containing a chemical group or molecule that can be readily and specifically detected. The incorporated group may be detected by a direct assay in which the incorporated group directly serves as a label that is measured by the assay. Often a **fluorophore** is used, a chemical group that can readily be detected because it absorbs energy of a specific wavelength (**excitation wavelength**) and re-emits the energy at a longer, but equally specific, wavelength (**emission wavelength**) (**Box 7.3**).

Alternatively, an indirect assay is used, in which the incorporated chemical group serves as a **reporter** that is specifically recognized and bound by some affinity molecule, such as a dedicated antibody. The affinity molecule has a **marker** group bound to it, a chemical group or molecule that can be assayed in some way (**Figure 7.7**).

BOX 7.3 FLUORESCENCE LABELING OF NUCLEIC ACIDS

Fluorescence labeling of nucleic acids was developed in the 1980s and has proved to be extremely valuable in many different applications, including chromosome *in situ* hybridization, tissue *in situ* hybridization, and automated DNA sequencing.

A **fluorophore** is a chemical group that absorbs energy when exposed to a specific wavelength of light (excitation) and then re-emits it at a specific but longer wavelength (**Table 1** and **Figure 1**). Direct labeling of nucleic acids with fluorophores is achieved by incorporating a modified nucleotide (often 2′ deoxyuridine 5′ triphosphate) containing an appropriate fluorophore.

Indirect labeling systems are also used. In this technique, the fluorophore is used as a marker and is attached to an affinity molecule (such as streptavidin or a digoxigenin-specific antibody) that binds specifically to modified nucleotides containing a reporter group (such as biotin or digoxigenin) (see Figure 7.7).

Figure 1 Structure of two common fluorophores. TRITC and a variety of other fluorophores have been derived from rhodamine.

fluorescein

rhodamine

TABLE 1 FLUOROPHORES FOR LABELING NUCLEIC ACIDS

Fluorophore	Maximum wavelength (nm)	
	Excitation	Emission
Blue		
AMCA	350	450
DAPI	358	461
Green		
FITC	492	520
Fluorescein (see Figure 1)	494	523
Red		
CY3	550	570
TRITC	554	575
Rhodamine (see Figure 1)	570	590
Texas red	596	620
CY5	650	670

AMCA, aminomethylcoumarin; DAPI, 4′, 6-diamidino-2-phenylindole; FITC, fluorescein isothiocyanate; CY3, indocarbocyanine; TRITC, tetramethylrhodamine isothiocyanate; CY5, indodicarbocyanine.

Detection of fluorophore-labeled nucleic acids

Fluorophore-labeled nucleic acids can be detected with laser scanners or by fluorescence microscopy. The fluorophores are detected by passing a beam of light from a suitable light source (an argon laser is used in automated DNA sequencing, a mercury vapor lamp is used in fluorescence microscopy) through an appropriate color filter. The filter is designed to transmit light at the desired *excitation wavelength*. In fluorescence microscopy systems, this light is reflected onto the fluorescently labeled sample on a microscope slide by using a dichroic mirror that reflects light of certain wavelengths while allowing light of other wavelengths to pass straight through (**Figure 2**). The light then excites the fluorophore to fluoresce; as it does so, it emits light at a

slightly longer wavelength, the *emission wavelength*. The light emitted by the fluorophore passes back up and straight through the dichroic mirror, through an appropriate barrier filter and into the microscope eyepiece. A second beam-splitting device can also permit the light to be recorded in a CCD (charge-coupled device) camera.

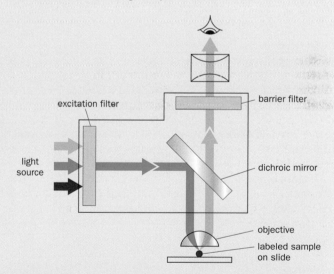

Figure 2 Fluorescence microscopy. The excitation filter allows light of only an appropriate wavelength (blue in this example) to pass through. The transmitted blue light is reflected by the dichroic (beam-splitting) mirror onto the labeled sample, which then fluoresces and emits light of a longer wavelength (green light in this case). The emitted green light passes straight through the dichroic mirror and then through a second barrier filter, which blocks unwanted fluorescent signals, leaving the desired green fluorescence emission to pass through to the eyepiece of the microscope.

There are two widely used indirect label detection systems. The biotin–streptavidin system depends on the extremely high affinity between two naturally occurring ligands. Biotin (a vitamin) acts as the reporter, and is specifically bound by the bacterial protein streptavidin with an affinity constant (also known as the dissociation constant) of 10^{-14}, one of the strongest known in biology. Biotinylated

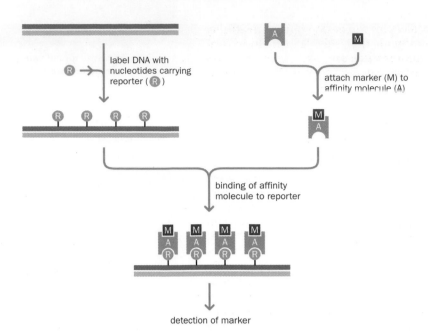

Figure 7.7 Indirect detection of labeled groups in nucleic acids. Nucleic acids can be labeled with chemical groups that are not detected directly. Instead, the incorporated groups serve as *reporter* groups that are bound with high specificity by an affinity molecule carrying a detectable marker. The marker can be detected in various ways. If it carries a specific fluorescent dye, it can be detected by fluorescence microscopy. A common alternative involves using an enzyme such as alkaline phosphatase to convert a substrate to give a colored product that is measured colorimetrically.

probes can be made easily by including a suitable biotinylated nucleotide in the labeling reaction (**Figure 7.8**). Streptavidin then serves as the affinity molecule. Another widely used reporter is digoxigenin, a steroid obtained from *Digitalis* plants (see Figure 7.8). A specific antibody raised against digoxigenin acts as the affinity molecule.

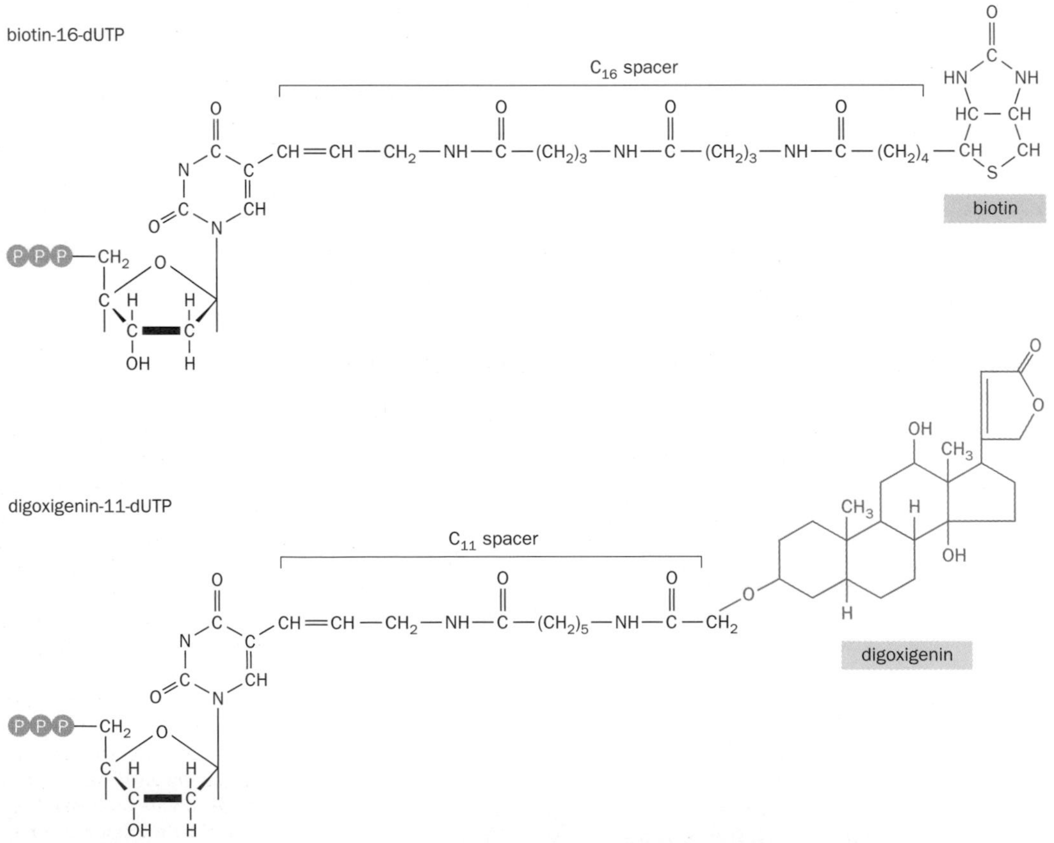

Figure 7.8 Structure of biotin- and digoxigenin-modified nucleotides. The biotin and digoxigenin reporter groups shown here are linked to the 5′ carbon atom of the uridine of dUTP by spacer groups consisting, respectively, of a total of 16 carbon atoms (biotin-16-dUTP) or 11 carbon atoms (digoxigenin-11-dUTP). The spacer groups are needed to ensure physical separation of the reporter group from the nucleic acid backbone, so that the reporter group protrudes sufficiently far to allow the affinity molecule to bind to it.

A variety of different marker groups or molecules can be conjugated to affinity molecules such as streptavidin and the digoxigenin antibody. They include various fluorophores that can be detected by fluorescence microscopy, or enzymes such as alkaline phosphatase and peroxidase that can permit detection by means of colorimetric assays or chemical luminescence assays.

7.3   HYBRIDIZATION TO IMMOBILIZED TARGET NUCLEIC ACIDS

Many different types of nucleic acid hybridization involve binding a target nucleic acid population to a solid support and then exposing it to a solution of labeled probe. Typically the probe is homogeneous and the object is to identify target sequences within a complex nucleic acid population in a test sample of interest. The nucleic acid population in the test sample may have been purified and deposited directly on the support or after previous size fractionation. Alternatively, cells or chromosome preparations may be fixed to a solid support and then treated to make their DNA or RNA available for hybridization.

After probe–target heteroduplexes have formed, unbound labeled probe is washed off, and the support is further washed and dried. After this treatment, any label detected on the support is expected to derive from heteroduplexes containing a target strand that is fixed to the support and also hydrogen-bonded to a labeled probe strand.

Dot-blot hybridization offers rapid screening and often employs allele-specific oligonucleotide probes

In dot-blotting the test sample is an aqueous solution of purified DNA or RNA, for example total human genomic DNA, and is simply spotted onto a nitrocellulose or nylon membrane and then allowed to dry. The variant technique of slot-blotting involves pipetting the DNA through an individual slot in a suitable template. In both methods the DNA sequences are then denatured before hybridization with a denatured labeled probe.

A useful application of dot-blotting involves distinguishing between alleles that differ by even a single nucleotide substitution. To do this, allele-specific oligonucleotide (ASO) probes are constructed from sequences spanning the variant nucleotide site. ASO probes are typically 15–20 nucleotides long and are normally employed under hybridization conditions in which the DNA duplex between probe and target is stable only if there is *perfect* base complementarity between them: a single mismatch between probe and target sequence is sufficient to render the short heteroduplex unstable.

Typically, two ASOs are designed to represent the two alleles, so that the single nucleotide difference between alleles occurs in a central segment of the oligonucleotide sequence (to maximize the thermodynamic instability of a mismatched duplex). Such discrimination can be employed for a variety of research and diagnostic purposes, for example in distinguishing the sickle-cell allele from normal β-globin alleles (**Figure 7.9**).

Figure 7.9 Testing for the sickle-cell mutation with dot-blot hybridization. The sickle-cell mutation is a single nucleotide substitution (A→T) at codon 6 in the β-globin gene, resulting in a GAG (Glu) →GTG (Val) substitution. Allele-specific oligonucleotides (ASOs) were designed to match the nucleotide sequence flanking the substitution: βA-ASO is specific for the normal β-globin allele and βS-ASO for the mutant β-globin allele. The schematic dot-blots show the results of probing with these two ASOs under stringent hybridization conditions. With the normal probe (βA-ASO), the results are positive (filled circle) for normal individuals and for heterozygotes, but negative for sickle-cell homozygotes (unfilled circle). With the mutant probe (βS-ASO), the results are positive for the sickle-cell homozygotes and heterozygotes, but negative for normal individuals.

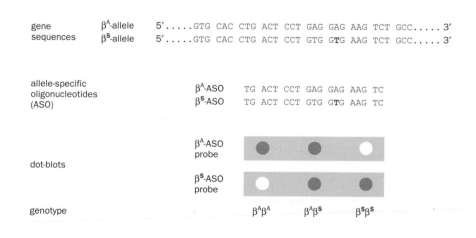

Southern and northern blot hybridizations detect size-fractionated DNA and RNA

Southern blot hybridization

A sample population of purified DNA is digested with one or more restriction endonucleases, generating fragments that are several hundred to thousands of base pairs in length. The restriction fragments are separated according to size by agarose gel electrophoresis (**Box 7.4**), denatured, and transferred to a nitrocellulose or nylon membrane. Labeled probes are hybridized to the membrane-bound target DNA, and the positions of the labeled heteroduplexes are revealed by autoradiography (**Figure 7.10**).

An important application of Southern blot hybridization is to identify target sequences that are similar but not identical to the gene used as a probe. The targets may be members of a family of evolutionarily related genes or DNA sequences within the same genome or a direct equivalent to the probe within another genome. In the latter case, genomes can be sampled from different individuals of the same species, or they can be from different species (a *zooblot*). Once a newly isolated probe is demonstrated to be related to other uncharacterized sequences, attempts can then be made to isolate the other members of the family by screening appropriate DNA libraries.

Northern blot hybridization

Northern blot hybridization is a variant of Southern blotting in which the samples contain undigested RNA instead of DNA. The principal use of this method is to obtain information on the expression patterns of specific genes. Once a gene has been cloned, it can be used as a probe and hybridized against a northern blot containing samples of RNA isolated from a variety of tissues (**Figure 7.11**). The data obtained can provide information on the range of cell types in which the gene is expressed, and the relative abundance of transcripts, as measured by the relative intensity of the hybridization band. Additionally, by revealing transcripts of different sizes, it may provide evidence for different isoforms, for example those resulting from alternative promoters, splice sites, or polyadenylation sites.

BOX 7.4 ELECTROPHORESIS OF NUCLEIC ACIDS

The phosphodiester backbone of nucleic acids means that they carry numerous negatively charged phosphate groups, so that when present in an electric field they will migrate in the direction of the positive electrode. By migrating through a porous gel, nucleic acid molecules can be separated according to size. This happens because the porous gel acts as a sieve, and small nucleic acid molecules pass easily through the pores of the gel but larger nucleic acid molecules are impeded by frictional forces.

Traditionally, quite small to moderately large nucleic acid molecules have been separated by using agarose slab gels with individual samples loaded into cut-out wells so that different lanes are used by the migrating samples (**Figure 1**). After electrophoresis, the gels are stained with chemicals, such as ethidium bromide or SYBR green, that bind to nucleic acids and that can fluoresce when exposed to ultraviolet radiation.

For fragments between 0.1 kb and 30 kb long, the migration speed depends on fragment length, but scarcely at all on the base composition. However, conventional agarose gel electrophoresis is not well suited to separating rather small or very large nucleic acid molecules. The percentage of agarose can be varied to increase resolution—decreasing the percentage helps to separate larger fragments, and increasing it makes it easier to separate smaller fragments. To get superior resolution of smaller nucleic acids, however, polyacrylamide gels are normally used, and they are used in DNA sequencing to separate fragments that differ in length by just a single nucleotide.

Very large DNA fragments are very poorly separated by conventional agarose gel electrophoresis. Instead, specialized equipment is used in which a discontinuous electric field is used. During electrophoresis the direction or polarity of the electric field is intermittently changed. The electric field may be pulsed so that a particular negative–positive orientation is established for a short time before being reversed for another short interval and the switching of the field is performed recurrently. In *pulsed-field gel electrophoresis*, the intermittent switching of the electric field forces the DNA to reorient to be able to migrate in the new electric field; the time taken to reorient depends on the length of the molecule. As a result, it is possible to separate fragments of DNA by size up to several megabases in length.

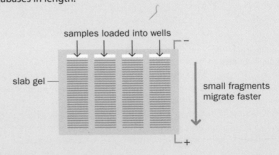

Figure 1 Gel electrophoresis.

Figure 7.10 Southern blot hybridization. A complex DNA sample is digested with restriction endonucleases. The resulting fragments are applied to an agarose gel and separated by size using electrophoresis (see Box 7.4). The gel is treated with alkali to denature the DNA fragments and is then placed against a nitrocellulose or nylon membrane. DNA will be transferred from the gel to the membrane, which is then soaked in a solution containing a radiolabeled single-stranded DNA probe. After hybridization, the membrane is washed to remove excess probe and then dried. The membrane is placed against an X-ray film and the position of the labeled probe will cause a latent image on the film that can be revealed by development of the autoradiograph (see Box 7.2) as a hybridization band.

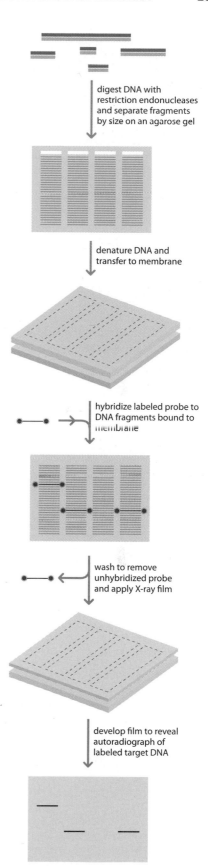

digest DNA with restriction endonucleases and separate fragments by size on an agarose gel

denature DNA and transfer to membrane

hybridize labeled probe to DNA fragments bound to membrane

wash to remove unhybridized probe and apply X-ray film

develop film to reveal autoradiograph of labeled target DNA

In an *in situ* hybridization test, sample DNA or RNA is immobilized within fixed chromosome or cell preparations

Some hybridization assays rely on hybridization to target nucleic acid within fixed chromosomes or cells trapped on microscope slides. Because the nucleic acid is immobilized within native structures the hybridization is said to occur *in situ.*

Chromosome *in situ* hybridization

A simple procedure for mapping genes and other DNA sequences is to hybridize a suitable labeled DNA probe against chromosomal DNA that has been denatured *in situ.* To do this, an air-dried microscope slide chromosome preparation is made, typically using metaphase or prometaphase chromosomes from peripheral blood lymphocytes or lymphoblastoid cell lines. RNA and protein are removed from the sample by treatment with RNase and proteinase K, and the remaining chromosomal DNA is denatured by exposure to formamide. The denatured DNA is then available for *in situ* hybridization with an added solution containing a labeled nucleic acid probe, overlaid with a coverslip.

Depending on the particular technique, chromosome banding of the chromosomes (see Chapter 2) can be arranged either before or after the hybridization step. As a result, the signal obtained after the removal of excess probe can be correlated with the chromosome band pattern to identify a map location for the

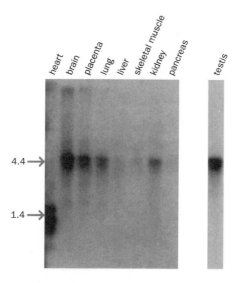

Figure 7.11 Northern blot hybridization. Northern blotting uses samples of total RNA or purified poly(A)$^+$ mRNA from tissues or cells of interest. The RNA is size-fractionated by electrophoresis, transferred to a membrane, and hybridized with a suitable labeled nucleic acid probe. In the example shown here, the probe was a labeled cDNA from the *FMR1* (fragile X mental retardation syndrome) gene. The results show comparative expression of *FMR1* in different tissues: strongest expression was found in the brain and testis (4.4 kb), with decreasing gene expression in the placenta, lung, and kidney, respectively, and almost undetectable expression in liver, skeletal muscle, and pancreas. Although no 4.4 kb transcript was evident in heart tissue, smaller transcripts (about 1.4 kb) were evident that could have been generated by alternative splicing or aberrant transcription. [From Hinds HL, Ashley CT, Sutcliffe JS et al. (1993) *Nat. Genet.* 3, 36–43. With permission from Macmillan Publishers Ltd.]

DNA sequences recognized by the probe. Chromosome *in situ* hybridization has been revolutionized by the use of fluorescently labeled probes in fluorescence *in situ* hybridization (FISH) techniques (see Chapter 2).

Tissue *in situ* hybridization

Labeled probe can also be hybridized against RNA in tissue sections. Very thin tissue sections are cut with a cryostat either from tissue blocks embedded in paraffin wax or from frozen tissue, and then mounted on glass slides. A hybridization mix including the probe is applied to the section on the slide and covered with a glass coverslip.

Single-stranded complementary RNA (cRNA) probes are preferred for this method. Because they must be complementary to the mRNA of a gene, antisense riboprobes are obtained by cloning a gene in the reverse orientation in a suitable vector such as pSP64 (see Figure 7.6). In such cases, the phage polymerase is designed to synthesize labeled transcripts from the opposite DNA strand to that normally transcribed *in vivo*.

The riboprobes can be labeled with a radioisotope such as ^{33}P or ^{35}S, and the hybridized probes are detected with autoradiographic procedures. The localization of the silver grains is often detected with only *dark-field microscopy*. Here, direct light is not allowed to reach the objective; instead, the illuminating rays of light are directed from the side so that only scattered light enters the microscope lenses and the signal appears as an illuminated object against a black background. However, better signal detection is possible with *bright-field microscopy*, in which the image is obtained by direct transmission of light through the sample (**Figure 7.12**). Alternatively, probes are subjected to fluorescence labeling, and detection is accomplished by fluorescence microscopy (see Box 7.3).

Hybridization can be used to screen bacterial colonies containing recombinant DNA

Bacterial colonies containing recombinant DNA are often screened to identify those that contain sequences related to a labeled probe of interest. To do this, the cell colonies are allowed to grow on an agar surface and then transferred by surface contact to a nitrocellulose or nylon membrane, a process known as *colony hybridization* (**Figure 7.13**). Alternatively, the cell mixture is spread out on a nitrocellulose or nylon membrane placed on top of a nutrient agar surface, and colonies are allowed to form directly on top of the membrane. In either approach, the membrane is then exposed to alkali to denature the DNA before it is hybridized with a labeled nucleic acid probe.

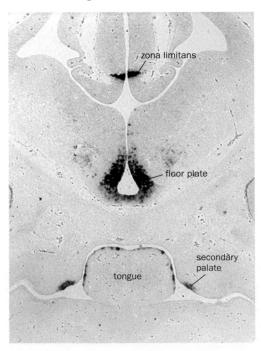

Figure 7.12 Tissue *in situ* hybridization. The example shows the expression pattern produced by hybridizing a digoxigenin-labeled Sonic hedgehog antisense riboprobe against a section of tissue from a mouse embryo at 13.5 embryonic days. The section is at the level of a part of the forebrain (top and middle) and the upper and lower jaws (bottom). Detection of the probe was by an alkaline phosphatase-coupled blue color assay. Strong expression was seen in very specific regions of the developing central nervous system, as indicated, and also in the secondary palate. (Courtesy of Mitsushiro Nakatomi & Heiko Peters, Newcastle University.)

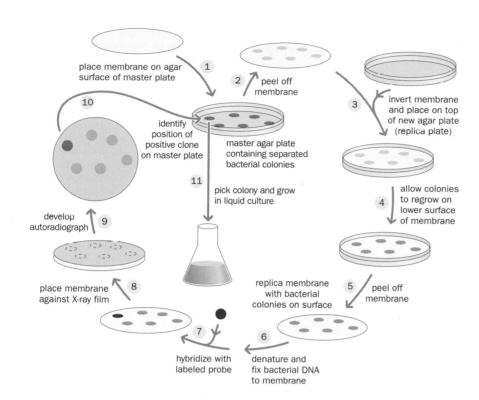

Figure 7.13 Colony blot hybridization. When a nylon or nitrocellulose filter is placed on top of discrete bacterial colonies on an agar plate, some cells from the colonies adhere to the membrane. The membrane is peeled off and then transferred to a new culture dish to allow the transferred cells to grow into colonies. The resulting replica membrane will have the same physical representation of bacterial colonies as the original culture dish (the master agar plate). The replica membrane is treated to denature the attached DNA and fix it to the surface of the membrane. The dried membrane is then hybridized with a labeled nucleic acid probe. After a wash to remove any unbound labeled probe, X-ray film is used to create an autoradiograph that reveals the position of labeled colonies on the replica membrane that can then be used to identify the equivalent colony on the original master plate. The positive colonies can be picked and grown individually in liquid culture.

After hybridization, the probe solution is removed, and the filter is washed extensively, dried, and processed to detect the bound labeled probe. The position of strong signals from bound probe is related back to a master plate containing the original pattern of colonies, to identify colonies containing DNA related to the probe. These can then be individually picked and grown in liquid culture to obtain sufficient quantities for extraction and purification of the recombinant DNA.

Once it became possible to create complex DNA libraries, more efficient automated methods of clone screening were required. Now robotic gridding devices can be used to pipette from clones arranged in microtiter dishes into predetermined linear coordinates on a membrane. The resulting high-density clone filters can permit rapid and efficient library screening, as described in Chapter 8.

7.4   MICROARRAY-BASED HYBRIDIZATION ASSAYS

Microarray hybridization allows highly parallel hybridization assays using thousands of different immobilized probes

Innovative and powerful microarray hybridization technologies were developed in the early 1990s to permit massively parallel assays, in which thousands of single hybridization assays are simultaneously conducted under the same conditions. Microarray hybridization was first envisaged as a tool for large-scale mapping and for sequencing DNA. However, the major driving force for the technology was the need for simultaneous analysis of the expression of very large numbers of genes. As cDNA libraries were developed and large numbers of characterized cDNA clones became available, researchers began to plan genome-wide analyses of gene expression. Multiplex hybridizations were needed using large numbers of gene-specific DNA probes simultaneously. Subsequently, microarray hybridization was used to assay DNA variation as well as for profiling gene expression.

When used in a hybridization assay, a **microarray** consists of many thousands of unlabeled DNA or oligonucleotide probe populations fixed to a glass or other suitable surface at precisely defined coordinates in a high-density grid format. A test sample, an aqueous solution containing a complex population of fluorescently labeled denatured DNA or RNA, is hybridized to the probe molecules on the microarray. After washing and drying steps, bound fluorescent label is detected with a high-resolution laser scanner, and the signal emitted from each

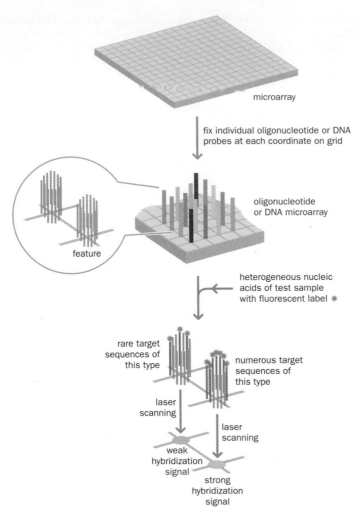

Figure 7.14 Principle of microarray hybridization. A microarray is a solid surface on which molecules can be fixed at specific coordinates in a high-density grid format. Microarray hybridization uses oligonucleotide or DNA microarrays that are prepared by fixing thousands of synthetic single-stranded oligonucleotide or DNA probes at specific predetermined positions in the grid. As shown by the expanded item, each specific position will have many thousands of identical copies of a single type of oligonucleotide or DNA probe (a *feature*). A test sample containing a heterogeneous collection of labeled DNA fragments or RNA transcripts is added to allow hybridization with the probes on the array. The target sequences bound by some probes may be numerous, resulting in a strong hybridization signal; for other probes there may be few target sequences in the test sample, resulting in a weak hybridization signal. After a washing and drying step, the hybridization signals for the thousands of different probes in the array are detected by laser scanning, giving huge amounts of data from a single experiment.

spot on the array is analyzed with digital imaging software that converts the signal into one of a palette of colors according to its intensity.

We will consider practical examples of microarray use in subsequent chapters, such as in Chapter 8 where we consider how gene expression is analyzed. Here we are concerned with the basic principles and technologies. The object of microarray hybridization is for each of the thousands of different probes to form heteroduplexes with target sequences in the labeled nucleic acids of the test sample. At any *one* position on the array numerous identical copies of just *one* type of oligonucleotide or DNA will be available for hybridization (**Figure 7.14**).

Target sequences will be present at different amounts in the sample under study. The amount of specific target DNA bound by each probe will depend on the frequency of that specific target DNA in the test sample. A probe that binds a relatively rare target nucleic acid will have less fluorescent label bound to it, resulting in a weak hybridization signal, whereas positions on the array that are strongly fluorescent after washing indicate probes that have bound an abundant target nucleic acid. Quantitating the amount of fluorescent label bound across all the positions in a microarray gives a huge data set that can reflect the relative frequencies of numerous target sequences in the sample population.

High-density oligonucleotide microarrays offer enormously powerful tools for analyzing complex RNA and DNA samples

The construction of first-generation DNA microarrays involved robotic spotting of DNA clones or synthetic oligonucleotides onto chemically treated glass microscope slides. However, modern DNA microarrays often use oligodeoxynucleotide probes because they are versatile. Oligonucleotides can be designed to be specific for just one target, and the synthesis of oligonucleotides to order means that high-density microarrays can be produced rapidly and efficiently. Data reliability can also be more efficiently assessed in many cases; for example, the expression of a single gene can be followed simultaneously with many different oligonucleotide probes, each of which is specific for that gene.

Most modern applications use microarrays produced by alternative ingenious technologies developed by two commercial companies. Affymetrix's approach was modeled on silicon chip manufacture, prompting the term **DNA chip** (although the term DNA chip can now mean any high-density DNA or oligonucleotide microarray). Illumina's technology depends on a random assembly of arrays of beads in wells.

Affymetrix oligonucleotide microarrays

Affymetrix chip technology involves light-directed *in situ* synthesis of oligonucleotide probes on the microarray, and can produce more than 6 million different oligonucleotide populations on a 1.7 cm^2 surface. Each type of oligonucleotide is assembled by adding mononucleotides sequentially to a linker molecule that terminates with a photolabile protecting group. A photomask (photolithograph) is used to determine which positions on the microarray will be exposed to an external light source, resulting in the destruction of photolabile groups. A chemical coupling reaction is then used to add a particular type of nucleotide to newly deprotected sites. The process is repeated sequentially with different masks for each of the other three nucleotides to achieve the addition of a single nucleotide—A, C, G, or T—to each linker molecule. Then the process is repeated to extend oligonucleotide synthesis by a further 24 nucleotides or so. Depending on the arrangement of holes cut in the masks used at each step, specific predetermined oligonucleotide sequences can thus be assembled at predetermined locations on the microarray (**Figure 7.15**).

Illumina oligonucleotide microarrays

Illumina's BeadArray™ technology depends on a random assembly of arrays of beads in wells. The beads are silica microspheres 3 μm in diameter and form the array elements. The oligonucleotides are pre-synthesized to be more than 70 nucleotides long (consisting of a 25-nucleotide address code and a 50-nucleotide gene-specific sequence). Each oligonucleotide is coupled to a particular batch of beads, and individual beads carry more than 100,000 identical oligonucleotides. The same principle underlies a new sequencing technology marketed by the Illumina company as will be described in Chapter 8.

Beads carrying different oligonucleotides are pooled and then spread across prefabricated microarrays with regularly spaced microwells designed to accept the bead size. The beads are immobilized within the cavities, and the 25-nucleotide addresses are decoded, allowing each bead to be identified. Thus, the arrangement of probes is random and positions are determined later. Hybridization of labeled target nucleic acids to complementary oligonucleotides in the bead arrays can then be tracked by scanning the array for the presence of bound label.

Microarray hybridization is used mostly in transcript profiling or assaying DNA variation

Although the technology for establishing DNA microarrays was developed only very recently, numerous important applications have already been developed, and their impact on future biomedical research and diagnostic approaches is expected to be profound. Most applications involve tracking gene transcripts or assaying DNA variation.

The first application—and still a very important one—was in simultaneously assessing the abundance of transcripts of multiple genes. Initially, small subsets of the total number of genes in complex genomes, such as the human genome, were assayed, but whole genome gene sets can now be assayed. This application will be explored further in Chapter 8. Transcript profiling has found many applications, notably in comparing expression profiles of closely related cell types, cells that are otherwise identical but differ in the presence or absence of a pathogenic mutation, and identical cells that are differently exposed to defined environments, such as the presence of a drug. Other aspects of transcript expression can also be assayed, such as the use of oligonucleotide probes to assay splice variants.

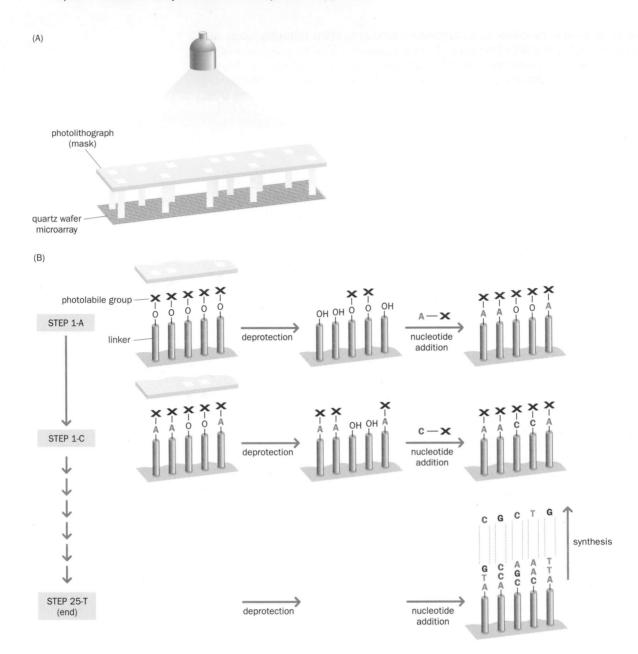

Figure 7.15 Construction of oligonucleotide microarrays by using light-directed oligonucleotide synthesis *in situ*.
(A) Affymetrix oligonucleotide microarrays are constructed by synthesizing thousands of oligonucleotides at predetermined positions on a quartz wafer microarray, using a photolithograph (mask) to determine the position of each added nucleotide. (B) Each oligonucleotide is assembled starting from a linker molecule that is fixed to the array and whose free end carries a protective photolabile group. A mask with holes cut into it at specific positions is placed above the array and light from an external source is shone through the holes. At the positions exposed to light the photolabile groups are destroyed (deprotection). A specific mononucleotide coupled to a photolabile group is added to the array and will combine with deprotected strands. The process is repeated three times with another three masks with holes cut in different positions, introducing in turn one of the other three nucleotides. The net result is that each linker molecule will have a single nucleotide attached, either A, C, G, or T, in a specific pre-programmed way. The sequential use of four different masks to allow coupling of the four different nucleotides is repeated for a further 24 cycles. The end result is an assembly of oligonucleotides 25 nucleotides long, each with a precisely determined sequence at a specific coordinate in the microarray.

Increasingly, pathway-specific microarrays are being developed to monitor the individual expression profiles for sets of closely related genes that participate in the same biological pathway. Diagnostic applications are also becoming important. For example, transcript profiles for key genes provide molecular fingerprints for defining different types and grades of tumors.

Several applications have been devised for tracking DNA variation. In addition to assaying mutations in known disease genes, vigorous efforts have been made to identify and catalogue human **single nucleotide polymorphism** (**SNP**) markers. As we will describe later, large-scale DNA variation analyses often use microarrays that have bacterial artificial chromosome DNA clones as probes that seek to identify large subchromosomal deletions and duplications.

In addition to microarray hybridization, it should be noted that microarrays have many other applications. Applications include the use of protein arrays, arrays that assay binding between nucleic acids and proteins, and many others.

FURTHER READING

General

Sambrook J & Russell DW (2001) Molecular Cloning. A Laboratory Manual. Cold Spring Harbor Laboratory Press.

Fluorescence labeling

Invitrogen Molecular Probes—The Handbook: A Guide To Fluorescent Probes And Labeling Technologies. http://www.invitrogen.com/site/us/en/home/References/Molecular-Probes-The-Handbook.html

In situ hybridization

Wilkinson D (1998) *In Situ* Hybridization: A Practical Approach, 2nd ed. IRL Press.

Microarray hybridization principles and basic technology

Various authors (1999) The Chipping Forecast. *Nat. Genet.* 21 (Suppl.), 1–60.
Various authors (2002) The Chipping Forecast II. *Nat. Genet.* 32 (Suppl.), 465–551.

Affymetrix and Illumina oligonucleotide microarray technology

Affymetrix GeneChip™. http://www.affymetrix.com/technology/index.affx
Gunderson KL, Kruglyak S, Graige MS et al. (2004) Decoding randomly ordered DNA arrays. *Genome Res.* 14, 870–877.
Illumina BeadArray™. http://www.illumina.com/pages.ilmn?ID=5

Microarray hybridization applications

Falciani F (ed.) (2007) Microarray Technology Through Applications. Taylor and Francis.
Hoheisel JD (2006) Microarray technology: beyond transcript profiling and genotype analysis. *Nat. Rev. Genet.* 7, 200–210.

Chapter 8

Analyzing the Structure and Expression of Genes and Genomes

8

KEY CONCEPTS

- DNA libraries are collections of cloned DNA fragments that collectively represent the genome of an organism (genomic DNA libraries) or genome subfractions.

- cDNA libraries are derived from reverse-transcribed RNA and so contain only the expressed part of the genome.

- DNA sequencing usually relies on synthesizing new DNA strands from single-stranded templates. The most modern methods sequence huge numbers of DNA samples in parallel without electrophoresis.

- Genome resequencing entails using an initial genome sequence as a reference for rapid resequencing of the genomes or genome subfractions of other individuals from the same species.

- DNA markers are DNA sequences that have a unique subchromosomal location and can be conveniently assayed. Marker maps provided important framework maps for the sequencing of complex genomes that have numerous repetitive DNA elements.

- Polymorphic DNA markers are needed to make genetic maps. The different markers are genotyped in individuals spanning different generations of a common pedigree.

- Both non-polymorphic and polymorphic DNA markers can be used to make physical maps of chromosomes, often by genotyping panels of cells that have different fragments of individual chromosomes.

- A clone contig is a series of cloned genomic DNA fragments, arranged in the same linear order as the subchromosomal region from which they were derived. Overlaps between each DNA fragment allow them to be placed in order, providing important framework maps for genome sequencing.

- Computer-based gene prediction often relies on identifying significant sequence similarity between a test DNA sequence (or derived protein sequence) and a known gene sequence or protein.

- The terms *transcriptome* and *proteome* describe, respectively, the complete set of RNA transcripts and the complete set of proteins produced by a cell. Whereas the different nucleated cells of an organism have stable, almost identical genomes, transcriptomes and proteomes are dynamic and each can vary very significantly from cell to cell.

- High-resolution expression analyses are designed to obtain detailed gene expression patterns in cells or tissues but are limited to analyzing one or a few genes at a time.

- Highly parallel expression analyses provide gene expression information for many thousands of genes at a time in two or more cellular sources.

The advent of DNA cloning and DNA sequencing technologies in the 1970s ushered in a new era in which it became possible to characterize genes in a systematic way. By sequencing a gene and its transcript(s) it became possible to work out exon–intron organization and to predict the sequence of any likely protein products.

Initial progress was slow but quickly picked up as technologies improved. By the early 1980s, researchers began to make plans for sequencing all the different DNA molecules in a cell (collectively known as a **genome**), and by the start of the 1990s, internationally coordinated projects had been launched to determine the complete sequence of the human genome and that of several model organisms. The genome sequences paved the way for comprehensive analyses of gene structure and gene expression.

8.1   DNA LIBRARIES

The first attempts at cloning human DNA fragments in bacterial cells took advantage of gene transcripts that were naturally found at high concentrations in some tissues. Much of the mRNA made in erythrocytes consists of α- and β-globin mRNA, for example. When reverse transcriptase is used to copy erythrocyte mRNA, the resulting erythrocyte cDNA is greatly enriched in globin cDNA, facilitating its isolation.

To enable a more general method of cloning genes and indeed all DNA sequences, technologies were subsequently developed with the aim of cloning all of the constituent DNA sequences in a starting population. The resulting comprehensive collections of DNA clones became known as **DNA libraries**. Although PCR can be used to generate DNA libraries, cell-based DNA cloning has traditionally been used, partly because it is more suited to cloning large DNA fragments.

Genomic DNA libraries comprise fragmented copies of all the different DNA molecules in a cell

For any one organism, the cells that contain a nucleus have essentially the same DNA content. To make a genomic DNA library, the starting material can be DNA from any representative nucleated cell, such as easily accessible white blood cells. The isolated genomic DNA is fragmented, often by cutting with a restriction endonuclease that recognizes a 4 bp sequence. For example, *Mbo*I recognizes the sequence GATC.

By controlling the digestion, it is possible to get the restriction enzyme to cleave DNA at only a small fraction of the available restriction sites; typically the enzyme is used at low concentration and incubation is performed for only a short time. Such *partial digestion* is designed to ensure that the average size of fragments produced is the optimum size for the cloning system that will be used (see Table 6.2).

In addition, because only a very small percentage of the possible restriction sites are cut by the enzyme, the DNA is *randomly* fragmented. Because the starting material usually consists of millions of identical cells, there are millions of copies of each of the different DNA molecules. Random DNA fragmentation means that for any given location on a starting DNA molecule, the pattern of cutting will vary randomly on the different copies of that DNA molecule (**Figure 8.1**).

The random fragmentation ensures that the library will contain as much representation as possible of the starting DNA. It also results in clones with overlapping inserts (see Figure 8.1). This means that any selected DNA clone from the library can be used to retrieve clones with inserts that overlap with it. As a result, the clones can be arranged in an order that corresponds to the chromosomal order of the cloned DNA sequences. This will be discussed further in the context of genome projects in Section 8.3.

cDNA libraries comprise DNA copies of the different RNA molecules in a cell

Unlike genomic DNA, the RNA content of an individual's cells can vary enormously from one cell type to another, and from one developmental stage to

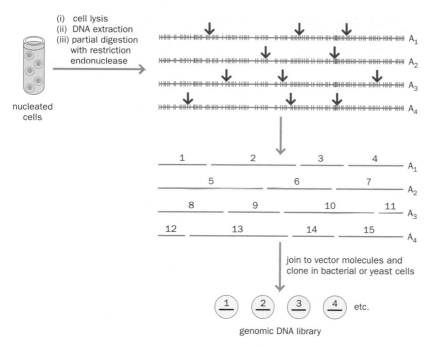

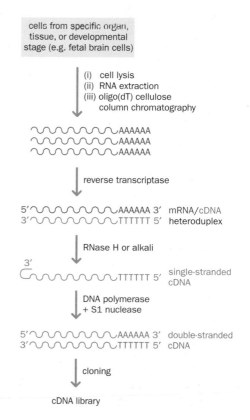

Figure 8.1 Making a genomic DNA library. All nucleated cells of an individual will have essentially the same genomic DNA content so that any easily accessible cells (such as blood cells) can be used as source material. Because DNA is extracted from numerous cells with identical DNA molecules, the isolated DNA will contain large numbers of identical DNA sequences, such as A_1, A_2, A_3, and A_4 shown here. However, partial digestion with a restriction endonuclease will cleave the DNA at only a fraction of the available restriction sites (short vertical blue bars indicate all restriction sites the enzyme could cut if the enzyme were used under normal conditions; vertical red arrows mark the positions of the small number of restriction sites that are cut). As a result, the pattern will differ between identical copies of the same chromosomal DNA molecules, resulting in almost random cleavage. This will generate a series of restriction fragments that, if they derive from the same subchromosomal region, may share some common DNA sequence (e.g. fragment 6 from original molecule A_2 partly overlaps fragments 2 and 3 from A_1, fragments 9 and 10 from A_3, and fragments 13 and 14 from A_4).

another. The starting material is usually total RNA from a desired tissue, from a cell line, or from an embryo at a specific developmental stage.

Until recently, the polypeptide-encoding mRNA fraction was considered to be virtually the only RNA class of interest, and because almost all mRNA is poly-adenylated, poly(A)+ mRNA would be selected by specific binding to a complementary oligo(dT) or poly(U) sequence connected to a solid matrix. Reverse transcriptase would then be used to convert the isolated poly(A)+ mRNA to a double-stranded cDNA copy (**Figure 8.2**). More recently, it has become clear that a substantial fraction of functional RNA is not polyadenylated, and so total cellular RNA is prepared and random hexanucleotide primers are used with reverse transcriptase to make cDNA.

An initial rationale for cDNA libraries was to provide a quick route toward gene clones (human coding DNA sequences typically account for only just over 1% of the DNA in a human cell). Because of differences in gene expression patterns, a wide variety of cDNA libraries have been constructed from different tissues, cell lines, and developmental stages for many organisms.

To be useful, DNA libraries need to be conveniently screened and disseminated

Like any other cell-based DNA cloning system, DNA library screening requires the initial selection of transformed cells, often using antibiotic resistance conferred by the vector molecule. As described in Section 6.1, recombinants are often also identified by insertional inactivation; the insertion site is designed to interrupt a marker gene within the cloning vector so that any sizeable insert causes this marker gene to be inactivated.

Library screening

To identify a clone containing a specific recombinant DNA, the library is screened by two methods. In **colony hybridization**, aliquots of cell clones from the DNA library are spotted in a high-density gridded formation onto a suitable nitrocellulose or nylon membrane and subjected to a cell lysis and DNA denaturation

Figure 8.2 Making a cDNA library. As shown here, the reverse transcriptase step often uses an oligo(dT) primer to prime the synthesis of the cDNA strand. More recently, mixtures of random hexanucleotide primers have been used instead to provide a more normal representation of sequences. RNase H will specifically digest RNA that is bound to DNA in an RNA–DNA hybrid. The 3′ end of the resulting single-stranded cDNA has a tendency to loop back to form a short hairpin. This can be used to prime second-strand synthesis by DNA polymerase, and the resulting short loop connecting the two strands can then be cleaved by the S1 nuclease, which specifically cleaves regions of DNA that are single-stranded.

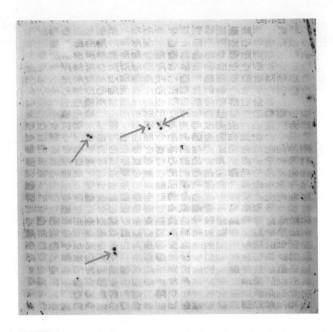

Figure 8.3 Screening a DNA library by colony hybridization. The autoradiograph is of a membrane containing 17,664 human YAC clones (total DNA from individual yeast clones containing human yeast artificial chromosomes) gridded in 6 × 6 clone units. The hybridization signals include weak signals from all clones obtained with a ^{35}S-labeled probe of total yeast DNA plus strongly hybridizing signals obtained with a ^{32}P-labeled unique human X chromosome probe (*DXYS646*). [Courtesy of Mark Ross, Sanger Institute, from Ross MT & Stanton VP (1995) Current Protocols in Human Genetics, vol. 1. With permission of Wiley-Liss, Inc., a subsidiary of John Wiley & Sons, Inc.]

treatment (see Figure 7.13). Addition of a labeled hybridization probe of interest can then identify cells on the membrane that have a related DNA insert (**Figure 8.3**), allowing the parent clone to be identified. Alternatively, PCR-based library screening is efficiently accomplished by screening different combinations of clones from microtiter dishes (**Figure 8.4**).

Library amplification and dissemination

Newly constructed libraries are said to be unamplified, although the initially transformed cells have been amplified to form visible cell colonies. To propagate the libraries, individually picked cell colonies are spotted in gridded arrays onto suitable membranes or into the wells of microtiter dishes, where they can be stored for long periods at about –70°C in the presence of a cell-stabilizing agent such as glycerol.

Some libraries are in high demand and are therefore distributed on a large scale. To do this, the original library needs to be amplified. For transport and long-term storage, small aliquots of the original libraries are diluted in cell medium containing a cell-stabilizing agent—the libraries can later be regenerated by plating out, to allow the growth of all cells containing recombinant DNA.

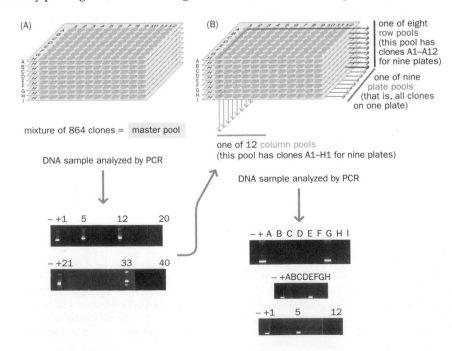

Figure 8.4 Screening a DNA library by PCR. Here, about 35,000 individual YAC clones were individually deposited into the 96 wells of 360 microtiter dishes. To facilitate screening, a total of 40 master pools were generated by combining all 864 clones in sets of nine microtiter dishes (plates A–I). (A) Primary screening. This involves PCR assay of the 40 master pools. In this example, three master pools were positive when referenced against positive (+) and negative (–) controls: pools 5, 12, and 33. (B) Secondary screening. Here, single YACs were identified by assaying different subsets of the 864 YACs in a positive master pool, in this case master pool 12. Three-dimensional screening of each of nine plate pools (96 YACs each), eight row pools (106 YACs each), and 12 column pools (72 YACs each) identified a positive YAC in plate 12G (top panel), row E (middle panel), column 5 (bottom panel). [Courtesy of Sandie Jones, Newcastle University and adapted from Jones MH, Khwaja OS, Briggs H et al. (1994) *Genomics* 24, 266–275. With permission from Academic Press Inc.]

During the amplification stage, different colonies may grow at different rates, however. For example, the growth of particular cells may be retarded because they contain a recombinant DNA that affects their metabolism in some way. Amplification may therefore result in a distortion of the original representation of cell clones in the unamplified library.

8.2 SEQUENCING DNA

The development of DNA libraries made it possible in principle to perform comprehensive DNA sequencing. Until very recently, much of modern DNA sequencing was based on an enzymatic sequencing method first developed in the 1970s, in which a DNA polymerase was used to synthesize new DNA chains by using a cloned single-stranded DNA template, consisting of millions of identical copies of a specific DNA sequence. Initially, DNA sequencing was slow and laborious and output was limited, but the need to scale up to conduct large-scale sequencing of library clones drove technological improvements.

Dideoxy DNA sequencing involves enzymatic DNA synthesis using base-specific chain terminators

The DNA sequencing method that was used to sequence the human genome and many other genomes was an enzymatic sequencing method pioneered by Fred Sanger in the mid-1970s. It relies on random inhibition of chain elongation, creating newly synthesized DNA strands of various lengths that can be separated by size. The DNA needs to be in a single-stranded form that will act as a template for making a new complementary DNA strand *in vitro* by using a suitable DNA polymerase.

The substrate for DNA sequencing was often a recombinant DNA that would be denatured so that a strand-specific sequencing primer could be used to direct new strand synthesis, or DNA fragments would be cloned into phagemid vectors that were manipulated to produce single-stranded recombinant DNA. Alternatively, and increasingly commonly, DNA produced by PCR amplification is used and converted to a single-stranded form to act as a sequencing template. The final product of either method is a population of many identical copies of the DNA to be sequenced.

Sequencing is conducted in four parallel reactions, each containing the four dNTPs (dATP, dCTP, dGTP, and dTTP) plus a small proportion of one of the four analogous dideoxynucleotides (ddNTPs) that will serve as a base-specific chain terminator. A ddNTP is closely related to its dNTP counterpart but lacks a hydroxyl group at the 3′ carbon position and also at the 2′ carbon (**Figure 8.5**). It can be incorporated into the growing DNA chain by forming a phosphodiester bond between its 5′ carbon atom and the 3′ carbon of the previously incorporated nucleotide. However, because it lacks a 3′ hydroxyl group, any dideoxynucleotide that is incorporated into a growing DNA chain cannot participate in phosphodiester bonding at its 3′ carbon atom. Once a dideoxynucleotide has been incorporated, therefore, it causes the abrupt termination of chain elongation.

By ensuring that one of the four dNTPs or the primer is labeled, the growing DNA strand becomes labeled. By setting the concentration of the dideoxynucleotide to be very much lower than that of the corresponding deoxynucleotide analog, there will be competition between a specific dideoxynucleotide and its deoxynucleotide counterpart for inclusion in the growing DNA chain. The deoxynucleotide is present in excess; when it is incorporated, chain elongation continues, but occasionally the dideoxynucleotide will be incorporated in the growing chain, ending polymerization and so causing chain termination. Each reaction is therefore a *partial* reaction because chain termination will occur randomly at one of the possible choices for a specific type of base in any one DNA strand.

Because the DNA sample is a population of identical molecules, each of the four base-specific reactions will generate a collection of labeled DNA fragments of different lengths. Each of the fragments in one reaction will have a common 5′ end (defined by the sequencing primer). However, the 3′ ends are variable because the insertion of the selected dideoxynucleotide occurs randomly at one of the many different positions that will accept that specific base (**Figure 8.6**).

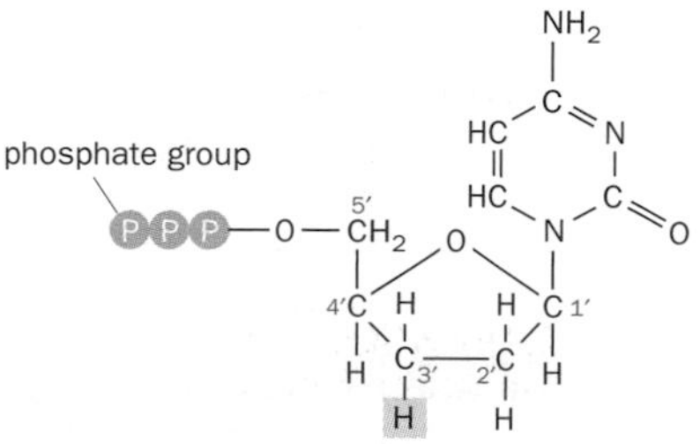

Figure 8.5 Structure of a dideoxynucleotide: 2′, 3′-dideoxy CTP (ddCTP). The hydroxyl group attached to carbon 3′ in normal nucleotides is replaced by a hydrogen atom (shown by shading).

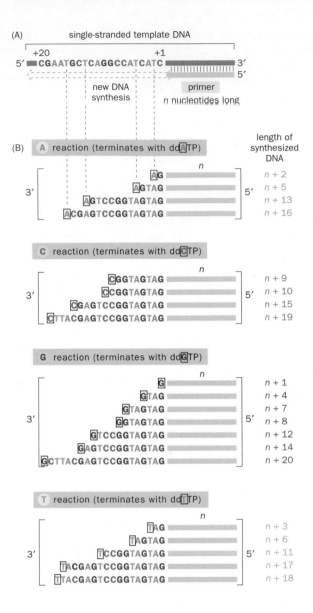

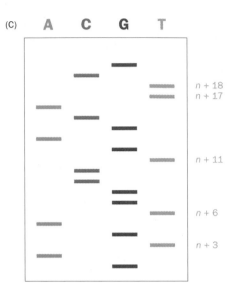

Figure 8.6 Dideoxy DNA sequencing.
(A) With the use of a primer about 20 nucleotides long, a complementary sequence is synthesized from a single-stranded DNA template. (B) Four parallel base-specific reactions are performed, each with all four dNTPs plus one ddNTP that competes with its dNTP counterpart for insertion into the growing chain. For example, the A reaction uses dATP, dCTP, dGTP, dTTP, and also ddATP, which competes with dATP and causes chain termination when it is incorporated. Because there are many identical copies of the template, a range of different fragments is produced, depending on the point at which the ddATP has been inserted. The fragments will have a common 5′ end (defined by the sequencing primer) but variable 3′ ends, depending on where the ddA (shown as A enclosed within a box) has been inserted. For example, in the A-specific reaction chain, extension occurs until a ddA nucleotide is incorporated. This will lead to a population of DNA fragments of lengths $n + 2$, $n + 5$, $n + 13$, $n + 16$ nucleotides, etc. (C) Size fractionation on a polyacrylamide gel enables the sequence to be ascertained, giving for $n + 1$ to $n + 20$ the sequence GATGATGGCCTGAGCATTCG, the reverse complement of the sequence shown in (A).

Fragments that differ in size by even a single nucleotide can be size-fractionated on a denaturing polyacrylamide gel, a gel that contains a high concentration of urea or other denaturing agent so that the migrating DNA remains single-stranded.

Automation of dideoxy DNA sequencing increased its efficiency

Automated DNA sequencing machines that used fluorescence labeling of DNA were developed in the early 1990s. Four different fluorescent dyes are used in the four base-specific reactions. By selecting dyes with different emission wavelengths, all four reactions could be loaded into a single sample well on the gel. During electrophoresis, the DNA fragments pass by an excitation source such as a laser, while a monitor detects and records the fluorescence signal as the DNA passes through a fixed point in the gel (**Figure 8.7**). This allows an output in the form of intensity profiles for each of the differently colored fluorophores while simultaneously storing the information electronically.

Early automated DNA sequencers used slab polyacrylamide gels, but greater DNA sequencing capacity became possible with capillary sequencing. In this technique, DNA samples migrate through long and extremely thin (0.1 mm in diameter) glass capillary tubes containing a polyacrylamide gel. Just like the slab-gel machines, capillary machines read the base sequence as DNA moves through the gel, but a higher degree of automation can be achieved by avoiding the need to cast large slab gels.

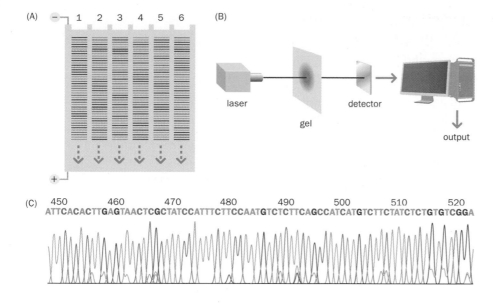

(C)
450   460   470   480   490   500   510   520
ATTCACACTTGAGTAACTCGCTATCCATTTCTTCCAATGTCTCTTCAGCCATCATGTCTTCTATCTCTGTGTCGGA

Figure 8.7 Auto...
with fluorescen...
fluorescent dye... base-specific re... incorporated b... specific ddNTF... the primer and having four set... corresponding to the four reactions). Samples containing mixtures of all four base-sequencing reactions are size-fractionated by polyacrylamide gel electrophoresis (here, fragments are shown migrating downward through a slab gel). (B) While the fragments are migrating during the electrophoresis run, a laser beam is focused at a specific constant position on the gel. As the individual DNA fragments migrate past this position, the laser causes the dyes to fluoresce. Maximum fluorescence occurs at different wavelengths for the four dyes; the information is recorded electronically and the interpreted sequence is stored in a computer database. (C) Example of DNA sequence output, showing a succession of dye-specific (and therefore base-specific) intensity profiles. The example illustrated shows a cDNA sequence from the human polyhomeotic gene *PHC3*.

Iterative pyrosequencing records DNA sequences while DNA molecules are being synthesized

With minor changes, the dideoxysequencing method underpinned molecular genetics for three decades. The method is, however, disadvantaged by relying on gel electrophoresis to fractionate newly synthesized DNA fragments. Not only does this make the method a laborious one, but more importantly it makes it difficult to sequence large numbers of DNA fragments at a time—the most advanced dideoxy DNA sequencing machines could sequence only up to 96 samples at a time. As a result, sequencing output is limited to ~30–60 kb of sequence per 3–4 hour electrophoresis run.

Fundamentally different DNA sequencing technologies that did not require gel electrophoresis were not developed until the early to mid-2000s. An important breakthrough was to develop methods that could record the DNA sequence while a DNA strand was being synthesized by a DNA polymerase from a single-stranded DNA template. That is, the sequencing method was able to monitor the incorporation of each nucleotide in the growing DNA chain and to identify which nucleotide was being incorporated at each step.

At the heart of the first such approach was a method known as pyrosequencing, which had been developed initially to assay single nucleotide polymorphisms. For use in sequencing, a series of successive pyrosequencing reactions (*iterative pyrosequencing*) are performed to record the DNA sequence as it is synthesized. DNA chains are synthesized from dNTP (deoxynucleoside triphosphate) precursors and the DNA polymerase reaction naturally causes cleavage between the α and β phosphates so that a dNMP (containing the α phosphate) is incorporated into DNA, leaving behind a pyrophosphate (PPi) residue made up of the β and γ phosphates. Pyrosequencing exploits the release of pyrophosphate each time a nucleotide is successfully incorporated into a growing DNA chain.

Sequential enzyme reactions are used to detect released PPi. First, ATP sulfurylase quantitatively converts PPi to ATP in the presence of adenosine 5′ phosphosulfate. Then, the released ATP drives a reaction in which luciferase converts luciferin to oxyluciferin, a product that generates visible light in amounts that are proportional to the amount of ATP. Thus, each time a nucleotide is incorporated, a light signal is detected.

In iterative pyrosequencing the individual dNTPs are provided sequentially (unlike in dideoxysequencing, in which a mix of dNTPs is used). If the selected dNTP is the one that can provide the required dNMP to continue chain elongation, PPi is released, and light is produced and recorded by a CCD camera. Any unused dNTPs and excess ATP are degraded by the enzyme apyrase, which is included in the reaction mixture. Thus, if the selected dNTP is not the one needed for the next synthesis step, no light is produced and apyrase will degrade the dNTP (**Figure 8.8**).

Massively parallel DNA sequencing enables the simultaneous sequencing of huge numbers of different DNA fragments

Many different massively parallel sequencing methods have recently been developed that can carry out up to billions of sequencing reactions in parallel. The first wave of the next-generation sequencing technologies to be developed use PCR to amplify target DNA and became widely available in 2007. Subsequently, methods were developed to sequence single DNA molecules that had not been amplified in any way, and the first commercial machine for this purpose became available in 2008.

Massively parallel DNA sequencing is transforming molecular genetics. Such methods are being extensively used in rapid genome resequencing for various purposes, including personal genome sequencing and the identification of mutations. In addition, they are providing comprehensive analyses of the **transcriptome**, the collection of different RNA molecules in cells. Furthermore, as we will see in Chapter 12, they are providing rapid endpoints for various analyses that seek to identify DNA–protein interactions and binding sites for regulatory proteins.

Massively parallel sequencing of amplified DNA

The first massively parallel sequencing methods to be developed used PCR amplification and involved sequencing-by-synthesis. Massively parallel pyrosequencing was developed in the 2000s by the 454 Life Sciences company and subsequently commercialized by Roche. It involves breaking the sample DNA into short fragments (300–500 bp) and preparing single-stranded templates. Two different types of oligonucleotide adaptor are ligated to the ends of the DNA fragments to provide universal priming sequences for amplification. After denaturation, those single-stranded fragments that have the different adaptors ligated at the different ends are selected. The next step is to attach the DNA molecules to beads (**Figure 8.9A**) and takes advantage of biotin–streptavidin binding. As described in Section 7.2, the bacterial protein streptavidin has an extraordinarily high binding affinity for the vitamin biotin, and so by designing one of the adaptors to contain a biotin tag, the DNA molecules will bind to beads coated with biotin's binding partner, streptavidin.

Single-stranded DNA templates are immobilized on beads, and the beads are separated from each other by creating an oil–water emulsion, with each droplet containing a single bead and the reagents needed for PCR. The droplets are known as microreactors (Figure 8.9B). After PCR amplification there are 10 million copies of one DNA fragment immobilized on one bead. The emulsion is then broken and the beads are deposited into picoliter wells on a slide (one bead per well) that are then layered with smaller beads that have ATP sulfurylase and luciferase attached to their surfaces (see Figure 8.9B). A fixed sequence of the dNTP precursors (T, then A, then C, then G) is washed over the beads and chemiluminescent light is emitted each time a nucleotide is incorporated; this is recorded for the individual positions.

The massively parallel pyrosequencing method developed by 454 Life Sciences (and marketed by Roche) produces reasonably long individual DNA sequence readouts, but many hundreds of thousands of sequences can be read in parallel. As a result, the throughput is close to ten thousand times greater than that of dideoxysequencing (**Table 8.1**). The related Solexa sequencing technology was developed more recently by the Illumina company and has even greater sequencing throughput, although accompanied by smaller individual sequence readouts.

An alternative technology, massively parallel sequencing-by-ligation has been developed by the Applied Biosystems company. Their SOLiD™ (Sequencing by Oligonucleotide Ligation Detection) technology also uses an emulsion-based PCR strategy to amplify individual DNA fragments. The amplification products are then randomly deposited onto an array for probing with fluorescently labeled oligonucleotides. The SOLiD method also has high sequencing throughput but with comparatively short individual sequence readouts.

Single-molecule sequencing

A new wave of technologies, sometimes known as third-generation sequencing, permit the sequencing of single DNA molecules that are not amplified in any way.

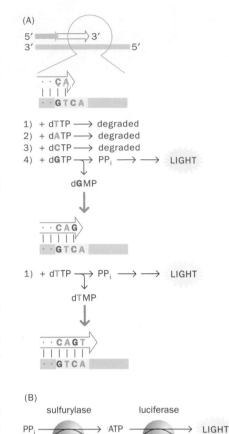

Figure 8.8 The principle underlying pyrosequencing. (A) Base insertion in pyrosequencing. In pyrosequencing, a DNA polymerase synthesizes a DNA chain by using a single-stranded DNA template and the four normal dNTPs. Instead of having a mixture of the four dNTPs, the individual dNTPs are provided sequentially. When the correct dNTP is provided, the incorporation of the dNMP nucleotide is tracked by the simultaneous production of a pyrophosphate (PP_i) group that is used to produce light. If an incorrect dNTP is provided it is degraded by the enzyme apyrase. In this example, the first dNMP to be incorporated is G, as indicated by the production of light in the G reaction only, to be followed by T. (B) The insertion of the correct base is monitored by light production in a two-step reaction. The released PP_i is used by the enzyme ATP sulfurylase to generate ATP, which in turn drives a luciferase reaction to produce light, as detected by a charge-coupled device (CCD) camera.

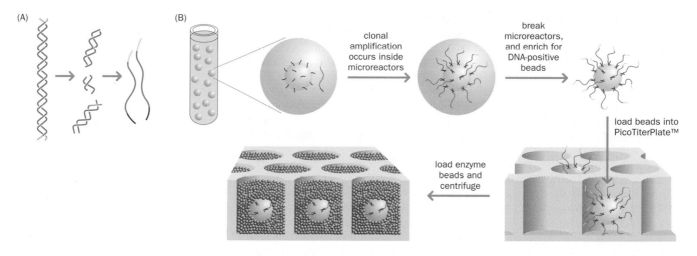

Figure 8.9 Sample preparation for massively parallel pyrosequencing. (A) Genomic DNA is isolated, fragmented, ligated to oligonucleotide adaptors, and separated into single strands. (B) Fragments are bound to beads under conditions that favor one fragment per bead. Thereafter the beads are captured in the droplets of a microreactor (an emulsion containing a PCR reaction mixture in oil). Clonal PCR amplification occurs within each droplet, resulting in beads each carrying 10 million copies of a unique DNA template. The emulsion is broken to release the beads, and the DNA strands are denatured. Single beads carrying single-stranded DNA clones are deposited into individual wells of a fiber-optic slide (PicoTiterPlate™) that contains 1.6 million wells. Each well will contain a picoliter sequencing reaction. Smaller beads carrying immobilized enzymes required for pyrophosphate sequencing (Figure 8.8B) are deposited into each well to enable the pyrosequencing reaction to take place. [Adapted from Margulies M, Egholm M, Altman WE et al. (2005) *Nature* 437, 376–380. With permission from Macmillan Publishers Ltd.]

Sequencing errors that occasionally arise as a result of amplifying template DNA can be avoided or minimized, and the technologies are both simpler than those using amplified templates and have much lower projected sequencing costs. The first single-molecule sequencing technology to be commercialized was developed by Helicos Biosciences, whose HeliScope™ machine became available in 2008. Because the HeliScope sequencer detects single molecules, sequencing templates can be densely packed on the Helicos flow cell surface (about 100 million single molecule templates per square centimeter, or billions per run). Sequencing occurs by a sequencing-by-synthesis reaction that involves cycles of single nucleotide addition steps with washes in between. As a result, the individual sequence reads are small, but the total sequence output is already very respectable (see Table 8.1) and may become greater still in the future, and the cost of reagents is substantially less than for sequencing of PCR-amplified DNA.

TABLE 8.1 EXAMPLES OF MASSIVELY PARALLEL DNA SEQUENCING METHODOLOGIES				
Methodology	DNA sequencing platform	Read length per reaction (nucleotides)	Output	Comments
Sequencing-by-synthesis of PCR-amplified DNA	Roche GS FLX sequencer	>300	>0.45 Gb per run of 7 hours	Long read lengths but high reagent costs and limited output
	Illumina Genome Analyzer IIx	~100	18 Gb for a run of 4 days	Currently most widely used but high reagents cost
		2 × ~100[a]	35 Gb for a run of 9 days	
Sequencing-by-ligation of PCR- amplified DNA	Applied Biosystems SOLiD 3	~50	Up to 30 Gb for a run of 7 days	Short read lengths, long runs and high reagent costs
		2 × ~50[a]	Up to 50 Gb for a run of 14 days	
Single molecule sequencing	Helicos Biosciences HeliScope	~30	~40 Gb per run of 8 days	Low reagent costs but short read lengths and comparatively high error rates
	Pacific Biosciences	~1000	Not currently available	Not expected to be released until late 2010/2011. Great potential (see text)

[a]When using a mate-pair run that involves obtaining sequences from both ends of the DNA molecules.

Other more powerful single-molecule technologies are being developed and are likely to transform DNA sequencing, both in terms of rapidity of sequencing and in reduced costs. A front runner is a revolutionary new single-molecule DNA sequencing method called SMRT (**single-molecule real-time** sequencing) that is being developed to sequence DNA at speeds of 20,000 times faster than second-generation sequencers currently on the market. It also entails sequencing-by-synthesis, but unlike the methods above the synthesis occurs in *real time* so that the sequencing reactions are extremely fast. Within cells, DNA polymerases naturally perform new DNA synthesis to duplicate whole genomes within minutes. The SMRT method visually tracks individual DNA polymerases as they go about synthesizing DNA molecules in real time. It records which nucleotides get incorporated at each position by using nucleotides tagged with one of four different fluorophores according to the base specificity.

SMRT exploits two key innovations. First, it uses unconventional fluorophore-labeled dNTPs. In normal DNA labeling, fluorophores are attached to the bases; as each nucleotide is incorporated its fluorophore becomes a permanent part of the DNA strand. Once multiple nucleotides have been incorporated into the DNA, however, the physical bulk of the added fluorophores sterically inhibits further synthesis by the DNA polymerase. In most sequencing-by-synthesis methods, therefore, the DNA is synthesized a single nucleotide at a time, starting and stopping the reaction after each incorporation. Large reagent volumes are required and the *processivity* of the polymerase, its ability to continue its catalytic function repetitively without dissociating from its substrate, is severely limited. In SMRT, by contrast, fluorophores are attached to the external (γ) phosphate group of the triphosphate residue of nucleotides (**Figure 8.10**). When such a *phospholinked* labeled dNTP pairs with a complementary base on the template strand, the fluorescence signal can be recorded before the polymerase cleaves the triphosphate group (**Figure 8.11B**). Cleavage of the dNTP generates an unlabeled dNMP that is incorporated into the growing DNA chain so that the polymerase can keep on inserting nucleotides without steric problems and can generate long read lengths. To maximize read lengths, the highly processive φ29 DNA polymerase is used.

The second key innovation is the SMRT chip, a very thin metal film deposited on a glass substrate. The metal film contains thousands of tiny sub-wavelength holes known as nanophotonic zero-mode waveguides (ZMWs; Figure 8.11A). An individual ZMW has a volume of just 20 zeptoliters (10^{-21} liters) and ZMWs constitute nanophotonic visualization chambers for directly watching a DNA polymerase through the glass support as it performs sequencing by synthesis on a single DNA molecule template. At this volume, the technology can detect the activity of a single molecule that is briefly immobilized when held at the active site of the enzyme among a background of thousands of constantly moving labeled nucleotides. With refinements, the technology is projected to be capable of sequencing a human genome in 1 hour for only US$100 or so.

Another major area of technological development in single molecule sequencing is *nanopore sequencing*. Materials such as silicon can be fabricated in such a way that they contain very tiny holes (nanopores). In principle, successive nucleotides on an individual single-stranded DNA molecule are induced to pass through a nanopore. While passing through the pore, each nucleotide partially blocks the nanopore to a different and characteristic degree, according to whether it is an A, a C, a G or a T, and the amount of electrical current that can pass through the nanopore varies according to the type of nucleotide that is obstructing it. The

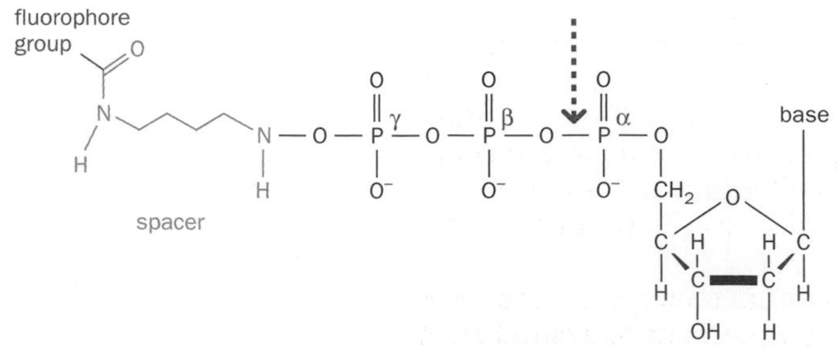

Figure 8.10 A phospholinked fluorescent labeled dNTP. The dashed arrow indicates cleavage that occurs when the nucleotide is incorporated into DNA. A pyrophosphate group containing the β- and γ-phosphate groups and an attached fluorophore is cleaved off from the dNMP residue that will be incorporated into DNA.

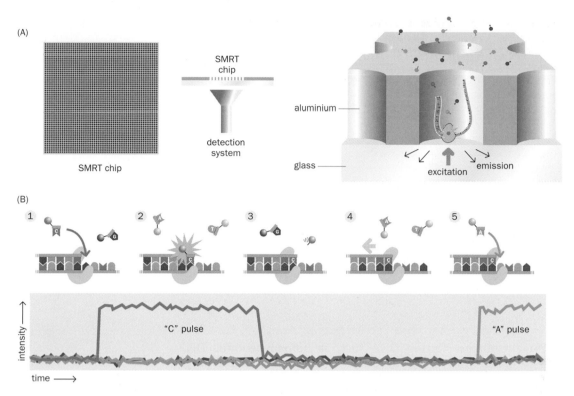

Figure 8.11 Principles of SMRT (single-molecule real-time sequencing) technology. (A) The SMRT chip and ZMW nanophotonic visualization chambers. The SMRT chip shown on the left is an aluminum film 100 nm thick deposited on a silicon dioxide substrate that extends just over 10 μm × 30 μm in size. The metal film contains thousands of perforations known as zero-mode waveguides (ZMWs) that can act as tiny visualization chambers, in each of which a DNA polymerase molecule performs DNA sequencing with a single DNA molecule template. The chip is mounted next to a fluorescence detection system. By viewing through the glass support it can visually record real-time DNA sequencing reactions in each of the individual ZMWs (right). (B) The phospholinked dNTP incorporation cycle plus the corresponding expected time trace of detected fluorescence intensity. SMRT is a sequencing-by-synthesis method that uses dNTPs labeled with one of four different fluorophores, depending on which base is present. However, because the fluorophore is attached to the γ-phosphate of dNTPs (see Figure 8.10), it does not get incorporated into the growing DNA chain. Nevertheless, a DNA polymerase molecule will hold an incoming dNTP steady at its active site for a brief time before the triphosphate group is cleaved and the dNMP is inserted. During this time, the engaged fluorophore emits fluorescent light whose color corresponds to the base identity. The steps shown are: (1) the phospholinked nucleotide begins to associate with the template in the active site of the polymerase, (2) causing an elevation of the fluorescence output on the corresponding color channel. (3) Phosphodiester bond formation liberates the dye–linker–pyrophosphate product, which diffuses out of the ZMW, thus ending the fluorescence pulse. (4) The polymerase translocates to the next position, and (5) the next nucleotide to be inserted binds the active site beginning the subsequent pulse. [Adapted from Eid J, Fehr A, Gray J et al. (2009) *Science* 323, 133–138. With permission from the American Association for the Advancement of Science.]

change in the current represents a direct reading of the DNA sequence. See http://www.nanoporetech.com/sequences for a video explanation of one such method. This is a fast moving area, but like SMRT has great potential for offering fast very high throughput DNA sequencing at low cost.

Microarray-based DNA capture methods allow efficient resequencing

Notwithstanding the huge recent progress in genome sequencing, many current applications of massively parallel DNA sequencing are focused on target sequences that collectively constitute a small fraction of a genome. For example, screening for cancer gene susceptibility could involve sequencing all exons, exon–intron boundaries, and known regulatory elements for all known cancer genes. There are many hundreds of known cancer genes, including many new genes identified by international programs such as the Cancer Genome Atlas. PCR amplification of what may be hundreds of sequence elements in each cancer gene is both tedious and time-consuming. As an alternative, it is possible to use microarray hybridization as a tool to enrich for the desired sequences that are then submitted for high-throughput sequencing.

The NimbleGen™ sequence capture system is a commercial system that uses DNA microarrays to permit hybridization-mediated enrichment of desired DNA

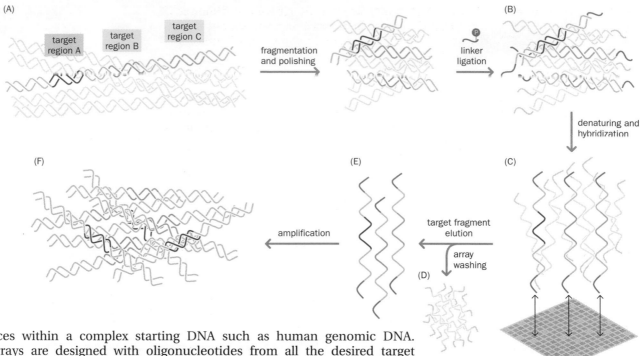

sequences within a complex starting DNA such as human genomic DNA. Microarrays are designed with oligonucleotides from all the desired target sequences and are hybridized with the randomly fragmented starting DNA that has been treated so that specific linker oligonucleotides are attached at both ends, and then denatured. The desired fragments within the starting DNA should selectively hybridize to the microarray and can later be recovered by elution, whereupon linker-specific primers can be used to PCR amplify the DNA in preparation for massively parallel sequencing (**Figure 8.12**). Genome-wide capture of essentially all human protein-coding exons, the human **exome**, for resequencing is now possible: the NimbleGen 2.1M human exome array can capture about 180,000 human exons and about 550 miRNA sequences (miRNA is a type of functional noncoding RNA that regulates the expression of certain target genes).

8.3 GENOME STRUCTURE ANALYSIS AND GENOME PROJECTS

Unlike bacterial genomes, which often consist of a single type of comparatively small DNA molecule, the genomes of complex multicellular organisms are composed of many large DNA molecules. The human genome, for example, consists of 25 different DNA molecules. There is one small mitochondrial DNA molecule, which is present in multiple copies per cell, and 24 different nuclear DNA molecules, with an average length of about 135 Mb, that correspond to the 24 different chromosomes. To begin to understand the structure and functions of the human genome, a necessary starting point was to obtain its sequence. Until the early 1980s, however, the idea of sequencing the human genome had seemed impossibly remote.

The full sequence of the circular human mitochondrial DNA (mtDNA) molecule was, nevertheless, published as early as April 1981, but this was very much the exception. The mtDNA could easily be purified because it was uniquely found in the cytoplasm, and it was comparatively tiny—somewhat less than 16.6 kb, accounting for 1/200,000 of the total human genome size. Given the substantial effort that had been needed to sequence the tiny mtDNA molecule, the task of sequencing the chromosomal DNA molecules had seemed hopelessly daunting. Not only were they known to be very large DNA molecules, but there was no easy way of separating one type of chromosomal DNA from another.

The advent of DNA libraries had offered, in principle, the possibility of shotgun sequencing of large genomes, whereby randomly selected clones in a genomic DNA library are sequenced until the full genome is covered. Clones with overlapping inserts are normally generated during the construction of genomic DNA libraries as a result of the random fragmentation of the DNA. Identi-cal

Figure 8.12 Array-based DNA capture with NimbleGen™ sequence capture technology. Resequencing specific fractions of a genome can be done conveniently by first capturing the desired target sequences by selective hybridization. To do this custom microarrays are designed to have DNA probes that represent the desired target sequence regions only. (A) A complex starting DNA is randomly fragmented and ends are filled in to generate a population of small (about 500 bp) blunt-ended fragments. (B) Common oligonucleotide linkers are ligated to the ends of the fragments. (C) The DNA is denatured and hybridized to a custom microarray; after hybridization, unbound sequences are removed by washing (D). The only DNA fragments that are left bound to the microarray should be the desired target DNA sequences, and they can be retrieved by elution (E). The retrieved target sequences are now amplified by using linker-specific primers (F) to provide an enriched amplified pool of target sequences ready for massively parallel sequencing. For alternative methods of capturing DNA for massively parallel DNA sequencing, see Mamanova et al. in Further Reading. (Figure based with permission on an image provided by Roche Inc.)

copies of the same chromosomal DNA molecules will be cleaved at different locations on the DNA, so any unique short sequence will be represented by a series of overlapping DNA fragments of different sizes produced by differential DNA cleavage.

The different fragments are then attached to identical vector molecules and randomly introduced into cells (see Figure 8.1). If the average insert size of a genomic DNA library is selected to be a few hundred base pairs, standard dideoxy DNA sequencing can be used to determine the inserts in all clones. If there is an accurate way of identifying sequences with overlaps and enough clones are sequenced, it becomes possible to sequence entire genomes.

The above approach of whole genome shotgun sequencing (**Figure 8.13A**) is most successfully applied to small genomes; there are major difficulties in applying it to the sequencing of large metazoan genomes for the first time. The human genome, for example, has huge numbers of repetitive DNA sequences. Because members of a repetitive DNA family can be very similar in sequence, it is difficult to map the individual sequences.

To circumvent the problem of repetitive DNA, a different strategy was required for complex metazoan genomes. What was needed was some kind of initial scaffold, a series of **framework maps**, to anchor sequences to defined subchromosomal regions as a prelude to shotgun sequencing (Figure 8.13B). The burning question was this: what kind of framework maps could be established to aid in the sequencing of the human genome?

Framework maps are needed for first time sequencing of complex genomes

For decades, human geneticists had envied geneticists working on model organisms in which high-resolution classical **genetic maps** could be established readily. Such maps were based on mutant genes: by crossing mutants, the inheritance of individual phenotypes could be tracked through generations. If two mutant phenotypes showed a tendency to be inherited together, the underlying genes could be expected to be reasonably closely linked on the same chromosome. Recombination between linked loci could provide a measure of the physical distance separating the two genes.

Very occasionally, high-resolution **physical maps** could also be available. *Drosophila* was the prime example: not only was genetic mapping simple, but high-resolution physical maps could also be constructed by using the giant polytene chromosomes found in the salivary glands of *Drosophila* larvae. The latter chromosomes are extremely unusual because they can be viewed under the microscope during interphase when the chromosomes are highly extended. This is possible because they have undergone numerous rounds of DNA replication without cell division (endomitosis). As a result, each chromosome consists of 1000 or more chromatids that remain bound together along their lengths, like tightly packed drinking straws in a box.

For ethical and practical reasons, classical genetic mapping could never be contemplated in humans. The only physical map that had been available was

Figure 8.13 Two strategies for sequencing a genome. (A) Whole genome shotgun sequencing involves indiscriminate fragmentation of the genome into small pieces of DNA that are readily sequenced. It quickly generates large amounts of sequence data, but anchoring sequences to specific locations in the genome may be problematic. Large amounts of repetitive DNA in complex genomes makes it difficult to locate sequences unambiguously to specific subchromosomal regions. (B) For first-time sequencing of complex genomes it is more efficient to assemble contigs of large insert clones for each chromosome and then to fragment individual clones into pieces that are sequenced to reconstruct the sequence of the parent clone. [Adapted from Waterston RH, Lander ES & Sulston JE (2002) *Proc. Natl Acad. Sci. USA* 99, 3712–3716. With permission from the National Academy of Sciences, USA.]

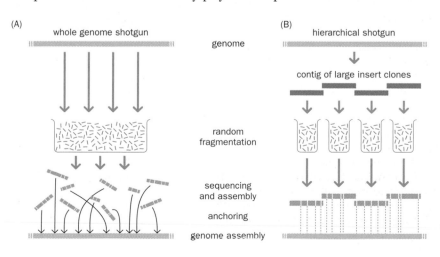

(A) whole genome shotgun

(B) hierarchical shotgun

genome

contig of large insert clones

random fragmentation

sequencing and assembly

anchoring

genome assembly

based on distinguishing chromosomes according to size and shape and using certain stains to produce subchromosomal banding patterns (see Figure 2.15). DNA library clones could be mapped to subchromosomal regions, for example, by hybridization to denatured chromosomal DNA (*in situ* hybridization), but the mapping process was laborious and the subchromosomal resolution had not been particularly high.

The breakthrough that paved the way to mapping the human genome came with the realization that genetic maps did not have to be based on genes. Mutation is essentially a random process. For any one type of human chromosome, the DNA sequence varies from one chromosome copy to another at about one nucleotide in a thousand on average. Only a tiny fraction of this variation causes mutant phenotypes, and the great bulk of the changes are found in regions between genes or within introns. Once assays had been developed to track this general type of *DNA polymorphism*, framework maps based on DNA markers could be established.

After the creation of initial framework maps based on genetic markers, high-density DNA marker maps were developed in which many of the markers were not polymorphic but were simply chosen because they had a unique sequence that could be assayed by PCR. Once framework *marker maps* with a suitably high density had been developed, it was possible to build framework *clone maps*. As described in the next section, DNA clone maps involve identifying and arranging DNA clones in a linear order that corresponds to the original linear order within the chromosomes of the cloned DNA sequences. Comprehensive clone maps for each chromosome provided the final substrate for genome sequencing that delivers the ultimate physical map at a resolution of 1 bp.

The linear order of genomic DNA clones in a contig matches their original subchromosomal locations

The perfect substrates for sequencing complex genomes are sets of genomic DNA clones containing large inserts that have been linearly ordered according to the subchromosomal origins of their inserts. A series of such clones where the insert of each clone partly overlaps that of its neighbors is known as a **clone contig**: it represents a contiguous (continuous) DNA sequence from a whole chromosome or subchromosomal region (**Box 8.1**).

Clones with overlapping inserts are normally generated during the construction of genomic DNA libraries because the DNA is first subjected to *random*

BOX 8.1 FRAMEWORK MAPS BASED ON DNA MARKERS AND CLONE CONTIGS

First time sequencing of complex genomes (which invariably have large amounts of repetitive DNA) is assisted by first constructing **framework maps** for each chromosomal DNA molecule. The framework maps best suited for DNA sequencing are based on **clone contigs**, linearly organized series of cloned overlapping DNA fragments that collectively represent chromosomal DNA sequences (**Figure 1**).

Assembling clone contigs depends on being able to identify overlaps between cloned DNA fragments. There are various ways in which this can be done but often clones are assayed for the presence of certain DNA markers known to map in the approximate subchromosomal region. A DNA marker is any short DNA sequence that can be assayed in some way, either by a hybridization assay or, more conveniently, by a PCR assay.

DNA markers may or may not be polymorphic (**Table 1**), but to be useful for mapping purposes a marker needs to have a unique chromosomal location. Different methods have been used to map a human DNA marker to a specific subchromosomal localization. A common approach is to type panels of artificially constructed **hybrid cells** that contain a full set of rodent chromosomes plus one or more specific human chromosomes or multiple human chromosome fragments. The latter are generated by exposing human cells to controlled radiation, causing chromosome breakage, and then fusing the human cells with rodent cells so that variable sets of human chromosome fragments can insert into the rodent chromosomes (*radiation hybrids*). Alternatively, DNA markers have been mapped on chromosomes by labeling clones known to contain the marker with a fluorophore and hybridizing them to suitably treated fixed human

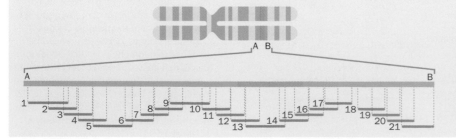

Figure 1 A schematic clone contig. The chromosomal DNA sequence from positions A to B is represented by overlapping DNA inserts in a linear series of genomic DNA clones. Clones with overlapping inserts are generated by the random fragmentation of DNA when a genomic DNA library is constructed (see Figure 8.1).

fragmentation (see Figure 8.1). Identical copies of the same chromosomal DNA molecules will be cleaved at different locations on the DNA, and so any unique short sequence will be represented by a series of overlapping DNA fragments of different sizes produced by differential DNA cleavage. The different fragments are then attached to identical vector molecules and randomly introduced into cells.

Early attempts to assemble a series of human genomic DNA clones with overlapping inserts would proceed from specific starting clones. A hybridization probe prepared from the insert DNA of a chosen genomic clone of interest would be used to screen all the cells of the DNA library to identify those that contained a strongly hybridizing DNA fragment, which could indicate an overlapping DNA fragment. Newly identified clones could then be used to prepare new hybridization probes to identify more distally overlapping clones. This procedure, known as *chromosome walking*, was extremely laborious.

More efficient, genome-wide methods of identifying clones with overlapping inserts were subsequently developed using **clone fingerprinting**. The idea was to submit *all* clones in the library to some assay that could help identify clones with

chromosome preparations on a microscope slide (chromosome FISH; see Figures 2.16A and 2.17A).

DNA marker maps can be built in different ways. Polymorphic markers can simply be assayed in members of multigenerational pedigrees to identify groups of linked markers that must map to the same chromosome. Non-polymorphic markers can be placed on marker maps by using PCR to assay the markers in panels of radiation hybrid cells, or in panels of YAC clones or other genomic DNA clones with large inserts. Common markers are **sequence-tagged site (STS)** markers originating from clones that have been at least partly sequenced, including previously identified and sequenced DNA clones, such as gene clones or other DNA clones of interest that have already been well characterized and mapped. More generally,

it is possible to sequence the ends of numerous large insert clones from a genomic DNA library, thereby generating thousands of markers.

If large numbers of STS markers are available, they can be used to build framework marker maps that allow clones to be assembled into clone contig maps. An early example of a short clone contig map that was assembled from STS marker mapping is shown in **Figure 2**. The idea is to arrange clones in a linear order according to whether they type positively or negatively for STS markers. Yeast artificial chromosome clones were initially a popular choice for assembling clone contig maps, but YACs containing large human inserts were found to be unstable, being prone to internal rearrangements and deletions (see Figure 2).

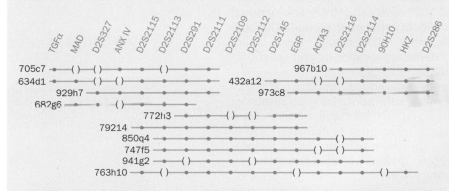

Figure 2 A practical example of an early human clone contig assembled by STS (sequence-tagged site) mapping. This example shows a YAC clone contig map representing a portion of human chromosome 2. Clones are shown as horizontal lines with clone names at the left. STS markers at the top include some derived from genes (TGFα, MAD, ANX IV, etc.) and anonymous chromosome 2 markers (D2S327, D2S2115, etc.). Positive typing for an STS marker is indicated by a closed circle; brackets indicate the absence of an expected STS (most probably as a result of YAC instability).

TABLE 1 COMMON MARKERS USED IN CONSTRUCTING FRAMEWORK DNA MAPS OF COMPLEX GENOMES		
Marker type	**Marker**	**Definition**
Polymorphic	Restriction fragment length polymorphism (RFLP)	Any DNA polymorphism that results in the creation or destruction of one recognition sequence for a specific restriction endonuclease. Used to be assayed by hybridization but now assayed by PCR
	Microsatellite	A sequence containing several (usually 10 or more) tandem repeats of a sequence of 1–4 nucleotides. Frequently found in complex genomes and often highly polymorphic
	Single nucleotide polymorphism (SNP)	A single nucleotide change that is polymorphic. The polymorphism is usually limited (typically, there are just two significantly frequent alleles), but SNPs are very suited to automated genotyping such as the use of pyrosequencing (see Figure 8.8)
Non-polymorphic	Sequence-tagged site (STS)	Any DNA sequence that has been mapped to a specific subchromosomal location and can be assayed by PCR amplification
	Expressed sequence tag (EST)	A subset of sequence-tagged sites that happen to be located within DNA sequences known to be transcribed

overlapping DNA inserts. For example, clone insert DNAs were submitted to restriction mapping and assays of repetitive DNA content to look for characteristic patterns of spacing of sites for specific restriction endonucleases or of particular repetitive sequences.

Clone fingerprinting based on restriction mapping and/or repetitive DNA content is still quite laborious. A yet more efficient method depends on the availability of a high-density DNA marker map. Various DNA markers can be used. They may be polymorphic (and may be localized on a genetic map) or nonpolymorphic. The only requirements are that they should have a unique subchromosomal localization and be able to be assayed in some way (see Box 8.1).

The Human Genome Project was an international endeavor and biology's first Big Project

The realization in the early 80s that even large genomes, such as the human genome, could be sequenced sparked serious planning efforts to sequence the human genome. The official Human Genome Project (HGP) envisaged a 15-year time-scale, from 1990 to 2005. In addition to sequencing the human genome, the HGP had three other goals: to develop mapping and sequencing technologies; to conduct genome projects for five model organisms; and to investigate the ethical, legal, and societal implications.

The prioritized model organisms were the bacterium *Escherichia coli*, the yeast *Saccharomyces cerevisiae*, the roundworm *Caenorhabditis elegans*, the fruit fly *Drosophila melanogaster*, and the mouse *Mus musculus*. The first four were known to have substantially smaller genomes than the human genome and so were expected to be test beds for evaluating and then refining genome sequencing strategies and methodologies. The bulk of the sequencing for the more complex human and mouse genomes was expected to be performed in the later stages, after learning from the smaller genome projects and taking maximum benefit from technological improvements.

Because of the large scale involved, the HGP was biology's first Big Project. The ultimate aim was to achieve a periodic table for biology, based on genes rather than elements. The publicly funded HGP was also a truly international endeavor. It came to be represented by the International Human Genome Sequencing Consortium of 20 different centers in the USA, the UK, France, Japan, Germany, and China, and was marked by extensive data sharing and collaboration on strategy, methodology, and data analysis.

Much of the genome sequencing technology was concentrated in a few very large genome mapping and sequencing centers with industrial-scale resources and massive data analysis capabilities. Interacting with these centers was a worldwide network of small laboratories mostly attempting to map and identify disease genes and typically focusing on very specific subchromosomal regions. In addition to publicly funded research efforts, privately funded research programs pursued similar but parallel sequencing projects.

The first framework maps for the human genome were genetic maps for individual chromosomes. The genetic maps were of low resolution, but they provided the backbone on which to build a series of ever more detailed physical maps, culminating in clone contigs for each chromosome that were then used for the final sequencing stage.

The first human genetic maps were of low resolution and were constructed with mostly anonymous DNA markers

In genetic mapping different polymorphic markers are assayed in all members of a variety of multigeneration families. The resulting genotypes are then analyzed by computer to work out how alleles of the different markers segregate at each meiotic event connecting parents to their offspring, and to identify markers where specific alleles co-segregate. The discovery of apparently random DNA polymorphisms in the human genome prompted the idea of constructing a comprehensive, nonclassical human genetic map, one that was based predominantly on anonymous DNA markers rather than on genes.

The first genetic linkage map of the human genome was published in 1987 and was based on **restriction fragment length polymorphisms** (**RFLPs**). An

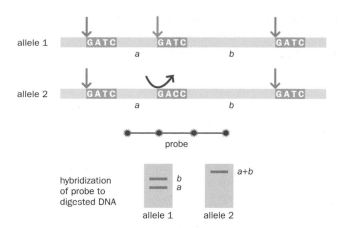

Figure 8.14 Restriction fragment length polymorphism (RFLP). An RFLP is any DNA polymorphism that causes a change in the recognition sequence for a restriction endonuclease. Often it results from a single nucleotide change such as the T→C transition shown here, which causes allele 1 to have an additional *Mbo*I recognition sequence (GATC) that is lacking in allele 2. The hybridization probe shown would detect sequences *a* and *b*, and so if genomic DNA is digested with *Mbo*I before hybridization, the probe can distinguish allele 1 (cutting produces two fragments *a* and *b*) and allele 2 (*a* and *b* are on the same fragment). Alternatively, but not shown here, the alleles can be conveniently typed by PCR by using an upstream primer derived from the *a* sequence and a downstream primer from the *b* sequence. Amplification products are digested with *Mbo*I and size-fractionated to distinguish between the alleles.

RFLP is any DNA polymorphism, often a single nucleotide change, that results in the creation or destruction of one recognition sequence for a specific restriction endonuclease (**Figure 8.14**). It can be assayed by digesting DNA with the restriction endonuclease and using a hybridization probe spanning the variable sequence (or immediately adjacent to it), or more conveniently by using primers flanking the variable sequence to amplify the required sequence and then digesting with the restriction endonuclease.

Although an outstanding achievement, the first human genetic map was of limited use. There were too few markers—only 393 RFLPs were used. In addition, the markers were not very polymorphic because RFLPs have only two alleles—either the restriction site is present or it is absent. To create a useful genetic map, a higher-density map was needed with markers that were more polymorphic.

Second-generation human genetic maps were based on a different class of DNA marker. **Microsatellite DNA** is the name given to short tandemly repeated DNA sequences, typically with a repeat unit one to four nucleotides long. They are common in the human genome. Individual microsatellites quite often have a large number of repeats, making them unstable so that the number of repeat units varies.

Microsatellite instability arises because during DNA replication the DNA polymerase often makes mistakes in copying long tandem repeats. Sometimes the polymerase skips past an individual repeat unit and fails to copy it; it can also go back by mistake to copy the same individual repeat unit twice. As a result, microsatellites are often highly polymorphic and the multiple alleles can be distinguished by using a PCR-based assay (**Figure 8.15**). The first human linkage map based on microsatellite markers was published in 1992 using 814 markers (corresponding to roughly one marker per 3.5 Mb). By 1994 an integrated genetic map (mostly based on microsatellites but containing some other markers) had a marker density of close to one per megabase.

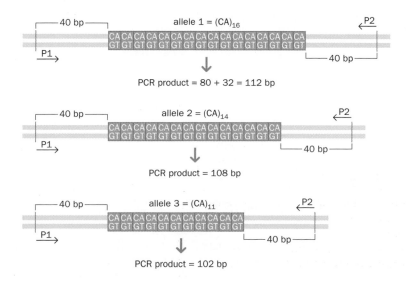

Figure 8.15 Designing a PCR assay for a microsatellite DNA polymorphism. A microsatellite DNA marker is a type of DNA polymorphism in which variation in the number of short tandem repeats [in this case, a (CA)/(TG) dinucleotide repeat] produces length polymorphism. In this example the 5′ ends of upstream primer P1 and downstream primer P2 are imagined to be each located 40 bp from the microsatellite array, and so the length of amplified fragment will be 80 + the array length (32, 28, or 22 bp in this example). The different alleles can be separated by polyacrylamide gel electrophoresis.

The 1994 map was of sufficiently high resolution to achieve the HGP's goals for genetic mapping, and from then onward the major focus was on developing and refining physical maps leading to the ultimate physical map, the complete DNA sequence of each chromosome. However, genetic mapping of the human genome has continued, in two major ways. First, microsatellite maps of even higher resolution have been developed. Second, high-density maps have been developed for single nucleotide polymorphisms (SNPs) by the International HapMap (haplotype mapping) Consortium. The major motivation has been to aid the identification of common disease genes. SNPs are not particularly polymorphic (they mostly have two alleles) but they are well suited to automated typing, by methods such as pyrosequencing (see Figure 8.8). The most recent such SNP map has over 3 million SNP markers, or roughly one SNP per kilobase.

Physical maps of the human genome progressed from marker maps to clone contig maps

Many different types of physical map have been constructed for the human genome (Table 8.2), but a major goal of the HGP was to construct clone contig maps for each chromosome as a prelude to genome sequencing. To do this, genomic libraries containing large insert DNAs were preferred, and initially yeast artificial chromosome (YAC) libraries were selected because of their large insert sizes.

In 1993, Daniel Cohen and colleagues at the Centre d'Études du Polymorphisme Humaine (CEPH) laboratory in Paris reported a human YAC library with more than 30,000 independent clones and an average insert size of just less than 1 Mb. Establishing clone overlaps by clone fingerprinting, they were eventually able to assemble a YAC contig map covering about 75% of the genome, with an average contig size of about 10 Mb.

Another important physical mapping aim was to build maps based on **sequence-tagged site** (**STS**) markers. An STS marker is any known unique DNA sequence that can be easily assayed by PCR. The STS markers included both polymorphic markers, such as microsatellites that can be easily genotyped by PCR, and a potentially huge number of non-polymorphic markers. Sequencing genomic DNA clones at random provided many non-polymorphic STS markers; others were obtained from sequenced genes and cDNA clones, and these were known as **expressed sequence tag** (**EST**) markers.

In 1995 a Whitehead Institute–Massachusetts Institute of Technology (MIT) collaboration published a human STS map containing more than 15,000 STS

TABLE 8.2 DIFFERENT TYPES OF PHYSICAL MAP USED TO MAP THE HUMAN NUCLEAR GENOME

Type of map	Examples/methodology	Resolution
Cytogenetic	chromosome banding maps	an average band has several Mb of DNA
Chromosome breakpoint maps	somatic cell hybrid panels with human chromosome fragments derived from natural, translocation, or deletion chromosomes	distance between adjacent chromosomal breakpoints on a chromosome is usually several Mb
	monochromosomal radiation hybrid (RH) maps	distance between breakpoints is often many Mb
	whole genome RH maps	resolution can be as high as 0.5 Mb
Restriction maps	created with restriction endonucleases that cut only rarely	several hundred kb
Clone contig maps	overlapping YAC clones	average YAC insert has several hundred kb
	overlapping BAC clones	average BAC insert is 160 kb
STS maps	requires previous sequence information from clones that have been mapped to subchromosomal regions	less than 1 kb is possible, but standard STS maps have resolutions in tens of kb
EST maps	requires cDNA sequencing, then mapping of cDNAs back to other physical maps	average resolution in the human nuclear genome is ~90 kb
DNA sequence maps	complete nucleotide sequence of DNA of each chromosome	1 bp

STS, sequence-tagged site; EST, expressed sequence tag.

markers with an average spacing of just less than 200 kb. In this landmark achievement, the chromosomal locations of the non-polymorphic STS markers had been obtained by two approaches. One approach used STS content mapping. YAC clones from the CEPH YAC library were assayed for large numbers of STS markers to identify clones sharing certain markers, and to assign them to YAC contigs with a known subchromosomal location. The second approach used hybrid cell mapping. Different human STS markers were assayed in panels of human–rodent **hybrid cells**. The hybrid cells each contained a full complement of rodent chromosomes but differed in that they contained different whole human chromosomes or different sets of fragments from human chromosomes that had integrated into the rodent chromosomes.

Although the YAC mapping achievements were impressive, there was a problem. YACs containing large human DNA inserts (which will contain many repetitive DNA sequences) proved to be prone to rearrangements and deletions (see Figure 2 in Box 8.1). Because the YAC inserts were often not faithful representations of the original starting human DNA, YAC clones could not be the template for the final genome sequencing effort.

Alternative large insert cloning systems were developed. Bacterial artificial chromosome (BAC) and phage P1 artificial chromosome (PAC) libraries have smaller inserts (80–250 kb) than YACs but, crucially, human inserts are comparatively stable in BACs and PACs. To provide even more STS markers, clones in the BAC libraries were routinely subjected to end-sequencing in which a few hundred nucleotides were sequenced at either end of the insert. Eventually, large BAC/PAC clone contigs were established for each human chromosome, paving the way for the final genome sequencing phase (see overview in **Figure 8.16**).

The final sequencing phase of the Human Genome Project was a race to an early finish

The progress of the Human Genome Project was faster than expected (see **Box 8.2** for a timeline). Genetic maps were developed ahead of the original schedule, and the final stage of large-scale DNA sequencing was facilitated by developments in automated fluorescence-based DNA sequencing. Competition between publicly and privately funded sequencing programs also drove a rapid final sequencing phase.

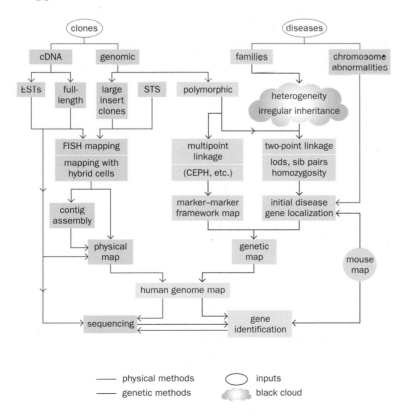

Figure 8.16 Major scientific strategies and approaches used in the Human Genome Project (HGP). The HGP required the isolation of human genomic and cDNA clones. The clones were used to construct high-resolution genetic and physical maps that paved the way for genome sequencing. Inevitably, the HGP interacted with research on mapping and identifying human disease genes. The data produced were channeled into mapping and sequence databases, permitting rapid electronic access and data analysis. Ancillary projects (not shown here) included studying genetic variation, genome projects for model organisms, and research on ethical, legal, and social implications. EST, expressed sequence tag; CEPH, Centre d'Études du Polymorphisme Humaine; lods, lod (logarithm of the odds) scores; STS, sequence-tagged site; FISH, fluorescence *in situ* hybridization.

BOX 8.2 MAJOR MILESTONES IN MAPPING AND SEQUENCING THE HUMAN GENOME

1956   The first physical map of the human genome is determined. Using light microscopy of stained tissue, JH Tjio and A Levan (*Hereditas* 42, 1 6) reveal that our cells normally contain 46 chromosomes and that there are 24 different types of human chromosome.

1977   Fred Sanger and colleagues publish the dideoxy DNA sequencing method (*Proc. Natl Acad. Sci. USA* 74, 5463–5467). With some further refinements (fluorescence labeling and automation) it will be the method used to sequence the human genome and many other genomes.

1980   David Botstein et al. (*Am. J. Hum. Genet.* 32, 314–331) propose that a human genetic map can be constructed by using a set of random DNA markers such as restriction fragment length polymorphisms (RFLPs).

1981   Sanger and colleagues publish the complete sequence of human mitochondrial DNA (Anderson S et al. *Nature* 290, 457–465).

1984   A workshop is held in Alta, Utah, to evaluate methods of mutation detection and characterization and to project future technologies. A principal conclusion is that an enormously large, complex, and expensive sequencing program is required to permit high-efficiency mutation detection.

1987   The US Department of Energy publishes a report on a Human Genome Initiative, which is the first of its kind.

1987   Helen Donis-Keller and colleagues report the first genetic linkage map of the human genome (*Cell* 51, 319–337). The map is based on RFLPs and has a low resolution.

1988   The US National Institutes of Health (NIH) sets up a dedicated Office of Human Genome Research (later renamed the National Center for Human Genome Research).

1988   The Human Genome Organization (HUGO) is established to coordinate international efforts by facilitating the exchange of research resources, encouraging public debate, and advising on the implications of human genome research.

1990   The Human Genome Project (HGP) is launched officially after implementation of a $3 billion 15-year project in the USA.

1992   The first comprehensive human genetic linkage map, based on microsatellite markers (Weissenbach J et al. *Nature* 359, 794–801).

1993   A first-generation physical map of the human genome is reported, based on YAC clones (Cohen D et al. *Nature* 366, 698–701).

1994   An improved genetic map is published, based mostly on microsatellite markers and with a spacing of one marker per centimorgan (Murray JC et al. *Science* 265, 2049–2054).

1995   The first detailed physical map of the human genome is published, based on sequence-tagged sites (Hudson TJ et al. *Science* 270, 1945–1954).

1996   A high-density human BAC library is published (Kim UJ et al. *Genomics* 34, 213–218).

1998   GeneMap '98, the first reasonably comprehensive map of gene-based markers, is published (Deloukas P et al. *Science* 282, 744–746).

1999   The first essentially complete DNA sequence for a human chromosome is reported, for chromosome 22 (Dunham I et al. *Nature* 402, 489–495).

2001   Draft sequences of the human nuclear genome, comprising roughly 90% of the total euchromatic component, are published by the International Human Genome Sequencing Consortium (IHGSC) (*Nature* 409, 860–921) and by Celera (*Science* 291, 1304–1351).

2001/2   Publication of a draft sequence of the mouse nuclear genome (Waterson B et al. *Nature* 420, 520–562). Human–mouse comparisons help human gene identification/characterization.

2003/4   The essentially completed sequence (about 99%) of the euchromatic component of the human genome is reported, comprising 2.85 Gb of DNA, or roughly 90% of the total of 3.2 Gb (analyses published in 2004 by the IHGSC, *Nature* 431, 931–945).

2004   Over 21,000 human genes are validated by full-length cDNA clones (Imanishi T et al. *PLoS Biol.* 2, e162).

2005–7   The International HapMap Consortium reports increasingly detailed single nucleotide polymorphism (SNP) maps for the human genome in 2005 (*Nature* 437, 1299–1320) and 2007 (*Nature* 449, 851–861). The latter map has more than 3.1 million SNPs, roughly one SNP per kilobase.

2007–8   The age of personal genome sequencing begins with delivery of euchromatic genome sequences for James Watson and Craig Venter and the launch of the 1000 Genomes Project (http://www.1000genomes.org/page.php).

An important rationale of the HGP—and a major motivation for privately funded genome sequencing—was to be able to study human genes. Starting in the early 1990s, attempts were made to obtain partial sequences from the 3′ untranslated regions of as many different human cDNA clones as possible, generating a huge number of ESTs. Introns are rarely found in the 3′ untranslated region of human genes and so a PCR assay based on the EST sequence could usually be used to type genomic DNA.

Subsequently, systematic large-scale mapping of ESTs against panels of radiation hybrids produced the first comprehensive human gene maps. The resulting gene map was published in 1998 (see Box 8.2) and seemed to identify the positions of 30,000 human genes. However, the full extent of our genes could not be known with more precision until the genome sequence was delivered.

From an early stage it was clear that human genes were not uniformly distributed along or between chromosomes. Some chromosomes were rich in genes; others were gene-poor (**Figure 8.17**). The heterochromatic regions of the genome—including most of the Y chromosome, and substantial regions on chromosomes 1, 9, 16, and so on—were known to be essentially devoid of genes and extraordinarily rich in repetitive DNA, which would make mapping extremely difficult. As a result, the HGP was focused almost exclusively on the remaining euchromatic regions that collectively accounted for about 90% of the human genome.

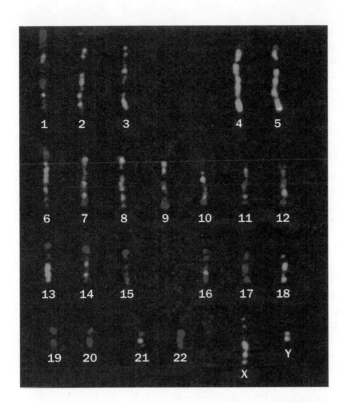

Figure 8.17 An early gene map for humans. The 5′ ends of most human (and vertebrate) genes have CpG islands, sequences about 1 kb long that differ from the bulk of the DNA in having many unmethylated CpG dinucleotides. The image shows the result of hybridizing a purified human CpG island fraction (labeled with a Texas Red stain) to human metaphase chromosomes. Late-replicating chromosomal regions (mostly transcriptionally inactive) are distinguished by the incorporation of FITC-labeled bromodeoxyuridine (green signal). Yellow regions (overlap of red and green signals) denote late-replicating regions rich in genes (or, strictly, CpG islands). Because CpG islands are gene markers, chromosomal regions that show a strong red signal have a high gene density (e.g. chromosome 22). Other chromosomes have very weak red signals and are gene-poor, such as chromosomes 4, 18, X, and Y. [Adapted from Craig JM & Bickmore WA (1994) *Nat. Genet.* 7, 376–381. With permission from Macmillan Publishers Ltd.]

For the publicly funded HGP, most of the sequence was contributed by large genome centers, notably the Wellcome Trust Sanger Institute, Washington University, Baylor College of Medicine, MIT/Harvard, the DoE Joint Genome Institute, the Japanese RIKEN Institute, and the French Genoscope center. To ensure efficiency, it was agreed that specific centers would take primary responsibility for assembling clone contigs and subsequent sequencing of individual chromosomes: for example, the Sanger Institute for chromosome 1, and Washington University for chromosome 2.

The publicly funded HGP met with aggressive competition from privately funded genome sequencing. In 1999 the Celera company announced that it intended to produce a draft human genome sequence in 2 years and that it would do this by whole genome shotgun sequencing (see Figure 8.13A) rather than the slower large contig assembly approach of the HGP.

The ensuing race between the publicly funded International Human Genome Sequencing Consortium (IHGSC) and the privately funded Celera accelerated the HGP timetable. In 2001, both sides published a draft sequence of the human genome that covered about 90% of the euchromatic genome sequence. The euchromatic component is about 90% of the total genome, and so the draft sequences actually represented about 80% of the total genome but 90% of the total gene sequence.

Although the race was perceived to have ended in a draw in 2001, it had not been a fair race, and the finishing line had not been reached. The IHGSC had made their data freely available, posting sequence data updates on the Web every 24 hours. Celera made use of huge blocks of the IHGSC's sequence data, reprocessed it, and fed the data back into its own sequence compilation. The Celera sequence was therefore not an independently obtained human genome sequence. Unlike the IHGSC, Celera denied external access to their sequence data. Long after they had published their analyses, Celera required expensive subscription charges to view their sequence data.

To complete the sequence of the human genome, the IHGSC continued alone with the hard work of filling in the gaps in the sequence. By 2003, the IHGSC had produced an essentially complete sequence of the euchromatic component of the human genome that was available on the Web and followed up by publishing analyses of the finalized sequence in 2004.

The 2.85 Gb of sequence reported in 2003 was extremely accurate and arranged in contigs of close to 9 Mb on average. The sequence was interrupted by 341 gaps that remained. The gaps are of two types. Structural gaps result from problems with ambiguous map assignment due to repetitive sequences (often duplications with near-identical sequences), but can be solved by standard (clone-based) DNA sequencing methods. Non-structural gaps are ones where the missing sequence can not be cloned in bacterial cells. In such cases, the sequence alternates between pyrimidine and purine nucleotides, making the DNA twist in a reverse orientation to form a left-handed helix that is toxic to bacteria. They can be closed only by using sequencing-by-synthesis methods. The number of gaps in the euchromatic genome has been progressively reduced to less than 200 in the most current sequence (NCBI build 37; released 2009).

Because a variety of DNA libraries had been used to generate the sequence, the final sequence was a composite one, representing various donor cells of different genotypes. This sequence acted as a reference sequence that hugely facilitated subsequent sequencing of the euchromatic genomes of individual people. An early demonstration was a 2007 report of the sequencing of Jim Watson's genome in a few months using massively parallel pyrosequencing. The age of personal genome sequencing began to take off with the international 1000 Genomes Project that was launched in early 2008. The aim is to obtain a detailed catalog of human genetic variation by sequencing the genomes of at least 1000 individuals representing a variety of different ethnic groups (see Box 8.2).

Once the initial draft genome sequence had been obtained, another major international effort focused on functionally important sequences with a priority of identifying and characterizing all human genes and regulatory elements. By 2004 more than 21,000 human genes were reported to have been validated by determining full-length cDNA sequences. However, as described in the sections below, there are still considerable uncertainties about exactly how many genes we have.

Genome projects have also been conducted for a variety of model organisms

From the outset, the goals of the HGP included sequencing of five model organisms, partly as technology test beds. The genome sequences for four of the five model organisms prioritized by the HGP—*E. coli*, *S. cerevisiae*, *C. elegans*, and *D. melanogaster*—were obtained at an early stage in the project and were helpful in guiding the mapping and sequencing strategies that would be applied to the more complex human genome.

Draft mouse genome sequences were obtained in 2001/2002. Celera produced a mouse genome sequence that was a composite of multiple mouse strains, and the publicly funded Mouse Genome Sequencing Consortium derived the genome sequence of the widely used C57BL/6J mouse strain. The latest C57BL/6J sequence update represents more than 90% of the total mouse genome. As detailed in Chapter 10, comparison of the human and mouse sequences was to prove extremely important in identifying genes and in establishing exon–intron organizations.

At an early stage, additional genome projects were launched for other organisms in addition to those that the HGP focused on. The first cellular genome was sequenced in 1995 (from the bacterium *Haemophilus influenzae*), and since then a large number of genomes from the three kingdoms of life have been sequenced. By late 2009, close to 1100 complete genome sequences had been published, and an additional 4500 genome projects were ongoing. Various archaeal and bacterial genomes have been sequenced to understand general and evolutionary aspects of prokaryotes. The principal motivation for genome sequencing of many bacteria has been to understand their involvement in pathogenesis or applications in biotechnology. For eukaryotes, varied motivations prompted genome sequencing: they include general research models, models of disease and development, models for evolutionary and comparative genomic studies, farm animals and crops, and pathogenic protozoa and nematodes. Relevant genome data can be accessed at certain websites that provide compilations of genome databases (see the next section).

Analysis of the sequences has shown that the number of genes in a genome is correlated with complexity of an organism, but the correlation is a poor one. For example it was a surprise that *C. elegans*, a 1 mm roundworm that has only

TABLE 8.3 MAJOR ELECTRONIC DATABASES THAT SERVE AS GENERAL NUCLEOTIDE OR PROTEIN SEQUENCE REPOSITORIES

Database type	Database	Originator/host	URL
Nucleotide sequence	GenBank	US National Center for Biotechnology Information (NCBI)	http://www.ncbi.nlm.nih.gov
	EMBL	European Bioinformatics Institute (EBI)	http://www.ebi.ac.uk
	DDBJ	National Institute of Genetics, Japan	http://www.ddbj.nig.ac.jp/
	dbEST	NCBI (division of GenBank that stores single-pass cDNA/ESTs)	http://www.ncbi.nlm.nih.gov/dbEST/
Protein sequence	SWISS-PROT	Swiss Institute of Bioinformatics, Geneva	http://ca.expasy.org/sprot
		European Bioinformatics Institute	http://www.ebi.ac.uk/swissprot/
	TREMBL	EBI (translations of coding sequences from the EMBL database that have not yet been deposited in SWISS-PROT)	http://www.ebi.ac.uk/trembl/ http://ca.expasy.org/sprot
	PIR	US National Biomedical Research Foundation (NBRF)	http://pir.georgetown.edu/
		Japan International Protein Information Database (JIPID)	http://www.ddbj.nig.ac.jp/
		Munich Information Center for Protein Sequences (MIPS)	http://mips.gsf.de/

Note that the databases have subsets that are devoted to particular sequence classes. For transcribed sequences and genes, the dbEST database (http://www.ncbi.nlm.nih.gov/dbEST) and the UniGene database (http://www.ncbi.nlm.nih.gov/unigene) have been widely used.

959 cells or 1031 cells depending on its gender, should have more than 20,000 protein-coding genes, about as many as a human. The correlation between gene content and organism complexity will be considered more fully in Chapter 10.

Powerful genome databases and browsers help to store and analyze genome data

From the earliest stages of genome and gene analysis, central computer repositories were established for storing mapping data and sequence data produced in laboratories throughout the world (Table 8.3). After major genome mapping and sequencing centers developed, a parallel data storage effort began when the individual genome centers developed dedicated in-house databases to store mapping and sequencing data produced in their own laboratories. The data were made freely available through the Web.

As genome data began to be produced in very large quantities, strenuous efforts were devoted to developing new genome databases (Table 8.4) and designing new software that would permit the huge amounts of mapping and sequence data and associated information to be searched in a systematic and user-friendly way. As we will see, a major new focus on *in silico* (computer-based) analyses made vital contributions to our understanding of the structure of genes and genomes.

An important advance was the development of **genome browsers** with graphical user interfaces to portray genome information for individual chromosomes and subchromosomal regions. Users of genome browsers can quickly navigate the sequence of a selected human chromosome moving from large scale to nucleotide scale, identifying genes and associated exons, RNAs, and proteins (Figure 8.18). As more and more information is obtained for genes and other functional units, more informative and precise *gene annotation* will be available in frequent, periodical updates of the genome browsers and databases.

Different computer programs are designed to predict and annotate genes within genome sequences

Genome sequence data can be interrogated by a suite of computer programs to seek novel genes by screening for particular characteristics of genes. For example, genes are transcribed into RNA, and they contain sequences that show strong evolutionary conservation (because of the need to conserve important gene functions). Genes are also associated with certain sequence motifs and sequence characteristics.

TABLE 8.4 MAJOR EUKARYOTIC GENOME BROWSERS AND DATABASES

Resource	Originator/host	URL
GENOME BROWSERS		
Ensembl	Wellcome Trust Sanger Institute/European Bioinformatics Institute (EBI)	http://www.ensembl.org
NCBI map viewer	US National Center for Biotechnology Information (NCBI)	http://www.ncbi.nlm.nih.gov/mapview/
UCSC genome browser	University of California at Santa Cruz	http://genome.ucsc.edu
GENOME COMPILATIONS		
EBI Genomes	European Bioinformatics Institute (EBI)	http://www.ebi.ac.uk/genomes
GOLD	Genomes Online Database	http://www.genomesonline.org/index.htm
ORGANISM GENOME DATABASES		
Flybase (*Drosophila*)	Flybase Consortium	http://flybase.org/
MGI (mouse genome informatics)	Jackson Laboratory	http://www.informatics.jax.org/
NCBI Human Genome Resources	US National Center for Biotechnology Information (NCBI)	http://www.ncbi.nlm.nih.gov/projects/genome/guide/human/
SGD (*Saccharomyces* genome database)	Stanford University	http://www.yeastgenome.org/
Wormbase (*C. elegans*)	Wormbase Consortium	http://www.wormbase.org/
ZFIN (Zebrafish Information Network)	University of Oregon	http://zfin.org/

Gene prediction programs depend heavily on *homology searching* to identify evolutionarily conserved sequences. A test DNA sequence can first be used to search for homologous RNA sequences from any organism by using programs such as BLAST (**Box 8.3**) to query nucleotide and EST databases. Thereafter, if the DNA is suspected to be transcribed, it can then be translated in all six reading frames (three for each DNA strand).

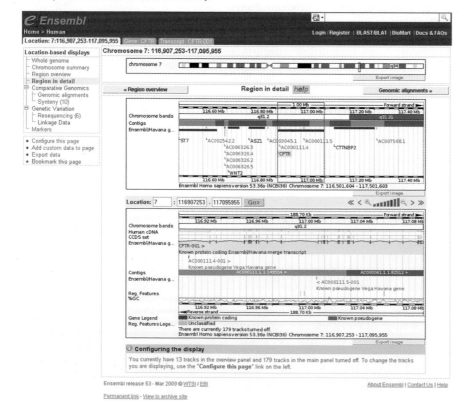

Figure 8.18 The Ensembl Genome Browser. Genome browsers such as Ensembl (http://www.ensembl.org) allow users to explore selected subchromosomal regions by using a graphical interface. Here the subject of the query was the human *CFTR* (cystic fibrosis transmembrane regulator) gene; the overview at the top shows the position of the *CFTR* gene (highlighted in pale green) in relation to neighboring genes and DNA markers on chromosome 7q31. The detailed view at the bottom shows a few of very many features that can be switched on or off according to the user's needs. Central to the navigation are click-over tools that enable numerous Internet connections to other databases and programs (including direct links to other assembled genome sequences), allowing users to follow through on requested information.

The predicted translation products can in turn be used to search for homology against all known protein sequences (using BLASTP), and against the predicted translation products of all known nucleotide sequences (using TBLASTN). Suspected candidate protein sequences can also be used as queries in homology searching against databases of protein domains and short motifs, as detailed in Chapter 10.

BOX 8.3 SEQUENCE HOMOLOGY SEARCHING

Powerful computer programs have been devised to permit the rapid searching of nucleic acid and protein sequence databases (and of dedicated genome sequence databases) for significant sequence matching (homology) with a test sequence. The different BLAST and FASTA programs are the most popular (**Table 1**).

The designs of comparable FASTA and BLAST programs are different and so they may give different results. The BLAST and FASTA programs are widely available, for example from the US National Center for Biotechnology Information (which also provides tutorials in how to use the different BLAST programs at http://www.ncbi.nlm.nih.gov/Education/BLASTinfo/information3 .html) and the European Bioinformatics Institute (http://www. ebi.ac.uk/Tools/similarityandanalysis.html); BLAT is hosted at the University of California at Santa Cruz (at http://genome.ucsc.edu/), but is available at many other locations too.

Programs such as BLAST and FASTA use algorithms to identify optimal sequence alignments and typically display the output as a series of pairwise comparisons between the test sequence (*query sequence*) and each related sequence that the program identifies in the database (*subject sequences*).

Different approaches can be taken to calculate the optimal sequence alignments. For example, in nucleotide sequence alignments the Needleman–Wunsch algorithm seeks to maximize the number of matched nucleotides, but the Waterman algorithm seeks to minimize the number of mismatches.

Pairwise comparisons of sequence alignments are comparatively simple when the test sequences are very closely matched and have similar, preferably identical, lengths. When the two sequences that are being matched are significantly different from each other, and especially when there are clear differences in length resulting from deletions or insertions, considerable effort may be necessary to calculate the optimal alignment (**Figure 1A**).

For coding sequences, nucleotide sequence alignments can be aided by parallel amino acid sequence alignments using the assumed translational reading frame for the coding sequence. This is so because there are 20 different amino acids but only four different nucleotides. Pairwise alignments of amino acid sequences may also be aided by taking into account the chemical subclasses of amino acids. Conservative nucleotide substitutions replace an amino acid with one that is chemically related to it, typically belonging to the same subclass.

TABLE 1 COMMONLY USED PROGRAMS FOR BASIC SEQUENCE HOMOLOGY SEARCHING

Program	Features
FASTA	Compares a nucleotide sequence against a nucleotide sequence database, or an amino acid sequence against a protein sequence database
TFASTA	Compares an amino acid sequence against a nucleotide sequence database translated in all six reading frames
BLASTN	Compares a nucleotide sequence against a nucleotide sequence database
BLASTX	Compares a nucleotide sequence translated in all six reading frames against a protein sequence database
BLASTP	Compares an amino acid sequence against a protein sequence database
TBLASTN	Compares an amino acid sequence against a nucleotide sequence database translated in all six reading frames
BLAT	BLAST-like program that delivers extremely rapid searching at nucleotide or protein levels against a defined sequenced genome

As a result, algorithms used to compare amino acid sequences typically use a scoring matrix in which pairs of scores are arranged in a 20×20 matrix in which higher scores are accorded to identical amino acids and to ones that are similar in character (e.g. isoleucine and leucine) and lower scores are given to amino acids that are different in character (e.g. isoleucine and aspartate).

The typical output gives two overall results for percent sequence relatedness, often termed **percentage sequence identity** (matching of identical residues only) and **percentage sequence similarity** (matching of both identical residues and ones that are chemically related; **Figure 1B**).

(A)
```
GATATTATCACTGGAGCCTGGCAGGAGCT          GATATTATCACTGGAGCCTGGCAGGAGCT
**** **** ********** * ******   OR    **** **** ********** - ******
GATTTATGACTGGAGCCTGA-AGGAGCT          GATTTATGACTGGAGCCT-GAAGGAGCT
```

(B)
```
Score = 52.8 bits (125), Expect = 9c-08
Identities = 39/120 (32%), Positives = 57/120 (47%), Gaps = 9/120 (7%)


QUERY:   1   AKLLIKHDSNIGIPDVEGKIPLHWAANHKDPSAVHTVRCILDAAPTESLLNWQDYEGRTP  60
             A+LL++HD+       G  PLH A +H +    + V+ +L +        W  Y   TP
SBJCT: 548   AELLLEHDAHPNAAGKNGLTPLHVAVHHNN---LDIVKLLLPRGGSPHSPAWNGY---TP 601


QUERY:  61   LHFAVADGNLTVVDVLTSY-ESCNITSYDNLFRTPLHWAALLGHAQIVHLLLERNKSGTI 119
             LH A   +V   L Y  S N S   + TPLH AA  GH ++V LLL +  +G +
SBJCT: 602   LHIAAKONOIEVARSLLOYGGSANAESVOGV--TPLHLAAOEGHTEMVALLLSKOANGNL 659
```

Figure 1 Sequence alignment, sequence identity, and sequence similarity. (A) Sequence alignment ambiguity. The two nucleotide sequences are clearly related, but there is ambiguity in how to align the sequence **GGC** with the corresponding sequence **GA** in the bottom sequence. * signifies identity; – signifies absence. (B) Sequence identity and sequence similarity. The BLASTP output here resulted from querying the SWISS-PROT protein database with an inversin protein query sequence. The subject sequence shown is an erythrocyte ankyrin sequence. Amino acids shown in red indicate sequence identity (39 of the 120 positions, or 32.5%). An additional 17 positions have chemically similar amino acids (shown as +) giving a total of 56 out of 120 positions (46.7%) that are identical or chemically similar.

The detection of novel genes can also be aided by computer algorithms that recognize certain elements commonly found in genes. Exon prediction programs such as GENSCAN, for example, test for conserved consensus sequences at splice junctions (see Figure 1.17) and assign high probability to a predicted exon if there is a large open reading frame that differentiates the sequence from noncoding DNA (where on average one of the three termination codon sequences will occur every 60 bp or so).

Integrated gene-finding software packages have been developed that combine programs designed to identify exons and gene-associated motifs with general sequence-homology-based database searching programs. One example is the Genotator program (http://www.fruitfly.org/~nomi/genotator/).

Another characteristic that helps identify genes in vertebrates is base composition. Vertebrate genes are very often associated with small regions, often about 1 kb long, that are both GC-rich and have a significantly higher frequency of the dinucleotide CpG than the bulk of the genome, where the CpG dinucleotide is statistically under-represented. Because these regions can be viewed as islands of normal CpG frequency in a sea of genomic DNA that is otherwise CpG-deficient, they are referred to as **CpG islands**.

As we will see in Chapter 9, the CpG dinucleotide is a target for DNA methylation that can cause local condensation of chromatin and inhibit gene expression. The upstream regions and other important sites that regulate gene expression need to be able to adopt an open chromatin conformation for gene expression, and so high frequencies of CpG are not tolerated in these regions. CpG islands can be identified both experimentally and by computer programs that scan the sequence for variation in base composition.

Obtaining accurate estimates for the number of human genes is surprisingly difficult

Analysis of the human genome sequence has led to the identification of many thousands of previously unstudied genes. As we will see in later chapters, the functions of predicted genes are being intensively studied in the post-genome-sequencing era, and systematic and hierarchical vocabularies are being developed to define gene function (Box 8.4). Surprisingly, however, many years after the complete euchromatic sequence was obtained and despite intense investigations, the total number of human genes is still not accurately known, and it may be some time before a precise figure can be obtained.

Before the human genome was sequenced, the number of human genes had been expected to be in the range 70,000–100,000, or close to three to four times the number of genes in *C. elegans*. By 2001, experimental evidence was available for about 11,000 human genes, but analyses of the draft genome sequence reported in that year raised the predicted number to only about 30,000–40,000 mostly protein-coding genes. The essentially complete sequence of the euchromatic component of the human genome was obtained by 2003. After analysis of the new sequence, the number of predicted protein-coding genes had fallen to somewhat less than 25,000, and current estimates suggest close to 21,500 human protein-coding genes.

The difficulty in establishing the precise number of genes is not confined to the human genome but to all moderately complex genomes. As more and more genomes were sequenced, however, comparative genomics has been extremely helpful in gene identification. In particular, comparative genomics has helped in the identification of protein-coding genes, and the recent estimate of about 21,500 human protein-coding genes is likely to be quite accurate. RNA genes are, however, much more difficult to identify for two major reasons. First, they lack any open reading frame, and second, their sequences tend to be much less conserved than protein-coding sequences.

During the course of the human genome project, little attention was devoted to RNA genes, and the various estimates of gene number were dominated by expectations of the number of protein-coding genes. Remarkably, the 2001 *Science* paper in which Celera reported their analyses of the draft human genome sequence does not mention RNA genes at all! Since then, there has been a revolution in our understanding of the importance of RNA genes. As detailed in Chapter 9, there are close to 1000 human miRNA genes, and many entirely new classes of

BOX 8.4 GENE ONTOLOGY AND THE GO CONSORTIUM

As the trickle of genome sequence data began to become a flood, biologists started to grapple with the difficult challenge of integrating sequence data with the vast and rapidly growing body of data from gene function analyses. Searching across the broad scientific literature and databases has been hampered by wide variations in terminology. What was needed was a standardized system of **gene ontology** to represent gene function across genomes and species. The Gene Ontology (GO) Consortium was formed in 1998 and the GO project began to develop a controlled vocabulary to describe the attributes of genes and gene products in any organism, to allow searching across multiple gene products and species for common characteristics.

To describe gene products, the GO Consortium developed three separate ontologies—biological process, cellular component, and molecular function. These ontologies permit the annotation of molecular characteristics across species and may be broad or more focused. For example, the biological process can be as broad as signal transduction or more restricted, for example α-glucoside transport. Each vocabulary is structured so that any term may have more than one parent as well as zero, one, or more children. This makes attempts to describe biology much richer than would be possible with a hierarchical graph. Currently the GO vocabulary consists of more than 17,000 terms that will, in time, all have strict definitions for their usage. See http://www.geneontology.org for more information.

TABLE 8.5 DIFFICULTIES IN ESTIMATING GENE NUMBER IN COMPLEX GENOMES

• RNA gene prediction. Genes encoding untranslated RNAs may be difficult to identify in the absence of a sizeable open reading frame, especially if they are small. RNA gene sequences are often comparatively poorly conserved during evolution. It is often difficult to distinguish between functional noncoding RNA and related pseudogene sequences.
• Genes that are expressed at low levels and/or at unusual cellular locations and stages of development may not be well represented in available cDNA libraries, and may not be evident if they have small exons.
• Very large genes have widely dispersed exons and may be misinterpreted as a cluster of two or more smaller genes. This type of overestimate of gene number arises because many of the cDNA libraries used to validate genes and their exon–intron organization have comparatively short inserts.
• Some nonfunctional gene copies (pseudogenes) are transcribed and may initially be mistaken for true genes.

RNA have recently been identified, many with a regulatory function. It may take some time before we have a clear idea of the number of human RNA genes. Table 8.5 lists some of the difficulties in estimating gene numbers.

8.4   BASIC GENE EXPRESSION ANALYSES

Principles of expression screening

Gene expression can be monitored at either the transcript level or the protein level with the use of a variety of technologies. Important parameters in gene expression are the source and nature of the product, the expression resolution, and the number of genes that are analyzed at any one time (Table 8.6).

Frequently, crude RNA/cDNA or protein extracts are used as source material. Sometimes, however, expression is sampled in tissue sections or even whole embryos that have been fixed so as to preserve the original *in vivo* morphology. Expression can also be studied in live cells in tissue culture. The expression of genes tagged with a fluorescent group can also be followed in living experimental organisms with optically transparent tissues.

TABLE 8.6 DIFFERENT LEVELS OF EXPRESSION MAPPING

Study material	Resolution	Gene expression throughput[a]	Examples
RNA	high	low	tissue *in situ* hybridization (Figures 7.12 and 8.20)
			cellular *in situ* hybridization
	low	low to medium	northern blot hybridization (Figure 7.11)
			RNA dot-blot hybridization
			ribonuclease protection assay
			RT-PCR/qPCR
	low	high	DNA microarray hybridization (Figure 8.24)
Protein	high	low	immunocytochemistry (Figure 8.22)
			fluorescence microscopy (Figure 8.23)
	low	low	immunoblotting (western blotting) (Figure 8.21)
	low	high	two-dimensional gel electrophoresis (Figure 8.25)
			mass spectrometry (Figures 8.26 and 8.27)

[a]Number of genes or proteins studied at one time.

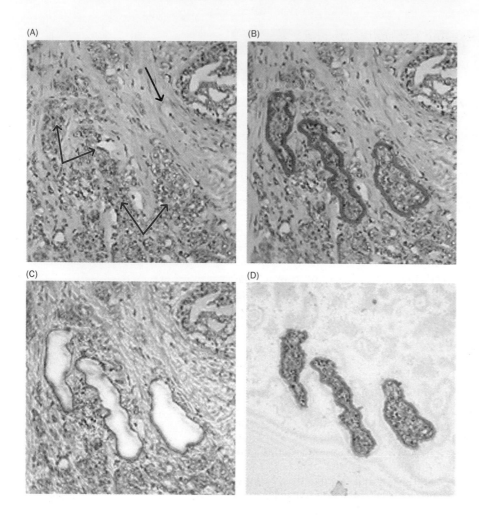

Figure 8.19 Laser capture microdissection of tissue sections. (A) Prostate section stained with hematoxylin and eosin. The black arrow indicates stromal (connective tissue) cells; red arrows indicate epithelial cells. (B) Laser outline of the cells to be collected. (C) The remaining cells after laser capture. (D) The cells collected from the outlined area. [Courtesy of JR Vielkind from Garnis C, Buys TP & Lam WL (2004) *Mol. Cancer* 3, 9. With permission from BioMed Central Ltd.]

Laser capture microdissection uses a laser to dissect out microscopic portions of a tissue to produce pure cell populations from sources such as tissue biopsies and stained tissue, and even single cells (**Figure 8.19**). As a result, gene expression analyses can be focused on single cells, or on homogeneous cell populations that will be more representative of the *in vivo* state than cell lines.

Low-resolution expression patterns are initially sought from genes by tracking gross expression in RNA extracts or protein extracts. In addition to being able to sample expression in different tissues, these patterns may provide useful information on the level of expression and on expression product variants that can differ in size (isoforms). Interesting expression patterns can be followed up by using methods to track expression within a cell, or within groups of cells and tissues that are spatially organized in a manner representative of the normal *in vivo* organization.

Low-throughput expression screening follows the expression of only one gene or a very small number of genes at a time. High-throughput methods can simultaneously track the expression of very many genes—often thousands of them—at a time, and can offer whole genome expression screening.

Hybridization-based methods allow semi-quantitative and high-resolution screening of transcripts of individual genes

A variety of hybridization-based procedures have been developed to study in detail the expression characteristics of individual genes. Initial studies may concentrate on establishing the sizes of the transcripts for individual genes in different cells, and their approximate abundances. More advanced studies will seek to conduct detailed analyses of where the genes are expressed in tissues and cells.

Hybridization-based methods for assaying transcript size and abundance

In **northern blot hybridization**, total RNA or poly(A)$^+$ RNA extracts prepared from different tissues or cell lines are size-fractionated by gel electrophoresis, transferred by blotting to a plastic membrane, then hybridized to a gene-specific nucleic acid probe. This method provides information on the size and comparative abundance of transcripts in multiple tissues or cell lines. It can also identify tissue-specific isoforms that may arise through alternative splicing, alternative polyadenylation, or usage of alternative promoters (see Figure 7.11).

The **ribonuclease protection assay** seeks to quantitate specific RNA transcripts in a complex mixture of total RNA or mRNA. A labeled single-stranded antisense RNA probe for the gene of interest is incubated with a sample of total RNA or mRNA to facilitate hybridization of the complementary region of interest to the labeled probe. After hybridization, the mixture is treated with ribonuclease (RNase), which digests all single-stranded RNA but no double-stranded RNA molecules. The only surviving labeled RNA molecules are those that have hybridized to the specific RNA transcripts to form double-stranded RNA. The labeled RNA is separated from degraded RNA by electrophoresis on a denaturing polyacrylamide gel. The amount of label detected gives a measure of the amount of the specific transcript in the sample.

Tissue *in situ* hybridization

High-resolution spatial expression patterns of RNA in tissues and groups of cells are normally obtained by **tissue *in situ* hybridization**. Usually, tissues are frozen or embedded in wax, then sliced with a microtome to give very thin sections (e.g. 5 μm or less), which are mounted on a microscope slide. Hybridization of a suitable gene-specific probe to the tissue on the slide can then give detailed expression images representative of the distribution of the RNA in the tissue of origin. Embryonic tissues are often used: their miniature size allows screening of many tissues in a single section (see Figure 7.12).

An extension of tissue *in situ* hybridization is to study expression in an intact embryo. Whole-mount *in situ* hybridization is a popular method for tracking expression during development in whole embryos from model vertebrate organisms. Because of the ethical and practical difficulties in conducting equivalent human gene analyses, there has been considerable reliance on extrapolating from analyses conducted on mouse embryos or those of other model vertebrates (**Figure 8.20**). The relatively large amount of tissue available means that the method is a relatively sensitive one, and automation of the technique has enhanced its popularity.

With the use of suitably labeled probes, specific RNA sequences can also be tracked within *single cells* to identify sites of RNA processing, transport, and cytoplasmic localization. By using quantitative fluorescence *in situ* hybridization (FISH) and digital imaging microscopy, it has even been possible to detect single RNA transcripts *in situ*. A further refinement uses combinations of different types of oligonucleotide probe labeled with spectrally distinct fluorophores. This has allowed transcripts from multiple genes to be tracked simultaneously.

Quantitative PCR methods are widely used for expression screening

The hybridization-based expression screening methods described above are designed to detect transcripts that have not been amplified. As a result, they are not well suited to the detection of low-copy-number transcripts. PCR-based methods are much more sensitive and can track gene expression in cell types or tissues that are not easy to access in great quantity, and even in single cells. There is a variety of such methods, but all of them require that the starting RNA be first copied into a cDNA sequence by using a reverse transcriptase.

In the basic **reverse transcriptase PCR** (**RT-PCR**) method, a cDNA copy is first made of RNA by using an oligo(dT) primer or random-sequence oligonucleotide primers, and this is then used to initiate a PCR reaction. After the PCR reaction is complete, an aliquot is taken for analysis by standard gel electrophoresis to fractionate the cDNA according to size, whereupon the separated products are

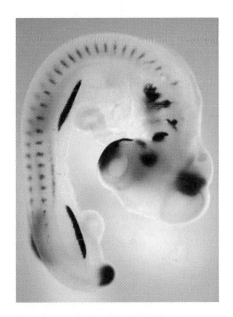

Figure 8.20 Whole-mount *in situ* hybridization. The example shows expression of the *Fgf8* fibroblast growth factor gene in a chick embryo at Hamburger Hamilton stage 20 (about 3 days of embryonic development after egg-laying). Transcripts were labeled with an antisense digoxigenin-labeled *Fgf8* probe and were detected with anti-digoxigenin antibodies coupled to alkaline phosphatase. The alkaline phosphatase assay used a combination of BCIP (5-bromo-4-chloro-3-indolyl phosphate) and NBT (Nitro Blue Tetrazolium), resulting in deep blue expression signals. Expression was evident in the developing eye, isthmus, branchial arches, somites, limb buds, and tailbud. (Courtesy of Terence Gordon Smith, Newcastle University.)

exposed to binding by ethidium bromide and revealed under ultraviolet radiation. The basic RT–PCR method has been useful for identifying and studying different RNA isoforms of an RNA transcript. However, the detection of DNA by ethidium bromide fluorescence is not very sensitive. It can detect products that have gone through a full 30–40 PCR cycles, but by this stage the exponential stage of amplification (the regular doubling of product per PCR cycle) has long since passed and it is not possible to obtain accurate RNA quantitation.

To obtain accurate quantitation of RNA transcripts, a variant method is used that was initially called **real-time PCR** but is often now simply called **qPCR** (quantitative PCR). Unlike the basic RT-PCR method, qPCR does not involve fractionating the final amplified products by electrophoresis. Instead, qPCR is a *kinetic* reaction that requires specialized PCR equipment. In qPCR the amplification products are continuously quantitated during the PCR reaction. To obtain the most accurate data, the quantitation measurements are made only during the early stages in the PCR reaction, when the amplification is still exponential.

To detect amplification products during a qPCR reaction, some measurable signal must be generated that is proportional to the amount of the amplified product. All current detection methods use fluorescence technologies, and the detection method may be nonspecific or specific (**Box 8.5**).

Specific antibodies can be used to track proteins expressed by individual genes

Because of their exquisite diversity, selectivity, and sensitivity in detecting proteins, antibodies are ideally placed to track gene expression at the protein level. The traditional way to obtain an antibody is to inject a suitable animal repeatedly with a specific **immunogen**, a molecule that is detected as being foreign by the host's immune system. To raise an antibody against a specific protein, the immunogen used is a synthetic peptide or a fusion protein that contains part of the sequence of the protein to be tracked (**Box 8.6**).

Antibodies can be labeled in different ways. In direct detection methods, the purified antibody is labeled by attaching a reporter molecule, such as a fluorophore or biotin, allowing the labeled antibody to bind directly to the target protein. Alternatively, the target protein is first bound by an unlabeled *primary antibody* that binds to the target protein and is then specifically bound in turn by a suitably labeled secondary molecule that may be a *secondary antibody*, a specific antibody raised against the primary antibody. Sometimes a general secondary molecule is used, such as protein A, a protein found in the cell wall of *Staphylococcus aureus*. For unknown reasons, protein A binds strongly to a common core region on antibody molecules (in the second and third constant regions of the Fc portion of immunoglobulin heavy chains).

Labeled antibodies can be used to track specific proteins in cell-free extracts, fixed tissues, and cells by using various methods. For cell-free extracts, a common application is **immunoblotting** (**western blotting**). In this method, proteins from cell-free extracts are first dissolved in a solution of sodium dodecyl sulfate (SDS), an anionic detergent that disrupts nearly all noncovalent interactions in native proteins, then separated according to size by SDS–polyacrylamide gel electrophoresis (SDS-PAGE). The fractionated proteins are transferred (blotted) to a sheet of nitrocellulose and then exposed to a specific antibody (**Figure 8.21**). As described in Section 8.5, two-dimensional PAGE gels may also be used.

In immunocytochemistry (= immunohistochemistry), an antibody is used to obtain an overall expression pattern for a protein within a tissue or other multicellular structure. As in tissue *in situ* hybridization, the tissues are typically either frozen or embedded in wax and then cut into very thin sections with a microtome before being mounted on a slide. A suitably specific antibody is allowed to bind to the protein in the tissue section and can produce expression data that can be related to histological staining of neighboring tissue sections (**Figure 8.22**).

Electron microscopy provides higher resolution than immunocytochemistry and may be used to investigate the intracellular localization of a protein in cells within fixed tissue that has been sliced into ultra-thin sections. The antibody is typically labeled with an electron-dense particle, such as colloidal gold spheres. As described in the next section, different fluorescence microscopy methods are used to track subcellular protein expression in cultured cells.

BOX 8.5 DETECTION METHODS USED IN QUANTITATIVE (REAL-TIME) PCR

Quantitation in qPCR depends on detecting a fluorescence signal that is generated after the binding of some reagent to amplified product at the early stages of the PCR reaction. Nonspecific detection methods are limited to detecting only one type of target amplicon at a time, whereas specific detection methods can distinguish between different targets and allow multiplex assays.

Specific detection is possible with single-stranded hybridization probes that are designed to bind to a specific type of amplicon during the annealing stage of the qPCR reaction. As detailed in the table and figures below, popular specific detection methods often use a hybridization probe that carries a fluorophore at the 5′ end and a quencher group at the 3′ end that absorbs photons emitted by the fluorophore and then dissipates the energy absorbed either in the form of heat or as light of a different wavelength.

Method	Basis of method
Nonspecific detection	
SYBR Green I dye	When free in solution, this dye shows relatively little fluorescence, but when it binds to double-stranded DNA, its fluorescence increases more than 1000-fold. It is, however, nonspecific in its binding of double-stranded DNA and so can bind to primer dimers or nonspecific amplification products (which is why it is not used in standard PCR). To control for this, a *melting curve* is conducted at the end of the run by increasing the temperature slowly from 60°C to 95°C while continuously monitoring the fluorescence. At a certain temperature, the whole amplified product will dissociate fully, resulting in a decrease in fluorescence as the dye dissociates from the DNA. The temperature of dissociation is dependent on the length and composition of the amplicon, allowing different DNA fragments to be distinguished (should there be nonspecific amplification or significant primer dimers).
Specific detection by hybridization probes	
Molecular Beacon probes	Molecular Beacon probes are designed to have a stem–loop structure that brings a fluorophore at the 5′ end in very close proximity to a quencher at the 3′ end, severely limiting the fluorescence emitted by the fluorophore. However, in the presence of a complementary target sequence, the probe unfolds and hybridizes to the target, causing the fluorophore to be displaced from the quencher (**Figure 1**). As a result, the quencher no longer absorbs the photons emitted by the fluorophore and the probe starts to fluoresce. **Figure 1 Molecular Beacon probes fluoresce only after they have hybridized to a target DNA.**
TaqMan® double-dye probes	Here the probe is an oligonucleotide that has a fluorophore such as FAM at the 5′ end, and a quencher group such as TAMRA or the Black Hole Quencher at the 3′ end. In this case the quenching mechanism is based on fluorescence resonance energy transfer (FRET) and can occur over a relatively long distance—as much as 10 nm or more. As above, the oligonucleotide probe can bind to a complementary sequence in the qPCR amplicon during an annealing step in the PCR reaction, but in this case the fluorophore of the intact hybridized probe does not fluoresce because it continues to pass on its energy to the quencher (**Figure 2**). After the probe has bound to an amplicon target, however, the advancing Taq polymerase displaces the 5′ end of the probe, which is then degraded by the 5′→3′ exonuclease activity of the Taq polymerase. While the polymerase continues to push aside the rest of the probe, cleavage continues and results in the release of the fluorophore and quencher into solution (**Figure 3**). Now they are physically separated from each other, the fluorophore FAM exhibits strong fluorescence but the quencher emits much less energy (at a different wavelength from that of the fluorophore).

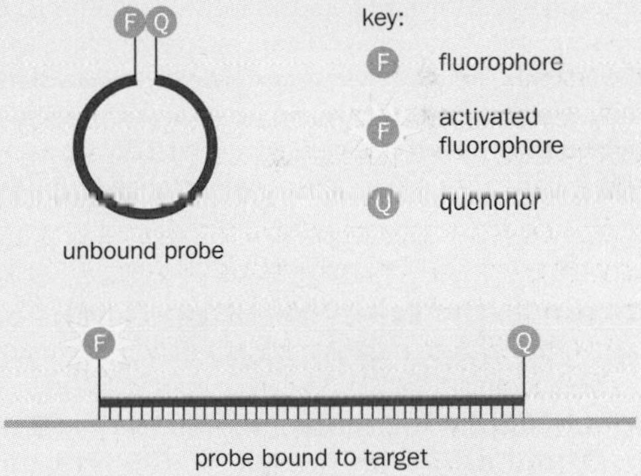

Figure 2 Hybridization of a TaqMan oligonucleotide probe to a PCR amplicon. F, fluorophore; Q, quencher.

Figure 3 The hybridized TaqMan probe is displaced by Taq polymerase and degraded by the associated exonuclease, thereby activating the probe's fluorophore.

key:
- F fluorophore
- F activated fluorophore
- Q quencher

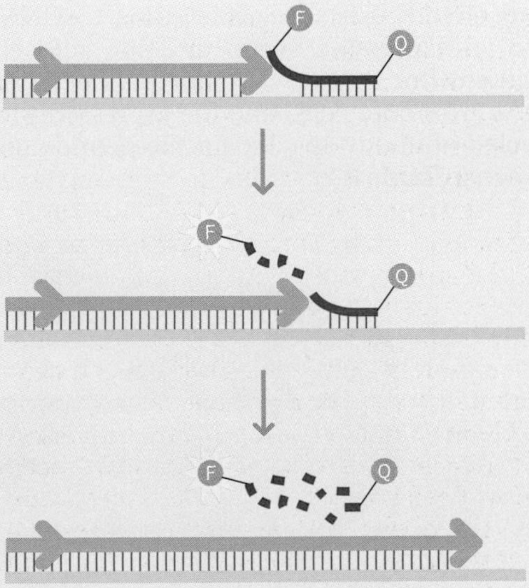

BOX 8.6 OBTAINING ANTIBODIES

Antibodies that can specifically detect a protein of interest can be raised in different ways. A comparatively recent approach has been to use *phage display*, an expression cloning system in which recombinant phages are used to display heterologous proteins on bacterial cell surfaces (see Section 6.3). However, the main method of obtaining antibodies has been the traditional route of repeatedly injecting animals (such as rodents, rabbits, or goats) with a suitable immunogen that represents a protein under investigation. One approach is to design a synthetic peptide, often 20–50 amino acids long, that is then conjugated to a suitable molecule (such as keyhole limpet hemocyanin) that will help maximize the immunogenicity. The hope is that the peptide will adopt a conformation resembling that of the native polypeptide sequence, but success is not guaranteed and several different peptides may need to be designed.

An alternative type of immunogen is a **fusion protein** that contains most or a large portion of the target protein sequence fused to another protein that confers some advantages, notably in assisting protein purification. The fusion protein is synthesized by cloning a cDNA for the desired target protein into a plasmid expression vector that also contains a cDNA for the companion protein and selecting for recombinants in which the two cDNAs are in frame (see Figure 6.12).

If the animal's immune system has responded, specific antibodies should be secreted into the serum. The antibody-rich serum (*antiserum*) contains a heterogeneous mixture of antibodies, each produced by a different B lymphocyte. The different antibodies recognize different parts (*epitopes*) of the immunogen and are known as **polyclonal antibodies**.

A homogeneous preparation of antibodies with a defined specificity can be prepared, however, by propagating a clone of cells (originally derived from a single B lymphocyte). Because B cells have a limited lifespan in culture, it is preferable to establish an immortal cell line: antibody-producing cells are fused with cells derived from an immortal B-cell tumor. From the resulting heterogeneous mixture of hybrid cells, those hybrids that have both the ability to make a particular antibody and the ability to multiply indefinitely in culture are selected. Such **hybridomas** are propagated as individual clones, each of which can provide a permanent and stable source of a single type of **monoclonal antibody (mAb)**.

Protein expression in cultured cells is often analyzed by using different types of fluorescence microscopy

The subcellular location of a protein of interest can also be tracked by antibodies in cultured cells with the use of different types of fluorescence microscopy. The protein may be followed directly by using a specific antibody raised against it. Alternatively, a peptide tag or other protein is coupled to the protein of interest, and expression of the protein is followed indirectly. In immunofluorescence microscopy, antibodies track protein expression directly. The antibodies are labeled with a suitable fluorescent dye, such as fluorescein or rhodamine, and after antibody has bound to the protein of interest, the expression of the protein is directly monitored within cells by using fluorescence microscopy (see Box 7.3 Figure 2 for the principle of fluorescence microscopy).

In **epitope tagging**, the relevant protein is tagged by attaching to its N- or C-terminus an immunogenic marker peptide sequence for which a specific antibody already exists. To achieve this, a cDNA for the protein of interest is cloned into an expression vector at a site adjacent to, and in frame with, a coding sequence for the relevant peptide. Recombinant plasmids are transfected into the appropriate cells, and the expressed protein is tracked with a fluorescently labeled antibody specific for the peptide tag. Commonly used epitope tags are shown in **Table 8.7**.

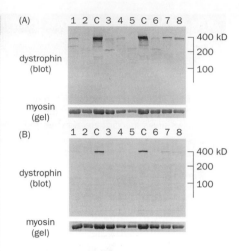

Figure 8.21 Immunoblotting (western blotting). Immunoblotting involves the detection of polypeptides after size fractionation in a polyacrylamide gel and transfer (blotting) to a membrane. Here, the blots show dystrophin detection in muscle samples from individual patients with Duchenne or Becker muscular dystrophy (lanes 1–8) and in normal control muscle (lanes C). (A) Blot using the Dy4/6D3 antibody, which is specific for the dystrophin rod domain. (B) Blot using the Dy6/C5 antibody, which is specific for the C-terminal region of dystrophin. Myosin was used as a loading control to provide comparative quantitation of the amount of muscle samples in the different lanes. [Courtesy of the late Louise Anderson (formerly Nicholson) from Nicholson LV, Johnson MA, Bushby KM et al. (1993) *J. Med. Genet.* 30, 737–744. With permission from the BMJ Publishing Group.]

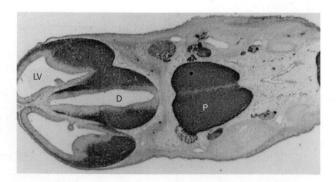

Figure 8.22 Immunocytochemistry. In this example, β-tubulin expression was screened in a transverse section of a mouse brain at embryonic day 12.5. The antibody detection system used identifies β-tubulin expression ultimately as a brown color reaction (based on horseradish peroxidase/ 3, 3'-diaminobenzidine). The underlying histology was revealed by counterstaining with a toluidine blue stain. D, diencephalon; LV, lateral ventricle; P, pons. (Courtesy of Steve Lisgo, Newcastle University.)

TABLE 8.7 COMMONLY USED EPITOPE TAGS FOR TRACKING PROTEIN LOCALIZATION

Sequence of tag	Peptide origin	Linked to protein at	Monoclonal antibody
DYKDDDDK	synthetic	N- or C-terminus	anti-Flag M1
EQKLISEEDL	human c-Myc protein	N- or C-terminus	9E10
MASMTGGQQMG	phage T7 gene 10 protein	N-terminus	T7.Tag antibody
QPELAPEDPED	herpes simplex virus protein D	C-terminus	HSV.Tag antibody
RPKPQQFFGLM	substance P	C-terminus	NC1/34
YPYDVPDYA	influenza HA1 protein	N- or C-terminus	12CA5

More recently, fluorescence microscopy has been used to track protein expression in cells without the use of antibodies. Instead, a naturally fluorescent protein is coupled to the protein of interest to provide a marker of its expression pattern. One such protein is the green fluorescent protein (GFP), a 238-amino acid protein originally identified in the jellyfish *Aequoria victoria*. Similar proteins are expressed in many jellyfish and seem to be responsible for the green light that they emit, being stimulated by energy obtained after the oxidation of luciferin or another photoprotein.

When the gene encoding GFP was cloned and transfected into target cells in culture, expression of GFP in heterologous cells was also marked by emission of the green fluorescent light. This means that GFP is an *autofluorescent* protein. Because it is a natural functional fluorophore, it can serve directly as a reporter that can be readily followed by conventional and confocal fluorescence microscopy, making it a popular tool for tracking gene expression in cell lines and even in some whole organisms.

Various genetically engineered mutants of GFP have improved its efficiency in different ways, and various color mutants have been generated, notably blue fluorescent protein, cyan fluorescent protein, and yellow fluorescent protein. Another autofluorescent protein, the red fluorescent protein, has been derived from the *Discosoma* coral.

To track the expression of a protein within living cells, the GFP-coding DNA sequence has been inserted at the beginning or end of the cDNA for the protein of interest within a suitable expression vector. Transfection and expression of the resulting hybrid cDNA sequence produces a fusion protein with GFP linked to the N- or C-terminus of the protein of interest. In many cases, the fusion protein is expressed in the same way as the original protein, and so helps to reveal its location (Figure 8.23).

8.5 HIGHLY PARALLEL ANALYSES OF GENE EXPRESSION

Highly parallel analyses of gene expression allow simultaneous screening of the expression of many hundreds or thousands of genes, even the full set of different transcripts (transcriptome) or of the different proteins (**proteome**) expressed by a cell. For large-scale transcript profiling, microarray hybridization has been the most widely used method; however, alternative sequencing-based methods have been developed, and massively parallel sequencing is now beginning to revolutionize transcriptome analysis. Global protein profiling has popularly been achieved with two-dimensional gel electrophoresis and mass spectrometry.

DNA and oligonucleotide microarrays permit rapid global transcript profiling

The targets for transcript profiling are complex RNA populations from cellular sources of interest, often cultured cells, surgically excised tissues or tumors, or isolated portions of these. Typical microarrays use many hundreds or thousands of gene-specific probes. cDNA probes have been used, but oligonucleotide probes are increasingly being used. As detailed in Section 7.4, two popular systems are Affymetrix GeneChip microarrays, in which oligonucleotides about

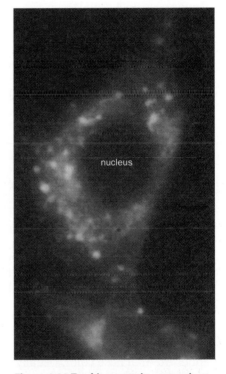

nucleus

Figure 8.23 Tracking protein expression with the use of green fluorescent protein tags. This example shows a live transiently transfected HeLa cell expressing a GFP-tagged Batten disease protein. A cDNA sequence for the Batten disease gene *CLN3* was cloned into the GFP expression vector pEGFP-N1 in order to express the CLN3 protein with a GFP sequence coupled to its C-terminus (a *GFP tag*). This cell is an example of a small proportion of HeLa cells expressing CLN3/GFP in a vesicular punctate pattern distributed throughout the cytoplasm. These and other analyses indicated that the Batten disease protein is a Golgi integral membrane protein. [From Kremmidiotis G, Lensink IL, Bilton RL et al. (1999) *Hum. Mol. Genet.* 8, 523–531. With permission from Oxford University Press.]

25 nucleotides long are synthesized *in situ* on an array, and Illumina microarrays, in which pre-synthesized oligonucleotides with a gene-specific component about 50 nucleotides long are attached to beads. More recently, high-density arrays of longer oligonucleotides up to 60 nucleotides long have also become available.

Microarray-based expression analyses are typically organized so as to compare two or more highly related cellular or tissue sources that differ in an informative way. For example, gene expression can be tracked in specific embryonic structures at a series of developmental time points, or expression can be monitored in cultured cells at various times after they have been exposed to a range of drugs and other chemicals. The same type of tumor tissue can be profiled at different stages of malignancy. Expression profiles for diseased and normal phenotypes can also be compared, and in mouse models the two cellular sources under comparison may be genotypically identical except for the presence of a defined pathogenic mutation.

To perform transcript profiling, the cellular RNA sample is normally reverse transcribed *en masse* to form a representative complex cDNA population. The cDNA can be labeled as it is synthesized by the inclusion of a fluorophore-conjugated nucleotide in the reaction mix. Alternatively, a two-step procedure is used—first, unlabeled cDNA is made, and then it is converted into a labeled complementary RNA (cRNA) by the incorporation of biotin (which will later be detected with fluorophore-conjugated streptavidin).

The labeled target cDNA or cRNA is then applied to the array and allowed to hybridize. Each individual feature or spot on the array contains millions to billions of copies of the same DNA sequence and is therefore unlikely to be completely saturated in the hybridization reaction. Under these conditions, the intensity of the hybridizing signal at each feature on the array is proportional to the relative abundance of that particular cDNA or cRNA in the target population, which in turn reflects the abundance of the corresponding mRNA in the original source population. The relative abundance of thousands of different transcripts can therefore be monitored in one experiment. Multiple oligonucleotides are also used to help distinguish between closely related transcripts from individual genes. It is possible to monitor splice variants, for example, and to design oligonucleotides specific for every single known exon.

The huge amount of expression data generated by the microarray data requires careful statistical analyses (**Box 8.7**). Stringent controls are also required to normalize expression data for cross-experiment variation. One way of avoiding such problems is to hybridize cDNA populations labeled with different fluorophores to the same array simultaneously. Under nonsaturating conditions, the signal at each feature will represent the relative abundance of each transcript in the sample. If two samples are used, the ratio of the signals from each fluorophore provides a direct comparison of expression levels between samples fully normalized for variations in signal-to-noise ratio even within the array. The array is scanned at two emission wavelengths, and a computer is used to combine the images and render them in false color. Usually, one fluorophore is represented as green and the other as red. Features representing differentially expressed genes show up as either green or red, whereas those representing equivalently expressed genes show up as yellow—see the example of using spotted cDNA arrays in **Figure 8.24A.**

Expression analyses with microarrays that have short oligonucleotide probes are similar in principle to those in which cDNA probes or long (more than 50-nucleotide) oligonucleotides are used. However, when using oligonucleotides that are only 25 nucleotides long, the hybridization specificity is not so great. There is a greater tendency for probes to hybridize to other sequences in addition to their expected target sequences, and so additional controls are needed. Accordingly, in Affymetrix GeneChip arrays, each gene is represented by 20 or so different oligonucleotide probes that are selected from different regions along the transcribed sequence. In addition to 20 perfect match (PM) oligonucleotides per gene, a corresponding series of 20 mismatch (MM) oligonucleotides are designed to control for nonspecific hybridization by changing a single base in each of the PM sequences (Figure 8.24B). To determine the signal for a particular gene, the signals of all 20 PM oligonucleotides are added together and the signals from all 20 MM oligonucleotides are subtracted from the total.

BOX 8.7 ANALYZING MICROARRAY EXPRESSION DATA

Microarray expression data are revolutionizing biological and biomedical research, but data interpretation remains a challenge. The raw expression data from microarray experiments are signal intensities that must be corrected for background effects and inter-experimental variation (*normalized*) and checked for errors caused by contaminants and extreme outlying values.

The data are summarized as a table of normalized signal intensities: rows on the table represent individual genes and the columns represent different conditions under which gene expression has been measured. In the simplest cases, the table has two columns (e.g. control and disease samples) and these may represent the signal intensities from two samples hybridized simultaneously to the array. However, there is no theoretical limit to the number of conditions that can be used.

Next, genes with similar expression profiles are grouped. Generally, the more conditions under which gene expression is tested, the more rigorous is the analysis. Two types of algorithm are used to mine the gene expression data, one in which similar data are clustered in a hierarchy and one in which the clusters are defined in a nonhierarchical manner.

Hierarchical clustering

The general approach in hierarchical clustering is to establish a *distance matrix* that lists the differences in expression levels between each pair of features on the array. Those showing the smallest differences, expressed as the distance function *d*, are then clustered in a progressive manner. *Agglomerative clustering* methods begin with the classification of each gene represented on the array as a singleton cluster (a cluster containing one gene). The distance matrix is searched, and the two genes with the most similar expression levels (the smallest distance function) are defined as neighbors; these are then merged into a single cluster. The process is repeated until there is only one cluster left. There are variations in how the expression value of the merged cluster is calculated for the purpose of further comparisons.

In the *nearest-neighbor (single linkage) method*, the distance is minimized. That is, where two genes i and j are merged into a single cluster ij, the distance between ij and the next nearest gene k is defined as the lower of the two values $d(i,k)$ and $d(j,k)$. In the *average linkage method*, the average between $d(i,k)$ and $d(j,k)$ is used. In the *farthest-neighbor (complete linkage) method*, the distance is maximized. These methods generate dendrograms with different structures (**Figure 1**). Less frequently, a divisive clustering algorithm may be used in which a single cluster representing all the genes on the array is progressively split into separate clusters.

Hierarchical clustering of microarray data is often represented as a heatmap—**Figure 2**.

Nonhierarchical clustering

A disadvantage of hierarchical clustering is that it is time-consuming and resource hungry. As an alternative, nonhierarchical methods partition the expression data into a certain predefined number of clusters. As a result, the analysis is speeded up considerably, especially when the data set is very large. In the *k-means clustering method*, several points known as cluster centers are defined at the beginning of the analysis, and each gene is assigned to the most appropriate cluster center.

On the basis of the membership of each cluster, the means are recalculated (the cluster centers are repositioned). The analysis is then repeated so that all the genes are assigned to the new cluster centers. This process is reiterated until the membership of the various clusters no longer changes. *Self-organizing maps* are similar in concept, but the algorithm is refined through the use of a neural network.

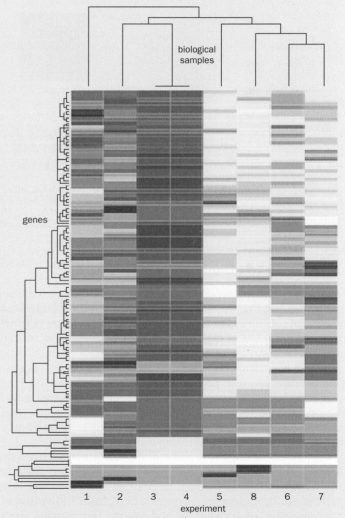

Figure 2 Heatmaps as a tool for visualizing microarray analysis. *Heatmaps* give a quick overview of clusters of genes that show similar expression values. They consist of small cells, each consisting of a color, which represent relative expression values. Heatmaps are often generated from hierarchical cluster analyses of different biological samples (typically portrayed in columns, as here) and genes (usually in rows and grouped together according to similarity of expression). [Adapted from Allison DB, Cui X, Page GP & Sabripour M (2006) *Nat. Rev. Genet.* 7, 55–65. With permission from Macmillan Publishers Ltd. See the same reference for alternative ways of visualizing microarray analysis.]

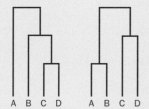

Figure 1 Microarray data analysis using the hypothetical expression profiles of four genes, A–D. Hierarchical clustering methods produce branching diagrams (*dendrograms*) in which genes with the most similar expression profiles are grouped together, but alternative clustering methods produce dendrograms with different topologies. The pattern on the left is typical of the topology produced by nearest-neighbor (single linkage) clustering; the pattern on the right is typical of the topology produced by farthest-neighbor (complete linkage) clustering.

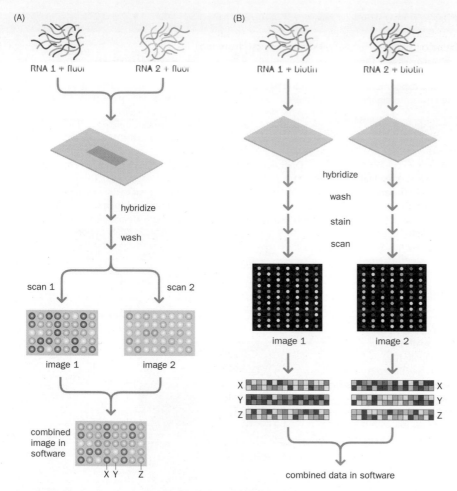

Figure 8.24 Comparative expression analysis with DNA microarrays. (A) *Using spotted cDNA arrays.* Here, comparative expression assays are usually performed by differentially labeling two mRNA or cDNA samples with different fluorophores that are hybridized to the arrayed cDNAs and then scanned to detect both fluorophores independently. Colored dots labeled X, Y, and Z at the bottom of the image illustrate three different gene expression patterns: elevated expression in sample 1 (X, red), elevated expression in sample 2 (Y, green), and similar expression in samples 1 and 2 (Z, yellow). (B) *Using Affymetrix GeneChips.* Here RNA is labeled in a two-step process to produce biotinylated cRNA. After hybridization and washing, biotin-cRNA bound to the array is stained by binding a streptavidin-conjugated fluorophore, and the bound fluorophore is detected by laser scanning. Each gene is represented by 15–20 different oligonucleotide probe pairs (16 are shown in this example). One member of each pair is a perfectly matched oligonucleotide probe; the other is a control oligonucleotide with a deliberate mismatch. The example shows expression data for three hypothetical genes, representing genes that are preferentially expressed in sample 1 (X), preferentially expressed in sample 2 (Y), or showing equivalent expression in samples 1 and 2 (Z). [Adapted from Harrington CA, Rosenow C & Retief J (2000) *Curr. Opin. Microbiol.* 3, 285–291. With permission from Elsevier.]

Modern global gene expression profiling increasingly uses sequencing to quantitate transcripts

Microarray hybridization provided an exciting new technology that in the late 1990s was the first to offer the ability to conduct large-scale and genome-wide transcript profiling. Global gene expression profiling is, however, also possible by direct sequencing of cDNA copies of the transcripts, so that transcript frequencies can be quantitated for individual genes.

cDNA libraries have been characterized by partial sequencing of clone inserts, generating numerous EST sequences, or sometimes full-length inserts have been sequenced. But because of the labor involved in generating libraries and retrieving recombinant clones, and the small number of samples that could be sequenced at a time, standard dideoxy sequencing of library clones did not seem an attractive way of quantitating transcripts.

As a result of the above difficulties, various sequence sampling methods were developed to reduce the effort involved in retrieving cDNA sequences. The methods would obtain RNA from cellular sources, convert it into cDNA, and then retrieve short sequences (*sequence tags*) from individual cDNAs that would be representative of transcripts and whose frequencies would be a quantitative measure of the expression of that transcript.

A popular, and ingenious, sequence sampling method is known as serial analysis of gene expression (SAGE). In this technique, sequence tags are collected from numerous cDNA copies, and ligated in a long series to form concatemers that are then sequenced. In the alternative method, massively parallel signature sequencing (MPSS), very large numbers of cDNAs are attached to microbeads in a flow cell and sequencing of short sequence tags is performed in parallel rather than in series as in SAGE.

The recent advent of massively parallel DNA sequencing means that both microarray-based expression profiling and serial sample sequencing (SAGE) will be eclipsed in the near future. The ability to sequence hundreds of millions of samples in parallel offers huge power in quantitating transcripts. Because it does not require PCR-based sequence amplification, single-molecule sequencing may ultimately provide the most accurate quantitation of transcripts.

Global protein expression is often profiled with two-dimensional gel electrophoresis and mass spectrometry

Transcriptome analysis is clearly very useful for following gene expression, but the ultimate products of many genes are proteins. Protein levels do not simply reflect the transcript levels of a protein-coding gene because there are differences in the rates of protein turnover among different protein-coding transcripts. Protein activity also often depends on post-translational modifications that are not predictable from the level of the corresponding transcript.

Like transcriptomics, proteomics can be used to monitor the abundance of different gene products. The expression of all the proteins in the cell can be compared between related samples, allowing proteins with similar expression patterns to be identified, and highlighting important changes in the proteome that occur, for example during disease or the response to particular external stimuli. This is sometimes termed expression proteomics.

Unlike the genome, the proteome varies widely between different cell types in an organism. Human cells typically contain tens of thousands of proteins differing in abundance over four or more orders of magnitude. Like nucleic acids, proteins can be detected and identified by specific molecular interactions, in most cases by using antibodies or other ligands as probes. However, unlike nucleic acids, there is no procedure for cloning or amplifying rare proteins. Furthermore, the physical and chemical properties of proteins are so diverse that no single, universal methodology analogous to hybridization can be used to study the entire proteome in a single experiment.

Currently, expression proteomics is largely based on 'separation and display' technology: complex protein mixtures are separated into their components so that interesting features (such as proteins present in a disease sample but absent from a matching healthy sample) can be selected for further characterization. Separation is usually performed with two-dimensional gel electrophoresis, a long-standing technique that has the power to resolve up to 10,000 proteins on a single gel. Thereafter, separated proteins of interest can be interrogated by mass spectrometry.

Two-dimensional polyacrylamide gel electrophoresis (2D-PAGE)

The principle of 2D-PAGE is to separate proteins according to their charge (first dimension) and then according to their mass (second dimension, at right angles to the first). A complex protein sample is loaded onto a denaturing polyacrylamide gel and separated in the first dimension by isoelectric focusing. In this technique, the proteins migrate in a pH gradient until they reach their isoelectric point, the position at which their charge is neutral with respect to the local pH.

The standard procedure is to prepare an immobilized pH gradient gel, in which the buffering groups are attached to the polyacrylamide matrix (to prevent them from drifting and becoming unstable during long gel runs). The gel is then equilibrated in the detergent SDS, which is used to disrupt noncovalent bonding in proteins so that all the proteins assume similar elongated conformations; by binding stoichiometrically to the backbone of denatured proteins it confers a massive negative charge that effectively cancels out any charge differences between individual proteins. Separation in the second dimension is therefore dependent on the mass of the protein, with smaller proteins moving more readily through the pores of the gel. The gel is then stained and the proteins are revealed as a complex pattern of spots (**Figure 8.25**).

Although 2D-PAGE has a high resolution and is the most widely used technique for protein separation in proteomics, there are several limitations to its usefulness in terms of representation, sensitivity, reproducibility, and convenience. Several classes of protein are under-represented on standard gels, including very basic proteins, proteins with poor solubility in aqueous buffers, and membrane proteins.

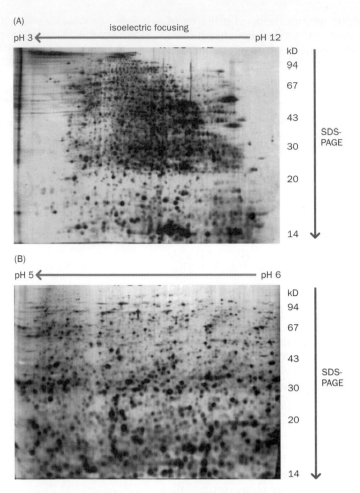

(A)

isoelectric focusing

pH 3 ← → pH 12

kD
94
67
43
30
20
14

SDS-PAGE

(B)

pH 5 ← → pH 6

kD
94
67
43
30
20
14

SDS-PAGE

Figure 8.25 Two-dimensional gel electrophoresis. In two-dimensional gel electrophoresis protein samples are initially fractionated by isoelectric focusing (often in a tube gel, or pre-cast strip, comprising a polyacrylamide matrix with an immobilized pH gradient). The separated fractions are then subjected to SDS-polyacrylamide gel electrophoresis in a second dimension at right angles to the first. Both images represent mouse liver proteins separated by two-dimensional gel electrophoresis and then stained with silver to reveal individual protein spots. (A) In a gel with a wide pH range (pH 3–12), the proteins are clustered in the middle because most proteins have isoelectric points in the range pH 4–7. (B) A gel with a narrow pH range (pH 5–6) has a higher resolving power. [From Orengo CA, Jones DT & Thornton JM et al. (2003) Bioinformatics: Genes, Proteins And Computers. BIOS Scientific Publishers.]

The sensitivity of 2D-PAGE is dependent on the detection limit for very scarce proteins, but the class of SYPRO® dyes can detect protein spots in the nanogram range. Sensitivity is also influenced by the resolution of the gel, because spots representing scarce proteins can be obscured by those representing abundant proteins. Pre-fractionation can remove abundant proteins and so simplify the initial sample loaded onto the gel, and the resolution can be improved by using gels with a very narrow pH range (see Figure 8.25).

Important developments to improve the efficiency of 2D-PAGE include software for spot recognition and quantitation, algorithms that can compare protein spots across multiple gels, and robots that can pick interesting spots and process them for later identification by mass spectrometry as described below. However, a remaining major limitation of 2D-PAGE is that it is not highly suited to automation, making it difficult to perform high-throughput analyses of many samples. Alternative separation methods based on multidimensional high-performance liquid chromatography (HPLC) may eventually displace 2D-PAGE as the major platform for protein separation.

Mass spectrometry

Mass spectrometry (MS) is used to determine the accurate masses of molecules in a particular sample. In expression proteomics it helps identify the proteins in selected spots that have been resolved by 2D-PAGE. It does this by accurate determination of molecular mass, a procedure that can be completed more quickly than direct sequencing of proteins by Edman degradation. MS is also easy to automate for high-throughput sample analysis, and so it can conveniently process thousands of spots from a two-dimensional gel or hundreds of HPLC fractions.

Until recently, MS could not be applied to large molecules such as proteins and nucleic acids because these were broken into random fragments during the ionization process. This limitation has been overcome using soft-ionization methods such as matrix-assisted laser desorption/ionization (MALDI) (**Figure 8.26** and **Box 8.8**) and electrospray ionization (ESI).

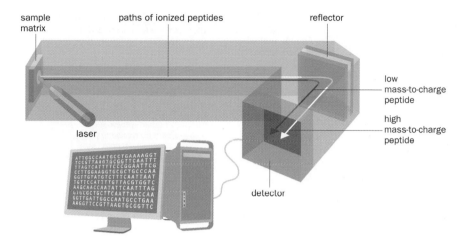

Figure 8.26 Principle of MALDI-TOF mass spectrometry. The *analyte* (the sample under study—usually a collection of tryptic peptide fragments) is mixed with a matrix compound and placed near the source of a laser. The laser heats up the analyte–matrix crystals, causing the analyte to expand into the gas phase without significant fragmentation. Ions then travel down a flight tube to a reflector, which focuses the ions onto a detector. The time of flight (the time taken for ions to reach the detector) is dependent on the mass/charge ratio and allows the mass of each molecule in the analyte to be recorded.

The masses of proteins, or more usually the peptide fragments derived from them by digestion with proteases, can be used to identify the proteins by correlating the experimentally determined masses with those predicted from database sequences. As described below there are three different ways to annotate a protein by MS (**Figure 8.27**):

- **Peptide mass fingerprinting** (**PMF**) is best suited for simple proteomes. A simple protein mixture (typically a single spot from a two-dimensional gel) is digested with trypsin to generate a collection of tryptic peptides. These are subject to MALDI-MS using a time-of-flight (TOF) analyzer (see Figure 8.26 and Box 8.8), which returns a set of mass spectra. The spectra are used as a search query against protein sequence databases. The search algorithm performs virtual trypsin digests of all the proteins in the database and calculates the masses of the predicted tryptic peptides. It then attempts to match these predicted masses against those determined experimentally.

BOX 8.8 MASS SPECTROMETRY IN PROTEOMICS

The mass spectrometer

A mass spectrometer has three components. An *ionizer* converts the sample to be analyzed (the analyte) into gas-phase ions and accelerates them toward the *mass analyzer*. The latter separates the ions according to their mass/charge ratio on their way to the *ion detector*, which records the impact of individual ions, presenting these as a mass spectrum of the analyte.

The ionization of large molecules without fragmentation and degradation is known as *soft ionization*. Two soft ionization methods are widely used in proteomics.

- **Matrix-assisted laser desorption/ionization** (**MALDI**) involves mixing the analyte (e.g. the tryptic peptides derived from a particular protein sample) with a light-absorbing matrix compound in an organic solvent. Evaporation of the solvent produces analyte/matrix crystals, which are heated by a short pulse of laser energy. The desorption of laser energy as heat causes expansion of the matrix and analyte into the gas phase. The analyte is then ionized and accelerated toward the detector.

- In **electrospray ionization** (**ESI**), the analyte is dissolved and the solution is pushed through a narrow capillary. A potential difference, applied across the aperture, causes the analyte to emerge as a fine spray of charged particles. The droplets evaporate as the ions enter the mass analyzer.

Mass analyzers

A **quadrupole mass analyzer** comprises four metal rods, pairs of which are connected electrically and carry opposing voltages that can be controlled by the operator. Mass spectra are obtained by varying the potential difference applied across the ion stream, allowing ions of different mass/charge ratios to be directed toward the detector. A **time-of-flight** (**TOF**) **analyzer** measures the time taken by ions to travel down a flight tube to the detector, a factor that depends on the mass/charge ratio.

Tandem mass spectrometry (MS/MS)

Two or more mass analyzers are operated in series. Various MS/MS instruments have been described, including triple quadrupole and hybrid quadrupole/time-of-flight instruments. The mass analyzers are separated by a collision cell that contains inert gas and causes ions to dissociate into fragments. The first analyzer selects a particular peptide ion and directs it into the collision cell, where it is fragmented. A mass spectrum for the fragments is then obtained by the second analyzer. These two functions may be combined in more sophisticated instruments, such as the ion-trap and Fourier-transform ion cyclotron analyzers.

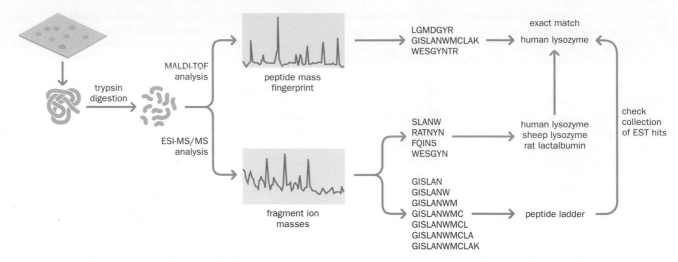

Figure 8.27 Protein annotation by mass spectrometry. Individual protein samples (such as spots from two-dimensional gels) are digested with trypsin, which cleaves at the C-terminal side of lysine (K) or arginine (R) residues as long as the next residue is not proline. The tryptic peptides can be analyzed as intact molecules by MALDI-TOF, and the masses used as search queries against protein databases. Algorithms are used that take protein sequences, cut them with the same cleavage specificity as trypsin, and compare the theoretical masses of these peptides with the experimental masses obtained by MS. Ideally, the masses of several peptides should identify the same parent protein, in this case human lysozyme. There may be no hits if the protein is not in the database or, more probably, that it has been subject to post-translational modification or artifactual modification during the experiment. In these circumstances, electrospray ionization coupled with tandem mass spectrometry (ESI-MS/MS) can be used to fragment the ions. The fragment ion masses can be used to search EST databases and obtain partial matches, which may lead eventually to the correct annotation. Alternatively, the masses of peptide ladders can be used to determine protein sequences *de novo*.

- **Fragment ion searching** is more suited to the analysis of complex proteomes, and the algorithm can be modified to take into account the masses of known post-translational modifications. The tryptic peptide fragments are analyzed by tandem mass spectrometry (MS/MS; see Box 8.8), during which the peptides are broken into random fragments. The mass spectra from these fragments can be used to search against EST databases (which cannot be searched with PMF data because the intact peptides are generally too large). Any EST hits can then be used in a BLAST search to identify putative full-length homologs. A dedicated algorithm called MS-BLAST is useful for handling the short sequence signatures obtained from peptide-fragment ions.

- *De novo* **sequencing of peptide ladders** is also performed because it is impossible to account for all variants, either at the sequence level (e.g. polymorphisms) or at the protein modification level (e.g. complex glycans). The sequencing of peptide ladders may provide sequence signatures that can be used as search queries to identify homologous sequences in the databases. In this technique, the peptide fragments generated by MS/MS are arranged into a nested set differing in length by a single amino acid. By comparing the masses of these fragments with standard tables of amino acids, it is possible to deduce the sequence of the peptide fragment *de novo*, even when a precise sequence match is not available in the database. In practice the *de novo* sequencing approach is complicated by the presence of two fragment series, one nested at the N-terminus and one nested at the C-terminus. The two series can be distinguished by attaching diagnostic mass tags to either end of the protein.

Comparative protein expression analyses have many applications

The aim of many proteomics experiments is to identify proteins whose abundance differs significantly across two or more samples. The varying abundance of specific proteins in healthy versus disease samples, or in samples representing the progression of a disease, can help to reveal useful markers and novel drug targets. Drugs typically target proteins by binding to them and modifying their behavior, and so many drug responses will not affect mRNA abundance and therefore cannot be assayed at the level of the transcriptome.

One way to compare protein expression in different samples is to examine two-dimensional gels and identify spots that show quantitative variation (several software packages are available to aid such comparative investigations). An alternative is to label proteins from two different samples, for example by conjugating them with Cy3 and Cy5 and then separating them on the same gel. This approach, known as difference gel electrophoresis (DIGE), exploits the same principle as differential gene expression with DNA microarrays.

Another approach is to label proteins from different sources with isotope-coded affinity tags (ICATs). A mass spectrometer can easily distinguish and quantitate two isotopically labeled forms of the same compound that are chemically identical and can be co-purified. An ICAT method has therefore been developed using a biotinylated iodoacetamide derivative to label protein mixtures selectively at their cysteine residues. By binding to streptavidin, the biotin tag allows affinity purification of the tagged cysteine peptides after proteolysis with trypsin.

The ICAT reagent is available in heavy and light isotopically labeled forms, which can be used to label cell pools differentially under different conditions (such as health versus disease). After labeling, the cells are combined and lysed, and the proteins are isolated so that purification losses occur equally in both samples. Isotope intensities are compared for peptides as they enter the mass spectrometer. If they are equivalent, no upregulation or downregulation has occurred, and the protein is of no immediate interest. If the intensities differ, a change in protein expression has taken place, and the protein is of interest. The amount of the two forms is measured, and the light peptide form is fragmented and identified by database searching.

The 2000s will be seen to have been a revolutionary decade for genome analysis. Technologies developed to allow rapid mapping and sequencing of complex genomes and whole genome sequences were obtained for various complex genomes, including the human genome. The advent of massively parallel DNA sequencing methods means that genome sequencing will be comparatively rapid in the future; perhaps by the year 2015–2020, personal genome sequencing will be routine and inexpensive. In addition, large-scale analyses have begun to identify all the genes in genomes and catalog all their expression products, namely the different RNA transcripts and translated proteins. Each type of expression product can be detected within intact cells (by hybridization or immunochemistry) and quantitated by techniques such as real-time PCR, western blotting, or mass spectrometry. Present refinements have moved analysis to greater and greater numbers of samples, so that now very large numbers of genes are studied in parallel in a single experiment. Regardless of the numbers of genes studied and the methodology employed, the major goal remains the same: to understand how genes function and how disease alters the cellular production of transcripts and proteins.

FURTHER READING

DNA sequencing

Clarke J, Wu HC, Jayasinghe L et al. (2009) Continuous base identification for single-molecule nanopore DNA sequencing. *Nat. Nanotechnol.* 4, 265–270.

Eid J, Fehr A, Gray J et al. (2009) Real-time DNA sequencing from single polymerase molecules. *Science* 323, 133–138.

Gupta P (2008) Single-molecule DNA sequencing technologies for future genomics research. *Trends Biotechnol.* 26, 602–611.

Hodges E, Xuan Z, Balija V et al. (2007) Genome-wide *in situ* exon capture for selective resequencing. *Nat. Genet.* 39, 1522–1527.

Mamanova L, Coffey AJ, Scott CE et al. (2010) Target-enrichment strategies for next generation sequencing. *Nat. Meth.* 7, 111–118.

Mardis ER (2008) The impact of next-generation sequencing technology on genetics. *Trends Genet.* 24, 133–141.

Metzker M (2010) Sequencing technologies—the next generation. *Nat. Rev. Genet.* 11, 31–46.

Tucker T, Marra M, Friedman JM (2009) Massively parallel DNA sequencing: the next big thing in genetic medicine. *Am. J. Hum. Genet.* 85, 142–154.

Human Genome Project and human genetic maps (see also Box 8.2 for additional historical references)

All About The Human Genome Project (HGP). http://www .genome.gov/10001772 [An educational resource maintained by the US National Human Genome Research Institute.]

Garber M, Zody MC, Arachchi HM et al. (2009) Closing gaps in the human genome using sequencing by synthesis. *Genome Biol.* 10, R60.

Genome Reference Consortium. http://www.ncbi.nlm.nih .gov/projects/genome/assembly/grc/ [Gives updates and graphical overviews of the human and mouse genomes.]

Imanishi T, Itoh T, Suzuki Y et al. (2004) Integrative annotation of 21,037 human genes validated by full-length cDNA clones. *PLoS Biol.* 2, e162.

International HapMap Consortium (2005) A haplotype map of the human genome. *Nature* 437, 1299–1320.

International HapMap Consortium (2007) A second generation human haplotype map of over 3.1 million SNPs. *Nature* 449, 851–861.

Various authors (2001) Human Genome Issue. *Nature* 409, 813–958. [Available at Nature Network OmicsGateway portal, http://www.nature.com/omics/subjects/ genomesequenceandanalysis/archive/2001/index.html]

Various authors (2001) Human Genome Issue. *Science* 291, 1177–1351 [Papers available at http://www.sciencemag.org/ content/vol291/issue5507/index.dtl]

Various authors (2003) User's Guide to the Human Genome. *Nature Genet.* 35 (1s), 1–79. [Available at http://www.nature .com/ng/journal/v35/n1s/index.html]

Various authors (2006) Human Genome Collection. *Nature* S1, 1–305. [Contains papers analyzing the sequences of all 24 human chromosomes; available at http://www.nature.com/ nature/supplements/collections/humangenome/index.html]

Waterston RH, Lander ES & Sulston JE (2002) On the sequencing of the human genome. *Proc. Natl Acad. Sci. USA* 99, 3712–3716.

Model Organism Genome Projects

Genome News Network. A Quick Guide To Sequenced Genomes. http://www.genomenewsnetwork.org/resources/ sequenced_genomes/genome_guide_index.shtml. [Gives a brief account of each model organism that has been sequenced and an associated literature cross-reference.]

Liolios K, Mavrommatis K, Tavernarakis N & Kyrpides NC (2008) The Genomes On Line Database (GOLD) in 2007: status of genomic and metagenomic projects and their associated metadata. *Nucleic Acids Res.* 36 (Database issue), D475–D479.

Genome databases, browsers, gene annotation, and associated bioinformatics

Barnes MR (ed.) (2007) Bioinformatics for Geneticists: A Bioinformatics Primer for the Analysis of Genetic Data, 2nd ed. John Wiley and Sons.

Bina M (2006) Use of genome browsers to locate your favorite genes. *Methods Mol. Biol.* 338, 1–7.

Reeves GA, Talavera D & Thornton JM (2009) Genome and proteome annotation: organization, interpretation and integration. *J. R. Soc. Interface* 6, 129–147.

Spudich G, Fernández-Suárez XM & Birney E (2007) Genome browsing with Ensembl: a practical overview. *Brief. Funct. Genomics Proteomics* 6, 202–219.

The Gene Ontology Consortium (2008) The Gene Ontology Project in 2008. *Nucleic Acids Res.* 36, D440–D444. [See also http://www.geneontology.org/]

Wheeler DL, Barrett T, Benson DA et al. (2008) Database resources of the National Center for Biotechnology Information. *Nucleic Acids Res.* 36 (Database issue), D13–D21.

Basic gene expression analyses

Applied Biosystems (undated) Real-Time PCR Vs. Traditional PCR. http://www.appliedbiosystems.com/support/tutorials/pdf/ rtpcr_vs_tradpcr.pdf [Real-time PCR tutorial.]

VanGuilder HD, Vrana KE & Freeman WM (2008) Twenty-five years of quantitative PCR for gene expression analysis. *Biotechniques* 44, 619–626.

Ward TH & Lippincott-Schwartz J (2006) The uses of green fluorescent protein in mammalian cells. *Methods Biochem. Anal.* 47, 305–337.

Highly parallel gene expression analyses: transcript profiling

Allison DB, Cui X, Page GP & Sabripour M (2006) Microarray data analysis: from disarray to consolidation and consensus. *Nat. Rev. Genet.* 7, 55–65.

Belacel N, Wang Q & Cuperlovic-Culf M (2006) Clustering methods for microarray gene expression data. *Omics* 10, 507–531.

Jongeneel CV, Delorenzi M, Iseli C et al. (2005) An atlas of human gene expression from massively parallel signature sequencing (MPSS). *Genome Res.* 15, 1007–1014.

Murray D, Doran P, MacMathuna P & Moss AC (2007) *In silico* gene expression analysis—an overview. *Mol. Cancer* 6, 50.

Various authors (2002) The Chipping Forecast II. *Nat. Genet.* 32 (Suppl.), 465–551.

Highly parallel gene expression analyses: protein profiling

Hamdan M & Righetti PG (2003) Assessment of protein expression by means of 2-D gel electrophoresis with and without mass spectrometry. *Mass Spectrom. Rev.* 22, 272–284.

Kolker E, Higdon R & Hogan JM (2006) Protein identification and expression analysis using mass spectrometry. *Trends Microbiol.* 14, 229–235.

Chapter 9

Organization of the Human Genome

9

KEY CONCEPTS

- The human genome is subdivided into a large nuclear genome with more than 26,000 genes, and a very small circular mitochondrial genome with only 37 genes. The nuclear genome is distributed between 24 linear DNA molecules, one for each of the 24 different types of human chromosome.

- Human genes are usually not discrete entities: their transcripts frequently overlap those from other genes, sometimes on both strands.

- Duplication of single genes, subchromosomal regions, or whole genomes has given rise to families of related genes.

- Genes are traditionally viewed as encoding RNA for the eventual synthesis of proteins, but many thousands of RNA genes make functional noncoding RNAs that can be involved in diverse functions.

- Noncoding RNAs often regulate the expression of specific target genes by base pairing with their RNA transcripts.

- Some copies of a functional gene come to acquire mutations that prevent their expression. These pseudogenes originate either by copying genomic DNA or by copying a processed RNA transcript into a cDNA sequence that reintegrates into the genome (retrotransposition).

- Occasionally, gene copies that originate by retrotransposition retain their function because of selection pressure. These are known as retrogenes.

- Transposons are sequences that move from one genomic location to another by a cut-and-paste or copy-and-paste mechanism. Retrotransposons make a cDNA copy of an RNA transcript that then integrates into a new genomic location.

- Very large arrays of high-copy-number tandem repeats, known as satellite DNA, are associated with highly condensed, transcriptionally inactive heterochromatin in human chromosomes.

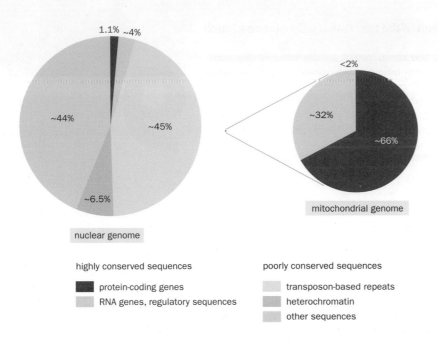

Figure 9.1 Sequence conservation and sequence classes in the human nuclear and mitochondrial genomes. To get an idea of the vast difference in scale between the nuclear (left) and mitochondrial (right) genomes, the tiny red dot in the center represents the equivalent of 25 mitochondrial DNA (mtDNA) genomes on the same scale as the single nuclear genome on the left. Note also the profound difference between the two genomes in the fractions of highly conserved DNA and also in the fraction of highly repetitive noncoding DNA.

nuclear genome

mitochondrial genome

highly conserved sequences

- protein-coding genes
- RNA genes, regulatory sequences

poorly conserved sequences

- transposon-based repeats
- heterochromatin
- other sequences

The human genome comprises two parts: a complex *nuclear genome* with more than 26,000 genes, and a very simple *mitochondrial genome* with only 37 genes (**Figure 9.1**). The nuclear genome provides the great bulk of essential genetic information and is partitioned between either 23 or 24 different types of chromosomal DNA molecule (22 autosomes plus an X chromosome in females, and an additional Y chromosome in males).

Mitochondria possess their own genome—a single type of small circular DNA—encoding some of the components needed for mitochondrial protein synthesis on mitochondrial ribosomes. However, most mitochondrial proteins are encoded by nuclear genes and are synthesized on cytoplasmic ribosomes before being imported into the mitochondria.

As detailed in Chapter 10, sequence comparisons with other mammalian genomes and vertebrate genomes indicate that about 5% of the human genome has been strongly conserved during evolution and is presumably functionally important. Protein-coding DNA sequences account for just 1.1% of the genome. The other 4% or so of strongly conserved genome sequences consists of non-protein-coding DNA sequences, including genes whose final products are functionally important RNA molecules, and a variety of *cis*-acting sequences that regulate gene expression at DNA or RNA levels. Although sequences that make non-protein-coding RNA have not generally been so well conserved during evolution, some of the regulatory sequences are much more strongly conserved than protein-coding sequences.

Protein-coding sequences frequently belong to families of related sequences that may be organized into clusters on one or more chromosomes or be dispersed throughout the genome. Such families have arisen by gene duplication during evolution. The mechanisms giving rise to duplicated genes also give rise to non-functional gene-related sequences (*pseudogenes*).

One of the big surprises in the past few years has been the discovery that the human genome is transcribed to give tens of thousands of different *noncoding RNA* transcripts, including whole new classes of tiny regulatory RNAs not previously identified in the draft human genome sequences published in 2001. Although we are close to obtaining a definitive inventory of human protein-coding genes, our knowledge of RNA genes remains undeveloped. It is abundantly clear, however, that RNA is functionally much more versatile than we previously suspected. In addition to a rapidly increasing list of human RNA genes, we have also become aware of huge numbers of pseudogene copies of RNA genes.

A very large fraction of the human genome, and other complex genomes, is made up of highly repetitive noncoding DNA sequences. A sizeable component is organized in tandem head-to-tail repeats, but the majority consists of interspersed repeats that have been copied from RNA transcripts in the cell by reverse transcriptase. There is a growing realization of the functional importance of such repeats.

In this chapter we primarily consider the *architecture* of the human genome. We outline the different classes of DNA sequence, describe briefly what their function is, and consider how they are organized in the human genome. In later chapters we describe other aspects of the human genome: how it compares with other genomes, and how evolution has shaped it (Chapter 10), DNA sequence variation and polymorphism (Chapter 13), and aspects of human gene expression (Chapter 11).

9.1 GENERAL ORGANIZATION OF THE HUMAN GENOME

The DNA sequence of the human mitochondrial genome was published in 1981, and a detailed understanding of how mitochondrial DNA (mtDNA) works has been built up since then. The more complex nuclear genome has been a much more formidable challenge. Comprehensive sequencing of the nuclear genome began in the latter part of the 1990s, and by 2004 essentially all of the euchromatic portion of the genome had been sequenced. Our knowledge of the nuclear genome remains fragmentary, however. As we see below, we still do not know how many genes there are in the nuclear genome, and recently obtained data are radically changing our perspective on how it is organized and expressed.

The mitochondrial genome is densely packed with genetic information

The human mitochondrial genome consists of a single type of circular double-stranded DNA that is 16.6 kilobases in length. The overall base composition is 44% (G+C), but the two mtDNA strands have significantly different base compositions: the heavy (H) strand is rich in guanines, but the light (L) strand is rich in cytosines. Cells typically contain thousands of copies of the double-stranded mtDNA molecule, but the number can vary considerably in different cell types.

During zygote formation, a sperm cell contributes its nuclear genome, but not its mitochondrial genome, to the egg cell. Consequently, the mitochondrial genome of the zygote is usually determined exclusively by that originally found in the unfertilized egg. The mitochondrial genome is therefore maternally inherited: males and females both inherit their mitochondria from their mother, but males do not transmit their mitochondria to subsequent generations. During mitotic cell division, the multiple mtDNA molecules in a dividing cell segregate in a purely random way to the two daughter cells.

Replication of mitochondrial DNA

The replication of both the H and L strands is unidirectional and starts at specific origins. Although the mitochondrial DNA is principally double-stranded, repeat synthesis of a small segment of the H-strand DNA produces a short third DNA strand called 7S DNA. The 7S DNA strand can base-pair with the L strand and displace the H strand, resulting in a triple-stranded structure (**Figure 9.2**). This region contains many of the mtDNA control sequences (including the major promoter regions) and so it is referred to as the *CR/D-loop region* (where CR denotes control region, and D-loop stands for displacement loop).

The origin of replication for the H strand lies in the CR/D-loop region, and that of the L strand is sandwiched between two tRNA genes (**Figure 9.3**). Only after about two-thirds of the daughter H strand has been synthesized (by using the L strand as a template and displacing the old H strand) does the origin for L-strand replication become exposed. Thereafter, replication of the L strand proceeds in the opposite direction, using the H strand as a template.

Mitochondrial genes and their transcription

The human mitochondrial genome contains 37 genes, 28 of which are encoded by the H strand and the other nine by the L strand (see Figure 9.3). Whereas nuclear genes often have their own dedicated promoters, the transcription of mitochondrial genes resembles that of bacterial genes. Transcription of mtDNA

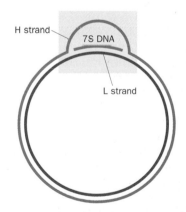

Figure 9.2 D-loop formation in mitochondrial DNA. The mitochondrial genome is not a simple double-stranded circular DNA. Repeat synthesis of a small segment of the H (heavy) strand results in a short third strand (7S DNA), which can base-pair with the L (light) strand and so displace the H strand to form a local triple-stranded structure that contains many important regulatory sequences and is known as the *CR/D-loop region* (shown by shading and in enlarged form for clarity).

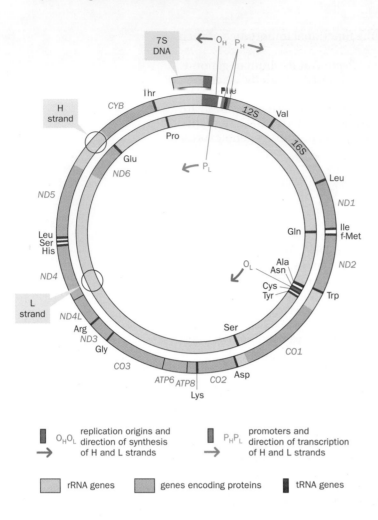

Figure 9.3 The organization of the human mitochondrial genome. The H strand is transcribed from two closely spaced promoter regions flanking the tRNA^Phe gene (grouped here as P_H); the L strand is transcribed from the P_L promoter in the opposite direction. In both cases, large primary transcripts are produced and cleaved to generate RNAs for individual genes. All genes lack introns and are closely clustered. The symbols for protein-coding genes are shown here without the prefix *MT-* that signifies mitochondrial gene. The genes that encode subunits 6 and 8 of the ATP synthase (*ATP6* and *ATP8*) are partly overlapping. Other polypeptide-encoding genes specify seven NADH dehydrogenase subunits (*ND4L* and *ND1–ND6*), three cytochrome *c* oxidase subunits (*CO1–CO3*), and cytochrome *b* (*CYB*). tRNA genes are represented with the name of the amino acid that they bind. The short 7S DNA strand is produced by repeat synthesis of a short segment of the H strand (see Figure 9.2). For further information, see the MITOMAP database at http://www.mitomap.org/.

starts from common promoters in the CR/D-loop region and continues round the circle (in opposing directions for the two different strands), to generate large multigenic transcripts. The mature RNAs are subsequently generated by cleavage of the multigenic transcripts.

Almost two-thirds (24 out of 37) of the mitochondrial genes specify a functional noncoding RNA as their final product. There are 22 tRNA genes, one for each of the 22 types of mitochondrial tRNA. In addition, two rRNA genes are dedicated to making 16S rRNA and 12S rRNA (components of the large and small subunits, respectively, of mitochondrial ribosomes). The remaining 13 genes encode polypeptides, which are synthesized on mitochondrial ribosomes. These 13 polypeptides form part of the mitochondrial respiratory complexes, the enzymes of oxidative phosphorylation that are engaged in the production of ATP. However, the great majority of the polypeptides that make up the mitochondrial oxidative phosphorylation system plus all other mitochondrial proteins are encoded by nuclear genes (**Table 9.1**). These proteins are translated on cytoplasmic ribosomes before being imported into the mitochondria.

Unlike its nuclear counterpart, the human mitochondrial genome is extremely compact: all 37 mitochondrial genes lack introns and are tightly packed (on average, there is one gene per 0.45 kb). The coding sequences of some genes (notably those encoding the sixth and eighth subunits of the mitochondrial ATP synthase) show some overlap (**Figure 9.4**) and, in most other cases, the coding sequences of neighboring genes are contiguous or separated by one or two noncoding bases. Some genes even lack termination codons; to overcome this deficiency, UAA codons have to be introduced at the post-transcriptional level (see Figure 9.4).

The mitochondrial genetic code

Prokaryotic genomes and the nuclear genomes of eukaryotes encode many hundreds to usually many thousands of different proteins. They are subject to a universal genetic code that is kept invariant: mutations that could potentially change

TABLE 9.1 THE LIMITED AUTONOMY OF THE MITOCHONDRIAL GENOME

Mitochondrial component		Encoded by	
		Mitochondrial genome	Nuclear genome
Components of oxidative phosphorylation system		**13 subunits**	**80 subunits**
I	NADH dehydrogenase	7	42
II	Succinate CoQ reductase	0	4
III	Cytochrome b–c_1 complex	1	10
IV	Cytochrome c oxidase complex	3	10
V	ATP synthase complex	2	14
Components of protein synthesis apparatus		**24 RNAs**	**79 proteins**
rRNA		2	0
tRNA		22	0
Ribosomal proteins		0	79
Other mitochondrial proteins		**0**	**All[a]**

[a]Includes mitochondrial DNA and RNA polymerases plus numerous other enzymes, structural and transport proteins, etc.

the genetic code are likely to produce at least some critically misfunctional proteins and so are strongly selected against. However, the much smaller mitochondrial genomes make very few polypeptides. As a result, the mitochondrial genetic code has been able to drift by mutation to be slightly different from the universal genetic code.

In the mitochondrial genetic code there are 60 codons that specify amino acids, one fewer than in the nuclear genetic code. There are four stop codons: UAA and UAG (which also serve as stop codons in the nuclear genetic code) and AGA and AGG (which specify arginine in the nuclear genetic code; see Figure 1.25). The nuclear stop codon UGA encodes tryptophan in mitochondria, and AUA specifies methionine not isoleucine.

The mitochondrial genome specifies all the rRNA and tRNA molecules needed for synthesizing proteins on mitochondrial ribosomes, but it relies on nuclear-encoded genes to provide all other components, such as the protein components of mitochondrial ribosomes and aminoacyl tRNA synthetases. Because there are only 22 different types of human mitochondrial tRNA, individual tRNA molecules need to be able to interpret several different codons. This is possible because of *third-base wobble* in codon interpretation. Eight of the 22 tRNA molecules have anticodons that each recognize families of four codons differing only at the third base. The other 14 tRNAs recognize pairs of codons that are identical at the first two base positions and share either a purine or a pyrimidine at the third base. Between them, therefore, the 22 mitochondrial tRNA molecules can recognize a total of 60 codons $[(8 \times 4) + (14 \times 2)]$.

Figure 9.4 The *MT-ATP6* and *MT-ATP8* genes are transcribed in different reading frames from overlapping segments of the mitochondrial H strand. *MT-ATP8* is transcribed from nucleotides 8366 to 8569, and *MT-ATP6* from 8527 to 9204. After transcription, the RNA encoding the ATP synthase 6 subunit is cleaved after position 9206 and polyadenylated, resulting in a C-terminal UAA codon where the first two nucleotides are derived ultimately from the TA at positions 9205–9206 and the third nucleotide is the first A of the poly(A) tail.

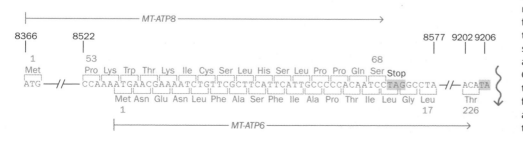

TABLE 9.2 THE HUMAN NUCLEAR AND MITOCHONDRIAL GENOMES

	Nuclear genome	Mitochondrial genome
Size	3.1 Gb	16.6 kb
Number of different DNA molecules	23 (in XX cells) or 24 (in XY cells); all linear	one circular DNA molecule
Total number of DNA molecules per cell	varies according to ploidy; 46 in diploid cells	often several thousand copies (but copy number varies in different cell types)
Associated protein	several classes of histone and nonhistone protein	largely free of protein
Number of protein-coding genes	~21,000	13
Number of RNA genes	uncertain, but >6000	24
Gene density	~1/120 kb, but great uncertainty	1/0.45 kb
Repetitive DNA	more than 50% of genome; see Figure 9.1	very little
Transcription	genes are often independently transcribed	multigenic transcripts are produced from both the heavy and light strands
Introns	found in most genes	absent
Percentage of protein-coding DNA	~1.1%	~66%
Codon usage	61 amino acid codons plus three stop codons[a]	60 amino acid codons plus four stop codons[a]
Recombination	at least once for each pair of homologs at meiosis	not evident
Inheritance	Mendelian for X chromosome and autosomes; paternal for Y chromosome	exclusively maternal

[a]For details see Figure 1.25.

In addition to their differences in genetic capacity and different genetic codes, the mitochondrial and nuclear genomes differ in many other aspects of their organization and expression (**Table 9.2**).

The human nuclear genome consists of 24 widely different chromosomal DNA molecules

The human nuclear genome is 3.1 Gb (3100 Mb) in size. It is distributed between 24 different types of linear double-stranded DNA molecule, each of which has histones and nonhistone proteins bound to it, constituting a chromosome. There are 22 types of autosome and two sex chromosomes, X and Y. Human chromosomes can easily be differentiated by chromosome banding (see Figure 2.15), and have been classified into groups largely according to size and, to some extent, centromere position (see Table 2.3).

There is a single nuclear genome in sperm and egg cells and just two copies in most somatic cells, in contrast to the hundreds or even thousands of copies of the mitochondrial genome. Because the size of the nuclear genome is about 186,000 times the size of a mtDNA molecule, however, the nucleus of a human cell typically contains more than 99% of the DNA in the cell; the oocyte is a notable exception because it contains as many as 100,000 mtDNA molecules.

Not all of the human nuclear genome has been sequenced. The Human Genome Project focused primarily on sequencing *euchromatin*, the gene-rich, transcriptionally active regions of the nuclear genome that account for 2.9 Gb. The other 200 Mb is made up of permanently condensed and transcriptionally inactive (constitutive) heterochromatin. The heterochromatin is composed of long arrays of highly repetitive DNA that are very difficult to sequence accurately. For a similar reason, the long arrays of tandemly repeated transcription units encoding 28S, 18S, and 5.8S rRNA were also not sequenced.

The DNA of human chromosomes varies considerably in length and also in the proportions of underlying euchromatin and constitutive heterochromatin (**Table 9.3**). Each chromosome has some constitutive heterochromatin at the

TABLE 9.3 DNA CONTENT OF HUMAN CHROMOSOMES

Chromosome	Total DNA (Mb)	Euchromatin (Mb)	Heterochromatin (Mb)	Chromosome	Total DNA (Mb)	Euchromatin (Mb)	Heterochromatin (Mb)
1	249	224	19.5	13	115	96.3	17.2
2	243	240	2.9	14	107	88.3	17.2
3	198	197	1.5	15	103	82.1	18.3
4	191	188	3.0	16	90	79.0	10.0
5	181	178	0.3	17	81	78.7	7.5
6	171	168	2.3	18	78	74.6	1.4
7	159	156	4.6	19	59	60.8	0.3
8	146	143	2.2	20	63	60.6	1.8
9	141	120	18.0	21	48	34.2	11.6
10	136	133	2.5	22	51	35.1	14.3
11	135	131	4.8	X	155	151	3.0
12	134	131	4.3	Y	59	26.4	31.6

Chromosome sizes are taken from the ENSEMBL Human Map View (http://www.ensembl.org/Homo_sapiens/Location/Genome). Heterochromatin figures are estimates abstracted from International Human Genome Sequencing Consortium (2004) *Nature* 431, 931–945. The size of the total human genome is estimated to be about 3.1 Gb, with euchromatin accounting for close to 2.9 Gb and heterochromatin accounting for 200 Mb.

centromere. Certain chromosomes, notably 1, 9, 16, and 19, also have significant amounts of heterochromatin in the euchromatic region close to the centromere (*pericentromere*), and the acrocentric chromosomes each have two sizeable heterochromatic regions. But the most significant representation is in the Y chromosome, where most of the DNA is organized as heterochromatin.

The base composition of the euchromatic component of the human genome averages out at 41% (G+C), but there is considerable variation between chromosomes, from 38% (G+C) for chromosomes 4 and 13 up to 49% (for chromosome 19). It also varies considerably along the lengths of chromosomes. For example, the average (G+C) content on chromosome 17q is 50% for the distal 10.3 Mb but drops to 38% for the adjacent 3.9 Mb. There are regions of less than 300 kb with even wider swings, for example from 33.1% to 59.3% (G+C).

The proportion of some combinations of nucleotides can vary considerably. Like other vertebrate nuclear genomes, the human nuclear genome has a conspicuous shortage of the dinucleotide CpG. However, certain small regions of transcriptionally active DNA have the expected CpG density and, significantly, are unmethylated or hypomethylated (*CpG islands*; Box 9.1).

The human genome contains at least 26,000 genes, but the exact gene number is difficult to determine

Several years after the Human Genome Project delivered the first reference genome sequence, there is still very considerable uncertainty about the total human gene number. When the early analyses of the genome were reported in 2001, the gene catalog generated by the International Human Genome Sequencing Consortium was very much oriented toward protein-coding genes. Original estimates suggested more than 30,000 human protein-coding genes, most of which were gene predictions without any supportive experimental evidence. This number was an overestimate because of errors that were made in defining genes (see Box 8.5).

To validate gene predictions supportive evidence was sought, mostly by evolutionary comparisons. Comparison with other mammalian genomes, such as

BOX 9.1 ANIMAL DNA METHYLATION AND VERTEBRATE CpG ISLANDS

DNA methylation in multicellular animals often involves methylation of a proportion of cytosine residues, giving 5-methylcytosine (mC). In most animals (but not *Drosophila melanogaster*), the dinucleotide CpG is a common target for cytosine methylation by specific cytosine methyltransferases, forming mCpG (**Figure 1A**).

DNA methylation has important consequences for gene expression and allows particular gene expression patterns to be stably transmitted to daughter cells. It has also been implicated in systems of host defense against transposons. Vertebrates have the highest levels of 5-methylcytosine in the animal kingdom, and methylation is dispersed throughout vertebrate genomes. However, only a small percentage of cytosines are methylated (about 3% in human DNA, mostly as mCpG but with a small percentage as mCpNpG, where N is any nucleotide).

5-Methylcytosine is chemically unstable and is prone to deamination (see Figure 1A). Other deaminated bases produce derivatives that are identified as abnormal and are removed by the DNA repair machinery (e.g. unmethylated cytosine produces uracil when deaminated). However, 5-methyl cytosine is deaminated to give thymine, a natural base in DNA that is not recognized as being abnormal by cellular DNA repair systems. Over evolutionarily long periods, therefore, the number of CpG dinucleotides in vertebrate DNA has gradually fallen because of the slow but steady conversion of **C**pG to **T**pG (and to **C**pA on the complementary strand; Figure 1B).

Although the overall frequency of CpG in the vertebrate genome is low, there are small stretches of unmethylated or hypomethylated DNA that are characterized by having the *normal*, expected CpG frequency. Such islands of normal CpG density (**CpG islands**) are comparatively GC-rich (typically more than 50% GC) and extend over hundreds of nucleotides. CpG islands are gene markers because they are associated with transcriptionally active regions. Highly methylated DNA regions are prone to adopting a condensed chromatin conformation, but for actively transcribing DNA the chromatin needs to be in a more extended, open unmethylated conformation that allows various regulatory proteins to bind more readily to promoters and other gene control regions.

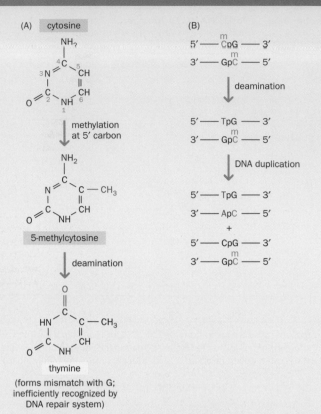

Figure 1 Instability of vertebrate CpG dinucleotides. (A) The cytosine in CpG dinucleotides is a target for methylation at the 5′ carbon atom. The resulting 5-methylcytosine is deaminated to give thymine (T), which is inefficiently recognized by the DNA repair system and so tends to persist (however, deamination of unmethylated cytosine gives uracil which is readily recognized by the DNA repair system). (B) The vertebrate CpG dinucleotide is gradually being replaced by TpG and CpA.

those of the mouse and the dog, failed to identify counterparts of many of the originally predicted human genes. By late 2009 the estimated number of human protein-coding genes appeared to be stabilizing somewhere around 20,000 to 21,000, but huge uncertainty remained about the number of human RNA genes. RNA genes are difficult to identify by using computer programs to analyze genome sequences: there are no open reading frames to screen for, and many RNA genes are very small and often not well conserved during evolution. There is also the problem of how to define an RNA gene. As we detail in Chapter 12, comprehensive analyses have recently suggested that the great majority of the genome—and probably at least 85% of nucleotides—is transcribed. It is currently unknown how much of the transcriptional activity is background noise and how much is functionally significant.

By mid-2009, evidence for at least 6000 human RNA genes had been obtained, including thousands of genes encoding long noncoding RNAs that are thought to be important in gene regulation. In addition, there is evidence for tens of thousands of different tiny human RNAs, but in many such cases quite large numbers of different tiny RNAs are obtained by the processing of single RNA transcripts. We look at noncoding RNAs in detail in Section 9.3.

The combination of about 20,000 protein-coding genes and at least 6000 RNA genes gives a total of at least 26,000 human genes. This remains a provisional total gene number; defining RNA genes is challenging and it will be some time before we obtain an accurate human gene number.

Human genes are unevenly distributed between and within chromosomes

Human genes are unevenly distributed on the nuclear DNA molecules. The constitutive heterochromatin regions are devoid of genes and, even within the euchromatic portion of the genome, gene density can vary substantially between chromosomal regions and also between whole chromosomes.

The first general insight into how genes are distributed across the human genome was obtained when purified CpG island fractions were hybridized to metaphase chromosomes. CpG islands have long been known to be strongly associated with genes (see Box 9.1). On this basis, it was concluded that gene density must be high in subtelomeric regions, and that some chromosomes (e.g. 19 and 22) are gene-rich whereas others (e.g. X and 18) are gene-poor (see Figure 8.17). The predictions of differential CpG island density and differential gene density were subsequently confirmed by analyzing the human genome sequence.

This difference in gene density can also be seen with Giemsa staining (G banding) of chromosomes. Regions with a low (G+C) content correlate with the darkest G bands, and those with a high (G+C) content with pale bands. GC-rich chromosomes (e.g. chromosome 19) and regions (e.g. pale G bands) are also comparatively rich in genes. For example, the gene-rich human leukocyte antigen (HLA) complex (180 protein-coding genes in a span of 4 Mb) is located within the pale 6p21.3 band. In striking contrast, the mammoth dystrophin gene extends over 2.4 Mb of DNA in a dark G band at Xp21.2 without evidence for any other protein-coding gene in this region.

Duplication of DNA segments has resulted in copy-number variation and gene families

Small genomes, such as those of bacteria and mitochondria, are typically tightly packed with genetic information that is presented in extremely economical forms. Large genomes, such as the nuclear genomes of eukaryotes, and especially vertebrate genomes, have the luxury of not being so constrained. Repetitive DNA is one striking feature of large genomes, in both abundance and importance.

Different types of DNA sequence can be repeated. Some are short noncoding sequences that are present in a few copies to millions of copies. These will be discussed further in Section 9.4. Many others are moderately long to large DNA sequences that often contain genes or parts of genes. Such duplicated sequences are prone to various genetic mechanisms that result in *copy-number variation* (*CNV*) in which the number of copies of specific moderately long sequences—often from many kilobases to several megabases long—varies between different haplotypes. Copy-number variation generates a type of *structural variation* that we consider more fully in Chapter 13, but we will consider some of the mechanisms below in the context of how genes become duplicated. It is clear, however, that CNV is quite extensive in the human genome. For example, when James Watson's genome was sequenced, 1.4% of the total sequencing data obtained did not map with the reference human genome sequence. As personal genome sequencing accelerates, new CNV regions are being identified with important implications for gene expression and disease.

Repeated duplication of a gene-containing sequence gives rise to a **gene family**. As we will see in Sections 9.2 and 9.3, many human genes are members of multigene families that can vary enormously in terms of copy number and distribution. They arise by one or more of a variety of different mechanisms that result in gene duplication. Gene families may also contain evolutionarily related sequences that may no longer function as working genes (*pseudogenes*).

Gene duplication mechanisms

Gene duplication has been a common event in the evolution of the large nuclear genomes found in complex eukaryotes. The resulting multigene families have from two to very many gene copies. The gene copies may be clustered together in one subchromosomal location or they may be dispersed over several chromosomal locations. Several different types of gene duplication can occur:

- *Tandem gene duplication* typically arises by crossover between unequally aligned chromatids, either on homologous chromosomes (*unequal crossover*) or on the same chromosome (*unequal sister chromatid exchange*). **Figure 9.5** shows the general mechanism. The repeated segment may be just a few kilobases long or may be quite large and contain from one to several genes. Two such repeats are said to be *direct repeats* if the head of one repeat joins the tail of its neighbor (→→) or *inverted repeats* if there is joining of heads (→←) or tails (←→). Over a long (evolutionary) time-scale the duplicated sequences can be separated on the same chromosome (by a DNA insertion or inversion) or become distributed on different chromosomes by translocations.

- *Duplicative transposition* describes the process by which a duplicated DNA copy integrates into a new subchromosomal location. Typically this involves *retrotransposition*: cellular reverse transcriptases make a cDNA copy of an RNA transcript, whereupon the cDNA copy integrates into a new chromosomal location. The same type of mechanism can often lead to defective gene copies and will be detailed in Section 9.2.

- *Gene duplication by ancestral cell fusion.* Aerobic eukaryotic cells are thought to have evolved through the endocytosis of a type of bacterial cell by a eukaryotic precursor cell. The current mitochondrial genome is thought to have derived from the bacterial cell's genome but is now a very small remnant because many of the original bacterial genes were subsequently excised and transferred to what is now the nuclear genome. As a result, the nuclear genome contains duplicated genes that encode cytoplasm-specific and mitochondrion-specific isoforms for certain enzymes and certain other key proteins.

- *Large-scale subgenomic duplications* can arise as a result of chromosome translocations. Euchromatic regions close to human centromeres and to telomeres (pericentromeric and subtelomeric regions, respectively) are comparatively unstable and are prone to recombination with other chromosomes. As a result, large segments of DNA containing multiple genes have been duplicated. Within the past 40 million years of primate evolution, this process has led to the duplication of about 400 large (several megabases long) DNA segments, accounting for more than 5% of the euchromatic genome. This type of duplication, known as **segmental duplication**, results in very high (often more than 95%) sequence identity between the DNA copies and can involve both intrachromosomal duplications and also interchromosomal duplications (**Figure 9.6**). Segmental duplications are important contributors to copy-number variation and to chromosomal rearrangements leading to disease and rapid gene innovation. We consider how they originate in Chapter 10.

- *Whole genome duplication.* It is now clear from comparative genomics studies that whole genome duplication has occurred at several times during evolution on a variety of different eukaryotic lineages. For example, there is compelling evidence that whole genome duplication occurred in the early evolution of chordates. This type of event could explain why vertebrates have four HOX clusters (see Figure 5.5). Whole genome duplication is detailed in Chapter 10 in the context of genome evolution.

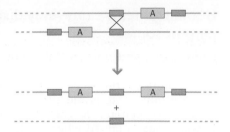

Figure 9.5 Tandem gene duplication. Crossover between misaligned chromatids can result in one chromosome with a tandem duplication of a sequence containing a gene (such as gene A shown here by a green box) and one in which the gene is lost. The mispairing of chromatids may be stabilized by closely related members of an interspersed repeated DNA family such as Alu repeats (as shown here by orange boxes). The crossover event leading to tandem gene duplication may result from unequal crossover (crossover between misaligned chromatids on homologous chromosomes) or unequal sister chromatid exchange (the analogous process by which sister chromatids are misaligned; see Figure 13.3 for an illustration).

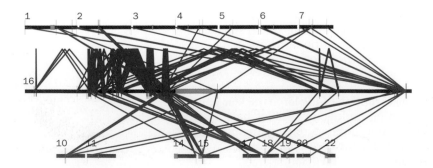

Figure 9.6 Segmental duplication. The horizontal bar in the center is a linear map of the DNA of human chromosome 16 (the central green segment represents heterochromatin). The black horizontal bars at the top and bottom represent linear maps of 16 other chromosomes containing large segments that are shared with chromosome 16, with red connecting lines marking the positions of homologous sequences. Intrachromosomal duplications are shown by blue chevrons (^) linking the positions of large duplicated sequences on chromosome 16. [From Martin J, Han C, Gordon LA et al. (2004) *Nature* 432, 988–994. With permission from Macmillan Publishers Ltd. For chromosomal coordinates and further information on segmental duplications in the human genome, dedicated segmental duplication databases can be accessed at http://humanparalogy.gs.washington.edu/ and http://projects.tcag.ca/humandup/]

9.2 PROTEIN-CODING GENES

For many years, molecular geneticists believed that the major functional end-point of DNA was protein. Studies of prokaryotic genomes supported this belief, partly because these genomes are rich in protein-coding DNA. It came as a surprise to find that the much larger genomes of complex eukaryotes have comparatively little protein-coding DNA. For example, protein-coding DNA sequences account for close to 90% of the *E. coli* genome but just 1.1% of the human genome.

Human protein-coding genes show enormous variation in size and internal organization

Size variation

Genes in simple organisms such as bacteria are comparatively similar in size and are usually very short (typically about 1 kb long). In complex eukaryotes, genes can show huge variation in size. Although there is generally a direct correlation between gene and product sizes, there are some striking anomalies. For example, the giant 2.4 Mb dystrophin gene is more than 50 times the size of the apolipoprotein B gene but the dystrophin protein has a linear length (total amino acid number) that is about 80% of that of apolipoprotein B (Table 9.4).

A small minority of human protein-coding genes lack introns and are generally small (see note to Table 9.4 for some examples). For those that do possess

TABLE 9.4 STRUCTURAL VARIATION IN SIZE AND ORGANIZATION OF HUMAN PROTEIN-CODING GENES

Human protein	Size of protein (no. of amino acids)	Size of gene (kb)	No. of exons	Coding DNA (%)	Average size of exon (bp)	Average size of intron (bp)
SRY	204	0.9	1	94	850	–
β-Globin	146	1.6	3	38	150	490
p16	156	7.4	3	17	406	3064
Serum albumin	609	18	14	12	137	1100
Type VII collagen	2928	31	118	29	77	190
p53	393	39	10	6.0	236	3076
Complement C3	1641	41	29	8.6	122	900
Apolipoprotein B	4563	45	29	31	487	1103
Phenylalanine hydroxylase	452	90	26	3	96	3500
Factor VIII	2351	186	26	3	375	7100
Huntingtin	3144	189	67	8.0	201	2361
RB1 retinoblastoma protein	928	198	27	2.4	179	6668
CFTR (cystic fibrosis transmembrane receptor)	1480	250	27	2.4	227	9100
Titin	34,350	283	363	40	315	466
Utrophin	3433	567	74	2.2	168	7464
Dystrophin	3685	2400	79	0.6	180	30,770

Where isoforms are evident, the given figures represent the largest isoforms. As genes get larger, exon size remains fairly constant but intron sizes can become very large. Internal exons tend to be fairly uniform in size, but the terminal exon or some exons near the 3′ end can be many kilobases long; for example, exon 26 of the *APOB* gene is 7.5 kb long. Note the extraordinarily high exon content and comparatively small intron sizes in the genes encoding type VII collagen and titin. In addition to *SRY*, other single-exon protein-coding genes in the nuclear genome include retrogenes (see Table 9.8) and genes encoding other SOX proteins, interferons, histones, many G-protein-coupled receptors, heat shock proteins, many ribonucleases, and various neurotransmitter receptors and hormone receptors.

introns, there is an inverse correlation between gene size and fraction of coding DNA (see Table 9.4). This does not arise because exons in large genes are smaller than those in small genes. The average exon size in human genes is close to 300 bp, and exon size is comparatively independent of gene length. Instead, there is huge variation in intron lengths, and large genes tend to have very large introns (see Table 9.4). Transcription of long introns is, however, costly in time and energy; transcription of the 2.4 Mb dystrophin gene takes about 16 hours. Thus, very highly expressed genes often have short introns or no introns at all.

Repetitive sequences within coding DNA

Highly repetitive DNA sequences are often found within introns and flanking sequences of genes. They will be detailed in Section 9.4. In addition, repetitive DNA sequences are found to different extents in exons. Tandem repetition of very short oligonucleotide sequences (1–4 bp) is frequent and may simply reflect statistically expected frequencies for certain base compositions. Tandem repetition of sequences encoding known or assumed protein domains is also quite common, and it may be functionally advantageous by providing a more available biological target.

The sequence identities between the repeated protein domains are often quite low but can sometimes be high. Lipoprotein Lp (a), encoded by the *LPA* gene on chromosome 6q26, provides a classical example. It contains multiple tandemly repeated kringle domains, which are each about 114 amino acids long and form disulfide-bonded loops. The different kringle domains are often nearly identical in amino acid sequence. Even at the nucleotide sequence level the DNA repeats that encode the kringle domains show very high levels of sequence identity, making them prone to unequal crossover. As a result, the *LPA* gene is subject to length polymorphism, and the number of kringle domains in lipoprotein Lp (a) varies but is usually 15 or more.

Different proteins can be specified by overlapping transcription units

Overlapping genes and genes-within-genes

Simple genomes have high gene densities (roughly one per 0.5, 1, and 2 kb for the genomes of human mitochondria, *Escherichia coli*, and *Saccharomyces cerevisiae*, respectively) and often show examples of partly overlapping genes. Different reading frames may be used, sometimes from the same sense strand. In complex organisms, such as humans, genes are much bigger, and there is less clustering of protein-coding sequences (Table 9.5).

Gene density varies enormously from chromosome to chromosome and within different regions of the same chromosome. In chromosomal regions with high gene density, overlapping genes may be found; they are typically transcribed from opposing DNA strands. For example, the class III region of the HLA complex at 6p21.3 has an average gene density of about one gene per 15 kb and is known to contain several examples of partly overlapping genes (Figure 9.7A).

Whole genome analyses show that about 9% of human protein-coding genes overlap another such gene. More than 90% of the overlaps involve genes transcribed from opposing strands. Sometimes the overlaps are partial, but in other cases small protein-coding genes are located within the introns of larger genes. The *NF1* (neurofibromatosis type I) gene, for example, has three small internal genes transcribed from the opposite strand (Figure 9.7B).

Recent analyses have also shown that RNA genes can frequently overlap protein-coding genes. The positioning of RNA genes will be covered in Section 9.3.

Genes divergently transcribed or co-transcribed from a common promoter

Some protein-coding genes share a promoter. In many cases the 5′ ends of the two genes are often separated by just a few hundred nucleotides and the genes are transcribed in opposite directions from the common promoter. This type of bidirectional gene organization may provide for common regulation of the gene pair.

Alternatively, genes with a common promoter are transcribed in the same direction to produce multigenic transcripts that are then cleaved to produce a

TABLE 9.5 HUMAN GENOME AND HUMAN GENE STATISTICS

SIZE OF GENOME COMPONENTS	
Mitochondrial genome	16.6 kb
Nuclear genome	3.1 Gb[a]
Euchromatic component	2.9 Gb (~93%)
Highly conserved fraction	~150 Mb (~5%)
Protein-coding DNA sequences	~35 Mb (~1.1%)
Other highly conserved DNA	~115 Mb (~3.9%)
Segmentally duplicated DNA	~160 Mb (~5.5%)
Highly repetitive DNA	~1.6 Gb (~50%)
Constitutive heterochromatin	~ 200 Mb (~7%; Table 9.3)
Transposon-based repeats	~1.4 Gb (~45%; Table 9.12)
DNA per chromosome	48 Mb—249 Mb (Table 9.3)
GENE NUMBER	
Nuclear genome	> 26,000
Mitochondrial genome	37
Protein-coding genes	~ 20,000–21,000
RNA genes	> 6000 (exact figure not known)
Pseudogenes related to protein-coding genes	> 12,000
GENE DENSITY	
Nuclear genome	>1 per 120 kb (but considerable uncertainty)
Mitochondrial genome	1 per 0.45 kb
LENGTH OF PROTEIN-CODING GENES	
Average length	53.6 kb
Smallest	a few hundred base pairs long (several examples)
Largest	2.4 Mb (dystrophin)
EXON NUMBER IN PROTEIN-CODING GENES	
Average number of exons in one gene[b]	9.8
Largest number in one gene	363 (in the titin gene)
Smallest number in one gene	1 (no introns—see Tables 9.4 and 9.7 for example)
EXON SIZE IN PROTEIN-CODING GENES	
Average exon size	288 bp (but exons at 3' end of genes tend to be large)
Smallest	< 10 bp (various; e.g. exon 3 of the troponin I gene *TNNI1* is just 4 bp long)
Largest	18.2 kb (exon 6 in *MUC16* isoform-201)
INTRON SIZE IN PROTEIN-CODING GENES	
Smallest	< 30 bp (various)
Largest	1.1 Mb (intron 5 in *KCNIP4*)
RNA SIZE	
Smallest noncoding RNA	< 20 nucleotides (e.g. many transcriptional start site-associated RNAs are 18 nucleotides)
Largest noncoding RNA	Several hundred thousand nucleotides; e.g. *UBE3A* antisense RNA is likely to be close to 1 Mb
Largest mRNA	> 103 kb (titin mRNA, NF-2A isoform)
POLYPEPTIDE SIZE	
Smallest	tens of amino acids (various neuropeptides)
Largest	34,350 amino acids (titin, NF-2A isoform)

[a]Note that the total size can vary between haplotypes because of copy-number variation. [b]For the longest isoform. Data were obtained largely from ENSEMBL release 55 datasets.

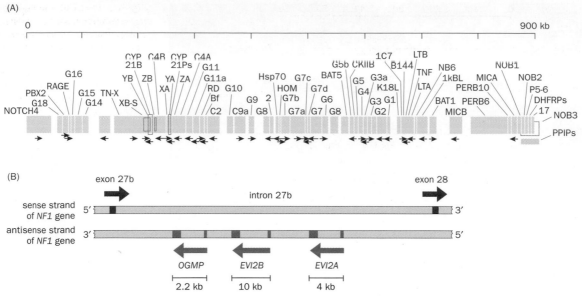

Figure 9.7 Overlapping genes and genes-within-genes. (A) Genes in the class III region of the HLA complex are tightly packed and overlapping in some cases. Arrows show the direction of transcription. (B) Intron 27b of the *NF1* (neurofibromatosis type I) gene is 60.5 kb long and contains three small internal genes, each with two exons, which are transcribed from the opposing strand. The internal genes (not drawn to scale) are *OGMP* (oligodendrocyte myelin glycoprotein) and *EVI2A* and *EVI2B* (human homologs of murine genes thought to be involved in leukemogenesis and located at ecotropic viral integration sites).

separate transcript for each gene. Such genes are said to form part of a *polycistronic* (= multigenic) transcription unit. Polycistronic transcription units are common in simple genomes such as those of bacteria and the mitochondrial genome (see Figure 9.3). Within the nuclear genome, some examples are known of different proteins being produced from a common transcription unit. Typically, they are produced by cleavage of a hybrid precursor protein that is translated from a common transcript. The A and B chains of insulin, which are intimately related functionally, are produced in this way (see Figure 1.26), as are the related peptide hormones somatostatin and neuronostatin. Sometimes, however, functionally distinct proteins are produced from a common protein precursor. The *UBA52* and *UBA80* genes, for example, both generate ubiquitin and an unrelated ribosomal protein (S27a and L40, respectively).

More recent analyses have shown that the long-standing idea that most human genes are independent transcription units is not true, and so the definition of a gene will need to be radically revised. Multigenic transcription is now known to be rather frequent in the human genome, and specific proteins and functional noncoding RNAs can be made by common RNA precursors. This will be explored further in Section 9.3.

Human protein-coding genes often belong to families of genes that may be clustered or dispersed on multiple chromosomes

Duplicated genes and duplicated coding sequence components are a common feature of animal genomes, especially large vertebrate genomes. As we will see in Chapter 10, gene duplication has been an important driver in the evolution of functional complexity and the origin of increasingly complex organisms. Genes that operate in the same or similar functional pathways but produce proteins with little evidence of sequence similarity are distantly related in evolution, and they tend to be dispersed at different chromosomal locations. Examples include genes encoding insulin (on chromosome 11p) and the insulin receptor (19p); ferritin heavy chain (11q) and ferritin light chain (22q); steroid 11-hydroxylase (8q) and steroid 21-hydroxylase (6p); and JAK1 (1p) and STAT1 (2q). However, genes that produce proteins with both structural and functional similarity are often organized in gene clusters.

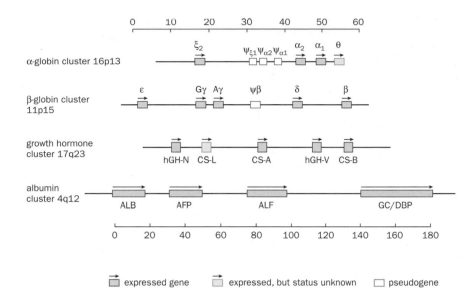

Figure 9.8 Examples of human clustered gene families. Genes in a cluster are often closely related in sequence and are typically transcribed from the same strand. Gene clusters often contain a mixture of expressed genes and nonfunctional pseudogenes. The functional status of the θ-globin and *CS-L* genes is uncertain. The scales at the top (globin and growth hormone clusters) and the bottom (albumin cluster) are in kilobases.

Different classes of human gene families can be recognized according to the degree of sequence similarity and structural similarity of their protein products. If two different genes make very similar protein products, they are most likely to have originated by an evolutionarily very recent gene duplication, most probably some kind of tandem gene duplication event, and they tend to be clustered together at a specific subchromosomal location. If they make proteins that are more distantly related in sequence, they most probably arose by a more ancient gene duplication. They may originally have been clustered together, but over long evolutionary time-scales the genes could have been separated by translocations or inversions, and they tend to be located at different chromosomal locations.

Some gene families are organized in multiple clusters. The β-, γ-, δ-, and ε-globin genes are located in a gene cluster on 11p and are more closely related to each other than they are to the genes in the α-globin gene cluster on 16p (**Figure 9.8**). The genes in the β-globin gene cluster on 11p originated by gene duplication events that were much more recent in evolution than the early gene duplication event that gave rise to ancestors of the α- and β-globin genes. An outstanding example of a gene family organized as multiple gene clusters is the olfactory receptor gene family. The genes encode a diverse repertoire of receptors that allow us to discriminate thousands of different odors; the genes are located in large clusters at multiple different chromosomal locations (**Table 9.6**).

Some gene families have individual gene copies at two or more chromosomal locations without gene clustering (see Table 9.6). The genes at the different locations are usually quite divergent in sequence unless gene duplication occurred relatively recently or there has been considerable selection pressure to maintain sequence conservation. The family members are expected to have originated from ancient gene duplications.

Different classes of gene family can be recognized according to the extent of sequence and structural similarity of the protein products

As listed below, various classes of gene family can be distinguished according to the level of sequence identity between the individual gene members.

- In gene families with closely related members, the genes have a high degree of sequence homology over most of the length of the gene or coding sequence. Examples include histone gene families (histones are strongly conserved, and subfamily members are virtually identical), and the α-globin and β-globin gene families.

TABLE 9.6 EXAMPLES OF CLUSTERED AND INTERSPERSED MULTIGENE FAMILIES

Family	Copy no.	Organization	Chromosome location(s)
CLUSTERED GENE FAMILIES			
Growth hormone gene cluster	5	clustered within 67 kb; one pseudogene (Figure 9.8)	17q24
α-Globin gene cluster	7	clustered over ~50 kb (Figure 9.8)	16p13
Class I HLA heavy chain genes	~20	clustered over 2 Mb (Figure 9.10)	6p21
HOX genes	38	organized in four clusters (Figure 5.5)	2q31, 7p15, 12q13, 17q21
Histone gene family	61	modest-sized clusters at a few locations; two large clusters on chromosome 6	many
Olfactory receptor gene family	> 900	about 25 large clusters scattered throughout the genome	many
INTERSPERSED GENE FAMILIES			
Aldolase	5	three functional genes and two pseudogenes on five different chromosomes	many
PAX	9	all nine are functional genes	many
NF1 (neurofibromatosis type I)	> 12	one functional gene at 22q11; others are nonprocessed pseudogenes or gene fragments (Figure 9.11)	many, mostly pericentromeric
Ferritin heavy chain	20	one functional gene on chromosome 11; most are processed pseudogenes	many

- In gene families defined by a common protein domain, the members may have very low sequence homology but they possess certain sequences that specify one or more specific protein domains. Examples include the PAX gene family and SOX gene family (**Table 9.7**).

- Examples of gene families defined by functionally similar short protein motifs are families of genes that encode functionally related proteins with a DEAD box motif (Asp-Glu-Ala-Asp) or the WD repeat (**Figure 9.9**).

Some genes encode products that are functionally related in a general sense but show only very weak sequence homology over a large segment, without very significant conserved amino acid motifs. Nevertheless, there may be some evidence for common general structural features. Such genes can be grouped into an evolutionarily ancient **gene superfamily** with very many gene members. Because multiple different gene duplication events have occurred periodically during the long evolution of a gene superfamily, some of the gene members make proteins that are very divergent in sequence from those of some other family members, but genes resulting from more recent duplications are more readily seen to be related in sequence.

TABLE 9.7 EXAMPLES OF HUMAN GENES WITH SEQUENCE MOTIFS THAT ENCODE HIGHLY CONSERVED DOMAINS

Gene family	Number of genes	Sequence motif/domain
Homeobox genes	38 *HOX* genes plus 197 orphan homeobox genes	homeobox specifies a homeodomain of ~60 amino acids; a wide variety of different subclasses have been defined
PAX genes	9	paired box encodes a paired domain of ~124 amino acids; *PAX* genes often also have a type of homeodomain known as a paired-type homeodomain
SOX genes	19	SRY-like HMG box which encodes a domain of 70–80 amino acids
TBX genes	14	T-Box encodes a domain of ~170 amino acids
Forkhead domain genes	50	the forkhead domain is ~110 amino acids long
POU domain genes	16	the POU domain is ~150 amino acids long

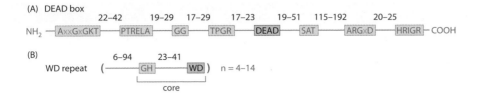

Figure 9.9 Some gene families encode functionally related proteins with short conserved amino acid motifs. (A) DEAD box family motifs. This gene family encodes products implicated in cellular processes involving the alteration of RNA secondary structure, such as translation initiation and splicing. Eight very highly conserved amino acid motifs are evident, including the DEAD box (Asp-Glu-Ala-Asp). Numbers refer to frequently found size ranges for intervening amino acid sequences; X represents any amino acid. (B) WD repeat family motifs. This gene family encodes products that are involved in a variety of regulatory functions, such as regulation of cell division, transcription, transmembrane signaling, and mRNA modification. The gene products are characterized by 4–16 tandem WD repeats that each contain a core sequence of fixed length beginning with a GH (Gly-His) dipeptide and terminating in the dipeptide WD (Trp-Asp), preceded by a sequence of variable length.

Two important examples of gene superfamilies are the Ig (immunoglobulin) and GPCR (G-protein-coupled receptor) superfamilies. Members of the Ig superfamily all have globular domains resembling those found in immunoglobulins, and in addition to immunoglobulins they include a variety of cell surface proteins and soluble proteins involved in the recognition, binding, or adhesion processes of cells (see Figure 4.22 for some examples). The GPCR superfamily is very large, with at least 799 unique full-length members distributed throughout the human genome. All the GPCR proteins have a common structure of seven α-helix transmembrane segments, but they typically have low (less than 40%) sequence similarity to each other. They mediate ligand-induced cell signaling via interaction with intracellular G proteins, and most work as rhodopsin receptors.

Gene duplication events that give rise to multigene families also create pseudogenes and gene fragments

Gene families frequently have defective gene copies in addition to functional genes. A defective gene copy that contains at least multiple exons of a functional gene is known as a **pseudogene** (Box 9.2). Other defective gene copies may have only limited parts of the gene sequence, sometimes a single exon, and so are sometimes described as *gene fragments*.

Clustered gene families often have defective gene copies that have arisen by tandem duplication. These are examples of *nonprocessed pseudogenes*. Copying can be seen to have been performed at the level of genomic DNA because non-processed pseudogenes contain counterparts of both exons and introns and sometimes also of upstream promoter regions. However, even if the copy has sequences that correspond to the full length of the functional gene, closer examination will identify inappropriate termination codons in exons, aberrant splice junctions, and so on. Classical examples of nonprocessed pseudogenes are found in the α-globin and β-globin gene clusters (see Figure 9.8). Sometimes, smaller truncated gene copies and gene fragment copies are also evident, as in the class I HLA gene family (**Figure 9.10**).

A few gene families that are distributed at different chromosomal locations can also have nonprocessed pseudogene copies of a single functional gene. Certain types of subchromosomal region, notably pericentromeric and subtelomeric regions, are comparatively unstable. They are prone to recombination events that can result in duplicated gene segments (containing both exons and introns) being distributed to other chromosomal locations. The gene copies are typically defective because they lack some of the functional gene sequence. Two illustrative examples are sequences related to the *NF1* (neurofibromatosis type I) and the *PKD1* (adult polycystic kidney disease) genes.

The *NF1* gene is located at 17q11.2. Because of pericentromeric instability, multiple nonprocessed *NF1* pseudogene/gene fragment copies are distributed over seven different chromosomes, nine being located at pericentromeric regions (**Figure 9.11A**). The *PKD1* gene has over 46 exons spanning 50 kb and is located in a subtelomeric region at 16p13.3. Six nonprocessed *PKD1* pseudogenes have been generated by segmental duplications during primate evolution and have inserted into locations within a region that is about 13–16 Mb proximal to the *PKD1* gene, corresponding to part of band 16p13.1 (Figure 9.11B). The pseudogenes lack sequences at the 3' end of the *PKD1* gene but have sequence counterparts of much of the genomic sequence spanning exons 1–32, showing 97.6% to 97.8% sequence identity to the PKD1 sequence.

Processed pseudogenes are defective copies of a gene that contain only exonic sequences and lack an intronic sequence or upstream promoter sequences. They arise by *retrotransposition*: cellular reverse transcriptases can use processed gene

BOX 9.2 THE ORIGINS, PREVALENCE, AND FUNCTIONALITY OF PSEUDOGENES

Pseudogenes are usually thought of as defective copies of a functional gene to which they show significant sequence homology. They typically arise by some kind of gene duplication event that produces two gene copies. Selection pressure to conserve gene function need only be imposed on one gene copy; the other copy can be allowed to mutate more freely (*genetic drift*) and can pick up inactivating mutations, producing a pseudogene. However, some sequences are referred to as pseudogenes even though they have not originated by DNA copying. For example, as we will see in Chapter 10, humans have rare *solitary pseudogenes* that are clearly orthologs of functional genes in the great apes and became defective after acquiring harmful mutations in the human lineage.

Different gene duplication mechanisms can give rise to multiple functional gene copies and defective pseudogenes. Either the genomic DNA sequence is copied, or a cDNA copy is made (after reverse transcription of a processed RNA transcript) that integrates into genomic DNA. For a protein-coding gene, copying at the genomic DNA level can result in duplication of the promoter and upstream regulatory sequences as well as of all exons and introns. A defective gene that derives from a copy of a genomic DNA sequence is known as a *nonprocessed pseudogene* (**Figure 1A**). Such pseudogenes usually arise by tandem duplication so that they are located close to functional gene counterparts (see Figures 9.8 and 9.10B for examples), but some are dispersed as a result of recombination (see Figure 9.11 for examples).

Copying at the cDNA level produces a gene copy that typically lacks introns, promoter elements, and upstream regulatory elements. Very occasionally, a processed gene copy can retain some function (a *retrogene*; see Table 9.7). However, because they lack important sequences needed for expression, most processed gene copies degenerate into *processed pseudogenes* (sometimes called *retrotransposed pseudogenes*; Figure 1B).

Prevalence and functionality of pseudogenes

Eukaryotic genomes typically have many pseudogenes. A long-standing rationale for their abundance is that gene duplication is evolutionarily advantageous. New functional gene variants can be created by gene duplication, and pseudogenes have long been viewed as unsuccessful by-products of the duplication mechanisms. Although some prokaryotic genomes seem to have many pseudogenes, pseudogenes are generally rare in prokaryotes because their genomes are generally designed to be compact.

The great majority of what are conventionally recognized as human pseudogenes are copies of protein-coding genes simply because it is relatively easy to identify them (by looking for frameshifting, splice site mutations, and so on). There are more than 8000 different processed pseudogene copies of protein-coding genes in the human genome, plus more than 4000 nonprocessed pseudogenes (see the pseudogene database at

http://www.pseudogene.org). Only about 10% of the 21,000 human protein-coding genes have at least one processed pseudogene, but highly expressed genes tend to have multiple processed pseudogenes. For example, the cytoplasmic ribosomal protein contains 95 functional genes encoding 79 different proteins (16 genes are duplicated) and 2090 processed pseudogenes.

RNA pseudogenes are often difficult to identify as pseudogenes (there is no reading frame to inspect, and RNA genes often lack introns). Nevertheless, pseudogene copies of many small RNAs are common (see the table below), notably if they are transcribed by RNA polymerase III (genes transcribed by RNA polymerase III often have internal promoters).

RNA family	Number of human genes	Number of related pseudogenes
U6 snRNA	49	~800
U7 snRNA	1	85
Y RNA	4	~1000

As will be described in Section 12.4, the Alu repeat, the most abundant sequence in the human genome, seems to have originated by the copying of 7SL RNA transcripts, and many other highly repeated interspersed DNA families in mammals are copies of tRNA. So, in a sense, RNA pseudogenes have come to be the most common sequence elements of mammalian genomes.

All the pseudogenes are located in the nuclear genome, but they do include defective copies of genes that reside in the mitochondrial genome (*mitochondrial pseudogenes*). The mitochondrial genome originated from a much larger bacterial genome and over a long evolutionary time-scale much of the DNA of the large precursor mitochondrial genome migrated in a series of independent integration events into what is now the nuclear genome. mtDNA pseudogenes now account for at least 0.016% of nuclear DNA (or about 30 times the content of the mitochondrial genome).

The functionality of pseudogenes has been an enduring debate, and different pseudogene classes have been envisaged. A significant number of pseudogenes (mostly processed pseudogenes) are transcribed, and antisense pseudogene transcripts may regulate parent genes. Pseudogenes have also been directly implicated in the production of endogenous siRNAs that regulate transposons, as described in Section 9.3. Finally, some pseudogene sequences may be co-opted for a different function. They have been described as *exapted pseudogenes*. An example is provided by the *XIST* gene. It makes a noncoding RNA that regulates X-chromosome inactivation, and two of its six exons are known to have originated from a pseudogene copy of a protein-coding gene.

Figure 1 Origins of nonprocessed and processed pseudogenes. (A) Copying of genomic DNA sequence containing gene A can produce duplicate copies of gene A. Strong selection pressure needs to be applied to one of the copies to maintain gene function (bold arrow), but the other copy can be allowed to mutate (dashed arrow). If it picks up inactivating mutations (red circles), a nonprocessed pseudogene (ψA) can arise.
(B) A processed pseudogene arises after cellular reverse transcriptases convert a transcript of a gene into a cDNA that then is able to integrate back into the genome (see Figure 9.12 for details). The lack of important sequences such as a promoter usually results in an inactive gene copy.

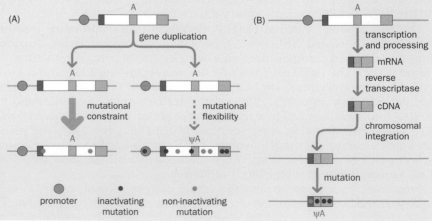

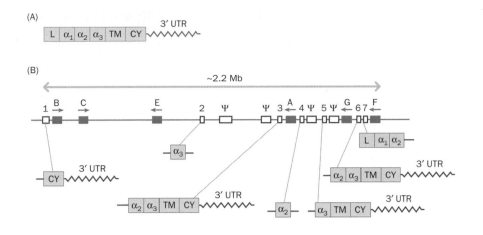

Figure 9.10 The class I HLA gene family: a clustered gene family with nonprocessed pseudogenes and gene fragments.
(A) Structure of a class I HLA heavy-chain mRNA. The full-length mRNA contains a polypeptide-encoding sequence with a leader sequence (L), three extracellular domains (α_1, α_2, and α_3), a transmembrane sequence (TM), a cytoplasmic tail (CY), and a 3′ untranslated region (3′ UTR). The three extracellular domains are each encoded essentially by a single exon. The very small 5′ UTR is not shown. (B) The class I HLA heavy chain gene cluster is located at 6p21.3 and comprises about 20 genes. They include six expressed genes (filled blue boxes), four full-length nonprocessed pseudogenes (long red open boxes labeled ψ), and a variety of partial gene copies (short open red boxes labeled 1–7). Some of the latter are truncated at the 5′ end (e.g. 1, 3, 5, and 6), some are truncated at the 3′ end (e.g. 7), and some contain single exons (e.g. 2 and 4).

transcripts such as mRNA to make cDNA that can then integrate into chromosomal DNA (**Figure 9.12**). Processed pseudogenes are common in interspersed gene families (see Table 9.5).

Processed pseudogenes lack a promoter sequence and so are typically not expressed. Sometimes, however, the cDNA copy integrates into a chromosomal DNA site that happens, by chance, to be adjacent to a promoter that can drive expression of the processed gene copy. Selection pressure may ensure that the processed gene copy continues to make a functional gene product, in which case it is described as a **retrogene**. A variety of intronless retrogenes are known to have testis-specific expression patterns and are typically autosomal homologs of an intron-containing X-linked gene (**Table 9.8**).

One rationale for retrogenes may be a critical requirement to overcome the lack of expression of certain X-linked sequences in the testis during male meiosis. During male meiosis, the paired X and Y chromosomes are converted to heterochromatin, forming the highly condensed and transcriptionally inactive **XY body**. Autosomal retrogenes can provide the continued synthesis in testis cells of certain crucially important products that are no longer synthesized by genes in the highly condensed XY body.

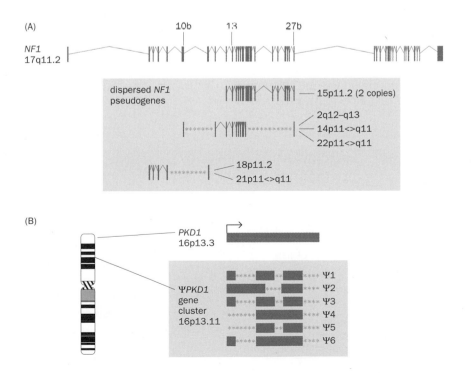

Figure 9.11 Dispersal of nonprocessed *NF1* and *PKD1* pseudogenes as a result of pericentromeric or subtelomeric instability. (A) The *NF1* neurofibromatosis type I gene is located close to the centromere of human chromosome 17. It spans 283 kb and has 58 exons. Exons are represented by thin vertical boxes; introns are shown by connecting chevrons (^). Highly homologous defective copies of the *NF1* gene are found in nine or more other genome locations, mostly in pericentromeric regions. Each copy has a portion of the full-length gene, with both exons and introns. Seven examples are shown here, such as two copies on 15p that have intact genomic sequences spanning exons 13 and 27b. Rearrangements have sometimes caused the deletion of exons and introns (shown by asterisks). (B) The 46 kb *PKD1* polycystic kidney disease gene is located close to the telomere of 16p and has over 40 exons. As a result of segmental duplication events during primate evolution, large components of this gene have been duplicated and six *PKD1* pseudogenes are located at 16p13.11, with large blocks of sequence (shown as blue boxes) copied from the *PKD1* gene (asterisks represent the absence of counterparts to *PKD1* sequences).

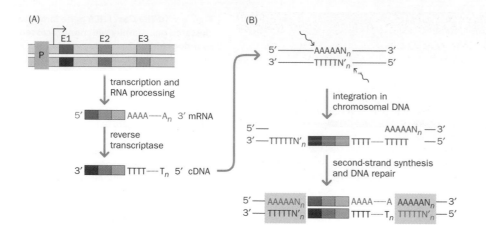

Figure 9.12 Processed pseudogenes and retrogenes originate by reverse transcription from RNA transcripts. (A) In this example, a protein-coding gene with three exons (E1–E3) is transcribed from an upstream promoter (P), and introns are excised from the transcript to yield an mRNA. The mRNA can then be converted naturally into an antisense single-stranded cDNA by using cellular reverse transcriptase function (provided by LINE-1 repeats). (B) Integration of the cDNA is envisaged at staggered breaks (indicated by curly arrows) in A-rich sequences, but could be assisted by the LINE-1 endonuclease. If the A-rich sequence is included in a 5′ overhang, it could form a hybrid with the distal end of the poly(T) of the cDNA, facilitating second-strand synthesis. Because of the staggered breaks during integration, the inserted sequence will be flanked by short direct repeats (boxed sequences).

9.3   RNA GENES

Much of the attention paid to human genes has focused on protein-coding genes because they were long considered to be by far the functionally most important part of our genome. By comparison, genes whose final products are functional **noncoding RNA (ncRNA)** molecules have been so underappreciated that one of the two draft human genome sequences reported in 2001 contained no analyses at all of human RNA genes! RNA was seen to be important in very early evolution (**Box 9.3**) but its functions were imagined to have been very largely overtaken by DNA and proteins. In recent times, the vast majority of RNA molecules were imagined to serve as accessory molecules in the making of proteins.

The last few years have witnessed a revolution in our understanding of the importance of RNA and, although the number of protein-encoding genes has been steadily revised downward since draft human genome sequences were reported in 2001, the number of RNA genes is constantly being revised upward. The tiny mitochondrial genome was always considered to be exceptional because 65% (24 out of 37) of its genes are RNA genes. Now we are beginning to realize that the RNA transcribed from the nucleus is not so uniformly dedicated to protein synthesis as we once thought; instead, it shows great functional diversity.

What has changed our thinking? First, completely unsuspected classes of ncRNA have recently been discovered, including several prolific classes of tiny regulatory RNAs. Second, recent whole genome analyses using microarrays and high-throughput transcript sequencing have shown that at least 85% and possibly more than 90% of the human genome is transcribed. That is, more than 85% of the nucleotide positions in the euchromatic genome are represented in primary transcripts produced from at least one of the two DNA strands. Two other major surprises were the extent of multigenic transcription and the pervasiveness of bidirectional transcription. The recent data challenge the distinction between genes and intergenic space and have forced a radical rethink of the concept of a gene (**Box 9.4**).

TABLE 9.8 EXAMPLES OF HUMAN INTRONLESS RETROGENES AND THEIR PARENTAL INTRON-CONTAINING HOMOLOGS		
Retrogene	Intron-containing homolog	Product
GK2 at 4q13	*GK1* at Xp21	glycerol kinase
PDHA2 at 4q22	*PDHA1* at Xp22	pyruvate dehydrogenase
PGK2 at 6p12	*PGK* at Xq13	phosphoglycerate kinase
TAF1L at 9p13	*TAF1* at Xq13	TATA box binding protein associated factor, 250 kD
MYCL1 at 1p34	*MYCL2* at Xq22	homolog of v-Myc oncogene
GLUD1 at 10q23	*GLUD2* at Xq25	glutamate dehydrogenase
RBMXL at 9p13	*RBMX* at Xq26	RNA-binding protein important in brain development

BOX 9.3 THE RNA WORLD HYPOTHESIS

Proteins cannot self-replicate, and so many evolutionary geneticists consider that *autocatalytic* nucleic acids must have pre-dated proteins and were able to replicate without the help of proteins. The *RNA world hypothesis* developed from ideas proposed by Alexander Rich and Carl Woese in the 1960s. It imagines that RNA had a dual role in the earliest stages of life, acting as both the genetic material (with the capacity for self-replication) and also as effector molecules. Both roles are still evident today: some viruses have RNA genomes, and noncoding RNA molecules can work as effector molecules with catalytic activity. RNAse P, for example, is a ribozyme that can cleave substrate RNA without any requirement for protein, and certain types of intron are autocatalytic and able to splice themselves out of RNA transcripts without any help from proteins (see text). Another observation consistent with RNA's being the first nucleic acid is that deoxyribonucleotides are synthesized from ribonucleotides in cellular pathways.

As well as storing genetic information, RNA has been imagined to have been used subsequently to synthesize proteins from amino acids. Different RNAs, including rRNA and tRNA, are central in assisting polypeptide synthesis. Many ribosomal proteins can be deleted without affecting ribosome function, and the crucial peptidyl transferase activity—the enzyme that catalyzes the formation of peptide bonds—is a ribozyme. However, RNA has a rather rigid backbone and so is not very well suited as an effector molecule. Proteins are much more flexible and also offer more functional variety because the 20 amino acids can have widely different structures and offer more possible sequence combinations (a decapeptide provides 20^{10} or about 10^{13} different possible amino acid sequences, whereas a decanucleotide has 4^{10} or about 10^6 different possible sequences).

The replacement of RNA with DNA as an information storage molecule provided significant advantages. DNA is much more stable than RNA, and so is better suited for this task. Its sugar residues lack the 2′ OH group on ribose sugars that makes RNA prone to hydrolytic cleavage. Greater efficiency could be achieved by separating the storage and transmission of genetic information (DNA) from protein synthesis (RNA). All that was needed was the development of a reverse transcriptase so that DNA could be synthesized from deoxynucleotides by using an RNA template.

We have known for many decades that various ubiquitous ncRNA classes are essential for cell function. Until recently, however, we have largely been accustomed to thinking of ncRNA as not much more than a series of *accessories* that are needed to process genes to make proteins. Transfer RNAs are needed at the very end of the pathway, serving to decode the codons in mRNA and provide amino acids in the order they are needed for insertion into growing polypeptide chains. Ribosomal RNAs are essential components of the ribosomes, the complex ribonucleoprotein factories of protein synthesis.

Other ubiquitous ncRNAs were known to function higher up the pathway to ensure the correct processing of mRNA, rRNA, and tRNA precursors. Various small RNAs are components of complex ribonucleoproteins involved in different processing reactions, including splicing, cleavage of rRNA and tRNA precursors, and base modifications that are required for RNA maturation. Typically these RNAs work as *guide RNAs*, by base pairing with complementary sequences in the precursor RNA.

We have also long been aware of a few ncRNAs that have other functions, such as RNAs implicated in X-inactivation and imprinting, and the RNA component of the telomerase ribonucleoprotein needed for synthesis of the DNA of telomeres (see Figure 2.13). But these RNAs seemed to be quirky exceptions.

In the past decade or so, however, there has been a revolution in how we view RNA. Many thousands of different ncRNAs have recently been identified in animal cells. Many of them are developmentally regulated and have been shown to have crucial roles in a whole variety of different processes that occur in specialized tissues or specific stages of development. Several ncRNAs have already been implicated in cancer and genetic disease.

Now that the human genome is known to have close to 20,000 protein-coding genes, about the same as in the 1 mm nematode *Caenorhabditis elegans*, which has only about 1000 cells—the question now is whether RNA-based regulation is the key feature in explaining our complexity. Certainly, the complexity of RNA-based gene regulation increases markedly in complex organisms, as is described in Chapter 10. Maybe it is time to view the genome as more of an RNA machine than just a protein machine.

BOX 9.4 REVISING THE CONCEPT OF A GENE IN THE POST-GENOME ERA

Until genome-wide analyses were performed, a typical human protein-coding gene was imagined to be well defined and separated from its neighbors by identifiable intergenic spaces. The gene would typically be split into several exons. Directed from an upstream promoter, a primary transcript complementary to just one DNA strand would undergo splicing. The functionally unimportant (junk) intronic sequences would be discarded, allowing fusion of the important exonic sequences to make an mRNA. Expression would be regulated by nearby regulatory sequences typically located close to the promoter.

This neat and cosy image of a gene has been buffeted by a series of complications. It has long been known that some nuclear genes partly overlap others or are entirely embedded within much larger genes. Different products were known to be produced from a single gene by using alternative promoters, alternative splicing, and RNA editing. Very occasionally, sequences from different genes could be spliced together at the RNA level, sometimes even in the case of genes on different chromosomes (*trans-splicing*). Occasionally, natural antisense transcripts were observed that could be seen to regulate the expression of the sense transcript of a gene. Some genes were known to have regulatory elements located many hundreds of kilobases away, sometimes within another gene.

Despite the above complications, scientists were not quite ready to give up on the simple idea of a gene described in the first paragraph above, until whole genome analyses shattered previous misconceptions. The significant findings that forced a reappraisal of gene organization were:

- *Transcription is pervasive.* More than 85% of the euchromatic human genome is transcribed, and multigenic transcription is common, so that the distinction between genes and intergenic space is now much less apparent (**Figure 1**). About 70% of human genes are transcribed from both strands.
- *Coding DNA accounts for less than one-quarter of the highly conserved (and presumably functionally important) fraction of the genome.* This meant that there must be many more functionally important

noncoding DNA sequences than had been expected. And so it proved. An unexpectedly high number of conserved regulatory sequences and numerous novel noncoding RNAs (ncRNAs) have been, and continue to be, identified, often within or spanning introns of known protein-coding genes.

As a result of intensive analyses of mammalian transcriptomes, many thousands of transcripts of unknown function have been uncovered. Typically, ncRNA and protein-coding transcripts overlap, creating complicated patterns of transcription (**Figure 2**). In some cases, such as the imprinted *SNURF–SNRPN* transcript at 15q12, a single transcript contains a coding RNA plus a noncoding RNA that are separated by RNA cleavage.

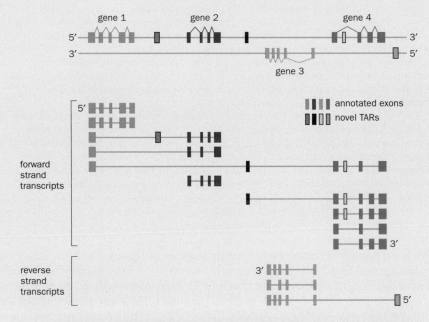

Figure 1 Blurring of gene boundaries at the transcript level. In the past, the four genes at the top would be expected to behave as discrete non-overlapping transcription units. As shown by recent analyses, the reality is more complicated. A variety of transcripts often links exons in neighboring genes. The transcripts frequently include sequences from previously unsuspected transcriptionally active regions (TARs). [From Gerstein MB, Bruce C, Rozowsky JS et al. (2007) *Genome Res.* 17, 669–681. With permission from Cold Spring Harbor Laboratory Press.]

Figure 2 Extensive transcriptional complexity of human genes. (A) Human genes are frequently transcribed on both strands, as shown in this hypothetical gene cluster. (B) A single gene can have multiple transcriptional start sites (right-angled arrows) as well as many interleaved coding and noncoding transcripts. Exons are shown as blue boxes. Known short RNAs such as small nucleolar RNAs (snoRNAs) and microRNAs (miRNAs) can be processed from intronic sequences, and novel species of short RNAs that cluster around the beginning and end of genes have recently been discovered (see the text). [From Gingeras TR (2007) *Genome Res.* 17, 682–690. With permission from Cold Spring Harbor Laboratory Press.]

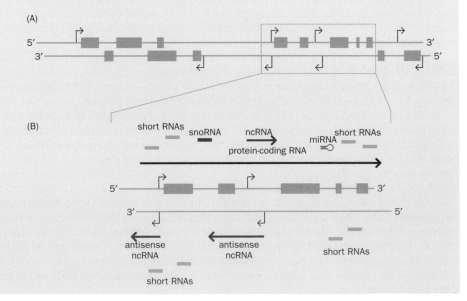

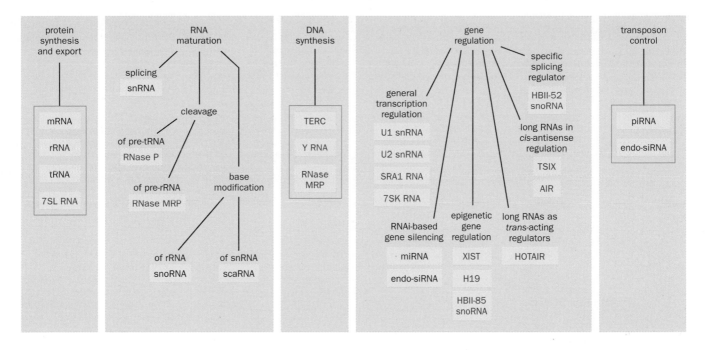

Figure 9.13 Functional diversity of RNA. Various ubiquitous RNAs function in housekeeping roles in cells, and in protein synthesis and protein export from cells (using 7SL RNA, the RNA component of the signal recognition particle). RNAs involved in RNA maturation include spliceosomal small nuclear RNAs (snRNAs), small nucleolar RNAs (snoRNAs), small Cajal body RNAs (scaRNAs), and two RNA ribonucleases. Telomere DNA synthesis is performed by a ribonucleoprotein that consists of TERC (the telomerase RNA component) and a reverse transcriptase (see Figure 2.13). The Y RNA family are involved in chromosomal DNA replication, and RNase MRP has a crucial role in initiating mtDNA replication as well as in cleaving pre-rRNA precursors in the nucleolus. Diverse classes of RNA serve a regulatory function in gene expression. Although some do have general accessory roles in transcription, regulatory RNAs are often restricted in their expression to certain cell types and/or developmental stages. Three classes of tiny RNA use RNA interference (RNAi) pathways to act as regulators: Piwi protein-interacting RNAs (piRNAs) regulate the activity of transposons in germ-line cells; microRNAs (miRNAs) regulate the expression of target genes; and endogenous short interfering RNAs (endo-siRNAs) act as gene regulators and also regulate some types of transposon. Some members of the snoRNA family, such as HBII-85 snoRNA, are involved in epigenetic gene regulation. A large number of long RNAs are involved in regulating various genes, often at the transcriptional level. Some are known to be involved in epigenetic gene regulation in imprinting, X-inactivation, and so on. RNA families are shown in black type; individual RNAs are shown in red.

Figure 9.13 gives a modern perspective on the functional diversity of RNA. In this section we consider the functions and gene organizations of the different human RNA classes (**Table 9.9**). Numerous databases have been developed recently to document data on ncRNAs (**Table 9.10**).

More than 1000 human genes encode an rRNA or tRNA, mostly within large gene clusters

Ribosomal RNA genes

In addition to the two mitochondrial rRNA molecules (12S and 16S rRNA), there are four types of cytoplasmic rRNA, three associated with the large ribosome subunit (28S, 5.8S, and 5S rRNAs) and one with the small ribosome subunit (18S rRNA). The 5S rRNA genes occur in small gene clusters, the largest being a cluster of 16 genes on chromosome 1q42, close to the telomere. Only a few 5S rRNA genes have been validated as functional, and there are many dispersed pseudogenes.

The 28S, 5.8S, and 18S rRNAs are encoded by a single multigenic transcription unit (see Figure 1.22) that is tandemly repeated to form megabase-sized *ribosomal DNA* arrays (about 30–40 tandem repeats, or roughly 100 rRNA genes) on the short arms of each of the acrocentric human chromosomes 13, 14, 15, 21, and 22. We do not know the precise gene numbers because the ribosomal DNA arrays were excluded from the Human Genome Project, as a result of the technical difficulties in obtaining unambiguous ordering of overlapping DNA clones for long regions composed of very similar tandem repeats.

TABLE 9.9 MAJOR CLASSES OF HUMAN NONCODING RNA

RNA class	Subclass or evolutionary/ functional subfamily	No. of different types	Function	Gene organization, biogenesis, etc.
Ribosomal RNA (rRNA), ~120–5000 nucleotides	12S rRNA, 16S rRNA	1 of each	components of mitochondrial ribosomes	cleaved from multigenic transcripts produced by H strand of mtDNA (Figure 9.3)
	5S rRNA, 5.8S rRNA, 18S rRNA, 28S rRNA	1 of each	components of cytoplasmic ribosomes	5S rRNA is encoded by multiple genes in various gene clusters; 5.8S, 18S, and 28S rRNA are cleaved from multigenic transcripts (Figure 1.22); the multigenic 5.8S–18S–28S transcription units are tandemly repeated on each of 13p,14p,15p, 21p, and 22p (= rDNA clusters)
Transfer RNA (tRNA), ~70–80 nucleotides	mitochondrial family	22	decode mitochondrial mRNA to make 13 proteins on mitochondrial ribosomes	single-copy genes. tRNAs are cleaved from multigenic mtDNA transcripts (Figure 9.3)
	cytoplasmic family	49	decode mRNA produced by nuclear genes (Figure 9.13)	700 tRNA genes and pseudogenes dispersed at multiple chromosomal locations with some large gene clusters
Small nuclear RNA (snRNA), ~60–360 nucleotides	spliceosomal family with subclasses Sm and Lsm (Table 9.10)	9	U1, U2, U4, U5, and U6 snRNAs process standard GU–AG introns (Figure 1.19); U4atac, U6atac, U11, and U12 snRNAs process rare AU–AC introns	about 200 spliceosomal snRNA genes are found at multiple locations but there are moderately large clusters of U1 and U2 snRNA genes; most are transcribed by RNA pol II
	non-spliceosomal snRNAs	several	U7 snRNA: 3′ processing of histone mRNA; 7SK RNA: general transcription regulator; Y RNA family: involved in chromosomal DNA replication and regulators of cell proliferation	mostly single-copy functional genes
Small nucleolar RNA (snoRNA), ~60–300 nucleotides	C/D box class (Figure 9.15A)	246	maturation of rRNA, mostly nucleotide site-specific 2′-O-ribose methylations	usually within introns of protein-coding genes; multiple chromosomal locations, but some genes are found in multiple copies in gene clusters (such as the HBII-52 and HBII-85 clusters—Figure 11.22)
	H/ACA class (Figure 9.15B)	94	maturation of rRNA by modifying uridines at specific positions to give pseudouridine	
Small Cajal body RNA (scaRNA)		25	maturation of certain snRNA classes in Cajal bodies (coiled bodies) in the nucleus	usually within introns of protein-coding genes
RNA ribonucleases, ~260–320 nucleotides		2	RNase P cleaves pre-tRNA in nucleus + mitochondria; RNase MRP cleaves rRNA in nucleolus and is involved in initiating mtDNA replication	single-copy genes
Miscellaneous small cytoplasmic RNAs, ~80–500 nucleotides	BC200	1	neural RNA that regulates dendritic protein biosynthesis; originated from Alu repeat	1 gene, *BCYRN1*, at 2p16
	7SL RNA	3	component of the signal recognition particle (SRP) that mediates insertion of secretory proteins into the lumen of the endoplasmic reticulum	three closely related genes clustered on 14q22
	TERC (telomerase RNA component)	1	component of telomerase, the ribonucleoprotein that synthesizes telomeric DNA, using TERC as a template (Figure 2.13)	single-copy gene at 3q26
	Vault RNA	3	components of cytoplasmic vault RNPs that have been thought to function in drug resistance	*VAULTRC1, VAULTRC2,* and *VAULTRC3* are clustered at 5q31 and share ~84% sequence identity
	Y RNA	4	components of the 60 kD Ro ribonucleoprotein, an important target of humoral autoimmune responses	*RNY1, RNY3, RNY4,* and *RNY5* are clustered at 7q36

TABLE 9.9 (*cont.*) MAJOR CLASSES OF HUMAN NONCODING RNA

RNA class	Subclass or evolutionary/ functional subfamily	No. of different types	Function	Gene organization, biogenesis, etc.
MicroRNA (miRNA), ~22 nucleotides	> 70 families of related miRNAs	~1000	multiple important roles in gene regulation, notably in development, and implicated in some cancers	see Figure 9.17 for examples of genome organization, and Figure 9.16 for how they are synthesized
Piwi-binding RNA (piRNA), ~24–31 nucleotides	89 individual clusters	> 15,000	often derived from repeats; expressed only in germ-line cells, where they limit excess transposon activity	89 large clusters distributed across the genome; individual clusters span from 10 kb to 75 kb with an average of 170 piRNAs per cluster
Endogenous short interfering RNA (endo-siRNA), ~21–22 nucleotides	many	probably more than 10,000[a]	often derived from pseudogenes, inverted repeats, etc.; involved in gene regulation in somatic cells and may also be involved in regulating some types of transposon	clusters at many locations in the genome
Long noncoding regulatory RNA, often > 1 kb	many	> 3000	involved in regulating gene expression; some are involved in monoallelic expression (X-inactivation, imprinting), and/or as antisense regulators (Table 9.11)	usually individual gene copies; transcripts often undergo capping, splicing, and polyadenylation but antisense regulatory RNAs are typically long transcripts that do not undergo splicing

[a]Based on extrapolation of studies in mouse cells.

Transfer RNA genes

The 22 different mitochondrial tRNAs are made by 22 tRNA genes in mtDNA. The Genomic tRNA Database lists over 500 human tRNA genes that make a cytoplasmic tRNA with a defined anticodon specificity. The genes can be classified into 49 families on the basis of anticodon specificity (Box 9.5). There is only a rough correlation of human tRNA gene number with amino acid frequency. For example, 30 tRNA genes specify the comparatively rare amino acid cysteine (which accounts for 2.25% of all amino acids in human proteins), but only 21 tRNA genes specify the more abundant proline (which has a frequency of 6.10%).

Although the tRNA genes seem to be dispersed throughout the human genome, more than half of human tRNA genes (273 out of 516) reside on either chromosome 6 (with many clustered in a 4 Mb region at 6p2) or chromosome 1. In addition, 18 of the 30 Cys tRNAs are found in a 0.5 Mb stretch of chromosome 7.

TABLE 9.10 MAJOR NONCODING RNA DATABASES

Database	Description	URL
NONCODE	integrated database of all ncRNAs except rRNA and tRNA	http://www.noncode.org
Noncoding RNA database	sequences and functions of noncoding transcripts	http://biobases.ibch.poznan.pl/ncRNA/
RNAdb	comprehensive mammalian noncoding RNA database	http://research.imb.uq.edu.au/rnadb/
Rfam	noncoding RNA families and sequence alignments	http://rfam.sanger.ac.uk/
antiCODE	natural antisense transcripts database	http://www.anticode.org
sno/scaRNAbase	small nucleolar RNAs and small Cajal body-specific RNAs	http://gene.fudan.sh.cn/snoRNAbase.nsf
snoRNA-LBME-db	comprehensive human snoRNAs	http://www-snorna.biotoul.fr/
Genomic tRNA Database	comprehensive tRNA sequences	http://lowelab.ucsc.edu/GtRNAdb/
Compilation of tRNA sequences and sequences of tRNA genes	just as its name suggests	http://www.tRNA.uni-bayreuth.de
miRBase	miRNA sequences and target genes	http://microrna.sanger.ac.uk/
piRNAbank	empirically known sequences and other related information on piRNAs reported in various organisms, including human, mouse, rat, and *Drosophila*	http://pirnabank.ibab.ac.in/

BOX 9.5 ANTICODON SPECIFICITY OF EUKARYOTIC CYTOPLASMIC tRNAs

There is no one-to-one correspondence between codons in cytoplasmic mRNA and the tRNA anticodons that recognize them. The 64 possible codons are shown in **Figure 1**, alongside the (unmodified) anticodons. Horizontal lines join codon–anticodon pairs. Alternative codons that differ in having a C or a U at the third base position can be recognized by a single anticodon (*third-base wobble*, shown by chevrons). There are three rules for decoding cytoplasmic mRNA codons:

- *Codons in two-codon boxes.* Those codons ending with U/C that encode a different amino acid from those ending with A/G are known as two-codon boxes. Here the U/C wobble position is typically decoded by a G at the 5′ base position in the tRNA anticodon. For example, at the top left for Phe, there is no tRNA with an **A**AA anticodon to match the UUU codon, but the **G**AA anticodon can recognize both UU**U** and UU**C** codons in the mRNA (see Figure 1).

- *Non-glycine codons in four-codon boxes.* Four-codon boxes are those in which U, C, A, and G in the third, wobble, position all encode the same amino acid. Here the U/C wobble position is decoded by a chemically modified adenosine, known as inosine, at the 5′ position in the anticodon (blue shaded boxes; see Figure 11.31 for the structure of inosine). Inosine can base pair with A, C, or U. For example, at the bottom left, the GU**U** and GU**C** codons of the four-codon valine box are decoded by a tRNA with an anticodon of **A**AC, which is no doubt modified to **I**AC. The IAC anticodon can recognize each of GU**U**, GU**C**, and GU**A**. To avoid possible translational misreading, tRNAs with inosine at the 5′ base of the anticodon cannot be used in two-codon boxes.

- *Glycine codons.* The four-codon glycine box provides the one exception to the rule above: GGU and GGC codons are decoded by a **G**CC anticodon, rather than the expected **I**CC anticodon.

Thus, only 16 anticodons are required to decode the 32 codons ending in a U/C. The minimum set of anticodons is therefore 45 (64 minus 3 stop codons, minus 16). On this basis, one would predict a total of 45 different classes of human tRNA. However, despite the generality of third-base wobble, three pairs of codons ending in U/C are served by two anticodons each (see Figure 1), and so there are an extra three tRNA classes. In addition, a specialized tRNA carries an anticodon to the codon UGA (which normally functions as a stop codon). At high selenium concentrations, this tRNA will very occasionally decode UGA to insert the 21st amino acid, selenocysteine, in a select group of selenoproteins. Thus, there are 45 + 3 + 1 = 49 different classes of human tRNA, encoded by several hundred genes (see Figure 1).

	codon	anticodon			codon	anticodon			codon	anticodon			codon	anticodon	
Phe	UUU	AAA	–	Ser	UCU	**A**GA	11	Tyr	UAU	AUA	1	Cys	UGU	ACA	–
	UUC	GAA	12		UCC	GGA	–		UAC	GUA	14		UGC	GCA	30
Leu	UUA	UAA	7		UCA	UGA	5	stop	UAA	UUA	–	stop	UGA	UCA	– (3)
	UUG	CAA	7		UCG	CGA	4	stop	UAG	CUA	–	Trp	UGG	CCA	9
Leu	CUU	**A**AG	12	Pro	CCU	**A**GG	10	His	CAU	AUG	–	Arg	CGU	**A**CG	7
	CUC	GAG	–		CCC	GGG	–		CAC	GUG	11		CGC	GCG	–
	CUA	UAG	3		CCA	UGG	7	Gln	CAA	UUG	11		CGA	UCG	6
	CUG	CAG	10		CCG	CGG	4		CAG	CUG	20		CGG	CCG	4
Ile	AUU	**A**AU	14	Thr	ACU	**A**GU	10	Asn	AAU	AUU	2	Ser	AGU	ACU	–
	AUC	GAU	3		ACC	GGU	–		AAC	GUU	32		AGC	GCU	8
	AUA	UAU	5		ACA	UGU	6	Lys	AAA	UUU	16	Arg	AGA	UCU	6
Met	AUG	CAU	20		ACG	CGU	6		AAG	CUU	17		AGG	CCU	5
Val	GUU	**A**AC	11	Ala	GCU	**A**GC	29	Asp	GAU	AUC	–	Gly	GGU	ACC	–
	GUC	GAC	–		GCC	GGC	–		GAC	GUC	19		GGC	GCC	15
	GUA	UAC	5		GCA	UGC	9	Glu	GAA	UUC	13		GGA	UCC	9
	GUG	CAC	16		GCG	CGC	5		GAG	CUC	13		GGG	CCC	7

Figure 1 Over 500 different human cytoplasmic tRNAs decode the 61 codons that specify the standard 20 amino acids. The relationships between the 64 possible codons (positioned next to amino acids on the left of the four major columns) and the corresponding anticodons (to the right of the four columns) are shown. The number next to each anticodon is the number of different human tRNAs that are documented in the Genomic tRNA Database (see Table 9.9) as carrying that anticodon. Note that 12 of the 61 anticodons that could recognize the codons that specify the standard 20 amino acids are not represented in the tRNAs (shown by dashes). This happens because of wobble at the third base position of most codons where the third base is U or C (exceptions are for codon pairs AU**U**/AU**C**, AA**U**/AA**C**, and UA**U**/UA**C**). The (3) indicated by an asterisk signifies that there are three different selenocysteine tRNAs with an anticodon that can recognize the codon UGA, which normally serves as a stop codon. The shaded adenines are most probably a modified form of adenine known as inosine, in which the amino group attached to carbon 6 is replaced by a C=O carbonyl group.

Dispersed gene families make various small nuclear RNAs that facilitate general gene expression

Various families of rather small RNA molecules (60–360 nucleotides long) are known to have a role in the nucleus in assisting general gene expression, mostly at the level of post-transcriptional processing. Initially, such RNAs were simply labeled as *small nuclear RNAs* (*snRNAs*) to distinguish them from pre-mRNA. Many of them were known to be uridine-rich and they were named accordingly (U2 snRNA, for example, does not honor a famous Irish rock band but simply indicates the second uridine-rich small nuclear RNA to be classified). Like rRNA,

TABLE 9.11 THE Sm AND Lsm CLASSES OF SPLICEOSOMAL snRNA

	Sm class[a]	Lsm class
Component of major spliceosome	U1, U2, U4, and U5 snRNAs	U6 snRNA
Component of minor spliceosome	U11, U12, U4atac, and U5 snRNAs	U6atac snRNA
Structure	see Figure 9.14A	see Figure 9.14B
Transcribed by	RNA polymerase II	RNA polymerase III
Bound core proteins	Sm proteins (SmB, SmD1, SmD2, SmD3, SmE, SmF, SmG)	seven Lsm proteins (LSM2–LSM8)
Location	synthesized in the nucleus and then exported to the cytoplasm, where they each associate with seven Sm proteins and undergo 5' and 3' end processing; then re-imported into the nucleus to undergo more RNA processing in Cajal bodies, before accumulating in speckles to perform spliceosomal function	never leave the nucleus; undergo maturation in the nucleolus and then accumulate in speckles to perform spliceosomal function
Site-specific nucleotide modification	performed by scaRNAs in the Cajal bodies of the nucleus	performed by snoRNAs in the nucleolus

[a]Although not a spliceosomal snRNA, U7 snRNA has a similar structure to the Sm class of snRNAs and five of its seven core proteins are identical to core proteins bound by the Sm snRNAs.

snRNA molecules bind various proteins and function as ribonucleoproteins (snRNPs).

Subsequently, various snRNAs, including some of the first to be classified, were found to be involved in post-transcriptional processing of rRNA precursors in the nucleolus; they were therefore re-classified as *small nucleolar RNAs* (*snoRNAs*), for example U3 and U8 snoRNAs. More recently, membership of the classes has been based on structural and functional classification.

A third group of small RNAs have been identified that resemble snoRNAs but are confined to Cajal bodies (also called *coiled bodies*), discrete nuclear structures in the nucleus that are closely associated with the maturation of snRNPs. They have been termed *small Cajal body RNAs* (*scaRNAs*). Hundreds of mostly dispersed human genes are devoted to making snRNA and snoRNA, and there are many hundreds of associated pseudogenes.

Spliceosomal small nuclear RNA (snRNA) genes

The nine human spliceosomal snRNAs vary in length from 106 to 186 nucleotides and bind a ring of seven core proteins. U1, U2, U4, U5, and U6 snRNAs operate within the major spliceosome to process conventional GU–AG introns (see Figure 1.19). U4atac, U6atac, U11, and U12 snRNAs form part of the minor spliceosome that excises rare AU–AC introns. Each of the spliceosomal snRNPs contains seven core proteins that are identical within a subclass and a unique set of snRNP-specific proteins. The Lsm subclass is made up of just U6 and U6atac snRNAs; the other spliceosomal snRNAs belong to the Sm subclass (**Table 9.11** and **Figure 9.14**).

More than 70 genes specify snRNAs used in the major spliceosome. They include 44 identified genes specifying U6 snRNA and 16 specifying U1 snRNA. There is some evidence for clustering. Multiple U2 snRNA genes are found at 17q21–q22 but the copy number varies; a cluster of about 30 U1 snRNA genes is located at 1p36.1.

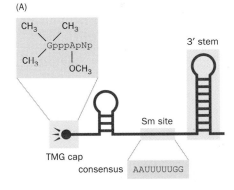

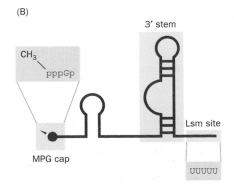

Figure 9.14 Structures of Sm-type and Lsm-type spliceosomal snRNAs.
(A) Sm-type snRNAs contain three important recognition elements: a 5'-trimethylguanosine (TMG) cap, an Sm-protein-binding site (Sm site), and a 3' stem–loop structure. The Sm site and the 3' stem elements are required for recognition by the survival motor neuron (SMN) complex for assembly into stable core ribonucleoproteins (RNPs). The consensus Sm site directs the assembly of a ring of the seven Sm core proteins (see Table 9.10). The TMG cap and the assembled Sm core proteins are required for recognition by the nuclear import machinery. (B) Lsm-type snRNAs contain a 5'-monomethylphosphate guanosine (MPG) cap and a 3' stem, and terminate in a stretch of uridine residues (the Lsm site) that is bound by the seven Lsm core proteins.

Non-spliceosomal small nuclear RNA genes

Not all snRNAs within the nucleoplasm function as part of spliceosomes. Both U1 and U2 snRNAs also have non-spliceosomal functions. U1 snRNA is required to stimulate transcription by RNA polymerase II. U2 snRNA is known to stimulate transcriptional elongation by RNA polymerase II. Several other small nuclear RNAs with a non-spliceosomal function have been well studied. They tend to be single-copy genes but there are many associated pseudogenes. Three examples are given below.

- U7 snRNA is a 63-nucleotide snRNA that is dedicated to the specialized 3′ processing undergone by histone mRNA which, exceptionally, is not polyadenylated.

- 7SK RNA is a 331-nucleotide RNA that functions as a negative regulator of the RNA polymerase II elongation factor P-TEFb.

- The Y RNA family consists of three small RNAs (less than 100 nucleotides long) that are involved in chromosomal DNA replication and function as regulators of cell proliferation.

Small nucleolar RNA (snoRNA) genes

SnoRNAs are between 60 and 300 nucleotides long and were initially identified in the nucleolus, where they guide nucleotide modifications in rRNA at specific positions. They do this by forming short duplexes with a sequence of the rRNA that contains the target nucleotide. There are two large subfamilies. H/ACA snoRNAs guide site-specific pseudouridylations (uridine is isomerized to give pseudouridine at 95 different positions in the pre-rRNA). C/D box snoRNAs guide site-specific 2′-O-ribose methylations (there are 105–107 varieties of this methylation in rRNA). Single snoRNAs specify one, or at most two, such base modifications (**Figure 9.15**).

At least 340 human snoRNA genes have been found so far, but there may be many more because snoRNAs are surprisingly difficult to identify with the use of bioinformatic approaches. The vast majority are found within the introns of larger genes that are transcribed by RNA polymerase II. These snoRNAs are produced by processing of the intronic RNA, and so the regulation of their synthesis is coupled to that of the host gene. Many snoRNA genes are dispersed single-copy genes. Others occur in clusters. For example, the large imprinted *SNURF–SNRPN* transcription unit at 15q12 contains six different types of C/D box snoRNA gene, two of which are present in large gene clusters: one contains about 45 almost identical HBII-52 snoRNA genes and the other 29 HBII-85 snoRNA genes (see Figure 11.22).

Most snoRNAs are ubiquitously expressed, but some are tissue-specific. For example, the six types of snoRNA gene within the *SNURF–SNRPN* transcription unit are predominantly expressed in the brain from only the paternal chromo-

Figure 9.15 Structure and function of snoRNAs. (A) C/D box snoRNAs guide 2′-O-methylation modifications. The box C and D motifs and a short 5′, 3′-terminal stem formed by intrastrand base pairing (shown as a series of short horizontal red lines) constitute a kink-turn structural motif that is specifically recognized by the 15.5 kD snoRNP protein. The C′ and D′ boxes represent internal, frequently imperfect copies of the C and D boxes. C/D box snoRNAs and their substrate RNAs form a 10–21 bp double helix in which the target residue to be methylated (shown here by the letter m in a circle) is positioned exactly five nucleotides upstream of the D or D′ box. R represents purine. (B) H/ACA box snoRNAs guide the conversion of uridines to pseudouridine. These RNAs fold into a hairpin–hinge–hairpin–tail structure. One or both of the hairpins contains an internal loop, called the pseudouridylation pocket, that forms two short (3–10 bp) duplexes with nucleotides flanking the unpaired substrate uridine (ψ) located about 15 nucleotides from the H or ACA box of the snoRNA. Although each box C/D and H/ACA snoRNA could potentially direct two modification reactions, apart from a few exceptions, most snoRNAs possess only one functional 2′-O-methylation or pseudouridylation domain.

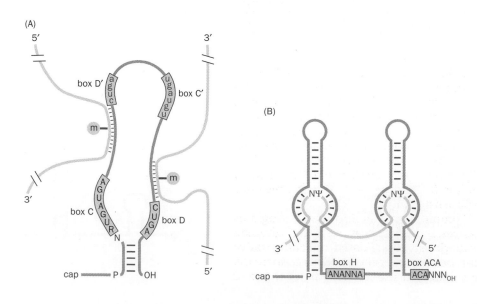

some 15. Nonstandard functions are known or expected for some snoRNA genes that do not have sequences complementary to rRNA sequences. For example, the HBII-52 snoRNA has an 18-nucleotide sequence that is perfectly complementary to a sequence within the *HTR2C* (serotonin receptor 2c) gene at Xp24, and regulates alternative splicing of this gene. The neighboring HBII-85 snoRNAs have recently been implicated in the pathogenesis of Prader–Willi syndrome (OMIM 176270).

Small Cajal body RNA genes

The scaRNAs resemble snoRNAs and perform a similar role in RNA maturation, but their targets are spliceosomal snRNAs and they perform site-specific modifications of spliceosomal snRNA precursors in the Cajal bodies of the nucleus. There are at least 25 human genes, each specifying one type of scaRNA. Like snoRNA genes, the scaRNA genes are typically located within the introns of genes transcribed by RNA polymerase II.

Close to 1000 different human microRNAs regulate complex sets of target genes by base pairing to the RNA transcripts

In addition to tRNA, we have known for some time about a variety of other moderately small (80–500-nucleotide) cytoplasmic RNAs. For example, the enzyme telomerase that synthesizes DNA at telomeres (see Figure 2.13) has both a protein component, TERT (telomerase reverse transcriptase), and also an RNA component, TERC, that is synthesized by a single-copy gene at 3q26.2 (see Table 9.8 for other examples). In the early 2000s it became clear that a novel family of tiny regulatory RNAs known as **microRNA** (**miRNA**) also operated in the cytoplasm.

MicroRNAs are only about 21–22 nucleotides long on average, and they were initially missed in analyses of the human genome. The first animal miRNAs to be reported were identified in model organisms such as the nematode (*C. elegans*) and the fruit fly (*Drosophila melanogaster*) by investigators studying phenomena relating to **RNA interference (Box 9.6)**, a natural form of gene regulation that protects cells from harmful propagation of viruses and transposons.

Because many miRNAs are strongly conserved during evolution, vertebrate miRNAs were quickly identified and the first human miRNAs were reported in the early 2000s. miRNAs regulate the expression of selected sets of target genes by base pairing with their transcripts. Usually, the binding sites are in the 3′ untranslated region of target mRNA sequences, and the bound miRNA inhibits translation so as to down-regulate expression of the target gene.

Synthesis of miRNAs involves the cleavage of RNA precursors by nucleus-specific and cytoplasm-specific RNase III ribonucleases, nucleases that specifically bind to and cleave double-stranded RNAs. The primary transcript, the *pri-miRNA*, has closely positioned inverted repeats that base-pair to form a hairpin RNA that is initially cleaved from the primary transcript by a nuclear RNase III (known as Rnasen or Drosha) to make a short double-stranded pre-miRNA that is transported out of the nucleus (**Figure 9.16**). A cytoplasmic RNase III called dicer cleaves the pre-miRNA to generate a miRNA duplex with overhanging 3′ dinucleotides.

A specific RNA-induced silencing complex (RISC) that contains the endoribonuclease argonaute binds the miRNA duplex and acts so as to unwind the double-stranded miRNA. The argonaute protein then degrades one of the RNA strands (the *passenger strand*) to leave the mature single-stranded miRNA (known as the *guide strand*) bound to argonaute. The mature miRNP associates with RNA transcripts that have sequences complementary to the guide strand. The binding of miRNA to target transcript normally involves a significant number of base mismatches. As a result, a typical miRNA can silence the expression of hundreds of target genes in much the same way that a tissue-specific protein transcription factor can affect the expression of multiple target genes at the same time—see the targets section of the miRBase database listed in Table 9.9.

To identify more miRNA genes, new computational bioinformatics programs were developed to screen genome sequences. By mid-2009, more than 700 human miRNA genes had been identified and experimentally validated, but comparative genomics analyses indicate that the number of such genes is likely to increase. Some of the miRNA genes have their own individual promoters; others are part of

BOX 9.6 RNA INTERFERENCE AS A CELL DEFENSE MECHANISM

RNA interference (RNAi) is an evolutionarily ancient mechanism that is used in animals, plants, and even single-celled fungi to protect cells against viruses and transposable elements. Both viruses and active transposable elements can produce long double-stranded RNA, at least transiently during their life cycles. Long double-stranded RNA is not normally found in cells and, for many organisms, it triggers an RNA interference pathway. A cytoplasmic endoribonuclease called dicer cuts the long RNA into a series of short double-stranded RNA pieces known as **short interfering RNA (siRNA)**. The siRNA produced is on average 21 bp long, but asymmetric cutting produces two-nucleotide overhangs at their 3′ ends (**Figure 1**).

The siRNA duplexes are bound by different complexes that contain an argonaute-type endoribonuclease (Ago) and some other proteins. Thereafter, the two RNA strands are unwound and one of the RNA strands is degraded by argonaute, leaving a single-stranded RNA bound to the argonaute complex. The argonaute complex is now activated; the single-stranded RNA will guide the argonaute complex to its target by base-pairing with complementary RNA sequences in the cells.

One type of argonaute complex is known as the *RNA-induced silencing complex (RISC)*. In this case, after the single-stranded guide RNA binds to a complementary long single-stranded RNA, the argonaute enzyme will cleave the RNA, causing it to be degraded. Viral and transposon RNA can be inactivated in this way.

Another class of argonaute complex is the *RNA induced transcriptional silencing (RITS)* complex. Here, the single-stranded RNA guide binds to complementary RNA transcripts as they emerge during transcription by RNA polymerase II. This allows the RITS complex to position itself on a specific part of the genome and then attract proteins such as histone methyltransferases (HMTs) and sometimes DNA methyltransferases (DNMTs), which covalently modify histones in the immediate region. This process eventually causes heterochromatin formation and spreading; in some cases, the RITS complex can induce DNA methylation. As a result, gene expression can be silenced over long periods to limit, for example, the activities of transposons.

Although mammalian cells have RNA interference pathways, the presence of double-stranded RNA triggers an interferon response that causes *nonspecific* gene silencing and cell death. This is described in Chapter 12 when we consider using RNA interference as an experimental tool to produce the specific silencing of pre-selected target genes. In such cases, artificially synthesized short double-stranded RNA is used to trigger RNAi-based gene silencing.

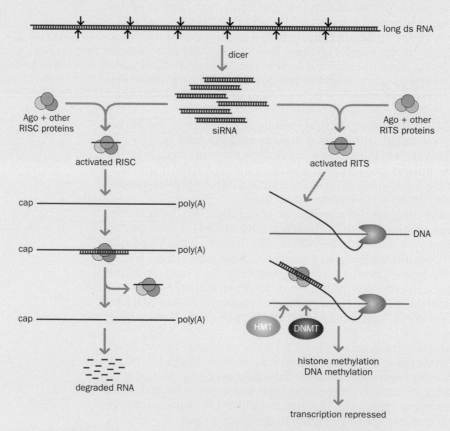

Figure 1 RNA interference. Long double-stranded (ds) RNA is cleaved by cytoplasmic dicer to give siRNA. siRNA duplexes are bound by argonaute complexes that unwind the duplex and degrade one strand to give an activated complex with a single RNA strand. By base pairing with complementary RNA sequences, the siRNA guides argonaute complexes to recognize target sequences. Activated RISC complexes cleave any RNA strand that is complementary to their bound siRNA. The cleaved RNA is rapidly degraded. Activated RITS complexes use their siRNA to bind to any newly synthesized complementary RNA and then attract proteins, such as histone methyltransferases (HMT) and sometimes DNA methyltransferases (DNMT), that can modify the chromatin to repress transcription.

a miRNA cluster and are cleaved from a common multi-miRNA transcription unit (**Figure 9.17A**). Another class of miRNA genes form part of a compound transcription unit that is dedicated to making other products in addition to miRNA, either another type of ncRNA (Figure 9.17B) or a protein (Figure 9.17C).

Many thousands of different piRNAs and endogenous siRNAs suppress transposition and regulate gene expression

The discovery of miRNAs was unexpected, but later it became clear that miRNAs represent a small component of what are a huge number of different tiny regulatory RNAs made in animal cells. In mammals, two additional classes of tiny regulatory RNA were first reported in 2006, and these are being intensively studied. Because huge numbers of different varieties of these RNAs are generated from multiple different locations in the genome, large-scale sequencing has been required to differentiate them.

Piwi-protein-interacting RNA

Piwi-protein-interacting RNAs (piRNAs) have been found in a wide variety of eukaryotes. They are expressed in germ-line cells in mammals and are typically 24–31 nucleotides long; they are thought to have a major role in limiting transposition by retrotransposons in mammalian germ-line cells, but they may also regulate gene expression in some organisms. Control of transposon activity is required because by integrating into new locations in the genome, active transposons can interfere with gene function, causing genetic diseases and cancer.

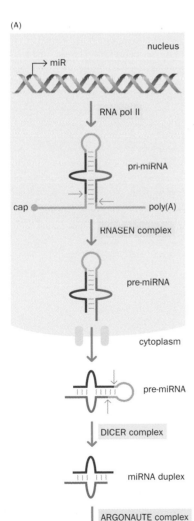

(A)

Figure 9.16 Human miRNA synthesis. (A) General scheme. The primary transcript, pri-miRNA, has a 5′ cap (m⁷GpppG) and a 3′ poly(A) tail. miRNA precursors have a prominent double-stranded RNA structure (RNA hairpin), and processing occurs through the actions of a series of ribonuclease complexes. In the nucleus, Rnasen, the human homolog of Drosha, cleaves the pri-miRNA to release the hairpin RNA (pre-miRNA); this is then exported to the cytoplasm, where it is cleaved by the enzyme dicer to produce a miRNA duplex. The duplex RNA is bound by an argonaute complex and the helix is unwound, whereupon one strand (the *passenger*) is degraded by the argonaute ribonuclease, leaving the mature miRNA (the *guide strand*) bound to argonaute. miR, miRNA gene. (B) A specific example: the synthesis of human miR-26a1. Inverted repeats (shown as highlighted sequences overlined by long arrows) in the pri-miRNA undergo base pairing to form a hairpin, usually with a few mismatches. The sequences that will form the mature guide strand are shown in red; those of the passenger strand are shown in blue. Cleavage by both the human Drosha and dicer (green arrows) is typically asymmetric, leaving an RNA duplex with overhanging 3′ dinucleotides.

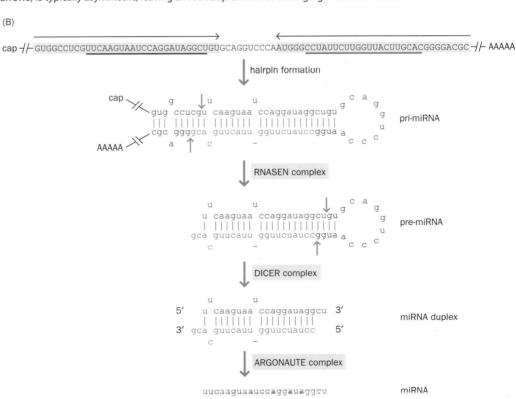

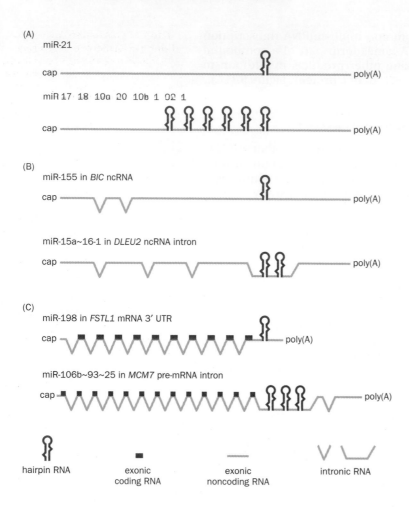

Figure 9.17 The structure of human pri-miRNAs. (A) Examples of transcripts that are used exclusively to make miRNAs: miR-21 is produced from a single hairpin within a dedicated primary transcript RNA; a single multigenic transcript with six hairpins that will eventually be cleaved to give six miRNAs, namely miR-17, miR-18, miR-19a, and so on. (B, C) Examples of miRNAs that are co-transcribed with a gene encoding either (B) a long noncoding RNA (ncRNA) or (C) a polypeptide. In each part, the upper example shows single miRNAs located within (B) an exon of an ncRNA (miR-155) and (C) in the 3′ untranslated region (UTR) within a terminal exon of an mRNA (miR-198). The lower examples show multiple miRNAs located within intronic sequences of (B) an ncRNA (miR-15a and miR-16-1) and (C) a pre-mRNA (miR-106b, miR-93, and miR-25). Cap, m^{7}G(5′)ppp(5′) G. [Adapted from Du T & Zamore PD (2005) *Development* 132, 4645–4652. With permission from the Company of Biologists.]

More than 15,000 different human piRNAs have been identified and so the piRNA family is among the most diverse RNA family in human cells (see Table 9.8). The piRNAs map back to 89 genomic intervals of about 10–75 kb long (for more information see the piRNAbank database; Table 9.9). They are thought to be cleaved from large multigenic transcripts. In humans, the multi-piRNA transcripts contain sequences for up to many hundreds of different piRNAs.

piRNAs are thought to repress transposition by transposons through an RNA interference pathway, by association with piwi proteins, which are evolutionarily related to argonaute proteins (**Figure 9.18**).

Endogenous siRNAs

Long double-stranded RNA in mammalian cells triggers nonspecific gene silencing through interferon pathways, but transfection of exogenous synthetic siRNA duplexes or short hairpin RNAs induces RNAi-mediated silencing of specific genes with sequence elements in common with the exogenous RNA. As we will see in Chapter 12, this is an extremely important experimental tool that can give valuable information on the cellular functions of a gene. Very recently, it has become clear that human cells also naturally produce *endogenous siRNAs* (*endo-siRNAs*).

In mammals, the most comprehensive endo-siRNA analyses have been performed in mouse oocytes. Like piRNAs, endo-siRNAs are among the most varied RNA population in the cell (many tens of thousands of different endo-siRNAs have been identified in mouse oocytes). They arise as a result of the production of limited amounts of natural double-stranded RNA in the cell. One way in which this happens involves the occasional transcription of some pseudogenes (**Figure 9.19**).

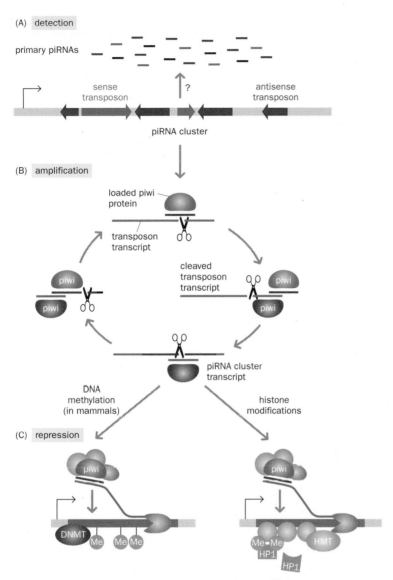

(A) detection

primary piRNAs

sense transposon   ?   antisense transposon

piRNA cluster

(B) amplification

loaded piwi protein

transposon transcript

cleaved transposon transcript

piwi

piwi

piwi

piwi

piRNA cluster transcript

DNA methylation (in mammals)

histone modifications

(C) repression

piwi

piwi

DNMT

Me Me Me

Me—Me

HP1

HP1

HMT

Figure 9.18 piRNA-based transposon silencing in animal cells. (A) Primary piRNAs (piwi-protein-interacting RNAs) are 24–31 nucleotides long and are processed from long RNA precursors transcribed from defined loci called piRNA clusters. Any transposon inserted in the reverse orientation in the piRNA cluster can give rise to antisense piRNAs (shown in red). (B) Antisense piRNAs are incorporated into a piwi protein and direct its slicer activity on sense transposon transcripts. The 3′ cleavage product is bound by another piwi protein and trimmed to piRNA size. This sense piRNA is, in turn, used to cleave piRNA cluster transcripts and to generate more antisense piRNAs. (C) Antisense piRNAs target the piwi complexes to cDNA for DNA methylation (left) and/or histone modification (right). DNMT, DNA methyltransferase; HMT, histone methyltransferase; HP1, heterochromatin protein 1. [From Girard A & Hannon GJ (2007) *Trends Cell Biol.* 18, 136–148. With permission from Elsevier.]

More than 3000 human genes synthesize a wide variety of medium-sized to large regulatory RNAs

Many thousands of different long ncRNAs, often many kilobases in length, are also thought to have regulatory roles in animal cells. They include antisense transcripts that usually do not undergo splicing and that can regulate overlapping sense transcripts, plus a wide variety of long mRNA-like ncRNAs that undergo capping, splicing, and polyadenylation but do not seem to encode any sizeable polypeptide, although some contain internal ncRNAs such as snoRNAs and piRNAs. The functions of the great majority of the mRNA-like ncRNAs are unknown. Some, however, are known to be tissue-specific and involved in gene regulation. Recently, in a systematic effort to identify long ncRNAs, 3300 different human long ncRNAs were identified as associating with chromatin-modifying complexes, thereby affecting gene expression.

Some long mRNA-like ncRNAs that are involved in epigenetic regulation have been extensively studied. The *XIST* gene encodes a long ncRNA that regulates X-chromosome inactivation, the process by which one of the two X chromosomes is randomly selected to be condensed in female mammals, with large regions becoming transcriptionally inactive. Many other long ncRNAs, such as the *H19* RNA, are implicated in repressing the transcription of either the paternal or maternal allele of many autosomal regions (*imprinting*). These mRNA-like ncRNAs are often regulated by genes that produce what can be very long antisense ncRNA transcripts that usually do not undergo splicing (**Table 9.12** gives examples).

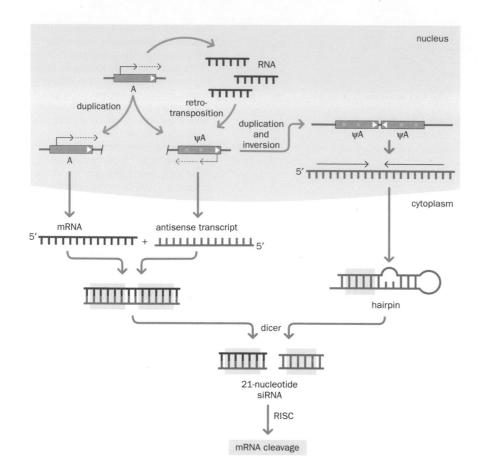

Figure 9.19 Pseudogenes can regulate the expression of their parent gene by endogenous siRNA pathways. Pseudogenes arise through the copying of a parent gene. Some pseudogenes are transcribed and, depending on the genomic context, can produce an RNA that is the antisense equivalent of the mRNA produced by the parent gene. An mRNA transcript of the parent gene (A) and an antisense transcript of a corresponding pseudogene (ψA) can then form a double-stranded RNA that is cleaved by dicer to give siRNA. Endogenous siRNAs can also be produced from duplicated inverted sequences such as the example shown here of an inverted duplication of the pseudogene (ψA ψA) at the right. Transcription through both copies of the pseudogene results in a long RNA with inverted repeats (blue, overlined arrows) causing the RNA to fold into a hairpin that is cleaved by dicer to give siRNA. In either case, the endogenous siRNAs are guided by RISC to interact with, and degrade, the parent gene's remaining mRNA transcripts. Green arrows indicate DNA rearrangements. [Adapted from Sasidharan R & Gerstein M (2008) *Nature* 453, 729–731. With permission from Macmillan Publishers Ltd.]

The pervasive involvement of long ncRNAs in regulating developmental processes is illustrated by a comprehensive (5 bp resolution) analysis of the transcriptional output at the four human *HOX* gene clusters. Although there are only 39 *HOX* genes, the transcriptional output of the *HOX* clusters was also found to include a total of 231 different long ncRNAs. Many of these are *cis*-acting regulators, but one of them, HOTAIR, was found to be a *trans*-acting regulator (see Table 9.12).

Some of the functional RNAs, such as XIST and AIR, have not been so well conserved during evolution. The fastest-evolving functional sequences in the human genome include components of primate-specific long ncRNAs that are strongly expressed in brain. We consider the evolutionary implications of such genes in Chapter 10.

TABLE 9.12 EXAMPLES OF LONG REGULATORY HUMAN RNAs				
RNA	**Size**	**Gene location**	**Gene organization**	**Function**
XIST	19.3 kb	Xq13	6 exons spanning 32 kb	regulator of X-chromosome inactivation
TSIX	37.0 kb	Xq13	1 exon	antisense regulator of XIST
H19	2.3 kb	11p15	5 exons spanning 2.67 kb	involved in imprinting at the 11p15 imprinted cluster associated with Beckwith–Wiedemann syndrome
KCNQTOT1 (= LIT1)	59.5 kb	11p15	1 exon	antisense regulator at the imprinted cluster at 11p15
PEG3	1.8 kb[a]	19q13	variable number of exons but up to 9 exons spanning 25 kb	maternally imprinted and known to function in tumor suppression by activating p53
HOTAIR	2.2 kb	12q13	6 exons spanning 6.3 kb	*trans*-acting gene regulator; although part of a regulatory region in the *HOX-C* cluster on 12q13, HOTAIR RNA represses transcription of a 40 kb region on the *HOX-D* cluster on chromosome 2q31

[a]Largest isoforms.

9.4   HIGHLY REPETITIVE DNA: HETEROCHROMATIN AND TRANSPOSON REPEATS

Genes contain some repetitive DNA sequences, including repetitive coding DNA. However, the majority of highly repetitive DNA sequences occur outside genes. Some of the sequences are present at certain subchromosomal regions as large arrays of tandem repeats. This type of DNA, known as heterochromatin, remains highly condensed throughout the cell cycle and does not generally contain genes.

Other highly repetitive DNA sequences are interspersed throughout the human genome and were derived by *duplicative transposition* (see Section 9.1). Sequences like this are sometimes described as *transposon repeats* and they account for more than 40% of the total DNA sequence in the human genome. In addition to residing in extragenic regions, they are often found in introns and untranslated sequences and sometimes even in coding sequences.

Constitutive heterochromatin is largely defined by long arrays of high-copy-number tandem DNA repeats

The DNA of constitutive heterochromatin accounts for 200 Mb or 6.5% of the human genome (see Table 9.3). It encompasses megabase regions at the centromeres and comparatively short lengths of DNA at the telomeres of all chromosomes. Most of the Y chromosome and most of the short arms of the acrocentric chromosomes (13, 14, 15, 21, and 22) consist of heterochromatin. In addition, there are very substantial heterochromatic regions close to the centromeres of certain chromosomes, notably chromomosomes 1, 9, 16, and 19.

The DNA of constitutive heterochromatin mostly consists of long arrays of high-copy-number tandemly repeated DNA sequences, known as *satellite DNA* (Table 9.13). Shorter arrays of tandem repeats are known as minisatellites and

TABLE 9.13 MAJOR CLASSES OF HIGH-COPY-NUMBER TANDEMLY REPEATED HUMAN DNA

Class[a]	Total array size unit	Size or sequence of repeat unit	Major chromosomal location(s)
Satellite DNA[b]	often hundreds of kilobases		associated with heterochromatin
α (alphoid DNA)		171 bp	centromeric heterochromatin of all chromosomes
β (*Sau*3A family)		68 bp	notably the centromeric heterochromatin of 1, 9, 13, 14, 15, 21, 22, and Y
Satellite 1		25–48 bp (AT-rich)	centromeric heterochromatin of most chromosomes and other heterochromatic regions
Satellite 2		diverged forms of ATTCC/GGAAT	most, possibly all, chromosomes
Satellite 3		ATTCC/GGAAT	13p, 14p, 15p, 21p, 22p, and heterochromatin on 1q, 9q, and Yq12
DYZ19		125 bp	~400 kb at Yq11
DYZ2		AT-rich	Yq12; higher periodicity of ~2470 bp
Minisatellite DNA	0.1–20 kb		at or close to telomeres of all chromosomes
Telomeric minisatellite		TTAGGG	all telomeres
Hypervariable minisatellites		9–64 bp	all chromosomes, associated with euchromatin, notably in sub-telomeric regions
Microsatellite DNA	< 100 bp	often 1–4 bp	widely dispersed throughout all chromosomes

[a]The distinction between satellite, minisatellite, and microsatellite is made on the basis of the total array length, not the size of the repeat unit.
[b]Satellite DNA arrays that consist of simple repeat units often have base compositions that are radically different from the average 41% G+C (and so could be isolated by buoyant density gradient centrifugation, when they would be differentiated from the main DNA and appear as *satellite bands*—hence the name).

microsatellites, respectively. Large tracts of heterochromatin are typically composed of a mosaic of different satellite DNA sequences that are occasionally interrupted by transposon repeats but are devoid of genes. Transposon repeats are also widely distributed in euchromatin and will be described below.

The vast majority of human heterochromatic DNA has not been sequenced, because of technical difficulties in obtaining unambiguous ordering of overlapping DNA clones. Thus, for example, only short representative components of centromeric DNA have been sequenced so far. However, Y-chromosome heterochromatin is an exception and has been well characterized. There are different satellite DNA organizations, and the repeated unit may be a very simple sequence (less than 10 nucleotides long) or a moderately complex one that can extend to over 100 nucleotides long; see Table 9.13.

At the sequence level, satellite DNA is often extremely poorly conserved between species. Its precise function remains unclear, although some human satellite DNAs are implicated in the function of centromeres whose DNA consists very largely of various families of satellite DNA.

The centromere is an epigenetically defined domain. Its function is independent of the underlying DNA sequence; instead, its function depends on its particular chromatin organization, which, once established, has to be stably maintained through multiple cell divisions. Of the various satellite DNA families associated with human centromeres, only the α-satellite is known to be present at all human centromeres, and its repeat units often contain a binding site for a specific centromere protein, CENPB. Cloned α-satellite arrays have been shown to seed *de novo* centromeres in human cells, indicating that α-satellite must have an important role in centromere function.

The specialized telomeric DNA consists of medium-sized arrays just a few kilobases long and constitutes a form of *minisatellite DNA*. Unlike satellite DNA, telomeric minisatellite DNA has been extraordinarily conserved during vertebrate evolution and has an integral role in telomere function. It consists of arrays of tandem repeats of the hexanucleotide TTAGGG that are synthesized by the telomerase ribonucleoprotein (see Figure 2.13).

Transposon-derived repeats make up more than 40% of the human genome and arose mostly through RNA intermediates

Almost all of the interspersed repetitive noncoding DNA in the human genome is derived from **transposons** (also called *transposable elements*), mobile DNA sequences that can migrate to different regions of the genome. Close to 45% of the genome can be recognized as belonging to this class, but much of the remaining unique DNA must also be derived from ancient transposon copies that have diverged extensively over long evolutionary time-scales.

In humans and other mammals there are four major classes of transposon repeat, but only a tiny minority of transposon repeats are actively transposing. According to the method of transposition, the repeats can be organized into two groups:

- *Retrotransposons* (also abbreviated to *retroposons*). Here the copying mechanism resembles the way in which processed pseudogenes and retrogenes are generated (see Figure 9.12): reverse transcriptase converts an RNA transcript of the retrotransposon into a cDNA copy that then integrates into the genomic DNAs at a different location. Three major mammalian transposon classes use this copy-and-paste mechanism: long interspersed nuclear elements (LINEs), short interspersed nuclear elements (SINEs), and retrovirus-like elements containing long terminal repeats.

- *DNA transposons.* Members of this fourth class of transposon migrate directly without any copying of the sequence; the sequence is excised and then re-inserted elsewhere in the genome (a cut-and-paste mechanism).

Transposable elements that can transpose independently are described as *autonomous*; those that cannot are known as *nonautonomous* (**Figure 9.20**). Of the four classes of transposable element, LINEs and SINEs predominate; we describe them more fully below. The other two classes are briefly described here.

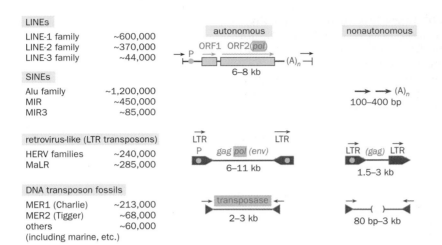

Figure 9.20 Mammalian transposon families. Only a small proportion of members of any of the illustrated transposon families may be capable of transposing; many have lost such a capacity after acquiring inactivating mutations, and many are short truncated copies. Subclasses of the four main families are listed, along with sizes in base pairs. ORF, open reading frame. [Adapted from International Human Genome Sequencing Consortium (2001) *Nature* 409, 860–921. With permission from Macmillan Publishers Ltd.]

Human LTR transposons

LTR transposons include autonomous and nonautonomous retrovirus-like elements that are flanked by long terminal repeats (LTRs) containing necessary transcriptional regulatory elements. *Endogenous retroviral sequences* contain *gag* and *pol* genes, which encode a protease, reverse transcriptase, RNAse H, and integrase. They are thus able to transpose independently. There are three major classes of human endogenous retroviral sequence (HERV), with a cumulative copy number of about 240,000, accounting for a total of about 4.6% of the human genome (see Figure 9.20).

Very many HERVs are defective, and transposition has been extremely rare during the last several million years. However, the very small HERV-K group shows conservation of intact retroviral genes, and some members of the HERV-K10 subfamily have undergone transposition comparatively recently during evolution. Nonautonomous retrovirus-like elements lack the *pol* gene and often also the *gag* gene (the internal sequence having been lost by homologous recombination between the flanking LTRs). The MaLR family of such elements accounts for almost 4% or so of the genome.

Human DNA transposon fossils

DNA transposons have terminal inverted repeats and encode a transposase that regulates transposition. They account for close to 3% of the human genome and can be grouped into different classes that can be subdivided into many families with independent origins (see the Repbase database of repeat sequences at http://www.girinst.org/repbase/index.html). There are two major human families, MER1 and MER2, plus a variety of less frequent families (see Figure 9.20).

Virtually all the resident human DNA transposon sequences are no longer active; they are therefore transposon fossils. DNA transposons tend to have short lifespans within a species, unlike some of the other transposable elements such as LINEs. However, quite a few functional human genes seem to have originated from DNA transposons, notably genes encoding the RAG1 and RAG2 recombinases and the major centromere-binding protein CENPB.

A few human LINE-1 elements are active transposons and enable the transposition of other types of DNA sequence

LINEs (long interspersed nuclear elements) have been very successful transposons. They have a comparatively long evolutionary history, occurring in other mammals, including mice. As autonomous transposons, they can make all the products needed for retrotransposition, including the essential reverse transcriptase. Human LINEs consist of three distantly related families: LINE-1, LINE-2, and LINE-3, collectively comprising about 20% of the genome (see Figure 9.20). They are located primarily in euchromatic regions and are located preferentially in the dark AT-rich G bands (Giemsa-positive) of metaphase chromosomes.

Of the three human LINE families, LINE-1 (or L1) is the only family that continues to have actively transposing members. LINE-1 is the most important human transposable element and accounts for a higher fraction of genomic DNA (17%) than any other class of sequence in the genome.

Full-length LINE-1 elements are more than 6 kb long and encode two proteins: an RNA-binding protein and a protein with both endonuclease and reverse transcriptase activities (**Figure 9.21A**). Unusually, an internal promoter is located within the 5′ untranslated region. Full-length copies therefore bring with them their own promoter that can be used after integration in a permissive region of the genome. After translation, the LINE-1 RNA assembles with its own encoded proteins and moves to the nucleus.

To integrate into genomic DNA, the LINE-1 endonuclease cuts a DNA duplex on one strand, leaving a free 3′ OH group that serves as a primer for reverse transcription from the 3′ end of the LINE RNA. The endonuclease's preferred cleavage site is TTTT↓A; hence the preference for integrating into AT-rich regions. AT-rich DNA is comparatively gene-poor, and so because LINEs tend to integrate into AT-rich DNA they impose a lower mutational burden, making it easier for their host to accommodate them. During integration, the reverse transcription often fails to proceed to the 5′ end, resulting in truncated, nonfunctional insertions. Accordingly, most LINE-derived repeats are short, with an average size of 900 bp for all LINE-1 copies, and only about 1 in 100 copies are full length.

The LINE-1 machinery is responsible for most of the reverse transcription in the genome, allowing retrotransposition of the nonautonomous SINEs and also of copies of mRNA, giving rise to processed pseudogenes and retrogenes. Of the 6000 or so full-length LINE-1 sequences, about 80–100 are still capable of transposing, and they occasionally cause disease by disrupting gene function after insertion into an important conserved sequence.

Alu repeats are the most numerous human DNA elements and originated as copies of 7SL RNA

SINEs (short interspersed nuclear elements) are retrotransposons about 100–400 bp in length. They have been very successful in colonizing mammalian genomes, resulting in various interspersed DNA families, some with extremely high copy numbers. Unlike LINEs, SINEs do not encode any proteins and they cannot transpose independently. However, SINEs and LINEs share sequences at their 3′ end, and SINEs have been shown to be mobilized by neighboring LINEs. By parasitizing on the LINE element transposition machinery, SINEs can attain very high copy numbers.

The human Alu family is the most prominent SINE family in terms of copy number, and is the most abundant sequence in the human genome, occurring on average more than once every 3 kb. The full-length Alu repeat is about 280 bp long and consists of two tandem repeats, each about 120 bp in length followed by a short A_n/T_n sequence. The tandem repeats are asymmetric: one contains an internal 32 bp sequence that is lacking in the other (Figure 9.21B). Monomers,

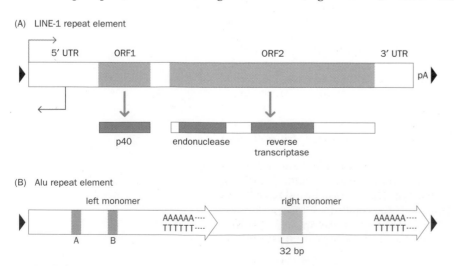

(A) LINE-1 repeat element

(B) Alu repeat element

Figure 9.21 The human LINE-1 and Alu repeat elements. (A) The 6.1 kb LINE-1 element has two open reading frames: ORF1, a 1 kb open reading frame, encodes p40, an RNA-binding protein that has a nucleic acid chaperone activity; the 4 kb ORF2 specifies a protein with both endonuclease and reverse transcriptase activities. A bidirectional internal promoter lies within the 5′ untranslated region (UTR). At the other end, there is an A_n/T_n sequence, often described as the 3′ poly(A) tail (pA). The LINE-1 endonuclease cuts one strand of a DNA duplex, preferably within the sequence TTTT↓A, and the reverse transcriptase uses the released 3′-OH end to prime cDNA synthesis. New insertion sites are flanked by a small target site duplication of 2–20 bp (flanking black arrowheads). (B) An Alu dimer. The two monomers have similar sequences that terminate in an A_n/T_n sequence but differ in size because of the insertion of a 32 bp element within the larger repeat. Alu monomers also exist in the human genome, as do various truncated copies of both monomers and dimers.

containing only one of the two tandem repeats, and various truncated versions of dimers and monomers are also common, giving a genomewide average of 230 bp.

Whereas SINEs such as the MIR (mammalian-wide interspersed repeat) families are found in a wide range of mammals, the Alu family is of comparatively recent evolutionary origin and is found only in primates. However, Alu subfamilies of different evolutionary ages can be identified. In the past 5 million or so years since the divergence of humans and African apes, only about 5000 copies of the Alu repeat have undergone transposition; the most mobile Alu sequences are members of the Y and S subfamilies.

Like other mammalian SINEs, Alu repeats originated from cDNA copies of small RNAs transcribed by RNA polymerase III. Genes transcribed by RNA polymerase III often have internal promoters, and so cDNA copies of transcripts carry with them their own promoter sequences. Both the Alu repeat and, independently, the mouse B1 repeat originated from cDNA copies of 7SL RNA, the short RNA that is a component of the signal recognition particle, using a retrotransposition mechanism like that shown in Figure 9.12. Other SINEs, such as the mouse B2 repeat, are retrotransposed copies of tRNA sequences.

Alu repeats have a relatively high GC content and, although dispersed mainly throughout the euchromatic regions of the genome, are preferentially located in the GC-rich and gene-rich R chromosome bands, in striking contrast to the preferential location of LINEs in AT-rich DNA. However, when located within genes they are, like LINE-1 elements, confined to introns and the untranslated regions. Despite the tendency to be located in GC-rich DNA, newly transposing Alu repeats show a preference for AT-rich DNA, but progressively older Alu repeats show a progressively stronger bias toward GC-rich DNA.

The bias in the overall distribution of Alu repeats toward GC-rich and, accordingly, gene-rich regions must result from strong selection pressure. It suggests that Alu repeats are not just genome parasites but are making a useful contribution to cells containing them. Some Alu sequences are known to be actively transcribed and may have been recruited to a useful function. The *BCYRN1* gene, which encodes the BC200 neural cytoplasmic RNA, arose from an Alu monomer and is one of the few Alu sequences that are transcriptionally active under normal circumstances. In addition, the Alu repeat has recently been shown to act as a *trans*-acting transcriptional repressor during the cellular heat shock response.

CONCLUSION

In this chapter, we have looked at the architecture of the human genome. Each human cell contains many copies of a small, circular mitochondrial genome and just one copy of the much larger nuclear genome. Whereas the mitochondrial genome bears some similarities to the compact genomes of prokaryotes, the human nuclear genome is much more complex in its organization, with only 1.1% of the genome encoding proteins and 95% comprising nonconserved, and often highly repetitive, DNA sequences.

Sequencing of the human genome has revealed that, contrary to expectation, there are comparatively few protein-coding genes—about 20,000–21,000 according to the most recent estimates. These genes vary widely in size and internal organization, with the coding exons often separated by large introns, which often contain highly repetitive DNA sequences. The distribution of genes across the genome is uneven, with some functionally and structurally related genes found in clusters, suggesting that they arose by duplication of individual genes or larger segments of DNA. Pseudogenes can be formed when a gene is duplicated and then one of the pair accumulates deleterious mutations, preventing its expression. Other pseudogenes arise when an RNA transcript is reverse transcribed and the cDNA is re-inserted into the genome.

The biggest surprise of the post-genome era is the number and variety of non-protein-coding RNAs transcribed from the human genome. At least 85% of the euchromatic genome is now known to be transcribed. The familiar ncRNAs known to have a role in protein synthesis have been joined by others that have roles in gene regulation, including several prolific classes of tiny regulatory RNAs and thousands of different long ncRNAs. Our traditional view of the genome is being radically revised.

In Chapter 10 we describe how the human genome compares with other genomes, and how evolution has shaped it. Aspects of human gene expression are elaborated in Chapter 11. Within Chapter 13 we also consider human genome variation.

FURTHER READING

Human mitochondrial genome

Anderson S, Bankier AT, Barrell BG et al. (1981) Sequence and organization of the human mitochondrial genome. *Nature* 290, 457–465.

Chen XJ & Butow RA (2005) The organization and inheritance of the mitochondrial genome. *Nat. Rev. Genet.* 6, 815–825.

Falkenberg M, Larsson NG & Gustafsson CM (2007) DNA replication and transcription in mammalian mitochondria. *Annu. Rev. Biochem.* 76, 679–699.

MITOMAP: human mitochondrial genome database. http://www .mitomap.org

Wallace DC (2007) Why do we still have a maternally inherited mitochondrial DNA? Insights from evolutionary medicine. *Annu. Rev. Biochem.* 76, 781–821.

Human nuclear genome

Clamp M, Fry B, Kamal M et al. (2007) Distinguishing protein-coding and noncoding genes in the human genome. *Proc. Natl Acad. Sci. USA* 104, 19428–19433.

Ensembl human database. http://www.ensembl.org/Homo _sapiens/index.html

GeneCards human gene database. http://www.genecards.org

International Human Genome Sequencing Consortium (2001) Initial sequencing and analysis of the human genome. *Nature* 409, 860–921.

International Human Genome Sequencing Consortium (2004) Finishing the euchromatic sequence of the human genome. *Nature* 431, 931–945.

Nature Collections: Human Genome Supplement, 1 June 2006 issue. [A collation that includes papers analyzing the sequence of each chromosome plus reprints of the papers reporting the 2001 draft sequence and the 2004 finished euchromatic sequence, available electronically at http://www.nature.com/nature/supplements/collections /humangenome/]

NCBI Human Genome Resources. http://www.ncbi.nlm.nih.gov /projects/genome/guide/human/

UCSC Genome Browser, Human (*Homo sapiens*) Genome Browser Gateway. http://genome.ucsc.edu/cgi-bin/hgGateway

Organization of protein-coding genes

Adachi N & Lieber MR (2002) Bidirectional gene organization: a common architectural feature of the human genome. *Cell* 109, 807–809.

Li YY, Yu H, Guo ZM et al. (2006) Systematic analysis of head-to-head gene organization: evolutionary conservation and potential biological relevance. *PLoS Comput. Biol.* 2, e74.

Sanna CR, Li W-H & Zhang L (2008) Overlapping genes in the human and mouse genomes. *BMC Genomics* 9, 169.

Soldà G, Suyama M, Pelucchi P et al. (2008) Non-random retention of protein-coding overlapping genes in Metazoa. *BMC Genomics* 9, 174.

Gene duplication, segmental duplication, and copy-number variation

Bailey JA, Gu Z, Clark RA et al. (2002) Recent segmental duplications in the human genome. *Science* 297, 1003–1007.

Conrad B & Antonarakis SE (2007) Gene duplication: a drive for phenotypic diversity and cause of human disease. *Annu. Rev. Genomics Hum. Genet.* 8, 17–35.

Kaessmann H, Vinckenbosch N & Long M (2009) RNA-based gene duplication: mechanistic and evolutionary insights. *Nat. Rev. Genet.* 10, 19–31.

Linardopoulou EV, Williams EM, Fan Y et al. (2005) Human subtelomeres are hot spots of interchromosomal recombination and segmental duplication. *Nature* 437, 94–100.

Redon R, Ishikawa S, Fitch KR et al. (2006) Global variation in copy number in the human genome. *Nature* 444, 444–454.

Tuzun E, Sharp AJ, Bailey JA et al. (2005) Fine-scale structural variation of the human genome. *Nat. Genet.* 37, 727–732.

The complexity of the mammalian transcriptome and the need to redefine genes in the post-genome sequencing era

Gerstein MB, Bruce C, Rozowsky JS et al. (2007) What is a gene, post-ENCODE? History and updated definition. *Genome Res.* 17, 669–681.

Gingeras T (2007) Origin of phenotypes: genes and transcripts. *Genome Res.* 17, 682–690.

Jacquier A (2009) The complex eukaryotic transcriptome: unexpected pervasive transcription and novel small RNAs. *Nat. Rev. Genet.* 10, 833–844.

Kapranov P, Cheng J, Dike S et al. (2007) RNA maps reveal new RNA classes and a possible function for pervasive transcription. *Science* 316, 1484–1488.

General reviews of noncoding RNA

Amaral PP, Dinger ME, Mercer TR & Mattick JS (2008) The eukaryotic genome as an RNA machine. *Science* 319, 1787–1789.

Carninci P, Yasuda J & Hayashizaki Y (2008) Multifaceted mammalian transcriptome. *Curr. Opin. Cell Biol.* 20, 274–280.

Griffiths-Jones S (2007) Annotating non-coding RNA genes. *Annu. Rev. Genomics Hum. Genet.* 8, 279–298.

Marakova JA & Kramerov DA (2007) Non-coding RNAs. *Biochemistry (Moscow)* 72, 1161–1178.

Mattick JS (2009) The genetic signatures of noncoding RNAs. *PLoS Genet.* 5, e1000459.

Prasanth KV & Spector DL (2007) Eukaryotic regulatory RNAs: an answer to the genome complexity conundrum. *Genes Dev.* 21, 11–42.

Small nuclear RNA and small nucleolar RNAs

Kishore S & Stamm S (2006) The snoRNA HBII-52 regulates alternative splicing of the serotonin receptor 2C. *Science* 311, 230–232.

Matera AG, Terns RM & Terns MP (2007) Non-coding RNAs: lessons from the small nuclear and small nucleolar RNAs. *Nat. Rev. Mol. Cell Biol.* 8, 209–220.

Sahoo T, del Gaudio D, German JR et al. (2008) Prader–Willi phenotype caused by paternal deficiency for the HBII-85 C/D box small nuclear RNA cluster. *Nat. Genet.* 40, 719–721.

MicroRNAs and ncRNAs as developmental regulators

Bushati N & Cohen SM (2007) microRNA functions. *Annu. Rev. Cell Dev. Biol.* 23, 175–205.

Chang T-C & Mendell JT (2007) microRNAs in vertebrate physiology and disease. *Annu. Rev. Genomics Hum. Genet.* 8, 215–239.

Makeyev EV & Maniatis T (2008) Multilevel regulation of gene expression by microRNAs. *Science* 319, 1789–1790.

Rinn JL, Kertesz M, Wang JK et al. (2007) Functional demarcation of active and silent chromatin domains in human *HOX* loci by non-coding RNAs. *Cell* 129, 1311–1323.

Stefani G & Slack F (2008) Small non-coding RNAs in animal development. *Nat. Rev. Mol. Cell Biol.* 9, 219–230.

piRNAs and endogenous siRNAs

Aravin AA, Sachidanandam R, Girard A et al. (2007) Developmentally regulated piRNA clusters implicate MILI in transposon control. *Science* 316, 744–747.

Girard A & Hannon GJ (2007) Conserved themes in small RNA-mediated transposon control. *Trends Cell Biol.* 18, 136–148.

Tam OH, Aravin AA, Stein P et al. (2008) Pseudogene-derived small interfering RNAs regulate gene expression in mouse oocytes. *Nature* 453, 534–538.

Watanabe T, Totoki Y, Toyoda A et al. (2008) Endogenous siRNAs from naturally formed dsRNAs regulate transcripts in mouse oocytes. *Nature* 453, 539–543.

Antisense and long noncoding regulatory RNAs

He Y, Vogelstein B, Velculescu VE et al. (2008) The antisense transcriptomes of human cells. *Science* 322, 1855–1858.

Khalil AM, Guttman M, Huarte M et al. (2009) Many human large intergenic noncoding RNAs associate with chromatin-modifying complexes and affect gene expression. *Proc. Natl Acad. Sci. USA* 106, 11667–11672.

Ponting CP, Oliver PL & Reik W (2009) Evolution and function of long noncoding RNAs. *Cell* 136, 629–641.

Wilusz JE, Sunwoo H & Spector DL (2009) Long noncoding RNAs: functional surprises from the RNA world. *Genes Dev.* 23, 1494–1504.

Promoter- and termini-associated RNAs

Affymetrix/Cold Spring Harbor Laboratory ENCODE Transcriptome Project (2009) Post-transcriptional processing generates a diversity of 5′-modified long and short RNAs. *Nature* 457, 1028–1042.

Pseudogenes and retrogenes

D'Errico L, Gadaleta G & Saccone C (2004) Pseudogenes in metazoa: origins and features. *Brief. Funct. Genomic. Proteomic.* 3, 157–167.

Duret L, Chureau C, Samain S et al. (2006) The *Xist* RNA gene evolved in eutherians by pseudogenization of a protein-coding gene. *Science* 312, 1653–1655.

Sasidharan R & Gerstein M (2008) Protein fossils live on as RNA. *Nature* 453, 729–731.

Zhang D & Gerstein MB (2004) Large-scale analysis of pseudogenes in the human genome. *Curr. Opin. Genet. Dev.* 14, 328–335.

Zheng D & Gerstein MB (2007) The ambiguous boundary between genes and pseudogenes: the dead rise up, or do they? *Trends Genet.* 23, 219–224.

Heterochromatin and transposon-based repeats

Choo KH, Vissel B, Nagy A et al. (1991) A survey of the genomic distribution of alpha satellite DNA on all the human chromosomes, and derivation of a new consensus sequence. *Nucleic Acids Res.* 19, 1179–1182.

Faulkner GJ, Kimura Y, Daub CO et al. (2009) The regulated retrotransposon transcriptome of mammalian cells. *Nat. Genet.* 41, 563–571.

Henikoff S, Ahmad K & Malik HS (2001) The centromere paradox: stable inheritance with rapidly evolving DNA. *Science* 293, 1098–1102.

Mariner PD, Walters RD, Espinoza CA et al. (2008) Human Alu RNA is a modular transacting repressor of mRNA transcription during heat shock. *Mol. Cell* 29, 499–509.

Mills RE, Bennett EA, Iskow RC & Devine SE (2007) Which transposable elements are active in the human genome? *Trends Genet.* 23, 183–191.

Muotri AR, Marchetto MCN, Coufal NG & Gage FH (2007) The necessary junk: new functions for transposable elements. *Hum. Mol. Genet.* 16, R159–R167.

Repbase: database of repeat sequences. http://www.girinst.org/repbase/index.html

Wicker T, Sabot F, Hua-Van A et al. (2007) A unified classification system for eukaryotic transposable elements. *Nat. Rev. Genet.* 8, 973–983.

Yang N & Kazazian HH Jr (2006) L1 retrotransposition is suppressed by endogenously encoded small interfering RNAs in human cultured cells. *Nat. Struct. Mol. Biol.* 13, 763–771.

Chapter 10

Model Organisms, Comparative Genomics, and Evolution

10

KEY CONCEPTS

- Model organisms have long been used for understanding facets of basic biology and for applied research, with important roles in modeling disease and helping evaluate potential therapies. Genome sequences for many model organisms are now available.

- Sequences from different genomes can be aligned. Comparing genome sequences provides genomewide information on the relatedness of DNA sequences and inferred protein sequences, and gives an insight into the processes shaping genome organization and genome evolution.

- Many functionally important sequences are subject to purifying (negative) selection whereby harmful mutations are selectively eliminated; these sequences are said to show evolutionary constraint and seem to be strongly conserved.

- Some mutations in functionally important sequences are subject to positive selection because they are beneficial. Such sequences underlie adaptive evolution.

- Neutral mutations, which are neither harmful nor beneficial, can become fixed in the population through random genetic drift.

- Evolutionary novelty often arises through duplicated genes; after duplication, genes can diverge in sequence and ultimately in function.

- Genes that originated by duplication of a gene within a single genome are known as paralogs; orthologs are homologous genes in different species that descended from the same gene in a common ancestor.

- One or both members of a duplicated gene pair can acquire a new function. Changes in gene function can be due to mutations in the *cis*-regulatory elements that regulate the gene's expression.

- Comparative genomics has contributed to the assessment of evolutionary relationships between organisms (phylogenetics).

- Morphological evolution is largely driven by mutations in *cis*-regulatory elements rather than differences in protein sequences.

- Fragments from transposable elements can be co-opted to make new functionally important sequences (exaptation); they can give rise to new exons (exonization) or to new *cis*-regulatory elements.

- Closely related species such as humans and chimpanzees have almost identical gene repertoires, but there are important differences between them in copy number of some gene families, in differential inactivation or modification of some orthologs, and in regulation of gene expression.

Now that we have the sequence of the human genome, biologists ultimately hope to get detailed answers to two big questions. First, at the molecular level, how does genome information program the normal functioning of cells and organisms? The corollary to this, of course, is: How does genome information become altered or misinterpreted to lead to hereditary diseases? Second, what makes individuals and species different? Getting detailed answers to these two fundamental questions will take a long time.

Obtaining genome sequences seemed to be an extremely arduous enterprise in the pre-genome era of the 1990s and early 2000s. In the post-genome era, as next-generation sequencing technologies become routine, we will soon be inundated with genome sequences from a vast array of organisms and from multiple individuals within species. With hindsight, we can now recognize that genome sequencing was the easy part, and just the beginning of a fascinating voyage of discovery.

To start to answer the question of how our own genes operate, we need to identify all the genes and gene products, all the regulatory elements, and all the other functional elements. In Chapter 12, we look at how gene function and gene regulation are studied in cells and model organisms. Genome sequences also give us the opportunity to compare representative individuals within and between species. In Chapter 13, we consider variation between individual humans. In this chapter, we address variation between species. We will consider how humans relate to other organisms in the tree of life and the forces that have shaped genome evolution.

Sequence comparisons across whole genomes from different organisms are now the mainstay of a new discipline, comparative genomics, and they are beginning to provide powerful new insights into our relationship to other organisms. We have a major interest in a variety of model organisms from both basic and applied research perspectives.

10.1 MODEL ORGANISMS

Our planet teems with countless organisms, but only a very few have been studied in the laboratory. Certain species that are amenable to experimental investigation have been particularly well investigated. A major motivation has been to increase our basic knowledge of cells and organisms as we seek to understand facets of biochemistry, cell biology, genetics, development, physiology, and evolution. Often there is strong evolutionary conservation of gene function and so we can glean insights from model organisms that help us understand how human genes operate.

In addition to helping us understand diverse aspects of basic biology, model organisms are extensively used in applied research—in agriculture, in industry, and also in medicine, where they are used to model and understand disease and to test new systems of treating disease.

The range of model organisms is large, extending from microbes to primates. Inevitably, if a model organism is evolutionarily distant from us, the amount that we can infer about human biological processes is limited to the most highly conserved aspects of cell function and to fundamental cellular processes.

Unicellular model organisms aid understanding of basic cell biology and microbial pathogens

Although very distantly related to us in evolutionary terms, unicellular organisms are useful to study for a variety of scientific and medical reasons. Various microbes are particularly suited to genetic and biochemical analyses, and offer important advantages such as extremely rapid generation times and easy large-scale culture. Species studied include representatives of the two prokaryotic kingdoms, bacteria and archaea, plus yeasts (unicellular fungi), protozoa (unicellular animals), and unicellular algae.

A variety of normally nonpathogenic bacteria have been long-standing and popular model organisms, notably *Escherichia coli* (**Box 10.1**). Different bacteria have also been studied because of their economic importance (for example, many are used in the preparation of fermented foods, in waste processing, in biological pest control, and in the manufacture of antibiotics and other chemi-

BOX 10.1 CHARACTERISTICS OF PRINCIPAL UNICELLULAR MODEL ORGANISMS

Escherichia coli is a rod-shaped prokaryotic microbe that lives in the gut of humans (and other vertebrates) in a symbiotic relationship (they synthesize vitamin K and B-complex vitamins, which their hosts gratefully absorb). Through intensive studies over decades we have built up more knowledge of *E. coli* than of any other type of cell, and most of our understanding of the fundamental mechanisms of life, including DNA replication, DNA transcription, and protein synthesis, has come from studies of this organism.

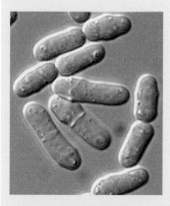

(Courtesy of Yoshifumi Jigami, National Institute of Advanced Industrial Science and Technology.)

Saccharomyces cerevisiae is a yeast that reproduces asexually by budding and has long been important in baking and brewing. Partly because of a high frequency of nonhomologous recombination, it has been very amenable to genetic analyses. It has been used as a model to dissect various aspects of cell biology, including cell cycle control, protein trafficking, and transcriptional regulation.

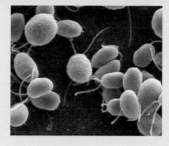

Chlamydomonas reinhardtii is a unicellular green alga about 10 μm in diameter. It has one large chloroplast and multiple mitochondria, and uses two anterior flagella for propulsion and mating. It is amenable to genetic analyses, and many mutants have been characterized; it has provided a useful model for understanding how eukaryotic flagella and basal bodies function. Cilia and flagella both originate from basal bodies, so *Chlamydomonas* has been valuable in helping identify and study genes involved in human ciliary disease.

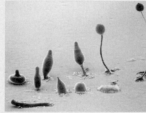

(Courtesy of Grimson MJ & Blanton RL, Texas Tech University.)

Dictyostelium discoideum is a social ameba. Its cells can grow independently, but when challenged by adverse conditions such as starvation they can interact to form multicellular structures, like the slugs and fruiting bodies illustrated here. This organism is ideally suited for studies of cytokinesis, motility, phagocytosis, chemotaxis, signal transduction, and aspects of development such as cell sorting, pattern formation, and cell-type determination. Many of these cellular behaviors and biochemical mechanisms are either absent or less accessible in other model organisms.

Schizosaccharomyces pombe is a yeast that divides asexually by fission and has been well studied, primarily as a model of cell cycle control. In some aspects of chromosome structure and RNA processing it more closely resembles higher eukaryotes than the distantly related yeast *S. cerevisiae*.

(Courtesy of Nicholas Rhind, University of Massachusetts Medical School.)

[From Robinson R (2006) *PLoS Biol* 4: e304 (doi:10.1371/journal.pbio.0040304).]

Tetrahymena thermophila is a ciliated protozoan and a well-established model for cellular and developmental biology, especially cell motility, developmentally programmed DNA rearrangements, regulated secretion, phagocytosis, and telomere maintenance and function. Like other ciliates, *Tetrahymena* has a striking variety of highly complex and specialized cell structures. It possesses hundreds of cilia and complicated microtubule structures and so is a good model with which to investigate the diversity and function of microtubule organizations. It is unusual in that two structurally and functionally differentiated types of nucleus coexist in the cell: a diploid *micronucleus* with five pairs of chromosomes serves as the germ line, and a large somatic *macronucleus* is actively expressed during vegetative multiplication but none of its DNA is transmitted to the sexual progeny.

cals). Other unicellular organisms have been studied because they are patho-gens, notably diverse bacterial species.

Archaea are not known to be associated with disease, and the major interest in studying them is to know more about how they evolved to be so different. They were initially found in unusual, often extreme environments, such as at very high temperatures, in waters of extreme pH or salinity, or in oxygen-deficient muds of marshes. They are now known to inhabit more familiar environments such as in soils and lakes and have been found thriving inside the digestive tracts of cows, termites, and marine life, where they produce methane. The metabolic and energy conversion systems of archaea resemble those of bacteria but the systems used to handle and process genetic information (DNA replication, transcription, and translation) are more closely aligned to those of eukaryotes.

To model eukaryotic cell functions, most reliance has been placed on yeasts because they are easy to culture and genetically very amenable, and because cer-tain key molecules are known to have been strongly conserved in eukaryotes, from yeasts to mammals. Yeasts are common on plant leaves and flowers, in soil, and in salt water, and are also found on the skin surface and in the intestinal tracts of warm-blooded animals, where they may live symbiotically or as para-sites. They typically replicate asexually by budding rather than by binary fission: the cytoplasm and dividing nucleus from the parent cell are initially a continuum with the *bud*, or daughter yeast, before a new cell wall is deposited to separate the two. Usually, yeasts are not associated with disease, but some *Candida* species can cause common health problems, including candidiasis (thrush).

Two genetically amenable yeasts, *Saccharomyces cerevisiae* and *Schizosaccharomyces pombe*, have been particularly well studied (see Box 10.1). Genetic screens have provided a host of valuable mutants that have been partic-ularly valuable in understanding aspects of the cell cycle and DNA repair that are very relevant to our understanding of human cells and cancer and birth defects.

Various protozoan models are studied, including amebae, flagellates, and cili-ates that move using, respectively, pseudopodia, flagella, and cilia. They are of interest to biomedical researchers as models of various facets of cell and develop-mental biology. Many are also parasites associated with disease, acting as hosts for pathogenic bacteria that cause diseases such as legionnaire's disease, salmo-nellosis, and tuberculosis, or they may cause disease directly as in some trypano-somes and malaria-causing *Plasmodium* species. As models of cell and develop-mental biology, most interest has been focused on the ameba *Dictyostelium discoideum* and the ciliate *Tetrahymena thermophila*.

Algae are plant-like organisms that exist in unicellular and multicellular forms. The most relevant unicellular model is *Chlamydomonas* (see Box 10.1).

With the advent of massively parallel DNA sequencing (see Chapter 8, p. 220), unicellular genome sequencing is now routine. As a result, it is now possible to conduct a comprehensive investigation of the human microbiome, the collective term for the microorganisms that live and interact inside and on humans (resid-ing mostly on the skin and in the digestive tract, and outnumbering human cells by a factor of 10 or so). As an extension to the Human Genome Project, an inter-national effort of linked projects called the Human Microbiome Project has been set up to sequence diverse genomes within the human microbiome.

There is also the expectation that *synthetic* model organisms will be created in the near future. The first cellular genome to be completely synthesized—that of *Mycoplasma genitalium*—was reported in early 2008. The 0.53 Mb synthetic genome was assembled by a variety of genetic engineering procedures, but ulti-mately derived from approximately 10,000 synthetic oligonucleotides about 50 nucleotides in length. This achievement came one year after another important breakthrough: the first successful case of cellular genome transplantation. One species, *Mycoplasma capricolum*, was changed into another, *Mycoplasma mycoides*, by removing the host chromosome from *M. capricolum* and replacing it with an equivalent chromosome isolated from *M. mycoides*.

Together with the ability to mutate genes, the above breakthroughs herald the start of a new era of **synthetic biology** that will lead to the production of artificial life. Synthetic unicellular organisms hold great promise for biomedical research and the possibility of developing novel solutions to environmental problems, such as the production of green biofuels and breaking down toxic waste. However, even with inbuilt protection systems, there are important safety issues and the threat of bioterrorism on a large scale.

Some invertebrate models offer cheap high-throughput genetic screening, and can sometimes model disease

To understand complex cell–cell interactions we need to study multicellular organisms. Metazoan (multicellular animal) models include various invertebrates and vertebrates. Invertebrate models are often easy and inexpensive to maintain, and can offer very large numbers of offspring and rapid generation times. These characteristics make them ideally suited to high-throughput genetic screening.

Invertebrates, however, are evolutionarily distant to humans, and thus may not have a counterpart of some human genes. They also lack some organs and functional adaptation found in vertebrates, such as a central nervous system, an adaptive immune system, and skeletal muscle. They are thus less suited to modeling disease. Nevertheless, many human diseases are caused by mutations in very highly conserved genes, and invertebrates have often been used to model disease processes and understand gene function at the *cellular* level.

The roundworm *Caenorhabditis elegans* and the fruit fly *Drosophila melanogaster* are the two most widely studied invertebrates, and they are particularly amenable to experimentation and high-throughput genetic screening. They have been studied largely as developmental and behavioral models. Because many human genes have homologs in *D. melanogaster* and/or *C. elegans*, these models have provided valuable insights into how human genes function. They have been useful in modeling cellular aspects of various human disorders, and they have also been used more recently as *in vivo* systems for high-throughput drug screening.

Roundworms (as opposed to flatworms and segmented worms) outnumber all other complex creatures on the planet and are found almost everywhere in the temperate world, flourishing in soil. *C. elegans* is a free-living form but others are symbiotic or parasitic. They devour crops and infect a billion humans, spreading diseases including river blindness and elephantiasis.

In a quest to understand a simple metazoan, Sydney Brenner initiated the study of *C. elegans* in the 1970s. Like other nematodes, it possesses very simple digestive, nervous, excretory, and reproductive systems but lacks a discrete circulatory or respiratory system. Its small size and short life cycle make it easy to culture in the laboratory, and its transparent body facilitates the study of cell lineage and its nervous system. These and other features make it a useful model organism (**Box 10.2**).

Although considerably more complex than a nematode, the fruit fly *D. melanogaster* also has a short life cycle and is particularly amenable to sophisticated genetic analyses (see Box 10.2). Studies over many decades have generated and systematically analyzed a large number of mutants, including many developmental mutants. As a result, we know more about how the fruit fly develops than for any other multicellular organism. It is also a principal model for studying behavior and neuroscience.

A variety of other invertebrate animal models are also being studied to understand aspects of evolution and invertebrate development, or to understand disease, either because they are human parasites or because they transmit pathogenic protozoan parasites or viruses (**Table 10.1**).

Various fish, frog, and bird models offer accessible routes to the study of vertebrate development

Mammalian development is not easy to analyze. First, egg cells are very small and difficult to manipulate. Second, developing embryos implant into the uterine epithelium; development then proceeds inside the mother, making it difficult to access material for study. Fish, frogs, and birds produce eggs that can be large and easy to manipulate, and the subsequent development proceeds externally to the mother. For these reasons, various non-mammalian vertebrate models are routinely studied to illuminate our understanding of vertebrate development.

BOX 10.2 CHARACTERISTICS OF THE TWO PRINCIPAL INVERTEBRATE ANIMAL MODELS

Caenorhabditis elegans is a roundworm 1 mm long. It can be cultured easily in the laboratory, on agar plates (feeding on bacteria) or in liquid culture, and is very amenable to genetic analyses. The predominant sex is hermaphrodite (XX), a modified female. By producing both sperm and eggs it can self-fertilize, resulting in increased homozygosity. Rarely, a male (X0) form develops through occasional loss of one of the two hermaphrodite X chromosomes, and hermaphrodites will mate preferentially with males when available. Several areas of study take advantage of the unique features of *C. elegans*:

(Courtesy of Bob Goldstein under the Creative Commons Attribution ShareAlike 3.0 license.)

- *Lineage studies. C. elegans* is transparent throughout its life cycle, and so every cell can be seen and followed during development. As a result, it is the only multicellular organism for which we know all the cells—there are precisely 959 somatic cells in the adult hermaphrodite and 1031 in the adult male—and the exact lineage of every cell.
- *Nervous system.* The transparency of *C. elegans* has allowed the observation of the connections between the cells (neurons) that comprise its nervous system. By the 1980s, a complete wiring diagram of the *C. elegans* nervous system was established, identifying all 302 neurons and all the interneural connections. There are significant similarities to vertebrate nervous systems. For example, the neurotransmitters are largely the same as in vertebrates, and *C. elegans* also possesses genes for most of the known molecular components of vertebrate brains.
- *Apoptosis.* Apoptosis forms a key part of *C. elegans* development: 1090 cells develop initially in the hermaphrodite, but 131 cells are programmed to die. Genes involved in apoptosis are often highly conserved.
- *Aging.* Because *C. elegans* develops from one cell into the fully grown form within 3 days and survives for only 2 weeks, the aging process is readily studied.
- *Gene expression and functional studies.* The transparency of *C. elegans* allows protein expression patterns to be followed by linking the green fluorescent protein gene to any *C. elegans* gene or cDNA. Transient inactivation of expression can be achieved for specific genes by *RNA interference (RNAi)*, and various large-scale RNAi screens have been performed to understand gene function.

The fruit fly *Drosophila melanogaster* has been studied extensively for decades, during which a vast amount of information has been built up about gene function. Some of the major features and approaches that have assisted genetic mapping and functional analyses are listed below.

(Courtesy of André Karwath under the Creative Commons Attribution ShareAlike 2.5 license.)

- *Polytene chromosomes.* Present in the salivary gland cells in the larval stages, these interphase chromosomes arise through repeated DNA replication without separation into daughter nuclei. Eventually, 1024 copies of a normally single DNA duplex are arranged side by side, like drinking straws in a box. As a result, these exceptional interphase chromosomes are visible under the light microscope, and so chromosome breakpoints and hybridized DNA clones can be mapped on chromosomes at high resolution.

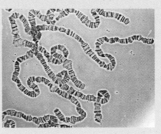

[From Beckingham KM, Armstrong JD, Texada MJ et al. (2005) *Gravitational Space Biol.* 18, 17–29. With permission from the American Society for Gravitational and Space Biology.]

- *Spatially and temporally restricted expression of transgenes* is possible with the GAL4–UAS system of conditional gene expression. Large-scale mutagenesis screens are possible. In many cases, loss of function does not result in a mutant phenotype, but transgene misexpression often gives clues to gene function by producing dominant/dominant-negative phenotypes. One-generation screens for suppressors/enhancers of dominant mutant phenotypes can identify interacting genes. The yeast Flp–FRT recombinase system (described in Chapter 20) can be used to induce mitotic clones and so form homozygous patches, permitting the observation of phenotypes of lethal recessive mutations at late stages of development. Mitotic recombination can also be used in a one-generation screen to score mutant phenotypes in clones and recover lethal mutations that affect late development.
- *The P element.* This *Drosophila* transposable element permits several types of experimental manipulation, including mutagenesis and *transgenesis*. Unequal recombination between adjacent P-element inserts can also produce precise deletions.

Two types of small freshwater fish have been popular developmental models: the zebrafish, which originated in India, and more recently the medaka (Japanese killifish—**Box 10.3**). In addition, the genomes of pufferfish are remarkably free of repetitive DNA, and two such species—*Takifugu rubripes* and *Tetraodon nigroviridis*—have been useful in aiding the discovery of genes and regulatory elements by comparative genomics.

Frogs of the genus *Xenopus* (African clawed frog) have been particularly important models for investigating both embryonic development and cell biology. The name *Xenopus* means strange foot and originates from the sharp claws on the toes of the large, strong, webbed hind feet. *Xenopus* has been an important

TABLE 10.1 PRINCIPAL REASONS FOR STUDYING MAJOR INVERTEBRATE MODEL ORGANISMS

Organism class		Principal models	Major reasons why studied
Cnidarians		*Hydra*	models of regeneration in a diploblast (having two germ layers)
Planaria (flatworms)		*Lineus ruber*	model of regeneration in a triploblast
Molluscs		*Aplysia californica* (sea-slug)	model of cell behavior and development
Nematodes	nonparasitic	*Caenorhabditis elegans*	model of development, behavior, and aging
	parasitic	*Wuchereria bancrofti* (causes elephantiasis) *Onchocerca volvulus* (causes river blindness)	to understand pathogenesis and for economic reasons (some nematodes attack crops)
Arthropods	nonpathogenic	*Drosophila melanogaster* (fruit fly)	model of development, behavior, and neuroscience; also for gaining insights into gene function, modeling some human diseases at the cellular level, and rapid drug screening
		Danaus plexippus (monarch butterfly)	to study the cellular and molecular mechanisms underlying a sophisticated circadian clock
		Apis mellifera (honeybee)	model of learning, memory, and social behavior
	pathogenic	*Anopheles gambiae* (transmits *Plasmodium falciparum*, which causes malaria) *Aedes aegypti* (transmits yellow fever and Dengue fever viruses)	these mosquitoes can transmit a variety of different parasites and viruses
Echinoderms		sea urchins, notably *Strongylocentrotus purpuratus*	models of development, and as basic deuterostomes
Ascidians		*Ciona intestinalis* (sea squirt) (Figure 10.7B)	model of development, and as a primitive chordate
Cephalochordates		*Amphioxus* (Figure 10.7A)	model of development, and because evolutionarily closely related to vertebrates

model for establishing the mechanisms for early fate decisions, patterning of the basic body plan, and organogenesis. There has also been seminal work on chromosome replication, chromatin and nuclear assembly, cell cycle components, cytoskeletal elements, and signaling pathways.

Birds have amniotic membranes, and avian development closely resembles that of mammals. However, whereas a mammalian embryo depends on the mother for its nutrition (with exchange occurring across the placenta), avian embryos do not have a placenta and so are self-developing systems. Because a bird embryo develops outside the body, it is accessible at all stages in development. The chick is a good model largely because its embryo is easily obtained (see Box 10.3).

Mammalian models are disadvantaged by practical limitations and ethical concerns

If the aim is to model some human characteristic, the most representative models are ourselves, followed by other mammals, which are very closely related to us at the biochemical, developmental, and physiological levels. Almost all human genes have an easily identifiable counterpart in other mammals. Nonhuman primates, and especially the great apes—chimpanzee, bonobo, gorilla, and orangutan—are the most closely related to us, and theoretically should be the best models.

Mammalian models are, however, disadvantaged by various practical considerations. Accommodating and breeding mammals is expensive, especially when they are large. Mammalian generation times are often long and there are usually few offspring, making genetic studies difficult. Rodents, however, have relatively short generation times and large numbers of offspring and so are more suited to genetic studies.

There are also ethical concerns. Some people are opposed to all animal experimentation. For those who feel that it is justified, there is often an understandable reluctance to accept the need for experimentation on our closest evolutionary

BOX 10.3 CHARACTERISTICS OF PRINCIPAL MODEL VERTEBRATES

The **zebrafish** (*Danio rerio*) has a short generation time, and large numbers of eggs are produced at each mating. Fertilization is external, so that all aspects of development are accessible; the embryo is transparent, facilitating the identification of developmental mutants. The high conservation of genes that are important in vertebrate development means that human developmental control genes normally have easily identifiable orthologs in zebrafish. Large-scale mutagenesis screens have produced many valuable developmental mutants, and some of these have been used to model human disorders. Antisense technology and/or protein-based translation blocking are also widely used to inactivate specific genes.

(Courtesy of Seotaro under the Creative Commons Attribution ShareAlike 3.0 license.)

The **medaka** (*Oryzias latipes*) is a distant relative of the zebrafish (the two most likely separated from a common ancestor more than 110 million years ago) and is also well suited to genetic and embryological analyses, with a short generation time (2–3 months), inbred strains, genetic maps, transgenesis, enhancer trapping, and availability of stem cells.

[From Amaya E, Offield MF & Grainger RM (1998) *Trends Genet.* 14, 253–255. With permission from Elsevier.]

Xenopus is a genus of African clawed frogs of which two species—*X. laevis* (pictured on the left) and the much smaller *X. tropicalis* (right)—have been widely used in studying development. All developmental stages are accessible, and the comparatively large size of *Xenopus* eggs and embryos facilitates micromanipulation, including microinjections (of mRNA, antibodies, and antisense oligonucleotides), cell grafting, and labeling experiments. Each cell of an amphibian embryo contains its own yolk supply, and so the cells are well able to cope with being transplanted and to continue to differentiate in an explant (where they can be incubated with selected protein factors) or as an individual cell in simple salt solutions. Powerful methods for generating transgenic embryos were developed in the mid-1990s. *X. laevis* has a longish generation time of 1–2 years, and can be induced to produce 300–1000 large (1.0–1.3 mm) eggs at a time. *X. tropicalis* has a comparatively short generation time (less than 5 months) and can be induced to produce 1000–3000 eggs at a time, 0.6–0.7 mm in size. Because of a recent genome duplication, *X. laevis* has a *pseudotetraploid* genome and is not suited to genetic analyses. *X. tropicalis* has a diploid genome, however, and is increasingly being used as a model.

(Courtesy of Joe M500 under the Creative Commons Attribution 2.0 Generic license.)

The **chick** (*Gallus gallus*) embryo is easily obtained and also has the advantage of being very large and relatively translucent, making delicate microsurgical manipulations easy. It thus offers an excellent system in which molecular studies can be combined with classical embryology. Popular experimental manipulations of chick embryos include surgical manipulations and tissue grafting, retrovirus-mediated gene transfer, electroporation of developing embryos, and embryo culture. Rapid advances are being made in chicken transgenics, embryonic stem (ES) cell technology, and cryopreservation of sperm, blastodisc cells, primordial germ cells, and ES cells.

(Courtesy of Rana under the Creative Commons Attribution ShareAlike 2.0 France.)

The **mouse** (*Mus musculus*) is particularly well suited to genetic studies and is an extensively used model of mammalian development. Its small size and short generation time have allowed large-scale mutagenesis programs and extensive genetic crosses, and various features aid in mapping genes and phenotypes. The ability to construct mice with predetermined genetic modifications to the germ line (by transgenic technology and gene targeting in embryonic stem cells) has been a powerful tool in studying gene function and in creating models of human disease, as will be detailed in Chapter 20.

The **rat**, being considerably larger than the mouse, has for many years been the mammal of choice for physiological, neurological, pharmacological, and biochemical analyses. Rat models have helped advance medical research in cardiovascular diseases (hypertension), psychiatric disorders (studies of behavioral intervention and addiction), neural regeneration, diabetes, surgery, transplantation, autoimmune disorders (rheumatoid arthritis), cancer, wound and bone healing, and space motion sickness. In drug development, the rat is routinely employed to demonstrate therapeutic efficacy and assess the toxicity of drug compounds in advance of human clinical trials. Genetic analysis in laboratory rats, however, is much less advanced than in mice, partly because of the relatively high cost of rat breeding programs and because until recently it has been much more difficult to modify the rat germ line by gene targeting.

cousins. Livestock animals, such as sheep and pigs, have some useful applications, but rodents are by a long way the most studied mammalian models, partly because of their greater amenability to genetic studies and partly because of financial and ethical considerations. Of these, the mouse is the premier model for a variety of reasons, although an increasing number of studies involve rat models (see Box 10.3). We consider mouse and rat models, largely in the context of genetic models of disease, in Chapter 20.

Rodents are not good models for understanding some human systems, notably how our brain functions. Nonhuman primates resemble humans in physiology, cognitive capabilities, detailed brain organization, social complexity, reproduction, and development, and have been particularly important in neuroscience research. Being large and normally living in complex social groups, great apes are expensive to maintain and their use in animal experimentation has been contentious. As a result, most primate studies have been conducted on smaller primates with shorter generation times, notably the rhesus macaque and marmoset (**Table 10.2**).

Humans are the ultimate model organism and will probably be a principal one some time soon

Although animal models can provide valuable clues to how human systems work, they inevitably have limitations. Such is the pace of change, too, that the incremental gain in knowledge of some microbial models might soon reach a plateau. Stanley Fields and Mark Johnston, for example, estimate that we will know essentially all that is worth knowing about *S. cerevisiae* somewhere around the years 2025–2035.

As the value of studying some of the more simple model organisms declines, humans are set to become much more thoroughly investigated. Human phenotypes are already the most intensively studied on the planet—a continuous and global screen of abnormal function and disease is maintained by individuals, families, and health professionals. As we move further into the age of personal genome sequencing, the number of individual human genome sequences will rapidly escalate—some predictions suggest that by 2020 personal genome sequencing will be so routine that most individuals in many regions of the planet will have had their genomes decoded.

TABLE 10.2 PRINCIPAL REASONS FOR STUDYING MAJOR MAMMALIAN MODEL ORGANISMS	
Organism	**Major reasons why studied**
NON-PRIMATE MODELS	
Cat	interactions between host gene and infectious pathogens, notably involving feline immunodeficiency virus; as a model of genetic disease, and testing treatment regimens; model of reproductive physiology, endocrinology, and behavior
Cow, sheep, pig	as models of human endocrinology and physiology; host–pathogen interactions in relation to food safety. Transgenic sheep have also been used to produce valued human proteins in their milk
Dog	as a model of genetic diseases (a long history of selective breeding means that many types of dog are prone to genetic diseases such as cancer, heart disease, deafness, blindness, and autoimmune disorders); an important model for the genetics of behavior, and used extensively in pharmaceutical research
Mouse	principal model for understanding gene function and mammalian development, and for modeling human disease and testing treatment regimens; as detailed in Chapter 20, sophisticated genetic tools and large numbers of mutants have long been available
Rat	large size compared with mouse makes it more suited to physiological, neurological, pharmacological, and biochemical analyses. Has provided useful models for various complex human disorders, but genetics not as well developed as mouse genetics; routinely used in testing drug toxicity and therapeutic efficiency
NONHUMAN PRIMATE MODELS	
Chimpanzee	for evolutionary reasons, being the species most closely related to humans; to identify the basis of its resistance to developing *Plasmodium falciparum*-induced malaria and its comparative resistance to the HIV progression to AIDS
Marmoset	used in studying brain function, drug sensitivity, immunity, and autoimmune disease. Also used for studying cognitive and behavioral disorders
Rhesus macaque	the most widely used primate model. The preferred model for vaccine development. Important model in neuroanatomy and neurophysiology, and better suited than marmosets for studying higher brain function

A series of white paper discussion documents on individual model organisms and proposed genome projects can be accessed after clicking on individual genome sequencing projects listed on the US National Human Genome Research Institute's list of approved sequencing targets (http://www.genome.gov/10002154).

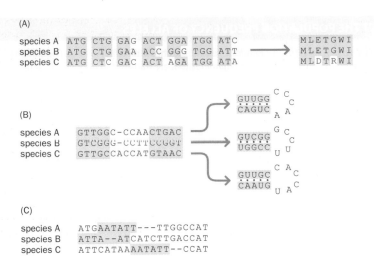

(A)

```
species A   ATG CTG GAG ACT GGA TGG ATC          M L E T G W I
species B   ATG CTG GAA ACC GGG TGG ATT    →     M L E T G W I
species C   ATG CTC GAC ACT AGA TGG ATA          M L D T R W I
```

(B)

```
species A   GTTGGC-CCAACTGAC
species B   GTCGGG-CCTTCCGGT
species C   GTTGCCACCATGTAAC
```

(C)

```
species A   ATGAATATT---TTGGCCAT
species B   ATTA--ATCATCTTGACCAT
species C   ATTCATAAAATATT--CCAT
```

Figure 10.1 Purifying selection results in evolutionarily constrained sequences. (A) In a protein-coding gene, the pattern of mutation is constrained because of the need to conserve functionally important amino acids. Base triplets that will specify codons at the RNA level are marked by the boxes; yellow shading indicates the predominant base at each position. Mutations at the third base position of codons are less likely to alter the encoded amino acid and so are more tolerated. The encoded heptapeptide sequence (shown on the right) seems well conserved, but the fifth amino acid position seems to be less constrained and may be less critical for the function of the protein (because of the nonconservative G to R substitution). (B) In this hypothetical RNA gene, the shaded sequences base-pair at the RNA level, forming a stem–loop structure, as shown on the right (note that G can base-pair with U as well as with C in double-stranded RNA). Bases in the loops are less well conserved, but base pairing within the 5 bp stem region is conserved and is maintained by compensatory double substitutions (as seen in the first base positions of the stem, immediately adjacent to the loop). (C) The sequence of the short elements (highlighted) that form *cis*-acting regulatory sequences are conserved but their positions vary in the aligned sequences from different species.

By comparing genome sequences, highly conserved functionally important sequences can be identified that are not readily apparent from analyzing nucleotide sequences in just one organism. Comparative genomics is particularly helpful in identifying functionally important noncoding DNA sequences, including regulatory elements, which are very short and extremely difficult to identify when a single genomic sequence is analyzed.

Rapidly evolving sequences and positive selection

During evolution, new functions can be acquired as a result of **positive selection** (sometimes called *Darwinian selection*), whereby new mutations that benefit an organism in some way are selectively retained. Positive selection is most readily identified in coding DNA as rapidly evolving codons within a background of evolutionarily constrained sequence. The standard way to do this involves calculating the relative frequencies of non-synonymous and synonymous substitutions—see **Box 10.5**. Signatures of positive selection may also be found in RNA genes (**Figure 10.2**).

A variety of computer programs allow automated genome sequence alignments

Underpinning comparative genomics is the need to perform extensive genome sequence comparisons. Basic sequence alignment tools such as the standard BLAST programs rely on simple queries; using them to compare large genome

---

BOX 10.5 ASSAYING THE EVOLUTION OF CODING SEQUENCES BY USING K_a/K_s RATIOS

Coding sequences are, by definition, functionally important and so they are subject to selection pressure. Most coding sequences are subject to purifying selection to conserve functionally important amino acid sequences, and they seem to evolve slowly. However, some coding sequences are subject to positive selection and diverge rapidly in sequence in a way that is beneficial to the organism. They include many immune system proteins that need to be able to adapt quickly to be able to recognize rapidly evolving microbial antigens, many proteins involved in reproductive processes, and many others.

The standard way of assessing whether a sequence is subject to positive or purifying selection is to calculate the ratio of the number of non-synonymous (amino acid-changing) substitutions per non-synonymous site (K_a or d_N) to the number of synonymous (silent) substitutions per synonymous site (K_s or d_S). This ratio is variously known as the K_a/K_s ratio or the d_N/d_S ratio, or ω. When comparing orthologs from different species, the overall K_a/K_s ratio is often very much less than 1.0 because a substitution that changes an amino acid is much less likely to be different between two species than one

that is silent. However, for immune system genes that co-evolve with genes of invading microorganisms in a battle to recognize and kill the invaders, the K_a/K_s ratio may be much greater than 1.0, which is very strong evidence that positive selection is involved.

Working out a K_a/K_s ratio over a full-length coding sequence can be misleading because one part of a protein may be subject to purifying selection while another domain is subject to positive selection. Modern methods can use multiple alignments of orthologous sequences from several species within a phylogeny to work out a K_a/K_s ratio for each codon.

Various potential pitfalls may affect the accuracy of the calculations. One assumption—that synonymous substitutions are neutral substitutions—may not always be true. For example, the silent site may be part of a regulatory sequence within an exon, such as an exonic splice enhancer. Also, if the compared sequences are from distantly related organisms, multiple consecutive substitutions may have taken place at a specific site but gone unnoticed.

For a Web-based server for calculating K_a/K_s ratios, see Selecton (http://selecton.tau.ac.il/).

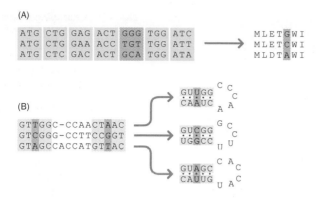

(A)

```
ATG CTG GAG ACT GGG TGG ATC        M L E T G W I
ATG CTG GAA ACC TGT TGG ATT        M L E T C W I
ATG CTC GAC ACT GCA TGG ATA        M L D T A W I
```

(B)

```
GTTGGC-CCAACTAAC
GTCGGG-CCTTCCGGT
GTAGCCACCATGTTAC
```

Figure 10.2 Sequence diversification by positive selection. (A) Positive selection within a coding sequence. In this example, the fifth base triplet is envisaged to be a target of positive selection and varies in all three species shown, whereas the surrounding sequence is constrained by purifying selection. (B) Suggestive positive selection within an RNA gene. The hypothetical sequences shown here have inverted repeats that permit base pairing to form hairpins with a 5 bp stem. The middle nucleotides in the 5 bp stem are highly variable, but compensatory double substitutions at the DNA level maintain the stem. The middle nucleotides of the stem seem to be selected to be diversified within a highly conserved RNA structure.

sequences, such as those of vertebrates, is remarkably inefficient. A significant difficulty is the extent of large-scale genome rearrangements (involving regions larger than 50 kb) that have occurred during evolution.

For the comparison of large genome sequences, more powerful programs are needed. For a collection of *extant* (currently existing) genomes, the challenge is to find the minimum set of subsequence groups so that inside each group the sequences to be compared are both homologous and collinear (**Figure 10.3**). This collinearity reconstruction problem seeks answers to two basic questions. First, which segments of extant genomes arose from the genome of a common ancestor? Second, what was the likely evolutionary path of events that generated the different segments? Various computer programs have been developed to reconstruct collinearity for use in large-scale alignment (**Table 10.3**).

Once collinear relationships between homologous genomic regions have been established, computer programs take the unaligned sequences and determine which bases in each sequence are orthologous. Multiple alignment of very large sequences consumes huge computing power and inevitably the programs use *heuristic* methods; that is, they apply workable solutions that are not formally correct so as to reduce computational time. Nevertheless, by 2007, megabase chunks of the human genome sequence could be aligned with sequences from 27 other vertebrates (see the TBA/MULTIZ program entry in Table 10.3). Additional programs are used to inspect the aligned sequence to identify sequences that are evolutionarily constrained, suggesting a conserved function.

Comparative genomics helps validate predicted genes and identifies novel genes

When the sequences of metazoan genomes were first reported, large numbers of novel genes were predicted for which there was little or no supportive experimental evidence. Often gene prediction relied rather heavily on the identification of open reading frames (ORFs). However, apparent ORFs can occur by chance within noncoding DNA. Confidence that an ORF is functionally significant increases according to its length, but smaller ORFs are more ambiguous. Gene predictions can also be made by using ESTs (expressed sequence tags), but these could reflect occasional artifacts in making cDNA libraries. A more robust way to validate a predicted gene is by supportive data from clearly recognizable sequence counterparts in related species.

(A)

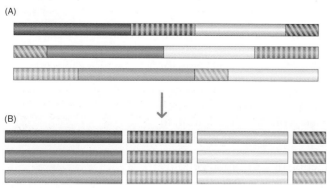

(B)

Figure 10.3 Aligning whole genome sequences begins by reconstructing homologous collinearity relationships. (A) Homologous genomic sequences (shown by similar shading patterns) from three organisms are located in nonhomologous positions as a result of intrachromosomal rearrangements since they diverged from a common ancestor (interchromosomal rearrangements are not shown for the sake of simplicity). (B) The homologous sequences need to be grouped into sets that are then arranged in a linear sequence so as to simplify subsequent whole genome alignments. [From Margulies EH & Birney E (2008) *Nat. Rev. Genet.* 9, 303–313. With permission from Macmillan Publishers Ltd.]

TABLE 10.3 SOME COMPUTER PROGRAMS USED TO ALIGN COMPLEX GENOME SEQUENCES AND DETECT EVOLUTIONARILY CONSTRAINED SEQUENCES

Task	Program	Characteristics	URL/reference
Reconstructing homologous collinearity	MERCATOR	uses only coding exons as initial anchoring points	http://www.biostat.wisc.edu/~cdewey/mercator/
	GRIMM	analyzes rearrangements in pairs of genomes	http://grimm.bioprojects.org/GRIMM/index.html
	MLAGAN	multiple global alignment of genomic sequences	Brudno et al. (2003) *Genome Res.* 13, 721–731
	PECAN	a consistency-based multiple-alignment program	http://www.ebi.ac.uk/~bjp/pecan
Multiple alignment	TBA/MULTIZ	TBA is designed to be able to align many megabase-sized regions of multiple mammalian genomes; to perform the dynamic-programming step, TBA runs a stand-alone program called MULTIZ, which has been used to align much of the human genome sequence with the sequences of 27 other vertebrate genomes	Miller et al. (2007) *Genome Res.* 17, 1797–1808; the 28-way vertebrate alignment can be viewed with the UCSC genome browser at http://genome.ucsc.edu, downloaded in bulk by anonymous FTP from http://hgdownload.cse.ucsc.edu/goldenPath/hg18/multiz28way, or analyzed with the Galaxy server at http://g2.trac.bx.psu.edu
Constraint detection	PHASTCONS	identifies evolutionarily conserved elements in a multiple alignment, given a phylogenetic tree	Siepel et al. (2005) *Genome Res.* 15, 1034–1050; http://compgen.bscb.cornell.edu/~acs/phastCons-HOWTO.html
	SCONE	gives a sequence conservation score after estimating the rate at which each nucleotide is evolving, and then computes the probability of neutrality given the estimated rate	Asthana et al. (2007) *PLoS Comp. Biol.* 3, 2559–2568

For further information on these and additional programs, see Margulies & Birney (2008) in Further Reading.

Comparative genomics has been used to great advantage in re-annotating genomes. Early examples included genome sequencing of multiple different *Saccharomyces* and *Drosophila* species to help annotate the previously studied *S. cerevisiae* and *D. melanogaster* genomes. Many originally predicted genes in the latter two genomes were discounted when equivalent ORFs or similar codon substitution frequencies were not consistently found in orthologous sequences in the other *Saccharomyces* or *Drosophila* species (**Figure 10.4**). Comparative genomics also helps uncover novel genes by identifying sufficient sequence conservation. For example, 1275 novel *C. elegans* genes were predicted only after comparison with the nematode *C. briggsae*.

Comparative genomics has also been of great help in identifying novel genes that produce functional noncoding RNA. Short RNA genes can easily be overlooked in bioinformatics analyses of a single genome sequence, but comparative genome analyses can identify signatures of purifying selection in noncoding sequences as well as protein-coding DNA (see Figure 10.1), and so they can focus attention on functionally important noncoding DNA regions that have subsequently been shown to be genes. Comparative genomics has also helped in the prediction of conserved small RNA genes, notably miRNA genes.

Comparative genomics reveals a surprisingly large amount of functional noncoding DNA in mammals

A driving force in sequencing the human genome was the desire to understand functionally important gene sequences. In particular, much of the focus has been on protein-coding DNA and on the noncoding RNA classes (such as rRNA, tRNA, and snRNA) known to be important in guiding the expression of genes that ultimately make polypeptides. More recently, various other classes of functional noncoding RNA have been identified, including various long noncoding RNA species and short RNAs such as miRNA and piRNA (see Table 9.9). In addition to RNA genes, *cis*-acting regulatory elements such as enhancers are an important class of functional noncoding DNA.

We have known for some time that a very small fraction of the human genome comprises protein-coding DNA—the most current estimates indicate 1.1%.

However, the amount devoted to functionally important noncoding DNA was quite unknown until the newly sequenced mouse genome was compared with the human genome. Although humans and mice diverged from a common ancestor around 80–90 million years ago, about 40% of the sequence in the human genome could be aligned with the mouse genome at the nucleotide level. The aligned sequences included some with no obvious function, such as many transposon repeats that had last been active in the common ancestor to humans and mice and that had been expected to be free from any constraints on mutation (*neutral mutation*).

When the extent of genomewide sequence conservation was compared with the neutral rate, about 5% of small segments (50–100 bp long) seemed to be significantly conserved (as a result of purifying selection). Follow-up analyses by the ENCODE (Encyclopedia of DNA Elements) project arrived at similar conclusions: about 3.9% of the genome is functionally important noncoding DNA, significantly more than the 1.1% that is coding DNA.

The above estimates of functionally important noncoding DNA may be on the low side. First, they assume that functionally important sequences are highly conserved; however, as we describe later in this chapter, some important sequences are evolving rapidly. Second, the figure is based on a neutral mutation reference that may be wrong. The relatively non-conserved 95% of our genome used to be considered junk DNA, because it seemed to lack any function. More recently, the possibility that it may yet be important has gained greater recognition, and it is sometimes even called the dark matter of our genome.

The new respect for what is the predominant component of our genome is founded in part on a reappraisal of its transcriptional activity. Until recently, the great bulk of the euchromatic genome had been thought to be transcriptionally

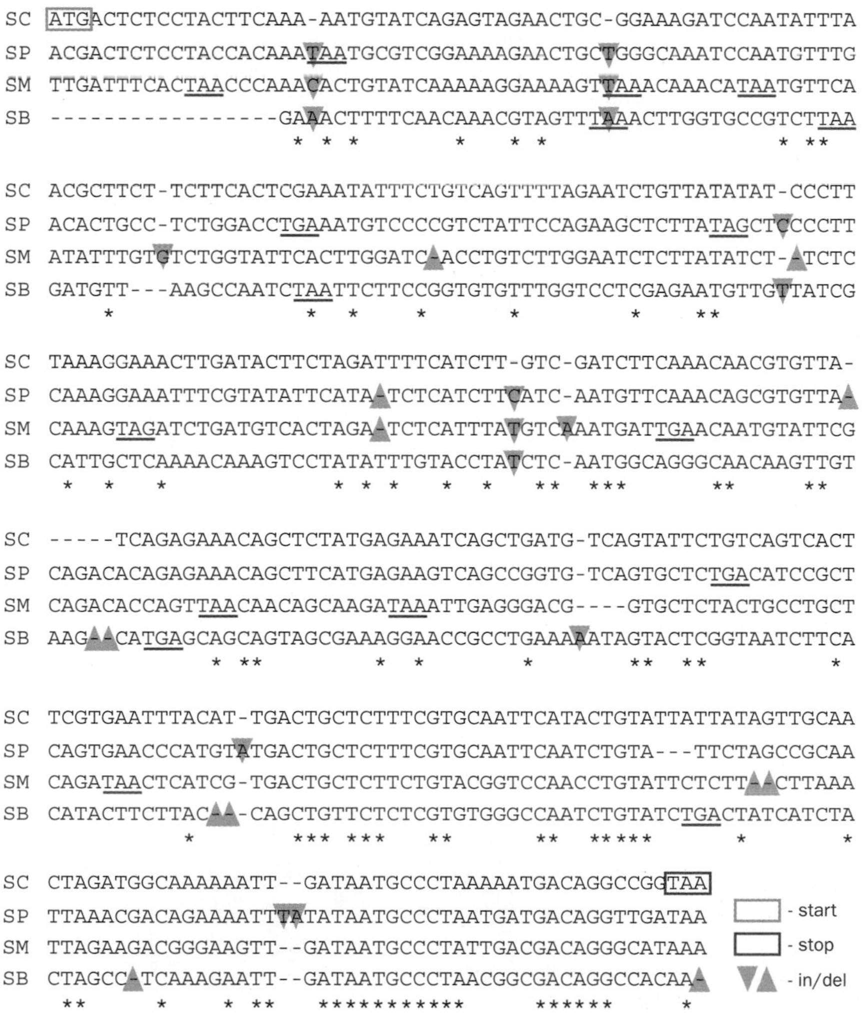

Figure 10.4 Using comparative genomics to test the validity of a predicted gene. The *S. cerevisiae* YDR102C sequence (SC) is aligned here with homologs from *S. paradoxus* (SP), *S. mikatae* (SM), and *S. bayanus* (SB). YDR102C was originally identified as an open reading frame (ORF) that was predicted to encode a 111 amino acid protein. Asterisks indicate nucleotide positions where there is identity in the four sequences. Trinucleotides underlined in red indicate potential premature termination codons. Not only is the start codon (green box) not conserved but the alignments require frequent insertions and deletions (arrows). Thus, the homologs in the other *Saccharomyces* species could not make a protein like the one that had been predicted for the YDR102C ORF, raising doubts about whether the YDR102C sequence is really a coding sequence. [From supplementary information in Kellis M, Patterson N, Endrizzi M et al. (2003) *Nature* 423, 241–254. With permission from Macmillan Publishers Ltd.]

inert, but high-throughput sequence sampling now suggests that at least 85% of the human genome is transcribed. In addition, as we describe later in this chapter, highly repetitive retrotransposons can give rise to functionally important sequences including enhancer sequences, exons, and even some genes.

Comparative genomics has been particularly important in identifying regulatory sequences

After all the intensive analyses of the human genome in recent years, one important area that is still poorly understood is how our genes are regulated. The binding specificities of various transcription factors have been analyzed over many years. In each case these factors consist of very short DNA sequence elements.

Because they are very short, novel elements of this type are extremely difficult to identify by bioinformatics analyses of a single genome. As a result, most previous investigations would start by scanning genomic sequences in the immediate vicinity of a gene of interest. Identifying distantly located regulatory elements was a challenge, and whole genome analyses were unthinkable at that time.

Comparative genomics has dramatically changed the landscape. Regulatory sequences are often quite strongly conserved. When genomes of related organisms are aligned, therefore, large numbers of conserved noncoding sequence elements can be identified across whole genomes. Depending on the evolutionary distance, there is a trade-off between the sensitivity of detection and the specificity of detection. The number of conserved noncoding elements detected by comparing the relatively evolutionarily distant human and fish genomes is low—many genuine regulatory elements are not detected (**Figure 10.5**)—but the specificity is very high and so virtually all of the sequences detected will be worth investigating. For more closely related genomes, such as humans and mice, there is a much higher detection rate, but there is also a higher false positive rate.

Cis-acting regulatory elements are made up of modules containing short sequence elements to which *trans*-acting factors bind. The functional short elements are sometimes separated by very poorly constrained sequences (*diffuse enhancers*), making them very difficult to detect by bioinformatics methods. In other cases, there may be very tight packing of binding sites for a range of different transcription factors, in which case sequence conservation is extended over long distances, making their detection easier.

Inspection of aligned genome sequences from humans, mice, and rats has led to the identification of more than 480 *ultraconserved elements* (*UCEs*), sequences longer than 200 bp that show three-way 100% orthologous sequence identity over the three genomes. They are also highly conserved in other mammals and birds (the average human–dog and human–chicken sequence identities are, respectively, 99% and 95%). They are not, however, well conserved over large evolutionary distances—a typical human UCE does not have recognizable invertebrate homologs. Just over 20% of UCEs overlap known protein-coding exons, but most are noncoding. Often, they are located within introns or close to genes that are involved in regulating transcription and development, and in many cases they have been shown to function as enhancers.

A variety of genome browser tools are available to access details of conserved noncoding sequences (**Table 10.4**). **Figure 10.6** shows an example in which by using the VISTA Genome Browser it is possible to identify highly conserved noncoding DNA sequences that in this case were clearly shown to be tissue-specific enhancers.

Figure 10.5 The sensitivity–specificity trade-off in detecting conserved noncoding elements by comparative genomics. (A) Many apparently conserved noncoding elements (CNEs) are identified by aligning the human and mouse genomes. Few functionally important CNEs are missed because of the high sensitivity, but many apparent CNEs may simply reflect the overall high human–mouse sequence similarity. The human and pufferfish (fugu) genome sequences are much more distantly related. Few CNEs will be identified, but the great majority are very likely to be functionally important. (B) Vertebrate conservation of 12 known mouse *cis*-regulatory elements that are important in regulating the expression of genes in the heart. Most of the elements are minimally conserved beyond mammals and would have been missed by human–pufferfish comparisons. [Adapted from Visel A, Bristow J & Pennacchio LA (2007) *Semin. Cell Dev. Biol.* 18, 140–152. With permission from Elsevier.]

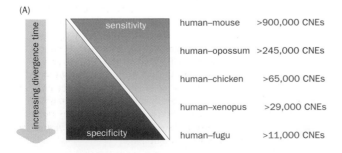

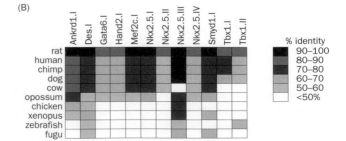

TABLE 10.4 PRINCIPAL GENOME BROWSERS FOR IDENTIFYING CONSERVED NONCODING SEQUENCES IN VERTEBRATES

Genome browser	Conserved elements identified with	Sequence alignment based on	URL
Dcode ECR Browser	PiP (percentage identity plot)	BLASTZ	http://ecrbrowser.dcode.org
UCSC Genome Browser	PHASTCONS	MULTIZ (multiple, local[a])	http://genome.ucsc.edu
VISTA Genome Browser	PiP	SLAGAN (pairwise, glocal[a])	http://pipeline.lbl.gov
	GUMBY	MLAGAN (pairwise, glocal[a])	
VISTA Enhancer Browser	GUMBY	MLAGAN (multiple, global[a])	http://enhancer.lbl.gov

[a]In *global* alignment, one sequence string is transformed into the other; in *local* alignment, all locations of similarity between the sequence strings under comparison are returned. Global alignments are less prone to demonstrating false homology because each letter of one sequence is constrained to being aligned with only one letter of the other. However, local alignments can cope with rearrangements between non-syntenic, orthologous sequences by identifying similar regions in sequences; this, however, comes at the expense of a higher false positive rate. *Glocal* alignment is a combination of global and local alignment that allows the user to create a map that transforms one sequence into another while allowing for arrangement events.

10.3 GENE AND GENOME EVOLUTION

Comparative genomics has been valuable in illuminating many aspects of genome evolution. In this section, we focus very largely on developments that have affected how vertebrate and especially mammalian genomes evolved. We also consider vertebrate lineages with respect to the lineages of the invertebrates that are evolutionarily most closely related to vertebrates. They are grouped along with vertebrates within the phylum Chordata (chordates). Other well-studied invertebrates such as *Drosophila melanogaster* and *Caenorhabditis elegans* are more distantly related to us, being members of different phyla, respectively Arthropoda and Nematoda.

Chordates are all animals that go through an embryonic stage in which they possess a notochord, nerve cord, and gill slits. They comprise three major groups:

- Craniates (all animals with skulls, including vertebrates and also hagfish, which lack a backbone);
- Cephalochordates (also called lancelets, or amphioxus; exclusively marine animals that resemble small slender fishes but without eyes or a definite head or brain—**Figure 10.7A**);
- Urochordates (also called tunicates or ascidians; underwater saclike filter feeders such as the sea squirt, *Ciona*—Figure 10.7B).

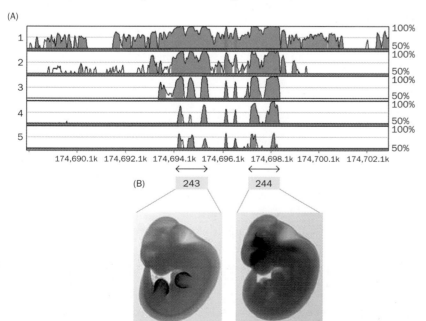

Figure 10.6 Using the VISTA Genome Browser to identify regions of highly conserved noncoding DNA. (A) The query here was a 15 kb region of human chromosome 2 (chr2: 174,688,000–174,703,000 in the hg18 reference sequence) that is located within the human *OLA1* gene (also called *PTD004*). This region was selected to be compared with five other vertebrate genomes and the outputs for individual genome comparisons are shown in the horizontal tracks as graphs that display percentage identity from 50% (bottom) to 100% (top). Tracks show comparison with individual genomes thus: 1, dog; 2, mouse; 3, chicken; 4, frog (*X. tropicalis*); 5, *Fugu* (pufferfish). Pale blue signifies exons; pink signifies noncoding DNA. Two highly conserved *OLA1* coding exons are shown near the center of the graphs. Note that most of the noncoding DNA is poorly conserved, but two regions of highly conserved noncoding DNA can be identified, spanning the regions shown by lines with double arrowheads. (B) The VISTA Enhancer Browser stores experimental data for conserved noncoding DNA sequences identified as enhancers. The enhancer assay involves cloning the test DNA into a reporter vector with a *lacZ* gene driven by a minimal promoter. The construct is injected into a one-cell mouse embryo and after 11.5 days of development the mouse embryo is analyzed to see where the *lacZ* gene is expressed (blue signal). Here, we see that the conserved noncoding DNA regions identified in panel (A) are tissue-specific enhancers: enhancer 243 regulates limb bud expression (left image), whereas enhancer 244 regulates forebrain expression (right image). See Table 10.4 for Web addresses for the VISTA browsers.

Gene complexity is increased by exon duplication and exon shuffling

The discovery in the mid- to late 1970s that most eukaryotic genes are split by introns was a surprise. Such introns are known as spliceosomal introns because the splicing reaction requires dedicated spliceosomes, unlike autocatalytic (self-splicing) introns. The latter are found in both bacteria and eukaryotes, but their distribution is very restricted, being primarily found in rRNA and tRNA genes, and in a few protein-coding genes found in some types of mitochondria and chloroplasts.

Spliceosomal introns are believed to have originated from a subclass of auto-catalytic introns known as group II introns that subsequently lost their capacity for self-splicing. They are thought to have arisen in eukaryotes and were probably inserted into genes at an early stage in eukaryote evolution, but there is evidence for some later integrations too. However, the great majority of introns pre-dated chordate evolution. Although humans and the primitive chordate, *Amphioxus*, diverged from a common ancestor approximately 550 million years ago, comparison of the human and *Amphioxus* genomes reveals that 85% of the introns are conserved at precisely analogous locations.

During the evolution of complex genomes, spliceosomal introns expanded in size by the insertion of different types of DNA sequence, notably repetitive DNA sequences originating by retrotransposition. As a result, exons came to be well separated in genomic DNA. Splitting functionally important sequences as exons in genomic DNA introns facilitated the copying of exons in evolutionarily advantageous ways, allowing duplication and new combinations of important sequences.

Exon duplication

Tandem duplication of exons is evident in about 10% of genes in humans, *D. melanogaster*, and *C. elegans*, and is often responsible for generating repeated protein domains (**Figure 10.8**). Unequal crossover (or unequal sister chromatid exchange) can result in intragenic duplication so that a segment of genomic DNA spanning one or more exons is duplicated.

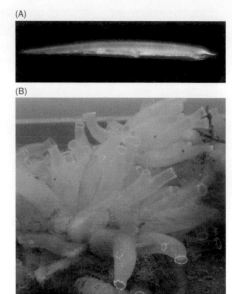

Figure 10.7 Amphioxus and tunicates represent two subphyla of chordates. (A) An example of an amphioxus (lancelet), *Branchiostoma lanceolata*. (B) A tunicate, *Ciona intestinalis*. [(A) courtesy of Hans Hillewaert under the Creative Commons Attribution ShareAlike 2.5 License. (B) courtesy of Clément Lamy under the Creative Commons Attribution ShareAlike 3.0 License.]

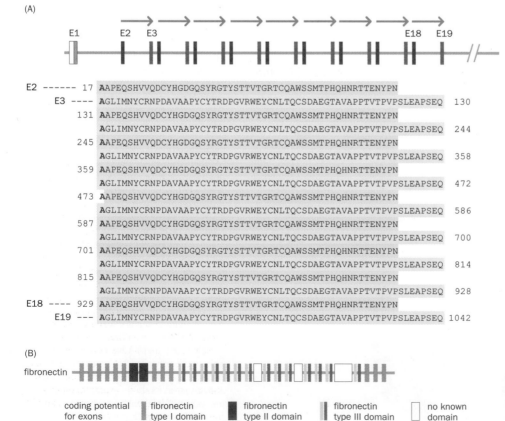

(A)

```
E2 ------ 17 AAPEQSHVVQDCYHGDGQSYRGTYSTTVTGRTCQAWSSMTPHQHNRTTENYPN
   E3 ---- AGLIMNYCRNPDAVAAPYCYTRDPGVRWEYCNLTQCSDAEGTAVAPPTVTPVPSLEAPSEQ 130
      131 AAPEQSHVVQDCYHGDGQSYRGTYSTTVTGRTCQAWSSMTPHQHNRTTENYPN
          AGLIMNYCRNPDAVAAPYCYTRDPGVRWEYCNLTQCSDAEGTAVAPPTVTPVPSLEAPSEQ 244
      245 AAPEQSHVVQDCYHGDGQSYRGTYSTTVTGRTCQAWSSMTPHQHNRTTENYPN
          AGLIMNYCRNPDAVAAPYCYTRDPGVRWEYCNLTQCSDAEGTAVAPPTVTPVPSLEAPSEQ 358
      359 AAPEQSHVVQDCYHGDGQSYRGTYSTTVTGRTCQAWSSMTPHQHNRTTENYPN
          AGLIMNYCRNPDAVAAPYCYTRDPGVRWEYCNLTQCSDAEGTAVAPPTVTPVPSLEAPSEQ 472
      473 AAPEQSHVVQDCYHGDGQSYRGTYSTTVTGRTCQAWSSMTPHQHNRTTENYPN
          AGLIMNYCRNPDAVAAPYCYTRDPGVRWEYCNLTQCSDAEGTAVAPPTVTPVPSLEAPSEQ 586
      587 AAPEQSHVVQDCYHGDGQSYRGTYSTTVTGRTCQAWSSMTPHQHNRTTENYPN
          AGLIMNYCRNPDAVAAPYCYTRDPGVRWEYCNLTQCSDAEGTAVAPPTVTPVPSLEAPSEQ 700
      701 AAPEQSHVVQDCYHGDGQSYRGTYSTTVTGRTCQAWSSMTPHQHNRTTENYPN
          AGLIMNYCRNPDAVAAPYCYTRDPGVRWEYCNLTQCSDAEGTAVAPPTVTPVPSLEAPSEQ 814
      815 AAPEQSHVVQDCYHGDGQSYRGTYSTTVTGRTCQAWSSMTPHQHNRTTENYPN
          AGLIMNYCRNPDAVAAPYCYTRDPGVRWEYCNLTQCSDAEGTAVAPPTVTPVPSLEAPSEQ 928
E18 ---- 929 AAPEQSHVVQDCYHGDGQSYRGTYSTTVTGRTCQAWSSMTPHQHNRTTENYPN
  E19 --- AGLIMNYCRNPDAVAAPYCYTRDPGVRWEYCNLTQCSDAEGTAVAPPTVTPVPSLEAPSEQ 1042
```

(B)

fibronectin

coding potential for exons | fibronectin type I domain | fibronectin type II domain | fibronectin type III domain | no known domain

Figure 10.8 Exon duplication. (A) The human *LPA* gene encodes apolipoprotein(a) and is implicated in coronary heart disease and stroke. There is extensive size polymorphism, and the N-terminus of the protein has a large but variable number of repeats of a 114 amino acid kringle IV domain. Each kringle repeat is encoded by a tandemly repeated pair of exons. The first nine kringle domains (spanning amino acids 17–1042) are shown here. They have identical amino acid sequences and are encoded by exons 2–19 of the *LPA* gene; grey arrows represent tandemly repeated exon pairs, starting with exons 2 and 3 (E2 and E3) and ending in exons 18 and 19 (E18 and E19). The 18 alanines shown in bold red letters are encoded by codons that are each interrupted by introns. (B) Fibronectin, a large extracellular matrix protein, has multiple repeats of three different domains. In the fibronectin gene, the type I and II domains are encoded by individual exons (orange and purple boxes, respectively), but type III domains are encoded by a tandemly repeated pair of exons (blue boxes).

Exon duplication offers several types of advantage. It can result in extension of a structural domain, as occurs in collagens. For example, 41 exons of the *COL1A1* gene encode the part of α1(I) collagen that forms a triple helix. Each exon encodes essentially one to three copies of the 18 amino acid motif, which consists of six tandem repeats of tripeptides beginning with glycine, $(Gly-X-Y)_6$.

Alternative splicing can produce different isoforms by selecting one exon sequence from a group of duplicated exons that have diverged in sequence. Many human genes undergo this type of splicing, but the *Drosophila Dscam* gene provides the most celebrated example.

Over evolutionarily long time-scales, intragenic repeats often diverge in sequence and acquire different, although related, functions. Sometimes extensive sequence divergence means that the repeated structure may not be obvious at the sequence level, as with the different immunoglobulin domains.

Exon shuffling

Protein domains are rarely restricted to one type of protein. For example, the protein-binding kringle domains of lipoprotein(a) shown in Figure 10.8 are widely found in blood-clotting factors and fibrinolytic proteins, and the type II domains of fibronectin are found in many cell surface receptors and extracellular matrix proteins. Different exon shuffling mechanisms can give rise to spreading of protein domains to different proteins. Non-allelic recombination is one possibility, but transposons are likely to have been important contributors. In particular, retrotransposons offer the possibility of a copy-and-paste mechanism such that domains are retained in the donor gene and copied into an acceptor gene (**Figure 10.9**).

Gene duplication can permit increased gene dosage but its major value is to permit functional complexity

Human genes have been duplicated by various duplication mechanisms (see Chapter 9, p. 263). Gene families are a major general feature of metazoan genomes, and one gene in a hundred is estimated to be duplicated and fixed in the population every million years. When a gene duplicates within a genome, the two resulting gene copies are initially identical. In rare cases, the duplicated genes may be maintained to produce identical, or essentially identical, gene products because increased amounts of gene product are advantageous in some way. The large numbers of genes that make essentially identical copies of different ribosomal RNA and histone proteins come into this category.

In response to an altered environment, other genes may undergo duplication to provide a selective advantage through increased gene dosage, as illustrated in the recent adaptation of human populations to starch diets. Starch is digested using the enzyme salivary amylase. The chimpanzee has a single salivary amylase

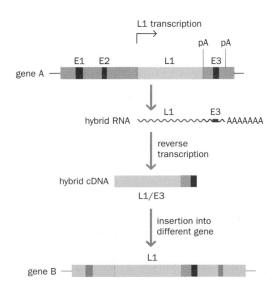

Figure 10.9 Exon shuffling between genes can be mediated by transposable elements. Exon shuffling can be performed with retrotransposons such as actively transposing members of the LINE-1 (L1) sequence family as shown here. LINE-1 elements have weak poly(A) signals, and so transcription often continues past such a signal until another nearby poly(A) signal is reached (e.g. after exon 3 in gene A). The resulting RNA copy contains a transcript not just of LINE-1 sequences but also of a downstream exon (in this case E3). The LINE-1 reverse transcriptase machinery can then act on the extended poly(A) sequence to produce a hybrid cDNA copy that contains both LINE-1 and E3 sequences. Subsequent transposition into a new chromosomal location may lead to the insertion of exon 3 into a different gene (gene B).

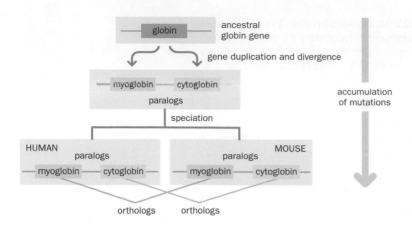

Figure 10.10 Orthologs and paralogs. Homologs, genes that have significant sequence identity suggesting a close evolutionary relationship, can be one of two types. Paralogs are closely related genes present in a single genome as a result of gene duplication. Paralogs are often identical in sequence immediately after gene duplication. Orthologs are genes present in the genomes of different species that are directly related through descent from a common ancestor.

gene, but in humans a series of evolutionarily very recent tandem gene duplications has produced multiple copies of such genes at the *AMY1* gene cluster at 1p21. *AMY1* copy number varies significantly between haplotypes, and *AMY1* copy number positively correlates with the expression of salivary amylase. Human populations that historically consume a high-starch diet have significantly more *AMY1* copies than populations that traditionally consume a low-starch diet; increased *AMY1* copy number seems to be an adaptive response to increased starch in the diet.

Gene duplication to provide increased gene dosage is comparatively rare. More frequently, after gene duplication the sequences of the two genes diverge quite extensively. One of the duplicated gene copies may become a pseudogene (see Box 9.2). Alternatively, two divergent functional gene copies are retained. The divergent gene copies may acquire different properties and often they come to be expressed in different ways that are functionally advantageous. Gene duplication is thus thought to be a major motor that drives increasing functional complexity. The two homologous gene copies are described as **paralogs** (as opposed to orthologs, which are present in different species; **Figure 10.10**).

Retrogenes offer an immediate possibility for functional divergence: because the gene copy is made at the cDNA level, it lacks the promoter and both neighboring and intronic regulatory elements of the parent gene (see Figure 9.12). Continued expression of a retrogene is therefore dependent on regulatory sequences in the neighborhood of its integration site that are usually rather different from the sequences that regulate the parent gene. As a result, the parent gene and retrogene copy can be expressed in different ways, often in different cell types. After being exposed to a different cellular (and molecular) environment, the retrogene can come to acquire different functions.

Genes duplicated at the genomic DNA level can also acquire different functions (**Figure 10.11**). One possibility is that one of the duplicate genes acquires a distinctive new function (*neofunctionalization*). Pure neofunctionalization is rare, however. Usually both genes undergo expression changes over evolutionary time. Sometimes the duplicated genes acquire complementary mutations in different *cis*-regulatory regions so that they diverge in expression while between them initially maintaining the expression domains of the ancestral gene. As a result, different subsets of the functions of the ancestral gene can be partitioned

Figure 10.11 Functional divergence of duplicated genes. (A) *Neofunctionalization.* Here, one of the duplicated genes retains the function of the ancestral gene. The other paralog undergoes mutations (green arrow) that result in its having altered characteristics (green box, but could be in regulatory sequences unlike as shown here), leading it to acquire a new function. (B) *Subfunctionalization.* The duplicated gene copies undergo complementary deleterious mutations, often at the level of regulatory elements. In this example, the ancestral gene is imagined to be regulated by two sets of upstream control elements (blue and red symbols) so that it is expressed in two different cell types, or different tissue types, or at different developmental stages (blue and red arrows). Mutations (jagged arrows) inactivate one set of control elements in one paralog and the other set in the second paralog so that the expression patterns and functions of the ancestral gene are partitioned between the two daughter genes.

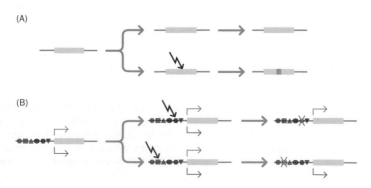

between the duplicated daughter genes (**subfunctionalization**). As we describe below, studies of whole genome duplication indicate that many duplicated genes can be retained over long time-scales.

The globin superfamily illustrates divergence in gene regulation and function after gene duplication

Vertebrate globin genes originated by a series of gene duplications from an ancestral globin gene, perhaps as much as 800 million years ago. At least some were tandem gene duplications, producing gene clusters on human chromosomes 11 and 16 (**Figure 10.12**). Some of the duplications have been unproductive, giving rise to pseudogenes, but the overall value has been to enable globins to adapt to various cellular and tissue environments.

The hemoglobins and myoglobin have long been known to transport both O_2 and CO_2 in blood and muscle, respectively. Recently, myoglobin has also been shown to have a crucial role as an intrinsic nitrite reductase that regulates responses to cellular hypoxia and reoxygenation, and other globins may have similar functions. Whereas hemoglobins work as tetramers with two identical polypeptide chains encoded by a member of the α-globin gene family and two other identical polypeptides encoded by a member of the β-globin family, other globins work as monomers. Neuroglobin works in the nervous system, and cytoglobin is widely distributed in different tissue types, but their exact functions are unknown.

By partitioning regulatory control sequences, subfunctionalization could have resulted in some globins being restricted to specialized tissues with the possibility of acquiring modified functions, maybe even additional or divergent

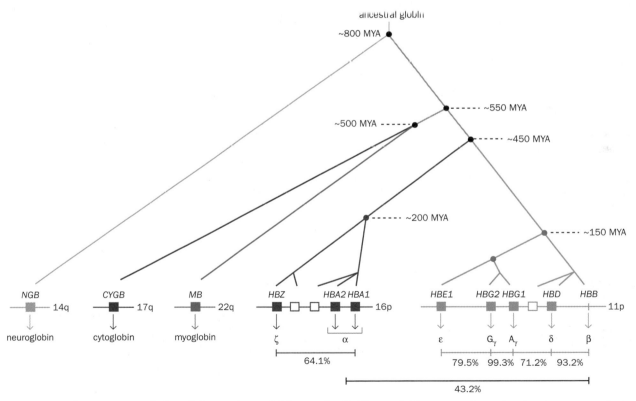

Figure 10.12 Evolution of the vertebrate globin superfamily. Human globin genes are located on five chromosomes (bottom). Those resulting from evolutionarily ancient gene duplications have subsequently been separated by chromosomal rearrangements. The more recent duplications have resulted in gene clusters on chromosomes 11 and 16. Globins encoded within a gene cluster show a greater degree of sequence identity (64.1–99.3%) than globins on different chromosomes (e.g. 43.2% between α- and β-globin; sequence identities of the α- and β-globin genes to globins on the other chromosomes are significantly less than 30%). Some of the duplications have been recent: the *HBA1* and *HBA2* genes encode identical α-globins, and the *HBG1* and *HBG2* genes encode γ-globins that differ by a single amino acid. Other vertebrates show some differences. Sometimes there are copy number differences (for example, mice have a single α-globin and two β-globin genes, goats have a triplicated β-globin gene cluster, and many fish have two cytoglobin genes). In addition, there are qualitative differences, such as a globin gene that is expressed specifically in the eye of birds. MYA, million years ago.

functions in the case of neuroglobin. Fish have two cytoglobin genes that are expressed widely but have diverged in gene expression: the *cygb-2* gene is expressed very strongly in neuronal tissues, unlike the *cygb-1* gene.

The blood globins have subsequently undergone further specialization, but this time differences in gene regulation result in differences in developmental timing. Thus, in early development, ζ-globin is expressed instead of α-globin, which will take its place at later stages, and ε-globin (embryonic period) and γ-globins (fetal period) are used instead of β-globin, which begins to be synthesized at three months' gestation and gradually accumulates as γ-globin production declines. One rationale is that the globins used in hemoglobin during embryonic and fetal periods are better suited to the more hypoxic environment at these stages.

As we shall see later on in this chapter, different metazoan species can have very different copy numbers for a specific gene family and, in some cases, the differences can clearly be seen to be evolutionarily advantageous.

Two or three major whole genome duplication events have occurred in vertebrate lineages since the split from tunicates

Whole genome duplication (WGD) offers a powerful way of increasing genetic complexity. At a stroke, the number of genes is doubled. However, in most gene pairs, one of the genes is eventually lost from the genome, so that WGD is followed by a period of *diploidization*. Over time the net result is that the total gene content is only slightly increased in comparison with that before WGD. Nevertheless, many gene pairs are retained, often because of dosage constraints (for example, where gene products work as part of a protein complex in which precise ratios of the interacting proteins are important). Over long evolutionary time-scales the retained duplicate genes can eventually diverge in function (**Figure 10.13**).

Although a change in polyploidy could be expected to cause problems at meiosis and thus be deleterious and difficult to establish, WGD seems to have been prominent in various evolutionary lineages, especially plants. An evolutionarily recent WGD event can be detected by identifying an organism that has twice as many chromosomes as in closely related species. An ancient WGD event is more difficult to detect because during the ensuing period of diploidization there is not only massive gene loss but also major genome rearrangements (such as chromosome inversions and translocations, which occur naturally over long periods).

Plants are adept at asexual reproduction, and they have provided the most evidence for WGD (most plants belonging to the eudicot clade of angiosperms are descended from a hexaploid ancestor). WGD has also been well studied in

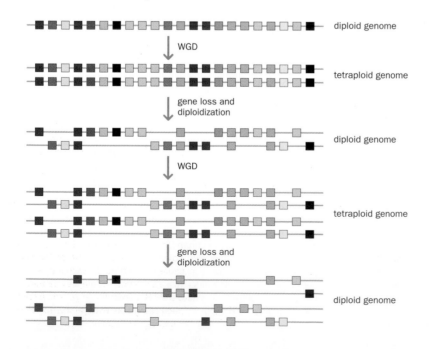

Figure 10.13 Paralogous genes originating from whole genome duplication followed by large-scale gene loss. This hypothetical genome has 22 different genes; only one haploid set is shown but the genome is diploid. Whole genome duplication (WGD) results in a complete series of paralogs in identical order and, initially, a tetraploid genome. In most of the paralogous gene pairs, one gene acquires disabling mutations to become a pseudogene and is eventually lost, restoring diploidy. A subsequent round of WGD followed by gene loss would result in paralogous sets of two, three, or very occasionally four gene copies. [Adapted from Dehal P & Boore JL (2006) *PLoS Biol.* 3, e314 (doi:10.1371/journal.pbio.0030314).]

yeasts, in which genome comparisons have confirmed a WGD in the evolutionary lineage leading to *S. cerevisiae*. The most convincing example of WGD has been documented in species of *Paramecium*, in which, as a result of recent WGD, more than 50% of the genes in the genome of *P. tetraurelia* are clearly seen to be duplicated. The *Paramecium* studies show that after WGD, gene loss does not occur initially on a massive scale but instead occurs over a very long time-scale, and that highly expressed genes and genes encoding proteins that function as part of complexes are more likely to be retained after WGD.

Within vertebrates, polyploidy is comparatively common in amphibians and reptiles, and some bony fish are polyploid. Constitutional polyploidy in mammals is expected to be extremely rare, however, because of gene dosage difficulties arising from the X–Y sex chromosome system.

Comparative genomics has defined time points for some of the major WGD events. For most land-dwelling vertebrates, the most recent WGD occurred early in chordate evolution after the split of the lineage giving rise to tunicates such as *Ciona* from the vertebrate lineage, when there seems to have been two rounds of genome duplication. A major WGD event also occurred in the lineage leading to teleost fishes (the bony fishes, including zebrafish, pufferfish, and medaka), subsequent to divergence from mammals (**Figure 10.14**).

Major chromosome rearrangements have occurred during mammalian genome evolution

Large-scale chromosome rearrangements have been frequent during mammalian genome evolution. Sometimes chromosome evolution can be seen to be uncoupled from phenotype evolution. A classic example is provided by the Chinese and Indian species of muntjac, a type of small deer. The two species are so closely related that they can mate and give birth, although the offspring are not viable. The Chinese muntjac has 46 chromosomes, but as a result of various chromosome fusion events, the Indian muntjac has far fewer, but much larger, chromosomes (**Figure 10.15**).

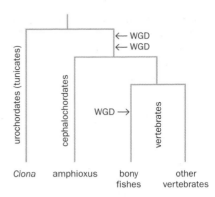

Figure 10.14 Major whole genome duplication events arising in vertebrate lineages. The evolutionary lineages leading to most of the current vertebrates have undergone two major whole genome duplication (WGD) events that took place after the split from tunicates some time within the approximate interval of 525–875 million years ago. In addition, a major WGD event occurred about 340–350 million years ago in the lineage leading to bony fishes.

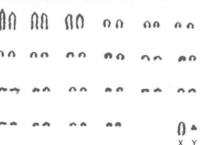

Figure 10.15 The Chinese and Indian muntjacs are very closely related phenotypically but have very different karyotypes. (A) The Chinese muntjac (*Muntiacus reevesi*) has 2n = 46 chromosomes. (B) Because of a series of chromosome fusions, the Indian muntjac (*Muntiacus muntjak*) has two pairs of very large autosomes plus two large X chromosomes in females (2n = 6) or, as shown here, an X chromosome and two different Y chromosomes in males (2n = 7). [Karyotype images adapted from Austin CR & Short RV (1976) The Evolution Of Reproduction. With permission from Cambridge University Press.]

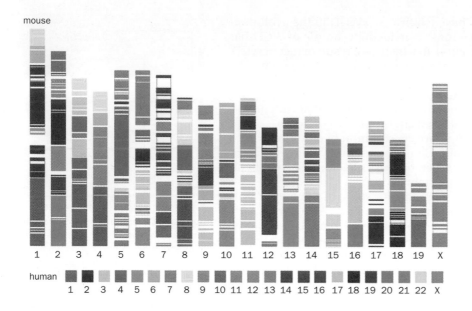

mouse

1 2 3 4 5 6 7 8 9 10 11 12 13 14 15 16 17 18 19 X

human

1 2 3 4 5 6 7 8 9 10 11 12 13 14 15 16 17 18 19 20 21 22 X

Figure 10.16 Human–mouse conservation of synteny is generally limited to small chromosome segments. Segments and blocks larger than 300 kb with conserved synteny in humans are superimposed on the 20 mouse chromosomes. Each color corresponds to a particular human chromosome. The 342 segments are separated from each other by thin white lines within the 217 blocks of consistent color. Human X chromosome segments are all found on the mouse X chromosome as single, reciprocal syntenic blocks. Human chromosome 20 corresponds entirely to a portion of mouse chromosome 2 and human chromosome 17 corresponds to mouse chromosome 11, but with extensive rearrangements into at least 16 segments. Other chromosomes show evidence of much more extensive interchromosomal rearrangement. [From Mouse Genome Sequencing Consortium (2002) *Nature* 420, 520–562. With permission from Macmillan Publishers Ltd.]

This example of karyotype divergence in closely related muntjacs is an extraordinary one, but various types of chromosome rearrangement occur regularly during mammalian evolution. Inversions seem to be particularly frequent; translocations somewhat less so. Centromeres (which are composed of rapidly evolving DNA sequences) can also change positions. During evolution, new centromeres periodically form in euchromatin. Although the underlying mechanisms are poorly understood, studies of such constitutional *neocentromeres* in humans suggest that they tend to form at certain chromosomal hotspots and that chromosome rearrangements can induce neocentromere formation.

Comparison of mammalian chromosome sets shows that, in general, conservation of gene order (synteny) is limited to small chromosome segments. Human–mouse comparisons show that conservation of synteny extends to somewhat less than 10 Mb on average. Genes on one human or mouse chromosome usually have orthologs on various different chromosomes in the other species, but the X chromosome is a notable exception (**Figure 10.16**).

Large-scale conservation of synteny on mammalian X chromosomes is a result of the X-chromosome inactivation that evolved to ensure an effective 2:1 gene dosage ratio for autosomal:X-linked genes. Even in the X chromosome, however, there have been numerous inversions, scrambling gene order between humans and mice. Analysis of the lengths and numbers of conserved synteny segments on the autosomes fits with a process of random chromosome breakage.

When human chromosomes are compared with those of our nearest living relatives, the great apes, there are very strong similarities in chromosome banding patterns. The most frequent rearrangements have been inversions, but some translocations have also occurred; however, the most obvious difference is that the haploid chromosome number is 24 in great apes as opposed to 23 in humans. After the human lineage diverged from those of the great apes, two chromosomes fused to give human chromosome 2.

In heteromorphic sex chromosomes, the smaller chromosome is limited to one sex and is mostly non-recombining with few genes

Different sex-determining systems have evolved in vertebrates. For some fish and reptiles, environmental factors—notably, the temperature at which an egg is incubated—are the primary determinants of the sex of the developing organism. However, genetic sex determination is prevalent in mammals, birds, and amphibians, and also occurs in some fish, reptiles, and invertebrates (**Table 10.5**).

Genetic sex determination involves the development of an unmatched pair of chromosomes. In some cases of genetic sex determination, the sex chromosomes seem, like autosomal homologs, to be virtually identical at the cytological level

TABLE 10.5 EXAMPLES OF DIFFERENT GENETIC SEX DETERMINATION SYSTEMS

System	Sex chromosome constitutions	Distribution/examples	Comments
X–Y	XY (male); XX (female)	prevalent in mammals, but also in some amphibians and in some fish[b]	Y chromosome carries male-determining factor; sperm are X or Y and so determine the sex
Z–W	ZW (female); ZZ (male)	prevalent in birds and reptiles, but also in some amphibians and in some fish[b]	eggs are Z or W and so determine the sex
X-autosome ratio	XX (female/hermaphrodite); X0[a]/XY (male)	in many types of insect and some nematodes	*C. elegans* is an example in which XX individuals are hermaphrodites; *D. melanogaster* has a Y chromosome but it does not confer maleness
Haplo-diploid	relies on ploidy only	in many types of insect	unfertilized eggs produce haploid fertile males; fertilized eggs are diploid and give rise to females or sterile males

[a]The 0 denotes zero, so that X0 means a single X. [b]There can be different systems in the same kind of organism. For example, some medaka species have the X–Y system and others have the Z–W system.

(*homomorphic*) but are presumed to differ at the gene level. In other cases, the sex chromosomes can clearly be seen to be structurally different and are said to be *heteromorphic*.

In heteromorphic sex chromosomes, one sex has two different sex chromosomes and is said to be *heterogametic*, being able to produce gametes with either one of the two sex chromosomes. The other sex is *homogametic* and produces gametes with the same sex chromosome. For some species, the male is heterogametic, whereupon the sex chromosomes are designated X and Y. For other species, the female is heterogametic and the sex chromosomes are designated Z and W (see Table 10.5).

In both the X–Y and Z–W sex chromosome systems, the chromosome that is exclusively associated with the heterogametic sex (Y and W) is comparatively small and poorly conserved with very few genes and with large amounts of repetitive DNA. For example, the human X chromosome contains many genes (including close to 900 protein-coding genes) within 155 Mb of DNA. The Y chromosome has just 59 Mb of DNA and much of it is genetically inert constitutive heterochromatin. There are more than 80 protein-coding genes on the Y chromosome, but several genes are repeated and effectively the human Y chromosome encodes only close to 30 different types of protein.

In female meiosis, the two X chromosomes pair up to form bivalents much like any pair of autosomal homologs, but X–Y pairing in male meiosis is more problematic because of major differences between the X and Y chromosomes in size and sequence composition. Recombination between the human X and Y chromosomes occurs exclusively in short terminal regions that, as a result of engaging in frequent X–Y crossovers, are identical on the X and Y chromosomes. Because of recombination these sequences can move between the X and Y chromosomes; they are neither X-linked nor Y-linked and are known as **pseudoautosomal regions**.

The remaining regions of the sex chromosomes, containing the great majority of their sequences, do not recombine in male meiosis and so are *sex-specific regions*. Because the Y chromosome is exclusively male, the non-recombining region on the Y chromosome never engages in recombination and is sometimes referred to as the *male-specific region*. The equivalent region on the X chromosome does, however, recombine with its partner X chromosome in female meiosis.

The Xp/Yp and Xq/Yq pseudoautosomal regions are allelic sequences, but the X-specific and Y-specific components outside these areas are rather different in sequence. Nevertheless, there are multiple regions of X–Y homology in the X-specific and Y-specific regions that indicate a common evolutionary origin for the X and Y chromosomes. They include 25 gene pairs with functional homologs on the X and Y, known as *gametologs* (**Figure 10.17**).

Figure 10.17 The human X and Y chromosomes show multiple regions of homology that indicate a common evolutionary origin. The pseudoautosomal regions at the tips of Xp and Yp are identical, as are those at the tips of Xq and Yq. The non-recombining male-specific region on the Y chromosome (MSY) and the equivalent, X-specific, region on the X chromosome are rather different in sequence but nevertheless show multiple homologous XY gene pairs (*gametologs*). The latter are generally given the same gene symbols followed by an X or a Y such as *SMCX* on proximal Xp and its equivalent *SMCY* on Yq. In some cases, however, the Y-chromosome homologs have degenerated into pseudogenes (with symbols terminating in a P; see examples in the gene clusters labeled a, b, and c). As a result of positive selection, the sequence of the male-determinant *SRY* is now rather different from its original gene partner on the X chromosome, the *SOX3* gene (highlighted in yellow). [Adapted from Lahn BT & Page DC (1999) *Science* 286, 964–967. With permission from the American Association for the Advancement of Science.]

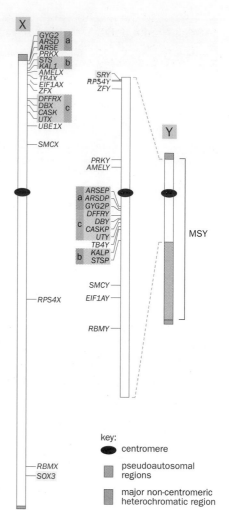

The pseudoautosomal regions have changed rapidly during evolution

Humans have two pseudoautosomal regions. The major one, PAR1, extends over 2.6 Mb at the extreme tips of Xp and Yp. The minor one, PAR2, spans 330 kb at the extreme tips of Xq and Yq. PAR1 is the site of an *obligate* crossover during male meiosis that ensures correct meiotic segregation. As a result, the sex-averaged recombination frequency is 28%, about 10 times the normal recombination frequency for a region of this size. X–Y crossovers also occur in PAR2, but they are not so frequent as in PAR1, and they are neither necessary nor sufficient for successful male meiosis.

The human PAR2 region has been an evolutionarily very recent addition to the sex chromosomes. There is no PAR2 equivalent in mouse or even in some primates, and known mouse orthologs of PAR2 genes are either autosomal or map close to the centromere of the X. There are also very significant species differences in PAR1. Mice do have a counterpart of PAR1 but it is much smaller (only 0.7 Mb long), and in the mouse X chromosome it is located at the tip of the *long* arm.

The mouse PAR shows very little sequence homology to human PAR1. Most of the human PAR1 genes do not have an equivalent mouse ortholog, but some do have identifiable autosomal mouse orthologs although the sequence is very divergent from the human gene (**Figure 10.18**). Other human PAR1 genes have autosomal orthologs in some other mammals. The PAR1 region is thought to have evolved by repeated addition of autosomal segments onto the pseudoautosomal region of one of the sex chromosomes before being recombined onto the other sex chromosome.

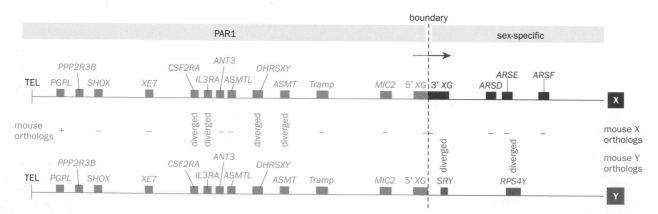

Figure 10.18 Organization and evolutionary instability of the major pseudoautosomal region. The 2.6 Mb human PAR1 region is common to the tips of Xp and Yp and contains multiple genes, of which 13 protein-coding genes are shown here. Adjoining PAR1 are the sex-specific parts of the X and Y chromosomes, with only the first few genes shown here. The PAR1 boundary occurs within the *XG* blood group gene, so that the 5′ end of *XG* is in the pseudoautosomal region whereas the 3′ end of *XG* is X-specific. On the Y chromosome, there is a truncated 5′ *XG* gene homolog containing just the promoter and the first few exons. The adjacent Y-specific DNA contains unrelated sequences including *SRY* and *RPS4Y*. Genes within PAR1 and in the neighboring sex-specific regions (which were part of a former pseudoautosomal region) have been poorly conserved during evolution, and mouse orthologs are often undetectable (–) or have extremely diverged sequences with low similarity to the human ones. TEL, telomere.

PAR1 and the neighboring regions are thought to be comparatively unstable regions. Frequent DNA exchanges result in a high incidence of gene fusions, exon duplications, and exon shuffling. Many of the PAR1 genes and genes in the adjacent part of the sex-specific regions (which were pseudoautosomal regions in the recent evolutionary past) do not seem to have detectable orthologs in the mouse. Those that do have orthologs are often rapidly diverging in sequence, as in the case of the major male determinant, *SRY* (located just 5 kb from the PAR1 boundary) and the neighboring steroid sulfatase gene, *STS*.

Human sex chromosomes evolved after a sex-determining locus developed on one autosome, causing it to diverge from its homolog

The X chromosome is highly conserved in mammals, as is the Z chromosome in birds. But the X–Y pair of sex chromosomes is not homologous to the Z–W sex chromosomes of birds. Mammalian X-linked genes have autosomal orthologs in birds; avian Z-linked genes have autosomal counterparts in mammals. In each case, the different sex chromosomes (X and Y; Z and W) are believed to have evolved from a pair of autosomal homologs that diverged after one of them happened to evolve a major sex-determining locus, such as a testis-determining factor.

The major pseudoautosomal region is known to be of very recent autosomal origin (see above) and, in addition, much of human Xp is also known to be of recent autosomal origin. Many X-linked genes in placental mammals are found to be also X-linked in marsupials, but genes mapping distal to Xp11.3 have orthologs on autosomes of both marsupials and monotremes. The simplest explanation is that at least one large autosomal region was translocated to the X chromosome early in the eutherian lineage that led to placental mammals.

The emergence of a sex-determining locus happens by modification of a preexisting gene. For example, the male determinant gene *SRY* (**S**ox-**r**elated gene on the **Y** chromosome) and the X-linked *SOX3* gene are thought to be an X–Y gene pair that evolved from a common autosomal gene. *SRY* is found in both placental mammals and marsupials such as the kangaroo, but it is absent in monotremes (egg-laying mammals, such as the platypus), and the monotreme *SOX3* gene is autosomal. It is likely that *SRY* developed from a modification of what was an autosomal *SOX3* gene some time after monotremes split off from the main mammalian lineage but before the divergence of marsupials and placental mammals (**Figure 10.19**). *SRY* has been subject to positive selection and so over millions of years its sequence has diverged very significantly from the *SOX3* sequence.

Once a sex-determining locus has been established, recombination needs to be suppressed in the region containing it to ensure that the sex-determining locus remains on just the one chromosome. This can be achieved through chromosomal inversions, which are known to be able to suppress recombination over broad regions in mammals, and which seem to have occurred on the Y chromosome. For example, a Y-specific inversion would explain why the PAR1 boundary crosses a gene that is intact on the X chromosome but disrupted on the Y (see Figure 10.18).

In the X–Y system, natural selection will favor the emergence of other alleles on the Y that confer an advantage on males; as we will see below, many genes on the Y are involved in testis functions. As more alleles are selected with a male advantage, the area in which recombination is suppressed is extended (**Figure 10.20**).

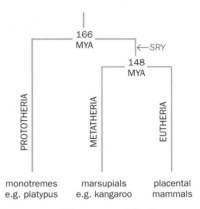

Figure 10.19 Divergence of the three major mammalian subclasses and the emergence of *SRY*. *SRY*, the major male determinant, and *SOX3* are an evolutionarily related X–Y gene pair in placental mammals and marsupials, but monotremes lack *SRY* and have autosomal *SOX3* genes. *SRY* probably developed from an autosomal *SOX3*-like gene on the main therian lineage leading to both placental mammals and marsupials, shortly after the split from prototherian lineages. MYA, million years ago.

Figure 10.20 The X and Y chromosomes evolved from a pair of autosomes. The progression from a homomorphic pair of ancestral autosomes to heteromorphic sex chromosomes is shown here in three steps. (1) The ancestral autosomes diverge into proto-X and proto-Y after a gene on the proto-Y chromosome is modified so as to become a sex-determining locus, such as a testis-determining factor (TDF). (2) Recombination is suppressed in the neighborhood of the male sex-determining locus to ensure that it remains on just the one chromosome, and natural selection promotes the acquisition of additional male-specific alleles, causing further suppression of recombination. The regions where recombination is suppressed become sex-specific. (3) The X-specific region can engage in recombination with its partner X chromosome in female meiosis, but the Y-specific region can never recombine with another chromosome. It will gradually pick up harmful mutations, causing genes to become inactive pseudogenes that are eventually lost by deletions because there is no selective constraint to maintain inactive genes. As a result, there is progressive loss of genes so that only a few remain, mostly ones that confer a male advantage, and the Y chromosome will contract. Eventually, recombination between the X chromosome and the small Y chromosome is limited to a small pseudoautosomal region (PAR), shown in white. [Adapted from Graves JA (2006) *Cell* 124, 901–914. With permission from Elsevier.]

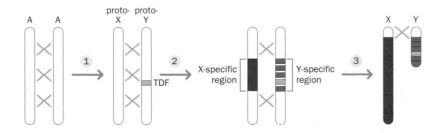

Unlike the human X chromosome, which can recombine throughout its length with a partner X chromosome in female meiosis, the human Y chromosome can be viewed as a mostly asexual (non-recombining) chromosome. Population genetics predicts that a non-recombining chromosome will acquire harmful mutations. This happens because natural selection acts through recombination to eliminate deleterious mutations.

Imagine a deleterious mutation arising in some gene on a chromosome that has another gene with a beneficial allele. Natural selection can act through recombination to uncouple the beneficial and harmful mutations, retaining the beneficial mutation and eliminating the harmful one. But if they cannot be uncoupled—because recombination between them is not possible—deleterious mutations can hitchhike along with strongly beneficial ones, or if strongly deleterious mutations are eliminated by selection, linked beneficial mutations will also be discarded.

Once harmful mutations accumulate in the non-recombining Y and cause loss of gene function, inactive pseudogenes are formed. There is no selective pressure to retain the relevant DNA segment, and genes will be progressively lost. Normal DNA turnover mechanisms will ensure that the chromosome gradually, but inexorably, contracts by a series of deletions.

The X chromosome is thought to have much the same DNA and gene content as the ancestral autosome. By contrast, the human Y chromosome has lost the great majority of its genes and much of the DNA content and has been envisaged to be heading for extinction. On the basis that the Y chromosome had the same number of genes as the X several hundred million years ago but is now reduced to a very small number, one calculation has suggested that the Y chromosome would be driven to extinction within only 10 million years from now. As we will see in the following section, that calculation did not foresee that *gene conversion*, a mechanism involving non-reciprocal sequence transfer, can occur in parts of the Y chromosome.

Abundant testis-expressed genes on the Y chromosome are mostly maintained by intrachromosomal gene conversion

As described above, during differentiation of the sex chromosomes the Y chromosome has lost numerous genes and much of the original DNA sequence. The euchromatic region of the male-specific region on the Y chromosome reveals about 80 protein-coding genes, the great majority of which are predominantly expressed in the testis, plus more than 200 pseudogenes. The Y chromosome seems to be a mosaic of three major sequence classes (**Figure 10.21**).

The sequences in the X-transposed class account for a total of 3.4 Mb in distal Yp and exhibit 99% sequence identity to sequences in Xq21. They originated from a massive X–Y transposition that occurred only about 3–4 million years ago, after the divergence of the human and chimpanzee lineages. Only two protein-coding genes have been identified (**Table 10.6**).

Figure 10.21 Sequence classes in the male-specific region of the human Y chromosome. Only the terminal pseudoautosomal regions (PAR1 and PAR2) of the Y chromosome recombine; the rest of the chromosome is non-recombining and is described as the male-specific region (MSY). The roughly 23 Mb euchromatic region spans Yp and proximal Yq, and consists of three major sequence classes: X-transposed sequences (transposed from Xq21 only 3–4 million years ago), X-degenerate sequences (surviving relics of ancient autosomes from which the X and Y chromosomes co-evolved), and ampliconic segments (sequences that show marked similarity to sequences elsewhere on the MSY). The ampliconic sequences include eight palindromes (P1 to P8) that have arms (9 kb–1.45 Mb in length) with essentially identical sequences separated by shorter (2–170 kb) unique spacer sequences. The structure of the P6 palindrome is shown at the bottom as an example. The rest of the MSY comprises heterochromatin. There is also a 0.4 Mb internal block of heterochromatin in the euchromatic region. [Adapted from Skaletsky H, Kuroda-Kawaguchi T, Minx PJ et al. (2003) *Nature* 423, 825–837. With permission from Macmillan Publishers Ltd.]

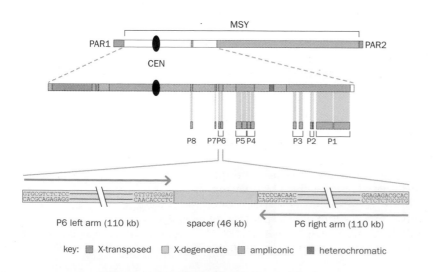

TABLE 10.6 ABOUT 75% OF PROTEIN-CODING GENES ON THE HUMAN MSY REGION FUNCTION IN THE TESTIS

MSY sequence class[a]	Gene number	Gene(s)	Copy number	Expression	Function
X-transposed	2	TGIF2LY	1	testis	transcription factor
		PCDH11Y	1	fetal brain, brain	protocadherin
X-degenerate	18	SRY	1	testis	major male determinant
		RPS4Y1, ZFY	1 each	ubiquitous	housekeeping
		AMELY	1	teeth	tooth development
		TBL1Y, PRKY, USP9Y, DDX3Y, DBY, UTY, TMSB4Y	1 each	ubiquitous	housekeeping
		NLGN4Y	1	fetal brain, brain, prostate, testis	neuroligin
		CYorf14, CYorf15A, CYorf15B, SMCY, EIF1AY, RPS4Y2	1 each	ubiquitous	housekeeping
Ampliconic	60	TSPY	35	testis	spermatogenesis factors, etc.
		VCY, XKRY, CDY2, CSPG4LY, HSFY	2 each		
		RBMY	6		
		GOLGA2LY, PRY	2 each		
		BPY2	3		
		DAZ	4		

[a]See Figure 10.21 for the location of the different gene classes in the human MSY (male-specific region of the Y chromosome). Data adapted from Skaletsky et al. (2003) *Nature* 423, 825–837.

A second class of euchromatic Y chromosome sequences is described as X-degenerate; it has several protein-coding genes and also many pseudogenes that are more distantly related by sequence to X-linked genes. They are thought to be surviving relics of the ancient autosomes from which the X and Y chromosomes evolved. Most of the functional genes are housekeeping genes and make proteins that are closely related to those made by partner genes on the X-chromosome, which all escape X-inactivation.

The third sequence class is made up of *ampliconic* segments. They are largely composed of sequences with pronounced similarity to other sequences in the male-specific region of the Y (MSY). These intrachromosomal repeat units can be large, and ampliconic sequences contain most of the protein-coding genes on the Y chromosome. The genes are members of families of testis-expressed genes whose copy number shows considerable variation (see Table 10.6). In addition, many testis-expressed noncoding transcription units have been identified within the ampliconic sequence class.

The amplicons in this third sequence class include eight very large palindromes (sequence elements that have the same 5′→3′ sequence on both DNA strands). The palindromes are imperfect because they consist of two paired arms with the same sequence separated by a short stretch of unique sequence. The sequence identity between the paired arms of individual palindromes is extraordinarily high. For example, the P1 palindrome consists of two arms that are 1.45 Mb long but show 99.97% sequence identity over their lengths.

The very high degree of sequence identity does not reflect very recent duplication, because the palindromes mostly originated before the chimpanzee–human split 5 million years ago. Instead, the palindrome arms seem to have undergone concerted evolution by a mechanism that resembles *gene conversion*, a recombination-associated mechanism involving non-reciprocal transfer of DNA sequence from a donor DNA sequence to a highly homologous acceptor

DNA sequence (a copy is made of the donor DNA sequence that is then used to replace the original acceptor sequence—see Box 14.1 for the mechanism). In the context considered here, the outcome is that one arm of a palindrome acts as a gene conversion donor sequence and the other arm as an acceptor so that the sequence of one arm is periodically replaced by a perfect copy of the sequence of the other arm. Essentially, this is a type of intrachromosomal recombination and may act so as to preserve the testis-expressed genes from the extinction that they might otherwise have faced in the absence of recombination.

X-chromosome inactivation developed in response to gene depletion from the Y chromosome

As a result of large-scale destruction of Y chromosome sequences, there is an imbalance in the copy number of X-linked genes between the two sexes. Females have two gene copies, but males mostly have just one copy of X-linked genes because there is almost always no partner gene on the Y chromosome (see Figure 10.17 for some exceptions). For many genes, dosage needs to be tightly controlled and because X-linked gene products interact with those made by autosomal genes, a form of gene dosage compensation evolved whereby a single X chromosome was selected to be inactivated in female cells (**X-chromosome inactivation**).

A small minority of human X-linked genes do have functional homologs on the Y chromosome and could be expected to escape X-inactivation. All PAR1 genes tested escape X-inactivation. PAR2 genes are different. The two most telomeric genes, *IL9R* and *CXYorf1*, escape inactivation but the two proximal genes, *SYBL1* and *HSPRY3*, are subject to X-inactivation in females. The copies of *SYBL1* and *HSPRY3* on the Y chromosome are methylated and not expressed, a compensatory form of local *Y-chromosome inactivation*, so that males and females each have one functioning copy for both these genes.

In addition to the genes in the pseudoautosomal regions, about one-fifth of the total genes on the X chromosome escape inactivation. Escaping genes tend to be concentrated in clusters mainly on Xp, and many of them seem to derive from recent autosomal additions to the sex chromosomes. Genes that do not have a functional homolog on the Y chromosome but yet escape inactivation may be ones for which 2:1 dosage differences are not a problem.

The expression pattern of non-pseudoautosomal genes varies between species. For example, the human non-pseudoautosomal genes *ZFX*, *RPS4X*, and *UBE1* all escape inactivation, but the murine homologs *Zfx*, *Rps4*, and *Ube1X* (which, unlike the human *UBE1* gene, has a homolog on the Y chromosome) are all subject to X-inactivation. We will explore the mechanism of X-inactivation in Chapter 11.

10.4 OUR PLACE IN THE TREE OF LIFE

Phylogenetics seeks to assess evolutionary relationships (**phylogenies**) between organisms or populations. Classical phylogenetic approaches have been based on anatomical and morphological features of living organisms and on information gleaned from the fossil record. With such approaches it has been possible to classify organisms at different hierarchical levels known as taxonomic units or **taxa** (**Figure 10.22**, and **Box 10.6** for a glossary).

More recently, molecular genetics has made a huge contribution to phylogenetics. Here, we consider the basis of molecular phylogenetics and how molecular genetics is transforming our understanding of the evolutionary relationships of metazoans and where we fit in within the tree of life. Finally, we consider what is being uncovered that can help us understand what makes us unique.

Molecular phylogenetics uses sequence alignments to construct evolutionary trees

Molecular phylogenetics uses nucleic acid or protein comparisons as the basis of classifying evolutionary relationships between organisms. Until recently, sequences from at most a very few genomic locations were used. However, phylogenies based on very small sequence data sets can be misleading and can give

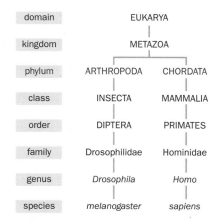

domain	EUKARYA	
kingdom	METAZOA	
phylum	ARTHROPODA	CHORDATA
class	INSECTA	MAMMALIA
order	DIPTERA	PRIMATES
family	Drosophilidae	Hominidae
genus	*Drosophila*	*Homo*
species	*melanogaster*	*sapiens*

Figure 10.22 Eight major taxonomic ranks are traditionally used to classify living organisms. A taxonomic unit or *taxon* (plural = taxa) is a name that either identifies a specific type of organism (*species*) or a group of related organisms (*genus, family, order*, etc.). There are eight major taxonomic ranks, ranging from domain (the most basic division of life, comprising bacteria, archaea, and eukarya) to genus. More detailed taxonomy information can be accessed at the US National Center for Biotechnology Information's Taxonomy Browser (http://www.ncbi.nlm.nih.gov/Taxonomy/CommonTree/wwwcmt.cgi).

BOX 10.6 A GLOSSARY OF COMMON METAZOAN PHYLOGENETIC GROUPS AND TERMS

Amniotes—vertebrates that develop an amnion. Includes reptiles, birds, and mammals but not fishes and amphibians.

Anthropoids—*hominoids* and monkeys.

Ascidians—sea squirts (e.g. *Ciona intestinalis*), a type of *tunicate*.

Bilaterians—bilaterally symmetrical metazoans, including the majority of animal phyla. They mostly have three germ layers and so are now almost synonymous with *triploblasts*.

Catarrhine primates—*hominoids* and *Old World monkeys* (see Figure 10.26).

Cephalochordates—*chordates* that do not have a skull or vertebral column (e.g. lancelets, *Amphioxus*).

Chordates—animals that undergo an embryonic stage in which they possess a notochord, a dorsal tubular nerve cord, and pharyngeal gill pouches; includes *urochordates* (*tunicates*), *cephalochordates,* and *craniates*.

Cnidarians (= coelenterates)—eukaryotes that have two germ layers and are radially symmetrical (e.g. jellyfish, corals, and sea anemones).

Coelomates—organisms with a coelom, or fluid-filled internal body cavity that is completely lined with mesoderm, as opposed to acoelomates (e.g. flatworms) and pseudocoelomates (e.g. nematodes that have a fluid-filled body cavity but one that is not completely lined with mesoderm). Divided into two groups, *protostomes* and *deuterostomes*.

Clade—a group of organisms including all the descendants of a last common ancestor. A clade is a monophyletic taxon.

Craniates—*chordates* with skulls; that is, vertebrates plus hagfish.

Deuterostomes—from the Greek, meaning *second mouth*. Animals in which the blastopore (the opening of the primitive digestive cavity to the exterior of the embryo) is located to the posterior of the embryo and becomes the anus. Later on, the mouth opens opposite the anus. Includes *chordates* and *echinoderms*. Compare with *protostomes*.

Diploblasts—metazoans with only two germ layers: ectoderm and endoderm. This group is often classified as *radiata*, because its members are radially symmetrical. It comprises *cnidarians* and ctenophora (comb jellies).

Echinoderms—radially symmetrical marine adult animals that develop from bilaterally symmetrical larvae. They have three germ layers and so are *triploblasts*.

Eutherian mammals—placental mammals.

Gnathostomes—jawed vertebrates. The great majority of vertebrates are gnathostomes, but lampreys are jawless vertebrates.

Hagfish—a type of fish that have skulls but are classed as invertebrates because the notochord does not get converted into a vertebral column. They are closely related to vertebrates.

Hominids—humans plus great apes (gorilla, chimpanzee, and bonobo) plus orangutans (see Figure 10.26).

Hominins—humans plus great apes (see Figure 10.26).

Hominoids—humans plus great apes plus orangutans plus gibbons (see Figure 10.26).

Metatherian mammals—marsupials (e.g. kangaroo).

Monotremes—*prototherian* egg-laying mammals (e.g. platypus).

New World monkeys (Platyrrhini)—monkeys that are limited to tropical forest environments of southern Mexico, and Central and South America.

Old World monkeys (Cercopithecidae)—monkeys that are found in a wide variety of environments in South and East Asia, the Middle East, and Africa.

Phylum—a group of species sharing a common body organization.

Platyrrhine primates—*New World monkeys*.

Prototherian mammals—*monotremes*.

Protostomes—from the Greek meaning *first mouth*. Organisms in which the mouth originates from the blastopore (the opening of the primitive digestive cavity to the exterior of the embryo). Later on, the anus will open opposite the mouth. Includes molluscs, annelids, and arthropods. Compare with *deuterostomes*.

Radiata—radially symmetrical animals with only two germ layers. Also known as *diploblasts*.

Taxon—a group of organisms recognized at any level of the classification.

Teleost fish—bony fish (as opposed to cartilaginous fish), characterized by a fully moveable upper jaw, rayed fins, and a swim bladder (e.g. zebrafish and pufferfish).

Triploblasts—metazoans with three germ layers; they are bilaterally symmetrical and hence can also be described as *bilaterians*.

Tunicates—a primitive chordate group (sea squirts, e.g. *Ciona intestinalis*).

Urochordates—*chordates* that have a notochord limited to the caudal region. There is only one major grouping, the *tunicates*.

For more information on phylogenetic groupings, see the NCBI Taxonomy Browser (http://www.ncbi.nlm.nih.gov/Taxonomy/).

differing results. Modern molecular phylogenetics can take advantage of whole genome comparisons as described in Section 10.2, and the discipline is transforming into *phylogenomics*.

To construct an evolutionary tree it is necessary to compare sequences from different species; nucleic acid sequences are often used, but if the sequences are from distantly related organisms they may be difficult to align; where available, protein sequences are used instead. If two or more sequences show a sufficient degree of similarity (*sequence homology*) they can be assumed to be derived from a common ancestral sequence. Sequence alignments can then be used to derive quantitative scores describing the extent of relationship between the sequences.

Comparing sequences of equal fixed length is usually straightforward if there is a reasonably high sequence homology. Often, however, the nucleic acid sequences that are compared will have previously undergone deletions or insertions, so that rigorous mathematical approaches to sequence alignment are needed.

Once the sequences have been aligned, evolutionary trees can be constructed. They are most commonly represented as diagrams that use combinations of lines (*branches*) and nodes. The different organisms (sequences) under comparison are placed at external nodes, connected via branches to interior nodes (intersections between the branches) that represent ancestral forms for two or more organisms.

A *rooted tree* (or *cladogram*) infers the existence of a common ancestor (represented as the trunk or *root* of the tree) and indicates the direction of the evolutionary process (**Figure 10.23B**). The root of the tree may be determined by comparing sequences against an *outgroup* sequence (one that is clearly but distantly related from the sequences under study). An *unrooted tree* (Figure 10.23A) does not infer a common ancestor and shows only the evolutionary relationships between the organisms. Note that the number of possible rooted trees is usually much higher than the number of unrooted trees.

Evolutionary trees can be constructed in different ways, and their reliability is tested by statistical methods

Different approaches are used for constructing evolutionary trees, but many employ a *distance matrix* method. The first step calculates the evolutionary distance between all pairs of sequences in the data set and arranges them in a table (matrix). This may be expressed as the number of nucleotide/amino acid differences between the two sequences, or the number of nucleotide/amino acid differences per nucleotide/amino acid site (**Figure 10.24**).

Having calculated the matrix of pairwise differences, the next step is to link the sequences according to the evolutionary distance between them. For example, in one approach the pair of sequences that have the smallest distance score are connected with a root between them. The average of the distances from each member of this pair to a third node is used for the next step of the distance matrix and the process is repeated until all sequences have been placed in the tree. This always results in a rooted tree, but variant methods such as the *neighbor relation* method can create unrooted trees.

The methods used to construct evolutionary trees usually make some assumptions that may not always be true. Thus, they usually assume that all changes are independent, which can cause false interpretations if a single event led to multiple changes. The mutation rate is also usually assumed to be constant in different lineages, which may not be correct. However, the *neighbor joining* method is a variant that does not require all lineages to have diverged by equal amounts per unit time. It is especially suited for data sets comprising lineages with greatly varying rates of evolution.

Alternatives to distance matrix methods include maximum-parsimony and maximum-likelihood methods. *Maximum-parsimony* methods seek to use the minimum number of evolutionary steps. They consider all the possible evolutionary trees that could explain the observed sequence relationships but then select those that require the fewest changes. *Maximum-likelihood* methods create all possible trees and then use statistics to evaluate which tree is likely. This

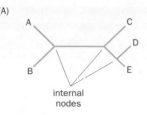

(A)

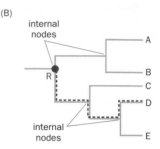

(B)

Figure 10.23 Unrooted and rooted evolutionary trees. (A) This unrooted tree has five external nodes (A, B, C, D, and E) linked by lines (branches) that intersect at three internal nodes. Such a tree specifies only the relationships between the sequences (or other characters) under study but does not define the evolutionary path. (B) In this rooted tree, from the internal node known as the root (R), there is a unique evolutionary path that leads to any other node, such as the path to D (dashed red line). With both trees, a close evolutionary relationship is indicated when the cumulative length of the lines that join two external nodes is short. Thus, both trees suggest that D and E are much more closely related to each other than either is to A.

(A)

(C)
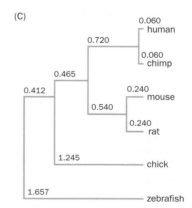

(B)
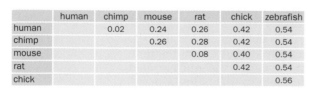

Figure 10.24 Constructing a phylogenetic tree with UPGMA. In this example, the N-terminal 50 amino acids of ζ-globins from human and six other vertebrate species were aligned, and a rooted tree was constructed by using the UPGMA (unweighted pair-group method with arithmetic mean) hierarchical clustering method, which first calculates a *distance matrix*. (A) The human ζ-globin sequence is used as a reference for the alignment. Sequence identity is shown by a dot. (B) A distance matrix is constructed in which pairwise comparisons are used to establish the fraction of amino acid sites that are different between two sequences. For example, the human and chimp sequences differ at one amino acid out of 50, and so the fraction is 0.02. (C) The distance matrix values are used to construct the phylogenetic tree. Note that the lengths of the horizontal branches are proportional to the distances between the nodes. Phylogeny software packages such as PHYLIP are available at various Web servers (e.g. at the Institut Pasteur in Paris, http://bioweb2.pasteur.fr/phylogeny/intro-en.html) and typically offer neighbor joining (unrooted trees) and UPGMA (rooted trees) as alternatives.

may be possible for a small number of sequences. For a large number of sequences, the number of trees generated becomes extremely large and the computational time needed to identify the best evolutionary tree becomes impossibly long. In that case, a subset of possible trees is created by using *heuristics*: methods are used that will produce an answer in a computable length of time, even if it may not be the optimal one.

Once an evolutionary tree has been derived, statistical methods can be used to gain a measure of its reliability. A popular method is **bootstrapping**, a form of Monte Carlo simulation. Typically, a subsample of the data is removed and replaced by a randomly generated equivalent data set; the resulting *pseudosequence* is analyzed to see whether the suggested evolutionary pattern is still favored. If there is a clear relationship between two sequences, the randomization introduced by the resampling will not erase it. If, however, a node connecting the two sequences in the original tree is spurious, it may disappear on randomization because the randomization process changes the frequencies of the individual sites.

Bootstrapping often involves resampling subsets of data 1000 times. The *bootstrap value* is typically given as a percentage. A value of 100 means that the simulations fully support the original interpretation; values of 95–100 indicate a high level of confidence in a predicted node; values less than 95 do not mean that the original grouping of sequences is wrong, but that the available data do not provide convincing support.

As a result of a combination of classical and molecular phylogenetic approaches, humans have been classified as belonging to a range of different groups of organisms. **Figure 10.25** shows simplified phylogenies for eukaryotes and metazoans, and **Figure 10.26** shows a vertebrate phylogeny.

The G-value paradox: organism complexity is not simply related to the number of (protein-coding) genes

Whereas the genomes of single-celled organisms are rather compact, genome (and gene) sizes tend to have increased in the lineages leading to metazoans, and vertebrate genomes tolerate many large introns and also large amounts of highly repetitive DNA. However, there is an inconstant relationship between the complexity of an organism and the amount of DNA in its cells (the *C-value paradox*), and some plants and even some types of ameba are reported to have considerably more DNA per cell than humans (see Chapter 4, p. 96). The paradox is partly explained by extensive polyploidy in some organisms.

The relationship between the total number of different genes and organism complexity is also complex. Because RNA genes are not easy to identify, the great majority of the identified genes in published whole genome sequences are protein-coding genes. Bacteria and archaea tend to have small genomes with small numbers of genes (often about 4000–5000 protein-coding genes), whereas vertebrate genomes tend to have about 20,000 protein-coding genes. However, the 1 mm long *C. elegans* has only 959 or 1031 cells depending on its sex but has virtually the same number of genes as humans. More recently, the genomes of a tiny water flea, *Daphnia pulex*, and an aphid, *Acyrthosiphon pisum*, have been reported to have many thousand more genes than the human genome (**Table 10.7**).

Part of the *G-value paradox* (where G value denotes gene number) relates to recent whole genome duplication—this is clearly evident in *Paramecium tetraurelia* for example, which seems to have undergone several recent whole genome duplications, with gene loss occurring gradually over time.

In some species, there seem to be constraints on the amount of genomic DNA that can be tolerated. Thus, although the fruit fly *D. melanogaster* is much more complex than *C. elegans* it has 6000 fewer genes. For some reason, there is a high genomewide intrinsic rate of DNA loss in *Drosophila* that can be imagined to constrain the extent of gene duplication. Rampant DNA deletion has been thought to explain the comparative lack of repeats (all except 8% of the *D. melanogaster* genome is composed of unique sequences) and pseudogenes (only about 100 in the *D. melanogaster* genome, in contrast with more than 12,000 in the human genome).

(A)

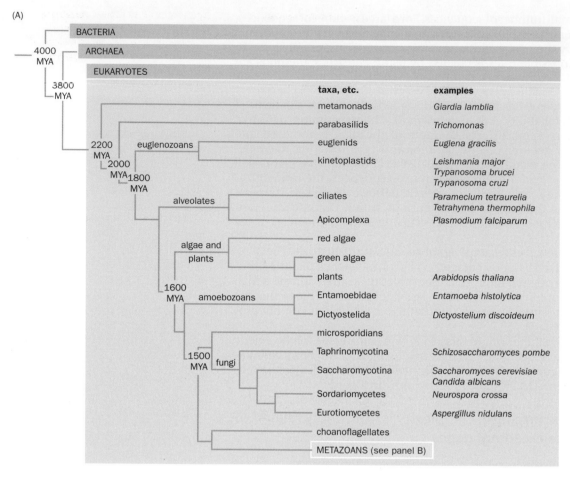

(B)

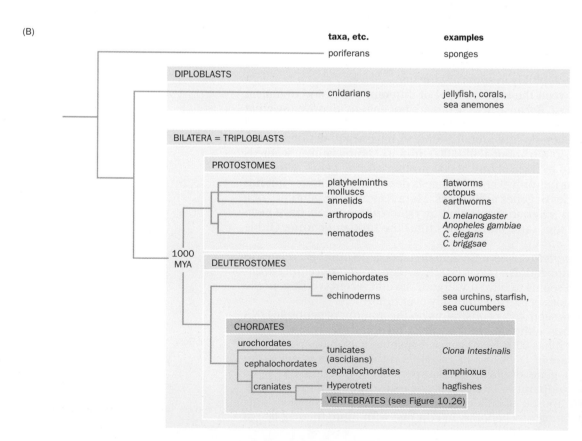

Figure 10.25 Simplified eukaryotic and metazoan phylogenies. (A) Life is divided into three domains—bacteria, archaea, and eukarya—and the eukaryotic phylogeny is shown here. The choanoflagellates (also called collar flagellates) are considered to be the closest living protist relatives of the sponges, the most primitive metazoans. (B) In the metazoan phylogeny, diploblasts have two germ layers whereas triploblasts have three germ layers and are bilaterally symmetrical. Note that the fundamental protostome–deuterostome split, which occurred about 1000 million years ago (MYA), means that *C. elegans* and *D. melanogaster* are more distantly related from humans than are some other invertebrates such as the sea urchin *Strongylocentrotus purpuratus*, and notably other chordates including amphioxus and the tunicate *Ciona intestinalis*.

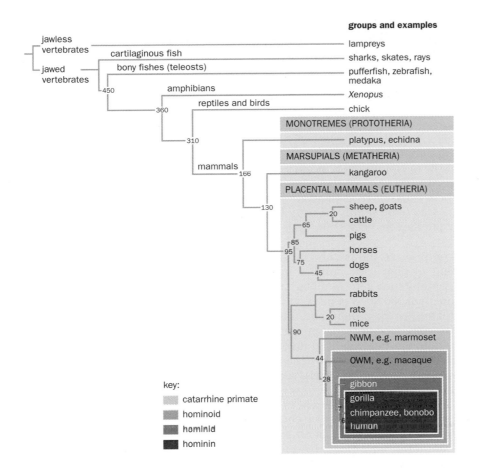

Figure 10.26 A simplified vertebrate phylogeny. Numbers at nodes show estimated divergence times in millions of years. OWM, Old World monkeys; NWM, New World monkeys.

As we will see in the next section, another explanation is lineage-specific expansion of the copy number of certain gene families.

The inconsistent relationship between gene number and biological complexity has been called the G-value paradox. It raises two important questions. First, if the total number of genes is not the primary determinant of organism complexity, what is? We consider alternative determinants later in this chapter. Second, why should such very simple organisms such as a water flea need so many genes?

One possible explanation for an unexpectedly high gene number in some simple organisms is pressure to adapt to a range of rather different environments and predators. For example, the water flea, *Daphnia*, can live in diverse aquatic environments that may be acidic, salty, or hot; sometimes it lives in water that is

TABLE 10.7 THE INCONSISTENT RELATIONSHIP BETWEEN GENE NUMBER AND ORGANISMAL COMPLEXITY			
Species	**Description**	**Genome size (Mb)**	**Number of protein-coding genes**
H. sapiens	humans	3100	~20,000
D. melanogaster	fruit fly	169	~14,150
C. elegans	nematode	100	~20,200
T. thermophila	*Tetrahymena*, single-celled protozoan (Box 10.1)	104[a]	>27,000[a]
A. pisum	pea aphid	517	>34,600
D. pulex	water flea		>39,000
P. tetraurelia	*Paramecium*, single-celled protozoan	72[a]	~39,600[a]

[a]This refers to the macronuclear (somatic) genome. Ciliated protozoans are unusual because they often have two nuclei—a large macronucleus with a genome that has a somatic, non-reproductive function, and a smaller micronucleus that is required for reproduction.

so shallow that it is extremely exposed to powerful sunlight. Depending on the predator, it can also develop tail spines, helmets, or ridges to protect itself. Additional gene duplication may have provided *Daphnia* with additional genetic capacity that helps it adapt to diverse environments.

Striking lineage-specific gene family expansion often involves environmental genes

The copy number of individual gene families can also vary very significantly between organisms and is the most striking difference between closely related species. The great majority of human genes have counterparts in mice, but the human and mouse genomes have undergone extensive remodeling in the 80–90 million years since human–mouse divergence. Thus, whereas there are just over 15,000 human and mouse genes that can be seen to be simply 1:1 orthologs, the remaining 5000 or so genes belong to gene families that can show major differences in copy number in humans and mice.

Comparison of the completed human, chimpanzee, mouse, and rat genomes reveals a few gene families that are present in high copy number in rodent genomes but absent from primate genomes. Many such gene families have roles in reproduction. In mouse, they include more than 100 *Speer* genes that specify a type of spermatogenesis-associated glutamate (E)-rich protein and more than 200 genes that encode vomeronasal type 1 or type 2 receptors that produce pheromones. Similar major differences in copy number are evident in invertebrate genomes: different *Caenorhabditis* species have many hundreds of genes encoding nuclear hormone receptors and chemoreceptors of the 7TM class (with seven transmembrane domains), whereas *Drosophila* species have a few tens of each.

Extended comparative genomics analyses show that the differences in gene copy number do not simply indicate rapid gene family expansion on certain lineages, but instead a rather complicated process of gene duplication and gene loss. Vomeronasal receptors, for example, are found in a range of vertebrates (**Table 10.8**); they are expressed in the vomeronasal organ, an auxiliary olfactory sense organ found in many animals but absent in primates, presumably as a result of evolutionarily recent gene loss.

Lineage-specific gene family expansions typically involve what have been called environmental genes; that is, genes involved in responding to the external environment. Their products may be chemoreceptors that can be involved in various functions such as sensing odors or pheromones (see Table 10.8), or they may be involved in immunity, in the response to infection, and in degrading toxins (such as the cytochrome P450 family).

A rationale for some gene family expansions may be straightforward. It is not surprising that rodents have many more olfactory receptor genes than we do (see Table 10.8). In mammals, olfactory neurons express just one type of olfactory receptor. Mice and rats rely much more on the sense of smell than we do, and so by having many hundred more olfactory receptor genes than us they are much better than we are at discriminating between a multitude of smells. Brain-expressed miRNA genes seem to have undergone significant expansion in primate lineages, which may have contributed to primate brain development. But it

TABLE 10.8 HIGHLY VARIABLE COPY NUMBER FOR CHEMOSENSORY RECEPTOR GENE FAMILIES IN DIFFERENT VERTEBRATES											
Gene family	Number of functional gene copies										
	Human	Chimp	Mouse	Rat	Dog	Cow	Opossum	Platypus	Chicken	*Xenopus*	Zebrafish
OR	388	399	1063	1259	822	1152	1198	348	300	1024	155
V1R	–	–	187	106	8	40	98	270	–	21	2
V2R	–	–	121	79	–	–	86	15	–	249	44
TAAR	6	3	15	17	2	17	22	4	3	6	109

OR, olfactory receptor; *V1R*, vomeronasal type 1 receptor; *V2R*, vomeronasal type 2 receptor; *TAAR*, trace amine-associated receptor. Data abstracted from Nei et al. (2008) *Nat. Rev. Genet.* 9, 951–963.

is less clear why Kruppel-associated box (KRAB)-zinc finger gene families have expanded as they have in various mammalian lineages.

Although gene family expansions can therefore be an adaptive change, there also seems to be a significant random element in the way in which gene family copy number can vary between species, a process that has been called *genomic drift*.

Regulatory DNA sequences and other noncoding DNA have significantly expanded in complex metazoans

If increased gene number does not account for organism complexity, what does? Alternative splicing has been put forward as a possible explanation to the paradox. It is comparatively rare in simple metazoans such as *C. elegans*, but the great majority of human genes undergo alternative splicing and there are significant human–chimpanzee differences in splicing of orthologous exons.

Another characteristic that clearly distinguishes between the genomes of complex and simple metazoans and may well underlie organism complexity is the proportion of noncoding DNA. Simple and complex metazoans can have much the same amount of coding DNA (e.g. 25 Mb in *C. elegans* and 32 Mb in humans) but the amount of noncoding DNA increases hugely in complex metazoans (75 Mb in *C. elegans*, but 3070 Mb in humans).

The increase in noncoding DNA includes a significant expansion in the amount of untranslated regions of protein-coding genes, which are known to contain a wide variety of regulatory DNA sequences. The amount of highly conserved (and presumably functionally important) noncoding DNA has also expanded very significantly during metazoan evolution. The most simple metazoans have very little highly conserved noncoding DNA. Thus, when the *C. elegans* genome is compared with that of another nematode, *C. briggsae*, almost all highly conserved sequence elements are found to overlap coding exons. In *D. melanogaster*, however, the amount of conserved noncoding DNA is about the same as the amount of coding DNA, and in humans the amount of highly conserved noncoding DNA is close to four times that of coding DNA.

The great majority of the conserved functionally important noncoding DNA sequences in the human genome are believed to be involved in gene regulation. They include *cis*-acting regulatory DNA sequences (such as enhancers and splicing regulators), and sequences transcribed to give regulatory RNA. As detailed in Section 9.3, recent analyses have uncovered a huge number of functional noncoding RNAs in complex genomes whose existence was not even suspected a few years ago.

Gene regulation systems in vertebrate cells are much more complex than in the cells of simple metazoans. As described below, there is considerable evidence that ever increasing sophistication in gene regulation and lineage-specific innovations lie at the heart of organism complexity.

Mutations in *cis*-regulatory sequences lead to gene expression differences that underlie morphological divergence

Genomewide comparisons reveal an average 99% identity between orthologous human and chimpanzee protein sequences. Phenotype evolution may be more attributable to changing gene regulation than changing protein sequences. Consistent with this view is the fact that extremely closely related species such as sibling species in the *D. melanogaster* subgroup show divergence in the expression of orthologous genes that is much greater than the corresponding sequence divergence. Transcription factor binding sites are also, in general, known to evolve rapidly—in the 80–90 million years after human–mouse divergence, more than one-third of human transcription factor binding sites are nonfunctional in rodents, and vice versa.

Analyses in various organisms have confirmed the importance of *cis-regulatory elements* (*CREs*) in dictating morphological evolution. In close to 30 case studies, precise changes in CREs that control single genes have been shown to underlie morphological differences. In each case, the divergence in traits arose between closely related populations or species, and CRE evolution is considered sufficient to account for the changes in gene regulation within and between

Figure 10.27 Humans have a gene that makes flies grow wings.
(A) *Drosophila* with a mutation in the *apterous* gene lack wings. The normal phenotype can be rescued after inserting the wild-type fly gene (B) or its human ortholog, *LHX2* (C), into the mutant fly embryo. [Adapted from Rincón-Limas DE, Lu CH, Canal I et al. (1999) *Proc. Natl Acad. Sci. USA* 96, 2165–2170. With permission from the National Academy of Sciences.]

(A)

(B)

(C)

closely related species. The cumulative data led Sean Carroll to propose a new genetic theory of morphological evolution in 2008.

Key developmental regulators typically exhibit *mosaic pleiotropism*: they serve multiple roles (and so are pleiotropic), but they also function independently in different cell types, germ layers, body parts, and developmental stages. Because the same protein can shape the development of many different body parts, the coding sequences of such regulators are under very considerable evolutionary constraint, and so are extremely highly conserved. At the protein level, their functions can be extraordinarily conserved; sometimes, their functions can be maintained after they have been artificially replaced by very distantly related orthologs, and sometimes even paralogs.

As shown in **Figure 10.27**, a human ortholog of the *Drosophila apterous* gene can regulate the formation of fly wings. The implication is that it is not just the gene regulators that are highly conserved, but also the proteins that they interact with to perform their functions, and that in some cases the conservation has been maintained over as much as the billion years of evolution that separate humans and flies (Figure 10.25B). Although there was a rapid expansion in the number of fundamentally different genes encoding transcription factors, signaling molecules, and receptors at the beginning of metazoan evolution, there has been comparatively little expansion since the period just before the origin of bilaterians (see Figure 10.25B). For example, 11 of the 12 vertebrate *Wnt* gene families, which encode proteins that regulate cell–cell interactions in embryogenesis, have counterparts in cnidarians such as jellyfish and, remarkably, six of the major bilaterian signaling pathways are represented in sponges that diverged from other metazoans at the very beginning of metazoan evolution (see Figure 10.25B).

The Hox genes, which specify the anterior–posterior axis and segment identity in embryogenesis, have also undergone very few changes over many hundreds of millions of years. Coelacanths, cartilaginous fish that are very distantly related to us, have much the same classical Hox genes as we do. More accurately, they have four more Hox genes than us because of a loss of Hox genes in vertebrate lineages leading to mammals (**Figure 10.28**).

Thus, although many other protein families were diversifying over the last several hundred million years, there has been comparatively little diversification of key developmental regulators, possibly because of constraints imposed by gene dosage. Instead, expansion of gene function without gene duplication has often been achieved by using the same gene to shape several entirely different traits. By appropriately regulating gene expression, the same gene product can be sent to different but specific tissues (*heterotopy*) and at different developmental stages (*heterochrony*) where, according to its environment, it can interact with

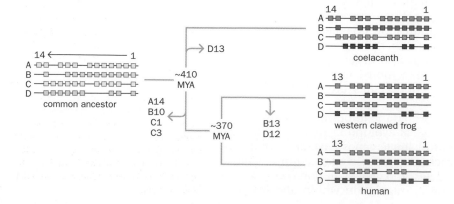

Figure 10.28 Gene duplication has not occurred over hundreds of millions of years of Hox cluster evolution. The coelacanth is a cartilaginous fish that descended from the earliest diverging lineage of jawed vertebrates (see Figure 10.26). Coelacanths, humans, and frogs have remarkably similar Hox gene organizations, and there does not seem to have been any gene duplication in the 410 million years or so since they diverged from a last common ancestor. Rather, there have been instances of occasional Hox gene loss during this time, shown by curved arrows. For simplicity, only the classical Hox genes are shown within the Hox cluster. MYA, million years ago. [Adapted from Hoegg S & Meyer A (2005) *Trends Genet.* 21, 421–424. With permission from Elsevier.]

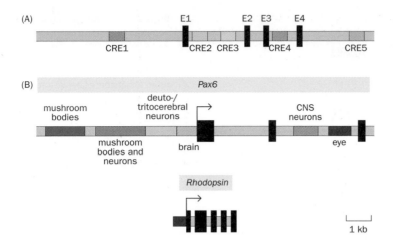

Figure 10.29 Complex *cis*-regulatory regions in pleiotropic developmental regulators. (A) In this generalized scheme, a gene with four exons (E1 to E4) is controlled by five *cis*-regulatory elements (CRE1 to CRE5). Each regulatory element consists of a variety of modules that contain binding sites for *trans*-acting proteins. Clustered binding sites have extended sequence conservation, which makes them easier to detect by comparative genomics. (B) The *Drosophila Pax6* gene, also called *eyeless* (*ey*), has three exons (black bars) and encodes a transcription factor that is expressed in specific parts of the developing brain, central nervous system (CNS), and eyes. Expression is regulated by six CREs (colored blocks), most being 1 kb in length or longer. The rhodopsin gene at the bottom is one of the target genes of Pax6. It has a single function that involves expression in the photoreceptor cells of the eye, and it has a single CRE, as is commonly found in genes that are restricted in expression. [Adapted from Carroll SB (2008) *Cell* 134, 25–36. With permission from Elsevier.]

different sets of molecules. For example, the *Drosophila* Decapentaplegic signaling protein shapes embryonic dorsoventral axis polarity, epidermal patterning, gut morphogenesis, and the patterning of wings, legs, and other appendages. To achieve this type of highly specific regulation, pleiotropic gene regulators typically have several large, modular CREs that *independently* regulate a specific pattern of expression (**Figure 10.29**).

Independent regulation of a developmental regulator by different CREs allows multifunctionality, and harmful mutations in one CRE will not affect the functions of other CREs that regulate the expression of the same protein or the protein itself. Each CRE consists of a series of modules with short binding sites (often about 6–12 nucleotides long) for *trans*-acting regulators, notably transcription factors. In some regulatory elements, individual binding sequences recognized by different transcription factors may be very tightly packed; in others, such as diffuse enhancers, they may be much more spread out along the genomic DNA.

Lineage-specific exons and *cis*-regulatory elements can originate from transposable elements

Complex metazoans have complex genetic regulatory networks, in which pleiotropic transcription factors control hundreds of target genes. *Cis*-regulatory elements are at the heart of these networks, and although some such sequences seem remarkably conserved within a taxon such as a class or an order, the sequence conservation often does not extend over broader evolutionary taxa. Note, for example, the restricted range of evolutionary conservation of many enhancers that bind heart-specific transcription factors (Figure 10.5B). New lineage-specific regulatory elements evolve from time to time, and co-evolution may result in *trans*-acting factors that are lineage-specific (**Figure 10.30**).

If CRE evolution is central to morphological evolution and phenotype divergence, how do CREs evolve? Existing CREs can be modified in different ways. Mutations in an existing CRE can result in new binding sites for transcription

Figure 10.30 The DNA-binding domain of brinker is highly conserved but insect-specific. Brinker is a transcriptional repressor that is important in dorsoventral patterning in insects. The alignment shows the conserved core from a selection of insect species, including the pea aphid *Acyrthosiphon pisum*, the silkworm *Bombyx mori*, the honeybee *Apis mellifera*, the wasp *Nasonia vitripennis*, the head louse *Pediculus humanus*, 10 *Drosophila* species, the flour beetle *Tribolium castaneum*, and three mosquito species *Culex pipiens*, *Aedes aegypti*, and *Anopheles gambiae*. The strong sequence conservation within insects supports the functional importance of this sequence to insect lineages. However, sequences with significant similarity cannot be identified in other organisms. [Adapted from Copley RR (2008) *Phil. Trans. R. Soc. B* 363, 1453–1461. With permission from the Royal Society.]

```
      A. pisum   CCLHKTYHAHSLLSVLDSYRQDSDCQGNQRATARKYGIHRRQIQKWLQTE
       B. mori   AGSRRIFPPQFKLQVLEAYRRDSQCRGNQRATARKFGIHRRQIQKWLQAE
   A. mellifera   MGSRRIFAPAFKLKVLDSYRNDIDCRGNQRATARKYGIHRRQIQKWLQCE
  N. vitripennis  MGSRRIFAPAFKLKVLDSYRNDIDCRGNQRATARKYGIHRRQIQKWLQCE
     P. humanus   VGSRRIFSPHFKLQVLDSYRYDADCRGNQRATARKNIHRRQIQKWLQCE
   D. mojavensis  MGSRRIFTPQFKLQVLESYRHDNDCKGNQRATARKYNIHRRQIQKWLQCE
 D. melanogaster  MGSRRIFTPQFKLQVLESYRNDNDCKGNQRATARKYNIHRRQIQKWLQCE
  D. pseudoobscura MGSRRIFTPQFKLQVLESYRNDNDCKGNQRATARKYNIHRRQIQKWLQCE
    D. ananassae   MGSRRIFTPQFKLQVLESYRNDNDCKGNQRATARKYNIHRRQIQKWLQCE
      D. erecta    MGSRRIFTPQFKLQVLESYRNDNDCKGNQRATARKYNIHRRQIQKWLQCE
      D. yakuba    MGSRRIFTPQFKLQVLESYRNDNDCKGNQRATARKYNIHRRQIQKWLQCE
    D. sechellia   MGSRRIFTPQFKLQVLESYRNDNDCKGNQRATARKYNIHRRQIQKWLQCE
     D. simulans   MGSRRIFTPQFKLQVLESYRNDNDCKGNQRATARKYNIHRRQIQKWLQCE
    D. grimshawi   MGSRRIFTPQFKLQVLESYRNDNDCKGNQRATARKYNIHRRQIQKWLQCE
      D. virilis   MGSRRIFTPQFKLQVLESYRNDNDCKGNQRATARKYNIHRRQIQKWLQCE
   T. castaneum   IGSRRIFAPQFKLQVLDSYRNDIDCKGNQRATARKYGIHRRQIQKWLQVE
    C. pipiens    MGSRRIFTPQFKLQVLDSYRNDSDCKGNQRATARKYGIHRRQIQKWLQVE
    A. aegypti    MGSRRIFTPQFKLQVLDSYRNDSDCKGNQRATARKYGIHRRQIQKWLQVE
    A. gambiae    MGSRRIFTAQFKLQVLDSYRNDGDCKGNQRATARKYGIHRRQIQKWLQVE
```

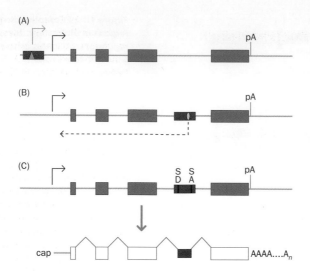

Figure 10.31 Transposable elements can influence gene expression in several ways. (A) When inserted upstream of a gene, a transposable element (red box) may provide new regulatory sequences, such as an enhancer sequence (green triangle) that alters the expression of a downstream gene. (B) A transposable element inserted within an intron may contain promoter elements (yellow oval) that drive antisense transcription. The antisense transcript may interfere with sense transcription and potentially regulate transcription. (C) An inserted transposable element may carry cryptic splice donor (SD) and splice acceptor (SA) sequences that allow it to be incorporated as an alternative exon (*exonization*). For various other possibilities, see Feschotte C (2008) *Nat. Rev. Genet.* 9, 397–405. See Figure 10.32 for specific examples of (A) and (C). pA, polyadenylation site.

factors, for example, or in the loss or modification of existing binding sites. But new CREs may evolve by random mutation and also as a result of sequence insertion events involving transposable elements.

Transposable elements have been viewed as genome parasites that owe their survival to their ability to replicate faster than the host that carries them; RNA interference mechanisms are needed to limit their spread within a genome. Transposable element repeats account for about 45% of the DNA in the human genome, although only a small percentage is active in transposition (see Chapter 9). By integrating into new locations in the DNA, they can influence the expression of nearby genes by a variety of possible mechanisms (**Figure 10.31**).

By disrupting the normal patterns of gene expression, transposable elements are known to cause disease. In addition to being a potential threat at the level of the genome and the individual, however, transposable elements can also be beneficial. Comparative genomic studies have shown that at least 10,000 human transposable element fragments have evolved under strong purifying selection, indicating that they are functionally important. Many of them are located close to genes encoding developmental regulators, and many have been strongly conserved over hundreds of millions of years.

The functional importance of transposable elements arises from their occasional ability to donate sequences that are then used by the host genome for a novel function, a process known as **exaptation**. Thus, the sequences of a small number of genes, sometimes known as *neogenes*, are thought to have originated in large part from transposons. They include *TERT* (telomerase reverse transcriptase), *CENPB* (encoding a centromere protein), and the recombination activating genes *RAG1* and *RAG2*.

More commonly, transposable element sequences have contributed smaller gene components. One way involves donating sequence fragments that have appropriate splice sites so they can be included as novel exons (*exonization*—Figure 10.31C). Such exons have contributed to both functional noncoding RNA (such as the BC200 RNA—see Table 9.9) and various proteins. In addition, transposable element sequences that contain binding sites for *trans*-acting factors have been exapted to provide novel CREs.

As an example, the LF-SINE family of retrotransposons are known to have contributed sequences to generate both a novel exon and a novel enhancer. The starting point for this discovery was uc.338, a 223 bp ultraconserved element located at 12q13 whose sequence is 100% conserved in human, mouse, and rat. The uc.338 sequence lies within the *PCBP2* gene, which encodes an RNA-binding regulatory protein, and spans an alternatively spliced exon that encodes part of the PCBP2 protein. Comparative genomics also showed that the sequence of uc.338 showed a high degree of sequence identity to members of the LF-SINE retrotransposon family in the coelacanth, a fish that has descended from one of the most evolutionary ancient lineages of jawed fishes (**Figure 10.32**).

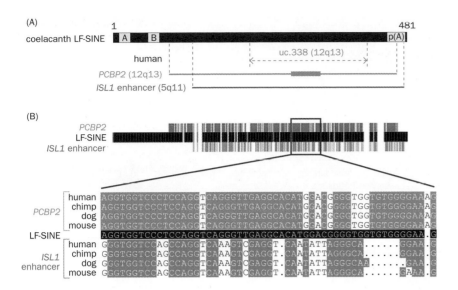

Figure 10.32 Extensive sequence homology indicates a transposon origin for an alternatively spliced *PCBP2* exon and an *ISL1* enhancer. (A) The consensus 481 bp coelacanth LF-SINE retrotransposon has tRNA-like A-box and B-box sequences and ends in a poly(A) tail [p(A)]. Remarkably, it shows more than 80% sequence identity over a 200 bp ultraconserved region on human chromosome 12 (the uc.338 element) that is located within a longer homology region in the *PCBP2* gene (the green box indicates an alternatively spliced exon). Significant homology is also apparent with an enhancer sequence that regulates the *ISL1* gene. (B) In the top panel, vertical bars denote regions of sequence homology of the coelacanth LF-SINE consensus with human *PCBP2* and proximal *ISL1* enhancer sequences, corresponding to panel (A). The expanded red open box in the lower panel shows detailed comparative sequence homology. Shaded *PCBP2* and *ISL1* enhancer sequences indicate sequence identity to the consensus coelacanth LF-SINE element; dots indicate an absence of corresponding nucleotides. [Adapted from Bejerano G, Lowe CB, Ahituv N et al. (2006) *Nature* 441, 87–90. With permission from Macmillan Publishers Ltd.]

There are about 100,000 LF-SINEs in the coelacanth, but, as in other SINE families, evolutionary conservation of LF-SINEs is limited. Nevertheless, a few diverged non-transposing copies are found in other vertebrates, including 244 human copies in addition to the sequence within *PCBP2*. Most have evolved more slowly than would be expected if they were neutral sequences, suggesting that they could have been exapted into cellular roles that benefited the host and then became subject to purifying selection. They also seem to be found preferentially near genes involved in transcriptional regulation and neuronal development, suggesting that they may function as enhancers. Testing of one such candidate showed that it acted as an enhancer that regulated the *ISL1* gene, which encodes a transcription factor needed for motor neuron differentiation. A conserved sequence in this enhancer is clearly related to the LF-SINE consensus sequence (see Figure 10.32).

The LF-SINE family members had been active in transposition in a common ancestor hundreds of millions of years ago and long before the appearance of mammals, and so the alternatively spliced *PCBP2* exon and the proximal *ISL1* enhancer sequence were generated by evolutionarily ancient retrotransposition events.

Gene family expansion and gene loss/inactivation have occurred recently in human lineages, but human-specific genes are very rare

What makes us human? To seek answers to this question, it is unsurprising that we should concentrate on comparisons with our closest living relatives, *Pan troglodytes*, the common chimpanzee, and *Pan paniscus*, the pygmy chimpanzee or bonobo. The human and chimpanzee lineages are thought to have diverged about 5–7 million years ago, about a million or so years after divergence of the lineage leading to present-day gorillas (see Figure 10.26).

Human–chimpanzee genome comparisons have identified 35 million single nucleotide substitutions (just over 1.2% sequence divergence), plus insertions and deletions that account for about 90 Mb or about another 3% sequence divergence. In addition, there are structural differences, with more than 1500 inversions ranging from cytogenetically invisible segments as small as 23 bp to clear chromosomal inversions of up to 62 Mb.

Several hundred regions show significant differences in copy number between the two genomes, often as a result of segmental duplication in one genome, and so gene copy number can vary significantly. For example, as described in Section 10.3, duplication of the salivary amylase gene *AMY1* has occurred repeatedly in human lineages since the human–chimpanzee split, but chimpanzees have a single such gene. The copy number of repetitive elements can also show major differences. The human Alu family, for example, has three times as many members as the chimpanzee Alu family.

TABLE 10.9 EXAMPLES OF PRIMATE-SPECIFIC GENES

Gene	Type of product	Comments
GLUD2	glutamate metabolism enzyme	hominoid-specific retrogene; evolved by retrotransposition from the housekeeping gene GLUD1; believed to have undergone positive selection resulting in specific targeting to mitochondria with novel, potentially brain-specific functions
CDC14Bretro	cell cycle protein	hominoid-specific retrogene that evolved by copying from a microtubule-expressed isoform of CDC14; believed to have undergone positive selection and now preferentially associates with endoplasmic reticulum
TCP10L	likely spermatogenesis factor	also expressed in hepatocytes, where it may be involved in maintaining the differentiation state
USP6	ubiquitin-specific peptidase and oncogene	implicated in a variety of cancers
ZNF674	zinc-finger-containing transcription factor (ZNF)	one of a group of primate-specific genes generated by lineage-specific expansion of a ZNF gene family; implicated in X-linked mental retardation and may be important in cognitive functioning
snaR gene family	~117-nucleotide noncoding RNAs that bind nuclear factor 90	hominid-specific and rapidly evolving; predominantly expressed in testis
Various miRNA genes	various microRNAs	generated by primate-specific expansions in miRNA gene families; they notably include brain-expressed miRNAs, some of which have been claimed to be human-specific

Various primate-specific genes have been identified, and during the past 60 million years or so of primate evolution there seems to have been a burst of retrogene formation—some estimates suggest that at least one new retrogene has formed on average every one million years. Other primate-specific genes have originated through the duplication of genomic DNA sequences, such as by segmental duplication. Some of the primate-specific genes are known to have evolved recently and may be present in hominoids but not monkeys, and some others are restricted to hominids (**Table 10.9**). Truly human-specific genes seem to be extremely rare; however, families of genes encoding brain-expressed miRNAs have undergone significant expansion in primate lineages, and some of the miRNA genes have been reported to be human-specific.

Lineage-specific differential expression of a single gene also plays a part. Inappropriate ectopic expression of a gene may have important consequences, as strikingly revealed with the *FGF4* gene in dogs (**Figure 10.33**). Human–chimpanzee gene differences resulting from lineage-specific gene inactivation are clearly evident. Since the divergence of humans and chimpanzees from a

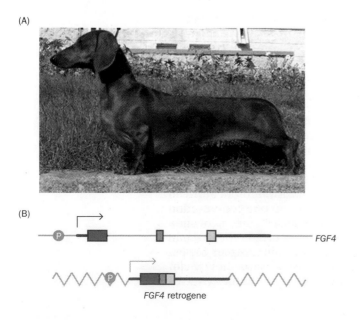

Figure 10.33 The short-legged phenotype of various dog breeds is due to duplication and ectopic expression of a single gene. (A) The dachshund, like many other breeds of dog such as the Welsh corgi and basset hound, has chondrodysplasia, with disproportionately short legs. (B) The chondrodysplasia in these dogs has resulted from the very recent formation of an *FGF4* retrogene that has inserted into a new genomic location. The *FGF4* retrogene is regulated by a different promoter (P) that causes it to be expressed in chondrocytes (mature cartilage cells), unlike the parent *FGF4* gene. [(A) courtesy of Igor Bredikhin under the Creative Commons Attribution 3.0 Unported License.]

TABLE 10.10 EXAMPLES OF EVOLUTIONARY INACTIVATION OR MODIFICATION OF GENES IN THE HUMAN LINEAGE AFTER DIVERGENCE FROM CHIMPANZEES

Gene	Gene product	Nature of change	Consequence
CMAH	CMP-N-acetyl-neuraminic acid hydrolyase	frameshifting deletion	unlike other primates (and mammals), humans do not make N-glycolylneuraminic acid (Neu5Gc), which is one of the two most common sialic acids in mammals and is known to be expressed in a developmentally regulated and tissue-specific fashion
MYH16	type of myosin heavy chain	inactivation	loss of this myosin protein in humans has been suggested to correlate with the considerably smaller masticatory muscles used to chew food, making larger brain sizes possible; see text and Stedman et al. (2004) *Nature* 428, 373–374
SIGLEC11	type of sialic acid	5′ end and first five exons replaced by sequence copied from a pseudogene	new and strong expression in human microglia

common ancestor about 5–7 million years ago, about 80 genes have been inactivated on the human lineage and now appear as nonprocessed pseudogenes, and lineage-specific gene modification is also apparent in some cases (**Table 10.10** shows some examples).

Differential inactivation of *MYH16*, a gene expressed in the jaw muscles, has been considered to be one of the most biologically meaningful differences between humans and chimpanzees. Humans cannot make a MYH16 protein because *MYH16* was inactivated by mutation in the human lineage about 2.4 million years ago, just before the modern hominid cranium evolved. Inactivation of *MYH16* in humans may have led to a decrease in the size of the jaw muscle, allowing the necessary room for the hominid cranium to be remodeled to accommodate a larger brain.

Comparative genomic and phenotype-led studies seek to identify DNA sequences important in defining humans

Identifying genes and noncoding DNA sequences that have been important in adaptive evolution in the human lineage is a challenging task. DNA sequence comparisons can be used to look for conserved but rapidly evolving sequences, and evidence can be sought for positive selection. As we will see, however, there have been difficulties in both detecting positive selection and correlating putative positive selection on human DNA sequences with specific changes in phenotype.

To detect very recent positive selection in the human genome, polymorphism-based searches are used; these involve a comparison of orthologous sequences from different human populations. To detect earlier positive selection that may have contributed to our unique abilities, human DNA sequences can be compared with orthologs from nonhuman primates, and also from *Homo neanderthalensis* by using the recently obtained Neanderthal genome sequence (modern humans and Neanderthals diverged from a common ancestor about 600,000 years ago). Phylogenetic analyses using maximum likelihood are then applied. Data on positive selection of human genes, obtained with phylogenetic methods, can be accessed at the human PAML browser (see Further Reading).

Genes that function in reproduction or in the immune system are often among the most rapidly evolving human genes and are more likely to be targets of positive selection. Positive selection is not, however, a distinguishing feature of human genes—many human genes that have been subject to positive selection in the human lineage may also be subject to positive selection in other lineages. For example, the HLA genes are celebrated examples of positively selected human genes, but their equivalents in many other animals are also known targets of positive selection. To understand why we are unique, positive selection acting *differentially* in the human lineage could more convincingly explain our most distinguishing characteristics.

TABLE 10.11 EXAMPLES OF DNA SEQUENCES THAT HAVE BEEN POSTULATED TO BE SUBJECT TO DIFFERENTIAL POSITIVE SELECTION IN THE HUMAN LINEAGE

DNA sequence	Type of sequence	Properties	References
FOXP2	protein-coding gene	encodes a transcription factor that is important in speech; see the text	Enard et al. (2002) *Nature* 418, 869–872; Enard et al. (2009) *Cell* 137, 961–971
MCPH	protein-coding gene	encodes microcephalin, a protein that regulates brain size	Evans et al. (2005) *Science* 309, 1717–1720
HAR1A (HAR1F)	RNA gene	expressed specifically in Cajal–Retzius neurons in the developing neocortex; co-expressed with it is reelin, a protein that is fundamentally important in specifying the six-layer structure of the brain	Pollard et al. (2006) *Nature* 443, 167–172
HACNS1	enhancer	important in early development; in comparison with orthologous sequences in chimpanzee and rhesus macaque it has gained a strong expression domain in the developing limb that has been postulated to result in altered human limb anatomy, such as specialization of the hand to facilitate using tools or the foot to allow walking on two legs	Prabhakar et al. (2008) *Science* 321, 1346–1350

In several cases of putative positive selection in the human lineage, the great majority of nucleotides altered in the human lineage arose by replacement of an A or T with a G or C. This AT→GC bias has prompted the suggestion that the nucleotide changes simply reflect biased gene conversion rather than positive selection; for a recent discussion, see Hurst LD (2009) *Nature* 457, 543–544.

We show in **Table 10.11** some examples of DNA sequences that have been believed to be positively selected in the human lineage, and we outline different approaches to identify them below. To identify genes that confer human characteristics, one major approach starts with phenotypes that relate to our special cognitive abilities and brain capacity. Studies of human microcephaly, a neurodevelopmental disorder characterized by decreased brain growth, have identified genes that regulate brain growth such as *MCPH1* for which there is evidence for positive selection in the human lineage. However, the ongoing adaptive evolution of such genes is not explained by increased intelligence. Various language and reading disorders with a significant genetic component have also been studied, including developmental dyslexia (reading disability), speech language disorder (SLD), and also specific language impairment (SLI), where the heritability is particularly high.

The SLD gene *FOXP2* has been extensively studied. It encodes a transcription factor that has been very well conserved during evolution—out of a total of 715 amino acids, only 3 are changed in the mouse. Of these, two amino acids, N303 and S325, seem to be human-specific when compared with various nonhuman primates and are thought to have been positively selected during recent human evolution (**Figure 10.34**).

Evidence that the apparently human-specific N303 and S325 residues are important in brain function comes from the *Foxp2^{hum/hum}* mouse, which is homozygous for a *humanized* version of the *Foxp2* gene (genetically engineered to express the human-specific N303 and S325). *Foxp2^{hum/hum}* mice have altered vocalization and also changes in certain neurons that are expected to enhance the efficiency of brain mechanisms that in humans are known to regulate motor control, including speech, word recognition, and several other cognitive properties.

```
                                          303                      325
modern human   DNGIKHGGLD LTTNNSSSTT SSNTSKASPP ITHHSIVNGQ SSVLSARRDS SSHEETGASH TLYGHGVCKW
 Neanderthal   DNGIKHGGLD LTTNNSSSTT SSNTSKASPP ITHHSIVNGQ SSVLSARRDS SSHEETGASH TLYGHGVCKW
       chimp   DNGIKHGGLD LTTNNSSSTT SSTTSKASPP ITHHSIVNGQ SSVLNARRDS SSHEETGASH TLYGHGVCKW
     gorilla   DNGIKHGGLD LTTNNSSSTT SSTTSKASPP ITHHSIVNGQ SSVLNARRDS SSHEETGASH TLYGHGVCKW
   orangutan   DNGIKHGGLD LTTNNSSSTT SSTTSKASPP ITHHSIVNGQ SSVLNARRDS SSHEETGASH TLYGHGVCKW
      rhesus   DNGIKHGGLD LTTNNSSSTT SSTTSKASPP ITHHSIVNGQ SSVLNARRDS SSHEETGASH TLYGHGVCKW
    marmoset   DNGIKHGGLD LTTNNSSSTT SSTTSKASPP ITHHSIVNGQ SSVLNARRDS SSHEETGASH TLYGHGVCKW
       mouse   DNGIKHGGLD LTTNNSSSTT SSTTSKASPP ITHHSIVNGQ SSVLNARRDS SSHEETGASH TLYGHGVCKW
```

Figure 10.34 Human-specific amino acids in the highly conserved FOXP2 protein. The FOXP2 protein has been extremely highly conserved during evolution, showing 99.6% identity to mouse Foxp2. Alignment with nonhuman primate orthologs suggests that amino acids N303 and S325 result from non-synonymous substitutions in the human lineage after human–chimp divergence but before the lineages leading to modern humans and Neanderthals diverged about 600,000 years ago.

Systematic genomewide screens for positive selection operating on protein-coding genes have also sought genes important in recent adaptive evolution in humans. However, the top candidate genes typically show a non-random pattern of nucleotide changes: the great majority of the altered nucleotides changed from A or T and became G or C. This AT→GC bias is typically found not just in non-synonymous substitutions but also in nearby synonymous substitutions and noncoding intron sequence. Rather than resulting from positive selection, nucleotide changes of this kind may result from a bias in the mechanism of gene conversion (see Box 14.1 for the normal mechanism of gene conversion). GC-biased gene conversion may even promote the fixation of deleterious amino acid changes in primates (see the footnote to Table 10.11).

Conserved noncoding sequences are also being extensively investigated to assess possible roles in adaptive evolution. Variation in gene regulation has long been viewed as being responsible for many of the phenotype differences between species. As a direct test, comparative genomewide gene expression screens have been performed on equivalent tissues from humans and nonhuman primates. Although the data are not easy to interpret, genes encoding transcription factors are particularly likely to show differences in expression between humans and chimpanzees.

Much of the effort to identify positively selected noncoding DNA has focused on short regions of the human genome that are known to have been highly conserved over vertebrate evolution but yet show remarkably high sequence divergence in the human lineage since human–chimpanzee divergence. The vast majority of these *human accelerated regions* (HAR) consist of noncoding DNA. The most remarkable is the HAR1 sequence at 20q13. Although the chimpanzee and chicken lineages diverged about 300 million years ago, their HAR1 sequences show more than 98% sequence identity (only 2 differences out of 118 nucleotides), but the human and chimpanzee HAR1 sequences differ at 18 out of the 118 nucleotides. An RNA gene, *HAR1A*, that overlaps the HAR1 sequence has been proposed to have an important role in human brain development (see Table 10.11). Another such region, variously known as *HACNS1* or HAR2, acts as an enhancer that seems to have acquired a human-specific gain of function (see Table 10.11). However, in the human lineage, nucleotide changes in the HAR regions also show emphatic AT→GC biases, raising the question of the contributions of biased gene conversion and positive selection, as above.

CONCLUSION

Model organisms have long been useful in research, to understand molecular and physiological processes and to study disease and evaluate potential therapies for human diseases. Now that we have the genome sequences for many of these model organisms as well as humans, the detection of differences and similarities between genomes is enabling us to learn more about the processes that shape a genome and allow it to evolve.

Comparative genomics has allowed the construction of molecular phylogenetic trees of extant species, deepening our understanding of our position in the tree of life, and revealing information about how genomes, including our own, have evolved.

As well as helping to identify and validate predicted protein-coding genes, comparative genomics has also been helpful in identifying functional noncoding DNA. The latter is thought to account for about 80% of the highly conserved DNA in the human genome, being made up of RNA genes and short regulatory elements that control how genes are expressed.

Functionally important sequences tend to be strongly conserved and are subject to purifying (negative) selection, with harmful mutations being selectively eliminated. Those mutations that are beneficial tend to be positively selected. These mutations underlie adaptive evolution.

Shuffling and duplication of the exons within protein-coding genes increases gene complexity. Transposable elements can also be copied and used to make coding sequences or to introduce new functionally important sequences, such as novel exons and novel regulatory elements.

Evolutionary novelty can also arise through the duplication of genes or even whole genomes. Once a gene has been duplicated, the two copies can diverge in sequence and ultimately in function. The mutations can be in the coding sequence or in *cis*-regulatory elements that determine when and where the gene is expressed.

Large-scale chromosomal rearrangements have occurred during mammalian genome evolution. Gene order is only conserved over small segments (synteny), and syntenic genes are often found on many different chromosomes when species are compared. The sex chromosomes are the exceptions. Comparative genomics has provided insights into how the sex chromosomes developed from autosomal ancestral chromosomes, and into the mechanisms that have led to the divergence of the X and Y chromosomes, with the loss of sequences on the Y chromosome.

Comparative genomics has also thrown up the G-value paradox, that organism complexity does not correlate with the number of protein-coding genes. Whole genome duplication and subsequent DNA loss can partly explain this paradox. In addition, there are lineage-specific expansions of certain gene families.

Variation in regulatory sequences and other noncoding DNA may also be important. Ever increasing sophistication in gene regulation and lineage-specific innovations may lie at the heart of the complexity of organisms.

The differences between the human genome and those of our closest relatives, the gorilla and the chimpanzee, are small in percentage terms, and subtle. There are some primate-specific genes, including many microRNA (miRNA) genes. There are very few human-specific genes, with only weak evidence for accelerated evolution of genes, particularly miRNAs, expressed in the brain. The rapidly evolving regions of the human genome are noncoding, including some *cis*-regulatory regions and noncoding RNAs, suggesting that gene regulation may have a significant role in distinguishing us from our phylogenetic cousins.

In Chapter 11 we take a closer look at the regulation of gene expression in humans, and in Chapter 12 we consider how gene function and gene regulation are studied in cells and model organisms.

FURTHER READING

Model organisms

Butte AJ (2008) The ultimate model organism. *Science* 320, 325–327.

Davis RH (2004) The age of model organisms. *Nat. Rev. Genet.* 5, 69–76.

Fields S & Johnston M (2005) Whither model organism research? *Science* 307, 1885–1886.

Spradling A, Ganetsky B, Hieter P et al. (2006) New roles for model genetic organisms in understanding and treating human disease: report from the 2006 Genetics Society of America Meeting. *Genetics* 172, 2025–2032.

Unicellular model organisms and synthetic biology

Gibson DG, Benders GA, Andrews-Pfamkoch C et al. (2008) Complete chemical synthesis, assembly, and cloning of a *Mycoplasma genitalium* genome. *Science* 319, 1215–1220.

Keller LC, Romijn EP & Zamora I (2005) Proteomic analysis of isolated *Chlamydomonas* centrioles reveals orthologs of ciliary-disease genes. *Curr. Biol.* 15, 1090–1098.

Mager WH & Winderickx J (2005) Yeast as a model for medical and medicinal research. *Trends Pharmacol. Sci.* 26, 265–273.

Turnbaugh PJ, Ley RE, Hamady M et al. (2007) The Human Microbiome Project. *Nature* 449, 804–810.

Invertebrate animal models

Beckingham KM, Armstrong JD, Texada MJ et al. (2005) *Drosophila melanogaster*—the model organism of choice for the complex biology of multi-cellular organisms. *Gravit. Space Biol. Bull.* 18, 17–29.

Davidson B & Christiaen L (2006) Linking chordate gene networks to cellular behavior in ascidians. *Cell* 124, 247–250.

Kaletta T & Hengartner MO (2006) Finding function in novel targets: *C. elegans* as a model organism. *Nat. Rev. Drug Discov.* 5, 387–398.

Segalat L (2007) Invertebrate animal models of disease as screening tools in drug discovery. *ACS Chem. Biol.* 2, 231–236.

WormBook. http://www.wormbook.org [An online review of *C. elegans* biology.]

Vertebrate models (rodents and zebrafish are referenced in Chapter 20)

Beck CW & Slack JMW (2001) An amphibian with ambition: a new role for *Xenopus* in the 21st century. *Genome Biol.* 2, reviews1029.1–1029.5 (doi:10.1186/gb-2001-2-10-reviews1029).

Capitanio JP & Emborg ME (2008) Contributions of non-human primates to neuroscience research. *Lancet* 371, 1126–1135.

Carlsson HE, Schapiro SJ, Farah I & Hau J (2004) Use of primates in research: a global view. *Am. J. Primatol.* 63, 225–237.

Cyranoski D (2009) Marmoset model takes centre stage. *Nature* 459, 492.

Pennisi E (2007) Boom time for monkey research. *Science* 316, 216–218.

Stern CD (2005) The chick: a great model system becomes even greater. *Dev. Cell* 8, 9–17.

Comparative genomics

Blanchette M (2007) Computation and analysis of multi-sequence alignments. *Annu. Rev. Genomics Hum. Genet.* 8, 193–213.

Dolinski K & Botstein D (2007) Orthology and functional conservation in eukaryotes. *Annu. Rev. Genet.* 41, 465–507.

Margulies EH & Birney E (2008) Approaches to comparative sequence analysis: towards a functional view of vertebrate genomes. *Nat. Rev. Genet.* 9, 303–313.

Miller W, Makova KD, Nekrutenko A & Hardison RC (2004) Comparative genomics. *Annu. Rev. Genomics Hum. Genet.* 5, 15–56.

Stern A, Doron-Faigenboim A, Erez E et al. (2007) Selecton 2007: advanced models for detecting positive and purifying selection using a Bayesian inference approach. *Nucleic Acids Res.* 35, W506–W511.

Ureta-Vidal A, Ettwiller L & Birney E (2003) Comparative genomics: genome-wide analysis in metazoan eukaryotes. *Nat. Rev. Genet.* 4, 251–262.

Comparative genomics: invertebrates

Drosophila 12 Genomes Consortium (2007) Evolution of genes and genomes on the *Drosophila* phylogeny. *Nature* 450, 203–218.

Stark A, Lin MF, Kheradpour P et al. (2007) Discovery of functional elements in 12 *Drosophila* genomes using evolutionary signatures. *Nature* 450, 219–232.

Stein LD, Bao Z, Blasiar D et al. (2003) The genome sequence of *Caenorhabditis briggsae*: a platform for comparative genomics. *PLoS Biol.* 1, e45.

Vertebrate comparative genomics and functional noncoding DNA

Bejerano G, Pheasant M, Makunin I et al. (2004) Ultraconserved elements in the human genome. *Science* 304, 1321–1325.

Elgar G & Vavouri T (2008) Tuning in to the signals: non-coding sequence conservation in vertebrate genomes. *Trends Genet.* 24, 344–352.

Gibbs RA, Weinstock GM, Metzker ML et al. (2004) Genome sequence of the Brown Norway rat yields insights into mammalian evolution. *Nature* 428, 493–521.

Margulies EH, Cooper GM, Asimenos G et al. (2007) Analyses of deep mammalian alignments and constraint predictions for 1% of the human genome. *Genome Res.* 17, 760–774.

Nickel GC, Tefft D & Adams MD (2008) Human PAML browser: a database of positive selection on human genes using phylogenetic methods. *Nucleic Acids Res.* 36, D800–D808.

Pang KC, Frith MC & Mattick JS (2006) Rapid evolution of non-coding RNAs: lack of conservation does not mean lack of function. *Trends Genet.* 22, 1–5.

Pheasant M & Mattick JS (2007) Raising the estimate of functional human sequences. *Genome Res.* 17, 1245–1253.

Visel A, Bristow J & Pennacchio LA (2007) Enhancer identification through comparative genomics. *Semin. Cell Dev. Biol.* 18, 140–152.

General molecular evolution

Bromham L (2008) Reading The Story In DNA: A Beginner's Guide To Molecular Evolution. Oxford University Press.

Li W-H (2007) Molecular Evolution. Sinauer Associates, Inc.

Gene and genome evolution

Aury J-M, Jaillon O, Duret L et al. (2006) Global trends of whole-genome duplication revealed by the ciliate *Paramecium tetraurelia*. *Nature* 444, 171–178.

Babushok DV, Ostertag EM & Kazazian HH Jr (2007) Current topics in genome evolution: molecular mechanisms of new gene formation. *Cell. Mol. Life Sci.* 64, 542–554.

Blomme T, Vandepoele K, De Bodt S et al. (2006) The gain and loss of genes during 600 million years of vertebrate evolution. *Genome Biol.* 7, R43.

Comai L (2005) The advantages and disadvantages of being polyploid. *Nat. Rev. Genet.* 6, 836–846.

Dehal P & Boore JL (2006) Two rounds of whole genome duplication in the ancestral vertebrate. *PLoS Biol.* 10, e314.

Force A, Lynch M, Pickett FB et al. (1999) Preservation of duplicate genes by complementary degenerative mutations. *Genetics* 151, 1531–1545.

Guth SIE & Wegner M (2008) Having it both ways: Sox protein function between conservation and innovation. *Cell. Mol. Life Sci.* 65, 3000–3018.

Hurles M (2004) Gene duplication: the genomic trade in spare parts. *PLoS Biol.* 2, e206.

Perry GH, Dominy NJ, Claw KG et al. (2007) Diet and the evolution of human amylase gene copy number. *Nat. Genet.* 39, 1256–1260.

Rodríguez-Trelles F, Tarrío R & Ayala FJ (2006) Origins and evolution of spliceosomal introns. *Annu. Rev. Genet.* 40, 47–76.

Semon M & Wolfe KH (2007) Consequences of genome duplication. *Curr. Opin. Genet. Dev.* 17, 505–512.

Tvrdik P & Capecchi MR (2006) Reversal of *Hox1* gene subfunctionalization in the mouse. *Dev. Cell* 11, 239–250.

Chromosome evolution

Ferguson-Smith MA & Trifonov V (2007) Mammalian karyotype evolution. *Nat. Rev. Genet.* 8, 950–962.

Graves JA (2006) Sex chromosome specialization and degeneration in mammals. *Cell* 124, 901–914.

Graves JA (2008) Weird animal genomes and the evolution of vertebrate sex and sex chromosomes. *Annu. Rev. Genet.* 42, 565–586.

Marshall OJ, Chueh AC, Wong LH & Choo KHA (2008) Neocentromeres: new insights into centromere structure, disease development, and karyotype evolution. *Am. J. Hum. Genet.* 82, 261–282.

Ross MT, Graham DV, Coffey AJ et al. (2005) The DNA sequence of the human X chromosome. *Nature* 434, 325–337.

Rosser ZH, Balaresque P & Jobling M (2009) Gene conversion between the X chromosome and the male-specific region on the Y chromosome at a translocation hotspot. *Am. J. Hum. Genet.* 85, 130–134.

Rozen S, Skaletsky H, Marszalek JD et al. (2003) Abundant gene conversion between arms of palindromes in human and ape Y chromosomes. *Nature* 423, 873–876.

Skaletsky H, Kuroda-Kawaguchi T, Minx PJ et al. (2003) The male-specific region of the human Y chromosome is a mosaic of discrete sequence classes. *Nature* 423, 825–837.

Molecular phylogenetics and taxonomy

Delsuc F, Brinkmann H & Philippe H (2005) Phylogenomics and the reconstruction of the tree of life. *Nat. Rev. Genet.* 6, 361–375.

Hall BG (2007) Phylogenetic Trees Made Easy: A How-To Manual, 3rd ed. Sinauer Associates, Inc.

NCBI Taxonomy Browser. http://www.ncbi.nlm.nih.gov/taxonomy

Nei M & Kumar S (2000) Molecular Evolution And Phylogenetics. Oxford University Press.

Phylogeny programs. http://evolution.genetics.washington.edu/phylip/software.serv.html [One of several sites collating phylogeny programs available through the Web.]

Recent mammalian genome sequence papers with major evolutionary implications

Mikkelsen TS, Wakefield MJ, Wakefield J et al. (2007) Genome of the marsupial *Monodelphis domestica* reveals innovation in non-coding sequences. *Nature* 447, 167–178.

Rhesus Macaque Genome Sequencing and Analysis Consortium (2007) Evolutionary and biomedical insights from the rhesus macaque genome. *Science* 316, 222–234.

Warren WC, Hillier LW, Graves JAM et al. (2008) Genome analysis of the platypus reveals unique signatures of evolution. *Nature* 453, 175–184.

Interspecific homology, lineage-specific gene family expansion, and gene inactivation/loss

Church DM, Goodstadt L, Hillier LW et al. (2009) Lineage-specific biology revealed by a finished genome assembly of the mouse. *PLoS Biol.* 7, e1000112.

Clusters of Orthologous Groups database. http://www.ncbi.nlm .nih.gov/COG/ [A phylogenetic classification of proteins in complete genomes.]

HomoloGene Database (NCBI). http://www.ncbi.nlm.nih.gov/ homologene [For automated detection of homologs in other species.]

Matsuya A, Sakate R, Kawahara Y et al. (2008) Evola: ortholog database of all human genes in H-InvDB with manual curation of phylogenetic trees. *Nucleic Acids Res.* 36, D787–D792.

Nei M, Niimura Y & Nozawa M (2008) Evolution of animal chemosensory receptor gene repertoires: roles of chance and necessity. *Nat. Rev. Genet.* 9, 951–964.

Ponting CP (2008) The functional repertoires of metazoan genomes. *Nat. Rev. Genet.* 9, 689–698.

Ponting CP & Goodstadt L (2009) Separating derived from ancestral features of human and mouse genomes. *Biochem. Soc. Trans.* 37, 734–739.

Wang X, Grus WE & Zhang J (2006) Gene losses during human origins. *PLoS Biol.* 4, e52.

Noncoding DNA, *cis*-acting regulatory sequences, and morphological evolution

Carroll SB (2008) Evo-Devo and an expanding evolutionary synthesis: a genetic theory of morphological evolution. *Cell* 134, 25–36.

Copley RR (2008) The animal in the genome: comparative genomics and evolution. *Phil. Trans. R. Soc. B* 363, 1453–1461.

Jeong S, Rebeiz M, Andolfatto P et al. (2008) The evolution of gene regulation underlies a morphological difference between two *Drosophila* sister species. *Cell* 132, 783–793.

Taft RJ, Pheasant M & Mattick JS (2007) The relationship between non-protein-coding DNA and eukaryotic complexity. *BioEssays* 29, 288–299.

Visel A, Prabhakar S, Akiyama JA et al. (2008) Ultraconservation identifies a small subset of extremely constrained developmental enhancers. *Nat. Genet.* 40, 158–160.

Wray GA (2007) The evolutionary significance of *cis*-regulatory mutations. *Nat. Rev. Genet.* 8, 206–216.

Transposon-mediated birth of genes, exons, and *cis*-regulatory elements

Bejerano G, Lowe CB, Ahituv N et al. (2006) A distal enhancer and an ultraconserved exon are derived from a novel retroposon. *Nature* 441, 87–90.

Feschotte C (2008) Transposable elements and the evolution of regulatory networks. *Nat. Rev. Genet.* 9, 397–405.

Oliver KK & Greene WK (2009) Transposable elements: powerful facilitators of evolution. *BioEssays* 31, 703–714.

Human–primate differences and the genetic basis of our uniqueness (see also Table 10.10)

Berezikov E, Thuemmler F, van Laake LW et al. (2006) Diversity of microRNAs in human and chimpanzee brain. *Nat. Genet.* 38, 1375–1377.

Bird CP, Stranger BE, Liu M et al. (2007) Fast-evolving non-coding sequences in the human genome. *Genome Biol.* 8, R118.

Galtier N & Duret L (2007) Adaptation or biased gene conversion? Extending the null hypothesis of molecular evolution. *Trends Genet.* 23, 273–277.

Gilad Y, Oshlack A, Smyth GK et al. (2006) Expression profiling in primates reveals a rapid evolution of human transcription factors. *Nature* 440, 242–245.

Kelley JL & Swanson WJ (2008) Positive selection in the human genome: from genome scans to biological significance. *Annu. Rev. Genomics Hum. Genet.* 9, 143–160.

Konopka G, Bomar JM, Winden K et al. (2009) Human-specific transcriptional regulation of CNS development genes by FOXP2. *Nature* 462, 213–218.

Kosiol C, Vinar T, da Fonseca RR et al. (2008) Patterns of positive selection in six mammalian genomes. *PLoS Genet.* 4, e1000144.

Noonan JP (2009) Regulatory RNAs and the evolution of human development. *Curr. Opin. Genet. Dev.* 19, 557–564.

Pollard KS, Salama SR, King B et al. (2006). Forces shaping the fastest evolving regions of the human genome. *PLoS Genet.* 2, e168.

Prabhakar S, Noonan JP, Paabo S & Rubin M (2006) Accelerated evolution of conserved non-coding sequences in humans. *Science* 314, 786.

Sabeti PC, Schaffner SF, Fry B et al. (2006) Positive natural selection in the human lineage. *Science* 312, 1614–1620.

Yang Z (2007) PAML 4: phylogenetic analysis by maximum likelihood. *Mol. Biol. Evol.* 24, 1586–1591.

Chapter 11

Human Gene Expression

KEY CONCEPTS

- The characteristic forms and behaviors of different cell types result from different patterns of expression of the same set of genes. Expression is regulated at both the transcriptional and translational levels.

- RNA polymerase II transcribes all protein-coding genes, and many that encode noncoding RNAs. A large multiprotein complex must be assembled at the gene promoter before transcription can be initiated.

- A fixed set of general transcription factors forms the basal transcription apparatus, with a large cast of gene-specific or tissue-specific transcription factors, co-activators, and co-repressors modulating promoter activity.

- Promoter activity is affected by sequence-specific DNA-binding proteins that bind either directly to the promoter or at more distant positions (enhancers, repressors, and locus control regions). DNA looping brings distant sites close to the promoter. Co-activators and co-repressors do not bind DNA but are recruited to promoters by protein–protein interactions.

- Large-scale studies of human transcripts have shown that the majority of human genomic DNA is transcribed. The function (if any) of most of the transcripts is unknown. The implications of this are currently not clear, but a substantial revision of the traditional view of discrete genes sparsely scattered along genomic DNA may be required.

- The local chromatin conformation may be more important than the local DNA sequence in defining promoters. It is governed by an interplay between histone-modifying enzymes, ATP-driven chromatin remodeling complexes, and DNA methyltransferases that methylate cytosine in CpG dinucleotides. RNA molecules may also have a role.

- Histones in nucleosomes can undergo many different covalent modifications. The pattern of these constitutes an additional layer of heritable information (a histone code) that has a major, but not exclusive, role in determining chromatin conformation.

- Heritable effects on gene expression that are not caused by a change in the DNA sequence are called epigenetic changes. The heritability is due to heritable patterns of methylation of CpG sequences.

- For a few dozen human genes, expression depends on parental origin because of imprinting. This is an epigenetic process whereby alleles at a locus carry an imprint of their parental origin that governs the pattern of expression.

- Most protein-coding genes encode multiple polypeptides because of the use of alternative promoters and/or alternative splicing. In many cases, the resulting protein isoforms have different tissue specificities, biological properties, and functions.

- Gene expression is also regulated by controlling whether or not an mRNA will be translated. This often depends on proteins and microRNAs (miRNAs) that bind to sequences in the 3′ untranslated region of the mRNA.

- The role of the myriad different miRNAs is still uncertain, but it is thought that they fine-tune the expression of many genes. Major changes in gene expression are more often due to controls that affect transcription.

All the cells of our body, apart from gametes, are derived by repeated mitosis from the original fertilized egg. All those cells, with a few minor exceptions, therefore contain exactly the same set of genes. The differences in anatomy, physiology, and behavior between cells, described in Chapters 4 and 5, are the result of differing patterns of gene expression. During normal development, cells follow branching trajectories in which successively more specialized patterns of gene expression define the transitions from totipotency through pluripotency and onward to terminal differentiation. Most of these developmental decisions are irreversible in the context of normal human physiology, although there is great interest in reversing them by artificial manipulation, to produce induced pluripotent cells (see Chapter 21) or to promote the regeneration of damaged tissues or organs. In addition to these fixed patterns, cells have a repertoire of short-term flexible changes in gene expression that enable them to respond to their fluctuating environments.

Controls on gene expression operate at several levels. On the largest scale, megabase regions of chromatin around centromeres and elsewhere form highly condensed and genetically inert heterochromatin that can be seen by Giemsa (G-band) staining of metaphase chromosomes. The dark G bands do contain some expressed sequences, but most expressed genes are located in regions that stain pale in a standard G-banded preparation (see Figure 2.14). The location of a chromosome within the interphase cell nucleus can also have a profound effect on its level of expression.

This chapter explores how gene expression is controlled at the single gene level, from the selection of which genes to transcribe, through processing of the transcript, to translation to produce the initial protein product.

As described in Chapter 8, techniques for studying gene expression have evolved rapidly. Early studies used RT-PCR to document the expression or otherwise of single genes of interest in different tissues or at different stages of development. Gene expression was quantitated by real-time RT-PCR. As with other aspects of genetics, the emphasis moved on to genomewide studies, using microarrays for expression profiling. Typically, the experiments compare expression profiles in two or more tissues or cell states. Now the focus is moving on to direct sequencing of cDNA on a massive scale. In some ways, this is just a refinement of the large-scale generation of expressed sequence tags (ESTs) by partial sequencing of cDNA libraries. However, the new massively parallel sequencing machines allow an unprecedented depth of sequencing, and hence the unbiased quantitation of even very rare transcripts and detection of rare splice isoforms.

Efforts to understand the factors controlling human gene expression have been focused by the ENCODE (Encyclopedia of DNA Elements) project. This is a systematic attempt to identify all functional sequence elements in human DNA (see http://genome.ucsc.edu/ENCODE). A pilot phase, launched in 2003, focused on 44 genomic regions, totaling 30 Mb or 1% of the human genome. In 2007, the project moved into the production phase, studying the whole genome. At the same time, parallel efforts (ModENCODE; http://www.modencode.org) targeted two important model organisms, the fruit fly *Drosophila melanogaster* and the nematode worm *Caenorhabditis elegans*. Data from the ENCODE project inform much of the discussion in this chapter.

There is a growing awareness that the way in which our genome works is much more complex than it once appeared. Early understanding naturally focused on the sparsely scattered protein-coding sequences and was guided by the idea that each gene encoded a single polypeptide. It now turns out that most protein-coding genes encode multiple polypeptides. Moreover, although less than 1.5% of our DNA codes for protein, much of the rest is transcribed, and cells are awash with newly recognized RNA molecules, of whose function (if any) we still have only a very limited understanding. The genome projects showed that we humans have scarcely more protein-coding genes than the 1 mm long *Caenorhabditis elegans* worm. It seems inevitable that our greater complexity must be the result of more complex organization and regulation of a broadly similar set of genes. Some of the new vision was presented in Box 9.5, which discussed revising the concept of the gene. We are still at an early stage of unraveling the mechanisms that control human gene expression, but this chapter outlines current concepts and sets the scene for emerging new data.

11.1 PROMOTERS AND THE PRIMARY TRANSCRIPT

Gene expression depends on transcription. As noted in Chapter 1, humans have three nuclear RNA polymerases (and in fact a fourth, used only in mitochondria). The three nuclear RNA polymerases have distinct functions and use different types of promoter:

- Polymerase I (pol I) specifically transcribes the ribosomal RNA genes that exist as tandem arrays on the short arms of the 10 acrocentric chromosomes (see Figure 1.22).

- Polymerase II (pol II) transcribes all protein-coding genes and many of the genes encoding functional RNAs, including the snoRNAs (small nucleolar RNAs) that modify ribosomal RNA and the microRNAs that regulate translation. This chapter is concerned with the expression of genes that are transcribed by pol II.

- Polymerase III (pol III) is used to transcribe the genes for transfer RNA and the 5S ribosomal RNA, together with some other small RNAs (see Figure 1.15).

- Mitochondria have an entirely different RNA polymerase, encoded by the nuclear *POLRMT* gene. Surprisingly, in the light of the supposed bacterial origin of mitochondria, this is not a bacterial-type polymerase but a single-polypeptide polymerase related to RNA polymerases of bacteriophages.

The molecular events that occur at active promoters are complex in their details, but the principle is straightforward enough. We consider these events first and then go on to the much more difficult question of why one DNA sequence rather than another should be active as a promoter.

Transcription by RNA polymerase II is a multi-step process

Although RNA polymerase is the enzyme that synthesizes the transcript, it does not act alone. Transcription is a highly regulated process. Decisions on which parts of the genome to transcribe, and how actively to do so, are crucial in allowing a cell to respond appropriately to its environment and the demands of the whole organism. These decisions are governed by a large cast of regulatory factors acting in a complex, multi-stage process.

Defining the core promoter and transcription start site

In terms of DNA sequence, pol II core promoters are highly heterogeneous. Several consensus elements have been identified, including the TATA, GC, and CCAAT boxes described in Chapter 1 (see Figure 1.14), but none of them is essential and many promoters lack all of them. The spacing of the various elements is important (**Figure 11.1**), but much depends on binding of locus-specific activators, either at the core promoter or at more distant sites. Characteristic histone modifications (see below) are better than the DNA sequence at predicting promoter activity, but how those modifications are targeted to promoter sequences is not well understood. One survey of 1031 human pol II core promoters found that only 32% of genes carried a TATA box and that such promoters are typically used by genes with highly tissue-specific expression. By contrast, promoters of housekeeping genes, or those with complex patterns of expression, tend to have CpG islands (see below) but not TATA boxes. Transcription start sites are tightly constrained at TATA box promoters but are distributed broadly over a 50–100-bp region at CpG-island-associated promoters. The actual start site is almost always the purine of a pyrimidine–purine dinucleotide. **Box 11.1** describes two of the main methods used to define the start sites of genes.

Figure 11.1 Sequence elements found in RNA polymerase II core promoters. The boxes show DNA sequence elements that are often found in RNA polymerase II promoters, together with their consensus sequences and positions relative to the transcription start site. The TATA box binds the TATA-binding protein subunit of the general transcription factor TFIID. The initiator (Inr) element defines the transcription start site (the highlighted A) when located 25–30 bp from a TATA box. The downstream core promoter (DPE) element is functional only when placed precisely at +28 to +32 bp relative to the highlighted A of an Inr element. The general transcription factor TFIIB binds to the BRE (TFIIB recognition element). However, none of these elements is either necessary or sufficient for promoter activity, and many active polymerase II promoters lack all of them. Py, pyrimidine (C or T) nucleotide; N, any nucleotide. [Adapted from Smale ST & Kadonaga JT (2003) *Annu. Rev. Biochem.* 72, 449–479. With permission from Annual Reviews.]

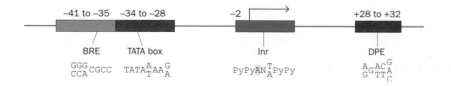

BOX 11.1 TECHNIQUES FOR IDENTIFYING THE TRANSCRIPTION START SITE

5′ RACE (**R**apid **a**mplification of **c**DNA **e**nds) is a form of RT-PCR that allows amplification of the whole mRNA sequence between an internal primer and the 5′ end of the RNA (**Figure 1**). This is a low-throughput method but is particularly useful for identifying unsuspected upstream exons of a gene, the result of initiation at a novel promoter. In the ENCODE project, 5′ RACE was systematically used to examine transcripts from 399 loci in 12 different tissues. RACE products were identified with the use of oligonucleotide microarrays. Novel fragments were found for 90% of all loci tested.

CAGE (**C**ap **a**nalysis of **g**ene **e**xpression) is a high-throughput method that sequences just the first 18 nucleotides adjacent to the 5′ cap of all mRNAs in a sample. The sequence allows the transcription start site to be identified, and the number of occurrences of that sequence is a direct count of the number of molecules of that mRNA in the original sample. The method depends on three basic steps:

*   cDNAs that correspond to the capped 5′ end of mRNAs are selected. This is achieved by a chemical reaction that biotinylates the mRNA cap, followed by selection with streptavidin-coated magnetic beads.
*   An adaptor containing the recognition site for the type IIs enzyme *Mme*I is ligated to the cDNA. Type IIs restriction enzymes recognize a specific DNA sequence but cut the DNA elsewhere, a defined number of base pairs away. Digestion with *Mme*I releases an 18-nucleotide fragment of the cDNA with a two-base overhang (a sticky end).
*   The 18-base fragments with their sticky ends are ligated together to form long concatemers, which are then sequenced.

The method is a variant of the **SAGE** (**S**erial **a**nalysis of **g**ene **e**xpression) technique that first introduced the use of type IIs restriction enzymes and sequencing of concatemerized tags. The short tags are sufficient to identify the molecule, and the concatemerization allows huge numbers of tags to be sequenced, permitting unprecedentedly accurate quantification of the relative abundances of different mRNAs. The purpose of using the type IIs enzyme to generate short tags is to reduce the amount of sequencing needed. With the introduction of massively parallel sequencing technologies, this may no longer be an important consideration, and it may be simpler just to sequence cap-selected cDNAs on a massive scale.

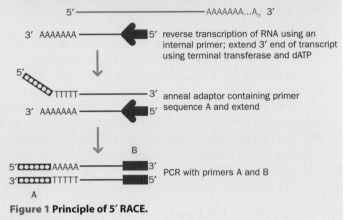

Figure 1 Principle of 5′ RACE.

Assembling the basal transcription apparatus

Given a favorable local chromatin environment, the basal transcription machinery will assemble on the DNA at a promoter. This comprises 27 polypeptide subunits: RNA polymerase II and five multisubunit **general transcription factors** called TFIIB, D, E, F, and H (**Table 11.1**). The machinery assembles sequentially (**Figure 11.2**). The first component to bind is TFIID, a multiprotein complex comprising the TATA-binding protein (TBP) and about 11 TAFs (TBP-associated factors). Promoters that include a TATA box are bound by the TBP component of TFIID; some of the other components presumably direct binding to promoters that lack a TATA box. Once TFIID has been located on the promoter, the remaining TFII factors and RNA polymerase II assemble to form the initiation complex.

TABLE 11.1 THE GENERAL TRANSCRIPTION FACTORS NEEDED FOR INITIATION OF TRANSCRIPTION BY RNA POLYMERASE II

Name	No. of subunits	Roles in initiation of transcription
TFIID		
TBP subunit	1	recognizes TATA box
TAF subunits	~11	recognize other DNA sequences near the start point; regulate DNA binding by TBP
TFIIB	1	recognizes BRE element in promoters; accurately positions RNA polymerase at the start site of transcription
TFIIF	3	stabilizes RNA polymerase interaction with TBP and TFIIB; helps attract TFIIE and TFIIH
TFIIE	2	attracts and regulates TFIIH
TFIIH	9	unwinds DNA at the transcription start point; phosphorylates Ser 5 of the RNA polymerase C-terminal domain; releases RNA polymerase from the promoter

TBP, TATA-binding protein; TAF, TBP-associated factors; BRE, TFIIB recognition element.

Figure 11.2 Initiation of transcription at a pol II promoter. Before RNA polymerase II can initiate transcription, several general transcription factors—TFIID, TFIIB, TFIIE, TFIIF, and TFIIH—are required to assemble sequentially on the promoter. The promoter illustrated here includes a TATA box, to which TFIID binds. Many promoters lack a TATA box and use other sequence elements to assemble the transcription factors and RNA polymerase II. When this basal transcription apparatus has fully assembled, TFIIH opens up the DNA double helix at the transcription start site and initiates a conformational change in the polymerase by phosphorylating its protruding C-terminal domain. This allows the polymerase to commence transcription. Additional DNA-binding transcription factors and protein-binding co-activators (not shown) control the level of activity of a promoter. TBP, TATA-box binding protein; CTD, C-terminal domain of RNA pol II. [From Alberts B, Johnson A, Lewis J et al. (2007) Molecular Biology of the Cell, 5th ed. Garland Science/Taylor & Francis LLC.]

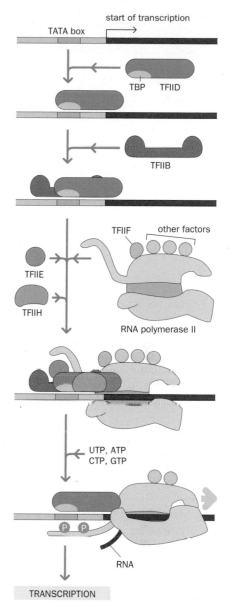

Elongation of the transcript

Unlike DNA polymerases, RNA polymerases do not need a primer to get started, but further actions of the TFII factors are needed before transcription can start. One subunit of TFIIH is a DNA helicase. This uses energy from the hydrolysis of ATP to open up the DNA double helix, giving pol II access to the template strand. The polymerase then starts RNA synthesis but initially fails to progress away from the promoter. Short oligoribonucleotides are produced in a process of abortive initiation until a set of conformational changes occur that allow the pol II to move into elongation mode. These changes are triggered by TFIIH-dependent phosphorylation of serine residues in the C-terminal domain of pol II. Some data suggest that this transition is the most critical part of transcription control. Most promoters, whether active or inactive, are said to be occupied by a pre-initiation complex, but only those on active genes are able to move into elongation mode.

The polymerase then moves along the template strand, leaving behind most of the general transcription factors but picking up new proteins, the **elongation factors** such as the Elongin complex. The average speed is about 20 nucleotides per second, although there are pauses and spurts. At this speed it would take more than 24 hours to transcribe the 2.4 Mb dystrophin gene!

Termination of transcription

There seem to be no specific sequences that act as stop signals for RNA polymerase; there are no transcriptional analogs of the UGA, UAG, and UAA translational stop codons. For many genes, termination occurs at various positions in different individual transcripts. It turns out that termination of transcription is linked to cleavage of the transcript. A well-known feature of mRNAs is the poly(A) addition site—AAUAAA or a closely similar sequence. As described in Chapter 1 (see Figure 1.21), the mRNA is cut a few nucleotides downstream of this site. This cleavage also initiates the events leading to the termination of transcription. The cut is made by a protein complex that is physically tethered to the RNA polymerase. The complex includes an exonuclease. Once the cleavage has produced a free RNA end, the exonuclease moves along the RNA that is still emerging from the polymerase, degrading it in a 5′→3′ direction. A sort of race ensues between the polymerase, continuing along the DNA and elongating the tail of the transcript, and the exonuclease coming up behind the polymerase and eating up the transcript (**Figure 11.3**). The exonuclease is the faster of the two, and when it catches up with the polymerase, transcription terminates.

Many other proteins modulate the activity of the basal transcription apparatus

The efficiency of transcription initiation is greatly influenced by further proteins that bind to the DNA or to the transcription initiation complex. They can be divided into two classes according to whether or not they bind DNA. **Transcription factors** are sequence-specific DNA-binding proteins that bind at or close to the core promoter. Many are tissue-specific or developmentally regulated. Promoters are often inactive in the absence of appropriate transcription factors. Other DNA-binding proteins bind to specific sequences (enhancers and silencers) located some way away from the promoter. Looping of the DNA brings them into close

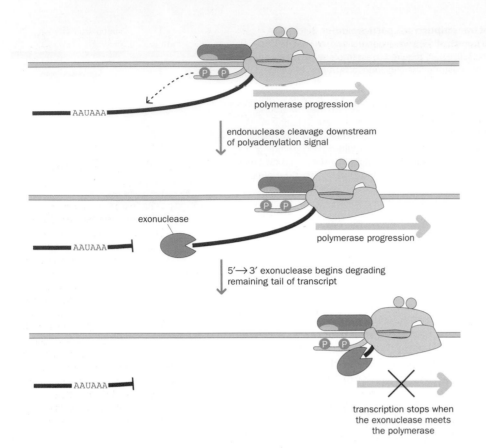

Figure 11.3 Termination of transcription.
After cleavage (dotted arrow) of the transcript (red line) downstream of the polyadenylation signal (AAUAAA) an exonuclease tethered to the RNA polymerase works its way along the tail of the nascent transcript. Transcription ceases when the exonuclease reaches the body of the polymerase.

proximity to the promoter. **Co-activators** and **co-repressors**, in contrast, do not bind DNA: they rely on protein–protein interactions to associate with the transcription initiation complex and modulate its activity.

Sequence-specific DNA-binding proteins can bind close to a promoter or at more remote locations

Transcription factors contain a DNA-binding domain that confers specificity, and an activation domain that mediates the functional effect. The main classes of DNA-binding protein motifs were described in Chapter 4: helix–loop–helix, helix–turn–helix, leucine zipper, and zinc finger motifs (see Box 4.2). These bind to DNA in a sequence-specific manner. Consensus binding sequences can be identified by experiments that systematically modify oligonucleotides to find the one that is most strongly bound. However, the actual sequences in the genome that are bound by a given transcription factor are not readily predictable. Chromatin immunoprecipitation (see below) can be used to identify the sequences actually bound *in vivo*, and these do not always correspond to the optimum binding sequence identified by the first method.

Binding of a transcription factor to DNA is not determined simply by the presence of a suitable binding sequence anywhere in the genome. Binding sites for most DNA-binding proteins are quite short, typically four to eight nucleotides, and some variation in sequence is tolerated. Possible binding sites for any one factor are present in thousands of copies across the genome, but only a tiny minority of them are used. Two factors govern the selection of actual sites from among the large number of potential sites. First, binding is combinatorial. Assembly of the transcription initiation complex depends on co-operative interactions between RNA polymerase and a whole set of transcription factors and co-activators. A typical promoter has multiple potential binding sites, usually for several different proteins. Second, proteins can only bind to the DNA if the relevant sequence is accessible. Access to the DNA is strongly restricted when it is wrapped around nucleosomes. Active promoters need to have a relaxed, open chromatin structure in which the DNA is relatively free of nucleosomes. We consider chromatin structure below.

Binding sites for transcription factors have traditionally been thought of as lying upstream of the transcription start site. However, data from the ENCODE project show that, in fact, averaged across all the genes studied, binding is symmetrically distributed about the transcription start site. Transcription factors are just as likely to bind downstream of the start site as upstream (**Figure 11.4**).

Distal DNA sequences where sequence-specific binding of proteins affects transcription are labeled **enhancers** or **silencers**, depending on whether they activate or repress transcription. They can lie either side of the promoter. The chromatin loops round to provide direct physical contact with proteins bound to the promoter (**Figure 11.5**). Developmental genes, in particular, often have complex sets of very distant enhancers. These have usually been discovered by chance, when a chromosomal translocation separates a gene from an enhancer. This produces the phenotype associated with loss-of-function mutations of the gene, although the coding sequence is intact. Typical examples include the following:

- Heterozygous loss-of-function mutations in the *PAX6* gene on chromosome 11p15 cause aniridia (absence of the iris of the eye; OMIM 106210). Some patients with aniridia have no mutation in *PAX6* but have translocations with breakpoints up to 125 kb downstream of the last *PAX6* exon. Experiments with transgenic mice located a cluster of essential regulatory elements in this region (**Figure 11.6A**). There are several different tissue-specific enhancers in the cluster.

- The role of the *SHH* (sonic hedgehog) gene in limb development was described in Chapter 5. Another gene, *LMBR1*, is located 1 Mb away from the *SHH* gene on chromosome 7q36. Several deletions, insertions, or translocation breakpoints in the *LMBR1* gene in humans or mice cause various limb abnormalities (Figure 11.6B). It was natural to assume that these were the result of a loss of function of *LMBR1*—but point mutations in *LMBR1* had no effect on limb development. Instead, these changes disrupt a series of enhancers of *SHH* that happen to be located in introns of *LMBR1*.

When a chromosomal rearrangement abolishes the expression of an apparently intact gene by separating it from a distant enhancer, as in the *PAX6* cases described above, this is often but incorrectly described as a *position effect*. That term was coined to describe the gene silencing observed in *Drosophila* when a chromosomal rearrangement relocated a gene to a position close to a heterochromatic region (see Chapter 16). Position effects reflect the tendency of heterochromatinization to spread along a chromosome. The only undoubted position effect known in humans is the occasional silencing of autosomal loci when an X-autosome translocation relocates them to an inactivated X chromosome (see below).

When clusters of distal regulatory elements were first discovered that were both necessary and sufficient to ensure the correct tissue-specific expression of a gene, they were called **locus control regions** (LCRs). In experimental systems, locus control regions can confer strong position-independent expression on transgenes, being required if the transgene is to function as a fully independent transcriptional unit. The best-known human examples control expression of the

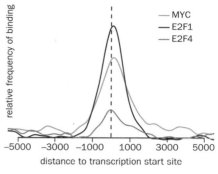

Figure 11.4 Transcription factors may bind either upstream or downstream of the transcription start site. The curves show the distribution of locations at which the transcription factors MYC, E2F1, and E2F4 bound to the promoters of a set of different actively expressed genes, relative to the transcription start site (vertical dashed line). The curves are roughly symmetrical about the transcription start sites, showing that the binding site for one of these transcription factors in a particular gene is just as likely to be downstream of the transcription start site as upstream in the conventional promoter location. [Adapted from the ENCODE Project Consortium (2007) *Nature* 447, 799–816. With permission from Macmillan Publishers Ltd.]

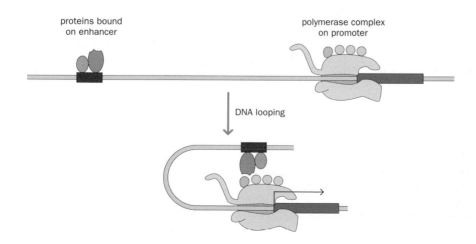

Figure 11.5 DNA looping brings proteins bound to enhancers or silencers into direct contact with proteins bound to the promoter. Regulatory elements such as enhancers (red box) may be located hundreds of kilobases upstream or downstream of the gene they control (blue box). DNA looping allows direct physical interactions between proteins bound to these distal elements and some of the many proteins bound to the promoter. For clarity, only the RNA polymerase is shown at the promoter.

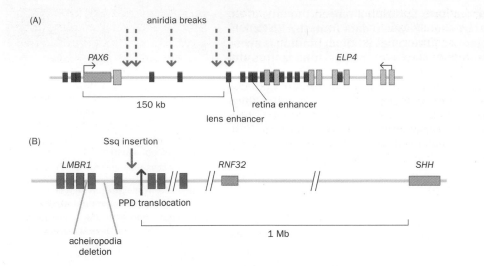

Figure 11.6 Long-range controls on gene expression. (A) Translocation breakpoints (vertical dashed arrows) up to 125 kb downstream of the 3′ end of an intact *PAX6* gene cause a loss of *PAX6* function in patients with aniridia. Regulatory elements essential for the expression of *PAX6* must lie distal of all these breakpoints. DNase hypersensitive sites (red boxes) mark stretches of DNA where nucleosomes are absent or unstable, which are therefore available for interaction with DNA-binding proteins. Two of these sites have been identified as retina-specific and lens-specific enhancers. The *PAX6* regulatory sequences lie within introns of *ELP4* (yellow boxes mark exons of this gene), an unrelated gene that is transcribed in the opposite orientation. (B) Function of the *SHH* (sonic hedgehog) gene in limb development depends on enhancers located 1 Mb away, within introns of the *LMBR1* gene (blue boxes show exons of this gene). An additional gene, *RNF32*, lies between *SHH* and these regulatory elements. Positions of a deletion, an insertion, and a translocation breakpoint are shown, all of which cause phenotypes due to abnormal control of *SHH* expression. Ssq, Sasquatch mouse mutant; PPD, human preaxial polydactyly (OMIM 174500); acheiropodia, human syndrome of bilateral absence of hands and feet (OMIM 200500). [From Kleinjan DA & van Heyningen V (2005) *Am. J. Hum. Genet.* 76, 8–32. With permission from Elsevier.]

two clusters of globin genes. In the α-globin cluster on chromosome 16p and the β-globin cluster on 11p, distant LCRs allow tissue-specific expression and developmental switching of genes in the cluster (**Figure 11.7**). The LCRs include enhancer and **insulator** elements—the latter preventing more distal sequences from affecting the expression of the genes controlled by the LCR. Deletion of the LCR prevents expression of the genes it controls, even though the gene sequences, including the promoters, are intact. Many examples are now known of genes in which the regulatory sequences are more scattered, and so the idea of discrete locus control regions has become less useful.

Co-activators and co-repressors influence promoters without binding to DNA

The components of transcription initiation complexes are held together by protein–protein interactions. Many proteins help glue together the complex without themselves binding DNA. Thus, some co-activators and co-repressors function simply as platforms for the assembly of a complex; others, however, have enzymatic activities. The CoREST co-repressor complex illustrates this. CoREST itself is recruited to promoters by the DNA-binding protein REST, where it forms an assembly platform for the other components of the CoREST co-repressor complex. These include multiple histone-modifying enzymes that collectively repress transcription.

The nuclear receptors exemplify the way in which co-activators and co-repressors build up elaborate regulatory networks. The 48 human nuclear receptors are transcription factors whose activity is inducible. In Chapter 4, we described their role in allowing cells to respond to external signals. In many cases, the activity is induced by binding of a steroid, thyroid, retinoid, or other fat-soluble hormone that can pass through the outer cell membrane. Alternatively, a nuclear receptor may be activated indirectly as a downstream consequence of the binding of an extracellular ligand to a cell surface receptor.

Once activated, the nuclear receptors associate with response elements in the promoters of the genes they regulate (see Figure 4.8). However, they do not act alone: they recruit a large cast of co-activators and co-repressors. These are multiprotein complexes incorporating a range of enzymatic activities relevant to the control of transcription. These enzymatic activities are in turn regulated by a

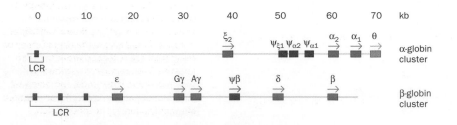

Figure 11.7 Locus control regions for the α- **and** β-**globin clusters.** The locus control regions (LCRs) are marked by one or more DNase I-hypersensitive sites (red boxes) located upstream of the cluster, where nucleosomes are absent or unstable, and the DNA is therefore available for interaction with DNA-binding proteins. These sites are hypersensitive in erythroid cells, where the globin genes are expressed, but not in cells of other lineages. Blue boxes show expressed genes, purple boxes pseudogenes. The functional status of the θ-globin gene (green box) is uncertain. Arrows mark the direction of transcription of expressed genes.

series of reversible covalent modifications (phosphorylation, methylation, acetylation, and so on) of proteins in the co-activator and co-repressor complexes—modifications that may themselves be performed by other co-activators or co-repressors. These interactions build very extensive networks to control and coordinate multiple cellular responses to external signals.

Between them, transcription factors, co-activators, and co-repressors control the activity of a promoter. They do not, however, explain why one DNA sequence rather than another is chosen to function as a promoter. As we mentioned above, there are thousands of potential binding sites for transcription factors across the genome, but only a limited subset are actually used. The choice depends mainly on the local chromatin configuration, which is described in the next section.

11.2 CHROMATIN CONFORMATION: DNA METHYLATION AND THE HISTONE CODE

If fully extended, the DNA in a diploid human cell would stretch for 2 meters. As described in Chapter 2, the bare double helix is subject to several levels of packaging. The most basic level is the nucleosome, in which 147 bp of DNA are wrapped round a complex of eight core histone molecules—normally two each of H2A, H2B, H3, and H4. A variable-length stretch of free DNA separates adjacent nucleosomes, and this is stabilized by one molecule of the linker histone H1. Histones are small (typically 130 amino acids long) highly basic proteins, rich in lysine and arginine. The isoelectric point of histones is 11 or greater, so that at the typical intracellular pH they carry a strong positive charge. This gives them an affinity for the negatively charged DNA. A histone molecule has a globular body as well as N- and C-terminal tails that protrude from the body of the nucleosome. Covalent modifications of amino acids in the tails govern nucleosome behavior.

Modifications of histones in nucleosomes may comprise a histone code

Histones in nucleosomes are subject to many different modifications that affect specific amino acid residues in the tails. The nomenclature used to describe them is explained in **Box 11.2**. Common modifications include acetylation and monomethylation, dimethylation, or trimethylation of lysines (**Figure 11.8**), and phosphorylation of serines. These are effected by large families of enzymes: histone acetyltransferases (HATs), histone methyltransferases (HMTs), and histone kinases. Histone deacetylases (HDACs), histone demethylases, and histone phos-

> **BOX 11.2 NOMENCLATURE FOR HISTONE MODIFICATIONS**
>
> A common shorthand is used for histone modifications. Specific amino acid residues are identified by the type of histone, the one-letter amino acid code, and the position of the residue, counting from the N-terminus. Thus, H3K9 is lysine-9 in histone H3.
>
> Modifications are then described using ac for acetylation, me for methylation, ph for phosphorylation, ub for ubiquitylation, and su for sumoylation. For example, H4K12ac is acetylated lysine-12 of histone H4 and H3K4me3 is trimethylated lysine-4 of histone H3.

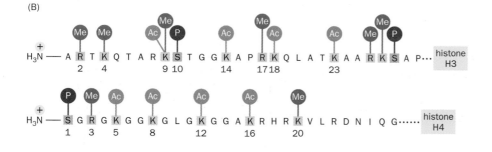

Figure 11.8 Histone modifications.
(A) The ε-amino group of lysine residues can be modified by acetylation or the addition of one to three methyl groups. The standard nomenclature of the modified lysines is shown. (B) The N-terminal tails of histones H3 and H4 are the sites of many of the modifications that control chromatin structure. The amino acid sequence is shown in single-letter code, and potential modifications are indicated. Ac, acetylation; Me, methylation; P, phosphorylation.

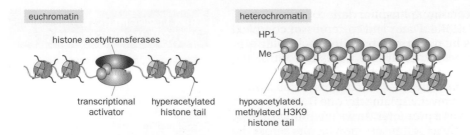

Figure 11.9 Euchromatin and heterochromatin. In euchromatin, nucleosomes are loosely packed, with nucleosome-free regions that can bind regulatory proteins. In heterochromatin, the nucleosomes are densely packed and associated with heterochromatin protein 1 (HP1). Different histone modifications mark the two states. In comparison with euchromatin, heterochromatin is gene-poor, contains much repetitious DNA sequence, and replicates late during S phase of the cell cycle. [From Grewal SI & Elgin SC (2007) *Nature* 447, 399–406. With permission from Macmillan Publishers Ltd.]

phatases reverse these effects. Less frequent modifications include monomethylation, dimethylation, or trimethylation of arginines, phosphorylation of threonines H3T3 and H3T11, and ubiquitylation of lysines H2AK119 and H2BK120.

Open and closed chromatin

Packaging of nucleosomes into higher-order structures is crucial in determining the activity of genes. There are two basic variants. Heterochromatin is a closed, inactive conformation; euchromatin is open and potentially active (**Figure 11.9**). Heterochromatin may be constitutive—that is, it maintains that structure throughout the cell cycle—or facultative. Facultative heterochromatin forms reversibly during the life of the cell as part of the controls on gene expression, as illustrated by the status of the X chromosome in females (see below). As mentioned in Chapter 9, the DNA of constitutive heterochromatin consists mainly of repeats and is largely devoid of genes. Specific histone modifications differentiate the main types of chromatin (**Table 11.2**). Several techniques allow chromatin conformation to be investigated on a local or genomewide scale (**Box 11.3**).

The **histone code** concept implies that particular combinations of histone modifications define the conformation of chromatin and hence the activity of the DNA contained therein. Although there is undoubtedly a strong general correlation, no single histone modification is completely predictive of chromatin state or DNA activity. The various histone modifications influence chromatin structure and function by acting as binding sites for a wide range of nonhistone effector proteins (chromatin typically contains roughly equal weights of histones and nonhistone proteins). These proteins contain domains that recognize specific histone modifications: **bromodomains** recognize acetylated lysines, and **chromodomains** recognize lysines that are methylated. Particular proteins in each class recognize particular specific lysine residues. Other frequent domains such as the PHD (plant homeodomain) interact in a more general way with chromatin. Many chromatin-binding proteins carry several of these domains, enabling them to read histone modifications in a combinatorial manner.

Amino acid	Constitutive heterochromatin	Facultative heterochromatin	Euchromatin
TABLE 11.2 CHARACTERISTIC HISTONE MODIFICATIONS IN DIFFERENT TYPES OF CHROMATIN			
H3K4			trimethylated[a]; monomethylated[b]
H3K9	trimethylated	dimethylated	acetylated
H3K27		trimethylated	
H4R3			methylated
H4K5			acetylated
H4K12	acetylated	acetylated	
H4K20	trimethylated		monomethylated

Within each category of chromatin there are sub-varieties and variant patterns of modification, but those shown are the most frequent modifications in each broad category of chromatin. See Box 11.2 for nomenclature of amino acids in histones. Blank boxes signify that there is no clear pattern. [a]At promoters. [b]At enhancers.

A good example of the importance of histone modifications for gene regulation is provided by the methylation of H3K4. Dimethylated and trimethylated H3K4 appear in discrete peaks in the genome that overlap precisely with promoter regions. This modification is in turn recognized by a PHD domain in the TAF3 subunit of the TFIID basal transcription machinery. Thus, H3K4 methylation constitutes a specific *chromatin landscape* that coincides with promoters, and this modification contributes directly to the recruitment of the RNA pol II transcription machinery. DNA methyltransferases, which methylate cytosines in CpG sites and operate as potent silencing complexes (see below), also sense the status of H3K4 in chromatin through a dedicated histone-binding module. In

BOX 11.3 TECHNIQUES FOR STUDYING CHROMATIN CONFORMATION

In **chromatin immunoprecipitation (ChIP)** (**Figure 1**), cells are treated with formaldehyde to form covalent bonds between the DNA and its associated proteins. The cells are lysed, the chromatin is fragmented, and an antibody against some protein of interest is used to precipitate the chromatin fragments that include that protein. Typical examples would be transcription factors or histones that carry some specific modification (antibodies are available against many modified histones). Incubation at 65°C reverses the protein–DNA crosslinking, and protein is removed by digestion with proteinase. The DNA that was associated with the protein is then identified. PCR with gene-specific primers can be used to check what proteins are bound near a gene of interest. More usually the recovered DNA is hybridized to a microarray to identify genomewide associations—the so-called ChIP-chip technique.

ChIP-chip can give a genomewide overview and has been widely used, but it has several limitations. It requires several micrograms of DNA to get a good signal, so the results are an average of the state of millions of cells. Repeated sequences and allelic variation are hard to study because of cross-hybridization, and the results are only semi-quantitative because of possible bias in the extensive PCR amplification needed to obtain enough DNA. An alternative approach is large-scale sequencing of the recovered DNA without amplification. Occurrences of each individual gene in the immunoprecipitate are simply counted. Sometimes just short tags from each end are sequenced [ChIP-PET (paired end tags) or STAGE (sequence tag analysis of genomic enrichment)]. Increasingly, the whole fragment is sequenced (ChIP-seq). The new generation of massively parallel sequencers is making this an attractive method for obtaining unbiased data to any desired level of resolution.

Chromosome conformation capture (3C, 4C) (**Figure 2, overleaf**) is used to identify DNA sequences that may be widely separated in the genome sequence but lie physically close together within the cell nucleus. Living cells or isolated nuclei are treated with formaldehyde to crosslink regions of chromatin that lie physically close to one another. The crosslinked chromatin is solubilized and digested with a restriction enzyme. Distant interacting sequences will be represented by crosslinked chromatin containing a DNA fragment from each of the interacting sequences. As a result of the restriction digestion, the two fragments will have compatible sticky ends. DNA ligase is used to try to ligate together the ends from the two different regions of DNA. The ligated DNA fragments are released from the chromatin as before. They can then be identified by any of several methods.

The original 3C method used quantitative PCR amplification of the ligation product to test the frequency with which two specified DNA sequences associated. One primer of each pair was specific for each of the two chosen sequences. 4C extends this idea to produce an unbiased list of sequences (interactors) with which a given bait sequence associates. Primers from the bait sequence are used in

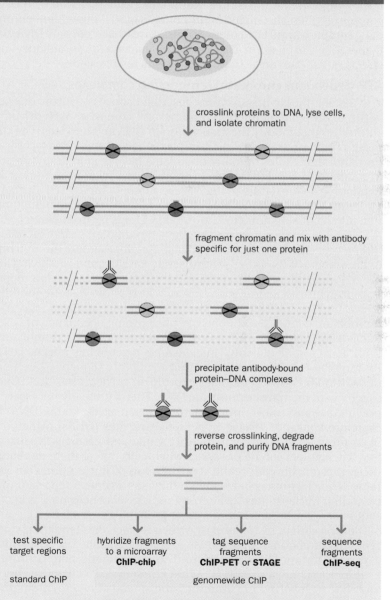

crosslink proteins to DNA, lyse cells, and isolate chromatin

fragment chromatin and mix with antibody specific for just one protein

precipitate antibody-bound protein–DNA complexes

reverse crosslinking, degrade protein, and purify DNA fragments

test specific target regions	hybridize fragments to a microarray **ChIP-chip**	tag sequence fragments **ChIP-PET** or **STAGE**	sequence fragments **ChIP-seq**

standard ChIP | genomewide ChIP

Figure 1 Chromatin immunoprecipitation.

inverse PCR to amplify interactor sequences in the circular ligation product. These are then identified by hybridization to microarrays or, increasingly, by mass sequencing. 3C and 4C are quantitative techniques: the aim is to identify interactions that are present more frequently than the many random events identified by these techniques.

contrast with TAF3, DNA methyltransferases can bind only to nucleosomes that are unmethylated at H3K4. Thus, H3K4 methylation contributes not only to recruitment of the transcription machinery but also to the protection of promoter regions against undesired silencing. This is particularly important for housekeeping genes transcribed from CpG island promoters, the natural targets for DNA methyltransferases.

A web of interactions affects the final outcome, and no one factor is fully determinative. The proteins that recognize specifically modified histones may themselves have histone-modifying activity, producing positive feedback loops—for example, some chromodomain proteins have histone deacetylase activity. Thus, histone modifications can be interdependent. Certain modifications follow on from one another. Phosphorylation of serine at H3S10 promotes acetylation of the adjacent H3K9 and inhibits its methylation. Ubiquitylation of H2AK119 is a prerequisite for the dimethylation and trimethylation of H3K4. The protein complexes often also include components of ATP-dependent chromatin remodeling complexes and DNA methyltransferases that methylate DNA, both described below. Gene expression depends on a balance of stimulatory and inhibitory effects rather than on a simple one-to-one histone code.

ATP-dependent chromatin remodeling complexes

The histone-modifying enzymes described above constitute one set of determinants of chromatin function; another set comprises ATP-driven multiprotein complexes that modify the association of DNA and histones. The components of these chromatin remodeling complexes are strongly conserved from yeast to humans, and studies in many organisms have demonstrated their involvement in numerous developmental switches. They can be grouped into families, depending on the nature of the ATPase subunit (**Table 11.3**). Each ATPase associates with a variety of partners to form large mix-and-match complexes with a rich and confusing nomenclature. Table 11.3 is far from being a comprehensive list. Some subunits are tissue-specific. The complexes often include proteins that interact specifically with modified histones, for example through bromodomains or chromodomains. Other subunits may have histone-modifying activity, so that the ATP-dependent and histone-modifying factors can be interdependent.

Chromatin structure is dynamic. Whereas the histone-modifying enzymes affect nucleosome function through covalent changes, the ATP-dependent remodeling complexes change the nucleosome occupancy of DNA. They move nucleosomes along the DNA or promote nucleosome assembly or disassembly. Both locally and globally, nucleosome positioning affects gene expression. Promoters of active genes and other regulatory sequences are relatively free of nucleosomes, with a characteristic dip in nucleosome occupancy coinciding exactly with the transcription start site. This is probably necessary to allow RNA polymerase and transcription factors to gain access to the DNA. These regions therefore appear as **DNase-hypersensitive sites** (**DHSs**), which therefore often mark promoters and other regulatory sequences. Changes in sensitivity to DNase at particular loci can be observed when cells differentiate or change their state; these presumably reflect the repositioning of nucleosomes. In yeast, the *Isw2* remodeling factor has been shown to act as a general repressor of transcription by sliding nucleosomes away from coding sequences onto adjacent regulatory sequences. The human counterparts probably act similarly.

Several variant histones exist, which replace the standard histones in specific types of chromatin or in response to specific signals. The best known is CENP-A, a variant histone H3 that is found in centromeric heterochromatin. Another example is H2A.Z, a variant H2A that is associated with active chromatin regions and may be involved in containing the spread of heterochromatin. The relative depletion of nucleosomes at active promoters and regulatory sequences, mentioned above, reflects the fact that nucleosomes at these locations tend to contain H2A.Z and another variant histone, H3.3. This makes them unstable, thus allowing DNA-binding proteins better access to these sequences. Another variant of H2A, macro-H2A, is characteristic of nucleosomes on the inactive X chromosome. Incorporation of these variant histones depends on the ability of some chromatin remodeling complexes to loosen the structure of nucleosomes and promote histone exchange. For example, the TIP60 complex (see Table 11.3) has a function in replacing H2A by H2A.Z.

DNA binding proteins

treat living cells or cell nuclei with formaldehyde

crosslink formed

extract and fragment chromatin; digest with restriction enzyme

DNA ligation

remove crosslinks by heat treatment and proteolysis

inverse PCR with primers from bait sequence

interactor sequences

identify interactor sequences on microarrays or by sequencing

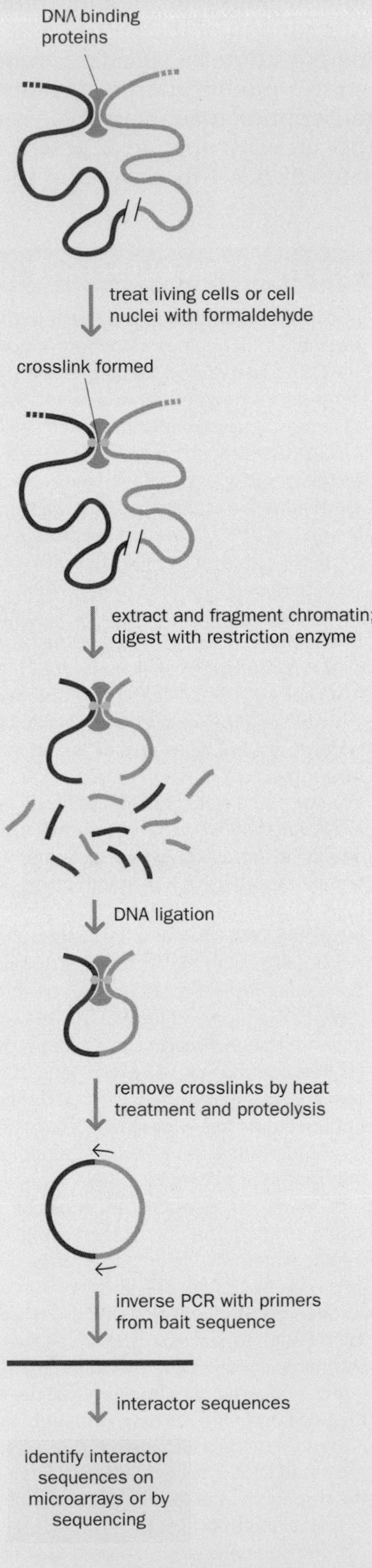

Figure 2 Chromosome conformation capture (4C). [Adapted from Alberts B, Johnson A, Lewis J et al. (2007) Molecular Biology of the Cell, 5th ed. Garland Science/Taylor & Francis LLC.]

TABLE 11.3 EXAMPLES OF ATP-DEPENDENT CHROMATIN REMODELING COMPLEXES

Family	Complex	No. of units	ATPase subunit	Histone-interacting subunit
SWI/SNF	BAF	10	BRG1 or BRM	BAF155 (chromodomain)
ISWI	NuRF	4	SNF2L	BPTF (bromodomain)
	CHRAC	3	SNF2H	–
	ACF	2	SNF2H	ACF1 (bromodomain)
CHD	NuRD	6	CHD3/4	HDAC1/2 (chromodomain)
IN080	TIP60	15	DOMINO	TIP60 (chromodomain)

BAF, Brg- or Brm-associated factor; NuRF, nucleosome remodeling factor; CHRAC, chromatin accessibility complex; ACF, ATP-dependent chromatin assembly factor; NuRD, nucleosome remodeling and histone deacetylase; TIP, Tat-interactive protein.

DNA methylation is an important control in gene expression

As mentioned in Box 9.1, DNA is sometimes modified by methylation of the 5-position of cytosine bases. Only cytosines whose downstream neighbor is guanine—so-called CpG sequences—are subject to methylation. The methyl group lies in the major groove on the outside of the DNA double helix; it does not interfere with base pairing. Thus, 5-methyl cytosine base-pairs with guanine in just the same way as unmodified cytosine (**Figure 11.10**), but the methyl group acts as a signal that is recognized by specific meCpG-binding proteins. These have a role in regulating chromatin structure and gene expression. MeCpG also has an important role in epigenetic memory, which is discussed below. The methylation state of individual cytosines can be investigated by examining the result of treating the DNA with sodium bisulfite (**Box 11.4**).

Cytosine methylation is accomplished by DNA methyltransferases (DNMTs). Humans have three functional DNMTs (**Table 11.4**); a fourth protein, DNMT3L, helps target the methylases to appropriate sequences, while a fifth, DNMT2, turned out to be an RNA-methylating enzyme, despite having a structure similar to DNA methyltransferases.

In mammals the large majority (about 70%) of all CpG sequences are methylated. Just as in plants and many invertebrates, CpG methylation is concentrated on repetitive sequences, including satellite repeats characteristic of pericentric heterochromatin and dispersed transposons. DNA methylation at repeated sequences probably serves to repress transcription, illustrating its role as a genome defense mechanism. Methylation is also sporadically distributed in the body of genes, in both introns and exons, and in intergenic sequences. As discussed in Box 9.1, the CpG dinucleotide is the least frequent dinucleotide in the human genome because of the tendency of methylated cytosines to deaminate to thymines. CpG islands, which are found at the promoters of many human genes, represent an exception to this rule, as they retain a relatively high proportion of CpG dinucleotides.

(A)

(B)

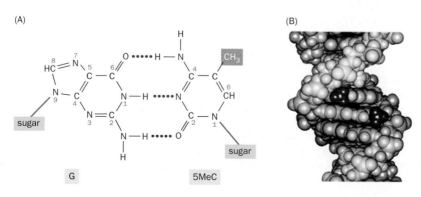

Figure 11.10 5-Methylcytosine.
(A) 5-Methylcytosine base-pairs with guanine in the same way as unmodified cytosine.
(B) There is a symmetrically methylated CpG sequence in the center of this molecule. The methyl groups (red) lie in the major groove of the double helix. [(B) designed by Mark Sherman. Courtesy of Arthur Riggs and Craig Cooney.]

BOX 11.4 BISULFITE MODIFICATION OF DNA

Bisulfite modification is a method for identifying the methylation status of cytosines in genomic DNA. When single-stranded DNA is treated with sodium bisulfite (Na_2SO_3) or metabisulfite ($Na_2S_2O_5$) under carefully controlled conditions, cytosine is deaminated to uracil but 5-methylcytosine remains unchanged. When the treated DNA is sequenced or PCR-amplified, uracil is read as thymine. By comparing the sequence before and after treatment with bisulfite, it is possible to identify which cytosines were methylated.

Several methods have been used to identify any changes induced by the treatment. The status of specific CpG sequences can be examined by sequencing, restriction digestion, or PCR. A C→T change may create or destroy a restriction site (in **Figure 1**, the treatment destroys a *Taq*I TCGA site if the C is unmethylated). Methylation-specific PCR uses primers that are specific for either modified (MSP) or unmodified (UMP) cytosines in the primer-binding site. Genomewide methylation profiles can be obtained by analyzing the bisulfite-treated DNA on specially designed oligonucleotide microarrays that contain probes to match either each normal sequence or its bisulfite-modified counterpart.

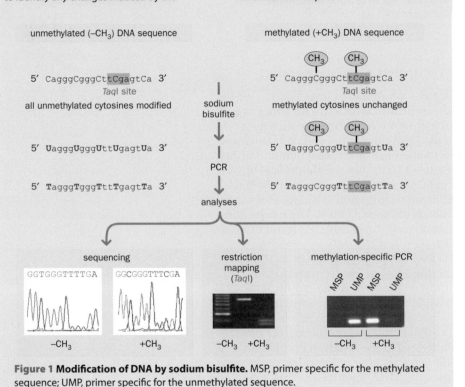

Figure 1 Modification of DNA by sodium bisulfite. MSP, primer specific for the methylated sequence; UMP, primer specific for the unmethylated sequence.

These dense clusters of CpG sequences tend to be protected from DNA methylation regardless of whether the associated gene is active or inactive, and they are likely to be under strong selective pressure to retain promoter activity. On occasion, however, CpG islands can acquire DNA methylation, which inevitably results in the shutting down of transcription, as observed on the inactive X chromosome or at imprinted genes (see below). Aberrant methylation of CpG islands, particularly those associated with tumor suppressor genes, is a general characteristic of cancer cells (this is discussed further in Chapter 17). A minority of CpG islands do show variable methylation that correlates with gene silencing, often as part of a developmental switch. **Figure 11.11** shows examples of the distributions of CpG methylation in different tissues and at different loci.

Methyl-CpG-binding proteins

The effect of DNA methylation on gene expression can be observed directly, in that some transcription factors such as YY1 fail to bind to methylated DNA. DNA methylation can also be read by proteins that contain a methyl-CpG-binding domain (MBD). These can then recruit other proteins associated with repressive

TABLE 11.4 HUMAN DNA METHYLTRANSFERASES

Enzyme	OMIM no.	Major functions[a]	Associated proteins
DNMT1	126375	maintenance methylase	PCNA (replication forks); histone methyltransferases; histone deacetylases; HP1 (heterochromatin); methyl-DNA-binding proteins
DNMT2[b]	602478	methylation of cytosine-38 in tRNA^{Asp}; no DNA-methylation activity	
DNMT3A	602769	*de novo* methylase	histone methyltransferases; histone deacetylases
DNMT3B	602900	*de novo* methylase	histone methyltransferases; histone deacetylases
DNMT3L	606588	binds to chromatin with unmethylated H3K4 and stimulates activity of DNMT3A/3B	DNMT3A, DNMT3B; histone deacetylases

PCNA, proliferating cell nuclear antigen; HP1, heterochromatin protein 1. [a]For the distinction between *de novo* and maintenance methylases, see Section 11.3 on p. 365. [b]DNMT2 turned out to have RNA rather than DNA as its substrate, but its structure is that of a DNA methyltransferase, and it is included here because it is listed as such in many publications.

structures, such as HDACs. Humans have five MeCpG-binding proteins, MBD1–4 and MECP2. MECP2 has been closely studied because loss of function causes Rett syndrome (OMIM 312750), a strange condition in which heterozygous girls develop normally for their first year but then regress in a very characteristic way. The gene encoding MECP2 is X-linked, and *MECP2* mutations are normally lethal in males. DNA methylation proceeds normally in patients with Rett syndrome but, as a result of the absence of MECP2 protein from cells that have inactivated the normal X chromosome, some signals are not read correctly. MECP2 function is particularly needed in mature neurons. Gene expression profiling in the hypothalamus of mice that either lacked or overexpressed MECP2 showed that the expression of many genes was affected but, unexpectedly, in 85% of affected genes MECP2 apparently acted to upregulate rather than repress expression.

DNA methylation in development

DNA methylation shows striking changes during embryonic development (**Figure 11.12**). Egg cells and, especially, sperm cells have heavily methylated DNA; the methylation profiles of germ cells are very different from those of any other somatic lineage. The genome of the fertilized oocyte is an aggregate of the sperm and egg genomes, with methylation differences at paternal and maternal alleles of many genes. Even before the fusion of the two parental genomes, the paternal pronucleus is subject to an active DNA demethylation process through an as yet uncharacterized enzymatic activity residing in the cytoplasm of the oocyte. Once

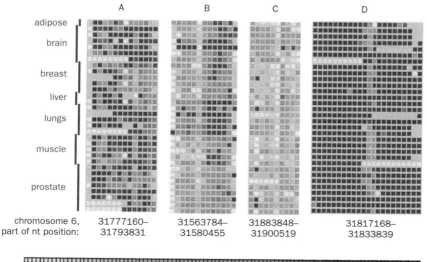

	A	B	C	D
chromosome 6, part of nt position:	31777160–31793831	31563784–31580455	31883848–31900519	31817168–31833839

0%   20%   40%   60%   80%   100%

Figure 11.11 Patterns of CpG methylation may be person-specific, tissue-specific, or locus-specific. Each square represents a CpG sequence within the major histocompatibility complex on chromosome 6, with the intervening nucleotides omitted. Sequences are from four subregions, as indicated. Squares are color-coded to indicate the extent of methylation (see scale at bottom). Rows show the methylation pattern seen in different tissue samples. Region A shows diverse patterns of methylation both between different tissues and between different samples of the same tissue, whereas region B shows more tissue-specific methylation. Region C is largely unmethylated, and region D is almost completely methylated. [Data from www.sanger.ac.uk/PostGenomics/epigenome, as depicted by Hermann A, Gowher H & Jeltsch A (2004) *Cell. Mol. Life Sci.* 61, 2571–2587. With permission from Springer.]

These regions bind the insulator protein CTCF when unmethylated, but not when methylated. CTCF can prevent a promoter from having access to an enhancer if, but only if, it lies between the two (**Figure 11.23**).

Two questions arise about imprinting: how is it done and why is it done?

The phenomenon of imprinting raises two difficult questions. The first is: How does it work? It must be reversible: a man may inherit a particular sequence with a maternal imprint from his mother, but if he passes it on to his child it will then carry a paternal imprint. The imprint probably consists of differential methylation of CpG dinucleotides in imprinting control centers in the paternal and maternal genomes. These are imposed by sex-specific *de novo* methylation during gametogenesis. How the specific regions to be imprinted are chosen is not known. Although methylation patterns are quite different in sperm and eggs, most of these differences are erased in the wave of demethylation in the early zygote (see Figure 11.12). However, for these special regions the imprint is erased only in primordial germ cells, not in somatic cells. Interestingly, at least in the mouse the *Dnmt1* DNA methyltransferase uses different promoters in spermatocytes, oocytes, and somatic cells, so that each contains a different isoform of the enzyme. Maybe gamete-specific isoforms allow this differential control of methylation.

The second question concerns the function of imprinting. One popular theory is based on a conflict of evolutionary interest between fathers and mothers. Selfish gene theory suggests that paternal genes might be best propagated by ensuring that offspring are born as robust as possible, even at the expense of the mother—a man can father children by many different mothers. Maternal genes, in contrast, are best propagated if the mother remains capable of further pregnancies. So, the theory goes, paternal genes program the fetus to extract nutrients at the greatest possible rate from the mother through the placenta, whereas maternal genes act to limit the depredations of a parasitic fetus. This does fit many imprinted loci, but is unlikely to be the full explanation. Why should some imprinting be tissue-specific? And not all imprinted genes have functions related to intrauterine growth, or are imprinted in the direction predicted by the parental conflict hypothesis.

Paramutations are a type of transgenerational epigenetic change

In maize and other plants, there are several well-studied examples of transgenerational epigenetic effects. These start with a heterozygous plant. One allele, the paramutagenic allele, somehow modifies the other (the paramutable allele) so that offspring that inherit the paramutable allele nevertheless express the phenotype of the paramutagenic allele. The paramutable allele itself becomes paramutagenic in the offspring, and the modification may persist through many generations, although, depending on the locus, it may eventually die away. Where mechanisms have been identified, they involve chromatin modification and noncoding RNAs.

A few similar cases have been reported in mice. Perhaps the clearest involves the *Kit* locus (**Figure 11.24**). Heterozygotes for a loss-of-function allele have white spotting. When these heterozygotes were intercrossed or backcrossed with wild-type homozygotes, a proportion of the homozygous wild-type offspring showed the white spotting characteristic of heterozygotes, despite having the wild-type genotype. The effect persisted for a few generations, but with diminishing intensity. Only some loss-of-function alleles showed the effect. A point mutation had no effect, but two mutants in which the gene was inactivated by insertion of different cassettes at different positions did, suggesting that some degree of meiotic mispairing was part of the mechanism. There is evidence that RNA contained in the gametes was involved—even sperm contain significant amounts of RNA.

How widely such mechanisms operate in mammals (including humans) is completely unknown—although the general success of Mendelian genetics suggests that these cannot be major effects. In general, the effects of reported paramutations are quantitative rather than qualitative (a rheostat rather than a switch), which makes them still harder to pin down in humans.

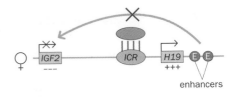

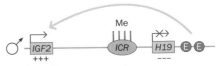

Figure 11.23 The *H19* and *IGF2* genes on chromosome 11p compete for an enhancer. Binding of the CTCF insulator protein to a differentially methylated region determines the outcome of the competition. On the maternal chromosome, the imprinting control region (ICR) is unmethylated, allowing it to bind CTCF (orange oval). This prevents the *IGF2* gene from accessing the enhancer, and so the enhancer drives *H19* expression. On the paternal chromosome, methylation of the imprinting control region prevents the binding of CTCF, thereby allowing *IGF2* to outcompete *H19* for access to the enhancer. [Adapted from Wallace JA & Felsenfeld G (2007) *Curr. Opin. Genet. Dev.* 17, 400–407. With permission from Elsevier.]

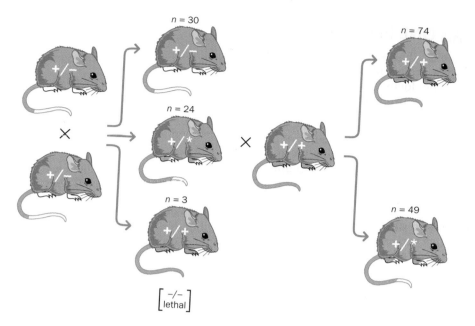

Figure 11.24 Paramutation at the Kit locus in mice. Mice heterozygous for a *Kit* mutation (+/–) show white spotting. When heterozygous mice were crossed, most (24 out of 27) of the genotypically wild-type (+/+) offspring nevertheless showed the white spotting characteristic of heterozygous animals. The wild-type allele inherited from the heterozygous parent has been somehow changed (paramutated, asterisk). The change is unstable: when +/* mice were crossed with wild-type animals, fewer than the predicted 50% of offspring showed the expected +/* phenotype. [Data from Rassoulzadegan M, Grandjean V, Gounon P et al. (2006) *Nature* 441, 469–474. With permission from Macmillan Publishers Ltd.]

Some genes are expressed from only one allele but independently of parental origin

X-inactivation ensures the monoallelic expression of most X-linked genes. However, monoallelic expression, independently of parental origin, is also a feature of a surprisingly high proportion (maybe as high as 5–10%) of autosomal genes. Three processes can be distinguished.

First, the immunoglobulin and T-cell receptor genes show monoallelic expression after programmed DNA rearrangements. Each B or T lymphocyte expresses only a single allele of the relevant gene. As described in Chapter 4, functional immunoglobulin and T-cell receptor genes are assembled from large batteries of potential gene segments by a complicated series of DNA rearrangements. Some random elements within the process mean that the chance of a productive rearrangement is quite low; however, once a functional protein has been produced, a feedback mechanism inhibits further DNA rearrangements.

Competition for a single-copy enhancer provides a second mechanism. The olfactory receptor genes are a well-studied example (**Figure 11.25**). As mentioned in Chapter 9, olfactory receptor genes are the largest gene family in humans and mice. Mice have some 1300 of them, and humans at least 900. Nevertheless, each olfactory neuron expresses just a single allele of a single receptor gene, so that it fires only in response to one specific odorant. In mice, and presumably also in humans, expression of any olfactory receptor gene depends absolutely on an enhancer sequence (H) present as only a single copy in each genome. Chromosome conformation capture experiments show that the single copy of H can associate with any one of the 1300 receptor genes, regardless of their chromosomal location. Although a diploid olfactory neuron contains two copies of H, only one is active; the other is inactivated by methylation (interestingly, apparently at CpA rather than CpG sequences).

Finally, the great majority of cases are random. In independent cell clones, a gene is monoallelically expressed in some clones, but both alleles are expressed in others. In cell cultures, the effects are stable within a clonal lineage. For a given gene, some clones may show paternal and others maternal expression. In an individual there is no consistency of parental origin between different monoallelically expressed genes, even when they lie on the same chromosome. Polymorphic variation in the expression level of alleles is common and heritable (expression quantitative trait loci or e-QTLs), so monoallelic expression may be the tail end of the distribution. The cause and significance of all this are unknown, but probably e-QTLs and random monoallelic expression explain much of the phenotypic diversity of our species.

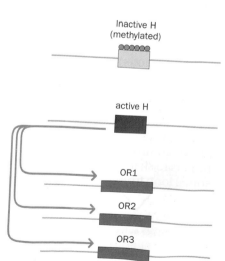

Figure 11.25 Competition for a single enhancer leads to monoallelic expression of olfactory receptors. Expression of any of the hundreds of olfactory receptor genes (OR1, OR2, OR3, ...) distributed around the genome depends on association with a single-copy enhancer, H. One of the two alleles of H in a diploid cell is inactivated by methylation (green circles), so that there is only a single active copy of H in each cell.

α-promoter   β-promoter

Figure 11.29 **Alternative splicing of the neurexin-3 transcript in the nervous system.** There are two alternative promoters, α and β (red bars). Exons 3, 4, 5, 12, and 20 (blue) can each be either included or skipped. Exon 7 (light green) can be included, using either of two alternative 5′ splice acceptor sites, or it can be completely skipped. Exon 22 (purple) has two alternative 3′ splice donor sites. Exon 23 (pink) has two alternative 5′ splice acceptor sites that use different reading frames, one of which leads to an in-frame stop codon within this exon. The protein produced from this variant lacks the transmembrane and cytoplasmic domains encoded by exon 24. Exon 24 (dark green) has three alternative 5′ splice acceptor sites. By using different combinations of variants, this single gene could potentially encode about 1000 different proteins.

As the most complex part of the human body, the central nervous system may need the most complex proteome, and alternative splicing is particularly marked in neurons. Many widely expressed genes have neuron-specific splice isoforms. As long ago as 1994 a compilation by Stamm and colleagues listed almost 100 examples of neuron-specific splicing, including every type shown in Figure 11.28. Many of the variant isoforms of ion channels and receptors are known to be functionally important. Some genes encode an extraordinary number of different transcripts. The example of protocadherins α and γ, with their batteries of alternative promoters, was mentioned earlier. **Figure 11.29** shows the neurexin 3 (*NRXN3*) gene. This large gene on chromosome 14q encodes a cell adhesion and receptor molecule that is present at synapses in the nervous system. It has two promoters and 24 downstream exons. Seven of the exons can each individually be included or excluded in transcripts; one exon has alternative splice donor sites, and two others have alternative splice acceptors. One of the alternatively spliced exons includes a stop codon, the use of which would produce a protein lacking transmembrane and cytoplasmic domains. Potentially this one gene could encode 1000 different proteins.

In addition to generating different protein isoforms, alternative splicing may also result in different 5′ or 3′ untranslated sequences. These may well be functionally significant, especially because the 3′ untranslated region (UTR) is the main site through which microRNAs affect gene expression, as described later in this chapter. The use of alternative polyadenylation signals is also quite common in human mRNA, and different types of alternative polyadenylation have been identified. In many genes, two or more polyadenylation signals are found in the 3′ UTR, and the alternatively polyadenylated transcripts can show tissue specificity. In other cases, alternative polyadenylation signals may be brought into play as a consequence of alternative splicing.

RNA editing can change the sequence of the mRNA after transcription

In contravention of the Central Dogma, there are examples in which the DNA sequence of a gene does not fully determine the sequence of its transcript. RNA editing involves the insertion, deletion, or modification of specific nucleotides in the primary transcript. Nucleotides may be modified by enzymatic deamination or transamination to effect C→U, A→I, or U→C conversions (I is inosine; see below). This happens on a large scale in the mitochondria and chloroplasts of vascular plants. In mammals, there is no evidence for insertion or deletion RNA editing, but modification of nucleotides has been observed in a limited number of genes.

- **C→U editing** occurs in the human apolipoprotein gene *APOB*. In the liver, this gene encodes the large ApoB100 protein. In the intestine, however, C→U editing at nucleotide position 6666 of the mRNA causes replacement of the CAA glutamine codon by a UAA stop codon. The mRNA now encodes a shorter polypeptide, ApoB48 (**Figure 11.30**). C→U editing similarly produces a stop codon in the *NF1* (neurofibromin) gene.

- **A→I editing** is performed by members of the ADAR (**a**denosine **d**eaminase **a**cting on **R**NA) family of deaminases (**Figure 11.31**). Inosine base-pairs with cytosine rather than thymine. ADAR editing often converts CAG codons, encoding glutamine, into CIG codons which, like CGG, encode arginine (Q/R editing). In the nervous system, Q/R editing occurs in several genes encoding neurotransmitter receptors or ion channels (*GABRA3*, *GRIA2*, and *GRIK2*). In the *HTR2C* serotonin receptor gene, A→I editing at splice sites modulates alternative splicing

- **U→C editing** has been observed in the transcript of exon 6 of the Wilms tumor gene.

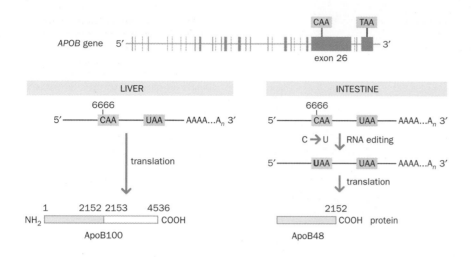

Figure 11.30 RNA editing: the two products of the *APOB* gene. In the liver, the *APOB* mRNA encodes a 14.1 kb 4536-residue protein product, ApoB100. However, in the intestine a cytosine deaminase, APOBEC1, specifically converts cytosine 6666 to uridine, changing the CAA glutamine codon 2153 into a UAA stop codon. The mRNA now encodes a product, ApoB48, consisting of just the first 2152 amino acids of ApoB100.

RNA editing has the appearance of unintelligent design—but it evidently satisfies an important need, because mouse embryos that are even heterozygous for a knockout of *Adar1* die in mid-gestation. Curiously, humans heterozygous for loss-of-function mutations in the orthologous gene, *DSRAD*, have only a relatively trivial skin condition, dyschromatosis symmetrica hereditaria (OMIM 127400).

11.5   CONTROL OF GENE EXPRESSION AT THE LEVEL OF TRANSLATION

Traditionally, controls at the transcriptional level have been the main focus of work to understand gene regulation—but effects at the level of translation have long been known and have recently received greatly increased attention with the discovery of short interfering RNAs (siRNAs) and microRNAs (miRNAs) (see Chapter 9, p. 283). Translational control of gene expression can permit a more rapid response to altered environmental stimuli than altering the level of transcription.

Further controls govern when and where a mRNA is translated

Newly synthesized mRNAs do not simply emerge from a nuclear pore and engage with the nearest ribosome. They may be transported to specific locations within the cell, and they may be translated immediately or may be held in reserve for future use.

mRNAs may be transported as ribonucleoprotein particles to particular locations within some types of cell. In neurons, for example, tau mRNA is localized to the proximal portions of axons rather than to dendrites, where many mRNA molecules are located. Myelin basic protein mRNA is transported with the aid of kinesin to the processes of oligodendrocytes. Transporting RNAs rather than their protein products may provide a more efficient way of localizing proteins, because a single mRNA molecule can give rise to many protein molecules. Specific RNA-binding proteins, such as the ELAV/Hu group, recognize and bind AU-rich sequences in the 3′ UTRs of some mRNAs. Functionally related mRNAs may be bound by the same factor, which might coordinate their localization and/or expression.

Not all mRNA molecules are immediately used to direct protein synthesis. Extrapolation from studies in model organisms suggests that a variety of mRNAs are stored in oocytes in an inactive form, characterized by their having short oligo(A) tails. The short tail means that they cannot be translated. At fertilization or later in development, the stored inactive mRNAs can be activated by cytoplasmic polyadenylation, which restores the normal-sized poly(A) tail.

External factors can also regulate the translation of mRNAs, as shown by two examples in iron metabolism. Increased iron levels stimulate the synthesis of the iron-binding protein ferritin without any corresponding increase in the amount of ferritin mRNA. Conversely, decreased iron levels stimulate the production of

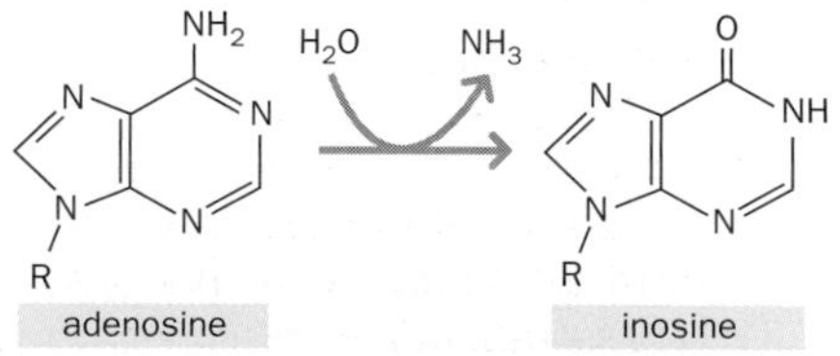

Figure 11.31 Deamination of adenosine. Enzymes of the ADAR family deaminate the amino group at carbon 6 of adenosine to produce inosine.

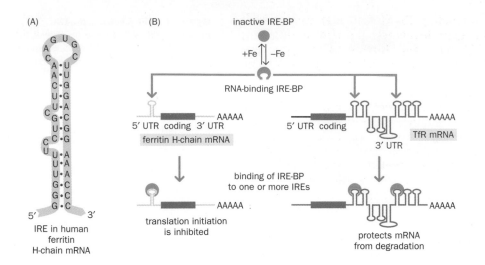

(A)

IRE in human
ferritin
H-chain mRNA

(B) inactive IRE-BP

+Fe ⇅ −Fe

RNA-binding IRE-BP

5′ UTR coding 3′ UTR — AAAAA
ferritin H-chain mRNA

5′ UTR coding — AAAAA
TfR mRNA
3′ UTR

binding of IRE-BP
to one or more IREs

— AAAAA
translation initiation
is inhibited

— AAAAA
protects mRNA
from degradation

Figure 11.32 Iron-response elements in the ferritin and transferrin receptor mRNAs. (A) Stem–loop structure of an iron-response element (IRE) in the 5′ untranslated region (UTR) of the ferritin heavy (H)-chain mRNA. (B) In the presence of iron, a specific IRE-binding protein (IRE-BP) is activated, enabling it to bind the IRE in the ferritin heavy-chain gene and also IREs in the 3′ untranslated region of the transferrin receptor mRNA. Binding inhibits the translation of ferritin but protects the transferrin receptor mRNA from degradation.

the transferrin receptor (TfR), part of the machinery for absorbing dietary iron, again without any effect on the production of the mRNA. The 5′ UTRs of both ferritin heavy-chain mRNA and light-chain mRNA contain a single **iron-response element (IRE)**. This is a specific *cis*-acting regulatory sequence that forms a hairpin structure. Several such IRE sequences are also found in the 3′ UTR of the transferrin receptor mRNA. Regulation is effected by a specific IRE-binding protein that is activated at low iron levels to bind IREs (**Figure 11.32**). Like many translational controls, the effects depend on the secondary structure of the mRNAs.

Until a few years ago, these scattered examples of RNA-based controls on gene expression were considered interesting but rare events. That view was changed with the discovery of RNAi and miRNAs.

The discovery of many small RNAs that regulate gene expression caused a paradigm shift in cell biology

Small RNAs are not all rare. Some are indeed rare but, for example, an adult *C. elegans* cell contains more than 50,000 molecules each of miR-2, miR-52, and miR-58. Why were they not discovered earlier? When researchers ran RNA gels, they assumed—usually with good reason—that the smear of very low-molecular-weight RNA at the bottom of the gel consisted of degradation products and was of no interest. In addition, such short molecules are difficult to study: standard bioinformatics approaches overlook them, and such tiny targets were seldom hit in mutagenesis experiments. Thus, although they must have been observed in many laboratory experiments, they were not recognized.

Two seemingly disparate lines of research in the 1990s, both involving the *C. elegans* worm, alerted biologists to the unsuspected importance of very small RNAs. Andrew Fire and Craig Mello received the 2006 Nobel Prize in Physiology or Medicine for their roles in understanding RNAi—the specific inhibition of gene expression by short double-stranded RNA molecules. Their Nobel lectures, describing the process of discovery, can be read at http://nobelprize.org/nobel _prizes/medicine/laureates/2006/. In addition, Victor Ambros and Gary Ruvkun, among others, opened up the world of miRNAs in development. Their accounts of how they came to make these discoveries can be read in two Commentaries in *Cell* (see Further Reading).

MicroRNAs as regulators of translation

The basic mechanisms through which small RNAs exert their effects were described in Chapter 9 (see Box 9.7 and Figures 9.16–9.18). As described there, several different types of small RNA are important in gene regulation. Synthetic siRNAs are hugely important as tools for selectively knocking down the expression of genes in the laboratory (see Chapters 12 and 20), whereas endogenous siRNAs are probably involved in the establishment of heterochromatin, as described above. The Piwi-protein-interacting RNAs (piRNAs) were initially described in germ cells, in which their major role seems to be in preventing transposons from becoming active. MicroRNAs have much broader roles in gene regulation in all types of cell and are considered in more detail here. Most, though not all, of their reported effects are at the translational level.

Most of the target sequences recognized by miRNAs and siRNAs are in the 3' UTR of an mRNA. The targeted mRNA may be degraded, or it may remain intact but not be translated. Which of these fates is chosen depends on the degree of matching between the two RNA species. Short interfering RNAs have perfect matches and degrade their target. MicroRNAs have imperfect matches and usually repress translation, although they can also trigger the degradation of their target. In either case, nucleotides 2–7 function as a seed for initial recognition of the target. If the rest of the small RNA can then make a perfect match, the target is cleaved between the nucleotides paired with residues 10 and 11 of the small RNA. If not, the target mRNA may remain intact but translation is repressed. The precise mechanism of repression is uncertain. It may work through recognition of the mRNA cap or by an effect on assembly of the ribosome. A protein, EIF6, that complexes with the 60S ribosomal subunit and inhibits assembly of the complete ribosome has been described as a component of a modified RISC. In one way or another, miRNAs usually operate by preventing the initiation of translation. When cells are transfected with a miRNA there is usually also a reduction in the quantity of target mRNA in the cell. It is still not certain whether this loss is because the miRNA degrades its target as well as inhibiting translation, or whether it is secondary to the repression of translation. Several studies support the view that the loss is secondary, and not a direct effect of the miRNA.

The imperfect matching of miRNAs to their targets makes it harder to identify the targets by bioinformatic analysis. The results of laboratory experiments to identify targets are surprising. When mass spectrometry was used to examine the effects on the proteome, transfection or knockdown of a single miRNA could be seen to affect the synthesis of hundreds, or even thousands, of proteins (**Figure 11.33**). Some proportion of these effects will be secondary: the miRNA represses the production of a protein, and the reduced availability of that protein, in turn, affects the production of other proteins. However, many changes were correlated with the presence of the seed sequence for the relevant miRNA in the 3' UTR of the mRNA, and so were likely to be direct. The other noteworthy observation from these experiments is how modest the effects were. Few proteins showed an increase as great as twofold when an inhibitory miRNA was knocked down, or repression greater than 50% when the cell was transfected with a miRNA. The general picture that emerges is of very wide-ranging but small effects.

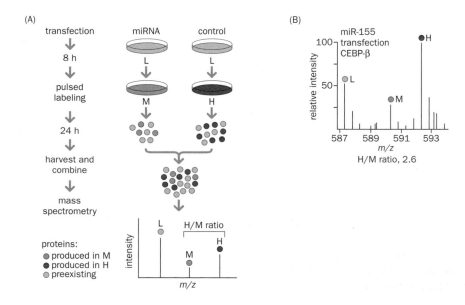

Figure 11.33 Proteomic analysis of the effect of changing the level of a miRNA. (A) An outline of the experimental procedure. HeLa cells were transfected with a microRNA in a medium containing light isotopic label (L); after 8 hours they were transferred to medium containing a medium-weight isotopic label (M). Control HeLa cells were transferred to a medium containing heavy isotopic labels (H). Newly synthesized proteins carry an M or H label. Extracts are combined and analyzed by mass spectrometry. (B) An example result: transfection with miR-155 resulted in a 2.6-fold decrease in synthesis of CEBP-β protein [compare the heights of the H (control) and M (experimental) peaks]. [From Selbach M, Schwanhäusser B, Thierfelder N et al. (2008) *Nature* 455, 58–63. With permission from Macmillan Publishers Ltd.]

MicroRNAs and cancer

As will be described in Chapter 17, cancer cells show changed expression of many miRNAs. There is a general depression of miRNA levels, consistent with the generally dedifferentiated state of cancer cells (miRNA expression is greater in differentiated cells than in the various types of stem cell). Some of these effects can be reproduced by targeted knockdown of the Dicer or Drosha enzymes that are essential for processing all miRNAs from their primary transcripts (see Figure 9.16). When this is done, cells show enhanced tumorigenic behavior. However, certain miRNAs show the opposite pattern, being upregulated in many tumors. The evidence described above suggests that these changes could affect the expression of thousands of genes, and might be significant in tilting a cell's behavior toward tumorigenesis.

Some unresolved questions

MicroRNAs exert their effects primarily through interactions with mRNA molecules within the cytoplasm. A number of reports have suggested that miRNAs may also have functions in the nucleus, maybe directly affecting transcription. These remain currently highly speculative. There is rather more solid evidence that miRNAs can sometimes stimulate, rather than inhibit, translation. Some of the same miRNAs that inhibit translation during normal cell growth may become positive regulators in conditions of cell stress.

CONCLUSION

Gene expression is controlled by decisions at all levels: on whether or not to transcribe a sequence; on the choice of transcription start site; on how to process an alternatively spliced primary transcript; on how, when, and where to translate the message in the mature mRNA; and on the modification and transportation, within or outside the cell, of the initial polypeptide.

Protein-coding genes, and many genes encoding functional RNAs, are transcribed by RNA polymerase II. Before transcription can start, a large complex of proteins must be assembled around the transcription start site. Apart from the RNA polymerase, these include the TFII transcription factors (a complex of at least 27 individual polypeptides) that comprise the basal transcription apparatus, plus many tissue-specific proteins. Some of these proteins recognize specific DNA sequence elements near the transcription start site, such as the TATA box. Others join the complex through protein–protein interactions. These include proteins bound to distant regulatory elements that are brought into close proximity to the promoter by DNA looping.

Transcription start sites are defined more by the local chromatin conformation than by the local DNA sequence. The conformation depends on a set of mutually reinforcing processes: histones in the core nucleosomes undergo numerous covalent modifications or are sometimes replaced by variant molecules; cytosines in CpG dinucleotides may be methylated, and ATP-dependent chromatin remodeling complexes alter the distribution of nucleosomes. The many nonhistone proteins in chromatin include the enzymes that add or remove modifying groups to or from the histones and DNA. They are themselves differentially bound to chromatin, depending on its conformation and state of modification.

Chromatin conformation controls gene expression, both in the short term as cells respond to fluctuating metabolic and environmental conditions, and in the longer term as developmental programs unfold. Enduring changes in conformation are the basis of epigenetic effects—changes in gene expression that do not depend on changes in the DNA sequence. For some genes, epigenetic effects depend on the parental origin—the phenomenon of genetic imprinting. Transcription requires an open chromatin conformation, whereas heterochromatin has a closed, tightly packed conformation and represses gene expression. Small noncoding RNA molecules are involved in the establishment of stable heterochromatin.

Primary transcripts are often subject to alternative splicing. Exons can be skipped or included, or variant donor or acceptor sites can be used. Some genes can potentially encode hundreds of different polypeptides because of extensive alternative splicing. Splicing is controlled by splicing enhancers and suppressors that bind proteins such as the SR and hnRNP proteins. Alternative splicing is often tissue-specific. In many cases, the different splice isoforms have clearly different functions, although some alternative splicing may reflect random inefficiencies in the selection of splice sites.

Controls at the level of translation fine-tune gene expression. These are less well understood than transcriptional controls; for the most part they seem to involve controlling the stability of mRNAs. The 3′ untranslated regions of many mRNAs contain important determinants of stability, including most miRNA-binding sites. MicroRNAs have very broad but not very deep effects on cell metabolism—they function as rheostats rather than switches. However, we should not underestimate the importance of rheostats. One of the lessons from research on complex diseases is that the cumulative effect of such minor changes can be very important. This may be especially true of miRNA-based effects because they affect very broad aspects of cell behavior. Nevertheless, this may still underestimate the importance of RNA-based controls. Small RNAs may have substantial roles in major decisions through their role in modifying chromatin structure.

A view is emerging of cellular metabolism comprising tightly coupled networks rather than a set of linear pathways. In their review, Maniatis and Reed (see Further Reading) quote a host of evidence that transcription, RNA processing, and export of the mature mRNA are tightly coupled. Rather than a linear production line, the gene expression machine is a network in which components whose main role is in later stages of the process also influence the efficiency of earlier stages. In Chapter 12, we move from gene expression to gene function, and it will become clear that most gene products function as part of very extensive networks of interacting proteins. The emerging science of systems biology studies such networks and is helping to bridge the gap between gene expression, as described in this chapter, and cell biology.

FURTHER READING

Promoters and the primary transcript

Fuda NJ, Ardehali B & Lis JT (2009) Defining mechanisms that regulate RNA polymerase II transcription *in vivo*. *Nature* 461, 186–192. [A nice review of controls on initiation of transcription, with good (but nonhuman) examples.]

Kleinjan DA & van Heyningen V (2005) Long-range control of gene expression: emerging mechanisms and disruption in disease. *Am. J. Hum. Genet.* 76, 8–32. [Gives detail on *PAX6*, *SHH*, and other clinically important examples.]

Lonard DM & O'Malley BW (2007) Nuclear receptor coregulators: judges, juries, and executioners of cellular regulation. *Mol. Cell* 27, 691–700. [Examples of regulation by networks of co-activators and co-repressors.]

Maniatis T & Reed R (2002) An extensive network of coupling among gene expression machines. *Nature* 416, 499–506. [An extra insight into control of gene expression.]

Sandelin A, Carnici P, Lenhard B et al. (2007) Mammalian RNA polymerase II core promoters: insights from genomewide studies. *Nat. Rev. Genet.* 8, 424–436. [Promoter structure and transcription start sites; useful boxes on techniques.]

Shiraki T, Kondo S, Katayama S et al. (1999) Cap analysis gene expression for high-throughput analysis of transcriptional starting point and identification of promoter usage. *Methods Enzymol.* 303, 19–44. [Gives details of the CAGE technique outlined in Box 11.1.]

Smale ST & Kadonaga JT (2003) The RNA polymerase II core promoter. *Annu. Rev. Biochem.* 72, 449–479. [Summarizes much work on the start sites of genes.]

Visel A, Rubin EM & Pennachio LA (2009) Genomic views of distant-acting enhancers. *Nature* 461, 199–205. [Review of the nature and function of enhancers, with emphasis on human examples.]

Chromatin conformation: DNA methylation and the histone code

Barski A, Cuddapah S, Cui K et al. (2007) High-resolution profiling of histone methylations in the human genome. *Cell* 129, 823–837. [Genomewide mapping of histone methylations across promoters, enhancers, insulators, and transcribed regions.]

ENCODE Project Consortium (2007) Identification and analysis of functional elements in 1% of the human genome by the ENCODE pilot project. *Nature* 447, 799–816. [Findings of the pilot project; details are in a series of papers in *Genome Research*.]

Koch CM, Andrews RM, Flicek P et al. (2007) The landscape of histone modifications across 1% of the human genome in five human cell lines. *Genome Res.* 17, 691–707. [Detail from the ENCODE pilot project.]

Lall S (2007) Primers on chromatin. *Nat. Struct. Mol. Biol.* 14, 1110–1115. [A good general introduction.]

Ooi L & Wood IC (2007) Chromatin cross-talk in development and disease: lessons from REST. *Nat. Rev. Genet.* 8, 544–554. [Function of the CoREST complex.]

Wang Z, Zang C, Rosenfeld JA et al. (2008) Combinatorial patterns of histone acetylations and methylations in the human genome. *Nat. Genet.* 40, 897–903. [A genomewide study of the distribution of over 30 specific histone modifications, that identifies specific combinations that are characteristic of promoters, enhancers, etc.]

Heterochromatin

Bühler M & Moazed D (2007) Transcription and RNAi in heterochromatic gene silencing. *Nat. Struct. Mol. Biol.* 14, 1041–1048. [Involvement of small RNAs in heterochromatin formation.]

Gaszner M & Felsenfeld G (2006) Insulators: exploiting transcriptional and epigenetic mechanisms. *Nat. Rev. Genet.* 7, 703–713. [General review of insulators.]

Hediger F & Gasser SM (2006) Heterochromatin protein 1: don't judge the book by its cover! *Curr. Opin. Genet. Dev.* 16, 143–150. [Functions of HP1.]

Trojer P & Reinberg D (2007) Facultative heterochromatin: is there a distinctive molecular signature? *Mol. Cell* 28, 1–13. [Features of various types of facultative heterochromatin.]

DNA methylation

de la Serna IL, Ohkawa Y & Imbalzano AN (2006) Chromatin remodelling in mammalian differentiation: lessons from ATP-dependent remodellers. *Nat. Rev. Genet.* 7, 461–473.

Eckhardt F, Lewin J, Cortese R et al. (2006) DNA methylation profiling of human chromosomes 6, 20 and 22. *Nat. Genet.* 38, 1378–1385. [An example of large-scale mapping of epigenetic signatures.]

Goll MG & Bestor TH (2005) Eukaryotic DNA cytosine methyltransferases. *Annu. Rev. Biochem.* 74, 481–514. [Review of the DNA methyltransferases.]

Effects of nuclear localization on gene expression

Akhtar A & Gasser SM (2007) The nuclear envelope and transcriptional control. *Nat. Rev. Genet.* 8, 507–517.

Dekker J, Rippe K, Dekker M & Kleckner N (2002) Capturing chromatin conformation. *Science* 295, 1306–1311. [The original 3C paper.]

Fraser P & Bickmore W (2007) Nuclear organization of the genome and the potential for gene regulation. *Nature* 447, 413–417.

Lanctôt C, Cheutin T, Cremer M et al. (2008) Dynamic genome architecture in the nuclear space: regulation of gene expression in three dimensions. *Nat. Rev. Genet.* 8, 104–115.

Linnemann AK, Platts AE & Krawetz SA (2009) Differential nuclear scaffold/matrix attachment marks expressed genes. *Hum. Mol. Genet.* 18, 645–654. [Underlines the difficulty of defining matrix/scaffold attachment regions; a good source of references.]

Epigenetic memory and imprinting

Cuzin F, Grandjean V & Rassoulzadegan M (2008) Inherited variation at the epigenetic level: paramutation from the plant to the mouse. *Curr. Opin. Genet. Dev.* 18, 193–196. [Discusses their work on the *Kit* paramutation.]

Geneimprint. Jirtle Laboratory at Duke University. www.geneimprint.com [A list of imprinted genes in humans and other organisms.]

Gimelbrant A, Hutchinson JN, Thompson BR & Chess A (2007) Widespread monoallelic expression on human autosomes. *Science* 318, 1136–1140. [Surprising data showing widespread monoallelic expression.]

Heard E & Disteche CM (2006) Dosage compensation in mammals: fine-tuning the expression of the X-chromosome. *Genes Dev.* 20, 1848–1867.

Horsthemke B & Wagstaff J (2008) Mechanisms of imprinting of the Prader–Willi/Angelman region. *Am. J. Med. Genet.* 146A, 2041–2052.

Lomvardas S, Barnea G, Pisapia DJ et al. (2006) Interchromosomal interactions and olfactory receptor choice. *Cell* 126, 403–413. [How the H enhancer ensures that each olfactory neuron expresses only a single olfactory receptor.]

Ringrose L & Paro R (2007) Epigenetic regulation of cellular memory by the polycomb and trithorax group proteins. *Annu. Rev. Genet.* 38, 413–443.

Wutz A & Gribnau J (2007) X-inactivation Xplained. *Curr. Opin. Genet. Dev.* 17, 387–393.

One gene—more than one protein

Maniatis T & Tasic B (2002) Alternative splicing and proteome expansion in metazoans. *Nature* 418, 236–243. [A good source of examples.]

Stamm S, Zhang MQ, Marr TG & Helfman DM (1994) A sequence compilation and comparison of exons that are alternatively spliced in neurons. *Nucleic Acids Res.* 22, 1515–1526. [Lists many examples.]

Südhof TC (2008) Neuroligins and neurexins link synaptic function to cognitive disease. *Nature* 455, 903–911. [Discusses alternative splicing of the neurexin 3 (*NRXN3*) gene shown in Figure 11.29.]

Control of gene expression at the level of translation

Baek D, Villén J, Shin C et al. (2008) The impact of microRNAs on protein output. *Nature* 455, 64–71.

Buchan JR & Parker R (2007) The two faces of microRNA. *Science* 318, 1877–1878. [Reviews the evidence that miRNAs can sometimes stimulate, rather than inhibit, translation.]

Chendrimada TP, Finn KJ & Ji X (2007) MicroRNA silencing through recruitment of EIF6. *Nature* 447, 823–828.

Klausner RD, Rouault TA & Harford JB (1993) Regulating the fate of messenger RNA: the control of cellular iron metabolism. *Cell* 72, 19–28. [Source of the data on iron-response elements in Figure 11.32.]

Lee R, Feinbaum R & Ambros V (2004) A short history of a short RNA. *Cell* 116 (2 Suppl.), S89–S92. [Personal recollections of groundbreaking work.]

Mathonnet G, Fabian MR, Svitkin YV et al. (2007) MicroRNA inhibition of translation initiation *in vitro* by targeting of the cap-binding complex eIF4F. *Science* 317, 1764–1767. [One way in which miRNAs inhibit translation; this may not be the whole answer.]

Ruvkun G, Wightman B & Ha Z (2004) The 20 years it took to recognize the importance of tiny RNAs. *Cell* 116 (2 Suppl.), S93–S96. [A personal account of this important work.]

Selbach M, Schwanhäusser B, Thierfelder N et al. (2008) Widespread changes in protein synthesis induced by microRNAs. *Nature* 455, 58–63. [Source of the data in Figure 11.33; illustrates an important experimental approach.]

Chapter 12

Studying Gene Function in the Post-Genome Era

12

KEY CONCEPTS

- Clues to the function of a gene can be inferred by comparing the sequence of the gene and any protein products with all previously investigated nucleotide and protein sequences to find related sequences. The related sequences may be closely related genes or proteins, or they may be short functional sequence components such as protein domains or peptide/nucleotide sequence motifs.

- Inactivating an individual gene often results in an altered phenotype that provides important insights into the function of the gene, but useful information can also sometimes be obtained by overexpressing genes or expressing them ectopically.

- Gene function can be conveniently studied in cultured cells by silencing the expression of a specific target gene using RNA interference. More extensive information about how a gene works can often be inferred by selectively inactivating a gene in whole model organisms.

- Yeast two-hybrid screens identify interactions between proteins by splitting a transcription factor into two functionally complementary domains that are then joined onto test proteins; interaction between two such test proteins can reconstitute the transcription factor and switch on a reporter gene.

- Large-scale protein interaction studies seek to define *interactomes*, global functional protein networks in cells.

- Protein–protein and protein–DNA interactions can be identified with the help of chemical crosslinking agents such as formaldehyde that can covalently bond proteins to other proteins or to DNA within cells.

- Chromatin immunoprecipitation identifies DNA-binding sites for a protein *in vivo* by using a specific antibody to capture a protein after it has been chemically crosslinked to DNA within cells.

Sequencing of the human genome was, with hindsight, the easy part. Now comes the hard part: how do we unravel what the sequence means and discover how it helps construct a functioning human being and differentiates us from other species?

In the post-genome era, work will continue for some time to get a complete inventory of human genes. As was described in Chapter 10, comparative genomics has been particularly helpful in identifying regulatory elements and in validating predicted protein-coding genes. The vast majority of protein-coding genes have been identified in the human genome, and we now know that there are about 20,000 genes of this kind. However, for the reasons described in Chapter 9, RNA genes have been much more difficult to identify within genomic DNA, and our knowledge of human RNA genes, although growing rapidly, remains rather incomplete.

As the effort to identify all the genes and other functional components in the human genome intensifies, researchers seek to understand our genome in different ways. One major priority, as will be detailed in Chapter 13, is prompted by the biological and medical need to understand genetic variation at the genome level. In this chapter we consider another priority, the need to understand the functions of genes. In the post-genome era, whole genome analyses are increasingly being applied to understand the function and regulation of genes. In many cases, such analyses have been developed from methods that were used to study one or a few genes at a time. However, getting a whole-genome perspective allows us to develop comprehensive networks of interacting genes and gene products, and so obtain a fuller picture of how genes and cells work.

12.1 STUDYING GENE FUNCTION: AN OVERVIEW

The function of a gene is actually the function of its product(s). About 20,000 human genes ultimately specify proteins, and many thousand additional genes produce functional noncoding RNA. The functions of human gene products can be conveniently classified at three levels:

- *The biochemical level.* For example, a gene product may be described as a kinase or a calcium-binding protein. This reveals little about its wider role in the organism.

- *The cellular level.* This builds in information about intracellular localization and biological pathways. For example, it may be possible to establish that a protein is located in the nucleus and is required for DNA repair, even if its precise biochemical function is unknown.

- *The organismal level.* This may include information about where and when a gene is expressed in different tissues and its role in the processes of development or physiology.

To obtain a complete picture of gene function, information is required at all three levels. Defined vocabularies have been developed to describe gene function across all genomes, such as the Gene Ontology system (http://www.geneontology.org/). Other useful systems include those employed by the Kyoto Encyclopedia of Genes and Genomes (KEGG; http://www.genome.ad.jp/kegg/) and the Enzyme Commission system for enzyme classification (http://www.chem.qmul.ac.uk/iubmb/enzyme/).

Investigations into the functions of human genes and the genes of various model organisms began long before genome projects were developed. In this pre-genome era, gene function analyses were very much done on a gene-by-gene basis. Individual small research laboratories throughout the world would focus on trying to understand the function of just one gene or a small number of genes that worked together in some common pathway of biological and/or medical interest.

The gene-by-gene approach is still being used today, but what genome sequences brought was the additional possibility to perform **functional genomics**—global, genomewide analyses of gene function. In the post-genome era, the big genome centers that were established to sequence genomes, including the human genome, are now major contributors to the new quest of studying gene function. They are at the forefront of internationally coordinated projects such as the ENCODE (Encyclopedia of DNA Elements) project. As detailed in

Chapter 11 (p. 362), the ENCODE project seeks to identify and analyze all the functional components of our genome by a combination of experimental approaches and bioinformatics analyses. The widespread availability of many experimental approaches to perform global gene analyses has meant that a network of small laboratories throughout the world are also making invaluable contributions in addition to large, well-equipped genome centers.

Gene function can be studied at a variety of different levels

Let us first consider the simple case of a single previously uncharacterized gene that has been predicted by using gene-finding programs and then validated by comparative genomics. How can we study its function?

First of all, we can use bioinformatic approaches to compare the sequence of a gene and its products with all other known nucleotide and protein sequences. Additionally, the sequences can be scanned to look for significant matching to short nucleotide and peptide sequence components that have previously been shown to be important contributors to gene function. The other major avenue to understanding gene function requires laboratory investigations, and a whole range of different types of experimental approach can be used, including those listed below.

Gene expression studies

Some studies look to see where the gene is expressed at the RNA level by using a complementary antisense RNA probe, and/or at the protein level (usually by using a suitably specific antibody). The expression studies often begin by analyzing expression in tissue sections of early embryos, or by probing RNA extracted from adult tissues, such as in northern blots or using RT-PCR. In which tissues is the gene of interest expressed? Ubiquitous expression would suggest some very general (housekeeping) cellular function. Alternatively, if the gene is expressed in a specific tissue—such as the testis—then a more defined role is indicated, in this case possibly a role in reproduction. Intracellular localization is another aspect of gene expression that can be studied in intact cells or by analyzing subcellular fractions. See Chapter 8 (p. 239) for a detailed description of expression screening techniques.

Gene Inactivation and inhibition of gene expression

To obtain more precise information on gene function, follow-up studies are usually designed to specifically inactivate the gene *in vivo* or to specifically block its expression. This can be done at two rather different levels.

One way is to perform the required genetic manipulation in cultured cells, often standard human cell lines. Conducting these manipulations in cultured cells can be done quickly and easily, but inevitably the results obtained only give information about the role of the gene within the cell type that is being studied. We describe this type of approach later in this chapter.

At another level, genetically modified animals can be produced. Early embryos can be manipulated to introduce specific changes in germ-line DNA or to specifically block the expression of a predetermined gene of interest, and the effects on the developing phenotype can be tracked at different developmental stages and in the postnatal period. Such studies can be time-consuming and laborious, but they offer the big advantage that the effect of inactivating or inhibiting a specific gene can be tracked in the whole organism, providing valuable information about genes that have roles in the functions of organs, tissues, and other sets of interacting cells. The technology for genetically manipulating animals is complex, and because it is also used extensively to produce animal models of disease we will consider the details in Chapter 20.

Defining molecular partners for gene products

To perform their functions, gene products need to interact with other molecules in the cells. Proteins, for example, often bind to other proteins, sometimes transiently, and sometimes as components of stable multisubunit complexes. Many regulatory proteins and functional noncoding RNAs bind to specific nucleic acid sequences, as in the case of transcription factors. To identify molecular partners for a specific gene product, various screens can be employed to test protein–protein and protein–nucleic acid interactions.

Genomewide analyses aim to integrate analyses of gene function

Arguably, the grand purpose of molecular biology over the past few decades has been to determine gene functions and link genes into pathways and networks to explain how living things work. What started as a reductionist approach of studying genes and their products one at a time has increasingly been overtaken by a shift in focus to a global, genomewide approach in which many or all gene products are studied simultaneously.

According to the class of gene product, global analyses have been named with an -*omics* suffix, following the long established use of *genomics* for global DNA analyses. Commonly used examples include *transcriptomics* (RNA), *proteomics* (protein), *lipidomics* (lipids), *glycomics* (carbohydrates), and *metabolomics* (metabolites). Such is the popularity of -*omics* or -*ome* that subclasses are increasingly being derived, such as *kinome* for the totality of kinases, and a new scientific journal *OMICS* is devoted to publishing global molecular analyses.

The transcriptome and the proteome have been particularly well studied, and techniques for analyzing them have been described in Chapter 8 (p. 245). The transcriptome represents the combined output of transcription, RNA processing, and RNA turnover; similarly, the proteome is the combined output of translation, post-translational processing including protein cleavage and protein modification, and protein turnover. Because processes such as transcription, RNA processing, protein synthesis, and post-translational processing are all regulated, the transcriptome and proteome differ significantly between different cell types, and even between cells of the same type in response to changes in the cell's environment.

Unlike the genome, which is virtually identical in all nucleated cells of an organism, the transcriptome and proteome are therefore highly variable. While acknowledging DNA variation between individuals, we can talk about the human genome, but the term human transcriptome or human proteome is essentially meaningless unless we reference it with respect to the origin of the cell (such as cell type or developmental stage) and, for cultured cells, the cell's environment. Analysis at the level of the transcriptome or proteome provides a snapshot of the cell in action, showing the abundance of all the RNAs and proteins in a *particular* set of circumstances.

In an effort to document mammalian gene expression, transcriptome analyses have been performed in many different human and animal tissues. The data can be analyzed to identify genes that show similar expression patterns that may indicate that they belong to common functional pathways. Quantitative transcriptomics uses microarray hybridization or DNA sequencing. In the latter case, the data used to be provided by methods such as the ingenious SAGE (serial analysis of gene expression) technique that captured and then sequenced short sequence components (*sequence tags*) from individual RNA molecules. These methods were subsequently overtaken by massively parallel DNA sequencing of larger sequence components of all the individual RNA molecules in the starting samples.

Large-scale tissue *in situ* hybridization studies are also offering high-resolution gene expression patterns during mouse development and in adult mouse brain. Proteomes are being studied in different tissues with the use of mass spectrometry and two-dimensional gel electrophoresis. The resulting huge expression data sets are being stored in dedicated databases (**Table 12.1**).

As we describe later in this chapter, additional global gene function analyses are being deployed in many different areas, including RNA interference-mediated gene inactivation in cultured cells, gene knockout in embryos, and protein–protein and protein–DNA interactions.

12.2 BIOINFORMATIC APPROACHES TO STUDYING GENE FUNCTION

Bioinformatics has been vitally important in gene prediction (Chapter 8, p. 235) and in gene validation, notably in comparative genomics analyses (Section 10.2, p. 306). Bioinformatics is also essential to support the massive data sets coming out of global analyses of gene function. It is no exaggeration to say that the functional genomics revolution has been driven as much by the development of new

TABLE 12.1 EXAMPLES OF PUBLIC TRANSCRIPTOMIC AND PROTEOMIC DATABASES			
Database class	Database or Web compilation	Contents	Website address
Transcriptomic (microarray hybridization)	Microarray World	compilation of microarray expression databases	http://www.microarrayworld.com/DatabasePage.html
	ArrayExpress	comprehensive microarray data	http://www.ebi.ac.uk/microarray-as/aer/entry
	Gene Expression Omnibus	archive of high-throughput data (microarray, SAGE[a], RNA-seq[b])	http://www.ncbi.nlm.nih.gov/geo/
Transcriptomic (DNA sequencing)	SAGE Genie	SAGE[a] data from normal and malignant human tissues	http://cgap.nci.nih.gov/SAGE
	Mouse Atlas of Gene Expression	SAGE[a] and sequence tag data from many mouse tissues during development	http://www.mouseatlas.org/mouseatlas_index_html
Transcriptomic (tissue hybridization)	Allen Brain Atlas	high-resolution expression patterns from sections of adult mouse brain	http://www.brain-map.org/
	EMAGE	high-resolution gene expression during mouse development	http://genex.hgu.mrc.ac.uk/Emage/database/emageIntro.html
Proteomics	Proteomic World	compilation of proteomics databases	http://www.proteomicworld.org/DatabasePage.html

[a]SAGE (serial analysis of gene expression) is a method that is designed to make double-stranded DNA copies of RNA populations. These are then processed enzymatically so that from individual RNA molecules short sequence fragments (tags) are recovered and then sequenced. [b]RNA-seq involves shotgun sequencing of the whole transcriptome, using massively parallel DNA sequencing.

databases and new algorithms—for example, to search and analyze the data or to model structures and molecular interactions—as it has by experimental advances. However, in addition, as described in this section, bioinformatics can be used directly to infer gene function simply by comparing sequences from previously uninvestigated genes with other nucleotide and protein sequences about which some function is already known.

Sequence homology searches can provide valuable clues to gene function

The function of a recently identified gene may be unknown, but if it can be shown to be evolutionarily related to another gene with a defined function, it is reasonable to suspect that the unknown gene has the same or at least a related function. Follow-up experimental analyses are then used to seek support for the function predicted by bioinformatic analyses.

The evolutionary relationships are initially determined by using *sequence homology searches* (see Box 8.3, p. 237). For genes that make proteins, it is most efficient to conduct sequence homology searches at the protein level. The inferred protein product is used to search protein sequence databases with programs such as BLASTP, or translated nucleotide sequences with TBLASTN. Protein comparisons are preferred to nucleotide sequence comparisons, partly because amino acid sequences are more conserved than nucleotide sequences. There is also greater confidence in identifying evolutionarily significant sequence matches given that there are 20 different amino acids but only 4 different nucleotides.

The robustness of functional annotation based on homology searching depends on many factors, including the degree of similarity between the query sequence and any database hits, the reliability of functional information already in the databases, and the degree to which conserved sequence and structure corresponds to conserved function. A particular query sequence may return a number of matches with different degrees of similarity, suggesting different evolutionary relationships.

Sometimes the matching sequences may be very similar and can be aligned over all or most of their lengths. In such cases, the query sequence and the matching sequence are usually *orthologs* that can be expected to perform identical or closely related functions in different species. If orthologs can be identified in well-studied model organisms, it may be possible to infer a large amount about the function of an unstudied human gene.

A lower degree of similarity is often found in the case of *paralogs*, homologous genes that have arisen by gene duplication and divergence within a genome. In such cases, functional predictions may be reliable at the biochemical level but their specific cellular and organism-level functions might be very different. Greater structural similarity generally implies greater functional similarity, and biochemical function is generally more highly conserved than cellular and organism-level function.

Identifying paralogous relationships can also lead to the identification of novel genes. For example, the vertebrate globin gene family was long thought to be limited to genes encoding blood hemoglobin subunits and the muscle myoglobin, but homology searching of expressed sequence tag (EST) databases revealed more distantly related sequences from previously unidentified genes, encoding a globin expressed in nervous tissue, neuroglobin, and the widely expressed cytoglobin (see Figure 10.12).

Database searching is often performed with a model of an evolutionarily conserved sequence

Sequence differences between closely conserved sequences can be displayed by multiple sequence alignment with programs such as CLUSTAL. A representative model is often derived from protein sequences aligned in this way and then used to interrogate databases, rather than simply relying on individual sequence queries.

The simplest approach is to establish a *consensus sequence*. In protein sequence alignments the majority amino acid is used unless there is too much variation. A *pattern* is a slightly more informative model that gives more weighting to a significant minority of amino acids. More informative still is a *position-specific scoring matrix* (*PSSM*). This model takes into account all the variation at each position, with highly conserved positions receiving high scores and weakly conserved positions scoring close to zero (**Figure 12.1**).

Extended PSSM models can be used to detect remote protein homologies that cannot be detected by standard BLAST searching. A popular program that uses this approach is PSI-BLAST (position-specific iterative BLAST). The underlying principle is to use a reiterative search based on a representative PSSM-like model or *profile* that is built up from previous sequence matches. The process begins by using a protein query in a normal BLAST search, but the initial hits are combined into a representative profile that is then used in a second round of searching. Any additional hits are combined with the profile and the process is repeated either for a predetermined number of iterations or until no new hits are identified.

PSI-BLAST has been shown to identify three times as many evolutionary relationships as standard homology searching (see Further Reading for access to a Web tutorial, and **Box 12.1** for a case study in which it was successfully used).

(A)
multiple sequence alignment

```
K R K K M G P C R G G G I Y G Y E
K R K G R G Y C T K R G S Y A V E
K R K F D G K C F P L A A S P E E
K R K G K G G C K T G G C E L K E
K R K L C G T C V Y P A A G L R E
```

list of amino acids at each position

```
1 2 3 4 5 6 7 8 9 1 1 1 1 1 1 1 1 1
                  0 1 2 3 4 5 6 7
K R K K M G P C R G G G I Y G Y E
      G R   Y   T K R A S S A V
      F D . K   F P L   A E P E
      L K   G   K T P   C G L K
          C   T   V Y           R
```

consensus

```
K R K * * G * C * * * G * * * * E
```

pattern

```
K-R-K-X(2)-G-X-C-X(3)-(GA)-X(4)-E
```

(B)
matrix of amino acid frequencies

```
  1 2 3 4 5 6 7 8 9 1 1 1 1 1 1 1 1 1
                    0 1 2 3 4 5 6 7
A 0 0 0 0 0 0 0 0 0 0 0 2 2 0 1 0 0
C 0 0 0 0 1 0 0 5 0 0 0 0 1 0 0 0 0
D 0 0 0 0 1 0 0 0 0 0 0 0 0 0 0 0 0
E 0 0 0 0 0 0 0 0 0 0 0 0 0 1 0 1 5
F 0 0 0 1 0 0 0 0 1 0 0 0 0 0 0 0 0
G 0 0 0 2 0 5 1 0 0 2 3 0 1 1 0 0
H 0 0 0 0 0 0 0 0 0 0 0 0 0 0 0 0 0
I 0 0 0 0 0 0 0 0 0 0 0 0 1 0 0 0 0
K 5 0 5 1 1 0 1 0 1 0 1 0 0 0 0 1 0
L 0 0 0 1 0 0 0 0 0 0 1 0 0 0 2 0 0
M 0 0 0 0 1 0 0 0 0 0 0 0 0 0 0 0 0
N 0 0 0 0 0 0 0 0 0 0 0 0 0 0 0 0 0
P 0 0 0 0 0 0 1 0 0 0 0 0 0 0 1 0 0
Q 0 0 0 0 0 0 0 0 0 0 0 0 0 0 0 0 0
R 0 5 0 0 1 0 0 0 1 0 1 0 0 0 0 1 0
S 0 0 0 0 0 0 0 0 0 0 0 0 0 1 1 0 0
T 0 0 0 0 0 0 1 0 1 0 0 0 0 0 0 0 0
V 0 0 0 0 0 0 0 0 1 0 1 0 0 0 0 1 0
W 0 0 0 0 0 0 0 0 0 0 0 0 0 0 0 0 0
Y 0 0 0 0 0 0 1 0 0 0 0 0 2 0 1 0
```

Figure 12.1 Modeling protein sequence conservation in preparation for database searching. It is often advantageous to interrogate databases by using an artificially constructed model of a conserved sequence. (A) Five related sequences each containing 17 amino acids are aligned (top); the majority amino acid at each amino acid position is highlighted in yellow. Sequence variation can be described by listing the alternative amino acids at each of the 17 positions. Sequence conservation can be described by a *consensus sequence* that gives the majority amino acid at each position unless there is too much variation, whereupon an asterisk (*) is used. Alternatively, sequence conservation is represented by a *pattern* that represents positions with too much variation as X but recognizes a significant minority amino acid; for example, [GA] denotes that G and A are alternatives at position number 12. (B) A more comprehensive *position-specific scoring matrix* (*PSSM*) initially calculates the frequency of every amino acid residue (vertical axis) at each amino acid position (horizontal axis), as shown here. Subsequently, the observed frequency is compared with the frequency at which a residue can be expected in a *random* sequence so as to provide a score for each position in the matrix (as a logarithm of the ratio of the observed and expected frequencies—not shown here).

BOX 12.1 HOW BIOINFORMATICS CAN HAVE A VITAL ROLE IN ELUCIDATING GENE FUNCTION: THE EXAMPLE OF THE *NIPBL* GENE

Positional cloning identified a previously unstudied gene at 5p13 as a major locus for a severe developmental malformation, Cornelia de Lange syndrome (CdLS; OMIM 122470). The sequence of the underlying gene provided few clues to its function, but sequence homology searching with the predicted 2804 amino acid protein revealed that it had well-studied orthologs in flies and fungi.

The *Drosophila* ortholog, Nipped-B, was known to be a developmental regulator that seemed to facilitate long-range interactions between promoters and remote enhancers in certain target genes, and so the human gene was named *NIPBL* (Nipped-B-like). Fungal orthologs of the NIPBL protein, such as budding yeast SCC2, were known to load cohesins onto chromatin and to be important in regulating chromosome segregation. Individuals with CdLS possessing a causative *NIPBL* mutation are heterozygotes and do not have a significantly increased incidence of chromosome segregation abnormalities. However, when both *NIPBL* alleles were selectively inactivated in human cells by RNA interference, clear defects in sister chromatid cohesion were found. Parallel studies in which the *Nipped-B* gene was inactivated in *Drosophila* cells also produced sister chromatid cohesion defects.

In budding yeast, SCC2 was known to bind a smaller protein SCC4, close to 600 amino acid residues long, and it had been shown that the SCC2–SCC4 complex was responsible for loading cohesins, multisubunit proteins that regulate chromosome segregation, on to chromatin. What had been intriguing, however, was that whereas SCC2 had been highly conserved during evolution with clear orthologs in all genomes, including NIPBL and Nipped-B, SCC4 seemed so poorly conserved that standard BLAST searching could find no matches other than homologs in closely related species of yeast that were confined to the *Saccharomyces* genus.

Different protein interaction methods were used to show that the fly Nipped-B protein and the human NIPBL protein each bound a small protein roughly 600 amino acid residues long, which turned out to be closely related orthologs of each other and unexpectedly also of the well-studied *C. elegans* Mau-2 protein. Mau-2 was known to be a regulator of cell and axon migration in *C. elegans*, but nothing was known about any of its orthologs in other species.

Standard BLAST searches did not find any evidence to suggest that Mau-2 and its orthologs were related to SCC4 in any way. However, subsequent analyses with PSI-BLAST, a specialized sequence homology search program that is useful for detecting similarities between distantly related sequences, showed that SCC4, Mau-2, and the proteins binding to NIPBL and Nipped-B were orthologs. Subsequent experimental analyses confirmed the idea that vertebrate orthologs functioned in loading cohesin onto chromatin as partners of vertebrate SCC2 orthologs. As for the SCC2/Nipped-B/NIPBL family, selective inhibition of expression of genes encoding Mau-2 and its orthologs also resulted in chromosome segregation defects (**Figure 1**).

It was concluded that the yeast SCC2–SCC4 cohesin-loading complex and its chromosome segregation function has been conserved during evolution but that, in vertebrates, the same proteins, including the NIPBL protein, are involved in various aspects of gene regulation during embryonic development. Subsequent work showed that both NIPBL and human Mau-2 were involved in regulating axon migration. Genes encoding certain cohesin subunits were also found to be mutated in CdLS, and cohesins were found to have roles in long-distance gene regulation in various cells including non-dividing neurons.

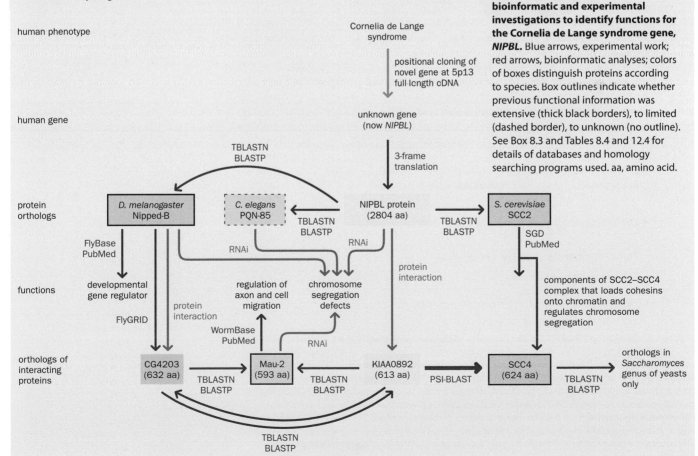

Figure 1 Synergy between bioinformatic and experimental investigations to identify functions for the Cornelia de Lange syndrome gene, *NIPBL*. Blue arrows, experimental work; red arrows, bioinformatic analyses; colors of boxes distinguish proteins according to species. Box outlines indicate whether previous functional information was extensive (thick black borders), to limited (dashed border), to unknown (no outline). See Box 8.3 and Tables 8.4 and 12.4 for details of databases and homology searching programs used. aa, amino acid.

TABLE 12.2 SECONDARY DATABASES OF PROTEIN SEQUENCES USED TO IDENTIFY CONSERVED ELEMENTS AND PROTEIN DOMAINS		
Database	**Contents**	**URL**
PROSITE	consensus sequences (*sequence patterns*) associated with protein families and longer *sequence profiles* representing full protein domains	http://ca.expasy.org/prosite
Pfam, SMART, ProDom	collections of protein domains	http://pfam.sanger.ac.uk/; http://smart.embl-heidelberg.de/
InterPro	a search facility that integrates the information from other secondary databases	http://www.ebi.ac.uk/interpro/

Comparison with documented protein domains and motifs can provide additional clues to gene function

In some cases, database searches also return hits that match the query sequence over only a relatively small part of its length, often signifying a specific protein domain. This reflects the modular nature of many proteins, with distinct functions performed by different protein domains. The matching genes have diverged not merely by the accumulation of point mutations but also by more complex events such as recombination between genes and gene segments, leading to *exon shuffling*.

Several secondary sequence databases have been created that contain protein domain motifs and profiles derived from the primary sequence databases (**Table 12.2**), and various programs allow a given protein sequence query to be interrogated to see whether it contains any of the known domains and sequence motifs. **Figure 12.2** gives an example.

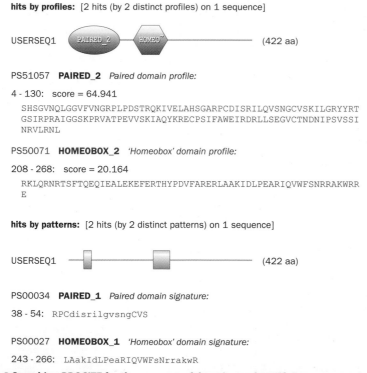

hits by profiles: [2 hits (by 2 distinct profiles) on 1 sequence]

USERSEQ1   PAIRED_2   HOMEO   (422 aa)

PS51057   **PAIRED_2**   *Paired domain profile:*

4 - 130:   score = 64.941

    SHSGVNQLGGVFVNGRPLPDSTRQKIVELAHSGARPCDISRILQVSNGCVSKILGRYYRT
    GSIRPRAIGGSKPRVATPEVVSKIAQYKRECPSIFAWEIRDRLLSEGVCTNDNIPSVSSI
    NRVLRNL

PS50071   **HOMEOBOX_2**   *'Homeobox' domain profile:*

208 - 268:   score = 20.164

    RKLQRNRTSFTQEQIEALEKEFERTHYPDVFARERLAAKIDLPEARIQVWFSNRRAKWRR
    E

hits by patterns: [2 hits (by 2 distinct patterns) on 1 sequence]

USERSEQ1   (422 aa)

PS00034   **PAIRED_1**   *Paired domain signature:*

38 - 54:   RPCdisrilgvsngCVS

PS00027   **HOMEOBOX_1**   *'Homeobox' domain signature:*

243 - 266:   LAakIdLPeaRIQVWFsNrrakwR

Figure 12.2 Searching PROSITE for the presence of domains and motifs in a test protein. The PROSITE database at http://www.expasy.org/prosite/ stores a large number of documented sequences for protein domains and motifs. In this example, searching with a 422 amino acid (aa) protein query returned two hits when compared against protein domain profiles with significant matching to both a paired domain (found in the PAX family of transcription factors) at amino acids 4–130 and a homeodomain and a homeobox variant at positions 208–268. A further two hits are evident when compared against patterns of protein motifs: a paired domain type motif [R-P-C-x(11)-C-V-S] spanning amino acids 38–54 and a motif characteristic of some homeodomains spanning amino acids 243–266. The motifs identified suggest strongly that the test protein is a DNA-binding protein and is most probably a transcription factor.

12.3 STUDYING GENE FUNCTION BY SELECTIVE GENE INACTIVATION OR MODIFICATION

Experimental methods to establish how genes function often use genetic manipulations that produce either altered gene sequences or altered gene expression, resulting in altered phenotypes. By correlating an altered phenotype, which may be at the level of a cell or early embryo or organism, with changes in the properties of a specific gene one can gain important clues about how that gene normally functions.

Clues to gene function can be inferred from different types of genetic manipulation

Model systems are often used to infer the functions of human genes. They include well-studied cultured cells and genetically amenable model organisms whose properties have been described in Chapter 10 (Section 10.1, p. 298). In principle, two major types of genetic analysis have been applied in model systems:

- **Classical or forward genetics** (phenotype→gene). The starting point is a phenotype of interest that may have arisen spontaneously but is usually generated by random mutagenesis using, for example, chemical mutagens or ionizing radiation. The aim is to identify the gene whose expression has been altered by mutation, causing the observed phenotype.

- **Reverse genetics** (gene→phenotype). A specific gene is genetically manipulated in some way in an effort to produce a phenotype that illuminates the role of the gene.

As described later in this section, screens for genetic function have largely been performed by genetically manipulating whole model organisms in which both forward and reverse genetic approaches can be used. The mouse has been the primary model organism for understanding how human genes function because, being a mammal, it has counterparts of almost all human genes, and because until very recently it was the only mammalian model organism in which it was possible to inactivate a specific predetermined gene. To do so, the gene of interest is inactivated in the mouse germ line to produce mutant descendants with a specific gene *knockout* that results in a loss of gene function. Other types of genetic modification can produce gain-of-function, dominant-negative, or modifier mutations (see below). By studying the phenotype of the mutant animals, valuable clues to how the genes work are frequently obtained.

The sophisticated technology required to make gene knockouts in mice has been the gold standard for assaying mammalian gene function, but, as described in Chapter 20, very recent advances in genetic technologies mean that construction of gene knockouts in other mammalian and vertebrate species will become much more common. However, the procedures involved are laborious and time-consuming. In recent years, a variety of methods have been systematically applied to the rapid investigation of gene function with cultured cells.

Cultured cells have the disadvantage that their use is inevitably restricted. They cannot directly provide information on gene functions that involve interaction between different cell types. They cannot easily replicate the natural *in vivo* environment of cells that are often components of tissues and organs. In addition, not all cell types of the body are easy to grow in cell culture. However, it has been possible to grow many different cell types in culture, cultured cell lines are very convenient to study, and rapid assays can often permit valuable insights to be gained into functions of genes within the limited context of a single cell type. Additionally, certain types of cultured human cells—such as HeLa cells and HEK (human embryonic kidney) 293 cells—have been extensively studied, allowing data to be related to an already extensive literature on these cell lines. Importantly, global analyses in some kinds of cultured cells have the potential to enable mapping of the entire range of gene functions and molecular interactions within a cell and so allow a kind of **systems biology**.

Different genetic screens can be conducted in cultured cells. However, the forward genetics approach is difficult to perform because meiotic recombination does not occur in cultured mammalian cells, and homozygous mutants cannot simply be generated. Genetic screens in cultured mammalian cells are therefore

TABLE 12.3 DIFFERENT TYPES OF REVERSE GENETIC SCREEN IN CULTURED MAMMALIAN CELLS	
Type of screen	Basis of method
Loss-of-function	usually, the RNA interference (RNAi) pathway is induced to selectively degrade RNA transcripts of a specific target gene (see Figures 12.3 and 12.4)
Gain-of-function	involves transfection of an exogenous gene copy driven by a suitable promoter to overexpress or misexpress that gene in a desired cell type
Dominant-negative	relies on producing a mutant gene product that interferes with the normal product of a specific target gene; similar to loss-of-function screen but suppresses gene activity more efficiently than RNA; often the mutant product is designed to be a truncated version of the wild type that competes with wild-type protein in some way, or the mutant protein becomes incorporated with wild-type proteins in multisubunit complexes, thereby inactivating the complex
Modifier	seeks to identify genes that enhance or suppress a specific phenotype; the initial phenotype is produced by one of the three methods above; thereafter, the mutant cells are subjected to a second genetic modification (which can again be any of the three methods above); if altered expression of a second gene enhances or suppresses the initial phenotype, the second gene is considered to be a modifier of the gene that caused the initial phenotype; it then remains to be worked out how the two genes interact

For further details see the review by Tochitani & Hayashizaki (2007).

typically reverse genetic screens. There are four different functional classes of mutation—gain-of-function, loss-of-function, dominant-negative, and modifier—and different screens can be conducted depending on the desired effect on the target gene (Table 12.3).

Of the different approaches listed in Table 12.3, by far the most widely used involves selective gene inactivation to cause a loss of function. An early general approach used the specificity of base pairing to selectively inhibit the expression of a predetermined gene. In **antisense technology**, RNA or oligonucleotide constructs are designed to have a sequence that is complementary to that of RNA transcripts from a gene of interest. By hybridizing to the sense transcripts, the antisense constructs block gene expression (gene *knockdown*).

Antisense technology was first developed in the 1980s. Specific antisense RNA molecules were initially used to *knock down* the expression of a target gene, but this was very much a hit-or-miss affair depending on which gene was targeted. Subsequently, antisense oligodeoxynucleotides were used because they were more stable than RNA. More recently, morpholino antisense oligonucleotides are used. They have a vastly more stable and robust structure than conventional oligonucleotides, and they are more consistent in producing significant inhibition of gene expression. As will be described in Chapter 20, morpholino oligonucleotides are now routinely used to knock down the expression of specific genes in various vertebrate models, notably zebrafish embryos.

As explained in the next section, the natural phenomenon of RNA interference (see Box 9.7) has also been widely applied as an experimental tool to knock down the expression of specific target genes in cultured cells and in the embryos of some invertebrates, notably the nematode *Caenorhabditis elegans*.

RNA interference is the primary method for evaluating gene function in cultured mammalian cells

The most widely used method to assess gene function in cultured mammalian cells exploits *RNA interference* (*RNAi*), a natural cellular pathway in which the formation of double-stranded RNA (dsRNA) induces specific gene inactivation. RNA interference is thought to have evolved to protect cells against the accumulation of potentially dangerous nucleic acid sequences, notably viruses (see Box 9.7).

To knock down a predetermined gene transcript, a suitably specific dsRNA needs to be provided. In some animal systems, such as the nematode *C. elegans*, long dsRNA is used experimentally to inactivate genes. After it has been transferred into the worm's cells, the dsRNA is chopped up by an endogenous cytoplasmic ribonuclease called dicer into uniformly small pieces of dsRNA (close to 20 bp in length and with 3′ dinucleotide overhangs) called *short interfering RNA*

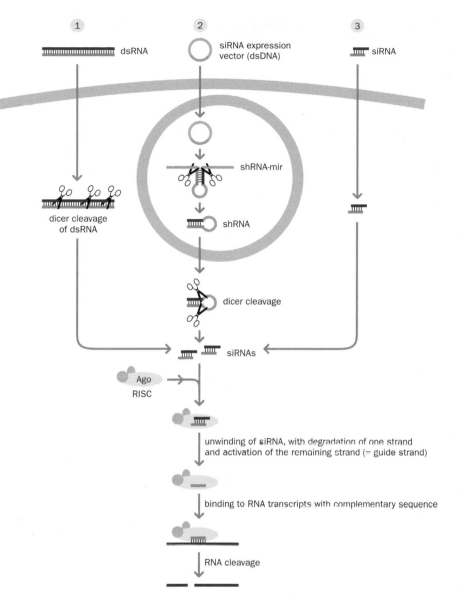

Figure 12.3 Three routes for experimentally inducing RNA interference (RNAi) in animal cells. To induce RNAi, cells are provided with double-stranded RNA (dsRNA) whose nucleotide sequence is complementary to a sequence in a predetermined target gene. The RNA is either transfected into cells (pathways 1 and 3) or produced from a transfected expression vector (pathway 2). Pathway 1 shows the use of long double-stranded RNA (typically longer than 200 bp) that can be used in invertebrate systems such as that in *C. elegans*. In mammalian systems, however, long dsRNA triggers *nonspecific* RNA cleavage (alternative pathways 2 and 3 are used). Short double-stranded RNA can be produced from an expression vector (pathway 2) transcribed in the nucleus to give a single RNA molecule with inverted repeats. The repeats base-pair to each other to produce a short *hairpin RNA* that resembles initial miRNA transcripts (shRNA-mir). The 5′ and 3′ tails of this primary transcript are then cleaved in the nucleus to give a more compact short hairpin RNA (shRNA) that is exported to the cytoplasm where it is processed by the dicer endonuclease to give double-stranded short interfering RNA (siRNA). Alternatively, two short oligoribonucleotides are chemically synthesized to have complementary sequences and are allowed to anneal to form double-stranded siRNA (pathway 3). Whichever pathway is used, the end result is to provide a siRNA duplex that is bound by the argonaute ribonuclease subunit (Ago) of the RNA-induced silencing complex (RISC). Unwinding of the siRNA and degradation of one of its strands activates the RISC, which then binds to a mRNA with a complementary sequence that is then cleaved by the argonaute subunit.

(*siRNA*) (**Figure 12.3**, pathway 1). The siRNA associates with a multisubunit protein complex, the RNA-induced silencing complex (RISC), and the duplex siRNA is unwound and one strand degraded by a RISC ribonuclease called argonaute. The RISC with one siRNA strand attached to it is now activated, and the siRNA strand will guide it to bind RNA transcripts containing complementary sequences. Any transcript that is bound by the specific antisense siRNA is then targeted for destruction by the argonaute ribonuclease.

The addition of long dsRNA to mammalian cells does not inhibit the expression of specific genes as in *C. elegans*. Instead, long dsRNA induces a general antiviral pathway that produces global, *nonspecific* gene silencing. It does this by activating the protein kinase PKR (which is also induced by interferon). Activated PKR phosphorylates the translation initiation factor 2 (eIF2α), thereby inactivating it and causing a generalized inhibition of translation. PKR also phosphorylates 2′,5′ oligoadenylate synthetase (2′,5′AS), which, when so activated, induces a ribonuclease that causes nonspecific degradation of mRNA.

As an alternative to adding long dsRNA to mammalian cells, two approaches are used (see Figure 12.3). In one method (Figure 12.3, pathway 2; see also next section), transgenes are added that can be expressed to give a partly double-stranded hairpin RNA (like the structure of miRNA). Another widely used method involves chemically synthesizing two short complementary oligoribonucleotides and allowing them to anneal to form a synthetic equivalent of siRNA. Often, the synthesized oligonucleotides are chosen to be about 20–25 nucleotides long and the sequences are chosen to form a duplex that has two deoxythymidine bases added as 3′ overhangs (duplexes with 3′ overhangs are more potent than blunt ones in inducing RNAi).

Because the RNAi efficiency can vary depending on the sequence, it is usual to design three or more complementary pairs of siRNA oligonucleotides corresponding to different sequences for any transcript. The efficiency of gene knockdown is assessed by real-time PCR or, where antibodies are available for the relevant protein product, by western blotting.

Global RNAi screens provide a systems-level approach to studying gene function in cells

RNA interference studies began by attempting to assess the function of individual genes. In the post-genome era, it became possible to perform large-scale, and eventually genomewide, analyses of gene function. As long as there is a functional assay that can be used in cultured cells, then parallel analyses of all available genes permit a systems-level approach to studying gene function in cells. That is, entire functional pathways can be dissected, such as regulation of the cell cycle (see Further Reading for examples).

To perform global RNAi screens, suitably large nucleic acid libraries need to be made. One way is to make *siRNA libraries*. Pairs of complementary siRNAs are synthesized for each gene in the genome and deposited as individual pairs in wells of huge microtiter plate arrays.

Another approach is to make libraries of genes encoding short hairpin RNA (*shRNA libraries*). The idea is to mimic the way in which miRNA is naturally made in cells, being produced from a short transcript that spontaneously forms a hairpin RNA, which then undergoes cleavage (see Figure 9.16). Pairs of long complementary oligonucleotides are designed in such a way that, when annealed and cloned into a suitable expression vector, a transcript can be produced with inverted repeats that base-pair (**Figure 12.4**).

In many shRNA libraries, the transcript is simply designed to be a short hairpin RNA that is exported to the cytoplasm, where it is cleaved by dicer to give siRNA. So-called second-generation shRNA libraries take this a step further by encoding sequences similar to the primary miRNA transcripts (pri-miRNA) that give rise to a short dsRNA through two successive types of cleavage (first by Rnasen/Drosha in the nucleus to generate pre-miRNA, and then by dicer in the cytoplasm; see Figure 9.16).

Inactivation of genes in the germ line provides the most detailed information on gene function

Assaying gene function in cultured cells or using cell extracts has its limitations. In particular, no detailed information can be obtained about how a gene functions in regulating processes that involve an interaction between different types of cell in the body, whether it be in the context of physiology (e.g. the nervous system) or embryonic development. Defining gene function in this wider context requires the genetic manipulation of model organisms.

Figure 12.4 Producing siRNA with the use of an shRNA expression vector. An shRNA expression vector seeks to mimic the way in which miRNA is produced naturally by cells. The example shown here illustrates the cloning of a DNA insert about 60 bp long that was prepared by synthesizing complementary oligonucleotides and allowing them to anneal. The sequence of the DNA insert is designed to have inverted repeats (letters in red overlined with red arrows). After the DNA insert has been transcribed by a vector promoter sequence (P) located close to the cloning site and subjected to post-transcriptional processing, the resulting RNA will have complementary nucleotide sequences (19 nucleotides long in this example) that can base-pair to form a short hairpin RNA (shRNA) with an unpaired loop. The shRNA will be cleaved asymmetrically by cytoplasmic dicer at the positions marked by the green arrows, generating a functional siRNA consisting of a central 19 bp dsRNA with 3′ di-uridine (UU) overhangs. The black arrow indicates the direction of transcription. [Adapted from Bernards R, Brummelkamp TR & Beijersbergen RL (2006) *Nat. Methods* 3, 701–706. With permission from Macmillan Publishers Ltd.]

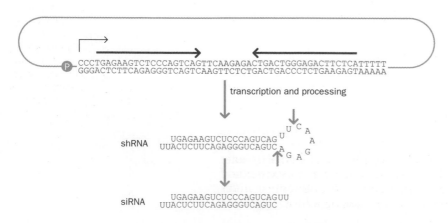

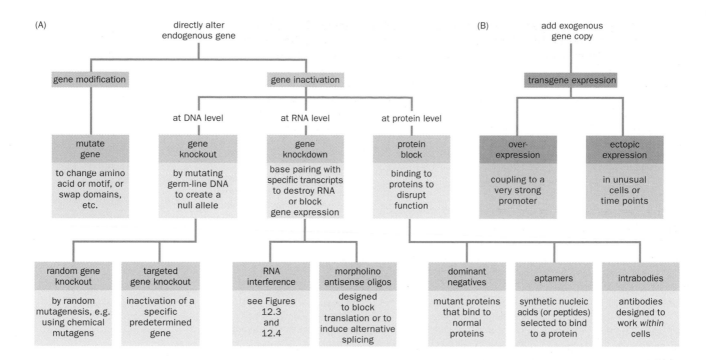

Although all four types of screen listed in Table 12.3 can also be performed *in vivo*, the most common approach to defining the function of a predetermined gene in whole model organisms is to inactivate a gene and then assess the resulting phenotype. This can be done by mutating the gene to produce a null allele, a *gene knockout,* or the gene can be effectively inactivated by inhibiting its expression at the RNA or protein level (**Figure 12.5**).

Valuable clues to gene function are usually obtained by this method, but not all genes produce a phenotype when knocked out because of genetic redundancy (sometimes two or more genes have identical or very similar roles and so double or triple gene knockouts may be needed to elucidate a phenotype). Other genes may be difficult to study because their functions are so crucial that inactivating them causes cell death.

The technologies involved in designing and studying genetically modified animals to understand gene function are highly sophisticated. We examine these in Chapter 20, where animal models of disease are described. Different methods are popular in different model eukaryotes.

Table 12.4 outlines the principal eukaryote models that are currently used to infer how human genes work. The extent to which they are valuable depends on the degree to which functions of genes have been conserved during evolution. Thus, although it may seem evident that mammalian models would generally be the most useful models (because of the high biochemical and physiological similarity to humans), even unicellular yeasts have also provided many insights into human gene function.

12.4 PROTEOMICS, PROTEIN–PROTEIN INTERACTIONS, AND PROTEIN–DNA INTERACTIONS

The majority of known human genes encode polypeptides that are incorporated in proteins. Proteins serve numerous functions in cells, and large-scale studies of how human proteins function have begun.

Proteomics is largely concerned with identifying and characterizing proteins at the biochemical and functional levels

In its traditional, broadest sense, **proteomics** encompasses all global, or at least large-scale, studies of proteins. However, it is now usual to use a more restricted meaning that excludes protein structure determination (*structural genomics*) and homology studies that compare sets of protein sequences and protein

Figure 12.5 Principal ways of studying gene function *in vivo* **by genetically modifying model eukaryotes.** (A) Most methods seek to inactivate a gene to produce a mutant phenotype that can be correlated with a specific gene (peach boxes). Usually this is done at the level of DNA, where large-scale random mutagenesis screens have been performed in many models, including both invertebrates (such as *D. melanogaster*) and vertebrates (notably zebrafish and mice). Specific knockouts of predetermined genes have also been conducted, notably by using homologous recombination in mice. The expression of a specific gene can also be selectively inhibited by targeting its transcripts—large-scale RNA interference-based genetic screens have been conducted in *C. elegans* and *D. melanogaster* and gene knockdown with antisense morpholino oligonucleotides is commonly performed in zebrafish embryos. Blocking proteins with dominant-negative mutant proteins, aptamers, or intrabodies also have potential roles in treating disease (by specifically downregulating harmful genes) and are discussed in more detail in Chapter 21. (B) An alternative approach is to use a transgene to overexpress a specific gene or to express it in tissues or developmental stages in which it is not normally expressed (*ectopic expression*) in an attempt to produce a phenotype that will provide clues to the function of the gene.

TABLE 12.4 PRINCIPAL MODEL EUKARYOTIC ORGANISMS USED TO ASSESS GENE FUNCTION

Organism	Class	Methods of studying gene function	Comments	Genes and mutants databases
Saccharomyces cerevisiae	yeast (unicellular fungus)	highly sophisticated genetics and mutant screens aided by high frequency of homologous recombination	despite being unicellular, yeasts have been useful in modeling how some genes work to perform some highly conserved cellular functions such as in DNA repair and cell cycle control	SGD (*Saccharomyces* Genome Database) at http://www.yeastgenome.org/
Caenorhabditis elegans	nematode (roundworm)	large-scale genetic screens using RNA interference	extensively studied; invariant and fully mapped cell lineage; the only model in which the complete wiring system is known for the nervous system (there are 302 neurons, and all their connections are known)	WormBase at http://www.wormbase.org/
Drosophila melanogaster	fruit fly	highly sophisticated genetics and mutant screens, aided by P1 transposon mutagenesis	extensively studied and huge numbers of well-characterized mutants	FlyBase at http://flybase.bio.indiana.edu/
Dario rerio	zebrafish	selective gene inactivation by using morpholino oligonucleotides to inhibit expression at level of translation; also, use of transgenes to express dominant-negative mutant proteins	principal uses are for understanding how genes function in early development and in offering a rapid assessment of how a selected gene functions in a vertebrate	ZFIN at http://zfin.org/
Mus musculus	mouse	large-scale random mutagenesis screens (using chemical mutagens or transgenes); large programs using homologous recombination to knock out specific genes; expression of mutant transgenes and overexpression/ectopic expression of normal transgenes; transgenic RNA interference	the no. 1 model for understanding human gene function because of the combination of being a mammal and the capacity for sophisticated genetic manipulation	MGI (Mouse Genome Informatics) at http://www.informatics.jax.org/

Nature Reviews Genetics has published a series of review articles on the Art and Design of Genetic Screens in model organisms that includes the above models as follows: yeast: Forsburg SL (2001) *Nat. Rev. Genet.* 2, 659–668; *C. elegans*: Jorgensen EM & Mango SE (2002) *Nat. Rev. Genet.* 3, 356–369; *Drosophila*: St Johnston D (2002) *Nat. Rev. Genet.* 3, 176–188; zebrafish: Patton EE & Zon LI (2001) *Nat. Rev. Genet.* 2, 956–966; mouse: Kile BT & Hilton DJ (2005) *Nat. Rev. Genet.* 6, 557–567. Many of the technologies are explained in detail in Chapter 20.

structures. Nevertheless, in the post-genome era proteomics has multiple facets, as illustrated in **Figure 12.6**.

Much of the effort in proteomics is devoted to identifying proteins and analyzing them at biochemical and functional levels. Analysis of the sequences of the human and other vertebrate genomes has led to the identification of a large number of novel open reading frames that are used in analyzing and referencing sequences recovered from mass spectrometry analyses. Certain types of post-translational modification are also being characterized at the proteome level with the use of proteome chip technology, notably phosphorylation (see Further Reading).

Expression analyses include profiling subcellular protein localization, and comparative two-dimensional gel electrophoresis (see Figure 8.25) where samples representing two conditions (e.g. with and without drug treatment) are compared. Large-scale assays of phenotypes generated by gene knockout or RNAi-induced gene knockdowns can be conducted at the proteome level to illuminate understanding of gene function.

Protein microarrays (or *protein chips*) have also been employed to understand protein function. In principle, they can be made in much the same way as DNA and oligonucleotide microarrays, with the proteins printed on glass slides. However, high throughput has been difficult to achieve in practice. Protein microarrays have been used mostly in protein identification and also to identify the

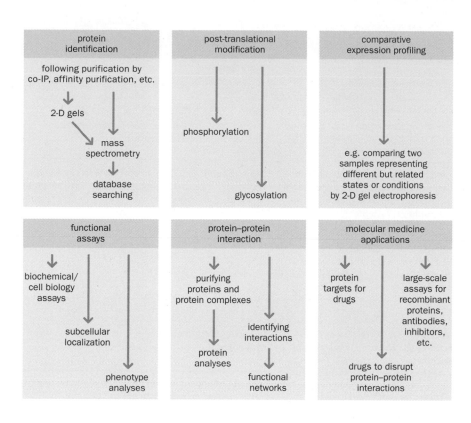

Figure 12.6 Major facets of proteomics. Protein–protein interactions are detailed in this chapter. Molecular medicine applications are described in Chapter 21. 2-D, two-dimensional.

substrates of enzymes such as protein kinases. Antibody arrays have been comparatively widely used.

Much of the focus of proteomics is currently being devoted to identifying protein–protein interactions and protein complexes. The ultimate aim is to define protein networks that will aid our understanding of how cells work.

In addition to binding to other proteins and to nucleic acids, proteins interact with a large number of small molecules that act as ligands, substrates, cargoes (in the case of transport proteins), cofactors, and allosteric modulators. In all cases, these interactions occur because the surface of the protein and the interacting molecule are complementary in shape and chemical properties. Protein–ligand interactions provide information on how genes function, and they are also important for therapeutic purposes: the basis of drug design is to identify molecules that interact specifically with target proteins and change their activities in the cell. Most drugs work by interacting with a protein and altering its activity in a manner that is physiologically beneficial. We will consider such interactions in the context of therapeutic applications in Chapter 21.

Large-scale protein–protein interaction studies seek to define functional protein networks

The function of a predicted protein may not be obvious even after a comprehensive study of its sequence and expression profile. One way of gaining insights into function is to identify those proteins with which a test protein interacts—previous information on the interacting proteins can provide strong clues to the function of the test protein. For example, if a previously uncharacterized protein is shown to interact with proteins required for RNA splicing, it is likely to have a role in the same process. In this way, proteins can be linked into *functional networks* in the cell.

A wide range of genetic, biochemical, and physical methods can be used to study protein–protein interactions. Genetic methods can be applied only in model organisms such as *Drosophila* and yeast, but biochemical and physical methods can be applied to human cells directly. High-throughput mapping methods seek partners for proteins by the global screening of proteins for their

TABLE 12.5 COMMONLY USED METHODS TO DETECT AND VALIDATE PROTEIN–PROTEIN INTERACTIONS

Method	Description	Comments
(A) PROTEIN INTERACTION SCREENING METHODS		
Yeast two-hybrid (Y2H) screening (Figure 12.7)	relies on mating of haploid yeast cells expressing hybrid proteins with complementary transcription factor domains; positive interactions bring together a novel transcription factor within the resulting diploid yeast cells	the specificity of interactions is often low; yeast cells are not ideal environments for testing interaction between mammalian proteins
Affinity purification–mass spectrometry (AP–MS) (Figure 12.8A)	a specific protein is covalently linked to an insoluble matrix; this bound protein is then used to capture any proteins that it interacts with from a cell lysate; captured protein can be eluted, purified, and analyzed by mass spectrometry	affinity purification is more specific than Y2H but does not readily detect transient interactions; like Y2H, it has been used extensively to work out the detail of protein interactomes in model organisms
TAP (tandem affinity purification) tagging (Figure 12.8B)	relies on two sequential rounds of affinity purification; this is achieved by transfecting cells with a cDNA of interest coupled with a sequence that will add a terminal tag to express two different proteins/peptides known to be bound by other proteins	
(B) METHODS FOR TESTING/VALIDATING INTERACTIONS		
Fluorescence resonance energy transfer (FRET) (Figure 12.9)	microscopy method that can monitor interactions in real time; it tests interaction of suggested interacting proteins after labeling them with different fluorophores	provides circumstantial supportive evidence only for a supposed protein–protein interaction
Co-immunoprecipitation (co-IP; Figure 12.10)	a specific antibody is used to precipitate a desired protein from a cell lysate; if the protein is complexed to other proteins, the interacting proteins should be co-precipitated	the gold standard for validating protein interaction, especially when using endogenous proteins
Pull-down assay	a variant of co-IP that is identical in all aspects except that it uses a specific protein ligand other than an antibody to capture a protein complex	

For more detail on these and other methods including protein microarrays, see the review by Lalonde et al. (2008) under Further Reading. Note that the division into (A) and (B) is somewhat arbitrary. For example, a directed Y2H assay can be designed to test the interaction between two specific proteins. On the other hand, although co-IP is often used for validating presumed interactions, it can also be used to isolate entire protein complexes.

ability to bind to individual test proteins, and various assays can be performed to validate interactions between specific proteins. We summarize the methods in **Table 12.5** and provide details of different methods in the next sections.

Yeast two-hybrid screening relies on reconstituting a functional transcription factor

In *yeast two-hybrid assays*, specific protein–protein interactions are detected by generating a functional transcription factor that is not normally made in the host yeast cell. The newly generated transcription factor then activates a reporter gene and/or selectable marker.

The method relies on the observation that transcription factors have two key domains that can maintain their function when separated: a DNA-binding domain (BD) and a transcription activation domain (AD). Coding sequences for individual BD and AD domains of a *specific* transcription factor can be joined to other cDNAs to produce hybrid (fusion) proteins carrying either a BD or AD domain. If a protein that is fused to the BD domain of the transcription factor interacts with another that is coupled to the AD domain of the same kind of transcription factor, the close association of BD and AD domains can produce a functional transcription factor even though the two domains are located in *different* proteins (**Figure 12.7A**).

The object of a standard two-hybrid screen is to use a specific protein of interest as a *bait* for specific recognition by an interacting protein selected from a large library. A haploid yeast strain is designed to express the bait protein of interest coupled to a DNA-binding domain of a transcription factor. It is mated with yeast strains from a library of haploid yeast cells that each express a single type of

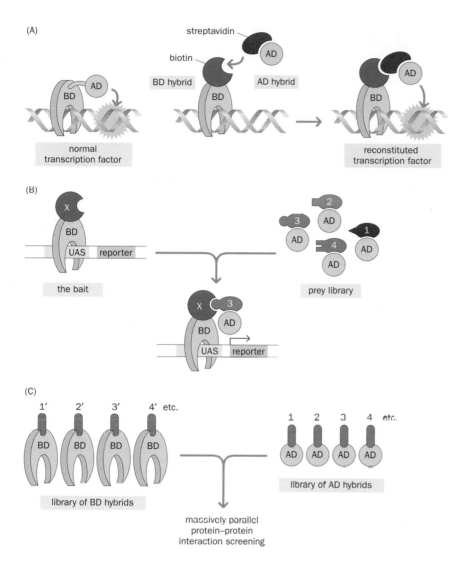

Figure 12.7 Yeast two-hybrid (Y2H) screening. (A) An active transcription factor comprises complementary DNA-binding (BD) and transcription activation (AD) domains. A cDNA for either domain can be spliced to cDNA for other proteins, producing AD and BD hybrid (fusion) proteins. Binding of the attached proteins, as for biotin and streptavidin as shown here, will associate the AD and BD hybrid proteins, reconstituting the transcription factor. (B) In a standard Y2H screen, one yeast strain expresses the *bait*, a BD hybrid protein with a specific test protein (X) fused to a binding domain for a specific transcription factor. The BD hybrid protein binds to a specific upstream activating sequence (UAS) positioned before a reporter gene but is inactive in the absence of the relevant AD. Possible protein partners for protein X are sought from within a *prey library* of numerous yeast strains containing different proteins linked to the relevant AD. Crossing of the bait strain with a prey strain that makes a protein which binds to protein X (protein 3 in this example) can produce a diploid cell in which AD and BD domains associate. As a result, the transcription factor is reconstituted and drives expression of the reporter gene. (C) Crossing of yeast cells from a library of cells expressing BD hybrid proteins with cells from a library expressing AD hybrid proteins permits massively parallel protein–protein interaction screening.

protein fused to the appropriate transcription factor activation domain but collectively express thousands of different hybrid proteins (the *prey library*; Figure 12.7B).

The target cells are also engineered to carry a reporter gene and/or a selectable marker gene that cannot be activated until the intended transcription factor has been assembled. Interactions are tested in the resulting diploid yeast cells. If the bait and prey do not interact, the two transcription factor domains remain separate; if they do interact, a new transcription factor is assembled and the marker genes is then activated, facilitating the visual identification and/or selective propagation of yeast cells containing interacting proteins. From these cells, the cDNA sequence of the prey construct can be identified.

Affinity purification–mass spectrometry is widely used to screen for protein partners of a test protein

In standard affinity purification–mass spectrometry (AP–MS), a particular bait protein is first immobilized by being covalently coupled to an insoluble supporting matrix. The matrix may be loaded on a chromatography column (affinity chromatography) or used to coat the surface of paramagnetic beads. A cell lysate is then added, and proteins that interact with the bait are retained on the supporting matrix. The non-binding proteins are then separated from the matrix by washing the matrix in a chromatography column as shown in **Figure 12.8A**, or by using a magnet to purify matrix-coated beads selectively. The interacting proteins can then be eluted from the matrix by increasing the salt concentration of the buffer, purified by gel electrophoresis, and submitted for analysis by mass spectrometry (see Figure 8.27).

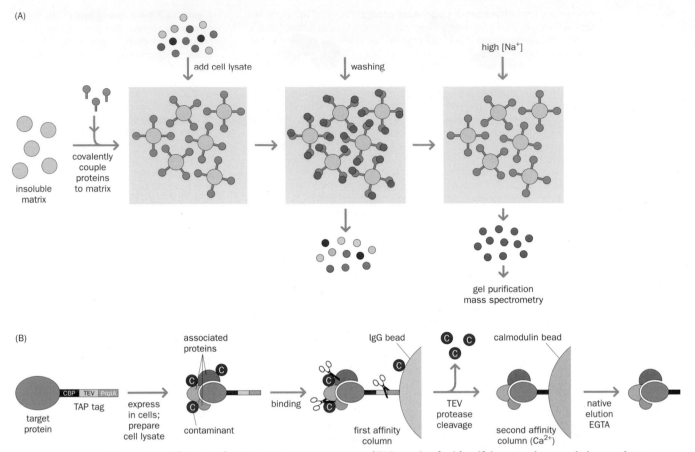

Figure 12.8 Affinity purification–mass spectrometry and TAP tagging for identifying protein–protein interactions.
(A) *Standard affinity purification–mass spectrometry*. Affinity purification starts with covalently coupling a protein of interest to beads of an insoluble matrix that is supported in some way, such as in a chromatography column. A cell lysate is passed over the beads to allow possible binding of proteins to the immobilized protein on the column. Proteins that are nonspecifically bound can be removed by washing; any interacting protein can then be eluted with buffers of high salt concentration and subjected to mass spectrometry analysis (see Figure 8.27) to identify constituent peptides and, ultimately, the protein. (B) *TAP (tandem affinity purification) tagging*. An expression vector containing the coding sequence for a TAP-tagged protein is transfected into suitable cells to allow the target protein to bind to proteins in the cells. The C-terminal TAP tag shown here has a calmodulin-binding peptide (CBP) and two immunoglobulin (IgG)-binding domains from *Staphylococcus aureus* protein A (ProtA) separated by a peptide sequence recognized by the tobacco etch virus (TEV) protease. The protein complex is recovered in a two-step affinity purification procedure. The complex will be separated from some contaminants by binding to IgG beads. The TEV protease is then used to cleave the TAP tag. The released protein complex now with just a CBP tag can bind specifically to calmodulin beads and is subsequently eluted and processed for analysis by mass spectrometry. EGTA, ethylene glycol tetracetic acid. [Adapted from Puig O, Caspary F, Rigaut G et al. (2001) *Methods* 24, 218–229. With permission from Elsevier.]

TAP (*tandem affinity purification*) *tagging* is a highly efficient variant of AP–MS. Here, a fusion protein is created by linking the target protein to an engineered affinity tag composed of two protein/peptide components known to bind strongly to different proteins, separated by a cleavage site for a specific protease. After expressing the construct in suitable cells, the target protein with interacting proteins attached can be recovered by a two-step affinity purification. The use of the protease to remove the protein complex from the matrix minimizes nonspecific binding (Figure 12.8B).

Suggested protein–protein interactions are often validated by co-immunoprecipitation or pull-down assays

Different methods are popularly used to test and validate protein interactions that have been suggested from the screening methods described above. For example, fluorescence resonance energy transfer (FRET) is a fluorescence microscopy method that monitors in real time any very close interaction between different protein molecules that have been labeled with different fluorophores.

In FRET, one of the two fluorophores acts as a potential energy donor. It is excited by light of a suitable wavelength, and part of the resulting energy can be

Figure 12.9 Using fluorescence resonance energy transfer (FRET) to test a protein interaction. Fluorescence microscopy using FRET can be used to test protein interactions within cells in real time. (A) Two proteins of interest (1 and 2) are expressed as hybrid proteins that have different fluorescent protein tags, such as cyan fluorescent protein (CFP) and yellow fluorescent protein (YFP) shown here. The cells are then exposed to light of a defined wavelength that will excite only one of the two fluorophores, in this case CFP. When excited by exposure to light of wavelength 436 nm, CFP emits blue light with a slightly longer wavelength of 480 nm and continues to do so if proteins 1 and 2 do not interact. (B) If, however, proteins 1 and 2 do interact, their attached fluorescent proteins come into very close proximity, and fluorescence energy is transferred from the donor CFP to the acceptor YFP in a non-radiative manner via dipole–dipole coupling (*resonance*). After excitation of CFP, only a small residual amount of the absorbed light is now emitted at 480 nm; most of the energy is transferred from CFP to the adjacent YFP. The acceptor YFP then emits light at a different wavelength (535 nm) that is readily distinguished.

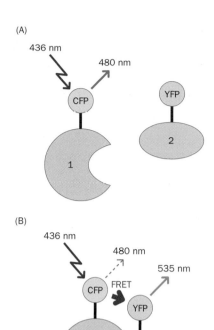

transferred to the other type of fluorophore—the acceptor—if the two proteins to which the donor and acceptor fluorophores are bound come into extremely close contact (within 1–10 nm). As a result, a significant amount of the energy absorbed by the excited donor is emitted at a much longer wavelength than if the donor and acceptor were distant from each other (**Figure 12.9**).

FRET only offers suggestive evidence of an interaction. To validate protein–protein interaction, more robust evidence is required, such as that obtained by using **co-immunoprecipitation** (**co-IP**). In this method, antibodies specific for a particular protein are added directly to a cell lysate. If the protein of interest is bound to one or more other proteins, the antibody can be used to immunoprecipitate the target protein while it is still bound to interacting proteins, often as part of a complex (**Figure 12.10**). A variant co-IP technique, called a *pull-down assay*, uses a specific ligand in place of an antibody to capture a protein of interest as part of a complex. In both methods, the bound proteins can be identified after peptide cleavage and mass spectrometry analysis.

The protein interactome provides an important gateway to systems biology

Affinity purification can be used to isolate entire protein complexes, allowing the different components to be identified by mass spectrometry. This is one of the major technology platforms in *interaction proteomics*. It has been widely used for

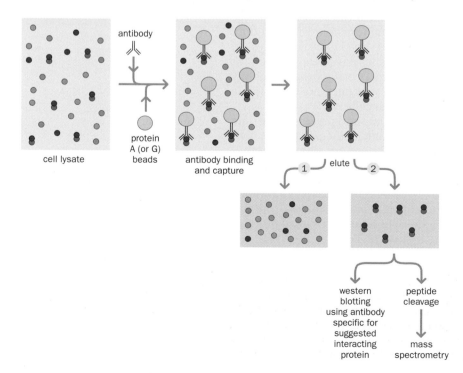

Figure 12.10 Co-immunoprecipitation (co-IP). This method is popularly used to validate a suggested interaction between two proteins, but it can also be used as a screening method to directly identify novel proteins that bind to a protein of interest. A cell lysate (or other protein mix) is incubated with an antibody specific for a protein of interest to form antigen–antibody complexes. The resulting immune complexes are then captured by binding to agarose or Sepharose beads to which protein A or protein G has been coupled (protein A and protein G are bacterial proteins that happen to bind strongly to immunoglobulins). As a result, the immune complexes are precipitated out of solution and, after centrifugation and washing, the unbound protein is eluted (1), leaving the target protein bound via its antibody to the beads. If the target protein was already bound to an interacting protein it too would be *co-precipitated*. The beads can then be treated to elute the target protein and any interacting proteins (2). To validate a suspected protein interaction, western blotting is then conducted with an antibody specific for the presumed protein partner. Alternatively, the method can be used as a screening tool to identify novel interacting proteins by treating them with proteolytic enzymes to generate peptides that can be analyzed and identified by mass spectrometry.

TABLE 12.6 MAJOR PUBLICLY ACCESSIBLE PROTEIN INTERACTION DATABASES		
Database	**Description**	**URL**
BioGRID (General Repository for Interaction Datasets)	curated set of physical and genetic interactions for more than 20 species; searchable by species	http://www.thebiogrid.org/
DIP (Database of Interacting Proteins)	curated experimentally determined interactions between proteins	http://dip.doe-mbi.ucla.edu/
DroID (*Drosophila* Interactions Database)	a comprehensive gene and protein interactions database designed specifically for the model organism *Drosophila* but with useful links to protein interaction data from other organisms	http://www.droidb.org/
Human Proteinpedia	community portal for sharing and integration of human protein data; as well as protein–protein interaction (PPI) data, it also holds data on post-translational modification, tissue and cell line expression, subcellular localization, etc.	http://www.humanproteinpedia.org/
HPRD (Human Protein Reference Database)	integrates data deposited in the Human Proteinpedia along with literature information curated in the context of individual proteins	http://www.hprd.org/
MINT (Molecular INTeraction database)	experimentally verified protein interactions mined from the scientific literature by expert curators	http://mint.bio.uniroma2.it/mint/Welcome.do
MIPS mammalian protein–protein interaction database	a collection of manually curated high-quality PPI data collected from the scientific literature by expert curators	http://mips.helmholtz-muenchen.de/proj/ppi

the characterization of protein complexes such as ribosomes, the anaphase-promoting complex, the nuclear pore complex, and signaling complexes, and has been applied on a genomic scale in some model organisms, notably yeast. Specific interactions within each complex can be studied by *chemical crosslinking* whereby interacting proteins are covalently bound to each other with a crosslinking agent (see below).

Systematic large-scale approaches to define the network of all protein interactions within a cell, the protein interactome, largely use yeast two-hybrid screening and AP–MS. Individual screens often suffer from a high rate of false positives, but confidence in a suggested interaction is greatly increased if different methods have independently converged on the same result, or if there is supporting evidence from genetic interactions. Additionally, gene expression data can be used to refine the networks (interacting proteins need to be co-expressed in the same tissue or cell). Extensive protein interactomes have been built up for various model eukaryotes (*S. cerevisiae*, *C. elegans*, and *D. melanogaster*) mainly by using the approaches described above, and concerted efforts are being made to define major protein interaction networks in human cells.

Various databases have been established to assimilate and present these data (**Table 12.6**). However, although the goal remains to link all the proteins in cells into functional pathways, protein interaction networks are not simple to represent, and clear data presentation is a significant challenge (**Figure 12.11**). However, as might be expected, proteins with a similar general function (e.g. membrane transport, DNA repair, or amino acid metabolism) tend to interact with each other rather more than with functionally unrelated proteins. Complex maps can therefore be simplified by grouping those functionally related proteins into foci representing basic cellular processes (**Figure 12.12**).

The processes themselves are linked in various ways. For example, proteins that control chromatin structure have more interactions with those involved in DNA repair and recombination than with those involved in amino acid metabolism and protein degradation. This type of presentation allows protein interactions to be presented in a hierarchical manner and also provides a benchmark allowing the plausibility of novel interactions to be judged. Three-quarters of all protein interactions occur within the same functional protein group, and most others occur with related functional groups. An unexpected interaction between

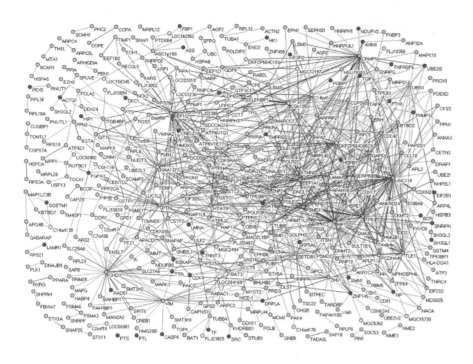

Figure 12.11 A small subset of the human protein interactome. Shown here is an interaction network involving 401 human proteins linked via 911 interactions. Orange, disease proteins (according to OMIM Morbid Map; NCBI); light blue, proteins with GO (gene ontology) annotation; yellow, proteins without GO and disease annotation. Interactions connecting the nodes are represented by color-coded lines according to their confidence scores: green, 3 quality points; blue, 4 quality points; red, 5 quality points; purple, 6 quality points. [From Stelzl U, Worm U, Lalowski M et al. (2005) *Cell* 122, 957–968. With permission from Elsevier.]

proteins involved in unrelated processes could be regarded with suspicion and tested by rigorous genetic, biochemical, and physical assays.

Defining nucleic acid–protein interactions is critical to understanding how genes function

A variety of proteins, notably transcription factors, bind to specific sequences in DNA and are important in gene regulation. Different methods can map where a protein binds to DNA. Some methods were designed to define protein–DNA interactions *in vitro*, but powerful modern methods can map protein–DNA interactions in intact cells.

Mapping protein–DNA interactions *in vitro*

Various methods have been used to map protein-binding sites on defined DNA sequences after incubating a labeled cloned DNA sequence with a cell extract. In *DNA footprinting*, the sites on the cloned DNA that are bound by protein in the

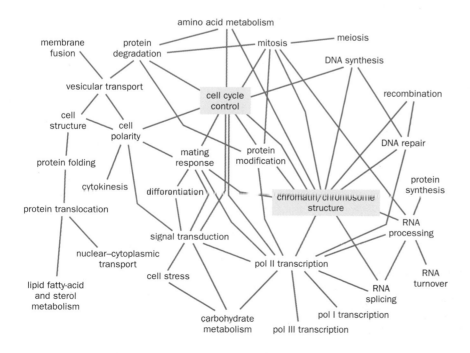

Figure 12.12 Simplified functional group interaction map of the yeast proteome. The data shown here were derived from a more complex map of individual interactions, obtained mostly from two-hybrid screens. Only connections for which there are 15 or more interactions between proteins of the connected groups are shown. A small number of interactions occur between almost all groups and often tend to be spurious, that is, based on false positive results. Only fully annotated yeast proteins are included, and many proteins are known to belong to several functional classes. [From Tucker CL, Gera JF & Uetz P (2001) *Trends Cell Biol.* 11, 102–106. With permission from Elsevier.]

Figure 12.13 DNase I footprinting. DNA molecules that are labeled (red) at one end are allowed to bind to a protein (green). Partial digestion of any unbound DNA with pancreatic DNase I can result in a ladder of DNA fragments of different sizes because the location of a cleavage site (gray vertical bars) on an *individual* DNA molecule is essentially random (left). If, however, the DNA contains a protein-binding site, incubation with a suitable protein extract may result in the binding of protein at a specific region (right). Subsequent partial digestion with pancreatic DNase I no longer produces random cleavage; instead, the bound protein protects the underlying DNA sequence from cleavage. As a result, the spectrum of fragment sizes is skewed against certain fragment sizes, leaving a gap when size-fractionated on a gel, a DNase I footprint.

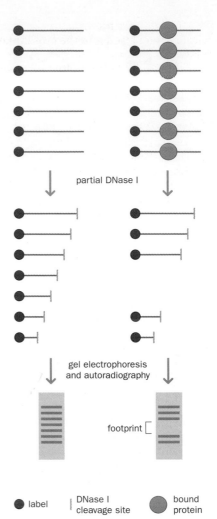

cell extract can be revealed by taking advantage of their comparative inaccessibility to nucleases, such as pancreatic DNase I (**Figure 12.13**).

In *gel mobility shift assays*, a suspected regulatory DNA sequence can be assayed for the ability to bind proteins, and the binding proteins can subsequently be purified. The basis of the method is that a labeled DNA clone bound to protein migrates more slowly in polyacrylamide gel electrophoresis than does unbound labeled DNA. Fractions of the gel that contain slowly migrating labeled DNA can be subjected to column chromatography to purify the bound protein.

To define DNA (or oligonucleotide) binding sites for a specific protein (or other ligand), an *in vitro* selection procedure can be used, such as *SELEX* (*systematic evolution of ligands by exponential enrichment*). Here, parallel oligonucleotide syntheses produce a huge library of oligonucleotides with constant sequences at their 5′ and 3′ ends for primer binding but with randomly generated internal sequences of a fixed length. Often about 10^{13}–10^{15} different sequences are designed. The oligonucleotide library is then mixed with the target protein to allow oligonucleotide–protein binding and the mixture is passed through an affinity chromatography column. After several rounds of column chromatography, oligonucleotide sequences are purified that can bind to the target protein (or other ligand) with extremely high affinity (**Figure 12.14**).

Mapping protein–DNA interactions *in vivo*

Methods such as those described above do not let us know where proteins bind to DNA *in vivo*. To obtain this information, proteins are covalently fixed to DNA

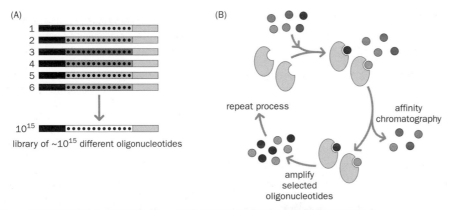

Figure 12.14 SELEX: *in vitro* selection of oligonucleotides for binding to a protein.
(A) Construction of a large library of different oligonucleotides by synthesizing billions of oligonucleotides (oligodeoxyribonucleotides or oligoribonucleotides) in parallel. Synthesis is designed so that the oligonucleotides have common end sequences (shown by black and gray boxes) but there is random incorporation of nucleotides at internal positions (colored boxes).
(B) The SELEX process selects for those rare oligonucleotides within the library that can bind a specific protein (or other ligand). In this example, oligonucleotide numbers 3 (red) and 5 (pink) are imagined to be able to bind to the protein for convenience, oligonucleotides are shown as filled circles here. Affinity chromatography with a specific antibody will capture the desired protein and any bound oligonucleotides; unbound oligonucleotides will be washed off. PCR primers corresponding to the common end sequences are used to amplify the selected oligonucleotides. The amplified products are made single-stranded usually by transcription (e.g. by having a T7 promoter sequence within one of the common flanking sequences and transcribing with phage T7 polymerase). The SELEX process is serially repeated to identify oligonucleotide sequences that have a strong binding capacity for the protein under study.

(*crosslinking*) within intact cells by using a reagent such as formaldehyde. The cells are then lysed and the crosslinked DNA–protein complexes are purified and analyzed.

Formaldehyde is a highly reactive dipolar molecule whose carbon atom acts as a nucleophilic center. It reacts readily with both amino and imino groups found in proteins and DNA. The most available reacting groups in proteins are the amino groups of lysine and arginine and the imino group of histidine. In DNA, the primary reacting groups are the amino groups of adenine and cytosine. **Figure 12.15** shows an example of how formaldehyde can cause crosslinking of macromolecules such as proteins and DNA.

The most popular method for mapping DNA–protein interactions *in vivo* is *chromatin immunoprecipitation* (*ChIP*; see Box 11.3). Proteins are cross-linked to DNA within cells. The cells are lysed, and the chromatin is extracted and fragmented by physical shearing or enzymatic treatment. DNA fragments that have a crosslinked protein of interest are isolated by using an antibody specific for that protein, and purified DNA fragments are analyzed to identify the location of the binding site for the protein.

The basic ChIP method is often devoted to investigating interactions between a specific protein and a few genomic regions that are known or suspected to have binding sites for that protein. More recent ChIP variants permit the identification of protein-binding sites on a genomewide scale. A major motivation is to identify all the binding sites for transcription factors within a genome. In the *ChIP-chip* method (also called ChIP-on-chip, ChIPArray, and genomewide *location analysis*), the ChIP-purified DNA is amplified, fluorescently labeled, and analyzed by hybridization to genomic DNA sequences distributed within a microarray. As an alternative, sequencing of the ChIP-purified DNA (*ChIP-seq*) is becoming widely used. In a test run of ChIP-seq by Johnson et al., a zinc finger repressor protein, the neuron-restrictive silencing factor (NRSF), was found to bind to 1946 locations in the human genome.

In addition to defining positions on DNA occupied by regulatory proteins, ChIP technologies are routinely being used to map the positions within genomes of various types of modified histone and methylated cytosine.

CONCLUSION

The study of gene function has been transformed in the post-genome era. Experimental studies on a gene-by-gene basis have been augmented by genomewide screens. In addition, the many databases that have arisen from genome projects and new analytical tools for their interrogation have driven much of the functional genomics revolution.

The function of a gene can often be inferred from sequence homology searches of increasing sophistication. Experimental approaches are then needed to verify this putative function. Whereas classical genetics began with an observed phenotype and searched to find a gene responsible for it, reverse genetics now allows us to alter a gene or its expression to create an altered phenotype. Model organisms, in particular the mouse, have enabled such studies at an organismal level. These are explored further in Chapter 20. More recently developed techniques such as RNA interference (RNAi) have been widely employed in genomewide studies of gene function in cultured human and animal cells.

Experimental techniques that determine the specific interactions between proteins and other ligands underpin the new field of proteomics. Yeast two-hybrid assays and affinity purification–mass spectrometry techniques can be used to screen for binding partners for specific proteins. Techniques such as fluorescence resonance energy transfer and co-immunoprecipitation can then be used to validate suggested protein–protein interactions.

The wealth of information from both genomewide experimental screens and bioinformatics is enabling the mapping of vast protein networks of interacting proteins in cells. Genes that encode functional noncoding RNA have not been explored so extensively, partly because identifying RNA genes has been a comparatively recent endeavor and is ongoing. Nevertheless, we are increasingly aware of the importance of many different noncoding RNAs in gene regulation and in disease; as we will see in Chapter 20, genetically manipulated model organisms are being used to dissect their functions.

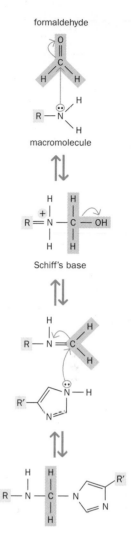

Figure 12.15 Formaldehyde-induced crosslinking. Formaldehyde treatment can result in covalent linking of proteins to DNA or other proteins, often through amino or imino groups. In this example, a nucleophilic attack causes covalent bonding between formaldehyde and an amino group carried by a macromolecule (R). A Schiff base intermediate can then form that can be covalently linked to a suitable chemical group (in this case, an imino group) on a different macromolecule (R′). The end result is that a –CH$_2$– bond contributed by a formaldehyde molecule is used to crosslink the macromolecules R and R′.

FURTHER READING

Sequence homology searching

Barnes MR (ed.) (2007) Bioinformatics For Geneticists, 2nd ed. John Wiley & Sons.

Durbin R, Eddy SR, Krogh A & Mitchison G (2005) Biological Sequence Analysis: Probabilistic Models Of Proteins And Nucleic Acids. Cambridge University Press.

Margulies EH & Birney E (2008) Approaches to comparative sequence analysis: towards a functional view of vertebrate genomes. *Nat. Rev. Genet.* 9, 303–313.

NCBI BLAST and PSI-BLAST tutorials are available at http://www.ncbi.nlm.nih.gov/Education/BLASTinfo/information3.html

Genetic screens in mammalian cells (see also the footnote to Table 12.4 for references to reviews on genetic screening in various model eukaryotes)

Bernards R, Brummelkamp TR & Beijersbergen RL (2006) shRNA libraries and their use in cancer genetics. *Nat. Methods* 3, 701–706.

Grimm S (2004) The art and design of genetic screens: mammalian culture cells. *Nat. Rev. Genet.* 5, 179–189.

Kittler R, Pelletier L, Heninger A-K et al. (2007) Genome-scale RNAi profiling of cell division in human tissue culture cells. *Nat. Cell Biol.* 9, 1401–1412.

Martin SE & Caplen NJ (2007) Applications of RNA interference in mammalian systems. *Annu. Rev. Genomics Hum. Genet.* 8, 81–108.

Silva JM, Li MZ, Chang K et al. (2005) Second-generation shRNA libraries covering the mouse and human genomes. *Nat. Genet.* 37, 1281–1288.

Tochitani S & Hayashizaki Y (2007) Functional screening revisited in the postgenomic era. *Mol. Biosyst.* 3, 195–207.

Proteomics

Bertone P & Snyder M (2005) Advances in functional protein microarray technology. *FEBS J.* 272, 5400–5411.

Gstaiger M & Aebersold R (2009) Applying mass spectrometry-based proteomics to genetics, genomics and network biology. *Nat. Rev. Genet.* 10, 617–627.

Nature Insight on Proteins to Proteomes (2007) *Nature* 450, 963–1009.

Pandey A & Mann M (2000) Proteomics to study genes and genomes. *Nature* 405, 837–846.

Ptacek J, Devgan G, Michaud G et al. (2005) Global analysis of protein phosphorylation in yeast. *Nature* 438, 679–684.

Protein–protein interaction

Cusick ME, Klitgord N, Vidal M & Hill DE (2005) Interactome: gateway into systems biology. *Hum. Mol. Genet.* 14, R171–R181.

Krogan NJ, Cagney G, Yu H et al. (2006) Global landscape of protein complexes in the yeast *Saccharomyces cerevisiae*. *Nature* 440, 637–643.

Lalonde S, Erhardt DW, Loque D et al. (2008) Molecular and cellular approaches for the detection of protein–protein interactions: latest techniques and current limitations. *Plant J.* 53, 610–635.

Rual JF, Venkatesan K, Hao T et al. (2005) Towards a proteome-scale map of the human protein–protein interaction network. *Nature* 437, 1173–1178.

Stelzl U, Worm U, Lalowski M et al. (2005) A human protein–protein interaction network: a resource for annotating the proteome. *Cell* 122, 957–968.

Identifying DNA–protein interactions

Johnson DS, Mortazavi A, Myers RM & Wold B (2007) Genome-wide mapping of *in vivo* protein–DNA interactions. *Science* 316, 1497–1502.

Kim TH & Ren B (2006) Genome-wide analysis of protein–DNA interactions. *Annu. Rev. Genomics Hum. Genet.* 7, 81–102.

Orlando V (2000) Mapping chromosomal proteins *in vivo* by formaldehyde-crosslinked-chromatin immunoprecipitation. *Trends Biochem. Sci.* 25, 99–104.

Stoltenburg R, Reinemann C & Strehlitz B (2007) SELEX—A (r)evolutionary method to generate high-affinity nucleic acid ligands. *Biomol. Eng.* 24, 381–403.

Wei C-L, Wu Q, Vega VB et al. (2006) A global map of p53 transcription-factor binding sites in the human genome. *Cell* 124, 207–219.

Chapter 13

Human Genetic Variability and Its Consequences

13

KEY CONCEPTS

- Human DNA variants can be classified as large scale versus small scale, common versus rare, and pathogenic versus nonpathogenic.

- Single nucleotide polymorphisms (SNPs) are the most numerous variants.

- Short tandem repeat polymorphisms (STRPs or microsatellites) are very common. The repeat units are most commonly 1, 2, or 4 bp long, and the string of tandem repeats usually extends over less than 100 bp. The number of repeats in a string often varies between people as a result of polymerase stuttering during DNA replication.

- Minisatellites are tandem repeats with longer repeat units, typically 10–50 bp. Again, the number of repeat units in a run often varies between people, but this is most often because of unequal recombination. Minisatellites are commoner toward the telomeres of chromosomes.

- Many larger sequences (between 1 kb and 1 Mb) show variation in copy number between normal healthy people. About 5% of the whole genome can vary in this way. A common cause of such variation is recombination between mispaired repeated sequences.

- Genomic DNA is subject to constant processes of damage and repair. Most damage passes unnoticed because it is efficiently repaired. Failures in repairing damage or correcting replication errors are the ultimate cause of most sequence variation.

- Most common variants are not pathogenic, although they might, in combination with other variants, increase or decrease susceptibility to multifactorial diseases. They are useful as markers (variants that can be used to recognize a chromosome or a person) in pedigree analysis, forensics, and studies of the origins and relationships of populations.

- Pathogenic effects can be mediated by either a loss of function or a gain of function of a gene product. Very many different changes in a gene can cause a loss of function. Usually only one or a few specific changes can cause a gain of function.

- Loss-of-function changes usually lead to recessive phenotypes, and gain-of-function changes to dominant phenotypes. Dominant phenotypes can also result from loss of function if a 50% level of the normal gene product is not sufficient to produce a normal phenotype (haploinsufficiency), or if the protein product of the mutant allele interferes with the function of the normal product (dominant-negative effects).

- Molecular pathology seeks to explain why a certain genetic change causes a particular clinical phenotype. However, well-defined genotype–phenotype correlations are rare in humans because humans differ greatly from one another in their genetic make-up and environments.

This chapter is about the differences that exist between different human individuals in the DNA sequence of their genomes (differences between humans and other species were covered in Chapter 10). Now that we have complete genome sequences for several named individuals, we can see that they differ from one another in millions of ways. Most of the many differences between individual human genomes seem to have no effect whatsoever. Other differences do affect the phenotype, producing the normal range of genetically determined variants in body build, pigmentation, metabolism, temperament, and so on, that make each of us individual. All these are normal variants. Some variants are pathogenic—that is, they either cause disease or make their bearer susceptible to a disease that they may or may not actually develop, depending on their other genes, their lifestyle, their environment, or pure luck. In this chapter, we consider first the normal variants and then those that are pathogenic. There is one further level of genetic variation that we do not consider here: epigenetic variation (variations in DNA methylation and chromatin conformation) between individuals. This was described in Chapter 11, but it is still a matter of speculation how much our individual phenotypes are due to genetic and how much to epigenetic differences between us.

13.1 TYPES OF VARIATION BETWEEN HUMAN GENOMES

Human genetic variation ranges from single nucleotide changes through to gains or losses of whole chromosomes. Variants can be usefully classified as large-scale or small-scale on the basis of whether or not they can be detected by sequencing a conventional PCR product of a few hundred base pairs. Small-scale variants normally have their primary effect, if they have any effect, on a single gene, whereas large-scale variants usually affect several or many genes. In reality, of course, there is a continuum across all these boundaries.

Single nucleotide polymorphisms are numerically the most abundant type of genetic variant

When the genomes of different humans are compared, the great majority of nucleotides are nearly always the same in everybody. That is why it is possible to talk in general terms about the human genome. Occasional variants may be seen at any nucleotide position, but they are nearly always rare. However, about one nucleotide in 300 is **polymorphic**—that is, more than one form is common in the population (**Box 13.1**). These **single nucleotide polymorphisms** or **SNPs** (pronounced snips) are cataloged in the public dbSNP database (http://www.ncbi.nlm.nih.gov/projects/SNP) and designated by numbers beginning with rs (reference SNP; **Box 13.2**).

In most cases, a SNP has two alternative forms (alleles)—say A or G at a certain position (**Figure 13.1A**). The frequency of the minor (rarer) allele may be

---

BOX 13.1 POLYMORPHISM AND MUTATION: WORDS WITH SEVERAL MEANINGS

The word **polymorphism** is used by human geneticists to mean several different things at different times. Needless to say, this can cause confusion.

- Molecular geneticists often describe a variant as a polymorphism if its frequency in the population is above some arbitrary value, often 0.01. For example, SNP rs1447295 (see Box 13.2 for an explanation of the notation) at 8q24 is a polymorphism with two alleles, C and A, with frequencies 0.93 and 0.07, respectively, in the European HapMap population. This is the meaning we use in this book, unless otherwise specified. Variants whose frequency is below the arbitrary threshold might be described as rare variants.
- Population geneticists define a polymorphism as the stable coexistence in a population of more than one genotype at frequencies such that the rarer type could not be maintained simply by recurrent mutation. On this definition, some pathogenic mutations would count as polymorphisms. For example, the

commonest mutation that causes cystic fibrosis in northern European populations (labeled p.F508del; see Box 13.2 for an explanation of the notation) has a frequency of 0.01–0.02 in northern European populations. As shown in Chapter 3, this could not be sustained by recurrent mutation in the face of the selective pressure against people with cystic fibrosis.
- Clinical geneticists often use polymorphism to mean a non-pathogenic variant, regardless of its frequency. Pathogenic variants, whether common or rare, would be described as mutations.

The word mutation can also be used with two different senses, referring to either the process or the product:

- an event that changes a DNA sequence—UV radiation produced a mutation in the DNA.
- a DNA sequence change that may have happened a long time ago—she had inherited a mutation from her father.

BOX 13.2 NOMENCLATURE FOR DESCRIBING DNA AND AMINO ACID VARIANTS

Each single nucleotide polymorphism (SNP) in the public database dbSNP (http://www.ncbi.nlm.nih.gov/projects/SNP) can be referred to by its unique identifier, such as rs212570, where rs stands for reference SNP and 212570 is a unique serial number. Short tandem repeat polymorphisms (STRPs) have identifiers such as *D6S282*, where D stands for DNA segment, 6 is the number of the chromosome on which the marker is located, S means a single copy sequence, and 282 is a unique serial number.

Any sequence change can be described using the conventions set out on the website of the Human Genome Variation Society, http://www.hgvs.org/mutnomen/. The commoner cases are described below.

All variants are prefixed with g. (genomic), c. (cDNA), r. (RNA), or p. (protein).

Nucleotide substitutions

For changes in a gene, the A of the initiator ATG codon is numbered +1; the base immediately preceding this is –1. There is no zero. The number of the altered nucleotide is followed by the change.

- **g.1162G>A**—in genomic DNA, replace guanine at position 1162 by adenine.
  For changes within introns, when only the cDNA sequence is known in full, specify the intron number by IVS (IVS stands for intervening sequence) or the number of the nearest exon position.
- **g.621+1G>T** or **IVS4+1G>T**—replace G by T at the first base of intron 4 (nucleotide 621 is the last base of exon 4).

Amino acid substitutions

Use either one-letter codes (X means a stop codon) or three-letter codes. Proteins are numbered with the initiator methionine as codon 1.

- **p.R117H** or **Arg117His**—replace arginine 117 by histidine.
- **p.G542X** or **Gly542Stop**—replace the codon for glycine 542 with a stop codon.

Deletions and insertions

Use del for deletions and ins for insertions, preceded by the nucleotide position or interval (for DNA changes) or the amino acid symbol (one-letter code; for amino acid changes).

- **p.F508del**—in a protein (p) delete phenylalanine (F) 508.
- **c.6232_6236del** or **c.6232_6236delATAAG**—delete five nucleotides starting with nucleotide 6232 of the cDNA. The identity of the deleted nucleotide(s) can be specified.
- **g.409_410insC**—insert C between nucleotides 409 and 410 of genomic DNA.
  A program, Mutalyzer, has been developed to generate the correct name for any sequence variant that the user inputs; see http://www.lovd.nl/mutalyzer/.

---

anything from 0.5 downward. We would not expect high-frequency SNP alleles to have any major phenotypic effect, because natural selection should have ensured that they were eliminated if harmful or fixed (present in everybody) if beneficial. However, those evolutionary processes might be incomplete at present for some particular allele. Additionally, heterozygote advantage is a mechanism that can produce stable polymorphisms even when one allele is harmful in homozygous form. This happens when asymptomatic carriers of a deleterious recessive condition have some selective advantage over normal homozygotes, as with sickle-cell anemia (see Chapter 3, p. 64). The argument from natural selection is less powerful with rare SNPs, but even so, these would also not generally be expected to have a phenotypic effect, if only because most SNPs are not located in coding or regulatory sequences.

Some polymorphic sites have more complex variants. A SNP may have three alleles, say A, G, or C, or two SNPs may be adjacent to one another. The great majority of more complex variants in the dbSNP database (about 2 million of the 12 million entries) are deletion/insertion polymorphisms (DIPs or **indels**). Typically one or two nonrepeated nucleotides are inserted or deleted (Figure 13.1B). Insertions or deletions of repeat units have different causes and dynamics, and are discussed below in the section on short tandem repeats (p. 408).

Sometimes a SNP will affect a nucleotide that is part of the recognition site for one or another restriction enzyme. Although many hundreds of restriction enzymes are known, most of them recognize a **palindromic** sequence such as GAATTC. The sequence is palindromic because the complementary strand, read in the 5′→3′ direction, also reads GAATTC:

```
5′-ACGATTGAATTCTAGGATC-3′
3′-TGCTAACTTAAGATCCTAG-5′
```

Only about 10% of all nucleotides fall in such palindromic sequences, so most SNPs do not affect any restriction site. When a SNP happens to lie within a restriction site, only one allele will retain the necessary recognition sequence. Thus, the polymorphism will create or abolish a restriction site. SNPs of this type (Figure 13.1C), known as **restriction fragment length polymorphisms** (**RFLPs**) or **restriction site polymorphisms** (**RSPs**), were the first widely used DNA markers and were used to create the first genetic linkage map of the human genome (see Chapter 8).

(A)

GCCTGTTTTATATTA**C/T**GATCCAATTTTTCA

(B)

GAGACAGAGTTTCGC**(T)**TCTTGTTGCCCAGGCT

(C)

CCAAGCCTGGA**GCTA/GGC**CGTGGGCCAGGCAAG

Figure 13.1 Types of single nucleotide polymorphism (SNP). (A) A simple SNP. rs212570 is a C/T variant on chromosome 20. (B) A deletion/insertion polymorphism. rs36126541 is insertion or deletion of a single T nucleotide in a sequence on chromosome 12. (C) A restriction fragment length polymorphism (RFLP). SNP rs36078338 creates or abolishes the sequence GCTAGC, which is the recognition site for the restriction enzyme *Nhe*I.

Why should certain nucleotides be polymorphic, when the surrounding nucleotides only rarely show variants? In general, it is not because something about that nucleotide makes it especially liable to mutate. Rather, the alternative variants mark alternative ancestral chromosome segments that are common in the present-day population. In comparison with short tandem repeats, SNPs are stable over evolutionary time. Very few SNPs are specific to one human race or ethnic group, although the allele frequencies may vary between groups—thus, they have mostly persisted from the earliest days of human evolution. The ability to define ancestral chromosome segments through their SNP content is hugely important in genetic research and will be a major preoccupation of Chapter 15.

Both interspersed and tandem repeated sequences can show polymorphic variation

As noted in Chapter 9, about 50% of the human genome comprises repeated sequences. Most are interspersed repeats—that is, repeated sequences that are scattered across all or part of the genome, rather than being clustered together. Most of the interspersed repeats in the human genome are derived from transposons (jumping genes that can spread within the genome like an intracellular infection—see Chapter 9). Several families of transposon-derived repeats are present in huge numbers—there are maybe 1.5 million copies of 100–300 bp SINEs (short interspersed nuclear elements) and 850,000 LINEs (long interspersed nuclear elements) of 6–8 kb, scattered throughout the genome. Some particular insertions are polymorphic in the human genome, being present on some chromosomes but absent from other copies of the same chromosome.

Tandem repeats are also common. Repeat units can be anything from a single nucleotide (e.g. a run of A nucleotides) to 100 or more nucleotides. Repeats with longer units are called **satellites**. The name comes from early studies in which whole genomic DNA was subjected to density gradient sedimentation in an ultracentrifuge. The repeated DNA had a different buoyant density from the bulk DNA and formed a small separate satellite band. A noted example is the α-satellite DNA found at chromosomal centromeres, which consists of tandem repeats of a 171 bp unit. Repeats with units of around 10–50 nucleotides are called **minisatellites**, and those with shorter units are **microsatellites**. Most microsatellites have repeat units of 1, 2, or 4 nucleotides (**Figure 13.2**).

Short tandem repeat polymorphisms: the workhorses of family and forensic studies

If one compares the genomes of different individuals, many tandem repeats are seen to vary in the number of repeat units. Unlike SNP variations, tandem repeat variations are the result of relatively recent events. These are of two types:

- Meiotic recombination between mispaired repeats produces step changes in unit number (**Figure 13.3**). This is thought to be the main mechanism generating minisatellite diversity.

- Polymerase stutter during DNA replication can change the repeat length by one or maybe two units. This is the main mechanism generating polymorphism of microsatellites (**Figure 13.4**).

Polymorphic microsatellites (**short tandem repeat polymorphisms**, or **STRPs**) have been the genetic markers of choice for family and forensic studies since the early 1990s. They are more informative than SNPs for distinguishing between individuals or for following a particular chromosomal segment through a pedigree, because there may be a large number of alleles in the population (for example, one microsatellite might have anything from 5 to 20 repeat units in different people). The early years of the Human Genome Project were largely devoted to

(A)

ACAGAGATAGA**CACACACACACACACACACACACACACACACA**AACAAGCATGCTC

(B)

ATCAATGGATGCATACGT**(AGAT)**$_{15}$GAGAGGGGATTTATTAGAGGAATTAGC

(C)

TGAATTGCCT**(TCTA)**$_4$**(TCTG)**$_6$**(TCTA)**$_3$TA**(TCTA)**$_3$TCA**(TCTA)**$_{10}$TCGTCTATC

Figure 13.2 Microsatellites. (A) *D6S282* is a (CA)$_n$ repeat on chromosome 6. (B) *D12S391* is a tetranucleotide repeat, (AGAT)$_n$ on chromosome 12. (C) *D21S11* is a complex tetranucleotide repeat on chromosome 21. Underlined sequences vary in repeat number; the figure shows examples of typical alleles.

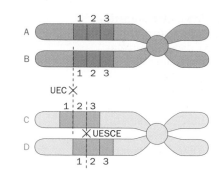

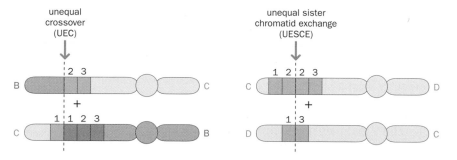

unequal
crossover
(UEC)

unequal sister
chromatid exchange
(UESCE)

Figure 13.3 Unequal crossovers and unequal sister chromatid exchanges cause insertions or deletions. The figure shows the result of recombination between misaligned repeats in an array of tandem repeats. The misaligned chromatids can be on homologous chromosomes, in which case the result is an unequal crossover (UEC), or they could be on sister chromatids, producing an unequal sister chromatid exchange (UESCE). In either case, the result is to produce two chromatids, one with one or more extra repeat units, the other with the corresponding units missing. For simplicity, the breaks are shown as located between repeat units, but they could equally occur at corresponding positions within units.

defining and mapping sufficient STRPs to construct a high-resolution marker framework map across the human genome; about 150,000 STRPs have been identified. Although STRPs still remain the tool of choice for forensic work, linkage work has moved to using SNPs because they can be genotyped on a huge scale with the use of microarrays.

Large-scale variations in copy number are surprisingly frequent in human genomes

Variants large enough to be seen by a cytogeneticist under the microscope are almost always the result of isolated pathogenic accidents and are not part of normal human variation. Cytogeneticists recognize just three types of relatively common normal variant:

- The size of the heterochromatic regions at the centromeres of chromosomes 1, 9, or 16 and the long arm of the Y chromosome may vary.

- The short arms of the acrocentric chromosomes (chromosomes 13, 14, 15, 21, and 22, which have the centromere close to one end—see Figure 2.15) vary considerably in size and morphology. Often the most distal part appears as a so-called satellite connected to the main body of the chromosome by a short, thin stalk. Note that this use of the word *satellite* is unrelated to its use to describe DNA tandem repeats.

- A variety of **fragile sites**—segments of uncoiled chromatin—may be seen when cells are cultured under conditions that make DNA replication difficult—for example, by starving the cells of thymidine or adding the DNA polymerase inhibitor aphidicolin (**Figure 13.5**). Most fragile sites are normal variants, although the *FRAXA* and *FRAXE* fragile sites discussed below (see p. 424) are pathogenic.

All these changes reflect varying copy numbers of tandemly repeated sequences. Until recently it was assumed that any deletion or insertion of a large stretch of non-repetitive DNA would be pathogenic. The advent of techniques such as array-comparative genomic hybridization (in which test and control DNAs compete to hybridize to a microarray; see Chapter 2) and whole genome SNP arrays made it relatively easy for the first time to scan whole genomes for extra or missing material, and it quickly became apparent that many large-scale changes can be found in the genomes of normal healthy people. A major study by Redon et al. of 269 healthy individuals from four geographically distinct populations (the HapMap sample; see Chapter 15) identified 1447 regions with variable numbers of copies of sequences of at least 1 kb in length. In total these variable

normal replication

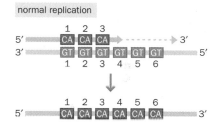

backward slippage causes insertion

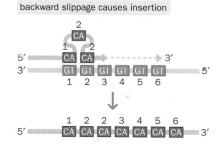

forward slippage causes deletion

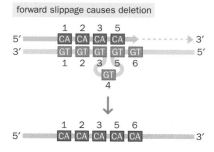

Figure 13.4 Slipped-strand mispairing during DNA replication. A new DNA strand (pink) is being synthesized, using the blue strand as template. During normal DNA replication the nascent strand often partly dissociates from the template and then reassociates. When there is a tandemly repeated sequence, the nascent strand may mispair with the template when it reassociates. This can result in the newly synthesized strand having fewer or more repeat units than the template strand.

regions were reported to cover 360 Mb, or 12% of the genome (although that estimate has since been revised downward), and included many known genes. The mean size of the variable segments was 250 kb. The techniques used were relatively insensitive to smaller, kilobase-sized variants. Many such smaller variable segments would have been missed. These smaller variants can be identified using the technique of **paired-end mapping**. This involves selecting random fragments of genomic DNA of a known size, and sequencing a few dozen nucleotides from each end of a fragment. One can then compare the distance apart of those end sequences in the selected fragment (whose size is known) with the distance apart of the same sequences in the reference human genome (**Figure 13.6**). If the separation is greater in the selected fragment than in the reference genome, the fragment must contain an insertion that is not present in the reference genome; conversely, a lesser distance indicates a deletion. Recent studies have shown that these smaller variants are even more frequent than larger ones (**Figure 13.7**).

Although SNPs are individually more numerous, copy-number variants (CNVs) are responsible for by far the greatest number of nucleotides that differ between two genomes. Thus, the human genome is considerably more variable between normal healthy people than had previously been assumed. The studies mentioned above did not reveal, when there is a gain in copy number, whether the multiple copies of the sequence are scattered across several chromosomal locations or clustered in tandem. Other studies show that they most commonly form tandem clusters. The Database of Genomic Variants (TCAG; http://projects.tcag.ca/variation/) is a database of variants seen in apparently healthy people, whereas the Decipher database (http://decipher.sanger.ac.uk/) records variants seen in people with phenotypic abnormalities (although whether any particular variant is the cause of the phenotypic abnormality is seldom known for certain).

The genome sequences of single individuals that are now becoming available give a striking picture of the overall variability present in ostensibly healthy individuals.

The pioneer genome scientist Craig Venter was the first individual whose full diploid genome sequence was determined. The reference human genome

Figure 13.5 A chromosomal fragile site. A stretch of relatively uncondensed chromatin in an otherwise highly condensed metaphase chromosome. There are about 120 locations in the human genome where a chromosome of a normal healthy person has some tendency to show a fragile site when cells are cultured under special conditions. Shown here is an X chromosome with the pathogenic fragile site *FRAXA* (gray arrow). (Courtesy of Graham Fews, University of Birmingham.)

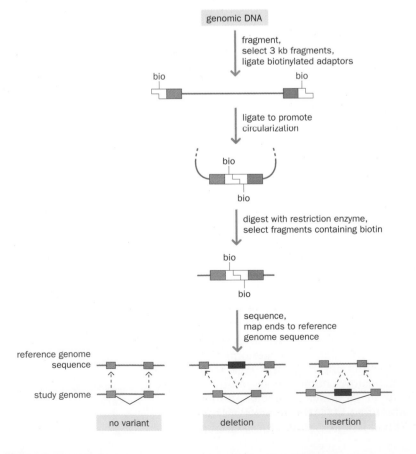

Figure 13.6 The principle of paired end mapping. In paired end mapping, small size-selected fragments of sheared genomic DNA from BAC or PAC clones are used, which are marked by ligation to biotinylated adaptors. In both cases, the aim is to identify a sequence representing the two ends of the original DNA joined by the circularization. In chromosome jumping this was to find sequences (green) that were 80–130 kb from a known sequence (blue), in order to speed up construction of a clone contig across a large genomic region. In paired end mapping, the aim is to identify structural variants, by finding sequences that are a different distance apart in the test genome than in the reference human genome sequence.

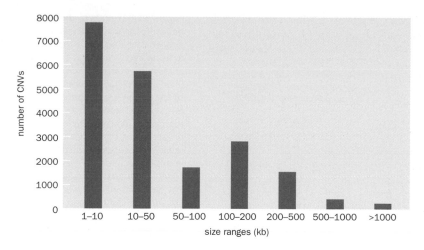

Figure 13.7 Size distribution of copy-number variants (CNVs) in the human genome. Only variants larger than 1 kb are shown. The peak in the size distribution at 100–200 kb is partly an artifact of the BAC array technology used in many studies; many variants reported in this size range are really smaller than this. [Data reproduced with permission from the Database of Genomic Variants http://projects.tcag.ca/variation, accessed January 2009.]

sequence was a composite of anonymous donors. Compared with this reference, the following variants were observed in Venter's genome:

- 3.2 million SNPs
- 290,000 heterozygous insertion/deletion variants (ranging from 1 to 571 bp in size)
- 559,000 homozygous insertion/deletion variants (ranging from 1 to 82,711 bp)
- 90 large inversions
- 62 large-copy-number variants

A total of 12,290,978 nucleotides differed from the reference sequence. Although the majority of this variation was in noncoding DNA, 44% of Craig Venter's genes had a sequence variant relative to the reference human genome sequence; 17% encoded an altered protein, and these included 317 genes in which some variants have been identified as pathogenic.

James Watson's diploid genome has also been published, and shows a similar extent of variation, with 3.3 million SNPs; these include 10,654 that caused amino acid changes in proteins.

13.2 DNA DAMAGE AND REPAIR MECHANISMS

Some DNA variants arise from errors in DNA replication or recombination, but a major source is a failure to repair DNA damage. The view of DNA as a stable genetic archive safely tucked away in the cell nucleus belies the difficulty of maintaining its stability. Chemical attack by exogenous or endogenous agents, and errors arising during its normal functioning, pose constant threats to the integrity of the genome (**Figure 13.8**).

DNA in cells requires constant maintenance to repair damage and correct errors

The agents that damage DNA can either be external to the cell or they may arise as undesired effects of internal cell chemistry.

There are three main external agents likely to cause DNA damage:

- *Ionizing radiation*—gamma rays and X-rays can cause single-strand or double-strand breaks in the sugar–phosphate backbone.
- *Ultraviolet radiation*—UV-C rays (with a wavelength of about 260 nm) are especially damaging, but the major source of UV damage in humans is from the UV-B rays (280–315 nm) in sunlight that can penetrate the ozone layer. UV radiation causes cross-linking between adjacent pyrimidines on a DNA strand to form cyclobutane pyrimidine dimers (**Figure 13.9D**) and other abnormal photoproducts.
- *Environmental chemicals*—these include hydrocarbons (for example, in cigarette smoke), some plant and microbial products such as the aflatoxins found on moldy peanuts, and chemicals used in cancer chemotherapy. Alkylating

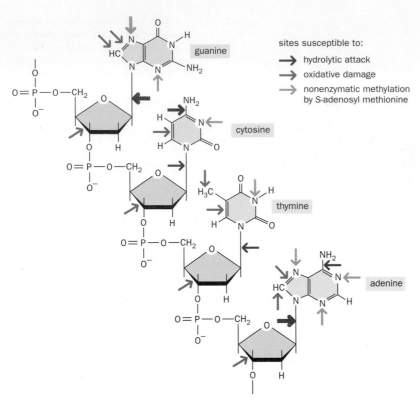

sites susceptible to:

→ hydrolytic attack
→ oxidative damage
→ nonenzymatic methylation by *S*-adenosyl methionine

Figure 13.8 Threats to the integrity of genomic DNA. The figure indicates the sites on each nucleotide that are known to be altered by hydrolytic attack (red arrows), spontaneous oxidative damage (blue arrows), and uncontrolled methylation by the methyl group donor *S*-adenosyl methionine (yellow arrows). The width of each arrow indicates the relative frequency of each event. Ultraviolet and ionizing radiation pose additional threats not shown in this figure. [From Alberts B, Johnson A, Lewis J et al. (2007) Molecular Biology of the Cell, 5th ed. Garland Science/Taylor & Francis LLC., after Lindahl T (1993) *Nature* 362, 709–715. With permission from Macmillan Publishers Ltd.]

agents can transfer a methyl or other alkyl group onto DNA bases and can cause cross-linking between bases within a strand or between different DNA strands.

Although external agents are more visible, the major threats to the stability of a cell's DNA come from internal chemical events, such as those illustrated in Figures 13.8 and 13.9:

- *Depurination*—about 5000 adenine or guanine bases are lost every day from each nucleated human cell by spontaneous hydrolysis of the base–sugar link (see Figures 13.8 and 13.9A).

- *Deamination*—at least 100 cytosines each day in each nucleated human cell are spontaneously deaminated to produce uracil (see Figures 13.8 and 13.9B). Less frequently, spontaneous deamination of adenine produces hypoxanthine.

- *Attack by reactive oxygen species*—highly reactive superoxide anions (O_2^-) and related molecules are generated as a by-product of oxidative metabolism in mitochondria. They can also be produced by the impact of ionizing radiation on cellular constituents. These reactive oxygen species attack purine and pyrimidine rings (see Figure 13.8).

- *Nonenzymatic methylation*—accidental nonenzymatic DNA methylation by *S*-adenosyl methionine produces about 300 molecules per cell per day of the cytotoxic base 3-methyl adenine, plus a quantity of the less harmful 7-methyl guanine (see Figure 13.8). This is quite distinct from the enzymatic methylation of cytosine to produce 5-methyl cytosine, which cells use as a major method of controlling gene expression (see Chapter 11). The methylated adenine or guanine bases distort the double helix and interfere with vital DNA–protein interactions.

In addition to damage of these various types, errors arise during normal DNA metabolism. A certain error rate (incorporation of the wrong nucleotide) is inevitable during DNA replication. Proofreading mechanisms correct the great majority of the resulting mismatches, but a few may persist, producing sequence variants. Failure of the proofreading mechanism in somatic cells is one cause of cancer (see Chapter 17). In addition, occasional errors in replication or recombination leave strand breaks in the DNA, which must be repaired if the cell is to survive.

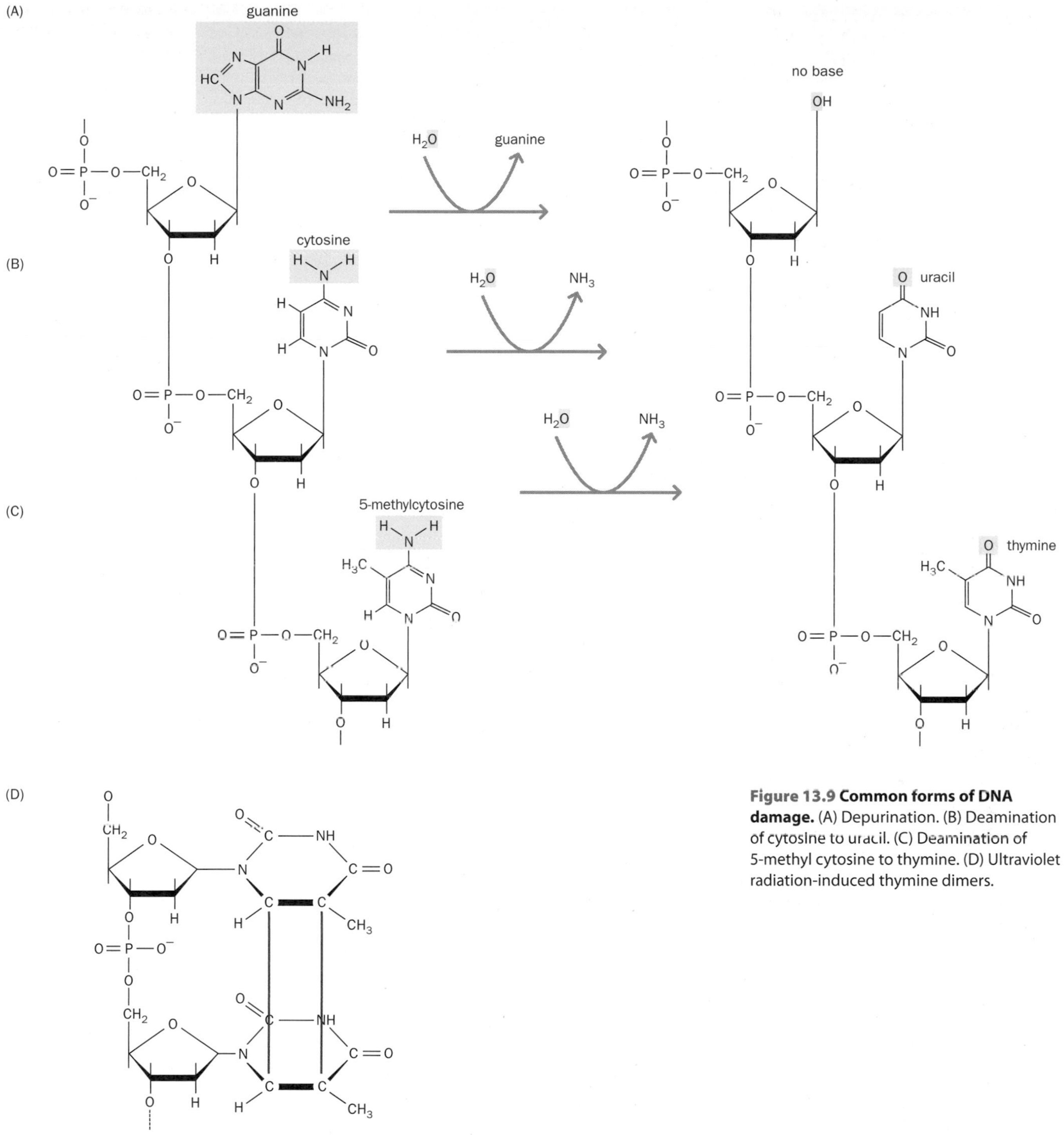

Figure 13.9 Common forms of DNA damage. (A) Depurination. (B) Deamination of cytosine to uracil. (C) Deamination of 5-methyl cytosine to thymine. (D) Ultraviolet radiation-induced thymine dimers.

The effects of DNA damage

DNA damage has two possible effects on a cell. Primarily it is cytotoxic. When the DNA is replicated, many types of damage will cause the replication fork to stall. Even outside S phase of the cell cycle, during transcription RNA polymerase stalls at lesions in the DNA, preventing expression of the damaged gene. These problems are potentially lethal to the cell. Special multiprotein machines constantly patrol the DNA of a cell, detecting and responding to damage. Progression through the cell cycle is delayed until the damage has been repaired, and irreparable damage triggers **apoptosis** (programmed cell death). Malfunctions of the systems for detecting DNA damage and coordinating the cellular response have major roles in the development of cancer, as described in Chapter 17.

BOX 13.3 MECHANISMS OF DNA REPAIR IN HUMAN CELLS

Human cells use at least six different DNA repair mechanisms. Three of them are used to correct abnormal modified bases when these are present in just one of the two strands of the DNA—that is, they are paired with a normal base. The damaged base can be repaired, or removed and replaced.

- **Base excision repair** (**BER**) corrects much the commonest type of DNA damage (of the order of 20,000 altered bases in each nucleated cell in our body each day). BER glycosylase enzymes remove abnormal bases by breaking the sugar–base bond (**Figure 1A**). Humans have at least eight genes encoding different DNA glycosylases, each responsible for identifying and removing a specific kind of base damage. After removal of the damaged base, an endonuclease and a phosphodiesterase cut the sugar–phosphate backbone at the position of the missing base and remove the sugar–phosphate residue. The gap is filled by resynthesis with a DNA polymerase, and the remaining nick is sealed by DNA ligase III.

- **Nucleotide excision repair** (**NER**) removes thymine dimers and large chemical adducts (Figure 1B). NER removes and resynthesizes a large patch around the damage, rather than just a single base as in BER. The sugar–phosphate backbone is cleaved at the site of the damage, and exonucleases remove a large stretch

of the surrounding DNA. As in BER, the gap is filled by resynthesis and sealed by DNA ligase. Nucleotide excision repair is also used to correct single-strand breaks in the sugar–phosphate backbone.

- **Direct reversal of the DNA damage** is an infrequently used mechanism of DNA repair in humans. Three human genes have been implicated in this mechanism, of which the best-characterized encodes the *O*-6-methylguanine-DNA methyltransferase, which is able to remove methyl groups from guanines that have been incorrectly methylated. In many organisms, UV radiation-produced thymine dimers can be directly resolved by the enzyme photolyase using the energy of visible light (photoreactivation). Although mammals possess enzymes related to photolyase, they use them for a quite different purpose, to control their circadian clock.

In the repair processes described above, the second, undamaged DNA strand acts as a template for accurate reconstruction of the damaged strand. Damage that affects both strands of DNA requires different mechanisms. There are two main processes.

- In **homologous recombination**, a single strand from the homologous chromosome invades the damaged DNA and acts as a template for accurate repair (**Figure 2A**). This type of repair normally occurs after DNA replication but before cell division,

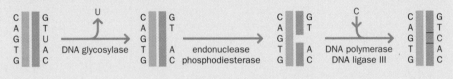

(A) base excision repair

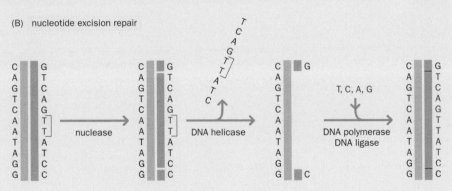

(B) nucleotide excision repair

Figure 1 DNA repair mechanisms. (A) Base excision repair. (B) Nucleotide excision repair.

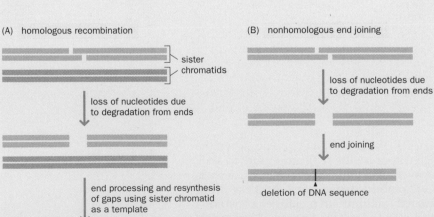

(A) homologous recombination

(B) nonhomologous end joining

Figure 2 Two different ways to repair DNA double-strand breaks.

(A) Homologous recombination. The correct sequence is reconstructed using the intact sister chromatid as a template (molecular details are not shown). This produces an accurate repair but is not always possible. (B) Nonhomologous end joining is always available, but the repaired strand has extra or missing nucleotides at the junction. [From Alberts B, Johnson A, Lewis J et al. (2007) Molecular Biology of the Cell, 5th ed. Garland Science/Taylor & Francis LLC.]

and involves the sister chromatid. The eukaryotic machinery for recombination repair is less well defined than the excision repair systems. Human genes involved in this pathway include *NBS* (Nijmegen breakage syndrome; OMIM 602667), *BLM* (Bloom syndrome; OMIM 604610), and the breast cancer susceptibility genes *BRCA2* (OMIM 600185) and *BRCA1* (OMIM 113705).

- In **nonhomologous end joining**, large multiprotein complexes are assembled at broken ends of DNA molecules, and DNA ligases rejoin the broken ends regardless of their sequence (Figure 2B). There is always some loss of DNA sequence from the broken ends. This is a desperate measure that is likely to cause mutations or

chromosomal rearrangements, but it is better than leaving breaks unrepaired. One reason why chromosomal telomeres require a special structure is to protect normal chromosome ends from this double-strand break response.

A final repair mechanism is concerned with correcting mismatches caused by replication errors. Cells deficient in mismatch repair have mutation rates 100–1000-fold normal, with a particular tendency to replication slippage in homopolymeric runs (see Figure 13.4). In humans, the mechanism involves at least five proteins, and defects cause Lynch syndrome (OMIM 120435 and OMIM 609310). Details of the mechanism are given in Chapter 17.

Even if a cell is able to survive with damaged DNA, unrepaired damage is likely to be mutagenic. If mutations occur in germ cells, they can give rise to new variants in the population that provide much of the raw material for evolution as well as the intraspecific variants considered in this chapter. A major source of mutations is error-prone trans-lesion DNA synthesis. The sloppy polymerases ζ (zeta) and ι (iota) are able to replicate damaged DNA, bypassing stalled replication forks—but at the cost of a high error rate. Additionally, modified bases may mispair during DNA replication, introducing permanent sequence changes. For example, uracil, produced by the deamination of cytosine, pairs with adenine, as does the 8-hydroxy guanine that is a product of oxidative attack on DNA.

A special case is 5-methyl cytosine. When cytosine is deaminated, the product is uracil. This is an unnatural base in DNA that is efficiently recognized and corrected. However, 5-methyl cytosine is deaminated to thymine, a natural base in DNA (Figure 13.9C). If the resulting G–T mispairing is not corrected before the next round of DNA replication, one daughter cell will have a permanent C→T mutation. Evidence from both evolutionary studies and human diseases shows the importance of deamination of 5-methyl cytosine as a source of sequence changes. Methylated cytosines are almost always found in CpG sequences (that is, the nucleotide 3′ of cytosine is guanosine), and CpG sequences are mutational hotspots (see also Chapters 8 and 11).

In response to these various forms of damage, cells deploy a whole range of repair mechanisms. Different mechanisms correct different types of lesion (**Box 13.3**). The importance of effective DNA repair systems is highlighted by the roughly 130 human genes participating in DNA repair, and by the severe diseases that affect people with deficient repair systems.

DNA replication, transcription, recombination, and repair use multiprotein complexes that share components

The different repair systems, except for direct repair, require exonucleases, endonucleases, helicases, polymerases, and ligases (see Box 13.3). Many of the same functions are also required for DNA replication, transcription, and recombination. In each case, large multiprotein complexes are assembled at the site of action and chromatin is remodeled. Although some components are specialized for different functions (such as the various DNA polymerases), many function in several different complexes. These include TFIIA (general transcription factor IIA), PCNA (proliferating cell nuclear antigen), and RPA (replication protein A). The transcription factor TFIIH is a multiprotein complex that exists in two forms, one concerned with general transcription and the other with repair, probably specifically repair of transcriptionally active DNA.

Defects in DNA repair are the cause of many human diseases

Mutation screens in *E. coli* and yeast have identified components of the repair pathways by looking for mutants that are hypersensitive to the damaging effects of radiation or chemical agents. In humans, similar investigations have used cell lines from patients. A confusing variety of human genetic diseases show partly overlapping phenotypes that suggest defects in some aspect of DNA repair. These include xeroderma pigmentosum (OMIM 278700), Cockayne syndrome (OMIM 216400), trichothiodystrophy (OMIM 601675), Fanconi anemia (OMIM 227650), ataxia-telangiectasia (OMIM 208900), Nijmegen breakage syndrome (OMIM

251260), and Bloom syndrome (OMIM 210900). Clinically, patients may suffer symptoms including hypersensitivity to sunlight, neurological and/or skeletal problems, anemia, and especially a high incidence of various cancers. In the laboratory, cells from patients with these disorders show hypersensitivity to various DNA-damaging agents.

Complementation groups

Many of these conditions can be divided into a series of complementation groups by using cell fusion tests (**Figure 13.10**). If two cells, A and B, lack function in different repair genes, then when the cells are fused the resulting hybrid cell will contain functioning copies of both genes. Cell A, with gene *A* defective, will provide a functioning copy of gene *B*, and cell B, in which gene *B* is defective, will provide a functional copy of gene *A*. Thus, the hybrid should recover the wild-type resistance to DNA damage. Using this technique, cells from patients with xeroderma pigmentosum, caused by defects in nucleotide excision repair (see Box 13.3), have been divided into seven different groups. Cells from any one group will complement (make good) the defect in a cell from any other group. Fanconi anemia, caused by a defective cellular response to DNA damage, has been divided into at least 12 groups. In general, a different gene is mutated in each separate complementation group, although the clinical phenotypes overlap. Details of these genetic diseases and their complementation groups can be found in the OMIM database (http://www.ncbi.nlm.nih.gov/omim).

Molecular studies of these various groups have defined a large number of genes involved in human DNA repair. Sorting out the individual pathways has been greatly aided by the very strong conservation of repair mechanisms across the whole spectrum of life. Not only the reaction mechanisms but also the protein structures and gene sequences are often conserved from *E. coli* to humans. Generally, eukaryotes have multiple systems corresponding to each single system in *E. coli*. For example, nucleotide excision repair requires six proteins in *E. coli* but at least 30 in mammals. A downside of the conservation is a confusing gene nomenclature, referring sometimes to human diseases (e.g. *xeroderma pigmentosum type D*, or *XPD*), sometimes to yeast mutants (*RAD* genes), and sometimes to mammalian cell complementation systems (ERCC—excision repair cross-complementing). So, for example, *XPD*, *ERCC2*, and *RAD3* are the same gene in human, mouse, and yeast.

Not all the diseases that involve hypersensitivity to DNA-damaging agents are caused by defects in the DNA repair systems themselves. Sometimes it is the broader cellular response to damage that is defective. Normal cells react to DNA damage by stalling progress through the cell cycle at a checkpoint until the damage has been repaired, or by triggering apoptosis if the damage is irreparable. Patients with ataxia-telangiectasia and Fanconi anemia have intact repair systems but are deficient in damage-sensing or response mechanisms. Defects in cell cycle control and in the apoptotic response are central to the development of cancer, and they will be discussed further in Chapter 17.

13.3 PATHOGENIC DNA VARIANTS

As mentioned above, the genomes of healthy individuals have huge numbers of sequence variants. The great majority of these are completely harmless and have no known effect on the phenotype. Even most of those that do affect the phenotype are part of the normal variation that makes us all individual. Special interest, however, naturally attaches to those variants that are pathogenic—that is, they either make us ill or make us susceptible to an illness.

Deciding whether a DNA sequence change is pathogenic can be difficult

Not every sequence variant seen in an affected person will be pathogenic. Just as perfectly healthy people carry innumerable sequence variants, the same will be true of a person with a genetic disease. How can we decide whether a sequence change we have discovered in such a person is the cause of their disease or a harmless variant? Only a functional test can give a definitive answer—but functional tests are often difficult to integrate into the work of a diagnostic laboratory.

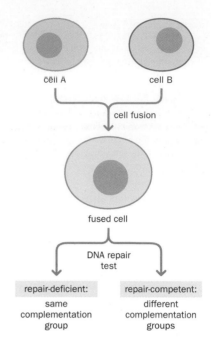

Figure 13.10 Assigning DNA repair defect genes to complementation groups. Repair-deficient cell lines from two unrelated patients (A and B) are fused in cell culture, and the fused cells are then tested for their ability to repair DNA. If the fused cells are competent in DNA repair, then cells A and B belong to different complementation groups and probably contain defects in different DNA repair genes. If the fused cells still lack the ability for DNA repair, then A and B belong to the same complementation group and may have defects in the same DNA repair gene.

In any case, for many gene products no laboratory test is available that checks all aspects of the gene's function *in vivo*. Some variants may be pathogenic only at times of environmental stress, and others may have subtle effects that manifest as susceptibility to a disease, perhaps only when in combination with certain other genetic variants.

In the absence of a definitive functional test, the nature of the sequence change often provides a clue. First we can ask whether the variant affects a sequence that is known to be functional. Such sequences would include the coding sequences of genes, sequences flanking exon–intron junctions (splice sites), the promoter sequence immediately upstream of a gene, and any other known regulatory sequences. The great majority of all known pathogenic variants affect sequences that were already known to be functional, and these comprise only a small percentage of our total DNA. However, it is always possible that a variant located outside any known functional sequence might lie in a currently unidentified functional element. As we saw in Chapter 11, the ENCODE project is revealing many previously unsuspected functional elements in the human genome. Such elements are suspected to be locations for variants that merely alter susceptibility to a disease, rather than directly causing any disease.

If a variant does affect a known functional sequence, we must try to predict its effect. A table of the genetic code (see Figure 1.25) can be used to identify the effect of a coding sequence variant on the protein product of a gene. As described below, nonsense mutations, frameshifts, and many deletions can be confidently predicted to wreck the protein. Similarly, changes to the invariant GT...AG sequences at splice sites are highly likely to be pathogenic. Changes that merely replace one amino acid with a different one (**missense changes**) are more difficult to interpret.

Another approach is to look for precedents. Maybe a variant is already documented in dbSNP, the database of single nucleotide polymorphisms (see above). Alternatively, it may be documented in one of the databases of pathogenic mutations listed in Further Reading. A different sort of precedent can be sought by checking the normal sequence of related genes. These may be in humans (paralogs) or other species (orthologs). If the variant is present as the normal, wild-type sequence of a related gene it is unlikely to be pathogenic.

Further aspects of this problem are considered in Chapter 18, where we discuss genetic testing. In the rest of this section we consider some of the many ways in which a change in a functional sequence can be pathogenic.

Single nucleotide and other small-scale changes are a common type of pathogenic change

Pathogenic changes are often caused by small-scale sequence changes in either the coding sequence or the regulatory region of a gene.

Missense mutations

A single nucleotide substitution within the coding sequence of a gene may or may not alter the sequence of the encoded protein. The genetic code is degenerate: the 64 codons encode only 20 different amino acids (plus three stop codons). Thus, some codon changes do not alter the amino acid—they are silent or synonymous. When the codon change does result in a changed amino acid (a non-synonymous change), the effect depends partly on the chemical differences between the old and new amino acids. As explained in Chapter 1, the 20 amino acids can be classified into acidic, basic, uncharged polar, and uncharged nonpolar types. Replacing an amino acid by one in the same class (a conservative substitution) has less effect on the protein structure than a nonconservative substitution. Adding or removing cysteine alters the potential for forming disulfide bridges, and so can cause major structural changes. Similarity matrices have been constructed that give a quantitative score for the likely disruptive effect of any substitution (see Further Reading).

Some amino acids are crucial to the functioning of a particular protein—for example, those at the active site of an enzyme. Others may be important for maintaining the protein structure. Globular proteins tend to have uncharged nonpolar amino acids in the interior and charged ones on the outside; any substitution that changes this may disrupt the three-dimensional folding. The sickle-

(A)

	V	H	L	T	P	**E**	E	K	S
normal *HBB* gene	GTG	CAT	CTG	ACT	CCT	G**A**G	GAG	AAG	TCT

	V	H	L	T	P	**V**	E	K	S
sickle-cell mutation	GTG	CAT	CTG	ACT	CCT	G**T**G	GAG	AAG	TCT

(B)

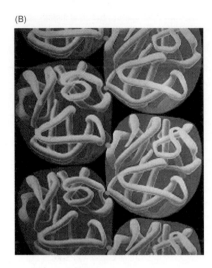

(C)

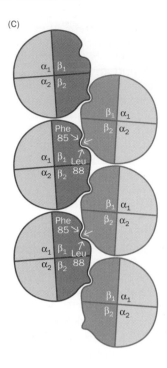

(D)

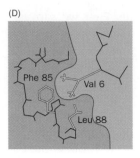

Figure 13.11 The sickle-cell mutation. (A) An A→T mutation in the β-globin (*HBB*) gene causes an amino acid change in the β-globin protein. The mutation replaces glutamic acid, a hydrophilic charged amino acid, with valine, a hydrophobic nonpolar amino acid. This change on the surface of the globin protein allows adhesive interactions between hemoglobin molecules. (B, C, and D) The result is sickle-cell anemia, in which hemoglobin molecules clump together, distorting red blood cells and preventing them from working efficiently. [Parts B, C, and D from Nelson DL & Cox M (2008) Lehninger Principles of Biochemistry, 4th ed. With permission from Palgrave Macmillan.]

cell mutation is pathogenic because it replaces a polar amino acid with a nonpolar one on the outside of the globin molecule (**Figure 13.11**). This makes the molecules tend to stick together. Protein aggregation, the result of abnormal proteins having sticky external areas, has emerged as a common pathogenic mechanism in a variety of diseases, especially progressive neurodegenerative conditions, and is discussed further on p. 425.

It is seldom possible to predict these effects with much confidence. It helps if the three-dimensional structure of the protein has been solved, so that one can model the likely structural effect of a substitution. If amino acid sequences of related proteins (from humans or other organisms) are known, we can see which amino acids are invariant and which seem free to vary widely between species. Most amino acid substitutions probably have no effect on the functioning of a protein.

Nonsense mutations

Three of the 64 codons in the genetic code are stop codons, and so it is quite common for a nucleotide substitution to convert the codon for an internal amino acid of a protein into a stop codon. When ribosomes encounter a stop codon they dissociate from the mRNA, and the nascent polypeptide is released (**Figure 13.12A**). However, genes containing premature termination codons seldom cause production of the truncated protein that might be predicted. Cells have a mechanism, **nonsense-mediated decay** (**NMD**), that detects mRNAs containing premature termination codons and degrades them. Thus, the usual result of a nonsense mutation is to prevent any expression of the gene.

NMD works because the spliced mRNA that travels from the nucleus to the ribosomes retains a memory of the positions of the introns. The splicing mechanism leaves proteins of the **exon junction complex** (**EJC**) attached to splice sites. During the first (pioneer) round of translation, as the ribosome passes each splice site it clears the EJC proteins attached to that site. If there is a premature termination codon, the ribosome will not have traversed every splice site before it detaches. Some EJC proteins will remain attached to the mRNA, and this marks the mRNA for destruction (Figure 13.12B).

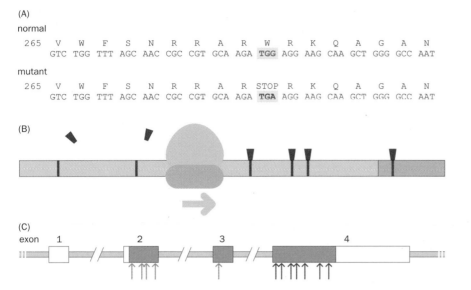

Figure 13.12 Nonsense mutations and nonsense-mediated decay. (A) A G→A change in exon 6 of the *PAX3* gene replaces the TGG codon for tryptophan 274 with a TGA stop codon. (B) Nonsense-mediated decay (NMD). This mature mRNA was transcribed from a gene that has seven exons. Splice junctions (red bars) retain proteins of the exon junction complex (EJC, red triangles). As the first ribosome moves along the mRNA, it displaces the EJC proteins. If it encounters a premature stop codon and detaches before displacing all EJCs, the mRNA is targeted for degradation. Stop codons in the last exon or less than 50 nucleotides upstream of the last splice junction (the green zone) do not trigger NMD. (C) Depending on whether or not a premature stop codon triggers NMD, the consequences of a nonsense mutation can be very different. Mutations in the *SOX10* gene that trigger decay (green arrows) result in Waardenburg syndrome type 4 (hearing loss, pigmentary abnormalities; Hirschsprung disease; OMIM 277580). Nonsense mutations in the 3′ region of the mRNA that escape NMD (red arrows) cause a much more severe neurological phenotype. Shaded areas are coding sequence, clear areas are the 5′ and 3′ untranslated regions of the gene.

Nonsense-mediated decay is not always fully effective. It does not apply to premature stop codons that are in the last exon of a gene, or less than about 50 nucleotides upstream of the last splice junction. In some cases, some quantity of truncated protein is produced even when the stop codon is not in this protected zone. Truncated proteins are potentially more pathogenic than a simple absence of the protein (Figure 13.12C) because they have the potential to interfere with the function of the normal product. Such dominant-negative effects will be discussed later in this chapter (see p. 431). It is assumed that NMD has arisen to protect against this problem.

Changes that affect splicing of the primary transcript

The positions of splice sites are marked by the (almost) invariant canonical GT... AG sequence, embedded within a less tightly defined consensus splice site recognition sequence (see Chapter 1). Mutations that change the canonical GT or AG will always prevent recognition of the site by the spliceosome and so will disrupt splicing at that site (**Figure 13.13A**), but a variety of other sequence changes may also affect it. Splicing is not an all-or-nothing process. As mentioned in Chapter 11, splice sites can be strong or weak. Because of variable use of weak splice sites,

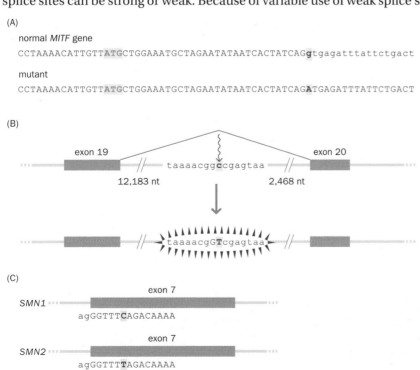

Figure 13.13 Mutations that affect splicing. (A) In the *MITF* gene, a G→A change in the canonical GT sequence that marks the position of the first intron will always disrupt splicing. The exon sequence is in capital letters, the intron in lower-case letters. The translation start codon is in green. (B) A single nucleotide change deep within intron 19 of the cystic fibrosis transmembrane conductance regulator encoded by the *CFTR* gene (called 3849+10 kb C→T, although actually the changed nucleotide is not 10 kb but 12,191 nucleotides from the 3′ end of exon 19) activates a cryptic splice site. The changed sequence is a strong donor splice site. (C) An apparently silent change that affects splicing. The *SMN1* gene is highly expressed, whereas *SMN2* produces almost no protein. The difference is due to a TTT→TTC change in exon 7. Although both TTT and TTC encode phenylalanine, the change inactivates a splicing enhancer and prevents correct splicing at the intron 6–exon 7 junction in *SMN2*.

(A)

ACAUUGUUAUGNOWYOUCANSEEHOWTHERNACANGETHIT

(B)

ACAUUGUUAUG NOW YOU CAN SEE HOW THE RNA CAN GET HIT

(C)

ACAUUGUUAUG NOW YOU CAN TSE EHO WTH ERN ACA NGE THI T

Figure 13.14 The reading frame. This continuous sequence of letters (A) uses the translation start signal AUG (green) to establish the correct reading frame (B). Inserting (or deleting) a letter (red) destroys the meaning (C).

most human genes produce a variety of alternatively spliced transcripts. Splicing enhancer or suppressor sequences modulate the strength of an adjacent splice site by binding proteins of the SR (serine and arginine-rich) and hRNP (heterogeneous ribonucleoprotein) families, possibly in a stage-specific or tissue-specific way. If one of these modulating sequences is mutated in a gene that naturally produces multiple splice isoforms, the effect may just be to alter the balance of isoforms. Depending on the functions of the various isoforms, this may abolish all gene function or make more subtle changes, such as affecting the pattern of tissue-specific isoforms.

Inactivating a splice site will usually destroy the function of a gene, or at least of all isoforms that use that site, but the precise molecular events are hard to predict. Sometimes an exon is skipped; sometimes intronic sequence is retained in the mature mRNA; often an adjacent cryptic splice site is used. **Cryptic splice sites** are sequences within a primary transcript (in exons or introns) that resemble true splice sites, but not sufficiently closely that they are normally recognized as such by the cell (Figure 13.13B). A nucleotide substitution within a cryptic site may increase the resemblance sufficiently to convert it into a functional site. This will disrupt the correct processing of the transcript. Alternatively, a sequence change may reduce the strength of a true splice site so that a nearby cryptic site is used preferentially.

It is often hard to predict from the DNA sequence whether or not a change will affect splicing. Apparent missense or silent mutations may actually be pathogenic through an effect on splicing. The difference between the two human *SMN* genes illustrates such an effect (Figure 13.13C). At least two duplicated but slightly divergent copies of the *SMN* gene are found on chromosome 5q13; the copy closest to the centromere is highly expressed, whereas the copy or copies nearer the telomere produce almost no protein. The difference is due to a TTT→TTC change in exon 7. Although this is apparently silent (both TTT and TTC encode phenylalanine), the change inactivates a splicing enhancer and prevents correct splicing at the intron 6–exon 7 junction. Individuals homozygous for a loss of function of the centromeric *SMN* gene suffer from spinal muscular atrophy (Werdnig–Hoffmann disease; OMIM 253300), but the severity is reduced in those patients who have multiple copies of the weakly expressed telomeric gene.

Computer programs are available that estimate the strength of a splice site, or check whether an apparent missense mutation might affect a splicing enhancer or suppressor, but there is no substitute for real experimental data obtained by RT-PCR. This is also the only method likely to identify a mutation that activates a cryptic splice site deep within an intron, as does the cystic fibrosis 3849+10 kb C→T mutation (see Figure 13.13B).

Frameshifts

The translational reading frame is set by the initiator AUG codon; any downstream change that adds or removes a non-integral number of codons (that is, a number of nucleotides that is not a multiple of three) will cause a frameshift (**Figure 13.14**). Two out of every three random changes to the length of a coding sequence would be expected to produce a frameshift. Because 3 out of 64 possible codons are stop codons, reading a message out of frame will usually quickly hit a premature termination codon. Nonsense-mediated decay will then most probably mean that no protein is produced. An example is found in the *GJB2* gene that encodes connexin 26, a component of gap junctions between cells. A run of six consecutive G nucleotides in the gene is prone to replication slippage. This introduces a premature stop codon (**Figure 13.15**). This mutation is the most frequent single cause of autosomal recessive congenital hearing loss (OMIM 220290) in most European populations.

Several different types of event can produce a frameshift. As well as small insertions or deletions within a single exon, abnormal splicing or whole exon

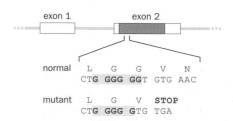

Figure 13.15 A simple frameshift. A frameshifting deletion of one nucleotide in the *GJB2* gene arises through replication slippage in a run of six G nucleotides. The mutation creates a premature stop codon in exon 2.

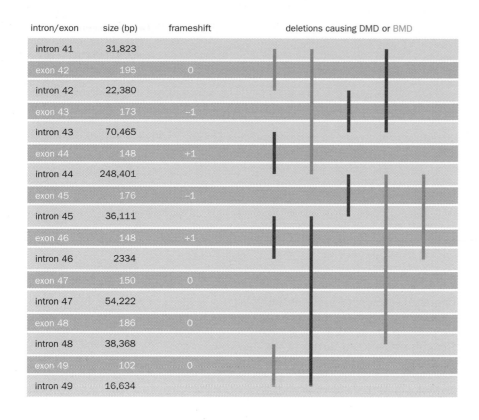

intron/exon	size (bp)	frameshift
intron 41	31,823	
exon 42	195	0
intron 42	22,380	
exon 43	173	–1
intron 43	70,465	
exon 44	148	+1
intron 44	248,401	
exon 45	176	–1
intron 45	36,111	
exon 46	148	+1
intron 46	2334	
exon 47	150	0
intron 47	54,222	
exon 48	186	0
intron 48	38,368	
exon 49	102	0
intron 49	16,634	

deletions causing DMD or BMD

Figure 13.16 Effect of deletions in the dystrophin gene. The gene consists of 79 small exons surrounded by large introns. About 65% of all mutations in this gene are intragenic deletions. Because the introns are much larger than the exons, the breakpoints nearly always lie within introns. The effect is to delete one or more whole exons. When this results in a frameshift (red bars) it causes the severe Duchenne muscular dystrophy (DMD; OMIM 310200). If the deleted exons are frame-neutral (green bars) the result is the milder Becker muscular dystrophy (BMD; OMIM 300376), even if the deletion is larger than one causing DMD.

deletions or duplications usually cause frameshifts. Most exons are not frame-neutral (that is, the number of nucleotides in the exon is not a multiple of three), so excluding one or more exons because of a deletion or splicing error will more often than not create a frameshift. Most introns are much larger than the exons they surround, so in most genes the breakpoints of random intragenic deletions or duplications will lie within introns. The result is to delete or duplicate one or more whole exons (**Figure 13.16**).

Changes that affect the level of gene expression

A variant in a control sequence might affect the level of transcription of a gene, so that although the gene product is entirely normal, too little (or too much) of it is produced. Most obviously this would happen if the variant changed the promoter sequence. In reality, few such variants have been described. This is partly because diagnostic laboratories do not routinely sequence promoters. If they did find a change, they would seldom know how to interpret its significance. As described in Chapter 1, promoters can include consensus binding sites for a variety of transcription factors, and the effect of a sequence change can usually only be determined by experiment.

Patients with α- or β-thalassemia have been thoroughly investigated in the search for mutations affecting transcription. Their diseases are forms of anemia caused by a quantitative deficiency of α- or β-globin, respectively, and so were prime candidates for mutations of this type. However, α-thalassemia is most commonly caused by reduced numbers of active α-globin genes (see below). Some mutations in the β-globin gene promoter have been identified (see OMIM 141900, listed mutations 370–381, and also **Figure 13.17A**), but the great majority of β-thalassemia mutations work by producing an unstable mRNA or an unstable globin protein, rather than directly repressing transcription. Three common types of event have been noted:

- Many thalassemia mutations are splicing errors or nonsense mutations that produce a premature termination codon, resulting in nonsense-mediated decay of the mRNA.

- Changes in the 3′ untranslated region may cause the mRNA to be unstable. A change in this region may create or abolish a critical binding site for a micro-RNA or a protein that regulates translation (see Chapter 11).

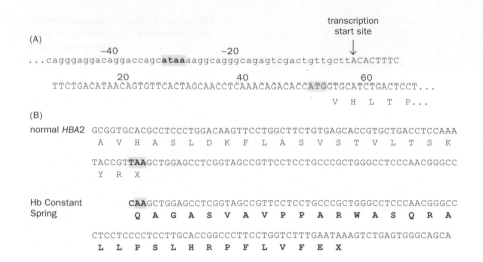

(A)

transcription
start site
↓

```
            −40              −20
...cagggaggacaggaccagcataaaaggcagggcagagtcgactgttgcttACACTTTC
            20               40               60
    TTCTGACATAACAGTGTTCACTAGCAACCTCAAACAGACACCATGGTGCATCTGACTCCT...
                                               V  H  L  T  P...
```

(B)

normal *HBA2*   GCGGTGCACGCCTCCCTGGACAAGTTCCTGGCTTCTGTGAGCACCGTGCTGACCTCCAAA
                  A V H A S L D K F L A S V S T V L T S K

                 TACCGT**TAA**GCTGGAGCCTCGGTAGCCGTTCCTCCTGCCCGCTGGGCCTCCCAACGGGCC
                  Y R X

Hb Constant   **CAA**GCTGGAGCCTCGGTAGCCGTTCCTCCTGCCCGCTGGGCCTCCCAACGGGCC
Spring   **Q A G A S V A V P P A R W A S Q R A**

                 CTCCTCCCCTCCTTGCACCGGCCCTTCCTGGTCTTTGAATAAAGTCTGAGTGGGCAGCA
                  L L P S L H R P F L V F E X
```

**Figure 13.17 Two globin gene mutations that cause thalassemia.** (A) Nucleotide substitutions at any of positions −31 to −28 (red) in the β-globin gene are associated with β-thalassemia. These nucleotides constitute the TATA box, which is an essential part of the promoter. Exon sequences are in capital letters, and the translation start codon is in green. (B) A variant of the major α-globin gene *HBA2* (Hb Constant Spring) has a TAA→CAA substitution at the normal stop codon. The resulting protein has 31 additional amino acids (shown in red), but it is unstable and so causes α-thalassemia. The sequence of exon 3 of the gene is shown here.

- Cells are very efficient at detecting and degrading abnormal proteins that fold incorrectly. For example, one form of α-thalassemia is caused by a change of the normal TAA stop codon of an α-globin gene to CAA. This encodes glutamine, and translation of the mRNA continues for a further 31 amino acids until another stop codon is encountered (Figure 13.17B). This variant hemoglobin (Hb Constant Spring) is unstable, and clinically the effect is α-thalassemia resulting from a quantitative deficiency of α-globin chains.

## Pathogenic synonymous (silent) changes

As mentioned above, synonymous changes are nucleotide substitutions that convert one codon for a particular amino acid into another that encodes the same amino acid. Such changes would not be expected to have any phenotypic effect. Sometimes, however, the altered nucleotide is part of a splicing enhancer or suppressor and the change affects splicing. As mentioned above, in spinal muscular atrophy a silent TTC→TTT change marks the difference between an active, correctly spliced gene and an inactive, incorrectly spliced gene (see Figure 13.13C). Even silent changes that do not affect splicing may not be entirely neutral. The *MDR1* gene encodes the P-glycoprotein, which is important in the transport of many drugs; its overexpression is an important determinant of multi-drug resistance in chemotherapy. A silent change, c.3435C→T, in exon 26 of the gene results in the production of a protein that has an unchanged amino acid sequence (the change converts one codon for isoleucine 1145 into another) but is subtly differently folded and has a different activity profile. This is thought to be because the alternative codon requires an alternative tRNA that is in short supply, and this therefore changes the kinetics of translation. How common such effects might be is not known.

## Variations at short tandem repeats are occasionally pathogenic

Whereas single nucleotide changes might affect any coding sequence, most short tandem repeats are located in noncoding DNA. Those that do occur in coding sequences are seldom polymorphic. However, tandem repeat variants located near promoters or splice sites of genes can sometimes affect gene expression. For example, different alleles of a 14 bp minisatellite near the promoter of the insulin gene on 11p15 are associated with differential risk of type 2 diabetes (see OMIM 176730). Another example occurs within the cystic fibrosis transmembrane conductance regulator (*CFTR*) gene, where a run of T nucleotides near the 3′ end of intron 8 affects the efficiency of the adjacent splice site. Alleles with five, seven, or nine T nucleotides are common. Whereas the splicing of 7-T or 9-T alleles is normal, 5-T alleles are often mis-spliced and exon 9 is skipped. 5-T alleles on their own do not reduce the output of correctly spliced mRNA so greatly as to be pathogenic, but in conjunction with other low-functioning variants they can be a cause of cystic fibrosis. Their effect is enhanced if a $(TG)_n$ repeat nearby in the intron has more than 11 repeats (**Figure 13.18**).
```

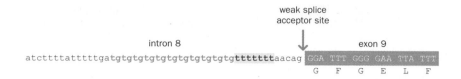

weak splice
acceptor site

intron 8 ↓ exon 9

atcttttatttttgatgtgtgtgtgtgtgtgtgtgtgtg**tttttttt**aacag GGA TTT GGG GAA TTA TTT
G   F   G   E   L   F

Figure 13.18 Variations in an intron of the *CFTR* gene affect the efficiency of splicing. Near the 3′ end of intron 8 (lower-case letters) are variable-length $(TG)_n$ (green) and T_n (red) sequences. Alleles with the 5T and $(TG)_{13}$ variants mis-splice a high proportion of transcripts at the nearby acceptor splice site, and so they produce only a low level of functioning CFTR protein. By themselves these changes are not pathogenic, but if that gene copy also contains a coding variant such as p.R117H that lowers the activity of the intact protein, the combined result of low protein level and low protein function makes it a pathogenic CF allele. Note that in the cystic fibrosis literature, for historic reasons, this intron/exon pair is usually called intron 8/exon 9, but in the genome databases they are intron 9/exon 10.

Tandem repeats within coding sequences are not normally polymorphic, but may be liable to pathogenic mutations because of polymerase stutter. Somatic mutations of this type are a major cause of disease in people with defects in the post-replicative mismatch repair system (see Chapter 17). Expanded polyalanine runs in certain proteins are responsible for several inherited diseases. Examples include the PHOX2B protein in people with congenital central hypoventilation syndrome (OMIM 209880) and the HOXD13 protein in patients with synpolydactyly 1 (OMIM 186000). These variants presumably originated through polymerase stutter, but within a family they are stably transmitted, just like any other STRP allele. In at least some cases, the expanded alanine run interferes with correct localization of the protein within the cell.

Dynamic mutations: a special class of pathogenic microsatellite variants

So-called **dynamic mutations** are STRPs that, above a certain size, become intensely unstable. The molecular causes are not well understood, but they may be a consequence of the way in which, when DNA is replicated, one strand (the lagging strand) is synthesized as a series of discontinuous fragments—the Okazaki fragments (see Chapter 1). A special endonuclease, FEN1, cuts off the overhangs in overlapping Okazaki fragments. One proposed mechanism for repeat expansion is that FEN1 fails to make the cuts, and overlapping fragments end up being joined end-to end. Repeats up to a certain size are stable, and it may be significant that, in most cases, the threshold of instability occurs when the repeat sequence reaches the typical size of an Okazaki fragment. Not all dynamic mutations are pathogenic, but several are (**Table 13.1**). Others are responsible for the nonpathogenic fragile sites seen by cytogeneticists when cells of some people are subjected to replicative stress (see Figure 13.5). For example the *FRA16A* fragile site on chromosome 16 is due to an expanded $(CCG)_n$ repeat, whereas *FRA16B* on the same chromosome is caused by an expanded 33 bp minisatellite.

The diseases in Table 13.1 are heterogeneous in many respects. There are different-sized repeat units, different degrees of expansion, different locations with respect to the affected gene, and different pathogenic mechanisms. Within these, the polyglutamine diseases form a well-defined group, of which Huntington disease (HD; OMIM 143100) is the prototype. In these conditions, modest expansions of a $(CAG)_n$ repeat in the coding sequence of a gene lead to an expanded polyglutamine run in the encoded protein (**Figure 13.19C**). This, in turn, predisposes the protein to form intracellular aggregates that are toxic to cells, especially neurons (**Box 13.4**). The result is a progressive late-onset neurodegenerative disease.

Other dynamic mutations affect gene sequences outside coding regions, and may involve much larger expansions. The expanded tandem repeat may be in the promoter, in the 5′ or 3′ untranslated region, or in an intron. Usually the effect is to prevent expression of the gene (Figure 13.19A), and the pathology comes from the resulting lack of gene function. When this occurs, the same disease can sometimes be caused by other loss-of-function mutations in the same gene. In myotonic dystrophy 1 (DM1; OMIM 160900) the mechanism is quite different. There is a gain of function involving a toxic mRNA. A massively expanded $(CTG)_n$ run in the 3′ untranslated region of the *DMPK* gene produces an mRNA that sequesters CUG-binding proteins in the nucleus. These are required for correct splicing of the primary transcripts of several unrelated genes, which therefore no longer function correctly. The result is a multisystem disease whose features bear no relation to the function of the *DMPK* gene product, a protein kinase, which is still produced in normal quantities. No other mutations in the *DMPK* gene produce myotonic dystrophy, but a very similar clinical disease (DM2; OMIM 116955) can be caused by massive expansion of a $(CCTG)_n$ sequence in the completely unrelated *ZNF9* gene.

TABLE 13.1 DISEASES CAUSED BY UNSTABLE EXPANDING NUCLEOTIDE REPEATS

Disease	OMIM no.	Mode of inheritance[a]	Name and location of gene	Location of repeat within gene	Repeat sequence	Stable repeat no.	Unstable repeat no.
1. VERY LARGE EXPANSIONS OF REPEATS OUTSIDE CODING SEQUENCE							
Fragile-X site A (FRAXA)	309550	X	*FMR1* Xq27.3	5′ untranslated region	$(CGG)_n$	6–54	200–1000+
Fragile-X site E (FRAXE)	309548	X	*FMR2* Xq28	5′ untranslated region	$(CCG)_n$	4–39	200–900
Friedreich ataxia (FRDA)	229300	AR	*FXN* 9q13	intron 1	$(GAA)_n$	6–32	200–1700
Myotonic dystrophy 1 (DM1)	160900	AD	*DMPK* 19q13	3′ untranslated region	$(CTG)_n$	5–37	50–10,000
Myotonic dystrophy 2 (DM2)	602668	AD	*ZNF9* 3q21.3	intron 1	$(CCTG)_n$	10–26	75–11,000
Spinocerebellar ataxia 10 (SCA10)	603516	AD	*ATXN10* 22q13.31	intron 9	$(ATTCT)_n$	10–20	500–4500
Myoclonic epilepsy of Unverricht and Lundborg	254800	AR	*CSTB* 21q22.3	promoter	$(CCCCGC\ CCCGCG)_n$	2–3	40–80
2. MODEST EXPANSIONS OF CAG REPEATS WITHIN CODING SEQUENCES							
Huntington disease (HD)	143100	AD	*HD* 4p16.3	coding	$(CAG)_n$	6–34	36–100+
Kennedy disease (SBMA)	313200	X	*AR* Xq12	coding	$(CAG)_n$	9–35	38–62
Spinocerebellar ataxia 1 (SCA1)	164400	AD	*ATXN1* 6p23	coding	$(CAG)_n$	6–38	39–82
Spinocerebellar ataxia 2 (SCA2)	183090	AD	*ATXN2* 12q24	coding	$(CAG)_n$	15–24	32–200
Machado-Joseph disease (SCA3)	109150	AD	*ATXN3* 14q32.1	coding	$(CAG)_n$	13–36	61–84
Spinocerebellar ataxia 6 (SCA6)	183086	AD	*CACNA1A* 19p13	coding	$(CAG)_n$	4–17	21–33
Spinocerebellar ataxia 7 (SCA7)	164500	AD	*ATXN7* 3p14.1	coding	$(CAG)_n$	4–35	37–306
Spinocerebellar ataxia 17 (SCA17)	607136	AD	*TBP* 6q27	coding	$(CAG)_n$	25–42	47–63
Dentatorubral–pallidoluysian atrophy (DRPLA)	125370	AD	*DRPLA* 12p13.31	coding	$(CAG)_n$	7–34	49–88
3. OTHER DYNAMIC EXPANSIONS							
Spinocerebellar ataxia 8 (SCA8)	608768	AD	? 13q21	untranslated RNA	$(CTG)_n$	16–34	74+
Spinocerebellar ataxia 12 (SCA12)	604326	AD	*PPP2R2B* 5q32	promoter	$(CAG)_n$	7–45	55–78
Huntington disease-like 2 (HDL2)	606438	AD	*JPH3* 16q24.3	variably spliced exon	$(CTG)_n$	7–28	66–78

[a]Modes of inheritance: X, X-linked; AD, autosomal dominant; AR, autosomal recessive.

A hallmark of diseases caused by dynamic repeats is *anticipation*—in successive generations the age of onset is lower and/or the severity worse because of successive expansions of the repeat. See Chapter 3 for a cautionary note about the way in which biases in the way in which family members are ascertained can produce a spurious appearance of anticipation. In some cases, the threshold for instability is lower than the threshold for pathogenic effects. In these cases intermediate-sized nonpathogenic but unstable **premutation alleles** are seen that readily expand to full mutation alleles when transmitted to a child (e.g. *FRAXA* repeats of 50–200 units in fragile X syndrome; OMIM 300624). In other cases, alleles below the pathogenic threshold only very occasionally expand (e.g. HD alleles with 29–35 repeats). In some cases, there is a sex effect such that large expansions are mainly seen in alleles inherited from a parent of one sex (the father in HD, the mother in myotonic dystrophy). These reflect a differential survival of gametes carrying large expansions, not an inherent tendency to expand in one sex rather than the other.

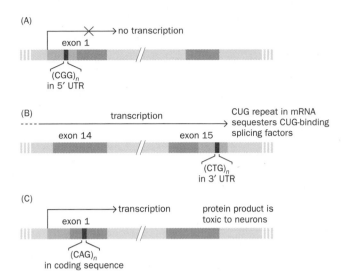

Figure 13.19 Three mechanisms by which dynamic mutations may be pathogenic. (A) In fragile X syndrome, the expanded repeat in the 5′ untranslated region (UTR) of the gene triggers methylation of the promoter and prevents transcription. (B) In myotonic dystrophy, the expanded repeat in the 3′ untranslated region causes the mRNA transcript to sequester splicing factors in the cell nucleus, preventing the correct splicing of several unrelated genes. (C) In Huntington disease, the gene containing the expanded repeat is transcribed and translated as normal, but the protein product has an expanded polyglutamine tract that renders it toxic. Expanded repeats are shown in red, and coding regions in dark blue.

Comparisons of repeat sizes in parent and child are hard to interpret mechanistically because they may reflect the mitotic or meiotic instability of the repeat in the parental germ line or, alternatively, the ability of gametes to transmit large repeats. Moreover, repeat sizes are usually studied in DNA extracted from peripheral blood lymphocytes, and if there is mitotic instability the lymphocytes may have very different repeat sizes from those in sperm or egg, or in the tissues involved in the disease pathology. Some repeats are unstable somatically, and so give a smeared band when blood DNA is analyzed by electrophoresis. Others are unstable between parent and child but stable in mitosis, and so blood DNA gives a sharp band, but of different sizes in parent and child.

Variants that affect dosage of one or more genes may be pathogenic

The pathogenic potential of abnormal gene dosage has long been appreciated because of the severe phenotypes produced by chromosomal trisomies. Nevertheless, only a minority of genes are sensitive to dosage. If a condition is recessive, that means that heterozygotes are phenotypically normal despite having only a single functional copy of the gene in question. For such genes, dosage evidently does not matter. The recent discovery of abundant and large-scale variations in copy number between apparently normal people reinforces the message that not all variants in copy number are harmful.

Chromosomal trisomies probably owe their characteristic phenotypes to just a few dosage-sensitive genes. For example, the characteristic features of Down syndrome are thought to be due largely to dosage effects of just two genes, *DSCR1* and *DYRK1A*. It is to be expected that more genes would produce phenotypic effects at half dosage than at 1.5-fold increased dosage. Thus, large deletions or monosomy of a whole chromosome are less well tolerated than duplications or trisomy in human development.

BOX 13.4 DISEASES CAUSED BY ABNORMAL PROTEIN AGGREGATION

The formation of protein aggregates is now recognized as a common feature of several adult-onset neurological diseases. These include the polyglutamine diseases of which Huntington disease is the prototype, Alzheimer disease, Parkinson disease, Creutzfeldt–Jakob disease, and the heterogeneous group known as amyloidoses.

Globular protein molecules resemble oil drops, with hydrophobic residues in the interior and polar groups on the outside. Correct folding is a critical and highly specific process, and naturally occurring proteins are selected from among all possible polypeptide sequences partly for their ability to fold correctly. Mutant proteins may be more prone to misfolding. Misfolded molecules with exposed hydrophobic groups can aggregate with each other or with other proteins, and this is somehow toxic to cells and, in particular, neurons. Sometimes

it seems that a conformational change can propagate through a population of protein molecules, converting them from a stable native conformation into a new form with different properties, in a process that is perhaps analogous to crystallization.

The behavior of prion proteins in diseases such as Creutzfeldt–Jakob disease is the most striking example. Misfolding might start with a chance misfolding of a newly synthesized structurally normal molecule (sporadic cases), a mutant sequence with a greater propensity to misfold (a genetic disease), or a misfolded molecule somehow acquired from the environment (infectious cases). Thus, this final common pathway brings together a set of diseases with very disparate origins. This topic is reviewed by Hardy & Orr (see Further Reading).

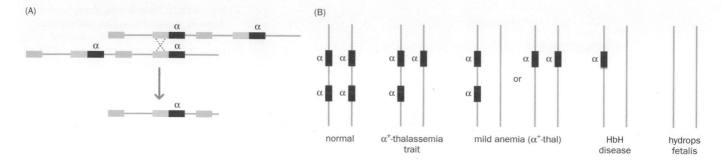

normal   α⁺-thalassemia trait   mild anemia (α⁺-thal)   HbH disease   hydrops fetalis

One common mechanism generating changes in gene dosage is **non-allelic homologous recombination** (NAHR). Segmental duplications (often defined as sequences 1 kb or longer with 95% or greater sequence identity) may misalign when homologous chromosomes pair in meiosis. NAHR then produces deletions or duplications. The misaligned repeats have the same sequence but not the same chromosomal location, so recombination is homologous but the sequences are not alleles. Many (but by no means all) of the common nonpathogenic variants in copy number seen in normal healthy people are generated by this mechanism. α-Thalassemia provides a good example of NAHR producing a pathogenic variation in gene dosage. Most people have four copies of the α-globin gene (αα/αα) as a result of an ancient tandem duplication. As shown in **Figure 13.20**, NAHR between low-copy repeat sequences flanking the α-globin genes can produce chromosomes carrying more or fewer α-globin genes. Reduced copy numbers of α-globin genes produce successively more severe effects. People with three copies (αα/α–) are healthy; those with two (whether the phase is α– /α– or αα/– –) suffer mild α-thalassemia; those with only one gene (α–/– –) have severe disease; and lack of all α genes (– –/– –) causes lethal hydrops fetalis (fluid accumulation in the fetus).

X-chromosome monosomy and trisomy are particularly interesting because X-inactivation (inactivation of all except one of the X chromosomes in a cell; see Chapter 3) ought to render them asymptomatic in somatic tissues. However, as noted in Chapter 11, a surprisingly large number of genes on the X chromosome escape inactivation. Some of these have a counterpart on the Y chromosome, but most do not. For those genes that escape X-inactivation but lack a Y-linked counterpart, normal females would have two functional copies and males only one. Turner (45,X) females would have the same single active copy as normal males—but perhaps in the context of female development, a single copy is not sufficient. The skeletal abnormalities of Turner syndrome are caused by haploinsufficiency for *SHOX* (50% of the normal gene product is not sufficient to produce a normal phenotype). This is a homeobox gene that is located in the Xp/Yp pseudoautosomal region, and so is present in two copies in both males and normal females.

Below the level of conventional cytogenetic resolution but above the single gene level, pathogenic variations in copy number are classified as microdeletions or microduplications (**Table 13.2**). Among these, three different molecular pathologies can be distinguished:

- **Single gene syndromes**, in which all the phenotypic effects are due to the deletion (or sometimes duplication) of a single gene. For example, Alagille syndrome (OMIM 118450) is seen in patients with a microdeletion at 20p11. However, 93% of Alagille patients have no deletion but instead are heterozygous for point mutations in the *JAG1* gene located at 20p12. The cause of the syndrome in all cases is a half dosage of the *JAG1* gene product.

- **Contiguous gene syndromes** are seen primarily in males with X-chromosome deletions (**Figure 13.21A**). The classic case was a boy BB who had Duchenne muscular dystrophy (DMD; OMIM 310200), chronic granulomatous disease (CGD; OMIM 306400), and retinitis pigmentosa (OMIM 312600), together with mental retardation. He had a chromosomal deletion in Xp21 that removed a contiguous set of genes and incidentally provided investigators with the means to clone the genes whose absence caused two of his diseases, DMD and CGD. Deletions of the tip of Xp are seen in another set of contiguous gene syndromes. Successively larger deletions remove more genes and add more

Figure 13.20 Deletions of α-globin genes in α-thalassemia. (A) Normal copies of chromosome 16 each carry two α-globin genes (red) arranged in tandem. Repeat blocks flanking the genes (shown in gray and yellow) may misalign, allowing unequal crossover. The figure shows one of several such misaligned recombinations that can produce a chromosome carrying only one active α gene. Unequal crossovers between other repeats (not shown) can produce chromosomes carrying no functional α gene. Thus, the number of α-globin genes may vary in individuals from none to four or more. (B) The consequences become more severe as the number of α genes diminishes. See Weatherall et al. (Further Reading) for detail.

normal amount. Thus, most loss-of-function mutations p...
notypes. For example, most inborn errors of metabolism a...
loss-of-function conditions may be dominant if there is h...
dominant-negative effect, as described below.

Allelic heterogeneity is a common feature of loss...
phenotypes

There are many ways of reducing or abolishing the funct...
When a clinical phenotype results from the loss of functic...
expect *any* change that inactivates the gene product to pro...
result. Gain of function, in contrast, is a rather specific p...
only a very specific change in a gene can cause a gain of fun...
of allelic heterogeneity is another strong, although far fro...
the underlying molecular pathology. For example, amor...
unstable trinucleotide repeats (see Table 13.1), fragile X syn...
and Friedreich ataxia (OMIM 229300) are occasionally ca...
mutation in their respective genes, pointing to a loss...
Huntington disease and myotonic dystrophy are never se...
of mutation, suggesting a gain of function.

When a clinical phenotype results from the loss of f...
should be able to find point mutations that have the sam...
disruption of the gene. For example, the ataxia-telangie...
gene on chromosome 11q23 encodes a protein 3056 am...
involved in the detection of DNA damage. When damage i...
phorylates downstream proteins, which sets the repair proc...
suffering from ataxia-telangiectasia are deficient in rej...
Different patients show a wide variety of *ATM* gene m...
frameshifting insertions or deletions, but also nonsense...
tions, some missense changes, and occasional large del...
Clearly, the cause of ataxia-telangiectasia is a loss of func...
the *ATM* gene.

Loss-of-function mutations produce dominant p...
there is haploinsufficiency

As mentioned above, for most gene products the precise...
Most loss-of-function mutations produce recessive phenc...
person who has only one functional gene copy will be phe...
some cases, however, 50% of the normal level is not suffi...
tion. People heterozygous for a loss-of-function mutatio...
which is therefore inherited as a dominant trait. This is call...
Waardenburg syndrome type 1 (OMIM 193500: hearing...
abnormalities) provides an example. As **Figure 13.23** show...
in the *PAX3* gene include amino acid substitutions, fram...
tions, and, in some patients, complete deletion of the g...
events produce the same clinical result, its cause must b...
PAX3. Nevertheless, Waardenburg syndrome is dominant,...
haploinsufficiency. **Table 13.3** lists some other examples.

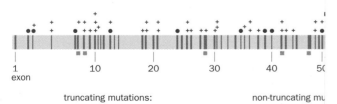

1   10   20   30   40   50
exon

truncating mutations:

● nonsense mutation

+ frameshifting insertion/deletion

■ splice site mutation

– deletions of all or part of the gene

non-truncating mu...

■ missense muta...

+ in-frame deletio...

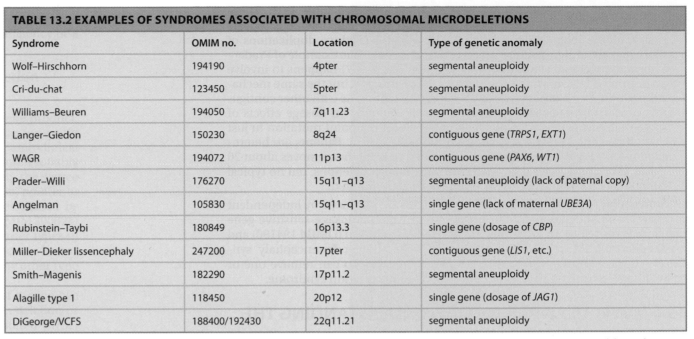

TABLE 13.2 EXAMPLES OF SYNDROMES ASSOCIATED WITH CHROMOSOMAL MICRODELETIONS

Syndrome	OMIM no.	Location	Type of genetic anomaly
Wolf–Hirschhorn	194190	4pter	segmental aneuploidy
Cri-du-chat	123450	5pter	segmental aneuploidy
Williams–Beuren	194050	7q11.23	segmental aneuploidy
Langer–Giedon	150230	8q24	contiguous gene (*TRPS1, EXT1*)
WAGR	194072	11p13	contiguous gene (*PAX6, WT1*)
Prader–Willi	176270	15q11–q13	segmental aneuploidy (lack of paternal copy)
Angelman	105830	15q11–q13	single gene (lack of maternal *UBE3A*)
Rubinstein–Taybi	180849	16p13.3	single gene (dosage of *CBP*)
Miller–Dieker lissencephaly	247200	17pter	contiguous gene (*LIS1*, etc.)
Smith–Magenis	182290	17p11.2	segmental aneuploidy
Alagille type 1	118450	20p12	single gene (dosage of *JAG1*)
DiGeorge/VCFS	188400/192430	22q11.21	segmental aneuploidy

Those syndromes caused by having only one functional copy of a single gene are often also seen in patients who do not have a microdeletion, but a point mutation in the gene. WAGR, Wilms tumor, aniridia, genital abnormalities, mental retardation; VCFS, velocardiofacial syndrome.

diseases to the syndrome. Microdeletions are relatively frequent in some parts of the X chromosome (such as Xp21 and proximal Xq) but are rare or unknown in others (such as Xp22.1–22.2 and Xq28). No doubt the deletion of certain individual genes, and visible deletions in gene-rich regions, would be lethal. Similar contiguous gene syndromes are much less common with autosomes because of the presence of the balancing normal chromosome (Figure 13.21B). Langer–Giedon syndrome (trichorhinophalangeal syndrome, type II; OMIM 150230) is a rare example.

- **Segmental aneuploidy syndromes** are a special type of contiguous gene syndrome that regularly recur with a well-recognized phenotype. Examples include Williams–Beuren (OMIM 194050), Prader–Willi (OMIM 176270), Angelman (OMIM 105830), Smith–Magenis (OMIM 182290), and DiGeorge/

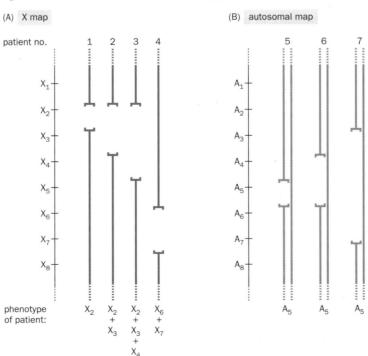

Figure 13.21 X-linked and autosomal microdeletion syndromes. (A) On the X chromosome, male patients 1–3 show a nested series of contiguous gene syndromes, whereas patient 4 has a different contiguous gene syndrome. Deletion of gene X₁ or X₅ is lethal in males. (B) On the autosome, contiguous gene syndromes are rare because of the balancing effect of the second chromosome. In this example, only gene A₅ is dosage sensitive. Patients 5–7, with different-sized deletions all encompassing gene A₅, all show the same phenotype as patient 8, who is heterozygous for a loss-of-function point mutation in the A₅ gene.

velocardiofacial (OMIM 188400/192430) syndromes (s
syndromes all have deletions produced by NAHR betw
that flank the region in question. NAHR will also pro
these regions, although these may not be pathogenic. Th
Willi and Angelman syndromes (see Figure 11.20, p. 36
an imprinted region, which complicates the phenotype,
nism produces the other syndromes mentioned above. A
ous gene syndromes, the phenotype usually depends
more than one gene and is not seen in people with a p
one of the genes. Williams–Beuren syndrome is typical
zygous for a 1.5 Mb deletion on chromosome 7q11.23 t
genes. Cases have been described who have smaller del
case has been found with just one gene deleted or muta

Some other recognizable recurrent syndromes are prod
random terminal deletions of chromosomes in which a c
lies close to the telomere. Examples are the Wolf–Hirschhor
cri-du-chat (OMIM 123450) syndromes. In Miller–Dieke
drome (OMIM 247200), random terminal deletions of 17
more dosage-sensitive genes, producing a contiguous dele

13.4 MOLECULAR PATHOLOGY: UNDERST
EFFECT OF VARIANTS

For genetic counseling and pedigree analysis, alleles are
nated *A* and *a*, with the capital letter representing the allele
nant. That is the appropriate level of description for those p
molecular pathology we need to look more closely. When v
type of a cystic fibrosis carrier as *Aa*, what we mean by
sequence that is mutated so that it does not produce a funct
nel. More than 1500 different such alleles have been describe
any sequence that functions sufficiently well not to cause
have to be optimally functional, just good enough to avoid
actual DNA sequence of *A* alleles in unrelated people will no
identical.

The biggest distinction in molecular pathology is
function and gain-of-function changes

For molecular pathology, the important thing is not the se
allele but its effect. Knowing the sequence of a mutant a
genetic testing, but for molecular pathology we need to know
not do. A mutated gene might have all sorts of subtle effects
a valuable first question to ask is whether it produces a loss

- In **loss-of-function mutations** the product has reduced

- In **gain-of-function mutations** the product does s
  abnormal.

Inevitably, some mutations cannot easily be classified as
function. Has a permanently open ion channel lost the fu
gained the function of inappropriate opening? A mutant gen
feres with the function of the remaining normal allele in a h
lost its normal function and gained a novel harmful funct
change the balance between several different functions o
Nevertheless, the distinction between loss of function and g
essential first tool for thinking about molecular pathology.

The mode of inheritance provides a clue to the underlyi
ogy. If there is a gain of function, the presence of a normal a
vent the mutant allele from behaving abnormally. The abno
be seen in a heterozygous person, and so we would expect
dominant. For example, both Huntington disease and my
dominant. For loss-of-function phenotypes, the picture is
gene products, the precise quantity is not crucial, and we c

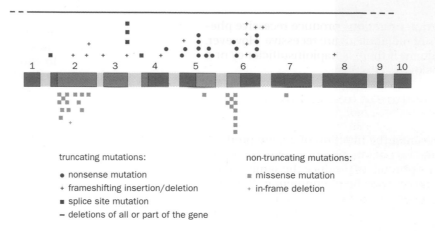

truncating mutations:
- ● nonsense mutation
- + frameshifting insertion/deletion
- ■ splice site mutation
- − deletions of all or part of the gene

non-truncating mutations:
- ■ missense mutation
- + in-frame deletion

Figure 13.23 Loss-of-function mutations in the *PAX3* gene in patients with type I Waardenburg syndrome. A wide range of both truncating and non-truncating mutations are seen, similar to the data for the *ATM* gene in Figure 13.21. Missense changes are concentrated in the two shaded areas that encode key functional DNA-binding domains (orange, paired domain; green, homeodomain) of the PAX3 protein. Although the disease is dominant, the data in the figure show that the cause must be a loss of function of the *PAX3* gene. This is therefore an example of haploinsufficiency.

One might reasonably ask why there should be haploinsufficiency for *any* gene product. Why has natural selection not managed things better? If a gene is expressed so that two copies make only a barely sufficient amount of product, selection for variants with higher levels of expression should lead to the evolution of a more robust organism, with no obvious price to be paid. The answer is that in most cases this has indeed happened—which is why relatively few genes are dosage-sensitive. There are a few cases in which a cell with only one working copy of a gene just cannot meet the demand for a gene product that is needed in large quantities. An example may be elastin. In people heterozygous for a deletion or loss-of-function mutation of the elastin gene, tissues that require only modest quantities of elastin (skin and lung, for example) are unaffected, but the aorta, where much more elastin is required, often shows an abnormality, supravalvular aortic stenosis (OMIM 185500). However, certain gene functions are inherently dosage-sensitive. These include:

- gene products that are part of a quantitative signaling system whose function depends on partial or variable occupancy of a receptor or a DNA-binding site, for example;

- gene products that compete with each other to determine a developmental or metabolic switch;

- gene products that cooperate in interactions with fixed stoichiometry (such as the α- and β-globins or many structural proteins).

In each case, the gene product is titrated against something else in the cell. What matters is not the correct absolute level of product, but the correct relative levels of interacting products. The effects are sensitive to changes in all the interacting partners; thus, these dominant conditions often show highly variable expression. Genes whose products act essentially alone, such as many soluble enzymes of metabolism, seldom show dosage effects.

TABLE 13.3 EXAMPLES OF PHENOTYPES PROBABLY CAUSED BY HAPLOINSUFFICIENCY		
Condition	**OMIM no.**	**Gene[a]**
Alagille syndrome type 1	118450	*JAG1*
Multiple exostoses type 1	133700	*EXT1*
Tomaculous neuropathy	162500	*PMP22*
Supravalvular aortic stenosis	185500	*ELN*
Trichorhinophalangeal syndrome type 1	190350	*TRPS1*

[a]Individuals with a single loss-of-function allele are affected, because the output of a single functional allele is insufficient for normal functioning or development.

Dominant-negative effects occur when a mutated gene product interferes with the function of the normal product

Dominant-negative effects provide another reason why some loss-of-function mutations produce dominant phenotypes. Dominant-negative effects are seen only in heterozygotes, where they cause more severe effects than simple null alleles of the same gene. A person heterozygous for a null mutation in a gene will still have 50% of the normal level of function, because they still have one normally functioning allele. But if the product of the mutant allele is not only nonfunctional in itself but also interferes with the function of the remaining normal allele, there will be less than 50% residual function. The mechanism of nonsense-mediated decay (see p. 418) probably evolved as a protection against possible dominant-negative effects of abnormal truncated proteins: it may be better to have no product from a mutant gene than to have an abnormal product. Proteins that build multimeric structures are particularly vulnerable to dominant-negative effects. Collagens provide a classic example.

Fibrillar collagens, the major structural proteins of connective tissue, are built of triple helices of polypeptide chains, sometimes homotrimers, sometimes heterotrimers, that are assembled into close-packed crosslinked arrays to form rigid fibrils. In newly synthesized polypeptide chains (preprocollagen), N- and C-terminal propeptides flank a regular repeating sequence $(Gly-X-Y)_n$, where X and Y are variable amino acids, at least one of which is often proline. Three preprocollagen chains associate and wind into a triple helix under control of the C-terminal propeptide. After formation of the triple helix, the N- and C-terminal propeptides are cleaved off. Mutations that replace glycine with any other amino acid usually have strong dominant-negative effects because they disrupt the tight packing of the triple helix. Missense mutations in type I collagen are responsible for the most severe forms of brittle bone disease (osteogenesis imperfecta type IIA; OMIM 166210) because of these dominant-negative effects. In heterozygotes the mutant collagen polypeptides associate with normal chains but then disrupt formation of the triple helix. This can reduce the yield of functional collagen to well below 50%. Null mutations in the same gene might be expected to produce more severe effects, but in fact the disease is milder. The simple absence of some collagen is less disruptive than the presence of abnormal chains (**Figure 13.24**).

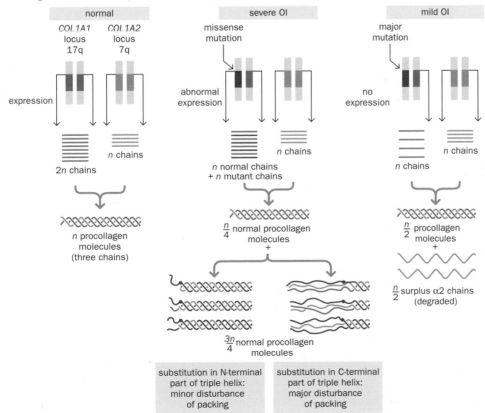

Figure 13.24 Dominant-negative effects of collagen gene mutations. Collagen fibrils are built of crosslinked arrays of triple-helical procollagen units. The type I procollagen comprises two chains encoded by the *COL1A1* gene and one encoded by *COL1A2*. In the triple helix, each polypeptide chain consists of repeating units, $(Gly-X-Y)_n$. In osteogenesis imperfecta type IIA (OI; OMIM 166210), mutations that replace glycine with any other amino acid usually have strong dominant-negative effects because they disrupt the packing. The helix is assembled starting at the C-terminus, and substitutions of glycines close to that end have a more severe effect than substitutions nearer the N-terminus. Null mutations in either gene result in fewer but otherwise normal triple helices forming (simple haploinsufficiency) and a less severe clinical phenotype.

Non-structural proteins that dimerize or oligomerize also show dominant-negative effects. For example, transcription factors of the bHLH-Zip family bind DNA as dimers. Mutants that cannot dimerize often cause recessive phenotypes, but mutants that are able to sequester functioning molecules into inactive dimers give dominant phenotypes. The ion channels in cell membranes provide another example of multimeric structures that are sensitive to dominant-negative effects. Connexin 26 provides an example. Six molecules of the connexin 26 protein associate to form a connexon, one half of a gap junction that allows small ions to move between cells. Figure 13.15 showed an example of a null mutation in the gene that encodes connexin 26. People homozygous for this mutation cannot make connexin 26; they lack functioning gap junctions in their inner ears, potassium ions cannot recirculate as they should, and the patients are deaf. Heterozygotes are entirely phenotypically normal (this is a problem if you wish to identify couples at risk of having deaf children). However, certain missense mutations produce structurally abnormal connexin 26 molecules. These disrupt the function of connexons, even if some of the six connexin molecules are normal; heterozygotes have hearing loss and the phenotype is dominant.

Gain-of-function mutations often affect the way in which a gene or its product reacts to regulatory signals

A mutation is unlikely to give a gene product a radically new function. The only mechanism that reliably generates novel gains of function is when a chromosomal rearrangement creates a novel chimeric gene by joining exons of two different genes. Exon shuffling of this type has been important in evolution, but not in inherited disease. The chromosomal translocations or inversions that produce these chimeric genes can propagate through successive mitotic divisions, when each chromosome behaves independently. However, in meiosis homologous chromosomes pair and recombine. As we saw in Chapter 2, pairing in carriers of chromosomal rearrangements involves complicated chromosomal conformations that can result in mis-segregation; in some cases, recombination will produce duplications and deletions. As a result, chimeric genes are often seen in tumors, which depend only on mitosis—many examples are described in Chapter 17—but not in inherited conditions, in which they would have to pass through meiosis. The evolutionarily successful examples must be rare exceptions to this.

Rather than producing a radically new function, most gain-of-function mutations affect the way in which a gene or its product reacts to regulatory signals. The mutated gene or its product may be unable to respond to negative regulatory signals. The gene may be expressed at the wrong time, in the wrong tissue, at the wrong level, or in response to the wrong signal. The product may show an abnormal and pathogenic interaction with other cellular components. **Table 13.4** lists some examples of ways in which a gene product can show a gain of function. A rare case of an inherited point mutation conferring a novel function on a protein is the Pittsburgh allele at the *PI* locus (OMIM 107400; **Figure 13.25**). An amino

---

TABLE 13.4 MECHANISMS OF GAIN-OF-FUNCTION MUTATIONS

Malfunction	Gene	OMIM no.	Disease	Comments
Overexpression	*NR0B1*	300018	male-to-female sex reversal	gene duplications cause overexpression and sex reversal
Acquire new substrate	*PI* (Pittsburgh allele)	107400	lethal bleeding disorder	a rare gain-of-function allele of the α_1-antitrypsin gene (Figure 13.25)
Ion channel open inappropriately	*SCN4A*	168300	paramyotonia congenita of von Eulenburg	specific mutations in this sodium channel gene delay closing
Protein aggregation	*HD*	143100	Huntington disease	proteins with expanded polyglutamine runs form toxic aggregates
Receptor permanently on	*GNAS1*	174800	McCune–Albright disease	somatic mutations only; constitutional form would probably be lethal
Chimeric gene	*BCR–ABL*	151410	chronic myeloid leukemia	somatic mutations only

acid substitution in the active site converts an elastase inhibitor into a novel, constitutively active, antithrombin, resulting in a lethal bleeding disorder.

Diseases caused by gain of function of G-protein-coupled hormone receptors

The G-protein-coupled hormone receptors provide good examples of activating mutations. Many hormones exert their effects on target cells by binding to the extracellular domains of transmembrane receptors. Binding of ligand to the receptor causes the cytoplasmic tail of the receptor to catalyze the conversion of an inactive (GDP-bound) G-protein into an active (GTP-bound) form, and this relays the signal further by stimulating ion channels or enzymes such as adenylyl cyclase. Some mutant receptors are constitutively active: they fire (signal to downstream effectors) even in the absence of ligand:

- Familial male precocious puberty (OMIM 176410: onset of puberty by age 4 years in affected boys) is found with a constitutively active luteinizing hormone receptor.

- Autosomal dominant thyroid hyperplasia can be caused by an activating mutation in the thyroid-stimulating hormone receptor (see OMIM 275200).

- The bone disorder Jansen metaphyseal chondrodystrophy (OMIM 156400) can be caused by a constitutively active parathyroid hormone receptor.

Allelic homogeneity is not always due to a gain of function

We stated above that allelic heterogeneity is normally a hallmark of loss-of-function phenotypes. However, it is not safe to assume that any condition that shows allelic homogeneity is caused by a gain of function. There are other possible explanations:

- In some diseases, the phenotype is very directly related to the gene product itself, rather than being a more remote consequence of the genetic change. The disease may then be defined in terms of a particular variant product, as in sickle-cell anemia.

- Some specific molecular mechanism may make a certain sequence change in a gene much more likely than any other change—for example the $(CGG)_n$ expansion in fragile X syndrome.

- There may be a **founder effect**. For example, certain disease mutations are common among Ashkenazi Jews. Present-day Ashkenazi Jews are descended from a fairly small number of founders. If one of those founders carried a recessive allele, it may be found at high frequency in the present Ashkenazi population.

- Selection favoring heterozygotes enhances founder effects and often results in one or a few specific mutations being common in a population.

Loss-of-function and gain-of-function mutations in the same gene will cause different phenotypes

We began this section on molecular pathology by emphasizing the distinction between loss-of-function and gain-of-function mutations. However, it is possible to see both types of mutation in the same gene. When this happens, the resulting phenotypes are often very different.

As described above, loss-of-function mutations in the *PAX3* gene cause the developmental abnormality Waardenburg syndrome type 1 (see Figure 13.23). A totally different phenotype is seen when an acquired chromosomal translocation creates a novel chimeric gene by fusing *PAX3* to another transcription factor gene, *FKHR*, in a somatic cell. The gain of function of this hybrid transcription factor causes the development of a tumor, alveolar rhabdomyosarcoma (OMIM 268220).

Another striking example concerns the *RET* gene. *RET* encodes a transmembrane receptor tyrosine kinase. Binding of ligand (glial cell line-derived neurotrophic factor, or GDNF) to the extracellular domain induces dimerization of the receptors. This activates tyrosine kinase modules in the cytoplasmic domain of the receptors, which then transmit the signal into the cell (see Figure 4.15, p. 110).

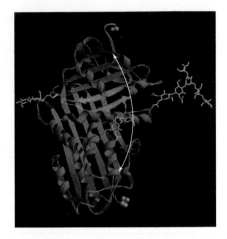

Figure 13.25 A mutation causing the α₁-antitrypsin molecule to gain a novel function. α₁-Antitrypsin is an inactivator of specific proteases. The protease cleaves the peptide bond between residues 358 and 359 in the α₁-antitrypsin molecule. As a result, the two residues (shown as green balls) spring 65 Å apart, as shown by the arrows here. The conformational change traps and inactivates the protease. In the wild-type allele, residues 358 and 359 are methionine and serine, respectively. This creates a substrate for elastase; thus, the wild-type α₁-antitrypsin functions as an anti-elastase agent. In the Pittsburgh variant a missense mutation, p.M358R, replaces the methionine with arginine. The structure, and the effect of cleavage of the peptide bond between residues 358 and 359, are unaltered, but now the specificity is for thrombin rather than elastase. Thus, the Pittsburgh variant has no anti-elastase activity, but has become a powerful antithrombin agent. The result is a lethal bleeding disorder. Orange and blue mark the parts of the α₁-antitrypsin molecule N-terminal and C-terminal of the cleavage site; carbohydrate residues are shown in gray. [Image S3D00427 from University of Geneva ExPASy molecular biology Web server, http://www.expasy.ch/]

A variety of loss-of-function mutations in the *RET* gene are one cause of Hirschsprung disease (OMIM 142623; absence of enteric ganglia in the bowel). These include frameshifts, nonsense mutations, and amino acid substitutions that interfere with the post-translational maturation of the RET protein. Certain very specific missense mutations in the *RET* gene are seen in a totally different set of diseases, familial medullary thyroid carcinoma (OMIM 155240) and the related but more extensive multiple endocrine neoplasia type II (OMIM 162300 and 171400). These are gain-of-function mutations, producing receptor molecules that react excessively to ligand or are constitutively active and dimerize even in the absence of ligand. Curiously, some people with missense mutations affecting cysteines 618 or 620, which are important for receptor dimerization, suffer from both thyroid cancer and Hirschsprung disease—simultaneous loss and gain of function. This reminds us that loss of function and gain of function are not always simple scalar quantities: mutations may have different effects in the different cell types in which a gene is expressed.

Table 13.5 lists several cases in which loss-of-function and gain-of-function mutations in a single gene result in different diseases. Usually, the gain-of-function mutant produces a qualitatively abnormal protein, but simple dosage effects can occasionally be pathogenic in both directions: increased and decreased gene dosage are both pathogenic but cause different diseases. The peripheral myelin protein gene *PMP22* is an example. Unequal crossovers between repeat sequences on chromosome 17p11 create duplications or deletions of a 1.5 Mb region that contains the *PMP22* gene (**Figure 13.26**). Heterozygous carriers of the deletion or duplication have one copy or three copies, respectively, of this gene. People who have only a single copy suffer from hereditary neuropathy with pressure palsies or tomaculous neuropathy (OMIM 162500), whereas people with three copies have a clinically different neuropathy, Charcot–Marie–Tooth disease 1A (CMT1A; OMIM 118220). These two diseases can also be caused by point mutations in the *PMP22* gene. Regardless of the molecular event, the cause of the pathology is changed activity of that one gene.

TABLE 13.5 EXAMPLES OF GENES IN WHICH LOSS-OF-FUNCTION AND GAIN-OF-FUNCTION MUTATIONS CAUSE DIFFERENT DISEASES					
Gene	Location	Loss (–) or gain (+)	Diseases	Symbol	OMIM no.
PAX3	2q35	–	Waardenburg syndrome type 1	WS1	193500
		+	Alveolar rhabdomyosarcoma	RMS2	268220
RET	10q11.2	+	Multiple endocrine neoplasia type IIA	MEN2A	171400
		+	Multiple endocrine neoplasia type IIB	MEN2B	162300
		+	Familial medullary thyroid carcinoma	FMTC	155240
		–	Hirschsprung disease	HSCR	142623
PMP22	17p11.2	–	Charcot–Marie–Tooth neuropathy type 1A	CMT1A	118220
		+	Tomaculous neuropathy	HNPP	162500
GNAS1	20q13.2	–	Albright hereditary osteodystrophy	PHP1A	103580
		+	McCune–Albright syndrome	MAS	174800
AR	Xq12	–	Testicular feminization syndrome	TFM	300068
		+	Spinobulbar muscular atrophy	SBMA	313200

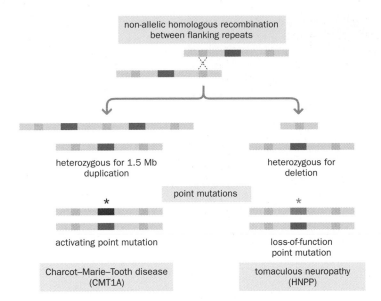

Unequal crossovers between flanking repeats (yellow boxes) generate duplications or deletions of a 1.5 Mb region containing the *PMP22* gene (blue box). Duplication causes Charcot–Marie–Tooth disease; deletion causes tomaculous neuropathy. The same two diseases can result from point mutations that either increase the activity of the *PMP22* gene or inactivate it.

13.5 THE QUEST FOR GENOTYPE–PHENOTYPE CORRELATIONS

Given the complexity of genetic interactions, it is not surprising that molecular pathology is a very imperfect science. The greatest successes so far have been in understanding cancer and hemoglobinopathies: for cancer, the phenotype to be explained—uncontrolled cell proliferation—is relatively simple, whereas hemoglobinopathies are a very direct result of abnormalities in an abundant and readily accessible protein. For most genetic diseases, the clinical features are the end result of a long chain of causation, and the holy grail of molecular pathology, genotype–phenotype correlation, will always be elusive. In reality even simple Mendelian diseases are not simple at all. The reviews on this topic by Scriver & Waters and by Weatherall are strongly recommended further reading.

The phenotypic effect of loss-of-function mutations depends on the residual level of gene function

DNA sequence changes can cause varying degrees of loss of function. Many amino acid substitutions have little or no effect, whereas some mutations totally abolish the function. A mutation may be present in one or both copies of a gene. People with autosomal recessive conditions are often **compound heterozygotes**—that is, they have different mutations in each allele. If both mutations cause a loss of function, but to differing degrees, the least severe allele will dictate the level of residual function. **Figure 13.27** represents four possible relations between the level of residual gene function and the clinical phenotype. Somebody heterozygous for a normal allele and one that is completely nonfunctional will have a 50% overall level of residual function. The result could be either a recessive

Figure 13.27 Four possible relationships between loss of gene function and clinical phenotype. The bars show the overall level of function produced by the combined effects of two alleles of the gene. (A) This condition, with 50% of residual gene function causing no effect, will be a simple recessive. (B) This condition will be dominant because of haploinsufficiency; loss of 50% of gene function causes the disease. (C) This condition is recessive, but the severity depends on the level of residual function, so that there is a genotype–phenotype correlation. (D) If the clinical consequences are very different, depending on the degree of residual function, the results may be described as different syndromes (A and A + B), and they may have different modes of inheritance, as here. Specific examples are discussed in the text.

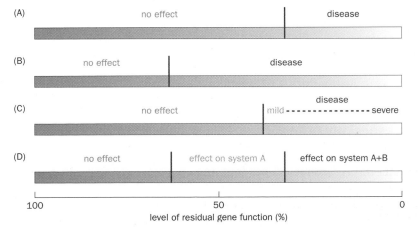

TABLE 13.6 CONSEQUENCES OF DECREASING FUNCTION OF HYPOXANTHINE GUANINE PHOSPHORIBOSYL TRANSFERASE

HPRT activity (% of normal)	Phenotype
>60	no disease
8–60	gout; no neurological problems (Kelley–Seegmiller syndrome)
1.6–8	gout plus variable neurological signs (clumsiness, choreoathetosis); normal intelligence
1.4–1.6	Lesch–Nyhan syndrome, but with normal intelligence
<1.4	Lesch–Nyhan syndrome (OMIM 300322): spasticity, choreoathetosis, self-mutilation, mental retardation

HPRT is an X-linked gene (OMIM 308000); the phenotype in males directly reflects the activity of their single *HPRT* gene.

(Figure 13.27A) or a dominant (Figure 13.27B) condition, depending on whether 50% activity is sufficient for normal function.

Mutations in the X-linked dystrophin gene provide an example of the situation shown in Figure 13.27C, in which the severity of disease correlates well with the extent of residual activity. Complete loss of gene function in affected males causes the severe Duchenne muscular dystrophy (DMD; OMIM 310200), whereas a smaller degree of loss causes a similar but milder condition, Becker muscular dystrophy (BMD; OMIM 300376). Despite their different names, both have the same cause and the same pathology; they differ only in severity. The reason for the difference was shown in Figure 13.16. Mutations in the X-linked hypoxanthine guanine phosphoribosyl transferase (*HPRT*) gene show a similar but clinically more marked series (**Table 13.6**). Thus, in these cases there is a clear genotype–phenotype correlation.

Sometimes different amounts of residual gene function can give rise to different clinical conditions that may even have different modes of inheritance (Figure 13.27D), for example:

- Mutations in the *DTDST* sulfate transporter gene that cause different degrees of loss of function cause autosomal recessive skeletal dysplasias that have been given different names depending on the severity: diastrophic dysplasia (OMIM 226600), multiple epiphyseal dysplasia 4 (OMIM 226900), atelosteogenesis II (OMIM 256050), and achondrogenesis type 1B (OMIM 600972). Extracellular matrix is rich in sulfated proteoglycans such as heparan sulfate and chon-droitin sulfate, and defects in sulfate transport interfere with skeletal development.

- Simple loss-of-function mutations in the *KVLQT1* K$^+$ channel have no effect in heterozygotes. However, a dominant-negative mutation in the same gene reduces the overall function to about 20% of normal and produces the dominantly inherited Romano–Ward syndrome (OMIM 192500: cardiac arrhythmia). Total loss of function in people homozygous for loss-of-function mutations causes the more severe recessive Jervell and Lange-Nielsen syndrome (OMIM 220400: heart problems and hearing loss).

- In the *COL1A1* or *COL1A2* genes that encode type I collagen, mutations in the same gene can produce two or more dominant conditions, the milder one by simple haploinsufficiency, and more severe forms through dominant-negative effects (see Figure 13.24).

Genotype–phenotype correlations are especially poor for conditions caused by mitochondrial mutations

Mutations in the mitochondrial DNA have rather unpredictable effects, both in terms of whether or not they will make somebody ill, and in terms of what disease they will cause. A given mtDNA sequence change may be seen in patients with several different diseases, and patients with the same mitochondrial disease may have different mutations in their mtDNA. In addition, because cells contain thousands of mtDNA molecules, they can be *homoplasmic* (every mtDNA molecule is the same) or *heteroplasmic* (a mixed population of normal and mutant mitochondrial DNA). Unlike mosaicism, heteroplasmy can be transmitted from mother to child through a heteroplasmic egg. Egg cells contain more than 100,000 mitochondria, so every child of an affected mother will inherit at least some of her mutant mitochondria even if the mother is heteroplasmic. However, mitochondrial diseases often show very reduced penetrance (see Figure 3.10, p. 68).

Leber hereditary optic neuropathy (OMIM 535000: sudden irreversible loss of vision) illustrates some of the problems. Eighteen different point mutations in mtDNA have been associated with this condition; 5 of these have a sufficiently serious effect to cause the disease on their own, and the other 13 have a contributory effect. Fifty per cent of affected people have a g.11778G→A substitution. Most of these patients are homoplasmic, but about 14%, no less severely affected, are heteroplasmic. Even in homoplasmic families the condition is highly variable; penetrance overall is 33–60%, and 82% of affected individuals are male.

There are several possible reasons for this poor correlation:

- Mitochondrial DNA is much more variable than nuclear DNA, and some syndromes may depend on the combination of the reported mutation with other unidentified variants.

- Some mitochondrial diseases seem to be of a quantitative nature: small mutational changes accumulate that reduce the energy-generating capacity of the mitochondrion and, at some threshold deficit, clinical symptoms appear.

- Heteroplasmy can be tissue-specific, and the tissue that is examined (typically blood or muscle) may not be the critical tissue in the pathogenesis.

- Early in germ-line development, cells go through a stage when they contain only a small number of mitochondria (the mitochondrial bottleneck). In a heteroplasmic mother, random sampling error at this stage means that the proportion of normal to mutant mitochondria may be significantly different before and after the bottleneck. Thus, the degree of heteroplasmy can vary widely between mother and child.

- Duplications and deletions of the mitochondrial genome often evolve over the years within a single affected individual; similarly, in heteroplasmic people, the proportion of mutant mitochondria may change over time.

- Many mitochondrial functions are encoded by nuclear genes, so that nuclear variation can be an important cause or modifier of mitochondrial phenotype.

The MITOMAP database of mitochondrial mutations (http://www.mitomap.org) has a good general discussion, and also extensive data tables showing just how great is the challenge of predicting phenotypes.

Variability within families is evidence of modifier genes or chance effects

Many Mendelian conditions are clinically variable even between affected members of the same family who carry exactly the same mutation. Intrafamilial variability must be caused by some combination of the effects of other unlinked genes (modifier genes) and environmental effects (including chance events). Humans are typical of a natural population, showing greater genetic diversity and a wider variety of environments than in laboratory animals, so it is not surprising that genotype–phenotype correlations are much looser in humans than in laboratory mice. Phenotypes depending on haploinsufficiency are especially sensitive to the effects of modifiers. Waardenburg syndrome is a typical example: Figure 13.23 shows the evidence that this dominant condition is caused by haploinsufficiency, and Figure 3.17 shows typical intrafamilial variation.

Genotype–phenotype correlations would become significantly more precise if the genotypes took account of modifiers as well as the primary mutation. However, identification and characterization of modifier genes in humans is extremely challenging. Given the inevitable limits of genetic analysis in humans, the most fruitful approach may come through the identification of all components of the biochemical pathway involved in a disease.

CONCLUSION

In this chapter, we have described the many ways in which the genomes of two humans can differ. Principally, these are through single nucleotide polymorphisms, variable numbers of tandem repeats, and large-scale copy number variants. Studies of these differences are directly relevant to understanding many features of human populations. The structure of populations in terms of inbreeding and reproductive isolation, the relationship of different populations to each other, the geographical and historical origins of populations—all these aspects are illuminated by studies of genetic variation.

When we come to studies of health and disease, it is not quite so straightforward. Connecting sequence variants with health and disease requires approaching the question from both ends. We have seen how one can try to decide whether a particular variant may be pathogenic or not. In some cases the answer is clear, but all too often simply looking at the DNA sequence does not give a clear answer. Even when we can be confident that a sequence change would be pathogenic, we can very rarely predict the actual clinical symptoms it would produce. To relate DNA to disease we need also to start from the disease end of the connection. Studying genetic diseases allows us to identify the underlying pathogenic sequence variants. This approach has been hugely successful with the Mendelian diseases, as the next chapters demonstrate. Given suitable families, it is relatively easy to identify the chromosomal location of the gene underlying a Mendelian character (Chapter 14), which usually allows the researcher to go on to identify the relevant gene (Chapter 16). Often it remains puzzling why a malfunction in a particular gene should cause that particular set of symptoms, but at least the pathogenic variants can usually be identified unambiguously.

Many genetic variants do not cause Mendelian conditions but may nevertheless have an influence on health. Genetic susceptibility factors are important determinants, though not the sole ones, of many common diseases. Linking specific variants to specific disease susceptibilities has been a very difficult task that is still far from fully accomplished. Chapters 15 and 16 review progress in that area. Hopefully we will eventually be in a position to derive useful predictive health information from analysis of a person's genome sequence—which will raise a whole series of issues about genetic testing and population screening. These are covered in Chapters 18 and 19.

FURTHER READING

Types of variation between human genomes

Bentley DR, Balasubramanian S, Swerdlow HP et al. (2008) Accurate whole-genome sequencing using reversible terminator chemistry. *Nature* 456, 53–59. [Mainly a technical account of using the Illumina technology to sequence an individual's genome.]

DbSNP. www.ncbi.nlm.nih.gov/projects/SNP [The 12 million entries in Build 126 include some rare pathogenic variants and some changes that are more complex than single nucleotide polymorphisms.]

Decipher database of pathogenic large-scale variants. http://decipher.sanger.ac.uk/

Den Dunnen JT & Antonarakis SE (2001) Nomenclature for the description of human sequence variations. *Hum. Genet.* 109, 121–124.

Horaitis O, Talbot CC Jr, Phommarinh M et al. (2007) A database of locus-specific databases. *Nat. Genet.* 39, 425. [A manifesto for locus-specific mutation databases.]

Jakobsson M, Scholz SW, Scheet P et al. (2008) Genotype, haplotype and copy-number variation in worldwide human populations. *Nature* 451, 998–1003. [Analysis of SNPs and CNVs in individuals from 29 populations.]

Kidd JM, Cooper GM, Donahue WF et al. (2008) Mapping and sequencing of structural variation from eight human genomes. *Nature* 453, 56–64. [A systematic study of variants 8 kb and larger in individuals from Asia, Europe, and Africa.]

Korbel JO, Urban AE, Affourtit JP et al. (2007) Paired-end mapping reveals extensive structural variation in the human genome. *Science* 318, 420–426. [Results of using a method of detecting CNVs down to 3 kb in size.]

Levy S, Sutton G, Ng PC et al. (2007) The diploid genome sequence of an individual human. *PLoS Biol.* 5, e254. [Craig Venter's sequence.]

Li JZ, Absher DM, Tang H et al. (2008) Worldwide human relationships inferred from genome-wide patterns of variation. *Science* 319, 1100–1104. [Using SNP data to infer population relationships.]

Lukusa T & Fryns JP (2008) Human chromosome fragility. *Biochim. Biophy. Acta* 1779, 3–16. [A review of chromosome fragile sites.]

Mutation databases. http://www.genomic.unimelb.edu.au/mdi/ The Human Genome Variation Society Mutation Database Initiative lists details and links for many databases, including http://www.hgvbaseg2p.org [Human Genome Variation

Database], and http://archive.uwcm.ac.uk/uwcm/mg/hgmd0.html [Human Gene Mutation Database], http://www.mitomap.org [database of mitochondrial mutations.]

Mutation nomenclature. http://www.hgvs.org/mutnomen/ [Guidelines on naming any DNA sequence change.]

Perry GH, Dominy NJ, Claw KG et al. (2007) Diet and the evolution of human amylase gene copy-number variation. *Nat. Genet.* 39, 1256–1260. [Example of a copy-number variant that affects the phenotype and has been subject to selection.]

Redon R, Ishikawa S, Fitch KR et al. (2006) Global copy number variation in the human genome. *Nature* 444, 444–454. [A major study of large-scale variation in the HapMap sample.]

TCAG Database of Genomic Variants. http://projects.tcag.ca/variation/ [Large-scale variants found in healthy people.]

Wang J, Wang W, Li R et al. (2008) The diploid genome sequence of an Asian individual. *Nature* 456, 60–65.

Wheeler DA, Srinivasan M, Ehgolm M et al. (2008) The complete genome of an individual by massively parallel DNA sequencing. *Nature* 452, 872–877. [James Watson's sequence.]

Wong KK, deLeeuw RJ, Dosanjh NS et al. (2007) A comprehensive analysis of common copy-number variations in the human genome. *Am. J. Hum. Genet.* 80, 91–104. [Variants 40 kb or larger among 95 individuals from 16 different ethnic groups.]

Failure to repair DNA damage is the ultimate cause of much genetic variation

Barnes DE & Lindahl T (2004) Repair and genetic consequences of endogenous DNA base damage in mammalian cells. *Annu. Rev. Genet.* 38, 445–476. [A comprehensive review.]

Buchwald M (1995) Complementation: one or more groups per gene? *Nat. Genet.* 11, 228–230. [Valuable discussion of the value and limitation of complementation testing, although the direct questions posed have long since been resolved.]

Friedberg EC (2003) DNA damage and repair. *Nature* 421, 436–440. [A readable historically oriented account of how our understanding of DNA damage and repair developed.]

Klein HL (2008) DNA endgames. *Nature* 455, 740–741. [A brief review of how DNA double-strand breaks are repaired.]

Kraemer KH, Patronas NJ, Schiffmann R et al. (2007) Xeroderma pigmentosum, trichothiodystrophy and Cockayne syndrome: a complex genotype-phenotype relationship. *Neuroscience* 145, 1388–1396. [The nucleotide excision repair pathway and its defects.]

McKinnon PJ & Caldecott KW (2007) DNA strand break repair and human genetic disease. *Annu. Rev. Genomics Hum. Genet.* 8, 37–55. [Reviews the repair of single-strand and double-strand breaks, and the diseases resulting from defects.]

Schumacher B, Garinis GA & Hoeijmakers JHJ (2007) Age to survive: DNA damage and ageing. *Trends Genet.* 24, 77–85.

Wang W (2007) Emergence of a DNA-damage response network consisting of Fanconi anaemia and BRCA proteins. *Nat. Rev. Genet.* 8, 735–748.

Pathogenic DNA variants

Bucciantini M, Giannoni E, Chiti F et al. (2002) Inherent cytotoxicity of polypeptide aggregates suggests a common origin of protein misfolding diseases. *Nature* 416, 507–511.

Grantham RF (1974) Amino acid difference formula to help explain protein evolution. *Science* 185, 862–864. [The Grantham Matrix quantifies the effect of amino acid substitutions.]

Groman JD, Hefferon TW, Casals T et al. (2004) Variation in a repeat sequence determines whether a common variant of the Cystic Fibrosis Transmembrane Conductance Regulator gene is pathogenic or benign. *Am. J. Hum. Genet.* 74, 176–179. [Effect of the 5T/7T/9T and (TG)$_n$ variants on splicing of the *CFTR* mRNA.]

Hardy J & Orr H (2006) The genetics of neurodegenerative disease. *J. Neurochem.* 97, 1690–1699.

Holbrook JA, Neu-Yilik G, Hentze MW & Kulozik AE (2004) Nonsense-mediated decay approaches the clinic. *Nat. Genet.* 36, 801–808.

Inoue K, Khajavi M, Ohyama T et al. (2004) Molecular mechanism for distinct neurological phenotypes conveyed by allelic truncating mutations. *Nat. Genet.* 36, 361–369. [SOX10 and MPZ as examples of variable molecular pathology due to NMD.]

Isken O & Maquat LE (2008) The multiple lives of NMD factors: balancing roles in gene and genome regulation. *Nat. Rev. Genet.* 9, 699–712. [A detailed review of the roles of nonsense-mediated decay in cells.]

Komar AA (2007) SNPs—silent but not invisible. *Science* 315, 466–467. [Discussion of data showing that a silent polymorphism in the *MDR1* gene changes substrate specificity.]

Mirkin SM (2007) Expandable DNA repeats and human disease. *Nature* 447, 932–940. [A review of possible mechanisms for the instability of dynamic mutations.]

Orr HT & Zoghbi HY (2007) Trinucleotide repeat disorders. *Annu. Rev. Neurosci.* 30, 575–621. [An authoritative overview of the diseases and their pathogenesis.]

Pearson CE, Edamura KN & Cleary JD (2005) Repeat instability: mechanisms of dynamic mutations. *Nat. Rev. Genet.* 6, 729–742. [Not easy reading, but a lot of data and ideas.]

Wang GS & Cooper TA (2007) Splicing in disease: disruption of the splicing code and the decoding machinery. *Nat. Rev. Genet.* 8, 749–761. [Many examples of pathogenic effects due to disturbances of splicing.]

Molecular pathology: understanding the effect of variants

Karniski LP (2001) Mutations in the diastrophic dysplasia sulfate transporter (DTDST) gene: correlation between sulfate transport activity and chondrodysplasia phenotype. *Hum. Mol. Genet.* 10, 1485–1490.

Lester HA & Karschin A (2000) Gain of function mutants: ion channels and G protein-coupled receptors. *Annu. Rev. Neurosci.* 23, 89–125. [Many detailed examples, not all from humans.]

Marini JC, Forlino A, Cabral WA et al. (2007) Consortium for osteogenesis imperfecta mutations in the helical domain of type I collagen: regions rich in lethal mutations align with binding sites for integrins and proteoglycans. *Hum. Mutat.* 28, 209–221. [Analysis of genotype–phenotype relationship in more than 800 mutations.]

Owen MC, Brennan SO, Lewis JH & Carrell RW (1983) Mutation of antitrypsin to antithrombin: α-1-antitrypsin Pittsburgh (358 met-to-arg), a fatal bleeding disorder. *N. Engl. J. Med.* 309, 694–698.

Rauch F & Glorieux FH (2004) Osteogenesis imperfecta. *Lancet* 363, 1377–1385. [A clinically oriented overview.]

Shenker A, Laue L, Kosugi S et al. (1993) A constitutively activating mutation of the luteinizing hormone receptor in familial male precocious puberty. *Nature* 365, 652–654.

Wollnik B, Schroeder BC, Kubisch C et al. (1997) Pathophysiological mechanisms of dominant and recessive *KVLQT1* K$^+$ channel mutations found in inherited cardiac arrhythmias. *Hum. Mol. Genet.* 6, 1943–1949.

Zlotogora J (2007) Multiple mutations responsible for frequent genetic diseases in isolated populations. *Eur. J. Hum. Genet.* 15, 272–278. [Limited allelic heterogeneity due to the combination of founder effects and selection.]

The quest for genotype–phenotype correlations

Carrell RW & Lomas DA (2002) α-1-Antitrypsin deficiency—a model for conformational diseases. *N. Engl. J. Med.* 346, 45–53.

Khrapko K (2008) Two ways to make an mtDNA bottleneck. *Nat. Genet.* 40, 134–135. [A brief review of the mitochondrial bottleneck.]

Manié S, Santoro M, Fusco A & Billaud M (2001) The RET receptor: function in development and dysfunction in congenital malformation. *Trends Genet.* 17, 580–589.

Pasini B, Ceccherini I & Romeo G (1996) RET mutations in human disease. *Trends Genet.* 12, 138–144.

Scriver CR & Waters PJ (1999) Monogenic traits are not simple: lessons from phenylketonuria. *Trends Genet.* 15, 267–272.

Wallace DC, Lott MT, Brown MD & Kerstann K (2001) Mitochondria and neuro-ophthalmologic diseases. In The Metabolic & Molecular Bases Of Inherited Disease, 8th ed. (CR Scriver, AL Beaudet, WS Sly, D Valle eds), pp 2425–2509. McGraw Hill.

Weatherall DJ (2001) Phenotype–genotype relationships in monogenic disease: lessons from the thalassaemias. *Nat. Rev. Genet.* 2, 245–255.

Weatherall DJ, Clegg JB, Higgs DR & Wood WG (2001) The Hemoglobinopathies. In The Metabolic & Molecular Bases Of Inherited Disease, 8th ed. (CR Scriver, AL Beaudet, WS Sly, D Valle eds), pp 4571–4636. McGraw Hill.

Chapter 14

Genetic Mapping of Mendelian Characters

KEY CONCEPTS

- Genetic mapping is needed to localize genes underlying observable phenotypes, which cannot be identified by searching the DNA sequence of the human genome.

- During meiosis there is normally at least one crossover in each arm of each pair of synapsed homologous chromosomes. Pedigree analysis reveals the result of these crossovers as genetic recombination.

- Genes that are physically close together on the same chromatid tend to travel together during meiosis; genes that are farther apart are liable to separate through crossovers between homologous chromosomes.

- The genetic distance between two loci is defined as the frequency with which recombination separates them during meiosis. This is often measured as the fraction of gametes that are recombinant. A recombinant fraction of 0.01 corresponds to 1% recombination and a genetic distance of 1 centimorgan (cM).

- Recombinant fractions never exceed 0.5, regardless of whether the two loci concerned are on the same or different chromosomes.

- Genetic maps are co-linear with physical maps, but distances do not necessarily correspond because the probability of recombination per megabase of DNA is not uniform along the length of a chromosome, and differs between males and females.

- To map human traits, it is necessary to assemble family pedigrees containing a sufficient number of informative meioses.

- Genes are mapped by checking a collection of pedigrees for co-segregation of the relevant phenotype and the alleles of genetic markers. The main markers used are short tandem repeat polymorphisms (STRPs or microsatellites) and single nucleotide polymorphisms (SNPs).

- The statistical criterion for assessing linkage is the logarithm of the odds or lod score. In simple situations, the threshold of significance for linkage is a lod score of +3 and against linkage −2.

- Human pedigrees often have complex structures in which recombinants cannot be identified through simple inspection, and so software programs are needed that calculate the relative likelihood of the observed data on the competing hypotheses of linkage or no linkage.

- Linkage analysis can be more efficient if data for more than two loci are analyzed simultaneously, an approach known as multipoint mapping.

- Multipoint linkage analysis has been used on large family pedigrees to construct marker framework maps, which can then be used to pinpoint the location of a disease gene.

- Autozygosity mapping is a powerful method of mapping genes for rare recessive conditions. It works by hunting for ancestral chromosome segments shared by affected people in large consanguineous families.

- The resolution of linkage analysis depends on the number of meioses analyzed. The ultimate resolution is obtained by typing individual sperm.

- The methods described above have been very successful for mapping Mendelian conditions, but loci conferring susceptibility to common complex diseases require different approaches, which are described in Chapter 15.

The finished sequence of the human genome is the ultimate physical map, allowing the chromosomal location of physical entities (such as DNA sequences and breakpoints) to be defined to the nearest nucleotide. However, nothing in the raw sequence would have allowed a researcher to locate the gene, *HD*, responsible for Huntington disease, or any other phenotypic character. Of course, the *HD* gene can now be located by using a genome browser, but this is only possible because the gene was previously localized and identified by other means, and that information was added to the annotation used by the browser. To locate genes responsible for a particular disease or any other phenotypic character, we still need a genetic map.

Genes defined through phenotypes can be identified in various ways (as will be described in Chapter 16), but the usual methods all involve first discovering the chromosomal location of the gene. Such genetic mapping can be performed for any phenotypic determinant. The first step is to collect a series of clinically well-phenotyped cases, which are then genotyped in the laboratory for a series of genetic markers. These genotypes are then analyzed to map the determinant. In this chapter we describe the basic principles of genetic mapping and their application to Mendelian characters. Mapping susceptibility factors for complex diseases, where there is genetic susceptibility but not a Mendelian inheritance pattern, is based on the same principles, but brings in extra complications, which are considered in Chapter 15.

14.1 THE ROLE OF RECOMBINATION IN GENETIC MAPPING

In principle, genetic mapping in humans is exactly the same as genetic mapping in any other sexually reproducing diploid organism. The aim is to discover how often two loci are separated by meiotic recombination.

Recombination is a normal part of every meiotic cell division. During prophase of meiosis I, pairs of homologous chromosomes synapse, and individual chromatids exchange segments (see Figure 2.6). This involves physically cutting and re-joining chromatids. The mechanisms that achieve this are quite complicated (**Box 14.1**). Recombination is a reciprocal process: equivalent segments are exchanged and nothing is gained or lost. **Gene conversion** is a non-reciprocal (one-way) process, in which a short (typically 1 kb) patch of a donor sequence replaces a recipient sequence, while the original sequence in the donor is retained. It can occur either as a concomitant of recombination, affecting sequences at the exact point of exchange, or independently of recombination, as an alternative outcome of the processes shown in Box 14.1.

Recombinants are identified by genotyping parents and offspring for pairs of loci

Consider a person who is heterozygous at two loci (*A* and *B*) and has the genotype $A_1A_2\,B_1B_2$. Suppose the alleles A_1 and B_1 in this person came from one parent, and A_2 and B_2 from the other. Any of that person's gametes that carries one of these parental combinations (A_1B_1 or A_2B_2) is **non-recombinant** for those two loci, whereas any gamete that carries A_1B_2 or A_2B_1 is **recombinant** (**Figure 14.1**). The proportion of gametes that are recombinant is the **recombination fraction** between the two loci *A* and *B*.

Recombination is normally observed by genotyping offspring rather than gametes (although some researchers have typed individual sperm by ultrasensitive PCR) and, in the context of human genetic mapping, it is normal to speak of a *person* being recombinant or non-recombinant. It is understood that we are really talking about one of the parental gametes that made the person. If there is any ambiguity about which parent's gamete is involved, it would be necessary to state this.

The recombination fraction is a measure of the genetic distance between two loci

If two loci are on different chromosomes, they will assort independently. In the pedigree shown in **Figure 14.2A**, the two loci are on different chromosomes.

BOX 14.1 RECOMBINATION AND GENE CONVERSION

Recombination is believed to involve the following steps (**Figure 1**):

- A double-strand break is made in one parental DNA molecule (yellow). Exonucleases strip back the 5′ ends of the broken strands, leaving single-stranded 3′ overhangs. The single-stranded ends are stabilized by cooperative binding of Rad51 or related proteins.
- The single strand invades the complementary double helix (red), initially forming a triple helix but eventually displacing one of the original complementary strands. DNA synthesis extends the invading strand for a kilobase or so (green). The displaced complementary strand pairs with the remaining strand of the cut molecule, which is extended by DNA synthesis (green).
- The two junctions linking the two double helices (known as Holliday junctions after Dr Robin Holliday, who first described their properties) are each resolved by cutting two strands and ligating the cut ends to the opposite partner.

Depending on the strands cut, the resulting double helices are recombinant or non-recombinant. However, within the region between the two Holliday junctions, things are more complicated. Recombination is a reciprocal process (an equal exchange between the two double helices), but in the region between the two junctions two mechanisms may cause a *non-reciprocal* transfer of sequence, known as **gene conversion**. First, both stretches of newly synthesized DNA (green) have been made by using the red strand as a template, rather than one using the red and the other the yellow strand. Second, if there were any sequence differences between the two original double helices, the hatched regions in the figure may contain mismatches. Mismatch repair enzymes will correct these by stripping back and resynthesizing one of the strands, chosen at random. The result of either of these processes is to patch short (typically 1 kb) stretches of sequence derived from one parental double helix into the other, without making any reciprocal replacement.

Recombination generally takes place only in prophase of the first division of meiosis, and normally involves truly allelic sequences. However, there are exceptions:

- Recombination happens in mitosis at maybe 10^{-4} times the frequency of meiotic recombination. Mitotic recombination is a significant mechanism in carcinogenesis and is discussed in Chapter 17.
- Many sequences occur more than once in the genome and so homologous but non-allelic recombination is not infrequent. In Chapter 13 we saw how recurrent microdeletions and

microduplications are produced by *non-allelic homologous recombination* between low-copy-number repeats spaced along the same chromosome.
- Nonhomologous recombination can sometimes also occur, presumably through chance malfunction of the recombination machinery. This can be a source of chromosomal rearrangements.
- Special recombination mechanisms rearrange immunoglobulin and T-cell receptor genes during the maturation of lymphocytes. These generate the diversity that allows an immune response to be mounted against virtually any antigen (see Chapter 4).

However, these exceptions do not affect the use of normal meiotic recombination for genetic mapping.

Figure 1 The stages of recombination and gene conversion. Note that here we show the individual strands of each double helix; Figure 14.3 shows chromatids, each of which contains an intact double helix. [From Alberts B, Johnson A, Lewis J et al. (2007) Molecular Biology of the Cell, 5th ed. Garland Science/Taylor & Francis LLC.]

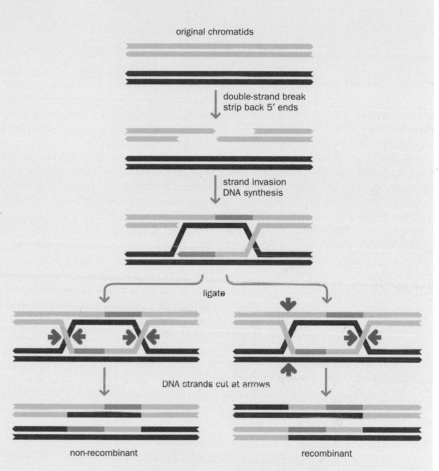

Consider spermatogenesis in individual II$_1$. At the end of meiosis, if a sperm receives the grandpaternal (red) allele for the upper locus, there is a 50% chance that it will receive the red allele and a 50% chance it will receive the blue (grandmaternal) allele at the lower locus. Thus, on average, 50% of the children will be recombinant and 50% non-recombinant for the paternal gamete. The recombination fraction is 0.5. In the pedigree shown in Figure 14.2B, the loci are **syntenic**; that is, they lie on the same chromosome. If there were no crossing over, every gamete would carry the same combination of alleles at the two loci as one of the parental chromosomes. Given that crossing over is a feature of every meiotic division, there is always the possibility that two syntenic loci will be separated by a crossover occurring in the interval between them. Thus, some proportion of the

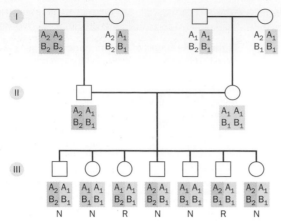

Figure 14.1 Recombinants and non-recombinants. In this family, there are two loci (A and B) at which alleles (A_1 and A_2, B_1 and B_2) are segregating. Colored boxes mark combinations of alleles that can be traced through the pedigree. In generation III, we can distinguish people who received non-recombinant (N; A_1B_1 or A_2B_2) or recombinant (R; A_1B_2 or A_2B_1) sperm from their father (II$_1$). Their mother (II$_2$) is homozygous at these two loci, and so we cannot identify which individuals in generation III developed from non-recombinant or recombinant oocytes.

offspring may receive a recombinant gamete, but unless the two loci are very far apart on the chromosome, we would expect only a minority of gametes to be recombinant. In this pedigree the recombination fraction is 0.1.

At the stage when crossing over occurs, the DNA has already been replicated and each chromosome consists of two sister chromatids. Only two of the four chromatids are involved in any particular crossover. A crossover that occurs between the positions of the two loci will create two recombinant chromatids and leave the two non-involved chromatids non-recombinant. Thus, one crossover generates 50% recombinants between loci flanking it.

Recombination will rarely separate loci that lie very close together on a chromosome, because only a crossover located precisely in the small interval between the two loci will create recombinants. Therefore, sets of alleles on the same small chromosomal segment tend to be transmitted as a block through a pedigree. Such a block of alleles is known as a **haplotype**. Haplotypes mark recognizable chromosomal segments that can be tracked through pedigrees and through populations when not broken up by recombination.

The farther apart two loci are on a chromosome, the more likely it is that a crossover will separate them. Thus, the recombination fraction is a measure of the distance between two loci. Recombination fractions define **genetic distance**. Two loci with a recombination fraction of 0.01, and thus 1% recombination between them, are defined as being 1 **centimorgan** (**cM**) apart on a genetic map (the M is capitalized because it refers to the name of TH Morgan).

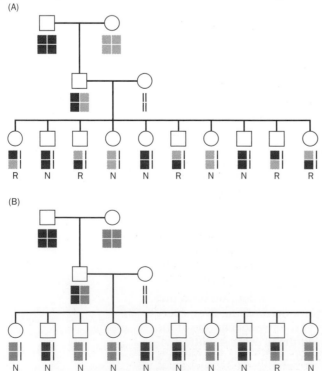

Figure 14.2 Independent assortment and linkage. In each pedigree, the upper and lower squares represent alleles at two loci, color-coded according to their parental origin. In each pedigree, individual II$_1$ is doubly heterozygous, and his partner is homozygous for a different allele (shown as a thin line). Using the logic of Figure 14.1, we can identify which individuals in generation III received a recombinant (R) or non-recombinant (N) gamete from II$_1$. (A) The two loci in this pedigree are on different chromosomes. They assort independently. There were five non-recombinant and five recombinant gametes, giving a recombination fraction of 0.5. (B) The two loci in this pedigree are a few megabases apart on the same chromosome. They show linkage. There were nine non-recombinant gametes and only one recombinant gamete. The recombination fraction is 0.1. We assume here that we already know the two loci to be linked. If this pedigree were being used to decide whether or not the two loci were linked, this pedigree would provide suggestive but not statistically significant evidence of linkage.

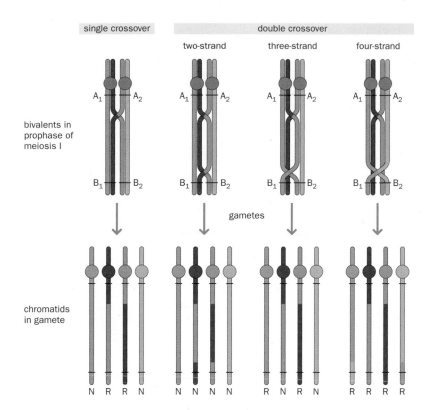

single crossover | double crossover

two-strand | three-strand | four-strand

bivalents in prophase of meiosis I

A_1 | A_2

B_1 | B_2

gametes

chromatids in gamete

N R R N | N N N N | R N R N | R R R R

Figure 14.3 Single and double crossovers. The figure shows the chromosomal events that determine whether a gamete is recombinant or non-recombinant for the two loci A and B. Each crossover involves two of the four chromatids of the two synapsed homologous chromosomes. One chromosome carries alleles A_1 and B_1, the other alleles A_2 and B_2. Chromatids in the gametes labeled N carry a parental combination of alleles (A_1B_1 or A_2B_2). Chromatids labeled R carry a recombinant combination (A_1B_2 or A_2B_1). Note that recombinant and non-recombinant are defined only in relation to these two loci. For example, in the result of the three-strand double crossover, the second chromatid from the left has been involved in crossovers—but it is non-recombinant for loci A and B because it carries alleles A_1 and B_1, a parental combination. A single crossover generates two recombinant and two non-recombinant chromatids (50% recombinants). The three types of double crossover occur in random proportions, so the average effect of a double crossover is to give 50% recombinants.

Recombination fractions do not exceed 0.5, however great the distance between two loci

A single recombination event produces two recombinant and two non-recombinant chromatids. When loci are well separated, there may be more than one crossover between them. Double crossovers can involve two, three, or all four chromatids, but the overall effect, averaged over all double crossovers, is to give 50% recombinants (**Figure 14.3**). Loci very far apart on the same chromosome might be separated by three or more crossovers. Again, the overall effect is to give 50% recombinants. Recombination fractions never exceed 0.5, however far apart the loci are.

Mapping functions define the relationship between recombination fraction and genetic distance

Because recombination fractions never exceed 0.5, they are not simply additive across a genetic map. If a series of loci, A, B, C, ..., are located at 5 cM intervals on a map, locus N may be 65 cM from locus A, but the recombination fraction between A and N will not be 65%. The mathematical relationship between recombination fraction and genetic map distance is described by the **mapping function**.

If crossovers occurred at random along a bivalent and had no influence on one another, the appropriate mapping function would be *Haldane's function*:

$$w = -\tfrac{1}{2}\ln(1 - 2\theta)$$
$$\text{or} \quad \theta = \tfrac{1}{2}[1 - \exp(-2w)]$$

where w is the map distance and θ the recombination fraction; as usual ln means logarithm to the base e, and exp means e to the power of. However, we know that the presence of one chiasma strongly inhibits the formation of a second chiasma nearby. This phenomenon is called **interference**. A variety of mapping functions exist that allow for varying degrees of interference. A widely used function for human mapping is *Kosambi's function*:

$$w = \tfrac{1}{4}\ln[(1 + 2\theta)/(1 - 2\theta)]$$
$$\text{or} \quad \theta = \tfrac{1}{2}[\exp(4w) - 1]/[\exp(4w) + 1].$$

On the basis of a very large set of linkage data, Broman & Weber in 2000 estimated the probability of a true double crossover between markers d cM apart as roughly $(0.0114d - 0.0154)^4$. For $d = 10$ cM this works out at only 0.001%, which is much lower than the theoretical chance of a double crossover between two markers 10 cM apart if there were no interference (10% × 10% = 1%) . The interested reader should consult Ott's book (see Further Reading) for a fuller discussion of mapping functions.

Interference makes true close double recombinants very unlikely, but gene conversion (see Box 14.1) can produce results that are indistinguishable from very close double recombinants. Only very short sequences are involved in gene conversion—of the order of 1 kb. This is far below the resolution of almost all genetic mapping, and so most gene conversion events have no influence on genetic maps. However, they could pose a problem for identifying conserved ancestral chromosome segments, as explained in Chapter 15 (see Section 15.4).

Chiasma counts give an estimate of the total map length

Each crossover during meiosis produces two recombinant and two non-recombinant chromatids, giving 50% recombination between flanking markers. As 1% recombination equates to 1 cM, one crossover therefore contributes 50 cM to the overall genetic map length. By counting chiasmata under the microscope and multiplying the number per cell by 50, we can estimate the total map length in centimorgans. Male meiosis can be studied relatively easily in testicular biopsies (**Figure 14.4A**). The average chiasma count is reported as 50.6 per cell, giving a map length of 2530 cM. Female meiosis I in humans takes place at 16–24 weeks of fetal life and so can only be observed in the ovaries of aborted female fetuses. Accurate chiasma counts were difficult to achieve until the application of fluorescent antibodies against MLH1, a mismatch repair protein that is part of the recombination machinery (Figure 14.4B). Chiasmata are more frequent in female meiosis, exemplifying Haldane's rule that the heterogametic sex (males in mammals and *Drosophila*, females in birds) has the lower chiasma count. The average autosomal count in ovaries from one aborted female fetus was 70.3, giving a map length of 3515 cM. The actual count varies both between individuals and between different meiotic cells in the same individual.

These cytologically derived map lengths can be compared with the estimates of 2590 cM (male) and 4281 cM (female, excluding the X chromosome) in the best current linkage map, obtained by Kong and colleagues by typing 5136 microsatellites in 1257 meioses in a collection of 146 Icelandic families. Estimates from genetic mapping necessarily represent an average over many individuals.

Recombination events are distributed non-randomly along chromosomes, and so genetic map distances may not correspond to physical distances

Physical maps show the order of features along the chromosome and their distance in kilobases or megabases; genetic maps show their order and the probability that they will be separated by recombination. Although the order of features should be the same on both maps, the distances will only correspond if the probability of recombination per megabase of DNA is constant for all chromosomal locations. In fact, recombination probabilities vary considerably according to sex and chromosomal location.

The distribution of recombination along a chromosome can be estimated either directly under the microscope (see Figure 14.4) or, with potentially much higher resolution, by relating genetic map distances to physical locations of markers. With the completion of the Human Genome Project, the physical location of microsatellite or SNP markers can be easily found in the human genome sequence. Both methods show a similar non-random pattern of recombination along chromosomes. There is more recombination toward the telomeres of chromosomes, especially in males, whereas centromeric regions have recombinants in females but not in males. **Figure 14.5** shows some examples.

The general trends shown in Figure 14.5 hide substantial local variations. The overall sex-averaged recombination rate is 1.22 cM per megabase (0.88 cM/Mb in males, 1.55 cM/Mb in females). This gives a useful rough rule of thumb of about 1 cM per megabase—but there are recombination deserts up to 5 Mb in length with sex-averaged recombination less than 0.3 cM/Mb, and recombination jungles with more than 3 cM/Mb. The most extreme deviation is shown by the

(A)

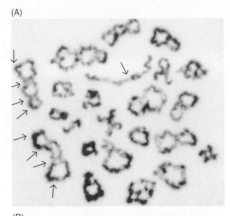

(B)

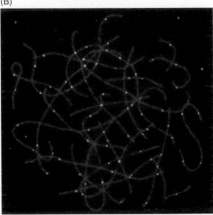

Figure 14.4 Crossovers in male and female meiosis. (A) Male meiosis late in prophase I. Chiasmata (arrows) mark the positions of crossovers within each bivalent as visualized under the microscope. (B) Female meiosis at mid-prophase I (pachytene stage). The bright dots of yellow fluorescent MLH1 antibody mark positions of crossovers. The same method can also be used to study male meiosis. [(A) courtesy of Maj Hultén, University of Warwick. (B) from Tease C, Hartshorne GM & Hultén MA (2002) Patterns of meiotic recombination in human fetal oocytes. *Am. J. Hum. Genet.* 70, 1469–1479. With permission from Elsevier.]

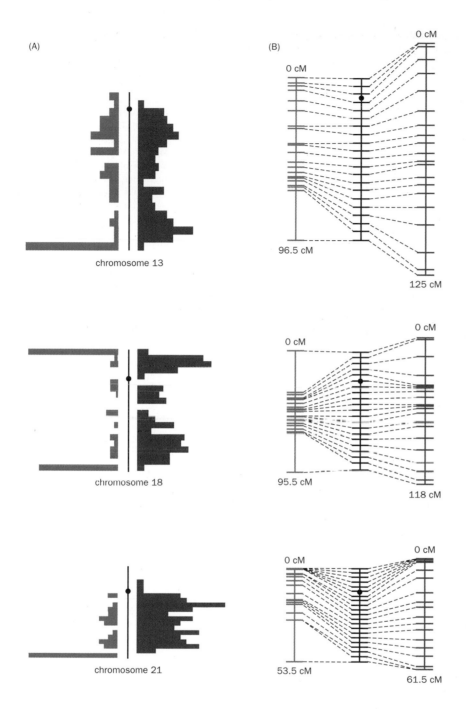

(A)

chromosome 13

chromosome 18

chromosome 21

(B)

0 cM

0 cM

96.5 cM

125 cM

0 cM

0 cM

95.5 cM

118 cM

0 cM

0 cM

53.5 cM

61.5 cM

Figure 14.5 The distribution of recombinants along the length of a chromosome is non-random and sex-specific. (A) Histograms illustrating the distributions of chiasmata in spermatocytes (blue) and MLH1 foci in oocytes (red) on chromosomes 13, 18, and 21. Each chromosome pair is divided into 5% length intervals. (B) The same crossover distribution patterns shown for each chromosome as recombination maps. These maps highlight the different patterns of crossover numbers and distributions in male and female germ cells and the consequent effect on the recombination map. [From Hultén MA & Tease C (2003) Genetic maps: direct meiotic analysis. In Encyclopedia Of The Human Genome (DN Cooper ed.). With permission from Macmillan Publishers Ltd.]

pseudoautosomal region at the tip of the short arms of the X and Y chromosomes. Males have an obligatory crossover within this 2.6 Mb region, so that it is 50 cM long. Thus, for this region in males 1 Mb = 19 cM, whereas in females 1 Mb = 2.7 cM. Uniquely, the Y chromosome, outside the pseudoautosomal region, has no genetic map because it is not subject to synapsis and crossing over in meiosis.

Two lines of evidence show that recombination is also highly non-random at a much finer, kilobase, level of resolution:

- When individuals are genotyped for a dense map of SNP markers, and the results from many different individuals are compared, it seems that our genome consists of a set of conserved ancestral segments separated by fairly sharp boundaries. This structure implies that, historically, very few recombination events have taken place within the ancestral blocks, whereas block boundaries are hotspots for recombination. Typical block sizes are 5–15 kb. These findings, from the International HapMap Project, are discussed in more detail in Chapter 15.

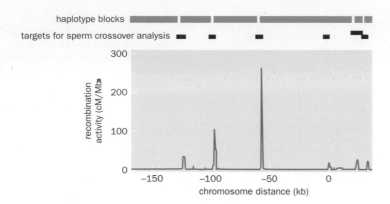

Figure 14.6 At the highest resolution, recombination is concentrated in small hotspots. DNA-level distribution of recombination in males in a 206 kb region of chromosome 1q42.3; 95% of all recombination occurs at highly localized hotspots, with average recombination frequencies up to 70 cM/Mb. Outside the hotspots, recombination is no more than 0.04 cM/Mb. Green bars mark relatively stable haplotype blocks where crossovers are infrequent. Black bars mark the short sequences that were selected for analysis of crossovers in individual sperm. [Adapted from Jeffreys AJ, Neumann R, Panayi M et al. (2005) *Nat. Genet.* 37, 601–606. With permission from Macmillan Publishers Ltd.]

- Recombination events can be observed directly at 1 kb or higher resolution by genotyping single sperm from suitably heterozygous men. Such work by Jeffreys and colleagues on two different chromosomal regions has clearly shown that almost all recombination occurs in scattered 1–2 kb hotspots (**Figure 14.6**). By and large (although not entirely), these hotspots correspond to the gaps between ancestral conserved segments. It is not possible to study female meiosis in the same way, but the general agreement with the block structure defined through population studies suggests that recombination in females must also be largely confined to the same small hotspots.

The restriction of recombination events to such very localized hotspots has major implications for studies of the mechanism of recombination. Tantalizingly, the DNA sequence at these points does not have any obvious special feature to explain this behavior. Identical DNA sequences can be hotspots in humans but not in chimpanzees. However, genetic mapping in general is not affected by this very fine-scale irregularity. The best current human genetic map has a resolution of only about 500 kb.

Individuals also vary in the average number of crossovers per meiosis, and if they do, then they will also have different genetic map lengths (**Figure 14.7**). Thus, genetic maps are always probabilistic and should not be over-interpreted. In fact, the same is true of physical maps. Although physical distances in any one individual's DNA can be stated with complete precision, we saw in Chapter 13 that human populations are polymorphic for numerous large deletions, insertions, and inversions. In the end, both genetic and physical maps are tools for achieving certain purposes, not absolute descriptions of human beings.

14.2 MAPPING A DISEASE LOCUS

The main purpose of genetic mapping is to locate the determinants of phenotypes. As mentioned above, these cannot be inferred by any inspection of the unannotated human genome sequence. Genetic mapping depends on estimating the recombination fraction between pairs of loci. This requires an individual who is heterozygous for both loci, as shown in Figure 14.1. People heterozygous for two different genetic diseases are extremely rare. Even if they can be found, they will probably have no children or be unsuitable for genetic analysis in some other way. Thus, it is seldom possible in humans to map diseases or other phenotypes against one another, as one would in *Drosophila*.

Mapping human disease genes depends on genetic markers

To map a human disease (or the locus determining any other uncommon phenotype), we need to collect families in which the character of interest is segregating, and then find some other Mendelian character that is also segregating in these same families, so that individuals can be scored as recombinant or nonrecombinant. A suitable character would have the following characteristics:

- It should show a clean pattern of Mendelian inheritance, preferably codominant so that the genotype can always be inferred from the phenotype.

- It should be scored easily and cheaply using readily available material (e.g. a mouthwash rather than a brain biopsy).

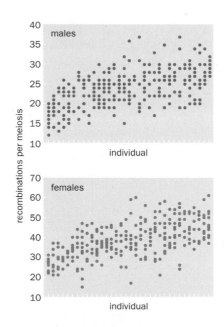

Figure 14.7 Individual differences in numbers of recombinations per meiosis. Cheung and colleagues studied an average of eight meioses per individual in 34 women and 33 men by genotyping members of the Centre d'Étude du Polymorphisme Humain (CEPH) Utah families for 6324 SNP markers. Each vertical line of dots shows the number of recombinations identified in the different meioses in one person. The scatter within and between individuals shows that there is no one correct human map length. [From Cheung VG, Burdick JT, Hirschmann D & Morley M (2007) *Am. J. Hum. Genet.* 80, 526–530. With permission from Elsevier.]

TABLE 14.1 THE DEVELOPMENT OF MARKERS FOR HUMAN GENETIC MAPPING

Type of marker	Dates used	No. of loci	Features
Blood groups	1910–1960	~20	May need fresh blood, and rare antisera. Genotype cannot always be inferred from phenotype because of dominance. No easy physical localization
Electrophoretic mobility variants of serum proteins	1960–1985	~30	May need fresh serum, and specialized assays. No easy physical localization. Often limited numbers of alleles and low frequency polymorphism
HLA tissue types	1970–	1	One set of closely linked loci, usually scored as a haplotype. Highly informative; many alleles with high or moderate frequency. Can only test for linkage to 6p21.3
DNA RFLPs	1975–	>10^5	Two-allele markers, maximum heterozygosity 0.5. Assayed previously by Southern blotting, now by PCR. Easy physical localization
DNA VNTRs (minisatellites)	1985–1990	~10^4	Many alleles, highly informative. Assayed by Southern blotting. Easy physical localization. Tend to cluster near ends of chromosomes
DNA VNTRs (microsatellites)	1990–	~10^5	Many alleles, highly informative. Can be assayed by automated multiplex PCR. Easy physical localization. Distributed throughout genome
DNA SNPs	2000–	~10^7	Two-allele markers. Densely distributed across the genome. Massively parallel technologies (microarrays, etc.) allow a sample to be assayed for thousands of SNPs in a single operation. More stable over evolutionary time than microsatellites

RFLP, restriction fragment length polymorphism; SNP, single nucleotide polymorphism; VNTR, variable number of tandem repeats.

- The locus that determines the character should be highly polymorphic, so that a randomly selected person has a good chance of being heterozygous.

Any character that satisfies these criteria can be used as a **genetic marker**. Originally, protein variants such as blood groups and tissue types were used as human genetic markers, but these have been superseded by DNA markers such as microsatellites and SNPs (**Table 14.1**).

Disease-marker mapping, if it is not to be a purely blind exercise, requires the markers themselves to have already been mapped, so that their chromosomal locations and distances apart are known. This avoids the frustrating situation that arose in 1985, when the long-sought first linkage was established to the cystic fibrosis locus. The linkage was to a protein polymorphism of the enzyme paraoxonase—but the chromosomal location of the paraoxonase gene was not known. Framework maps of markers, developed as part of the effort to sequence the human genome (see Chapter 8), are also an important tool for human genetic mapping. They are generated by marker–marker mapping, in which the segregation of multiple markers is followed in large, multigenerational pedigrees.

For linkage analysis we need informative meioses

Linkage analysis requires us to be able to score meioses as recombinant or non-recombinant for the loci in question. A meiosis is said to be informative for linkage when we can do this. The minimum requirement for this is that the person in whom the meiosis takes place should be heterozygous at both loci. The mean heterozygosity of a marker (the chance that a randomly selected person will be heterozygous) is used as the measure of its usefulness for linkage analysis. If there are marker alleles $A_1, A_2, A_3, ..., A_n$ with gene frequencies $p_1, p_2, p_3, ..., p_n$, then the proportion of people who are heterozygous is $1 - (p_1^2 + p_2^2 + p_3^2 + ... + p_n^2)$.

Being heterozygous is necessary but is not always sufficient for a meiosis to be informative. In the pedigree shown in **Figure 14.8B** the meiosis is uninformative even though it occurs in a doubly heterozygous person, because both parents are heterozygous for the same alleles. A more sophisticated but seldom used measure, the **polymorphism information content** (**PIC**), allows for such cases.

Suitable markers need to be spread throughout the genome

To be able to map a locus that might be anywhere in the genome, markers need to be available for every part of every chromosome. How densely the markers need to cover each chromosome depends on the amount of pedigree data

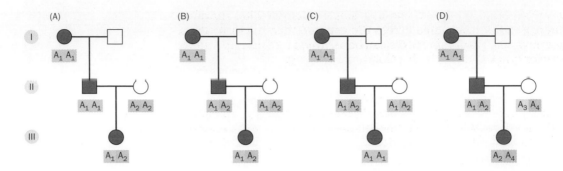

available. In theory, linkage could be detected between loci that are 40 cM apart, but the amount of data required to do this is prohibitive (**Table 14.2**).

Obtaining enough family material to test more than 20–30 meioses can be seriously difficult for a rare disease. Thus, as can be seen in Table 14.2, mapping requires markers spaced at intervals no greater than 10–20 cM across the genome. Given the genome lengths calculated above, and allowing for imperfect informativeness, we need a minimum of several hundred markers to cover the genome. Human genetic mapping only became generally feasible with the identification of DNA markers, starting with restriction fragment length polymorphisms (RFLPs) in the 1970s. As well as being more numerous than earlier protein-based markers, DNA markers have the advantage that they can all be assayed with the same techniques. Moreover, their chromosomal location can be easily determined by reference to the human genome sequence. This allows DNA-based genetic maps to be cross-referenced to physical maps.

Linkage analysis normally uses either fluorescently labeled microsatellites or SNPs as markers

The first DNA-based linkage studies used RFLPs that had to be assayed by the very laborious Southern blotting technique, which made a genomewide search for linkage extremely onerous. Subsequent developments have steadily made linkage analysis quicker and easier. Microsatellite (STRP) markers were introduced that were highly informative because they had many possible alleles, and could be assayed using PCR. In the early days, each microsatellite was amplified individually and run individually on a manually poured slab gel. Later, compatible sets of microsatellite markers were developed that could be amplified together in a multiplex PCR reaction to give non-overlapping allele sizes, so that they could be run in the same gel lane. With fluorescent labeling in several colors, it became possible to genotype a sample for 10 or more markers in a single gel lane or capillary of a sequencing machine.

A major achievement of the early stages of the Human Genome Project was to identify upward of 10,000 highly polymorphic microsatellite markers and place them on framework maps (see Chapter 8). These maps made possible the spectacular progress of the 1990s in mapping Mendelian diseases. The original microsatellites were mostly repeats of the dinucleotide sequence CA. Trinucleotide and tetranucleotide repeats have gradually replaced dinucleotide repeats as the markers of choice because they give cleaner results.

With the widespread adoption of microarray technology, SNPs have come to the fore as genetic markers. SNPs normally have only two alleles, which means that a higher proportion of meioses are uninformative for linkage using a SNP than using a microsatellite. Two to three SNPs will generally be required to

Figure 14.8 Informative and uninformative meioses. In each pedigree, individual II$_1$ has a dominant condition that we know he inherited along with marker allele A_1 because he inherited the condition from his mother, who is homozygous A_1A_1. Is the sperm that produced his daughter recombinant or non-recombinant between the loci for the condition and the marker? (A) This meiosis is uninformative: the father (II$_1$) is homozygous and so his marker alleles cannot be distinguished. (B) This meiosis is uninformative because both parents are heterozygous for alleles A_1 and A_2: the child could have inherited A_1 from the father and A_2 from the mother, or vice versa. (C) This meiosis is informative and non-recombinant: the child inherited A_1 from the father. (D) This meiosis is informative and recombinant: the child inherited A_2 from the father.

TABLE 14.2 THE NUMBER OF INFORMATIVE MEIOSES NEEDED TO OBTAIN EVIDENCE OF LINKAGE BETWEEN TWO LOCI							
Recombination fraction	0	0.05	0.10	0.15	0.20	0.30	0.40
Minimum no. of informative meioses	10	14	19	26	36	84	343

The higher the recombination fraction is between two linked loci, the more meioses are needed to obtain evidence that they are linked. See Box 14.2 for an explanation of how these figures were calculated.

produce the same amount of linkage data as a single microsatellite. However, SNPs are far more densely distributed across the genome than microsatellites, and with microarrays 500,000 or more SNPs can be assayed in a single operation, opening the way for quick and very high-resolution mapping.

14.3 TWO-POINT MAPPING

Having collected families where a Mendelian disease is segregating, and genotyped them with an informative marker, how do we know when we have found linkage? There are two aspects to this question:

- How can we work out the recombination fraction?

- What statistical test should we use to see whether the recombination fraction is significantly different from 0.5, the value expected on the null hypothesis of no linkage?

Scoring recombinants in human pedigrees is not always simple

In some families, the first question can be answered very simply by counting recombinants and non-recombinants. The family shown at the start of this chapter in Figure 14.1 is one example. There are two recombinants in seven meioses, and the recombination fraction is 0.28. **Figure 14.9A** shows another example. Individual II_1 is a double heterozygote and is also **phase-known**: we know which combination of alleles was inherited from each parent, and so we can unambiguously score each meiosis as recombinant or non-recombinant. In Figure 14.9B, individual II_1 is again doubly heterozygous, but this time **phase-unknown** (both parents are dead and therefore not available for genotyping). Among her children, either there are five non-recombinants and one recombinant, or else there are five recombinants and one non-recombinant. We can no longer identify

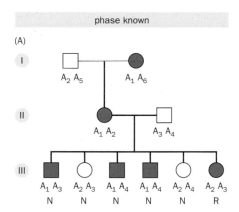
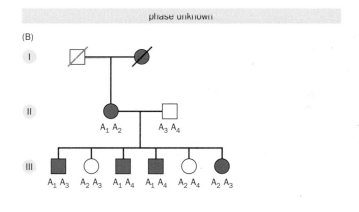
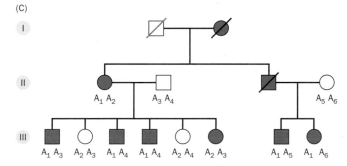

Figure 14.9 Recognizing recombinants. Three versions of a family with an autosomal dominant disease, typed for a marker with alleles A_1–A_6. (A) All meioses are phase-known. We can identify III_1–III_5 unambiguously as non-recombinant (N) and III_6 as recombinant (R). (B) The same family, but phase-unknown. The mother, II_1, could have inherited either marker allele A_1 or A_2 with the disease; thus, her phase is unknown. Either III_1–III_5 are non-recombinant and III_6 is recombinant; or III_1–III_5 are recombinant and III_6 is non-recombinant. (C) The same family after further tracing of relatives. III_7 and III_8 have also inherited marker allele A_1 along with the disease from their father—but we cannot be sure whether their father's allele A_1 is identical by descent to the allele A_1 in his sister II_1. Maybe there are two copies of allele A_1 among the four grandparental marker alleles. The likelihood of this depends on the gene frequency of allele A_1. Thus, although this pedigree contains linkage information, extracting it is problematic.

recombinants unambiguously, even if the first alternative seems much more likely than the second. Figure 14.9C adds yet more complications: further relatives have been traced who have also inherited marker allele A_1 along with the disease from their father II_3—but we cannot be sure whether this A_1 allele is identical by descent to the A_1 allele in their aunt II_1. However, if this were a family with a rare disease, no researcher would be willing to ignore it. Some method is needed to extract the linkage information from a family with only the incomplete or ambiguous identification of recombinants and non-recombinants.

Computerized lod score analysis is the best way to analyze complex pedigrees for linkage between Mendelian characters

In the pedigree shown in Figure 14.9B, it is not possible to identify recombinants unambiguously and count them. It is possible, however, to calculate the overall likelihood of the pedigree, on the alternative assumptions that the loci are linked (recombination fraction = θ) or not linked (recombination fraction = 0.5). The ratio of these two likelihoods gives the odds of linkage, and the logarithm of the odds (lod) is the **lod score**. Lod scores are symbolized as Z. Newton Morton demonstrated in 1955 that lod scores represent the most efficient statistic for evaluating pedigrees for linkage, and derived formulae to give the lod score (as a function of the recombination fraction θ) for various standard pedigree structures. **Box 14.2** shows how this is done for simple structures such as those in Figure 14.9A and B. Being a function of the recombination fraction, lod scores are calculated for a range of θ values. The most likely recombination fraction is the value at which the lod score is maximal. In a set of families, the overall probability of linkage is the product of the probabilities in each individual family; therefore lod scores (being logarithms) can be added up across families.

Calculating the full lod score for the family in Figure 14.9C is difficult. To calculate the likelihood that III_7 and III_8 are recombinant or non-recombinant, we

---

BOX 14.2 CALCULATION OF LOD SCORES

Let us take as an example a calculation of lod scores for the families in Figure 14.9A and B.

- Given that the loci are truly linked, with recombination fraction θ, the likelihood of a meiosis being recombinant is θ and the likelihood of its being non-recombinant is $1 - \theta$.
- If the loci are, in fact, unlinked, the likelihood of a meiosis being either recombinant or non-recombinant is 0.5.

Family A

There are five non-recombinants $(1 - \theta)$ and one recombinant (θ).
The overall likelihood, given linkage, is $(1 - \theta)^5 \times \theta$.
The likelihood, given no linkage, is $(0.5)^6$.
The likelihood ratio is $(1 - \theta)^5 \times \theta / (0.5)^6$.

θ	0	0.1	0.2	0.3	0.4	0.5
Z	$-\infty$	0.577	0.623	0.509	0.299	0

The lod score, Z, is the logarithm of the likelihood ratio.

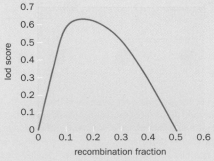

Figure 1 Lod score curve for family A.

Family B

The mother (II_1 in Figure 14.9B) is phase-unknown.
If she inherited A_1 with the disease, there are five non-recombinants and one recombinant.
If she inherited A_2 with the disease, there are five recombinants and one non-recombinant.
The overall likelihood is $\frac{1}{2}[(1 - \theta)^5 \times \theta/(0.5)^6] + \frac{1}{2}[(1 - \theta) \times \theta^5/(0.5)^6]$.
This allows for either possible phase, with equal prior probability.
The lod score, Z, is the logarithm of the likelihood ratio.

θ	0	0.1	0.2	0.3	0.4	0.5
Z	$-\infty$	0.276	0.323	0.222	0.076	0

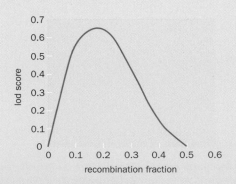

Figure 2 Lod score curve for family B.

Family C

At this point non-masochists turn to the computer.

must take likelihoods calculated for each possible genotype of I_1, I_2, and II_3, weighted by the probability of that genotype. For I_1 and I_2, the genotype probabilities depend on both the gene frequencies and the observed genotypes of II_1, III_7, and III_8. Genotype probabilities for II_3 are then calculated by simple Mendelian rules. Human linkage analysis, except in the very simplest cases, is entirely dependent on computer programs that implement algorithms for handling these branching trees of genotype probabilities, given the pedigree data and a table of gene frequencies.

The result of linkage analysis is a table of lod scores at various recombination fractions, like the two tables in Box 14.2. Positive lod scores give evidence in favor of linkage, and negative lods give evidence against linkage. Note that only recombination fractions between 0 and 0.5 are meaningful, and that all lod scores are zero at $\theta = 0.5$ [because they are then measuring the ratio of two identical probabilities, and $\log_{10}(1) = 0$]. The results can be plotted to give curves (see Box 14.2).

Returning to the two questions posed at the start of this section, we now see that *the most likely recombination fraction is the one at which the lod score is highest*. If there are no recombinants, the lod score will be maximum at $\theta = 0$. If there are recombinants, Z will peak at the most likely recombination fraction ($0.167 = 1/6$ for family A in Figure 14.9, but it is harder to predict for family B).

Lod scores of +3 and –2 are the criteria for linkage and exclusion (for a single test)

The second question posed at the start of this section concerned the threshold of significance. Here the answer is at first sight surprising. $Z = 3.0$ is the threshold for accepting linkage, with a 5% chance of a type 1 error (falsely rejecting the null hypothesis). For most statistics, $p < 0.05$ is used as the threshold of significance, but $Z = 3.0$ corresponds to 1000:1 odds [$\log_{10}(1000) = 3.0$]. The reason why such a stringent threshold is chosen lies in the inherent improbability that two loci, chosen at random, should be linked. With 22 pairs of autosomes to choose from, it is not likely the two loci would be located on the same chromosome (syntenic) and, even if they were, loci well separated on a chromosome segregate independently. Common sense tells us that if something is inherently improbable, we require strong evidence to convince us that it is true. This common sense can be quantified in a Bayesian calculation (**Box 14.3**), which shows that odds of 1000:1 in fact correspond precisely to the conventional $p = 0.05$ threshold of significance. Thus, $Z = 3.0$ is the threshold for accepting linkage with a 5% chance of falsely rejecting the null hypothesis. Linkage can be rejected if $Z < -2.0$. Values of Z between –2

BOX 14.3 BAYESIAN CALCULATION OF LINKAGE THRESHOLD

The likelihood that two loci should be linked (the prior probability of linkage) has been argued over, but estimates of about 1 in 50 are widely accepted.

Hypothesis	Loci are linked (recombination fraction = θ)	Loci are not linked (recombination fraction = 0.5)
Prior probability	1/50	49/50
Conditional likelihood: 1000:1 odds of linkage [lod score $Z(\theta) = 3.0$]	1000	1
Joint probability (prior × conditional)	20	~1

Because of the low *prior probability* that two randomly chosen loci should be linked, evidence giving 1000:1 odds in favor of linkage is required in order to give overall 20:1 odds in favor of linkage. This corresponds to the conventional $p = 0.05$ threshold of statistical significance. The calculation is an example of the use of **Bayesian statistics** to combine probabilities. See the text for a description of the lod score.

Figure 14.10 Lod score curves. Graphs of lod score against recombination fraction (θ) from a hypothetical set of linkage experiments. The green curve shows evidence of linkage (Z > 3) with no recombinants. The red curve shows evidence of linkage (Z > 3), with the most likely recombination fraction being 0.23. The purple curve shows linkage excluded (Z < –2) for recombination fractions below 0.12, inconclusive for larger recombination fractions. The blue curve is inconclusive at all recombination fractions.

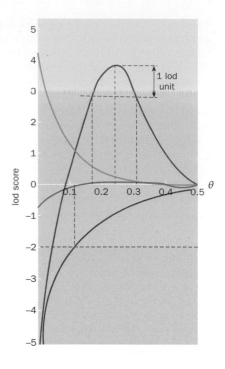

and +3 are inconclusive. The same logic suggests a threshold lod score of 2.3 for establishing linkage between an X-linked character and an X-chromosome marker (prior probability of linkage ≈ 1/10).

Confidence intervals are hard to deduce analytically, but a widely accepted support interval extends to recombination fractions at which the lod score is 1 unit below the peak value (the *lod – 1 rule*). Thus, the red curve in **Figure 14.10** gives acceptable evidence of linkage (Z > 3), with the most likely recombination fraction 0.23 and support interval 0.17–0.32. The curve will be more sharply peaked for greater amounts of data but, in general, peaks are quite broad. It is important to remember that distances on human genetic maps are often very imprecise estimates.

Negative lod scores exclude linkage for the region where Z < –2. The purple curve in Figure 14.10 excludes the disease from 12 cM (= 0.12 recombination fraction) either side of the marker. Although gene mappers hope for a positive lod score, exclusions are not without value. They tell us where the disease is not (**exclusion mapping**). This can exclude a possible candidate gene, and if enough of the genome is excluded, only a few possible locations may remain.

Table 14.3 shows a real example of two-point lod scores from a study of families with migraine with aura. There are several points worth noting:

- There is evidence of linkage—some lod scores are well above the 3.0 threshold (green shaded rows).

TABLE 14.3 TWO-POINT LOD SCORES FOR A SERIES OF MARKERS IN FAMILIES SHOWING AUTOSOMAL DOMINANT INHERITANCE OF MIGRAINE WITH AURA

Marker	lod (Z) at θ =						
	0.001	0.01	0.05	0.1	0.2	0.3	0.4
D15S817	–1.95	–0.96	–0.30	–0.06	0.11	**0.17**	0.14
D15S128	1.73	2.65	**3.02**	2.88	2.27	1.47	0.59
D15S10	0.78	1.57	**1.98**	1.93	1.50	0.93	0.35
D15S986	–0.41	1.36	2.40	**2.56**	2.23	1.58	0.78
D15S113	–0.05	1.72	2.76	**2.92**	2.56	1.84	0.91
GABRB3	**5.56**	5.47	5.04	4.48	3.32	2.10	0.87
D15S97	**5.31**	5.22	4.80	4.26	3.12	1.94	0.78
GABRA5	3.16	4.07	**4.34**	4.08	3.20	2.10	0.92
D15S975	0.74	1.67	**2.02**	1.88	1.31	0.80	0.36
D15S1019	**4.48**	4.41	4.07	3.63	2.71	1.71	0.69
D15S1010	–7.44	–3.59	–0.88	0.13	**0.75**	0.71	0.35
D15S971	–13.62	–7.64	–3.59	–1.99	–0.69	–0.22	**–0.06**
D15S118	–11.80	–5.88	–2.02	–0.65	0.26	**0.37**	0.16
D15S1012	–14.11	–8.14	–4.08	–2.46	–1.05	–0.43	**–0.14**

Markers from chromosome 15q11–q13 are listed in chromosomal order. For each marker, the maximum lod score is shown in bold. The markers highlighted in green show significant evidence of linkage to the locus (i.e. Z > 3.0) determining migraine in these families. A multipoint analysis of the same data is shown in Figure 14.10. [Data from Russo L, Mariotti P, Sangiorgi E et al. (2005) *Am. J. Hum. Genet.* 76, 312–326.]

- For some markers, the maximum lod score is at zero recombination (like the green curve in Figure 14.10; actually $\theta = 0.001$ is used to avoid problems with minus infinity scores for unlinked markers). For some others (*D15S986* and *D15S113*) it is slightly negative at $\theta = 0.001$ but rises almost to the threshold of significance at $\theta = 0.1$, like the red curve in Figure 14.10 although less extreme; for others (*D15S1012* and *D15S971*) it is always negative, like the purple curve in Figure 14.10.

Tables like this should not be over-interpreted. The precise lod score depends on the vagaries of which meioses were informative with which markers. The thing to note is the overall trend, showing strongly positive scores for markers toward the center of the region covered, with negative scores at each end of the region.

For whole genome searches a genomewide threshold of significance must be used

In disease studies, families are usually genotyped for hundreds or thousands of markers, each of which is checked for evidence of linkage. The appropriate threshold for significance is a lod score such that there is only a 0.05 chance of a false positive result occurring *anywhere* during a search of the whole genome. As shown above, a lod score of 3.0 corresponds to a significance of 0.05 at a single point. But if 500 markers have been used, the chance of a spurious positive result is greater than if only one marker has been used. A stringent procedure (Bonferroni correction) would multiply the p value by the number of markers used before testing its significance. The threshold lod score for a study using n markers would be $3 + \log(n)$; that is, a lod score of 4 for 10 markers, 5 for 100, and so on. However, this is over-stringent. Linkage data are not independent. If one location is excluded, then the prior probability that the character maps to another location is raised. The threshold for a *genomewide significance level* of 0.05 has been much argued over, but a widely accepted answer for Mendelian characters is 3.3. In practice, claims of linkage based on lod scores below 5, whether with one marker or many, should be regarded as provisional.

14.4 MULTIPOINT MAPPING

Linkage analysis can be more efficient if data for more than two loci are analyzed simultaneously. Multilocus analysis or **multipoint mapping** is particularly useful for establishing the chromosomal order of a set of linked loci. Experimental geneticists have long used *three-point crosses* for this purpose. The most usual three-point cross has one parent heterozygous at each of three linked loci (*A, B,* and *C*) and the other a triply recessive homozygote (*abc/abc*). Among the offspring of this cross, the rarest recombinant class is that which requires a double recombination. In **Table 14.4**, the gene order *A–C–B* is easily seen. This procedure is more efficient than estimating the recombination fractions for the intervals *A–B, A–C,* and *B–C* separately in three separate two-point crosses. Ideally, in any linkage analysis the whole genome would be screened for linkage, and the full data set would be used to calculate the likelihood at each location across the genome.

A second advantage of multipoint mapping in humans is that it helps overcome problems caused by the limited informativeness of markers. Some meioses in a family might be informative with marker *A,* and others uninformative for *A* but informative with the nearby marker *B.* Only simultaneous linkage analysis of the disease with markers *A* and *B* extracts the full information. This is less important for mapping with highly informative microsatellite markers rather than two-allele RFLPs, but it surfaces again when SNPs are used.

The CEPH families were used to construct marker framework maps

The power of multipoint mapping in ordering loci is particularly useful for constructing marker framework maps. However, ordering the loci in such maps is not a trivial problem. For n markers, there are $n!/2$ possible orders [i.e. for $n = 100$, $n!/2 = (100 \times 99 \times 98 \times \ldots \times 1)/2$], and current maps have hundreds of markers per chromosome. Before the human genome sequence was completed, this was a

TABLE 14.4 GENE ORDERING IN *DROSOPHILA* BY A THREE-POINT CROSS		
Class of offspring	Position of recombination (×)	Number of offspring
ABC/abc *abc/abc*	Non-recombinant	853
ABc/abc *abC/abc*	(A, B)–×–C	5
Abc/abc *aBC/abc*	A–×–(B, C)	47
AbC/abc *aBc/abc*	B–×–(A, C)	95

One thousand offspring have been scored from crosses between female *Drosophila* fruit flies heterozygous for recessive characters at three linked loci (*ABC/abc*) and triply homozygous males (*abc/abc*). This design allows the genotype of the eggs to be inferred by inspection of the phenotype of the offspring. The rarest class of offspring will be those whose production requires two crossovers. Of the 1000 flies, 142 (95 + 47) are recombinant between *A* and *B*, 52 (47 + 5) between *A* and *C*, and 100 (95 + 5) between *B* and *C*. Only five flies are recombinant between *A* and *C* but not between *A* and *B*, so these must have double crossovers, *A*–×–*C*–×–*B*. Therefore the map order is *A*–*C*–*B* and the genetic distances are approximately *A*–(5 cM)–*C*–(10 cM)–*B*.

serious difficulty for map makers. Something more intelligent than brute-force computing had to be used to work out the correct order. Even without large-scale sequence data, physical mapping information was immensely helpful. Markers that can be assayed by PCR can be used as sequence-tagged sites (STS) and physically localized, either by database searching or experimentally by using radiation hybrids. The result is a physically anchored marker framework.

Disease–marker mapping suffers from the necessity of using whatever families can be found in which the disease of interest is segregating. Such families will rarely have ideal structures. All too often the number of meioses is undesirably small, and some are phase-unknown. Marker–marker mapping can avoid these problems. Markers can be studied in any family, so families can be chosen that have plenty of children and ideal structures for linkage, like the family in Figure 14.1. The construction of marker framework maps has benefited greatly from a collection of families each with eight or more children and all four grandparents, assembled specifically for the purpose by the Centre d'Étude du Polymorphisme Humain (CEPH; now the Fondation Jean Dausset) in Paris. Immortalized cell lines from every individual of these **CEPH families** ensure a permanent supply of DNA, and sample mix-ups and non-paternity have long since been ruled out by typing with many markers. As an example, the widely used 1998 CHLC (Cooperative Human Linkage Center) map was based on the results of scoring eight CEPH families with 8325 microsatellites, resulting in more than 1 million genotypes.

Multipoint mapping can locate a disease locus on a framework of markers

For disease–marker mapping, the starting point is the framework map of markers. This is taken as given, and the aim is to locate the disease gene in one of the intervals of the framework. Programs such as Linkmap (part of the Linkage package) or Genehunter can move the disease locus across the marker framework, calculating the overall likelihood of the pedigree data with the disease locus in each possible location, taking into account the genotypes at more than just the immediately flanking markers—ideally of every marker used, although this may not be computationally feasible. The result (**Figure 14.11**) is a curve of lod score against map location. This method is also useful for exclusion mapping: if the curve stays below a lod score of –2 across the region, the disease locus is excluded from that region.

The apparently quantitative nature of Figure 14.11 is largely spurious. Peak heights depend crucially on the precise genetic distances between markers,

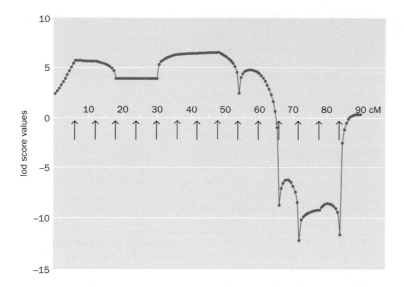

Figure 14.11 Multipoint mapping in humans. The data from Table 14.3 on linkage between migraine with aura and markers from chromosome 15, analyzed by a multipoint mapping program (Simwalk). The genetic map of markers is taken as fixed; the disease locus is placed successively at each point on the genetic marker framework (marker positions are shown by black arrows; colored arrows indicate the position of two candidate genes) and the multipoint lod score is calculated for each possible location of the disease locus. Lod scores dip to strongly negative values near the position of loci that show recombinants with the disease. The highest peak marks the most likely location; odds in favor of this location are measured by the degree to which the highest peak overtops its rivals. The peak lod score is 6.54; that is, a tenfold stronger likelihood than the maximum of 5.56 obtained by two-point analysis. There were no recombinants within this region in this data set, so the candidate region could not be narrowed down further. [From Russo L, Mariotti P, Sangiorgi E et al. (2005) *Am. J. Hum. Genet.* 76, 327–333. With permission from Elsevier.]

which are usually only very roughly known. The physical distances may be known precisely, but for this purpose we need the genetic distances. For markers 1 cM apart on the highest resolution map currently available (the Icelandic map of Kong et al.), the 95% confidence interval for their true distance is 0.5–1.8 cM. Moreover, those figures relate only to the individuals whose data were used in constructing the map, who will not be the same individuals as those used in any particular disease mapping study. As Figure 14.7 shows, total map length varies between individuals. Additionally, none of the mapping functions in linkage programs even approximates to the real complexities of chiasma distribution (see Figure 14.5). However, unless distances on the marker map are radically wrong, it remains true that the highest peak marks the most likely location. If the marker framework is physically anchored, as described above, the stage is then set to search the DNA of the candidate interval and identify the disease gene.

14.5 FINE-MAPPING USING EXTENDED PEDIGREES AND ANCESTRAL HAPLOTYPES

The resolution of mapping depends on the number of meioses—the more meioses that are analyzed, the greater is the chance of a recombination event narrowing down the linked region. The small size of most human families severely limits the resolution attainable in family studies. However, extended family structures can sometimes be used for high-resolution mapping. In some societies, people are very aware of their clan membership and may see themselves as part of highly extended families. Even in societies in which people limit their family feeling to close relatives, ultimately everybody is related, and sometimes one can identify shared **ancestral chromosome segments** among ostensibly unrelated people. Autosomal recessive diseases lend themselves to such analyses, because a mutated allele can be transmitted for many generations; for most dominant or X-linked diseases, the turnover of mutant alleles is too fast to allow sharing over extended families. For a recessive disease, the potential resolution of mapping is set by the number of meioses linking distantly related affected people through their common ancestor.

Autozygosity mapping can map recessive conditions efficiently in extended inbred families

Autozygosity is a term used to mean homozygosity for markers **identical by descent**—that is, the two alleles are both copies of one specific allele that was present in a recent ancestor. In a consanguineous marriage, both partners share some of their ancestors and so may have each inherited a copy of the same ancestral allele at a locus. Their child may then be autozygous for that allele. Suppose that the parents are second cousins (i.e. offspring of first cousins): they would be expected to share 1/32 of all their genes because of their common ancestry, and a child would be autozygous at only 1/64 of all loci (**Figure 14.12**). If that child is homozygous for a particular marker allele, this could be because of autozygosity or it could be because a second, independent example of the same allele has entered the family at some stage (these alleles can be described as **identical by state**, or **IBS**). The rarer the allele is in the population, the greater is the likelihood that homozygosity represents autozygosity. In practice, the requisite rarity is achieved by using a haplotype of closely linked markers. For an infinitely rare allele or haplotype, a single homozygous affected child born to second cousins generates a lod score of $\log_{10}(64) = 1.8$. If there are two other affected sibs who are both also homozygous for the same rare allele, the lod score is $\log_{10}(64 \times 4 \times 4) = 3.0$; the chance that a sibling would have inherited the same pair of parental haplotypes even if they are unrelated to the disease is 1 in 4.

Thus, quite small consanguineous families can generate significant lod scores. Autozygosity mapping is an especially powerful tool if families can be found with multiple people affected by the same recessive condition in two or more sibships linked by consanguinity. Suitable families may be found in Middle Eastern countries and parts of the Indian subcontinent where first-cousin and other consanguineous marriages are common. The method has been applied with great success to locating genes for autosomal recessive hearing loss (**Figure 14.13**). As mentioned in Chapter 3, extensive locus heterogeneity makes recessive hearing loss impossible to analyze by combining data from several small nuclear families.

A bold application of autozygosity in a Northern European population enabled Houwen and colleagues in 1994 to map a rare recessive condition, benign recurrent intrahepatic cholestasis (OMIM 243300), using only four affected individuals (two sibs and two supposedly unrelated people) from an isolated Dutch

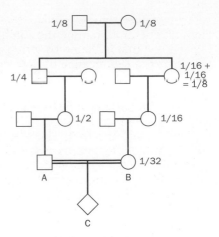

Figure 14.12 Second cousins share 1/32 of their genes, on average, by virtue of their relationship. Individuals A and B are second cousins (their mothers are first cousins). The figures show the proportion of A's genes that each linking relative is expected to share. C, the child of A and B, would be expected to be autozygous at 1/64 of all loci.

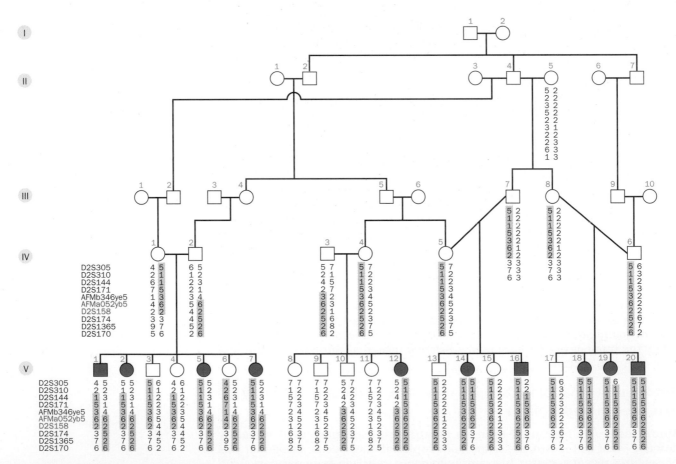

Figure 14.13 Autozygosity mapping. A large, multiply consanguineous family in which several members suffer from autosomal recessive profound congenital deafness (filled symbols). Shaded blocks mark a haplotype of microsatellite markers from chromosome 2 that segregates with the deafness. Markers *AFMa052yb5* and *D2S158* (blue) are homozygous in all affected people and no unaffected people. The deafness gene should lie somewhere between the two markers that flank these (*AFMb346ye5* and *D2S174*). [Adapted from Chaib H, Place C, Salem N et al. (1996) *Hum. Mol. Genet.* 5, 155–158. With permission from Oxford University Press.]

village. Similar virtuoso applications of autozygosity have been reported from Finland. The more remote the shared ancestor, the smaller is the proportion of the genome that is shared by virtue of that common ancestry, and therefore the greater is the significance of demonstrating that the patients share a segment identical by descent. However, at the same time, the remoter the common ancestor, the more chances there are for a second independent allele to enter the family from outside, and so the less likely it is that homozygosity represents autozygosity, either for the disease or for the markers. With remote common ancestry, everything depends on finding people with a very rare recessive condition who are homozygous for a rare marker allele or (more likely) haplotype. The statistical power of the study of Houwen et al. seems almost miraculous, but it is important to remember that this methodology applies only to diseases and populations in which most affected people are descended from a common ancestor who was a carrier.

Identifying shared ancestral chromosome segments allows high-resolution genetic mapping

Cystic fibrosis (CF; OMIM 219700) is uncommon in the non-European countries where family structures are more likely to allow autozygosity mapping, and so mapping CF depended on rare unfortunate nuclear families with more than one affected child. Using these, CF was mapped in 1985 to chromosome 7q31.2; however, after all available recombinants had been used, the candidate region was still dauntingly large for the gene identification techniques available at the time. Arguing that CF mutations might be mostly very old (not only is there no selection against heterozygotes, there is probably positive selection in their favor; see Section 3.5 on factors affecting gene frequency), researchers set out to identify shared ancestral chromosomal segments on CF chromosomes from unrelated patients. Sharing would be indicated by repeatedly finding the same haplotype of marker alleles. The phenomenon is called **linkage disequilibrium** (LD); see Chapter 15 for a fuller description. Table 14.5 shows typical data for two markers from within the CF candidate region. Non-CF chromosomes show a random selection of haplotypes, but CF chromosomes tend to carry the alleles X_1, K_2. The significance of this is that LD is a very short-range phenomenon—shared ancestral segments are short because of recurrent recombination—and so it pointed researchers to the exact location of the elusive CF gene.

A more recently cloned gene, governing Nijmegen breakage syndrome (NBS; OMIM 251260), shows a more detailed application of the same principle. NBS is a very rare autosomal recessive disease characterized by chromosome breakage, growth retardation, microcephaly, immunodeficiency, and a predisposition to cancer. This condition was mentioned in Chapter 13 because its suspected cause is a defect in DNA repair. Conventional linkage analysis in small nuclear families located the *NBS* locus to chromosome 8q21, but, after all recombinants were used, the target region still spanned 8 Mb between markers *D8S271* and *D8S270*. Fifty-one apparently unrelated patients and their parents were then genotyped for a series of microsatellite markers spaced across the candidate region. This

TABLE 14.5 ALLELIC ASSOCIATION IN CYSTIC FIBROSIS		
Marker alleles	Chromosomes with CF mutation	Normal chromosomes
X_1, K_1	3	49
X_1, K_2	147	19
X_2, K_1	8	70
X_2, K_2	8	25

Data from typing for two RFLP markers, the first with alleles X_1 and X_2 and the second with alleles K_1 and K_2 in 114 British families with a child with cystic fibrosis (CF). Chromosomes carrying the CF disease mutation tend to carry allele X_1 and allele K_2. [Data from Ivinson AJ, Read AP, Harris R et al. (1989) *J. Med. Genet.* 26, 426–430.]

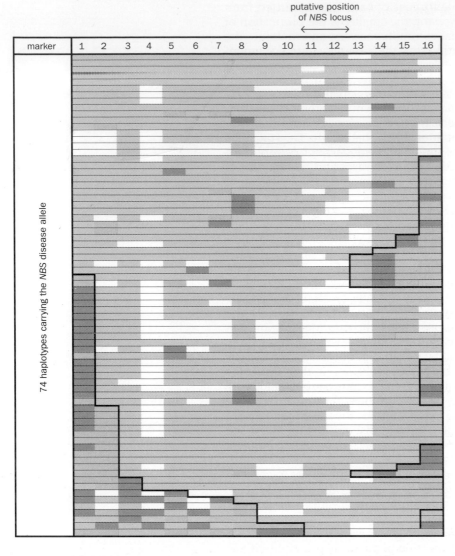

putative position
of *NBS* locus

74 haplotypes carrying the *NBS* disease allele

Figure 14.14 An ancestral haplotype in European patients with Nijmegen breakage syndrome. In a set of 51 apparently unrelated patients with this recessive condition, 74 out of 102 haplotypes around the disease location on chromosome 8q21 seemed to be derived from a common ancestor. The 74 haplotypes (rows) were defined using 16 markers, shown in chromosomal order across the top of the table. The beige color marks locations with alleles identical to those of the inferred ancestral haplotype. Alleles colored orange differ from the putative ancestral version but might be the result of a mutation of the ancestral allele. Alleles colored blue differ substantially from the ancestral allele, and are probably the result of recombination. Blanks mark loci for which there are no data. Only at loci 11 and 12 are there no recombinant (blue) alleles, suggesting that the *NBS* gene maps to this position. [Data reported in Varon R, Vissinga C, Platzer M et al. (1998) *Cell* 93, 467–476. With permission from Elsevier.]

generated 102 haplotypes of chromosomes that carried an NBS disease muta-tion. Of these, 74 looked like derivatives of a common ancestral haplotype, most probably of Slav origin. The most highly conserved region lay between markers 11 and 12 (**Figure 14.14**), which therefore marked the likely location of the *NBS* gene. Subsequently, a gene encoding a novel protein was cloned from this loca-tion and was shown to carry mutations in NBS patients. As predicted, patients with the common haplotype all had the same mutation, whereas those with inde-pendent haplotypes had independent mutations.

Linkage disequilibrium is a central tool in efforts to identify susceptibility genes for complex diseases, and is discussed in detail in Chapter 15.

14.6 DIFFICULTIES WITH STANDARD LOD SCORE ANALYSIS

Standard lod score analysis is a tremendously powerful method for scanning the genome in 20 Mb segments to locate a disease gene, but it can run into difficul-ties. These include:

• vulnerability to errors

• computational limits on what pedigrees can be analyzed

• problems with locus heterogeneity

• limits on the ultimate resolution achievable

• the need to specify a precise genetic model, detailing the mode of inheritance, gene frequencies, and penetrance of each genotype.

Errors in genotyping and misdiagnoses can generate spurious recombinants

With highly polymorphic markers, common errors such as misread gels, switched samples, or non-paternity will often lead to a situation in which an offspring's stated genotype is incompatible with the parental genotypes. The linkage analysis program will stall until such errors have been corrected. During data checking, markers with allelic distributions that deviate significantly from Hardy–Weinberg proportions (see Chapter 3) are flagged. Significant deviation generally indicates systematic errors in genotyping (usually missing heterozygotes)—although in a sample of people selected because they have a certain disease, such a deviation might indicate a true connection between the marker genotype and disease status.

Errors that introduce theoretically possible but incorrect genotypes are difficult to identify because they are not flagged during data checking. The most serious problems are introduced by misassignment of disease status (an individual who is affected is scored as unaffected, or vice versa). Such individuals will seem to be recombinant between the disease locus and any marker that is, in truth, linked to the disease locus. Multilocus analysis can help, because such spurious recombinants appear as close double recombinants. As we saw previously, interference makes close double recombinants very unlikely. In **Figure 14.15**, the probability of a true double recombinant with markers 5 cM apart is extremely small, well below $0.05 \times 0.05 = 0.0025$ (the probability if there were no interference). Apparent double recombinants usually signal an error in genotyping the markers, a clinical misdiagnosis, or locus heterogeneity. Mutation in one of the genes or germinal mosaicism are rarer causes. When marker framework maps are made, error-checking routines test the extent to which the map can be shortened by omitting any single test result. Individual genotypes that significantly lengthen the map (i.e. add recombinants) are suspect.

Errors in the order of markers on marker framework maps used to cause problems (single recombinants could seem to be double), but these have receded as genetic maps are cross-checked against physical sequence.

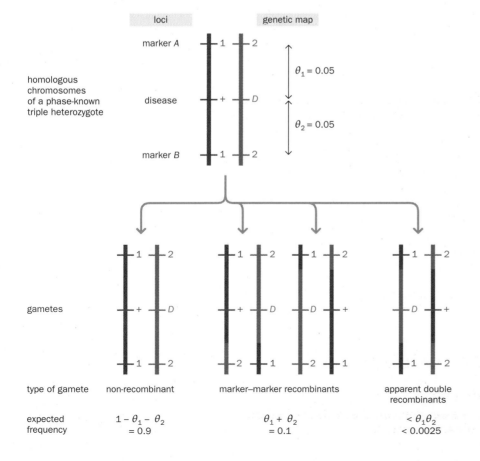

Figure 14.15 Apparent double recombinants suggest errors in the data. In this phase-known triple heterozygote, two markers (*A* and *B*) flank the disease locus (*D*). On the genetic map, the markers are each only 5 cM from the disease locus, giving a recombination frequency for each single crossover of 0.05. The recombination frequencies of the possible gametes are shown below. Because of interference, the probability of true double recombinants would be substantially below the theoretical value of $0.05 \times 0.05 = 0.0025$.

Computational difficulties limit the pedigrees that can be analyzed

As we saw in the discussion of the pedigree in Figure 14.9C, human linkage analysis depends on computer programs that implement algorithms for handling branching trees of genotype probabilities, given the pedigree data and gene frequencies. These algorithms are computationally extremely intensive, and large analyses can stretch the capacity of even mainframe machines. Liped was the first generally useful program, and Mlink (part of a package called Linkage) used the same basic algorithm, the Elston–Stewart algorithm, but extended it to multipoint data. The Elston–Stewart algorithm can handle arbitrarily large pedigrees, but the computing time increases exponentially with increasing numbers of possible haplotypes (more alleles and/or more loci). This limits its utility for analyzing multipoint data. An alternative algorithm, the Lander–Green algorithm, is able to handle any number of loci (the computing time increases linearly with the number of loci) but has memory problems with large pedigrees. This algorithm is implemented in the Genehunter and Merlin programs. These programs are particularly good for analyzing whole genome searches of modest-sized pedigrees.

Locus heterogeneity is always a pitfall in human gene mapping

As we saw in Chapter 3, it is common for mutations in several unlinked genes to produce the same clinical phenotype. Even a dominant condition with large families can be hard to map if there is **locus heterogeneity** within the collection of families studied. It took years of collaborative work to show that tuberous sclerosis was caused by mutations at either of two loci, *TSC1* (OMIM 191100) at 9q34 and *TSC2* (OMIM 191092) at 16p13. With recessive conditions, the difficulty is multiplied by the need to combine many small families. Autozygosity mapping (see above) is the main solution for recessive conditions.

Software programs such as Genehunter or Homog (see Further Reading) can compare the likelihood of the data on the alternative assumptions of locus homogeneity (all families map to the location under test) and heterogeneity (a proportion α of unlinked families), and give a maximum-likelihood estimate of α. However, because this introduces an extra degree of freedom into the hypothesis, more data are required to produce definitive answers.

Pedigree-based mapping has limited resolution

The resolution of mapping depends on the number of meioses analyzed. Human families are quite limiting for this purpose—for example, the CEPH family collection mentioned above can provide an average resolution of only about 3 Mb. One solution is to genotype sperm. Men may have too few children for high-resolution mapping, but they produce untold millions of sperm. The protocol illustrated in **Figure 14.16** allows the identification of individual sperm that are recombinant across a predefined region of a few kilobases. The recombination fraction is estimated by comparing the number of recombinant sperm recovered with the total number of sperm tested, as measured by the total amount of sperm DNA used in the experiment. The exact position of each crossover can be determined, allowing the identification of recombination hotspots. This method cannot be used to map diseases, but it does allow a researcher with the necessary

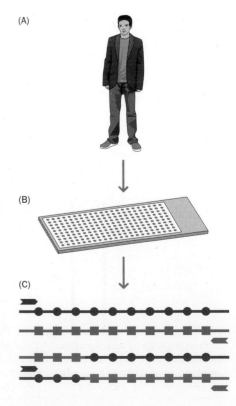

(A)

(B)

(C)

Figure 14.16 Scoring recombinant sperm. (A) A man is selected who is heterozygous for many SNPs across the region of interest (typically a few kilobases). (B) His sperm DNA is divided into aliquots sufficiently small that each is estimated to contain, on average, less than one molecule that is recombinant across the region of interest, along with several thousand non-recombinant molecules. (C) PCR is performed on each aliquot by using allele-specific primers (red and blue arrows); red circles and blue squares represent the alternative alleles at each heterozygous SNP. Only recombinant molecules (the bottom molecule in the figure) will hybridize to both primers and be amplified. Any PCR product can then be genotyped for the individual SNPs to identify the exact position of the crossover that produced each recombinant sperm.

high degree of technical skill to identify extremely rare recombinants between very closely linked markers. In 2005, Jeffreys and colleagues identified the existence of very localized recombination hotspots (see Figure 14.6) by measuring recombination fractions as low as 0.00001. Of course, sperm analysis only gives information on male recombination.

Characters with non-Mendelian inheritance cannot be mapped by the methods described in this chapter

The methods of lod score analysis described in this chapter require a precise genetic model. Before the computer can start the analysis, it requires a series of parameters to be specified. These include the mode of inheritance, the gene frequencies, and the penetrance of each genotype. Penetrance is the most difficult of these to specify. Ideally, there would be an exact relationship between somebody's disease status and their genotype at the disease locus. But two factors limit this. First, as we saw in Chapter 3, there may be reduced penetrance of the disease genotype. Sometimes a person has the genotype that would normally cause disease but for one reason or another (such as other genes, lifestyle, or simple luck) is unaffected. Second, somebody may be affected despite not having the disease genotype—that is, they are a **phenocopy**. For example, in a family being studied to map a gene causing deafness, one family member might be deaf because of head trauma or meningitis rather than because he has inherited the family mutation. If no allowance were made for possible non-penetrance and phenocopies, such persons would be wrongly scored as recombinant, following the logic shown in Figure 14.15. However, if the penetrance is set too low, there is a reduction in the power to detect linkage, because a less precise hypothesis is being tested. Families collected for study are usually selected because they include multiple affected people, and this systematically biases the estimate of penetrance upward.

Despite these complications, for Mendelian characters it is not usually a major problem to provide plausible values for the parameters. However, for common complex diseases such as diabetes or schizophrenia, we have no real idea of the gene frequencies or penetrance of any susceptibility alleles, or even the mode of inheritance. The problem of specifying an appropriate genetic model is far more intractable. As a result, the methods described in this chapter are not appropriate for mapping the genetic factors that govern susceptibility or resistance to complex diseases; this task requires different approaches, which are described in Chapter 15.

CONCLUSION

The human genome sequence provides the ultimate physical map, enabling physical entities such as genes, DNA sequences, and chromosomal breakpoints to be localized to the nearest nucleotide. However, localizing the determinants of phenotypes such as genetic diseases requires an alternative approach, namely *genetic* mapping. Genetic mapping depends on identifying a physical entity that consistently travels together (co-segregates) with the phenotypic determinant through multiple meioses in a collection of families. Because of recombination during meiosis, only sequences that lie very close together on the same chromosomal segment will consistently co-segregate.

The usual physical entities used in genetic mapping are microsatellites or single nucleotide polymorphisms (SNPs). Array-based technologies allow hundreds of thousands of SNPs, distributed across the whole genome, to be tested quickly and easily for co-segregation with a phenotype of interest. If there is significant but imperfect co-segregation, the fraction of meioses in which the two fail to co-segregate (the recombination fraction) is a measure of their genetic distance apart. Genetic distances are not identical to physical distances, measured in kilobases, because the probability of a crossover is not uniform along the whole length of every chromosome. It varies systematically between the sexes and between individuals. The relationship between the genetic distance, measured in centimorgans, and the recombination fraction is not linear, but must be expressed mathematically by a mapping function. This is necessary because recombination fractions never exceed 0.5, no matter how far apart physically two loci are.

Additionally, the presence of one crossover inhibits the formation of further crossovers close to it (interference).

The consistency with which a genetic marker co-segregates with a phenotype is assessed by calculating the lod score (the logarithm of the odds of linkage). For most pedigrees, elaborate computer programs are needed to do this. There is a trade-off between the number of meioses studied and the number of markers that would be necessary to have the possibility of achieving a statistically significant lod score. Fewer markers can be used if more meioses are studied, and vice versa. It is more efficient to calculate lod scores by using multiple markers (multipoint mapping) than to calculate the score with each individual marker in turn. Framework marker maps have been established to assist in multipoint mapping.

The resolution of genetic mapping—that is, the precision with which a genetic entity can be localized—is determined by the number of meioses studied. Very high resolution can be achieved by studying affected people who are related through a remote common ancestor (autozygosity mapping). For mapping DNA sequences, even higher resolution is achievable by genotyping individual spermatozoa produced by a doubly heterozygous male. This allows the exact positions of crossovers to be identified in studies of the mechanism of recombination.

The methods of lod score analysis described in this chapter have been very successful with Mendelian characters, but they have some limitations. These methods are called **parametric** because a series of parameters needs to be specified before analysis can begin. These include the mode of inheritance, the gene frequencies, and the penetrance of each genotype. For Mendelian characters it is not usually a major problem to provide plausible values for these parameters. However, for common complex diseases such as diabetes or schizophrenia, the problems are far more intractable. Any genetic model is no more than a hypothesis—we have no real idea of the gene frequencies or penetrance of any susceptibility alleles, or even the mode of inheritance. This makes it very unwise to apply the methods described in this chapter to such diseases. Nevertheless, identifying the genetic components of susceptibility to complex diseases is now a major part of human genetics research. The ways in which one can attempt to do this are the subject of Chapter 15.

FURTHER READING

General

Laboratory of Statistical Genetics, Rockefeller University. http://linkage.rockefeller.edu/ [A very useful website, giving access to all the programs mentioned in this chapter, and a wealth of general information about human linkage analysis.]

Ott J (1999) Analysis Of Human Genetic Linkage, 3rd ed. Johns Hopkins University Press. [Gives a full discussion of linkage analysis and mapping functions.]

Terwilliger J & Ott J (1994) Handbook For Human Genetic Linkage. Johns Hopkins University Press. [Full of practical advice indispensable to anybody undertaking human linkage analysis.]

The role of recombination in genetic mapping

Broman KW & Weber JL (2000) Characterization of human crossover interference. *Am. J. Hum. Genet.* 66, 1911–1926. [An attempt to define the best mapping function for human linkage analysis.]

Cheung VG, Burdick JT, Hirschmann D & Morley M (2007) Polymorphic variation in human meiotic recombination. *Am. J. Hum. Genet.* 80, 526–530. [A study of an average of eight meioses in each of 67 individuals using 6324 SNPs, giving detail of inter-individual and chromosomal variations in recombination.]

Holliday R (1990) The history of the DNA heteroduplex. *BioEssays* 12, 133–142. [An interesting personal review of the history of ideas in genetic recombination, from the man who proposed the Holliday junction. The full text is available electronically.]

Hultén MA & Lindsten J (1973) Cytogenetic aspects of human male meiosis. *Adv. Hum. Genet.* 4, 327–387. [Classical microscope study of male meiosis.]

Jeffreys AJ, Neumann R, Panayi M et al. (2005) Human recombination hotspots hidden in regions of strong marker association. *Nat. Genet.* 37, 601–606. [An elegant and comprehensive exploration of recombination hotspots in a region of human chromosome 1; also a useful source of references to earlier work.]

Keeney S (2001) Mechanism and control of meiotic recombination initiation. *Curr. Top. Dev. Biol.* 52, 1–53. [A detailed review of the molecules involved in meiotic recombination, focusing on yeast.]

Mapping a disease locus

Botstein D, White RL, Skolnick M & Davis RW (1980) Construction of a genetic linkage map in man using restriction fragment length polymorphisms. *Am. J. Hum. Genet.* 32, 314–331. [A classic in the history of developing markers for linkage.]

Shete S, Tiwari H & Elston RC (2000) On estimating the heterozygosity and polymorphism information content value. *Theor. Popul. Biol.* 57, 265–271. [For those interested in a thorough mathematical analysis of informativeness of markers.]

Two-point mapping

Morton NE (1955) Sequential tests for the detection of linkage. *Am. J. Hum. Genet.* 7, 277–318. [The blockbuster paper that established the lod score method for human linkage. Not easy reading.]

Russo L, Mariotti P, Sangiorgi E et al. (2005) A new susceptibility locus for migraine with aura in the 15q11–q13 genomic region containing three GABA$_A$ receptor genes. *Am. J. Hum. Genet.* 76, 312–326. [The study described in Table 14.3.]

Multipoint mapping

Abecasis GR, Cherny SS, Cookson WO & Cardon LR (2002) Merlin—rapid analysis of dense genetic maps using sparse gene flow trees. *Nat. Genet.* 30, 97–101.

Gudbjartsson DF, Jonasson K, Frigge ML & Kong A (2000) Allegro, a new computer program for multipoint linkage analysis. *Nat. Genet.* 25, 12–13.

Kong A, Gudbjartsson DF, Sainz J et al. (2002) A high-resolution recombination map of the human genome. *Nat. Genet.* 31, 241–247. [The most comprehensive human marker–marker genetic map, following 5136 microsatellites through 1257 meioses in a set of Icelandic families.]

Lathrop GM, Lalouel JM, Julier C & Ott J (1985) Multilocus linkage analysis in humans: detection of linkage and estimation of recombination. *Am. J. Hum. Genet.* 37, 482–498. [Discusses the advantages of three-point over two-point linkage, and introduces the Linkage program package.]

Fine-mapping using extended pedigrees and ancestral haplotypes

Brooks AS, Bertoli-Avella AM, Burzynski GM et al. (2005) Homozygous nonsense mutations in *KIAA1279* are associated with malformations of the central and enteric nervous systems. *Am. J. Hum. Genet.* 77, 120–126. [Includes a very neat example of autozygosity mapping.]

Houwen RHJ, Baharloo S, Blankenship K et al. (1994) Genome screening by searching for shared segments: mapping a gene for benign recurrent intrahepatic cholestasis. *Nat. Genet.* 8, 380–386. [An extreme example of the power of autozygosity mapping.]

Varon R, Vissinga C, Platzer M et al. (1998) Nibrin, a novel DNA double strand break repair protein, is mutated in Nijmegen breakage syndrome. *Cell* 93, 467–476. [The work illustrated in Figure 14.14.]

Difficulties with standard lod score analysis

Elston RC & Stewart J (1971) A general model for the analysis of pedigree data. *Hum. Hered.* 21, 523–542. [The Elston–Stewart algorithm.]

Kruglyak L, Daly MJ & Lander ES (1995) Rapid multipoint linkage analysis of recessive traits in nuclear families, including homozygosity mapping. *Am. J. Hum. Genet.* 56, 519–527. [Description of the Lander–Green algorithm and its use in homozygosity mapping.]

Chapter 15

Mapping Genes Conferring Susceptibility to Complex Diseases

KEY CONCEPTS

- Many common diseases are multifactorial, with a variety of environmental and genetic factors each reducing or increasing an individual's susceptibility to that disease. They are also usually complex, having many different possible causes.

- The major genetic effects on human health and disease come from genetic factors that affect susceptibility to common complex diseases, rather than those that cause Mendelian diseases. A main aim of much current human genetic research is to identify such factors.

- Evidence for the role of genetic factors in many common complex diseases comes from studies of families, twins, and adopted people. Such studies need careful interpretation to disentangle genetic effects from the effects of a shared family environment.

- Linkage analysis has shown only very limited success in mapping susceptibility factors for common complex diseases. Standard lod score analysis cannot be used because it is not possible to specify certain parameters for the susceptibility factors, such as the mode of inheritance, gene frequencies, or penetrances. Instead, non-parametric methods are used, which do not require a detailed genetic model to be provided. Non-parametric linkage methods take affected relatives and look for chromosomal segments that they share more often than expected by chance. Affected sib pairs are the most frequently used sets of relatives.

- An alternative approach to finding susceptibility factors is to seek populationwide statistical associations between a certain genotype and a disease. Such associations may arise because many supposedly unrelated people share a chromosome segment inherited from a distant common ancestor who carried a susceptibility factor. However, not all populationwide associations have a genetic cause.

- The International HapMap Project has defined the ancestral chromosome segments in four human populations, and cataloged markers (tagging SNPs) that can be used to identify them.

- Shared ancestral chromosome segments are extremely small (typically a few kilobases). Thus, seeking associations requires the use of a dense array of closely spaced markers. It has only recently become feasible to conduct genomewide association studies.

- Identifying susceptibility factors, either by linkage studies or by association, has proved unexpectedly difficult. Only recently have studies become sufficiently powerful to identify susceptibility factors reliably.

- Association studies can only identify factors that are present on chromosome segments that are shared by many individuals in the study group. Thus, the ability of association studies to identify susceptibility factors for common complex diseases depends on the common disease–common variant hypothesis, which supposes that most susceptibility factors are ancient variants found on shared ancestral chromosome segments.

- An alternative hypothesis, the mutation–selection hypothesis, supposes most factors to be the result of a highly heterogeneous set of individually rare recent mutations.

- If the mutation–selection hypothesis is true, identifying susceptibility factors will require large-scale resequencing of individual genomes from cases and controls. New sequencing technologies make this feasible.

The main genetic contribution to morbidity and mortality is through the genetic component of common complex diseases. These diseases have no one single cause, but result from the cumulative effects of a variety of genetic and environmental factors, often different in different affected individuals. Identifying the genes concerned is a central task for medical research. However, this is a much more formidable task than identifying the mutations that cause monogenic diseases. For Mendelian diseases, given a sufficient collection of affected families, linkage analysis as described in Chapter 14 can usually localize the causative gene to a small chromosomal segment containing only a handful of candidate genes. Similar studies in complex diseases have been much less successful.

A meta-analysis by Altmüller and colleagues in 2001 reviewed 101 linkage studies in 31 complex diseases. The result was sobering. Candidate regions defined in different linkage studies of the same disease rarely coincided. There were some real successes, reviewed by Lohmueller and colleagues (see Further Reading) but, despite huge efforts by leading research teams, overall the studies had made only limited progress in localizing susceptibility genes. Recently this has changed. A combination of new technology (high-resolution SNP chips) and new understanding of the structure of human genomes (the HapMap project) is finally allowing susceptibility factors to be reliably identified. In this chapter, we describe the ways in which this difficult problem has been approached.

15.1 FAMILY STUDIES OF COMPLEX DISEASES

Before work can begin on uncovering the genetic factors involved in common complex diseases, it is necessary to establish the criteria by which people are to be labeled as affected or unaffected. With Mendelian syndromes it is usually fairly obvious which features of a patient form part of the syndrome and which are coincidental. Different features may have different penetrances, but basically the components of the syndrome are those that co-segregate in a Mendelian pattern. Things are much less clear cut for non-Mendelian conditions. Even physically obvious conditions such as the major birth defects are very variable in severity— where is the line to be drawn? Should we lump together or split apart the various types of congenital heart malformation for example?

Psychiatric and behavioral conditions are especially difficult. A diagnostic label can be valid, in the sense that two independent investigators will agree whether or not it applies to a given patient, without being biologically meaningful. Great efforts have been made to establish valid diagnostic categories. These are codified in the successive versions of the Diagnostic and Statistical Manual of Mental Disorders (DSM) published by the American Psychiatric Association. Adhering to such conventions helps make different studies comparable, but it does not guarantee that the right genetic question is being asked.

Having established clear diagnostic criteria for a complex disease, it is necessary to show whether or not genetics has a role in the etiology of the condition. People who share more of their DNA should be more likely to share the phenotype under investigation. The obvious way to approach this is to show that the character runs in families. This involves family, twin, or adoption studies. The difficulty of distinguishing the effects of shared family environment from those of heredity has often made such studies controversial, especially for psychiatric conditions.

The risk ratio (λ) is a measure of familial clustering

Nobody would dispute the involvement of genes in a character that consistently gives Mendelian pedigree patterns or that is associated with a chromosomal abnormality. However, with non-Mendelian characters, whether continuous (quantitative) or discontinuous (dichotomous), there is no such reality check. It is necessary to prove claims of genetic determination. The obvious way to approach this is to show that the character runs in families.

The degree of family clustering of a disease can be expressed by the **risk ratio** ($\lambda_{\mathbf{R}}$), the risk to a relative (R) of an affected proband compared with the risk in the general population. A risk ratio of 1 implies no additional risk above that of the general population. Separate values can be calculated for each type of relative, for example λ_S for sibs. The mathematical properties of λ_R are derived in the 1990

TABLE 15.1 RISK OF SCHIZOPHRENIA AMONG RELATIVES OF SCHIZOPHRENICS: POOLED RESULTS OF SEVERAL STUDIES

Relative	No. at risk	Risk (%)	λ
Parents	8020	5.6	7
Sibs	9920.7	10.1	12.6
Sibs, one parent affected	623.5	16.7	20.8
Offspring	1577.3	12.8	16
Offspring, both parents affected	134	46.3	58
Half-sib	499.5	4.2	5.2
Uncles, aunts, nephews, nieces	6386.5	2.8	3.5
Grandchildren	739.5	3.7	4.6
Cousins	1600.5	2.4	3

Numbers at risk are corrected to allow for the fact that some at-risk relatives were below or only just within the age of risk for schizophrenia (say, 15–35 years). λ values are calculated assuming a population incidence of 0.8%. [Data from McGuffin P, Shanks MF & Hodgson RJ (eds) (1984) The Scientific Principles of Psychopathology. Grune & Stratton.]

papers by Risch (see Further Reading). As an example, Table 15.1 shows pooled data from several studies of schizophrenia. Family clustering is evident from the raised λ values, for example a sevenfold increased risk for somebody, one of whose parents is schizophrenic. As expected, λ values drop back toward 1 for more distant relationships such as nephews, nieces, and cousins.

Shared family environment is an alternative explanation for familial clustering

Geneticists must never forget that humans give their children their environment as well as their genes. Many characters run in families because of the shared family environment—whether one's native language is English or Chinese, for example. One always has to ask whether shared environment might be the explanation for familial clustering of a character. This is especially important for behavioral attributes such as IQ or schizophrenia, which depend at least partly on upbringing. Even for physical characters or birth defects it cannot be ignored: a family might share an unusual diet or some traditional medicine that could cause developmental defects. Among the Fore people of New Guinea, a degenerative brain disease, kuru, ran in families because, as part of funerary rituals, close relatives ate infectious brain material from deceased affected people. Something more than a familial tendency is necessary to prove that a non-Mendelian character is under genetic control. These reservations are not always as clearly stated in the medical literature as perhaps they should be. Table 15.5 on p. 472 shows what can happen if shared family environment is ignored.

Twin studies suffer from many limitations

Francis Galton, the brilliant but eccentric cousin of Charles Darwin, who laid so much of the foundation of quantitative genetics, pointed out the value of twins for human genetics. Monozygotic (MZ) twins are genetically identical clones and should always be **concordant** (both the same) for any genetically determined character. This is true regardless of the mode of inheritance or number of genes involved; the only exceptions are for characters dependent on post-zygotic somatic genetic changes (the pattern of X-inactivation in females, the repertoire of functional immunoglobulin and T-cell receptor genes, and random post-zygotic somatic mutations leading to mosaicism). Dizygotic (DZ) twins share half their genes on average, the same as any pair of sibs.

Concordance can be calculated in two ways. Pairwise concordance counts the number of pairs of twins, ascertained through an affected proband, in which

TABLE 15.2 TWIN STUDIES IN SCHIZOPHRENIA			
Study	Country	Concordant pairs	
		MZ	DZ
Kringlen et al. (1968)	Norway	14/50 (0.28)	6/94 (0.06)
Fischer et al. (1969)	Denmark	5/21 (0.23)	4/41 (0.10)
Tienari et al. (1975)	Finland	3/20 (0.15)	3/42 (0.07)
Farmer et al. (1987)	UK	6/17 (0.35)	1/20 (0.05)
Onstad et al. (1991)	Norway	8/24 (0.33)	1/28 (0.04)

The numbers show the total number of twin pairs ascertained and the number that were concordant (both twins diagnosed as schizophrenic). Diagnostic and inclusion criteria varied between studies; despite the heterogeneity there is a clear tendency for more monozygotic (MZ) than dizygotic (DZ) pairs to be concordant. [For references, see Onstad S, Skre I, Torgersen S & Kringlen E (1991) *Acta Psychiatr. Scand.* 83, 395–401.]

both twins have the condition (concordant, +/+) and the number of pairs in which only one twin is affected (discordant, +/–). Probandwise concordance is obtained by counting a pair twice if both were probands, and thus gives higher values for the concordance. For example, in the study by Onstad et al. cited in **Table 15.2**, in 7 of the 24 MZ twin pairs, both twins were independently ascertained as affected probands. Thus, the probandwise concordance was 0.48, calculated as $[(8 + 7)/(24 + 7)]$, in comparison with the pairwise concordance of 0.33 (8/24). Probandwise concordances are thought to be more comparable with other measures of family clustering.

Genetic characters should show a higher concordance in MZ than DZ twins, and many characters do. However, a higher concordance in MZ twins than in DZ twins does not prove a genetic effect. For a start, half of DZ twins are of different sexes, whereas all MZ twins are the same sex. Even if the comparison is restricted to same-sex DZ twins (as it is in the studies shown in Table 15.2), at least for behavioral traits the argument can be made that MZ twins are more likely to look very similar, to be dressed and treated the same, and thus to share more of their environment than DZ twins.

Separated monozygotic twins

Monozygotic twins separated at birth and brought up in entirely separate environments seem to provide an ideal experiment for separating the effects of shared genes and shared environment. Francis Crick once made the tongue-in-cheek suggestion that one of each pair of twins born should be donated to science for this purpose. Such separations happened in the past more often than one might expect—the birth of twins was sometimes the last straw for an overburdened mother. Fascinating television programs can be made about twins reunited after 40 years of separation, who discover they have similar jobs, wear similar clothes, and like the same music. As research material, however, separated twins have many drawbacks:

- Any research is necessarily based on small numbers of arguably exceptional people.
- The separation was often not total—often the twins were separated some time after birth, and were brought up by relatives.
- There is a bias of ascertainment—everybody wants to know about strikingly similar separated twins, but separated twins who are very different are not newsworthy.
- Research on separated twins cannot distinguish intrauterine environmental causes from genetic causes. This may be important, for example in studies of sexual orientation (the mapping gay gene), in which some people have suggested that maternal hormones may affect the fetus *in utero* so as to influence its future sexual orientation.

Thus, for all their anecdotal fascination, separated twins have contributed relatively little to human genetic research.

Adoption studies are the gold standard for disentangling genetic and environmental factors

If separating twins is an impractical way of disentangling heredity from family environment, studying adopted people is much more promising. Two study designs are possible:

- Find adopted people who suffer from a particular disease known to run in families, and ask whether it runs in their biological family or their adoptive family.
- Find affected parents whose children have been adopted away from the family, and ask whether being adopted saved the children from the family disease.

A celebrated but controversial study by Rosenthal & Kety (see Further Reading) used the first of these designs to test for genetic factors in schizophrenia. The diagnostic criteria used in this study have been criticized, and there have also been claims (disputed) that not all diagnoses were made truly blind. However, an independent re-analysis using DSM-III diagnostic criteria reached substantially the same conclusion: it was the genes rather than the environmental influence of a schizophrenic parent that increased the risk for the offspring. Table 15.3 shows the results of a later extension of this study.

The main obstacle in adoption studies is lack of information about the biological family, frequently made worse by the undesirability of approaching them with questions. Efficient adoption registers exist in only a few countries. A secondary problem is selective placement, in which the adoption agency, in the interests of the child, chooses a family likely to resemble the biological family. Adoption studies are unquestionably the gold standard for checking how far a character is genetically determined, but because they are so difficult, they have in the main been performed only for psychiatric conditions, for which the nature–nurture arguments are particularly contentious.

15.2 SEGREGATION ANALYSIS

Pure Mendelian and pure polygenic characters represent the opposite ends of a continuum. In between are *oligogenic traits* governed by a few major susceptibility loci, possibly operating against a polygenic background, and possibly subject to major environmental influences. **Segregation analysis** is a statistical tool for analyzing the inheritance of any character. It can provide evidence for or against a major susceptibility locus and can at least partly define its properties. The results can help guide future linkage or association studies.

Complex segregation analysis estimates the most likely mix of genetic factors in pooled family data

Analyzing data on the relatives of a large collection of people affected by a familial but non-Mendelian disease is not a simple task. There could be both genetic and environmental factors at work; the genetic factors could be polygenic, oligogenic, or monogenic (Mendelian) with any mode of inheritance, or any mixture

TABLE 15.3 AN ADOPTION STUDY IN SCHIZOPHRENIA		
Case types	Schizophrenia cases among biological relatives	Schizophrenia cases among adoptive relatives
Index cases (47 chronic schizophrenic adoptees)	44/279 (15.8%)	2/111 (1.8%)
Control adoptees (matched for age, sex, social status of adoptive family, and number of years in institutional care before adoption)	5/234 (2.1%)	2/117 (1.7%)

The study involved 14,427 adopted persons aged 20–40 years in Denmark; 47 of them were diagnosed as chronic schizophrenic. The 47 were matched with 47 non-schizophrenic control subjects from the same set of adoptees. [Data from Kety SS, Wender PH, Jacobsen B et al. (1994) *Arch. Gen. Psychiatry* 51, 442–455.]

TABLE 15.4 COMPLEX SEGREGATION ANALYSIS OF HIRSCHSPRUNG DISEASE

Model	d	t	q	H	z	x	χ^2	p
Mixed	1.00	7.51	9.6×10^{-6}		0.01	0.15		
Sporadic							334	$< 10^{-5}$
Polygenic				1.00	1.00		78	$< 10^{-5}$
Major recessive locus	0.00	8.22	3.8×10^{-3}				35	$< 10^{-5}$
Major dominant locus	1.00	7.56	1.2×10^{-5}			0.19	2.8	0.42

Data are for families ascertained through a proband with long-segment Hirschsprung disease (OMIM 142623). Parameters that can be varied are as follows: d, the degree of dominance of any major disease allele; t, the difference in liability between people homozygous for the low-susceptibility and the high-susceptibility alleles of a major susceptibility gene, measured in units of standard deviation of liability; q, the gene frequency of any major disease allele; H, the proportion of total variance in liability that is due to polygenic inheritance, in adults; z, the ratio of heritability in children to heritability in adults; x, the proportion of cases due to new mutation. The values shown are those that best account for the family data using the stated model. The χ^2 statistic is a standard test that compares the performance of each model with the mixed model, in which a mix of all mechanisms is allowed. A single major locus encoding dominant susceptibility explains the data as well as the mixed model. [Data from Badner JA, Sieber WK, Garver KL & Chakravarti A (1990) *Am. J. Hum. Genet.* 46, 568–580.]

of these, and the environmental factors may include both familial and non-familial variables. In complex segregation analysis, a whole range of possible inheritance patterns, gene frequencies, penetrances, and so on, are modeled by computer analysis to find the mix of scenarios that gives the greatest overall likelihood for the observed data. Factors in this mixed model are then omitted or constrained to identify the minimum that must be included so as to avoid a significant loss of explanatory power. As with lod score analysis (see Chapter 14), the question asked is how much more likely the observations are when one hypothesis is compared with another.

In the example of **Table 15.4**, the ability of specific models (sporadic, polygenic, recessive, dominant) to explain the data was compared with the likelihood calculated by a general mixed model, in which the computer program could freely optimize the mixture of single-gene, polygenic, and random environmental causes. All models were constrained by the overall epidemiology, sex ratios, and probabilities of ascertainment estimated from the collected data. A single-locus dominant model is not significantly worse than the mixed model at explaining the data ($\chi^2 = 2.8$, $p = 0.42$). In contrast, models assuming no genetic factors (sporadic), pure polygenic inheritance, or pure recessive inheritance perform very badly. On the argument that simple explanations are preferable to complicated explanations, the analysis suggests the existence of a major dominant susceptibility to Hirschsprung disease. One such factor has now been identified, the *RET* (rearranged during transfection) oncogene (OMIM 164761).

However clever the segregation analysis program is, it can do no more than maximize the likelihood across the parameters it was given. If a major factor is omitted, the result can be misleading. This was well illustrated by the data in **Table 15.5**. McGuffin & Huckle asked their classes of medical students which of their relatives had attended medical school. When they fed the results through a

TABLE 15.5 A RECESSIVE GENE FOR ATTENDING MEDICAL SCHOOL?

Model	d	t	q	H	χ^2	p
Mixed	0.087	4.04	0.089	0.008		
Sporadic					163	$< 10^{-5}$
Polygenic				0.845	14.4	< 0.005
Major recessive locus	0.00	7.62	0.88		0.11	n.s.

The data are taken from a survey of medical students and their families. The meaning of the symbols is explained in the foornote to Table 15.4. Affected is defined as attending medical school. The analysis seems to support recessive inheritance, because this accounts for the data equally well as the unrestricted model (but see the text). n.s., not significant. [Data from McGuffin P & Huckle P (1990) *Am. J. Hum. Genet.* 46, 994–999.]

segregation analysis program, it came up with results apparently favoring the existence of a recessive gene for attending medical school. Though amusing, this was not done as a joke, nor to discredit segregation analysis. The authors did not allow the segregation analysis program to consider the likely true mechanism, namely a shared family environment. The program's next best alternative was mathematically valid but biologically unrealistic. The serious point that McGuffin & Huckle were making was that there are many pitfalls in segregation analysis of human behavioral traits, and incautious analyses can generate spurious genetic effects.

Although complex segregation analysis can provide a valuable framework for detailed genetic studies of a condition, in recent years its use in research has declined. This is partly because most of the major complex diseases have already been investigated. Another key reason is that segregation analysis is not able to decompose the heterogeneity that is characteristic of complex traits. Segregation analysis necessarily provides a top-down, bird's-eye view of a condition. Given the extreme genetic heterogeneity of most if not all multifactorial diseases, the value of such a view can be questioned. Provided there is evidence that genetic factors are involved somewhere in the etiology, it may be more productive to dive in and use the tools of molecular genetics to hunt for the factors directly, rather than worry about their overall statistical properties. Thus, the focus has moved on, first to linkage analysis and more recently to association studies.

15.3 LINKAGE ANALYSIS OF COMPLEX CHARACTERS

Linkage studies for complex diseases use rather different methods from those described in Chapter 14. Rather than test a detailed hypothesis about recombination fractions, the analysis seeks to identify chromosomal segments shared by affected family members, without having to specify exactly how any susceptibility factors carried on those shared segments contribute to the disease. These methods are very robust for detecting susceptibility factors that are neither necessary nor sufficient for the disease to develop.

Standard lod score analysis is usually inappropriate for non-Mendelian characters

Standard lod score analysis is called **parametric** because it requires a precise genetic model, detailing a series of parameters: the mode of inheritance, gene frequencies, and information about the penetrance of each genotype. As long as a valid model is available, parametric linkage provides a wonderfully powerful method for scanning the genome in 5–10 Mb segments to locate a disease gene. For Mendelian characters, specifying an adequate model should be no great problem. Non-Mendelian conditions, however, are much less tractable. Although complex segregation analysis can provide parameters for the overall genetic susceptibility, the parameters for any one susceptibility factor are impossible to guess in advance. Thus, the type of linkage analysis that was described in Chapter 14 cannot be used for complex diseases.

Near-Mendelian families

One approach to this problem is to look for a subset of families in which the condition segregates in a near-Mendelian manner. Segregation analysis is used to define the parameters of a genetic model in those families, which are then used in a standard (parametric) linkage analysis. Such families may arise in three ways:

- Any complex disease is likely to be heterogeneous, so the family collection may well include some with Mendelian conditions phenotypically indistinguishable from the non-Mendelian majority.
- The near-Mendelian families may represent cases in which, by chance, many determinants of the disease are already present in most people, so that the balance is tipped by the Mendelian segregation of just one of the many susceptibility factors.
- The near-Mendelian pattern may be spurious—just chance aggregations of affected people within one family.

In the first case, identifying the Mendelian subset is intrinsically valuable, but it does not necessarily cast any light on the causes of the non-Mendelian disease. That was the case with breast cancer and Alzheimer disease. In breast cancer this led to the discovery of the *BRCA1* and *BRCA2* genes, as described in Chapter 16 (see Case 7, p. 526). Mutations in the *PSEN1*, *PSEN2*, and *APP* genes cause rare Mendelian forms of early-onset Alzheimer disease. However, the common non-Mendelian forms of breast cancer and Alzheimer disease do not usually involve any of these genes. In the second case, the loci mapped are also susceptibility factors for the common non-Mendelian disease—Hirschsprung disease is an example. Finally, an early study of schizophrenia exemplified the third case, producing a lod score of 6 that is now generally agreed to have been spurious. This debacle was enough to persuade most investigators to switch to alternative methods of analysis.

Non-parametric linkage analysis does not require a genetic model

Model-free or **non-parametric** methods of linkage analysis look for alleles or chromosomal segments that are shared by affected individuals more often than random Mendelian segregation would predict. Some of the basic ideas underlying these approaches were set out in 1990 in the three papers by Risch mentioned previously.

Identity by descent versus identity by state

It is important to distinguish segments *identical by descent* (IBD) from those *identical by state* (IBS). Alleles that are IBD are *demonstrably* copies of the same ancestral (usually parental) allele. IBS alleles look identical, and may indeed be copies, but their common ancestry is not demonstrable. This may be because of a lack of information about any common ancestor, or because the genotypes do not permit unambiguous tracing of the ancestral origin of the alleles. **Figure 15.1** illustrates the difference. In non-parametric analysis, IBD alleles are treated mathematically in terms of the Mendelian probability of inheritance from the defined common ancestor; for IBS alleles, the population frequency is used. For very rare alleles, two independent origins are unlikely, so IBS generally implies IBD. With common alleles, no such inference can be made. Multiallele microsatellites are more efficient than two-allele markers such as SNPs for defining IBD, and multilocus multiallele haplotypes are better still, because any one haplotype is likely to be rare. Shared segment analysis can be conducted with either IBS or IBD data, provided that the appropriate analysis is used. IBD is the more powerful of the two, but it requires samples from more relatives because it is necessary to work out the likelihood that sharing is IBD rather than IBS.

Affected sib pair analysis

Picking a chromosomal segment at random, pairs of sibs are expected to share 0, 1, or 2 parental haplotypes with frequencies of ¼, ½, and ¼, respectively (ratios of 1:2:1). However, if both sibs are affected by a genetic disease, then any segment of chromosome that carries the disease locus is likely to be shared. For a fully penetrant Mendelian dominant disease, affected sibs would always share the parental haplotype that carried the disease allele; if the disease is recessive they would always share *both* the relevant parental haplotypes (**Figure 15.2**). If a susceptibility factor is neither necessary nor sufficient for disease, then not all affected sib pairs will share the relevant chromosomal segment, but there will still be a statistical tendency to share that segment more often than just by chance. This allows a simple form of linkage analysis. **Affected sib pairs** (**ASPs**) are typed for markers,

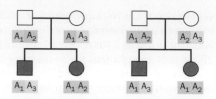

Figure 15.1 Identity by state and identity by descent. Both sib pairs share allele A_1. The first sib pair have two independent copies of A_1 (colored red and blue), indicating identity by state but not by descent. The second sib pair share copies of the same paternal A_1 allele (red), showing identity by descent. The difference is only apparent if the parental genotypes are known.

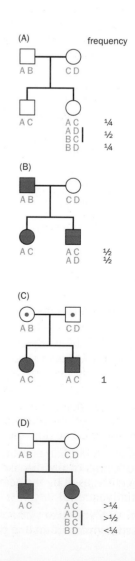

Figure 15.2 Affected sib pair analysis. (A) By random segregation, sib pairs share 2 (both *AC*), 1 (*AC* and either *AD* or *BC*), or 0 (*AC* and *BD*) parental haplotypes ¼, ½, and ¼ of the time, respectively. (B) Pairs of sibs who are both affected by a Mendelian dominant condition must share the segment that carries the disease allele, and they may or may not (a 50:50 chance) share a haplotype from the unaffected parent. (C) Pairs of sibs who are both affected by a Mendelian recessive condition necessarily share the same two parental haplotypes for the relevant chromosomal segment. (D) For complex conditions, haplotype sharing above that expected to occur by chance (as in panel A) identifies chromosomal segments containing susceptibility genes.

and chromosomal regions are sought where the sharing is above the random 1:2:1 ratios of sharing 2, 1, or 0 haplotypes IBD. If the sib pairs are tested only for IBS, the expected sharing on the null hypothesis must be calculated as a function of the gene frequencies.

Shared segment analysis can be performed using any set of affected relatives and without making any assumptions about the genetics of the disease. Affected sib pairs are especially favored because they are relatively easy to collect. Multilocus analysis is preferable to single-locus analysis because it more efficiently extracts the information about IBD sharing across the chromosomal region. The Mapmaker/Sibs computer program is widely used to analyze multipoint ASP data. Programs such as Genehunter extend shared segment analysis to other relationships. The programs calculate the extent to which affected relatives share alleles IBD. The result across all affected pedigree members is compared with the null hypothesis of simple Mendelian segregation (markers will segregate according to Mendelian ratios unless the segregation among affected people is distorted by linkage or association). The comparison can be used to compute a **non-parametric lod** (**NPL**) **score**. Theoretical arguments by Lander & Kruglyak suggest genomewide lod score thresholds of 3.6 for IBD testing of affected sib pairs, and 4.0 for IBS testing.

Linkage analysis of complex diseases has several weaknesses

One drawback of ASP analysis is that the candidate regions it identifies are large and likely to contain many genes. Few recombinants separate sibs, so shared parental chromosome segments are large. Identifying the actual disease gene or susceptibility factor in such a large region will be challenging. Crucially, complex disease analysis has no process analogous to the end-game of Mendelian mapping, in which closer and closer markers are tested until there are no more recombinants. Moreover, sib pairs share many segments by chance. Nevertheless, because of its simplicity and robustness, ASP mapping has been one of the main tools for seeking genes conferring susceptibility to common complex diseases.

Any individual susceptibility factor is neither necessary nor sufficient for a person to develop a complex disease. This means that any genetic hypothesis being tested is necessarily much looser than the hypotheses involved in Mendelian linkage analysis. In addition, it is often supposed that many susceptibility factors are of ancient origin and common in the population, unlike the variants that cause Mendelian diseases. This means that, in extended families, even if two affected relatives both have a certain factor, they may have inherited it through two different recent ancestors, maybe in association with different alleles of the marker. As a result of these problems, lod scores are typically modest, and only occasionally reach the threshold of significance.

Significance thresholds

For a genomewide study (whether of a Mendelian character or a complex one), the threshold of significance is a lod score at which the probability of finding a false positive anywhere in the genome is 0.05. In Chapter 14, we noted the distinction between pointwise (or nominal) and genomewide significance:

- The **pointwise p value** of a linkage statistic is the probability of exceeding the observed value at a specified position in the genome, assuming the null hypothesis of no linkage at that location.

- The **genomewide p value** is the probability that the observed value will be exceeded anywhere in the genome, assuming the null hypothesis of no linkage for each individual location.

Lander & Kruglyak proposed the terminology shown in **Table 15.6** for reporting the strength of linkage data. This is now widely accepted for use in reports. Note that a p value of 10^{-3} is *not* equivalent to a lod score of 3.0—the two measures are not the same:

- A lod score of 3 means that the data are 10^3 times more likely on the given linkage hypothesis than on the null hypothesis. This is a measure of likelihood, or the relative probability of the data on two competing hypotheses.

- A p value of 10^{-3} means that the stated lod score will be exceeded only once in 10^3 times, given the null hypothesis. This is a statement about the probability

TABLE 15.6 CRITERIA FOR REPORTING LINKAGE IN GENOMEWIDE STUDIES

Category of linkage	Expected number of occurrences by chance in a genomewide scan	Range of approximate p values	Range of approximate lod scores
Suggestive	1	7×10^{-4} to 3×10^{-5}	2.2–3.5
Significant	0.05	2×10^{-5} to 4×10^{-7}	3.6–5.3
Highly significant	0.001	$\leq 3 \times 10^{-7}$	≥ 5.4
Confirmed	0.01 in a search of a candidate region that gave significant linkage in a previous independent study		

Criteria suggested by Lander E & Kruglyak L (1995) *Nat. Genet.* 11, 241–247. The figures for *p* values and lod scores are for ASP studies as given by Altmüller J, Palmer LJ, Fischer G et al. (2001) *Am. J. Hum. Genet.* 69, 936–950.

of the data on the given hypothesis. For example, as explained in Chapter 14, in two-point analysis of a Mendelian condition, a lod score of 3 corresponds to an absolute probability of 0.05.

Complex disease studies often use simulation to estimate their significance thresholds. In a typical *permutation test*, 1000 replicates of the family collection are generated by a computer program with random marker genotypes, but based on correct allele frequencies, recombination fractions, and so on. A genomewide search is conducted in each simulated data set and the maximum lod score is noted. The genomewide threshold of significance is set at a score that is exceeded in less than 5% of the replicates. This is taken to be a lod score that has no more than a 5% probability of having arisen purely by chance in the real data set.

Striking lucky

The actual lod score for any particular factor depends on how it happened to segregate in the particular families studied. But there are many susceptibility factors, and a genomewide scan tests for all of them. If there are a dozen or more susceptibility loci, but the studies are only marginally powerful enough to detect the effect of any of them, then there is a large and unavoidable element of chance. In one set of families, a certain factor may happen to segregate in a way that gives a significant lod score, whereas in an independent set of families used in a replication study, that factor may not happen to show such a favorable pattern (**Figure 15.3**). In fact, targeted replication requires a study design with a much higher statistical power than the original study. Thus, failure to replicate the conclusion of a study does not necessarily mean that the original report was a false positive. Nevertheless, for a long time the paucity of well-replicated findings cast a shadow over linkage studies of complex disease.

An example: linkage analysis in schizophrenia

As an example of these problems, **Table 15.7** gives a rough summary of major findings from the 10 genomewide linkage studies of schizophrenia that formed part of the meta-analysis by Altmüller et al. that was cited in the introduction to this chapter. That paper should be consulted for references to the 10 studies, and the original papers should be consulted for more details—the studies often used sophisticated multi-stage designs and produced large amounts of interesting information. Nevertheless, the conclusion is clear: suggestive linkages found in one study were almost never confirmed in any of the others, and overall there was no evidence that any true susceptibility loci had been identified.

Figure 15.4 shows the results of one of these studies in more detail. Because it is unclear what definition of affected might be the most biologically relevant, these authors analyzed their data using three alternative criteria for affected status (for details see the figure legend). This is an entirely reasonable strategy, but it does introduce extra degrees of freedom into the analysis, and hence requires a more stringent threshold of significance to be used. In this case, no suggestive or significant linkages were found.

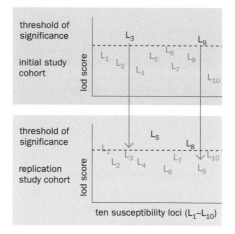

Figure 15.3 The role of chance in detecting weak effects. In a genomewide linkage study of a disease, alleles at each of ten loci (L_1–L_{10}) confer susceptibility to the disease under study, but each has such a weak effect that a linkage study of a given collection of families is not predicted to give a significant lod score. By chance, however, genotypes at two of the loci, L_3 and L_9, happen to give a significant lod score in this collection of families. In a replication study using an independent sample, family members do not happen to have this chance favorable set of genotypes for L_3 and L_9, and therefore the previous significant lod scores are not confirmed. Instead, other loci, L_5 and L_8, happen to have genotypes that give significant lod scores.

TABLE 15.7 POSITIVE FINDINGS FROM 10 GENOMEWIDE LINKAGE STUDIES OF SCHIZOPHRENIA

Study	Sample	No. of individuals genotyped	Significance level	Chromosomal region reported
Coon et al. (1994)	families (mixed)	126[a]	suggestive	4q, 22q
Moises et al. (1995)	families (mixed)	213[a]	none	
Blouin et al. (1998)	families (mixed)	363	significant	13q
			suggestive	8p
Faraone et al. (1998)	families (European American)	146	suggestive	10p
Kaufman et al. (1998)	families (African American)	98	none	
Levinson et al. (1998)	families (mixed)	269	none	
Shaw et al. (1998)	ASPs (European descent)	171[a]	none	
Hovatta et al. (1999)	families (Finnish isolate)	20 families	suggestive	1q
Williams et al. (1999)	ASPs (UK or Irish Caucasian)	196 ASPs	suggestive	4p, 18q, Xcen
Ekelund et al. (2000)	ASPs (Finnish)	134 ASPs	suggestive	1q, 7q

[a]Number of affected individuals. ASPs, affected sib pairs. Significance levels follow the Lander–Kruglyak criteria shown in Table 15.6. These studies formed part of the meta-analysis by Altmüller J, Palmer LJ, Fischer G et al. (2001) *Am. J. Hum. Genet.* 69, 936–950. Full references are given in that paper.

15.4 ASSOCIATION STUDIES AND LINKAGE DISEQUILIBRIUM

The low success rate of linkage studies for complex traits in the 1990s suggested that many, if not most, of the susceptibility factors must be relatively weak, highly heterogeneous, or both. More statistical power would be needed to detect them reliably. Some researchers responded to the challenge by starting studies of much larger samples. However, a seminal paper by Risch & Merikangas in 1996 showed that power to detect weak effects should be achieved more easily by studying associations rather than through linkage analysis. This prompted a general move to association studies. Rather than studying affected relatives, association studies seek populationwide associations between a particular condition and a particular allele or haplotype somewhere in the genome. In this section, we consider how populationwide associations between a specific susceptibility allele and a disease arise, and how they might be detected.

Association is not a specifically genetic phenomenon; it is simply a statistical statement about the co-occurrence of alleles or phenotypes. Allele *A* is associated with disease D if people who have D also have *A* significantly more often (or maybe less often) than would be predicted from the individual frequencies of D and *A* in the population. For example, *HLA-DR4* is found in 36% of the general UK population but in 78% of people with rheumatoid arthritis (RA). Thus *HLA-DR4* is *associated* with RA. The strength of the association is measured by the **relative risk**. This is the amount by which being DR4-positive multiplies the baseline risk of RA. It would be calculated by ascertaining the incidence of RA in *DR4*-positive and *DR4*-negative people. A relative risk of 1 means that being *DR4*-positive confers no additional risk of RA. An alternative measure that is often used is the **odds ratio**. This is explained in Box 19.3 (p. 609). The odds ratio has the advantage that it can be calculated directly from the results of a case-control study, without the need for any information about the population incidence. Again, an odds ratio of 1 means that the factor has no effect on risk.

Associations have many possible causes

A population association can have many possible causes, not all of which are genetic.

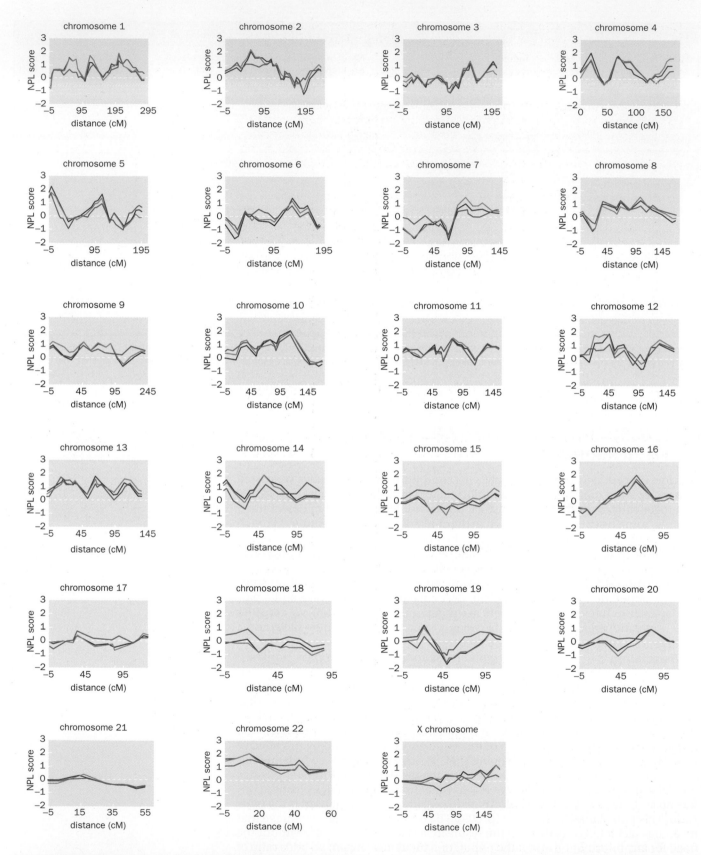

Figure 15.4 Results of a genomewide association study of schizophrenia. A total of 171 affected individuals from 70 multiply affected sibships were genotyped for 338 microsatellite markers spaced across the genome. The blue line shows results when only people with schizophrenia (DSM-III criteria) were counted as affected. For the red line, individuals with DSM schizoaffective disorder were also included. The green line uses a broad definition of affected (schizophrenia, schizoaffective disorder, paranoid or schizotypal personality disorder, delusional disorder, or brief reactive psychosis). In the event, no significant or suggestive lod scores were observed. NPL, nonparametric lod score. [From Shaw SH, Kelly M, Smith AB et al. (1998) *Am. J. Med. Genet. (Neuropsychiatr. Genet.)* 81, 364–378. With permission of Wiley-Liss, Inc., a subsidiary of John Wiley & Sons, Inc.]

- **Direct causation:** having allele *A* makes you susceptible to disease D. Possession of *A* may be neither necessary nor sufficient for somebody to develop D, but it increases the likelihood.

- **An epistatic effect:** people who have disease D might be more likely to survive and have children if they also have allele *A*.

- **Population stratification:** the population contains several genetically distinct subsets, and both the disease and allele *A* happen to be particularly frequent in one subset. Lander & Schork give the example of the association in the population of the San Francisco Bay area between carrying the *A1* allele at the HLA locus and being able to eat with chopsticks. *HLA*A1* is more frequent among Chinese than among Caucasians.

- **Type I error:** association studies normally test a large number of markers for association with a disease. Even without any true effect, 5% of results will be significant at the $p = 0.05$ level and 1% at the $p = 0.01$ level. These are false positives, or type I errors. The raw p values need correcting for the number of questions asked. In the past, researchers often applied inadequate corrections. Even after adequate correction, there will remain a certain probability of a false positive result.

- **Linkage disequilibrium (LD):** the disease-associated allele *A* marks an ancestral chromosome segment that carries a sequence variant causing susceptibility to the disease, as described below. Most disease association studies aim to discover associations caused by linkage disequilibrium; it is then an additional step to identify the actual causative sequence variant.

Association is quite distinct from linkage, except where the family and the population merge

In principle, linkage and association are totally different phenomena. Linkage is a relation between *loci* (physical sites on the chromosome), but association is a relation between specific *alleles* and/or *phenotypes*. Linkage is a specifically genetic relationship, whereas association is simply a statistical observation that might have various causes. However, where the family and the population merge, so do linkage and association.

Linkage does not of itself produce any association in the general population. For example, the *STR45* microsatellite locus is linked to the dystrophin locus. Nevertheless, the distribution of *STR45* alleles among a set of unrelated Duchenne muscular dystrophy patients (OMIM 310200), all of whom carry dystrophin mutations, is just the same as in the general population. The mutations arose independently, on a set of chromosomes whose distribution of *STR45* alleles was typical of the general population. However, *within a family* where a particular dystrophin mutation is segregating, we would expect affected people to share the *same* allele of *STR45*, because the loci are tightly linked. Thus, linkage creates associations within families, but not between unrelated people. But how far does a family extend?

All humans are related, if we go back far enough. A rough calculation suggests that, in the UK, two unrelated people would typically have common ancestors not more than 22 generations ago. If fully outbred, they would have $2^{22} = 4$ million ancestors each at that time. Assuming a generational time of 25 years, 22 generations is about 500 years, and in the year 1500 the population of Britain was only a little over 4 million (**Figure 15.5**). It is nevertheless useful to draw a distinction between people who know they are part of one family and those who simply have unknown common ancestors. We will use the word **unrelated** to describe people who do not have a *demonstrable* common ancestor, normally in the past four or so generations.

Association studies depend on linkage disequilibrium

If two supposedly unrelated people with disease D have actually inherited their disease susceptibility from a distant common ancestor, they may well tend also to share particular ancestral alleles at loci closely linked to the susceptibility locus. Thus, in so far as a population is one extended family, population-level associations should exist between alleles of ancestral disease susceptibility genes

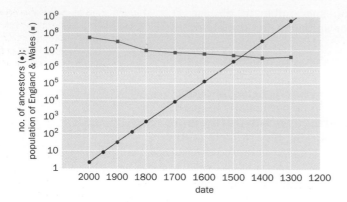

Figure 15.5 Merging into the gene pool. The number of a person's ancestors (red circles) compared with the population of England and Wales (blue squares). This model assumes that there are no consanguineous marriages and that there is a 25-year generation time. On this model two unrelated present-day people would share all their ancestors in 1470. In reality, of course, the population is not fully outbred, and the two people would have strongly overlapping but not identical pools of ancestors.

and closely linked markers. Particular combinations of alleles at closely linked loci occur more often (or less often) than the individual allele frequencies would predict. This phenomenon is called **linkage disequilibrium** (**LD**). Strictly, it would be better called allelic association, but the use of the term LD is firmly established in human genetics.

Linkage disequilibrium can only point to susceptibility factors that have been inherited from ancient common ancestors. A set of unrelated Duchenne muscular dystrophy patients would not be expected to share an ancient disease allele. As mentioned in Chapter 3, there is a rapid turnover of mutant dystrophin alleles, a high rate of loss through natural selection being balanced by a high rate of fresh mutations. Unrelated patients are likely to carry unrelated mutations. Linkage disequilibrium can only cause populationwide disease associations if the disease allele has persisted for many generations. The hope behind disease association studies is that many susceptibility factors are in themselves quite benign, at least in the environments in which our genomes evolved (or alternatively, any deleterious effect in relation to one disease might be balanced by an advantageous effect for a different condition). It is only when they get into combination with other factors that they cause susceptibility. Moreover, many common diseases affect mainly older people well past childbearing age, by which time natural selection has far less relevance for the survival of their genes.

The size of shared ancestral chromosome segments

Suppose that two unrelated people each inherit a disease susceptibility allele from their common ancestor. During the many generations and many meioses that separate them from their common ancestor, repeated recombination will have reduced the shared chromosomal segment to a very small region (**Figure 15.6**). Only alleles at loci tightly linked to the disease susceptibility locus will still be shared. For a marker showing recombination fraction θ with the shared disease locus, a proportion θ of ancestral chromosomes will lose the association each generation, and a proportion $(1 - \theta)$ will retain it. After n meioses, a fraction $(1 - \theta)^n$ of chromosomes will retain the association. The half-life of LD between loci 1 cM ($\theta = 0.01$) and 2 cM ($\theta = 0.02$) apart is 69 and 34 meioses respectively, since $(0.99)^{69} \approx (0.98)^{34} \approx 0.5$.

We calculated above that the ancestry of two unrelated British people merges completely 22 generations back. That calculation was grossly simplified because it assumed that the entire British population has been one freely interbreeding unit over the past 500 years. However, it provides a first crude estimate that the ancestral segments shared between two unrelated people from a large population might be of the order of a few megabases long (using the rule of thumb that 1cM = 1 Mb). For a population association, we require ancestral segments that are shared not just by two particular individuals but by a significant proportion of all descendants of that ancestor. The locations of crossovers will be different in each lineage leading down from the common ancestor. Thus, segments that are shared by enough people to produce population associations will, on average, be much shorter than the segments shared by any two particular individuals (in reality this will be a few kilobases—see Figure 15.6). More sophisticated calculations use a Poisson distribution of recombination events and incorporate

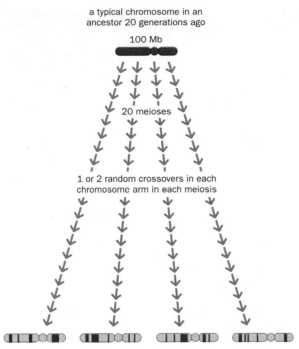

a typical chromosome in an
ancestor 20 generations ago

100 Mb

20 meioses

1 or 2 random crossovers in each
chromosome arm in each meiosis

the same chromosome in four present-day descendants

Figure 15.6 The size of shared ancestral chromosome segments. A typical chromosome is shown in a common ancestor, 20 generations ago, of four present-day individuals. There will be one or two random crossovers in each chromosome arm in each of the 20 meioses linking each present-day person to their common ancestor. Only a small proportion of the sequence of the ancestor's chromosome will be inherited by descendants after 20 generations (red segments). The ancestral segments that are shared by a significant proportion of all descendants are very small, typically 5–15 kb.

assumptions about population structure and history to estimate the size distribution of shared ancestral segments. However, the wide stochastic variance and reliance on unknowable details of population history make even the most elaborate calculations unreliable. What we need is data, and recently increasing quantities of real data have become available.

Studying linkage disequilibrium

LD is a statistic about populations, not individuals. To study it, a sample of individuals from the chosen population is genotyped for a series of linked polymorphic markers. SNPs are the markers of choice, for three reasons:

- They are sufficiently abundant that they can be used to check very short chromosome segments for disequilibrium.

- In comparison with microsatellites, they have a far lower rate of mutation. This is important when the aim is to identify ancient conserved chromosomal segments.

- SNPs are easy to genotype on a large scale, up to 1 million at once, spread across the genome. This is very hard with other polymorphisms.

Unless genotyping is done on individual chromosomes isolated by laboratory manipulation (an expensive option), the raw data will consist of the genotypes at each individual locus, rather than haplotypes of alleles across multiple linked loci. The raw genotype data must be *phased*—that is, converted into haplotypes—before any useful analysis can be done. This can be done by genotyping other family members to infer haplotypes; alternatively, computer programs can be used to convert genotypes into haplotypes. These programs use a maximum-likelihood procedure: they guess possible haplotypes, and keep trying until they find the guess that best explains the whole data set with the minimum number of plausibly related haplotypes (haplotypes within a population should be related to one another by a minimum number of recombinations).

Various statistics have been used as measures of LD (**Box 15.1**).

The HapMap project is the definitive study of linkage disequilibrium across the human genome

Early studies of specific chromosome regions showed that LD is common but unpredictable. The data were in striking contradiction to the naive expectation that LD between any two loci would fall off as a smooth function of their physical

BOX 15.1 MEASURES OF LINKAGE DISEQUILIBRIUM

If two loci have alleles A,a and B,b with frequencies p_A, p_a, p_B, and p_b, there are four possible haplotypes: AB, Ab, aB, and ab. Suppose that the frequencies of the four haplotypes are p_{AB}, p_{Ab}, p_{aB}, and p_{ab}. If there is no LD, $p_{AB} = p_A p_B$ and so on, because the haplotype will be constructed just by random assortment of the constituent alleles. The degree of departure, D, from this random association can be measured by $D = p_{AB} p_{ab} - p_{Ab} p_{aB}$. As a measure of LD, D suffers from the property that its maximum absolute value depends on the gene frequencies at the two loci, as well as on the extent of disequilibrium. Among preferred measures are:

- $D' = (p_{AB} - p_A p_B)/D_{max}$, where D_{max} is the maximum value of $|p_{AB} - p_A p_B|$ possible with the given allele frequencies; the vertical lines indicate the absolute value or modulus of the expression.
- $\Delta^2 = (p_{AB} - p_A p_B)^2/(p_A p_a p_B p_b)$.

D' is the most widely used. It varies between 0 (no LD) and ± 1 (complete association) and is less dependent than D on the allele frequencies. As a rule of thumb, $D' > 0.33$ is often taken as the threshold level of LD above which associations will be apparent in the usual size of data set. The proliferation of alternative measures suggests that none is ideal.

distance apart. Closely spaced SNPs often showed little or no LD, whereas sometimes there was strong LD between more widely separated SNPs. **Box 15.2** shows how the patterns of disequilibrium can be represented graphically, and also illustrates just how complex and irregular the patterns can be. These irregularities reflect the combined effects of several factors:

- Recombination, the force that breaks up ancestral segments, occurs mainly at a limited number of discrete hotspots (see Chapter 14). SNPs that are close together but separated by a hotspot show little or no LD, whereas, conversely, LD may be strong across even a large chromosomal region if it is devoid of hotspots.

- Gene conversion (see Box 14.1) may replace a small internal part of a conserved segment, producing a localized breakdown of LD, whereas markers either side of the replaced segment continue to show LD with each other.

- Population history is important. The older a population is, the shorter are the conserved segments. LD is more extensive and of longer range in populations derived from recent founders, as often occurs with small, genetically isolated populations. LD will have a shorter range in populations that have remained constant in size than in populations that have undergone a recent expansion. Superimposed on this regularity are many stochastic effects. Chromosome segments may carry a mixture of marker alleles that are IBD and alleles that are only IBS (through independent mutations, back mutation, and so on), but LD statistics do not distinguish between these two cases.

Efficient disease association studies depend on a detailed knowledge of the patterns of LD across the genome. It is important to know how big and how diverse the ancestral chromosome segments are, so that SNPs can be chosen to test each conserved ancestral segment. The International HapMap Project was established to provide this detailed knowledge. A consortium of academic institutions and pharmaceutical companies typed several million SNPs in 269 individuals drawn from four human populations: 30 white American parent–child trios from Utah (CEU), 30 Yoruba parent–child trios from Ibadan, Nigeria (YRI), 45 individual Han Chinese from Beijing (CHB), and 44 individual Japanese from Tokyo (JPT). All were ostensibly healthy and, although not a formal population sample, were hopefully typical of the population from which they were drawn. The Phase I report in 2005 (see Further Reading) summarized results from 1,007,329 SNPs; Phase II has added a further 3 million. The same individuals have been used in several other studies, for example in surveys of copy-number variants and of gene expression levels, adding value and depth to the HapMap data.

The CEU and YRI samples, which each consisted of parent–child trios, could be phased directly. For each trio, the genotype of the child could be used to infer phase in the parental genotypes. Thus, each set of 30 trios yielded 120 phased parental haplotypes. For the CHB and JPT samples, computer programs were used to convert genotypes into haplotypes.

The key findings can be summarized as follows (**Table 15.8**):

- All four populations show a similar structure of blocks of strong linkage disequilibrium, with little or no LD between markers in adjacent blocks. Block boundaries are generally at similar positions across the four populations.

BOX 15.2 GRAPHICAL REPRESENTATIONS OF LINKAGE DISEQUILIBRIUM

In HapMap and similar studies, hundreds of SNPs are genotyped across a fairly small chromosomal region, and the extent of linkage disequilibrium (LD) is calculated for every pair of markers. To help the human mind grasp patterns in the data, the results are usually displayed graphically. The markers, in chromosomal order, are set out on both the horizontal and vertical axes of a grid. Each axis may be a physical map, with markers spaced according to their actual physical distance apart, or markers may simply be listed in order along each axis. The linkage disequilibrium between each pair of markers is indicated by color coding of the point on the graph at which the vertical and horizontal coordinates of the two markers intersect. **Figure 1** and **Figure 2** show two examples.

In Figure 1 a physical map has been used, and the computer has interpolated colors to produce a continuous distribution of colors. The distribution is necessarily symmetrical about the diagonal because coordinate (*m,n*) contains the same data as coordinate (*n,m*). In Figure 2 this redundancy is removed by showing only one half of the square. Tilted over so that what was the diagonal is now the horizontal axis, this axis can be envisaged as a map of the chromosome, with the colored triangles giving an impression of the size of haplotype blocks. In this case, the actual physical distance between markers is not represented; the axes simply list them in order.

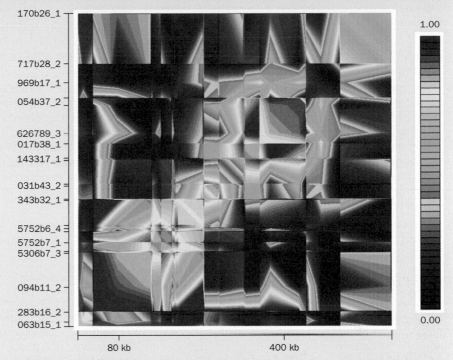

Figure 1 The pattern of linkage disequilibrium across a 500 kb section of chromosome 13, as displayed by the GOLD program. (Courtesy of William Cookson, University of Oxford.)

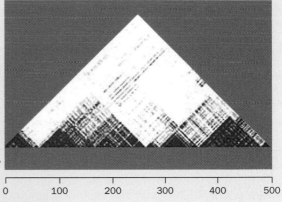

Figure 2 The pattern of linkage disequilibrium at 7q31.33 in the HapMap CEU sample.
Key: red, *D′* = 1, lod score ≥ 2; pink, *D′* < 1, lod score ≥ 2; blue, *D′* = 1, lod score < 2; white, *D′* < 1, lod score < 2. [From The International HapMap Consortium (2005) *Nature* 437, 1299–1320. With permission from Macmillan Publishers Ltd.]

TABLE 15.8 HAPLOTYPE BLOCK STRUCTURES IN FOUR HUMAN POPULATIONS AS REPORTED BY PHASE I OF THE HAPMAP PROJECT

Parameter	YRI	CEU	CHB+JPT
Average number of SNPs per block	30.3	70.1	54.4
Average length per block (kb)	7.3	16.3	13.2
Percentage of genome spanned by blocks	67	87	81
Average number of haplotypes per block	5.57	4.66	4.01
Percentage of chromosomes accounted for by these haplotypes	94	93	95

Only haplotypes having a frequency of 0.05 or greater in the relevant population are reported here. The CHB and JPT samples have been combined because they show very similar patterns. A different statistical method of defining blocks gave somewhat different detailed figures but a similar overall pattern. YRI, Yoruba from Ibadan, Nigeria; CEU, white Americans from Utah; CHB, Han Chinese from Beijing; JPT, Japanese from Tokyo. [Data from The International HapMap Consortium (2005) *Nature* 437, 1299–1320.]

- Blocks vary in size, but are typically 5–15 kb. Much longer blocks can be found in chromosomal regions where there is very little recombination, such as centromeres, or that have been subject to recent strong selection.

- The blocks, as defined here, cover only 67–87% (depending on the statistic used to define a block) of the genomic regions examined; in other words, 13–33% of the regions analyzed show a more chaotic and diverse pattern of variation between individuals.

- The general size of blocks is similar in the CEU, CHB, and JPT samples (13.2–16.3 kb), but smaller in the YRI samples (7.3 kb). This accords with the Out of Africa model of human origins that suggests that humans originated in Africa and evolved there for a considerable time before the progenitors of modern non-African populations migrated out. Sub-Saharan African populations are older and more variable than all others. Their founders lie further in the past, and the present structure reflects a greater number of generations during which meiotic recombination has had the chance to fragment the founder chromosomes.

- Human genetic variability is much more limited than the number of SNPs might suggest. In each population studied, on average 4.0–5.6 haplotypes account for 93–95% of copies of any given block.

The use of tag-SNPs

Given the average in Table 15.8 of 30–70 SNPs per block, each with two alleles, each block has 2^{30}–2^{70} possible haplotypes. In reality, four to six haplotypes per block account for the great majority of all chromosomes. This is a key finding for studies of complex disease that rely on linkage disequilibrium. For each block, a small number of SNPs (**tag-SNPs**) can be defined that identify which of the four to six common alternative blocks a chromosome carries (**Figure 15.7**). Thus, a disease association study no longer needs to genotype all 30–70 SNPs in a block; genotyping as few as two to four tag-SNPs may serve to capture most of the genetic variability.

15.5 ASSOCIATION STUDIES IN PRACTICE

Searching for population associations is an attractive option for identifying disease susceptibility genes. Association studies are easier to conduct than linkage analysis, because neither multi-case families nor special family structures are needed. Under some circumstances, association can also be more powerful than linkage for detecting weak susceptibility alleles (see below). However, association depends on linkage disequilibrium, which is a very short-range phenomenon in comparison with linkage. Linkage disequilibrium with a susceptibility factor can only be detected with markers located on the same haplotype block—that is, within a few kilobases of the factor. A genomewide association study would therefore require samples to be genotyped for hundreds of thousands of markers (in comparison with the few hundred needed for a genomewide linkage scan).

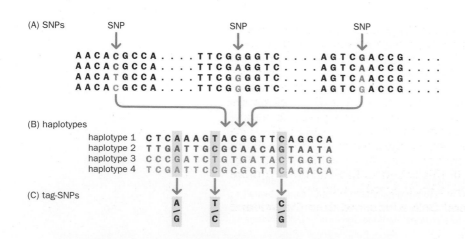

Figure 15.7 SNPs, haplotypes, and tag-SNPs. (A) A short segment of four individual copies of the same chromosome shows three biallelic SNPs. (B) Haplotypes from a larger region on these four chromosomes containing 20 SNPs, showing which allele of each SNP the chromosome carries. Although there are 2^{20} possible combinations of 20 biallelic SNPs, a population survey shows that most copies of this chromosome have one of these four haplotypes. (C) Genotyping just three of the 20 SNPs serves to identify each of these four haplotypes. [From The International HapMap Consortium (2003) *Nature* 426, 789–796. With permission from Macmillan Publishers Ltd.]

Until recently, association studies had to be focused on small candidate chromosomal regions, and reported associations were seldom replicated in follow-up studies. New technical developments, and the availability from the HapMap project of a genomewide catalog of ancestral chromosome segments, have led to a new generation of genomewide association studies that are at last producing robust, replicable results. The background to these studies is discussed in the rest of this section.

Early studies suffered from several systematic weaknesses

Disease association studies in human genetics have a very long but distinctly checkered history. In the 1960s, long before the identification of common DNA variants made human linkage analysis generally useful, many studies looked for associations between particular HLA tissue types and various diseases. Mostly these were autoimmune diseases, in which an HLA effect was *a priori* plausible, but many other diseases were studied, and many associations were reported. Few were replicated. There were three main reasons for this poor record:

- **Inadequate matching of controls.** These were all case-control studies, and often insufficient attention was paid to matching cases and controls. This is in fact a major worry, even in the most carefully designed studies.

- **Insufficient correction for multiple testing.** As we mentioned in connection with genomewide linkage scans, in any set of tests 5% of random observations will be significant at the $p = 0.05$ level and 1% at the $p = 0.01$ level. Each time the data are checked for another possible association, there is another chance of a type I error. The overall threshold of significance must be adjusted for the number of independent questions asked. If the frequencies in cases and controls of alleles at three HLA loci (HLA A, HLA B, and HLA-DR) are compared, it is not sufficient to say that this represents three independent questions. Each allele that was checked was a separate, although not fully independent, question. A full (Bonferroni) correction divides the threshold p value by N, the total number of questions asked. To maintain an overall 5% chance of a false positive result, the threshold p value for a single question is 0.05, for 10 questions 0.005, and for 1,000,000 questions (typical of studies using high-density SNP arrays) 5×10^{-8}. For large values of N this is overconservative, and the preferred threshold of significance is $p' = 1 - (1 - p)^N$. The report from the Wellcome Trust Case-Control Consortium (see Further Reading) has a good discussion of this problem.

- **Striking lucky in underpowered studies.** We noted that chance is a factor causing low rates of replication in linkage studies of complex disease (see Figure 15.3). The same applies to association studies. An underpowered study may occasionally get lucky, but this luck is unlikely to be repeated in a similarly powered study. Targeted replication studies need much more power than the initial trawl.

The transmission disequilibrium test avoids the problem of matching controls

In any association study, the choice of the control group is crucial. One way of avoiding the matching problem altogether is to use internal controls—that is, the control data come from the same people as the case data. The **transmission disequilibrium test** (**TDT**) implements this idea. The TDT starts with couples who have one or more affected offspring. It is irrelevant whether or not either parent is affected. Suppose we wish to test the hypothesis that allele *I* of marker M is associated with the disease. We identify cases in which a parent is heterozygous for *I* and any other allele of the marker (denoted *X*—**Box 15.3**). *I* is probably not necessary for the disease to develop, so the affected offspring will not necessarily have inherited that allele; however, if *I* has any role in susceptibility, we would expect the affected children in this cohort to have inherited *I* more often than the parent's other allele. We therefore proceed as follows:

- Affected probands are ascertained and DNA is obtained from the proband and both parents.

BOX 15.3 THE TRANSMISSION DISEQUILIBRIUM TEST

The transmission disequilibrium test (TDT) statistic, which is based on the standard χ^2 statistic, is

$$(a - b)^2/(a + b)$$

where a is the number of times that a heterozygous parent transmits I to the affected offspring, and b is the number of times that the other allele is transmitted. The mathematically inclined can find the justification for this procedure in the paper by Spielman, McGinnis, and Ewens (see Further Reading).

Figure 1 shows the transmission disequilibrium test applied to type 1 diabetes. Ninety-four families were investigated for an association between type 1 diabetes and a particular allele at a repetitive sequence upstream of the insulin gene. Among the 94 families were 57 parents who were heterozygous for the allele under investigation (denoted by I) and some other allele (denoted collectively by X). These 57 parents transmitted 124 alleles to diabetic offspring (some had more than one affected child). Of these, 78 were allele I and 46 allele X.

The TDT statistic had a value of $(78 - 46)^2/(78 + 46) = 32^2/124 = 8.26$, corresponding to $p = 0.004$. Thus, the data demonstrated an association between the I allele and type 1 diabetes.

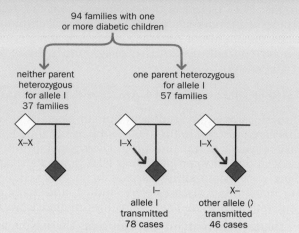

Figure 1 An example of TDT data. Only one parent is shown in each case. [Data reported in Spielman RS, McGinnis RE & Ewens WJ (1993) *Am. J. Hum. Genet.* 52, 506–516. With permission from Elsevier.]

- The probands and their parents are genotyped for marker M.
- Only data from those parents who are heterozygous for marker allele I are considered. It does not matter what their other allele is, provided it is not I.

Box 15.3 shows the TDT statistic applied to data from families with one or more diabetic children.

The result is unaffected by population stratification because the non-inherited parental allele serves as an internal control. An extended TDT test (ETDT) has been developed to handle data from multiallelic markers such as microsatellites. The TDT can be used when only one parent is available, but this may bias the result. When there is no parent available (a common problem with late-onset diseases) an alternative variant, sib-TDT, looks at differences in marker allele frequencies between affected and unaffected sibs.

There has been some argument about whether the TDT is a test of linkage or association. Because it asks questions about alleles and not loci, it is fundamentally a test of association—but association in the presence of linkage. The associated allele may itself be a susceptibility factor, or it may be in linkage disequilibrium with a susceptibility allele at a nearby locus. The TDT cannot detect linkage if there is no linkage disequilibrium—this is a point to remember when considering schemes to use the TDT for whole-genome scans.

Association can be more powerful than linkage studies for detecting weak susceptibility alleles

In 1996, Risch & Merikangas (see Further Reading) compared the power of linkage (affected sib pair, or ASP) and association (TDT) testing to identify a marker tightly linked to a disease susceptibility locus. They calculated the number of ASPs or TDT trios (affected child and both parents) required to distinguish a genetic effect from the null hypothesis, with a given power and significance level. **Box 15.4** illustrates their method (consult the original paper for more detail), and Table 1 in Box 15.4 shows typical results of applying their formulae. The conclusion is clear. ASP analysis would require unfeasibly large samples to detect susceptibility loci that confer a relative risk of less than about 3, whereas TDT might detect alleles giving a relative risk below 2 with manageable sample sizes. Note, however, that the calculation incorporates various assumptions. In particular, it assumes a single ancestral susceptibility allele at the disease locus. Any allelic heterogeneity would rapidly degrade the performance of an association test such as TDT, while not affecting the power of a linkage test.

Although the specific calculations used by Risch & Merikangas applied to the TDT, the implication is more general: association tests are more powerful than linkage tests to detect weak susceptibility factors. Moreover, it is easier to collect

1000 isolated cases and their parents than it is to collect 1000 affected sib pairs, at least for early-onset diseases, for which both parents are likely to be still alive. The paper helped trigger a widespread move away from linkage studies and toward studies of association.

Case-control designs are a feasible alternative to the TDT for association studies

As an alternative to TDT, conventional case-control studies are now coming back into favor. Case-control studies have advantages, provided that the matching of cases and controls is not problematic. They need 50% fewer samples than TDT (two rather than three per comparison) and are more feasible for late-onset diseases, for which parents are seldom available.

How closely the controls need to be matched depends on the population. In the UK, the large Wellcome Trust Case-Control Consortium study excluded individuals who reported non-European or non-Caucasian ancestry, but having done that, dividing the UK population into 12 geographic regions identified only a small number of chromosomal locations where gene frequencies showed significant differences between the 12 regions. By comparing allele frequencies at a range of unlinked loci in the cases and controls, the data can be checked for stratification. However, caution must remain: a recent study by Campbell and colleagues (see Further Reading) showed how subtle population stratification explained an association between tall stature and persistence of intestinal lactase (which allows adults to digest milk) among European Americans. The frequency

---

BOX 15.4 SAMPLE SIZES NEEDED TO FIND A DISEASE SUSCEPTIBILITY LOCUS BY A GENOMEWIDE SCAN

Risch & Merikangas (1996) calculated the sample sizes needed to find a disease susceptibility locus by a genomewide scan by using either affected sib pairs (ASPs) or the transmission disequilibrium test (TDT). This Box summarizes their formulae and equations, but the original paper (see Further Reading) should be consulted for the derivations and for details.

A standard piece of statistics tells us that the sample size M required to distinguish a genetic effect from the null hypothesis with power $(1 - \beta)$ and significance level α is given by $(Z_\alpha - \sigma Z_{1-\beta})^2/\mu^2$, where Z refers to the standard normal deviate. The mean μ and variance σ^2 are calculated as functions of the susceptibility allele frequency (p) and the relative risk γ conferred by one copy of the susceptibility allele. The model assumes that the relative risk for a person carrying two susceptibility alleles is γ^2, that the marker used is always informative, and that there is no recombination with the susceptibility locus.

For ASP, the expected allele sharing at the susceptibility locus is given by $Y = (1 + w)/(2 + w)$, where $w = [pq(\gamma - 1)^2]/(p\gamma + q)$. $\mu = 2Y - 1$ and $\sigma^2 = 4Y(1 - Y)$. The genomewide threshold of significance (probability 0.05 of a false positive anywhere in the genome; testing for sharing IBD) requires a lod score of 3.6, corresponding to $\alpha = 3 \times 10^{-5}$, and $Z_\alpha = 4.014$. For 80% power to detect an effect, $1 - \beta = 0.2$ and $Z_{1-\beta} = -0.84$.

For the TDT, the probability that a parent will be heterozygous for the allele in question is $h = pq(\gamma + 1)/(p\gamma + q)$. $P(trA)$, the probability that such a heterozygous parent will transmit the high-risk allele to the affected child, is $\gamma/(1 + \gamma)$. $\mu = \sqrt{h(\gamma - 1)/(\gamma + 1)}$, and $\sigma^2 = 1 - [h(\gamma - 1)^2/(\gamma + 1)^2]$. As discussed above, for a genomewide screen involving 1,000,000 tests, $\alpha = 5 \times 10^{-8}$, $Z_\alpha = 5.33$, and, as before, $Z_{1-\beta} = -0.84$.

In **Table 1**, the Z_α, $Z_{1-\beta}$, μ, and σ^2 values are used to calculate sample sizes by substitution in the formula $M = (Z_\alpha - \sigma Z_{1-\beta})^2/\mu^2$. For the TDT, the answer is halved because each parent–child trio allows two tests, one on each parent.

Table 1 Comparison of the power of linkage and association studies

Susceptibility factor		ASP analysis		TDT analysis	
γ	p	Y	No. of pairs	$P(trA)$	No. of trios
5	0.01	0.534	2530	0.830	747
	0.1	0.634	161	0.830	108
	0.5	0.591	355	0.830	83
3	0.01	0.509	33,797	0.750	1960
	0.1	0.556	953	0.750	251
	0.5	0.556	953	0.750	150
2	0.1	0.518	9167	0.667	696
	0.5	0.526	4254	0.667	340
1.5	0.1	0.505	115,537	0.600	2219
	0.5	0.510	30,660	0.600	950
1.2	0.1	0.501	3,951,997	0.545	11,868
	0.5	0.502	696,099	0.545	4606

The table shows the number of affected sib pairs (ASPs) or parent–child trios needed for 80% power to detect significant linkage or association in a genomewide search. γ is the relative risk for individuals of genotype Aa compared with aa; p is the frequency of the susceptibility allele, A. For affected sib pair analysis, Y is the expected allele sharing; for the trios analyzed by the transmission disequilibrium test, $P(trA)$ is the probability that a parent heterozygous for allele A will transmit A to an affected child. For low values of γ, unfeasibly large numbers of affected sib pairs are needed to detect an effect. [Data from Risch N & Merikangas K (1996) *Science* 273, 1516–1517.]

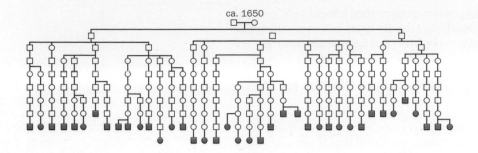

Figure 15.8 Patterns of recent common ancestry in an isolated Finnish population. The people in the youngest generation were recruited because they each had multiple offspring affected by schizophrenia. Note, however, that the ancestry shown is highly selective; multiple ancestors have been removed from each generation for clarity. Ten generations ago, in 1650, an outbred person has 1024 ancestors, but only two are shown here. [From Hovatta I, Varilo T, Suvisaari J et al. (1999) *Am. J. Hum. Genet.* 65, 1114–1124. With permission from Elsevier.]

of each of these two characters varies considerably between populations, and rematching of individuals on the basis of European ancestry greatly decreased the association.

Special populations can offer advantages in association studies

Populations derived from a small number of relatively recent founders are expected to show limited haplotype diversity and more extensive linkage disequilibrium than older populations. The HapMap data summarized in Table 15.8 illustrate this. In comparison with the Yoruba subjects from Nigeria, the non-African study subjects come from populations that were established more recently, and on average their haplotype blocks are longer and less diverse. Ancestral disease-bearing haplotypes should be easier to identify in populations that derive from a small number of founders and have remained relatively isolated since that time. This belief lies behind the DeCode project in Iceland, and similar projects in Quebec and elsewhere. **Figure 15.8** shows an example from an isolated Finnish population. The 39 shaded individuals each have two or more offspring affected by schizophrenia, and were ascertained as part of the study by Hovatta et al. listed in Table 15.7. All could be traced back to at least one common ancestral couple 7–10 generations back. Because the common ancestors are relatively recent (dating to about 1650), any segments of their chromosomes that are shared by several of the 39 shaded individuals are likely to be quite large in comparison with ancestral segments in an older population.

Populations derived by recent admixture are a second group that might offer special advantages. For example, consistent long-range linkage disequilibrium has been documented in the Lemba, a Bantu–Semitic hybrid population in Africa. If the two source populations had widely different incidences of a common disease, the mixed population could be used to map the determinants rather efficiently. This is the human analog of an interspecific mouse cross, although of course all humans are the same species. In reality, the availability of large numbers of potential subjects with good medical records may be more important than population structure—certainly, that is the thinking behind the BioBank project in the UK, which seeks to collect medical and lifestyle data and DNA from 500,000 British people aged 45–69 years and follow their health prospectively.

A new generation of genomewide association (GWA) studies has finally broken the logjam in complex disease research

It would be wrong to paint a wholly negative picture of the search for complex disease susceptibility factors over the decade up to 2005. Although the frequency of successful replication has been low, it has not been negligible. In 2003, Lohmueller and colleagues made a meta-analysis of 301 publications on 25 frequently studied associations in 11 different diseases. They concluded that at least 8 of the 25 associations had been adequately replicated. **Table 15.9** lists some associations that were well established before the current generation of GWA studies.

It remains true that the successes listed in Table 15.9 have been hard won. The cost, in money and person-hours, of identifying and thoroughly validating a common disease association has been orders of magnitude higher than the cost of identifying a Mendelian disease locus. But at least we have now put the history of irreproducible results behind us. Several key developments have made this possible:

TABLE 15.9 SOME CONFIRMED DISEASE ASSOCIATIONS

Disease	Gene containing associated variant	Odds ratio	Frequency of risk allele or haplotype
Type 1 diabetes	IF1H1	1.17[a]	0.65
	CTLA4	1.31[b]	0.53
Type 2 diabetes	ABCC8	2.23[b]	0.02
	PPARG	1.21[b]	0.85
	SLC2A1	1.82[b]	0.30
	TCF7L2	1.49[a]	0.26
Age-related macular degeneration	CFH	2.45[a]	0.46
Systemic lupus erythematosus	IRF5	1.78[a]	0.12
Schizophrenia	DRD3	1.13[b]	0.01
	HTR2A	1.07[b]	0.39

Different studies of the same disease and candidate gene may report associations with different markers, or with multilocus haplotypes. [a]These odds ratios and all allele or haplotype frequencies are representative examples of data from one out of several reports on each association. [b]These odds ratios are from the meta-analysis by Lohmueller KE, Pearce CL, Pike M et al. (2003) *Nat. Genet.* 33, 177–182.

- Consortia have been established to perform case-control studies with 1000 or more subjects in each arm—and funding bodies have recognized that work on this scale is necessary. In the USA a private–public partnership, the Genetic Association Information partnership (GAIN), and in the UK the Wellcome Trust Case-Control Consortium, are conducting or coordinating GWA studies on large cohorts of patients with several different complex diseases. Similar efforts are underway in other countries.

- Efficient whole-genome amplification methods (using φ29 DNA polymerase) have been developed that allow very extensive analysis of even small DNA samples.

- Massively parallel microarray or bead-based methods are available, by which a sample can be genotyped for 500,000 or more SNPs in parallel.

- The HapMap project has provided data that permit a rational choice of tag-SNPs to capture a significant proportion of all the genetic variation in a population.

A good flavor of the new clutch of studies is provided by the report from the Wellcome Trust Case-Control Consortium (WTCCC), published in June 2007. Large consortia of British researchers assembled 2000 well-phenotyped cases of each of seven diseases: bipolar disorder (manic-depressive psychosis), coronary artery disease, Crohn disease, hypertension, rheumatoid arthritis, type 1 diabetes, and type 2 diabetes. In addition, 3000 healthy controls were collected. All samples were typed for more than 500,000 SNPs. The estimated power of the study to detect risk factors was 43% for a factor conferring a relative risk of 1.3, and 80% for a relative risk of 1.5 (risks are expressed on a per-allele basis using a multiplicative model—that is, if allele A confers a relative risk of x, the risks for genotypes aa, Aa, and AA are $1{:}x{:}x^2$).

Across the seven patient groups, 25 independent disease-association signals were identified with p values beyond the threshold of significance, calculated as $p < 5 \times 10^{-7}$ (**Figure 15.9**). The key question is: How many of these could be replicated? Previous studies in the seven diseases were considered to have identified 15 variants with strong, replicated evidence of association with one or other of the seven diseases. Thirteen of the 15 were unambiguously identified in the

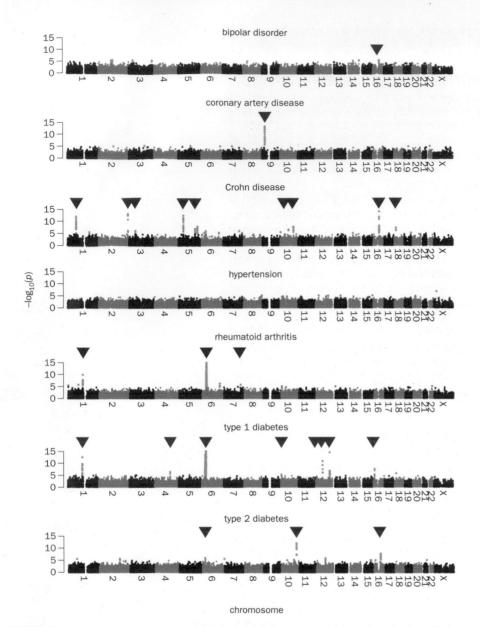

Figure 15.9 Results of the Wellcome Trust Case-Control Consortium (WTCCC) genomewide association study. For each of the seven diseases studied, the distribution of *p* values (as $-\log_{10}p$) for the association of each SNP with the disease is shown, at the appropriate chromosomal position. Most of the 469,557 SNPs that passed all the quality control checks showed weak or no association with the respective disease (blue dots, merged together). Those showing stronger evidence of association ($p < 10^{-5}$) are marked with green dots. The 25 most strongly associated SNPs or clusters of SNPs ($p < 5 \times 10^{-7}$, the threshold of significance in this study) are marked with red triangles. [Adapted from The Wellcome Trust Case-Control Consortium (2007) *Nature* 447, 661–676. With permission from Macmillan Publishers Ltd.]

WTCCC study, and one of the remaining two showed positive signals that did not quite reach significance. Almost all of the novel susceptibility factors have subsequently been confirmed in independent follow-up studies.

The WTCCC study is not the only one to have exploited the new tools and knowledge noted above. Between them, these studies clearly show that the era of reliable GWA studies has finally arrived.

The size of the relative risk

Considering the 25 SNPs showing the strongest evidence of association in the WTCCC study, the odds ratios (risk for a heterozygous carrier of the risk allele compared with a homozygous non-carrier) varied between 1.09 and 5.49. However, only the previously well-known associations of certain HLA types with autoimmune disease (type 1 diabetes and rheumatoid arthritis) produced high odds ratios. For all other cases, the highest ratio observed was 2.08, and in only five cases did it even exceed 1.5. As already mentioned, the study was predicted to have 80% power to detect a factor with an odds ratio of 1.5. The clear conclusion is that there are no individually strong ancient susceptibility factors for any of these typical common complex diseases.

A common story is emerging from this and other similar studies: the ancient susceptibility factors can be identified by current methods—but the relative risks they confer are quite small.

15.6 THE LIMITATIONS OF ASSOCIATION STUDIES

The new GWA studies on large cohorts of subjects have unprecedented power to detect susceptibility factors conferring relative risks as low as 1.2, and a proven track record of success. The question remains, however, whether these studies will unravel the complete genetic architecture of complex disease susceptibility.

One complicating factor is the existence of large-scale copy-number variations (CNVs) in the genomes of normal healthy persons. As described in Chapter 13, these CNVs actually account for a greater number of variable nucleotides than do all the SNPs. Individually, they are less numerous than SNPs, but each one covers many nucleotides. Many involve variable copy numbers of one or more genes, and in some cases (e.g. salivary amylase) there are demonstrable phenotypic effects of differing gene dosage. It is entirely likely that some CNVs will be susceptibility factors for common diseases. The question therefore arises: How well would current SNP association studies pick up associations between a disease and a CNV?

This question was addressed in the study by Redon et al. described in Chapter 13 of CNV in the HapMap sample. Their data showed that the HapMap Phase I SNPs would act as reasonably good tags for about 51% of the biallelic CNV in the non-African subjects, but only about 22% in the Africans. In other words, in 51% or 22% of cases, respectively, a particular SNP allele was strongly associated with a particular form of an insertion–deletion CNV, so that the tag-SNP would be associated with any disease risk caused by variation at the CNV. Multiallelic CNVs were not well tagged by any SNPs. Thus, older SNP association studies would be quite poor at detecting any disease susceptibility caused by a CNV. However, this is a passing phase. Newer generations of SNP genotyping arrays include assays for common CNVs.

The common disease–common variant hypothesis proposes that susceptibility factors have ancient origins

A more fundamental problem lies in the basic premise of testing for associations: the assumption that susceptibility factors are ancient common variants. This is the **common disease–common variant hypothesis**. But this hypothesis is controversial. Evidence both for and against it has been presented.

The fact that GWA studies are identifying associations that can be replicated tells us that some susceptibility factors are indeed ancient common variants. But the modest size of their effects leaves open the question of how much of the total disease susceptibility they will explain. Studies of breast cancer provide interesting figures. Breast cancer is about twice as common in first-degree relatives of affected women as in the general population. All known susceptibility factors identified before 2007 (*BRCA1*, *BRCA2*, *BRIP1*, *PALB2*, *ATM*, *CHEK2*, and *CASP8*) collectively accounted for less than 25% of the familial tendency of breast cancer. A whole-genome association study by Easton and colleagues (described in Chapter 16) identified five new factors. Despite the very large scale of the study, involving more than 20,000 patients, the new factors together accounted for only a further 3.6% of the excess familial risk.

Indeed, the underlying premise of the common disease–common variant hypothesis seems quite a tall order. The variants in question must have been able to persist through thousands of generations of natural selection to be associated with ancient haplotype blocks and remain in the population at high frequency. Yet, at the same time they must be sufficiently pathogenic, at any rate against certain relatively common genetic backgrounds and under certain relatively common environmental conditions, to be significant risk factors for serious diseases. It is hardly surprising that the common susceptibility variants that are now being identified through association studies have such weak effects. In defense, we can note that many of the diseases are of late onset, and so are relatively immune from natural selection and relatively unimportant in overall health terms until our current aging population made such diseases more prominent. Additionally, the precipitating environments may be features of modern life that have existed only recently. Genetically, we are all adapted to life as cavemen.

The mutation–selection hypothesis suggests that a heterogeneous collection of recent mutations accounts for most disease susceptibility

The alternative view is that common diseases are common because of mutation–selection balance. On this view, many or most susceptibility factors are deleterious enough to be removed by natural selection. These are replaced by new deleterious variants generated by recurrent random mutation. There is a simple relationship between the deleterious effect of a variant genotype and its likely persistence in a population:

- The variants that cause Mendelian diseases have very strong effects, so that most people with the susceptible genotype have the relevant disease. Such strongly disadvantageous variants, such as the dystrophin mutations that cause Duchenne muscular dystrophy, have a very rapid turnover.

- At the other end of the spectrum of pathogenicity, completely neutral variants can persist indefinitely in a large population (in a small population, sooner or later by random chance a variant will be either lost or fixed—that is, it will either be lost or entirely replace the alternative form).

- Very mildly deleterious variants may persist long enough to be present on ancient haplotype blocks, and so may be identifiable through association studies.

- Variants with a rather stronger effect will be removed from the population too fast to persist on common ancient haplotype blocks. The removal is balanced by *de novo* mutation creating fresh mildly deleterious variants.

The **mutation–selection hypothesis** supposes that this last class of variants make up the major susceptibility factors for common complex diseases. Calculations by Bodmer & Bonilla (see Further Reading) show that a susceptibility factor could have a penetrance as high as 10–20%—that is, that factor alone could have a 10–20% likelihood of causing disease, independently of other genetic or environmental factors—and yet would not give rise to a Mendelian pedigree pattern of disease. Typically, the physiological basis of disease susceptibility might be a partial loss of function in some complex pathway. This loss could be caused by any number of possible mutations in any of the genes involved in that pathway. There would be heterogeneity at the locus level (different affected people could have deleterious variants in different genes that are involved in the affected pathway) and very great heterogeneity at the allelic level (even if two people have variants in the same gene, the actual sequence change would most probably be different in each case). Each individual mutation may be rare, but deficiencies in the pathway may overall be quite common in the population.

Detecting such a heterogeneous set of rare variants, and proving a convincing role for them in pathogenesis of the disease, is quite a challenge. Each individual variant is directly pathogenic, but rare in the population. Association studies have little power to detect variants present in less than 5% of subjects, and so would be unable to pick up the postulated variants. Linkage analysis is not sensitive to allelic heterogeneity, and the individual effects might well be strong enough to show up in ASP analysis—if only all the sib pairs in the study sample had variants in the same gene. But the extensive locus heterogeneity would prevent success, because different affected sib pairs would often have variants at different unlinked loci.

Instead, it would be necessary to resequence candidate genes in large collections of cases and controls. Relevant genes would carry a higher overall frequency of rare, mildly deleterious variants in affected people than in controls. This excess would have to be picked out against a background of rare neutral variants that would be similarly diverse but equally common in the cases and controls. How easily this could be done would depend on the proportion of *de novo* mutations that are deleterious rather than neutral.

If the proportion of deleterious variants is low, they would probably be obscured among all the neutral variants. It could be difficult or impossible to find convincing differences between cases and controls. Inspection of the sequence would not usually reveal whether a change is neutral or very mildly deleterious. However, studies of naturally occurring missense changes in many different genes suggest that maybe half of all such changes may be mildly deleterious (with a quarter being seriously deleterious and a quarter neutral). For silent changes and changes in noncoding sequences, the proportion of neutral variants is likely to be much higher.

Several preliminary resequencing studies have indeed found an excess of rare missense variants in candidate genes in disease cohorts, but such studies need to be conducted on a wider selection of genes and in more diseases to see whether the effect is general. The new massively parallel sequencing technologies allow this to be done, so that definitive tests of the mutation–selection hypothesis are now possible.

A complete account of genetic susceptibility will require contributions from both the common disease–common variant and mutation–selection hypotheses

The common disease–common variant and mutation–selection hypotheses should not be seen as mutually exclusive (Table 15.10). Both may very well be true. *A priori*, the common disease–common variant hypothesis is less plausible than the mutation–selection hypothesis. We know for sure that mutation and selection happen, and that there is a whole spectrum of mutations, ranging from lethal, through varying degrees of deleterious effect, to neutrality. It was not equally obvious that ancestral haplotypes carrying variants that predispose to disease should be able to withstand natural selection over immense time-scales and remain common in present-day populations. However, the successful identification of some common susceptibility factors shows that this has indeed happened, at least in some cases. There is no reason to suppose that all factors have to be of one sort. Different experimental designs can reveal different factors, and a complete account of genetic susceptibility will require contributions of both types.

Under either hypothesis, the problem of distinguishing pathogenic from neutral variants looms large. Factors identified by large-scale resequencing are expected to be directly pathogenic, but it will still be necessary to pick them out against a background of neutral variants. Meanwhile, linkage and association studies only flag the chromosomal location of the causative factor—to a relatively large region by linkage, or to a haplotype block by association. Thus, there is always a problem of identifying the actual causal variant. In Chapter 16 we consider how one approaches this problem, starting with Mendelian disease genes, then moving on to susceptibility factors. Finally, the implications of this work for predictive testing and personalized medicine are the subject of Chapter 19.

TABLE 15.10 COMPARISON OF THE COMMON DISEASE–COMMON VARIANT HYPOTHESIS AND THE MUTATION–SELECTION HYPOTHESIS

Parameter	Common disease–common variant hypothesis	Mutation–selection hypothesis
Frequencies of susceptibility alleles	high	low
Effect sizes of susceptibility alleles	small	moderate
Locus heterogeneity (number of susceptibility loci for a given disease)	high	could be low
Allelic heterogeneity (number of different susceptibility alleles at a locus)	low	high
Origin of susceptibility alleles	ancient common ancestor	relatively recent mutations
Technology to detect susceptibility factors	association studies	resequencing

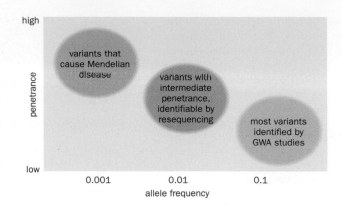

Figure 15.10 Frequency and effect size of pathogenic alleles. Variants causing Mendelian diseases have high penetrance but are individually rare. The variants conferring susceptibility to complex diseases that are identified through association studies may be common, but they have very weak effects (low penetrance). An intermediate class of individually rare susceptibility factors may exist that have stronger effects. These would not be expected to show populationwide associations with disease, and could be identified only by large-scale resequencing. [Adapted from McCarthy MI, Abecasis GR, Cardon LR et al. (2008) *Nat. Rev. Genet.* 9, 356–369. With permission from Macmillan Publishers Ltd.]

CONCLUSION

DNA sequence variants associated with disease can be divided into those that cause disease directly and those that merely modulate a person's susceptibility or resistance to environmental triggers of disease. These two types of sequence are not discrete, but lie on a continuum of disease penetrance. Variants that have a direct cause underlie Mendelian diseases, and are necessarily rare. As described in Chapter 14, they can be mapped by parametric linkage analysis to very small candidate chromosomal regions. Variants that confer susceptibility lie at the other end of the continuum—they may be common, but they have only very modest effects on susceptibility (**Figure 15.10**).

Parametric linkage analysis cannot be used to identify these low-penetrance factors, because it is not possible to specify the necessary detailed genetic model. Non-parametric linkage analysis, for example of affected sib pairs, has been widely used, but with limited success because of the low statistical power of the method. Large-scale successful identification has depended mainly on association studies. **Figure 15.11** summarizes the general protocol of genomewide association studies.

The HapMap data allow suitable SNPs to be identified for genomewide association studies and placed on high-density SNP arrays, and funding agencies have recognized the need for very large-scale collaborative studies. As a result, the trickle of validated associations reported before 2005 has turned into a flood. Association studies of feasible size (a few thousand cases and controls) can detect only factors that are common in populations, typically with allele frequencies of 0.05 or more. Factors that are common and associated with tag-SNPs are likely to be present on shared ancestral chromosome segments that have existed in the population for thousands of years. The common disease–common variant hypothesis proposes that such factors are largely responsible for the genetic susceptibility to complex diseases. It would be unusual for such an ancient variant to have a strongly pathogenic effect: natural selection should have long ago removed such variants from the population. It is therefore not surprising that almost all the susceptibility factors identified through association studies have very small individual effects.

The alternative mutation–selection hypothesis proposes that much susceptibility is due to individually rare variants that are likely to be of fairly recent origin. The hypothesis predicts extensive allelic heterogeneity, but maybe more limited locus heterogeneity—that is, many different mutations at a possibly limited number of loci. Such variants would be too rare to be identified by association studies, but may be detected by large-scale resequencing of candidate regions (or whole genomes) in cases and controls. Because the variants may not have survived in the population for very long periods, they may have stronger individual effects than the ancient common variants discussed above.

Most of the hundreds of well-validated susceptibility factors identified through association studies are probably not directly responsible for disease susceptibility, but rather are in linkage disequilibrium with the truly functional variants. Identifying the true causative variants would have many benefits. Chapter 16 discusses the various strategies that have been used to identify variants that function to cause or contribute to disease.

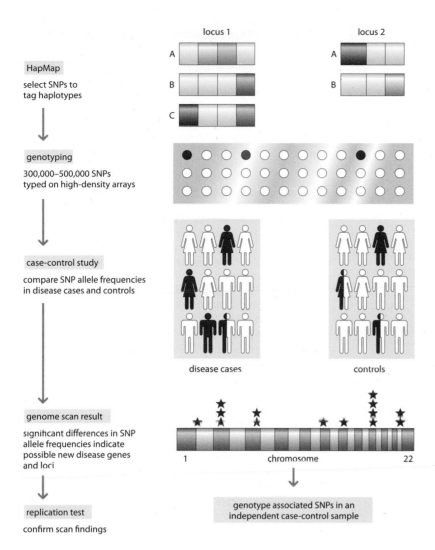

HapMap

select SNPs to
tag haplotypes

genotyping

300,000–500,000 SNPs
typed on high-density arrays

case-control study

compare SNP allele frequencies
in disease cases and controls

genome scan result

significant differences in SNP
allele frequencies indicate
possible new disease genes
and loci

replication test

confirm scan findings

locus 1

A
B
C

locus 2

A
B

disease cases

controls

1   chromosome   22

genotype associated SNPs in an
independent case-control sample

Figure 15.11 A genomewide association scan. Common haplotypes at each location in the genome are defined by SNPs (color versus white). Information from the HapMap project is used to select a subset of SNPs that will serve to identify (tag) each haplotype— purple and blue for locus 1, and either red or blue for locus 2. Disease cases and controls are genotyped with microarrays. SNPs that are associated with disease at an appropriate statistical threshold (stars) are then genotyped in a second independent sample of cases and controls to establish which of the associations from the primary scan are robust. [From Mathew CG (2008) *Nat. Rev. Genet.* 9, 9–14. With permission from Macmillan Publishers Ltd.]

FURTHER READING

Family studies of complex disease

Burmeister M, McInnis MG & Zollner S (2008) Psychiatric genetics: progress amid controversy. *Nat. Rev. Genet.* 9, 527–540. [An overview of progress since the pioneering studies used as examples in this section.]

McGuffin P, Shanks MF & Hodgson RJ (eds) (1984) The Scientific Principles Of Psychopathology. Grune & Stratton.

Risch N (1990) Linkage strategies for genetically complex traits. 1. Multilocus models. 2. The power of affected relative pairs. 3. The effect of marker polymorphism on analysis of affected relative pairs. *Am. J. Hum. Genet.* 46, 222–228; 229–241; 242–253. [Three key papers establishing the statistical basis of familial clustering and shared segment analysis.]

Rosenthal D & Kety SS (1968) The Transmission Of Schizophrenia. Pergamon Press.

Segregation analysis

Badner JA, Sieber WK, Garver KL & Chakravarti A (1990) A genetic study of Hirschsprung disease. *Am. J. Hum. Genet.* 46, 568–580. [A good example of segregation analysis applied to a non-Mendelian condition.]

McGuffin P & Huckle P (1990) Simulation of Mendelism revisited: the recessive gene for attending medical school. *Am. J. Hum. Genet.* 46, 994–999. [A warning about some pitfalls of segregation analysis.]

Linkage analysis of complex characters

Altmüller J, Palmer LJ, Fischer G et al. (2001) Genomewide scans of complex human diseases: true linkage is hard to find. *Am. J. Hum. Genet.* 69, 936–950. [A sobering meta-analysis of 101 linkage studies in 31 complex diseases.]

Lander ES & Kruglyak L (1995) Genetic dissection of complex traits: guidelines for interpreting and reporting linkage results. *Nat. Genet.* 11, 241–247. [Introducing the widely used categories of suggestive, significant, and highly significant results.]

Association studies and linkage disequilibrium

Cardon LR & Bell JI (2001) Association study designs for complex diseases. *Nat. Rev. Genet.* 2, 91–99. [A general non-mathematical discussion of designs for association studies.]

International HapMap Consortium (2003) The International HapMap Project. *Nature* 426, 789–796. [A description of the aims and methods of the project.]

International HapMap Consortium (2005) A haplotype map of the human genome. *Nature* 437, 1299–1320. [The primary report of Phase I of the HapMap project; available for download from http://www.hapmap.org]

International HapMap Consortium (2007) A second generation human haplotype map of over 3.1 million SNPs. *Nature* 449, 851–861. [Report of Phase II.]

Jobling MA, Hurles ME & Tyler-Smith C (2004) Human Evolutionary Genetics: Origins, People and Disease. Garland Science. [A unique and excellent textbook that, among many other things, sets out the whole background to linkage disequilibrium.]

Lohmueller KE, Pearce CL, Pike M et al. (2003) Meta-analysis of genetic association studies supports a contribution of common variants to susceptibility to common disease. *Nat. Genet.* 33, 177–182. [Showing that some early association studies did indeed produce replicable results.]

Risch N & Merikangas K (1996) The future of genetic studies of complex human diseases. *Science* 273, 1516–1517. [The power calculations in this paper helped trigger the move from linkage to association studies; see also *Science* 275, 1327–1330 (1997) for discussion.]

Slatkin M (2008) Linkage disequilibrium—understanding the evolutionary past and mapping the medical future. *Nat. Rev. Genet.* 9, 477–485. [A useful general introduction to LD, including definitions, measures, and origins.]

Association studies in practice

Altshuler D & Daly M (2007) Guilt beyond a reasonable doubt. *Nat. Genet.* 39, 813–815. [A brief review of the achievements of the first wave of successful genomewide association studies.]

Campbell CD, Ogburn EL, Lunetta KL et al. (2005) Demonstrating stratification in a European American population. *Nat. Genet.* 37, 868–872. [A cautionary tale about the need to match cases and controls very carefully.]

Database of Genotypes and Phenotypes (dbGaP). http://www.ncbi.nlm.nih.gov/entrez/query/Gap/gap_tmpl/about.html [Planned to be a central resource for raw data relating genotypes and phenotypes, including results of association studies.]

Schaid DJ (1998) Transmission disequilibrium, family controls and great expectations. *Am. J. Hum. Genet.* 63, 935–941. [A review of the strengths and weaknesses of the TDT.]

Spielman RS, McGinnis RE & Ewens WJ (1993) Transmission test for linkage disequilibrium: the insulin gene region and insulin-dependent diabetes mellitus (IDDM). *Am. J. Hum. Genet.* 52, 506–516. [Describes the statistical basis for the transmission disequilibrium test, and shows an example of its power.]

Wellcome Trust Case-Control Consortium (2007) Genome-wide association study of 14,000 cases of seven common diseases and 3,000 shared controls. *Nature* 447, 661–678. [An excellent overview of how to perform GWA studies and what they might show.]

Wilson JF & Goldstein DB (2001) Consistent long-range linkage disequilibrium generated by admixture in a Bantu–Semitic hybrid population. *Am. J. Hum. Genet.* 67, 926–935. [The advantages of an admixed population for association studies.]

Wright AF, Carothers AD & Pirastu M (1999) Population choice in mapping genes for complex diseases. *Nat. Genet.* 23, 397–404. [Despite its age, a good review of the options.]

The limitations of association studies

Bodmer W & Bonilla C (2008) Common and rare variants in multifactorial susceptibility to common disease. *Nat. Genet.* 40, 695–710. [An important contribution to the debate about the relative merits of the common disease–common variant and mutation–selection hypotheses.]

Helbig I, Mefford HC, Sharp AJ et al. (2009) 15q13.3 microdeletions increase risk of idiopathic generalized epilepsy. *Nat. Genet.* 41, 160–162. [One of several studies showing that the microdeletion acts as a typical low-penetrance susceptibility factor for several different neuropsychiatric disorders. This paper includes references to other studies.]

Jacobsson M, Scholz SW, Scheet P et al. (2008) Genotype, haplotype and copy number variation in worldwide human populations. *Nature* 451, 998–1003. [An extension of HapMap-type investigations to 485 individuals from 29 populations.]

Kryukov GV, Pennachio LA & Sunyaev SR (2007) Most rare missense alleles are deleterious in humans: implications for complex disease and association studies. *Am. J. Hum. Genet.* 80, 727–739. [Data and calculations suggesting that around 50% of rare coding-sequence variants are mildly deleterious, thus supporting the mutation–selection hypothesis.]

Pritchard JK & Cox NJ (2002) The allelic architecture of human disease genes: common disease–common variant or not? *Hum. Mol. Genet.* 11, 2417–2423. [Arguments against the common disease–common variant hypothesis.]

Reich DE & Lander ES (2001) On the allelic spectrum of human disease. *Trends Genet.* 17, 502–510. [Arguments in favor of the common disease–common variant hypothesis.]

Topol EJ & Frazer KA (2007) The resequencing imperative. *Nat. Genet.* 39, 439–440. [A brief review of the case for large-scale resequencing and results to date.]

Chapter 16

Identifying Human Disease Genes and Susceptibility Factors

16

KEY CONCEPTS

- Positional cloning has been the main route through which the genes underlying Mendelian diseases have been identified. In positional cloning, a gene is identified from its chromosomal location alone.

- Positional candidate genes are identified by drawing up a list of genes at that location, on the basis of their function, expression pattern, and homologies.

- Model organisms can provide clues to gene identification because the structure and function of many genes are conserved between humans and other organisms.

- Mice are particularly useful for identifying disease genes and studying pathogenic mechanisms because of the ease with which genes can be mapped and manipulated in mice, and because of the relatively close similarity of mice and humans.

- Patients with chromosome abnormalities can provide essential pointers to the location of a disease gene. Large abnormalities may be recognized by conventional cytogenetics, and small ones by comparative genomic hybridization or using SNP arrays. These approaches have been especially useful for severe dominant conditions, in which most cases are new mutations, making linkage analysis impossible.

- Some disease genes have been identified by alternative, position-independent approaches. These include functional studies of cells in culture, direct identification of altered proteins or changed gene expression, and extrapolation from animal models.

- Candidate genes are tested by seeking mutations in collections of unrelated affected cases. Functional studies may be needed to confirm an uncertain identification.

- Susceptibility factors for complex disease are mainly sought through association studies. These may be targeted at genes whose function makes them likely candidates, or studies may be conducted on a genomewide scale.

- Association studies identify haplotype blocks that are associated with susceptibility to a disease. Identifying the actual causal variant is difficult because of linkage disequilibrium between all the variants in the block. Both statistical and functional evidence are needed.

- Association studies have identified susceptibility factors for many complex diseases, but for most diseases the known factors only weakly influence susceptibility and collectively account for only a small part of the overall genetic determination, leaving a problem of hidden heritability.

- The hidden heritability in complex diseases may consist of the cumulative effects of very many extremely weak genetic factors, or of the effects of a highly heterogeneous collection of individually rare mutations, or possibly of epigenetic changes.

Few subjects have moved as fast as human disease gene identification. Before 1980, very few human genes had been identified as disease loci. The few early successes involved a handful of diseases with a known biochemical basis, for which it was possible to purify the gene product. In the 1980s, advances in recombinant DNA technology allowed a new purely genetic approach, sometimes given the rather meaningless label *reverse genetics*. The number of disease genes identified started to increase, but these early successes were hard-won, heroic efforts. With the advent of polymerase chain reaction (PCR) techniques for linkage studies and mutation screening, it all became much easier. Now that the human and other genome projects have made available a vast range of resources, the ability to identify a Mendelian disease gene depends almost entirely on having access to suitable families.

The 1990s saw triumphant progress in identifying the causes of Mendelian diseases, so that the genes underlying most common Mendelian diseases have now been identified. The exceptions tend to be those such as nonsyndromic mental retardation, in which the hunt for causative genes is bedeviled by irreducible genetic heterogeneity or, alternatively, very rare conditions for which it is difficult to find enough suitable families. However, identifying the factors conferring susceptibility to common complex diseases is much more challenging. As described in Chapter 15, only association studies have sufficient power to identify most susceptibility factors; linkage studies are seldom productive. For years, researchers were unable to make much progress. After many false dawns, the new generation of large-scale genomewide association studies, reported in increasing numbers from 2005 onward, has finally broken the logjam. Many susceptibility factors have now been mapped, but identifying the actual causal sequence change remains very difficult.

In this chapter, we first consider positional cloning, which has been the most successful general approach to the identification of Mendelian disease genes. Most commonly, the candidate chromosomal region is identified through linkage studies, but alternatives are to use chromosomal breakpoints or animal models. We next consider some other approaches: cloning based on knowledge of the gene product, and identifying causal variants through genomewide association studies. An important question is how one knows when one has succeeded. How can we know that a candidate sequence variant really is the cause of the phenotype? This connects with the discussion in Chapter 13 of pathogenic versus non-pathogenic sequence changes. Toward the end of the chapter is a series of case studies that show how the various methods have been used to identify important phenotypic determinants.

The approaches described are just as applicable to identifying determinants of normal variations such as red hair or red–green color blindness as they are to identifying disease genes. The determinants that we would wish to identify may not necessarily be genes, in the sense of protein-coding sequences. The primary determinant might affect noncoding RNAs or alter chromatin structure, and it might be located some distance from any protein-coding sequence. Whatever the mechanism, however, we are looking for DNA sequence variants that cause phenotypic variation. Epigenetic changes that do not alter the DNA sequence (such as changes in DNA methylation) also cause phenotypic variation, ultimately mediated by changes in the expression of protein-coding genes.

It is important not to be misled by phrases such as the gene for cystic fibrosis, the gene for diabetes, and so on. You would not describe your domestic freezer as a machine for ruining frozen food. Genes do a job in cells; if the job is not done, or is done wrongly, the result may be a disease. However, many human genes were first discovered through research on the diseases caused by mutations in them, so terminology such as the cystic fibrosis gene does reflect historical, if not biological, reality.

16.1 POSITIONAL CLONING

In positional cloning, a disease gene is identified knowing nothing except its approximate chromosomal location. This typically involves a combination of clinical, laboratory, and bioinformatic work, and maybe also the exploration of animal models (**Figure 16.1**).

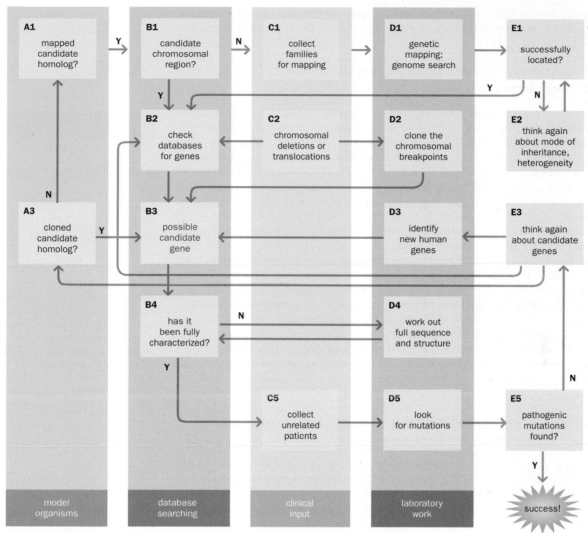

Positional cloning identifies a disease gene from its approximate chromosomal location

A major test bed for positional cloning methods was Duchenne muscular dystrophy (DMD; OMIM 310200). Although the first successful application of positional cloning in humans had come slightly earlier, with the identification of the gene for X-linked chronic granulomatous disease (OMIM 306400) by Royer-Pokora and colleagues in 1985, it was the identification of the gene underlying DMD that really caught the imagination. This was a relatively common (1 in 3000 boys) and devastating disease that was clearly caused by mutations in a gene somewhere on the X chromosome. Years of careful investigation of the pathological changes in affected muscle had failed to reveal the biochemical basis of DMD. In the mid-1980s, several groups competed to clone the DMD gene, using different approaches. The pioneering work of these groups, overcoming formidable technical difficulties to reveal a gene of a size and complexity never before seen, was probably the major inspiration for most subsequent positional cloning efforts. This work is the subject of the first of the case studies at the end of this chapter.

The successful conclusion of this work in 1987 marked the start of a triumphant new era for human molecular genetics. One after another, the genes underlying important Mendelian disorders such as cystic fibrosis (see case study 2 at the end of this chapter), Huntington disease, adult polycystic kidney disease, and familial colorectal cancer were isolated. The logic of positional cloning in this era followed the scheme of **Figure 16.2**. These early efforts could nevertheless be desperately hard work, because of the need to clone all the DNA and identify all the genes within the candidate region. This required the screening of genomic libraries to identify clones and construct a contig across the region, followed by

Figure 16.1 How to identify a human disease gene. As this figure emphasizes, there is no single pathway to success, but the key step is to arrive at a plausible candidate gene (box B3). This can then be tested for mutations in affected people (D5). The candidate gene might be identified through knowledge of a homolog in another species (A1 and A3), through a patient who has both the phenotype of interest and a chromosome structural abnormality (C2), or through linkage analysis (C1 and D1). Success depends on an interplay between clinical work, laboratory benchwork, and database analysis, with the latter becoming steadily more crucial as genome information accumulates.

major efforts in screening cDNA libraries to identify and characterize the genes. By 1995, only about 50 inherited disease genes had been identified by this approach.

The first step in positional cloning is to define the candidate region as tightly as possible

The difficulty of positional cloning depends very largely on the size of the candidate region, so the first priority is to narrow this down as far as possible. For Mendelian diseases, this is mainly a function of the number of meioses available for linkage analysis. The limit of resolution is reached when the last recombinant has been mapped between closely spaced markers. Using the generalization of 1 cM = 1 Mb, a family collection with 100 informative meioses might localize a Mendelian disease to a candidate region of about 1 Mb—depending on luck and on the frequency of recombination in the particular chromosomal region.

At this stage, the raw data should be scrutinized by inspecting haplotypes rather than relying on computer analysis (**Figure 16.3**). When single recombinants define the boundaries of the candidate region, it is important to consider possible sources of error. Meticulous clinical diagnoses are imperative. Key recombinants are more reliable if they occur in unambiguously affected people—an unaffected individual might be a non-penetrant carrier of the disease gene. Apparent close *double recombinants* are highly suspect; as explained in Chapter 14, they most probably reflect errors in diagnosis or genotyping. The possibility of *phenocopies* must be considered. Sometimes, despite good positive lod scores, there seem to be recombinants with every marker tried. This is usually an indication of locus heterogeneity—in one or more of the families in the study the disease does not map to the region under study.

The second step is to establish a list of genes in the candidate region

Before completion of the Human Genome Project, compiling a list of genes in the candidate region involved much laborious effort. Contigs of genomic clones had to be constructed across the candidate region by successive rounds of screening

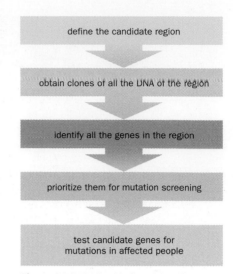

Figure 16.2 The logic of positional cloning. The figure illustrates the logical progression of positional cloning as originally practiced. Nowadays a list of genes in the candidate region can be downloaded from genome databases. Before the current sequence data and high-resolution marker maps were available, researchers tried all sorts of shortcuts to reduce the labor of pure positional cloning.

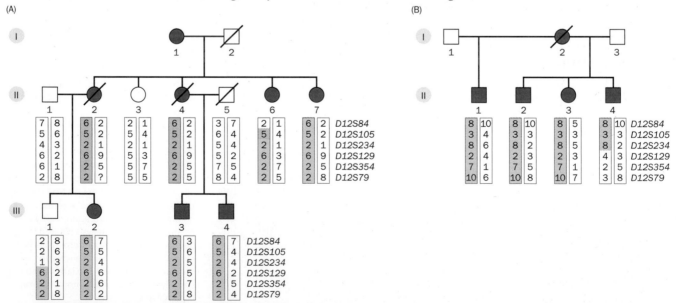

Figure 16.3 Defining the minimal candidate region by inspection of haplotypes. The two pedigrees show a dominantly inherited skin disorder, Darier–White disease (OMIM 124200), which had previously been mapped to chromosome 12q. The 12q marker haplotype that segregates with the disease is highlighted in yellow. Gray boxes mark inferred haplotypes in dead people. Numbers within the boxes identify different alleles of the relevant marker. (A) The recombination in individual II$_6$ in this pedigree shows that the disease must be located distal of marker *D12S84*. *D12S105* (green) is uninformative because I$_1$ was evidently homozygous for allele 5 (II$_3$ and II$_7$ evidently inherited opposite haplotypes from I$_1$, but both have *D12S105* allele 5). The recombination shown in III$_1$ suggests that the disease gene maps proximal to *D12S129*, but this requires confirmation because the interpretation requires the genotypes of II$_1$ and II$_2$ to be inferred correctly, and that III$_1$ not be a non-penetrant gene carrier. (B) The recombination in individual II$_4$ in this pedigree provides the confirmation. The combined data locate the Darier–White gene to the interval between *D12S84* and *D12S129*. [From Carter SA, Bryce SD, Munro CS et al. (1994) *Genomics* 24, 378–382. With permission from Elsevier.]

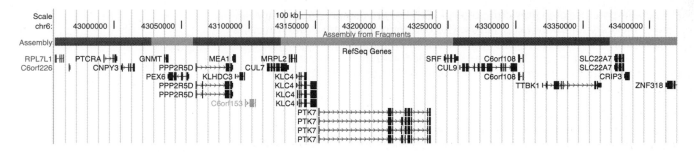

Figure 16.4 Using a genome browser to list the genes in a candidate region. This partial screendump from the UCSC genome browser (http://www.genome.ucsc.edu/) shows genes in a 500 kb region of chromosome 6p21.1. Exons of a gene (vertical bars) are linked by a horizontal line, and arrowheads show the direction of transcription.

genomic libraries (*chromosome walking*), and then expressed sequences from within the clone contig had to be identified by mapping transcripts. The campaign to clone the cystic fibrosis gene, described in case study 2 at the end of this chapter, gives an excellent insight into the difficulties researchers had to overcome in this early phase.

Nowadays a genome browser such as Ensembl (http://www.ensembl.org/) or the Santa Cruz browser (http://genome.cse.ucsc.edu/) would be used to display all the definite and possible genes in the candidate region (**Figure 16.4**). Deeply impressive though these displays are, it is important not to rely totally on them. These extremely sophisticated tools need to be used as adjuncts to, and not replacements for, thought. They must be supplemented by first-hand in-depth study of the region. The reference sequence used by the browsers is currently far from perfect. Going back to the raw data can identify missing or variable parts of the sequence. The ENCODE (Encyclopedia of DNA Elements) project (see Chapter 11) has shown that gene structures identified by the browsers are often not complete. The current annotations do include virtually all protein-coding genes, but they may not identify all the possible exons or alternative transcripts, especially those using small exons separated by large introns, and they certainly omit many noncoding transcripts. It can be useful to compare the predictions of several different gene-finding programs. In the laboratory, techniques such as reverse transcriptase (RT)-PCR (see Chapter 8) or 5′ RACE (see Box 11.1) can be used to verify claimed isoforms and search for new ones.

The third step is to prioritize genes from the candidate region for mutation testing

To identify mutations, samples from a collection of unrelated affected people need to be sequenced. Typically, a candidate region defined by linkage analysis would contain several genes with 100 or more exons in total. One approach, made increasingly possible by new sequencing technologies, is simply to sequence every exon across the candidate region. More usually, however, a candidate gene is selected, and each exon of that gene is individually amplified by PCR and sequenced. In choosing genes for sequencing it makes sense to start with the most promising ones, although ease of analysis is also a consideration. All things being equal, one would deal with a gene with 4 exons before tackling one with 65.

Appropriate expression

A good candidate gene should have an expression pattern consistent with the disease phenotype. Expression need not be restricted to the affected tissue, because there are many examples of widely expressed genes causing a tissue-specific disease, but the candidate gene should at least be expressed in the place where the pathology is seen, and at or before the time at which the pathology becomes evident. For example, genes responsible for neural tube defects should be expressed in the neural tube shortly before it closes (which happens during the third and fourth weeks of human embryonic development; see Chapter 5). The expression of candidate genes can be tested by RT-PCR, northern blotting (see Chapter 7), or serial analysis of gene expression (SAGE; see Box 11.1). Much of the preliminary work can be performed with databases (dbEST or the SAGE database, both accessible through the NCBI homepage at http://www.ncbi.nlm.nih.gov/) rather than in the laboratory. *In situ* hybridization against mRNA in tissue sections, or immunohistochemistry with labeled antibodies, provides the most detailed picture of expression patterns. Studies are usually done on mouse tissues, especially for embryonic stages. The common assumption that humans and mice will show similar expression patterns is not always justified, and centralized resources of staged human embryo sections have been established to allow the equivalent analyses to be performed, where necessary, on human embryos (see http://www.hdbr.org/).

Appropriate function

When the function of a gene in the candidate region is known, it may be obvious whether or not it is a good candidate for the disease. For example, rhodopsin was a good candidate for retinitis pigmentosum, and fibrillin for the connective tissue disorder Marfan syndrome (OMIM 154700). For a novel gene, sequence analysis will often provide clues to its function through the recognition of common sequence motifs such as transmembrane domains or tyrosine kinase motifs. These may be sufficient to prioritize a gene as a candidate, given the pathology of the disease. For example, ion transport in the inner ear is known to be critical for hearing, so an ion channel gene would be a natural candidate in positional cloning of a deafness gene.

Homologies and functional relationships

Sometimes a gene in the candidate region turns out to be a close homolog of a known gene (a paralog in humans, or an ortholog in other species). If mutations in the homologous gene cause a related phenotype, the new gene becomes a compelling candidate. For example, Marfan syndrome is caused by mutations in the fibrillin gene. A clinically overlapping condition, congenital contractural arachnodactyly (CCA; OMIM 121050), was mapped to chromosome 5q. The candidate region contained the *FBN2* gene, which is a paralog of fibrillin. *FBN2* mutations were soon demonstrated in CCA patients. Weak homologies may be missed by the programs used for automatic genome annotation. A more directed, hypothesis-driven search may come up with significant weak homologies that can point to the gene's function.

Candidate genes may also be suggested on the basis of a close functional relationship to a gene known to be involved in a similar disease. The genes could be related by encoding a receptor and its ligand, or other interacting components in the same metabolic or developmental pathway. For example, mutations in the *RET* gene were known to cause some but not all cases of Hirschsprung disease. *RET* encodes a cell-surface tyrosine kinase receptor. The genes encoding the *RET* ligands GDNF and NRTN were then obvious candidates for hunting for further Hirschsprung mutations.

Now that we have genome sequences of many different species, it has become clear how far structural and functional homologies extend across even very distantly related species. The great majority of mouse genes have an identifiable human counterpart, and the same is true of other mammalian species. As described in Chapter 10, extensive homologies can be detected between genes in humans and model organisms such as zebrafish, the fruit fly *Drosophila*, the nematode worm *Caenorhabditis elegans* and even the yeasts *Saccharomyces cerevisiae* and *Schizosaccharomyces pombe*. For these distantly related organisms, homology may be more evident in the amino acid sequence of the protein product than in the DNA sequence of the gene. Looking at the amino acid level allows synonymous coding sequence changes to be ignored.

A very powerful means of prioritizing candidates from among a set of human genes is therefore to see what is known about homologous genes in well-studied model organisms. Such data might include the pattern of expression and the phenotype of mutants. Papers listed in Further Reading show how data from *Drosophila* and yeast can be used to infer the function of human genes. Mice are especially useful for such investigations, and their use is considered in more detail below. Even more than gene sequences, pathways are often highly conserved, so that knowledge of a developmental or control pathway in *Drosophila* or yeast can be used to predict the likely working of human pathways—although mammals often have several parallel paths corresponding to a single path in lower organisms.

By contrast, mutant phenotypes are less likely to correspond closely. A striking example is the wingless *apterous* mutant of *Drosophila*. A human gene, *LHX2*,

BOX 16.1 MAPPING MOUSE GENES

Several methods are available for easy and rapid mapping of phenotypes or DNA clones in mice. Together with the ability to construct transgenic mice, they make the mouse especially useful for comparisons with humans.

Interspecific crosses (*Mus musculus* crossed with either *Mus spretus* or *Mus castaneus*)

Different mouse species have different alleles at many polymorphic loci, making it easy to recognize the origin of a marker allele. This is exploited in two ways:

- **Constructing marker framework maps.** Several laboratories have generated large sets of F2 backcrossed mice. Any marker or cloned gene can be assigned rapidly to a small chromosomal segment defined by two recombination breakpoints in the collection of backcrossed mice. For example, the Collaborative European backcross was produced from a *M. spretus* × *M. musculus* (C57BL) cross. Five hundred F2 mice were produced by backcrossing with *M. spretus*, and 500 by backcrossing with C57BL. All microsatellites in the framework map are scored in every mouse.

- **Mapping a new phenotype.** A cross must be set up specifically to do this but, unlike with humans, any number of F2 mice can be bred to map to the desired resolution. *M. musculus* × *M. castaneus* crosses are easier to breed than *M. musculus* × *M. spretus*.

Recombinant inbred strains

These are obtained by systematic inbreeding of the progeny of a cross. For example, the widely used BXD strains are a set of 26 lines derived by more than 60 generations of inbreeding from the progeny of a C57BL/6J × DBA/2J cross. They provide unlimited supplies of a panel of chromosomes with fixed recombination points. DNA is available as a public resource, analogous to the human CEPH families. Recombinant inbred strains are particularly suited to mapping quantitative traits, which can be defined in each parent strain and averaged over a number of animals of each recombinant type. In comparison with mice from interspecific crosses, it may be harder to find a marker in a given region that distinguishes the two original strains, and the resolution is lower because of the smaller numbers.

Congenic strains

These are identical except at a specific locus. They are produced by repeated backcrossing and can be used to explore the effect of changing just one genetic factor on a constant background.

---

is able to complement the deficient function of the mutant, so that the flies grow normal wings (see Figure 10.27). We humans must have a virtually identical developmental pathway to *Drosophila*, but clearly we use it for a different purpose. No human mutation has been described, but *Lhx2* knockout mice die before birth with defects in development of the eyes, forebrain, and erythrocytes. Branchio-oto-renal syndrome (OMIM 113650), the subject of case study 3 at the end of this chapter, provides another example of a conserved pathway used for a different purpose.

Mouse models have a special role in identifying human disease genes

Phenotypic homologies between humans and mice provide particularly valuable clues toward identifying human disease genes. Programs of systematic mutagenesis are generating very large numbers of mouse mutants. These include point mutations, knockouts, deletions, and conditional knockouts (in which the gene is inactivated only in specific cells or tissues). There are several programs for the systematic knockout of all genes and for the systematic generation of conditional mutants (see Further Reading).

Genetic mapping is quick and accurate in the mouse (**Box 16.1**). Thus, most mouse mutant phenotypes have been mapped or can easily be mapped. Cross-matching of human and mouse genome sequences has provided a very detailed picture of the relationship between human and mouse chromosomes (**Figure 16.5**). Once a chromosomal location for a phenotype of interest is known in the mouse, it is usually possible to predict the corresponding location in humans.

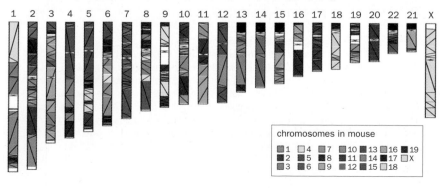

chromosomes in mouse

■1	□4	■7	■10	■13	■16	■19
■2	■5	■8	■11	■14	■17	□X
■3	■6	■9	■12	■15	□18	

Figure 16.5 Relationship between human and mouse chromosomes. The diagram shows the 22 human autosomes, plus the X chromosome. Segments are color coded to show where corresponding sequences occur on the 19 mouse autosomes plus the mouse X chromosome. Within each segment, oblique lines indicate whether the sequence is directly repeated (lines running from top left to bottom right) or inverted (lines running from top right to bottom left) relative to the human sequence. Red triangles mark the human centromeres. [From Church DM, Goodstadt L, Hillier LW et al. (2009) *PLoS Biol.* 7(5), e1000112. doi:10.1371/journal.pbio.1000112.]

Exon sequences and exon–intron structures are usually well conserved between orthologous human and mouse genes. Close counterparts of human mutations can be recognized or constructed in mice, and orthologous mutations are more likely to produce phenotypes resembling their human counterpart in mice than in flies or worms. Nevertheless, the similarities may not be as close as one might wish.

Techniques for the genetic manipulation of mice are very well developed. Once a candidate gene has been identified in humans, mouse mutants can be constructed to allow functional analysis. Our ability to make total or conditional knockouts and to engineer specific mutations in an organism fairly closely related to ourselves makes the mouse a very powerful tool for the exploration of human gene function.

16.2 THE VALUE OF PATIENTS WITH CHROMOSOMAL ABNORMALITIES

Chromosomal abnormalities can sometimes provide an alternative method of localizing a disease gene, in place of linkage analysis. For conditions that are normally sporadic, such as many severe dominant diseases, chromosome aberrations may provide the only method of arriving at a candidate gene. With luck, they may even point directly to the precise location, rather than defining a megabase-sized candidate region, as with linkage. Alert clinicians have a crucial role in identifying such patients (**Box 16.2**).

Patients with a balanced chromosomal abnormality and an unexplained phenotype provide valuable clues for research

A balanced translocation or inversion, with nothing extra or missing, would not be expected to have any phenotypic effect on the carrier. If a person with an apparently balanced chromosomal abnormality is phenotypically abnormal, there are three possible explanations:

- The finding is coincidental.
- The rearrangement is not in fact balanced—there is an unnoticed loss or gain of material.
- One of the chromosome breakpoints causes the disease by disrupting a critical gene.

A chromosomal break can cause loss of function of a gene if it disrupts the coding sequence of that gene or if it separates the coding sequence from a *cis*-acting regulatory region. Alternatively, it could cause a gain of function, for exam-

---

BOX 16.2 POINTERS TO THE PRESENCE OF CHROMOSOME ABNORMALITIES

Clinicians have made major contributions to identifying disease genes by identifying patients who have causative chromosome abnormalities.

A cytogenetic abnormality in a patient with the standard clinical presentation

If a disease gene has already been mapped to a certain chromosomal location, and then a patient with that disease is found who has a chromosome abnormality affecting that same location, the chromosome abnormality most probably caused the disease.

Patients with balanced translocations or inversions often have breakpoints located within the disease gene, or very close to it. Cloning these breakpoints can provide the quickest route to identifying the disease gene. With deletions, the breakpoints may be located some distance from the disease gene, but if the deleted segment is smaller than the current candidate region, defining the breakpoints helps localize the gene.

Most such patients will have *de novo* mutations. Some researchers feel that performing chromosome analysis on all patients with *de novo* mutations is a worthwhile expenditure of research effort.

Additional mental retardation

A patient may have a typical Mendelian disease but may in addition be mentally retarded. This may be coincidental, but such cases can be caused by deletions that eliminate the disease gene plus additional neighboring genes. Large chromosomal deletions almost always cause severe mental retardation, reflecting the involvement of a high proportion of our genes in fetal brain development. When the patient is also a *de novo* case of a condition that is normally familial, cytogenetic and molecular analysis is warranted.

Contiguous gene syndromes

Very rarely, a patient seems to suffer from several different genetic disorders simultaneously. This may be just very bad luck, but sometimes the cause is simultaneous deletion of a contiguous set of genes. Contiguous gene syndromes are described in Chapter 13; they are particularly well defined for X-linked diseases. One famous example, described at the end of this chapter, led directly to the identification of the gene mutated in Duchenne muscular dystrophy.

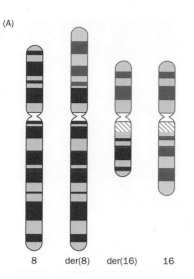

(A)

8   der(8)   der(16)   16

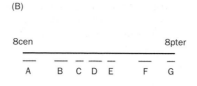

(B)

8cen   8pter

A   B   C   D   E   F   G

(C)

Test	FISH probe	Hybridization pattern
1	A	8, der(8)
2	G	8, der(16)
3	B	8, der(8)
4	F	8, der(16)
5	E	8, der(16)
6	D	8, der(8), der (16)

Figure 16.6 Using fluorescence *in situ* hybridization to define a translocation breakpoint. (A) A translocation t(8;16)(p22;q21) is defined by cytogenetics. (B) The physical map of part of the breakpoint region in a normal chromosome 8 shows the approximate locations of seven clones, A–G. (C) Results of successive FISH experiments reveal that the breakpoint is within the sequence represented in clone D. This result would normally be confirmed by using clones from chromosome 16.

ple by splicing together exons of two genes to create a novel chimeric gene—this is rare in inherited disease but common in tumorigenesis (see Chapter 17). In either case, the breakpoint provides a valuable clue to the exact physical location of the disease gene. The position of the breakpoint is most easily defined by using FISH (**Figure 16.6**). This will often be sufficient to identify the gene involved. If necessary, once the breakpoint has been localized to a single clone, its exact position could be defined by finding a sequence that can be amplified by PCR from DNA of the patient by using a pair of primers, one from each of the two chromosomes involved. An alternative approach would be to make a genomic library from the patient's DNA and then sequence appropriate clones. Breakpoints of deletions or duplications can be easily mapped on microarrays by looking for a change in the hybridization intensity.

An example of the power of this approach is the identification of the Sotos syndrome gene (**Figure 16.7**). This syndrome (OMIM 117550) involves childhood overgrowth, dysmorphic features, and mental retardation. Because it always occurs *de novo*, there were no multigeneration families that might allow the causative gene to be localized by linkage analysis. A single patient with a *de novo* balanced 5;8 translocation, 46,XX,t(5;8)(q35;q24.1), provided the vital clue. As shown in Figure 16.7, the translocation breakpoint disrupted the *NSD1* gene. This could have been purely coincidental. Proof that *NSD1* was the gene mutated in Sotos syndrome came from showing that 4 out of 38 independent patients with Sotos syndrome had point mutations in the *NSD1* gene, and 20 out of 30 patients had microdeletions involving this gene (**Figure 16.8**).

X–autosome translocations are a special case

Even if a translocation breakpoint disrupts a gene, we have lost the function of only one of the two copies of the gene. There will be no phenotypic effect unless

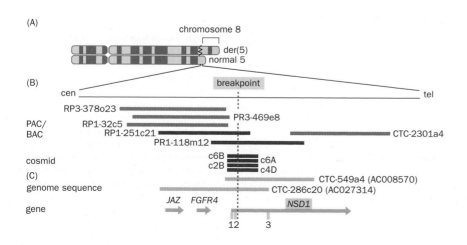

(A)

chromosome 8

der(5)
normal 5

(B)

cen   breakpoint   tel

RP3-378o23
                              PR3-469e8
PAC/   RP1-32c5
BAC   RP1-251c21   CTC-2301a4
            PR1-118m12

cosmid   c6B   c6A
                    c2B   c4D

(C)
genome sequence   CTC-549a4 (AC008570)
                              CTC-286c20 (AC027314)

gene   JAZ   FGFR4   NSD1

            12   3

Figure 16.7 A balanced 5;8 translocation disrupts the *NSD1* gene in a patient with Sotos syndrome. (A) The two copies of chromosome 5 are shown. One of them has part of 5qter missing and replaced by part of chromosome 8 (beige). (B) An expanded view of the region surrounding the breakpoint. A contig of PAC/BAC and cosmid clones was constructed around the breakpoint. Clones in red cross the breakpoint. (C) Sequencing revealed a partial genomic sequence homologous to mouse *Nsd1*. The human *NSD1* gene was isolated and characterized. Mutations and deletions were then demonstrated in unrelated patients with Sotos syndrome (not shown). [From Kurotaki N, Imaizumi K, Harada N et al. (2002) *Nat. Genet.* 30, 365–366. With permission from Macmillan Publishers Ltd.]

Figure 16.8 A microdeletion in a patient with Sotos syndrome demonstrated by fluorescence *in situ* hybridization. The two homologs of chromosome 5 are identified by the red FISH probe that recognizes a sequence on 5pter. The green FISH probe recognizes a sequence from 5qter containing the *NSD1* gene; this sequence is lacking on one copy of chromosome 5. [From Kurotaki N, Imaizumi K, Harada N et al. (2002) *Nat. Genet.* 30, 365–366. With permission from Macmillan Publishers Ltd.]

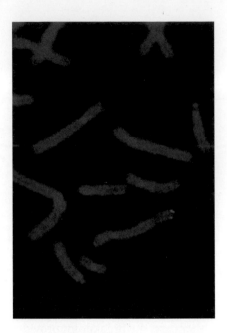

a 50% decrease in the level of the product causes problems (*haploinsufficiency*). X–autosome translocations in females are an exception to this, because of X-inactivation (see Chapter 3). One X chromosome, picked at random, in each cell is permanently inactivated. However, when one X is involved in a balanced X–autosome translocation, cells that inactivate the translocated X often suffer lethal genetic imbalances. Inactivation spreads *in cis*, and complete inactivation requires an intact X chromosome. If the imbalance kills all these cells in the early embryo (or at least prevents them contributing to further development), a female carrier of such a translocation will consist entirely of cells that have inactivated the normal X. If the X chromosome breakpoint disrupts a gene, the woman is left with no active functional copy of that gene. **Figure 16.9** shows how such a translocation, with a breakpoint in the dystrophin gene, can cause a woman to be affected by Duchenne muscular dystrophy.

Rearrangements that appear balanced under the microscope are not always balanced at the molecular level

If a patient has a *de novo* apparently balanced abnormality but is also dysmorphic or mentally retarded without any other known cause, it is quite likely that the chromosomal rearrangement is more complex than it appears under the microscope. Often there is a submicroscopic deletion. The loss of a few megabases of DNA would not be visible on standard cytogenetic preparations. Branchio-oto-renal syndrome (BOR; OMIM 113650), described as case study 3 at the end of this chapter, is an example. Large but still submicroscopic deletions at breakpoints can be detected by FISH, using as probes clones that map close to the breakpoints. However, it is the advent of array-comparative genomic hybridization (array-CGH; see below) that has made investigating these cases simple. Systematic study of patients with apparently balanced translocations but also a phenotype has resulted in the discovery of many microdeletions and some microduplications. From a research point of view, these are less valuable than truly balanced rearrangements (which would not be detectable by array-CGH) because they necessarily give a less precise positional clue. Nevertheless, such abnormalities have been instrumental in identifying several disease genes. CHARGE syndrome, described as case study 6 at the end of this chapter, is an example.

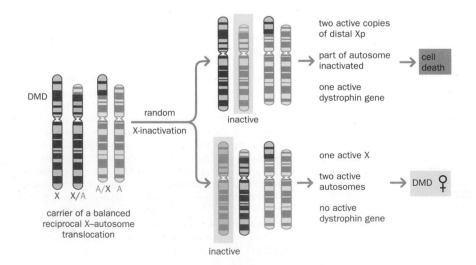

Figure 16.9 Non-random X-inactivation occurs in female Duchenne muscular dystrophy patients with Xp21–autosome translocations. The translocation between the X chromosome (X) and an autosome (A) is balanced, but the X chromosome breakpoint disrupts the dystrophin gene (red). X-inactivation is random, but cells that inactivate the translocated X die because of lethal genetic imbalance. The embryo develops entirely from cells in which the normal X is inactivated, leading to a woman with no functional dystrophin gene. The resulting failure to produce any dystrophin causes DMD.

Comparative genomic hybridization allows a systematic search for microdeletions and microduplications

Microdeletions were long suspected to be the cause of some unexplained recurrent genetic syndromes and many individual cases in which a patient had a unique pattern of abnormalities, but until recently there had been no way to search systematically for them. Their presence could be confirmed by FISH, but one had to know which probe to use. *Comparative genomic hybridization* (CGH) has removed this barrier. CGH uses competitive hybridization between DNA from a patient and from an unaffected control to identify chromosomal regions where there is a difference in copy number between the two. DNA samples from the patient and the control, labeled with different colored dyes, are mixed in equal amounts and hybridized to a sample from a normal individual. Originally, the hybridization target was a spread of metaphase chromosomes on a microscope slide, prepared as in FISH. A scan along each chromosome identified regions where there was differential hybridization of the patient and control DNAs. This has now been superseded by array-CGH (**Figure 16.10**), in which the hybridization target is a microarray of normal genomic DNA. The microarray might carry BAC clones or oligonucleotides, and it can be designed either to give complete, overlapping coverage (tiling) of the human genome or to give high-resolution cover of a specific region of interest.

Array-CGH is a flexible and very powerful technique. The resolution and coverage can be varied at will, because they depend entirely on the choice of probes used to construct the microarray. The location and size of any abnormality detected can be immediately identified, because we know the location in the human genome sequence of each probe on the array. An alternative route to similar results uses high-throughput SNP chips or bead arrays (see Chapter 8). In this case there is no control; labeled DNA from the patient is hybridized to the array, and deletions or duplications are observed as variations in hybridization

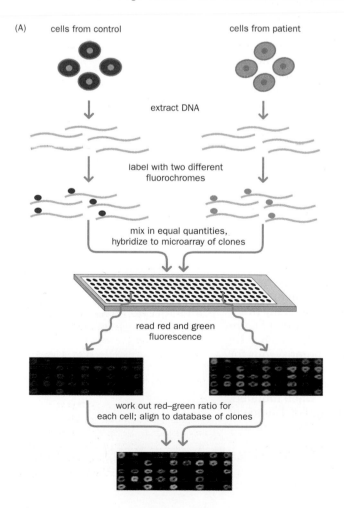

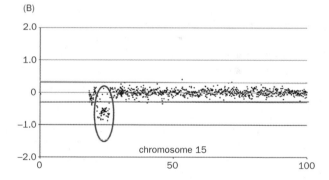

Figure 16.10 Comparative genomic hybridization. (A) Principle of array-CGH. DNA samples from a patient and an unaffected control are labeled with two different fluorochromes. Equal quantities of the two DNA samples are mixed and hybridized to a microarray of clones from a normal individual. After washing off unbound probe, the fluorescence at both wavelengths is measured and the red–green ratio for each clone is calculated. (B) Example of results. The microarray was made of cloned sequences from a normal chromosome 15. The red–green fluorescence ratio (*y* axis) is plotted for each clone on the microarray. Results are arranged along the *x* axis according to the chromosomal location of each clone. Red and green horizontal lines mark the limits of fluorescence ratios when the red-labeled and green-labeled DNAs are present in equal quantities. A series of contiguous clones can be seen (red oval) where the patient's DNA (green) has a lower copy number than that of the control DNA (red). The patient has a heterozygous deletion of the DNA represented by those clones. [Adapted from Read A & Donnai D (2006) New Clinical Genetics. With permission from Scion Publishing Ltd.]

intensity. The SNP genotypes can also be informative: SNPs from within a deleted region will show only a single allele (apparent homozygosity), and SNPs from a duplicated region may show three alleles.

Whichever technique is used, it is important to confirm the results by FISH. Balanced abnormalities would not be detected, and the results for a duplication or amplification would not distinguish tandem repeats from dispersed repeats. The main limitation of these techniques is in the interpretation of the results. As mentioned in Chapter 13, large-scale application of these methods has identified innumerable copy number variations in normal healthy people. Just because a microdeletion is seen in a patient with a clinical phenotype, it does not follow that the deletion caused the phenotype. Databases of normal variants, such as TCAG (see Further Reading), can be consulted to eliminate known nonpathogenic variants, but there always remains the possibility that the patient carries a novel but harmless variant. Thus, as always with chromosomal abnormalities, variants detected by CGH suggest hypotheses; these must then be checked by other means.

Long-range effects are a pitfall in disease gene identification

Chromosomal rearrangements can sometimes affect the expression of a gene located hundreds of kilobases away from the breakpoint, even though the coding sequence is not disrupted. There are two possible reasons why this may happen.

First, rearrangements can affect the structure of large-scale chromatin domains. Classic studies in the fruit fly *Drosophila melanogaster* showed that a structurally intact gene could be silenced if a chromosomal rearrangement placed it within or close to heterochromatin. This is known as a **position effect**. In both flies and humans, the histone changes that mark heterochromatin can propagate along a chromosome; this spreading is limited by insulator sequences. For example, when a gene affecting eye color in the fly was moved to a location close to heterochromatin, the resulting variable inactivation was seen as position-effect variegation in the eye. Position effects have been quite often invoked to explain findings in humans with chromosome abnormalities, but it is not clear how often this is the correct explanation.

A better authenticated cause of long-distance effects in humans is interference with distant *cis*-acting regulatory elements. Such elements are much more common than was formerly supposed. The ENCODE project has dramatically illustrated that we can no longer think of a gene as simply the coding sequence plus a kilobase or so of upstream promoter. *Cis*-acting regulatory elements or additional tissue-specific promoters may be located far away from the coding sequence. Several human examples are known of translocation breakpoints affecting the expression of a gene up to a megabase away. Aniridia (absence of the iris of the eye; OMIM 106210) is normally caused by heterozygosity for classical point mutations in the *PAX6* gene, but several patients with aniridia have balanced translocations with breakpoints up to 150 kb downstream of an intact *PAX6* gene (see Figure 11.6). Similarly, campomelic dysplasia (short limbs and other abnormalities; OMIM 114290) is caused by loss-of-function mutations in the *SOX9* gene, but translocation breakpoints up to 900 kb upstream of the coding sequence can have the same effect.

Sometimes the distal control elements are located within introns of another gene. Some limb abnormalities in humans are caused by mutations in introns of the *LMBR1* gene on chromosome 7—but the affected sequences actually function as tissue-specific enhancers of the *SHH* gene, located 1 Mb away (see Figure 11.6). Ellis–van Creveld syndrome (OMIM 225500) might be another such case. Patients with this autosomal recessive skeletal dysplasia can be homozygous for mutations in either of two adjoining genes on chromosome 4p16. The two genes are oppositely oriented and may have a common bidirectional promoter. Possibly each gene can independently cause exactly the same phenotype when mutated, but it is puzzling that the two genes share no evident homology. Persistence of intestinal lactase, described in case study 5 at the end of this chapter, is another example.

Thus, patients with a *de novo* balanced abnormality and a *de novo* Mendelian phenotype provide valuable pointers to the relevant gene, but their role is to suggest a hypothesis. This must then be confirmed by other studies—for example, by studying gene expression in cells from that patient, or by finding unrelated chromosomally normal cases with point mutations in the relevant gene.

16.3 POSITION-INDEPENDENT ROUTES TO IDENTIFYING DISEASE GENES

Human disease genes have mostly been identified by first defining their chromosomal location, as described in the previous two sections. Sometimes, however, positional information has not been used, either because it was not available or because an alternative approach made it unnecessary.

A disease gene may be identified through knowing the protein product

If we know at least a partial amino acid sequence of a gene product, we can use a table of the genetic code to infer the gene sequence. Because the code is degenerate (a given amino acid may be encoded by any one of several codons), the predicted gene sequence is also degenerate. A probe used to screen a cDNA library has to be a cocktail of all the possible sequences that could encode those amino acids. Because only one of the probes in the mix will correspond to the authentic sequence, it is important to keep the number of different oligonucleotides low so as to increase the chance of identifying the correct target. Tryptophan and methionine are very helpful here, because each has only a single codon. Arginine, leucine, and serine, with six codons each, are avoided as far as possible. **Figure 16.11** shows how degenerate probes were used by Gitschier and colleagues in 1984 to isolate the Factor VIII gene.

An alternative route, if the protein is available in even minute quantities, is to raise an antibody to the protein and use this to find the gene. Back in 1982 the mRNA encoding phenylalanine hydroxylase was recovered by immunoprecipitation of polysomes that were synthesizing the protein in a cell-free system. Nowadays an expression library could be made by cloning pooled cDNA into an expression vector. Host cells containing clones with the desired gene should produce the protein, or at least parts of it, and could be identified by screening colony filters from the library with an appropriate antibody. Everything here depends on the specificity of the antibody and on the hope that the protein is not toxic to the host cell. *Phage display* would give an alternative approach.

Most of this work can now be done computationally. A protein can usually be identified directly by mass spectrometric analysis of peptide fragments and chemical microsequencing. Databases can then be queried to identify the relevant gene. Case study 4 at the end of this chapter describes two ways by which the gene underlying multiple sulfatase deficiency was identified; one of them exemplifies this approach.

A disease gene may be identified through the function or interactions of its product

If the biochemical or functional defect that causes a disease can be observed in cultured cells, a search can be made for a protein or DNA fragment that will rectify the defect.

Figure 16.11 Cloning the Factor VIII gene. Degenerate probes were synthesized to correspond to a partial amino acid sequence of porcine Factor VIII, and used to screen a porcine cDNA library to recover a full-length cDNA. Screening was performed in two stages. The primary screen used a partly degenerate 45 nucleotide probe; this was long enough to tolerate occasional mismatched bases. A secondary screen used a mixture of 16 much shorter oligonucleotides, only 15 nt long, which would hybridize only to a cDNA that perfectly matched. The porcine Factor VIII genomic clone was then used to screen human DNA libraries to recover the human gene. See Gitschier J, Wood WI, Goralaka TM et al. (1984) *Nature* 312, 326–330.

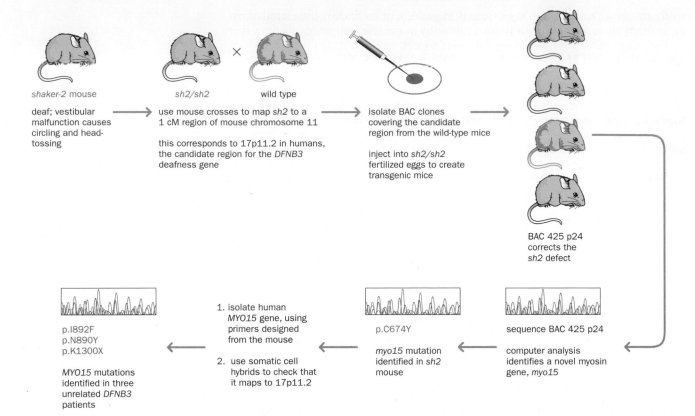

Figure 16.12 Functional complementation in transgenic mice as a tool for identifying a human disease gene. In mice, the *shaker-2* mutation causes deafness. This mutation was identified by finding a wild-type clone that corrected the defect. Members of a human family with a similar phenotype that mapped to the corresponding chromosomal location proved to have mutations in the orthologous gene, *MYO15*.

- The Fanconi anemia (see OMIM 227650) groups A, C, and G genes (see Chapter 13, p. 416) were identified by screening cDNA libraries for clones that could cure the DNA repair defect in cells from patients.

- The second of the two ways in which the gene responsible for multiple sulfatase deficiency was identified (see case study 4 at the end of this chapter) involved finding a chromosome fragment that was able to correct the defect in a cell line from a patient.

- A similar functional complementation approach, but in this case in a whole animal, was used to identify the mouse *shaker-2* and human *DFNB3* deafness genes (**Figure 16.12**). In fact, this approach did use some positional information, but it is described here because it is another example of functional complementation. A form of autosomal recessive hearing loss was mapped to a unique locus, *DFNB3*, in a very large extended Indonesian kindred. Because all the patients were likely to be homozygous for the same single mutation, it was unlikely that the pathogenic change could be unambiguously identified by sequencing DNA from the candidate region in family members. Comparative mapping showed that *DFNB3* and *shaker-2* were likely to be orthologs. Transgenic *shaker-2* mice were constructed that contained wild-type BAC clones from the *shaker-2* candidate region. A BAC that corrected the phenotype was identified and turned out to contain an unconventional myosin gene, *myo15*. The human *MYO15* gene was then isolated on the basis of its close homology with the mouse gene; its position within the *DFNB3* candidate region was confirmed, and the Indonesian patients were then shown to have a mutation in *MYO15*.

When a disease affects a biochemical pathway, each component of the pathway is a non-positional candidate for the disease. The components may be identified through protein or RNA studies.

- Protein–protein interactions can be identified through yeast two-hybrid experiments or by using mass spectrometry to identify the components of a multiprotein complex. For example, *BRIP1* and *PALB2* (see case study 7 at the end of this chapter) were identified as susceptibility factors for familial breast cancer because the proteins they encode interact with the proteins encoded by the known breast cancer susceptibility genes *BRCA1* and *BRCA2*.

- Microarrays are often used to compare mRNA profiles from patients and controls, or from cells before and after the expression of some protein is knocked down by RNAi (RNA interference; see Box 9.6, p. 284 and Chapter 12). This produces a list of genes whose expression is altered—but the list is usually very long. It is then necessary somehow to distinguish primary effects from all the downstream changes.

A disease gene may be identified through an animal model, even without positional information

Many human disease genes have been identified with the help of animal models—but nearly always this has been after checking positional information. Maybe a mouse mutant and a phenotypically similar human disease had been mapped to chromosomal locations that corresponded, as with *DFNB3* and the *shaker-2* mouse. Then if the mouse gene was cloned, its human homolog became a natural candidate. Alternatively, a disease gene may have been identified in the mouse, and the human homolog was then isolated; this could be mapped by FISH, and becomes a candidate gene for any relevant human disease mapping to that location. This is how the *MITF* gene was identified as a cause of type 2 Waardenburg syndrome (OMIM 193510).

Sometimes, however, a gene identified in an animal model is tested directly in human patients without first using positional information to confirm that these patients are the appropriate ones to test. One such case is *SOX10* (**Figure 16.13**). This gene was identified by laborious positional cloning of the mouse *Dominant megacolon* (*Dom*) mutation. *Dom* mice are a long-studied model of human Hirschsprung disease, but they also have white spotting or white coats. Patients with Waardenburg syndrome type IV (WS4; OMIM 277580) have a similar combination of Hirschsprung disease and pigmentary abnormalities. WS4 is very rare and normally occurs in families too small for mapping, so a panel of WS4 patients was tested for mutations in *SOX10* without any prior knowledge that their disease gene mapped to the corresponding location. The gamble paid off when mutations in *SOX10* were found, although not in all of the patients.

A disease gene may be identified by using characteristics of the DNA sequence

All the examples discussed so far go from a disease to the relevant gene. Sometimes the link is established in the opposite direction. A researcher might ask what diseases could be caused by mutations in a gene that he is already interested in for some other reason. A common approach is to use RNAi to knock down the expression of the gene in a model organism, and then check for phenotypic effects.

An interesting application of position-independent DNA sequence knowledge is the attempt to clone genes containing novel expanded trinucleotide

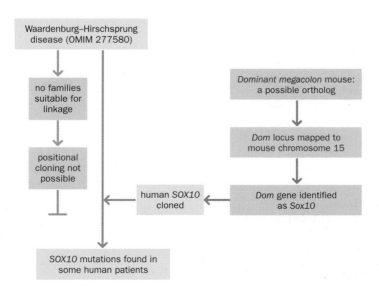

Figure 16.13 Identifying the gene causing Waardenburg–Hirschsprung disease. Individuals affected by this condition (OMIM 277580) have pigmentary anomalies, hearing loss, Hirschsprung disease, and sometimes other neurological problems. The condition is very rare, and affected people seldom reproduce, so it was not possible to identify the gene responsible by first mapping it in families and then testing positional candidates. The *Dominant megacolon* (*Dom*) mouse has a similar combination of pigmentary anomalies and lack of enteric ganglia. Breeding experiments in mice allowed the causative gene (*Sox10*) to be identified. The human homolog was then cloned, and DNA from patients was tested directly for mutations.

repeats. As mentioned in Chapter 13, expanded trinucleotide repeats cause several inherited neurological disorders. Often these disorders show *anticipation*—that is, the disease presents at an earlier age and/or with increased severity in successive generations. If a disease under investigation shows any of these features, it may be worth screening DNA from affected patients for triplet repeat expansions. This approach was used in a completely position-independent way to identify a novel repeat expansion involved in a form of spino-cerebellar ataxia (SCA8; OMIM 608768).

16.4 TESTING POSITIONAL CANDIDATE GENES

As we have seen above, there are many different ways of identifying a disease gene, but all paths converge on a candidate gene (box B3 in Figure 16.1). In one way or another, a candidate gene is identified; the researcher then tests the hypothesis that this is the disease gene.

For Mendelian conditions, a candidate gene is normally screened for mutations in a panel of unrelated affected patients

Screening for patient-specific mutations in the candidate gene is always the first approach. A panel of unrelated patients is assembled, and coding exons and splice sites of the candidate gene are sequenced. Mutation screening is often straightforward for diseases in which a good proportion of patients carry independent mutations, and for which the pathogenic mechanism is a loss of gene function. Typically, these are severe early-onset dominant or X-linked disorders. If the correct gene is tested, a panel of samples from unrelated patients will usually show a variety of different mutations. With luck, these will include some with an obviously deleterious effect on gene expression such as nonsense mutations or frameshifts. It may be difficult to identify missense mutations as being pathogenic, as opposed to being rare neutral variants with no major effect on gene expression. This problem was discussed in Chapter 13 and will be revisited in Chapter 18.

Normal controls need to be checked to prove that any changes are not non-pathogenic population variants. When one is deciding on the number of controls to screen, it is necessary to consider what degree of certainty one wishes to attain. If a variant exists in the population at a frequency q, and n chromosomes are tested to look for it, the chance that none of the n chromosomes will show the variant is $(1 - q)^n$. So, for example, to have a 95% chance of identifying a variant that has an allele frequency of 1 in 100 in the normal population, we would have to screen 298 chromosomes ($0.99^{298} = 0.05$). As a rule of thumb, $3n$ chromosomes must be tested with negative results to be 95% confident that a variant does not have a frequency in the normal population greater than 1 in n.

Normally, genomic DNA is sequenced. Where mRNA is available, it may be better to sequence cDNA, especially for genes with many small exons, each of which would otherwise have to be PCR amplified individually from genomic DNA. However, one must bear in mind the risk that the mutant transcript may be unstable because of nonsense-mediated decay, and that it may not be represented in a cDNA sample.

To save sequencing effort, an initial scan of a candidate gene may be performed by checking individual PCR-amplified exons from the panel of patients for the presence of nonstandard structures—heteroduplexes in double-stranded DNA or unusual internally base-paired conformations in single-stranded DNA. These methods are described in Chapter 18. They may allow quick identification of those exons that contain sequence variants; however, they do not identify the actual sequence change, and some variants will be missed by these methods. One frequent class of mutations, heterozygous deletions of whole exons, requires special detection methods such as MLPA (multiplex ligation-dependent probe amplification). Again, this technique is described in Chapter 18.

Sometimes the same sequence change is found in most patients, even though they are apparently unrelated. In Chapter 13 we discussed the likely reasons for such mutational homogeneity. Unless the observed change is unambiguously pathogenic (e.g. a nonsense mutation), some further evidence is needed to give confidence that the correct disease gene and mutation have been identified.

In some cases, the desired mutations may be hard to find. Apart from the practical problem of screening a large gene with many exons, some mutations are not readily discoverable by PCR testing of genomic DNA. For example, one frequent mutation of the cystic fibrosis transmembrane receptor gene (*CFTR*) that causes cystic fibrosis, c.3849+10kb C>T, is a single nucleotide change deep within an intron that activates a cryptic splice site. It would be missed by the standard exon-by-exon PCR and sequencing protocols. Initial detection of this mutation required RT-PCR, although once it had been identified a PCR assay could easily be designed to detect further cases by using genomic DNA.

In another example, most mutations that cause severe hemophilia A (OMIM 306700) could not be detected even after sequencing all 26 exons of the *F8A* gene. It turned out that a recurrent inversion disrupted the gene. The intragenic breakpoint was in a 32 kb intron, and so each exon of the gene remained intact and appeared normal on PCR or sequencing (**Figure 16.14**). The inversion was detected by Southern blotting. It is caused by non-allelic homologous recombination between a repeat located in intron 22 of the *F8A* gene and copies of the repeat located 360 kb and 435 kb away. Once the relevant sequences were known, a long PCR test could be designed to detect it.

Epigenetic changes might cause a disease without changing the DNA sequence

Epigenetic changes (epimutations) are a very common feature of tumor cells. As we saw in Chapter 11, patterns of methylation of cytosine in CpG sequences, and concomitant changes in modification of histones, can switch chromatin structure from an open to a closed conformation and abolish expression of a gene. Such patterns can be transmissible from cell to daughter cell. In many tumors, vital tumor suppressor genes are inactivated by epigenetic changes, without there being any underlying DNA sequence change. These matters are discussed in more detail in Chapter 17.

There is currently little evidence for epigenetic changes as *primary* causes of human hereditary disease. Heritable DNA sequence variants can exert their phenotypic effect by triggering epigenetic changes, as in fragile X syndrome, but heritable epigenetic changes without any underlying sequence change have not been unambiguously identified as causes of human hereditary disease. However, this is controversial. In mice there is at least one good example of a heritable **paramutation** that seems to depend on this mechanism (see Figure 11.24), and similar mechanisms are well documented in plants. Disentangling such effects from the effects of environment, especially intrauterine environment, is not easy, and it is possible that the role of epigenetic changes in determining hereditary human phenotypes has been underestimated. There is indeed some evidence for heritable epigenetic changes that may originally have arisen in response to environmental stresses—in other words, a molecular mechanism for inheritance of acquired characters (see Further Reading). The curious thing is that the proposed effects, if real, seem to adapt people to their grandparents' environment rather than their own, which does not seem very logical.

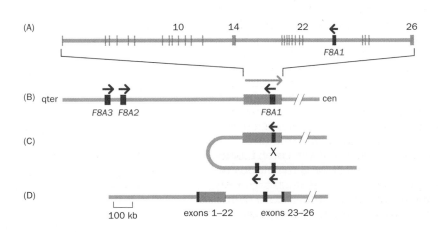

Figure 16.14 A common inversion in the Factor VIII gene causes severe hemophilia but is not detectable by sequencing each exon of the *F8A* gene. (A) A repetitive sequence in intron 22 of the *F8A* gene (*F8A1*, red bar) is present in two additional copies located 360 kb and 435 kb upstream of the *F8A* gene (B). Arrows indicate the relative orientations of the three copies. (C) During male meiosis, this part of the X chromosome has no homologous pairing partner. The *F8A* repeats may pair, forming a loop. (D) A crossover between paired *F8A* repeats causes inversion of a 500 kb segment. Although the *F8A* gene is disrupted and nonfunctional, each individual exon and its flanking intronic sequence is still intact.

The gene underlying a disease may not be an obvious one

Our ability to guess likely candidate genes is currently quite limited. Over and over again, when a disease gene is finally identified, it remains a complete mystery why mutations should cause that particular disease. Why should loss of function of the FMR1 protein, involved in transporting RNA from nucleus to cytoplasm, cause mental retardation and macro-orchidism (fragile X syndrome; OMIM 309550), whereas certain mutations in the TATA box-binding protein (part of the general transcription apparatus) cause SCA17 spino-cerebellar ataxia (OMIM 607136)?

Mutations leading to deficiency of a protein are not necessarily in the structural gene encoding the protein. For example, agammaglobulinemia (lack of immunoglobulins, leading to clinical immunodeficiency) is often Mendelian. It is natural to assume that the cause would be mutations in the immunoglobulin genes. But agammaglobulinemias do not map to chromosomes 2, 14, or 22, where the immunoglobulin genes are located; many forms are X-linked. Remembering the many steps needed to turn a newly synthesized polypeptide into a correctly functioning protein, and the special complexities of immunoglobulin gene rearrangements (Chapter 4), this lack of one-to-one correspondence between the mutation and the protein's structural gene should not come as any great surprise. Failures in immunoglobulin gene processing, in B-cell maturation, or in the overall development of the immune system will all produce immunodeficiency.

One gene defect can sometimes produce multiple enzyme defects. In I-cell disease or mucolipidosis II (OMIM 252500) there are deficiencies of multiple lysosomal enzymes. The primary defect is not in the structural gene encoding any of these enzymes but in an enzyme, N-acetylglucosamine-1-phosphotransferase, that phosphorylates mannose residues on the glycosylated enzyme molecules. Phosphomannose is a signal that targets the enzymes to lysosomes; in its absence, lysosomes lack a whole series of enzymes. Multiple sulfatase deficiency, described in case study 4 at the end of this chapter, is another example.

Mutations often affect only a subset of the tissues in which a gene is expressed. Thus, the pattern of tissue-specific expression of a gene is a poor predictor of the clinical effects of mutations. Tissues in which a gene is not expressed are unlikely to suffer primary pathology, but the converse is not true. Usually only a subset of expressing tissues are affected. The *HTT* (huntingtin) gene is widely expressed, but Huntington disease (OMIM 143100) affects only limited regions of the brain. The retinoblastoma gene (Chapter 17) is expressed ubiquitously, but only the retina is commonly affected by inherited mutations. This is also strikingly seen in the lysosomal disorders. Gene expression is required in a single cell type, the macrophage, which is found in many tissues. But not all macrophage-containing tissues are abnormal in affected patients. Explanations are not hard to find:

- Genes are not necessarily expressed only in the tissues in which they are needed. Provided it does no harm, there may be little selective pressure to switch off expression, even in tissues in which it confers no benefit.

- Loss of a gene's function will affect some tissues much more than others, because of the varying roles and metabolic requirements of different cell types and varying degrees of functional redundancy in the meshwork of interactions within a cell.

- A gain of function may be pathological for some cell types, harmless for others.

Locus heterogeneity is the rule rather than the exception

Locus heterogeneity describes the situation in which the same disease can be caused by mutations in any one of several different genes. It is important to think about the biological role of a gene product, and the molecules with which it interacts, rather than expecting a one-to-one relationship between genes and syndromes. Clinical syndromes often result from a failure or malfunction of a developmental or physiological pathway; equally, many cellular structures and functions depend on multicomponent protein aggregates. If the correct functioning of several genes is required, then mutations in any of the genes may cause the same, or a very similar, phenotype.

Not surprisingly, the most extreme example of locus heterogeneity comes from nonsyndromic mental retardation. More than 10,000 genes are expressed in the central nervous system, many more or less exclusively. All of them are candidate genes for mental retardation. Although mental retardation is an extreme example, almost all observable phenotypes depend on the action of more than one gene, and so locus heterogeneity is a very common and unsurprising observation. Indeed, at first sight it seems surprising that any condition should not be grossly heterogeneous. Locus homogeneity probably results in part from the way in which most conditions are defined by combinations of symptoms. For example, each individual feature of cystic fibrosis can have a variety of causes, but the combination is specific to people who have no functional copy of the *CFTR* gene. Additionally, we are good at seeing very subtle differences between people and labeling them accordingly. If mutations in two different genes cause very similar but subtly different phenotypes, we would probably label them differently in humans but not in mice, flies, or worms.

Further studies are often necessary to confirm that the correct gene has been identified

Sometimes the mutation evidence on its own is sufficient to make a convincing case that the correct gene has been identified. If most patients show mutations in the candidate gene, and the mutations include a reasonable selection of nonsense, frameshift, and splice site mutations, few would doubt that the mutations are the cause of the disease. In other cases, some further proof is needed. Maybe only a single mutation has been found. This is often true of highly heterogeneous conditions such as nonsyndromic mental retardation or nonsyndromic deafness, in which a mutation is found in just one family in a researcher's collection. It can also happen if only one specific sequence change will produce the specific pathogenic gain of function that characterizes the disease. Some conditions depend on a partial loss of function of a gene or a change in the ratio of splice isoforms—complete loss may be lethal, or it may produce a different phenotype. In such cases only missense or intronic mutations may be found, leaving some uncertainty whether these really are the cause of the phenotype.

Further proof is easiest if the mutant phenotype can be observed in cultured cells. RNA interference (RNAi) can be used to knock down expression of the gene in wild-type cells to see whether this produces the mutant phenotype. Alternatively, we can check whether transfection of a normal allele of the candidate gene, cloned into an expression vector, is able to rescue the mutant and restore the normal phenotype. For gain-of-function mutations, the effect of overexpression of the normal sequence, or expression of the mutant sequence, can be studied in transfected cells.

Once a putative disease gene has been identified, a transgenic mouse model can be constructed, if no relevant mutant already exists. Loss-of-function phenotypes can be modeled by knocking out the gene in the mouse germ line. For gain-of-function phenotypes, the disease allele must be introduced into the mouse germ line. The mutant mice are expected to show some resemblance to humans with the disease, although this expectation may not always be met even when the correct gene has been identified.

Identifying the gene involved in a genetic disease has often been the route for understanding aspects of normal function. For example, until the Duchenne muscular dystrophy gene was identified we knew nothing about the way in which the contractile machinery of muscle cells is anchored to the sarcolemma. For clinicians, the ability to identify mutations should immediately lead to improved diagnosis and counseling, while understanding the molecular pathology (why the mutated gene causes the disease) may eventually lead to more effective treatment, as well as providing insight into related diseases.

16.5 IDENTIFYING CAUSAL VARIANTS FROM ASSOCIATION STUDIES

The strategies outlined above have been used with great success for identifying the genes and mutations underlying Mendelian diseases, but they have not had much impact on the search for factors governing susceptibility to complex diseases. As explained in Chapter 15, linkage studies for these conditions have not been very successful and, even when a linkage is fully confirmed, the candidate region is usually too large to search for mutations. However, starting in about 2005, genomewide association studies in many different complex diseases have identified SNPs that are reproducibly associated with susceptibility. As described in Chapter 15, these studies typically genotype 500,000 or more tagging SNPs (as defined by the HapMap project) and copy-number variants in very large case-control designs. For example, the Wellcome Trust Case-Control Consortium (WTCCC) genotyped 500,568 SNPs in 14,000 patients with seven diseases and 3000 controls.

Identifying causal variants is not simple

Association depends on the short-range phenomenon of linkage disequilibrium. Thus, when these studies are successful they define candidate regions that are much smaller than those typically defined by linkage. For example, the 25 strongest associations in the WTCCC data defined candidate regions with an average size of 295 kb (range 40–670 kb). Because the regions identified by association studies are small, researchers seldom face the problem of selecting one gene for sequencing from a long list of candidate genes. They do, however, face a severe problem in moving from an associated SNP to the actual causal variant. Association studies use tagging SNPs to identify a haplotype block that is associated with disease susceptibility. All the SNPs in the block will be associated with susceptibility, but only one of them (in a simple situation) will be a true functional variant. Identifying the causal variant against such a background will be far from easy.

One problem is that the causal variant is only a susceptibility factor. It is neither necessary nor sufficient to cause the disease on its own. It will be absent from many patients and present in many controls. It will manifest as a difference in frequency in the two groups, rather than as a patient-specific variant as in Mendelian diseases. For example, susceptibility alleles for Crohn disease (the subject of case study 8 at the end of this chapter) showed frequencies in cases and controls, respectively, of 0.295 and 0.403 for a variant in the *IL23R* gene, and 0.364 and 0.453 for a variant in *ATG16L1*. Because of the large numbers in the study, these differences were highly significant. Nevertheless, such small differences can easily be produced by inadequate matching of cases and controls, even when there is no true association. Thus, an essential part of all large-scale association studies is meticulous filtering of the raw data. Any individual sample in which fewer than 90% of SNPs are successfully genotyped is usually excluded, as is any individual SNP that cannot be genotyped in at least 90% (95% is better) of samples. Genotype frequencies in controls are checked for conformity with the Hardy–Weinberg distribution. Systematic errors are suspected for any SNP that gives a non-Hardy–Weinberg distribution of genotypes, and all results for that SNP are excluded. A proportion of tests are repeated to check the overall error rate. As a result of all these checks, the number of samples and SNPs used in the analysis is usually only 70–80% of the number actually tested.

In addition, association studies detect variants that fulfill the *common disease–common variant hypothesis* (see Chapter 15). Because such variants are necessarily ancient, they cannot have been subject to strong negative selection. Thus, they are likely to have fairly subtle effects on gene expression or function, unlike the frameshift, splice-site, or nonsense mutations commonly seen in Mendelian conditions. Susceptibility factors may be polymorphisms in noncoding DNA that have some small effect on promoter activity, splicing, or mRNA stability. They may be located at a large distance from the gene whose activity they affect—the case study of persistence of intestinal lactase at the end of this chapter illustrates the sort of variant that may be expected. It is therefore unlikely that the true causal variant will be identifiable by simple inspection of the sequence, in the way that it often is for a Mendelian disease.

Causal variants are identified through a combination of statistical and functional studies

Figure 16.15 sketches the way in which one would hope to solve these problems. The original association study will most probably have used a limited number of tag-SNPs (described in Chapter 15); sequencing of the entire associated region in

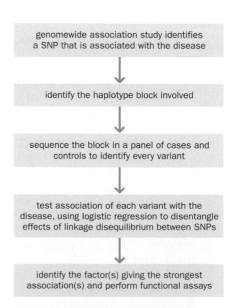

genomewide association study identifies a SNP that is associated with the disease

↓

identify the haplotype block involved

↓

sequence the block in a panel of cases and controls to identify every variant

↓

test association of each variant with the disease, using logistic regression to disentangle effects of linkage disequilibrium between SNPs

↓

identify the factor(s) giving the strongest association(s) and perform functional assays

Figure 16.15 Procedure for identifying a causal variant through association studies. Genomewide association studies use tagging SNPs to identify haplotype blocks that are associated with disease susceptibility. Any of the variants in that block might be the actual functional variant. Statistical and functional studies are needed to identify it.

a large panel of the cases and controls will reveal all variants in the study sample. These will include copy-number variants and polymorphisms with minor allele frequencies below the 5% threshold normally used in association studies.

Next, each individual variant will be tested for association with the disease, ideally in an independently ascertained set of patients and controls. Haplotype blocks, such as those in the HapMap data, are defined by an arbitrary cutoff of linkage disequilibrium that is always less than 100%, so that the degree of association is not necessarily identical for every variant within a block. Moreover, most variants are two-allele SNPs, but within the population there will usually be more than two alternative haplotype blocks at any location. Thus, particular variants will probably be present on more than one of the alternative blocks. If a variant that is present on two different blocks is causative, it will show a much stronger association with the disease than variants that are present on only one of the blocks. Conversely, a nonfunctional variant that is present on one block that also carries the functional variant but is also present on another block that does not do so will show a weaker association with susceptibility (**Figure 16.16** shows this in a simplified example). If the investigators are lucky, testing every variant for association will show one peak of especially strong association for some small cluster of SNPs, against a background of the general association of all the SNP alleles that are found on the relevant haplotype block.

Several different variants in a region may each contribute, independently or in combination, to susceptibility. Sorting out the causal relationships is extremely difficult when all the variants are in linkage disequilibrium with each other. The main tool is *logistic regression*: a principal variant is selected, and the effect of other variants is studied, conditioned on the effects of the principal variant. If this procedure still shows an effect, the second variant does indeed make an independent contribution. An additional problem is that the main determinants of susceptibility may be different in different populations. Apart from differences due to interaction of genetic factors with different environments, associations identify shared ancestral chromosome segments, which may be different in different populations. Readers interested in a more detailed discussion of the statistical methods used in association studies should consult the review by Balding (see Further Reading).

When a candidate causal variant has been identified, some sort of functional test must be used to confirm that it has a biological effect. Missense changes in coding sequences are the easiest to investigate, through direct tests of protein function. The balance of splice isoforms can be checked for variants located within gene sequences, and variants in promoters or enhancers can be studied with the use of appropriate expression systems. But laboratory studies seldom capture the full function of a complex gene in all cells and all tissues, and under all conditions. In a systematic study of gene knockouts in the budding yeast *Saccharomyces cerevisiae*, only a minority of all knockouts had any detectable effect. It is highly unlikely that most yeast genes really have no function—what the result tells us is that standard laboratory biochemical tests are not able to assess the full function of a gene, even in a simple organism. In a similar study in the mouse, 96% of all knockouts did show some abnormality, but the abnormalities could be subtle, often behavioral. If this is true of knockouts, it is likely to be

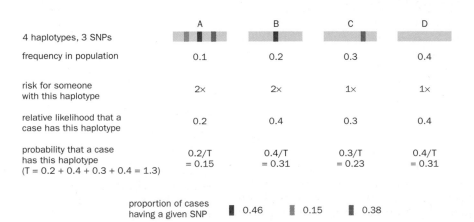

4 haplotypes, 3 SNPs	A	B	C	D
frequency in population	0.1	0.2	0.3	0.4
risk for someone with this haplotype	2×	2×	1×	1×
relative likelihood that a case has this haplotype	0.2	0.4	0.3	0.4
probability that a case has this haplotype (T = 0.2 + 0.4 + 0.3 + 0.4 = 1.3)	0.2/T = 0.15	0.4/T = 0.31	0.3/T = 0.23	0.4/T = 0.31

proportion of cases having a given SNP   ■ 0.46   ■ 0.15   ■ 0.38

Figure 16.16 SNP associations with disease susceptibility. In this imaginary population there are four alternative haplotypes (A–D) at a certain location. Three different SNPs are shown as three colored bars; the alternative allele in each case is blank. The red SNP is a true susceptibility factor for the disease under study, doubling the risk. The blue and green SNPs have no causal effect. When each SNP is tested for association with the disease, all will show an association because each is present on haplotype A, but it will be strongest for the true causal SNP.

doubly so for the more subtle functional changes that are suspected of causing susceptibility to complex diseases.

In a few cases, such investigations have convincingly explained the association of a variant with disease susceptibility or resistance. A likely example comes from the genomewide association study of Crohn disease described in the case studies. The p.T300A variant in the *ATG16L1* gene is an amino acid change in a protein that has convincing functional links to the disease, and variation at this amino acid accounts for the whole of the observed association.

Functional analysis of SNPs in sequences with no known function is particularly difficult

Often the candidate variant will be in intronic or intergenic DNA that has no known function. This makes it extremely difficult to produce the sort of functional data that would confirm a variant as being truly causal. In the current state of knowledge there are many variants that have statistical but not functional support for a role in disease susceptibility. How many of these will eventually be fully confirmed remains to be seen. The two cases below illustrate this frustrating situation.

Calpain-10 and type 2 diabetes

A pioneering study of type 2 diabetes (T2D) by Horikawa and colleagues in 2000 identified one susceptibility factor as heterozygosity for a haplotype of three intronic SNPs within the calpain-10 (*CAPN10*) gene on chromosome 2q37. Linkage analysis in Mexican-Americans had identified a 7 cM candidate region at 2q37. Physically, this corresponded to 1.7 Mb of DNA (recombination rates tend to be higher near telomeres, so there was a 7% chance of a crossover occurring in this 1.7 Mb region). Polymorphisms from the region were tested, not just for association with T2D but also for association with the evidence for linkage. The rationale was that only a subset of cases was linked to the 2q37 locus, but these should be the ones carrying the susceptibility determinant. Initial analyses suggested a 66 kb target region. This was sequenced in a panel of 10 Mexican diabetics and it turned out to contain three genes—*CAPN10*, *RNPEPL1*, and *GPR35*—and 179 sequence variants. Eventually a variant, UCSNP-43, was identified in which the homozygous G/G genotype showed association with the evidence for linkage and also probably with diabetes (odds ratio 1.54, confidence interval 0.88–2.41). This SNP lies deep within an intron of the *CAPN10* gene, and the G allele has a frequency of 0.75 in unaffected controls. A search was then initiated for haplotypes that were (a) increased in frequency in the patient groups with greatest evidence of 2q37 linkage, (b) shared by affected sib pairs more often than expected, and (c) associated with an increased risk of diabetes. A heterozygous combination of two haplotypes (defined by UCSNP-43 and two other SNPs within introns of the *CAPN10* gene) fulfilled all the criteria. Neither haplotype in homozygous form was a risk factor. *CAPN10* was an entirely unexpected gene to be involved in T2D. Later studies have suggested some reasons why this gene might be relevant, but nine years after the original publication, the role of the SNPs, and indeed their validity as susceptibility factors, remains unclear.

Chromosome 8q24 and susceptibility to prostate cancer

Several studies of susceptibility to prostate cancer have implicated a region on chromosome 8q24. In 2007, three groups independently reported large-scale association and resequencing studies across the candidate region (see Further Reading). Very strong associations were seen for several SNPs—but these came from three separate regions of 8q24, spread across 600 kb and not in linkage disequilibrium with each other (**Figure 16.17**). SNPs in the three regions seem to be independent risk factors that have a multiplicative overall effect. The three studies are mutually confirmatory—yet none of the SNPs lies near a gene and none has any obvious functional effect. The *MYC* oncogene lies only 260 kb telomeric of region 1. Clearly an effect on *MYC* expression would be a very attractive explanation of the data, but none of the groups was able to provide any data to support this.

The case studies of breast cancer and Crohn disease below illustrate successful genomewide association studies. In each case, novel susceptibility loci were

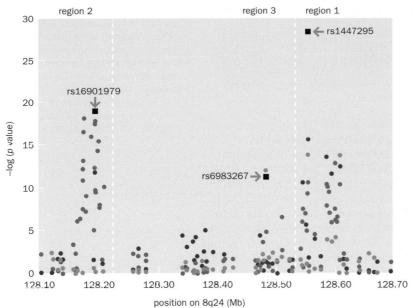

Figure 16.17 SNPs at 8q24 associated with susceptibility to prostate cancer. Round dots show results for individual SNPs across a 600 kb region of chromosome 8q24 in three different studies (colored red, green, and blue, respectively). Small black squares show the most significant results obtained by combining the raw data from the three studies. These results suggest that there are susceptibility factors in three separate regions of 8q24, which are not in linkage disequilibrium with each other. [From Witte JS (2007) *Nat. Genet.* 39, 579–580. With permission from Macmillan Publishers Ltd.]

identified, and for most of these new loci plausible candidate genes could be suggested. The odds ratios for the susceptibility alleles were low, mostly in the range 1.1–1.5 (an odds ratio of 1 means that the factor has no effect on risk). The studies demonstrate that it is possible to identify weak susceptibility factors while leaving open the question of whether doing so is worth the trouble.

16.6 EIGHT EXAMPLES OF DISEASE GENE IDENTIFICATION

The following series of case studies has been chosen to put some flesh on the bones of the methodological descriptions above. Duchenne muscular dystrophy (gene cloned in 1987) and cystic fibrosis (1989) were two pioneering studies that helped establish the feasibility of positional cloning. Branchio-oto-renal syndrome (1997) illustrates some of the approaches that were used after the pioneering phase but before the Human Genome Project had made it all easier. Multiple sulfatase deficiency (2003), lactase persistence (2002–2007), and CHARGE syndrome (2004) illustrate some more recent achievements. Finally the ongoing stories of breast cancer and Crohn disease show examples of tackling complex diseases.

Case study 1: Duchenne muscular dystrophy

To quote the excellent review by Worton and Thompson (see Further Reading):

> Duchenne muscular dystrophy (DMD, OMIM 310200) is a lethal X-linked genetic disease that for many years was one of the most frustrating and perplexing disorders in clinical genetics. Until the advent of molecular genetic approaches the nature of the primary defect remained elusive. All attempts to detect an altered protein in muscle tissue, cultured muscle cells or other tissues from patients had yielded negative results. Attempts to determine the basic biochemical or physiological defect were frustrated by the difficulty of distinguishing the primary defect from the numerous secondary manifestations of the disease. The same difficulty also handicapped attempts to identify carriers and carry out prenatal diagnosis in Duchenne families.

In 1983, the DMD locus was mapped to Xp21, making DMD a promising test bed for positional cloning. Being X-linked, affected males should show any genetic changes in a straightforward way without a homologous chromosome to confuse the picture. As the disease process was known to affect muscle, gene expression studies could be targeted to this fairly accessible tissue, from which good cDNA libraries could be prepared. Rare patients with chromosomal abnormalities enabled two groups of researchers to approach isolating the DMD gene by two different routes.

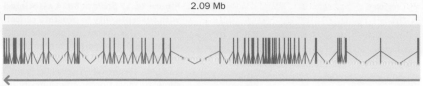

direction of transcription

Figure 16.18 The dystrophin gene. There are 79 exons (vertical bars) encoding a 14 kb mature transcript that produces a 3685-residue protein. Note that although the encoded RNA and protein are quite small, the gene is extremely large because it contains many big introns. Only 0.3% of the genomic sequence is present in the mature mRNA. [Download from Ensembl of Vega transcript OTTHUMT00000056182.]

Lou Kunkel in Boston started with a boy, BB, who had DMD and a cytogenetically visible Xp21 deletion. This unfortunate boy simultaneously suffered from DMD, chronic granulomatous disease (CGD; OMIM 306400), retinitis pigmentosa (OMIM 300389), McLeod phenotype (a red blood cell disorder; OMIM 314850), and mental retardation. He was mentioned in Chapter 13 as an example of a contiguous gene syndrome, and in fact his DNA was first used to clone the chronic granulomatous disease gene. A technically very difficult **subtraction cloning** procedure was used to isolate clones from normal DNA that corresponded to sequences deleted in BB. Individual DNA clones in the subtraction library were then used as probes in Southern blot hybridization against DNA from BB. Of the few hundred clones obtained, eight failed to hybridize to BB's DNA and therefore may have come from within the deletion.

One of the eight clones, pERT87-8, detected microdeletions in DNA from about 7% of cytogenetically normal patients with DMD. It also detected polymorphisms that were shown by family studies to be linked to the DMD locus. These results showed that pERT87-8 was located much closer to the DMD gene than any previously isolated clones (later studies showed that it was actually within the gene, in intron 13). Other nearby genomic probes were isolated by chromosome walking; conserved sequences were then sought by **zoo blotting** (hybridization to Southern blots of human, bovine, mouse, hamster, and chicken genomic DNA). One clone contained conserved sequence and identified a 14 kb transcript on northern blotting of RNA from normal muscle. This genomic clone was used to screen muscle cDNA libraries. Given the low abundance of dystrophin mRNA and, as we now know, the small size and widely scattered location of the exons, finding cDNA clones was far from easy, but eventually clones were identified, and subsequently the whole remarkable dystrophin gene was characterized (**Figure 16.18**).

Ron Worton in Toronto took a different approach. There are 20–30 women worldwide who suffer from Duchenne muscular dystrophy as a result of X–autosome translocations. As explained above (see Figure 16.9), because of X-inactivation even a balanced X–autosome translocation can cause a woman to experience an X-linked recessive disease, if the translocation breakpoint disrupts the disease gene. Each translocation in the women with DMD involved a different autosomal breakpoint, but the X-breakpoint was always in Xp21, the location identified by linkage analysis as harboring the DMD gene.

One woman had an Xp;21p translocation, and this provided Worton's group with a method of cloning the DMD gene. Knowing that 21p is occupied by arrays of repeated rRNA genes, Worton's group prepared a genomic library from this woman's DNA and set out to find clones that contained both rDNA and X-chromosome sequences. Any X-chromosome sequences identified in this way should come from within the DMD gene. This led to the isolation of XJ (X-junction) clones, which turned out to be located within the dystrophin gene, in intron 7. A clone contig was established around this initial sequence by extensive chromosome walking, and muscle cDNA libraries were screened with subclones from the contig. Eventually a 2 kb cDNA was isolated that contained exons 1–16 of the dystrophin gene.

Case study 2: cystic fibrosis

Cystic fibrosis (CF) illustrates positional cloning in its purest form, without any chromosomal abnormalities to assist the process. As one of the commonest severe Mendelian diseases in Europe and the USA, and one in which all attempts to identify the underlying defect through biochemical and physiological approaches had failed, cystic fibrosis (OMIM 219700) was a prime target in early attempts at positional cloning. This task presented major difficulties. The disease was autosomal recessive, and families were generally small in the countries where it was prevalent. Thus, mapping relied on combining large numbers of affected sib pairs and hoping that there was no locus heterogeneity. Frustratingly, the first linkage had been to the locus encoding the enzyme paraoxonase—whose chromosomal location was unknown. This was rectified when paraoxonase, and hence CF, were localized to chromosome 7q31.

Unlike with Duchenne muscular dystrophy, there were no patients with CF who had appropriately located chromosomal breakpoints to speed the discovery process. Instead, years of grinding effort were needed to clone as much as possible of the DNA from the candidate region. This became an intensely competitive race between three major groups. The winners, a US–Canadian group led by Lap-Chee Tsui, used 12 different genomic libraries, including one made from a human–hamster hybrid cell that contained only human chromosome 7. A major problem was to build up a clone contig across the candidate region. **Chromosome jumping** was used to bypass the size limitation of cosmids, which have a maximum cloning capacity of 45 kb. This was an early version of the **paired-end mapping** technique that is now used to scan the human genome for structural variants (see Figure 13.6). In this early manifestation, DNA fragments of 80–130 kb from a partial *Mbo*I digest of genomic DNA were recovered from preparative pulsed-field gels. The fragments were mixed in very dilute solution with an excess of a short marker sequence that had compatible sticky ends; they were then exposed to DNA ligase. The hoped-for ligation products were circles of 80–130 kb of genomic DNA with the marker fragment located at the position where the ends joined. Circles were fragmented with a restriction enzyme and the fragments were cloned. The desired clones contained the marker and a known 7q31 sequence, joined to an unknown sequence, which was hoped to come from a position in 7q31 that was 80–130 kb away from the known sequence (**Figure 16.19**).

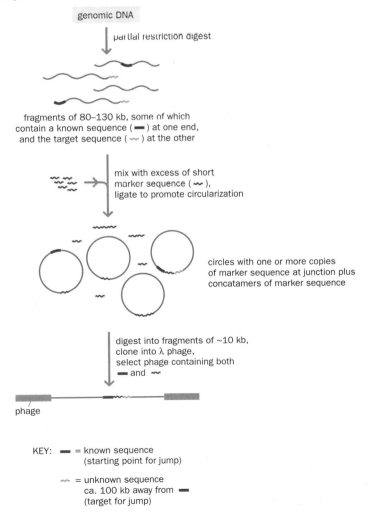

Figure 16.19 Chromosome jumping. The starting material was a mixture of restriction fragments 80–130 kb long from a partial *Mbo*I digest of genomic DNA, isolated from preparative pulsed-field gels. These were exposed to DNA ligase in very dilute solution in the presence of a large excess of a short marker sequence (*SupF*) that had compatible sticky ends. Some of the restriction fragments formed circles with the two original ends joined by a *SupF* sequence. The circles were cut and fragments isolated that contained a known sequence from 7q31 linked by a *SupF* sequence to a novel 7q31 sequence. The latter represented new loci 80–130 kb away from the original sequence.

As well as showing the persistence and detective work involved in identifying the cause of a phenotype, this example illustrates the way in which a clinically described condition may map imperfectly to a genetic description. CHARGE has now been redefined as a syndrome, rather than an association, and cases without *CHD7* mutations may need an alternative label.

Case study 7: breast cancer

Breast cancer is usually sporadic, but because it is so common—the lifetime risk for a British woman is 1 in 12—it often happens that several relatives have breast cancer entirely coincidentally. Nonetheless, there are two senses in which it may be genuinely familial. Statistically, having an affected first-degree relative approximately doubles a woman's risk of breast cancer. Some of this may be shared environment, but some is probably due to shared genetic predisposition. Superimposed on this weak familial tendency is a much stronger one in certain families. A strong family history of breast cancer is also associated with an unusually early age of onset, with breast plus ovarian cancer, with frequent bilateral tumors, and occasionally with affected males. Investigation of such families has led to the identification of the *BRCA1* and *BRCA2* genes.

In 1988, segregation analysis (a statistical method for comparing different genetic hypotheses; see Chapter 15) was used to study 1500 families with multiple cases of breast cancer. The results supported the view that 4–5% of cases, particularly those with early onset, might be heritable in a Mendelian manner. Families with near-Mendelian pedigree patterns were collected for linkage analysis (**Figure 16.25**). In 1990 a major susceptibility locus, named *BRCA1* (OMIM 113705), was mapped to 17q21. The mean age at diagnosis in 17q-linked families was below 45 years. Later-onset families gave negative lod scores. Subsequently, in 1994, a linkage search in 15 large families with breast cancer not linked to 17q identified a *BRCA2* locus at 13q12 (OMIM 600185). A hectic race ensued to identify the two mapped genes. *BRCA1* was cloned in 1994, and *BRCA2* in 1995. It

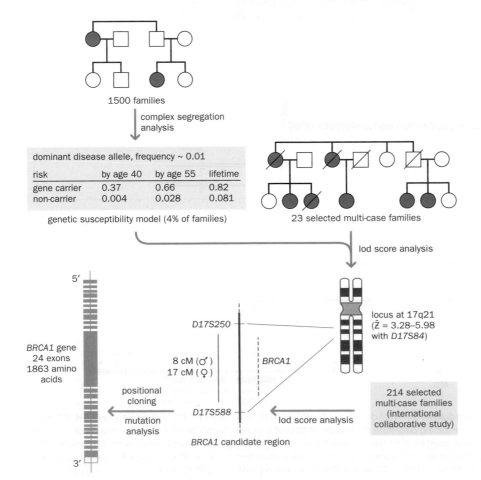

Figure 16.25 How the *BRCA1* gene was found. Initial segregation analysis of 1500 families with multiple cases of breast cancer suggested that some cases might be heritable in a Mendelian manner. Linkage analysis of 23 of these families identified a susceptibility locus at 17q21, which led to the identification of the *BRCA1* gene.

seemed that *BRCA1* might account for 80–90% of families with both breast and ovarian cancer but a much smaller proportion of families with breast cancer alone. Male breast cancer was seen mainly in *BRCA2* families. Evidence concerning the function of these two genes, and why mutations cause cancer, is discussed in Chapter 17.

Important though these discoveries were, it would be a mistake to suppose that we now understand the genetic basis of breast cancer, or even of familial breast cancer. Mutations at these two loci account for only 20–25% of the familial risk of breast cancer. A survey by Ford and colleagues of 257 families with four or more cases of breast cancer suggested that, of those with four or five cases of female breast cancer, but no ovarian or male breast cancer, 67% probably did not involve mutations in *BRCA1* or *BRCA2*. Despite intensive research, no third major gene has been discovered that could explain the remaining familial tendency. Segregation analysis is consistent with the remaining genetic susceptibility being polygenic—that is, due to a large number of genes, each having only a modest effect.

The *BRCA1/2* genes were identified by focusing on small numbers of special families. Identifying the other susceptibility factors requires large-scale case-control studies. Many candidate genes have been the subject of focused studies that have identified additional susceptibility factors.

- A specific variant, c.1100delC, in the *CHEK2* gene (encoding a protein important in repair of DNA damage, and hence a good candidate for involvement in cancer) was seen in 201 out of 10,890 cases and only 64 out of 9065 controls, giving an odds ratio of 2.34.

- By sequencing all 62 exons of the *ATM* gene (described in more detail in Chapter 17), mutations were found in 12 out of 443 cases but only 2 out of 521 controls. The odds ratio was 2.37. All cases were familial, but they did not have *BRCA1/2* mutations.

- By sequencing all 20 exons of the *BRIP1* gene (encoding a protein that interacts with the BRCA1 protein), mutations were found in 9 out of 1212 cases but only 2 out of 2081 controls, giving an odds ratio of 2.0. Again, the cases were chosen to be familial but negative for *BRCA1/2* mutations.

- By sequencing all 13 exons of the *PALB2* gene (encoding a protein that interacts with the BRCA2 protein), mutations were found in 10 out of 923 cases but none of 1084 controls, suggesting an odds ratio of 2.3 for mutation carriers. As before, the cases were familial and had been checked for *BRCA1/2* mutations.

These factors are relevant in the small minority of families in which they are segregating, but at a population level their effect is small. For example the *CHEK2* c.1100delC variant was seen in only 1.9% of cases. In fact, all these susceptibility factors together, including *BRCA1* and *BRCA2*, accounted for only about 25% of the familial tendency. A very large genomewide association study attempted to identify other factors. **Figure 16.26** outlines the study design.

SNPs at five novel loci gave convincing evidence of association with breast cancer susceptibility. Four of the loci contained genes that, on functional grounds, were considered plausible candidates (*FGFR2*, *TNRC9*, *MAP3K1*, and *LSP1*). The fifth did not fall close to any likely candidate gene but, interestingly, was close to the region on chromosome 8q24 that has been implicated in a complex pattern of susceptibility to prostate cancer (see Figure 16.17). The odds ratios for the five loci were 1.26 (*FGFR2*), 1.11 (*TNRC9*), 1.13 (*MAP3K1*), 1.07 (*LSP1*), and 1.08 (the anonymous 8q24 SNP). All the putative risk alleles were common in the population so, despite the low odds ratios, these factors may have an appreciable role in the overall incidence of breast cancer. Whether knowing one's genotype for these factors would be of any value to an individual is much more questionable—an issue to which we return in Chapter 19.

Perhaps more interesting than these specific identifications was the overall pattern of association. **Table 16.1** shows how many SNPs showed varying levels of association in the combined stage 1 and stage 2 data, compared with the number expected by chance. In every significance band, the number of observed SNPs was greater than the number predicted if these were all random statistical fluctuations. This pattern suggests that, although there are no major susceptibility

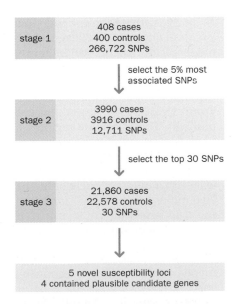

Figure 16.26 A three-stage genomewide association study of breast cancer susceptibility. This protocol was used by Easton and colleagues to identify five novel susceptibility factors. Four were in plausible candidate genes (*FGFR2*, *TNRC9*, *MAP3K1*, and *LSP1*), and the fifth was an anonymous SNP at 8q24.

TABLE 16.1 OBSERVED AND EXPECTED NUMBERS OF SNPs SHOWING ASSOCIATION WITH SUSCEPTIBILITY TO BREAST CANCER

Level of significance (p value)	Observed number of SNPs[a]	Expected number of SNPs[b]
<0.00001	13	0.96
0.00001–0.0001	12	7.0
0.0001–0.001	88	53
0.001–0.01	517	347
0.01–0.05	1162	934

Data are from stages 1 and 2 of the study outlined in Figure 16.26. [a]Observed data after certain statistical corrections. [b]Number expected on the null hypothesis of no true association.

factors like *BRCA1/2* waiting to be discovered, there may well be a considerable number of variants that make a definite but very minor contribution. This would support the conclusion from segregation analysis, that much of the residual causation is polygenic.

Case study 8: Crohn disease

Crohn disease (CD) and ulcerative colitis (UC) are forms of inflammatory bowel disease (IBD) that together affect 1–3 out of every 1000 people in the Western world. In terms of identifying susceptibility factors, Crohn disease has been one of the success stories of complex disease genetics. The results are shown here in some detail to give a feel for the complexity of this type of data.

Family and twin studies support a genetic susceptibility for each disease with some shared components. The risk to siblings of an affected individual (sib risk λ_s) is 20–30 for CD and 8–15 for UC. Sib-pair and family linkage analyses by many groups have defined several regions with at least suggestive linkage (**Table 16.2**). Linkage to *IBD1*, in particular, was confirmed with a maximum lod score of 5.79 in CD (but not UC) families in a very large pooled study of 613 families with 1298 CD and/or UC affected sibs.

In 2001, three groups independently identified the gene involved in *IBD1*. Yasunori Ogura and colleagues had cloned *CARD15* (*NOD2*) as a regulator of the nuclear factor κB (NF-κB) inflammatory response. They tested it in inflammatory bowel disease because it mapped to the *IBD1* locus and biologically was a promising IBD candidate. A German–British collaboration led by Stefan Schreiber and Christopher Mathew followed essentially the same strategy. By contrast, a French group (Jean-Pierre Hugot and colleagues) pursued a pure linkage-association pathway. The candidate region was narrowed by linkage, and then association studies were undertaken. Interestingly, the *transmission disequilibrium test* (TDT) result that led on to *CARD15* was only borderline significant ($p = 0.05$). This was replicated ($p < 0.01$) in a second set of families, but this time the association was with a different allele of the marker! Both these studies were small (108 and 76 families, respectively), and the chosen SNP was, in retrospect, not very close to the true susceptibility locus. Eventually, stronger though not marked associations highlighted a genomic region from which they cloned the then unpublished *CARD15* gene.

CARD15 makes excellent sense as a candidate gene: it encodes a protein involved in mounting a response to intracellular bacteria. However, the actual susceptibility variants are a heterogeneous collection of mutations, much as might be seen in a Mendelian disease. One survey identified homozygous or heterozygous mutations in 50% of 453 patients. Of these, three common mutations, p.R702W, p.G908R, and c.1007delC, made up, respectively, 32%, 18%, and 31% of all mutations. The remaining 19% included 27 different mutations. The *relative risk* of CD is about 3 for heterozygotes and 23 for homozygotes. The high relative risks and mutational heterogeneity are not typical of expectations under the common disease–common variant hypothesis. The association study of Hugot and colleagues would have failed if the relative risks had been lower.

TABLE 16.2 GENETICS OF INFLAMMATORY BOWEL DISEASE

Candidate region	OMIM number	Location (candidate gene)
IBD1[a]	266600	16q12 (NOD2/CARD15)
IBD2[a]	601458	12p13.2–q24.1
IBD3[a]	604519	6p21.3 (TNF)
IBD4[a]	606675	14q11–q12
IBD5[a]	606348	5q31
IBD6[a]	606674	19p13
IBD7[a]	605225	1p36
IBD8[a]	606668	16p
IBD9[a]	608448	3p26
IBD10	611081	2q37.1 (ATG16L1)
IBD11[a]	191390	7q22 (MUC3A)
IBD12	612241	3p21 (MST1)
IBD13[a]	612244	7q21.1 (ABCA1)
IBD14	612245	7q32 (IRF5)
IBD15	612255	10q21
IBD16	612259	9q32 (TNFSF15)
IBD17	612261	1p31.1 (IL23R)
IBD18	612262	5p13.1
IBD19	612278	5q33.1
IBD20[a]	612288	10q24
IBD21	612354	18p11
IBD22	612380	17q21
IBD23	612381	1q32
IBD24	612566	20q13
IBD25	612567	21q22
IBD26	612639	12q15
IBD27	612796	13p13.3

Linkage and association studies have suggested locations for a long and growing list of susceptibility factors. In many cases a candidate gene has been identified. [a]Regions supported by suggestive linkage data (lod scores ≥ 2.2). Other regions were defined by association data, although in some cases there was also weak linkage data.

CARD15 mutations account for only part of the susceptibility to Crohn disease. As usual in such cases, many other candidate regions were proposed in these early linkage studies, two of which are noteworthy. A region on 5q31 (*IBD5*) showed linkage and association in several studies, with an odds ratio of 1.5. The candidate region contains several cytokine genes that are plausible candidates, but attempts to identify the precise variant causing susceptibility have not so far been successful. A coding variant in the *IL23R* gene at chromosome 1p31 (p.R381Q) is strongly protective, with an odds ratio of 0.26—but this variant accounts for only a small proportion of overall susceptibility, because the arginine-381 allele that confers susceptibility is present in 93% of controls as well as in 98.1% of patients.

The era of large-scale genomewide association studies has brought new progress (see Table 16.2). The same German–British group that was involved in the original *CARD15* work typed 19,779 SNPs in 735 patients and 368 controls. The SNPs were chosen because they caused amino acid changes in a protein (cSNPs). SNPs showing signs of association were checked in independent samples of 498 patients, 1032 controls, and also 498 TDT trios (see Chapter 15). Strong confirmation was found for a variant (cSNP rs2241880; p.T300A) in the *ATG16L1* locus on chromosome 2q37, with odds ratios of 1.4 and 1.7 for individuals carrying, respectively, one and two copies of the susceptibility allele. It is worth noting that these strong results translated into only quite small differences in frequency of this allele between patients and controls (0.36 and 0.45, respectively, in the data of Table 16.3)—a point to bear in mind when discussing whether people could be screened for susceptibility (see Chapter 19).

A second study by US-based researchers identified the same variant in a case-control study involving 988 patients and 1007 controls. This was part of a genomewide association study using 317,503 SNPs. The results (see Table 16.3, Figure 16.27) show strong associations with the known susceptibility factors *IL23R* (chromosome 1) and *CARD15* (chromosome 16). Interestingly, the confirmed association with *IBD5* (5q31) does not show up, either on the figure or in the detailed statistics. A single study, even one as large and thorough as this one, gives only suggestions about susceptibility factors; nothing is conclusively accepted or rejected until several independent studies have concurred and functional evidence has been produced to support suggestions.

SNPs from the 23 most significantly associated chromosomal locations were tested in an independent replication sample of 650 TDT trios, using a different genotyping technology. This narrowed the candidates down to six that showed significant evidence of replication. These six were tested in a second independent series of 353 patients and 207 controls, using yet another genotyping method. The results from the top 10 of the 23 regions are shown in Table 16.3. *ATG16L1*, the gene identified by the German–British group, was strongly confirmed. Functional studies showed a role in the response of intestinal cells to intracellular microbes. The next strongest association was with a 70 kb region on chromosome 10q21 that did not contain any known gene.

TABLE 16.3 STATISTICAL ANALYSIS OF THE DATA FROM A GENOMEWIDE STUDY OF CROHN DISEASE

Chromosome no.	Initial GWA study		Replication cohort 1: TDT result (transmissions/non-transmissions)	Replication cohort 2: allele frequency (cases/controls)	Overall		Gene
	Allele frequency (cases/controls)	p			Odds ratio	p	
16	0.358/0.244	7×10^{-14}	N.D.	N.D.		N.D.	*CARD15*
1	0.295/0.403	3×10^{-12}	45/130[a]	N.D.	0.26[a]	6×10^{-19} [a]	*IL23R*
2	0.364/0.453	6×10^{-8}	220/306[b]	0.353/0.478	0.68	4×10^{-8}	*ATG16L1*
4	0.038/0.077	8×10^{-7}	39/75[b]	0.057/0.047	0.69	0.0084	*PHOX2B*
12	0.156/0.102	2×10^{-6}	121/136	N.D.		N.D.	
1	0.279/0.212	2×10^{-6}	failed	N.D.		N.D.	
18	0.054/0.094	3×10^{-6}	96/99	N.D.		N.D.	
3	0.218/0.160	6×10^{-6}	166/140	N.D.		N.D.	
10	0.134/0.191	8×10^{-6}	94/149[b]	0.140/0.230	0.60	3×10^{-7}	?
9	0.399/0.332	1×10^{-5}	274/252[b]	N.D.	N.D.		

The data analyzed here are also shown in Figure 16.27. The frequency of the minor SNP allele in cases and controls is shown for the initial scan and second replication set. TDT data show the transmission/nontransmission of the minor SNP allele from parent to affected offspring. GWA, genomewide associations; N.D., not done. [a]Data from an earlier report by the same group. [b]SNPs showing significant replication in the TDT series.

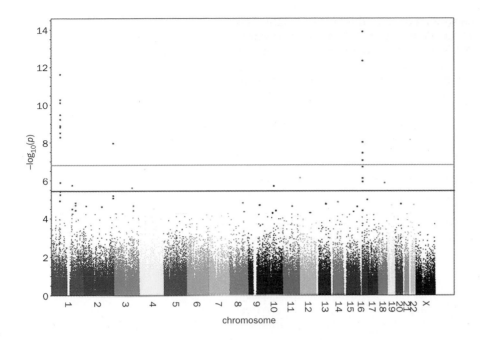

Figure 16.27 Results of a genomewide association study of Crohn disease. A total of 988 patients and 1007 controls were genotyped for 317,503 SNPs. In this so-called Manhattan plot, the y axis shows p values for association of each SNP with Crohn disease (on a logarithmic scale, after filtering out questionable data) and the x axis shows the SNPs in their position across the whole genome, color-coded by chromosome. The pink and green horizontal lines show the threshold p values for suggestive and significant associations (3.28×10^{-6} and 1.64×10^{-7}, respectively). [Adapted from Rioux JD, Xavier RJ, Taylor KD et al. (2007) *Nat. Genet.* 39, 596–604. With permission from Macmillan Publishers Ltd.]

This work has been chosen as a good example of the progress that has been made over the past few years. The list of susceptibility loci in Table 16.2 is long and still growing. The work has demonstrated that determinants of susceptibility to the two types of inflammatory bowel disease, Crohn disease and ulcerative colitis, are mostly different, and identifying the main Crohn disease susceptibility factors has produced a better understanding of the pathological process.

16.7 HOW WELL HAS DISEASE GENE IDENTIFICATION WORKED?

The goals of disease gene identification must be to understand the pathogenesis of diseases, with the hope that understanding will lead to therapy, and to be able to identify individuals at risk of disease, with the hope that onset of the disease can be prevented or its course ameliorated. Despite recent significant progress, there is still an obvious contrast between the great success with Mendelian diseases and the limited results achieved with complex diseases.

Most variants that cause Mendelian disease have been identified

There is still much work to be done in identifying the genes that cause rare or obscure Mendelian conditions, or those with nonspecific phenotypes, but the gene responsible for virtually every reasonably common condition with a distinct phenotype has been identified. Moreover, the majority of mutations in most Mendelian diseases can be detected by sequencing the exons and splice sites identified in current annotations, supplemented by RT-PCR to pick up unexpected splice variants, and other methods for finding whole exon deletions. Although there will certainly be exceptions, it looks as though the extreme complexity of transcripts detected by ENCODE researchers (see Chapter 11) does not invalidate the broad generalization that most Mendelian conditions depend on rather obvious changes in conventionally defined gene sequences. The causative variants can be identified in most patients, allowing accurate diagnosis and counseling.

Nonsyndromic mental retardation remains the major challenge. There is extreme genetic heterogeneity, individual families are usually too small for linkage analysis, and there are very large numbers of possible candidate genes. As previously mentioned, thousands of genes are expressed in the brain and there is seldom any way of choosing between them. Systematic resequencing looks like the only way to identify causative mutations. A pilot study by Tarpey and colleagues (see Further Reading) attempted to do this for X-linked mental retardation (XLMR). After excluding cases with mutations in known XLMR genes, 208 families were chosen for study, and an attempt was made to sequence every exon on the X chromosome in an affected male from each family. The study identified nine novel disease genes, but in 50–75% of families (depending on the degree of certainty required) the cause remained unknown. Worryingly, in several genes the study identified clearly inactivating mutations in normal male controls. Extending this to the much larger number of autosomal disease genes is clearly a formidable task.

Genomewide association studies have been very successful, but identifying the true functional variants remains difficult

Since 2005, genomewide association studies of a great variety of complex diseases have produced a torrent of reproducible data. Many disease-associated SNPs have been identified and confirmed in replication studies. These SNPs seldom cause the susceptibility directly, but they define haplotype blocks that contain the actual functional variant(s). To understand the pathogenesis of the disease it is necessary to pinpoint the true functional variant but, as discussed above, this can be very difficult. Crohn disease (see case study 8) is one of the clearest success stories, as described above. Significant progress has also been made in understanding the pathogenesis of many other complex diseases, for example both types of diabetes.

Psychiatric conditions remain the most intractable. Given their high incidence and immense costs, both in money and suffering, this is singularly unfortunate. The big problem with these conditions is that we have so little understanding of the biology. At the level of brain anatomy or physiology we simply do not know what has gone wrong in schizophrenia, bipolar disorder, or autism. Studies rely on clinical labels that probably lump together the consequences of a highly heterogeneous set of causes. Maybe the way forward is to identify **endophenotypes**—characters that are correlated with the clinical diagnosis but lie closer to biology, such as reaction times or eye movements. The problem here is to know whether the chosen character has any causal relevance or is just a downstream consequence of the clinical problem. An alternative line of attack is to look for common variants that, in different combinations, predispose to a variety of psychiatric conditions. Copy-number variants at 1q21 and 15q13 have each been associated with a variety of conditions, including schizophrenia, autism, seizures, and mental retardation. The same variants are also seen in normal individuals, often the parent of an affected child. Thus, they are neither necessary nor sufficient to cause disease, but evidently confer significant susceptibility to a range of conditions. Understanding these effects may help us understand the biology.

Clinically useful findings have been achieved in a few complex diseases

Most of the many risk factors that have been identified for complex diseases have individually very small effects. Even in combination they are not usefully predictive for individuals. There are, however, exceptions. Some complex diseases have a high-penetrance Mendelian subset. As described above, the identification of the *BRCA1* and *BRCA2* genes has allowed accurate diagnosis and risk prediction in the 5% of cases of breast cancer in which there are mutations in these genes. Alzheimer disease also has a rare Mendelian subset, described below. The other examples below show that some complex diseases are less complex than others. In some cases, mutations in one or two candidate genes, together with environmental effects, can explain much of the incidence of the disease and suggest clinically useful interventions.

Alzheimer disease

Early-onset disease (before age 65 years) is sometimes Mendelian, caused by mutations in the genes encoding presenilin 1 (*PSEN1*; OMIM 104311), presenilin 2 (*PSEN2*; OMIM 600759), or amyloid beta A4 precursor protein (*APP*; OMIM

104760). In affected families, the disease is inherited in a clear autosomal dominant manner. Accurate diagnosis and prediction are possible, raising the same kinds of ethical and personal dilemma as with Huntington disease. However, these Mendelian forms are extremely uncommon, even among families with early onset; the overwhelming majority of cases have a non-Mendelian pattern of inheritance, and usually a later onset.

As is usual for a common complex disease, many linkage and association studies have reported possible susceptibility loci for late-onset Alzheimer disease (see OMIM 104300 and the review by Bertram and Tanzi in Further Reading). Most remain unconfirmed but, unusually, one genetic variant has been unambiguously identified as a powerful risk factor. This is the *E4* allele of apolipoprotein E (*ApoE*; OMIM 107741). *ApoE* has three common variants: *E2*, *E3*, and *E4*. Frequencies of the three vary in different populations; in the UK they are 0.09, 0.76, and 0.15, respectively. There are reported effects on many clinical variables, but a major effect is on the risk of late-onset Alzheimer disease. *E4* is the risk allele, with odds ratios of 3 for heterozygotes and 14 for homozygotes, in contrast with *E3* homozygotes. This factor alone accounts for about 50% of the total genetic susceptibility to late-onset Alzheimer disease. The implications of this are discussed in Chapter 19—but meanwhile it is interesting to note that this is the only genotype that Jim Watson was unwilling to reveal when making his genome sequence public.

Age-related macular degeneration (ARMD)

This complex condition is the main cause of failing eyesight in the elderly and is a major cost to health providers. Family studies have established genetic susceptibility, and linkage studies have suggested several susceptibility loci, identified as *ARMD1-11* (see OMIM 603075). Variants in the *CFH* (complement factor H) gene at chromosome 1q32 are a major risk factor, accounting for maybe half of the overall risk of ARMD. The effect is strong enough to have been detected in an association study that used only 96 cases and 50 controls. The primary determinant seems to be a coding polymorphism p.Tyr402Iis, but other variants in this region may contribute independently to susceptibility or resistance. Variants in other genes in the complement system have also been implicated, although with much weaker effects. A second major risk locus is a coding SNP (p.Ala69Ser) in the *LOC38775* (*ARMS2*) gene at 10q26. Between them, these two susceptibility factors may account for three-quarters of the genetic susceptibility to age-related macular degeneration. Several environmental factors, particularly smoking, strongly modify the genetic risk. The genetic and environmental factors combine into a credible overview of the pathogenesis (**Figure 16.28**).

Eczema (atopic dermatitis)

Childhood eczema is a typical common complex disease, with familial aggregation but no clear Mendelian pattern. However, much of the susceptibility is due to mutations in a single gene, filaggrin (*FLG*; OMIM 135940). This is one of 25 functionally related genes forming the 2 Mb epidermal differentiation complex on chromosome 1q21.3. The gene encodes a large proprotein whose proteolysis releases multiple copies of the 342-residue protein filaggrin. Filaggrin causes the keratin cytoskeleton of epidermal keratinocytes to collapse, transforming these cells into the flattened squames that form the skin barrier.

FLG mutations are common. About 10% of Europeans carry one or other of five specific filaggrin mutations, and additional individuals may carry rarer or private mutations. Different mutations are common in different populations, suggesting that there has been widespread selection in favor of carrying a mutation. All the common mutations act as null alleles. Heterozygotes and homozygotes have greatly increased risks of eczema. Half or more of all children with moderate to severe eczema have a filaggrin mutation. Professor Marcus Pembrey has suggested that one explanation for the current epidemic of allergies could be

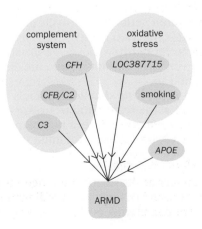

Figure 16.28 The pathogenesis of age-related macular degeneration (ARMD). The condition is caused by a combination of defects in the complement pathway and oxidative stress. [Adapted from Haines JL, Spencer KM & Pericak-Vance MA (2007) *Proc. Natl Acad. Sci. USA* 104, 16725–16726. With permission from the National Academy of Sciences.]

that modern living conditions (dry air and frequent washing) aggravate the skin barrier defects in filaggrin-deficient babies. These babies then first encounter antigens through a defective skin barrier rather than through the normal intestinal route, maybe during a short sensitive period.

The problem of hidden heritability

The examples discussed above are unusual. The great majority of all the SNPs identified through association studies have extremely modest effects on disease susceptibility. Odds ratios are seldom greater than 1.5, and are often closer to 1.1. This is a predictable consequence of the study designs. Genomewide association studies can detect only factors that conform to the common disease–common variant hypothesis. Such methods would probably not have discovered the filaggrin mutations described above, given the multiple risk alleles, each with an individually low population frequency. As discussed in Chapter 15, variants that are even modestly pathogenic would not be expected to survive long enough to give populationwide associations with tagging SNPs. Thus, although the genomewide studies have generated long and reproducible lists of susceptibility factors, they have not explained much of the familial tendency of most complex diseases. In most cases it is also questionable how much new light they have thrown on the pathogenic mechanisms. This has given rise to much discussion of the hidden heritability of complex diseases.

There are at least three possible explanations for the hidden heritability:

- Susceptibility may be due to large numbers of individually weak associated factors. Studies of a few thousand cases and controls have good power to detect factors giving odds ratios of 1.5 or greater, but many real factors conferring odds ratios of 1.05–1.2 will have gone undetected. The data on breast cancer (see Table 16.1) support this view.

- Much susceptibility may be due to a highly heterogeneous collection of rare variants that might individually give quite strong odds ratios. The filaggrin mutations that cause susceptibility to atopic dermatitis, described above, are a possible model. The new generation of ultra-high-throughput sequencing technologies provides a route to discover these.

- Epigenetic changes may be a major component of susceptibility. Thus far there has been little systematic search for such factors.

The overall conclusion is that, despite all the impressive progress in association studies, we are still a long way from identifying the factors that cause susceptibility to common complex diseases.

CONCLUSION

There are many ways in which a disease gene may be identified. Knowledge of the protein product or of a related gene (in humans or another species) can lead directly to identification of the gene. More often, first the chromosomal location of the gene is defined, and then a search of that region is made for a gene that carries mutations in patients but not in controls (positional cloning). Sometimes chromosomal abnormalities can suggest a location. Translocations, inversions, or deletions may cause clinical problems by disrupting or deleting genes. If two unrelated patients share a chromosomal breakpoint and also a particular clinical feature, maybe a gene responsible for that feature is located at or near the breakpoint. More often, however, disease genes are localized by linkage analysis in a collection of families where the disease is segregating, as described in Chapters 14 and 15.

Once a candidate location has been defined, it is necessary to identify all the genes in the region and prioritize them for mutation analysis. The genome databases are a powerful but not infallible tool for identifying candidate genes. It may be necessary to perform additional searches or laboratory experiments to identify all exons of a gene and other transcripts. Genes are prioritized for testing on the basis of their likely function and expression pattern, and also on practical considerations about the ease of testing. As sequencing has become faster and cheaper, it has become increasingly attractive simply to sequence all exons across the candidate region, without trying too hard to decide priorities.

This approach has been outstandingly successful at identifying the genes and mutations underlying Mendelian conditions, but less so for complex diseases. As described in Chapter 15, linkage analysis, even using non-parametric methods, has had very limited success with complex diseases, probably because of the great heterogeneity of causes. Association studies are much more successful. Genomewide association studies of many different complex diseases have used high-resolution SNP genotyping in large case-control studies to identify numerous SNPs that are associated with either susceptibility or resistance to the disease. Generally these SNPs will not directly affect susceptibility. Instead, they will be in linkage disequilibrium with the functional variant. Identifying the true functional variants is extremely challenging.

The question will remain: what is the practical value of having a list of susceptibility factors that confer relative risks of 1.2 or less? Funding agencies and pharmaceutical companies have invested heavily in research in these areas in the belief, or at least the hope, that such knowledge would revolutionize medicine. In Chapter 19 we ask how far these hopes look like being realized. Meanwhile, perhaps the area of medicine in which analyzing genomewide changes using microarrays is closest to clinical application is oncology. We explore this area in the next chapter.

FURTHER READING

Positional cloning

Church DM, Goodstadt L, Hillier LW et al. (2009) Lineage-specific biology revealed by a finished genome assembly of the mouse. PLoS Biol. 7(5), e1000112.

Comprehensive Mouse Knockout Consortium (2004) The knockout mouse project. Nat. Genet. 36, 921–924.

Database of Genomic Variants. http://projects.tcag.ca/variation/ [A database of copy-number and other structural variants found in apparently normal subjects.]

European Mouse Mutagenesis Consortium (2004) The European dimension for the mouse genome mutagenesis program. Nat. Genet. 36, 925–927.

Peters LL, Robledo RF, Bult CJ et al. (2007) The mouse as a resource for human biology: a resource guide for complex trait analysis. Nat. Rev. Genet. 8, 58–69. [An impressive detailing of the very extensive resources in mouse genetics that can help in the identification of human disease factors.]

Steinmetz LM, Scharfe C, Deutschbauer AM et al. (2002) Systematic screen for human disease genes in yeast. Nat. Genet. 31, 400–404.

The value of patients with chromosomal abnormalities

Kurotaki N, Imaizumi K, Harada N et al. (2002) Haploinsufficiency of NSD1 causes Sotos syndrome. Nat. Genet. 30, 365–366.

Vissers LELM, Veltman JA, Geurts van Kessel A & Brunner HG (2005) Identification of disease genes by whole genome CGH arrays. Hum. Mol. Genet. 14, R215–R223.

Position-independent routes to identifying disease genes

Koob MD, Benzow KA, Bird TD et al. (1998) Rapid cloning of expanded trinucleotide repeat sequences from genomic DNA. Nat. Genet. 18, 72–75. [Use of the repeat expansion detection method for position-independent identification of genes containing expanded repeats.]

Testing positional candidate genes

Bier E (2005) Drosophila, the golden bug, emerges as a tool for human genetics. Nat. Rev. Genet. 6, 9–23. [An overview of the ways in which Drosophila can be used to help understand human disease.]

Cuzin F, Grandjean V & Rassoulzadegan M (2008) Inherited variation at the epigenetic level: paramutation from the plant to the mouse. Curr. Opin. Genet. Dev. 18, 193–196. [A review of paramutation by the group that described the Kit paramutation in the mouse.]

Jirtle RL & Skinner MK (2007) Environmental epigenomics and disease susceptibility. Nat. Rev. Genet. 8, 253–262. [A review of possible cases in which epigenetic effects allow acquired characters to be heritable.]

Identifying causal variants from association studies

Balding DJ (2006) A tutorial on statistical methods for population association studies. Nat. Rev. Genet. 7, 781–790. [A detailed discussion of the statistical methods used in association studies.]

Horikawa Y, Oda N, Cox NJ et al. (2000) Genetic variation in the gene encoding calpain-10 is associated with type 2 diabetes mellitus. Nat. Genet. 26, 163–175. [A heroic early attempt to identify the precise variant causing susceptibility.]

Wellcome Trust Case Control Consortium (2007) Genome-wide association study of 14,000 cases of seven common diseases and 3,000 shared controls. Nature 447, 661–678. [A good overview of the state of the art in genomewide association studies.]

Witte JS (2007) Multiple prostate cancer risk variants on 8q24. Nat. Genet. 39, 579–580. [A commentary on three papers in this number of the journal that report association studies of susceptibility to prostate cancer.]

Eight examples of disease gene identification

Duchenne muscular dystrophy

Koenig M, Hoffman EP, Bertelson CJ et al. (1987) Complete cloning of the Duchenne muscular dystrophy (DMD) cDNA and preliminary genomic organization of the DMD gene in normal and affected individuals. Cell 50, 509–517. [The original report of the cloning of this gene.]

Royer-Pokora B, Kunkel LM, Monaco AP et al. (1985) Cloning the gene for an inherited human disorder—chronic granulomatous disease—on the basis of its chromosomal location. *Nature* 322, 32–38. [The first gene to be positionally cloned.]

Worton RG & Thompson MW (1988) Genetics of Duchenne muscular dystrophy. *Annu. Rev. Genet.* 22, 601–629. [An excellent review, giving a good feel for the obstacles overcome in cloning the dystrophin gene.]

Cystic fibrosis

Rommens JM, Januzzi MC, Kerem B-S et al. (1989) Identification of the cystic fibrosis gene: chromosome walking and jumping. *Science* 245, 1059–1065. [A humbling illustration of the huge effort needed in the pioneering days of positional cloning.]

Branchio-oto-renal syndrome

Abdelhak S, Kalatzis V, Heilig R et al. (1997) A human homologue of the *Drosophila eyes absent* gene underlies Branchio-Oto-Renal (BOR) syndrome and identifies a novel gene family. *Nat. Genet.* 15, 157–164.

Multiple sulfatase deficiency

Cosma M, Pepe S, Annunziata I et al. (2003) The Multiple Sulfatase Deficiency gene encodes an essential and limiting factor for the activity of sulfatases. *Cell* 113, 445–456. [Identifying the gene through microcell-mediated chromosome transfer.]

Dierks T, Schmidt B, Borissenko LV et al. (2003) Multiple sulfatase deficiency is caused by mutations in the gene encoding the human C_α-formylglycine generating enzyme. *Cell* 113, 435–444. [Identifying the gene through purification of the protein.]

Persistence of intestinal lactase

Tishkoff SA, Reed FA, Ranciaro A et al. (2007) Convergent adaptation of human lactase persistence in Africa and Europe. *Nat. Genet.* 39, 31–40. [Particularly interesting for the discussion of the evidence for strong selection.]

CHARGE syndrome

Vissers LE, van Ravenswaaij CM, Admiraal R et al. (2004) Mutations in a new member of the chromodomain gene family cause CHARGE syndrome. *Nat. Genet.* 36, 955–957. [A cautionary tale for anybody who believes that array-CGH makes it easy to discover new syndromes.]

Breast cancer

Easton DF, Pooley KA, Dunning AM et al. (2007) Genome-wide association study identifies novel breast cancer susceptibility loci. *Nature* 447, 1087–1093. [The study outlined in Figure 16.26; also a source of references to other work on breast cancer susceptibility.]

Ford D, Easton DF, Stratton M et al. (1998) Genetic heterogeneity and penetrance analysis of the *BRCA1* and *BRCA2* genes in breast cancer families. *Am. J. Hum. Genet.* 62, 676–689. [A large survey that shows how *BRCA1/2* mutations account for only a minor part of familial breast cancer.]

Crohn disease

Hampe J, Cuthbert A, Croucher PJP et al. (2001) Association between insertion mutation in *NOD2* gene and Crohn disease in German and British populations. *Lancet* 357, 1925–1928. [One of the three studies mentioned on p. 528.]

Hugot J-P, Chamaillard M, Zouali H et al. (2001) Association of *NOD2* leucine-rich repeat variants with susceptibility to Crohn disease. *Nature* 411, 599–603. [The linkage-association study mentioned on p. 528. This paper is well worth reading for the picture it gives of the methodology.]

Klionsky DJ (2009) Crohn disease, autophagy and the Paneth cell. *N. Engl. J. Med.* 360, 1785–1786. [A succinct account of our understanding of the pathogenesis.]

Ogura Y, Bonen DK, Inohara N et al. (2001) A frameshift mutation in *NOD2* associated with susceptibility to Crohn disease. *Nature* 411, 603–606. [One of the three studies mentioned on p. 528.]

Rioux JD, Xavier RJ, Taylor KD et al. (2007) Genome-wide association study identifies new susceptibility loci for Crohn disease and implicates autophagy in disease pathogenesis. *Nat. Genet.* 39, 596–604. [The study from which Figure 16.27 and Table 16.3 are taken.]

How well has disease gene identification worked?

Ben-Shachar S, Lanpher B, German JR et al. (2009) Microdeletion 15q13.3: a locus with incomplete penetrance for autism, mental retardation and psychiatric disorders. *J. Med. Genet.* 46, 382–388.

Brunetti-Pierri N, Berg JS, Scaglia F et al. (2008) Recurrent reciprocal 1q21.1 deletions and duplications associated with microcephaly or macrocephaly and developmental and behavioural abnormalities. *Nat. Genet.* 40, 1466–1471.

Mefford H, Sharp A, Baker C et al. (2008) Recurrent rearrangements of chromosome 1q21.1 and variable pediatric phenotypes. *N. Engl. J. Med.* 359, 1685–1699.

Tarpey PS, Smith R, Pleasance E et al. (2009) A systematic, large scale resequencing screen of X-chromosome coding exons in mental retardation. *Nat. Genet.* 41, 535–543. [A major attempt to identify all the X chromosome genes responsible for nonsyndromic mental retardation.]

Van Bon BWM, Mefford HC, Menten B et al. (2009) Further delineation of the 15q13 microdeletion and duplication syndromes: a clinical spectrum varying from non-pathogenic to a severe outcome. *J. Med. Genet.* 46, 511–523.

Alzheimer disease

Bertram L & Tanzi RE (2008) Thirty years of Alzheimer's disease genetics: the implications of systematic meta-analyses. *Nat. Rev. Neurosci.* 10, 768–778. [A review of likely genetic susceptibility factors and their possible modes of action.]

Age-related macular degeneration

Kanda A, Chen W, Othman M et al. (2007) A variant of mitochondrial protein LOC387715/ARMD2, not HTRA1, is strongly associated with age-related macular degeneration. *Proc. Natl Acad. Sci. USA* 104, 16227–16232. [An interesting example of identifying a functional variant from a set of variants in linkage disequilibrium.]

Klein RJ, Zeiss C, Chew EY et al. (2005) Complement factor H polymorphism in age-related macular degeneration. *Science* 308, 385–389. [One of three papers in the same issue of *Science* reporting similar findings; see also Haines et al. *Science* 308, 419–421 and Edwards et al. *Science* 308, 421–424.]

Schmidt S, Hauser MA, Scott WK et al. (2006) Cigarette smoking strongly modifies the association of LOC387715 and age-related macular degeneration. *Am. J. Hum. Genet.* 78, 852–864. [An example of disentangling gene–environment interaction in a complex disease.]

Filaggrin and eczema

Irvine AD & McLean WHI (2006) Breaking the (un)sound barrier: filaggrin is a major gene for atopic dermatitis. *J. Invest. Dermatol.* 126, 1200–1202.

Chapter 17

Cancer Genetics

17

KEY CONCEPTS

- Cancer is the result of somatic cells acquiring genetic changes that confer on them six general features: (1) independence of external growth signals, (2) insensitivity to external anti-growth signals, (3) the ability to avoid apoptosis, (4) the ability to replicate indefinitely, (5) the ability of a mass of such cells to trigger angiogenesis and vascularize, and (6) the ability to invade tissues and establish secondary tumors.

- These features are acquired through a normal Darwinian process of natural selection acting on random mutations. The microevolution occurs in several stages, with each successive change giving an extra selective advantage to descendants of a cell.

- Highly evolved and sophisticated defense mechanisms protect the body against the proliferation of such mutant cells, but tumor cells have mutations that disable these defenses.

- Stem cells, unlike most somatic cells, have the ability to replicate indefinitely, and so they already possess one of the features of cancer cells. Thus, stem cells are likely candidates to evolve into tumor cells.

- Oncogenes normally act to promote cell division, but a complex regulatory network limits this activity. In tumor cells, one copy of an oncogene is often abnormally activated—through point mutations, copy-number amplification, or chromosomal rearrangements—so that it escapes regulation.

- Chromosomal rearrangements in tumor cells often create novel chimeric oncogenes but may alternatively upregulate the expression of an oncogene by placing it under the influence of a powerful enhancer.

- The normal role of tumor suppressor genes is to limit cell division. In tumor cells, both copies of a tumor suppressor gene are often inactivated by deletions, point mutations, or methylation of the promoter.

- Many tumor suppressor genes have been identified through investigation of familial cancer predisposition syndromes. In these syndromes, individuals inherit a mutation that inactivates one allele of a tumor suppressor gene.

- Oncogenes and tumor suppressor genes normally function in the cell signaling that controls the cell cycle, or in the response to DNA damage. Understanding these processes is central to understanding what goes wrong in cancer.

- Genomic instability is a normal feature of tumor cells. Because of this instability, tumors contain large populations of cells carrying a great variety of mutations, on which natural selection can act. Driver mutations contribute to tumor development and are subject to positive selection; passenger mutations are chance by-products of the genomic instability of most tumor cells.

- Most instability is seen at the chromosomal level. However, some tumors are cytogenetically normal but have a high level of DNA replication errors.

- New sequencing technology allows the totality of acquired genetic changes in tumor cells to be cataloged. The results of these studies emphasize the individuality and large number of such changes in tumor cells.

- Histological and molecular profiling of a tumor can provide important prognostic information and a guide to treatment. Drugs can target specific acquired genetic changes. Genomewide studies using expression arrays are an important tool for this.

- Tumorigenesis is best understood by thinking in terms of altered pathways rather than individual mutated genes.

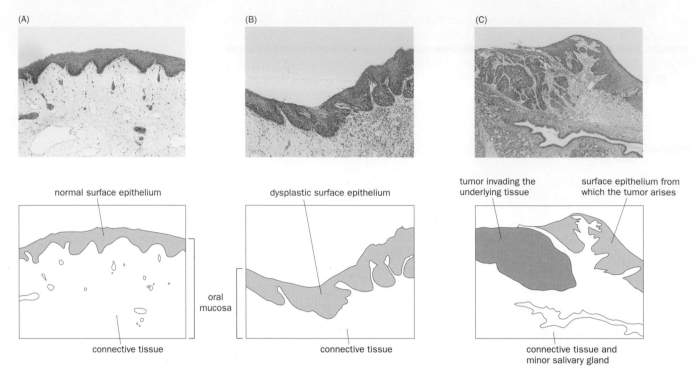

Figure 17.1 Stages in the development of a carcinoma. Three histological sections of oral mucosa, stained with hematoxylin and eosin, showing stages in the development of oral cancer. (A) Normal epithelium. (B) Dysplastic epithelium, which is a potentially premalignant change. The epithelium shows disordered growth and maturation, abnormal cells, and an increased mitotic index (the proportion of cells undergoing mitosis). (C) Cancer arising from the surface epithelium and invading the underlying connective tissues. The islands of the tumor (carcinoma) show disordered differentiation, abnormal cells, and increased and atypical mitoses. Pathologists use such changes in tissue architecture to identify and grade tumors. (Courtesy of Nalin Thakker, University of Manchester.)

Cancer—a condition in which cells divide without control—is not so much a disease as the natural end state of any multicellular organism. We are all familiar with the basic Darwinian idea that a population of organisms that show hereditary variation in reproductive capacity will evolve by natural selection. Genotypes that reproduce faster or more extensively will come to dominate later generations, only to be supplanted, in turn, by yet more efficient reproducers. Exactly the same applies to the population of cells that constitutes a multicellular organism. The determining factor can be an increased birth rate or a decreased death rate. Cellular birth and death are under genetic control, and if somatic mutation creates a variant that proliferates faster, the mutant clone will tend to take over the organism. Cancers are the result of a series of somatic mutations with, in some cases, also an inherited predisposition. Thus, cancer can be seen as a natural evolutionary process.

Tumors can be classified according to the tissue of origin; thus, carcinomas are derived from epithelial cells, sarcomas from bone or connective tissue, and leukemias and lymphomas from blood cell precursors. Solid tumors can be considered as organs, in that they consist of a variety of cell types and are thought to be maintained by a small population of (cancer) stem cells. An important distinction is between benign (noninvasive) and malignant (invasive) tumors. Pathologists classify tumors based on the histology as seen under the microscope (**Figure 17.1**). Genetic tests for specific chromosomal rearrangements or genomewide expression profiling (discussed later in this chapter) allow further refinements of the classification. These classifications are important for deciding prognosis and management, but they do not explain how the cancer evolved. The aim of cancer genetics is to understand the multi-step mutational and selective pathway that allowed a normal somatic cell (but maybe a stem cell) to found a population of proliferating and invasive cancer cells. As the key molecular events are revealed, new prognostic indicators become available to the pathologist. It is hoped that some of these events will also present new therapeutic targets to the oncologist.

The detail in cancer genetics can be overwhelming. Every tumor is individual. There are so many different genes that acquire mutations in one or another tumor, and they interact in such complex ways, that it is easy to get lost in a sea of detail. However, an important and highly recommended review by Hanahan and Weinberg points out that any invasive cancer is likely to depend on cells that have acquired six basic capabilities:

- Independence of external growth signals
- Insensitivity to external anti-growth signals
- Ability to avoid apoptosis
- Ability to replicate indefinitely
- Ability of a mass of such cells to trigger angiogenesis and vascularize
- Ability to invade tissues and establish secondary tumors

In this chapter, we focus on how a cell might acquire these capabilities, rather than on cataloging genes and mutations. By concentrating on the principles, as currently understood, rather than the individual details, it is hoped that a framework of understanding can be established.

17.1 THE EVOLUTION OF CANCER

As described above, cells are under strong selective pressure to evolve into tumor cells. However, although tumors are very successful as organs, as organisms they are hopeless failures. They leave no offspring beyond the life of their host. At the level of the whole organism, there is therefore powerful selection for mechanisms that prevent people from dying from tumors, at least until they have borne and brought up their children. Thus, we are ruled by two opposing sets of selective forces. But selection for tumorigenesis occurs over the short term, whereas selection for resistance occurs over the long term. The microevolution from a normal somatic cell to a malignant tumor takes place within the life of an individual and has to start afresh with each new individual. But an organism with a good anti-tumor mechanism transmits this to its offspring, where it continues to evolve. A billion years of evolution have endowed us with sophisticated interlocking and overlapping mechanisms to protect us against tumors, at least during our reproductive life. Potential tumor cells are either repaired and brought back into line or made to kill themselves (*apoptosis*). No single mutation can circumvent these defenses and convert a normal cell into a malignant one. As early as 50 years ago, studies of the age-dependence of cancer suggested that on average six or seven successive mutations are needed to convert a normal epithelial cell into an invasive carcinoma. In other words, only if six or seven independent defenses are disabled by mutation can a normal cell be converted into a malignant tumor.

The chance that a single cell will undergo six independent mutations is negligible, suggesting that cancer should be vanishingly rare. However, two general mechanisms exist that can allow the progression to happen (**Box 17.1**). An initial

---

BOX 17.1 TWO WAYS OF MAKING A SERIES OF SUCCESSIVE MUTATIONS MORE LIKELY

Turning a normal epithelial cell into an invasive cancer cell requires perhaps six specific mutations in the one cell. It would seem extremely unlikely that any one cell should suffer so many mutations (which is why most of us are alive). If a typical mutation rate is 10^{-7} per gene per cell generation, the probability of this happening to any one of the 10^{13} cells in a person is $10^{13} \times 10^{-42}$, or 1 in 10^{29}. Cancer nevertheless happens because of a combination of two mechanisms:

- Some mutations enhance cell proliferation, creating an expanded target population of cells for the next mutation (Figure 17.2). This may require a combination of two or more mutations.
- Some mutations affect the stability of the entire genome, at either the DNA or the chromosomal level, increasing the overall mutation rate. Malignant tumor cells usually advertise their genomic instability by their abnormal karyotypes.

Because cancers depend on these two mechanisms, they develop in stages, starting with tissue hyperplasia or benign growths. Within a stage, successive random mutations generate an increasingly diverse cell population until eventually, by chance, one cell acquires a change or combination of changes that gives it a growth advantage.

mutation can increase the likelihood that a cell will pick up subsequent mutations, either by conferring a growth advantage (**Figure 17.2**) or by inducing genomic instability. Accumulating all these mutations nevertheless takes time, so that cancer is mainly a disease of post-reproductive life, when there is little selective pressure to improve the defenses still further.

Cell types differ in the extent to which their normal capabilities approach these tumor cell capabilities—for example, some cell types divide rapidly, some are relatively resistant to apoptosis, and so on. Tumors arise most easily from cells in which that approach is closest. The existence of populations of rapidly dividing and relatively undifferentiated cells in fetuses and infants explains the special cancers seen in young children. Stem cells are thought to be important as tumor progenitors, because they already possess the capacity for indefinite proliferation. There is controversy over how far all tumors arise from mutated stem cells, but there is agreement that tumor precursor cells have stem cell-like properties, whether these are innate or are acquired by mutation.

The genes that are the targets of these mutations can be divided into two broad categories, although, as always in biology, these are more tools for thinking about cancer than watertight exclusive classifications.

- **Oncogenes** are genes whose normal activity promotes cell proliferation. Gain-of-function mutations in tumor cells create forms that are excessively or inappropriately active. A single mutant allele may affect the behavior of a cell. The nonmutant versions are properly called **proto-oncogenes**.

- **Tumor suppressor genes** are genes whose products act to limit normal cell proliferation. Mutant versions in cancer cells have lost their function. Some tumor suppressor gene products prevent inappropriate cell cycle progression, some steer deviant cells into apoptosis, and others keep the genome stable and mutation rates low by ensuring accurate replication, repair, and segregation of the cell's DNA. Both alleles of a tumor suppressor gene must be inactivated to change the behavior of a cell.

By analogy with a bus, one can picture the oncogenes as the accelerator and the tumor suppressor genes as the brake. Jamming the accelerator on (a dominant gain of function of an oncogene) or having all the brakes fail (a recessive loss of function of a tumor suppressor gene) will make the bus run out of control. Traditionally, both categories have been seen as protein-coding genes, but similar reasoning could be applied to other genetic control elements—in particular microRNAs (see Chapter 9, p. 283).

17.2 ONCOGENES

Oncogenes were discovered in the 1960s when it was realized that some animal cancers (especially leukemias and lymphomas) were caused by viruses. Some of the viruses had relatively complicated DNA genomes (SV40 virus and papilloma viruses, for example), but others were acute transforming **retroviruses** that had very simple RNA genomes. The genome of a standard retrovirus has just three transcription units: *gag*, encoding internal proteins; *pol*, encoding a reverse transcriptase and other proteins; and *env*, encoding envelope proteins. Post-transcriptional processing produces several proteins from single transcripts. Great excitement ensued when it was discovered that the ability of acute transforming retroviruses to *transform* cells in culture (that is, to change their growth pattern to one resembling that of tumor cells) was entirely due to their possession of one extra gene, the oncogene (**Figure 17.3**). For a short time, some enthusiasts hoped that the whole of cancer might be explained by infection with viruses carrying oncogenes, and that once these had been identified anti-cancer vaccines could be developed.

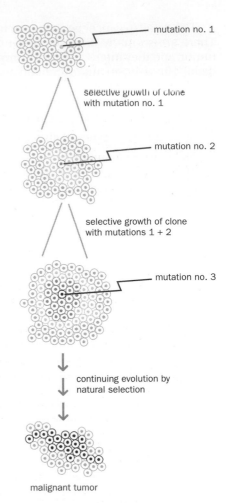

Figure 17.2 Multi-stage evolution of cancer. Each successive mutation gives the cell a growth advantage, so that it forms an expanded clone, thus presenting a larger target for the next mutation.

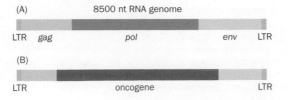

Figure 17.3 An acute transforming retrovirus. (A) The genome of a standard retrovirus has three transcription units: *gag*, encoding internal proteins; *pol*, encoding proteins needed for replication; and *env*, encoding the proteins needed for packaging new virus particles. There are also long terminal repeated sequences (LTR). (B) In a random processing error, part of the normal retroviral genome has been replaced by a cellular oncogene. The virus is no longer competent to replicate, but it can ferry the oncogene into a cell. nt, nucleotide.

Researchers soon discovered that the viral oncogenes were copies of normal cellular genes, proto-oncogenes, that had become accidentally incorporated into the retroviral particles. The viral versions were somehow activated, enabling them to transform infected cells.

Most human cancers do not depend on viruses, but nevertheless their resident proto-oncogenes have become activated. In the late 1970s an assay was developed to detect activated cellular oncogenes. NIH 3T3 mouse fibroblast cells were transfected with random DNA fragments from human tumors. Some of the cells were transformed, and the human DNA responsible could be identified by constructing a genomic library in bacteriophage from fragments of the DNA of the transformed 3T3 cells and then screening for recombinant phage that contained the human-specific *Alu* repeat. The oncogenes identified in the transforming fragments included many that were already known from the viral studies.

Oncogenes function in growth signaling pathways

Functional understanding of oncogenes began with the discovery in 1983 that the viral oncogene v-*sis* (the v- suffix denotes a viral oncogene) was derived from the normal cellular platelet-derived growth factor B (*PDGFB*) gene. Uncontrolled overexpression of a growth factor would be an obvious cause of cellular hyperproliferation. The roles of many cellular oncogenes have now been elucidated. The forms in normal cells are properly termed proto-oncogenes, but it is now common to ignore these distinctions and simply use the term oncogenes for the normal genes. The abnormal versions can be described as activated oncogenes (**Table 17.1**). Gratifyingly, they turn out to control exactly the sort of cellular

Left margin (Table 17.3, partial)

Tumor
CML
Ewing sarcoma
Ewing sarcoma
Malignant mela...
Desmoplastic s...
Liposarcoma
AML
Papillary thyroi...
Pre-B-cell ALL
ALL
Acute promyelo...
Alveolar rhabdo...

Note how the sa...
/Census. ALL, ac...

8q24) close to
2p12, or *IGL* at
locations do n
under the influ
of the immuno
cation, the *MY*
to head. Often
(which is nonc
normal upstre
expressed at a
malignancies

breakp...

C_α C_ε $C_{\gamma2}$

TABLE 17.1 VIRAL AND CELLULAR ONCOGENES

Function	Cellular proto-oncogene[a]	Location	Viral oncogene[a]	Viral disease
SECRETED GROWTH FACTORS				
Platelet-derived growth factor B subunit	*PDGFB*	22q13.1	v-*sis*	simian sarcoma
CELL SURFACE RECEPTORS				
Epidermal growth factor receptor	*EGFR*	7p11.2	v-*erbb*	chicken erythroleukemia
Macrophage colony-stimulating factor receptor	*CSF1R*	5q32	v-*fms*	McDonough feline sarcoma
SIGNAL TRANSDUCTION COMPONENTS				
Receptor tyrosine kinase	*HRAS*	11p15.5	v-*ras*	Harvey rat sarcoma
Protein tyrosine kinase	*ABL1*	9q34.1	v-*abl*	Abelson mouse leukemia
DNA-BINDING PROTEINS				
AP-1 transcription factor	*JUN*	1p32.1	v-*jun*	avian sarcoma 17
DNA-binding transcription factor	*MYC*	8q24.21	v-*myc*	avian myelocytomatosis
DNA-binding transcription factor	*FOS*	14q24.3	v-*fos*	mouse osteosarcoma
CELL CYCLE REGULATORS				
D-type cyclins:				
Cyclin D1 Cyclin D2 Cyclin D3	*CCND1* *CCND2* *CCND3*	11q13 12p13 6p21	K-cyclin of Kaposi sarcoma-associated herpesvirus[b]	Kaposi sarcoma[b]

[a]The viral genes are designated v-*sis*, v-*myc*, etc., their cellular counterparts can be described as c-*sis*, c-*myc*, etc. [b]Kaposi sarcoma-associated herpesvirus encodes a virus-specific D-type cyclin, which is an independent gene, not an activated version of one of the human cyclin D genes. However, oncogenic activated versions of all three human D-type cyclins, the result of somatic mutations, have been identified in certain leukemias.

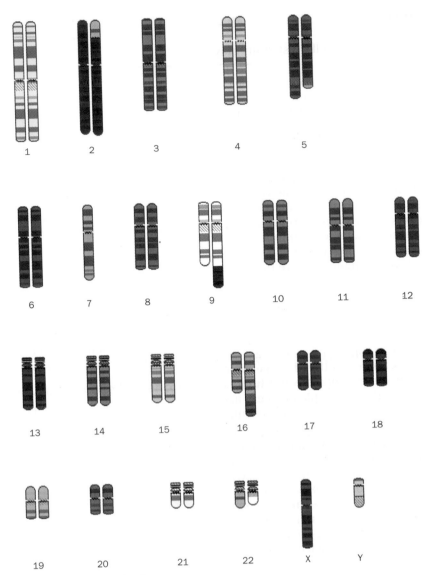

Figure 17.17 SKYgram of cells from a patient with acute myeloid leukemia. The diagram summarizes the result of using multicolor chromosome paints. There is a three-way variant of the standard 9;22 translocation (involving chromosome 2), plus an additional 5;16 translocation and the loss of one copy of chromosome 7. The karyotype is interpreted as 45,XY,t(2;9;22)(p21;q34;q11),t(5;16)(q31;q24),–7. [Case 6 of H Padilla-Nash, reproduced with permission from the NCBI SKY archive (http://www.ncbi.nlm.nih.gov/sky/).]

ABL1

BCR g

Ph¹ fus

translocation
the *BCR* (brea
fusion gene. T.
to the *ABL1* pr

Many othe
oncogenes ha
15–25% of leuk
1% or less of th
fusions in epit
difficulties. A
valuable discu
the Mitelman
nih.gov/Chron
involving 337
found in fusio
most extreme
partners. Fusic
sions or, occasi
two genes.

Clinically, t
an important b
by a targeted F
ently colored F
interphase cell
ABL signal, and
tion (see Figure

**Activation by
region**

Burkitt lympho
Papua New Gui
role in the etiol
acteristic chro
patients (Figur
Each of these t

They probably arise in three ways:

- In many tumor cells the spindle checkpoint is defective. Chromatids can start to separate before all of them are correctly attached to spindle fibers. There is then no way of ensuring that exactly one chromatid of each chromosome goes into each daughter cell. Cancer cells often contain extra centrosomes, which may produce abnormal spindles. Failure of the spindle checkpoint is probably the main source of the many numerical abnormalities seen in cancer cells.

- Tumor cells are able to progress through the cell cycle despite having unrepaired DNA damage. Normally unrepaired damage generates a signal that stalls the cell cycle until the damage is repaired, and irreparable damage triggers apoptosis. In cancer cells, either the damage signaling system or the apoptotic response is often defective. Structural chromosome abnormalities can be a by-product of attempts at DNA replication or mitosis with damaged DNA.

- Tumor cells may replicate to the point that telomeres become too short to protect chromosome ends, leading to all sorts of structural abnormalities.

Telomeres are essential for chromosomal stability

As we saw in Chapter 2, the ends of all human chromosomes carry a repeat sequence $(TTAGGG)_n$. This forms a special DNA structure that binds specific proteins and protects chromosome ends from being treated by the cell as DNA double-strand breaks. Telomere length declines by 50–100 bp with each cell generation because of the inability of DNA polymerase to use the extreme 3′ end of a DNA strand as a template for replication (see Figure 2.13). Telomere length is restored by a special RNA-containing enzyme system, telomerase, that is present in the human germ line but absent from most somatic tissues.

In prolonged culture, normal cells reach a point of senescence, at which they stop dividing. Cells with certain genetic defects (e.g. fibroblasts with deficiency of both p53 and pRb, or harboring viral oncoproteins) continue beyond senescence and hit *crisis*, which probably represents the point at which telomeres are so short that they can no longer protect chromosome ends. The cell then treats chromosome ends as double-strand DNA breaks, which it attempts to repair by homologous recombination or nonhomologous end joining. Random joining of chromosomes produces translocations, including chromosomes with two centromeres (dicentric chromosomes). These may be pulled in two directions at anaphase of mitosis, forming bridges. The bridges eventually break, creating new broken ends and triggering further rounds of fusion and breakage. The result is chromosomal chaos.

At crisis most cells die, but the 1 in 10^7 or thereabouts that survive have gross chromosomal abnormalities. Additionally, they have become immortal. In one way or another, tumor cells always acquire the ability to replicate indefinitely while maintaining their telomeres. Of full-blown metastatic cancers, 85–90% have contrived to re-express telomerase; the remainder use an alternative (ALT) mechanism, based on recombination.

DNA damage sends a signal to p53, which initiates protective responses

As described in Chapter 13, the DNA of every cell is constantly suffering damage from exogenous agents such as ionizing radiation and ultraviolet radiation, and from endogenous processes such as depurination, deamination of cytosines, and attack by reactive oxygen species. As mentioned above, the normal response of a cell to such damage is to stall the cell cycle until the damage is repaired, and to trigger apoptosis if the damage is irreparable. Proteins encoded by several well-known tumor suppressor genes are involved in the system that senses DNA damage and organizes the cellular response. Inherited loss-of-function mutations in these genes are associated with an elevated risk of cancer.

ATM: the initial detector of damage

The ATM protein may be the primary detector of damage. This 3056-residue protein kinase is somehow activated by DNA double-strand breaks—maybe by changes in the chromatin conformation. Once activated, the ATM kinase phosphorylates numerous substrates, including the p53, NBS1 (Nibrin), CHEK2, and BRCA1 proteins (**Figure 17.18**). A related kinase, ATR, performs a similar function in respect of stalled replication forks, which are a consequence of some types of DNA damage. Loss of function of the *ATM* gene causes ataxia telangiectasia (AT; OMIM 208900). Affected homozygotes have a predisposition to various cancers, plus immunodeficiency, chromosomal instability, and cerebellar ataxia. Heterozygous carriers of certain specific *ATM* mutations are at increased risk of breast cancer, presenting tricky problems of medical management because they are also sensitive to radiation, so X-ray mammography may increase the risk of breast cancer.

Nibrin and the MRN complex

Nibrin is phosphorylated by ATM and forms a complex (MRN) with the MRE11 and RAD50 proteins. The complex localizes to sites of damage and may help to recruit repair enzymes. Lack of Nibrin causes Nijmegen breakage syndrome (NBS; OMIM 251260). NBS is clinically rather similar to AT but includes microcephaly and growth retardation in place of ataxia. *MRE11* mutations also result in an AT-like disorder (ATLD; OMIM 604391).

CHEK2: a mediator kinase

CHEK2 is a mediator kinase. It is activated when it is phosphorylated by ATM. It then relays the ATM signal to other substrates including p53. In Chapter 16 a frameshift mutation in *CHEK2* was mentioned as an inherited factor in breast cancer susceptibility, giving carriers a relative risk of 2.34. Intriguingly, a missense variant that is found in about 5% of people in central Europe has been shown to *decrease* the risk of lung cancer in smokers by about half.

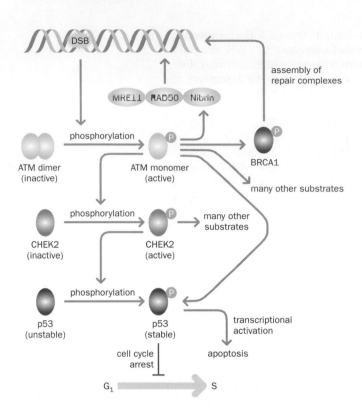

Figure 17.18 Part of the signaling mechanism by which DNA damage causes cell cycle arrest. The ATM kinase is activated by DNA double-strand breaks (DSB) and phosphorylates numerous substrates, some of which are involved in DNA repair (e.g. Nibrin and BRCA1). ATM also activates the CHEK2 protein, which in turn activates p53, the principal target of the DNA damage detection system. Phosphorylated p53 acts to block further progress through the cell cycle and also promotes apoptosis.

The role of BRCA1/2

BRCA1 protein, the product of the first gene implicated in familial breast cancer, localizes to sites of DNA damage. A multiprotein complex, the BASC (for BRCA1-associated genome surveillance complex), has been isolated that includes proteins involved in DNA damage response and repair. BRCA1 also has functions in recombination, chromatin remodeling, and control of transcription.

The *BRCA2* gene encodes another very large protein (3416 residues) with multiple functional domains, and a major role in familial breast cancer. It has no structural similarity to BRCA1 protein but shares many functions. One function is to load many molecules of the RAD51 protein onto single-stranded DNA at sites of damage, as part of the recombinational repair mechanism. As well as being the cause of some hereditary breast cancer, a particular class of homozygous *BRCA2* mutations cause the D1 form of Fanconi anemia (see OMIM 227650 and 605724). This is an autosomal recessive syndrome of congenital abnormalities, progressive bone marrow failure, cellular hypersensitivity to DNA damage, and predisposition to cancer.

p53 to the rescue

p53 is a principal downstream target of the DNA damage detection system. DNA damage causes p53 to be phosphorylated. As mentioned above (see Figure 17.14), this stabilizes p53, preventing Mdm2 from flagging it for degradation. The raised level of p53 leads to activation of the Cdk2 inhibitor p21$^{\text{WAF1/CIP1}}$, which prevents further progress toward S phase of the cell cycle. Raised levels of p53 can also lead to transcription of pro-apoptotic genes. When the signaling system is defective, the G$_1$/S checkpoint fails and cells may attempt to replicate damaged DNA. This leads to genomic instability, as described above. Cells with defects in the proteins described here seem to be particularly bad at repairing double-strand breaks. It may be that other types of damage do not rely so heavily on detection by this system for their repair. Details of the way in which DNA damage leads to delay of the cell cycle are reviewed by Kastan and Bartek (see Further Reading).

Defects in the repair machinery underlie a variety of cancer-prone genetic disorders

Tumor cells may fail to signal DNA damage, or alternatively the signal may be intact but the cell may have a defect that makes it unable to repair certain types of damage. Any cell that is unable to repair DNA damage should be eliminated by apoptosis, but in a whole population of such cells there must be selective pressure for variants that can escape death. The different types of damage and different repair mechanisms were described in Chapter 13 and are reviewed by Hoeijmakers (see Further Reading). Defects in each are associated with specific forms of cancer:

- **Nucleotide excision repair** is defective in several cancer-prone syndromes, particularly the various forms of xeroderma pigmentosum (XP; OMIM 278700). Patients with XP are homozygous for inherited loss-of-function mutations in one or other of the genes involved in nucleotide excision repair. They are un-able to repair DNA damage caused by ultraviolet radiation. They are exceedingly sensitive to sunlight and develop many tumors on exposed skin.

- **Base excision repair** defects are not common in human cancers, but a survey of patients with colon cancer by Farrington and colleagues (see Further Reading) showed that rare homozygous defects in the MUTYH repair enzyme were associated with a 93-fold increased risk of colon cancer. Overall, these accounted for about 0.5% of all cases of colon cancer.

- **Double-strand break repair** defects primarily affect homologous recombination rather than nonhomologous end joining, although both mechanisms require the ATM–NBS1–BRCA1–BRCA2 machinery mentioned above.

- **Replication error repair** defects cause a large generalized increase in mutation rates. These defects came to light when studies of familial colon cancer revealed the phenomenon of microsatellite instability, described below.

Microsatellite instability was discovered through research on familial colon cancer

Most colon cancer is sporadic. The rare familial cases fall into two main categories:

- **Familial adenomatous polyposis** or adenomatous polyposis coli (FAP or APC; OMIM 175100) is an autosomal dominant condition in which the colon is carpeted with hundreds or thousands of polyps. The polyps (adenomas) are not malignant, but if left in place one or more of them is virtually certain to evolve into invasive carcinoma. The cause is an inherited mutation in the *APC* tumor suppressor gene (see Table 17.4).

- **Hereditary non-polyposis colon cancer** (Lynch syndrome I; OMIM 120435 and 120436) is also autosomal dominant and highly penetrant but, unlike FAP, there is no preceding phase of polyposis. The causative genes were mapped by linkage analysis in affected families to two locations, 2p21 and 3p21.3.

Studies on FAP tumors and the polyps that precede them revealed losses of heterozygosity (LOHs) that led to a pioneering and influential description of the molecular events involved in the multi-stage evolution of colorectal tumors (see below). However, similar LOH studies on the tumors in patients with Lynch syndrome I produced unexpected, counterintuitive results. Rather than lacking alleles present in the constitutional DNA, tumor specimens seemed to contain extra, novel, alleles of the microsatellite markers used (**Figure 17.19**). LOH is a property of particular chromosomal regions, and may reveal the locations of tumor suppressor genes, but the microsatellite instability (MIN) in Lynch syndrome I is

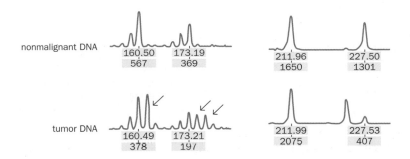

nonmalignant DNA

160.50   173.19
567   369

211.96   227.50
1650   1301

tumor DNA

160.49   173.21
378   197

211.99   227.53
2075   407

Figure 17.19 Microsatellite instability. Electropherograms of two PCR-amplified microsatellite markers. Upper traces: blood DNA. Lower traces: tumor DNA. Note the extra peaks in the tumor DNA. (Courtesy of Lise Hansen, University of Aarhus.)

TABLE 17.5 GENES INVOLVED IN DNA REPLICATION ERROR REPAIR

E. coli	Human	Chromosomal location	Frequency in Lynch syndrome I
MutS	MSH2	2p21	35%
	MSH3	5q14.1	0%[a]
	MSH6	2p16	5%[b]
MutL	MLH1	3p22.2	60%
	MLH3	14q24.3	0%[a]
	PMS2	7p22	few cases reported

[a]Data from only one group, and interpretations are ambiguous. [b]Atypical late-onset Lynch syndrome I and endometrial cancer. For details, see Jiricny J & Nystrom-Lahti M (2000) *Curr. Opin. Genes Dev.* 10, 157–162.

general. Many tumors show occasional instability of one or a few microsatellites, but high-frequency general instability (often defined as affecting 30% or more of all markers tested) defines a class of MIN[+] tumors with distinct clinicopathological features. Microsatellite instability is seen in 10–15% of colorectal, endometrial, and ovarian carcinomas, but only occasionally in other tumors.

In a wonderful example of lateral thinking, Fishel and colleagues in 1993 related the MIN[+] phenomenon to so-called mutator genes in *Escherichia coli* and yeast. These genes encode an error-correction system that checks newly synthesized DNA for mismatched base pairs or small insertion–deletion loops. Mutations in the *MutHLS* genes that encode the *E. coli* system led to a 100–1000-fold general increase in mutation rates. Fishel and colleagues cloned a human homolog of one of these genes, *MutS*, and showed that it mapped to the location on 2p of one of the Lynch syndrome I genes, and was constitutionally mutated in some families with the syndrome. In all, six homologs of the *E. coli* genes have been implicated in human mismatch repair (**Table 17.5** and **Figure 17.20**).

Patients with Lynch syndrome I are constitutionally heterozygous for a loss-of-function mutation, almost always in *MLH1* or *MSH2*. Their normal cells still have a functioning mismatch repair system and do not show the MIN[+] phenotype. In a tumor, the second copy is lost by one of the mechanisms described for retinoblastoma or, at least in the case of *MLH1*, it may be intact but silenced by promoter methylation.

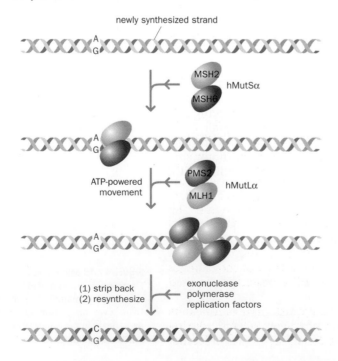

Figure 17.20 Mechanism of mismatch repair. Replication errors can produce mismatched base pairs or small insertion–deletion loops. These are recognized by hMutSα, a dimer of MSH2 and MSH6 proteins, or sometimes by the MSH2—MSH3 dimer hMutSβ. The proteins translocate along the DNA, bind the MLH1–PMS2 dimer hMutLα, then assemble the full repairosome, which strips back and resynthesizes the newly synthesized strand.

17.6 GENOMEWIDE VIEWS OF CANCER

The general move of genetic research away from specific genes and toward genomewide analysis has been particularly marked in cancer research. Several large collaborative projects are generating genomewide data on a huge scale. The goal of the Cancer Genome Anatomy Project (http://cgap.nci.nih.gov/) is 'to determine the gene expression profiles of normal, pre-cancer, and cancer cells, leading eventually to improved detection, diagnosis, and treatment for the patient.' The International Cancer Genome Consortium (http://www.icgc.org/home) has as its goal 'to obtain a comprehensive description of genomic, transcriptomic and epigenomic changes in 50 different tumor types and/or subtypes which are of clinical and societal importance across the globe.'

Cytogenetic and microarray analyses give genomewide views of structural changes

Multicolor FISH (see Figure 17.17) makes it possible to identify all cytogenetically visible chromosomal rearrangements in a tumor cell. Results from this and earlier techniques are accumulated in the Mitelman database (http://cgap.nci.nih.gov/Chromosomes/Mitelman). As at 16 December 2009 this contained results on 57,402 cases. On a finer scale, microarray-based methods (array-CGH and SNP chips) provide genomewide information on copy-number variations. For example, Weir et al. cataloged copy-number variations in 371 primary lung adenocarcinomas by comparing the hybridization intensity ratio between each tumor DNA and a panel of normal DNAs on a high-resolution SNP chip. These methods have superseded studies of loss of heterozygosity at individual gene loci and the use of individual FISH probes, although those still have important roles in diagnosis.

New sequencing technologies allow genomewide surveys of sequence changes

Massively parallel sequencing technologies are enabling researchers to generate unprecedented amounts of sequence data from tumor samples. Studies may be targeted at protein-coding sequences, or they may sequence the entire tumor genome. In the former category, a noteworthy pioneering example was a study by Greenman and colleagues of all the 518 known protein kinase genes in 210 tumors, representing 10 different types of cancer. Genes showing nonsynonymous somatic mutations were sequenced in a further series of 454 tumors. Two independent analyses of glioblastoma multiforme (a deadly brain tumor) reported in autumn 2008 show the increasing scale of work. Parsons and colleagues sequenced all coding exons of 20,661 protein-coding genes in 22 tumor samples, and the Cancer Genome Atlas Research Network sequenced exons of 601 selected genes in matched tumor and normal samples from 91 cases. Both studies supplemented the sequencing by studies of copy-number changes and gene expression. Many other such studies are being reported.

The first whole genome sequence of a tumor cell was reported by Ley and colleagues in 2008. This was of a rather unusual tumor type, an acute myeloid leukemia with no evidence of any acquired chromosomal or copy-number abnormalities, implying that tumor development was driven by small-scale genetic changes. To generate meaningful data, it is important to check each sequence many times (Ley and colleagues used a 32-fold coverage) and to sequence the normal DNA of the same patient so that tumor-specific somatic mutations can be distinguished from the many individual variants present in the patient's normal genome. The tumor genome had more than 3.8 million variants in comparison with the standard human genome reference sequence. The great majority of these were also present in the patient's normal genome, reflecting the typical individuality of normal genomes (as described in Chapter 13). However, 63,000 of the changes were not present in the patient's normal genome, and so represented acquired somatic mutations.

A major problem in all sequencing studies is to distinguish **driver mutations** (which contribute to carcinogenesis and have been positively selected during microevolution of the tumor) from **passenger mutations** that are an incidental consequence of the many cell divisions and genomic instability of cancer cells. Passenger mutations probably greatly outnumber driver mutations. For example, the glioblastoma multiforme study by Parsons et al. identified one or more non-silent coding sequence changes in 685 of the 20,661 genes studied, with a mean of 47 mutations per tumor. Combining data from sequencing (including limited testing of a further 83 tumor samples), copy-number changes, and expression data suggested that 42 of these genes might harbor driver mutations.

These and other similar results clearly show the complexity and heterogeneity of tumorigenesis. Individual tumors have large numbers of mutations and, even among tumors of the same type, there is only limited overlap. To extract significant patterns, the International Cancer Genome Consortium plan to sequence at least 2000 genes in 250 samples each from 50 tumor types, a total of 12,500 samples. The review by Stratton, Campbell, and Futreal gives a good picture of the achievements of this work and the challenges it faces.

Further techniques provide a genomewide view of epigenetic changes in tumors

Genomewide methylation patterns can be investigated by using specific antibodies to precipitate methylated DNA. The sequences present in the immunoprecipitate are identified by hybridization to microarrays or by large-scale sequencing (*ChIP-chip* or *ChIP-seq*; see Chapter 11). Comparison with the methylation pattern in the corresponding normal cells allows the cancer epigenome to be characterized. Array-CGH can be used to perform the comparison directly. Widespread hypomethylation is a feature of almost all tumor cells. Presumably, the result is to activate many genes whose expression is normally repressed by methylation. This is one of the earliest observable changes, and, as discussed in a review by Feinberg, Ohlsson, and Henikoff, some workers speculate that this epigenetic change is central to tumorigenesis.

Superimposed on the general *hypo*methylation is *hyper*methylation of CpG islands in the promoters of specific genes. This can be studied by hybridizing bisulfite-treated DNA to a custom microarray. As described previously (see Box 11.4), sodium bisulfite converts cytosine to uracil but leaves 5-methylcytosine unchanged. Oligonucleotide probes on the microarray hybridize specifically to either the converted or the unconverted sequence. With this technique, the Cancer Genome Atlas Research Network tested 2305 genes for tumor-specific promoter methylation in their study of glioblastoma multiforme described above. This revealed specific methylation at the promoter of *MGMT*, a gene encoding a DNA repair enzyme that removes alkyl groups from guanine residues. Tumors with this feature showed pervasive hypermutation, with a specific spectrum of nucleotide sequence changes.

Tumors in many organs can be divided into those showing methylation of many CpG islands (CpG island methylation phenotype; CIMP$^+$) and those that are CIMP$^-$. It is claimed that CIMP$^+$ cancers form a clinically and etiologically distinct subgroup, although the mechanism and implications of this are not currently understood. The factor underlying these patterns may be activities of the *polycomb* group of chromatin remodeling proteins (see Chapter 11).

Genomewide views of gene expression are used to generate expression signatures

Pathologists have always classified tumors by their appearance under the microscope (see Figure 17.1); however, from the mid-1990s onward, substantial efforts have been devoted to developing classifications based on patterns of gene expression. Most research has used microarrays. An alternative is to sequence total cDNA, either directly or in the form of concatenated short tags (SAGE, for Serial Analysis of Gene Expression; see Chapter 11). Sequencing produces data that are more quantitative, and the precision of quantification can be increased to any desired extent by sequencing more individual molecules.

In a typical experiment, a microarray containing probes for every human gene is used to compare RNA or cDNA from a tumor with RNA or cDNA from the corresponding normal tissue. This might be done directly in a competitive hybridization or by two sequential hybridizations. Alternatively, RNAs or cDNAs from a

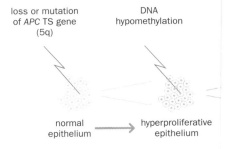

loss or mutation
of *APC* TS gene
(5q)

DNA
hypomethylation

normal
epithelium

hyperproliferative
epithelium

Figure 17.22 A model for the mul
genes is seen at particular histologi
rather than a firm description. Every
underlying genetic changes are mo

- Colorectal carcinomas, but not adenoma
  loss or mutations.

These observations have been summari
of **Figure 17.22**. This scheme should not b
thinking about the multi-stage evolution of
always happens. As the statistics quoted abo
show these changes. Indeed, only about 60
syndrome I tumors follow a different route.

Similar but more rudimentary schemes
which the data are not so rich. The early mu
a tumor are more critical than the later on
ways in which a small number of mutations
wise normal cell that has all its defenses in
stages allows a much wider range of possibi
act. According to the *gatekeeper hypothesis*
renewing cell population one particular ge
constant cell number. Mutation of a gateke
between cell division and cell death, where
long-term effect if the gatekeeper is functio
genes identified by studies of Mendelian ca
for the tissue involved, for example *RB1*
Schwann cells, and *VHL* for kidney cells. S
successive mutations in the gatekeeper—b
blastoma, this is compatible with the norm
relative rarity of each specific sporadic car
prostate or lung cancer do not have clean M
problem for the gatekeeper hypothesis beca
mutation substantially increases the likelih

Interestingly, inherited *APC* mutations a
or nonsense (30%) changes, whereas the s
familial cases, and both mutations in spo
gene fully. This fits well with the gatekeep
FAP remain normal and healthy until one
should be some difference between their i
mutations that are postulated to change c
seems specific to the *APC* gene. It is perha
regulates the level of β-catenin, a critical co
also has a role in ensuring chromosomal s
attachment of spindle microtubules to th
have already noted that even very early ade
It is likely that a single *APC* mutation (of th
advantage on colonic epithelium, wherea
gives no advantage to ductal epithelium. T
very common in sporadic colon cancer b
10–15% of sporadic breast cancers.

series of tumors can be used to search for common patterns or meaningful differences. Meticulous control of the experimental conditions and very careful filtering and processing of the raw data are necessary to produce reproducible results. The processed data are displayed as a matrix in which each column shows data for one sample and each row shows data for one gene. Cells of the matrix are colored to indicate raised, lowered, or unaltered levels of expression.

The challenge is to extract meaningful patterns from the matrix. The object may be to find gene expression signatures—patterns of overexpression or underexpression of a limited number of genes—that are correlated with known clinical variables and may serve to predict that variable in an unknown case. Alternatively, the object may be to identify unexpected between-sample similarities and differences that can define new classes of tumor. Hierarchical clustering methods rearrange the rows and columns of the matrix to maximize similarities.

MicroRNAs (miRNAs), small noncoding RNAs that are thought to have a regulatory role, have also been targets of genomewide expression profiling. The profiles show characteristic changes in different tumor types. Most miRNAs are downregulated in most tumors. This suggests a general dedifferentiation of the cells, a reversion to a more embryonic state. MiRNA expression tends to increase as cells differentiate.

Superimposed on an overall decrease in miRNA expression, cancer cells show abnormal miRNA expression profiles. Lu and colleagues reported profiles of 217 miRNAs in 334 samples of normal human tissue and various tumors (see Further Reading). The profiles were remarkably diverse and performed better than mRNA profiles at discriminating between different tumor types, and between tumor and normal cells. Other researchers showed that overexpression of one miRNA, miR-10b, in nonmetastatic breast tumors triggered invasion. Clinical progression of primary breast tumors correlated with miR-10b levels. Other miRNAs (miR-126 and miR-335) had the opposite effect: metastatic breast cancer cells lose expression of these two miRNAs, and restoring their expression decreased the metastatic potential of the cells.

The data are currently confusing, but some individual miRNA genes do fulfill the criteria for oncogenes or tumor suppressor genes. This may be because they control the expression of one of the well-known classical oncogenes or tumor suppressor genes, or it might reflect some independent action of the miRNA.

Expression profiling has been widely and successfully used for identifying distinct categories of tumor (class discovery). **Figure 17.21** shows typical examples. The eventual aim must be to develop prognostic and predictive indicators. Prognostic indicators would predict the likely survival of the patient, regardless of treatment. Predictive indicators would predict which treatments would be most likely to succeed with that particular tumor. How far expression profiling has moved toward such direct clinical relevance is discussed in Chapter 19.

17.7 UNRAVELING THE MULTI-STAGE EVOLUTION OF A TUMOR

As emphasized at the start of this chapter, all cancers are the result of a microevolutionary process that progresses through a series of stages. A full understanding of any cancer requires a description of the successive stages in its development. Our understanding of tumor evolution is best developed for colorectal cancers, because all the stages of tumor development can be studied in resected colon from patients with familial adenomatous polyposis (FAP).

The microevolution of colorectal cancer has been particularly well documented

The earliest lesions in otherwise normal colonic epithelium are microscopic aberrant crypt foci. These may develop into benign epithelial growths called polyps or adenomas. Adenomas evolve through early (less than 1 cm in size), intermediate (larger than 1 cm but without foci of carcinoma), and late (larger than 1 cm and with foci of carcinoma) stages, to become carcinomas, which eventually metastasize. For each stage, a sample of lesions can be used to assemble a picture of the molecular changes characteristic of that stage. Some changes are

(A)

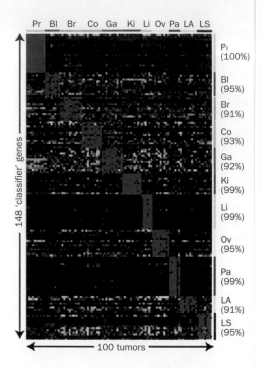

Figure 17.21 Examples of the use patterns of 148 genes (rows) classif gastroesophagus; Ki, kidney; Li, live indicates increased gene expressio Clustering of gene expression distir Five-year survival in the two group 6817 genes in 60 tumor cell lines w 30 cell lines (columns) that predict cancer. Expression patterns of 25,0 marker genes (rows) in 78 cancers (those predicted to require addition [(A) from Su AI, Welsh JB, Sapinoso for Cancer Research. (B) from Alizac Publishers Ltd. (C) from Staunton J from the National Academy of Scie permission from Macmillan Publish

frequent in very early stage lesions, wherea principal observations are as follows:

- The earliest detectable lesions, aberrant people with FAP, who have one inheritec every cell, about 1 epithelial cell in 10^6 d sistent with loss of the second *APC* allele

- Early stage adenomas, like most cancer of CpG sequences in the DNA. Despit islands in the promoters of specific gene previously.

- About 50% of intermediate and late ade adenomas, have mutations in the *KRAS* often be required for progression from e

- About 50% of late adenomas and carcino 18q. This is relatively uncommon in e seems likely that the relevant gene is *SM DCC.*

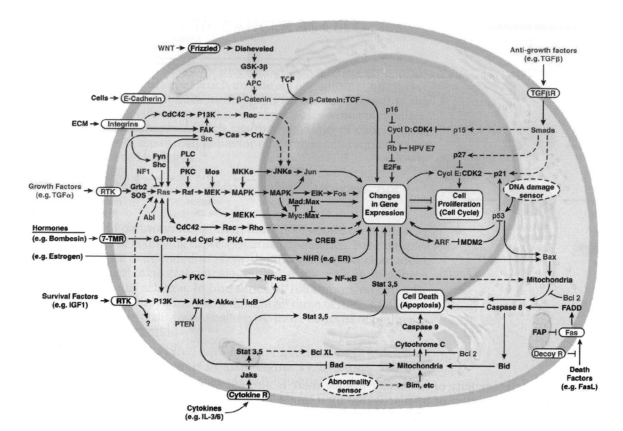

Systems biology may eventually allow a unified overview of tumor development

Each of the six capabilities listed by Hanahan and Weinberg may be acquired by a variety of different genetic changes, but in each case acquiring it requires certain specific pathways to be activated or inactivated. **Figure 17.25**, taken from the review by Hanahan and Weinberg, shows some of the target pathways. The capabilities are not necessarily acquired in the same order in different tumors, but the requirements for genomic instability and successive clonal expansions at intermediate stages impose a certain regularity on the process.

All these pathways are used to steer a cell in certain directions in particular circumstances (**Figure 17.26A**). Life would be very simple if signal and response were connected by a single unbranched linear pathway (Figure 17.26B) but this seems never to be the case. Rather, multiple branching, overlapping, and partially redundant pathways control the behavior of cells (Figure 17.26C). Such complicated networks are probably necessary to confer stability and resilience on the extraordinarily complex machinery of a cell. Experimentally, unraveling the precise genetic circuitry of the controls is exceedingly difficult, partly because of their complexity and partly because it is difficult to distinguish direct from indirect effects in transfection or knockout experiments. The science of systems biology is devoted to quantitative modeling of such networks, and one of its main applications is in cancer. When a full quantitative description of the control network in each different cell type has become available, we will be in a position to truly understand cancer.

CONCLUSION

Cancer is the result of a Darwinian microevolutionary process among the cells of our body. Just as in the evolution of a population of animals, individual cells acquire mutations at random, and natural selection favors any cell containing mutations that allow it to reproduce faster or resist death better than its neighbors. To become a cancer cell, a cell must be able to continue replicating, independently of external growth and anti-growth signals. This requires several successive mutations to disable multiple regulatory pathways.

Figure 17.25 Pathways important for the acquisition of cancer cell capabilities. This figure is an initial attempt to show the involvement of oncogenes and tumor suppressor genes (shown in red) in many pathways that affect a cell's choice of progressing through the cell cycle, exiting from the cycle, or undergoing apoptosis. Many specific examples are discussed by Hanahan and Weinberg, from whose paper this figure is taken. RTK, receptor tyrosine kinase; 7-TMR, 7-transmembrane segment cell surface receptor; ECM, extracellular matrix. [From Hanahan D & Weinberg RA (2000) *Cell* 100, 57–70. With permission from Elsevier.]

Genes that normally act to stimulate growth suffer activating mutations and are known as oncogenes; in contrast, those that normally act to restrain growth suffer inactivating mutations and are known as tumor suppressor genes. Genomic instability allows the more rapid accumulation of subsequent mutations. These consist of a small number of driver mutations and a much larger number of passenger mutations. Driver mutations contribute to the development of the tumor and are positively selected; passenger mutations are incidental by-products of tumor development. Through many rounds of ill-controlled replication with high mutation rates, tumors come to contain a heterogeneous collection of clones, probably maintained by a small population of cancer stem cells.

Whereas the evolution of a population of organisms is brought about by natural selection of germ-line mutations, the microevolution of a tumor depends on selection of somatic mutations. Nevertheless, certain predisposing mutations may be passed through the generations in the germ line, causing the various rare familial cancer syndromes. This can happen because mutations in most tumor suppressor genes have no phenotypic effect until both alleles are inactivated (as Knudson put it, they must suffer two hits). A single inactivating mutation can be transmitted through the germ line, so that it is present in every cell of the offspring. Such cells still function normally but are vulnerable to a single hit that inactivates the remaining functional allele. This is a much more frequent occurrence than two hits happening by chance on a single normal cell, so the result is an inherited high risk of cancer.

Every tumor is a unique product of a separate microevolutionary process, and contains a unique spectrum of driver and passenger mutations. However, all tumors need to have acquired a set of specific capabilities: in addition to independence from external growth and anti-growth signals, tumors need to avoid apoptosis, replicate indefinitely, develop their own blood supply (angiogenesis), and, in malignant tumors, invade other tissues to establish secondary tumors. These capabilities depend on subverting the same network of signaling pathways in a given cell type. Thus, considered in terms of pathways, tumors are much less heterogeneous than when the individual gene mutations they contain are listed. Understanding cancer requires a quantitative and genomewide view of how the signaling network functions in normal cells of each particular type. Only when we have obtained such a view will we then understand how specific genetic or epigenetic changes can destabilize the network and allow a tumor to develop.

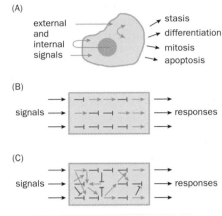

Figure 17.26 The options open to a cell, and how it chooses. (A) In response to internal and external signals, a cell chooses between stasis, mitosis, apoptosis, and sometimes differentiation. (B) An imaginary cell in which signals are linked to responses by linear unbranched pathways of stimulation (→) or inhibition (—|). Human cells do not function like this. (C) In real cells, signals feed into a complex network of partly redundant interactions whose outcome is not easy to predict analytically.

FURTHER READING

The evolution of cancer

Hanahan D & Weinberg RA (2000) The hallmarks of cancer. *Cell* 100, 57–70. [An important review, introducing the concept of the six essential capabilities of a tumor cell and giving many examples.]

Rosen JM & Jordan CT (2009) The increasing complexity of the cancer stem cell paradigm. *Science* 324, 1670–1673. [A review of the controversies and complexities surrounding the concept of cancer stem cells.]

Oncogenes

Bishop JM (1983) Cellular oncogenes and retroviruses. *Annu. Rev. Biochem.* 52, 301–354. [A nice review of early work on oncogenes.]

Boxer LM & Dang CV (2001) Translocations involving c-*myc* and c-*myc* function. *Oncogene* 20, 5595–5610. [An insight into the complexities of the Burkitt lymphoma and other *MYC* translocations.]

Cancer Gene Census. http://www.sanger.ac.uk/genetics/CGP/ Census [A database of genes for which mutations have been causally implicated in cancer.]

Meyer N & Penn LZ (2008) Reflecting on 25 years with MYC. *Nat. Rev. Cancer* 8, 976–990. [A historical review illustrating the many roles of one of the major oncogenes.]

Mitelman F, Johansson B & Mertens F (eds) (2002) Mitelman Database of Chromosome Aberrations in Cancer. http://cgap. nci.nih.gov/Chromosomes/Mitelman

Mitelman F, Johansson B & Mertens F (2007) The impact of translocations and gene fusions on cancer causation. *Nat. Rev. Cancer* 7, 233–244.

Tumor suppressor genes

Burkhart DL & Sage J (2008) Cellular mechanisms of tumour suppression by the retinoblastoma gene. *Nat. Rev. Cancer* 8, 671–682. [An in-depth review of the functions of the pRb protein.]

Cavenee WK, Dryja TP, Phillips RA et al. (1983) Expression of recessive alleles by chromosomal mechanisms in retinoblastoma. *Nature* 305, 779–784. [The paper that established Knudson's two-hit hypothesis. Some of the data are in an earlier paper, Godbout R, Dryja TP, Squire J et al. (1983) Somatic inactivation of genes on chromosome 13 is a common event in retinoblastoma. *Nature* 304, 451–453.]

Fodde R & Smits R (2002) One-hit carcinogenesis. *Science* 298, 761–763. [Examples of haploinsufficient tumor suppressor genes.]

Knudson AG (2001) Two genetic hits (more or less) to cancer. *Nat. Rev. Cancer* 1, 157–162. [A review of his two-hit hypothesis.]

Thiagalingam S, Laken S, Willson JK et al. (2001) Mechanisms underlying losses of heterozygosity in human colorectal cancers. *Proc. Natl Acad. Sci. USA* 98, 2698–2702. [A nice study using DNA markers and FISH to identify the mechanisms of loss in 62 tumors. The mechanisms often differed from those described by Cavenee and colleagues in retinoblastoma.]

Cell cycle dysregulation in cancer

Kastan MB & Bartek J (2004) Cell cycle checkpoints and cancer. *Nature* 432, 316–323. [One of a cluster of eight reviews of cell division and cancer in this issue of *Nature*. This review focuses on the response to DNA damage.]

Kops GJPL, Weaver BAA & Cleveland DW (2005) On the road to cancer: aneuploidy and the mitotic checkpoint. *Nat. Rev. Cancer* 5, 773–785. [Gives details of the mitotic checkpoint mechanism.]

Malumbres M & Barbacid M (2005) Mammalian cyclin-dependent kinases. *Trends Biochem. Sci.* 30, 630–641. [A general review of the cyclin-dependent kinases and their functions.]

Malumbres M & Barbacid M (2009) Cell cycle, CDKs and cancer: a changing paradigm. *Nat. Rev. Cancer* 9, 153–166. [Suggests that Cdk1 is the only generally essential cyclin-dependent kinase; others may have specialized roles in certain cell types.]

Massagué J (2004) G1 cell-cycle control and cancer. *Nature* 432, 298–306. [A detailed review of mechanisms controlling progression through G_1 and across the G_1/S threshold.]

Instability of the genome

Brennan P, McKay J, Lee M et al. (2007) Uncommon *CHEK2* missense variant and reduced risk of tobacco-related cancers: case-control study. *Hum. Mol. Genet.* 16, 1794–1801. [An interesting study and worth reading for the clear discussion of this unexpected finding—but not a comfort blanket for smokers!]

Farrington SM, Tenesa A, Barnetson R et al. (2005) Germ line susceptibility to colorectal cancer due to base-excision repair gene defects. *Am. J. Hum. Genet.* 77, 112–119.

Fishel R, Lescoe MK, Rao MRS et al. (1993) The human mutator gene homolog MSH2 and its association with hereditary nonpolyposis colon cancer. *Cell* 75, 1027–1038. [Identifying *MSH2* mutations as a cause of human mismatch repair defects.]

Hoeijmakers J (2009) DNA damage, aging and cancer. *New Engl. J. Med.* 361, 1475–1485.

Genomewide views of cancer

Cancer Genome Anatomy Project. http://cgap.nci.nih.gov/ [A project whose goal is 'to determine the gene expression profiles of normal, precancer, and cancer cells, leading eventually to improved detection, diagnosis, and treatment for the patient.']

Cancer Genome Atlas Research Network (2008) Comprehensive genomic characterization defines human glioblastoma genes and core pathways. *Nature* 455, 1061–1068.

Feinberg AP, Ohlsson R & Henikoff S (2006) The epigenetic progenitor origin of human cancer. *Nat. Rev. Genet.* 7, 21–32. [Argues for an early epigenetic event being crucial to tumorigenesis.]

Greenman C, Stephens P, Smith R et al. (2007) Patterns of somatic mutation in human cancer genomes. *Nature* 446, 153–158. [Sequence analysis of 518 protein tyrosine kinase genes in 210 diverse human cancers.]

International Cancer Genome Consortium. http://www.icgc.org/home [A project whose goal is to obtain a comprehensive description of genomic, transcriptomic and epigenomic changes in 50 different tumor types and/or subtypes which are of clinical and societal importance across the globe.]

Ley TJ, Mardis ER, Ding L et al. (2008) DNA sequencing of a cytogenetically normal acute myeloid leukaemia genome. *Nature* 456, 66–72. [The first full sequence of a tumor genome.]

Lu J, Getz G, Miska E et al. (2005) MicroRNA expression profiles classify human cancers. *Nature* 435, 834–838.

Ma L, Teruya-Feldstein J & Weinberg RA (2007) Tumor invasion and metastasis initiated by microRNA-10b in breast cancer. *Nature* 449, 682–688.

Parsons DW, Jones S, Zhang X et al. (2008) An integrated genomic analysis of human glioblastoma multiforme. *Science* 321, 1807–1812.

Quackenbush J (2006) Microarray analysis and tumor classification. *N. Engl. J. Med.* 354, 2463–2472. [A clear and readable account of how microarray data are processed and interpreted.]

Stratton MR, Campbell PJ & Futreal PA (2009) The cancer genome. *Nature* 458, 719–724. [A review of achievements and challenges in obtaining genomewide views of cancer.]

Tavasoie SF, Alarcón C, Oskarsson T et al. (2008) Endogenous human microRNAs that suppress breast cancer metastasis. *Nature* 451, 147–152.

Weir BA, Woo MS, Getz G et al. (2007) Characterizing the cancer genome in lung adenocarcinoma. *Nature* 450, 893–898. [Identifying copy-number variations in 371 tumors.]

Unraveling the multi-stage evolution of a tumor

Fodde R, Smits R & Clevers H (2001) *APC*, signal transduction and genetic instability in colorectal cancer. *Nat. Rev. Cancer* 1, 55–67. [A detailed look at APC function and the surprising regularities in the types of mutation found.]

Kinzler KW & Vogelstein B (1996) Lessons from hereditary colorectal cancer. *Cell* 87, 159–170. [A review of their important early work on the development of colorectal cancer, and an introduction to the gatekeeper hypothesis.]

Integrating the data: cancer as cell biology

Hahn WC & Weinberg RA (2002) Modelling the molecular circuitry of cancer. *Nat. Rev. Cancer* 2, 331–340. [An attempt to define a road map of carcinogenesis.]

Nguyen DX & Massagué J (2007) Genetic determinants of cancer metastasis. *Nat. Rev. Genet.* 8, 341–352. [A detailed discussion of how and when a tumor acquires metastatic potential.]

Rajagopalan H, Bardelli A, Lengauer C et al. (2002) RAF/RAS oncogenes and mismatch-repair status. *Nature* 418, 934. [Showing that *RAS* and *BRAF* mutations are alternative ways of achieving the same thing in tumorigenesis.]

Smith G, Carey FA, Beattie J et al. (2002) Mutations in APC, Kirsten-ras, and p53 –alternative genetic pathways to colorectal cancer. *Proc. Natl Acad. Sci. USA* 99, 9433–9438. [This study highlights the heterogeneity of molecular events in the evolution of colorectal cancer.]

Vogelstein B & Kinzler KW (2004) Cancer genes and the pathways they control. *Nat. Med.* 10, 789–799. [A good overview of pathways in tumor development.]

Chapter 18

Genetic Testing of Individuals

18

KEY CONCEPTS

- In considering what method is appropriate for a genetic test, it is important to distinguish cases in which it is necessary to scan one or more whole genes for any mutation that might be anywhere in their sequence from cases in which we wish to check for a specific predefined sequence change.

- Testing can be done with either DNA or RNA. RNA must be obtained from a tissue in which the gene in question is expressed, and such samples need much more careful handling than DNA samples. However, RNA analysis may be more economical for a gene that has many small exons, and it can reveal abnormal splicing that may not be apparent from DNA testing.

- The first step in testing is almost always to amplify the relevant sequences by using PCR, or RT-PCR if RNA is being analyzed.

- In scanning a gene for any possible mutation, the normal procedure is to sequence the DNA, exon by exon, or to sequence an RT-PCR product.

- There are several methods for quickly checking each exon of a multi-exon gene to see whether it is likely to harbor any sequence variant. Their purpose is to reduce the amount of sequencing that is necessary to detect a mutation. These methods usually exploit differences in properties of DNA heteroduplexes, or the conformation of single-stranded DNA under non-denaturing conditions.

- The protein truncation test can be used to scan a gene for mutations that introduce a premature termination codon.

- Novel missense or synonymous changes identified while scanning a candidate gene for pathogenic mutations pose great problems for interpretation. Many approaches are used to try to decide whether or not a change is pathogenic, but if these fail to give a definitive answer, a change has to be listed as an unclassified variant.

- Many methods exist for testing for a predefined sequence change. They often involve allele-specific hybridization, allele-specific DNA ligation, or extending a primer across the site in question. These methods are also used for genotyping single nucleotide polymorphisms.

- Special techniques to identify deletions or other copy-number changes include multiplex ligation-dependent probe amplification (MLPA; used to detect the duplication or deletion of one or a few exons) and comparative genomic hybridization (CGH; used to detect larger changes).

- Gene tracking is sometimes used in family studies to discover whether somebody has inherited a pathogenic allele, in cases in which it is impracticable or undesirable to check directly for the mutation.

- Genotypes at a limited number of highly polymorphic microsatellites can be used to produce a profile of a DNA sample. DNA profiling can be used to determine the zygosity of twins or the paternity of a child, or in criminal investigations to identify the source of DNA recovered from a crime scene.

Genetic testing is a routine tool for almost all biomedical scientists. PCR-based tests are the standard way of identifying pathogens and are the everyday working tools of microbiologists and virologists. Emerging diseases are identified by sequencing the pathogen's genome and are tracked by genetic tests. Animal and plant breeders use genetic tests to guide their work. Our modern understanding of evolution and development is based on genetic tests. However, this being a book on human molecular genetics, we will confine ourselves here to tests focused on the human genome. Even here the field is very extensive. Previous chapters have covered many applications of genetic tests in research, for example to map disease genes or susceptibility factors, or to identify the cause of a disease. Here we consider the methodology of testing in human diagnostics and forensics. Many of the principles have already been covered in Chapter 7.

When considering diagnostic tests, wherever possible we will use two of the most common Mendelian diseases, cystic fibrosis (CF) and Duchenne muscular dystrophy (DMD), to illustrate the various testing methods. Both CF and DMD involve large genes with extensive allelic heterogeneity, but beyond that they pose rather different sets of problems for DNA diagnosis (Table 18.1). Between them, they show many of the issues involved in testing for Mendelian diseases. As always in this book, we concentrate on the principles and not the practical details. Best practice guidelines for laboratory diagnosis of the commoner Mendelian diseases are available on the websites of the American College of Medical Genetics and the UK Clinical Molecular Genetics Society, among others. The reader interested in specific procedures or conditions should consult these (see Further Reading).

Genetic tests are unusual among clinical tests because a genetic test is normally performed just once and the result forms a permanent part of a person's health record. Thus, it is especially important to avoid mistakes in diagnostic testing. Any laboratory offering clinical (or forensic) testing must have adequate quality assurance measures in place. These are beyond the scope of this chapter—any interested reader should consult the principles and guidelines published by the Organisation for Economic Co-operation and Development (OECD) (see Further Reading).

When a clinician brings a sample of a patient's DNA to a laboratory for diagnostic testing, there are three possible questions that the laboratory might be asked to answer:

- Does the patient have *any* mutation in *any* gene that would explain the disease? This is an unreasonable question to ask for routine diagnosis, even if it might very occasionally be possible in a research context. Gene testing has to be targeted. Even though the day may come when a person's entire genome sequence forms part of his or her basic health record, we would still need to know where to look among the several million specific variants present in a person's DNA. At best it might be reasonable to ask a laboratory to check

TABLE 18.1 THE CONTRASTING GENETICS OF CYSTIC FIBROSIS AND DUCHENNE MUSCULAR DYSTROPHY		
Condition	Cystic fibrosis	Duchenne muscular dystrophy
Mode of inheritance	autosomal recessive	X-linked recessive
Type of pathogenic mutation	loss-of-function mutations	loss-of-function mutations
Characteristics of gene	fairly large gene: 250 kb genomic DNA, 27 exons, 6.5 kb mRNA	giant gene: 2400 kb genomic DNA, 79 exons, 14 kb mRNA
Spectrum of mutations	almost all mutations are single-nucleotide changes; a few large deletions	65% of mutations are deletions encompassing one or more complete exons; 5% duplications; 30% nonsense, splice site, or small frameshifting mutations; missense mutations are very unusual
Frequency of new mutations	new mutations are extremely rare	new mutations are very frequent
Frequency of mosaicism	mosaicism is not a problem	mosaicism (see Chapter 3, p. 77) is common
Frequency of intragenic recombination	little intragenic recombination	recombination hotspot (12% between markers at either end of the gene)

maybe 100 candidate genes—and it is quite speculative to suppose that even this could become a routine diagnostic procedure. It is, however, practical to ask whether the patient's condition might be explained by a large-scale deletion or duplication anywhere in the genome, as described in Section 18.4.

- Does the patient have *any mutation* in *this particular gene* that might cause the disease? Standard ways of answering this question are considered in Section 18.2, and an alternative approach, gene tracking, is considered in Section 18.5.

- Does the patient have *a specific mutation* in *this particular gene*, for example, a three-base deletion of the codon for phenylalanine 508 in his or her *CFTR* gene? The circumstances in which this type of very specific question can be asked, and ways of answering it, are considered in Section 18.3.

The question might be asked not by a clinician about somebody's diagnosis or prognosis but by a policeman or lawyer about somebody's identity or the relationship between two people. Such questions are addressed by DNA profiling, which is described in Section 18.6. DNA profiling is also occasionally needed to answer clinical questions.

18.1 WHAT TO TEST AND WHY

DNA for genetic testing can be obtained from any sample containing nucleated cells, but clinical considerations may dictate the choice of sample. Sometimes it is more appropriate to test RNA or to perform a functional test, for example of enzyme activity; in those cases the sample must come from a tissue in which the gene is expressed and/or the gene product is normally functional.

Many different types of sample can be used for genetic testing

Genetic testing almost always begins with amplification of the DNA or RNA sample by PCR, applying the methods described in Chapter 6. The sensitivity of PCR makes it possible to use a wide range of tissue samples (Table 18.2). Blood is the most reliable general source of genomic DNA, but mouthwashes or buccal scrapes are used when sampling must be noninvasive. For prenatal diagnosis, chorionic villi or amniotic fluid are normally used. However, the biopsy procedure is invasive and carries a roughly 1% risk of causing a miscarriage. Many

TABLE 18.2 SOURCES OF MATERIAL FOR GENETIC TESTING	
Source	**Comments**
Peripheral blood	the best general source of DNA
Mouthwash or buccal scrape	noninvasive; usually enough DNA for several dozen PCR tests, but quality and quantity very variable
Skin, muscle, etc.	for RNA studies; needs to be a tissue where the gene of interest is expressed; requires a biopsy, which is invasive and often unpleasant
Hair roots, semen, cigarette butts, etc.	scene-of-crime samples for forensic analysis
Single cell from a blastocyst	for pre-implantation diagnosis; technically very demanding
Chorionic villi	for prenatal diagnosis at 9–14 weeks
Amniotic fluid	a relatively poor source of fetal DNA compared with chorionic villi, but used for testing at 15–20 weeks of pregnancy
Fetal DNA in maternal blood	a promising alternative to chorionic villi; detectable from 6 weeks; paternal alleles only
Pathological specimens	vital resource for genotyping dead people; also tumor, etc., biopsies; fixed paraffin-embedded tissue requires special procedures for DNA purification
Guthrie card	the cards used for neonatal screening have one or more spots of the baby's blood, not all of which are normally used for the screening; if cards are archived they are a possible source of DNA from a deceased baby

attempts have been made to use fetal cells in the maternal circulation for less invasive prenatal diagnosis. Unfortunately, the cells are present only in tiny numbers, and the results have never been reliable enough for routine use. Free fetal DNA is more promising, although only tests for paternally derived alleles are informative, because the fetal DNA comprises only maybe 3% of the free DNA in the maternal blood circulation; the rest is maternal. Archived pathological samples are very important sources of DNA from deceased persons (although consent laws in some countries make it difficult to use them). When samples have been formalin-fixed, special procedures are necessary, and only short sequences, typically 200 base pairs or less, can usually be recovered.

Southern blotting (see Figure 7.10) is still necessary for a few applications, including as an alternative to FISH (see Chapter 2, p. 47) for testing for major balanced rearrangements that would not be picked up by array-CGH (see Chapter 2, p. 49) and, in some circumstances, for fragile X full mutations (see below).

RNA or DNA?

If a gene has to be scanned for unknown mutations, testing RNA by reverse transcriptase PCR (RT-PCR; see Chapter 8) offers several advantages. Direct testing of DNA usually involves amplifying and testing each exon separately, and this can be a major chore in a gene with many exons. A gene with 15 exons that encodes a 2.5 kb transcript can be scanned by RT-PCR in perhaps four overlapping segments, whereas the same analysis on genomic DNA would require 15 PCR reactions and 30 sequencing runs (assuming that both strands of each PCR product are sequenced, which would be normal practice). In addition, only RT-PCR can reliably detect aberrant splicing. It is often hard to predict whether a DNA sequence change will affect splicing; moreover, aberrant splicing may result from activation of a cryptic splice site deep within an intron, which would not be detected by the usual exon-by-exon protocols for sequencing genomic DNA (see Chapter 8).

However, analyzing RNA has disadvantages. RNA is much less convenient to obtain and work with. Samples must be handled and processed with extreme care to avoid degrading mRNA. Importantly, the gene of interest may not be expressed in any readily accessible tissue. A very low level of illegitimate transcripts may sometimes be detectable in tissues in which the gene is not normally expressed, but reliable analysis of very low-level mRNAs can be difficult. In addition, truncating mutations usually result in unstable mRNA because of nonsense-mediated decay (see Chapter 13), so that the RT-PCR product from a heterozygous person may show only the normal allele. Treating cultured cells or whole blood with the translation inhibitor puromycin has been shown to inhibit nonsense-mediated decay, which may allow cDNA sequencing to detect transcripts that include premature termination codons.

Functional assays

When considering pedigree patterns in Chapter 3, we simply distinguished two classes of allele, normal and abnormal, or *A* and *a*. Assaying the function of a gene might allow a similar distinction to be made in the laboratory—which is, after all, the essential question in most diagnoses. Function might be tested at the DNA level, for example by a cell transfection assay. Alternatively, the protein product of a gene could be tested biochemically if its function can be adequately assayed in the laboratory. The problem with functional assays is that they are specific to a particular gene or protein; by contrast, DNA technology is generic. This has obvious advantages for the diagnostic laboratory, but in addition it encourages technical development, because any new technique could be used for a wide variety of problems.

18.2 SCANNING A GENE FOR MUTATIONS

For the great majority of diseases, as with CF and DMD, there is extensive *allelic heterogeneity* (see Chapter 13, p. 429). Diagnostic testing therefore usually involves searching for mutations that might be anywhere within or near the relevant gene or genes.

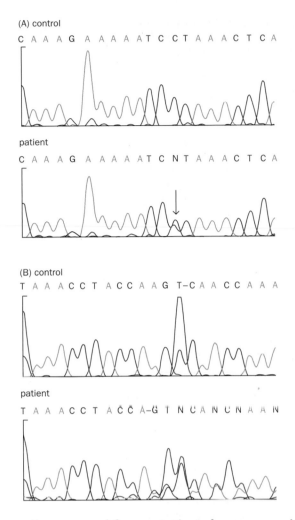

(A) control

C A A A G A A A A A T C C T A A A C T C A

patient

C A A A G A A A A A T C N T A A A C T C A

(B) control

T A A A C C T A C C A A G T–C A A C C A A A

patient

T A A A C C T A C C A–G T N C A N C N A A N

Figure 18.1 Detecting mutations in the *CFTR* gene by dideoxy sequencing. The illustrations show traces from a fluorescence sequencer. The nucleotide sequences above the traces (base calls) are produced automatically by the analysis software, but require manual checking before a sequence is reported. N means the software was unable to identify which base was present at that position. (A) A base substitution in exon 3. The double peak (arrow) shows a heterozygous mutation g.332C>T (p.P67L). (B) A single base deletion c.3659delC in exon 19. The sequence downstream of the deletion is confused, reflecting the overlapping sequences of the two alleles in this heterozygote. The change would be confirmed by sequencing the reverse strand. (Courtesy of Andrew Wallace, St Mary's Hospital, Manchester.)

A gene is normally scanned for mutations by sequencing

Increasingly, sequencing is the method of choice for mutation scanning: either sequencing each exon in genomic DNA or sequencing an RT-PCR product (**Figure 18.1**). The advent of ultra-fast alternatives to dideoxy (Sanger) sequencing (Chapter 8) can only accelerate this trend. Such testing is referred to as *resequencing*, meaning that one is checking a sequence against a known normal (or wild-type) sequence. Sequencing is the gold standard for mutation detection, but the DNA quality has to be high to produce unambiguous data. Problems can arise, especially when using tumor or archive material as DNA sources.

A PCR product containing a single exon plus a dozen or so base pairs of flanking intron would typically be less than 200 bp long, whereas a single dideoxy sequencing run on a standard capillary electrophoresis machine can produce 500–800 bp of sequence. It would be desirable to sequence more than one exon in a single sequencer run, but most introns are too big to allow this to be done directly from genomic DNA. An ingenious solution to this problem uses special PCR primers in a multiplexed reaction to produce concatenated amplicons. Primers for different exons are designed to carry complementary pairs of 5′ extensions. Because of their complementary ends, the PCR products (*amplicons*) anneal together to form *concatemers*; a second round of PCR using primers located at either end of the concatenated products will produce a single molecule containing several exons. Amplicons can be linked in any order to form a product of optimal size for sequencing. However, this meta-PCR method requires extremely careful primer design and optimization; the necessary development effort is only worthwhile for an analysis that will be performed very many times.

Sequencing generates more data than other methods, so the requirement for analysis is greater. In fact, analysis has become the bottleneck in sequence-based mutation scanning, because it needs to be performed carefully; before a clinical

TABLE 18.3 SOME METHODS FOR SCANNING A GENE FOR MUTATIONS BEFORE SEQUENCING

Method	Advantages	Disadvantages
Heteroduplex gel mobility	very simple to set up; cheap and simple to run	sequences <200 bp only, limited sensitivity; does not reveal position of change
dHPLC	quick, high throughput; quantitative	expensive equipment; does not reveal position of change
SSCP	simple, cheap; sensitive when formatted for capillary electrophoresis	sequences <200 bp only; does not reveal position of change
DGGE	high sensitivity	choice of primers is critical; expensive GC-clamped primers; does not reveal position of change
Melting curve analysis	high sensitivity	proprietary system using a Light-Cycler® machine
Protein truncation test	high sensitivity for chain-terminating mutations; shows position of change	chain-terminating mutations only; expensive, difficult technique; usually needs RNA

dHPLC, denaturing high-performance liquid chromatography; SSCP, single-strand conformation polymorphism; DGGE, denaturing gradient gel electrophoresis.

report is issued the data must be checked by two separate individuals, and the likely significance of any variant must be considered. Programs are available that automatically report differences between the test sequence and a standard sequence, provided that the sequence is of good quality. DNA quality is critical for avoiding artifacts and for reliably detecting base substitutions in heterozygotes.

A variety of techniques have been used to scan a gene rapidly for possible mutations

In the past, when sequencing was relatively expensive, various methods were devised to decrease the sequencing load by scanning exons quickly and cheaply to define which amplicons might contain features of interest. They are falling out of use in diagnostic laboratories, but they are briefly described below because they are by no means completely obsolete (Table 18.3).

Scanning methods based on detecting mismatches or heteroduplexes

Many tests use the properties of heteroduplexes to detect differences between two sequences. Most mutations occur in heterozygous form; even with autosomal recessive conditions, affected people born to non-consanguineous parents are often *compound heterozygotes*, with two different mutations. Heteroduplexes can be formed simply by heating the PCR product to denature it, and then cooling it slowly. For homozygous mutations, or X-linked mutations in males, it would be necessary to add some reference wild-type DNA. Several properties of heteroduplexes can be exploited:

- Heteroduplexes often have abnormal mobility on non-denaturing polyacrylamide gels (**Figure 18.2A**, lower panel). This is a particularly simple method to set up and use. If fragments no more than 200 bp long are tested, insertions, deletions, and most but not all single-base substitutions are detectable.

- Heteroduplexes have abnormal denaturing profiles. This is exploited in *denaturing high-performance liquid chromatography* (dHPLC, **Figure 18.3A**) and *denaturing gradient gel electrophoresis* (DGGE, Figure 18.2B). In both cases, the mobility of a fragment changes markedly when it denatures. *Melting curve analysis* studies the same phenomenon through changes in fluorescence of the test DNA hybridized to labeled probes. These methods require tailoring to the particular DNA sequence under test, and so are best suited to routine analysis of a given fragment in many samples. dHPLC allows high throughput, which makes it especially suitable for genotyping a large collection of samples for a small panel of single nucleotide polymorphisms. DGGE requires special primers with a 5′ extension of 30–50 nucleotides rich in G and C (a GC clamp). Once optimized, these methods have high sensitivity.

(A)

(B)

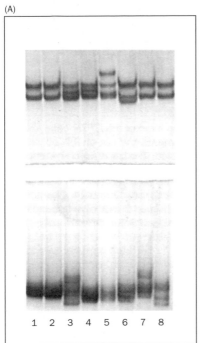

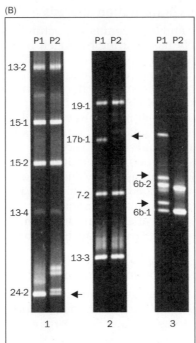

Figure 18.2 Scanning the *CFTR* gene for mutations. (A) Heteroduplex and SSCP analysis. Exon 3 was PCR amplified from the genomic DNA of eight unrelated CF patients. After denaturation and snap-cooling, the samples were loaded on a non-denaturing polyacrylamide gel. Some of each product re-annealed to give double-stranded DNA. This runs faster in the gel and gives the heteroduplex bands seen in the lower panel. The single-stranded DNA runs more slowly in the same gel (upper panel). Lanes 1 and 2 show the pattern typical of the wild-type sequence; the mutations in these patients must be elsewhere in the gene. Variant patterns can be seen in lanes 3–8; sequencing revealed mutations p.G85E, p.L88S, p.R75X, p.P67L, p.E60X, and p.R75Q, respectively.
(B) Denaturing gradient gel electrophoresis. Exons of the *CFTR* gene from the genomic DNA of subjects P1 and P2, two unrelated patients with CF, are PCR amplified in one or more segments and run on 9% polyacrylamide gels containing a gradient of urea-formaldehyde denaturant. The band from any amplicon that contains a heterozygous variant usually splits into four sub-bands (arrows). In the lanes shown, subject P1 (left lane in each panel) has a variant in amplicon 6 and subject P2 (right lanes) has variants in amplicons 17 and 24. Characterization of the variants showed that subject P2 was heterozygous for R1070Q (exon 17). Other variants were nonpathogenic. [(A) courtesy of Andrew Wallace, St Mary's Hospital, Manchester. (B) courtesy of Hans Scheffer, University of Groningen, The Netherlands.]

• Mismatched bases in heteroduplexes are sensitive to cleavage by chemicals or enzymes. The *chemical cleavage of mismatch* (CCM, Figure 18.3B) method is a sensitive method for mutation detection, with the advantages that quite large fragments (more than 1 kb in size) can be analyzed, the location of the mismatch is pinpointed by the size of the fragments generated, and variants present in only 5% or so of the sample can be detected. However, it never attained wide popularity because many of the protocols use toxic chemicals and all are experimentally quite difficult.

(A)

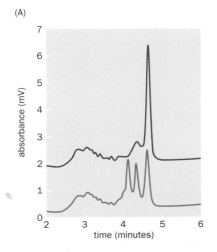

(B)

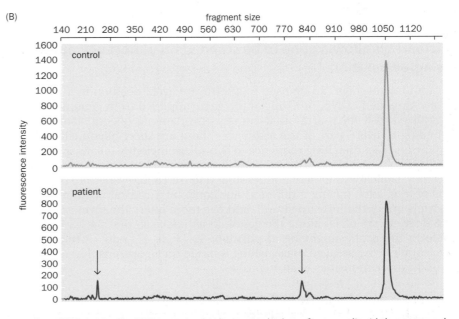

Figure 18.3 Scanning the *DMD* and *NF2* genes for mutations. (A) Scanning for DMD mutation by denaturing high-performance liquid chromatography (dHPLC). Exon 6 of the dystrophin gene gives different patterns in an affected male (blue trace) and a normal control (red trace). Sequencing revealed the splice site mutation c.738+1G>T. For males, because this is an X-linked condition, test DNA must be mixed with an equal amount of normal DNA to allow the formation of heteroduplexes. (B) Mutation scanning of the *NF2* (neurofibromatosis 2) gene by chemical cleavage of mismatches. A fluorescently labeled meta-PCR product contained exons 6–10 of the *NF2* gene. The lower track shows the sample from a patient; a heterozygous intron 6 splice site mutation 600-3c>g is revealed by hydroxylamine cleavage of the 1032 bp meta-PCR product to fragments of 813 + 239 bp (arrows). The upper track is a control sample. [(A) courtesy of Richard Bennett, Children's Hospital, Boston. (B) courtesy of Andrew Wallace, St Mary's Hospital, Manchester.]

Scanning methods based on single-strand conformation analysis

Single-stranded DNA has a tendency to fold up and form complex structures stabilized by weak intramolecular bonds, notably base-pairing hydrogen bonds. The electrophoretic mobility of such a structure in a non-denaturing gel will depend not only on its molecular weight but also on its conformation, which is dictated by the DNA sequence. Single-strand conformation polymorphisms (SSCPs) are detected by amplifying a genomic or cDNA sample, heating it to denature it, snap-cooling, and loading it on a non-denaturing polyacrylamide gel (see Figure 18.2A). Primers can be radiolabeled, or unlabeled products can be detected by silver staining. The precise pattern of bands seen is very dependent on details of the conditions. Control samples must be run, so that differences from the wild-type pattern can be noticed. SSCP analysis is very cheap and reasonably sensitive (about 80%) for fragments up to 200 bp long, so it still finds some uses. SSCP analyses can also be formatted to run on a DNA sequencer (conformation-sensitive capillary electrophoresis). SSCP analysis and heteroduplex analysis can be combined on a single gel, as in Figure 18.2A (some heteroduplex forms even in snap-cooled samples).

Scanning methods based on translation: the protein truncation test

The protein truncation test (PTT; **Figure 18.4**) is a specific test for frameshifts, or splice site or nonsense mutations that create a premature termination codon. The starting material is an RT-PCR product or, occasionally, a single large exon in genomic DNA such as the 6.5 kb exon 15 of the *APC* gene or the 3.4 kb exon 10 of the *BRCA1* gene. Nonsense-mediated mRNA decay (see Chapter 13, p. 418) would normally prevent the production of a truncated protein in an intact cell, but the relevant machinery is not present in this *in vitro* system and, in any case, there is no exon–exon splicing in the assay. The PTT detects only certain classes of mutation, which can be either a weakness or a strength. It would not be useful for CF, in which most mutations are non-truncating. But in DMD, adenomatous polyposis coli, or *BRCA1*-related breast cancer, missense mutations are infrequent, and any such change found may well be coincidental and nonpathogenic. For such diseases, the PTT has several advantages. It ignores silent or missense base substitutions, and (like mismatch cleavage methods, but unlike SSCP) it reveals the approximate location of any mutation. Several variants have been developed to give cleaner results, usually by incorporating an immunoprecipitation step—but PTT remains a demanding technique that is not easy to get working well.

Microarrays allow a gene to be scanned for almost any mutation in a single operation

Custom microarrays can be used to interrogate every position in a gene in one assay. Amplified cDNA, or exons of the gene amplified from genomic DNA, are hybridized to a microarray that contains overlapping oligonucleotides corresponding to every part of the sequence. These are short oligonucleotides that require an exact matching sequence to hybridize efficiently (allele-specific oligonucleotides; see below). The Affymetrix system uses 40 probes per nucleotide position, each about 25 nucleotides long (**Figure 18.5**). Probes are organized in sets of four, each having a different nucleotide at a central position. Five such quartets query the forward strand, and five more query the reverse strand. The five quartets are offset along the genomic sequence so that the variable nucleotide in the probe might be at position –2, –1, 0, +1, and +2 relative to the nucleotide being assayed. Base calling is based on algorithms that compare the hybridization intensities of all 40 probes.

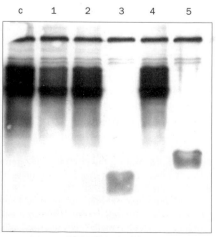

DMD segment 1EF

Figure 18.4 DMD mutation scanning with the protein truncation test (PTT). A coupled transcription–translation reaction is used to produce labeled polypeptide products encoded by a segment of a cDNA. Segments containing premature termination codons produce truncated polypeptides. The figure shows results for one segment of the dystrophin mRNA studied in five affected boys and one unaffected control (c). The samples in lanes 3 and 5 produce truncated, faster-running polypeptides. The position of the termination codon within a sample can be determined from the size of the truncated polypeptide. (Courtesy of Johan den Dunnen and D. Verbove, Leiden, The Netherlands.)

wild-type   AGGTCGTATCCATGCCTTACAGTC
A>C mutant   AGGTCGTATCCCTGCCTTACAGTC

cell	oligo	wild-type mismatch	hyb	mutant mismatch	hyb
−1					
11A	GGTCGTATCaaTGCCTTACA	1	+	2	−
11G	GGTCGTATCgaTGCCTTACA	1	+	2	−
11C	GGTCGTATCCaTGCCTTACA	0	++	1	+
11T	GGTCGTATCtaTGCCTTACA	1	+	2	−
0					
12A	GTCGTATCCaTGCCTTACAG	0	++	1	+
12G	GTCGTATCCgTGCCTTACAG	1	+	1	+
12C	GTCGTATCCCTGCCTTACAG	1	+	0	++
12T	GTCGTATCCtTGCCTTACAG	1	+	1	+
+1					
13A	TCGTATCCaaGCCTTACAGT	1	+	2	−
13G	TCGTATCCagGCCTTACAGT	1	+	2	−
13C	TCGTATCCacGCCTTACAGT	1	+	2	−
13T	TCGTATCCaTGCCTTACAGT	0	++	1	+

wild-type				mutant			
cell	11	12	13	cell	11	12	13
A				A			
G				G			
C				C			
T				T			

Figure 18.5 Principle of mutation detection with Affymetrix oligonucleotide arrays. For each nucleotide position, the labeled PCR product is checked for hybridization (hyb) to five quartets of oligonucleotides (oligos) on the array. A further five quartets interrogate the reverse strand, giving a total of 40 probes. The figure shows the quartets centered at positions −1, 0, and +1 relative to the mutation at position 12. The hybridization intensities for each probe are indicated as − (white), + (light green), or ++ (dark green).

Extra tests would be needed to check for deletions or other larger-scale changes. The range of mutations to be detected has to be defined in advance and designed into the chip. Thus, their main role in mutation detection may be for initial scanning of samples to pick up the common mutations, leaving the difficult cases to be sorted out by other methods. Given the high set-up costs of microarrays, they would only be used for genes such as the breast cancer genes *BRCA1/2*, where there is a very large demand for mutation analysis.

DNA methylation patterns can be detected by a variety of methods

In Chapter 11 we described the major role of DNA methylation in controlling gene expression. Methylation analysis is important in several clinical contexts:

- In the diagnosis of fragile X syndrome (the expanded repeat in the *FMR1* gene, described in Chapter 13, triggers methylation of the promoter, and it is the methylation that silences the gene and so causes the clinical syndrome).

- In testing for abnormal patterns of parent-specific methylation at imprinted gene loci, for example in Prader–Willi, Angelman, Beckwith–Wiedemann, and Russell–Silver syndromes (see Chapter 11, p. 368).

- In defining genetic changes in tumors, because tumor suppressor genes are often silenced by methylation (see Chapter 17, p. 549).

PCR products are always unmethylated, because they are made from the normal monomers supplied in the reaction mix. Thus, none of the methods described so far gives any information on the methylation patterns in a DNA sample. Genomewide studies of methylation use chromatin immunoprecipitation with an antibody against methylated DNA (see Chapter 11, p. 353), but to study the methylation status of an individual sequence, two main methods are used:

- *Restriction enzyme digestion. Hpa*II cuts only unmethylated CCGG, whereas *Msp*I cuts any CCGG, whether or not it is methylated. Thus, methylated genomic DNA gives different patterns of restriction fragments with the two enzymes. Alternatively, a PCR template containing a CCGG sequence can be digested with either *Hpa*II or *Msp*I before amplification. If the CCGG is methylated, the template will not be cleaved by *Hpa*II, and the PCR will yield a product (**Figure 18.6**).

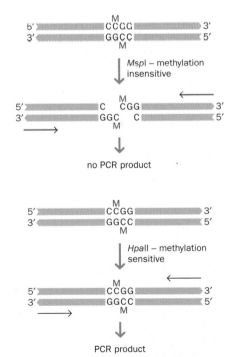

Figure 18.6 Using restriction enzymes to analyze the methylation status of a DNA sequence. The restriction enzyme *Hpa*II cuts CCGG, but only when the cytosines are not methylated. *Msp*I cuts the same sequence regardless of methylation. If genomic DNA is digested with *Hpa*II before PCR amplification, any amplicon that contains a CCGG sequence will give a product only if methylation of the CCGG site protects it from digestion.

- *Bisulfite modification.* As described in Chapter 11 (see Box 11.4), when DNA is treated with sodium bisulfite, cytosine (but not 5-methylcytosine) is converted to uracil. The changed nucleotides can be identified in various ways. The most obvious method is conventional sequencing of the bisulfite-treated product. Cytosine and 5-methylcytosine in the original genomic DNA show as thymine and cytosine, respectively, in the sequence trace. Alternatively, PCR primers can be designed to amplify specifically either the altered or unaltered sequence. Often, however, a quantitative assay is required. The original genomic DNA will be derived from many cells, and it is often relevant to ask how frequently a given CpG dinucleotide is methylated. Pyrosequencing (see Figure 8.8) or melting curve analysis can be used for this purpose. All these procedures require very careful optimization and control of conditions to yield reliable data.

Unclassified variants are a major problem

Sequencing the whole of a candidate gene in a large series of patients will inevitably reveal many variants. Deletions, frameshifts, nonsense mutations, and changes to the canonical GT...AG splice sites are highly likely to be pathogenic. But deciding whether or not a novel nucleotide substitution is pathogenic (and hence represents the sought-after mutation) can be very difficult. Guidelines developed by British and Dutch diagnostic laboratories suggest that some or all of the following checks could be useful:

- *Presence or absence on single nucleotide polymorphism (SNP) databases.* Clearly it is prudent to check dbSNP (http://www.ncbi.nlm.nih.gov/projects/SNP) to see whether the variant has previously been reported—but many rare nonpathogenic variants will not be in the database.

- *Co-segregation with the disease in the family.* This should always be checked where possible, because failure to co-segregate is powerful evidence that the variant is not pathogenic. However, co-segregation does not prove pathogenicity. Assuming that the variant is in the correct gene, there is a 50% chance that it will co-segregate with the disease even if the true cause is an undetected mutation elsewhere in the gene. The variant will co-segregate with the disease if it is in *cis* with the true mutation (i.e. in the same physical copy of the gene) but not if it is in *trans* (on the other allele).

- *Occurrence of a new variant concurrent with the (sporadic) incidence of the disease.* A *de novo* change (one not present in either parent) in a candidate gene in a *de novo* case of a dominant disease is highly suspect.

- *Testing ethnically matched controls.* It would be normal to perform this check, but it is less useful than people often imagine. Finding the same change in a healthy control would rule it out as a cause of a highly penetrant dominant or X-linked condition, but could not exclude the possibility that it causes a recessive or low-penetrance dominant condition. Not finding it provides a much weaker inference. As mentioned in Chapter 16 (see p. 512), a useful rule of thumb is that we can be 95% certain that a variant has a population frequency in healthy people of less than 1 in n if it has not been found in $3n$ control chromosomes. In practical terms, this means that testing controls is very poor at ruling out any variant with a population frequency under 0.01.

- *Functional studies*, where possible, would be important. However, it is often not possible to model *in vitro* all aspects of the function of a gene *in vivo*, and functional tests may be too complicated and gene-specific to fit readily into the workload of a diagnostic laboratory.

- *RNA studies* help identify effects on splicing. *In silico* splice site prediction should be attempted where a change is near an exon–intron boundary. Various programs are available to do this, ranging from quick and simple (www.fruitfly.org/seq_tools/splice.html—despite the name, it works with human sequences) to extremely comprehensive (www.umd.be/SSF), but none is infallible.

- *In silico predictions of pathogenic effect* can be attempted where there are good multiple sequence alignments, especially if there is also information on the three-dimensional structure of the gene product or of relevant domains within it. Web-based resources such as SIFT (http://sift.jcvi.org/) or Polyphen (http://genetics.bwh.harvard.edu/pph/) help perform these analyses.

- *Species conservation* provides some useful pointers. If the putative mutant sequence occurs as the wild type in some other species (or in a paralogous gene in humans), the change is unlikely to be pathogenic. Changes to highly conserved amino acids may well be pathogenic. For non-conserved amino acids, only extreme changes (introducing cysteine or proline in particular, which can have strong effects on the protein structure) have much likelihood of being pathogenic.

18.3 TESTING FOR A SPECIFIED SEQUENCE CHANGE

Testing for the presence or absence of a known sequence change is a different and much simpler problem than scanning a gene for the presence of *any* mutation. Samples can always be genotyped by sequencing, but conventional sequencing is not an efficient method if only a single nucleotide position is being checked. Some of the main genotyping methods are summarized in Table 18.4. Many variants of these and other methods have been developed as kits by biotechnology companies. Typical applications include the following:

- Diagnosis of diseases with limited allelic heterogeneity (Table 18.5).

- Diagnosis within a family. Mutation scanning methods may be needed to define the family mutation, but once it has been characterized, other family members normally need be tested only for that particular mutation.

- Testing control samples to see whether a change seen in a patient is actually a low-frequency population polymorphism.

- SNP genotyping. Although the aim is not to find a pathogenic mutation, the problem is identical: to test a DNA sample for a predefined sequence variant.

Testing for the presence or absence of a restriction site

When a base substitution mutation creates or abolishes the recognition site of a restriction enzyme, this allows a simple direct PCR test for the mutation. A suitable length sequence containing the potential mutation is PCR amplified. The PCR product is digested with the relevant restriction enzyme, and the products of

TABLE 18.4 METHODS OF TESTING FOR A SPECIFIED MUTATION OR SNP	
Method	**Comments**
Restriction digestion of PCR-amplified DNA; check size of products on a gel	only when the mutation creates or abolishes a natural restriction site or one engineered by the use of special PCR primers (see Figure 18.7); tests for a single mutation
Hybridize PCR-amplified DNA to allele-specific oligonucleotides (ASO) on a dot-blot or gene chip	general method for specified point mutations; large arrays allow massively parallel scanning for a very large number of mutations or SNPs
PCR using allele-specific primers (ARMS test)	general method for single point mutations; primer design is critical (see Figure 6.17)
Single-nucleotide primer extension	general method for specified point mutations; formatted for readout on DNA sequencer or on microarray for parallel assay of many sites
Oligonucleotide ligation assay (OLA)	general method for specified point mutations (see Figure 18.9); a few dozen tests can be multiplexed
Pyrosequencing	high-throughput method for identifying a few nucleotides at a specified position; quantitative result (see Figure 8.8)
Mass spectrometry	very fast (< 1 second) versatile general method; quantitative result (see Box 8.8 and Figure 8.26)
PCR with primers located either side of a translocation breakpoint	successful amplification shows the presence of the suspected deletion or specified rearrangement
Check size of expanded repeat	for dynamic repeat diseases (see Table 13.1) only; large expansions may require Southern blots

TABLE 18.5 EXAMPLES OF DISEASES THAT SHOW A LIMITED RANGE OF MUTATIONS

Disease	Cause	Comments
Sickle-cell disease	only this particular mutation produces the sickle cell phenotype	p.E6V in *HBB* gene (see Figure 7.9)
Achondroplasia	only G380R produces this particular phenotype; high frequency of new mutant cases	two distinct changes, both causing p.G380R in *FGFR3* gene
Huntington disease, myotonic dystrophy	gain-of-function mutations	unstable expanded repeats (see Table 13.1)
Fragile X	common molecular mechanism: expansion of an unstable repeat	see Table 13.1; other mutations occur, but are rare
Charcot–Marie–Tooth disease (HMSN1)	common molecular mechanism: recombination between misaligned repeats	duplication of 1.5 Mb at 17p11.2 (see Figure 13.25); point mutations also occur
α- and β- thalassemia	selection for heterozygotes leads to different ancestral mutations being common in different populations	see Figure 13.19 (α-thalassemia) and Table 18.6 (β-thalassemia)
Tay–Sachs disease	founder effect in Ashkenazi Jews; ancient heterozygote advantage	two common *HEXA* mutations in Ashkenazim: 4 bp insertion in exon 11 (73%); exon 11 donor splice site G>C (15%)
Cystic fibrosis	common ancestral mutations in northern European populations, ancient heterozygote advantage	see Table 18.7 and Section 3.5

See Section 13.4 for a further discussion of the reasons why some diseases show a limited range of mutations, whereas others have extensive allelic heterogeneity.

digestion are separated by electrophoresis to see whether or not a cut has occurred. Although hundreds of restriction enzymes are known, they almost all recognize symmetrical palindromic sites, and many point mutations will not happen to affect such sequences. In addition, sites for rare and obscure restriction enzymes are unsuitable for routine diagnostic use because the enzymes are expensive and often of poor quality. If a mutation does not change a suitable site, sometimes one can be introduced by a form of PCR mutagenesis using carefully designed primers. **Figure 18.7** shows an example.

Allele-specific oligonucleotide hybridization

Under suitably stringent hybridization conditions, a short (typically 15–20 nucleotides) oligonucleotide will hybridize only to a perfectly matched sequence. Allele-specific oligonucleotide (ASO) probes are made by chemical synthesis. Figure 7.9 demonstrated the use of dot-blot hybridization with ASO probes to detect the single base substitution that causes sickle-cell disease. For diagnostic purposes, a reverse dot-blot procedure has often been used. A screen for a series of defined cystic fibrosis mutations, for example, would use a series of ASOs specific for each mutant allele and for its wild-type counterpart, spotted onto a single membrane or anchored to some other solid support. This is then hybridized

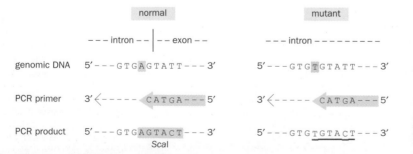
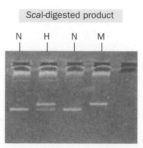

Figure 18.7 Introducing an artificial diagnostic restriction site. An A>T mutation in the intron 4 splice site of the *FACC* gene does not create or abolish a restriction site. The PCR primer stops short of this altered base but has a single base mismatch (red G) in a non-critical position that does not prevent it from hybridizing to and amplifying both the normal and mutant sequences. The mismatch in the primer introduces an AGTACT restriction site for *Sca*I into the PCR product from the normal sequence but not the mutant sequence. The *Sca*I-digested product from homozygous normal (N), heterozygous (H), and homozygous mutant (M) patients is shown. (Courtesy of Rachel Gibson, Guy's Hospital, London.)

Figure 18.8 Multiplex ARMS test to detect 29 cystic fibrosis mutations. Each sample is tested with four multiplex mutation-specific PCR reactions (lanes A–D). Within a multiplex, each product is a different size. If one of the 29 mutations is present there is an extra band. The identity of the mutation is revealed by the position of the band in the gel, and by which track contains it. Each tube also amplifies two control sequences—these are the top and bottom bands in each gel track; they differ between tracks so that each multiplex carries its own signature pattern. Note that the normal alleles of mutations other than p.F508del (band in lane B) are not tested for. Samples 1, 2, and 3 show no mutation-specific bands. In sample 4, the extra bands in lanes A and D show the presence of p.F508del and c.1898+1G>A, respectively, showing that this DNA comes from a compound heterozygote. (Courtesy of Michelle Coleman, St Mary's Hospital, Manchester; data obtained using the Elucigene™ kit from Orchid Biosciences.)

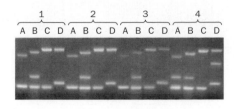

to labeled PCR-amplified test DNA. The same reverse dot-blot principle is applied on a massively parallel scale in DNA microarrays (see above and Chapter 7, p. 207).

Allele-specific PCR amplification

The principle of the ARMS (amplification refractory mutation system) method was shown in Figure 6.17. Paired PCR reactions are performed. One primer (the conserved primer) is the same in both reactions; the other exists in two slightly different versions, one specific for the normal sequence and the other specific for the mutant sequence. Additional control primers are usually included, to amplify some unrelated sequence from every sample as a check that the PCR reaction has worked. The location of the common primer can be chosen to give different-sized products for different mutations, so that the PCR products of multiplexed reactions can be separated by gel electrophoresis. The pairs of mutation-specific primers can also be made to give distinguishable products. For example, they can be labeled with different fluorescent or other labels, or given 5′ extensions of different sizes. Multiplexed mutation-specific PCR is well suited to screening fairly large numbers of samples for a given panel of mutations (**Figure 18.8**).

A modified procedure, pyrophosphorolysis-activated polymerization, is effective for detecting variant alleles that are present at very low levels. Applications would include testing for residual disease in cancer, testing for rare somatic mutations, and testing for paternally derived alleles in fetal DNA extracted from maternal blood. The reaction uses a primer with a dideoxy 3′ end that pairs at the position of the mutant nucleotide. In the presence of pyrophosphate, DNA polymerase will remove the 3′ dideoxynucleotide, in a reversal of the normal polymerization reaction, but only if it is correctly paired with the test strand. The polymerase then extends the primer in the normal way, using normal dNTPs. The variant nucleotide is thus read twice: once by the reverse polymerase reaction, and again by the forward reaction. This dual selectivity results in an extremely low level of errors.

The oligonucleotide ligation assay

In the oligonucleotide ligation assay (OLA) for base substitution mutations, two oligonucleotides are constructed that hybridize to adjacent sequences in the target, with the join sited adjacent to the position of the mutation. DNA ligase will not covalently join the two oligonucleotides unless they are perfectly hybridized (**Figure 18.9**). When PCR is then performed with one primer that hybridizes to each oligonucleotide, the ligated oligonucleotides, but not the unligated ones, can be amplified. In a multiplexed OLA, all the probes carry the same PCR primer-binding sequences at their 5′ ends, allowing the use of a single pair of universal PCR primers. Stuffer sequences of different lengths, and differently colored dye labels in the oligonucleotides, ensure that the product of each reaction can be distinguished by size and color on a fluorescence sequencer (**Figure 18.10**).

Minisequencing by primer extension

Minisequencing uses the principle of dideoxy sequencing (see Figure 8.6) but adds just a single nucleotide. The 3′ end of the sequencing primer is immediately upstream of the nucleotide that is to be genotyped. The reaction mix contains

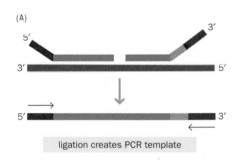

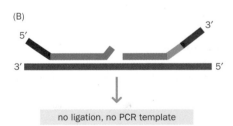

Figure 18.9 Principle of the oligonucleotide ligation assay (OLA). Two oligonucleotides (blue) hybridize to the target sequence (red) with their ends abutting immediately 3′ of the position of the nucleotide to be queried. (A) If the ends are correctly hybridized, DNA ligase will seal the gap, creating a single PCR-amplifiable molecule. (B) If the 3′ end nucleotide is mismatched, DNA ligase will not join the two oligonucleotides, and hence there is no substrate for the PCR reaction. Stuffer sequences (green) of different sizes allow individual amplicons to be identified in a multiplex OLA.

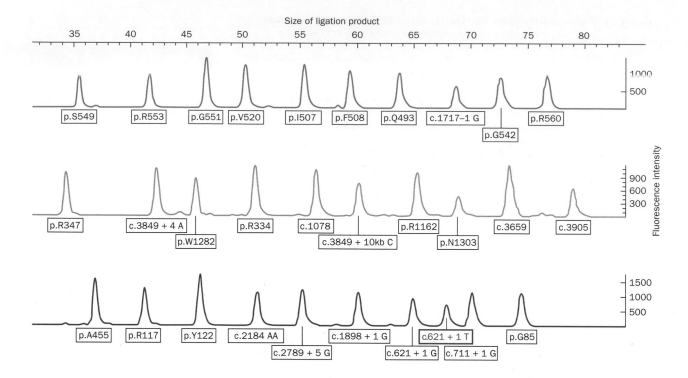

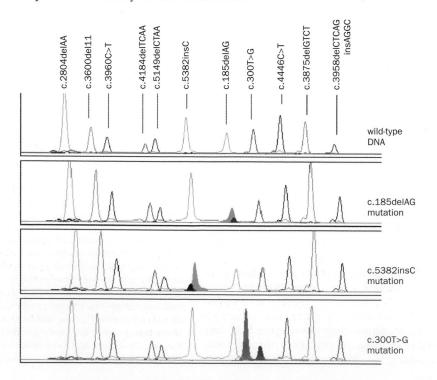

DNA polymerase plus four differently labeled ddNTPs. The test DNA acts as template for the addition of a single labeled dideoxynucleotide to the primer (**Figure 18.11**). The nucleotide added to the primer is identified, for example, by running the extended primer on a fluorescence sequencer, allowing the base at the position of interest to be identified. Minisequencing can readily be adapted to an array-based format (*APEX; arrayed primer extension*) to allow the entire sequence of a gene to be checked for base substitutions in a single operation. The same idea, but using reversibly blocked fluorescently labeled monomers, is the principle behind the Illumina/Solexa ultra-high-throughput sequencing technology (see Chapter 8, p. 220). After a single nucleotide has been added and identified, the blocking groups can be removed with a suitable flash of light, and the cycle can be repeated to identify the next nucleotide.

Figure 18.10 Using the oligonucleotide ligation assay to test for 29 known CF mutations. A multiplex OLA is performed and the products are amplified by PCR. Ligation oligonucleotides are designed so that products for each mutation and its normal counterpart can be distinguished by size and color of label. A ligation product from the splice site mutation 621+1g>t is seen (red box). The person may be a carrier or may be a compound heterozygote with a second mutation that is not one of the 29 detected by this kit. (Courtesy of Andrew Wallace, St Mary's Hospital, Manchester.)

Figure 18.11 A single-nucleotide primer extension assay for 11 specific mutations in the *BRCA1* gene. Eleven segments of the *BRCA1* gene were amplified from the test DNA in a multiplex PCR reaction. Eleven specific primers were then added, each with its 3' end adjacent to a nucleotide to be queried, together with four dye-labeled ddNTPs and DNA polymerase. The polymerase added a single colored ddNTP to each primer, the color depending on the relevant nucleotide in the test DNA. Primers were different lengths, so that the products of this multiplex reaction could be separated and identified by capillary electrophoresis. The figure shows results of four samples, three of which have mutations, each in heterozygous form. [From Révillion F, Verdière A, Fournier J et al. (2004) *Clin. Chem.* 50, 203–206. With permission from the American Association for Clinical Chemistry Inc.]

Pyrosequencing

As described in Chapter 8, pyrosequencing is a method of examining very short stretches of sequence—typically 1–5 nucleotides—adjacent to a defined start point. The main use is for SNP typing, in which only one or two bases are sequenced. Pyrosequencing uses an ingenious cocktail of enzymes to couple the release of pyrophosphate that occurs when a dNTP is added to a growing DNA chain to light emission by luciferase (see Figure 8.8). A primer is hybridized to the test DNA and offered each dNTP in turn. When the correct dNTP is present, so that the primer can be extended, a flash of light is emitted. The method has been developed into a machine that can automatically analyze 10,000 samples a day. Output is quantitative, so that allele frequencies of a SNP can be estimated in a single analysis of a large pooled sample. The same technology, applied on a massively parallel scale, is the basis of the Roche/454 technique for ultra-high-throughput sequencing (see Figure 8.9).

Genotyping by mass spectrometry

As described in Chapter 8, mass spectrometry (MS) techniques can be used to identify molecules from an accurate measurement of their mass. MS measures the mass–charge ratio of ions by accelerating them in a vacuum toward a target, and either timing their flight or measuring how far the ions are deflected by a magnetic field (see Box 8.8). The MALDI (matrix-assisted laser desorption/ionization) technique allows the MS analysis of large nonvolatile molecules such as DNA or proteins by embedding the macromolecule in a tiny spot of a light-absorbing substance, which is then vaporized by a brief laser pulse.

The combination of MALDI with time-of-flight analysis is known as MALDI-TOF (see Figure 8.26). Applied to DNA, the technique can measure a mass up to 20 kD with an accuracy of ±0.3%. MALDI-TOF MS can be used as a very much faster alternative to gel electrophoresis for sizing oligonucleotides up to about 100 nucleotides long. A test takes less than a second in the machine. For small oligonucleotides, the accuracy is sufficient to deduce the base composition directly from the exact mass. Alternatively, SNPs can be analyzed by primer extension using mass-labeled ddNTPs. The spots of DNA to be analyzed can be arrayed on a plate and the machine will automatically ionize each spot in turn. Current systems can genotype tens of thousands of SNPs per day and, as with pyrosequencing, samples can be pooled to measure allele frequencies directly.

Array-based massively parallel SNP genotyping

Microarrays, like those used for the genomewide association studies described in Chapter 15, genotype a sample for maybe half a million of the tagging SNPs defined by the HapMap project. Before this technology could be implemented, two problems had to be overcome. First, it is impracticable to multiplex PCR on such a scale when preparing a sample for analysis. The total primer concentration would be impracticable and also, as the number of primer pairs increases, the number of undesired primer–primer interactions increases exponentially. In highly multiplexed PCR reactions, much of the product consists of unwanted artifactual primer-dimers. Second, an extremely high level of allele discrimination is required if 500,000 SNPs are to be genotyped without producing thousands of erroneous results. Distinguishing homozygotes for the two SNP alleles may not be too difficult, but reliably identifying heterozygotes can be a serious challenge.

The multiplexing problem is solved by arranging for the sequences to be amplified to carry universal adaptors on their ends, so that all sequences can be amplified by using a single set of primers. Different strategies are used to combine single-primer amplification with high allelic discrimination:

- In the Affymetrix system, universal adaptors are ligated onto restriction fragments of whole genomic DNA. After PCR amplification using a single pair of primers, the PCR products are fragmented, labeled, and hybridized to oligonucleotides on the microarray (**Figure 18.12**). Genotyping uses the same system as mutation detection, described above. Forty probes per SNP are arranged in five quartets for the forward strand and five for the reverse strand. The principle was illustrated in Figure 18.5.

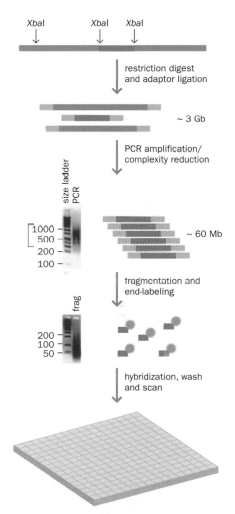

Figure 18.12 SNP genotyping using arrayed allele-specific oligonucleotides. In the Affymetrix system, genomic DNA is cut with a restriction enzyme. Universal adaptors (blue) are ligated to the fragments, allowing them to be amplified using a single pair of PCR primers. PCR products are fragmented, labeled, and hybridized to 25-mer oligonucleotides on the microarray, each of which specifically hybridizes to fragments containing one allele of one specific SNP. [From Matsuzaki H, Loi H, Dong S et al. (2004) *Genome Res.* 14, 414–425. With permission from Cold Spring Harbor Laboratory Press.]

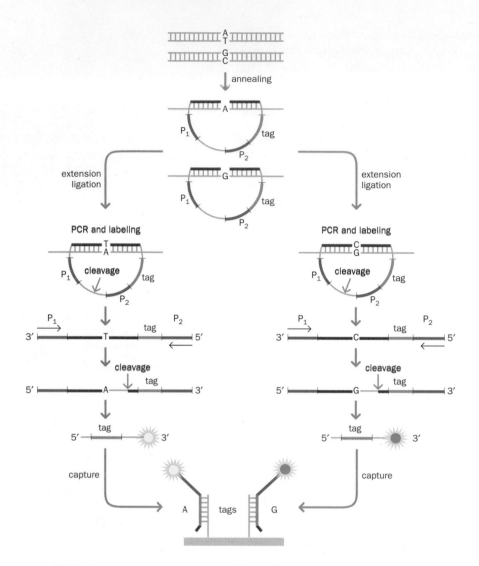

Figure 18.13 SNP genotyping using a molecular inversion probe (MIP). Each MIP includes locus-specific recognition sequences at its two ends (red), universal primer-binding sequences (P_1 and P_2, blue), and a locus-specific tag sequence (green). Two separate base-specific primer extension reactions are used, corresponding to the two alleles of the SNP being assayed (here, A or G). After circularization and then cleavage to release the MIP in linear form, the products of the two base-specific reactions are amplified with universal primers labeled with different dyes and hybridized to an oligonucleotide array that recognizes the tag sequences. [From Syvanen AC (2005) *Nat. Genet.* 37 (Suppl), S5–S10. With permission from Macmillan Publishers Ltd.]

- In the molecular inversion probe system, a single oligonucleotide probe with recognition sequences for the specific target at each end is used for each SNP. After hybridization to genomic DNA, the oligonucleotide forms a circular structure with ends separated by a single-base gap at the position of the SNP. Allele discrimination is achieved by a single-base primer extension in four separate reactions, each with one specific dNTP. The successful reaction creates a substrate for DNA ligase (**Figure 18.13**). Ligation circularizes the allele-specific product, which padlocks it to the genomic fragment, allowing all non-circularized material to be washed away. The circular oligonucleotide is then released by cleaving it, and amplified by PCR. Each oligonucleotide includes two universal primer-binding sequences; the products of the four base-specific reactions are amplified by using the same pair of primers but carrying a different dye label for each base-specific reaction product. The combined PCR products are then hybridized to an oligonucleotide array. The array probes recognize not the sequence surrounding the SNP but an individual tag sequence built into each SNP-specific oligonucleotide. This allows greater flexibility in assay design and greater control over the hybridization conditions.

- Illumina's GoldenGate® system is somewhat similar. Two allele-specific oligonucleotides (ASOs) and one locus-specific (but not allele-specific) oligonucleotide (LSO) are used to effect allele discrimination by a combination of allele-specific primer extension and ligation. For each SNP, one ASO becomes ligated to the LSO, creating a PCR substrate. All LSOs carry a common PCR primer-binding sequence. The ASOs carry one of two alternative primer-binding sequences, allowing the ligated product from each allele to be

Figure 18.14 SNP genotyping by the Illumina GoldenGate® method. Two allele-specific (P_1 and P_2) oligonucleotides and one locus-specific (P_3) oligonucleotide are used, with a combination of allele-specific primer extension and DNA ligation to provide reliable discrimination of alleles. As with the MIP system (see Figure 18.13), ligation creates a PCR substrate that is amplified with universal dye-labeled primers, and the products are identified through hybridization of the tag sequence to an array—in this case, a bead array. [From Syvanen AC (2005) *Nat. Genet.* 37 (Suppl), S5–S10. With permission from Macmillan Publishers Ltd.]

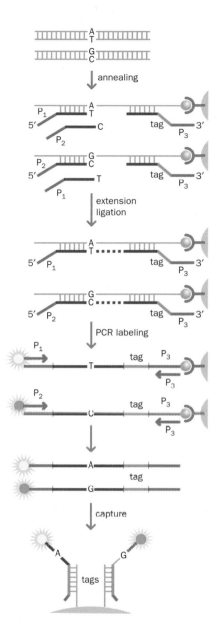

amplified with one of two universal dye-labeled primers (**Figure 18.14**). The LSO carries a tag sequence that is used to hybridize the PCR product to a specific bead in a bead array.

18.4 SOME SPECIAL TESTS

The techniques described previously are applicable to the detection of almost any variant in almost any DNA sequence. Some particular situations, however, either require special techniques or have special features that can be exploited to make testing easier or more effective. In this section, we describe several such cases.

Testing for whole-exon deletions and duplications requires special techniques

Generally, introns are much bigger than the exons they flank, and so random breakpoints in a gene will most probably lie within an intron. Thus, kilobase-scale partial deletions or duplications of a gene sequence, if they do not lie wholly within one intron, will most probably remove or duplicate one or more whole exons. Such changes will not be apparent when genomic DNA is amplified exon by exon. In heterozygous deletions of whole exons, the mutant allele will give no product. The test sees only the normal allele, and the result looks entirely normal. Similarly, duplications will not show up by these methods. PCR in its normal form is not quantitative, so any extra yield of product would not be noticed. RNA testing would solve the problem, but RNA may be difficult to obtain.

This is a problem that is particular to kilobase-scale deletions or duplications. Small deletions or duplications that lie wholly within an exon will be apparent when the exon is sequenced. Large-scale copy-number changes that might be anywhere in the genome are efficiently detected by array-CGH. However, this is an expensive technique for routine diagnosis, and the normal BAC arrays have low sensitivity for changes involving only a kilobase or so of DNA. When the candidate gene is already known, the technique of choice for detecting whole-exon deletions or duplications is **multiplex ligation-dependent probe amplification (MLPA)**.

The multiplex ligation-dependent probe amplification (MLPA) test

MLPA is a development of the oligonucleotide ligation assay (OLA) described above. As in the OLA, pairs of oligonucleotides hybridize to adjacent locations on the test DNA, and DNA ligase seals the gap between them to produce a single molecule. In the OLA the gap is positioned at a suspected variant nucleotide and ligation will occur only if there is an exact match; in MLPA the gap is positioned at a (hopefully) invariant nucleotide in the test DNA, so ligation will always take place if the template sequence is present. Ligation creates a PCR-amplifiable molecule (see Figure 18.9). The presence of a PCR product signals the presence of the appropriate matching sequence in the test DNA.

MLPA is a multiplex procedure with up to 45 probe pairs combined in some of the available kits. The individual PCR products are distinguished by length, so as to generate a series of peaks when the products are run on a DNA sequencer (**Figure 18.15**). Although the PCR is not truly quantitative (absolute amounts of product vary between probes), the relative amounts of a particular product in two samples should reflect the relative amounts of the template in the two samples. When results from test and control samples are compared, the relative peak heights give a direct readout of the dosage of each sequence in the test DNA relative to the control DNA. Normally, one ligation is used for each exon of a gene, allowing whole-exon deletions and duplications to be detected.

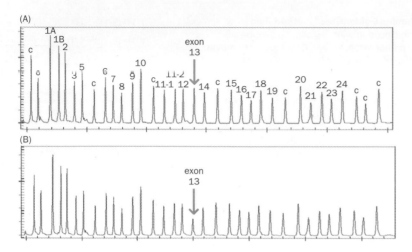

Figure 18.15 Using multiplex ligation-dependent probe amplification (MLPA) to identify an exon 13 deletion in the *BRCA1* gene. (A) Results from a control sample. Numbered peaks represent products from each exon; peaks labeled c are control probes. (B) The same analysis on DNA from a patient with breast cancer. In comparison with the control sample, the exon 13 peak is only half the size. (Courtesy of MRC Holland.)

Several other techniques can similarly identify whole-exon deletions and duplications in a candidate gene, but MLPA is the most widely used. The laboratory procedure is simple and the method has generally proved robust. Rare sequence variants that interfere with hybridization or ligation of the probes could lead to an exon being falsely scored as deleted. These could be checked by sequencing any suspect exon. MLPA probes are quite complex constructs (the stuffer fragment may be several hundred nucleotides long), and each individual probe needs to be carefully designed, so laboratories would normally buy commercial kits for their gene of interest. These are available for many of the genes in which deletions commonly cause clinical problems. Real-time PCR (see Box 8.5) would be an alternative for a gene for which no MLPA kit is available.

Dystrophin gene deletions in males

About 60% of DMD mutations are deletions of one or more exons of the dystrophin gene located on the X chromosome. MLPA is needed to detect female carriers of such deletions, but testing in affected males is simple: the deleted exon(s) will not amplify from the patient's DNA. In affected males, two multiplex PCR reactions (that shown in **Figure 18.16** and one testing exons in the 5′ part of

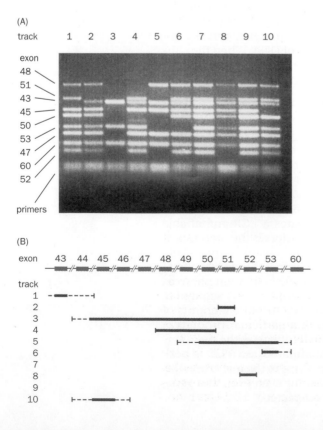

Figure 18.16 Multiplex screen for dystrophin deletions in males. (A) Products of multiplex PCR amplification of nine exons, using samples from 10 unrelated boys with Duchenne/Becker muscular dystrophy. PCR primers have been designed so that each exon, with some flanking intron sequence, gives a different-sized PCR product. (B) Interpretation of the results in (A). Solid lines show exons definitely deleted, dotted lines show the possible extent of deletion running into untested exons. No deletion is seen in samples 7 and 9; these patients may have point mutations, or deletions of exons not examined in this test. Exon sizes and spacing are not to scale. Compare with Figure 13.15. [(A) courtesy of R. Mountford, Liverpool Women's Hospital.]

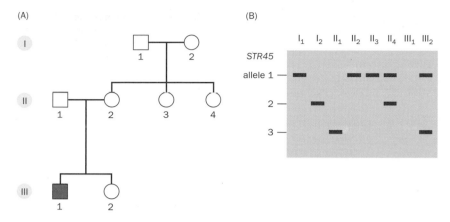

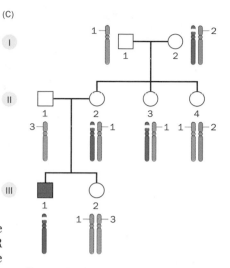

Figure 18.17 Deletion carriers in a DMD family revealed by apparent non-maternity. Pedigree (A) and results (B) of genotyping with the intragenic marker *STR45*. The affected boy, III$_1$, has a deletion that includes *STR45* (his lane on the gel is blank). His mother, II$_2$, and his aunt, II$_3$, inherited no allele of *STR45* from their mother, I$_2$, showing that the deletion is being transmitted in the family. I$_2$ is apparently homozygous for this highly polymorphic marker (lane 2), but in fact is hemizygous as a result of the deletion. The boy's other aunt, II$_4$, and his sister, III$_2$, are heterozygous for the marker and therefore do not carry the deletion. (C) Pedigree showing how the marker genotypes of individuals I$_2$, II$_2$, and II$_3$ are hemizygous as a result of the deletion.

the gene) will reveal 98% of all deletions. It is important to consider alternative explanations for failure to amplify: maybe there was a technical failure of the PCR or a base substitution in one of the primer-binding sites. Most deletions remove more than one exon. Deletions of just a single exon and deletions that seem to affect non-contiguous exons need confirming. Deletions can be confirmed by using alternative PCR primers or by MLPA. About 5% of dystrophin mutations are duplications of one or more exons, and detecting these requires MLPA in males as well as in females.

Apparent non-maternity in a family in which a deletion is segregating

If a deletion is segregating in a DMD family, genotyping females for microsatellites that map within the deletion may reveal *apparent non-maternity*, in which a mother has transmitted no marker allele to her daughter because of the deletion (**Figure 18.17**). In such families, non-maternity proves that a woman is a carrier, whereas heterozygosity (in the daughter or sister of a deletion carrier) proves that a woman is not a carrier. Several markers suitable for this purpose have been identified in the introns at deletion hotspots. The method works best in families in which there is an affected male in whom the deletion can first be defined. In principle, the same approach could be applied to tracking any other deletion, whether X-linked or autosomal, through a family. However, for any autosomal deletion, a quantitative method would be needed to identify any heterozygous deletion in the first place, and if such a method were available it would be simpler to use the same method to test other family members.

A quantitative PCR assay is used in prenatal testing for fetal chromosomal aneuploidy

For most diagnostic laboratories, a substantial part of the workload is prenatal testing for fetal chromosome anomalies. Women might request testing because they have had a previous chromosomally abnormal baby, because they (or their partner) have a balanced chromosomal abnormality that predisposes to fetal abnormality (see Chapter 2), or because of a suspect finding on an ultrasound scan. For all these cases, a full karyotype is usually prepared by standard cytogenetic techniques. Fetal cells are obtained by amniocentesis, usually at 14–20 weeks of gestation, or chorionic villus biopsy at 9–12 weeks (the exact date ranges vary from country to country).

However, the commonest single indication for fetal chromosome analysis is some combination of maternal age and maternal serum biochemistry that indicates an above-average risk of a fetal trisomy. The screening methods that identify these high-risk pregnancies are described in Chapter 19. For these women, fetal cells are not usually karyotyped. From the woman's point of view it takes far too long, and from the laboratory viewpoint it is far too expensive. The main risk is specifically for the three autosomal trisomies that are compatible with survival to term, namely trisomies 13, 18, and 21. These are usually checked by *QF-PCR*, a quantitative fluorescence-labeled multiplex PCR test. A multiplex PCR is performed with dye-labeled primers for several highly polymorphic microsatellite markers on each of chromosomes 13, 18, and 21. If a trisomy is present, the markers for that chromosome show either three peaks or two peaks in a 2:1 size ratio,

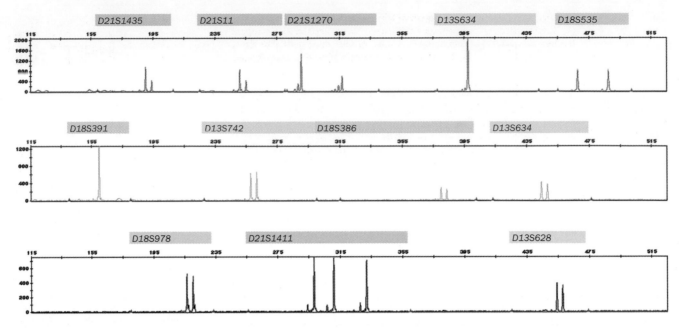

Figure 18.18 Prenatal diagnosis of trisomy 21 with QF-PCR. Fetal DNA, obtained by chorionic villus biopsy or amniocentesis, was amplified in a multiplex PCR reaction using dye-labeled primers for four highly polymorphic microsatellites from each of chromosomes 13, 18, and 21. The markers from chromosomes 13 and 18 each show two equal-sized peaks (if heterozygous) or one larger peak (if homozygous). Markers from chromosome 21 all show three peaks (*D21S1411*) or two peaks in a 2:1 size ratio (*D21S11*, *D21S1270*, and *D21S1435*). (Courtesy of Susan Hamilton, St Mary's Hospital, Manchester.)

depending on the informativeness of the marker (**Figure 18.18**). If there is any suspicion that the test DNA might be derived from contaminating maternal material, rather than from the fetus, this can be checked by comparing the genotypes with genotypes of DNA extracted from the mother's blood.

QF-PCR gives results in one day, in contrast with an average of 14 days for karyotyping cells from amniotic fluid. Mosaicism with the minor component less than about 15% is not reliably detected; nor, of course, are any chromosome abnormalities other than the specific trisomies targeted. There has been much debate about whether this specific focus on trisomies 13, 18, and 21 is an advantage or a disadvantage of the method. Some abnormalities will undoubtedly be missed; however, advocates of the QF-PCR approach point out that few of these are likely to result in a liveborn abnormal baby.

Some triplet repeat diseases require special tests

The expanded repeats that cause various neurological diseases (see Table 13.1) involve a special set of mutation-specific tests (**Figure 18.19**). The number of repeats is different in normal and in affected people, with a threshold for pathogenicity. For the polyglutamine repeat diseases such as Huntington disease, a single PCR reaction makes the diagnosis. Fragile X syndrome is more of a challenge. Normal (fewer than 50 repeats) and pre-mutation (50–200 repeats) alleles give clean PCR products, but full mutations have hundreds or thousands of CGG repeats and do not readily amplify by PCR, especially because of the high GC content. A full mutation in a male is easily recognized because the PCR product shows only a vague smear rather than a discrete band, but in a female the problem is more difficult. If a female sample gives only one band, this may be because she is homozygous for a particular repeat size in the normal range, or it may be because only one of her alleles amplifies, the other being a full expansion. This can only be settled by Southern blotting. Similarly, in people (male or female) who are mosaic for a full mutation and a pre-mutation, the full mutation would be missed without Southern blotting. Additionally, unlike the polyglutamine diseases, the fragile X expansion causes disease by loss of function, and occasional affected patients have deletions or point mutations that would be missed by testing just the repeat. Some of the other very large pathogenic expansions listed in Table 13.1 can be similarly difficult to identify reliably by PCR.

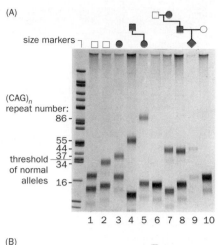

Figure 18.19 Laboratory diagnosis of trinucleotide repeat diseases.
(A) Huntington disease. A fragment of the gene containing the (CAG)$_n$ repeat has been amplified by PCR and run out on a polyacrylamide gel. Bands are revealed by silver staining. The scale shows numbers of repeats; those greater than 36 are pathogenic. Lanes 1, 2, 6, and 10 are from unaffected people; lanes 3, 4, 5, 7, and 8 are from affected people. Lane 5 is a juvenile-onset case; her father (lane 4) had 55 repeats but she has 86. Lane 9 is an affected fetus, diagnosed prenatally. (B) Myotonic dystrophy. Southern blot of genomic DNA digested with *Eco*RI and hybridized to a labeled probe consisting of part of the *DMPK* gene. Bands of 9 kb or 10 kb are normal variants, the result of a nonpathogenic insertion-deletion polymorphism. The grandfather (lane 4) has cataracts but no other sign of myotonic dystrophy. His 10 kb band seems to be very slightly expanded compared with the same band from the normal male in lane 3, but this is not definite on the evidence of this gel alone. His daughter (lane 1) has one normal and one definitely expanded 10 kb band; she has classical adult-onset myotonic dystrophy. Her son (lane 6) has a massive expansion and the severe congenital form of the disease. (C) Fragile X. The DNA of the inactivated X chromosome in a female, and of any X chromosome carrying the full mutation, is methylated. Genomic DNA is digested with a combination of *Eco*RI and the methylation-sensitive enzyme *Ecl*XI, Southern blotted, and hybridized to Ox1.9 or a similar probe. The X chromosome in a normal male (lane 1) and the active normal X chromosome in a female (lanes 2, 3, 4, and 6) give a small fragment (labeled N). Unmethylated pre-mutation alleles (P) give a slightly larger band in lanes 4 and 5 (female pre-mutation carriers) and lane 7 (a normal transmitting male). Methylated (inactive) X-chromosome sequences do not cut with *Ecl*XI and give a much larger band (NM), and the fully expanded and methylated sequence gives a very large smeared band (F) because of somatic mosaicism. [(A) courtesy of Alan Dodge, St Mary's Hospital, Manchester. (B) and (C) courtesy of Simon Ramsden, St Mary's Hospital, Manchester.]

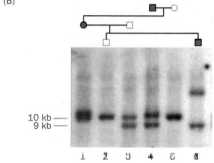

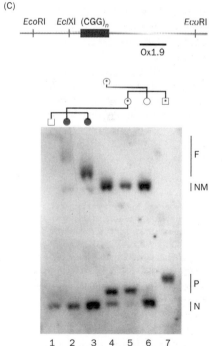

The mutation screen for some diseases must take account of geographical variation

The population genetics of recessive diseases is often dominated by founder effects or the effects of heterozygote advantage (see Chapter 3, p. 89). The resulting limited diversity of mutations in a population can make genetic testing much easier. β-Thalassemia and cystic fibrosis are good examples. For both these conditions, a very large number of different mutations in the relevant gene have been described, but in each case a handful of mutations account for most cases in any particular population. With β-thalassemia, DNA testing is not needed to diagnose carriers or affected people—orthodox hematology does this perfectly well—but it is the method of choice for prenatal diagnosis. Different mutations are predominant in different populations (**Table 18.6**). Provided that one has DNA samples from the parents and knows their ethnic origin, the parental mutations can often be found by using only a small cocktail of specific tests, after which the fetus can be readily checked.

In cystic fibrosis, the p.F508del mutation is the commonest in all European populations and is believed to be of ancient origin. However, the proportion of all mutations that are p.F508del varies, being generally high in the north and west of Europe and lower in the south. Testing for cystic fibrosis mutations divides into two phases. First, a limited number of specified mutations, always including p.F508del but otherwise population-specific, are sought with the methods described in Section 18.3. As **Table 18.7** shows, there is no obvious natural cutoff in terms of diminishing returns on testing for specific mutations. If this phase fails to reveal the mutations, then, if resources allow, a screen for unknown mutations may be instituted, using the methods described in Section 18.2. Alternatively, gene tracking (see below) may be used. The impact of this diversity on proposals for population screening is discussed in Chapter 19.

Surprisingly often, when a recessive disease is particularly common in a certain population, it turns out that more than one mutation is responsible. An example is Tay–Sachs disease among Ashkenazi Jews, where there are two common *HEXA* mutations (see Table 18.5). It is difficult to explain this situation except by assuming there has been a long-continuing heterozygote advantage, favoring the accumulation of mutations in the gene.

TABLE 18.6 THE MAIN β-THALASSEMIA MUTATIONS IN DIFFERENT COUNTRIES

Population	Mutation	Frequency (%)	Clinical effect
Sardinia	codon 39 (C>T)	95.7	β^0
	codon 6 (delA)	2.1	β^0
	codon 76 (delC)	0.7	β^0
	intron 1-110 (G>A)	0.5	β^+
	intron 2-745 (C>G)	0.4	β^+
Greece	intron 1-110 (G>A)	43.7	β^+
	codon 39 (C>T)	17.4	β^0
	intron 1-1 (G>A)	13.6	β^0
	intron 1-6 (T>C)	7.4	β^+
	intron 2-745 (C>G)	7.1	β^+
China	codon 41/42 (delTCTT)	38.6	β^0
	intron 2-654 (C>T)	15.7	β^0
	codon 71/72 (insA)	12.4	β^0
	−28 (A>G)	11.6	β^+
	codon 17 (A>T)	10.5	β^0
Pakistan	codon 8/9 (insG)	28.9	β^0
	intron 1-5 (G>C)	26.4	β^+
	619 bp deletion	23.3	β^+
	intron 1-1 (G>T)	8.2	β^0
	codon 41/42 (delTCTT)	7.9	β^0
US black African	−29 (A>G)	60.3	β^+
	−88 (C>T)	21.4	β^+
	codon 24 (T>A)	7.9	β^+
	codon 6 (delA)	0.8	β^0

In each country, certain mutations are frequent because of a combination of founder effects and selection favoring heterozygotes. Data courtesy of J Old, Institute of Molecular Medicine, Oxford. β^0, complete absence of β-globin chains; β^+, β-globin present but in insufficient quantity. The nomenclature of mutations used here is nonstandard (see Box 13.2, p. 407), but is widely used for thalassemia.

Testing for diseases with extensive locus heterogeneity is a challenge

As sequencing becomes cheaper and faster, diagnostic laboratories are able to do more and more individual tests on a patient. With careful optimization of protocols, even genes with many exons can be fully analyzed in a cost-effective way. The focus is now moving on to conditions that are often Mendelian but can be caused by mutations in any one of a large number of genes. Profound mental retardation and congenital profound hearing loss are examples. Identifying the causative mutation in a patient may involve testing dozens of genes. The key question here is whether certain specific mutations, or maybe defects in one particular gene, account for a significant proportion of cases. Hearing loss is one

TABLE 18.7 DISTRIBUTION OF *CFTR* MUTATIONS IN 300 CF CHROMOSOMES FROM THE NORTHWEST OF ENGLAND

Mutation	Exon	Frequency (%)	Cumulative frequency (%)
p.F508del	10	79.9	79.9
p.G551D	11	2.6	82.5
p.G542X	11	1.5	84.0
p.G85E	3	1.5	85.5
p.N1303K	21	1.2	86.7
c.621+1G>T	4	0.9	87.6
c.1898+1G>A	12	0.9	88.5
p.W1282X	21	0.9	89.4
p.Q493X	10	0.6	90.0
c.1154insTC	7	0.6	90.6
c.3849+10 kb (C>T)	intron 19	0.6	91.2
p.R553X	10	0.3	91.5
p.V520F	10	0.3	91.8
p.R117H	4	0.3	92.1
p.R1283M	20	0.3	92.4
p.R347P	7	0.3	92.7
p.E60X	3	0.3	93.0
Unknown/private	–	7.0	100

p.F508del and a few of the other relatively common mutations are probably ancient and spread through selection favoring heterozygotes; the other mutations are probably recent, rare, and highly heterogeneous. Cystic fibrosis is more homogeneous in this population than in most others. See Box 13.2 for nomenclature of mutations. Data courtesy of Andrew Wallace, St Mary's Hospital, Manchester.

case in which this does happen. Worldwide, 20–50% of children with autosomal recessive profound congenital hearing loss have mutations in the *GJB2* gene that encodes connexin 26. Different specific mutations are common in different populations—c.30delG in Europe, c.235delC in East Asia, c.167delT in Ashkenazi Jews. A simple PCR test therefore provides the answer in a good proportion of cases. For the remainder, it would be necessary first to sequence the whole *GJB2* gene and then, if resources allowed, to examine a large number of other genes.

In many other cases, no one gene accounts for a significant proportion of cases. Learning difficulties present the ultimate challenge in this respect. Custom chips are being developed to allow a large panel of genes to be screened; alternatively, the new exon capture and ultra-fast sequencing technologies may allow dozens of genes to be sequenced at a reasonable cost. Technically, the challenge is identical to the problem of screening a person's DNA for large numbers of variants that confer susceptibility or resistance to common complex diseases. However, data interpretation presents different problems. For susceptibility screening, the problem is to know the combined risk from many variants, each of which modifies risk to only a small degree. Risks cannot be simply added or multiplied; combining them requires a quantitative model of the effect of each variant on overall cell biology. For heterogeneous Mendelian conditions this is not a problem: one mutation is responsible for the condition. The problem is the large number of unclassified variants that will undoubtedly be identified.

18.5 GENE TRACKING

Gene tracking was historically the first type of DNA diagnostic method to be widely used. It uses knowledge of the map location of the disease locus, but not knowledge about the actual disease gene. Most of the Mendelian diseases that form the bulk of the work of diagnostic laboratories went through a phase of gene tracking when the disease gene had been mapped but not yet cloned. Once the gene had been identified, testing moved on to direct gene analysis. Huntington disease, cystic fibrosis, and myotonic dystrophy are familiar examples. However, gene tracking may still have a role even when a gene has been cloned. In the setting of a diagnostic laboratory, it is not always cost effective to search all through a large multi-exon gene to find every mutation. Moreover, there are always cases in which the mutation cannot be found. In these circumstances, gene tracking using linked markers is the method of choice. The prerequisites for gene tracking are:

- The disease should be well mapped, with no uncertainty about the map location, so that markers can be used that are known to be tightly linked to the disease locus.

- The pedigree structure and sample availability must allow the determination of phase (see below).

- There must be unequivocally confirmed clinical diagnoses and no uncertainty as to which locus is involved in cases in which there is locus heterogeneity.

Gene tracking involves three logical steps

Box 18.1 illustrates the essential logic of gene tracking. This logic can be applied to diseases with any mode of inheritance. There has to be at least one parent who

---

BOX 18.1 THE LOGIC OF GENE TRACKING

Shown here are three stages in the investigation of a late-onset autosomal dominant disease where, for one reason or another, direct testing for the mutation is not possible.

- Individual III$_2$ (arrow), who is pregnant, wishes to have a presymptomatic test to show whether she has inherited the disease allele. The first step is to tell her mother's two chromosomes apart. A marker, closely linked to the disease locus, is found for which II$_3$ is heterozygous (2–1).

- Next we must establish phase—that is, work out which marker allele in II$_3$ is segregating with the disease allele. The maternal grandmother, I$_2$, is typed for the marker (2–4). Thus, II$_3$ must have inherited marker allele 2 from her mother, which therefore marks

her unaffected chromosome. Her affected chromosome, inherited from her dead father, must be the one that carries marker allele 1.

- By typing III$_2$ and her father, we can work out which marker allele she received from her mother. If she is 2–1 or 2–3, it is good news: she inherited marker allele 2 from her mother, which is the grandmaternal allele. If she types as 1–1 or 1–3 it is bad news: she inherited the grandpaternal chromosome, which carries the disease allele.

Note that it is the segregation pattern in the family, and not the actual marker genotype, that is important: if III$_2$ has the same marker genotype, 2–1, as her affected mother, this is good news, not bad news, for her.

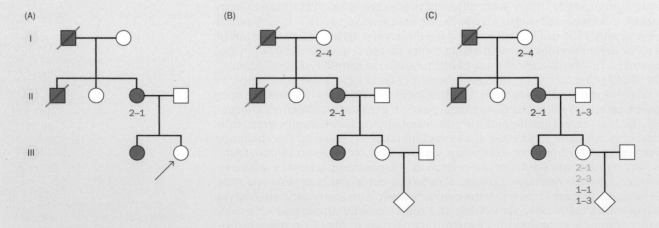

Figure 1 Using gene tracking to predict the risk of inheriting an autosomal dominant disease.

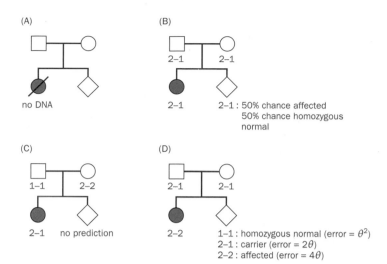

Figure 18.20 Gene tracking for prenatal diagnosis of an autosomal recessive disease. Four families each have a child affected with a recessive disease. Direct mutation testing is not possible (either because the gene has not been cloned or because the mutations could not be found). (A) No diagnosis is possible if there is no sample from the affected child. (B) If everybody has the same heterozygous genotype for the marker, the result is not clinically useful. (C) If the parents are homozygous for the marker, no prediction is possible with this marker. (D) Both parents are heterozygous carriers, and the genotype of the affected child shows that in each parent the pathogenic mutation is on the chromosome that carries allele 2 of the marker, allowing a successful prediction to be made. The error rates shown are the risk of predicting an unaffected pregnancy when the fetus is affected, or vice versa, if the marker used shows a recombination fraction θ with the disease locus. These examples emphasize the need for both an appropriate pedigree structure (DNA must be available from the affected child) and informative marker types.

could have passed on the disease allele to the proband, and who may or may not have actually done so. The process always follows the same three steps:

1. Distinguish the two chromosomes in the relevant parent(s)—that is, find a closely linked marker for which they are heterozygous.

2. Determine phase—that is, work out which chromosome carries the disease allele.

3. Work out which chromosome the proband received.

Figure 18.20 shows gene tracking for an autosomal recessive disease. The pedigrees emphasize the need for both an appropriate pedigree structure (DNA must be available from the affected child) and informative marker types. Even if the affected child is dead, if the Guthrie card used for neonatal screening can be retrieved, sufficient DNA for PCR typing can usually be extracted from the remnants of the dried blood spot. Informativeness of the marker should not be a big problem. With more than 20,000 highly polymorphic microsatellites mapped across the human genome, it should always be possible to find informative markers that map close to the disease locus.

Recombination sets a fundamental limit on the accuracy of gene tracking

Because the DNA marker used for gene tracking is not the sequence that causes the disease, there is always the possibility that recombination may separate the disease allele and the marker. This would lead to an erroneous prediction. The recombination fraction, and hence the error rate, can be estimated from family studies by standard linkage analysis (see Chapter 14). With almost any disease there should be a good choice of markers showing less than 1% recombination with the disease locus. This follows from the observations that 1 nucleotide in 300 is polymorphic, and that loci 1 Mb apart show roughly 1% recombination (see Chapter 14, p. 446). Ideally, one uses an intragenic marker, such as a microsatellite within an intron.

Recombination between marker and disease can never be completely ruled out, even for very tightly linked markers, but the error rate can be greatly reduced by using two marker loci situated on opposite sides of the disease locus. With such *flanking* or *bridging markers*, a recombination between either of the markers and the disease locus will also produce a marker–marker recombinant, which can be detected (e.g. III₁ in **Figure 18.21**). If a marker–marker recombinant is seen in the consultand, then no prediction can be made about inheritance of the disease, but at least a false prediction has been avoided. Provided that no marker–marker recombinant is seen, the only residual risk is that of double recombinants. As we saw in Chapter 14, the true probability of a double recombinant is very low because of interference (see p. 445). Thus, the risk of an error due to unnoticed recombination is much smaller than the risk of a wrong prediction due to human error in obtaining and processing the DNA samples. Perhaps

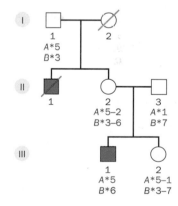

Figure 18.21 Gene tracking in Duchenne muscular dystrophy with flanking markers. The family has been typed for two polymorphisms, *A* and *B*, that flank the dystrophin locus. Individual III₁ has a recombination between marker locus *A* and DMD, but this does not confuse the prediction for III₂ because we know unambiguously that the disease allele in her mother is carried on a chromosome bearing marker alleles *A*2 and *B*6. III₂ can have inherited DMD only if she has a double recombination, one between marker *A* and DMD and another between DMD and marker *B*. If the recombination fractions are θ_A and θ_B, respectively, then the probability of a double recombinant is of the order $\theta_A\theta_B$, which typically will be well under 1%.

a greater risk is unexpected locus heterogeneity, so that the true disease locus in the family is actually different from the locus being tracked.

Calculating risks in gene tracking

Gene tracking can be used for Mendelian diseases with any mode of inheritance; however, unlike direct mutation testing, gene tracking always involves a calculation. Factors to be taken into account in assessing the final risk include:

- The probability of disease–marker and marker–marker recombination.
- Uncertainty due to imperfect pedigree structure or limited informativeness of the markers, about who transmitted which marker allele to whom (see Figure 14.9C for an example).
- Uncertainty as to whether somebody in the pedigree carries a newly mutant disease allele (see Figure 3.21B for an example of this problem in DMD).

Two alternative methods are available for performing the calculation: Bayesian calculations and linkage analysis.

Bayesian calculations

Bayes's theorem provides a general method for combining independent probabilities into a final overall probability. The theory, procedure, and a sample calculation are shown in **Box 18.2**.

BOX 18.2 USE OF BAYES'S THEOREM FOR COMBINING PROBABILITIES

A formal statement of Bayes's theorem is

$$P(H_i|E) = P(H_i) \times P(E|H_i)/\Sigma[P(H_i) \times P(E|H_i)]$$

$P(H_i)$ means the probability of the ith hypothesis, and the vertical line means *given*, so that $P(E|H_i)$ means the probability of the evidence (E), given hypothesis H_i. An example will probably make this clearer (**Figure 1**).

The steps in performing a Bayesian calculation are:
- Set up a table with one column for each of the alternative hypotheses. Cover all the alternatives.
- Assign a *prior probability* to each alternative. The prior probabilities of all the hypotheses must sum to 1. It is not important at this stage to worry about exactly what information you should use to decide the prior probability, as long as it is consistent across the columns. You will not be using all the information (otherwise there would be no point in doing the calculation because you would already have the answer), and any information not used in the prior probability can be used later.
- Using one item of information not included in the prior probabilities, calculate a *conditional probability* for each hypothesis. The conditional probability is the probability of the information, given the hypothesis, namely $P(E|H_i)$ [*not* the probability of the hypothesis given the information, $P(H_i|E)$]. The conditional probabilities for the different hypotheses do not necessarily sum to 1.
- The previous step can be repeated as many times as necessary until all information has been used once and once only. The end result is a number of lines of conditional probabilities in each column.
- Within each column, multiply together the prior and all the conditional probabilities. This gives a *joint probability*, $P(H_i) \times P(E|H_i)$. The joint probabilities do not necessarily sum to 1 across the columns.
- If there are just two columns, the joint probabilities can be used directly as odds. Alternatively, the joint probabilities can be scaled to give final probabilities, which do sum to 1. This is done by dividing each joint probability by the sum of all the joint probabilities, $\Sigma[P(H_i) \times P(E|H_i)]$.

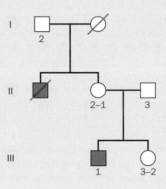

hypothesis: III$_2$ is	a carrier	not a carrier
prior probability	1/2	1/2
conditional (1): DNA result	0.05	0.95
conditional (2): CK data	0.7	1
joint probability	0.0175	0.475
final probability	0.0175/0.4925	0.475/0.4925
	= 0.036	= 0.964

Figure 1 Calculating the risk that III$_2$ is a carrier of DMD. Individual III$_2$ wishes to know her risk of being a carrier of DMD, which affected her brother III$_1$ and uncle II$_1$. Serum creatine kinase (CK) testing (an indicator of subclinical muscle damage common in DMD carriers) gave carrier–non-carrier odds of 0.7:1. A DNA marker that shows on average 5% recombination with DMD gave the types shown in red. These form two conditional probabilities and allow the risk to be calculated, following the guidelines in this Box, to give her overall risk of being a carrier as 3.6%.

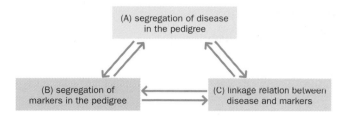

Figure 18.22 Use of linkage analysis programs for calculating genetic risks. Given information on any two of these subjects, the program can calculate the third. For linkage analysis, the program is given (A) and (B), and calculates (C). For calculating genetic risks, the program is given (B) and (C), and calculates (A).

Bayesian calculations give a quick answer for simple pedigrees, but the calculations can get very elaborate for more complex ones. Few people feel fully confident of their ability to work through a complex pedigree correctly, although the attempt is a valuable mental exercise for teasing out the factors contributing to the final risk. An alternative is to use a linkage analysis program.

Using linkage programs for calculating genetic risks

At first sight it may seem surprising that a program designed to calculate lod scores can also calculate genetic risks, but in fact the two are closely related (**Figure 18.22**). As described in Chapter 14, linkage analysis programs are general-purpose engines for calculating the likelihood of a pedigree, given certain data and assumptions. For calculating the likelihood of linkage we calculate the ratio

$$\frac{\text{likelihood of data} \mid \text{linkage, recombination fraction } \theta}{\text{likelihood of data} \mid \text{no linkage } (\theta = 0.5)}$$

For estimating the risk that a proband carries a disease gene, we calculate the ratio

$$\frac{\text{likelihood of data} \mid \text{proband is a carrier, recombination fraction } \theta}{\text{likelihood of data} \mid \text{proband is not a carrier, recombination fraction } \theta}$$

As in Box 18.2, the vertical line | means *given that*.

The special problems of Duchenne muscular dystrophy

Duchenne muscular dystrophy (DMD) poses a remarkably wide range of problems for the diagnostic laboratory. Fortunately, two-thirds of mutations in this X-linked disease are deletions, which are easily identified in males (see Figure 18.16), although more difficult in females. Duplications are hard to spot in either sex without using MLPA or some similar technique, and are undoubtedly underdiagnosed. The 30–35% of point mutations pose major problems. Scanning the DNA for point mutations requires all 79 exons to be individually amplified and sequenced, and therefore gene tracking is often used. However, DMD presents special problems for gene tracking because there is an extremely high recombination frequency across the gene. Even intragenic markers show an average of 5% recombination with the disease. It is therefore prudent to use flanking markers, as in Figure 18.21.

The problems do not end here. There is a high frequency of new mutations. The mutation-selection equilibrium calculations in Chapter 3 (p. 88) show that for any lethal X-linked recessive condition (fitness = 0), one-third of cases are fresh mutations. Therefore, the mother of an isolated DMD boy has only a two-thirds chance of being a carrier. This has two unfortunate consequences:

- It greatly complicates the risk calculations that are necessary for interpreting gene tracking results.

- As shown in Figure 3.21, the first mutation carrier in a DMD pedigree is very often a mosaic (male or female). This raises yet more problems, both for risk estimation and for interpretation of the results of direct testing.

These factors, together with the particularly distressing clinical course of the disease, the high recurrence risk within families, and the high frequency of DMD in the population, mean that DMD remains perhaps the most challenging of all diseases for genetic service providers.

18.6 DNA PROFILING

Here we move from problems of clinical diagnosis to problems of identifying individuals and determining relationships. We use the term **DNA profiling** to refer in a general way to the use of DNA tests to establish identity or relationships. **DNA fingerprinting** is reserved for a particular variant of profiling, the historically revolutionary but now obsolete technique published by Alec Jeffreys in 1985, using multilocus probes.

A variety of different DNA polymorphisms have been used for profiling

DNA profiling normally uses highly polymorphic microsatellites, but other types of variant (minisatellites, SNPs) also are, or have been, used.

DNA fingerprinting using minisatellite probes

The DNA fingerprinting technique pioneered by Alec Jeffreys used probes containing the common core of a hypervariable dispersed repetitive sequence GGGCAGGAXG (X is any nucleotide). The sequence is present in many minisatellites spread around the genome, at each of which the number of tandem repeats varies between individuals. When hybridized to Southern blots, the probes give an individual-specific fingerprint of bands (**Figure 18.23**). DNA fingerprinting revolutionized forensic practice but is now obsolete because of two problems:

- The Southern blot procedure requires several micrograms of DNA, corresponding to the content of perhaps a million cells. It is also very laborious and time consuming.

- It is not possible to tell which pairs of bands in a fingerprint represent alleles. Thus, when comparing two DNA fingerprints, the investigator matches each band individually by position and intensity. The continuously variable distance along the gel has to be divided into a number of bins. Bands falling within the same bin are deemed to match. Then if, say, 10 out of 10 bands match, the odds that the suspect, rather than a random person from the population, is the source of the sample are 1 in p^{10}, where p is the chance that a band in a random person would match a given band (we have simplified by

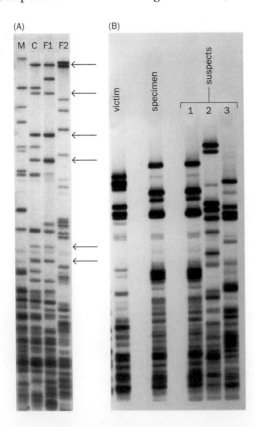

Figure 18.23 DNA fingerprinting.
(A) Resolving disputed paternity. DNA fingerprints from a mother (M), her child (C), and two possible fathers (F1 and F2). Arrows show bands present in the child but not in the mother. They could have come from F1 but not F2, thus ruling out F2 as the father. (B) A criminal investigation. A DNA fingerprint from a vaginal swab taken from a rape victim, together with DNA fingerprints from three suspects. Because the swab may contain DNA of the victim, her own fingerprint is also shown. The fingerprint of suspect 1 matches the specimen. (Images courtesy of Cellmark Diagnostics, Abingdon, Oxfordshire, UK.)

TABLE 18.8 PANELS OF MICROSATELLITE MARKERS USED FOR DNA PROFILING

Marker	Chromosome	CODIS (USA)	UK first panel	SGM (UK)	SGM+ (UK)
D2S1338	2				+
TPOX	2	+			
D3S1358	3	+			+
FGA	4	+		+	+
CSF1PO	5	+			
D5S818	5	+			
F13A1	6		+		
D7S820	7	+			
D8S1179	8	+		+	+
THO1	11	+	+	+	+
VWA	12	+	+	+	+
D13S317	13	+			
FES/FPS	15		+		
D16S539	16	+			+
D18S51	18	+		+	+
D19S433	19				+
D21S11	21	+		+	+
Amelogenin	X, Y	+		+	+
Average match probability		$1{:}10^{13}$	1:10,000	$1{\cdot}5 \times 10^7$	$1{:}10^9$

Current panels are CODIS and SGM+. The UK first panel and SGM panel are no longer used; they are included to illustrate how testing panels have evolved over time. The match probability of a profile is the probability that two unrelated individuals would have the same profile by random chance. It is calculated by multiplying together the chance of a match for each individual marker allele.

assuming p to be the same for every bin and ignoring the need to match on intensity as well as position). Even for $p = 0.2$, p^{10} is only 10^{-7}. It is imperative that the same binning criteria be used for judging matches between two profiles and for calculating p. The criteria can be arbitrary within certain limits, but they must be consistent.

DNA profiling using microsatellite markers

Single-locus profiling uses microsatellite polymorphisms, usually trinucleotide or tetranucleotide repeats, that can be genotyped by PCR. In theory, even a single cell left at the scene of a crime can be profiled. Alleles can be defined unambiguously by the precise repeat number, which avoids the binning problem. If the gene frequency of each allele in the population is known, an exact calculation can be made of the odds that a suspect, rather than an unrelated member of the population, is the source of a DNA sample. Standardized panels of microsatellites are used in different jurisdictions; the two main ones are CODIS in the USA and SGM+ in Europe (Table 18.8). It is usual also to include amelogenin as a sex marker. The X and Y chromosomes each have a copy of the amelogenin gene (*AMELX, AMELY*), but the copies differ so that each gives a different-sized PCR product. Figure 18.24 shows a typical SGM+ profile.

Y-chromosome and mitochondrial polymorphisms

For tracing relationships to dead persons, Y-chromosome and mitochondrial DNA markers are especially useful because, in each case, an individual inherits

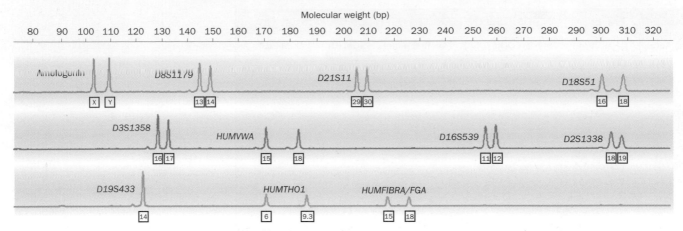

Figure 18.24 A DNA profile produced using the SGM+ set of markers. The SGM+ multiplex uses three different colored labels for primers to make the results clearer. Numbers in boxes are the number of repeat units in each microsatellite allele. [From Jobling MA & Gill P (2004) *Nat. Rev. Genet.* 5, 739–751. With permission from Macmillan Publishers Ltd.]

the complete genotype from a single definable ancestor. For the Y chromosome, a panel of short tandem repeat markers is used, whereas mitochondrial genotypes are defined by SNPs. An interesting example was the identification of the remains of the Russian tsar and his family, killed by the Bolsheviks in 1917, by comparing DNA profiles of excavated remains with those of living distant relatives (see Further Reading). These markers are also sometimes used in criminal investigations, particularly when the question arises of whether the criminal might be a relative of somebody whose profile is stored on a forensic database.

DNA profiling is used to disprove or establish paternity

Exclusion of paternity is simple: if the child has marker alleles that are not present in either the mother or alleged father, then the alleged father is not the biological father. Discrepancies at more than one locus are needed for certainty, because of the possibility that the child has a new mutant microsatellite allele. Proving paternity is, in principle, impossible—one can never prove that there is not another man in the world who could have given the child that particular set of marker alleles. All one can do is establish a probability of non-paternity that is low enough to satisfy the courts and, if possible, the putative father.

Single-locus microsatellites allow an explicit calculation of the odds (**Figure 18.25**). A series of 10 unlinked highly polymorphic single-locus markers gives overwhelming odds favoring paternity if all the bands fit.

DNA profiling can be used to identify the origin of clinical samples

The problem of maternal contamination of prenatal DNA samples has already been mentioned. If QF-PCR is to be used to check for fetal trisomies, the sample will in any case be genotyped for a series of polymorphic microsatellites. It is then simple to type a maternal sample for the same markers and compare the results. If the genotypes are insufficiently informative, other markers, for example a commercial DNA profiling kit, can be used. DNA profiling is also used to check the progress of a bone marrow transplant. If engraftment is successful, the recipient's blood DNA will take on the donor profile, whereas other tissues will retain the original profile of the recipient.

DNA profiling can be used to determine the zygosity of twins

In studying non-Mendelian characters (see Chapter 15), and sometimes in genetic counseling, it is important to know whether a pair of twins are monozygotic (MZ, identical) or dizygotic (DZ, fraternal). Traditional methods depended on an assessment of phenotypic resemblance or on the condition of the membranes at birth (see Box 5.3). Errors in zygosity determination systematically inflate heritability estimates for non-Mendelian characters, because very similar DZ twins are wrongly counted as MZ, whereas very different MZ twins are wrongly scored as DZ.

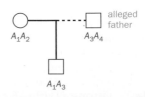

Figure 18.25 Using single-locus markers for a paternity test. The odds that the alleged father, rather than a random member of the population, is the true father are $\frac{1}{2}:q_3$, where q_3 is the allele frequency of A_3. A series of n unlinked markers would be used and, if paternity were not excluded, the odds would be $(\frac{1}{2})^n:q_A \times q_B \times q_C \times ... \times q_N$.

Genetic markers provide a much more reliable test of zygosity. The extensive but now obsolete literature on using blood groups for this purpose is summarized by Race & Sanger (see Further Reading). DNA profiling is nowadays the method of choice. The Jeffreys fingerprinting probe provided an immediate impression: samples from MZ twins look like the same sample loaded twice, and samples from DZ twins show differences. When single-locus markers are used, if twins give the same types, then for each locus the probability that DZ twins would type alike is calculated. If the parents have been typed, this follows from Mendelian principles; otherwise, the probability of DZ twins typing the same must be calculated for each possible parental mating and weighted by the probability of that mating calculated from population gene frequencies. The resultant probabilities for each (unlinked) locus are multiplied, to give an overall likelihood P_l that DZ twins would give the same results with all the markers used. The probability that the twins are MZ is then

$$P_m = m/[m + (1 - m)P_l],$$

where m is the proportion of twins in the population who are MZ (about 0.4 for same-sex pairs).

DNA profiling has revolutionized forensic investigations but raises issues of civil liberties

DNA profiling for forensic purposes follows the same principles as paternity testing. Scene-of-crime material (such as bloodstains, hairs, or a vaginal swab from a rape victim) is typed and matched to a DNA sample from the suspect. One of the most powerful applications of DNA profiling is for proving that a suspect is not the criminal. In the USA, the Innocence Project has demonstrated that many miscarriages of justice would have been avoided if DNA evidence had been used properly. If the samples do not match, the suspect is excluded, regardless of any circumstantial evidence to the contrary.

Technical Issues

The match probabilities in Table 18.8 are the probability that an unrelated individual would have a specific DNA profile—for example, the profile of DNA recovered from a crime scene. These are conservative estimates: some alleles are more common than others, so the exact match probability will vary between different profiles. For the current CODIS and SGM+ marker sets, the chances of a fortuitous match look so low as to make a DNA match irrefutable. However, three factors may reduce the degree of certainty.

First, the chances are much higher that relatives would give a full match. Because criminality often runs in families, this is not a purely academic point. Second, crime-scene samples often do not give a full profile. When the quantity of DNA is very small and/or is degraded, some alleles may fail to amplify (allele drop-out). Below a certain threshold quantity of DNA (typically 100 pg, which is the DNA content of about 17 cells), stochastic effects begin to be significant. Running the PCR for extra cycles or in a reduced volume may produce a profile when the standard protocols fail, but the profile is often only partial. This so-called Low Copy Number (LCN) technique has been quite controversial. Some major crimes have been solved with it, but there are questions about how reliable it is. Finally, crime-scene samples often contain a mixture of the DNA of several individuals. How can the individual profiles be teased out of a mixed profile? Provided that the component DNAs are present in different amounts, the peak sizes can be used to disentangle the individual contributions. With LCN data this is much more of a problem, because stochastic effects mean that the relative amounts of PCR products do not necessarily reflect the amounts of template. In addition, with very small amounts of DNA, even the most trivial contamination can be very significant.

The markers in the CODIS and SGM+ sets have been chosen, among other reasons, because they give no personal information; they merely serve to identify a person. This helps make profiling less contentious. However, the DNA sample potentially contains all sorts of personal information. One intriguing possibility concerns the link between Y-chromosome haplotypes and surnames. In societies in which the surname is taken from the father, such links should exist. Pilot studies have indeed demonstrated linkage, at least for unusual surnames. Much of our physical appearance is also encoded in our DNA. People dream of a DNA photofit—a portrait (and maybe surname) of the criminal revealed by the DNA left at the crime scene! Considerable development work is taking place toward that goal, but it remains far distant. Currently only eye and hair color can be even roughly predicted. It is also worth remembering that, whatever might be possible with a good blood sample, crime-scene DNA is often too limited in quantity and quality to allow extensive genotyping.

Courtroom issues

If the genotypes do all match, the court needs to know the odds that the criminal is the suspect rather than a random member of the population. Of course, if the alternative were the suspect's brother or the suspect's identical twin, the odds would look very different. It is also important to remember that a DNA match may prove beyond reasonable doubt that the suspect had been present at the crime scene, but (apart perhaps in rape cases) it cannot prove what he or she did there. By itself, a DNA match is not normally enough to secure a conviction.

The fate of DNA evidence in courts provides a fascinating insight into the difference between scientific and legal cultures. Suppose that a suspect's DNA profile matches the scene-of-crime sample. There are still at least three obstacles to a rational use of the DNA data.

First, the jury may simply not believe, or may perhaps choose to ignore, the DNA data, as evidently happened in the O.J. Simpson trial. Maybe they will decide the incriminating DNA was planted—a not unreasonable proposition.

Second, an unscrupulous lawyer may try to lead the jury into a false probability argument, the so-called Prosecutor's Fallacy. This consists of confusing the probability that the suspect is innocent, given the match, with the probability of a match, given that the suspect is innocent. The jury should consider the first probability, not the second. As **Box 18.3** shows, the two are very different.

Finally, objections may be raised to some of the principles by which DNA-based probabilities are calculated:

- The *multiplicative principle*, that the overall probabilities can be obtained by multiplying the individual probabilities for each allele or locus, depends on the assumption that genotypes are independent. If the population actually consisted of reproductively isolated groups, each of whom had genotypes that were fairly constant within a group but very different between groups,

---

BOX 18.3 THE PROSECUTOR'S FALLACY

Suppose that a court case has good evidence that the person whose DNA was found at the crime scene committed the crime. The suspect's DNA profile matches the crime-scene sample. Does that make the suspect guilty? Consider two different probabilities:

- The probability that the suspect is innocent, given the match.
- The probability of a match, given that the suspect is innocent.

Using Bayesian notation (see Box 18.2) with M = match, G = suspect is guilty, and I = suspect is innocent, the first probability is $P_I|M$ and the second is $P_M|I$. The Prosecutor's Fallacy consists of arguing that the relevant probability is $P_M|I$ when in fact it is $P_I|M$. The calculation below shows how different these two probabilities are.

If the suspect were guilty, the samples would necessarily match: $P_M|G = 1$. Let us suppose that population genetic arguments say there is a 1 in 10^6 chance that a randomly selected person would have the same profile as the crime-scene sample: $P_M|I = 10^{-6}$. Suppose that the guilty person could have been any one of 10^7 men in the population. If there is no other evidence to implicate him, he is simply a random member of the population, and the prior probability that he is guilty (before considering the DNA evidence) is $P_G = 10^{-7}$. The prior probability that he is innocent is $P_I = 1 - 10^{-7}$, which is very close to 1.

Bayes's theorem tells us that

$$P_I|M = (P_I \times P_M|I) / [(P_I \times P_M|I) + (P_G \times P_M|G)]$$
$$= 10^{-6}/(10^{-6} + 10^{-7})$$
$$= 1.0/1.1$$
$$= 0.9.$$

We already saw that $P_M|I = 10^{-6}$—quite a difference!

Courts make fools of themselves by ignoring compelling DNA evidence—but this calculation also shows that DNA evidence could not safely convict somebody unless $P_M|I$ was well below 10^{-6}. Plans to screen all men in a large town to find a rapist need to take account of this.

the calculation would be misleading. This is serious because it is the multiplicative principle that allows such exceedingly definite likelihoods to be given.

- The match probability depends on the gene frequencies. DNA profiling laboratories maintain databases of gene frequencies—but were these determined in an appropriate ethnic group for the case being considered?

Taken to extremes, the argument about the independence of genotypes implies that the DNA evidence might identify the criminal as belonging to a particular ethnic group, but would not show which member of the group it was who committed the crime. These issues have been debated at great length, especially in the American courts. The argument is valid in principle, but the question is whether it makes enough difference in practice to matter. It has become clear that, in general, it does not. It would be ironic if courts, seeing opposing expert witnesses giving odds of correct identification differing a millionfold (10^5:1 versus 10^{11}:1), were to decide that DNA evidence is hopelessly unreliable and were to rely instead on eye-witness identification (for which the odds of correct identification are less than 50:50).

Ethical and political issues

DNA profiling is a hugely powerful tool in the fight against crime, but its power raises a host of ethical and social issues. Most of these center on the associated databases. DNA databases are not inevitable concomitants of DNA profiling. A profile from a crime scene could be compared directly with profiles obtained from suspects in that particular case, without needing any database. However, most governments have established databases containing profiles of named individuals because these greatly increase the likelihood of identifying the source of a crime-scene sample. The UK has, for many years, had the largest forensic DNA database in the world per head of population, incorporating approximately six per cent of the population (but a much higher proportion of young adult males, and especially of men from certain ethnic minorities). At the end of March 2009, the UK National DNA Database held DNA profiles of 4,859,934 named individuals and 350,033 profiles from crime-scene samples. The US CODIS database is numerically slightly larger but covers a smaller percentage of the population.

Many ethical and political questions concern whose profile should be on the database. In the UK, successive modifications of the rules have led to ever wider criteria for inclusion. Anybody arrested in the UK for a recordable offence (anything except the most trivial offences) can be compelled to provide a sample, and the profile will remain on the database forever (or at least until the 100th anniversary of their birth). There is no requirement that they should be convicted of any crime, or even charged—the simple fact of being arrested triggers sampling. The rules in other countries are usually more restrictive, for example specifying that the person must be convicted, and maybe for only certain categories of serious crime. Civil liberties groups complain about the excessive use of DNA sampling in the UK; the police quote cases where the profile of somebody arrested for a minor offence has implicated them in a major unsolved crime. Following a 2008 ruling by the European Court of Human Rights, the policy in the UK is under review.

Further questions concern two additional uses of profiling. Because allele frequencies differ between ethnic groups, it is possible to produce an ethnic inference about the person who left DNA at a crime scene. Inferences from the standard CODIS or SGM+ profiles are usually quite weak; stronger inferences may be obtained by typing for panels of ancestry informative markers—markers whose allele frequencies differ widely between groups. Some people find ethnic inferencing ethically questionable. Another controversial technique is familial searching. If the crime-scene profile does not exactly match any in the database, it is possible to ask whether the person who left DNA at the crime scene could be a relative—say the son or a brother—of somebody whose profile is in the database. Familial searching generates very long lists of names, which must then be prioritized and eliminated by standard police methods. Because of this, it is only used for the most serious crimes—but some people worry about the possibility of improperly uncovering family secrets. These and other ethical concerns have been extensively reviewed in a report by the Nuffield Council on Bioethics (see Further Reading).

CONCLUSION

In this chapter we have reviewed some of the many ways in which a DNA sequence can be checked for the presence of known or unspecified variants.

For cases in which it is necessary to *scan one or more candidate genes for any variant*, DNA sequencing is the main tool. Normally, genomic DNA is sequenced, but cDNA may be preferable for large genes with numerous small exons, or if abnormal splicing is suspected. However, RNA requires more careful handling than DNA. DNA or RNA samples are generally amplified by PCR (RT-PCR in the case of RNA). If it is important to reduce the sequencing load, various techniques allow a quick scan of exons of a gene, allowing sequencing to be restricted to those that show some variant. These techniques exploit differences in the properties of DNA heteroduplexes or in the conformation of single-stranded DNA. Alternatively, microarrays can be used to scan a whole gene sequence for variants. The continuing rapid improvements in sequencing technology have made all these alternatives to sequencing less attractive; for most diagnostic laboratories the main bottleneck in mutation analysis is now in analyzing the sequence data rather than in generating it.

Another group of tests are designed to detect *specific mutations* in *a particular gene*. Commercial kits for such tests are increasingly available. Allele-specific oligonucleotide probes can be used in reverse dot-blot hybridization or, on a large scale, as probes on a DNA microarray. In ARMS (amplification refractory mutation system) testing, pairs of allele-specific primers are used in PCR to discriminate between normal and mutant sequences. In the oligonucleotide ligation assay (OLA), mutations prevent the DNA ligation of two paired oligonucleotides and thus PCR amplification of the test sequence. Specialized sequencing technologies such as pyrosequencing or minisequencing identify just one or a few nucleotides adjacent to a primer, thus allowing specific mutations to be detected. The same microarray techniques used for mutation detection can also be used to search whole genomes for specific mutations or SNPs.

Special tests are required to identify heterozygous deletions or copy-number changes in genomic DNA. Multiplex ligation-dependent probe amplification (MLPA), a variant of the OLA test, is a popular method for detecting deletions or duplications of exons within a gene. At the level of whole chromosomes, quantitative PCR of a panel of fluorescence-labeled microsatellites (QF-PCR) is often used as an alternative to standard karyotyping for prenatal diagnosis of the common fetal trisomies. For genomewide detection of copy-number changes, array-comparative genomic hybridization or high-resolution SNP chips can be used.

The older technique of gene tracking in a family pedigree is still useful in some situations, but care needs to be taken in the choice of DNA marker and calculation of the risk.

DNA profiles, consisting of genotypes for a panel of polymorphic markers, are of great use in identifying individuals and determining relationships. The older DNA fingerprinting technique, based on hypervariable minisatellites, has been superseded by single-locus profiling using standardized panels of polymorphic microsatellites. The main use of DNA profiling is in criminal investigations, in which profiling and the associated police DNA databases have revolutionized practice but have also raised concerns about possible infringements of civil liberties. Courts have not always found it easy to handle the scientific questions raised by DNA profiling within an adversarial legal framework. It is important that they should recognize the limitations, as well as the powers, of DNA profiling. Other uses for DNA profiling include resolving cases of disputed paternity, identifying the origin of clinical samples (e.g. between mother and fetus, or transplant donor and recipient), and determining the zygosity of twins.

The emphasis in this chapter has been on the techniques used to genotype a sample; except for the problem of unclassified variants there was little need to discuss the significance of the result. However, the same techniques can be used in situations where there is much more uncertainty about how useful the results might be. One such case is the use of genetic tests to predict somebody's response to a drug. Individual variations in response often have a genetic basis, and there are hopes that in the future genetic testing will lead to more efficient prescribing, with the ultimate goal of personalized medicine. How realistic is this goal? Further questions arise from the use of genetic tests in population screening. This raises a series of new issues, primarily about the circumstances in which screening would be useful, rather than about the technical procedures. All these matters are considered in Chapter 19.

FURTHER READING

Best practice guidelines

American College of Medical Genetics. http://www.acmg.net/
Clinical Molecular Genetics Society (UK). http://cmgsweb.shared
.hosting.zen.co.uk/
Cystic fibrosis mutation database. http://www.genet.sickkids
.on.ca/cftr/ [The best source of overview information about cystic fibrosis mutations.]
Leiden muscular dystrophy pages. http://www.dmd.nl/
[A comprehensive and well-maintained source of information on all molecular aspects of muscular dystrophy.]

RNA analysis

Andreutti-Zaugg C, Scott R & Iggo R (1997) Inhibition of nonsense-mediated messenger RNA decay in clinical samples facilitates detection of human *MSH2* mutations with an in vivo fusion protein assay and conventional techniques. *Cancer Res.* 57, 3288–3293. [Deals with the use of puromycin to identify mRNAs containing premature termination codons.]

Methods for scanning a gene for mutations

Cotton RGH, Edkins F & Forrest S (eds) (1998) Mutation Detection: A Practical Approach. IRL Press. [Individual chapters review a wide range of methods.]
Keen J, Lester D, Inglehearn C et al. (1991) Rapid detection of single base mismatches as heteroduplexes on Hydrolink gels. *Trends Genet.* 7, 5.
Sheffield VC, Beck JS, Kwitek AE et al. (1993) The sensitivity of single strand conformational polymorphism analysis for the detection of single base substitutions. *Genomics* 16, 325–332.
Sheffield VC, Beck JS, Nichols B et al. (1992) Detection of multiallele polymorphisms within gene sequences by GC-clamped denaturing gradient gel electrophoresis. *Am. J. Hum. Genet.* 50, 567–575.
van der Luijt R, Khan PM, Vasen H et al. (1994) Rapid detection of translation-terminating mutations at the adenomatous polyposis coli (*APC*) gene by direct protein truncation test. *Genomics* 20, 1–4.
Wallace AJ, Wu CL & Elles RG (1999) Meta-PCR: a novel method for creating chimeric DNA molecules and increasing the productivity of mutation scanning techniques. *Genet. Test.* 3, 173–183.

Analysis of DNA methylation

Ehrlich M, Nelson MR, Stanssens P et al. (2005) Quantitative high-throughput analysis of DNA methylation patterns by base-specific cleavage and mass spectrometry. *Proc. Natl Acad. Sci. USA* 102, 15785–15790.
Laird PW (2003) The power and the promise of DNA methylation markers. *Nat. Rev. Cancer* 3, 253–266.
Zeschnigk M, Lich C, Buiting K et al. (1997) A single-tube PCR test for the diagnosis of Angelman and Prader–Willi syndrome based on allelic methylation differences at the *SNRPN* locus. *Eur. J. Hum. Genet.* 5, 94–98. [An example of methylation-specific PCR.]

The problem of unclassified variants

International Agency for Research on Cancer: Align-GVGD. agvgd.iarc.fr/agvgd_input.php [A freely available, Web-based program that combines the biophysical characteristics of amino acids and protein multiple sequence alignments to predict where missense substitutions in genes of interest fall in a spectrum from deleterious to neutral.]
J. Craig Venter Institute: SIFT (Sorting Intolerant from Tolerant). http://sift.jcvi.org/ [Predicts whether an amino acid substitution affects protein function on the basis of sequence homology and the physical properties of amino acids.]
PolyPhen (Polymorphism Phenotyping). genetics.bwh.harvard .edu/pph/ [A tool that predicts the possible impact of an amino acid substitution on the structure and function of a human protein.]

Methods for testing for a specified sequence

Fakhrai-Rad H, Pourmand N & Ronaghi M (2002) Pyrosequencing: an accurate detection platform for single nucleotide polymorphisms. *Hum. Mutat.* 19, 479–485.
Liu Q & Sommer SS (2004) PAP: detection of ultra rare mutations depends on P* oligonucleotides: "sleeping beauties" awakened by the kiss of pyrophosphorolysis. *Hum. Mutat.* 23, 426–436.
Monforte JA & Becker CH (1997) High-throughput DNA analysis by time-of-flight mass spectrometry. *Nat. Med.* 3, 360–362.

Large-scale SNP genotyping

Affymetrix GeneChip. http://www.affymetrix.com/technology/ index.affx [Requires a log-in.]
Fan J-B, Oliphant A, Shen R et al. (2003) Highly parallel SNP genotyping. *Cold Spring Harbor Symp Quant Biol.* 68, 69–78.
Gut IG (2004) DNA analysis by MALDI-TOF mass spectrometry. *Hum. Mutat.* 23, 437–441.
Hardenbol P, Yu F, Belmont J et al. (2005) Highly multiplexed molecular inversion probe genotyping: over 10,000 targeted SNPs genotyped in a single tube assay. *Genome Res.* 15, 269–275.
Illumina BeadArray. http://www.illumina.com/pages.ilmn?ID=5
Stanssens P, Zabeau M, Meersseman G et al. (2004) High-throughput MALDI-TOF discovery of genomic sequence polymorphisms. *Genome Res.* 14, 126–133.
Tõnisson N, Zernant J, Kurg A et al. (2002) Evaluating the arrayed primer extension resequencing assay of TP53 tumor suppressor gene. *Proc. Natl Acad. Sci. USA* 99, 5503–5508.
Various authors (2002) The Chipping Forecast II. *Nat. Genet.* 32 (Suppl), 465–551. [Reviews of microarray-based technologies.]

Gene tracking

Bridge PJ (1997) The Calculation Of Genetic Risks: Worked Examples In DNA Diagnostics, 2nd ed. Johns Hopkins University Press. [A very detailed set of calculations covering almost every conceivable situation in DNA diagnostics.]
Young ID (2006) An Introduction To Risk Calculation In Genetic Counseling. Oxford University Press. [A clinically oriented set of risk calculations.]

DNA profiling

Gill P, Ivanov PL, Kimpton C et al. (1994) Identification of the remains of the Romanov family by DNA analysis. *Nat. Genet.* 6, 130–135.

Jeffreys AJ, Wilson V & Thein LS (1985) Individual-specific fingerprints of human DNA. *Nature* 316, 76–79.

Jobling MA & Gill P (2004) Encoded evidence: DNA in forensic science. *Nat. Rev. Genet.* 5, 739–751.

Nuffield Council on Bioethics (2007) The forensic use of bioinformation: ethical issues. http://www.nuffieldbioethics .org/go/ourwork/bioinformationuse/publication_441.html [An influential report on the ethical implications of the use of DNA profiles and other bioinformation by the police.]

Race RR & Sanger R (1975) Blood Groups In Man, 6th ed. Blackwell. [The classic source on these old techniques.]

Vogel F & Motulsky AG (1996) Human Genetics: Problems And Approaches, 3rd ed. pp 761–767. Springer. [Sample calculations for testing twins for monozygosity are given in Appendix 4 of that book.]

Weir BS (1995) DNA statistics in the Simpson matter. *Nat. Genet.* 11, 365–368. [A fascinating account of the O.J. Simpson trial.]

Quality assurance

Organisation for Economic Co-operation and Development. OECD Guidelines for quality assurance in molecular genetic testing. http://www.oecd.org/dataoecd/43/6/38839788.pdf

Chapter 19

Pharmacogenetics, Personalized Medicine, and Population Screening

KEY CONCEPTS

- Medicine in the post-genomic era holds the promise of testing of individuals to deliver personalized medicine and screening of populations to determine susceptibility to common diseases.

- Any clinical test, including genetic tests, should be evaluated by first considering its analytical validity, which includes its sensitivity, specificity, and predictive value.

- A broader evaluation of a clinical test should consider clinical validity, clinical utility, and ethical implications.

- The most crucial question about a proposed test is how often the result, in combination with other available data, will move participants across a threshold for possible further action.

- Pharmacogenetics is concerned with the way in which genetic variations between individuals cause different responses to a drug.

- Many pharmacogenetic variations affect the rate of absorption or metabolism of drugs (pharmacokinetics).

- Other pharmacogenetic variations affect the response of a drug target to a given level of the drug (pharmacodynamics).

- One aim of personalized medicine is to tailor the drug and dosage that an individual receives to his/her genotype. Obstacles to personalized prescribing include the inadequate predictive power of most genotypes and the need to develop rapid real-time bedside genotyping methods.

- Genotype-specific prescribing is closest to reality in cancer medicine, in which some drugs are already prescribed only after a tumor specimen has been genotyped.

- Expression profiling of tumors may eventually guide treatment, but its value has yet to be confirmed by large prospective studies.

- Testing for susceptibility to complex diseases is a growing but largely unregulated industry. Tests marketed directly to consumers over the Internet are often not supported by evidence of clinical utility.

- Population screening usually aims at defining a high-risk subset of the population. This will almost certainly include many individuals who do not, in fact, have the condition being screened for. People in the high-risk group are then offered some intervention, most usually a definitive diagnostic test.

- Screening tests usually involve a trade-off between sensitivity and specificity, with an arbitrary intervention level being defined to optimize the compromise. Screening programs need very careful consideration to ensure that they do more good than harm.

- A vision for the future role of genetics in medicine is to move from the current reactive 'diagnose and treat' system to a proactive 'predict and prevent' system. Whether this will be achievable, and if so when it will happen, are extremely contentious and unresolved questions.

In the previous chapter, we considered the technicalities of genetic testing and described many examples of the way in which laboratory tests can be used either to check for the presence of a specific predetermined DNA sequence change or to scan a gene sequence for any change. But many proposed applications of genetic testing raise issues that were not covered in that chapter.

As genotyping and DNA sequencing become cheaper and faster, more and more people will come to have more and more information about their own genomes. Some of this information will predict their responses to drugs and their susceptibilities to diseases. Will this lead to radical changes in the whole system of health care—a switch from the current 'diagnose and treat' mode to a 'predict and prevent' approach—over the next one or two decades? Or do these visions ignore the complexity of human life, the profound roles of environment and lifestyle in shaping health, and the fact that in a free society people can, and often do, choose to ignore good advice? These are major questions about future developments in the interlinked areas of personalized medicine and population screening.

We cannot give firm answers to the big questions, but we hope this chapter will provide readers with information and tools for forming their own opinions. The first tools for this task are those for evaluating any proposed test.

19.1 EVALUATION OF CLINICAL TESTS

All clinical tests need to be evaluated. This is as true of genetic tests as of any other. When a health care provider or insurance company is paying for the test, they need to know that they are getting value for money. There is never enough money to do everything that one might wish to do in a perfect world, so it is not enough that the test should offer some value: it should represent a better use of that money than any of the alternatives. Individuals choosing to buy a test over the Internet may perhaps not be so concerned with value for money, but they still need to understand what it is that they are buying, and to be confident that the advertising is not misleading. The test must do what it claims to do and produce realistic risk estimates.

A widely used framework for evaluation is the ACCE scheme (**Box 19.1**). Sometimes a distinction is made between an *assay* and a *test*. An assay is the actual laboratory result—a genotype or measurement of an analyte—whereas a test, in this context, is the overall result of deriving and assembling the information on which to base a clinical decision. The analytical validity would consider the assay, whereas the wider issues are part of the test. The distinction can be helpful in avoiding overemphasis on the assay in determining the usefulness of a test.

The analytical validity of a test is a measure of its accuracy

Analytical validity is the most basic measure of test performance. Measures of analytical validity relevant to genetic tests are set out in **Box 19.2**, and include sensitivity and specificity. These can involve wider or narrower considerations, depending on what the test claims to do. Where an assay claims simply to identify a single specific genotype, the definition of a positive or negative result is unambiguous—the target genotype is either present or absent. Some assays report the concentration of an analyte. In that case, the analytical validity would be measured by the variation in replicate measures on the same sample, and by the accuracy in reporting the result on test samples that had a known true result. In all

---

BOX 19.1 THE ACCE FRAMEWORK FOR EVALUATING TESTS

The ACCE framework considers four aspects of test performance:
- **Analytical validity:** how well does the test measure what it claims to measure?
- **Clinical validity:** how well does the test predict the health outcome that it claims to?
- **Clinical utility:** what use is the result?
- **Ethical aspects:** does the test conform to ethical standards?

This framework is applicable to any clinical test, including genetic tests.

BOX 19.2 MEASURING THE ANALYTICAL VALIDITY OF A TEST

The simplest measures of test performance are its sensitivity and specificity. Using the numbers defined in **Table 1**, the **sensitivity** is $a/(a + b)$; that is, the proportion of all people who have the condition who are identified by the test. The **specificity** is defined as $d/(c + d)$, and is the proportion of all people who do not have the condition where the test result correctly predicts absence of the condition. Other measures include:

- **False positive rate** = $c/(a + c)$: the proportion of positive test results that wrongly identify a person as having the condition.
- **False negative rate** = $b/(b + d)$: the proportion of negative test results that wrongly identify a person as not having the condition.
- **Positive predictive value** = $a/(a + c)$: the proportion of positive test results that correctly identify a person with the condition.
- **Negative predictive value** = $d/(b + d)$: the proportion of negative test results that correctly identify a person as not having the condition.

Table 1 **Performance of a test**

	Condition present	Condition absent
Test positive	a	c
Test negative	b	d

these cases the analytical validity covers just laboratory quality issues. It should not be taken for granted. In 2001, investigators in California sent hair samples from the same person to six of the leading US laboratories that offered hair mineral analysis to people concerned about nutritional deficiencies. The results from the different laboratories were grossly disparate and overall not far from random. Before relying on results from a laboratory it is advisable to check that it is enrolled in a reputable quality assurance scheme.

The measures become more interesting if the test is defined as the totality of what the laboratory does to try to answer a clinical question. In Chapter 18 we distinguished between testing for a specific predefined sequence change and scanning an entire gene for mutations. The clinically relevant question is usually whether any mutation has been detected anywhere in the relevant gene. The sensitivity of the test will depend on how hard we look. Will we sequence every exon, or will we rely on a method such as single strand conformation polymorphism or melting curve analysis? Will we use multiplex ligation-dependent probe amplification (MLPA) to check for whole exon deletions, in addition to sequencing exons? Scanning the entire gene also raises the problem of unclassified variants, discussed in Chapter 18. If a nonpathogenic variant is mistakenly reported as a positive test result, the specificity of the test is reduced.

The clinical validity of a test is measured by how well it predicts a clinical condition

Clinical validity is a separate issue from analytical validity. Even if the hair-testing laboratories cited above had correctly reported the mineral content of the samples, there is precious little evidence that the results would have any relevance to the health of the person whose hair they analyzed. Even some mainstream tests have limited clinical validity. Men concerned about their risk of prostate cancer are often offered a test of their blood level of prostate specific antigen (PSA). A laboratory may measure the PSA level with high analytical validity, but it is a very poor indicator of prostate cancer. The distributions do differ in cases and controls, and extremely high levels are a danger signal, but there is a large overlap and little predictive value.

In genotyping for Mendelian conditions, it is obvious whether a test result is positive or negative; the only question is what such a result should mean. But the clinical validity depends on the degree of correlation between genotype and phenotype. Genetic diseases are often variable, especially the many dominant

Mendelian diseases that are caused by haploinsufficiency. The severity of these conditions is usually extremely variable, even within families. Figure 3.17 illustrates this for Waardenburg syndrome. People occasionally request prenatal testing for this syndrome. A test could accurately tell whether a fetus had inherited the family mutation, but it could not predict how severely that individual would be affected. Hemochromatosis (OMIM 235200; an iron overload condition) is another example. It is desirable to identify people with this condition because very serious liver problems, including cirrhosis and cancer, can be avoided by the simple expedient of regular bloodletting before symptoms develop. Most people with clinical hemochromatosis are homozygous for one specific mutation, p.C282Y in the *HFE* gene. A simple test for that mutation could have high analytical validity—but the clinical validity would be low because most p.C282Y homozygotes remain healthy and do not suffer the clinical condition.

When the laboratory assay is of a quantitative variable, such as the concentration of some analyte or the level of expression of a gene, a decision must be made as to where to draw the line between a normal and an abnormal level, or a positive and a negative test result. These are analog tests, as compared with genotyping tests, which are digital. Here, as in other fields, digital data are easier to interpret and manage than analog data. Where the line is drawn can have a major bearing on the clinical validity of the test. Prenatal screening for Down syndrome, considered in Section 19.5, is a good example of a test in which an arbitrary threshold must be drawn across continuously variable analog test data. Such decisions involve a trade-off between sensitivity and specificity. The threshold can be drawn so as to catch virtually all cases (high sensitivity), but only at the cost of also catching many unaffected people (low specificity). Alternatively, a high threshold can be used so as to avoid catching unaffected people (high specificity), but this will usually be at the price of missing many affected cases (low sensitivity).

Clinical validity is of particular relevance when considering tests for genetic susceptibility to common diseases. Laboratories can genotype SNPs with high analytical validity, but the power of the results to predict disease susceptibility is much more limited. These factors are neither necessary nor sufficient for the disease to develop; instead, what matters is the relative risk. This is measured by the likelihood ratio or odds ratio, as set out in **Box 19.3**. The whole question of the clinical validity and utility of testing for genetic susceptibility to common diseases is discussed in Section 19.4.

Tests must also be evaluated for their clinical utility and ethical acceptability

Clinical utility, the third aspect of the ACCE framework (see Box 9.1), broadens the questions further. Even if the test makes an accurate prediction of disease risk, is this information useful to the patient? When the prediction leads to successful prevention or treatment the utility is obvious, but what about the test for Huntington disease? This test has high clinical validity. Finding a CAG repeat number of 45 in the *HTT* gene tells the patient that they will definitely eventually develop the disease (unless something else kills them first). But at present there is nothing they can do to modify that risk. So has any useful purpose been served by doing the test? Different patients will have different answers to that question. Anybody who takes the test will have a family history of the disease and will know they are at high risk. Some people want to know for certain. Of course they hope their test will be negative, but even if it is positive they can plan ahead for when the disease takes over their life. However, most people at risk of Huntington disease decide after careful counseling that they would rather not know. They would rather live with uncertainty than take the risk of discovering that they are doomed to develop the disease. For them there is no clinical utility.

Ethical issues would include how far the test is voluntary, and its potential to stigmatize people or adversely affect their prospects of insurance, employment, and so on. Other issues might concern the desirability of offering prenatal diagnosis for a relatively trivial or late-onset condition, or the implications of offering a screening program to people of just one of several ethnic groups in a country when the disease in question is found mainly just in that one group.

BOX 19.3 MEASURES OF RELATIVE RISK

Tests of disease susceptibility seek to compare an individual's risk of developing a disease with the general population risk. Available measures of the relative risk include the likelihood ratio and the odds ratio.

The likelihood ratio measures the likelihood that a person who has the susceptibility factor in question will develop the disease, compared with the likelihood of a person in the general population (who may or may not have the susceptibility factor). Alternatively, one could measure the likelihood relative to somebody who does not have the susceptibility factor. In either case, calculating the likelihood ratio requires knowledge of the epidemiology of the disease.

For case-control studies, the usual measure is the odds ratio (OR). In contrast with the likelihood ratio, this can be calculated directly from the results of the study, as shown in **Table 1**, without the need for wider population data. Its clinical validity depends, of course, on the control group's being representative of the wider population, as well as being matched to the cases for other factors that could independently produce an association with the disease.

Table 1 Categories in a case-control study

	Cases	Controls
Risk genotype present	a	c
Risk genotype absent	b	d

In this study, for a person who has the risk genotype, the odds of being affected (being a case rather than a control) are a:c or a/c:1. For a person without the risk genotype, the odds are b:d or b/d:1. The odds ratio is

$$\frac{(a/c)}{(b/d)} = ad/bc.$$

Imagine a study of 1000 cases and 1000 controls in which the risk factor is present in 80% of the cases but only 70% of controls (**Table 2**).

Table 2 Data from a hypothetical case-control study

	Cases	Controls
Risk genotype present	800	700
Risk genotype absent	200	300

The OR is $(800 \times 300)/(200 \times 700) = 1.7$. However, if the factor were present in only 8% of cases and 7% of controls, the OR in the same study would be $(80 \times 930)/(920 \times 70) = 1.15$. A more intuitive measure of the relative risk, the frequency of the risk genotype in cases compared with controls, is the same in both examples: 8/7 or 1.14. Odds ratios approach that intuitive risk ratio when the absolute risk is low (8% versus 80% in the examples above).

Odds ratios must be interpreted with great caution, especially for high-frequency risk alleles. It would be easy, but very wrong, to suppose that the odds ratio of 1.7 in our first example means that having the risk genotype gives a person a 70% greater chance of developing the disease. However, the intuitive interpretation of an OR of 1 is correct: it means that a factor has no effect on risk.

Despite being less intuitive, ORs are used in case-control studies because of their statistical properties, particularly in relation to logistic regression, the statistical technique used to tease out single effects from a complex set of factors. Note that the OR in Table 2 does not change if the number of cases or controls is changed, provided that the proportion of each who have the risk genotype remains unchanged.

If we consider a genetic risk allele, the OR will probably be different for people heterozygous or homozygous for that allele. Unless there is clear evidence of dominance or recessiveness, susceptibility alleles are usually assumed to have a multiplicative effect. The ORs for genotypes AA, Aa, and aa are assumed to scale in the ratio

1:r:r^2.

A per-allele OR is normally quoted.

The mathematically adept can find some extensions of these calculations in the paper by Wang and colleagues (see Further Reading).

In the end, the value of a test must be judged by the changes it brings about. Whatever the performance of a test according to the criteria outlined above, in the end what matters is what it achieves for patients. Does it change what people do? For most conditions there will be some threshold for action—for taking X-rays, doing a biopsy, or prescribing a drug. The result of any single test will be considered alongside all other relevant information—the age and sex of the patient, their general health and social situation, and the results of any other tests. A useful test is one that makes a substantial contribution to moving people across the action threshold, in either direction. In this connection, absolute risk is much more important than relative risk. A test may alter somebody's risk tenfold, but if the effect is only to change the risk from 1 in 10,000 to 1 in 1000 it will probably not make any practical difference. In contrast, a positive test for a *BRCA1* mutation has a relative risk of only about 7 (80% for a carrier versus 12% general population risk), but the test result is important because of the high absolute risk.

19.2 PHARMACOGENETICS AND PHARMACOGENOMICS

The core of any move to personalized medicine must be personalized prescribing. We all know that many drugs affect different people in different ways. Often there is differential sensitivity: some people need higher or lower doses to achieve the same therapeutic effect. Sometimes a drug simply has no therapeutic effect in certain individuals. A drug that most people take without any problems may cause an adverse reaction in a few individuals. Some of these differential effects are due to environmental causes: a person's ability to absorb or metabolize a drug may be changed by his illness or lifestyle. Sick people are often taking multiple drugs, and some combinations may interact. But many differences are due to genetic variation between people. **Pharmacogenetics** is the study of the roles of specific genes in these effects, whereas **pharmacogenomics** uses genomewide tools for the same purpose.

Pharmacokinetics studies the absorption, activation, metabolism, and excretion of drugs, whereas **pharmacodynamics** considers the actual target response. In other words, pharmacokinetics addresses what the body does to a drug, whereas pharmacodynamics addresses what the drug does to the body. Genetic factors are relevant to both. There are at least four stages at which genetic variations can affect a patient's response to a drug:

- *Absorption*—individuals may differ in their ability to transport an oral drug into the bloodstream.

- *Activation*—many drugs are given in the form of a prodrug that must undergo an enzymatic reaction, usually in the liver, to convert it to the active form.

- *Target response*—in different patients, the process or pathway targeted by the drug may respond differently to a given local concentration of the drug.

- *Catabolism and excretion*—individuals often differ in the rate at which they catabolize and dispose of the active drug molecule. Slow metabolizers will have a longer or stronger response to a given concentration of the drug than fast metabolizers will.

Adverse drug reactions are a serious problem. It has been estimated that they are responsible for about 100,000 deaths a year in the USA, and for 1 in 15 of all hospital admissions in the UK. Added to this is all the wasted time and money, and the extra suffering involved when a patient fails to respond to a drug—an especially severe problem with psychiatric drugs.

Many genetic differences affect the metabolism of drugs

The reactions of drug metabolism are traditionally divided into two phases. Phase 1 reactions (oxidation, hydroxylation, and hydrolysis) often produce the biologically active molecule, although sometimes the phase 1 product might be an intermediate in the inactivation and degradation of the drug. Phase 2 reactions (conjugation reactions such as acetylation, glucuronidation, or sulfation) produce a water-soluble compound that is more easily excreted. Not every drug is processed through both phases, but **Figure 19.1** shows a common sequence of events.

Enzymes involved in both phases often show polymorphic variations in activity that affect responses to many drugs. The natural function of these enzymes is in handling xenobiotics (foreign substances), maybe in particular plant alkaloids present in the diet. Interesting studies relate these polymorphisms to shifting selective pressures driven by variations in diet across time and location. Because the variations affect the handling not just of drugs but also of innumerable other chemicals in our diet and environment, the genes encoding these enzymes have been widely studied as candidates for susceptibility to cancer and other diseases, as well as for their effect on drug metabolism.

The P450 cytochromes are responsible for much of the phase 1 metabolism of drugs

P450 cytochromes constitute a large family of enzymes that have an iron–sulfur active site and a spectral absorption peak at 450 nm. They act by inserting a single oxygen atom derived from molecular oxygen into a very wide range of organic compounds. The end product is usually a polar, water-soluble hydroxylated

Figure 19.1 Stages in the metabolism of a drug. Phase 1 often involves oxidation, hydroxylation, or hydrolysis to produce a polar compound that is water soluble. This may be the active molecule itself or a degradation product. In phase 2, the molecule may be conjugated with an acetyl, glucuronosyl, or glutathionyl group to form a water-soluble molecule that can be excreted.

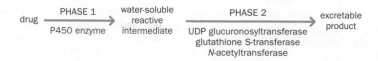

derivative of the substrate. The P450 cytochromes probably evolved to deal with toxic plant metabolites, and variants may be important in determining the response to chemicals in the diet and environment, as well as to prescribed drugs. The enzymes are found mainly in the liver, but some occur in the small intestine and elsewhere. Some drugs have an additional effect of inducing or inhibiting specific P450 enzymes, which can lead to unforeseen interactions between these drugs and those that are substrates of the enzyme. The Cytochrome P450 Drug Interaction Table (see Further Reading) maintains comprehensive lists.

Humans have about 60 P450 genes, encoding enzymes that, between them, are responsible for the phase 1 metabolism of maybe 60% of all prescribed drugs. They are grouped into several families. Genes have names such as *CYP2D6* (cytochrome P450 family 2, subfamily D, polypeptide 6). At least 10 different P450 enzymes have significant roles in phase 1 drug metabolism. Many of them can metabolize a considerable range of different drugs, and individual drugs are often substrates for more than one P450 enzyme. These wide and overlapping specificities mean that there are not always close correlations between any one P450 activity in a patient and his/her handling of a specific drug. Nevertheless, some drugs are metabolized by just one P450 enzyme, and in pharmacological research these drugs can be used as probes to identify individuals with unusual functional variants of that enzyme.

CYP2D6

Back in the 1970s, it was noted that some individuals showed markedly enhanced sensitivity to the antihypertensive drug debrisoquine and to the anti-arrhythmic drug sparteine. On investigation, these individuals showed high plasma levels of the drugs but low urinary levels of the catabolism products, implying that they were failing to metabolize and excrete the drugs. The cause was eventually identified as low activity of the CYP2D6 enzyme. Individuals can be classified into poor, intermediate, extensive, and ultra-rapid metabolizers. **Figure 19.2** shows how CYP2D6 activity governs the effective dose of the antidepressant drug nortriptyline, which is metabolized by this enzyme.

CYP2D6 is involved in the metabolism of perhaps 25% of all drugs. Variation in its activity has significant effects on the response to some beta-blockers used to treat hypertension and heart disease, and to several psychiatric drugs including tricyclic antidepressants. Poor metabolizers are at risk of overdose effects of these drugs. In contrast, CYP2D6 is also required to convert codeine into its active form, morphine. Codeine is ineffective in pain relief for poor metabolizers, whereas ultra-rapid metabolizers risk adverse effects such as sedation and impaired breathing.

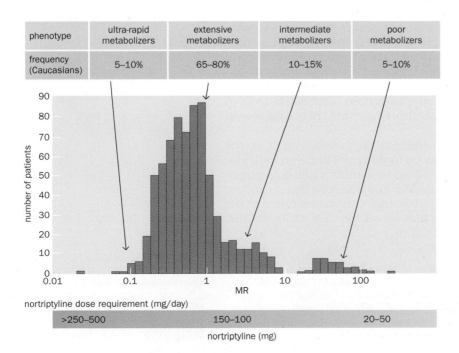

Figure 19.2 Pharmacogenetics of debrisoquine and nortriptyline. The activity of the CYP2D6 enzyme is measured by the metabolic ratio (MR), which is the ratio of amounts of a substrate drug, debrisoquine and its metabolic product in urine after a standard dose of the drug. High ratios show poor conversion due to low enzyme activity. The graph shows observed ratios for a range of patients. The same enzyme is largely responsible for phase 1 catabolism of the antidepressant drug nortriptyline. Depending on the CYP2D6 phenotype as measured by the metabolic ratio, patients require different doses of nortriptyline. [Adapted from Meyer UA (2004) *Nat. Rev. Genet.* 5, 669–676. With permission from Macmillan Publishers Ltd.]

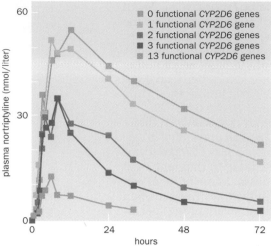

Figure 19.3 Effect of variable copy number of *CYP2D6* genes on the rate of metabolism of nortriptyline. Subjects with different numbers of *CYP2D6* genes were given a standard oral dose of 25 mg. The graphs show the plasma concentrations of nortriptyline over the following 72 hours. [Adapted from Dalén P, Dahl ML, Bernal Ruiz ML et al. (1998) *Clin. Pharmacol. Ther.* 63, 444–452. With permission from Macmillan Publishers Ltd.]

The *CYP2D6* gene has nine exons and is located on chromosome 22q13. Sequencing the gene in poor metabolizers has revealed a variety of point mutations, mainly nonsense, frameshift, or splice site changes, and occasional complete gene deletions. Interestingly, the Human Genome Project chromosome 22 reference sequence happened to be from a deletion carrier, so *CYP2D6* was not recorded in the original sequence. Ultra-rapid metabolizers have increased numbers of *CYP2D6* genes (**Figure 19.3**).

Other P450 enzymes

Another P450 cytochrome, **CYP2C9**, hydroxylates various drugs, including non-steroidal anti-inflammatory drugs, sulfonylureas, inhibitors of angiotensin-converting enzyme, and oral hypoglycemics. For example, rare poor metabolizers have an exaggerated response to tolbutamide, a hypoglycemic agent that is used to treat type 2 diabetes. The role of CYP2C9 variants in hypersensitivity to warfarin is discussed below in the section on personalized medicine.

CYP2C19 metabolizes a variety of drugs, including anti-convulsants such as mephenytoin, proton pump inhibitors such as omeprazole (used to treat stomach ulcers), proguanil (an antimalarial), and certain antidepressants. Subjects can be classified phenotypically into poor, intermediate, and extensive metabolizers (**Figure 19.4**). Intermediate metabolizers are compound heterozygotes for an active and an inactive allele; poor metabolizers have two inactive alleles. The frequency of poor metabolizers is 3–5% among Caucasians and African-Americans, but higher in Orientals and very high in Polynesians. Mephenytoin and omeprazole are degraded by CYP2C19, so poor metabolizers require a lower dose. A particular risk for poor metabolizers is unacceptably prolonged sedation with a standard dose of diazepam as a result of slow demethylation by CYP2C19. In contrast, proguanil is a prodrug that requires activation by CYP2C19 to form the active molecule, cycloguanil; poor metabolizers therefore show a decreased effect of that drug.

CYP3A4 is another P450 cytochrome involved in the metabolism of maybe 40% of all drugs. Its activity in the liver varies up to thirtyfold between individuals, but the cause of this variability has been hard to identify. Coding sequence variants are not common. The enzyme is inducible, and it may be that regulatory effects are the main cause of the variability. A further complication is that the activity of CYP3A4 strongly overlaps with that of yet another P450 enzyme, CYP3A5.

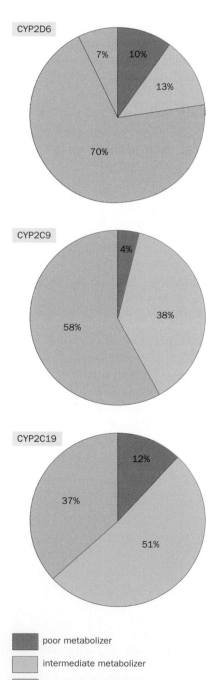

Figure 19.4 Population frequencies of P450 activity variants.
The distributions differ in different populations. [Data for CYP2D6 adapted from Meyer UA (2004) *Nat. Rev. Genet.* 5, 669–676. Data for CYP2C9 for northern Europeans from Service RF (2005) *Science* 308, 1858–1860. Data for CYP2C19 for Taiwanese from Liou YH, Lin CT, Wu YJ & Wu LS (2006) *J. Hum. Genet.* 51, 857–863.]

Other P450 enzymes with important roles in drug metabolism include CYP1A2 and maybe CYP2A6. Figure 19.4 shows the frequencies of variants for three of the most important P450 enzymes.

Another phase 1 enzyme variant causes a problem in surgery

Suxamethonium (succinyl choline) is a muscle relaxant used in surgery. About 1 in 3500 Europeans suffer prolonged apnea (failure to breathe spontaneously) after a standard dose. Spontaneous breathing resumes when the drug is inactivated by the enzyme butyrylcholinesterase (also known as pseudocholinesterase). The prolonged effect is seen in people who are homozygous for low-activity variants of the enzyme and who therefore inactivate suxamethonium abnormally slowly.

Phase 2 conjugation reactions produce excretable water-soluble derivatives of a drug

Phase 2 reactions can involve acetylation, glucuronidation, sulfation, or methylation of the drug. People with deficient phase 2 activity inactivate and excrete the relevant drugs abnormally slowly.

Fast and slow acetylators

Phase 2 metabolism of many drugs involves N-acetylation. Humans have two aryl-*N*-acetyltransferase enzymes, each involved in phase 2 metabolism but for different spectra of drugs. They are encoded by the highly homologous *NAT1* and *NAT2* genes that lie close together on chromosome 8p22. The NAT1 enzyme is relatively invariant, but all human populations show frequent polymorphism for NAT2 variants with different enzymatic activity. Rapid acetylation, the wild type, is dominant over slow acetylation. Slow acetylators eliminate drugs and other xenobiotics more slowly, and so show enhanced sensitivity to their effects. Slow acetylation has also been linked to increased susceptibility to bladder cancer. Despite these apparently deleterious effects, there is good evidence that the slow acetylator phenotype has been positively selected in Western Eurasians over the past 6500 years. About 50% of white Americans, but only 17% of Japanese, are slow acetylators. One suggestion is that slow acetylation is beneficial to meat-eaters because acetylation by NAT2 of compounds present in well-cooked meat may produce carcinogens.

Variable acetylation of the anti-tubercular drug isoniazid was one of the earliest observations in pharmacogenetics. Many years ago, it was noticed that individuals had greatly variable plasma concentrations of the drug after receiving a standard dose (**Figure 19.5**), and this had important clinical consequences. Slow acetylators are at increased risk of developing peripheral neuropathy, a known

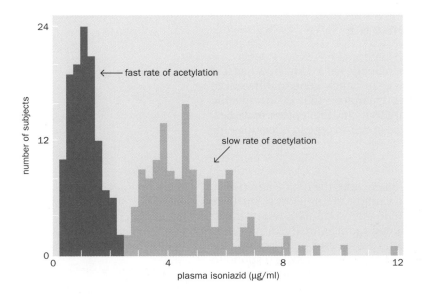

Figure 19.5 A bimodal distribution of plasma levels of isoniazid due to genetic polymorphism in the *NAT2* (N-acetyltransferase 2) gene. Plasma concentrations were measured in 267 subjects 6 hours after an oral dose of isoniazid. Fast acetylators remove the drug rapidly. [Adapted from Price Evans DA, Manley KA & McKusick VA (1960) *BMJ* 2, 485–491. With permission from BMJ Publishing Group Ltd.]

adverse effect of the drug. Other drugs for which variations in acetylation rate can be clinically important include procainamide (an anti-arrhythmic), hydralazine (an antihypertensive), dapsone (anti-leprosy), and several sulfa drugs.

UGT1A1 glucuronosyltransferase

Many drugs are excreted in the form of glucuronide conjugates. These are formed by the action of UDP glucuronosyltransferases. As mentioned in Chapter 11, the *UGT1A* locus on chromosome 2q37 has 13 alternative first exons, which are spliced onto invariable exons 2–5. Exon 1 determines the substrate specificity of the enzyme, whereas exons 2–5 encode the active site. One of the variants, UGT1A1, is responsible for catabolism of both bilirubin, the normal breakdown product of hemes from red blood cells, and the anti-cancer drug irinotecan. Many variants of UGT1A1 with decreased enzymatic activity have been described, primarily in connection with hyperbilirubinemias, but patients with low UGT1A1 activity also suffer severe side effects when treated with irinotecan.

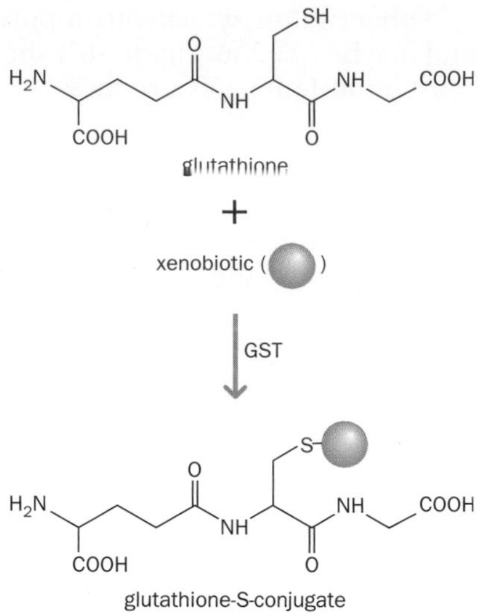

Glutathione S-transferase GSTM1, GSTT1

Glutathione S-transferases (GSTs) are a large family of enzymes that are involved in the detoxification of a variety of xenobiotics and carcinogens (**Figure 19.6**). There are six main classes—alpha (α), kappa (κ), mu (μ), pi (π), sigma (σ), and theta (θ), each encoded by several closely related genes. Different enzymes differ in their tissue and substrate specificities. The genes encoding the GSTM1 and GSTT1 enzymes have been the most extensively investigated. At each of these loci, gene deletions are common (about 50% and 38%, respectively, in white northern Europeans), probably as a result of unequal crossover in tandemly repeated gene clusters.

People can be classified into non-conjugators, low conjugators, and high conjugators with respect to any particular GST activity. Numerous studies have reported associations of low conjugation with susceptibility to genotoxic effects. People with low GST activity may be unable to cope with high doses of drugs whose phase 2 detoxification involves conjugation with glutathione. In some cases, the effect has been in the opposite direction—for example, low conjugators are less efficient at removing isothiocyanates that come from dietary cabbage and related cruciferous plants. These compounds are supposed to be protective against lung cancer, so low conjugators may be at increased risk.

Thiopurine methyltransferase

Thiopurine S-methyltransferase (TPMT) transfers a methyl group from S-adenosylmethionine onto the immunosuppressant drugs azathioprine and 6-mercaptopurine, leading to their inactivation. About 10% of Europeans are heterozygous, and 0.3% are homozygous, for low-activity variants of TPMT. These individuals require a lower dose of the drugs. Homozygotes can suffer life-threatening bone marrow toxicity when given a standard dose of either drug. Three relatively common variants account for about 90% of the low-activity alleles.

Genetic variation in its target can influence the pharmacodynamics of a drug

The effects considered above affect the pharmacokinetics of a drug: how fast it is activated, inactivated, and excreted. Another way in which genetic differences can affect drug responses is through pharmacodynamics—that is, the specific response of a drug target to a given drug. Drug targets include receptors, enzymes, and signal transduction systems. Genetic variants in the target can affect the efficacy of a drug.

Variants in beta-adrenergic receptors

The *ADRB2* gene located on chromosome 5q32–34 encodes the beta-2 adrenergic receptor, which is the target for many drugs. Two variants in this gene, p.Arg16Gly and p.Gln27Glu, are common (frequency 0.4–0.6) in many populations and are in strong linkage disequilibrium with each other. Beta-2 agonists are the most widely used drugs for treating asthma. Individuals homozygous and heterozygous for the Arg16 variant are 5.3-fold and 2.3-fold, respectively, more likely to respond to the anti-asthmatic drug albuterol than those who are homozygous for Gly16.

Figure 19.6 Glutathione conjugation.
Phase 2 metabolism of many drugs and other xenobiotics involves conjugation with glutathione. The reaction is catalyzed by glutathione S-transferase (GST).

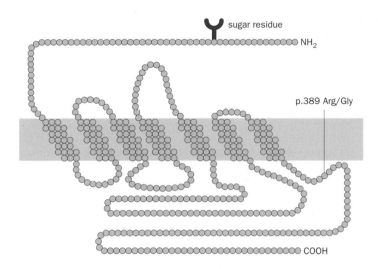

Figure 19.7 Beta-adrenergic receptor polymorphism. The beta-1 adrenergic receptor is a seven-transmembrane domain protein 477 amino acid residues long. Circulating catecholamines bind to the extracellular, N-terminal part of the receptor, activating an intracellular signaling cascade. Beta-1 receptors are located mainly in the heart and kidneys, and are the target of many drugs, both beta-blockers and beta-agonists. A common polymorphism, p.389 Arg/Gly, affects the response to many beta-blockers.

However, much depends on precisely what response is being studied. The review by Evans & McLeod (see Further Reading) gives details.

The *ADRB1* gene located on chromosome 10q24–26 encodes the beta-1 receptor. This has only limited homology to the beta-2 receptor and is targeted by different drugs (**Figure 19.7**). A common polymorphism, p.Arg389Gly, has been tested for pharmacodynamic effects. The Gly389 allele is associated with a decreased cardiovascular response to beta-blocker drugs. For example, Arg389 homozygotes had a much better response to the beta-blocker bucindolol than those who are heterozygous or homozygous for the Gly389 allele.

Variations in the angiotensin-converting enzyme

The angiotensin-converting enzyme (ACE) is a peptidase that converts angiotensin I into angiotensin II. The latter is an important regulator of blood pressure and has many other physiological functions. A polymorphic insertion/deletion of an Alu sequence in an intron of the *ACE* gene is associated with variations in enzyme activity (**Figure 19.8**). DD (deletion) homozygotes have about twice the level of circulating ACE than II (insertion) homozygotes.

ACE inhibitors such as enalapril and captopril are widely used drugs for treating heart failure. Several reports suggest that ACE inhibitors are more effective in DD than II patients. The insertion–deletion polymorphism has also been extensively studied for association with diseases. DD homozygotes are at increased risk of myocardial infarction and coronary artery disease, and possibly also for complications of type 2 diabetes, but at slightly decreased risk of late-onset Alzheimer disease.

Variations in the HT2RA serotonin receptor

The serotonin (5-hydroxytryptamine) receptor 2A has been a target for many investigations because of the central role of serotonin in many neurological processes. Response to psychiatric drugs is notoriously unpredictable, and several studies have looked for associations between *HT2RA* variants and variations in

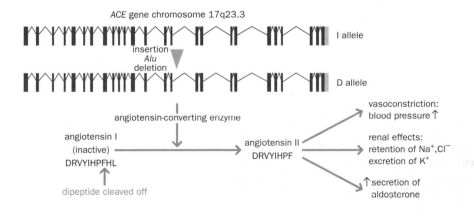

Figure 19.8 An insertion–deletion polymorphism in the *ACE* gene that encodes the angiotensin-converting enzyme. Insertion (I) alleles have an insertion of an *Alu* element into an intron of the gene. Homozygotes for the deletion allele (D) have higher levels of circulating enzyme than those with the I allele. Peptide sequences are shown with the single-letter code.

response. Two variants, p.Ile197Val and p.His452Tyr, have been associated with a poor response to the antipsychotic drug clozapine, whereas an intronic SNP, rs7997012, has been associated with variable response to the antidepressant citalopram. There are sometimes conflicting results regarding possible associations of *HT2RA* variants with conditions including schizophrenia, obsessive–compulsive disorder, and alcohol dependence.

Malignant hyperthermia and the ryanodine receptor

Rare individuals have a life-threatening response to inhalation anesthetics such as halothane or isoflurane. They suffer severe muscle damage (rhabdomyolysis) and an extreme rise in temperature. This condition, known as malignant hyperthermia (MH), is genetic but heterogeneous. Many, but not all, cases have mutations in the ryanodine receptor gene, *RYR1*, on chromosome 19q13. This gene encodes a calcium release channel in the muscle sarcoplasmic reticulum. It is thought that MH results when a mildly temperature-sensitive ryanodine receptor allows excessive calcium flow, which causes the temperature to rise and initiates a catastrophic positive feedback loop.

19.3 PERSONALIZED MEDICINE: PRESCRIBING THE BEST DRUG

The goal of personalized prescribing is to tailor the choice and dosage of a drug to the individual patient's genetic makeup. In the light of the evidence presented above, this would seem a straightforwardly rational thing to do; yet personalized medicine has been slow to develop. Some of the variations in phase 1 and phase 2 metabolism have been known for years, but clinicians still tend to take a reactive rather than a proactive approach. Generally, the patient is given a standard dose of a standard drug and monitored. If the drug has no effect, or if adverse effects appear, the drug or the dose is changed. Only a very few of the many drugs whose metabolism is affected by pharmacogenetic variation are labeled to draw attention to this. In the USA, the antipsychotic thioridazine is marked as contraindicated in patients who have low activity of CYP2D6. In other cases, for example the antipsychotic aripiprazole and the psychostimulant modafinil, the label draws attention to the importance of CYP2D6 variants, but does not mandate genotyping. There are several reasons, apart from simple resistance to change, to explain this apparent reluctance to embrace personalized medicine.

Without bedside genotyping it is difficult to put the ideal into practice

Seriously ill patients need immediate treatment, and even those who are not seriously ill often expect to leave the physician's office clutching a prescription. This makes genotyping difficult to integrate into much of clinical practice. A microarray (Roche AmpliChip®) is available that scores 27 variants of *CYP2D6* and 3 of *CYP2C19*. Nevertheless, this is still a laboratory-based test that imposes an inevitable delay between seeing the patient and prescribing the drug, so it is not suitable for all circumstances. If a simple instant dipstick test were available, physicians might be more inclined to investigate a patient's genotype. In countries with well-developed integrated national health services and electronic patient records, a realistic alternative could be to genotype everybody routinely at some young age for all relevant variants, and for the results to be part of their standard health record.

Drug effects are often polygenic

Probably the main single scientific obstacle to widespread personalized prescribing is that single genotypes are often not reliably predictive of a person's response to a drug. This is especially the case with P450 enzyme variants. An individual drug may be metabolized by any of several P450 enzymes, or successive stages in the metabolism may be handled by different enzymes. Where a single genotype does predict the response, as with butyrylcholinesterase or thiopurine methyltransferase, testing is much more widely used. The following example of warfarin illustrates some of the problems of polygenic effects.

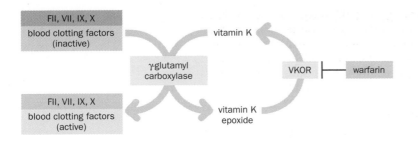

Figure 19.9 Mechanism of action of warfarin. Warfarin inhibits the enzyme vitamin K epoxide reductase (VKOR). This limits the supply of active vitamin K, which is an essential cofactor for the activation of blood clotting factors II, VII, IX, and X.

Warfarin

Warfarin is a powerful anticoagulant, used for patients at risk of embolism or thrombosis (and also to poison rats and mice by triggering internal bleeding). It works by decreasing the availability of vitamin K. This is an essential cofactor for the enzyme γ-glutamyl carboxylase that activates blood clotting factors II (prothrombin), VII, IX, and X. During these reactions, vitamin K is converted to the inactive vitamin K epoxide, and this is recycled by the action of vitamin K epoxide reductase (VKOR). Warfarin inhibits VKOR (**Figure 19.9**).

Achieving the correct level of anticoagulation is clinically very important. If the level is too low, the patient remains at risk of thrombosis or embolism; however, if it is too high, there is a risk of hemorrhage, which can be life-threatening. The therapeutic window (the range of doses that are efficacious and not harmful) is narrow. After insulin, warfarin is the most common prescription drug responsible for emergency hospital admissions. The average cost of a bleeding episode has been estimated as $16,000. But the effective warfarin dose varies up to twentyfold between individuals. This has made warfarin something of a test case for the general utility of pharmacogenetically informed prescribing.

Warfarin as normally used is a mixture of two stereoisomers, *R*-warfarin and *S*-warfarin. Both isomers are active, but the *S* form is about three times as potent as the *R* form. CYP2C9 is the principal enzyme that catalyzes the conversion of *S*-warfarin to inactive 6-hydroxy and 7-hydroxy metabolites, whereas the oxidative metabolism of *R*-warfarin is catalyzed mainly by CYP1A2 and CYP3A4. Variants in *CYP2C9* and *VKORC1* (the gene encoding subunit 1 of VKOR), together with a limited set of environmental factors, can explain about 50–60% of the variability of response to warfarin. However, as **Figure 19.10** shows, several other

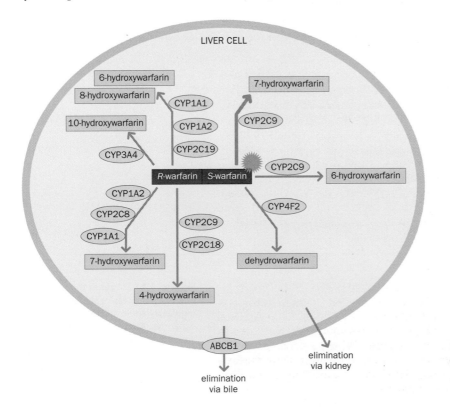

Figure 19.10 The catabolism of warfarin. Warfarin, as prescribed, is a mixture of two stereoisomers, *R*- and *S*-warfarin, each of which can be catabolized by several different P450 enzymes. As a result, the rate of catabolism is not wholly governed by the activity of any one enzyme, although CYP2C9 has the major effect. (Courtesy of PharmGKB. With permission from PharmGKB and Stanford University.)

P450 enzymes have a role in removing warfarin. In 2007, the US Food and Drug Administration put wording on the label of warfarin that recommended (but did not mandate) genotyping these two loci. Many clinicians, however, remain skeptical about the value of genotyping their patients, because *CYP2C9* and *VKORC1* genotypes do not fully predict the response. A major goal in pharmacogenetics is to identify the remaining factors and develop a simple test that will accurately predict the correct dose of warfarin for a patient. The review by Wadelius and Piromohamed (see Further Reading) describes 30 other genes in which variation might explain some of the variable effect, and discusses the prospects for better prediction.

Pharmaceutical companies have previously had little incentive to promote personalized medicine

Drug companies are in business to make money, and so they want to sell their drugs to the maximum possible number of people. They want their drug to be safe and effective in the widest possible range of patients, and to achieve that end they have embraced genotyping in the drug development process. But once a drug is on the market, there is less incentive for genotyping. A company would not necessarily welcome a genetic test that reduced sales of one of its leading drugs by identifying those people for whom it worked poorly. There would be a stronger incentive to use tests that predicted adverse effects, rather than simple efficacy. But only if genetically assisted prescribing gave a drug an advantage over its competitors would a company have a natural interest in promoting genotyping.

The stages in drug development

The total process, from initial laboratory identification of a promising compound to regulatory approval and launch of a new drug on the market, can take up to 15 years and cost $1 billion. The stages of evaluation and trial are illustrated in **Figure 19.11**. For pharmaceutical companies, it is very important to identify any problems at the earliest possible stage in this process, before too much time and money have been spent. Genotyping is important in the development process.

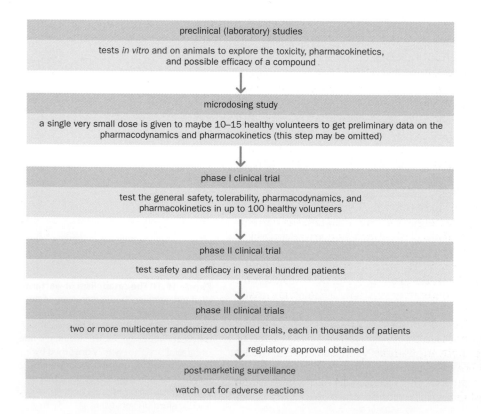

Figure 19.11 Stages in the development of a drug. Successive stages toward regulatory approval are increasingly expensive. Any effects of genetic variation among patients that might influence marketability need to be identified as early as possible in the process, to avoid unnecessary expenditure.

health. Screening, as defined here, is a top-
authority to whole populations (or maybe
tion), without reference to the health of th
with conventional genetic testing, which is ;
viduals or their physicians to answer specii
toms or family history.

Not all testing falls neatly into one or otl
programs, as defined here, raise particular i
individuals. Because these are large, centr;
them have to assess very carefully whether
ment any proposal for screening. Screening
icy, not individual whims. Individuals may b
or not to take a genetic test, but central autl
cies or insurance companies—have to make
screening. Within the ACCE framework (se
more significant in screening programs than

Screening tests are not diagnostic te

Screening programs normally aim at identify
can then be offered some intervention (Figu
test or preventive measures. For example, m;
natal screening for genetic disorders. The l
country to country, but it always includes ph
A screening test measures the concentratio
baby's blood. A high level does not make the
ther investigations. Detailed biochemical st
typing for mutations in the *PAH* (phenylalar
minority of those high-risk babies that have
low-phenylalanine diet. Provided that the fa
ing dietary restrictions, the baby's developı
intervention, the child would suffer profounc

Like disease susceptibility testing, screeni
analog scale, rather than pointing to a digita
individual may choose how to react to some
bility, a screening program has to define a t
screening for Down syndrome illustrates som

Prenatal screening for Down syndrom threshold for diagnostic testing

Pregnant women are naturally anxious that t
variety of prenatal tests for fetal abnormalit
abnormalities, the diagnostic test requires a s
This is done by chorionic villus biopsy at 10–
centesis at 16–20 weeks. Both of these procedu
carry an approximately 1% risk of triggering
probability of having a baby with a chromosc
tages may be worth disregarding, but how are

Some women are at high risk because they
chromosomal translocation (see Chapter 2), bı
risk factor. The probability of having a baby
numerical chromosome abnormality rises sh
(**Figure 19.18**). Until fairly recently, it was sta
offer amniocentesis or chorionic villus biopsy
age (usually in the range 35–38 years). But th
detecting fetuses with Down syndrome. Alth
higher for older mothers, it is not negligible f

population → screening test → high-risk subgroup

Nowadays companies try to avoid compounds that are metabolized by the most variable enzymes. Many of the drugs described above, for which pharmacogenetic effects are important, were developed years ago when these effects were less well understood. They are also now off-patent, so that the company has little financial incentive to develop a genetic test.

Predicting adverse effects is particularly important. It can be both a financial and a public relations disaster for a company if an adverse effect becomes apparent only after a drug has been brought to market. There will probably be litigation and claims for compensation. Vioxx® (rofecoxib), a nonsteroidal anti-inflammatory drug used to treat arthritis and other forms of pain, is a recent example. Vioxx was withdrawn from the market in 2004 by its manufacturer, Merck, because of evidence that high doses could increase the risk of heart attacks and stroke. Sales in the previous year had generated $2.5 billion in revenue in the USA alone. If it could be shown that only patients with certain genotypes were vulnerable to the adverse effect, the company would have a huge incentive to market the drug in combination with a genetic test. But identifying rare genotypes that predispose to adverse effects can be difficult. If only a few patients in a phase III clinical trial experience these effects, there will be little statistical power to identify the cause; moreover, obtaining the necessary DNA for investigation may be complicated and require extra consent.

Some drugs are designed or licensed for treating patients with specific genotypes

Despite the obstacles described above, a growing number of drugs are designed or marketed for patients with specific genotypes. A controversial example has been BiDil®, a specific combination of two long-established and generically available drugs, isosorbide dinitrate and hydralazine hydrochloride. BiDil was licensed in 2005 for treating congestive heart failure, but only in African-Americans. This surprising decision was the result of a successful clinical trial restricted to self-described African-Americans. This followed an earlier broader-based trial that did not provide convincing evidence of efficacy. Presumably, the real determinant is not skin color or self-described ethnicity but a low vascular level of nitric oxide. It happens that this is more common in African-Americans than in white Americans—but skin color is a very poor surrogate for this distinction. Better-founded examples of genotype-specific prescribing come from cancer therapy.

Trastuzumab for breast cancer with HER2 amplification

Herceptin® (trastuzumab) is a monoclonal antibody against HER2, a receptor tyrosine kinase. Amplification of the *ERBB2* gene that encodes HER2 (also known as Neu) is found in 20–30% of early-stage breast cancers. Herceptin is effective against those tumors, but not against those lacking *ERBB2* amplification. Because the drug is extremely expensive ($70,000 for a full course), an immunohistochemical assay for HER2, or a PCR test for *ERBB2*, is performed on a biopsied tumor sample before the drug is prescribed. Herceptin has been licensed for the treatment of advanced cancer, but there is evidence that it may also be effective as a first-line treatment.

Although Herceptin was a pioneering drug that helped establish the reality of personalized medicine, the results have not been stunning. The benefits of treatment have to be weighed against the risk of heart disease, a known side effect of the drug. One randomized controlled trial suggested that 56.7 women would need to be treated to save one extra life over a two-year follow-up, but one in 20 of the women treated would suffer some heart damage, and one in 51 would have symptomatic congestive heart failure. Alternative treatment protocols may lessen this risk. In the UK a government-appointed body, the National Institute for Health and Clinical Excellence (NICE), has been tasked with determining which drugs should be available on the National Health Service (NHS). Drugs such as Herceptin test this mechanism to the limit. Individuals who take to the media and the courts to try to force the NHS to fund their treatment are highly visible and attract public sympathy. The larger number of people who might gain much greater benefit from alternative uses of the money remain invisible.

to consumers through the Internet. Typica
health, Bone health, Immunogenomic profi
A Dutch–US group reported in 2008 on th
offered by seven such companies (see Furthe
identified the particular SNPs used for eac
these were significant risk factors for the clai
did little to bolster confidence. They concl
excess disease risk associated with many ge
profiles has not been investigated in meta-an
imal or not significant.... Although genom
enhance the effectiveness and efficiency of p
scientific evidence for most associations be
risk is insufficient to support useful applicati

Few would question the right of people t
constitution, as long as they do so at their ow
be able to trust the claims that are made foi
vided that the advertising is honest, people
according to personal whim. This may not ac
economists. A news article in *Nature* (vol. 45
dotal case of a young woman whose test resu
the Internet, claimed to change her risk of o
and sex, to 34%. She was a satisfied customer c
as saying, "It really does show that if I wasn't a
probably be overweight or diabetic."

Table 19.4 summarizes the situation in 20
Whether the situation will be very different :
debates in public health.

19.5 POPULATION SCREENING

The aim of population screening is not usually
define a high-risk subset of the population, w
ventive or diagnostic intervention. The word :
synonym for testing, but we will use it here in
large-scale testing performed as part of a pro

**TABLE 19.4 TESTING FOR MENDELIAN CONDITIC
SUSCEPTIBILITY COMPARED AND CONTRASTED**

Mendelian condition	Suscept
Environment and lifestyle do not usually have a major role	environi importa a test
A single genetic test is highly predictive	a single value
Genotype of one person may have implications for their relatives	no impli depends genotyp breaks th
Testing normally conducted within a framework of clinical responsibility	tests at p consume counselii package
Tests assessed for sensitivity, specificity, and predictive value before they are made available	tests asse they are c
Testing laboratories subject to external quality assurance	testing la any qualit

TABLE 19.7 A TEST THAT PERFORMS WELL IN THE LABORATORY MAY BE USELESS FOR POPULATION SCREENING

Prevalence of condition	True positives in population screened	True positives detected by screening	True negatives in population	False positives detected by screening	Positive predictive value
1/100,000	10	10	999,990	10,000	0.001
1/10,000	100	99	999,900	9999	0.0098
1/1000	1000	990	999,000	9990	0.09

In a laboratory trial of this hypothetical test, 99/100 cases and only 1/100 controls scored positive. The table shows the predicted result of screening 1 million people, for different prevalences of the condition tested.

different pathogenic mutations have been described. Even with the new ultra-high-throughput sequencing methods it would not be feasible to test for all of them as a population screen. Screening would test for some subset of the commoner mutations. People carrying rare mutations not included in the screening panel would be false negatives. The question is, what sensitivity would be acceptable? **Figure 19.19** shows a flowchart for the screening program.

It is clear that simply testing for the most common mutation, p.F508del, would not produce an acceptable program; more affected children would be born to couples who tested negative on the screening program than would be detected by the screening. Whether or not such a program were financially cost-effective, it would surely be socially unacceptable. What constitutes an acceptable program is harder to define. One suggestion focuses on +/− couples; that is, couples with one known carrier and the partner negative on all the tests. The partner still might be a carrier of an undetectable or unidentifiable rare mutation. Writing in the journal *Nature* (see Further Reading), Leo ten Kate suggested that an acceptable screening program is one in which the risk for such +/− couples is no higher than in the general population risk before screening. For cystic fibrosis screening of northern Europeans, that would require a sensitivity of about 96%. Rather than screen everybody with such high sensitivity, it might be more efficient to use an

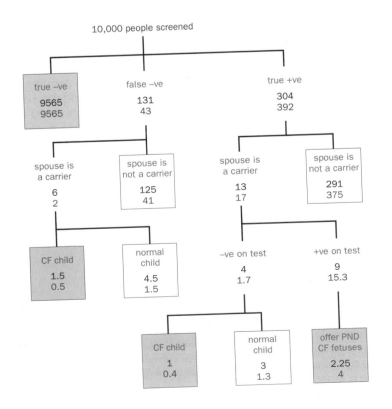

Figure 19.19 **Flowchart for cystic fibrosis carrier screening.** Results of screening 10,000 people, 1 in 23 of whom is a carrier. If a person tests positive, his or her spouse is then tested. Black figures show results using a test that detects 70% of CF mutations (i.e. testing for p.F508del only); red figures show results for a test with 90% sensitivity. Pink boxes represent cases that would be seen as successes for the screening program (regardless of what action they then take), pale blue boxes represent failures. PND, prenatal diagnosis.

initial screen of fairly low sensitivity and involving very large numbers of subjects, but then to do a very thorough screen of the much smaller number of partners of the carriers identified in the initial screen.

Choosing subjects for screening

As well as considering how many mutations to test, organizers of a screening program for CF carriers would have to decide whom to test. As Table 19.8 shows, there are several options, each of which has advantages and disadvantages.

An ethical framework for screening

Ethical issues in genetic population screening have been discussed by a committee of distinguished American geneticists, clinicians, lawyers, and theologians (see Further Reading). It is in the nature of ethical problems that they have no correct solutions, but certain principles emerge in their report:

- Any program must be voluntary, with participants taking the positive decision to opt in.

- Programs must respect the autonomy of the participants. People who score positive on a test must not be pressured into any particular course of action. For example, in countries with insurance-based health care systems, it would be unacceptable for insurance companies to put pressure on carrier couples to accept prenatal diagnosis and terminate affected pregnancies.

- Programs must respect the privacy of the participants. Information should be confidential. This may seem obvious, but it can be a difficult issue—we like to think that drivers of heavy trucks or jumbo jets have been tested for all possible risks.

In societies with insurance-based health care systems, the ethical principle of nondiscrimination conflicts with the principles of underwriting on which commercial insurance rests. Insurance companies could argue that they are penalizing low-risk people by not loading the premiums of high-risk people. In the USA, the 2008 Genetic Information Nondiscrimination Act (GINA) was passed only after a 10-year struggle.

Even if a screening program fulfills the ethical criteria listed above, it may still be controversial. Two issues that trouble some people are discussed below.

Some people worry that prenatal screening programs might devalue and stigmatize affected people

Some disability rights campaigners and patient groups argue that offering prenatal diagnosis for a condition, with the option of terminating the pregnancy if the fetus is affected, is equivalent to saying that affected people are worthless and should not be alive. On the other side, many people maintain that wanting a healthy baby is not incompatible with loving a child that is born with a disability.

TABLE 19.8 POSSIBLE WAYS OF ORGANIZING POPULATION SCREENING FOR CARRIERS OF CYSTIC FIBROSIS		
Group tested	**Advantages**	**Disadvantages**
Newborns	easily organized	no consequences for 20 years; many families would forget the result; unethical to test children
School leavers	easily organized; inform people before they start relationships	difficult to conduct ethically; risk of stigmatization of carriers
Couples from physician's lists	couple is unit of risk; stresses physician's role in preventive medicine; pre-conception screening allows a full range of reproductive options	difficult to control quality of counseling
Women in antenatal clinic	easily organized; rapid results	bombshell effect for carriers; partner may be unavailable; time pressure on laboratory
Adult volunteers (drop-in CF center)	few ethical problems	bad framework for counseling; no targeting to suitable users; inefficient use of resources?

This argument has special force for couples who already have one affected child. They often request prenatal testing, saying that however much they love that child, they could not cope with having a second affected child. Whatever their position on this argument, all civilized people must surely agree that society has an obligation to look after people born with disabilities and do whatever is possible to allow them a full life.

Some people worry that enabling people with genetic diseases to lead normal lives spells trouble for future generations

If neonatal screening detects a baby with PKU, and the family is able to keep to the low-phenylalanine diet, the baby will grow into a person who can lead a normal life. That may well include becoming a parent. Phenotypically such people are normal, but genetically they have PKU, and they will pass on one of their mutant alleles to any child. If they had not been treated, they could never have become parents and their mutant allele would have died with them. Is this not going against natural selection and storing up trouble for future generations?

It has to be true that helping those affected by a mutation to live normal lives has the effect of increasing the number of mutant alleles in the next generation—the question is, by how much? A little population genetics helps here:

- For recessive conditions, only a very small proportion of the disease genes are carried by affected people. The great majority are in healthy heterozygotes. The Hardy–Weinberg equation (see Chapter 3) gives the ratio of carriers to affected in a population as $2pq{:}q^2$, where q is the frequency of the disease allele and $p = 1 - q$. Because each carrier has one copy of the disease allele and each affected person has two, the proportion of disease alleles present in affected people is $2q^2/(2pq + 2q^2)$. This expression simplifies to just q when numerator and denominator are both divided by $2q$. So, for a recessive disease affecting one person in 10,000 ($q^2 = 1/10,000$; $q = 0.01$), only 1% of disease alleles are in affected people. Whether or not we allow affected people to transmit their disease alleles has very little effect on the frequency of the disease in future generations. We do not need to think of measures to prevent them from having children as the price of treatment.

- For dominant conditions, the effect on the gene frequency would be more marked. If a condition is sufficiently serious that, untreated, it is incompatible with reproduction, then it must be maintained by recurrent mutation. If affected people were treated so well that they had the same reproductive success as unaffected people, then the frequency of the condition would double in the next generation. On the other hand, if treatment were so successful, would an increase in the number of affected people be such a disaster?

19.6 THE NEW PARADIGM: PREDICT AND PREVENT?

Personalized medicine, susceptibility testing, and population screening are the constituents of a major debate about the place of genetics in twenty-first-century medicine. Nobody disputes that clinical genetics services are of great value here and now to the relatively small number of individuals and families who are affected by a Mendelian or chromosomal disease. The debate is about how much further this will go as the science develops. **Figure 19.20** caricatures the extreme positions in this debate.

Figure 19.20 Head in clouds or head in sand—contrasting views on the impact of new genetic discoveries on the future of medicine. (Cartoon by Maya Evans.)

Optimists believe that medicine will move from a reactive 'diagnose and treat' to a proactive 'predict and prevent' model. At present, we wait until we are unwell and then visit our physician for a cure. In the vision of the optimists, all this will change. Some time when we are young and healthy, we will donate a DNA sample that will be analyzed to identify our pattern of genetic disease susceptibility. Maybe our whole genome will be sequenced, or maybe the DNA will be typed for a few thousand SNPs that govern susceptibility to major diseases and reactions to most drugs. However, being able to predict is not useful unless we are also able to prevent. To fulfill the vision, the knowledge that we are susceptible to a certain disease must allow us to take effective measures to avoid it.

If all this works, the pattern of health care spending will be revolutionized (**Figure 19.21**). A small extra expenditure early in life will bring huge and steadily increasing savings later on. Even in the shorter term, there will also be savings because the pharmacogenetic profile will ensure that whenever we are ill, money is not wasted on ineffective drugs or unnecessarily high doses.

Pessimists (who like to call themselves realists) ask: where is the evidence for any of this? Some of the strongest disease susceptibility factors and many of the variants underlying drug response have been known for decades but have made very little impact on the practice of medicine. The main reason is that the knowledge does not lead to any useful action. It would be easy to screen everybody for the HLAB*27 tissue type, which is found in 6–8% of people in many populations. B27 has been known for decades to be a very powerful susceptibility factor for a serious disease, ankylosing spondylitis. The relative risk is about 80—vastly higher than any risk discovered by the current generation of genomewide association studies. B27 typing is indeed used when there is a question about diagnosis, but there is no point in testing healthy people because although the relative risk is high, the absolute risk is small and there is nothing that a B27-positive person can be told to do to avoid the risk. To give another example, the strongest association found by the Wellcome Trust Case Control Consortium (see Table 19.2) was between certain HLA types and type 1 diabetes. Again, this has been known for decades. This is a common severe disease that is hugely costly to health services. We have all the knowledge to screen for susceptibility—but nobody does so because no useful action would follow.

With other complex diseases we do know strategies to reduce the risk—but these do not depend on a person's genotype. Type 2 diabetes is a prime example. The incidence is rocketing in many countries as a direct consequence of unhealthy lifestyles. Getting some exercise and losing weight are strong protective factors. Intervention programs with high-risk people (defined by weight and lifestyle, not by genotype) have demonstrated the benefits of exercise and weight loss. But the advice to eat sensibly and get some exercise applies to everybody, regardless of genotype. That is the dilemma faced by companies offering lifestyle genetic profiling. Regardless of their results, they must offer the same generic healthy living advice to everybody. There is no genotype that says it is OK to live on a diet of French fries.

This argument cuts both ways. We all know that not smoking, eating sensibly, controlling our weight, and getting some exercise are the keys to a healthy life. All these options are available to everybody at zero or even negative cost—yet health care budgets are increasingly eaten up by the consequences of people choosing unhealthy lifestyles. Does this mean that we need genetic tests to identify those people on whom the healthy living advice should be concentrated? Or does it mean that many people will simply ignore any advice, regardless of how well-founded it is?

It may be that the pessimists simply lack vision. Given the staggering rate of progress of human molecular genetics, it is a brave person who claims dogmatically that nothing can be done. Perhaps a more prudent position is to accept that prediction will come long before prevention, but not to rule out the possibility of effective prevention for at least some common complex diseases. If only one or two of the currently intractable major complex diseases could be prevented through identifying and treating people at high genetic risk, it would be a major advance in public health. Maybe the correct place for our heads is neither in the sand nor in the clouds, but somewhere in between, where we have a clear vision.

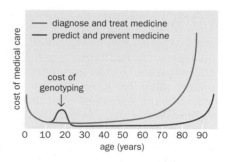

Figure 19.21 Changing profile of health care expenditure if medicine moves from a diagnose and treat to a predict and prevent model. The extra cost of genotyping everybody at an early age is more than offset by savings due to personalized prescribing throughout life and the avoidance of late-onset complex diseases. The curves are indicative only and do not represent the result of quantitative calculations.

CONCLUSION

Clinical tests must be evaluated against several criteria. The analytical validity measures how far the test measures the thing it seeks to measure. Typical considerations would be the sensitivity, specificity, false positive rate, and positive predictive value. For a test of a continuously variable quantity, suitable measures would be the test–retest reliability and the ability to report correctly the value in a known test sample. Reputable laboratories enroll in external quality assurance schemes to ensure that their results have good analytical validity.

Clinical validity measures the extent to which the test result predicts the clinical outcome that it is testing. For tests of genotypes, the clinical validity depends on the extent of genotype–phenotype correlation. Further criteria are the clinical utility—how useful is the result to the patient?—and considerations of the ethics of the test.

An important facet of individual genetic variation is the way in which individuals differ in their response to many drugs. Pharmacogenetics considers the effect of specific genotypes on pharmacokinetics (what the body does to a drug by way of absorption, metabolic activation, and elimination) and pharmacodynamics (the variable response of a target organ to a drug). There are many examples in which a drug is either ineffective or dangerous to patients who have some specific genetic variant. This has led to an interest in patient-specific prescribing, in which the choice and dose of a drug would be influenced by a genetic test on the patient. This is already a reality with some expensive anti-cancer drugs, but it has been slow to become part of wider medical practice. Reasons for the slow uptake include the limited predictive value of many pharmacogenetic tests, and the logistic problems of integrating genetic testing into normal clinical routines.

A second area where common individual genetic differences are of clinical interest is their role in modulating susceptibility to common complex diseases. Previous chapters have described the scientific and technical breakthroughs that have led to the identification of long lists of genetic susceptibility factors for many complex diseases. However, attempts to use this knowledge as a basis for personalized medicine seem premature. For most common diseases, the currently identified genetic factors account for only a modest part of the individual differences in susceptibility. Genotyping may be performed to satisfy individual curiosity, but for most susceptibility tests the clinical validity is low, and no clinical utility has been established.

Measures of susceptibility find current use in population screening programs. In contrast with diagnostic tests, population screening tests do not usually seek to identify cases directly. Instead, the usual aim is to define a high-risk group of people who can then be offered a definitive diagnostic test or some other useful intervention. The initial screening tests for genetic conditions are usually based on clinical or biochemical findings, rather than tests of genotypes. Usually, there is a trade-off between sensitivity and specificity in deciding where to draw the line between a normal and a high-risk result. In contrast with diagnostic tests, large-scale population screening programs target people who had no particular prior reason to be concerned about the condition being screened for. This makes it very important to consider the social and ethical implications of any proposed screening program. A poorly designed program could easily do more harm than good.

In the longer term, there are hopes that susceptibility testing could become much more predictive and could be combined with preventive measures. This would allow medicine to move from the current diagnose and treat model to a predict and prevent model. People would be genotyped at some early age for a whole range of susceptibility factors, their individual susceptibilities identified, and appropriate preventive measures recommended. How far this vision could ever be realized is a matter of considerable controversy. Progress will depend on identifying stronger determinants of susceptibility and better means of prevention. The chances of achieving these goals will differ from one disease to another.

FURTHER READING

Evaluation of tests

Burke W, Atkins D, Gwinn M et al. (2002) Genetic test evaluation: information needs of clinicians, policy makers and the public. *Am. J. Epidemiol.* 156, 311–318. [The ACCE framework and its implications.]

Seidel S, Kreutzer R, Smith D et al. (2001) Assessment of commercial laboratories performing hair mineral analysis. *JAMA* 285, 67–72.

Wang Q, Khoury MJ, Botto L et al. (2003) Improving the prediction of complex diseases by testing for multiple disease-susceptibility genes. *Am. J. Hum. Genet.* 72, 636–649. [The conclusions were largely refuted by Janssens and colleagues—see below—but the paper is a good source of equations relevant to evaluating tests.]

Pharmacogenetics

Arranz MJ, Munro J, Owen MJ et al. (1998) Evidence for association between polymorphisms in the promoter and coding regions of the 5-HT$_{2A}$ receptor gene and response to clozapine. *Mol. Psychiatry* 3, 61–66. [An example of a clinically important pharmacogenetic effect.]

Durham WJ, Aracena-Parks P, Long C et al. (2008) RyR1 S-nitrosylation underlies environmental heat stroke and sudden death in Y522S RyR1 knockin mice. *Cell* 133, 53–65. [The mechanism of malignant hyperthermia.]

Evans WE & McLeod HL (2003) Pharmacogenomics—drug disposition, drug targets and side effects. *N. Engl. J. Med.* 348, 538–549. [A general review.]

Flockhart DA (2007) Drug Interactions: Cytochrome P450 Drug Interaction Table. Indiana University School of Medicine, Division of Clinical Pharmacology. http://medicine.iupui.edu/clinpharm/ddis/table.asp [accessed 25 June 2008].

McMahon FJ, Buervenich S, Charney D et al. (2006) Variation in the gene encoding the serotonin 2A receptor is associated with outcome of antidepressant treatment. *Am. J. Hum. Genet.* 78, 804–814. [An example of the importance of genetic variants for treating psychiatric disease.]

Patin E, Barreiro LB, Sabeti PC et al. (2006) Deciphering the ancient and complex evolutionary history of human arylamine N-acetyltransferase genes. *Am. J. Hum. Genet.* 78, 423–436. [Natural selection and NAT gene polymorphisms.]

The Pharmacogenomics Knowledge Base [PharmGKB]. http://www.pharmgkb.org (accessed 25 June 2008).

Weinshilboum R (2003) Inheritance and drug response. *N. Engl. J. Med.* 348, 529–537. [A general review.]

Personalized medicine

Aspinall MG & Hamermesh RG (2007) Realizing the process of personalized medicine. *Harvard Business Rev.* 85, 108–117. [Discusses organization of services, not the science.]

International Warfarin Pharmacogenetics Consortium (2009) Estimation of the warfarin dose with clinical and pharmacogenetics data. *N. Engl. J. Med.* 360, 753–764. [Performance of an algorithm for warfarin dosage that incorporates both clinical and genetic data.]

Lyssenko V, Jonsson A, Almgren P et al. (2008) Clinical risk factors, DNA variants and the development of type 2 diabetes. *N. Engl. J. Med.* 359, 2220–2232. [The two studies summarized in Section 19.4.]

Meigs JB, Shrader MPH, Sullivan LM et al. (2008) Genotype score in addition to common risk factors for prediction of type 2 diabetes. *N. Engl. J. Med.* 359, 2208–2219.

Smith I, Procter M, Gelber RD et al. (2007) 2-year follow-up of trastuzumab after adjuvant chemotherapy in HER2-positive breast cancer: a randomised controlled trial. *Lancet* 369, 29–36. [Trastuzumab therapy for breast cancer is effective only for tumors with amplification of the *ERBB2* gene.]

Wadelius M & Piromohamed M (2007) Pharmacogenetics of warfarin: current status and future challenges. *Pharmacogenom. J.* 7, 99–111. [Reviews a long list of potentially relevant genetic variants.]

Wald NJ & Law MR (2003) A strategy to reduce cardiovascular disease by more than 80%. *Br. Med. J.* 326, 1419–1423. [A proposal for a universal polypill for all, in contrast with tailored, personalized medicine.]

Tumor expression profiling

Armstrong SA, Staunton JE, Silverman LB et al. (2002) *MLL* translocations specify a distinct gene expression profile that distinguishes a unique leukemia. *Nat. Genet.* 30, 41–47. [An example of successful molecular subtyping.]

Lakhani SR & Ashworth A (2001) Microarray and histopathological analysis of tumours: the future and the past? *Nat. Rev. Cancer* 1, 151–157.

Petricoin EF, Zoon KC, Kohn EC et al. (2002) Clinical proteomics: translating benchside promise into bedside reality. *Nat. Rev. Drug Discov.* 1, 683–695.

Sorlie T, Tibshirani R, Parker J et al. (2003) Repeated observation of breast tumor subtypes in independent gene expression data sets. *Proc. Natl Acad. Sci. USA* 100, 8418–8423. [Source of the data in Figure 19.13.]

Sotiriou C & Piccart MJ (2007) Taking gene-expression profiling to the clinic: when will molecular signatures become relevant to patient care? *Nat. Rev. Cancer* 7, 545–553.

Van de Vijver MJ, He YD, van 't Veer LJ et al. (2002) A gene-expression signature as a predictor of survival in breast cancer. *N. Engl. J. Med.* 347, 1999–2009. [Work subsequently commercialized as the MammaPrint system.]

Susceptibility testing

Easton DF, Pooley DA, Dunning AM et al. (2007) Genomewide association study identifies novel breast cancer susceptibility loci. *Nature* 447, 1087–1095. [Identification of five novel susceptibility factors using a three-stage protocol, as described in Chapter 16.]

Janssens AC, Pardo MC, Steyerberg EW & van Duijn CM (2004) Revisiting the clinical validity of multiplex genetic testing in common diseases. *Am. J. Hum. Genet.* 74, 585–588. [A riposte to the paper by Wang and colleagues; see above.]

Janssens AC, Gwinn M, Bradley LA et al. (2008) A critical appraisal of the scientific basis of commercial genomic profiles used to assess health risks and personalize health interventions. *Am. J. Hum. Genet.* 82, 593–598. [A very negative appraisal of some commercial lifestyle genotyping offers.]

Pharoah PDP, Antoniou AC, Easton DF & Ponder BA (2008) Polygenes, risk prediction and targeted prevention of breast cancer. *N. Engl. J. Med.* 358, 2796–2803. [A proposal to use susceptibility genotypes to modify mammographic screening criteria.]

Wellcome Trust Case Control Consortium (2007) Genome-wide association study of 14,000 cases of seven common diseases and 3,000 shared controls. *Nature* 447, 661–678. [One of the definitive genomewide association studies, involving 17,000 individuals and 7 diseases.]

Zheng SL, Sun J, Wiklund F et al. (2008) Cumulative association of five genetic variants with prostate cancer. *N. Engl. J. Med.* 358, 910–919. [The analysis of five susceptibility factors described in Section 19.4.]

Population screening

Andrews LB, Fullarton JE, Holtzman NA & Motulsky AG (1994) Assessing Genetic Risks: Implications For Health And Social Policy. The National Academies Press, Washington DC. http://www.nap.edu/catalog.php?record_id=2057#toc [Report on ethical issues in genetic population screening from a committee of distinguished American geneticists, clinicians, lawyers, and theologians.]

Malone FD, Canick JA, Ball RH et al. First- and Second-Trimester Evaluation of Risk (FASTER) Research Consortium (2005) First-trimester or second-trimester screening, or both, for Down syndrome. *N. Engl. J. Med.* 353, 2001–2011. [Data on the performance of American screening protocols.]

ten Kate L (1989) Carrier screening in CF. *Nature* 342, 131. [Discusses the sensitivity required for a CF carrier screening program.]

Wald NJ, Huttly WJ & Hackshaw AK (2003) Antenatal screening for Down syndrome with the quadruple test. *Lancet* 361, 835–836. [Data on the performance of a British screening protocol.]

Chapter 20

Genetic Manipulation of Animals for Modeling Disease and Investigating Gene Function

20

KEY CONCEPTS

- The study of gene function and human genetic disease can be greatly aided by genetically modified animal models, often known as transgenic animals.

- To make a transgenic animal, exogenous DNA (known as a transgene) is inserted into the germ line; the transgene may contain more than one gene and accompanying regulatory or marker sequences.

- Transgenes are often introduced into a fertilized oocyte, whereupon they can *randomly* integrate into the genome and be transmitted to all cells of the developing animal.

- Transgenes can also be introduced into embryos via cultured pluripotent stem cells that can ultimately give rise to all cells of the animal.

- In *gene targeting* a specific predetermined gene is genetically modified within cultured pluripotent stem cells and the genetically modified stem cells are used to introduce the mutation into the germ line.

- Gene targeting may be designed to inactivate a gene (a *gene knockout*) or to modify it; gene targeting frequently exploits homologous recombination, a natural cellular process, to replace specific DNA sequences within a gene by DNA sequences carried on an introduced vector DNA.

- Transgenes can also be directed to integrate within a specific target gene so that the target gene is inactivated while transgene sequences are expressed under the regulation of the promoter of the target gene (a *gene knock-in*).

- Larger genomic changes can be brought about by chromosome engineering. Short recognition sequences for a microbial recombinase are inserted by gene targeting into specific positions in the chromosomes of pluripotent stem cells. A specific microbial recombinase is then used to induce the recognition sequences to recombine to produce a chromosomal deletion, inversion, or translocation.

- Specific predetermined target genes can also be *knocked down* in animal cells at the RNA level by using RNA interference or synthetic antisense oligonucleotides to selectively target transcripts of the gene to be knocked down.

- Large-scale mutagenesis screens often use *insertional mutagenesis*, whereby specific transgenes are inserted into the genome, causing inactivation of genes in a random or semi-random way. The transgene acts as a marker to identify the inactivated gene.

- Disease resulting from a loss of gene function is often modeled by using gene targeting to inactivate the orthologous gene in a suitable animal model.

- Disease resulting from a gain of function can be modeled by making transgenic animals with mutant transgenes.

Much of our knowledge of gene function and human disease has been obtained or illuminated by studying model organisms. As outlined in Chapter 5, the conservation of basic developmental processes between species has enabled the study of model organisms to aid our understanding of human development. The characteristics of a wide variety of model organisms were detailed in Section 10.1. In this chapter, we explore how knowledge of the genome sequences of humans and of various animals has facilitated the genetic manipulation of animal models to understand gene function and to mimic human disease states.

20.1 OVERVIEW

Unicellular models—notably *Escherichia coli* and the yeast *Saccharomyces cerevisiae*—have been invaluable in shedding light on how genes function and on many general cellular functions (see Box 10.1 and Table 10.1). About 30% of human genes known to be involved in disease have functional homologs in yeast, and yeasts have also been of great help in modeling some kinds of human disease that are caused by defects in basic cellular functions, such as abnormalities of protein glycosylation.

As described in Chapter 12, studies of cultured invertebrate and mammalian cells have also been invaluable for investigating gene function and understanding pathogenesis. However, many types of cells are not easy to culture, and cultured cells can take us only so far—they cannot reveal much about interactions between different cell types. If we are to begin to understand our biology and to model disease we also need to study whole organisms.

Animal models are extremely valuable in providing insights into gene function and disease processes, allowing detailed physiological assessments and investigations of the cellular and molecular basis of disease. They are also crucially important in allowing us to test drugs and other novel therapies before conducting clinical trials on human subjects.

A wide range of species are used in animal modeling

Various animal species have been used as experimental models to help us understand the molecular basis of disease and how genes function. The details of the different models are presented in Section 10.1; here we provide an overview in **Table 20.1**. Animal models exist for some representatives of all the major human disease classes: genetically determined diseases, disease due to infectious agents, sporadic cancers, and autoimmune disorders.

Despite the large number of different animal models, many individual human disorders do not have a good animal model. Some classes of disorders are comparatively badly served by animal models, and disease modeling needs to take into account important differences—physiological, biochemical, and genetic—that distinguish the different experimental animals from humans.

Theoretically, nonhuman primates should be the best animal models of disease. The great apes have rarely been used as disease models because of ethical concerns, the large expense in maintaining them, and various practical difficulties; however, chimpanzees have been studied because of their resistance to certain human diseases. As detailed in Table 10.2, various small nonhuman primates are used for research but are not as well suited to genetic analyses. Instead, rodents (rats and particularly mice) have been widely used in modeling disease and investigating gene function. Being mammals, rodents are quite closely related to us at physiological, biochemical, and genetic levels, and there are many practical advantages in using them as experimental models. Mice are less expensive to maintain than rats and have been particularly amenable to genetic analyses (see Table 20.1).

Most animal models are generated by some kind of artificially designed genetic modification

Some animal models of human disease have originated spontaneously. They have come to our attention because they have appeared in livestock, pets, or laboratory animals, but most spontaneous mutations arise in wild animals and are not recorded. Spontaneous animal disease models include models of single-gene

TABLE 20.1 PRINCIPAL ANIMAL CLASSES USED TO MODEL HUMAN DISEASE AND INVESTIGATE GENE FUNCTION

Organism class	Advantages as models of disease and gene function	Principal disadvantages and practical limitations
Nonhuman primates (see Table 10.2)	closely resemble humans at physiological, biochemical, and genetic levels; high intelligence; great apes have similar brains to those of humans; important for neuroscience research (understanding brain function, etc.) and modeling resistance to certain diseases	primate experimentation is ethically unacceptable to many people; primate colonies are very expensive to maintain; generally long generation times and small numbers of offspring are a major limitation to genetic studies, but marmosets, which have a relatively short gestation period and are reasonably prolific, may be more amenable, especially after successful germ line transmission of transgenes first reported in 2009 (see the text)
Other mammals (see Table 10.2 and Box 10.3)	resemble humans quite closely at different levels; sheep and pigs, and to a lesser extent rats, are more suited to physiological studies than mice; dogs provide models of genetic disorders and genetics of behavior; cats are good models for studying host-gene–pathogen interactions; rodents have short generation times and large numbers of offspring, making them more suited to genetic analyses; mice have been especially amenable to genetic manipulation, with large numbers of disease models; rats have provided better models of complex disorders than mice	mammals other than rodents have comparatively long generation times and low numbers of offspring, and experimentation on them is not ethically acceptable to some people; rodents are not good models for studying normal human brain function and disorders of higher brain function, such as Alzheimer disease; they also show important genetic and biochemical differences from humans, and their short generation times may make them less suited to modeling late-onset disorders
Other vertebrates (see Box 10.3)	chickens, frogs (*Xenopus*), and zebrafish are good models for studying early developmental processes; the large eggs of *Xenopus* are useful for biochemical analyses; zebrafish are inexpensive to maintain and are particularly suited to genetic analyses, with short generation times and large numbers of offspring; they have a transparent embryo that facilitates mutant screening, and good genetic manipulation	genetic manipulation has been limited in non-mammalian vertebrates other than zebrafish; some models are relatively expensive to maintain and have moderately long generation times and moderate offspring numbers; more distantly related to humans than mammals are, and lack counterparts for some human genes
Invertebrates (see Box 10.2)	the nematode *C. elegans* and fruit fly *D. melanogaster* are inexpensive to maintain and are extremely well suited to genetic analyses, having very short generation times and very large numbers of offspring, and being amenable to sophisticated genetic manipulations; both organisms have been very important for exploring the functions of the sizeable number of human genes that have homologs in the worm or fruit fly, and they are often very useful in providing *cellular* models with which to understand human disease processes	invertebrates are evolutionarily very distant from humans and so are physiologically remote, even though some cellular processes are reasonably conserved; they lack counterparts for a considerable number of human genes

disorders such as the *mdx* muscular dystrophy mouse model, and of more complex disorders such as the NOD non-obese diabetic mouse.

The vast majority of animal disease models have been produced by human intervention at different levels. For some models, disease has been induced by treating animals with chemicals or infectious agents. In others, autoimmune disease has been induced by challenging the immune system with suitable antigens; for example, injection of pristane, a synthetic adjuvant oil, into the dermis of rats has produced a model of arthritis. Some other models have been obtained by selective breeding to obtain animal strains that are genetically susceptible to disease. However, most animal models are produced by some kind of genetic modification.

The techniques of genetic modification of animals will be described in this chapter. In Section 20.2 we describe strategies in which the aim is to insert an *exogenous* gene into an animal's germ line and then express it to help study gene function or to induce disease. Other strategies simply rely on modifying *endogenous* genes within the germ line to produce animals with different mutant phenotypes, and two main approaches are used. Sometimes, as detailed in Section 20.3, the aim is to specifically alter the expression of a single *predetermined* endogenous gene to produce a disease model and/or understand the function of that gene. Alternatively, as described in Section 20.4, endogenous genes can be mutated at *random* to generate a large series of mutants whose phenotypes are then analyzed and categorized.

Multiple types of phenotype analysis can be performed on animal models

Irrespective of the origins of an animal model, ideally the phenotype should be thoroughly investigated and documented to maximize the usefulness of the model. When an animal model has been generated by modification of a predetermined gene, phenotype analyses inevitably concentrate on particular aspects that are deemed particularly worthy of investigation. If, for example, a poorly understood human gene is suspected of having a role in the immune system, it is reasonable to focus on a detailed analysis of the immune system in animal models in which the aim is to understand the gene's function. Similarly, for disease models that arise from modification of a known gene, analysis of the phenotype is particularly directed to aspects of the phenotype that are considered most relevant to the human disease; **Figure 20.1** shows an example.

Phenotype analyses can include a whole range of possible investigations. Various anatomical studies can be conducted that may include histological staining of skeletal preparations (Figure 20.1B) and standard hematoxylin and eosin staining of tissue sections. Morphological studies may include radiography (Figure 20.1A), whole animal magnetic resonance imaging (for small animals such as mice—see http://www.bcm.edu/phenotyping/mri.htm for examples), and detailed scanning electron microscopy. According to need, detailed metabolic, physiological, and behavioral assays are performed.

In animal models that arise from modification of a predetermined gene, additional follow-up work will often include molecular genetic analyses to analyze expression patterns of genes involved in pathogenesis and/or genes that interact with a gene whose function is being analyzed. Labeled probes and antibodies are used to track respectively transcripts and proteins in tissue sections, while microarray analyses can provide global transcript profiling that can help elucidate molecular aspects of pathogenesis or gene–gene interactions. Particular cells

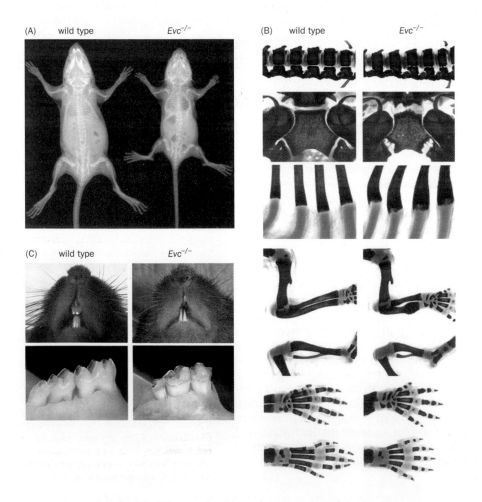

Figure 20.1 Anatomical and morphological phenotyping in a mouse disease model. The figure shows some of the phenotype analyses that can be applied, in this case to a mouse model of Ellis van Creveld syndrome (OMIM 225500). This autosomal recessive disorder results from mutations in the *EVC* gene or its close neighbor, *EVC2*, and affected individuals have skeletal dysplasia with short limbs, short ribs, postaxial polydactyly, and dysplastic nails and teeth. The figure shows a mouse model in which both alleles of the *Evc* gene have been inactivated, and illustrates many of the features of the human disease. Skeletal and dental abnormalities were recorded with the use of X-rays, histological staining, and photographic examination. (A) Radiographs at P18 (postnatal day 18). (B) Histological analysis of skeletal preparations; cartilage is stained by the dye alcian blue and bone by alizarin red. (C) $Evc^{-/-}$ mutants show small dysplastic incisors (upper panels) and conical lower molars, size reduction of the first molar, and an enamel defect (lower panels) that was not seen in the wild type. [Courtesy of Judith Goodship, Newcastle University; from Ruiz-Perez VL, Blair HJ, Rodriguez-Andres ME et al. (2007) *Development* 134, 2903–2912. With permission from The Company of Biologists.]

may be analyzed to gain insights into abnormal cellular functions, and relevant permanent cell lines may be produced for convenient analysis of the mutant phenotype at the cellular level.

In large-scale mutagenesis screens, large numbers of mutants are produced, and those that produce easily ascertained phenotypes tend to be over-represented. Mutants with major defects in limbs or craniofacial structures are easier to identify than those with minor alterations to some internal organs, but at least in zebrafish abnormalities of some internal organs such as the heart are more readily observed because the embryo is transparent. As we will see in Section 20.4, standardized phenotyping is important for large-scale phenotyping studies.

20.2 MAKING TRANSGENIC ANIMALS

As detailed in Chapter 12, different gene transfer technologies can be used to add new and functional genes to animal cells. The additional exogenous DNA sequence is generally called a **transgene**, although it may contain multiple genes. The general aim of introducing transgenes into cells is to get the recipient cells to express one or more new gene products (resulting in a *gain of function*) or to overexpress a gene product. The purpose of some transgenes is to provide extra functional copies of an endogenous gene, with the aim of overexpressing a specific gene product. Transgenes are also commonly designed to express mutant alleles or mutant genes from other species in an attempt to induce disease. Still other transgenes may contain very large human genes or chromosome segments with a view to understanding gene function or modeling disease.

Transgenic animals have exogenous DNA inserted into the germ line

To make a **transgenic animal**, an exogenous DNA sequence is transferred into the germ line of an animal. Genes within the exogenous DNA can therefore be transmitted to, and expressed, in the progeny. To produce a fully transgenic animal (an animal containing the same transgene in the same context in every cell), the animal must develop from a transgenic zygote.

Transgenes can be inserted into the germ line by direct transfer into the zygote, gametes, and embryonic or somatic cells that eventually contribute to the germ line (**Figure 20.2**). Of the possible routes described in Figure 20.2 and below, the fertilized oocyte and embryonic stem cell routes are the most popular.

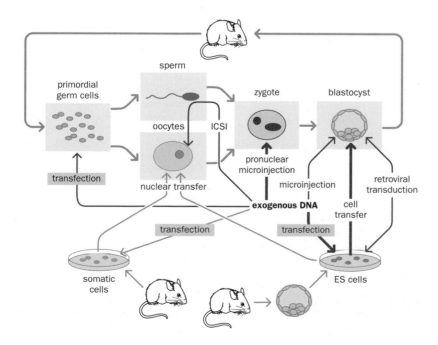

Figure 20.2 Genetically modified mice can be produced by a variety of routes. The boxes linked by gray arrows at the top show components of the mouse life cycle and represent the multiple stages at which, potentially, genetic modification can be performed. The lower part of the figure shows two mice as sources of donor cells for nuclear transfer procedures (process shown in green arrows). Red arrows show the input and transport of exogenous DNA for all other routes. The most widely used gene transfer methods—microinjection into the male pronucleus of the fertilized oocyte, and transfection of embryonic stem (ES) cells followed by injection of genetically modified ES cells into a blastocyst—are shown by thick red arrows. ICSI, intracytoplasmic sperm injection. For convenience, some methods described in the text are not shown.

Pronuclear microinjection is an established method for making some transgenic animals

Transgenic animal technologies came of age in the early 1980s, when the first transgenic mice and transgenic fruit flies were reported. Subsequently a wide variety of other transgenic animals—including worms, birds, frogs, fish, and many types of mammal—have been produced.

Transgenic mice are often produced by injecting DNA into a newly fertilized zygote. Sometimes the DNA is injected into the cytoplasm of the fertilized egg, but it is more efficient to introduce exogenous DNA directly into the large male pronucleus.

After pronuclear microinjection, the microinjected transgene randomly integrates into chromosomal DNA, usually at a single site, and typically as multiple tandemly repeated copies (**Figure 20.3**).

If the transgene integrates immediately, the resulting mouse will be fully transgenic. However, it is more common for the DNA to integrate after one or two cell divisions, in which case the resulting mouse is a *mosaic* that contains both transformed and nontransformed cells. When the transformed cells contribute to the germ line, the transgene is passed to the next generation (as verified by PCR or Southern blotting or some test for transgene expression), resulting in fully transgenic mice.

It is possible to achieve germ-line transmission in up to 40% of microinjected mouse eggs. In other mammals, the transmission rate is generally much lower (often less than 1%) partly because of the difficulty in handling eggs, and partly as a result of lower survival rates.

Microinjection can be used to introduce recombinant plasmid DNA into fish and amphibian eggs, but the introduced DNA tends to persist separately from the host cell's chromosomes and undergoes extensive independent replication. Injected plasmid DNA may increase 50–100-fold in the early stages of development. There is no transcription in early fish and amphibian development; instead, cell functions such as DNA replication are performed by an extensive stockpile of previously synthesized enzymes and proteins contributed by the oocyte.

The general lack of integration of exogenous DNA in fish and frog eggs means that gene expression is often transient (the transgenes are progressively diluted out after cell division). However, some of the exogenous DNA can integrate; if germ-line transmission is established, transgenic lines can be produced. This method is not used in frogs because of the long generation intervals, but it is the standard procedure for producing transgenic fish.

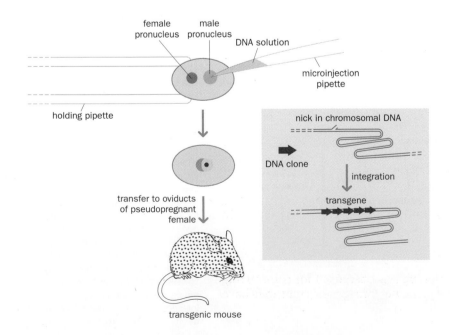

Figure 20.3 Construction of transgenic mice by pronuclear microinjection. A fertilized oocyte is held in place with a holding pipette. A microinjection pipette with a very fine point is then used to pierce first the oocyte and then the male pronucleus (which is bigger than the female pronucleus), delivering an aqueous solution of a desired DNA clone. The introduced DNA integrates at a nick (single-stranded DNA break) that has occurred randomly in the chromosomal DNA. The integrated transgene usually consists of multiple head-to-tail copies of the DNA clone. Surviving oocytes are re-implanted into the oviducts of *pseudopregnant* foster females (they will have been mated with a vasectomized male to initiate physiological changes in the female that stimulate the development of the implanted embryos). DNA analysis of tail biopsies from resulting newborn mice checks for the presence of the desired DNA sequence.

Transgenes can also be inserted into the germ line via germ cells, gametes, or pluripotent cells derived from the early embryo

Various other routes have also been used to introduce transgenes into the germ line, in addition to the fertilized oocyte (zygote) route. Some of them are less widely used but can be useful in some circumstances and can be popular with non-mammalian species.

Gene transfer into gametes and germ cell precursors

Germ cell precursors (see Section 5.7) are a favorite target for making transgenic fruit flies (*Drosophila melanogaster*). However, in mammals, although primordial germ cells are relatively easy to isolate, culture, and transfect, it is difficult to persuade the modified cells to contribute to the germ line when reintroduced into a host animal.

Transgenic frogs are traditionally made by a procedure that allows transgenes to integrate into the sperm genome. In restriction enzyme-mediated integration (REMI), sperm nuclei are isolated, decondensed, and mixed with plasmid DNA. They are then treated with limiting amounts of a restriction enzyme to introduce nicks in the DNA. The decondensed nuclei are then transplanted into unfertilized eggs, where the nicks in the DNA are repaired, resulting in the integration of plasmid DNA into the genome.

A less conventional way of genetically modifying zygotes is to use sperm-mediated transgene delivery into unfertilized eggs. Sperm heads bind spontaneously to DNA *in vitro*, and so sperm can be used simply as vehicles for delivering DNA—the sperm genome itself is not modified. *Intracytoplasmic sperm injection (ICSI)*, a method used in infertility treatment in which sperm heads are injected into the cytoplasm of the egg, has been adapted to produce transgenic mice. To do this, sperm heads are coated with plasmid DNA before being introduced into mouse eggs. However, the transgene often fails to integrate into the genome in this method.

More recently, the injection of recombinant retroviruses into the perivitelline space of isolated oocytes has allowed the production of transgenic cattle and the first transgenic primate, a rhesus monkey named ANDi.

Gene transfer into pluripotent cells of the early embryo or cultured pluripotent stem cells

Genes cloned into retroviral vectors have been transferred into unselected pluripotent cells of very early embryos, notably in mice and birds. In mice, recombinant retroviruses can be used to infect pre-implantation embryos or can be injected into early postimplantation embryos. After the stable transformation of some cells, mosaics are produced that may give rise to transgenic offspring.

A convenient alternative is to use cultured embryonic stem (ES) cells, pluripotent stem cells derived from cells selected from the early embryo. In mammals, ES cells are obtained by culturing cells taken from the *inner cell mass* of a blastocyst (see Figure 4.19 and Box 21.2). ES cells can be used to insert exogenous DNA into the germ line by first transfecting the DNA into cultured ES cells *in vitro*, selecting suitable transfectants, and then injecting them into isolated blastocysts that can be re-implanted into suitably treated female mice.

For technical reasons, certain strains of mouse ES cells have been particularly suitable for allowing gene transfer into the germ line. The *in vitro* step allows sophisticated genetic manipulations to be performed, and the ES cell route is the primary route for inactivating specific genes in the mouse germ line to make so-called *knockout mice*. The technology is especially important for modeling disease and understanding gene function and will be described in detail in Section 20.3.

Gene transfer into somatic cells (animal cloning)

As described in the next section, animal cloning has been used for transferring transgenes in some mammals in which the principal transgenesis methods have not been readily applicable.

Nuclear transfer has been used to produce genetically modified domestic mammals

The principal methods used to generate transgenic mice have not been readily applicable in some animals. However, **somatic cell nuclear transfer**, a technically very demanding method that allows animals to be cloned, has occasionally been used to make transgenic mammals other than mice.

Somatic cell nuclear transfer involves the replacement of an oocyte nucleus with the nucleus of a somatic cell, which is then reprogrammed by the oocyte. Despite the differentiated state of the donor cell, the reprogramming can make the nucleus *totipotent* so that it is able to recapitulate the whole of development.

The technology itself is not new. It has been used for more than 50 years to clone amphibians, and it has been possible to produce cloned mammals from embryonic cells since the 1980s. In 1995, it was shown for the first time that mammals could be cloned from the nuclei of cultured cells, and in 1997 the first mammal cloned from an adult cell was produced, a sheep called Dolly.

Dolly was produced by transferring a nucleus from a mammary gland cell from a Finn Dorset sheep into an enucleated oocyte taken from a Scottish blackface sheep. The resulting artificially fertilized oocyte was allowed to develop to the blastocyst stage before implantation into the uterus of a foster mother (**Figure 20.4**). Out of a total of 434 oocytes, only 29 developed to the transferable blastocyst stage, and of these only one developed to term, giving rise to Dolly. Dolly has the same nuclear genome as the Finn Dorset sheep donor of the mammary gland cell, so they are considered to be clones. However, their mitochondrial genomes are different because Dolly inherited her mtDNA from the Scottish blackface sheep from which the enucleated oocyte was derived. Subsequently, a variety of different mammals have been cloned by the nuclear transfer procedure, including mice, rats, cats, dogs, horses, mules, and cows.

We will consider here the nuclear transfer method as a way to generate transgenic animals. The essential point is that if the donor nucleus is taken from a somatic cell that has been genetically manipulated to contain a transgene of interest, the animal that develops from the manipulated oocyte will be transgenic. Transgenic livestock have been produced in this way with transgenes that produce therapeutic proteins, as described in Chapter 21. It is also possible to modify a specific predetermined endogenous gene in a somatic cell, such as a fibroblast, and then use somatic cell nuclear transfer to make an animal with the mutation in all its cells, as in a pig model of cystic fibrosis that will be described in Section 20.5. We consider the ethical implications of human cloning in Chapter 21.

Exogenous promoters provide a convenient way of regulating transgene expression

To make a useful transgenic animal, the transgene not only has to be inserted into the host's genome but also has to be expressed to produce functional products. Some regulatory elements that influence transgene expression are found within the host genome, but important elements are provided with the transgene, most notably a suitable promoter and a polyadenylation site.

In transgenic animals, it is often desirable to express the transgene in particular tissues or at particular developmental stages that are appropriate for the intended function of the transgene. To achieve a desired expression pattern, the transgene will have a suitable tissue-specific or stage-specific promoter. Maximum control over transgene expression is provided by *inducible promoters* that can be switched on and off according to need, usually by controlling the supply of a particular chemical ligand. Typically, the transcription factors that regulate such promoters are structurally modified by this ligand.

Several naturally inducible promoters have been used, but endogenous promoters are generally disadvantaged by leakiness (a high background expression level), and by relatively low levels of induction, unwanted co-activation of other endogenous genes that respond to the same ligand, and differential rates of uptake and elimination of the ligand in different organs. More robust inducible promoter systems have therefore been developed by using non-mammalian,

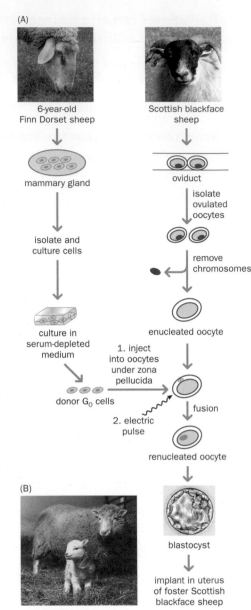

Figure 20.4 The first successful attempt at mammalian cloning from adult cells resulted in a sheep called Dolly. (A) The donor nuclei were derived from a cell line established from adult mammary gland cells of a Finn Dorset sheep. The donor cells were deprived of serum before use, forcing them to exit from the cell cycle and enter a quiescent state known as G_0, in which only minimal transcription occurs. Nuclear transfer was accomplished by fusing individual somatic cells to enucleated, metaphase II-arrested oocytes from a Scottish blackface sheep. Eggs are normally fertilized by transcriptionally inactive sperm whose nuclei are presumably programmed by transcription factors and other chromatin proteins available in the egg, and so the G_0 nucleus may represent the ideal basal state for reprogramming. (B) Dolly with her firstborn, Bonnie. [(B) courtesy of the Roslin Institute, Edinburgh.]

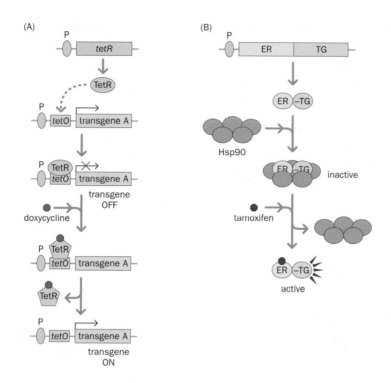

Figure 20.5 Inducible expression of transgenes. (A) Tetracycline-inducible expression. A *tetO* operator inserted just upstream of transgene A is a recognition sequence that can be specifically recognized and bound by the *E. coli* Tet repressor protein (TetR). When constitutively expressed from a separately introduced *tetR* transgene, the Tet repressor protein binds to the *tetO* operator and so prevents the expression of transgene A. However, if doxycycline (an analog of tetracycline) is subsequently provided, it binds to the Tet repressor protein, causing it to change its conformation so that it no longer binds to the *tetO* operator. As a result, expression of the previously silenced transgene A is switched on. Note that toxicity effects reflecting high-level constitutive expression of the repressor limit the use of this system in some cells. (B) Tamoxifen-inducible expression. Here coding sequences for the ligand-binding domain of a mutant mouse estrogen receptor (ER) are fused to a transgene of interest (TG) in an expression vector. The expressed ER–TG fusion protein is bound by the Hsp90 inhibitory protein complex that prevents it from performing its function. The mutant ER ligand-binding domain is not recognized by estrogen but will bind to tamoxifen, an estrogen analog. Following tamoxifen binding, the fusion protein is released from the Hsp90 complex and the protein of interest is activated, even though it is part of a fusion protein. P, promoter.

often bacterial elements, or by engineering mammalian proteins so that they are incapable of responding to endogenous inducers. Two widely used systems are described below.

Tetracycline-regulated inducible transgene expression

This system is based on the *E. coli tet* operon, and induction occurs at the level of transcription. Two transgenes are involved. One includes the *E. coli tetR* gene and is designed to express the Tet repressor protein constitutively. The other includes a gene of interest that has been modified so that the *tetO* operator sequence is inserted between the promoter and the coding sequence. The Tet repressor protein will bind to the *tetO* operator specifically and prevent expression of the downstream coding sequence. However, gene expression can be restored by providing doxycycline, a tetracycline analog, which binds to the Tet repressor protein, inducing a conformational change and dissociation of the Tet repressor protein (**Figure 20.5A**).

Tetracycline-regulated inducible transgenes are widely used, but because expression is induced at the transcriptional level there may be a significant delay (while the RNA is translated and the active protein is produced) before a response to induction is seen. A similar delay will occur between the removal of the inductive stimulus and return to the basal state after RNA transcripts and protein have been broken down.

Tamoxifen-regulated inducible transgene expression

Where rapid induction and decay are essential, an inducible system that works at the protein level is needed. One popular system is based on the post-translational regulation of the estrogen receptor. Like other nuclear hormone receptors, the estrogen receptor is bound by the Hsp90 protein inhibitory complex that keeps it in an inactivated state within the cytoplasm until its ligand, estrogen, enters the cell. Estrogen binds to the C-terminal ligand-binding domain of the estrogen receptor, causing a change of conformation. As a result, the Hsp90 inhibitor is released and the activated estrogen receptor translocates to the nucleus to activate certain target genes (see Figure 4.8 for the structure of nuclear hormone receptors, and Figure 4.9 for the mechanism of activation by ligands).

Tamoxifen-inducible expression uses a mutant ligand-binding domain of the mouse estrogen receptor Esr1. The mutant domain does not bind its natural ligand, estrogen, at physiological concentrations but does bind the synthetic ligand

4-hydroxytamoxifen. A cDNA for the mutant Esr1 ligand-binding domain is fused to the coding sequence of a gene of interest. Hsp90 binding prevents expression of the fusion protein until tamoxifen is provided (Figure 20.5B).

Transgene expression may be influenced by position effects and locus structure

Transgene integration is usually allowed to occur randomly. Identical transgene constructs can therefore produce differences in the level or pattern of transgene expression in independently derived transgenic animals. Transgene expression can be radically affected by the influence of local regulatory elements and chromatin structure (*position effect*; see Chapter 16, p. 508). If a transgene integrates next to an enhancer, its expression may be increased; if it integrates within a heterochromatin domain, its expression may be blocked (*transgene silencing*).

The structure of the transgenic locus can also influence expression. If two neighboring copies of a transgene happen to be arranged as an inverted repeat, then transcription through both copies can generate a hairpin RNA structure that can activate *RNA interference* pathways (see Box 9.7). Various other aspects of the locus structure may trigger cellular defenses against invasive DNA, leading to transgene silencing by *de novo* DNA methylation.

In many cases a transgene contains a cDNA copy of a gene plus its proximal regulatory elements. Inevitably, such copies lack some of the full complement of sequences required when endogenous genes are expressed naturally. Normally, master regulatory elements—often found within introns and sometimes located some distance from the genes that they regulate—establish open chromatin domains and protect endogenous genes from the influence of regulatory elements and chromatin structure in neighboring domains.

To avoid these position effects, additional regulatory elements need to be present in transgenes. To study long-range effects more accurately, and to investigate the expression and regulation of human genes in the context of their own *cis*-acting regulatory elements, large transgenes are used that are provided by yeast artificial chromosome (YAC) cloning (see Figure 6.7). YAC transgenes may contain a megabase or so of DNA, and YAC transgenic animals can be produced by using cell fusion, microinjection, or transfection with liposomes. *Transchromosomic mice* containing human chromosomes or large chromosome fragments have also been developed, as will be described in Section 20.5.

20.3 TARGETED GENOME MODIFICATION AND GENE INACTIVATION *IN VIVO*

The aim of the transgenic technology described in Section 20.2 is to express a gene so as to produce some desired product that may be a protein of interest or a mutant gene product. In this section, we consider a quite different approach in which the aim is to make an animal that carries a predetermined DNA change at a specific location in its genome. Essentially, any mutation can be introduced to order. Often the aim is to selectively inactivate or otherwise modify a *specific* predetermined target gene (*gene targeting*), but site-specific recombination can also be used to produce large DNA changes (*chromosome engineering*) that include subchromosomal deletions, chromosome inversions, and translocations. For some animal species, a precisely defined mutation can be introduced into the germ line, giving rise to animals with that mutation in their cells.

The mouse has long been the premier vertebrate species to be manipulated in this way, but the procedures are laborious and rather time-consuming. Recently, faster methods have been popularly used in some animals in which the function of a specific gene is selectively inhibited *in vivo* at the RNA level without necessarily abolishing all gene function.

The isolation of pluripotent ES cell lines was a landmark in mammalian genetics

The first mammalian embryonic stem (ES) cell lines were obtained in the early 1980s, after culturing cells retrieved from the inner cell mass of surgically excised mouse blastocysts; subsequently, ES cell lines were reported from a variety of other mammals. Human ES cell lines were first reported in 1998 and were obtained from surplus embryos after fertility treatment. In Box 21.2 we describe the procedures for isolating ES cells in the context of human ES cells.

ES cells are pluripotent, being able to give rise to all the different cell types present in the body, including germ cells. However, unlike the fertilized oocyte they are not totipotent: if implanted in a uterus they are unable to give rise to an embryo. The huge advantage of ES cell lines is that while in culture they can be subjected to sophisticated genetic manipulation, allowing specific mutations to be introduced at predetermined positions in the genome. As described in the next section, the genetically modified ES cells can then be used to introduce the mutations into the germ line to eventually produce an animal with a precisely defined mutation. Often, the mutation is designed to inactivate a specific gene, yielding valuable information on gene function. The same approach is widely used to make animal models of human disorders.

The usefulness of an ES cell line depends on various properties, including its ability to maintain pluripotency and to permit germ-line transmission. One particularly successful mouse ES cell line was derived from the mouse strain 129 and continues to be used extensively; however, as we describe in Section 20.4, other mouse ES cell lines are recently beginning to be used more widely. ES cell lines have also been established from other mammals but because they generally had far from ideal properties the mouse has been the premier mammalian model for making gene knockouts. As we will see later in this chapter, however, novel ES cell lines and other types of pluripotent cell lines are currently being developed that will allow gene targeting in various experimental organisms other than mice.

Gene targeting allows the production of animals carrying defined mutations in every cell

A powerful approach to modeling disease and understanding gene function is to make mutants in which the sequence of an individual predetermined gene is modified in a specific way (**gene targeting**). In vertebrates the technology is most developed for mice: since the pioneering work in the late 1980s, thousands of mutant mouse strains have been generated.

The origins of gene targeting lie in a natural phenomenon called *homologous recombination*, in which pairing of maternal and paternal homologous chromosomes is followed by sequence exchanges between the aligned homologs. Homologous recombination during meiosis permits shuffling of our genes, increasing genetic variation within the population. It is also used to repair DNA: after one allele suffers DNA damage, homologous recombination can be used to replace the damaged material with a copy of the normal (allelic) sequence from the other homologous chromosome (see Box 13.3).

In gene targeting by homologous recombination, the cell's natural recombination machinery is used to recombine endogenous sequences within the target gene and closely related exogenous DNA sequences contained in a targeting vector. If gene targeting is conducted in ES cells, the specific gene change can subsequently be introduced into the germ line by injecting genetically modified ES cells into a blastocyst that can develop into a mouse within a foster mother (**Figure 20.6**).

Homologous recombination is a very rare process in mammalian cells, occurring about 10^4–10^5 times less frequently than random integration. The frequency of homologous recombination depends on both the length of the homologous regions (shared by the endogenous gene and the targeting vector) and the degree of similarity with the target gene. Because of this, the exogenous DNA clone generally has long sequences that are *isogenic* (identical by sequence) with the genomic DNA of the host cell. Even then, the frequency of genuine homologous recombination events is normally very low and may be difficult to identify against a sizeable background of random integrations. The strategy used to identify cells in which gene targeting has taken place depends on the construct design (see below).

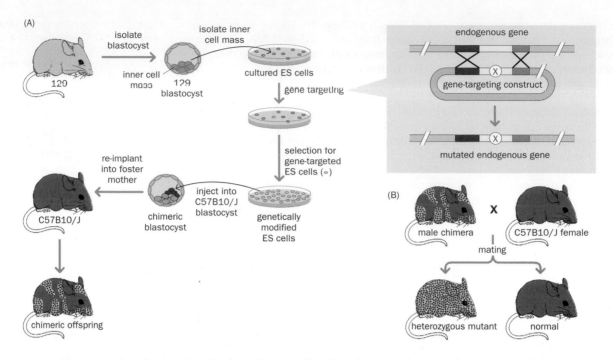

Figure 20.6 Genetic targeting of embryonic stem cells to introduce mutations into the mouse germ line. (A) An embryonic stem (ES) cell line is made by excising blastocysts from the oviducts of a suitable mouse strain (such as 129), and cells from the blastocyst's inner cell mass are cultured to eventually give an ES cell line (see Box 21.2 for the details). ES cells can be genetically modified while in culture by transfection of a linearized recombinant plasmid containing DNA sequences that are identical to certain sequences in a predetermined endogenous gene, except for a desired genetic modification. Homologous recombination allows sequence swapping so that the targeted gene is mutated in the ES cell's genome. Only a very few cells will undergo homologous recombination to produce the expected modification of the targeted gene, but they can be selected for and amplified. The modified ES cells are then injected into isolated blastocysts of another mouse strain such as C57B10/J, which has black coat coloration that is recessive to the agouti color of the 129 strain; the cells are then implanted into a pseudopregnant foster mother of the same strain as the blastocyst. Subsequent development of the introduced chimeric blastocyst results in a chimeric offspring containing two populations of cells that derive from different zygotes (here, 129 and C57B10/J, normally evident by the presence of differently colored coat patches). (B) Backcrossing of chimeras can produce mice that are heterozygous for the genetic modification (bottom left). Subsequent interbreeding of heterozygous mutants generates homozygotes.

Different gene-targeting approaches create null alleles or subtle point mutations

Gene knockouts

A common strategy in gene targeting is to completely inactivate a predetermined target gene, creating a *null allele* (a **gene knockout**). The objective may be to model a human disease caused by a loss of gene function, or simply to investigate the function of a gene. Homologous recombination is commonly used to insert an exogenous DNA sequence into the target gene in the place of some DNA sequence that is essential for the target gene's function. The net loss of important DNA sequence in the target gene causes gene inactivation. If the target gene is very small, the whole gene may be replaced. Usually, however, exogenous DNA is inserted to replace just one or a few exons at the 5′ end of the target gene, in order to induce a shift in the translational reading frame so that an early premature termination codon will be introduced.

To track its incorporation into the ES cell genome, the exogenous DNA carries some marker gene such as the *neo* gene, which makes neomycin phosphotransferase and can confer resistance to neomycin or its analog G418. To distinguish desired homologous recombination events from uninteresting random integration events, a positive–negative marker selection strategy is often used (**Figure 20.7**).

Gene knock-ins

Modern strategies for knocking out mouse genes are often designed to also introduce a reporter gene such as *lacZ* or a gene that encodes an autofluorescent protein such as GFP (green fluorescent protein), RFP (red), or YFP (yellow). The

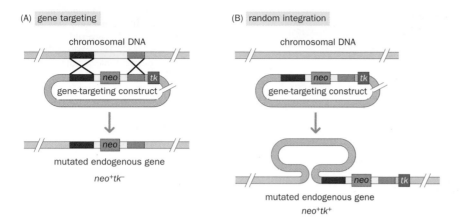

(A) gene targeting

chromosomal DNA

gene-targeting construct

mutated endogenous gene

neo⁺tk⁻

(B) random integration

chromosomal DNA

gene-targeting construct

mutated endogenous gene

neo⁺tk⁺

Figure 20.7 Positive–negative selection to enrich for ES cells containing a desired gene-targeting event. Here the linearized plasmid used for gene targeting contains two marker genes: the neomycin phosphotransferase gene (*neo*), which confers resistance to neomycin, and its analog G418; and the *Herpes simplex* thymidine kinase gene (*tk*). (A) In some gene-targeting events that occur by homologous recombination, double crossover leads to incorporation of the *neo* gene but not of the *tk* gene. (B) Random integration of the gene-targeting construct at a chromosome break leads to integration of both the *neo* and *tk* genes. Suitably modified ES cells can be identified by selecting for *neo⁺tk⁻* cells.

targeting construct, containing a reporter gene, is usually designed to integrate at the 5′ end of the endogenous gene. This inactivates (*knocks out*) the endogenous gene while using the endogenous gene's promoter and other expression controls to activate the reporter gene (*knocking in*). Because the expression of the knocked-in reporter gene is regulated by the endogenous target gene's regulatory sequences, it should faithfully mimic the expression of the target gene. The principle underlying a **gene knock-in** is illustrated in **Figure 20.8**; a practical example of knocking in a reporter gene is shown in **Figure 20.9**.

Knock-ins sometimes also examine the effect of placing a transgene of interest other than a simple reporter gene under the regulatory control of a specific endogenous promoter simultaneously with inactivating the endogenous gene's normal transcript. One application involves knocking in mutant transgenes into the endogenous gene locus to model disease, as in some models of Huntington disease that are described later in this chapter.

Creating point mutations

Mice have also been engineered to contain point mutations that alter one or a few nucleotides at a predetermined position within a target gene. The object may be to investigate how a small sequence, such as a specific codon, contributes to gene function or to replicate pathogenic human mutations. In either case, it is important that the end product is a mutant allele that does not have any marker gene (which could possibly affect its function).

Various strategies to introduce point mutations are possible and typically involve two rounds of recombination. In the tag-and-exchange strategy, for example, the desired gene is first tagged by inserting a selectable marker during one gene-targeting round. The ES cells containing the suitably tagged target gene are then identified so that a second round of targeting can be performed to introduce the desired point mutation with a high degree of specificity and simultaneously remove the marker gene.

Microbial site-specific recombination systems allow conditional gene inactivation and chromosome engineering in animals

Site-specific recombination systems are used naturally in microbes, including bacteria and yeast. In such systems, the recombinase specifically recognizes a short defined DNA sequence and will induce recombination between two copies

Figure 20.8 A gene knock-in replaces the activity of one chromosomal gene by that of an introduced transgene. In this example, the endogenous gene is envisaged to have multiple exons (E1, E2, ..., E*n*); coding sequences are in filled boxes and untranslated sequences are in open boxes. The gene-targeting vector contains two sequences from the endogenous target gene: **a**, a 5′ flanking sequence including the promoter, and a very small 5′ component of exon 1 that is mostly made up of the 5′ untranslated sequence; and **b**, an internal segment that spans much of intron 1. Between these two sequences is the coding sequence of the gene to be knocked in plus a marker gene whose expression is regulated by its own promoter. Arrows indicate expression driven by upstream promoters (P). The targeting procedure results in a double crossover (X) and in this case an inactivating deletion of a large part of exon 1 that is usually designed to result in a shift in the translational reading frame. The knocked-in gene of interest comes under the control of the endogenous promoter. The *neo* marker helps identify ES cells with the correct gene-targeting event, and a thymidine kinase (*tk*) gene marker helps select against random integration as illustrated in Figure 20.7.

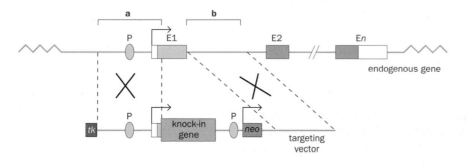

endogenous gene

knock-in gene

targeting vector

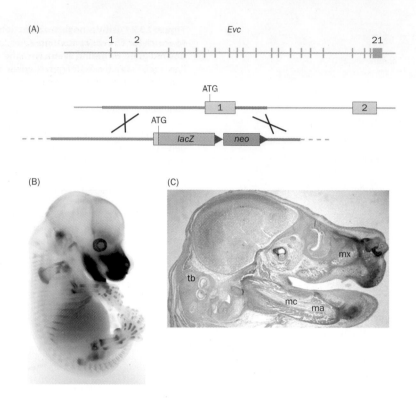

Figure 20.9 Knocking in a *lacZ* reporter transgene to simultaneously inactivate a gene and monitor its normal expression. This example shows a knock-in at the *Evc* gene, the mouse ortholog of the Ellis van Creveld syndrome gene, which has 21 exons. (A) Gene targeting involved homologous recombination with a transgene construct that included a *lacZ* reporter (encoding the enzyme β-galactosidase) and a *neo* marker gene plus target homology sequences. The latter corresponded to a 6 kb upstream region preceding exon 1 (green line) plus a 5′ region of exon 1 terminating in the ATG translation initiator trinucleotide and then a 1 kb region immediately following exon 1 (lilac line). Integration of the transgene inactivated the *Evc* gene by deleting the rest of the exon 1 coding sequence and also brought the *lacZ* gene under the regulation of the endogenous *Evc* promotor. (B, C) Expression images show β-galactosidase activity in *Evc*[+/−] embryos and represent expression in the whole E15.5 embryo (15.5 days *post coitum*) (B) and in a sagittal paraffin section of the head of an E15.5 embryo (C). Note the strong expression in the mouth and tooth-forming areas in addition to the developing skeleton. mx, maxilla; ma, mandible; mc, Meckel's cartilage; tb, temporal bone. [Courtesy of Judith Goodship, Newcastle University; from Ruiz-Perez VL, Blair HJ, Rodriguez-Andres ME et al. (2007) *Development* 134, 2903–2912. With permission from The Company of Biologists.]

of the recognition sequence. Recognition sites for microbial recombinases can be engineered easily into gene-targeting vectors and can be inserted into specific locations in an animal genome. The microbial recombinase can be supplied conditionally so that it is expressed only in certain tissues or at certain developmental stages in the animal, or so that it is inducible. To do this, another transgene is required in which the microbial gene coding for a recombinase is ligated to regulatory components such as a suitably regulated promoter or inducible gene expression system, such as the tetracycline-inducible system.

In genetically modified mice, the most widely used site-specific recombination systems are the Cre–*loxP* system originating from bacteriophage P1, and the FLP–FRT system from the 2 μm plasmid of *Saccharomyces cerevisiae*. Both the Cre recombinase and the FLP (flippase) recombinase recognize specific target sequences that are 34 base pairs long, respectively the *loxP* sequence and the FRT sequence (FLP recombinase target). The size of these microbial sequences is such that when introduced into an animal genome they will be quite distinctive (the odds that a sequence identical to the *loxP* recognition sequence will exist by chance in an animal genome of 3000 Mb and 40% GC are less than 1 in 10^{20}).

Although their sequences are different, *loxP* and FRT have essentially the same structure, comprising inverted 13 bp repeats separated by a central asymmetric 8 bp spacer (**Figure 20.10**). If two copies of the recombinase target site are located on the same DNA molecule and in the same orientation, recombination results in excision of the intervening sequence; if the *loxP* sites are in opposite orientations, recombination produces an inversion. Site-specific recombination between two target sequences on different DNA molecules is also possible.

The Cre–*loxP* system has been applied in many different ways in genetically modified mice. It can allow the site-specific integration of transgenes, the conditional activation and inactivation of transgenes, and the deletion of unwanted

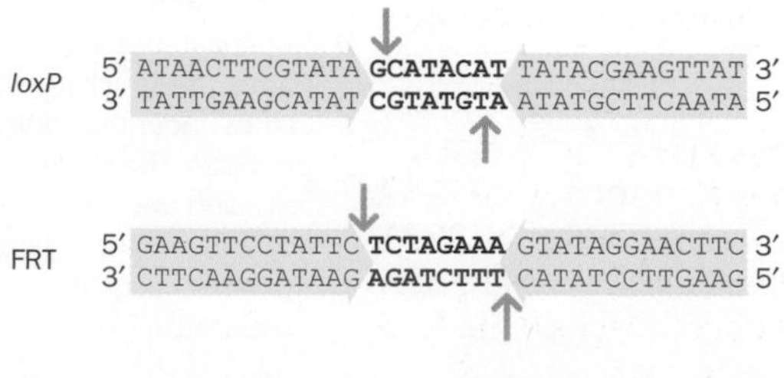

Figure 20.10 Structure of the *loxP* and FRT recognition sequences. Recombinases that catalyze site-specific recombination typically recognize and cleave a sequence that has inverted repeats flanking a central asymmetric core. The Cre recombinase from bacteriophage P1 recognizes the *loxP* (**l**ocus **o**f **X**-over, **P**1) and the FLP (flippase) recombinase from the *S. cerevisiae* 2 μm plasmid recognizes the FRT (flippase recognition target) sequence. Both the *loxP* and FRT recognition sequences are 34 bp long, with 13 bp inverted repeats (within arrows) flanking a central asymmetric 8 bp sequence (shown in bold). A recombinase monomer binds to both inverted repeats, and the two monomers form an active dimer that asymmetrically cleaves the central core sequence (breakpoints are shown by vertical green arrows) within. The central sequence confers orientation: for example, in one orientation *loxP* has the central sequence shown (GCATACAT); in the other orientation the central sequence is ATGTATGC.

loxP   5′ ATAACTTCGTATA **GCATACAT** TATACGAAGTTAT 3′
        3′ TATTGAAGCATAT **CGTATGTA** ATATGCTTCAATA 5′

FRT   5′ GAAGTTCCTATTC **TCTAGAAA** GTATAGGAACTTC 3′
        3′ CTTCAAGGATAAG **AGATCTTT** CATATCCTTGAAG 5′

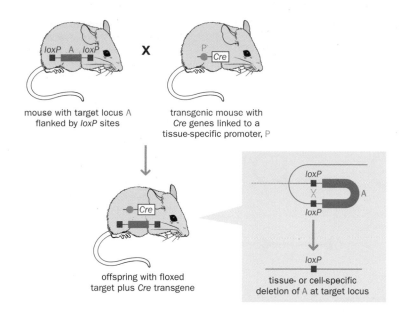

Figure 20.11 Conditional gene inactivation using Cre–*loxP* site-specific recombination. The mouse shown at the top left has been subjected to gene targeting so as to introduce two *loxP* sequences at positions flanking a specific genomic DNA sequence, A, that may be a small gene or an exon that, if deleted, could be expected to cause gene inactivation. The floxed A sequence will be specifically deleted in the presence of Cre recombinase. To introduce Cre recombinase, a mouse is required that has previously been constructed to have a *Cre* transgene ligated to a tissue-specific promoter P of interest (top right). By breeding the two types of mouse it is possible to obtain offspring containing both the *loxP*-flanked target locus plus the *Cre* transgene (bottom left). Cre recombinase will be produced in the desired type of tissue and will cause recombination between the two *loxP* sequences in cells of that tissue, leading to deletion of the A sequence and tissue-specific gene inactivation.

marker genes. The most important applications, however, are conditional gene inactivation and conditional recombination leading to major rearrangements in chromosomal DNA.

Conditional gene inactivation

Some genes have vitally important roles in cells or during development. A homozygous knockout for such a gene will typically result in embryonic lethality, but the heterozygous gene knockout may not show a phenotype (the presence of one functional allele may be enough for the organism to function normally). To study such a gene, a *conditional* gene knockout is often made. The gene is designed to be inactivated in only a selected tissue or group of cells or at a desired developmental stage so that the effect of homozygous gene inactivation can be studied.

Conditional gene inactivation typically involves replacing exons of an endogenous gene with a homologous gene segment flanked by *loxP* sequences (the segment is said to be *floxed*). Mice carrying the floxed target sequence are then mated with a strain of mice carrying a *Cre* transgene under the control of a tissue-specific or developmental stage-specific promoter (**Figure 20.11**).

Offspring of the cross that contain both a floxed target sequence and a *Cre* transgene are identified and analyzed. Depending on the promoter regulating the *Cre* transgene, the Cre recombinase should be expressed only in certain cells or at certain stages of development so that viable mutants can be analyzed.

Inducible Cre transgenics have also been generated. The *Cre* transgene is designed to have a Tet operator *tetO* and to be repressed by the binding of a Tet repressor protein (see Figure 20.5A), but it can be switched on by supplying doxycycline in the animal's drinking water. Alternatively, Cre is expressed constitutively as a fusion protein with a mutant estrogen receptor ligand-binding domain and is activated at the protein level by tamoxifen (Figure 20.5B).

Chromosome engineering

Inducible site-specific recombination has also been applied to produce deletions of specific mouse subchromosomal regions (from tens of kilobases up to tens of megabases) and to engineer specific chromosome translocations. Gene targeting is first used to integrate *loxP* sites at the desired chromosomal locations. Subsequently, transient expression of Cre recombinase induces a selected chromosomal rearrangement, such as a chromosomal translocation after *loxP* sites have been targeted to different chromosomal DNA molecules (**Figure 20.12A**).

Alternatively, engineering two *loxP* sites into the same chromosomal DNA can result in excision or inversion of the intervening sequence (Figure 20.12B). Excision occurs when the *loxP* sites are in the same orientation and is essentially irreversible because it generates a single *loxP* site. Inversion occurs when the *loxP*

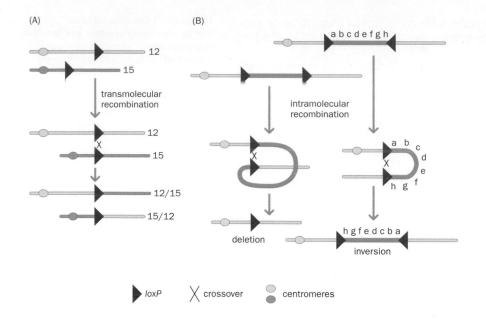

(A)

(B)

transmolecular recombination

intramolecular recombination

deletion

inversion

▶ loxP   ✕ crossover   ● centromeres

Figure 20.12 Chromosome engineering using Cre–*loxP* site-specific recombination. (A) To generate a chromosomal translocation, gene targeting is used to insert *loxP* sites (and accompanying marker genes not shown) at predetermined positions on the two chromosomes (chromosomes 12 and 15 in this example). Subsequent exposure of these chromosomes to Cre recombinase results in chromosome breakage at the *loxP* sites and re-joining to produce the desired translocation chromosome. (B) Targeted insertion of *loxP* sites permits intrachromosomal deletion (left) or an inversion (right) by intrachromosomal recombination. To generate a deletion, the flanking *loxP* sites need to be in the same orientation. To generate an inversion, the *loxP* sites must be in opposite orientations. Because two intact *loxP* sites remain, in this case there is the possibility of a secondary recombination between the *loxP* sequences causing the inverted segment to flip back again; to minimize this possibility, two different *loxP* sequences can be used (see the text).

sites are in opposite orientations and yields two *loxP* sites that are indistinguishable from the original *loxP* pair, so that the inversion could be reversed by a subsequent recombination. To avoid this, *loxP* sequences have been mutated to generate two slightly different mutant *loxP* sequences that can be used together to promote inversion in the desired direction.

Such **chromosome engineering** offers the exciting possibility of creating novel mouse lines with specific chromosomal abnormalities for genetic studies. A drawback has been that the frequency of such events has been very low. Recently, however, methods using transposons seem particularly suited to chromosome engineering, and these are described in Section 20.4.

Zinc finger nucleases offer an alternative way of performing gene targeting

Since the development of stable mouse ES cell lines in the 1980s, targeted modification of the genomes of vertebrates has focused almost exclusively on the mouse, in which homologous recombination in ES cells is used to engineer specific genome modifications into the germ line. By contrast, reported ES cell lines from other mammals have often been unstable, showing a rapid decline in proliferation potential and pluripotency. In the absence of stable authentic ES cell lines for some mammals, such as sheep and pigs, gene knockouts have in some cases been made by somatic cell nuclear transfer. This approach, however, is technically arduous and inefficient, and is rarely used.

Two types of recent technological development are now offering the prospect of rapidly extending gene targeting to many other mammals and vertebrates. One breakthrough involves novel pluripotent stem cells and includes recently reported non-murine ES cell lines and novel types of pluripotent cells derived originally from somatic cells. We consider these at the end of this chapter. The other breakthrough is described in this section and concerns **zinc finger nucleases**, artificially constructed endonucleases that are designed to make a double-strand DNA break at just one location in a genome, within a predetermined target gene. Natural cellular DNA repair pathways can repair the double-strand break, but often the repair is naturally inaccurate, or can be artificially induced to be inaccurate, resulting in specific inactivation of the target gene.

Zinc finger nucleases are engineered to have separate DNA-binding and DNA-cleaving domains and are produced as fusion proteins by ligating different protein-coding DNA sequences. The DNA-cleaving domain comes from *Fok*I, one of the few restriction nucleases that have separate DNA-binding and DNA-cleaving domains (in most restriction enzymes, a single domain is used for DNA binding and cleavage). The DNA-binding domain of a zinc finger nuclease is assembled by fusing coding sequences for different types of zinc finger to produce a particular combination of zinc fingers with the desired DNA-binding sequence specificity (**Figure 20.13A**).

Like restriction nucleases, zinc finger nucleases work as dimers. To make a double-strand DNA break, the two monomers bind to and cut complementary DNA strands. Standard restriction nucleases typically work as homodimers (with the result that both monomers recognize the same DNA sequence and the recognition sequence is palindromic). However, zinc finger nucleases are deliberately designed to work as heterodimers, and each monomer is designed to recognize different DNA sequences that are located within a very short distance from each other in the target gene (Figure 20.13B). A zinc finger nuclease heterodimer composed of monomers each with four zinc fingers (as shown in Figure 20.13) effectively binds a 24 (12 + 12) nucleotide recognition sequence that can be expected to occur just once in a genome.

The induced double-strand break can be repaired by different natural cellular dsDNA repair pathways (as described in Box 13.3). One such pathway, the non-homologous end joining (NHEJ) pathway, is naturally prone to error: because the broken ends are simply joined together, the joining process is frequently accompanied by a loss or gain of nucleotides (Figure 20.13C). For the vast majority of locations in the DNA sequence there would be no serious consequences, but when the double-strand break is designed to occur in a coding sequence, the result is often gene inactivation. Successful gene targeting with zinc finger nucleases was first conducted in zebrafish in 2008 and in rats in 2009. In both cases mRNAs encoding zinc finger nucleases were injected into one-cell embryos, and target genes were inactivated after inaccurate repair by the NHEJ DNA repair pathway.

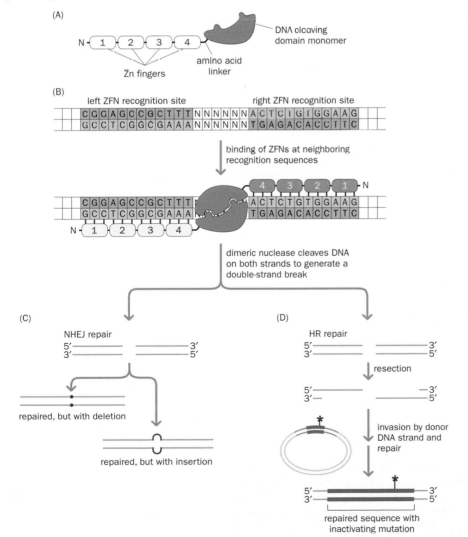

Figure 20.13 Targeted gene inactivation using genetically engineered zinc finger nucleases. (A) Zinc finger nucleases (ZFNs) contain a series of zinc fingers joined by an amino acid linker to a DNA-cleaving domain. Individual zinc fingers bind to *specific* triplet sequences in DNA, and combinations of different zinc finger modules can be assembled in a defined order to bind to specific DNA sequences of interest. (B) A double-strand break is introduced into functionally important sequence in a specific gene with a pair of different ZFNs. In this example the ZFNs each have four zinc fingers and are designed to specifically recognize neighboring 12 bp sequences at the target site. The effective recognition sequence is 12 + 12 = 24 nucleotides and so should occur just once in the genome. The targeting event is designed so that the DNA-cleaving domains of the two different ZFNs can come into contact and dimerize, activating cleavage of both DNA strands. The resulting double-strand break (DSB) activates cellular DNA repair pathways. The cleavage domains of the two ZFNs can be designed to be asymmetrical so that heterodimers form but not homodimers. (C) In the nonhomologous end joining (NHEJ) DNA repair pathway, the repair is naturally often an imperfect one, with nucleotides being deleted or inserted that may cause gene inactivation. (D) In the homologous recombination (HR) pathway, the 5′ ends of the DSB are first trimmed back (*resection*), allowing strand invasion by donor DNA strands. A plasmid with homologous sequence (bold lines) that is designed to have an inactivating mutation (red asterisk) may be used as the donor DNA, acting as a template for synthesizing new DNA from a *portion* of the homologous sequence (thick blue lines) to replace the previously existing sequence. As a result, the selected mutation is copied into the gene sequence, causing gene inactivation.

The alternative major pathway for repairing double-strand DNA breaks, the homologous recombination (HR) pathway, offers a general and more powerful way of artificially inducing faulty DNA repair. In this cellular pathway the broken DNA ends are trimmed back and new DNA, copied from a donor DNA strand with the homologous sequence, is inserted (Figure 20.13D). To induce an inactivating mutation, a plasmid can be provided that contains a portion of the target gene sequence spanning the location of the double-strand break but altered to include a specific inactivating mutation. The HR pathway would use the plasmid as a donor template DNA to direct repair and in so doing would be able to copy the inactivating mutation into the repaired sequence (see Figure 20.13D). This approach allows exquisite control over the mutagenesis procedure, unlike the random inactivation of the NHEJ pathway.

Targeted gene knockdown at the RNA level involves cleaving the gene transcripts or inhibiting their translation

Gene targeting has resulted in numerous informative gene knockouts, but the procedure is laborious and time-consuming. Alternative ways of inactivating genes *in vivo* have therefore been tried.

Quick and comparatively easy methods enable a gene's function to be selectively inhibited at the RNA level. The downside is that such methods rarely produce the complete abolition of gene function that is possible with gene targeting at the DNA level. Instead, the net effect is usually a major decrease in gene expression and is described as a **gene knockdown**, to distinguish it from a gene knockout. To knock down a gene, cellular RNA interference pathways (see Box 9.6) are exploited to selectively degrade transcripts from a gene, or an exogenous oligonucleotide or RNA base-pairs with transcripts from the target gene, inhibiting its expression.

In vivo gene knockdown by RNA interference

RNA interference (RNAi) is a very rapid, straightforward method that is used extensively to block expression of specific target genes in cultured cells from a variety of different animal species (see Chapter 12, p. 390). By cleaving transcripts, RNAi can cause a severe decrease in the expression of a target gene, but not normally the complete abolition of gene expression that is expected in gene knockouts. Despite this restriction, highly efficient *in vivo* RNAi knockdowns have produced a variety of informative loss-of-function phenotypes.

In vivo RNAi has been performed on a large scale in *D. melanogaster* and *Caenorhabditis elegans*. However, RNAi using dsRNA is often problematic in vertebrates. In zebrafish, dsRNA often triggers nonspecific interference and is often toxic to embryos. This has prompted the use of alternative systems for knocking down gene expression, as described in the next section.

In mammals, long dsRNA elicits a nonspecific antiviral pathway that causes a general inhibition of translation and subsequently programmed cell death (see Chapter 12, p. 391). Instead, specific small RNAs are used in RNA interference studies in mammals. Although small interfering RNA (siRNA) is often suitable for temporary gene knockdown in cultured cells, it is not suited to the maintenance of a sustained gene-silencing and germ-line transmission of RNA interference. Instead, transgene constructs designed to encode a specific shRNA (short hairpin RNA; see Figure 12.4) are injected into one-cell embryos; the shRNA can be naturally processed within cells of the transgenic animal to give siRNAs that activate endonuclease complexes to cleave transcripts from the target gene.

RNAi knockdown can result in developmental phenotypes in rodents, and successful RNAi-mediated gene knockdowns have also recently been performed in monkeys with the aim of testing the efficacy and safety of RNAi therapeutics. This area is covered in Chapter 21.

Gene knockdown with morpholino antisense oligonucleotides

Antisense technology seeks to inhibit the expression of a predetermined target gene by providing oligonucleotide or polynucleotide reagents that base-pair with the transcripts of the target gene. The most efficient approach is to make highly stable and robust types of oligonucleotides that are completely resistant to

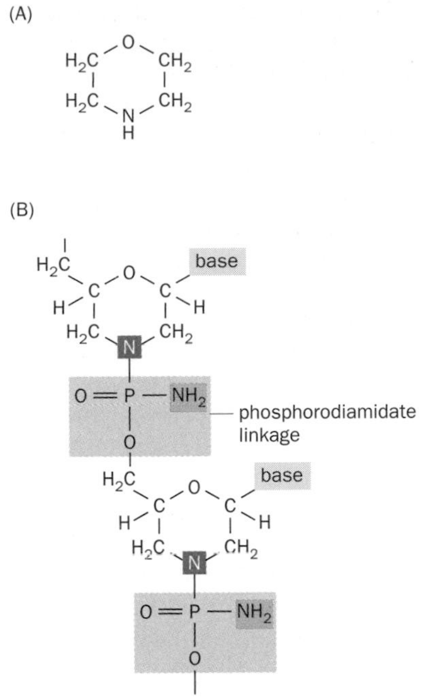

Figure 20.14 Structure of phosphorodiamidate morpholino oligonucleotides. (A) Structure of morpholine. (B) Repeating structure of morpholino oligonucleotides. Like conventional nucleotides (see Figure 1.5), the individual mononucleotide repeat units have a carbon-based ring to which is attached one of the four bases, and the repeat units are joined by phosphorus-containing bonds. However, the ribose or deoxyribose sugar is replaced by a ring that has an additional nitrogen atom (blue shading), and successive morpholine units are connected by phosphorodiamidate linkages (highlighted) instead of phosphodiester bonds. For more information on morpholino synthesis, see the Gene Tools website at http://www.gene-tools.com/.

nuclease attack, enabling them to remain intact and function much longer than standard oligonucleotides or polynucleotides.

Phosphorodiamidate morpholino oligonucleotides (often called *morpholinos*) are synthetic polymers of modified nucleotides in which bases are attached to morpholine rings instead of to a ribose or deoxyribose sugar. The morpholines of adjacent nucleotides are connected by non-ionic phosphorodiamidate linkages instead of the usual phosphodiester bonds (**Figure 20.14**; compare with Figure 1.5).

Morpholinos are very stable and are not degraded in cells. They are usually designed to have a sequence of about 25 nucleotides and hybridize to their targets under conditions that would prevent the equivalent normal oligonucleotides from binding. They may be directed to bind to a 5′ untranslated region (UTR) sequence close to the translation start site, so that they inhibit translation. Alternatively, they are designed to bind and blockade a splice site in the precursor RNA, causing alternative splicing.

Morpholinos are widely used to make gene knockdowns to investigate gene function in vertebrate embryos. Like other oligonucleotides, morpholinos do not easily cross membranes, so that delivery is not straightforward. Microinjection into embryos from a wide variety of species has been possible, but most work has been done in zebrafish embryos, which have the advantage of being transparent. One disadvantage is that morpholinos may have only a transient effect when introduced into the single-cell zygote and are diluted out after cell division and growth.

20.4 RANDOM MUTAGENESIS AND LARGE-SCALE ANIMAL MUTAGENESIS SCREENS

Gene targeting can allow precise mutations to be engineered into known genes in animals, but it can be laborious to conduct gene targeting on a large scale. Genomewide mutagenesis screens often involve rapid methods in which genes are mutated essentially at *random*. That is, some method is employed that allows significantly increased mutation (above background levels) to occur within the genome.

Random mutagenesis is not driven by genes—no attempt is made to direct the mutation to specific genes, and it is not possible to predict which genes will be affected. However, thousands of mutants can quickly be produced with a range of interesting phenotypes that can then be analyzed, and so random mutagenesis screens can be viewed as being phenotype-driven.

To correlate phenotypes with genotypes, it is desirable to identify the underlying mutations. Ionizing radiation or certain chemicals are very effective mutagens, but there is no easy way to identify underlying mutations that they induce. As an alternative, mutagenesis methods have been developed whereby known DNA sequences are allowed to integrate randomly into animal genomes. Such **insertional mutagenesis** involves the inactivation of an endogenous gene through the integration of either defective reporter transgenes (*gene trapping*) or certain transposons. Because the affected endogenous genes are tagged by a *known* sequence, they can be quickly identified by PCR.

Extensive mutagenesis screens have been performed in *D. melanogaster*, in *C. elegans*, and in zebrafish. As a major step toward understanding human genes and modeling disease, international consortia have very recently been using a combination of mutagenesis approaches to produce mutant phenotypes for all of the 20,000 or so mouse genes.

Random mutagenesis screens often use chemicals that mutate bases by adding ethyl groups

Traditionally, random mutagenesis screens have used ionizing radiation (e.g. X-rays) or chemical mutagens. In the latter case, ethylnitrosourea (ENU) and ethyl methanesulfonate (EMS) have been popular because of their extremely high efficiency. They act as alkylating agents, transferring their ethyl group (**Figure 20.15**) to bases in the DNA, causing base changes. ENU predominantly modifies A-T base pairs to form *N*-3-ethyladenine and *O*-4-ethylthymine, respectively, and EMS preferentially modifies guanine to form *O*-6-ethylguanine. For those mutations that by chance alter the coding sequence, the most common class consists of missense mutations that can result in loss of function or gain of function.

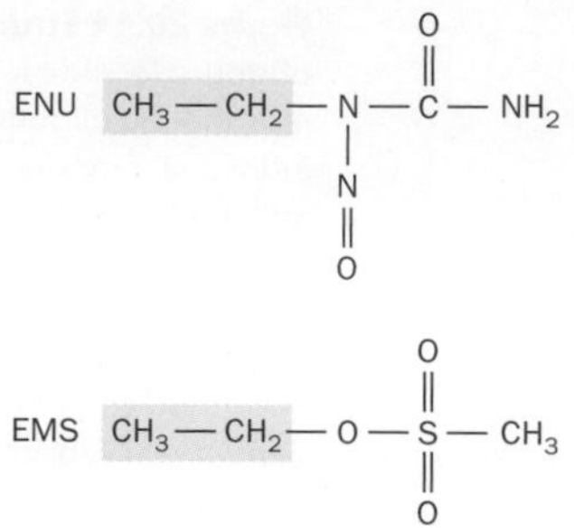

EMS has been a favorite mutagen in screens of *D. melanogaster* and *C. elegans*. Some screens have been confined to specific chromosome regions (by using strains with deletions or chromosome inversions in the crosses); others have been genomewide. Most of the screens have been phenotype-driven: investigators set out to identify a particular class of mutants that had some common phenotype of interest, such as wing abnormalities.

The first genomewide mutagenesis screens in any vertebrate were reported in 1996 when groups from Tübingen and Boston, MA, reported ENU mutagenesis of male zebrafish. Mutation in spermatogonial stem cells led to the identification of thousands of developmental phenotypes in the descendants of the mutagenized fish. More recently, large-scale mutagenesis programs have been developed for the African clawed frog, *Xenopus laevis*. In similar programs for mice, exposure to ENU can induce a mutation in any one gene in roughly 1 in 1000 sperm cells (see also below).

Identifying individual ENU-induced mutations can be laborious. One PCR-based approach called TILLING (**t**argeting **i**nduced **l**ocal **l**esions **in g**enomes) has been popularly used to screen sets of genes for mutations. It is based on enzymatic cleavage of wild-type–mutant heteroduplexes with the plant enzyme CEL-1. Future mutation screening may be performed by massively parallel DNA sequencing (Chapter 8, p. 220).

Figure 20.15 Structures of *N*-ethyl-*N*-nitrosourea (ENU) and ethyl methanesulfonate (EMS). These chemical mutagens act as alkylating agents, causing point mutations by transferring their ethyl groups (shaded) to bases in DNA.

Insertional mutagenesis can be performed in ES cells by using expression-defective transgenes as gene traps

Random insertional mutagenesis by a known DNA sequence has the advantage that the insert leaves a sequence tag, so mutated genes are comparatively easy to identify by PCR. One such method, which has been widely used in mice, is the **gene trap** approach and involves insertional mutagenesis of ES cells.

In gene trapping, the transgene is deliberately designed to lack some element needed for expression. The idea is that after it has randomly integrated into an endogenous gene, the defective reporter gene can be activated by gaining access to a complementary expression signal from the endogenous gene. For example, the reporter gene is often designed to lack a functional promoter but, after it integrates within a 5′ intron of an endogenous gene, the endogenous promoter may drive the expression of a fusion transcript containing the reporter coding sequence (**Figure 20.16**).

The principle of gene trapping is much the same as that of gene knock-ins, but it is applied randomly in the genome of ES cells rather than being targeted to one gene. An endogenous gene is inactivated because the insert effectively blocks its expression, and expression of the reporter gene allows the identification of ES cells in which a gene has been trapped in this way. Although insertion into a 5′ intron often causes gene inactivation, the outcome may not be complete gene inactivation. Instead, a *leaky mutation* may sometimes result, in which transcripts of the endogenous gene continue to be produced at low levels in addition to the fusion transcript.

Transposons cause random insertional gene inactivation by jumping within a genome

Transposons are mobile elements: they can move around to different positions within the genome of a single cell (see Chapter 9, p. 289). As they move to new

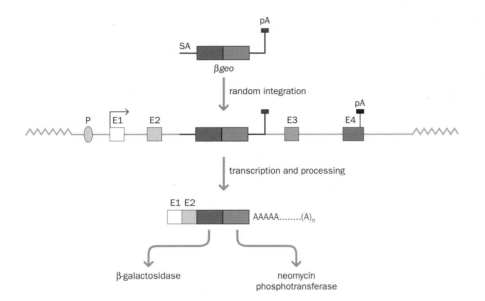

Figure 20.16 Gene trapping. In this example, the gene trap construct lacks a promoter and consists of a splice acceptor sequence (SA), a hybrid gene *βgeo* containing a *lacZ* cDNA encoding β-galactosidase fused to a *neo* cDNA specifying neomycin phosphotransferase, and a polyadenylation signal (pA). Random integration into a gene that was expressed in ES cells can result in insertional inactivation of that gene because of failure to make the normal transcripts. Instead, fusion transcripts are produced containing 5′ exons of the endogenous gene (here E1 and E2) fused to the βgeo coding sequence. Identification of ES cells containing an insertionally inactivated gene is possible by screening ES cells for resistance to the neomycin analog G418 and for β-galactosidase expression. The mutated genes can be identified by PCR assays such as 5′ RACE.

locations, they sometimes integrate into genes and can inactivate them. Retrotransposons encode a reverse transcriptase and they move via an RNA intermediate. DNA transposons move as DNA and encode a DNA transposase that binds to short inverted terminal repeats (see Figure 9.20).

DNA transposons have been widely used in germ-line mutagenesis in *D. melanogaster* and *C. elegans*. The 2907 bp P element of *D. melanogaster* is a celebrated example that has revolutionized the study of gene function in the fruit fly. It moves by a *cut-and-paste* mechanism: each element needs to be excised from the DNA before moving to a new location (as opposed to the *copy-and-paste* mechanisms used by retrotransposons; see Figure 10.9). Germ-line mutations caused by transposon mutagenesis are molecularly tagged, greatly facilitating their identification.

Until very recently, transposons were not available to mutagenize vertebrate genomes in a meaningful way. Transposable elements had been isolated from several vertebrate species, but nearly all of the elements were found to be inactive or had low frequencies of transposition. Transposons with broad host ranges were sought for use as effective mutagens in a wide range of vertebrate genomes.

Two recently developed transposon systems that offer high-frequency transposition in vertebrate genomes are described below. In addition to applications involving germ-line mutagenesis, vertebrate transposon systems can be used in the mutagenesis of somatic cells for cancer gene discovery and in modeling somatic cancers. They can also be used in standard transgenesis via pronuclear microinjection, offering both high efficiency and single-copy transgenes.

Insertional mutagenesis with the Sleeping Beauty transposon

A defective transposon in salmonid fish, presumed to have been inactive for millions of years, was resuscitated by molecular engineering and named *Sleeping Beauty* (**Figure 20.17**). It can transpose in cells of a wide variety of vertebrate species, but it is popularly used to produce genetically modified mice by inserting transgenes into the mouse genome, causing insertional inactivation of genes. Because its sequence is rather different from the endogenous mouse transposons, it is readily identifiable.

Although Sleeping Beauty shows no integration preference with respect to gene structure, it tends to reintegrate into sites located within a short distance (often less than 3 Mb) from the locus it transposes from. This phenomenon, called *local hopping*, is a feature of many transposons (the *D. melanogaster* P1 transposon usually reintegrates within 100 kb of its donor site). The local hopping tendency can be exploited by designing modified Sleeping Beauty transposons to permit region-specific mutagenesis and chromosome engineering (see the review by Carlson & Largaespada in Further Reading).

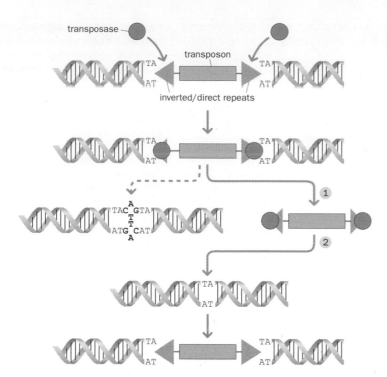

Figure 20.17 The Sleeping Beauty transposon system. The two key elements are the Sleeping Beauty transposase and its transposon substrate. The transposase recognizes inverted repeat or direct repeat elements that flank the DNA that will be transposed, and then excises the transposon from its original location (arrow 1), leaving behind a canonical three-base footprint, C(A/T)G. The excised transposon can subsequently re-insert (arrow 2) at a new location that can be anywhere that a TA dinucleotide is present. This TA is duplicated on re-insertion. [Adapted from Carlson CM & Largaespada DA (2005) *Nat. Rev. Genet.* 6, 568–580. With permission from Macmillan Publishers Ltd.]

piggyBac-mediated transposition

Many DNA transposons, including the P element of *D. melanogaster*, are nonfunctional outside their natural hosts—presumably they need some host cell factor to transpose. Others, however, have a broad host range. piggyBac (PB) is a DNA transposon originally derived from the cabbage looper moth, but PB element-like sequences have been found in diverse genomes from fungi to mammals. PB transposition is not so dependent on host cell factors, and PB transposes efficiently in human and mouse cell lines and also in mouse germ-line cells.

PB transposons insert into the tetranucleotide TTAA and regenerate the sequence on excision. Like Sleeping Beauty, piggyBac displays preferential integration into actively transcribed loci but has a higher transposition efficiency; there is no evidence for local hopping but a bias toward reintegration within intragenic regions.

The International Mouse Knockout Consortium seeks to knock out all mouse genes

Vertebrate gene knockouts have largely been performed in the mouse. Until very recently, knockout mice were made on a small scale in many different laboratories, mostly by gene targeting. There was a lack of coordination between the laboratories, and access to knockout mice was not straightforward.

After the human and mouse genomes had been (mostly) sequenced, serious preparations began toward developing publicly funded systematic and comprehensive approaches to making mouse knockouts. The International Mouse Knockout Consortium (http://www.knockoutmouse.org) was officially established in 2007. With the use of a combination of ES-cell-based gene-trapping and gene-targeting approaches, the consortium seeks to mutate all the mouse protein-coding genes and make knockouts widely available to the scientific community.

Although much progress has been made, there are perceived deficiencies of the 129 mouse strain that has been widely used as a source of ES cells for making gene knockouts. Knockout mice kept on a pure 129 background do not breed well and show abnormal anatomy, behavior, and immunology (**Box 20.1**). To increase fertility and facilitate phenotype analysis, knockout mouse lines are usually backcrossed with the more robust C57BL/6 strain. The resulting mice are therefore a mixture of the 129 and C57BL/6 strains, and lengthy times are required for repeated backcrossing with C57BL/6 to reduce the 129 contribution to a small region surrounding the mutant locus.

BOX 20.1 MOUSE GENETIC BACKGROUNDS AND MODIFIER GENES

The **genetic background** of a mouse describes the genetic constitution (all alleles at all loci) except for the mutated gene of interest and a very small amount of other genetic material (generally from one or two other mouse strains). The genetic background is important because it can influence the phenotype of a mutant allele in different ways.

Generally, a mutant allele of interest is maintained on one to a small number of backgrounds that are considered advantageous (e.g. they display strong phenotypes, and breed well). Thus, whereas ES cells used to make chimeras are traditionally of the 129 strain, knockout mice are not usually kept on a pure 129 background because they do not breed well, and they show abnormalities in their anatomy, immunology, and behavior. Consequently, knockout lines are backcrossed, often with the robust C57BL/6 strain to increase fertility and assist phenotype analysis.

When identical mutations are placed on different genetic backgrounds, large differences may be seen in the mutant phenotype. At the core of the differences between the different genetic backgrounds are a small number of *modifier* genes that somehow interact with the mutant allele.

A useful example of the importance of genetic background is the *Min* (multiple intestinal neoplasia) mouse that was generated by ENU mutagenesis and results from mutations in the mouse *Apc* gene. Mutations in the orthologous human gene, *APC*, cause adenomatous polyposis coli and related colon cancers (OMIM 175100) and the *Min* mouse has been regarded as a good model for such disorders.

The genetic background, however, drastically modifies the phenotype of the *Min* mouse. For example, the number of colonic polyps in mice carrying the Apc^{Min} mutation is strikingly dependent on the strain of mouse. Thus, B6 mice heterozygous for the Apc^{Min} mutation are very susceptible to developing intestinal polyps, but offspring of these mice mated with AKR/J, MA/MyJ, or CAST strains are significantly less susceptible. The latter three strains have strain-unique Apc^{Min} modifier loci, named *Mom1* (modifier of Min1). Similar phenotypic variability is found in human adenomatous polyposis coli families, in which different members of the same family may have strikingly different tumor phenotypes although they possess identical mutations in the *APC* gene.

C57BL/6 ES cells are generally less efficient than 129 ES cells in colonizing developing embryos and so usually give rise to low-level chimeras, but once they have done this there is no problem in achieving germ-line transmission. Recently, a C57BL/6-derived strain of mouse ES cells known as JM8 has been shown to be highly competent for germ-line transmission and has been genetically modified to show a dominant pale (agouti) coat color like the 129 strain. Because of these advantages, the JM8 strain is likely to be used widely in large-scale mouse knockouts.

Although the next few years will see a massive increase in the number of publicly available mouse knockouts, much work will be required to move from knockouts in ES cells to the time-consuming development and phenotyping of the resulting knockout mice. For the large mutagenesis screens, the challenge of analyzing the many facets of the phenotype—collectively called the **phenome**—is being addressed, and various phenotyping standardization protocols have been developed and are being applied.

20.5 USING GENETICALLY MODIFIED ANIMALS TO MODEL DISEASE AND DISSECT GENE FUNCTION

Genetic manipulation of animals has revolutionized our ability to understand how genes work in animal cells and has provided a huge and rapidly increasing number of disease models.

Genetically modified animals have furthered our knowledge of gene function

Powerful site-specific mutagenesis methods can be used with great precision and specificity to remove or swap functional modules in gene products or to change predetermined nucleotides and amino acids. There has been a long and successful tradition of using gene targeting to make mouse knockouts of mammalian protein-coding genes. More recently, various miRNA gene knockouts have also produced interesting and informative phenotypes. However, it is important to realize that, for some genes, knockouts do not always produce immediately discernible phenotypes. Gene duplication is common in vertebrate genomes and can lead to **genetic redundancy**, whereby closely related homologs perform the same type of function. Knocking out a single such gene may therefore yield a phenotype that appears normal.

Improvements in technology now allow us to make transgenic animals with very large transgenes. Yeast artificial chromosomes containing about 1 Mb of human DNA have been inserted into the mouse germ line by pronuclear injection. Such *YAC transgenics* have been useful in many ways. For example, regulatory elements controlling large genes can be investigated by manipulating the full-size gene and flanking regions. Further insights can be gained by overexpressing or ectopically expressing YAC transgenes. More recently, transgenic mice have been constructed with human artificial chromosomes. In one case, as described in Chapter 21, mice have been engineered to make human antibodies by the deletion of endogenous immunoglobulin loci and then insertion of a human artificial chromosome containing the full-length human IgH and IgK loci. In another extraordinary example, a mouse was constructed with almost a full-length human chromosome 21 to model Down syndrome.

Accompanying these advances have been striking technological improvements in studying how genes are expressed and in imaging biological processes. Genomewide expression analyses—at both the transcript and protein levels—are now widely available. In addition, multiple fluorophore labeling systems and powerful imaging methods are allowing new windows on how genes work in developing embryos and animals. **Figure 20.18** shows some extraordinary images in which adjacent neurons are distinguished in mice by using the delightfully named Brainbow transgenes (a fusion of *brain* and *rainbow*).

Inevitably, much of our knowledge on how genes work in human cells comes from studies of the genes in vertebrate models, notably the mouse. It is important to realize, however, that simple invertebrate animals that are particularly amenable to genetic analyses—such as the fruit fly *D. melanogaster* and the roundworm *C. elegans*—are more closely evolutionarily related to us than they seem. The majority of *D. melanogaster* genes and *C. elegans* genes have directly distinguishable human gene counterparts. Because of strong conservation of many developmental and cellular pathways studies of how genes work in flies and worms can provide crucially important information on how their human and vertebrate orthologs work.

Creating animal models of human disease

Various animals have also been genetically manipulated to model the major disease classes, including genetic disease, autoimmune disease, sporadic cancer, and disease due to infectious agents. Mouse disease models have predominated, but other vertebrates have also played a part. Invertebrate models have often been useful in illuminating the cellular or molecular basis of disease.

In genetic disease, many different models have been produced for single-gene disorders, and a few for chromosomal and multifactorial disorders. Various cancer models have also been made, but with varying degrees of success in recapitulating the human disease. The variable success in making models that accurately resemble the intended human disease reflects significant differences between humans and the animals used to model disease.

Loss-of-function mutations are modeled by selectively inactivating the orthologous mouse gene

Many single-gene disorders, including essentially all recessively inherited disorders and many dominantly inherited disorders, are thought to result from a loss of gene function. The usual approach to modeling disorders of this type is to use

Figure 20.18 Visualizing synaptic circuits in mice with the use of Brainbow transgenes to label neurons with multiple distinct colors. In Brainbow transgenes, Cre–*loxP* recombination is used to create a stochastic choice of expression between three or more fluorescent proteins (XFPs). Integration of tandem Brainbow copies in transgenic mice yielded combinatorial XFP expression, and thus many colors, thereby providing a way of distinguishing adjacent neurons and visualizing other cellular interactions. (A) With a *Brainbow* transgene expressing three XFPs, independent recombination of three transgene copies can, in principle, generate 10 distinct color combinations. (B) Oculomotor axons of a mouse line with a Brainbow transgene construct driven by regulatory elements from the *Thy1* gene, which permit transgenes to be strongly expressed in a variety of neurons. Boxes at the top right show sample regions from different axons. (C) Dentate gyrus from the same transgenic mouse line. Scale bars, 10 μm. [From Livet J, Weissman TA, Kang H et al. (2007) *Nature* 450, 56–62. With permission from Macmillan Publishers Ltd.]

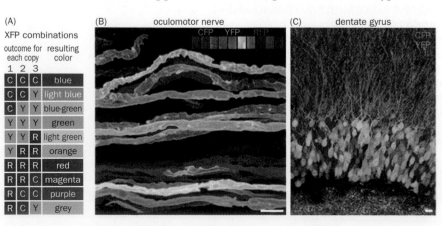

homologous recombination in mouse ES cells to engineer mutations in the orthologous mouse gene and thereby produce mice that are deficient in the gene product. As outlined in Chapter 12, the same approach has been used as a general method to investigate gene function, irrespective of whether the gene is known to be associated with disease.

For recessive disorders, the level of gene product is substantially decreased and in some cases can be zero. For those dominantly inherited disorders in which disease arises from haploinsufficiency, the presence of one functional allele often means that at least 50% of the gene product is made. However, extensive mutational heterogeneity in some disorders may result in considerable variation in the amounts of normal gene product. For some genes, a gradation of phenotypes may be apparent at different dosage thresholds (see Figure 13.26).

Depending on the extent of the gene product deficiency to be modeled, different gene-targeting strategies can be used. Initially a heterozygous mouse is obtained, and thereafter brother–sister heterozygotes are mated in an attempt to identify homozygous mutants that may or may not be viable.

Null alleles

To model a null allele, the mutation should result in complete inactivation of the targeted mouse gene. A general strategy for protein-coding genes is to introduce an early premature termination codon, as was described above. Sometimes mice with null alleles are not viable and so conditional expression systems such as Cre–*loxP* alleles are needed so that null mutations can be produced in just desired tissues or developmental stages.

Humanized alleles

Many single-gene disorders show considerable mutational heterogeneity. For disorders arising from a loss of function of a polypeptide-encoding gene, most chain-terminating mutations can be expected to be null alleles, but the effects of some splicing mutations and some other point mutations may be more difficult to predict. Disease alleles found at high frequency may warrant the effort needed to engineer the pathogenic human mutation into the sequence of the mouse allele, thereby creating a *humanized allele* (**Box 20.2** gives some examples).

Leaky mutations and hypomorphs

Although gene targeting may have been intended to produce a null mutation, sometimes the result is a *leaky mutation*: the mutant allele retains some gene expression. For example, alternative splicing can bypass the inserted DNA segment to produce a product with at least part of the coding sequence. As a result, the phenotype might be much milder than expected. Sometimes a cryptic splice site is activated within an integrated gene-targeting construct; an early attempt to make a mouse model of cystic fibrosis is an informative example (see Box 20.2).

Sometimes it may be desirable to engineer leaky mutations deliberately. The gene-targeting constructs can offer alternative splice sites that can permit partial gene activity, resulting in a **hypomorph**, a phenotype that has partial wild-type activity. Site-specific recombination may be used to produce a variety of different alleles ranging from complete inactivation to mild effects on gene expression (for an example, see the paper by Kist et al. in Further Reading).

As an alternative to gene targeting, mouse gene-trap databases may be screened to identify whether there is an ES cell strain with a gene-trap insertional mutation in the gene of interest. Gene traps not infrequently result in leaky mutations: altered wild-type transcripts sometimes continue to be made, but at a much lower level, in addition to the expected fusion transcripts caused by the splicing of 5′ exons to the gene-trap construct (see Figure 20.16).

Gain-of-function mutations are conveniently modeled by expressing a mutant transgene

For dominantly inherited gain-of-function mutations and also sporadic cancers caused by oncogenes, the presence—and the subsequent expression—of an introduced DNA sequence is itself sufficient to induce pathogenesis. To model

BOX 20.2 MODELING RECESSIVE DISORDERS IN MICE: THE EXAMPLE OF CYSTIC FIBROSIS

Cystic fibrosis (CF; OMIM 219700) is a recessive disorder that is particularly common in populations of Caucasian ancestry. It is caused by mutations in the *CFTR* (cystic fibrosis transmembrane regulator) gene, which has 27 exons, specifying a polypeptide of 1480 amino acid residues. The ΔF508 mutation, a trinucleotide deletion in exon 10, is very frequent (about 70% of CF alleles). Other relatively common alleles result in the amino acid substitutions G551D and G480C.

In the absence of a suitable animal model, the identification of *CFTR* as the CF locus prompted efforts to make a mouse knockout model. Different types of mutation were generated in the mouse *Cftr* gene, and the resulting knockout mice were produced on various genetic backgrounds. Not unexpectedly, the different CF models showed very considerable differences in phenotype (**Table 1**).

Initial attempts focused on mutating exon 10, either deleting it to cause a downstream shift in the translational reading frame or introducing an insertional mutation. Three such models failed to show evidence of lung pathology (see Table 1). Two mutations seemed to be null alleles, resulting in an absence of *Cftr* from mutant homozygotes, and did result in severe intestinal phenotypes; the other model had some residual *Cftr* mRNA because of a *leaky* mutation.

Other investigators deleted an early exon (exon 1 or exon 2) to cause early premature termination. Again, there was no lung pathology. However, in the deleted exon 1 model, the severity of the intestinal disease was observed to vary significantly depending on the genetic background. This prompted the originators of the first exon 10 deletion model to make an inbred **congenic strain** by repeatedly backcrossing with C57BL/6 until the contribution of the ES cell strain was limited to a small region including the *Cftr* allele. The resulting mouse now developed the hoped-for lung pathology.

Attempts to model human alleles with a single amino acid deletion or substitution produced milder phenotypes because the mutant protein had some residual functional activity. Two attempts to model the ΔF508 mutation produced different phenotypes because of differences in the amount of expression of the engineered *Cftr*^{ΔF508} allele. Other models have been produced for the missense mutations. A mouse with the G551D mutation did have some intestinal disease but less than for the null mice, and a mouse with the G480C allele had a very mild phenotype. The milder models can nevertheless provide valuable information on how the *Cftr* gene functions.

TABLE 1 SOME OF THE DIFFERENT GENE-TARGETING STRATEGIES USED TO MAKE MOUSE MODELS OF CYSTIC FIBROSIS

Mutant allele	Nature of mutation[a]	*Cftr* mRNA (percentage of wild type)	Genetic background (ES cell strain—embryo strain)	Phenotype
m1UNC	ΔEx10	0	E14TG2a—C57BL/6	less than 5% survive to maturity; no lung pathology even in adulthood, but severe intestinal disease
		0	inbred congenic C57BL/6 strain[b]	develops spontaneous and progressive lung disease at outset
m1HGU	Ins. Ex10	10	129—MF1	a *leaky mutation* by which some transcripts underwent exon skipping to produce some functional mRNA; the resulting mild phenotype included mild intestinal obstruction; about 95% of mice survive to maturity
m1CAM	ΔEx10	0	mixed: 129—MF1 and 129—C57BL/6	similar pathology to the *m1UNC* on a mixed genetic background, except has additional pathology in the lacrimal gland of the eye
m3BAY	ΔEx2	0	129—C57BL/6	similar pathology to the original *m1UNC* mouse
m1HSC	ΔEx1	0	129—CD1	severe phenotype with 30% survival but, after subsequent crossing with variety of different mouse strains, the survival rates and severity of phenotypes are seen to vary widely
m1EUR	ΔF508	100	129—FBV	disease mild (possibly because of high expression of the mutant protein that is known to have partial Cl⁻ channel function)
m2CAM	ΔF508	30	129—C57BL/6	about 35% of mice die by day 16 (compared with 80% for null mutants); higher survival is probably due to production of mutant *Cftr* mRNA with some functional activity
m1KTH	ΔF508	very low	129—C57BL/6	survival rate only about 40% because virtually no *Cftr* mRNA of any kind is produced in intestine

[a]ΔEx, deletion of exon; Ins., insertional mutation. For historical reasons, the 6th and 7th exons of the human *CFTR* and mouse *Cftr* genes are labeled as exons 6a and 6b, and so in this table exons 10 and 11 are in fact the 11th and 12th exons of the *Cftr* gene, respectively.
[b]As a result of extensive backcrossing with C57BL/6 and selection for the original *Cftr* allele, only a very small region (including the *Cftr* gene) originated from the E14TG2a ES cells, and the vast majority of the DNA is from the C57BL/6 strain.

such disorders, a cloned mutant gene should be available or, as required, be designed by *in vitro* mutagenesis.

The mutant gene is conveniently inserted as a transgene by microinjection into fertilized oocytes. The introduced mutant gene need not integrate at a specific location, and human mutant transgenes will suffice. Dominantly inherited disorders resulting from unstable expansion of short tandem repeats provide some illustrative examples. In such cases the disease often stems from the production of a mutant protein or RNA, representing a gain of function.

An illustrative example is provided by myotonic dystrophy, in which expansion of certain noncoding short tandem repeats containing CTG results in the production of a mutant RNA that interferes with the expression of other genes. Massive expansion of a CTG repeat in the 3′ UTR of the *DMPK* gene or of a CCTG repeat in the first intron of the completely unrelated *ZNF9* gene results in a mutant RNA that is retained in the nucleus. The mutant RNA causes defective splicing or aberrant transcriptional regulation of some unrelated genes, resulting in a multisystem disease. The hypothesis that a mutant RNA was involved was confirmed by producing mice that expressed human transgenes containing long CTG repeat arrays. Even when an expanded CTG array was placed in the context of a 3′ UTR of the unrelated human skeletal actin gene, the production of an actin mRNA with a large poly(CUG) tract was enough to induce myotonic dystrophy.

Huntington disease (OMIM 143100) is a prominent example of diseases caused by mutant proteins with abnormally long polyglutamine tracts. Repetition of the trinucleotide CAG in exon 1 of the *HD* gene is associated with pathogenesis when the number of repeats exceeds 36 (see Figure 18.19). Expansion of the encoded polyglutamine tract results in aggregation of the mutant protein and proves to be toxic to some cells, notably many types of neuron. Very large CAG expansions are associated with more severe disease and early onset of symptoms.

A variety of different animals have been genetically manipulated in an attempt to model this disorder and understand the function of the *HD* gene (**Table 20.2**). Initially, the significance of the CAG expansion was not well understood, and loss of function seemed a possibility. When gene targeting was used to inactivate the orthologous mouse gene, *Hdh* homozygous nulls lacking any huntingtin protein died at an early stage in embryonic development (*embryonic lethals*), and heterozygotes were normal. This suggested that Huntington disease could not be modeled by gene inactivation; instead, the disease was caused by one allele gaining some function after undergoing abnormal CAG expansion.

To model gain of function in mice, both standard transgenics (with transgenes randomly integrated into the genome by pronuclear microinjection) and knock-ins were made. The first mouse model of Huntington disease showing a neurological phenotype had a transgene that included exon 1 of a mutant *HD* gene under the control of a human *HD* promoter (**Figure 20.19**). Subsequent transgenic models were constructed that expressed gene fragments or full-length cDNAs with various expansion sizes for the CAG repeat. Various useful models were produced that exhibited brain atrophy and rapid onset of motor deficits.

Mice with a doxycycline-inducible transgene containing a mutant exon 1 provided a remarkable insight. After allowing symptoms to develop (with doxycycline added to the drinking water), the transgene was switched off by removing

TABLE 20.2 ANIMAL MODELS FOR MODELING HUNTINGTON DISEASE PATHOGENESIS AND EXPLORING THE FUNCTION OF THE *HD* GENE

Animal	Models and application(s)
Mouse (see also Figure 20.19)	gene knockout to investigate effects of loss of huntingtin protein function; conventional transgenics that express mutant huntingtin to study pathogenesis; knock-ins to study pathogenesis when mutant mouse huntingtin is regulated by the endogenous promoter
Rat	conventional transgenics that express mutant huntingtin; transgenics in which viral vectors were used to deliver mutant transgenes to brain
Rhesus macaque	transgenics with mutant *HTT* gene encoding expanded polyglutamine tract
Drosophila melanogaster	enhancer and suppressor screens to identify genes that modify polyglutamine toxicity; phenotype rescue by inhibition of transcriptional repressor complexes
Caenorhabditis elegans	RNAi screen to identify genes that modulate polyglutamine aggregation
Saccharomyces cerevisiae	enhancer and suppressor screens to identify genes that modify polyglutamine toxicity

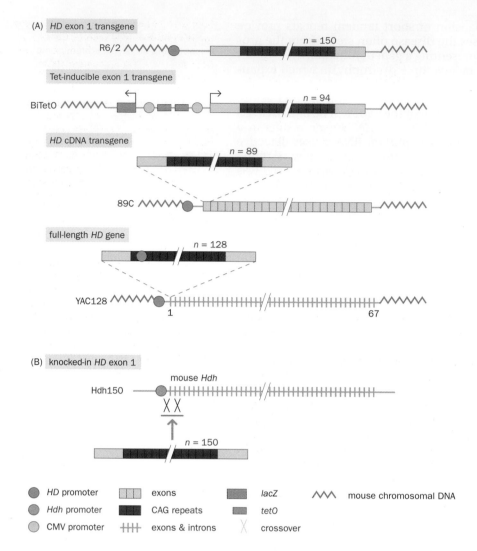

(A) HD exon 1 transgene

(B) knocked-in HD exon 1

HD promoter
Hdh promoter
CMV promoter
exons
CAG repeats
exons & introns
lacZ
tetO
crossover
mouse chromosomal DNA

Figure 20.19 Different approaches to the modeling of Huntington disease in mice with the use of mutant transgenes. (A) Mouse models of Huntington disease have often relied on transgenes containing mutant human *HD* alleles. Typically, the *HD* transgenes have been inserted randomly into the mouse genome after pronuclear microinjection. Early models used mutant *HD* exon 1 transgenes such as the widely used R6/2 strain, which has 150 CAG repeats. The BiTetO mouse model has an integrated inducible transgene containing a mutant exon 1 with 94 CAG repeats and a *lacZ* reporter gene, both of which are regulated by minimal cytomegalovirus (CMV) promoters. Other mouse strains were constructed with randomly integrated transgenes that had full-length mutant cDNAs or, in the case of the YAC128 model, a mutant full-length *HD* gene with all 67 exons, introns, and regulatory sequences. (B) Alternatively, knock-in models such as Hdh150 have been made in which a gene-targeting construct has included a mutant mouse *Hdh* transgene with flanking sequences that are homologous to the endogenous mouse *Hdh* gene. Homologous recombination in ES cells can then place the mutant transgene into the 5′ end of the mouse *Hdh* gene so that it comes under the regulation of the endogenous *Hdh* promoter. The more widely used transgenic models, such as R6/2 and YAC128, have particularly large CAG expansions that are generally associated with more severe phenotypes.

the doxycycline inducer, whereupon huntingtin aggregates were readily cleared from the mouse brains and the motor symptoms were reversed. It was the first indication that treatment of Huntington disease patients in the early stages might be able to reverse the clinical symptoms.

The rationale behind creating knock-in mice to model Huntington disease was to simulate the natural situation by placing an expanded CAG repeat within the natural context; that is, regulation by the endogenous *Hdh* promoter. The knock-in models, however, generally do not present with robust motor deficits or possess the brain atrophy and neuronal loss characteristic of the human disease.

In another approach, YAC transgenics were constructed with the full-length human *HD* gene including promoter, exons, introns, and upstream and downstream regulatory elements. Different strains were constructed to have different numbers of CAG repeats; one with 128 CAG repeats, the YAC128 strain, has become a widely used Huntington disease model (see Further Reading).

Modeling chromosomal disorders is a challenge

Existing mouse models for human chromosomal disorders are sparse, and chromosome engineering in mice is not a simple solution to providing badly needed models. Modeling chromosome microdeletions and microduplications is comparatively easy because of the comparatively small sizes of the genomic regions involved. In the former case, chromosome engineering using Cre–*loxP* recombination has been used to create deletions of the equivalent gene regions in mice, and YAC transgenes can be used to overexpress genes represented in chromosome microduplications. However, numerical chromosomal abnormalities are

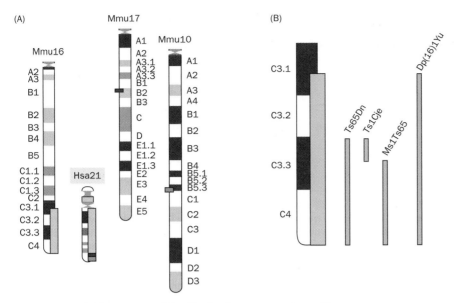

Figure 20.20 Segmental trisomy 16 mouse models of Down syndrome. (A) Genes on the long arm of human chromosome 21 (Hsa21) mostly have counterparts on the distal part of mouse chromosome 16 (Mmu16) (pale green boxes). Genes on the distal extremity of Hsa21 have orthologs on mouse chromosome 17 (Mmu17) (red boxes) or 10 (Mmu10) (yellow boxes). (B) Extent of trisomy in segmental trisomy 16 mouse models. The widely used Ts65Dn mouse model is trisomic for a distal region of mouse chromosome 16 that spans about 60% of the region showing conservation of synteny with Hsa21. However, it is also trisomic for a more than 5.8 Mb region on mouse chromosome 17. Complementary components of the trisomic region in Ts65Dn are represented in the Ts1Cje (small proximal component) and Ms1Ts65 (large distal component) models. The Dp(16)1Yu model was generated very recently by long-range chromosome engineering with Cre–*loxP* site-specific recombination; it carries an accurate duplication of the full 22.9 Mb region on mouse chromosome 16 that shows conserved synteny with Hsa21.

difficult to model because of the limited conservation of human–mouse synteny (gene order; see Figure 10.16). Nevertheless, considerable effort has been devoted to making mouse models of trisomy 21 (Down syndrome).

In trisomy 21, the pathogenesis results from the dosage sensitivity of a few genes on chromosome 21, for which the presence of three copies instead of two is sufficient to produce abnormal gene regulation. Because 21p is essentially made up of heterochromatin and of rRNA genes that are also present in multiple copies on four other chromosomes, the problem lies with genes on 21q. Most of these genes have counterparts on mouse chromosome 16 (**Figure 20.20A**). However, mouse trisomy 16 does not resemble human trisomy 21; trisomy 16 mice die *in utero* and are not useful models. Importantly, most genes on mouse chromosome 16 have orthologs on chromosomes other than human chromosome 21. Genes in the distal 4.5 Mb of human 21q have orthologs on mouse chromosome 17 or 10.

To obtain more accurate models, various mice with segmental trisomy 16 have been produced. Earlier models relied on using random mutagenesis and retrieving mice with partial trisomy 16 arising through translocations. The Ts65Dn mouse has been a very widely used model of trisomy 21 because of its learning and behavior deficits. However, it is trisomic for less than 60% of the human chromosome 21 syntenic region on mouse chromosome 16. Additionally, it is trisomic for a region of mouse chromosome 17. Recently, long-range chromosome engineering using Cre–*loxP* recombination has produced a promising model with a duplication of the entire 22.9 Mb region on mouse chromosome 16 that is syntenic with human chromosome 21 (see Figure 20.20).

As a radical alternative to duplicating mouse chromosome segments, **transchromosomic mice** have been generated that carry a single almost complete human chromosome 21. To do this, chromosomes from human fibroblast cells were transferred into mouse ES cells by using *microcell-mediated chromosome transfer* (see Chapter 16, p. 523, for a description of this technique). With the use of a marker gene, those ES cells that had picked up human chromosome 21 were selected. Some of the resulting ES cells had a nearly complete human chromosome 21 and were used to make an aneuploid transchromosomic mouse line, Tc1, that had various characteristics of trisomy 21 (**Figure 20.21**). Although the Tc1 transchromosomic mouse model appeared promising, it has not been easy to work with. Random loss of the Hsa21 fragment during development means that analyses are complicated by the variable levels of mosaicism in different tissues.

Modeling human cancers in mice is complex

Certain mouse strains develop cancer spontaneously or after exposure to radiation, chemicals, pathogenic viruses, and so on. Such models have been useful, but they are limited because they develop only certain types and grades of tumors

Genetic redundancy

Kafri R, Springer M & Pilpel Y (2009) Genetic redundancy: new tricks for old genes. *Cell* 6, 389–392.

Genetic screens and gene function analyses in vertebrates

Aitman TJ, Critser JK, Cuppen E et al. (2008) Progress and prospects in rat genetics: a community view. *Nat. Genet.* 5, 516–522.

Jorgensen EM & Mango SED (2002) The art and design of genetic screens: *Caenorhabditis elegans*. *Nat. Rev. Genet.* 3, 356–369.

Kile BT & Hilton DJ (2005) The art and design of genetic screens: mouse. *Nat. Rev. Genet.* 6, 557–567.

Mashimo T, Yanagihara K, Tokuda S et al. (2008) An ENU-induced mutant archive for gene targeting in rats. *Nat. Genet.* 5, 514–515.

Nguyen D & Xu T (2008) The expanding role of mouse genetics for understanding human biology and disease. *Dis. Model Mech.* 1, 56–66.

Skromne I & Prince VE (2008) Current perspectives in zebrafish reverse genetics: moving forwards. *Dev. Dyn.* 237, 861–882.

St Johnston D (2002) The art and design of genetic screens: *Drosophila melanogaster*. *Nat. Rev. Genet.* 3, 176–188.

Modeling single-gene disorders

Guilbault C, Saeed Z, Downey GP & Radzioch D (2007) Cystic fibrosis mouse models. *Am. J. Respir. Cell Mol. Biol.* 36, 1–7.

Hickey MA & Chesselet MF (2003) The use of transgenic and knock-in mice to study Huntington's disease. *Cytogenet. Genome Res.* 100, 276–286.

Kist R, Watson M, Wang X et al. (2005) Reduction of Pax9 gene dosage in an allelic series of mouse mutants causes hypodontia and oligodontia. *Hum. Mol. Genet.* 14, 3605–3617.

Mankodi A, Logigian E, Callahan L et al. (2000) Myotonic dystrophy in transgenic mice expressing an expanded CUG repeat. *Science* 289, 1769–1773.

Seznec H, Agbulut O, Sergeant N et al. (2001) Mice transgenic for the human myotonic dystrophy region with expanded CTG repeats display muscular and brain abnormalities. *Hum. Mol. Genet.* 10, 2717–2726.

Slow EJ, van Raamsdonk J, Rogers D et al. (2003) Selective striatal neuronal loss in a YAC128 mouse model of Huntington disease. *Hum. Mol. Genet.* 12, 1555–1567.

Yamamoto A, Lucas JJ & Hen R (2000) Reversal of neuropathology and motor dysfunction in a conditional model of Huntington's disease. *Cell* 101, 57–66.

Modeling chromosomal disorders

Li Z, Yu T, Morishima M et al. (2007) Duplication of the entire 22.9 Mb human chromosome 21 syntenic region on mouse chromosome 16 causes cardiovascular and gastrointestinal abnormalities. *Hum. Mol. Genet.* 16, 1359–1366.

O'Doherty A, Ruf S, Mulligan C et al. (2005) An aneuploidy mouse strain carrying human chromosome 21 with Down syndrome phenotypes. *Science* 309, 2033–2037.

Reeves RH & Gardner CC (2007) A year of unprecedented progress in Down syndrome basic research. *Ment. Retard. Dev. Disabil. Res. Rev.* 13, 215–220.

Modeling cancers

Evers B & Jonkers J (2006) Mouse models of BRCA1 and BRCA2 deficiency: past lessons, current understanding and future prospects. *Oncogene* 25, 5885–5897.

Frese KK & Tuveson DA (2007) Maximizing mouse cancer models. *Nat. Rev. Cancer* 7, 645–658.

Jonkers J & Berns A (2002) Conditional mouse models of sporadic cancer. *Nat. Rev. Cancer* 2, 251–265.

Rangarajan A & Weinberg RA (2003) Opinion: comparative biology of mouse versus human cells: modelling human cancer in mice. *Nat. Rev. Cancer* 3, 952–959.

Santos J, Fernandez-Navarro P, Villa-Morales M et al. (2008) Genetically modified mouse models in cancer studies. *Clin. Transl. Oncol.* 10, 794–803.

Stoletov K & Klemke R (2008) Catch of the day: zebrafish as a human cancer model. *Oncogene* 27, 4509–4520.

Mouse genetic background and modifier genes

Dietrich WF, Lander ES, Smith JS et al. (1993) Genetic identification of *Mom-1*, a major modifier locus affecting *Min*-induced intestinal neoplasia in the mouse. *Cell* 75, 631–639.

Jackson Laboratory Manual on Mouse Genetic Background. http://jaxmice.jax.org/manual/index.htmlgenetics

Kwong LN, Shedlovsky A, Biehl BS et al. (2007) Identification of *Mom7*, a novel modifier of *Apc*Min/+ on mouse chromosome 18. *Genetics* 176, 1237–1244.

Nadeau J (2003) Modifier genes and protective alleles in humans and mice. *Curr. Opin. Genet. Dev.* 13, 290–295.

General reviews on mouse disease models and difficulties in replicating human disease phenotypes

Bartlett NW, Walton RP, Edwards MR et al. (2008) Mouse models of rhinovirus-induced disease and exacerbation of allergic airway inflammation. *Nat. Med.* 14, 199–204.

Bedell MA, Largaespada DA, Jenkins NA & Copeland NG (1997) Mouse models of disease. Part II: recent progress and future directions. *Genes Dev.* 11, 11–43.

Dennis C (2006) Cancer: off by a whisker. *Nature* 442, 739–741.

Elsea SH & Lucas RE (2003) The mousetrap: what we can learn when the mouse model does not mimic the human disease. *ILAR J.* 43, 66–79.

Liao B-Y & Zhang J (2008) Null mutations in human and mouse orthologs frequently result in different phenotypes. *Proc. Natl Acad. Sci. USA* 105, 6987–6992.

Thyagarajan T, Totey S, Danton MJS & Kulkarni AB (2003) Genetically altered mouse models: the good, the bad and the ugly. *Crit. Rev. Oral Biol. Med.* 14, 154–174.

Genetically modified invertebrates for modeling disease and drug discovery

Bier E (2005) *Drosophila*, the golden bug, emerges as a tool for human genetics. *Nat. Rev. Genet.* 6, 9–23.

Bilen J & Bonini NM (2005) *Drosophila* as a model for human neurodegenerative disease. *Annu. Rev. Genet.* 39, 153–171.

Olsen A, Vantipalli MC & Lithgow GJ (2006) Using *Caenorhabditis elegans* as a model for aging and age-related diseases. *Ann. N.Y. Acad. Sci.* 1067, 120–128.

Genetic modification of non-murine vertebrates: progress and novel pluripotent stem cells

Buehr M, Meek S, Blair K et al. (2008) Capture of authentic embryonic stem cells from rat blastocysts. *Cell* 135, 1287–1298.

Bugos O, Bhide M & Zilka N (2009) Beyond the rat models of human neurodegenerative disease. *Cell. Mol. Neurobiol.* 29, 859–869.

Honda A, Hirose M, Inoue K et al. (2008) Stable embryonic stem cell lines in rabbits: potential small animal models for human research. *Reprod. Biomed. Online* 17, 706–715.

Ingham PW (2009) The power of the zebrafish for disease analysis. *Hum. Mol. Genet.* 18, R107–R112.

Kari G, Rodeck U & Dicker AP (2007) Zebrafish: an emerging model system for human disease and drug discovery. *Clin. Pharmacol. Ther.* 82, 70–80.

Li P, Tong C, Mehrian-Shai R et al. (2008) Germline competent embryonic stem cells derived from rat blastocysts. *Cell* 135, 1299–1310.

Lieschke GJ & Currie PD (2007) Animal models of human disease: zebrafish swim into view. *Nat. Rev. Genet.* 8, 353–367.

Liu H, Zhu F, Yong J et al. (2008) Generation of induced pluripotent stem cells from adult rhesus monkey fibroblasts. *Cell Stem Cell* 3, 587–590.

Rogers CS, Stoltz DA, Meyerholz DK et al. (2008) Disruption of the *CFTR* gene produces a model of cystic fibrosis in newborn pigs. *Science* 321, 1837–1841.

Sasaki E, Suemizu H, Shimada A et al. (2009) Generation of transgenic non-human primates with germline transmission. *Nature* 459, 523–527.

Yang S-H, Cheng PH, Banta H et al. (2008) Towards a transgenic model of Huntington's disease in a non-human primate. *Nature* 453, 921–924.

Despite much work, however, relatively few genetic diseases have wholly satisfactory treatments.

Genetic treatment of disease may be conducted at many different levels

Any disease, whether it has a genetic cause or not, is potentially treatable by using a range of different procedures that involve applying genetic knowledge or genetic manipulations in some way (**Figure 21.1**).

Sometimes genetic techniques form part of a treatment regime that also involves conventional small-molecule drugs or vaccines. For example, as described in Chapter 19, individuals may be genotyped to predict their pattern of favorable and adverse responses to specific drug treatments. Such genotyping may become routine as massively parallel DNA sequencing permits extensive screening of genes in vast numbers of people. Another approach involves the use of knowledge of genetics and cell biology to identify new targets for drug development. Genetic techniques can also be used directly in producing drugs and vaccines for treating disease.

Another active area concerns treating disease with therapeutic proteins that are produced or modified by genetic engineering. They include so-called *recombinant proteins*, which are produced by expression cloning and include hormones, blood factors, and enzymes, and also genetically engineered antibodies. We consider in Section 21.2 genetic inputs to treating disease with drugs, or therapeutic proteins and vaccines.

In addition, there are two other major areas where genetic interventions are being designed to treat disease. One involves using genetic techniques to assist *cell transplantation* methods for treating diseases in which the pathogenesis involves a loss of tissue and cells. Stem cells may offer important therapeutic possibilities in this area, and the principles and genetic inputs into stem cell therapies are discussed in Section 21.3.

Finally, we consider in Sections 21.4–21.6 a whole range of different therapeutic methods that involve the direct genetic modification of a patient's cells (*gene therapy*). Some of the methods involve transplanting genes into the cells of patients to overcome some genetic deficiency or to encourage the killing of harmful cells. A wide range of other gene therapy methods involve altering the expression of genes in the cells of patients in some way, by modifying the genome or by targeting gene expression products. Because they are needed to replenish tissue and blood cells, stem cells are important target cells for gene therapy.

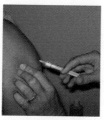

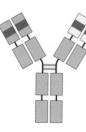

small-molecule drugs: screening and development of novel drug targets

production of genetically engineered antibodies

production of recombinant therapeutic proteins

production of genetically engineered vaccines

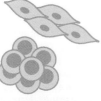

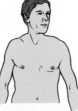

IL2R

production of reagents for gene therapies

genetic modification of patient or donor cells in gene and cell therapy

testing of treatment in genetically modified animals

genotyping of patients in personalized medicine

Figure 21.1 Some of the many different ways in which genetic technologies are used in the treatment of disease. See the text for detail. IL2R, interleukin type 2 receptor.

Chapter 21

Genetic Approaches to Treating Disease

21

KEY CONCEPTS

- Genetic techniques can be applied in various ways in treating disease, regardless of whether the disease has a genetic origin or not.

- Genetics is being used extensively in the drug discovery process to identify new drug targets.

- Therapeutic recombinant proteins produced by expressing cloned DNA are safer than those sourced from animal and human sources.

- Genetic engineering of vaccines may involve inserting genes for desired antigens into a viral vector and using either the modified vector or the purified antigen as the vaccine. Alternatively, DNA can be used as a vaccine by direct injection into muscle cells.

- Cells lost by disease or injury can be replaced by suitable stem cells that have been directed to differentiate into the correct specialized cell type.

- Nuclear reprogramming involves fundamental alteration of the chromatin structure and gene expression pattern so as to convert differentiated cells into less specialized cells (*dedifferentiation*) or to a different type of differentiated cell (*transdifferentiation*).

- Autologous cell therapy involves conversion of a patient's cells into a type of cell that is deficient through disease or injury. In principle this can be done by artificially stimulating the mobilization of existing stem cells in the patient or by nuclear reprogramming.

- Skin fibroblasts from a patient can be dedifferentiated to make patient-specific totipotent or pluripotent stem cells that can be differentiated to required cell types for use in autologous cell therapy and for studying disease. In *therapeutic cloning* the nucleus from a patient's cell is transplanted into an enucleated egg cell from a donor, giving rise to a totipotent cell. More conveniently, the skin cells can simply be induced to express key transcription factors that will convert them to pluripotent stem cells.

- Gene therapy involves transferring genes, RNA, or oligonucleotides into the cells of a patient so as to alter gene expression in some way that counteracts or alleviates disease. The patient's cells are often genetically modified in culture and then returned to the body.

- Gene therapy strategies often seek to compensate for underproduction of an important gene product. Other strategies are designed to block the expression of a mutant gene that makes a harmful gene product, or to induce alternative splicing so that a harmful mutation does not get included in a mRNA, or to kill harmful cells, for example in cancer gene therapy.

- Viral vectors are commonly used in gene therapy because they have good gene transfer rates, but they sometimes provoke strong immune responses and also abnormal activation of cellular genes such as proto-oncogenes.

- Zinc finger nucleases can be designed to make just one double-strand DNA break within the genome of intact cells. Cellular pathways that repair double-strand DNA breaks can then be induced to replace a sequence containing a harmful gene mutation with the normal allelic sequence.

In the preceding chapters in this part of the book, we have looked at how genetics can be used to test individuals and populations for genetic diseases and susceptibility for complex diseases. Chapter 20 explored genetic manipulation of animals to model human diseases. In this chapter, we look at how genetics can be employed to treat diseases. A chapter on disease treatment might reasonably consider two quite separate matters: the treatment of genetic disease, and the genetic treatment of disease. We therefore consider briefly what these two areas cover before describing how genetic techniques are being used directly to treat disease.

21.1 TREATMENT OF GENETIC DISEASE VERSUS GENETIC TREATMENT OF DISEASE

There is no connection at all between cause and treatability of a disease. Orthodox medical treatment aimed at alleviating the symptoms of a disease is just as applicable to genetic diseases as to any other disease. A profoundly deaf child should be offered hearing aids or a cochlear implant based solely on the child's symptoms and family situation, quite regardless of whether the hearing loss is due to a genetic disease or has another cause.

Treatment of genetic disease is most advanced for disorders whose biochemical basis is well understood

For many genetic conditions, existing treatments are unsatisfactory. A survey published by Costa and colleagues in 1985 estimated that treatment improved reproductive capacity in only 11% of Mendelian diseases, improved social adaptation in only 6%, and extended life span to normal in only 15% (of those that reduced longevity). No doubt the figures would be better now, but not dramatically so. Causative genes for many genetic disorders have been identified only quite recently, and it may take many years of research to identify how the underlying genes function normally in cells and tissues. Then it may be possible to design effective treatments.

Treatment has been possible, however, for some genetic disorders, for which we have detailed knowledge of the cellular and molecular basis of disease. For Mendelian disorders, inborn errors of metabolism are well understood at the biochemical level and are well suited to conventional treatments. There are many potential entry points for intervention in inborn errors of metabolism. Some examples are given below.

- *Substrate limitation.* Babies with phenylketonuria (OMIM 261600) have a genetic deficiency of phenylalanine hydroxylase. The resulting excess of phenylalanine has a neurotoxic effect that leads to impaired postnatal cognitive development (see Chapter 19, p. 629). Modifying the diet to reduce the intake of phenylalanine is a successful treatment. Even in successfully treated patients, however, cognitive development averages half a standard deviation below normal. Several other inborn errors respond equally well to dietary treatment.

- *Replacement of a deficient product.* One example is supplying thyroid hormone to infants with congenital hypothyroidism; another is enzyme replacement therapy, which is available for various metabolic disorders. Enzymes are often conjugated with polyethylene glycol to make them more active and more stable.

- *Using alternative pathways to remove toxic metabolites.* Treatments of this type range from simple bleeding as a very effective treatment for an iron overload condition, hemochromatosis (OMIM 235200), to using benzoate to increase nitrogen excretion in patients with disorders of the urea cycle.

- *Using metabolic inhibitors.* Babies with type 1 tyrosinemia (OMIM 276700) are unable to metabolize tyrosine effectively and therefore suffer liver damage from an excess of certain toxic intermediates. Their prognosis is markedly improved when the drug 2-(2-nitro-4-trifluoromethylbenzoyl)-1,3-cyclohexanedione (NTBC) is administered to block the enzyme 4-hydroxyphenylpyruvate dioxygenase.

21.2 GENETIC APPROACHES TO DISEASE TREATMENT USING DRUGS, RECOMBINANT PROTEINS, AND VACCINES

The pharmaceutical industry is responsible for developing most chemical treatments of disease. For decades, the drug discovery process has involved screening huge numbers of small molecules for evidence that they can reduce pathogenic effects. The drug screening process begins with laboratory studies involving cell culture and animal disease models (**Box 21.1**) before moving on to clinical trials. This process is costly and time-consuming (see Figure 19.11), and the drugs currently on the market were first developed when information about possible targets was scarce. In addition, the entire diversity of drugs currently on the market act through only a few hundred target molecules, and the declining number of new drug applications and approvals over the past few years has reflected a crisis in drug target identification and validation. Even when a drug receives regulatory approval, it is rarely effective in 100% of the patients to whom it is prescribed and may adversely affect some patients.

In Section 19.2 we considered how genetic variation between individuals affects the way in which people respond to drugs. In this section, we look at the various ways in which genetics is being applied in the drug discovery process. In addition to expanding the range of small-molecule drugs, genetics technologies have been applied to making genetically engineered therapeutic proteins and vaccines.

Drug companies have invested heavily in genomics to try to identify new drug targets

As described in Box 21.1, most drugs currently on the market bind to target molecules that belong to just a few protein classes. Novel drug targets are badly needed. One solution is to identify *generic drugs* that are not focused on a specific gene product, such as drugs that can suppress stop codons. The potential applications are huge because nonsense mutations promoting premature translational termination are responsible for causing anywhere from 5% to 70% of individual cases for most inherited diseases. Aminoglycoside antibiotics such as gentamycin were known to work in this way, but although gentamycin can cause readthrough of premature nonsense codons in mammalian cells it was found not to be clinically useful, partly because of a lack of potency and partly because of toxicity problems. Recently, however, extensive small-molecule screening has

---

BOX 21.1 DRUG TARGETS AND SMALL-MOLECULE DRUG SCREENING

Drugs are often small chemical compounds that are extracted from natural sources or chemically synthesized. Generally a *drug target* is some naturally existing molecule or cellular structure that is involved in a pathology of interest, and which the intended drug is designed to act on to treat the pathology. The molecular targets that have been considered most druggable have traditionally been proteins. Frequently they are receptors (notably G-protein-coupled receptors), enzymes (especially protein kinases, phosphatases, and proteases, or protein channels (voltage-gated or ion-gated). Small-molecule drugs fit tightly into crevices on the surface of their target proteins, thereby modulating their function.

Drug screening frequently involves *high-throughput screening*. With the use of liquid handling devices and robotics, hundreds of thousands to millions of different chemical compounds are individually assayed for their ability to react with some kind of biological material in specified wells of numerous microtiter dishes. The biological material has traditionally involved a cell-free system, microbes (bacteria, yeasts), or cultured cell lines, but more recently *in vivo* animal drug screening has also been possible. Searching for compounds may occur in a non-directed approach with normal cells or tissues in which the object is to identify any phenotypes produced

and correlate with the drugs that caused that effect. Often, however, the search for compounds is a directed one: the object is to search for drugs that react with a *defined* target or disease phenotype.

Various types of cultured cells, and immortalized aneuploid cell lines, can be used for drug screening. Recently isolated stem cell lines allow the possibility of controlled differentiation of euploid cells to a desired differentiated cell type. As described in Section 21.3, skin fibroblasts from any individual can now also be reprogrammed to make pluripotent stem cells, and so disease-specific pluripotent stem cells are being produced that can be directed to differentiate to give desired human disease cells for drug screening.

Animal disease models offer a physiological disease environment. Mice and rats are important in the testing of drug therapies before launching clinical trials, but rodent embryos are impractical for high-throughput drug screening. The culture conditions of small invertebrates such as *C. elegans* and *D. melanogaster* allow high-throughput drug screening in a physiological context, but they are limited to drugs that recognize evolutionarily very highly conserved targets. Zebrafish embryos offer the advantage of both being vertebrates and also transparent (allowing live tissue imaging) and are becoming increasingly important in drug screening.

TABLE 21.1
EXPRESSION

Recombinant
Insulin
Growth hormo
Blood clotting
Blood clotting
α-Interferon
β-Interferon
γ-Interferon
Tissue plasmin
Granulocyte/ stimulating fa
Leptin
Erythropoieti

For genetically

lines (or m because lar ways that si is secreted i sequence fo normally m animal's ge

donor

The process of using transgenic livestock (such as goats, pigs, or sheep) to make a therapeutic human protein has been called *pharming*, pharmaceutical-led farming. Milk proteins can be produced in large amounts, and a flock of about 150 goats could produce the world's supply of a specific protein. The cost of keeping flocks of transgenic animals is also significantly less than maintaining large-scale industrial bioreactors. In 2009, ATryn® became the first therapeutic protein produced by a transgenic animal to be approved by the US Food and Drug Administration (FDA). Produced by GTC Biotherapeutics, ATryn is an anti-thrombin expressed in the milk of goats and is designed to be used in anti-blood clotting therapy.

Other types of transgenic organism have been investigated to produce human proteins. Transgenic chickens are designed to lay eggs rich in a wanted product. Transgenic plants are also becoming popular. Plants do not replicate human-specific glycosylation patterns, but they have many advantages for expression cloning in terms of cost and safety. For example, there has been considerable interest in producing golden rice strains engineered to counter the vitamin A deficiency that is a serious public health problem in at least 26 countries in Africa, Asia, and Latin America.

Genetic engineering has produced novel antibodies with therapeutic potential

One class of therapeutic protein that has been widely used to treat disease comprises genetically engineered antibodies. As detailed in Section 4.6, each one of us has a huge repertoire of different antibodies that act as a defense system against innumerable foreign antigens. Antibody molecules function as adaptors: they have binding sites for foreign antigen at the variable end, and binding sites for effector molecules at the constant end. Binding of an antibody may by itself be sufficient to neutralize some toxins and viruses but, more usually, the bound antibody triggers the complement system and cell-mediated killing (see Figures 4.20 and 4.23).

Artificially produced therapeutic antibodies are designed to be specific for a single antigen (monospecific). Traditional **monoclonal antibodies** (**mAbs**) are secreted by **hybridomas**, immortalized cells produced by the fusion of antibody-producing B lymphocytes from an immunized mouse or rat with cells from an immortal mouse B-lymphocyte tumor (see Box 8.6). Hybridomas are propagated as individual clones, each of which can provide a permanent and stable source of a single mAb. Unfortunately, the therapeutic potential of mAbs produced in this way is limited. Although rodent mAbs can be raised against human pathogens and cells, they have a short half-life in human serum, often causing the recipient's immune system to make anti-rodent antibodies. In addition, only some of the different classes can trigger human effector functions.

Genetic engineering has been used to produce modified monoclonal antibodies in which some or all of the rodent protein sequence is replaced by equivalent human sequence. Often, this is achieved by swapping parts of the rodent sequence at the DNA level for equivalent human DNA sequences. *Chimeric antibodies* have been produced, for example, in which the rodent variable regions are retained but the constant (C) region sequences are replaced by human C region sequences. Ultimately *humanized antibodies* were constructed in which the only rodent sequences were from the complementarity determining regions (CDRs), the hypervariable sequences of the antigen-binding site (**Figure 21.3**). Phage display technology (see Figure 6.13) bypasses hybridoma construction altogether and allows innovative combinations of antibody domains to be constructed.

More recently, it has been possible to prepare *fully human antibodies* (see Figure 21.3D), by different routes. For example, mice have been genetically manipulated so that they can make nothing but fully human antibodies. To do this, the mouse immunoglobulin heavy- and light-chain loci were deleted in embryonic stem (ES) cells and then a human artificial chromosome containing the entire human heavy-chain and γ light-chain loci was inserted into the ES cells and used to give rise to transgenic strains with a human immunoglobulin (Ig) repertoire (**Figure 21.4**). Fully human antibodies can also be isolated by *mammalian cell display*, a mammalian equivalent of phage display. With this approach,

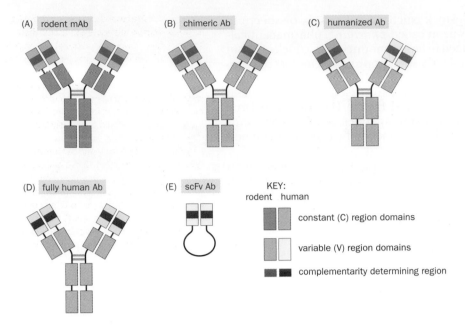

(A) rodent mAb   (B) chimeric Ab   (C) humanized Ab

(D) fully human Ab   (E) scFv Ab

KEY:
rodent human

constant (C) region domains

variable (V) region domains

complementarity determining region

Figure 21.3 Antibody engineering.
(A) Classic monoclonal antibodies (mAbs) are monospecific rodent antibodies synthesized by hybridomas. (B) Chimeric V/C antibodies (Abs) are genetically engineered to have rodent variable region sequences containing the critically important hypervariable complementarity determining region (CDR) joined to human constant region sequences. (C) Humanized antibodies can be engineered so that all the molecule except the CDR is of human origin. (D) More recently, it has been possible to obtain fully human antibodies by different routes (see the text and Figure 21.4). (E) Single-chain variable fragment (scFv) antibodies contain variable region domains jointed by a peptide linker. They are particularly well suited to working within the reducing environment of cells and serve as *intrabodies* (intracellular antibodies) by binding to specific antigens within cells. Depending on the length of the linker, they bind their target as monomers, dimers, or trimers. Multimers bind their target more strongly than monomers.

fully human antibodies have recently been obtained against nicotine. They have been shown to be effective in inhibiting the entry of nicotine into the brain in mice; if the same effect is replicated in clinical trials they could be a valuable aid in stopping smoking.

One very promising class of therapeutic antibody is designed to have a single polypeptide chain. *Single-chain variable fragment (scFv) antibodies* have almost all the binding specificity of a mAb but are restricted to a single non-glycosylated variable chain (see Figure 21.3E). scFV antibodies can be made on a large scale in bacterial cells, yeast cells, or even plant cells. They are particularly well suited to acting as intracellular antibodies (**intrabodies**). Instead of being secreted like normal antibodies, they are designed to bind specific target molecules within cells and can be directed as required to specific subcellular compartments. Unlike standard antibodies, which have four polypeptide chains linked by disulfide bridges, intrabodies are stable in the reducing environment within cells.

Intrabodies can be used to carry effector molecules that can perform specific functions when antigen binding occurs. However, for many therapeutic purposes, they are designed simply to block specific protein–protein associations within cells. As such, they complement conventional drugs. Protein–protein interactions usually occur across large, flat surfaces and are considered unsuitable targets for typical small-molecule drugs, which normally operate by fitting snugly into clefts on the surface of macromolecules. Promising therapeutic target proteins for intrabodies include mutant proteins that tend to misfold in a way that causes neurons to die, as in various neurodegenerative diseases including Alzheimer, Huntington and prion diseases.

From inauspicious beginnings in the 1980s, mAbs have become the most successful biotech drugs ever, and the market for mAbs is the fastest-growing component of the pharmaceutical industry. By the middle of 2008, a total of 21 mAb therapies had been approved by the US FDA. Of the therapeutic mAbs currently in use, the eight bestsellers together generate an annual income of close to $30 billion. Another 200 mAb products are in the pipeline. Of the FDA-approved mAbs, 19 are partly or fully human and most of them are aimed at treating cancers or autoimmune disease (**Table 21.2**). Only one of the approved antibodies is directed against infectious disease, being used to fend off respiratory syncytial virus in infants.

Safety issues can sometimes be a concern, however. In 2006 a clinical trial to assess a mAb called TGN1412 left six volunteers fighting for their lives. TGN1412 was hoped to be an effective agent in *cancer immunotherapy*, in which the strategy is to stimulate immune system cells so as to provoke a strong immune response against the cancer cells. However, TGN1412 activated immune cells to such an extent that the volunteers developed intense and nearly fatal allergic reactions to the mAb.

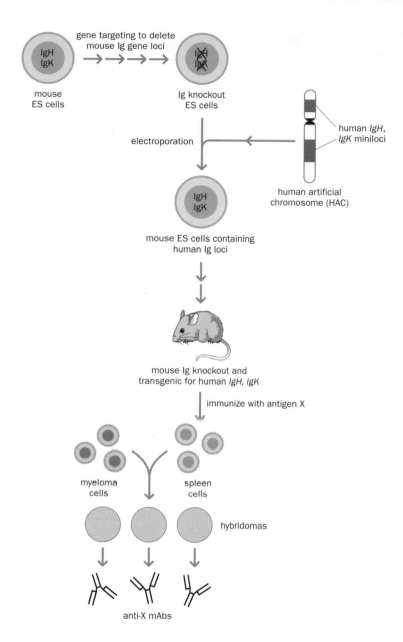

Figure 21.4 Making transgenic mice that express human monoclonal antibodies. To make fully human monoclonal antibodies (mAbs), Isao Ishida and colleagues generated a human artificial chromosome (HAC) containing the entire human Ig heavy-chain locus plus the entire human Igγ light-chain loci. This HAC was then introduced into mouse ES cells that had previously been subjected to rounds of Cre–*loxP* gene targeting to delete the endogenous mouse Ig loci. The resulting mouse can generate immunoglobulins by rearrangements of the introduced human Ig loci, and so can be used to produce human mAbs of any desired specificity. For full details see US patent 7041870, accessible by searching the patent database at http://www.patentstorm.us.

Aptamers are selected to bind to specific target proteins and inhibit their functions

Short oligonucleotide (or peptide) **aptamers** can also be identified that will bind to (and inhibit) specific target proteins, with affinities similar to those of antibodies. Starting from a large pool of random oligonucleotides, oligonucleotide aptamers are conveniently made by repeated rounds of selection and amplification, usually by the SELEX (**s**ystematic **e**volution of **l**igands by **ex**ponential enrichment) method that was described in Chapter 12 (see Figure 12.14).

Automated SELEX has generated an increasing number of aptamers for inhibiting protein function, and chemical modifications of the oligonucleotides make them more resistant to enzymatic degradation in body fluids. Like small-molecule drugs, aptamers can fit tightly into crevices on the surface of target macromolecules, but they can also fold to form clefts into which protruding parts of the target protein can bind. As a result, the potential number of contacts made with the target increases, allowing aptamers to form tighter, more specific interactions than small-molecule drugs can.

Potential therapeutic applications for aptamers include the inhibition of both intracellular target molecules, such as transcription factors, and extracellular targets. The latter can include targeting of specific viral components so as to inhibit

TABLE 21.2 PARTLY OR FULLY HUMAN THERAPEUTIC MONOCLONAL ANTIBODIES (mAbs) APPROVED BY THE US FOOD AND DRUG ADMINISTRATION AS OF MID-2008

Disease category	Target[a]	mAb generic name (trade name)	mAb class[b]	Disease treated
Autoimmune disease/ immunological	CD11a	efalizumab (Raptiva®)	humanized	psoriasis
	CD25 = IL2R	basiliximab (Simulect®)	chimeric	prevention of kidney transplant rejection
		daclizumab (Zenapax®)	humanized	
	complement-5	eculizumab (Soliris®)	humanized	paroxysmal nocturnal hemoglobinuria
	IgE	omalizumab (Xolair®)	humanized	asthma
	Integrin α4	natalizumab (Tysabri®)	humanized	multiple sclerosis
	TNF-α	infliximab (Remicade®)	chimeric	Crohn disease, rheumatoid arthritis
		certolizumab pegol (Cimzia®)	humanized	
		adalimumab (Humira®)	fully human	
Cancer	CD20	rituximab (Rituxan®, MabThera®)	chimeric	non-Hodgkin lymphoma
	CD33	gemtuzumab ozogamicin (Mylotarg®)	humanized	CD33-acute myeloid leukemia
	CD52	alemtuzumab (Campath®)	humanized	B-cell chronic lymphocytic leukemia
	EGFR	cetuximab (Erbitux®)	chimeric	colorectal cancer
		panitumumab (Vectibix®)	fully human	
	HER2	trastuzumab (Herceptin®)	humanized	metastatic breast cancer
	VEGF	bevacizumab (Avastin®)	humanized	colorectal, breast, renal, NSCL cancer
Other diseases	GPIIb/IIIa	abciximab (ReoPro®)	chimeric	adjunct to percutaneous transluminal coronary angioplasty
	RSV	palivizumab (Synagis®)	humanized	respiratory syncytial virus prophylaxis
	VEGF	ranibizumab (Lucentis®)	humanized	age-related macular degeneration (intravitreal injection)

[a]CD11a, CD20, CD25, CD33, and CD52, white blood cell antigens; IL2R, interleukin type 2 receptor; IgE, immunoglobulin E; TNF-α, tumor necrosis factor α; EGFR, epidermal growth factor receptor; HER2, human epidermal growth factor receptor 2; VEGF, vascular endothelial growth factor; GPIIb/IIIa, a platelet integrin; RSV, respiratory syncytial virus; NSCL, non-small-cell lung. [b]See Figure 21.3 for structures.

chronic viral infections. Several aptamers are currently being tested in preclinical and clinical trials, and recently the aptamer pegaptanib (Macugen®) received FDA approval for the treatment of neovascular (wet) age-related macular degeneration, a principal cause of blindness in adults. Loss of vision in this disorder arises by the formation of new blood vessels (angiogenesis) and leakage from blood vessels, both of which are promoted by vascular endothelial growth factor (VEGF). Pegaptanib aptamers specifically bind to VEGF and inhibit its angiogenesis functions.

Vaccines have been genetically engineered to improve their functions

Vaccines against infectious disease are made by killing or attenuating (weakening) the disease organism (such as virus, bacteria, or fungus) and injecting it into the patient to stimulate the patient's immune system to produce suitably specific antibodies. If the patient were subsequently to come into natural contact with the disease organism, the body's immune system would be expected to mount a strong response that would prevent illness.

Genetic engineering has been applied in different ways to make novel types of vaccine. Genetic modification can be used to disable pathogenic microorganisms, for example, so that an attenuated live vaccine can be used safely. Genetically modified plants might be used to produce edible vaccines. In addition, changes can be made to an antigen to improve its visibility to the immune system, so as to produce an enhanced response.

Genes for desired antigens can be inserted into a vector, usually a modified virus with very low virulence, to produce *recombinant vaccines*. The vaccine may consist of the antigen-expressing vector or purified antigen. The hepatitis B vaccine is an illustrative example. A cloned gene encoding a hepatitis B surface antigen was inserted into an expression vector and transfected into yeast cells. The protein product of the expressed recombinant is purified for injection as a vaccine. The recombinant vaccine is much safer than using the attenuated hepatitis B virus, which could potentially revert to its virulent phenotype, causing lethal hepatitis or liver cancer.

DNA vaccines are the most recently developed vaccines and, as in the case of recombinant vaccines, genes for the desired antigens are located and cloned. With DNA vaccines, however, the DNA is directly injected into the muscle of the recipient often by biolistics—a gene gun uses compressed gas to blow the DNA into the muscle cells. Some muscle and dendritic cells express the pathogen DNA to stimulate the immune system. DNA vaccines induce both humoral and cellular immunity.

Although clinical trials conducted so far have provided overwhelming evidence that DNA vaccines are well tolerated and have an excellent safety profile, the early designs of DNA vaccines failed to demonstrate sufficient immunogenicity in humans. More recent results have been more encouraging but the efficacy of many new and more complex DNA vaccines must await the outcomes of many clinical trials currently underway.

Cancer vaccines

Cancers constitute another class of disease that is potentially amenable to vaccination. Some types of cancer are caused by viruses, such as HBV (causing some forms of liver cancer) and human papillomavirus (causing about 70% of cases of cervical cancer). Traditional vaccines raised against those viruses can effectively *prevent* the development of the relevant cancer (*preventive cancer vaccines*). However, vaccines for treating *existing* cancers (*therapeutic cancer vaccines*) have been much less successful. Therapeutic cancer vaccines often involve immunotherapy strategies in which certain types of gene are transfected into patient cells, constituting a type of gene therapy that we will consider in Section 21.4.

21.3 PRINCIPLES AND APPLICATIONS OF CELL THERAPY

Present-day organ transplantation is based on replacing a defective organ by a healthy one from a suitable donor. Cell-based therapies can be seen as a natural extension of the same principle, and advances in our understanding of stem cells offer the hope of radically extending the present range of options.

Stem cell therapies promise to transform the potential of transplantation

Much of the focus in stem cell therapy involves simple cell replacement therapy. Stem cells would be used to differentiate into suitable cell types so as to compensate for a deficit in that type of cell. In theory, the possibilities are endless—any damaged or worn-out tissue or organ resulting from disease or injury might be renovated or re-created using appropriate stem cells, and these might be subjected to any sort of genetic manipulation in advance. As we will see later in this chapter, stem cells are also important targets in gene therapy.

As detailed in Section 4.5, stem cells are cells that can both self-renew and also give rise to differentiated progeny. They are probably present at all stages of development and in all tissues. There is a gradient in their ability to give rise to diverse cell types (*plasticity*), from pluripotent stem cells that can potentially produce germ-line and all somatic cell types of an organism, through to stem cells that are more restricted in their differentiation potential and that are described as multipotent, oligopotent, or unipotent (see Table 4.6).

Pluripotent embryonic stem cells are derived from the early embryo, a stage in development at which cells are naturally still unspecialized. Later, as tissues form in development, cells become more and more specialized but some stem cells are retained in the fetus and in adults. They are needed to regenerate tissue. Our blood, skin, and intestinal epithelial cells, for example, have limited lifetimes and are replaced regularly by new cells produced by differentiation from stem cells. Stem cells are also used in tissue repair. Although many stem cells derived from fetal and adult tissues are tissue-specific, mesenchymal stem cells are found in a wide range of tissues and may constitute a subset of cells associated with minor blood vessels (pericytes).

Embryonic stem cells

Cells of the inner cell mass of a blastocyst can be isolated and cultured to generate embryonic stem (ES) cell lines (**Box 21.2**). Mouse ES cells have been studied for many years but, after the isolation of human ES cell lines was first reported in 1998, stem cell therapy was propelled to the forefront of both medical research and ethical debate. ES cells have clear advantages for cell therapy because they can be grown comparatively readily in culture and propagated indefinitely and can give rise to cell lineages. Different ES cell lines show significant variation in some properties, however, such as their potential for differentiation. Some ES cell lines readily differentiate toward mesodermal lineages, for example, whereas others do not.

The use of ES cell lines is contentious because the methods used to isolate them have traditionally involved the destruction of a human embryo to obtain cells in the inner cell mass. Nevertheless, some human ES cell lines have been made without destroying an embryo. For example, individual cells (*blastomeres*) can be carefully isolated from very early embryos in such a way that the embryo remains viable, and a few human ES cell lines have been generated from individual blastomeres isolated in this way.

Tissue stem cells

Deriving stem cells from adult human tissues and adult and cord blood has not been contentious. Tissue stem cells were initially thought to be comparatively

---

BOX 21.2 MAKING AND VALIDATING EMBRYONIC STEM CELL LINES

Embryonic stem (ES) cell lines are cultured cells that are derived from cells from the inner cell mass (ICM) of the blastocyst. For laboratory animals, blastocysts are excised from oviducts. Blastocysts for making human ES cell lines are traditionally obtained as a by-product of *in vitro* fertilization (IVF) treatment for infertility. In IVF, eggs provided by donors are fertilized by sperm in culture dishes; surplus embryos produced in this way can be used for research purposes with the informed consent of the egg donors.

Isolated blastocysts are disaggregated to allow the outgrowth of cells from the ICM in culture dishes. Traditionally, the inner surface of the culture dish is coated with *feeder cells*, mouse embryonic fibroblasts that have been treated so that they do not divide. The feeder cells provide a sticky surface to which the ICM cells can attach and they secrete growth factors and other chemicals that will stimulate the growth of the ICM cells. Surviving ICM cells that divide and proliferate are collected and re-plated on fresh culture dishes and subsequently undergo many cycles of re-plating and subculturing (known as *passages*). Embryonic stem cells that have continued to proliferate after subculturing for a period of 6 months or longer and that are judged to be pluripotent and seem to be genetically normal are known as an **embryonic stem (ES) cell line**.

The pluripotency of ES cell lines can be validated in different ways. In animals, ES cells can be differentiated to give tissues from all three germ layers, and can also be shown to give rise to a whole new organism. For human ES cell lines, pluripotency is validated by demonstrating differentiation to give tissues from all three germ layers. One way is to grow the cells in liquid suspension culture instead of the usual growth on surfaces to form colonies. In suspension, the ES

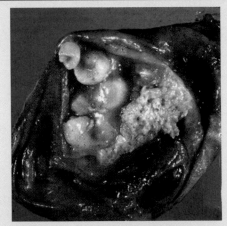

Figure 1 Human ovarian teratoma containing various tissues including teeth. [Image reproduced with permission from Virginia Commonwealth University.]

cells can spontaneously aggregate and differentiate to form *embryoid bodies* that consist of cells from all three germ layers. A more rigorous test is to inject the cells into an immunodeficient mouse to form a benign multilayered tumor known as a *teratoma*. Teratomas consist of different tissues derived from all three germ layers that can be associated in chaotic ways, giving a bizarre appearance. They also arise spontaneously in some individuals (**Figure 1**).

plastic under the right experimental conditions. For example, when bone marrow stem cells are placed in the environment of liver tissue they seem to be induced to give rise to new liver cells. However, when rigorously tested, there is little evidence to support claims of this type. Instead, tissue stem cells are now known to be almost always rather limited in their differentiation potential. An exception is provided by the adult germ line—pluripotent germ-line stem cells have recently been derived from spermatogonial cells of the adult human testis and show many similarities to ES cells. Many tissue stem cells are also difficult to grow in culture.

Practical difficulties in stem cell therapy

Stem cell therapy is also challenged by various practical limitations that depend on the effective expansion of stem cells in culture, controlled efficient differentiation, effective delivery of the desired replacement cells, and correct physical and functional integration within the desired tissue. Defining the molecular basis of the stemness that distinguishes a stem cell from other related cells has been an enduring problem, and so it has been extremely difficult to obtain highly purified stem cell populations. Our knowledge of the factors that control how stem cells differentiate, although rapidly advancing, remains primitive. As described below, some clinical trials have simply involved injecting crude preparations of stem cells directly into the damaged tissue and relying on the host tissue to send signals to guide the stem cells to differentiate into the desired cell type. It would clearly be more desirable to be able to induce efficient differentiation of stem cells by exogenously controlled factors once all the factors that regulate the relevant differentiation pathway are known.

Different diseases and tissue injuries may be less or more suited to cell therapy, according to tissue accessibility. Blood disorders and skin burns can be expected to pose fewer challenges than brain disorders. Another consideration is complexity of tissue architecture and regulation of cell function. Heart and liver cell replacement therapies would be facilitated by the simple tissue structures of these organs. Diabetes is more challenging because the pancreatic islets of Langerhans have a complex organization, and insulin-producing beta cells are just one of five different cell types in the islets and are subject to considerable regulation.

Important safety issues also need to be addressed. The pluripotency of cells such as ES cells carries the risk that some will give rise to tumors. Thus, differentiation would need to be maximized to reduce the chances of leaving residual undifferentiated stem cells. ES cells and many other stem cells can also show chromosomal instability if cultured for long periods.

Allogeneic and autologous cell therapy

In *allogeneic cell therapy* the transplanted cells come from a donor and are genetically different from those of the recipient. Except for a few *immunologically privileged* sites, such as the brain and the cornea and anterior chamber of the eye, there is a high risk of immune rejection of the donor cells. To counter this problem, stem cells can be genetically manipulated to make them less visible to the recipient's immune system. Stem cell banks containing large numbers of different ES cell lines could also offer individual cell lines that are closely HLA-matched to the cells of an intended recipient of cell therapy.

In *autologous cell therapy* the cells used to replace cells lost in the patient also originate from the patient, thereby avoiding immune rejection. One possible route relies on converting non-stem cells into stem cells by reprogramming their nuclei and gene expression patterns. This is a rapidly expanding field that will be described below. Another way is to artificially enhance the mobilization and differentiation of existing stem cells in the body. Our capacity for tissue repair depends on the degree to which stem cells are mobilized and effective in producing the necessary differentiated cells, and it is subject to developmental programming. Thus, in the fetus, wounds to skin tissue can be perfectly repaired by stem cells that mobilize to regenerate all the required cell types within the correct tissue architecture. With increasing age, this regenerative capacity is lost: adult skin heals through scar formation, with erratic vascular formation, loss of hair follicles, and disorganized collagen deposition.

While normal mobilization of an individual's stem cells can help make small repairs to different degrees, large injuries and serious disease pose too much of a challenge for the body's natural repair processes. But what if we could artificially enhance the normal stem cell response to repair? Bone marrow cells that are mobilized in response to disease or injury are an attractive target for artificially enhanced mobilization. They include mesenchymal stem cells, which can become bone or cartilage cells, and endothelial progenitor cells, which produce the cells that make up our blood vessels and help with tissue repair. A promising report published in early 2009 showed that administration of the drug Mozobil® and a growth factor was found to elicit a hundredfold increase in the release of endothelial and mesenchymal progenitor cells from the bone marrow in mice.

Nuclear reprogramming offers new approaches to disease treatment and human models of human disease

In principle, autologous cell therapy could be carried out by reprogramming cells from a patient so that they are converted to, or give rise to, the type of cells that the patient lacks. To do this, conveniently accessible cells, such as skin fibroblasts, need to be extracted from the patient and then manipulated in culture to change the differentiation status of the cell (**Box 21.3**). Until recently, the developmental progression that causes mammalian cells to gradually lose their plasticity and to become progressively more specialized, and more and more

---

BOX 21.3 EXPERIMENTAL INDUCTION OF NUCLEAR REPROGRAMMING IN MAMMALIAN CELLS

Nuclear reprogramming entails altering nuclear gene expression to bring about a profound change in the differentiation status of a specialized cell. It may cause a somatic cell to regress so that it acquires the properties of an unspecialized embryo or a pluripotent or progenitor cell (*dedifferentiation*), or it may cause a lineage switch so that one type of differentiated somatic cell changes into another type (*transdifferentiation*). The fertilized egg cell provides a natural example of nuclear reprogramming: the sperm is a highly differentiated cell and the nucleus is highly condensed until reprogrammed by factors in the oocyte. Different experimental procedures can bring about nuclear reprogramming (**Figure 1** for an overview).

Somatic cell nuclear transfer
The first experiments that caused reversal of vertebrate cell differentiation were performed in frogs more than 40 years ago. With the use of microsurgical techniques, nuclei were isolated from specialized cells of the embryo or terminally differentiated cells and transplanted into unfertilized eggs whose nuclei had either been removed or been inactivated by exposure to irradiation. This could result in reprogramming of a somatic cell nucleus to make it totipotent (Figure 1, path A). Frogs that were derived by nuclear transfer from somatic cells of a single individual were considered to be essentially genetically identical (by having the same nuclear genome) and were considered to be *clones*.

These experiments showed that it was possible to fully reverse differentiation in vertebrates. However, amphibians were considered to be exceptional; some amphibians—such as salamanders—are able to regenerate limbs and tails. The regenerative capacity of mammals is extremely limited by comparison. The prospects for reversing differentiation to clone mammals seemed bleak until the birth of a cloned sheep called Dolly in 1996 (see Figure 20.4). Although technically much more difficult than cloning frogs, the success with Dolly prompted the subsequent successful cloning of many other types of mammal, including cows, mice, goats, pigs, cats, rabbits, dogs, and horses. A fierce debate arose as to whether human cloning could ever be contemplated (see Box 21.4).

Cell fusion
Two different types of cell can be induced to fuse, forming a single cell with two different nuclei, a *heterokaryon*. By using an inhibitor of cell division, the two nuclei in the heterokaryon can be kept separate.

Each nucleus is potentially subject to regulation by factors that normally regulate gene expression in the other nucleus. One nucleus predominates—typically the nucleus of the larger and more actively dividing cell—and the other is reprogrammed to adopt the pattern of gene expression of the dominant nucleus. Thus, if actively dividing pluripotent ES cells are fused with somatic cells, the somatic cell nucleus will be reprogrammed to the pluripotent state (Figure 1, path B).

Specific transcription factors can induce pluripotency
The observation that the nuclei of somatic cells could be reprogrammed to reverse differentiation prompted intense research to identify the factors responsible. Transcription factors known to be involved in maintaining pluripotency in ES cells were likely candidates. Nevertheless, it was a huge surprise when in 2006 Shinya Yamanaka and colleagues reported that transfection of genes encoding just four transcription factors—Oct-3/4, Sox2, c-Myc, and Klf-4—could reprogram mouse fibroblasts to become pluripotent cells. Follow-up studies showed that human and rat fibroblasts could also be reprogrammed toward pluripotency with defined transcription factors, as could various other cell types, including lymphocytes and liver, stomach, and pancreatic beta cells.

Cells reprogrammed in this way are called *induced pluripotent stem cells* (*iPS cells*) because they closely resemble ES cells. They show telomerase activity and reactivation of pluripotency genes, and they have genomewide transcriptional and epigenetic patterns characteristic of ES cells. In addition, iPS cell chimeras can give rise to offspring, indicating their ability to contribute to the germ line. The mechanism underlying the induction of pluripotency is under intense investigation. The combination of transcription factors necessary for reprogramming is well conserved in mammals, but shows some variability according to cell type. A variety of human disease-specific iPS cell lines have already been made and have exciting potential for modeling disease and drug screening, as well as possibly providing therapeutic applications (see the main text).

Transdifferentiation
Nuclear reprogramming can also be induced so as to cause transdifferentiation. This may occur indirectly by *lineage switching*, when a cell is induced to regress along a differentiation pathway to a branch point that will allow some cells to then follow a different

committed to their fate, was thought to be irreversible. The cloning of a sheep called Dolly changed that view, because the nucleus of an adult somatic cell was successfully reprogrammed, giving rise to a totipotent cell.

As detailed in Section 20.2, the cloning of Dolly was made possible by isolating nuclei from mammary gland cells from a Finn-Dorset sheep and then transplanting individual nuclei into enucleated eggs from a Scottish blackface sheep (*somatic cell nuclear transfer*—see Figure 20.4). In one out of 434 attempts, factors in the egg cytoplasm somehow reprogrammed the somatic cell nucleus to return it to the totipotent state. The resulting cell was stimulated to develop to the blastocyst stage and was then implanted in the uterus of a foster mother to give rise to Dolly. As Dolly's nuclear genome was identical to that of the Finn-Dorset sheep, she was considered to be a clone of that sheep. A variety of other mammals have now been cloned by somatic cell nuclear transfer. However, cloned mammals have a very high incidence of abnormalities, most probably as a result of inappropriate epigenetic modifications.

Although natural human clones—identical twins and triplets, for example— are reasonably common, artificial human cloning is considered ethically unjustifiable (**Box 21.4**) and is banned by legislation in many countries. Nevertheless, somatic cell nuclear transfer offered for the first time the possibility of making an ES cell line from any patient willing to donate some easily accessible skin fibroblasts. Individual nuclei would be isolated from the skin cells and transferred into unfertilized eggs that had been donated by donors and subsequently enucleated.

BOX 21.3 *CONT.*

pathway to produce a different but related cell type (Figure 1, path C), or by direct cell conversion without an obvious intermediate dedifferentiation step (Figure 1, path D).

Ectopic expression of just a single type of transcription factor can cause transdifferentiation. Thus, for example, the transcription factors PU.1 and GATA-1 are mutually antagonistic in the development of blood cell lineages. If PU.1 predominates, differentiation occurs toward myeloid cells, but if GATA-1 dominates, cells go down the

megakaryocyte/erythroid pathway. As a result, transfection of a gene expressing the GATA-1 transcription factor into myeloid cells can cause transdifferentiation toward megakaryocyte/erythroid cells; ectopic expression of PU.1 in megakaryocyte/erythroid cells stimulates transdifferentiation into myeloid cells (Figure 1, path C). Efficient direct conversion of adult pancreatic exocrine cells to beta cells can be achieved by transfecting genes for three transcription factors that are essential for beta cell function: Ngn3, Pdx1, and Mafa (Figure 1, path D).

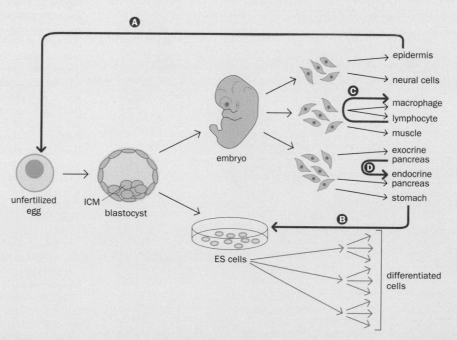

Figure 1 Experimental routes for nuclear reprogramming in mammals. Normal steps in development and tissue differentiation are shown by black arrows. Experimental procedures are shown as blue arrows. (A) Nuclear transfer into an unfertilized egg reprograms a somatic nucleus to dedifferentiate to give a totipotent cell. (B) A somatic cell is stimulated to dedifferentiate to become a pluripotent stem cell that closely resembles an ES cell. (C) Lineage switching is a form of transdifferentiation that involves dedifferentiation followed by differentiation down a different differentiation pathway. (D) Transdifferentiation by direct conversion of one somatic cell into another. ICM, inner cell mass.

BOX 21.4 THE ETHICS OF HUMAN REPRODUCTIVE CLONING

In principle, reproductive cloning to produce a human clone could provide the only option for a couple to have children in certain cases. For example, a woman with a serious mitochondrial disorder could avoid transmission of the disease by becoming pregnant with a donated oocyte into which a nucleus had been transplanted from one of her or her partner's somatic cells. The technology could also meet more general needs such as the production of cloned individuals to replace a dying child or other loved one. However, human reproductive cloning is banned by legislation in many countries because of powerful ethical arguments.

Of the two major ethical arguments, the practical argument is the more powerful one. It points to the low success rate of all mammalian cloning experiments, and to the fact that many of the cloned animals born have serious abnormalities, most probably because the epigenetic modifications of the donor cell are not reliably reprogrammed. Allowing human reproductive cloning to proceed with current imperfect technology would be grossly unethical because of the early deaths and major abnormalities that could be expected. Conceivably, advances in knowledge might remove this objection, although it is not clear how we could know without conducting unethical experiments.

Even if the technology were perfectly efficient and safe and had no health consequences, the second major ethical argument in principle says that humans should be valued for themselves and not treated as instruments to achieve a purpose.

The resulting cells would be stimulated to develop into blastocysts, and inner cell mass cells would be removed to make ES cells. This type of approach became known as *therapeutic cloning*, to distinguish it from the banned *reproductive cloning* (**Figure 21.5**).

Therapeutic cloning has been controversial. Not only does it involve the destruction of an unfertilized human egg, but also, by providing the technology for cloning human blastocysts, it has raised fears that human reproductive cloning could be facilitated. In addition, obtaining eggs from human donors involves an invasive and potentially dangerous procedure. The attraction of therapeutic cloning was that it offered the possibility of making *patient-specific ES* cells that could, in principle, be used for autologous cell therapy, and disease-specific ES cells that could be used for studying pathogenesis and in drug screening as described below. In practice, cloning human blastocysts by somatic cell nuclear transfer proved to be extraordinarily inefficient—thus far, there is no report of a human ES cell line made by this method.

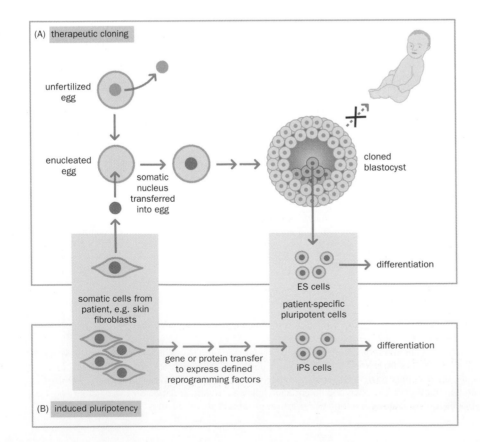

Figure 21.5 Nuclear reprogramming can generate patient-specific and disease-specific pluripotent stem cells.
(A) Therapeutic cloning. Here, a cell nucleus from suitably accessible cells from a patient donor is transplanted into an enucleated egg. Factors in the oocyte cytoplasm are able to reprogram the somatic nucleus so that it becomes totipotent. The resulting cell can then be stimulated to develop to the blastocyst stage. The nuclear genome of the resulting blastocyst will be genetically identical with that of the donating patient. Cells are removed from the inner cell mass of the blastocyst, which is then destroyed. The retrieved cells are cultured to establish a pluripotent ES cell line that can be induced to differentiate to different types of somatic cell. The large red X signifies that using cloned human blastocysts for reproductive purposes is regarded as ethically unacceptable. (B) Induced pluripotency does not involve the destruction of a human embryo. Somatic cells are transfected with genes encoding a few specific transcription factors that are important in maintaining pluripotency, and cells are selected that express the reprogramming factors. Alternatively, the reprogramming proteins can be added directly to the cells. The resulting induced pluripotent stem (iPS) cells are essentially equivalent to ES cells.

Induced pluripotency in somatic cells

An alternative way of generating patient-specific and disease-specific pluripotent stem cells does not raise the same ethical concerns. Pluripotency can be induced in somatic cells just by ectopically expressing a very few specific transcription factors. Shinya Yamanaka and colleagues reported an astounding breakthrough in 2006 (**Box 21.5**) when they were able to regress mouse fibroblasts to a pluripotent state by transfecting genes encoding just four types of transcription factor—Oct-3/4, Sox2, c-Myc, and Klf-4—known to be involved in maintaining pluripotency. By selecting for cells expressing the transfected genes they identified a small population of **induced pluripotent stem (iPS) cells**.

iPS cells were subsequently made from fibroblasts, or other types of somatic cell, from a variety of other mammals. iPS cells are highly similar to ES cells at both the molecular and functional levels, and, like ES cells, iPS cells can contribute to all tissues as well as to the germ line. In 2009 viable fertile mice were reported that originated exclusively from mouse iPS cells (see Box 21.5).

Because iPS cells are functionally equivalent to ES cells in pluripotency but are unhampered by the kind of ethical concerns that have hindered human ES cell research, research studies on iPS cells have developed rapidly. The ability to make patient-specific and disease-specific iPS cells and to allow them to differentiate into somatic cells affected in the relevant disease has important biomedical applications (**Figure 21.6**).

iPS cells can potentially provide a wide range of human models of human disease. For many important disorders, the affected tissues and cells are not readily accessible. In neurodegenerative conditions of neuronal tissue, such as Huntington, Alzheimer, and Parkinson diseases, the pathogenesis has had to be studied in post-mortem specimens or in animal models in which the phenotype is often somewhat different from the human disease phenotype. Differentiation of the appropriate disease-specific iPS cells will provide live human cells for studying the pathology.

BOX 21.5 SOME MILESTONES IN NUCLEAR REPROGRAMMING IN MAMMALIAN CELLS

1996	production of the first cloned mammal, a sheep called Dolly, by somatic cell nuclear transfer; see Figure 20.4 (*Nature* 380, 64–66; PMID 8598906)
2004	chemical screens identify small molecules that cause reversal of differentiation, such as reversine, which can cause myoblasts to dedifferentiate into progenitor cells (*J. Am. Chem. Soc.* 126, 410–411; PMID 14719906 and *Nat. Biotechnol.* 22, 833–840; PMID 15229546)
2004	reprogramming of nuclei transferred from olfactory sensory neurons to eggs to generate totipotent cells and subsequently cloned mice (*Nature* 428, 44–49; PMID 14990966)
2004	stepwise reprogramming of B cells into macrophages—an early example of a lineage switch (*Cell* 117, 663–676; PMID 15163413)
2005	nuclear reprogramming of somatic cells after fusion with human ES cells (*Science* 309, 1369–1373; PMID 16123299)
2006	reprogramming of cultured mouse embryonic and adult fibroblasts to pluripotent stem cells following transfection by genes encoding just four transcription factors (*Cell* 126, 663–676; PMID 16904174)
2007	induction of pluripotency in human somatic cells (*Cell* 131, 861–872; PMID 18035408 and *Science* 318, 1917–1920; PMID 18029452)
2008	transdifferentiation by *in vivo* reprogramming of adult pancreatic exocrine cells to beta cells (*Nature* 455, 627–632; PMID 18754011)
2008	nuclear reprogramming provides the first patient-specific and disease-specific pluripotent stem cells (*Cell* 134, 877–886; PMID 18691744 and *Science* 321, 1218–1221; PMID 18669821)
2009	generation of fertile viable mice derived exclusively from iPS cells (*Nature* 461, 86–90; PMID 19672241, *Nature* 461, 91–94; PMID 19672243, and *Cell Stem Cell* 5, 135–138; PMID 19631602)
2009	generation of iPS cells using recombinant proteins only (*Cell Stem Cell* 4, 381–384; PMID 19398399)
2010	reprogramming of fibroblasts to give functional neurons (Vierbuchen T et al., *Nature* 463, 1035–1041; PMID 20107439)

iPS cell, induced pluripotent stem cell; PMID, PubMed identifier number.

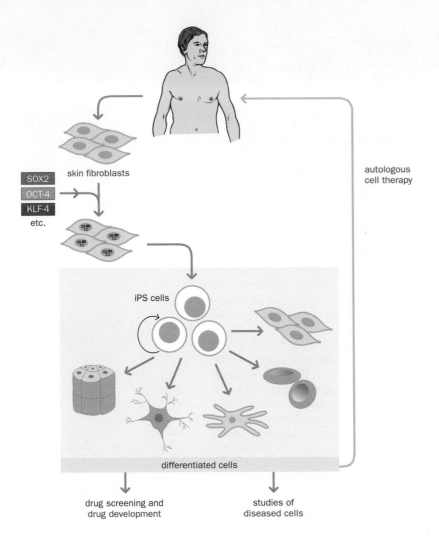

skin fibroblasts

SOX2
OCT-4
KLF-4
etc.

iPS cells

differentiated cells

autologous
cell therapy

drug screening and
drug development

studies of
diseased cells

Figure 21.6 Genesis and biomedical applications of human induced pluripotent stem cells. Skin fibroblasts from individuals with disease and from healthy individuals can be transfected with genes encoding certain pluripotency-associated transcription factors, such as SOX2 or OCT-4, or they can be transduced by a cocktail of the transcription factor proteins. As a result, the fibroblast nuclei are reprogrammed to produce self-regenerating ES-like cells known as induced pluripotent stem (iPS) cells. The iPS cells can be directed to differentiate along different pathways to give diverse cell types with multiple biomedical applications. Autologous cell therapy might be achieved by differentiating the iPS cells to produce cells of a type that a patient lacks, which are then returned to the patient. For genetic disorders, the genetic mutation could be repaired in the iPS cells from the individuals with disease before they are differentiated and returned to the patient. Differentiated cells produced from the patient's iPS cells could also be valuable in drug screening and in providing human cellular models of disease.

For some disorders, the use of iPS cell-derived cells also means that drugs can now be tested directly on the relevant human cells. Drugs can be tested for their ability to reduce the progression of the pathogenesis or even to counteract it in some way, and the toxicity levels can be determined directly, rather than relying on animal studies.

The potential of iPS cells for efficient autologous cell therapy remains unclear. Safety concerns were an issue with the original technology, which used retroviral vectors to transfect reprogramming genes, one of which was a known oncogene, c-*Myc*. However, new methods do not use c-*Myc*, and avoid genome integration by using non-integrating vectors or by transducing cells with the pluripotency-inducing transcription factor proteins themselves, or miRNAs and other small molecules involved in the same pathways. Performing efficient precisely directed differentiation remains a major challenge. Like ES cells, undifferentiated iPS cells can give rise to tumors, and there is considerable ignorance about the molecular details of the long, multi-step differentiation pathways required to convert pluripotent stem cells to differentiated somatic cells.

Transdifferentiation

An alternative to iPS cells is to use lineage-reprogramming methods that would result in limited dedifferentiation to produce oligopotent progenitors of a desired cell type, or a direct cell conversion (*transdifferentiation*) to produce the desired cell type (see Box 21.3 Figure 1, pathways C and D). For example, pancreatic exocrine cells could be reprogrammed to make pancreatic beta cells that could possibly be induced to join pancreatic islets to replace beta cells lost in type 1 diabetes. By just taking one step back in differentiation or a sideways step, differentiation toward producing the desired cell type would be much simpler or not even required. Additionally, disruption to the epigenetic marks that are set during development would be expected to be minimal. Reprogramming between closely related cell types should also require fewer proliferation steps and should presumably reduce the chance of mutations. Unlike iPS cell technology, however, in which much the same reprogramming factors are needed to dedifferentiate a wide range of different cell types, lineage-reprogramming factors can be expected to be cell type-specific and have thus far not been well characterized for many cell types (see Box 21.5 for two examples).

Stem cell therapy has been shown to work but is at an immature stage

Notwithstanding the difficulties described above, stem cell therapy can work. We consider here some illustrative examples of therapy in which the object has simply been to replace cells lost through disease or injury. This represents one component of a burgeoning field that is often described as *regenerative medicine* and also includes *tissue engineering*, the combinatorial use of cells, materials, and engineering principles toward producing functional replacement tissues and organs. For genetic disorders, future autologous cell therapies would be expected to include the use of genetic engineering to correct harmful mutations in stem cells in culture before returning cells to the patient. In later sections in this chapter we describe the use of progenitor cells and stem cells as preferred cellular targets for gene therapy.

The stem cells that have been used most effectively in cell replacement therapy have been classes of tissue-specific multipotent stem cells that are comparatively frequent and easy to access, notably bone marrow and umbilical cord blood, which are known to be enriched in stem cells. Bone marrow transplantation is a form of stem cell therapy that has been used successfully for several decades in treating blood cancers, notably leukemia and lymphoma. Because hematopoietic stem cells are formed in the bone marrow, a bone marrow transplant is effectively a form of enrichment of hematopoietic stem cells, although the percentage of stem cells in the transplants remains low. More recently, umbilical cord blood stem cells have also been used for the same purpose. In both cases, the therapy is designed to replace cells that are killed during previous chemotherapy treatment, which, as well as killing cancer cells, depletes dividing blood cells, including hematopoietic stem cells.

Because of the ready access to hematopoietic stem cells, blood disorders are comparatively easy targets for cell therapy (Table 21.3). More difficult to treat are diseases or injuries affecting organs for which stem cells have not been so readily accessible. In such cases, easily accessible stem cell populations, notably bone marrow cell preparations, have often been applied directly at the site of disease or injury in the hope that signals from the tissue environment would induce the introduced stem cells to transdifferentiate into the desired cell type. In the case of ischemic heart disease, for example, heart muscle cells (cardiomyocytes) that die after a heart attack need to be replaced, and some clinical trials have claimed a measure of success with injected bone marrow cells. However, there is some controversy over whether any improvement is due to cell therapy, and if so what the basis would be; there is little evidence for transdifferentiation—if it does occur it is a very infrequent event.

Pluripotent stem cells have very recently been considered in stem cell therapy. The first clinical trial with human ES cells was initially given FDA approval in early 2009, but was subsequently suspended pending further investigations. Researchers at Geron had hoped to expand ES cells and differentiate them into oligodendrocyte precursors for treating spinal cord injury. As we will see later in the chapter, new genetic manipulations make it easier to repair genetic defects at the DNA level in somatic cells. This may mean that cell therapy could be extended to monogenic genetic disorders by repairing genetic defects in iPS cells taken from patients.

However, formidable challenges lie ahead. Much basic research needs to be done on analyzing the properties of stem cells and differentiation pathways. And we are reminded of the difficulties of cell engraftment when, several years after transplantation of fetal neuronal cells to treat Parkinson disease, Lewy bodies—abnormal protein aggregates that develop in neurons—have appeared in what had originally been healthy grafts.

TABLE 21.3 SOME PRINCIPAL TARGETS FOR STEM CELL THERAPY AND EXAMPLES OF SUCCESSFUL CELL REPLACEMENT THERAPY IN RODENT DISEASE MODELS

Disease/injury	Cell affected	Stem cell therapy	Comments/references
Leukemia/lymphoma	hematopoietic stem cells killed by chemotherapy	allogeneic bone marrow transplantation from matched donor	successful over many decades
Eye injury/disease	limbal stem cells in one eye	limbal stem cells are taken from the healthy eye, expanded *ex vivo*, and transplanted back into the affected eye	Pellegrini et al. (1997) *Lancet* 349, 990–993; PMID 9100626 and Kolli et al. (2010) *Stem Cells* in press; PMID 20014040
Spinal cord injuries	neurons	transplantation of fetal neural stem cells or ES cell-derived oligodendrocyte progenitors	some evidence for successful treatment; first target for approved clinical trials using ES cells
Heart disease	cardiomyocytes (heart muscle cells)	direct injection of bone marrow stem cells (including hematopoietic stem cells and mesenchymal stem cells)	conflicting views on the degree of success, and uncertainty regarding the underlying cause of any clinical improvement
Stroke	neurons	human fetal neural stem cells used as cell source	first clinical trial underway
Parkinson disease	midbrain dopaminergic neurons	fetal neuronal cells used as cell source	difficulties in ensuring that grafted neurons do not themselves develop disease after a few years
Rat Parkinson disease model	midbrain dopaminergic neurons	iPS cells were differentiated into neural precursor cells that were transplanted into the fetal mouse brain; the transplanted cells were functionally integrated, causing improvements in symptoms	Wernig et al. (2008) *Proc. Natl Acad. Sci. USA* 105, 5856–5861; PMID 18391196
Brain disease (in mouse *shiverer* mutant)	glial cells as a result of myelin deficiency	human fetal glial progenitor cells were injected into the nervous system to correct abnormal brain development	Windrem et al. (2008) *Cell Stem Cell* 2, 553–565; PMID 18522848
Blindness (in *Crx*-deficient mice)	degeneration of photoreceptors	human ES cells were differentiated to give retinal cells that were transplanted into the retinas of the mouse *Crx* mutant and restored some visual function	Lamba et al. (2009) *Cell Stem Cell* 4, 73–79; PMID 19128794

PMID, PubMed identifier number.

21.4 PRINCIPLES OF GENE THERAPY AND MAMMALIAN GENE TRANSFECTION SYSTEMS

Gene therapy involves the direct genetic modification of cells of the patient to achieve a therapeutic goal. There are basic distinctions in the types of cell modified, and the type of modification effected. *Germ-line gene therapy* would produce a permanent transmissible modification and could be achieved by the modification of a gamete, a zygote, or an early embryo. It is widely banned for ethical reasons (**Box 21.6**). *Somatic gene therapy* seeks to modify specific cells or tissues of the patient in a way that is confined to that patient.

All current gene therapy trials and protocols are for somatic gene therapy. As gene therapy successes accumulate and the technologies become increasingly refined and safe, extending the technology to the germ-line to produce designer babies can be expected, with associated ethical concerns (**Box 21.7**).

In somatic cell gene therapy, the cells that are targeted are often those directly involved in the pathogenic process, but in some cancer gene therapies the object is to modify normal immune system cells genetically in the patient so as to provoke a powerful immune response against tumor cells. The somatic cells might be modified in several different ways (**Figure 21.7**):

- *Gene augmentation* (or *gene addition*). The aim is to supply a functioning gene copy that will *supplement* a defective gene. The obvious application is to treat diseases that are the result of a gene not functioning here and now. Cystic fibrosis would be a typical candidate. Gene augmentation would not be

BOX 21.6 THE ETHICS OF GERM-LINE GENE THERAPY

Germ-line gene therapy involves making a genetic change that can be transmitted down the generations. This would most probably be done by genetic manipulation of a pre-implantation embryo, but it might occur as a by-product of a treatment aimed at somatic cells that incidentally affected the patient's germ cells. Pure somatic gene therapy treats only certain body cells of the patient without having any effect on the germ line. Technical difficulties currently make germ-line therapy unrealistic. Even if these problems are solved, ethical concerns will remain: genetic manipulation of the germ line is prohibited by law in many countries.

- The main argument against germ-line therapy is that these treatments are necessarily experimental. We cannot foresee every consequence, and the risk is minimized by ensuring that its effects are confined to the patient we are treating. This would imply that once we have enough experience of somatic therapy, it might be ethical to proceed to germ-line therapy. However, even if the initial treatment were performed with informed consent, later generations are given no choice. This leads to the view that we have a responsibility not to inflict our ideas or products on future generations, and so germ-line therapy will always be unethical.

- The argument in favor of germ-line therapy is that it solves the problem once and for all. Why leave the patient's descendants at risk of a disease if you could equally well eliminate the risk?

    The argument in favor needs to be set against the population genetic background. The Hardy–Weinberg equations set out in Chapter 3 (p. 84) are highly relevant here; in addition there is a strong practical argument that germ-line therapy is unnecessary.

- For recessive conditions, only a very small proportion of the disease genes are carried by affected people; the great majority are in healthy heterozygotes. For a recessive disease affecting one person in 10,000 ($q^2 = 1/10,000$; $q = 0.01$) only 1% of disease alleles are in affected people. Whether or not we stop affected people from transmitting their disease genes (either by germ-line therapy or by the cruder option of sterilization as the price of treatment) has very little effect on the frequency of the disease in future generations.

- For fully penetrant dominant conditions, all the disease alleles are carried by affected people, and for X-linked recessives the proportion is one-third. But the dream of eliminating the disease once and for all from the population falls down because, as the equations in Box 3.7 show, most serious dominant or X-linked diseases are maintained in the population largely by recurrent mutation.

- A third, and cogent, objection is that germ-line therapy is simply not necessary. Candidate couples would most probably have dominant or recessive Mendelian disorders (recurrence risk 50% and 25%, respectively). Given a dish containing half a dozen IVF embryos from the couple, it would seem crazy to select the affected ones and subject them to an uncertain procedure, rather than simply to select the 50% or 75% of unaffected ones for re-implantation.

    The argument that somatic therapy is less risky than germ-line therapy seems incontrovertible. In particular, the safer non-integrating vectors could not be used for germ-line therapy. We are less convinced by the general argument that it is unethical to impose our choices on future generations.

suitable for loss-of-function conditions in which irreversible damage has already been done, for example through some failure in embryonic development. Cancer therapy could involve providing a normal gene copy to supplement a defective tumor suppressor gene.

- *Elimination of pathogenic mutations.* The object is to restore the function of a mutated gene. The gene could be repaired at the DNA level by replacing a sequence containing the pathogenic mutation with a normal equivalent sequence. Such gene correction would be required for gain-of-function diseases in which the resident mutant gene is doing something positively harmful to cells or tissues. Another possibility, not shown in Figure 21.7 but detailed later in this chapter, is to induce altered gene splicing causing skipping of an exon containing the harmful gene mutation.

BOX 21.7 THE ETHICS OF "DESIGNER BABIES"

The designer baby catchphrase encapsulates two sets of worries. First, people will use *in vitro* fertilization and pre-implantation diagnosis to select embryos with certain desired qualities and reject the rest even though they are normal. This contrasts with the use of the same procedure to avoid the birth of a baby with a serious disease. Second, people will use the therapeutic technologies described in this chapter, not to treat disease but for *genetic enhancement*; that is, endowing genetically normal people with superior qualities.

The first scenario is already with us in the form of pre-implantation sex selection and a few highly publicized cases in which a couple have sought to ensure that their next child can provide a perfectly matched transplant tissue to save the life of a sick child. Current cases involve a transplant of stem cells from cord blood. That would do no harm to the baby—it would be different if it were proposed to take a kidney. It is often suggested that these cases are the start of a slippery slope that leads inevitably to demands for very extensive specification of the genotype of the baby—as envisaged in the phrase designer baby. The reality is different. Simply selecting for HLA compatibility means that only one in four embryos is selected. Most IVF procedures produce only a handful of embryos, and usually two or three are implanted to maximize the chance of success. Selection on multiple criteria is simply not compatible with having enough embryos to implant. More generally, nature has endowed us with a simple and highly agreeable method of making babies that is very effective for the large majority of couples, and it is hard to imagine most people abandoning this in favor of a long-drawn-out, unpleasant, and highly invasive procedure that is very expensive and has a low success rate.

Genetic enhancement is a difficult question—or it will become one, once we have any idea which genes to enhance. On the one hand, parents are supposed to do what they can to give their children a good start in life; on the other hand, options available only to the rich are often seen as buying an unfair advantage. Perhaps fortunately, we are a long way from identifying suitable genes, even if the techniques for using them were available. Attempts to produce genetically enhanced animals have not been a success and in some cases have been spectacular failures. In the long term, the possibilities must be immense, and there will surely be very difficult ethical issues to confront.

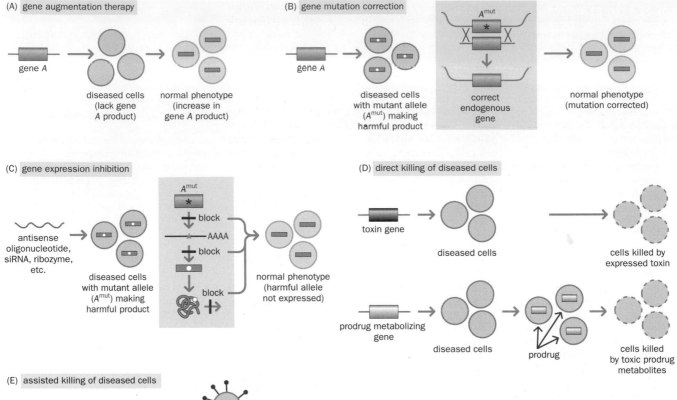

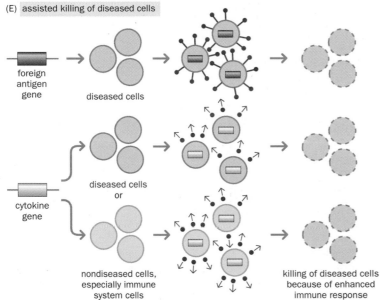

(A) gene augmentation therapy

gene A → diseased cells (lack gene A product) → normal phenotype (increase in gene A product)

(B) gene mutation correction

gene A → diseased cells with mutant allele (A^{mut}) making harmful product → A^{mut} correct endogenous gene → normal phenotype (mutation corrected)

(C) gene expression inhibition

antisense oligonucleotide, siRNA, ribozyme, etc. → diseased cells with mutant allele (A^{mut}) making harmful product → A^{mut} block / AAAA / block / block → normal phenotype (harmful allele not expressed)

(D) direct killing of diseased cells

toxin gene → diseased cells → cells killed by expressed toxin

prodrug metabolizing gene → diseased cells → prodrug → cells killed by toxic prodrug metabolites

(E) assisted killing of diseased cells

foreign antigen gene → diseased cells →

cytokine gene → diseased cells or nondiseased cells, especially immune system cells → killing of diseased cells because of enhanced immune response

Figure 21.7 Five strategies for gene therapy. (A) In some genetic disorders, the problem is deficiency for a normal gene product. Gene augmentation therapy involves supplying a functioning cloned gene to supplement the lack of product, shown here as the protein product (green horizontal bar) of gene A. In other cases shown in (B) and (C) a mutant gene (A^{mut}) makes a harmful product that is shown in (C) as a mutant protein (green bar with white circle signifying altered residues), but it could also be a harmful RNA product. Targeted correction of the gene mutation (B) seeks to replace the sequence of the mutation at the DNA level with wild-type sequence, causing it to revert to a normal allele. Alternatively, expression of the mutant gene can be selectively blocked at the levels of transcription and translation by using gene-specific antisense oligonucleotides or siRNA, or by using specific oligonucleotides to bind the mutant protein and so inhibit its effects. For treating cancer, the object is often simply to kill the cancer cells as shown in (D) and (E). This can be done directly (D), either by targeting a toxin gene to the cells or a prodrug-metabolizing gene that makes the cells vulnerable to the effects of some prodrug that can be added later (see Figure 21.17 for an example). Alternatively, in *immunotherapy* immune system cells can be incited in different ways to kill cancer cells (E).

- *Targeted inhibition of gene expression.* This is used especially in infectious disease, in which essential functions of the pathogen are targeted, and to silence activated oncogenes in cancer. Another application would be to damp down unwanted responses in autoimmune disease and maybe to silence a gain-of-function mutant allele in inherited disease.

- *Targeted killing of specific cells.* As will be described below, this approach is particularly applicable to cancer treatment. It can involve direct killing using genes encoding toxins, etc. (see Figure 21.7D). Alternatively, gene transfection is designed simply to provoke a very strong immune response that is designed to kill the cancer cells (immunotherapy; see Figure 21.7E). This is possible because our immune system has evolved to recognize nonself antigens, and although cancer cells share some antigens with other body cells they may sometimes express an antigen that is specific for the cancer cells or greatly overexpress a normal antigen in a way that can provoke weak immune responses. Immunotherapy is designed to amplify the naturally weak immune response to cancer cells.

Genes can be transferred to a patient's cells either in culture or within the patient's body

In gene therapy, cloned genes, RNA, or oligonucleotides are inserted into the cells of a patient with the use of some delivery method. Often vectors are used that are based on viruses because, over long evolutionary time-scales, viruses have become highly effective in infecting cells, inserting their genomes, and getting their genes to be expressed. Transfer of DNA into human (and other animal) cells by using viruses is known as **transduction**. In some cases the virus vectors can integrate into the genome and so provide the means for long-lasting transgene expression; however, as we describe below, integrating vectors currently pose safety risks. Other virus vectors do not integrate into the host genome. Instead, they remain in extrachromosomal locations in cells.

Non-viral methods can also be used to transfer DNA, RNA, or oligonucleotides into human (or other animal) cells (**transfection**). Because such methods do not involve integration of DNA into the genomes, they have the advantage of high safety in therapeutic applications, but they are generally disadvantaged by low transfection efficiencies.

In many gene therapy protocols, suitable target cells are excised from a patient, grown in culture, and then genetically modified by transfer of the desired nucleic acid or oligonucleotide. The cells can be analyzed at length to identify those cells in which the intended genetic modification has been successful. Selected cells can then be amplified in culture and injected back into the patient. Because the patient's cells are genetically modified outside the body, this approach is known as *ex vivo* gene therapy (**Figure 21.8**). It is most appropriate for disorders in which the target cells are accessible for initial removal and can be induced to engraft and survive for a long time after replacement. Examples include cells of the hematopoietic system and skin cells.

The alternative is *in vivo* gene therapy (see Figure 21.8), in which the gene transfer process is performed *in situ* within the patient, for example by injection into an organ (such as brain, muscle, or eye) or, for example, by using aerosols for the lung. *In vivo* gene therapy is the only option in disorders in which the target cells cannot be cultured *in vitro* in sufficient numbers (e.g. brain cells) or in which cultured cells cannot be re-implanted efficiently in patients.

Tissue targeting is an important consideration. As an alternative to placing the gene transfer construct directly into the target tissue, it may be introduced systemically. For example, injection into the portal vein can be used for delivery to the liver. Also some virus vectors naturally infect cells of a particular type. As there is no way of selecting and amplifying cells that have taken up and expressed the intended nucleic acid or oligonucleotide, the success of *in vivo* gene therapy is crucially dependent on the general efficiency of gene transfer and, where appropriate, expression in the correct tissue.

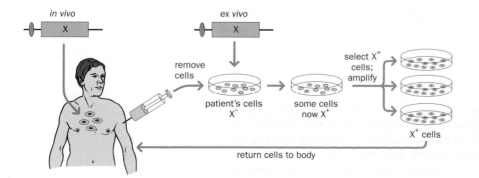

Figure 21.8 *In vivo* **and** *ex vivo* **gene therapy.** In *ex vivo* gene therapy, cells are removed from the patient, genetically modified in some way in the laboratory, and returned to the patient. This allows just the appropriate cells to be treated, and they can be checked to make sure they have the correct genetic modification before they are returned to the patient. For many tissues, this is not possible and the cells must be modified within the patient's body (*in vivo* gene therapy).

Integration of therapeutic genes into host chromosomes has significant advantages but raises major safety concerns

For achieving long-term gene expression it would seem desirable to integrate a therapeutic gene into a chromosome of the host cell—preferably a stem cell. Then the construct is replicated whenever the host cell or its daughters divide. Currently, certain virus vectors are used to ensure chromosomal integration, but integration typically occurs in a random or semi-random way so that inserts can be located at different sites in the genome in different cells of the patient. The local chromosomal environment can have unpredictable effects on expression of the construct—it may never be expressed, it may be expressed at an undesirably low level, or it may be expressed for a short time and then irreversibly silenced.

Lack of control over where virus vectors integrate within chromosomes carries very significant safety risks because integration may alter the expression of endogenous genes. The insertion point might be within the sequence of an endogenous gene, leading to insertional inactivation of that gene. The greatest worry is that insertion of a highly expressed construct may activate an adjacent oncogene, causing cancer in a manner similar to the activation of *MYC* in Burkitt lymphoma (see Figure 17.7). As we describe below, this has been a recurring issue when genes are inserted into the genomes of patients' cells, and it will remain so until genes can be inserted with high efficiency into specific safe locations within the genome, where they can be expressed without disturbing the functions of other genes.

Delivery systems in which the transferred DNA simply becomes an extrachromosomal episome within cells do not raise the same kind of safety concerns as integrating vectors. Their disadvantage is the limited duration of gene expression. If the target cells are actively dividing, the extrachromosomal gene will tend to be diluted out as the cell population grows. There is no possibility of achieving a permanent cure, and repeated treatments may be necessary. For some purposes, such as killing cancer cells or combating an acute infection, this is not a problem—long-term expression is not needed. Moreover, if something does go wrong, a non-inserted gene is self-limiting in a way that a gene inserted into a chromosome is not.

Viral vectors offer strong and sometimes long-term transgene expression, but many come with safety risks

Viruses are very efficient at delivering genes into cells. They attach to suitable host cells by recognizing and binding specific receptor proteins on the host cells. Some viruses infect a broad range of cells and are said to have a broad **tropism**. Other viruses have a narrow tropism because they bind to receptors expressed by only a few cell types. For example, herpes viruses are tropic for cells of the central nervous system. The natural tropism of viruses may be retained in vectors or genetically modified in some way, so as to target a particular tissue, for example.

Some enveloped viruses, such as HIV, enter cells by fusing with the host plasma membrane to release their genome and capsid (coat) proteins into the cytosol, but other enveloped viruses first bind to cell surface receptors and trigger receptor-mediated endocytosis, fusion-based transfer, or endocytosis-based transfer. Some viruses go on to access nuclear components of cells only after the nuclear envelope dissolves during mitosis. They are limited to infecting dividing cells. Other viruses have devised ways to transfer their genomes efficiently through nuclear membrane pores so that both dividing and non-dividing cells can be infected.

Different viral vector systems offer different advantages (Table 21.4). Viruses with large genomes can potentially accept proportionally large DNA inserts. Because the genome of a virus is constrained to be within a limited size range (to allow correct packaging into the viral protein coats) non-essential components of the viral genome must be eliminated to allow the insertion of a therapeutic gene. Packaging cell lines can be designed to contain stripped-down viral vectors containing therapeutic genes, and non-essential viral genes are added separately and can be expressed to provide *in trans* viral products required for the correct packaging of virus vectors within virus protein coats.

TABLE 21.4 VIRAL VECTORS OF USE IN MAMMALIAN GENE TRANSFECTION AND GENE THERAPY

Virus class	Viral genome	Cloning capacity	Interaction with host genome	Target cells	Transgene expression	Comments
γ-Retroviruses (oncoretroviruses)	ssRNA	7–8 kb	integrating	dividing cells only	long-lasting	moderate vector yield[a]; risk of activation of cellular oncogene
Lentiviruses	ssRNA; ~9 kb	up to 8 kb	integrating	dividing and non-dividing cells; tropism varies	long-lasting and high-level expression	high vector yield[a]; risk of oncogene activation
Adenoviruses	dsDNA; up to 38 kb	often 7.5 kb, but up to 35 kb for gutless vectors	non-integrating	dividing and non-dividing cells	transient but high-level expression	high vector yield[a]; immunogenicity can be a major problem
Adeno-associated viruses	ssDNA; 5 kb	<4.5 kb	non-integrating	dividing and non-dividing cells	high-level expression in medium to long term (year)	high vector yield[a]; small cloning capacity but immunogenicity is less significant than for adenovirus
Herpes simplex virus	dsDNA; 120–200 kb	>30 kb	non-integrating	central nervous system	potential for long-lasting expression	able to establish lifelong latent infections
Vaccinia virus	dsDNA; 130–280 kb	25 kb	non-integrating	dividing and non-dividing cells	transient	

ss, single-stranded; ds, double-stranded.

[a] High vector yield, 10^{13} transducing units/ml; moderate vector yield, 10^{10} transducing units/ml.

Some viruses that are used to make gene therapy vectors are naturally pathogenic, and some can potentially generate strong immune responses. For safety reasons, viral vectors are generally designed to be disabled so that they are *replication-defective*. However, replication-competent viruses have sometimes been used as therapeutic agents, notably oncolytic viruses for treating cancers. Where a virus is naturally immunogenic, the viral vectors are modified in an effort to reduce or eliminate the immunogenicity.

Retroviral vectors

Retroviruses—RNA viruses that possess a reverse transcriptase—deliver a nucleoprotein complex (*preintegration complex*) into the cytoplasm of infected cells. This complex reverse-transcribes the viral RNA genome and then integrates the resulting cDNA into a single, but random, site in a host cell's chromosome. Retroviruses offer high efficiencies of gene transfer but can be generated at titers that are not so high as some other viruses, and they can afford moderately high gene expression. Because they integrate into chromosomes, long-term stable transgene expression is possible, but uncontrolled chromosome integration constitutes a safety hazard because promoter/enhancer sequences in the recombinant DNA can inappropriately activate neighboring chromosomal genes.

γ-Retroviral vectors are derived from simple mouse and avian retroviruses that contain three transcription units: *gag*, *pol*, and *env* (**Figure 21.9**). In addition, a *cis*-acting RNA element, ψ, is important for packaging, being recognized by viral proteins that package the RNA into infectious particles. Because native γ-retroviruses transform cells, the vector systems need to be engineered to ensure that they can produce only permanently disabled viruses. γ-Retroviruses cannot get their genomes through nuclear pores and so infect dividing cells only. This limitation can, however, be turned to advantage in cancer treatment. Actively dividing cancer cells in a normally non-dividing tissue such as brain can be selectively infected and killed without major risk to the normal (non-dividing) cells.

Lentiviruses are complex retroviruses that have the useful attribute of infecting non-dividing as well as dividing cells, and can be produced in titers that are a hundredfold greater than is possible for γ-retroviruses. In addition to expressing late (post-replication) mRNAs from *gag*, *pol*, and *env* transcription units, six early

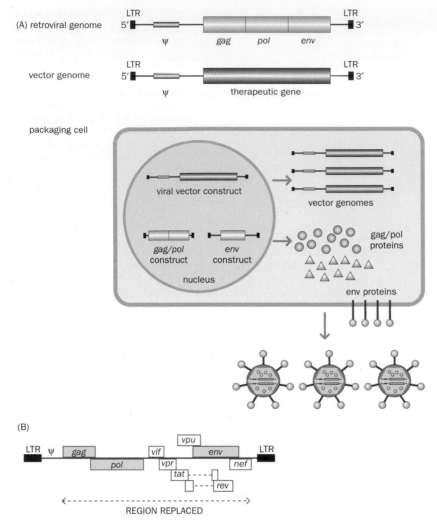

Figure 21.9 Simple and complex retroviral vectors. (A) A simple γ-retroviral genome contains three transcription units: *gag* (makes internal proteins), *pol* (makes reverse transcriptase and some other proteins), and *env* (makes viral envelope proteins), plus a ψ (psi) sequence that is recognized by viral proteins for assembly of the RNA into a virus particle. In vectors based on γ-retroviruses, *gag*, *pol*, and *env* are deleted and replaced by the therapeutic gene. The ψ sequence is retained. The packaging cell brings together viral vector recombinants and supplies the *gag*, *pol*, and *env* functions on physically separate molecules. The long terminal repeats (LTR) include promoter/ enhancer regions and sequences involved with integration. Recombinant viral genomes are packaged into infective but *replication-deficient* virus particles, which bud off from the cell and are recovered from the supernatant. (B) Lentiviruses, such as the human immunodeficiency virus (HIV), are complex retroviruses. In addition to a single-stranded RNA genome with flanking long terminal repeats (LTR) and characteristic *gag*, *pol*, and *env* genes, they possess a variety of other additional genes, such as *vif*, *vpr*, *vpu*, *tat*, *rev*, and *nef*, that encode proteins involved in regulating and processing of viral RNA and other replicative functions. In recombinant vectors, all of the genome except for the LTRs and ψ sequence can be replaced. [(A) adapted from Somia N & Verma IM (2000) *Nat. Rev. Genet.* 1, 91–99. With permission from Macmillan Publishers Ltd.]

viral proteins are produced before replication of the virus. The early proteins include two regulatory proteins, tat and rev, that bind specific sequences in the viral genome and are essential for viral replication. Like other retroviruses, they allow long-term gene expression. Results with marker genes have been promising, showing prolonged *in vivo* expression in muscle, liver, and neuronal tissue. Most lentiviral vectors are based on HIV, the human immunodeficiency virus, and much work has been devoted to eliminating unnecessary genes from the complex HIV genome and generating safe packaging lines while retaining the ability to infect non-dividing cells. Lentiviruses appear to have a safer chromosome integration profile than γ-retroviruses; self-inactivating lentivirus vectors provide an additional layer of safety.

Adenoviral and adeno-associated virus (AAV) vectors

Adenoviruses are DNA viruses that cause benign infections of the upper respiratory tract in humans. As with retroviral vectors, adenoviral vectors are disabled and rely on a packaging cell to provide vital functions. The adenovirus genome is relatively large, and gutless adenoviral vectors, which retain only the inverted terminal repeats and packaging sequence, can accommodate up to 35 kb of therapeutic DNA (**Figure 21.10**).

Adenovirus virus vectors can be produced at much higher titers than γ-retroviruses and so transgenes can be highly expressed. They can also efficiently transduce both dividing and non-dividing cells. A big disadvantage is their immunogenicity. Even though a live replication-competent adenovirus vaccine has been safely administered to several million US army recruits over several decades (for protection against natural adenoviral infections), unwanted immune reactions have been a problem in several gene therapy trials, as described below. Moreover, because these vectors are non-integrating, gene expression is short-

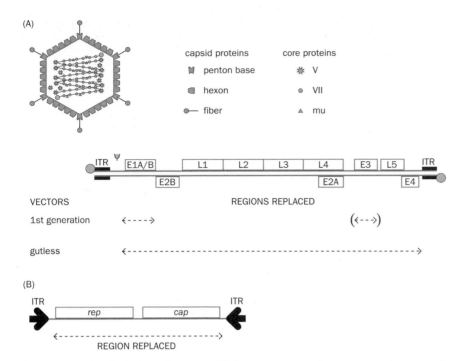

Figure 21.10 Adenovirus and adeno-associated virus as vehicles for gene delivery into mammalian cells. (A) Upper image: adenovirus structure. The adenovirus genome consists of a 36 to 38 kb double-stranded linear DNA that has a covalently attached protein molecule at the 5′ end (filled green circles). With various other core proteins, the adenoviral genome is encased in a capsid consisting of three types of capsid protein as shown (plus three types of cement protein, not shown). Lower image: adenoviral genome. The ends of the genome have inverted terminal repeats (ITRs) to which the terminal protein is covalently attached. Boxes above and below the double blue lines indicate transcription units transcribed from the opposing DNA strands. The early transcription units E1 to E4 encode mostly regulatory proteins; the late transcription units L1 to L5 are focused on producing structural proteins for packaging the genome. In the first series of adenovirus vectors, early transcription units, notably E1, were eliminated and could be replaced by insert DNA, but the more recent gutless vectors can replace all except the terminal repeats and the ψ sequence. (B) Adeno-associated virus (AAV) has a small, very simple 4.7 kb single-stranded DNA genome with terminal inverted repeats (ITRs), and just two open reading frames (ORFs). The *rep* ORF encodes various proteins required for the AAV life cycle, and the *cap* ORF makes the three capsid proteins. In a recombinant AAV vector, the *rep* and *cap* ORFs are replaced by a desired transgene. The functions necessary for production of virus particles (including *rep*) are provided by a packaging cell.

term; repeated administration would be necessary for sustained expression, which could only exacerbate the immune response.

Adeno-associated viruses (AAVs) are nonpathogenic single-stranded DNA viruses. They are completely unrelated to adenoviruses; their name comes from their reliance on co-infection by an adenovirus (or herpes) helper virus for replication. The genome contains only two genes: the *rep* gene makes proteins that control viral replication, structural gene expression, and chromosome integration; the *cap* gene encodes capsid structural proteins. Multiple different serotypes of AAV have been isolated and some have usefully narrow tropism. For example, the AAV9 variant is highly tropic for the spinal cord and brain astrocytes, and may be useful in future treatments of spinal cord injuries. Another important advantage of AAV vectors is that they can permit robust *in vivo* expression of transgenes in various tissues over several years while exhibiting little immunogenicity and little or no toxicity or inflammation.

Unmodified human AAV integrates into chromosomal DNA at a specific site on 19q13.3–qter. This highly desirable property would provide long-term expression without the safety risks of random insertional mutagenesis. Unfortunately, the specificity of integration is provided by the *rep* protein, and because the *rep* gene needs to be deleted in the constructs used for gene transfer, chromosomal integration of AAV vectors occurs randomly (AAV is disadvantaged by having a small genome; even vectors in which 96% of the AAV genome has been deleted accept inserts with a maximum size of just 4.5 kb).

Other viral vectors

Herpes simplex viruses are complex viruses containing at least 80 genes; they are tropic for the central nervous system (CNS). They can establish lifelong latent infections in sensory ganglia, in which they exist as non-integrated extrachromosomal elements. The latency mechanism might be exploited to allow the long-term expression of transferred genes, in the hope that they will spread through a synaptic network. Their major applications would be in delivering genes into neurons for the treatment of diseases such as Parkinson disease and CNS tumors. Vectors have a high insert capacity. Practical vectors are still at an early stage of development.

Modified vaccinia virus Ankara (MVA) is a highly attenuated vaccinia virus that has been used safely as a smallpox vaccine. Infection with MVA results in rapid replication of viral DNA, but it is largely non-propagative in human and mammalian cells. Replication-competent MVA vectors have been constructed and used largely for transferring suicide genes to kill tumor cells in cancer gene therapy.

Non-viral vector systems are safer, but gene transfer is less efficient and transgene expression is often relatively weak

In the laboratory, it is relatively easy to get foreign DNA into cells, and some of the methods have potential in gene therapy if the safety problems of viral systems prove intractable. However, whereas many viruses are able to insert their genomes through nuclear pores, transfers using non-viral delivery methods are generally much less efficient and they are often characterized by poor transfer rates and a low level of transgene expression.

Transport of plasmid DNA to the nucleus of non-dividing cells is very inefficient because the plasmid DNA often cannot enter nuclear membrane pores. Various methods can be used to facilitate nuclear entry such as conjugating specific DNA sequences or protein sequences (nuclear localization sequences) that are known to facilitate nuclear entry, or compacting the DNA to a small enough size to pass through the nuclear pores.

Although transfected DNA can not be stably integrated into the chromosomes of the host cell, this may be less of a problem in tissues such as muscle that do not regularly proliferate, and in which the injected DNA may continue to be expressed for many months.

Transfer of naked nucleic acid by direct injection or particle bombardment

In some cases, naked DNA has been injected directly with a syringe and needle into a target tissue such as muscle. An alternative direct injection approach uses particle bombardment (biolistic or gene gun) techniques: DNA is coated on to metal pellets and fired from a special gun into cells. Successful gene transfer into several different tissues has been obtained with this simple and comparatively safe method. However, gene transfer rates are low and transgene expression is generally weak.

Lipid-mediated gene transfer

Liposomes are synthetic vesicles that form spontaneously when certain lipids are mixed in aqueous solution. Phospholipids, for example, can form bilayered vesicles that mimic the structure of biological membranes, with the hydrophilic phosphate groups on the exterior and the hydrophobic lipid tails in the interior. Cationic liposomes have been the most commonly used in gene transfer experiments. The lipid coating allows the DNA to survive *in vivo*, bind to cells, and be endocytosed into the cells (**Figure 21.11**). Unlike viral vectors, the DNA–lipid complexes are easy to prepare and there is no limit to the size of DNA that is transferred. However, the efficiency of gene transfer is low, with comparatively weak transgene expression. Because the introduced DNA is not designed to integrate into chromosomal DNA, transgene expression may not be long-lasting.

As discussed in Section 21.5, short interfering RNA (siRNA) is often complexed with cholesterol for delivery *in vivo*.

Compacted DNA nanoparticles

Because of its phosphate groups, DNA is a polyanion. Polycations bind strongly to DNA and so cause the DNA to be significantly compacted. To form DNA nanoparticles, DNA is complexed with a polyethylene glycol (PEG)-substituted poly-L-lysine known as PEG-CK30 (because it contains 30 lysine residues and an N-terminal cysteine to which the PEG is covalently bound). Within this complex, the DNA forms a very condensed structure. Because of their much reduced size, compacted DNA nanoparticles are comparatively efficient at transferring genes to dividing and non-dividing cells and have a plasmid capacity of at least 20 kb.

21.5 RNA AND OLIGONUCLEOTIDE THERAPEUTICS AND THERAPEUTIC GENE REPAIR

For many gene therapy protocols, as outlined in the previous section, the object is simply to transfer genes into suitable target cells so that they are expressed at high levels. In this section we are mostly concerned with quite different gene therapy strategies in which the object is not to supply a missing gene product or

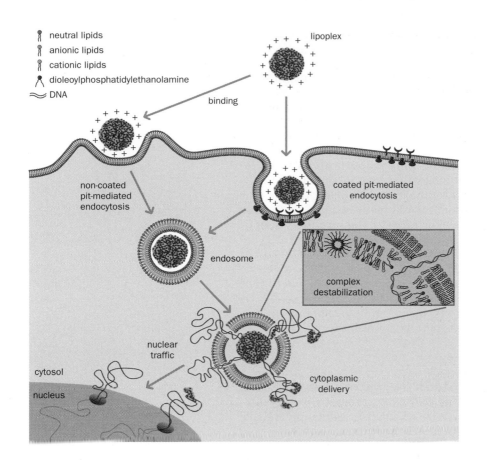

Figure 21.11 Cationic liposomes as vectors for gene delivery into mammalian cells. The gene to be transferred is complexed with cationic liposomes to form *lipoplexes* that can interact with cell membranes. The lipoplexes are taken up by cells through different endocytosis pathways in which the cell membrane evaginates to form a pit. Large lipoplexes are taken up by pits coated with clathrin complexes (top center); small lipoplexes are taken up by non-coated pits (top left). In either case the lipoplexes become trapped in endosomes that are targeted for destruction by lysosomes, where the DNA would be degraded if it were unable to escape. The inclusion within the liposomes of certain helper lipids—often electrically neutral lipids—helps to destabilize the endosomal membranes, causing the passenger DNA to escape to the cytoplasm. For the DNA to be transcribed it must pass to the nucleus. In dividing cells, the breakdown of the nuclear envelope during mitosis allows the DNA to gain access to the nucleus, but in non-dividing cells the precise mechanism of entry into the nucleus is unclear. [From Simões S, Filipe A, Faneca H et al. (2005) *Expert Opin. Drug Deliv.* 2, 237–254. Taylor & Francis Ltd.]

to kill harmful cells, but instead to counteract the harmful effects of genes within cells. Some conditions are caused by *gain-of-function* mutations (Chapter 13, p. 428) or a *dominant-negative effect* (Chapter 13, p. 431), and the problem is a resident gene that is doing something positively harmful. Diseases that might benefit from such strategies could include cancers that arise through activated oncogenes, dominant Mendelian conditions other than those caused by loss of gene function, and autoimmune diseases. Infectious diseases might also be treated by targeting a pathogen-specific gene or gene product.

In principle, two groups of strategies could be used to counteract a gene's harmful effects. One way is to specifically block expression of the harmful gene by downregulating transcription, by destroying the transcript, or by inhibiting the protein product. In Section 21.2 we outlined various methods that inhibit gene expression at the protein level; here, we will consider methods that inhibit or cleave specific RNA transcripts. A second type of strategy seeks to restore the normal gene function in some way. Either the DNA sequence is altered to correct a genetic defect, or alternative gene splicing is artificially induced so as to bypass a causative mutation at the RNA level.

Whatever method is used, some sort of agent or construct must be delivered to a target cell and made to function there. In that respect, the problems of efficient delivery and expression are similar to those described in the previous section. In the case of dominant diseases or activated oncogenes, there is the additional problem of designing an agent that will selectively attack the mutant allele without affecting the normal allele.

Therapeutic RNAs and oligonucleotides are often designed to selectively inactivate a mutant allele

Antisense RNA therapy, in which an antisense RNA is designed to base-pair with and selectively inhibit a target RNA, has been used in various clinical trials but the results have been mixed, and often disappointing. Introduced RNA is prone to attack by nucleases so that by the time the antisense RNA had reached its target tissue there was often little intact RNA. More stable antisense oligodeoxynucleotides were used, but successful knockdowns could not be guaranteed. To ensure that sufficient intact antisense oligonucleotides (AOs) arrived at the target tissue, clinical applications focused on applying the AOs directly to diseased tissue. In 1998, Vitravene® (fomivirsen) became the first FDA-approved therapeutic AO; it is applied directly to the eyes to treat cytomegalovirus retinitis in AIDS patients. The cornea (and anterior chamber of the eye) is an immunologically privileged site and so a high concentration of reagents can be administered directly by injection.

Subsequent technological developments produced a second generation of AOs based on nucleic acid analogs that are much more chemically stable than normal oligonucleotides, such as 2-O-methyl phosphorothioate oligonucleotides, morpholino phosphorothioate oligonucleotides, and locked nucleic acids (**Figure 21.12**). As we saw in Section 20.3, morpholino oligonucleotides are now widely used to knock down genes in some model organisms, but there have been few therapeutic applications that exploit morpholinos or other second-generation AOs to block gene expression. Although second-generation AOs are chemically stable, off-target effects can occur when the AO inhibits RNAs other than the target.

In addition, as for any other oligonucleotide, efficient delivery of AOs to cells has been a significant problem. In an effort to increase the *in vivo* uptake of AOs by cells, some second-generation AOs, such as the morpholino oligonucleotides, have been modified by conjugating peptides to them; however, this can sometimes provoke immune reactions.

Therapeutic ribozymes

A separate wave of RNA therapeutics focused on RNA enzymes (**ribozymes**), such as the plant hammerhead ribozyme, that naturally cleave other RNA target molecules. Genetically modified ribozymes have been made in which a catalytic, RNA-cleaving domain from a natural ribozyme is fused to an antisense RNA sequence designed to bind to transcripts from a specific target gene of interest. The modified ribozyme would bind to the transcripts of the target gene and cleave them. Ribozymes have, however, not been so effective in practice: clinical trials to treat cancer and other diseases have had rather disappointing outcomes.

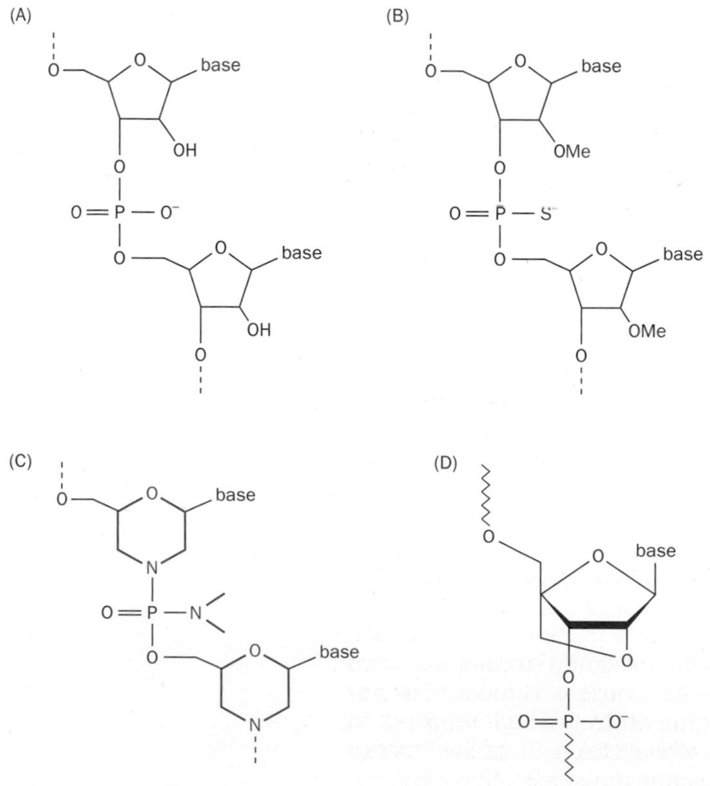

Figure 21.12 Second-generation antisense oligonucleotides. (A) The normal ribose phosphate backbone of RNA. (B) Structure of a 2-O-methyl phosphorothioate oligonucleotide. (C) Structure of a morpholino phosphorothioate oligonucleotide. (D) Structure of an oligonucleotide known as a locked nucleic acid.

Therapeutic siRNA

In the 1990s, the discovery of RNA interference (RNAi) transformed the potential for specific inhibition of gene expression and provided a stimulus to the development of RNA therapeutics. Experimentally, the object is to produce a *double-stranded* RNA with a base sequence corresponding to a transcribed part of the target gene, and so provoke natural cellular defense pathways to destroy transcripts of the specific gene. Long double-stranded RNA cannot be used for this purpose in human (or other mammalian) cells because it provokes interferon responses that result in *nonspecific* inhibition of gene expression (see Chapter 12, p. 391). However, short double-stranded RNA is highly efficient at inducing knockdown of the expression of *specific* genes.

Performing RNAi in mammalian cells often involves transfecting transgenes that express a short hairpin RNA. After transfection into cells, the expressed RNA is cleaved by the cytoplasmic enzyme dicer to give a double-stranded siRNA about 21–23 nucleotides long (see Figures 12.3 and 12.4). Other RNAi therapies deliver just the naked siRNA.

siRNA technology is highly efficient at gene knockdown in mammalian cells and is transforming our understanding of human gene function. There is therefore considerable excitement about its therapeutic potential. Promising therapeutic targets would include viral infections, cancers (targeting oncogenes), and neurodegenerative disease (targeting harmful mutant alleles). The method has been used with some success in treating various animal models of disease, but delivery is comparatively inefficient because of the reliance on non-viral vector systems.

Some early clinical trials using siRNA have focused on the immunologically privileged eye. One principal target is the *VEGF* gene that underlies macular degeneration (as described in the section on therapeutic aptamers on page 686) and has involved simply injecting naked siRNA. Applying RNAi to target diseases with less localized pathology is a much more formidable task. However, systemic and reasonably effective delivery has been possible in animal models by attaching siRNAs to cholesterol, to ferry the siRNA through the bloodstream.

Antisense oligonucleotides can induce exon skipping to bypass a harmful mutation

Antisense therapeutics is not always devoted to blocking the expression of a harmful gene. Sometimes the objective is quite different: to restore function to the mutated gene. AOs have been used to induce altered gene splicing so as to bypass a causative mutation. The mutation is typically a nonsense or frameshifting mutation in a coding DNA exon, which leads to the loss of functionally important downstream sequences.

Therapeutic AOs are normally designed to bind to relevant exon–intron junctions in the pre-mRNA; blocking of splicing at that junction may induce skipping of an adjacent exon containing the harmful mutation. For the strategy to be useful, the skipped exon(s) must not contain sequences that are vitally important for the function of the gene product, and exon skipping must not result in a shift in the translational reading frame. Thus, if a coding DNA exon with a harmful mutation contains a number of nucleotides that is not exactly divisible by three, it may be necessary to also induce skipping of neighboring exons to maintain the reading frame.

Antisense-mediated exon skipping strategies have been particularly well developed for Duchenne muscular dystrophy (DMD), a severe disorder that is typically caused by frameshifting or nonsense mutations in the giant dystrophin gene. Many of the causative mutations occur in internal exons specifying functionally non-essential internal protein sequence. After a proof of principle had been shown by using cultured cells, antisense-mediated exon skipping was used very successfully to restore dystrophin function in the mouse *mdx* model that has a nonsense mutation in exon 23. After encouraging early findings reported in 2007, clinical trials are now underway that seek to skip exon 51 of the human gene, the exon with the most reported DMD mutations (**Figure 21.13**).

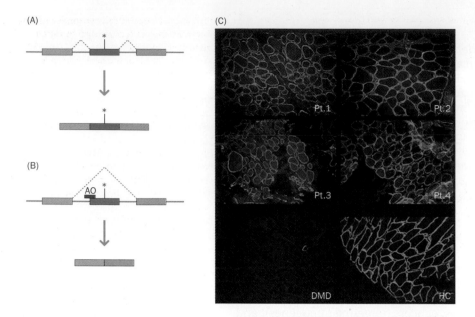

Figure 21.13 Antisense-mediated exon skipping can restore gene function. (A) In normal splicing (dashed blue lines), an exon containing a harmful mutation (red asterisk) is included in the coding sequence. (B) An antisense oligonucleotide (AO) is designed to base-pair with a splicing regulatory sequence, in this case the splice acceptor region of the intron preceding the exon that carries the harmful mutation. By shielding the splicing sequence from the spliceosomal machinery, the AO provokes exon skipping with the result that the harmful mutation is bypassed. (C) Antisense oligonucleotide PRO051 induces skipping of exon 51 of the dystrophin gene without causing a frameshift in the translational reading frame. After intramuscular injection with PRO051 four Duchenne muscular dystrophy (DMD) patients with a causative mutation in exon 51 (Pt.1 to Pt.4) show evidence of restoration of dystrophin in muscle fibers, as revealed by immunofluorescence analysis with a dystrophin-specific antibody. Bottom panels represent controls representing an untreated DMD patient and a healthy control (HC). [(B) from van Deutekom JC, Janson AA, Ginjaar IB et al. (2007) *N. Engl. J. Med.* 357, 2677–2686. With permission from the Massachusetts Medical Society.]

Gene targeting with zinc finger nucleases can repair a specific pathogenic mutation or specifically inactivate a target gene

Homologous recombination can replace one sequence with a closely related one; it is routinely used to replace endogenous sequences by exogenous sequences to selectively inactivate a target gene within intact cells (see Figure 20.6). Cellular homologous recombination pathways also repair double-strand DNA breaks (see Box 13.3). The possibility of using homologous recombination to repair mutant alleles within cells seemed remote, however, because it occurs at a very low frequency in mammalian cells (about one event per 10^6 cells). A new class of genetically engineered nucleases known as zinc finger nucleases has transformed the picture.

Zinc finger nucleases are effectively synthetic restriction nucleases that are designed to recognize long target sequences and cut at a *single* site in a complex genome. As explained in Chapter 20 (p. 654), the specificity comes from a series of zinc finger DNA-binding motifs each of which can bind to a specific triplet sequence in DNA. By cutting and ligating DNA sequences at the DNA level, it is possible to make a DNA construct that will be expressed to give four consecutive zinc fingers of a particular type that can bind to a chosen 12-nucleotide sequence. Attached to the zinc fingers is a DNA-cleaving domain that originates from the restriction endonuclease *Fok*I and which, as a dimer, cleaves double-stranded DNA. If two different zinc finger nucleases are designed to bind to two very closely positioned 12-nucleotide recognition sequences at non-overlapping sites within a target locus, the effective recognition sequence is 24 nucleotides long and may occur only once in the entire genome. The two DNA-cleaving domains come together to form a dimer and make just one double-strand break (**Figure 21.14**).

The occurrence of a double-strand break triggers cellular double-strand DNA repair pathways. One pathway involves homologous recombination; when activated, the frequency of homologous recombination is greatly increased (from 100-fold to 10,000-fold). The homologous recombination (HR) pathway trims back the DNA sequence at the cut site, and new DNA is copied from a donor strand of *homologous* DNA. If the original DNA segment that is repaired contained a pathogenic mutation, the HR pathway can use a donor wild-type homologous sequence to repair the mutation and restore gene function (see Figure 21.14). The donor sequence may come from a normal allele, but the repair is facilitated by introducing a plasmid with the correct sequence (see Figure 21.14).

The utility of this approach has been demonstrated in cellular models. For example, a pathogenic mutation in *IL2RG*, a locus for X-linked severe combined immunodeficiency, has been corrected with high efficiency in cultured human cells. Specific gene inactivation induced by zinc finger nucleases (ZFNs) can also

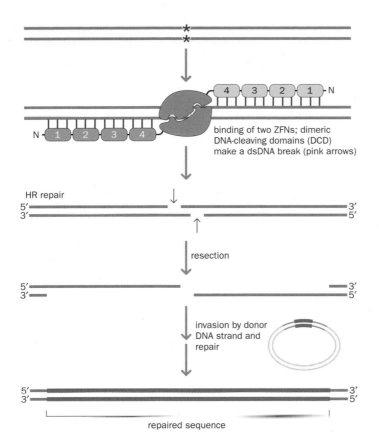

Figure 21.14 Genetically engineered zinc finger nucleases can be used to repair genes within cells. See Chapter 20 p. 654 and Figure 20.13 for the background to zinc finger nucleases. A pair of zinc finger nucleases (ZFNs) each containing four zinc fingers can be designed to specifically bind within a cell to a target sequence that includes a defective mutation (red asterisks) within a mutant gene. The two DNA cleaving domains (shown in green) work together to make a double-stranded DNA break (DSB) that activates cellular pathways that naturally repair such breaks. In the homologous recombination (HR) DNA repair pathway, the 5′ ends of the DSB are first trimmed back (resection), allowing strand invasion by endogenous or exogenous donor DNA strands that are used as templates for synthesizing new DNA to replace the previously existing sequence. A plasmid with a relevant segment of the normal gene sequence (bold blue lines) may be provided as a donor DNA. By acting as a template for synthesizing new DNA from a *portion* of the homologous sequence (thick blue lines), it facilitates replacement of the pathogenic point mutation by the normal sequence.

be used potentially to counteract harmful genes. For example, ZFN-induced gene inactivation has been used to make CD4+ helper T cells resistant to HIV-1 infection. HIV-1 is able to infect helper T cells by binding to the CCR5 chemokine receptor normally expressed by this type of cell. However, some individuals who are naturally resistant to HIV-1 infection are homozygous for a naturally occurring 32 bp deletion in the *CCR5* gene that renders it inactive without clinical consequences. By using ZFN-mediated gene inactivation it has been possible to inactivate the *CCR5* gene, leading to the hope that immunity to HIV-1 can be conferred on individuals.

21.6 GENE THERAPY IN PRACTICE

Expectations about gene therapy have followed a cyclical course over the past quarter of a century or so; periods of overoptimism were followed by bouts of excessive pessimism in response to significant setbacks (**Box 21.8**). Perhaps one cause of these exaggerated reactions is confusion over the natural time-scales of such work. Because diagnostic testing can often start within weeks of a gene being cloned, therapy could be expected to be not far behind, whereas really this is drug development and runs on a time-scale of decades.

Although practical gene therapy is a long-haul process, many academic and commercial laboratories are working hard in this area. By 2009 about 1500 trial protocols related to gene therapy had been approved (**Figure 21.15** gives a breakdown). However, only 3% of these are phase III trials, in which the efficacy of the therapy is tested on a large scale. Of the 1500 or so approved gene therapy protocols, nearly two-thirds have been for cancer; monogenic disease, infectious diseases, and cardiovascular diseases accounted for less than 10% each. We give brief overviews below of the progress of gene therapy in several important areas.

Despite the limited number of trials, monogenic diseases have always been high on the gene therapy agenda, and the first definitive successes have been in that area. In particular, recessive disorders, in which the problem arises because of deficiency of a single gene product, seemed comparatively easy targets; even an overall low-level expression of introduced genes might produce some clinical

BOX 21.8 SOME OF THE MAJOR UPS AND DOWNS IN THE PRACTICE OF GENE THERAPY

1990–	commencement of the first clinical trial involving gene therapy, using recombinant γ-retroviruses to overcome an autosomal recessive form of severe combined immunodeficiency (SCID) due to deficiency of adenosine deaminase (ADA); the trial was hailed as a success but patients had also been treated in parallel with standard enzyme replacement using polyethylene glycol (PEG)-ADA
1999	death of Jesse Gelsinger in September 1999, just a few days after receiving recombinant adenoviral particles by intra-hepatic injection in a clinical trial of gene therapy for ornithine transcarbamylase deficiency; a massive immune response to the adenovirus particles resulted in multiple organ failure
2000–	the first unambiguous gene therapy success involved using recombinant γ-retroviruses to treat an X-linked form of SCID (see Figure 21.16); however, several of the treated children went on to develop leukemia, with one death, as a result of insertional activation of cellular oncogenes (see the main text)
2006–	transient success in using γ-retroviral gene therapy to treat two adult patients with chronic granulomatous disease, a recessive disorder that affects phagocyte function causing immunodeficiency; despite initial success, both patients went on to show transgene silencing, and also myelodysplasia as a result of insertional activation of cellular genes (see the main text); 27 months after gene therapy one patient died from overwhelming sepsis; see Ott et al. (2006) *Nat. Med.* 12, 401–409; PMID 16582916 and Stein et al. (2010) *Nat. Med.* 16, 198–204; PMID 20098431
2006	report of limited success in gene therapy for hemophilia B: AAV2 vectors were used successfully to transduce hepatocytes and express a Factor IX transgene; however, an immune response led to destruction of the transduced cells; see Manno et al. *Nat. Med.* 12, 342–347; PMID 16474400
2009	successful gene therapy for ADA deficiency reported by Aiuti et al. *N. Engl. J. Med.* 360, 447–458; PMID 19179314
2009	first report of successful gene therapy for a central nervous system disorder, X-linked adrenoleukodystrophy, and the first using a lentiviral vector; see Cartier et al. *Science* 326, 818–823; PMID 19892975
2009	report of successful *in vivo* gene therapy using retinal injection of recombinant AAV to treat Leber congenital amaurosis, a form of childhood blindness; the significant post-treatment gain in vision was found to be maintained after one year; see Cideciyan et al. *N. Engl. J. Med.* 361, 725–727; PMID 19675341

PMID, PubMed identifier number.

benefits. By contrast, dominant disorders—in which one of the two alleles is normal and makes a gene product—present much greater challenges. Getting the correct gene dosage to treat haploinsufficiency could be problematic, and counteracting the effect of a mutant allele that makes a harmful product would require very efficient and selective silencing of the mutant allele, while leaving the normal allele unaffected.

The first gene therapy successes involved recessively inherited blood cell disorders

Blood cells are highly accessible and are suited to *ex vivo* gene therapy—the cells can be genetically modified in culture, and carefully screened before selecting and expanding suitably modified cells and returning them to the patient (see

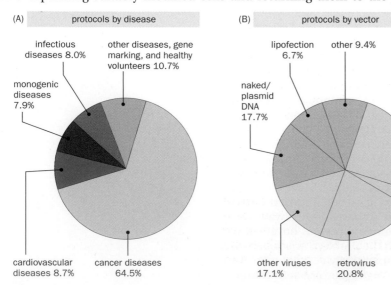

(A)   protocols by disease

infectious diseases 8.0%
other diseases, gene marking, and healthy volunteers 10.7%
monogenic diseases 7.9%
cardiovascular diseases 8.7%
cancer diseases 64.5%

(B)   protocols by vector

lipofection 6.7%
other 9.4%
adenovirus 23.9%
naked/plasmid DNA 17.7%
other viruses 17.1%
retrovirus 20.8%
AAV 4.4%

Figure 21.15 Approved gene therapy trial protocols. (A) Distribution by disease. (B) Distribution by gene delivery system. AAV, adeno-associated virus. The figures include all approved protocols (*n* =1481) for completed, ongoing, or pending trials listed by the end of 2008 in the Wiley database of worldwide gene therapy clinical trials (http://www.wiley.co.uk/genmed/clinical/).

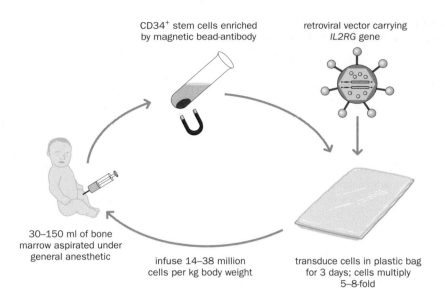

CD34$^+$ stem cells enriched by magnetic bead-antibody

retroviral vector carrying *IL2RG* gene

30–150 ml of bone marrow aspirated under general anesthetic

infuse 14–38 million cells per kg body weight

transduce cells in plastic bag for 3 days; cells multiply 5–8-fold

Figure 21.16 *Ex vivo* **gene therapy for X-linked severe combined immunodeficiency disease (X-SCID).** Bone marrow was removed from the patient and antibody affinity was used to enrich for cells expressing the CD34 antigen, a marker of hematopoietic stem cells. To do this, bone marrow cells were mixed with paramagnetic beads coated with a CD34-specific monoclonal antibody; beads containing bound cells were removed by using a magnet. The transduced stem cells were expanded in culture before being returned to the patient. For details, see Cavazzana-Calvo M et al. (2000) *Science* 288, 669–672; PMID 10784449 and Hacein-Bey-Abina S et al. (2002) *N. Engl. J. Med.* 346, 1185–1193; PMID 11961146.

Figure 21.8). Disorders resulting from recessively inherited defects in white blood cell function were among the first targets for gene therapy. White blood cells have key roles in the immune system (see Tables 4.7 and 4.8), and defects in their function can give rise to severe immunodeficiencies.

The first definitive gene therapy success came in treating severe combined immunodeficiency (SCID), in which both B and T lymphocyte function is defective. Patients are extremely vulnerable to infectious disease, and in some cases they have been obliged to live in a sterile environment. The most common form of SCID is due to inactivating mutations in the X-linked *IL2RG* gene that makes the γ_c common gamma subunit for multiple interleukin receptors, including interleukin receptor 2. Lymphocytes use interleukins as cytokines to signal to each other and to other immune system cells, and so lack of the γ_c cytokine receptor subunit had devastating effects on lymphocyte and immune system function. Another common form of SCID is due to adenosine deaminase (ADA) deficiency; the resulting build-up of toxic purine metabolites kills T cells. Because T cells regulate B-cell activity, both T- and B-cell function is affected.

SCID gene therapy has involved *ex vivo* γ-retroviral transfer of *IL2RG* or *ADA* coding sequences into autologous patient cells (see Box 21.8). To aid the chances of success, bone marrow cells were used because they are comparatively enriched in hematopoietic stem cells, the cells that give rise to all blood cells (see Figure 4.17); a further refinement was to select for cells expressing CD34, a marker of hematopoietic stem cells (**Figure 21.16**). By 2008 Fischer et al. reported that 17 out of 20 X-linked SCID patients and 11 out of 11 ADA-deficient SCID patients had been successfully treated and retained a functional immune system (for more than 9 years after treatment in the earliest patients).

With the use of retrovirus vectors, genes could be inserted into chromosomes; the therapeutic DNA would be stably transmitted after cell division, giving long-lasting transgene expression. This clear advantage was eroded by the lack of control over where the transgenes integrated. In one clinical trial for X-linked SCID, four patients developed T-acute lymphoblastic leukemia after retroviral integration because promoter/enhancer sequences in the inserted DNA inappropriately activated a neighboring *LMO2* proto-oncogene. Activation of *LMO2* is now known to promote the self-renewal of pre-leukemic thymocytes, so that committed T cells accumulate additional genetic mutations required for leukemic transformation.

Similar protocols were applied to treating patients with an X-linked form of chronic granulomatous disease (CGD). Patients with CGD have immunodeficiency arising from mutations in any of four genes that encode subunits of the NADPH oxidase complex. NADPH oxidase is involved in making free radicals and other toxic small molecules that phagocytes use to kill the microbes that they engulf; a defective NADPH oxidase results in defective phagocyte function.

Retroviral transfer of a suitable transgene restored functional NADPH oxidase, and the treatment was transiently successful but subsequently transgenes were silenced after insertional activation of cellular proto-oncogenes. In this case, gene-transduced hematopoiesis in both patients was dominated by cell clones containing integrations in the *MDS-EVI1* locus and resulted in overexpression of these proto-oncogenes. The neighboring inserted transgene was silenced by promoter methylation; see the paper by Stein et al. (2010) in the 2006 entry in Box 21.8.

Despite the above successes, some other recessively inherited blood disorders have proven to be less straightforward to treat. The limited success with hemophilia B (see Box 21.8) was derived from unexpected immune responses, but it is now realized that AAV is surprisingly prevalent, existing in perhaps 40% of all human livers; testing for preexisting AAV antibodies might have identified suitable patients for therapy. Gene therapy for the most prevalent blood disorders, the thalassemias, is a formidable challenge because of the tight regulation of globin gene expression that is designed to maintain a 1:1 production ratio of α-globin to β-globin.

Gene therapies for many other monogenic disorders have usually had limited success

Other than recessively inherited blood cell disorders, there have been few gene therapy successes. One preliminary success has involved an X-linked form of adrenoleukodystrophy caused by inactivating mutations in *ABCD1* (see Box 21.8). The gene defect results in the accumulation of saturated very long-chain fatty acids, leading to demyelination in oligodendrocytes and microglia, and consequent dysfunction of the nervous system. Even here, however, the successful therapy was based on lentiviral-mediated gene transfer into autologous hematopoietic stem cells. The transduced stem cells gave rise to myelomonocytic cells that migrated into the central nervous system to replace diseased microglial cells and relieve the lipid storage.

Two other promising targets are disorders of the eye and skin, both of which are highly accessible. The eye is a small and enclosed target organ, and because it is an immunologically privileged site immune responses tend to be weak. Recent advances include a report in 2009 of successful AAV-mediated gene transfer by retinal injection to treat Leber congenital amaurosis (Box 21.8).

Despite the above successes, gene therapy for many single-gene disorders has proved to be challenging, as in the case of *in vivo* gene therapies for cystic fibrosis (CF) and Duchenne muscular dystrophy (DMD). CF is caused by lack of the *CFTR*-encoded chloride channel, mainly in airway epithelium. Studies of patients with partly active *CFTR* alleles suggest that 5–10% of the normal level would be sufficient to produce a good clinical response, but numerous clinical gene therapy trials have been unsuccessful.

A significant problem in CF gene therapy is the physical barrier of mucus and polysaccharide that covers lung airway epithelial cells, especially in the infected lungs of patients with CF. Gene therapy agents can be delivered into the airways, but more sophisticated vehicles will be necessary to allow efficient transduction of epithelial cells. Ideally, stem cells should be targeted, because surface epithelial cells have a life span of only about 120 days, so that repeated administration will be necessary, with all the attendant problems of the immune response.

Patients with DMD suffer progressive wasting of first skeletal and then cardiac muscle. Studies of female DMD carriers and patients with milder Becker muscular dystrophy show that restoring about 20% of normal dystrophin gene expression in muscle would benefit DMD patients. However, the dystrophin coding sequence (14 kb) is too large for many vectors, and there is the challenge of getting efficient gene delivery into both skeletal and cardiac muscle cells. The size problem has been partly addressed by deleting less important coding sequences that encode a central domain in dystrophin to produce much smaller cDNAs, but efficient delivery remains a problem.

The prospects of antisense-mediated exon skipping may be good. The current clinical trials seek to skip exon 51 of the *DMD* gene by using an intramuscular injection of antisense oligonucleotides (see Figure 21.13C) and are applicable to just a small but significant percentage of DMD patients. However, another strategy under development seeks to use a cocktail of different antisense oligonucleotides to block splicing of each of exons 45–55; this would be applicable to most DMD patients. The loss of 11 consecutive exons does not change the reading frame and is associated with a mild phenotype.

Restoration of dystrophin gene function has also been possible by using autologous blood-derived progenitor cells expressing the CD133 antigen, which are known to be able to regenerate dystrophic fibers. The prospects of autologous stem cell-mediated restoration of gene function in different recessive disorders has been boosted by the comparative ease of preparing patient-specific induced pluripotent stem cells (see Section 21.3), and encouraging results are being obtained in animal models.

Cancer gene therapies usually involve selective killing of cancer cells, but tumors can grow again by proliferation of surviving cells

Nearly two-thirds of all approved gene therapy trial protocols have been for cancer (see Figure 21.15A). **Table 21.5** gives some examples, chosen to illustrate the range of approaches. They include:

- Gene addition to restore tumor suppressor gene function (e.g. *TP53* or *BRCA1*)
- Gene inactivation to prevent expression of an activated oncogene (e.g. *ERBB2*)
- Genetic manipulation of tumor cells to trigger apoptosis
- Modification of tumor cells to make them more antigenic, so that the immune system destroys the tumor
- Modification of dendritic cells to increase a tumor-specific immune reaction

TABLE 21.5 EXAMPLES OF CANCER GENE THERAPY TRIALS

Disorder	Cells altered	General strategy	Gene therapy protocol
Ovarian cancer	tumor cells	gene addition to restore tumor suppressor gene function	intraperitoneal injection of retrovirus or adenovirus encoding full-length cDNA encoding p53 or BRCA1, with the hope of restoring cell cycle control
Ovarian cancer	tumor cells	oncogene inactivation	inject adenovirus encoding a scFv antibody to ErbB2; hope to inactivate a growth signal
Malignant melanoma	tumor-infiltrating lymphocytes (TILs)	genetic manipulation of tumor cells to trigger apoptosis	extract TILs from surgically removed tumor and expand in culture; infect TILs *ex vivo* with a retroviral vector expressing TNF-α, infuse into patient; hope that TILs will target remaining tumor cells, and the TNF-α will kill them (see Figure 21.7E for the principle)
Various tumors	tumor cells	increase antigenicity of tumor cells so that immune system destroys tumor	transfect tumor cells with a retrovirus expressing a cell surface antigen such as HLA-B7 or a cytokine such as IL-12, IL-4, GM-CSF, or γ-interferon; hope that this enhances the immunogenicity of the tumor, so that the host immune system destroys it; often done *ex vivo* with lethally irradiated tumor cells (see Figure 21.7E for the principle)
Prostate cancer	dendritic cells	modify dendritic cells to enhance tumor-specific immune reaction	treat autologous dendritic cells with a tumor antigen or cDNA expressing the antigen, to prime them to mount an enhanced immune response to the tumor cells (see Figure 21.7E for the principle)
Malignant glioma (brain tumor)	tumor cells	genetic modification of tumor cells so that they convert a non-toxic prodrug into a toxic compound that kills them	inject a retrovirus expressing thymidine kinase (TK) or cytosine deaminase (CD) into the tumor; only the dividing tumor cells, not the surrounding non-dividing brain cells, are infected; then treat with ganciclovir (TK-positive cells convert this to the toxic ganciclovir triphosphate) or 5-fluorocytosine (CD-positive cells convert it to the toxic 5-fluorouracil); virally infected (dividing) cells are killed selectively (see Figure 21.17)
Head and neck tumors	tumor cells	use of oncolytic viruses that are engineered to selectively kill tumor cells	inject ONYX-015 engineered adenovirus into tumor; the virus can only replicate in p53-deficient cells, so it selectively lyses tumor cells; this treatment was effective when combined with systemic chemotherapy

TNF, tumor necrosis factor; IL, interleukin; GM-GSF, granulocyte/macrophage colony-stimulating factor. See the National Institute of Health clinical trials database at http://clinicaltrials.gov/ for a comprehensive survey.

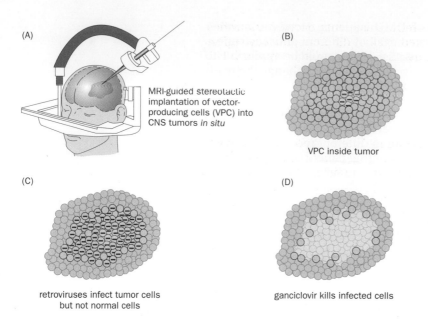

(A) MRI-guided stereotactic implantation of vector-producing cells (VPC) into CNS tumors *in situ*

(B) VPC inside tumor

(C) retroviruses infect tumor cells but not normal cells

(D) ganciclovir kills infected cells

Figure 21.17 Design of *in vivo* gene therapy for brain tumors. (A, B) A retrovirus is engineered to produce the herpes simplex virus thymidine kinase (HSV-TK). Vector-producing cells (VPC; blue) are injected into the brain tumor. (C) Because retroviruses infect only dividing cells, they infect the tumor cells (pink) but not the surrounding normal brain tissue (green). (D) The non-toxic prodrug ganciclovir is given intravenously. In TK$^+$ cells, ganciclovir is converted to the highly toxic ganciclovir triphosphate, causing cell killing and shrinkage of the tumor. Because retroviral transduction of the cancer cells is not 100% efficient, some tumor cells survive, so the tumor may grow again. MRI, magnetic resonance imaging.

- Use of oncolytic viruses that are engineered to kill tumor cells selectively

- Genetic modification of tumor cells so that they, but not surrounding non-tumor cells, convert a non-toxic prodrug into a toxic compound that kills them (**Figure 21.17**).

As at 2009, three cancer immunotherapies were at the phase III stage of clinical trials. Also at the phase III stage are adenoviral-based transfer of a herpes simplex virus thymidine kinase suicide gene for treating glioblastoma (see Figure 21.17) and adenoviral-based expression of p53 for the treatment of head and neck cancers and for Li–Fraumeni (OMIM 151623) tumors.

A general problem is that the methods for killing cancer cells are far from 100% efficient, and even when the efficiency is very high so that tumors shrink, they can grow again as surviving cancer cells proliferate. The idea of curing cancer is being replaced by a more realistic long-term management of cancer disorders.

Multiple HIV gene therapy strategies are being pursued, but progress toward effective treatment is slow

As the most important viral pathogen of humans, HIV-1 is the target of huge research efforts. Much work has gone into attempts to develop genetically engineered vaccines; in addition, many researchers have investigated the genetic manipulation of host cells to make them resistant to HIV. Much like the treatment for X-linked SCID described above, many of the gene therapy trials have used retroviral vectors to transfect hematopoietic stem cells *ex vivo*; treated cells are then returned to the patient. The transfected genes are designed in the hope that they will inhibit HIV replication and so prevent HIV infection from developing into AIDS.

To make lymphocytes resistant to HIV, the major targets for therapy have been the tat and rev regulatory proteins, and the viral RNA sequences to which they bind—TAR and RRE, respectively. A variety of reagents have been used:

- *Antisense RNAs.* Retroviruses have been constructed that encode antisense RNAs to TAR, to the overlapping *tat* and *rev* mRNAs, and to the *pol* and *env* mRNAs.

- *Decoy RNAs.* A retrovirus directing high-level expression of a transcript containing the RRE sequence might be able to sequester all the rev protein and prevent HIV replication.

- *Dominant-negative mutants.* Some retroviral constructs encode a mutant rev protein, RevM10. This binds the RRE but then will not assemble the multiprotein complex required to export the RNA from the nucleus.

- *Ribozymes.* Several groups have made retroviruses that encode ribozymes. RRz2 is a hammerhead ribozyme directed against the *tat* regulatory region; another type, the hairpin ribozyme, has also been used to cleave the HIV genome.

- *Intrabodies.* Retroviruses encoding scFv intracellular antibodies have been used to try to inactivate the tat or rev regulatory proteins, or the gp160 coat glycoprotein.

- *ZFN gene targeting.* As detailed in the section above on ZFNs, clinical trials will seek to use ZFNs to specifically inactivate the gene encoding the CCR5 chemokine receptor used by HIV-1 to gain entry into helper T cells.

In the early stages of HIV infection, there is a massive turnover of lymphocytes, as the immune system struggles to destroy infected cells. If the immune response were rather more effective, it might be possible to contain the virus at this stage. Apart from efforts at developing vaccines, work has also been devoted to modifying T cells so that they kill infected cells more efficiently. Retroviruses have been designed to make CD8$^+$ T cells express a chimeric T-cell receptor that targets their cytotoxic response to HIV-infected cells.

Details of all these approaches can be found in the NIH database of clinical trials. *In vitro*, the manipulated cells often show high resistance to HIV infection; *in vivo*, long-term (from several months to 1 or 2 years) bone marrow engraftment has been demonstrated in a few trials. The open question is whether engraftment can be made to occur at a sufficiently high level to provide a clinically useful pool of HIV-resistant lymphocytes. High-level engraftment might be achieved by first destroying the patient's existing marrow with cytotoxic chemicals and radiation—but doing this to a patient with AIDS would be a desperate measure.

CONCLUSION

Genetic technologies can be used in the treatment of disease, regardless of whether the disease has a genetic cause or not. This can be part of a treatment regime that involves conventional drugs or vaccines, but genetic manipulations can also be used in the production of drugs and vaccines. Genetic engineering has also been used to make therapeutic proteins. Expression cloning in microorganisms, cultured cells, or transgenic livestock (in which the protein of interest is expressed in milk or eggs) avoids the health risks associated with harvesting these proteins from animal or human sources. Genetic engineering has also been applied to make partly or fully human monoclonal antibodies. They are more stable in human serum than rodent monoclonal antibodies and are consequently better suited for therapeutic purposes.

Genetic interventions are also being used in two developing and interconnected therapeutic strategies: cell replacement therapy (to replace cells lost through disease or injury, enabling repair of damaged tissues) and gene therapy (involving genetic modification of a patient's cells). Stem cells are important in cell therapy because they are both self-renewing and also capable of being differentiated to make new tissue cells. Bone marrow transplants are an established form of stem cell therapy to treat blood disorders and hematopoietic stem cell depletion after chemotherapy.

To avoid immune rejection, autologous cells are prefered for cell therapy. In principle, they can be generated by artificially enhancing the mobilization and differentiation of stem cells in the patient or by using laboratory nuclear reprogramming methods to reverse the normal developmental progression of cells toward specialization, which has opened new research and therapeutic opportunities. The transplant of the nucleus from an adult somatic cell into an enucleated oocyte allowed the cloning of Dolly the sheep and various other types of animal species, and could in principle be used to make patient-specific embryonic stem cells; however, the procedures are technically challenging and ethically contentious.

Nuclear reprogramming of skin fibroblasts from patients by ectopic expression of specific transcription factors offers a new route to obtaining patient-specific and disease-specific pluripotent stem cells, with few ethical concerns.

Such induced pluripotent stem (iPS) cells can be differentiated to provide human cellular models of disease and are being used to test drugs; in principle they could provide sources of cells for autologous cell therapy. Transdifferentiation techniques hold the promise of making more subtle changes to cell lineage (e.g. altering pancreatic exocrine cells into beta cells to replace those lost in type 1 diabetes).

In gene therapy, the aim is to transfer a DNA, RNA, or oligonucleotide into a patient's cells to counteract or alleviate disease. Transfer into cells is often achieved with virus vectors, which offer high efficiency of gene transfer and high-level, sometimes long-lasting, transgene expression. Safety issues are a concern because integration into chromosomes is uncontrolled and can accidentally activate a neighboring proto-oncogene. Non-viral vectors are safer, but gene transfer rates are low and expression is often transient and comparatively weak.

The genetic modification could theoretically be directed to germ-line cells (a permanent, transmissible modification), but for ethical reasons germ-line gene therapy is not being attempted, and all current trials involve somatic gene therapy. The aim is often to add a gene that can functionally replace a defective gene, but some therapies seek to block expression of a mutant or ectopically expressed gene at the RNA level, usually using RNA interference, or at the protein level.

Another class of gene therapy strategy seeks to restore the function of a defective gene, either at the RNA level by inducing exon skipping so that a causative mutation is bypassed, or at the DNA level by using zinc finger nucleases. These enzymes are genetically engineered to act as restriction nucleases that cleave genomic DNA at just one site, such as at the location of an inactivating point mutation. The double-strand DNA break can trigger a cellular repair pathway that uses homologous recombination to replace the mutant sequence by a homologous wild-type sequence provided by a transfected plasmid.

Most clinical trials of gene therapy have sought to treat cancer, and here the strategy is to kill cancer cells selectively; however, success has been limited. The first clearly successful gene therapies have been for recessive blood cell disorders, which are more amenable because the cells are highly accessible and can be genetically modified outside the body before being returned to the patient, and because even a small amount of transgene expression can often bring about some clinical benefit.

FURTHER READING

General

Costa T, Scriver CR & Childs B (1985) The effect of Mendelian disease on human health: a measurement. *Am. J. Med. Genet.* 21, 231–242.

Treacy EP, Valle D & Scriver CR (2001) Treatment of genetic disease. In The Metabolic And Molecular Basis Of Inherited Disease, 8th ed. (Scriver CR, Beaudet AL, Sly WS, Valle MD eds). McGraw-Hill.

Genetic approaches to identifying novel drugs and drug targets

Caskey CT (2007) The drug development crisis: efficiency and safety. *Annu. Rev. Med.* 58, 1–16.

King A (2009) Researchers find their Nemo. *Cell* 139, 843–846. [A profile on zebrafish that includes its advantages for drug screening.]

Welch EM, Barton ER, Zhuo J et al. (2007) PTC124 targets genetic disorders caused by nonsense mutations. *Nature* 447, 87–91.

Antibody engineering and therapeutic antibodies and proteins

Beerli RR, Bauer M, Buser RB et al. (2008) Isolation of human monoclonal antibodies by mammalian cell display. *Proc. Natl Acad. Sci. USA* 105, 14336–14341.

Cardinale A & Biocca S (2008) The potential of intracellular antibodies for therapeutic targeting of protein-misfolding diseases. *Trends Mol. Med.* 14, 373–380.

Hoogenboom HR (2005) Selecting and screening human recombinant antibody libraries. *Nat. Biotechnol.* 23, 1105–1116.

Kim SJ, Park Y & Hong HJ (2005) Antibody engineering for the development of therapeutic antibodies. *Mol. Cell* 20, 17–29.

Lillico SG, Sherman A, McGrew MJ et al. (2007) Oviduct-specific expression of two therapeutic proteins in transgenic hens. *Proc. Natl Acad. Sci. USA* 104, 1771–1776.

Reichert JM (2008) Monoclonal antibodies as innovative therapeutics. *Curr. Pharm. Biotechnol.* 9, 423–430.

Stocks M (2005) Intrabodies as drug discovery tools and therapeutics. *Curr. Opin. Chem. Biol.* 9, 359–365.

Stöger E, Vaquero C, Torres E et al. (2000) Cereal crops as viable production and storage systems for pharmaceutical scFv antibodies. *Plant Mol. Biol.* 42, 583–590.

Genetic approaches to disease treatment using aptamers or vaccines

Bunka DH & Stockley PG (2006) Aptamers come of age—at last. *Nat. Rev. Microbiol.* 4, 588–596.

Dausse E, Gomes SDR & Toulme J-J (2009) Aptamers: a new class of oligonucleotides in the drug discovery pipeline. *Curr. Opin. Pharmacol.* 9, 602–607.

Ellington Lab Aptamer Database. http://aptamer.icmb.utexas .edu/index.php

Kutzler MA & Weiner DB (2008) DNA vaccines: ready for prime time? *Nat. Rev. Genet.* 9, 776–788.

Soler E & Houdebine LM (2007) Preparation of recombinant vaccines. *Biotechnol. Annu. Rev.* 13, 65–94.

General reviews on pluripotent stem cells and nuclear reprogramming

Conrad S, Renninger M, Hennenlotter J et al. (2008) Generation of pluripotent stem cells from adult human testis. *Nature* 456, 344–349.

Fenno LE, Ptaszek LM & Cowan CA (2008) Human embryonic stem cells: emerging technologies and practical applications. *Curr. Opin. Genet. Dev.* 18, 324–329.

Hochedlinger K & Plath K (2009) Epigenetic reprogramming and induced pluripotency. *Development* 136, 509–523.

Jaenisch R & Young R (2008) Stem cells, the molecular circuitry of pluripotency and nuclear reprogramming. *Cell* 132, 567–582.

Rossant J (2009) Reprogramming to pluripotency: from frogs to stem cells. *Cell* 138, 1047–1050.

Yamanaka S (2008) Pluripotency and nuclear reprogramming. *Philos. Trans. R. Soc. Lond. B* 363, 2079–2087.

Zhou Q & Melton DA (2008) Extreme makeover: converting one cell into another. *Cell Stem Cell* 3, 382–388.

General stem cell therapy and regenerative medicine (see also references in Table 21.3)

Ährlund-Richter L, De Luca M, Marshak DR et al. (2009) Isolation and production of cells suitable for human therapy: challenges ahead. *Cell Stem Cell* 4, 20–26.

Nature Insight on Regenerative Medicine (2008) *Nature* 453, 301–351.

Pitchford SC, Furze RC, Jones CP, Wengner AM & Rankin SM (2009) Differential mobilization of subsets of progenitor cells from the bone marrow. *Cell Stem Cell* 4, 62–72.

iPS cells and disease-specific pluripotent stem cells

Colman A & Dreesen O (2009) Pluripotent stem cells and disease modeling. *Cell Stem Cell* 5, 244–247.

Kiskinis E & Eggan K (2010) Progress towards the clinical application of patient-specific pluripotent stem cells. *J. Clin. Invest.* 120, 51–59.

Muller R & Lengerke C (2009) Patient-specific pluripotent stem cells: promises and challenges. *Nat. Rev. Endocrinol.* 5, 195–203.

Nature Methods Reviews on Induced Pluripotent Stem Cells (2010) *Nat. Methods* 7, 17–33.

Nishikawa S-I, Goldstein RA & Nierras CR (2008) The promise of human induced pluripotent stem cells for research and therapy. *Nat. Rev. Mol. Cell Biol.* 9, 725–729.

Saha K & Jaenisch R (2009) Technical challenges in using human induced pluripotent stem cells to model disease. *Cell Stem Cell* 5, 584–595.

Gene therapy: general background reviews and information

Gene Therapy Net. http://www.genetherapynet.com/ [An information resource for basic and clinical research in gene therapy, cell therapy, and genetic vaccines.]

Gordon JW (1999) Genetic enhancement in humans. *Science* 283, 2023–2024.

O'Connor TP & Crystal RG (2006) Genetic medicines: treatment strategies for hereditary disorders. *Nat. Rev. Genet.* 7, 261–276.

Verma IM & Weitzman MD (2005) Gene therapy: twenty-first century medicine. *Annu. Rev. Biochem.* 74, 711–738.

Gene delivery into mammalian cells

Bouard D, Alazard-Dany N & Cosset FL (2009) Viral vectors: from virology to transgene expression. *Br. J. Pharmacol.* 157, 153–165.

Chowdhury EH (2009) Nuclear targeting of viral and non-viral DNA. *Expert Opin. Drug Deliv.* 6, 697–703.

Farjo R, Skaggs J, Quiambao AB et al. (2006) Efficient non-viral ocular gene transfer with compacted DNA nanoparticles. *PLoS ONE* 1, e38.

Pringle, IA, Hyde SC & Gill DR (2009) Non-viral vectors in cystic fibrosis gene therapy: recent developments and future prospects. *Expert Opin. Biol. Ther.* 9, 991–1003.

Rolland A (2006) Nuclear gene delivery: the Trojan horse approach. *Expert Opin. Drug. Deliv.* 3, 1–10.

Simoes S, Filipe A, Faneca H et al. (2005) Cationic liposomes for gene delivery. *Expert Opin. Drug. Deliv.* 2, 237–254.

Wolfrum C, Shi S, Jayaprakash KN et al. (2007) Mechanisms and optimization of *in vivo* delivery of lipophilic siRNAs. *Nat. Biotechnol.* 25, 1149–1157.

RNA and antisense therapeutics

Castanotto D & Rossi JJ (2009) The promises and pitfalls of RNA interference-based therapeutics. *Nature* 457, 426–433.

Rayburn ER & Zhang R (2008) Antisense, RNAi, and gene silencing strategies for therapy: mission possible or impossible? *Drug Discov. Today* 13, 513–521.

Van Ommen G-J, van Deutekom J & Aartsma-Rus A (2008) The therapeutic potential of antisense-mediated exon skipping. *Curr. Opin. Mol. Ther.* 10, 140–149.

Wood M, Yin H & McClorey G (2007) Modulating the expression of disease genes with RNA-based therapy. *PLoS Genet.* 3, e109.

Yokota T, Takeda S, Lu QL et al. (2009) A renaissance for antisense oligonucleotide drugs in neurology: exon skipping breaks new ground. *Arch. Neurol.* 66, 32–38.

Zinc finger nucleases and potential therapeutic applications

Cathomen T & Joung JK (2008) Zinc-finger nucleases: the next generation emerges. *Mol. Ther.* 16, 1200–1207.

Perez EE, Wang J, Miller JC et al. (2008) Establishment of HIV-1 resistance in CD4$^+$ T cells by genome editing using zinc-finger nucleases. *Nat. Biotechnol.* 26, 808–816.

Urnov FD, Miller JC, Lee Y-L et al. (2005) Highly efficient endogenous human gene correction using designed zinc-finger nucleases. *Nature* 435, 646–651.

Wu J, Kandavelou K & Chandrasegaran S (2007) Custom-designed zinc finger nucleases: what is next? *Cell Mol. Life Sci.* 64, 2933–2944.

Gene therapy in practice (see also references in Box 21.8)

Baxevanis CN, Perez SA & Papmichail M (2009) Cancer immunotherapy. *Crit. Rev. Clin. Lab. Sci.* 46, 167–189.

Edelstein ML, Abedi MR & Wixon J (2007) Gene therapy clinical trials worldwide to 2007—an update. *J. Gene Med.* 9, 833–842.

Fischer A & Cavazzana-Calvo M (2008) Gene therapy of inherited diseases. *Lancet* 371, 2044–2047.

Frank NY, Schatton T & Frank MH (2010) The therapeutic promise of the cancer stem cell concept. *J. Clin. Invest.* 120, 41–50.

Gene Therapy Clinical Trials Worldwide. http://www.wiley.co.uk/ genmed/clinical/ [Wiley's database of worldwide gene therapy clinical trials.]

Griesenbach U & Alton EWFW (2009) Gene transfer to the lung: lessons learned from more than 2 decades of CF gene therapy. *Adv. Drug Deliv. Rev.* 61, 128–139.

NIH database of gene therapy trials. http://clinicaltrials.gov/

Rossi JJ, June CH & Kohn DB (2007) Genetic therapies against HIV. *Nat. Biotechnol.* 25, 1444–1454.

Glossary

3′ end The end of a DNA or RNA strand that is linked to the rest of the chain only by carbon 5 of the sugar, not carbon 3. (Figure 1.8)

3′, 5′- phosphodiester bond The link between adjacent nucleotides in DNA or RNA. (Figure 1.5)

5′ end The end of a DNA or RNA strand that is linked to the rest of the chain only by carbon 3 of the sugar, not carbon 5. (Figure 1.8)

5′ RACE (rapid amplification of cDNA ends) A technique for characterizing the ends of mRNAs, in this case the 5′ ends. (Box 11.1)

adaptive immune system The immune system of vertebrates that creates immunological memory.

affected sib pair (ASP) analysis A form of non-parametric linkage analysis based on measuring haplotype sharing by sibs who both have the same disease. (Figure 15.2)

affinity tag In genetic manipulation, a short peptide that is attached to a recombinant protein in order to allow the protein to be isolated by affinity chromatography.

alleles Alternative forms of the same gene.

allelic heterogeneity The existence of many different mutations, but all within the same gene, in unrelated people with the same phenotype.

amino acid The building blocks of proteins. (Figure 1.4)

amplification An increase in the copy number of a DNA sequence as a result of cloning, PCR or natural processes.

anaphase lag Loss of a chromosome because it moves too slowly at anaphase to get incorporated into a daughter nucleus.

ancestral chromosome segments, shared Chromosomal segments that are shared by apparently unrelated people because they are inherited from an unknown distant common ancestor. (Figures 14.14 and 15.6)

aneuploidy A chromosome constitution with one or more chromosomes extra or missing from a full euploid set.

anneal Allowing two complementary single-stranded nucleic acids to form a base-paired double strand. The reverse of denaturation.

antibody A protein produced by activated B-cells in response to a foreign molecule or microorganism. (Figure 4.21)

anticipation The tendency for the severity of a condition to increase in successive generations. Commonly due to bias of ascertainment (Section 3.3), but seen for real with dynamic mutations. (Section 13.3)

anticodon The 3-base sequence in a tRNA molecule that base-pairs with the codon in mRNA.

antigen A molecule that can induce an adaptive immune response or that can bind to an antibody or T-cell receptor.

antiparallel Of the strands in a double-stranded nucleic acid molecule, running in opposite directions so that where one strand has its 5′ end the complementary strand has its 3′ end.

antisense RNA A transcript complementary to a normal mRNA. Naturally occurring antisense RNAs, made using the nontemplate strand of a gene, are important regulators of gene expression.

antisense strand (template strand) The DNA strand of a gene, which, during transcription, is used as a template by RNA polymerase for synthesis of mRNA. (Figure 1.13)

antisense technology Experimental inhibition of expression of a gene by use of an RNA, an oligonucleotide or a morpholino derivative complementary to the mRNA of the gene.

apoptosis Programmed cell death.

aptamers Single-stranded nucleic acid molecules designed to bind specifically to an antigen, mimicking antibodies.

archaea Single-celled prokaryotes superficially resembling bacteria, but with molecular features indicative of a third kingdom of life.

ARMS (amplification refractory mutation system) test Allele-specific PCR. (Figures 6.17 and 18.8)

array CGH (comparative genomic hybridization) Competitive hybridization of a test and control sample to a microarray of mapped clones to detect copy number variations. (Figure 16.10)

ASOs (allele-specific oligonucleotides) Under stringent hybridization conditions, oligonucleotides 15–20 nt long will hybridize only to a perfectly matched target. This provides the basis for various methods of distinguishing alleles that differ by only a single nucleotide. (Figure 7.9)

association   A tendency of two characters (diseases, marker alleles etc.) to occur together at non-random frequencies. Association is a simple statistical observation, not a genetic phenomenon, but can sometimes be caused by linkage disequilibrium. (Section 15.4)

assortative mating   Mating where the partner is chosen on the basis of phenotypic or genotypic similarity (e.g. tall people tend to marry tall people, deaf people tend to marry deaf people; some people prefer to marry relatives). Assortative mating can produce a non-Hardy–Weinberg distribution of genotypes in a population.

autoimmunity   An abnormal state in which the distinction between self and nonself fails, so that the body mounts an adaptive immune response against one or more self molecules. (Box 4.5)

autophagy   Digestion of worn-out organelles by a cell's own lysosomes.

autoradiography   Using photographic film to make a radiolabeled molecule reveal its location on a gel or in a cell. (Box 7.2)

autosome   Any chromosome other than the sex chromosomes, X and Y.

autozygosity   In an inbred person, homozygosity for alleles identical by descent. (Section 14.5)

B cells (B lymphocytes)   The lymphocytes that make antibodies.

bacteria   Single-celled organisms that lack a membrane-bound nucleus or other organelles.

bacterial artificial chromosome (BAC)   A cloning vector in which inserts up to 300 kb long can be propagated in bacterial cells.

bacteriophage (phage)   A virus that infects bacteria. Modified phage are used as vectors for cloning in bacterial cells.

banding   In a preparation of chromosomes, treatments to make the chromosomes stain in a reproducible pattern of dark and light bands to aid identification of chromosomes and detection of structural abnormalities. (Figure 2.14)

Barr body   The chromatin of an inactive X chromosome, seen as a blob of condensed chromatin at the edge of the nucleus of interphase cells that contain one or more inactive X chromosomes. Also called sex chromatin. (Figure 3.8)

basal lamina   Thin mat of extracellular matrix that separates epithelial sheets and many types of cell from the underlying connective tissue.

basal transcription apparatus   The multiprotein complex that is required for RNA polymerase II to transcribe any gene. Additional, tissue-specific, proteins are usually also needed. (Figure 11.2)

base complementarity   The relationship between bases on opposite strands of a double-stranded nucleic acid: A always opposite T (or U in RNA) and G always opposite C.

base pair   The unit of length of a double-stranded nucleic acid. Also, more narrowly, two purine or pyrimidine bases on opposite strands of a double-stranded nucleic acid, hydrogen-bonded to each other. (Figure 1.6)

Bayesian statistics   A method of combining a number of independent probabilities. It forms the basis of much genetic risk estimation. (Box 18.2)

bias of ascertainment   Distorted proportions of phenotypes in a dataset caused by the way cases are collected. (Section 3.2)

biometrics   The statistical study of quantitative characters.

biotin-streptavidin system   A tool for isolating labeled molecules. The bacterial protein streptavidin binds the B-vitamin biotin with exceptionally high affinity. Biotinylated molecules can be isolated using streptavidin-coated magnetic beads.

bisulfite sequencing   A method for identifying methylated cytosines in a DNA sample. Sodium bisulfite converts unmethylated cytosines, but not methylated cytosines, to uracil. When the product is sequenced, cytosines that were originally methylated are still read as cytosines, but those that were unmethylated are read as thymine. (Box 11.4)

bivalent   The four-stranded structure seen in prophase I of meiosis, comprising two synapsed homologous chromosomes. (Figure 14.4)

blastocyst   An embryo at a very early stage of development when it consists of a hollow ball of cells with a fluid-filled internal compartment, the blastocele. (Figure 5.9)

blastomere   One of the many cells formed by cleavage of a fertilized egg.

bootstrapping   A statistical method designed to check the accuracy of an evolutionary tree constructed from comparative sequence analysis. (Section 10.4)

branch site   In mRNA processing, a rather poorly defined sequence (consensus YNCTRAY; R = purine, Y = pyrimidine, N = any nucleotide) located 10–50 bases upstream of the splice acceptor, containing the adenosine at which the lariat splicing intermediate is formed. (Figure 1.17)

bromodomain   A protein domain that stimulates binding to acetylated lysine residues, primarily in histones.

C value paradox   The lack of a direct relationship between the amount of DNA in the cells of an organism (the C value) and the complexity of the organism. (Table 4.2)

CAAT box   A short sequence, GGCCAATCT or a close variant that is found in the promoter of many genes that are transcribed by RNA polymerase II.

CAGE (cap analysis of gene expression)   A high-throughput technique for cataloging bulk mRNAs by isolating around 18 nucleotides adjacent to the 5′ cap for sequencing. (Box 11.1)

capping   A stage in RNA processing, addition of a special nucleotide, 7-methylguanosine triphosphate, by a 5′-5′ bond to the 5′ end of a primary transcript. Capping is important for the stability of the RNA. (Figure 1.20)

cell   The basic unit of all living things.

cell adhesion   The way cells recognize and bind to each other.

cell cycle The reproductive cycle of a cell, comprising mitosis (M phase), the first gap (G1 phase), DNA synthesis (S phase), a second gap (G2 phase), then mitosis of the next cycle. (Figure 2.1)

cell differentiation The process by which a less specialized cell becomes more specialized.

cell fate In developmental biology, the normal progeny of a given cell as development proceeds.

cell junctions Specialized structures that control the passage of molecules between cells. They may form totally impermeable barriers, or they may allow specific types or sizes of molecule to pass. (Figure 4.3)

cell lineage In development, the ancestry and descendants of a cell, as traced backwards or forwards through successive cell divisions.

cell polarity Asymmetry of a cell. In embryos, asymmetric cells often divide to form daughter cells that follow different fates in development. (Box 4.3)

centimorgan (cM) The unit of genetic distance. Loci 1 cM apart have a 1% probability of recombination during meiosis. Figure 14.5 shows the relationship between genetic and physical distances.

central dogma As formulated by Francis Crick, DNA→RNA→protein—that is, the DNA specifies the nucleotide sequence of an RNA which in turn specifies the amino acid sequence of a protein. Useful, but not always true.

centriole A cylinder of short microtubules located in the centrosome. (Box 2.1)

centromere The primary constriction of a chromosome, separating the short arm from the long arm, and the point at which spindle fibers attach to pull chromatids apart during cell division.

centrosome In cell division, the microtubule organizing center that forms a spindle pole. (Box 2.1)

CEPH families A set of families assembled by the Centre d'Etude du Polymorphisme Humain in Paris to assist the production of marker-marker framework maps.

CGH—*see* comparative genomic hybridization

character An observable property of an individual.

checkpoint Checkpoints prevent further progress through the cell cycle unless the genome and the cell are in a suitable state to proceed. (Figure 17.13)

chiasma (plural chiasmata) The physical manifestation of meiotic recombination, as seen under the microscope. (Figure 14.4)

chimera (1) An organism derived from more than one zygote. (Figure 3.23) (2) A chimeric gene is a gene created when a chromosomal rearrangement brings together parts of two different genes to create a novel functional gene – a frequent event in tumors. (Figure 17.6)

chromatin A general term for the packaged DNA in a cell nucleus. The basic conformation is a 30 nm coiled coil of DNA and histones. (Figure 2.8)

chromatin immunoprecipitation (ChIP) A technique for identifying the DNA sequences that bind a specific protein.

Protein and DNA are reversibly cross-linked, the chosen protein is precipitated with an antibody, and the associated DNA sequenced. (Box 11.3)

chromatin remodeling complexes Protein complexes that can move, dissociate or reconstitute nucleosomes in chromatin, as part of the systems controlling chromatin conformation. (Table 11.3)

chromodomain A protein domain that stimulates binding to methylated lysine residues, primarily in histones.

chromosome conformation capture A set of techniques for identifying DNA sequences that lie close together in interphase nuclei. (Box 11.3)

chromosome engineering Genetic engineering techniques to produce specific large-scale chromosomal deletions or rearrangements. (Figure 20.12)

chromosome jumping An obsolete technique used to obtain clones from a chromosomal location some distance (typically 100 kb) away from a previously isolated clone. (Figure 16.19)

chromosome painting Fluorescence labeling of a whole chromosome by a FISH procedure in which the probe is a cocktail of many different DNA sequences from that particular chromosome. (Figures 2.16, 2.18, and 17.17)

chromosome set The chromosomes of a haploid genome.

chromosome walking Isolating sequences adjacent on the chromosome to a characterized clone by screening genomic libraries for clones that partially overlap it.

cis-acting Controlling the activity of a gene only when it is part of the same DNA molecule or chromosome as the regulatory factor. Compare trans-acting regulatory factors which can control their target sequences irrespective of their chromosome location.

clonal selection and expansion The process responsible for immunological memory. Binding of an antigen to a B or T cell stimulates it to multiply, forming a clone of cells that react to the same antigen. However, clones that respond to self-antigens are eliminated.

clone fingerprinting Identifying independent clones that contain overlapping inserts by comparing the pattern of fragments produced by a series of restriction enzymes.

clones Identical copies (of a DNA sequence, a cell, an organism). In genetic research, often means cells containing identical recombinant DNA molecules (the cells themselves may or may not be identical).

cloning Production of many identical copies of a DNA sequence, a cell or a whole organism.

co-activators Proteins that enhance transcription of a gene through protein-protein rather than protein-DNA interactions.

coding RNA Messenger RNA that codes for protein.

codon A nucleotide triplet (strictly in mRNA, but by extension, in genomic coding DNA) that specifies an amino acid or a translation stop signal.

coefficient of inbreeding The proportion of loci at which a person is homozygous by virtue of the consanguinity of their parents. (Section 3.5)

coefficient of relationship   Of two people, the proportion of loci at which they share alleles identical by descent. (Box 3.6)

coefficient of selection   The chance of reproductive failure for a certain genotype, relative to the most successful genotype. (Section 3.5)

cofactor   A small molecule or metal ion that is required for the biological activity of a protein. More generally, any factor that assists the principal actor in a process.

co-immunoprecipitation (co-IP)   Using an antibody to precipitate a known molecule complexed with its binding partners, as a way of identifying the binding partners. (Figure 12.10)

colony hybridization   Screening a DNA library by hybridizing a labeled probe to gridded-out colonies of cells containing recombinant molecules. (Figure 8.3)

common disease–common variant hypothesis   The hypothesis that most factors conferring susceptibility to common non-Mendelian diseases are ancient variants that are common in the population (in distinction to the mutation–selection hypothesis, *q.v.*). (Section 15.6)

compaction   In an early embryo, the tightening of cell-cell contacts that converts the loosely bound products of the initial cleavage divisions of the zygote into a compact morula.

comparative genomic hybridization (CGH)   Competitive fluorescence *in situ* hybridization of a test and control sample, normally to a microarray of mapped clones but originally to a spread of normal chromosomes, to detect chromosomal regions in the test sample that are amplified or deleted compared to the control sample.

comparative genomics   A systematic and comprehensive comparison of the genomes of different organisms.

complement system   A system of serum proteins activated by antigen-antibody complexes or by microorganisms. (Figure 4.20)

complementary   Of two nucleic acid strands, having sequences such that they can anneal to form a double-stranded molecule.

complementary DNA (cDNA)   A DNA copy of an RNA, made by reverse transcriptase. (Figure 8.2)

complementation   Two alleles complement if in combination they restore the wild-type phenotype (Table 3.1). Normally alleles complement only if they are at different loci, although some cases of interallelic complementation occur.

complementation test   A breeding or cell culture experiment to establish the relationship between two recessive mutations by checking the phenotype of an organism or cell that inherits one of the mutations from each parent or each of two donor cells. If the mutations are allelic, the phenotype will be mutant; if not, it will be wild type. (Figures 3.9 and 3.12)

complete truncate ascertainment   Sampling a population by collecting every family where at least one child has a certain recessive condition. The collection will not include families where both parents are carriers but none of the children is affected. (Figure 3.11)

complex   Of a phenotype, one that can have a variety of different causes and modes of inheritance in different people.

compound heterozygote   A person with two different mutant alleles at a locus.

concatemers   Molecules joined end-to-end in a chain.

concordance   Of twins, the frequency with which co-twins have the same phenotype.

conformation   Of a complex molecule, the 3-dimensional shape – the result of the combined effects of many weak noncovalent bonds.

congenic strain   In experimental animals, two strains that have identical genetic backgrounds and differ only in some chosen gene or location.

consanguineous mating   A mating where both parties share one or more identifiable recent common ancestors.

consensus sequence   A representation of the main shared elements of a family of functionally related DNA sequences.

conservative change   In a protein, replacement of one amino acid by a chemically similar one.

conserved sequence   A sequence (of DNA or sometimes protein) that is identical or recognizably similar across a range of organisms.

constitutional abnormality   An abnormality that was present in the zygote, and so is present in every cell of a person.

contig   A set of overlapping clones. (Box 8.1)

contiguous gene syndrome   A syndrome that is the result of a deletion that inactivates two or more contiguous genes, each of which contributes to the phenotype. (Table 13.2)

continuous character   A character like height, which everybody has, but to differing degree – as compared with a dichotomous character like polydactyly, which some people have and others do not.

copy number variation (CNV)   Variation between individuals in the number of copies of a particular DNA sequence in their genomes. (Section 13.1)

co-repressors   Proteins that suppress transcription of a gene through protein-protein rather than protein-DNA interactions.

cosmid   A vector for cloning in *E. coli*. Cosmids have an insert capacity of up to 44 kb.

CpG island   Short stretch of DNA, often < 1 kb, containing frequent unmethylated CpG dinucleotides. CpG islands tend to mark the 5′ ends of genes.

Cre–*loxP* technique   A technique that allows tissue- or stage-specific knockout of a gene in an intact animal. (Figure 20.11)

crossover   An act of meiotic recombination, or the physical manifestation of that, as seen under the microscope. (Figures 14.3 and 14.4)

cryptic splice site   A sequence in pre-mRNA with some homology to a splice site. Cryptic splice sites may be used as splice sites when splicing is disturbed or after a base substitution mutation that increases the resemblance to a normal splice site. (Figure 13.13)

cytogenetics   The study of chromosomes.

cytokines   Extracellular signaling proteins or peptides that act as local mediators in cell-cell communication.

cytokinesis   The final event of cell division, when the cytoplasm of the cell divides. (Figure 2.3)

cytosol   The contents of the cytoplasm of a cell, excluding membrane-bound organelles such as mitochondria or lysosomes.

degenerate   Used in genetics to describe a many-to-one relation between structure and function. The genetic code is degenerate because most amino acids can be incorporated into a polypeptide in response to any of several different codons in the mRNA. An oligonucleotide is degenerate if it is a mixture of several related sequences.

denaturation   Dissociation of double-stranded nucleic acid to give single-strands. Also destruction of the 3-dimensional structure of a protein by heat or high pH.

dichotomous character   A character like polydactyly, which some people have and others do not have – as compared to a continuous character like height, which everybody has, but to differing degree.

dideoxy (Sanger) sequencing   The standard method of DNA sequencing, developed by Fred Sanger and using didexoynucleotide chain terminators. (Figure 8.6)

diploid   Having two copies of each type of chromosome; the normal constitution of most human somatic cells.

distal (of chromosome)   Positioned comparatively distant from the centromere.

disulfide bridge   In proteins, an intramolecular or intermolecular link between the SH groups of two cysteine residues. Disulfide bridges are important for maintaining the 3-dimensional folding of proteins. (Figure 1.29)

DNA chip   A high-density microarray carrying oligonucleotides or longer single-stranded DNA molecules.

DNA duplex   A double-stranded DNA molecule.

DNA fingerprinting   A now obsolete method of identifying a person for legal or forensic purposes based on probing Southern blots with a hypervariable minisatellite probe. (Figure 18.23)

DNA libraries   The result of cloning random DNA fragments or molecules. A collection of cells containing different recombinant vectors, which must then be screened to find any desired sequence. (Figures 8.1 to 8.4)

DNA methylation   Conversion of cytosine in DNA to 5-methyl cytosine, a signal that helps regulate gene expression. (Figures 11.10 to 11.13)

DNA polymerase   The family of enzymes that can add nucleotides to the 3′ end of a DNA molecule. (Table 1.3)

DNA primase   An enzyme that synthesizes a short RNA molecule that serves as a primer for DNA replication.

DNA profiling   Using genotypes at a series of polymorphic loci to recognize a person, usually for legal or forensic purposes. (Section 18.6)

DNA repair   Correcting lesions in DNA caused by mistakes during replication or by external agents such as radiation or chemicals.

DNA replication   The process of copying a DNA molecule to make two identical daughter molecules.

DNase-hypersensitive sites (DHSs)   Regions of chromatin that are rapidly digested by DNase I because the DNA is relatively exposed rather than being tightly packaged in nucleosomes. They are believed to mark important long-range control sequences. (Section 11.2)

dominant   In human genetics, any trait that is expressed in a heterozygote. See also semi-dominant.

dominant negative effect   The situation where a mutant protein interferes with the function of its normal counterpart in a heterozygous person. (Figure 13.24)

double helix   The normal structure of DNA; two antiparallel DNA strands wrapped round one another.

driver mutations   In cancer, mutations that are subject to positive selection during tumorigenesis because they assist development of the tumor. *Cf.* passenger mutations.

dynamic mutation   An unstable expanded repeat that changes size between parent and child. (Section 13.3)

ectoderm   One of the three germ layers of the embryo. It is formed during gastrulation from cells of the epiblast and gives rise to the nervous system and outer epithelia. (Figures 5.1 and 5.17)

electroporation   A method of transferring DNA into cells *in vitro* by use of a brief high-voltage pulse.

electrospray ionization   A mass spectrometric technique in which a solution containing the analyte is sprayed as very fine charged droplets into the machine. (Box 8.8)

elongation factors   Factors that assist progression of RNA polymerase along a DNA sequence once transcription has been initiated.

embryonic germ cells   Pluripotent cells derived from cultured primordial germ cells of embryos.

embryonic stem (ES) cell line   Embryonic stem cells that have continued to proliferate after subculturing for a period of 6 months or longer and that are judged to be pluripotent and genetically normal.

embryonic stem (ES) cells   Undifferentiatied, pluripotent cells derived from an embryo. A key tool for genetic manipulation. (Figure 21.5)

empiric risks   Risks calculated from survey data rather than from genetic theory. Genetic counseling in most non-Mendelian conditions is based on empiric risks. (Section 3.4)

endoderm   One of the three germ layers of the embryo. It is formed during gastrulation from cells migrating out of the epiblast layer. (Figures 5.1 and 5.17)

endonuclease   An enzyme that cuts DNA or RNA at an internal position in the chain.

endophenotype   A phenotype that marks one facet of the changes that occur during development of a complex disease. Hopefully endophenotypes are directly related to an underlying biological change.

endoplasmic reticulum   A meshwork of membranes in the cytoplasm of cells, forming a compartment where membrane-bound and secreted proteins are made.

enhancer   A set of short sequence elements which stimulate transcription of a gene and whose function is not critically dependent on their precise position or orientation. (Figures 11.5 and 11.6)

epiblast   The layer of cells in the pregastrulation embryo that will give rise to all three germ layers of the embryo proper, plus the extraembryonic ectoderm and mesoderm. *Cf.* hypoblast. (Figure 5.13)

epigenetic   Heritable (from mother cell to daughter cell, or sometimes from parent to child), but not produced by a change in DNA sequence. DNA methylation is the best understood epigenetic mechanism.

episome   Any DNA sequence that can exist in an autonomous extra-chromosomal form in the cell. Often used to describe self-replicating and extra-chromosomal forms of DNA.

epistasis   Literally 'standing above'. Gene A is epistatic to gene B if A functions upstream of B in a common pathway. Loss of function of A will cause all the effects of loss of function of B, and maybe other effects as well.

epitope   The part of an immunogenic molecule to which an antibody responds.

epitope tagging   A method for visualizing a specific protein in cells or tissues. A recombinant version of the protein is produced that has attached a marker peptide for which a fluorescently labeled antibody is available. (Table 8.7)

EST—*see* expressed sequence tag

euchromatin   The fraction of the nuclear genome which contains transcriptionally active DNA and which, unlike heterochromatin, adopts a relatively extended conformation. (Figure 11.9)

eukaryotes   Organisms made of cells with a membrane-bound nucleus and other organelles (Box 4.1). One of the three kingdoms of life.

exaptation   An unusual evolutionary process in which sequences derived from a transposable element are used by the host genome for a novel function. (Figure 10.32)

exclusion mapping   Genetic mapping with negative results, showing that the locus in question does not map to a particular location. Particularly useful for excluding a possible candidate gene without the labor of mutation screening.

exome   The totality of exons in a genome.

exon   A segment of a gene that is retained during splicing. Individual exons may contain coding and/or noncoding DNA (untranslated sequences). (Figure 1.16)

exon junction complex (EJC)   A set of proteins that are bound to mRNAs during splicing, at the positions where introns have been removed. Exon junction complexes are removed from the mRNA during a 'pioneer' round of translation by ribosomes. Incomplete removal, because of a premature termination codon, triggers nonsense-mediated decay of the mRNA. (Figure 13.12)

exonuclease   An enzyme that digests DNA or RNA from one end. May be a 3′ or 5′ exonuclease.

expressed sequence tag (EST)   Short partial sequences of cDNAs that can be used to follow gene expression or to isolate a full-length cDNA.

expression array   A microarray of probes for expressed sequences, used to analyze the pattern of gene expression in a given cell type or tissue by hybridizing to labeled cDNA from the cell or tissue. (Figures 8.24 and 17.21)

expression cloning   Cloning in vectors that are deigned to allow genes in the insert to be expressed. Used to make purified gene product. (Figures 6.11 and 6.15)

extracellular matrix   A meshwork of polysaccharide and protein molecules found within the extracellular space and in association with the basement membrane of the cell surface. It provides a scaffold to which cells adhere and serves to promote cellular proliferation.

first-degree relatives   Parents, children or sibs.

FISH—*see* fluorescent *in situ* hybridization

fitness (f)   In population genetics, a measure of the success in transmitting genotypes to the next generation, relative to the most successful genotype. Also called biological or reproductive fitness. f always lies between 0 and 1.

fluorescent *in situ* hybridization (FISH)   *In situ* hybridization using a fluorescently labeled DNA or RNA probe. A key technique in modern molecular genetics. (Figures 2.16 and 2.17)

fluorophore   A fluorescent chemical group, used for labeling nucleic acids or proteins. (Box 7.3)

founder effect   High frequency of a particular allele in a population because the population is derived from a small number of founders, one or more of whom carried that allele.

fragile sites   Locations on chromosomes where, under special culture conditions, the chromatin of metaphase chromosomes appears uncondensed. Most are nonpathogenic variants present at varying frequencies in normal healthy individuals, but a few are pathogenic. (Figure 13.5)

fragment ion searching   A method of identifying proteins in tandem mass spectrometry from the mass of fragments they produce. (Figure 8.27)

frameshift mutation   A mutation that alters the triplet reading frame of a mRNA (by inserting or deleting a number of nucleotides that is not a multiple of 3). (Figures 13.14 and 13.15)

framework map   A map of the locations of some physical entities – genetic markers, sequence-tagged sites or clones – across a genome or chromosome. Used as a step towards a full genome sequence. (Box 8.1)

functional genomics   Analysis of gene function on a large scale, by conducting parallel analyses of gene expression/function for large numbers of genes, even all genes in a genome.

fusion protein   The product of a natural or engineered fusion gene: a single polypeptide chain containing amino acid sequences that are normally part of two or more separate polypeptides. (Figures 6.12 and 17.6)

gain-of-function mutations   Mutations that cause the gene product to do something abnormal, rather than simply to lose function. Usually the gain is a change in the timing or level of expression. (Section 13.4)

gamete   Sperm or egg; a haploid cell formed when a primordial germ cell undergoes meiosis.

gastrulation Conversion of the two-layer embryo (consisting of epiblast plus hypoblast) to one that contains the three germ layers: ectoderm, mesoderm and endoderm.

GC box A short sequence, GGGCGG or a close variant, that is found in the promoters of many genes that are transcribed by RNA polymerase II.

gene (1) A functional DNA unit (but see Box 9.4). (2) A factor that controls a phenotype and segregates in pedigrees according to Mendel's laws.

gene conversion A naturally occurring nonreciprocal genetic exchange in which a sequence of one DNA strand is altered so as to become identical to the sequence of another DNA strand. (Box 14.1)

gene expression Production of the gene product (a protein or a functional RNA).

gene family A set of related genes with a presumed common ancestry. (Table 9.6)

gene frequency The proportion of all alleles at a locus that are the allele in question (Section 3.5). Really we mean allele frequency, but the use of gene frequency is too well established now to change.

gene knockdown Targeted inhibition of expression of a gene by, for example, using siRNA or a morpholino oligonucleotide to bind to RNA transcripts.

gene knock-in A targeted mutation that replaces activity of one gene by that of an introduced gene (usually an allele). (Figures 20.8 and 20.9)

gene knockout The targeted inactivation of a gene within an intact cell.

gene ontology A formal controlled vocabulary for describing the functions of genes, as an aid to automated cross-referencing.

gene pool All the genes (in the whole genome or at a specified locus) in a particular population. (Section 3.5)

gene superfamily A set of multiple genes and gene families that show signs of overall distant structural and functional relationships – for example the immunoglobulin and the G-protein coupled receptor superfamilies.

gene targeting Targeted modification of a gene in a cell or organism. (Figures 20.6 to 20.9)

gene therapy Treating disease by genetic modification. May involve adding a functional copy of a gene that has lost its function, inhibiting a gene showing a pathological gain of function, or more generally, replacing a defective gene.

gene tracking Following a disease gene through a pedigree by use of linked markers rather than a direct test for the pathogenic change. (Box 18.1)

gene trap Using random insertions of a reporter construct into genes of embryonic stem cells to generate embryos with random genes inactivated. The affected genes can be identified via the reporter. (Figure 20.16)

general transcription factors DNA-binding proteins that are always required to allow transcription to take place (as distinct from tissue-specific or stage-specific transcription factors). (Table 11.1)

genetic background The genotypes at all loci other than one under active investigation. Variations in genetic background are a major reason for imperfect genotype-phenotype correlations.

genetic code The relationship between a codon and the amino acid it specifies. (Figure 1.25)

genetic distance Distance on a genetic map, defined by recombination fractions and the mapping function, and measured in centimorgans. (Section 14.1)

genetic drift Random changes in gene frequencies over generations because of random fluctuations in the proportions of the alleles in the parental population that are transmitted to offspring. Only significant in small populations.

genetic map A map showing the sequence and recombination fractions between genes, based on breeding experiments or observation of human pedigrees (Table 14.4 and Figure 14.11). Figure 14.5 shows examples of the relation between genetic and physical maps. *Cf.* physical maps.

genetic marker Any character that can be used to follow the segregation of a particular chromosomal segment through a pedigree or in a population. Normally a DNA sequence polymorphism.

genetic redundancy Partially or completely overlapping function of genes at more than one locus, so that loss of function mutations at one locus do not cause overall loss of function.

genome The total set of different DNA molecules of an organelle, cell or organism. The human genome consists of 3×10^9 bp of DNA divided into 25 molecules, the mitochondrial DNA molecule plus the 24 different chromosomal DNA molecules. *Cf.* transcriptome, proteome.

genome browser A program that provides a graphical interface for interrogating genome databases.

genomewide association study (GWAS) The standard approach to identifying factors governing susceptibility to complex disease. (Figure 15.11)

genotype–phenotype correlation The extent to which a phenotype can be predicted from a genotype. Typically much higher in experimental animals, which are inbred and live under standard laboratory conditions, than in humans. Poor correlation is a major limitation in human genetic testing and counseling.

germ line The germ cells and those cells which give rise to them; other cells of the body constitute the soma.

germinal (gonadal, or gonosomal) mosaic An individual who has a subset of germ-line cells carrying a mutation that is not found in other germ-line cells. (Figure 3.22)

glycolipid A lipid molecule with a covalently attached sugar or oligosaccharide.

glycosaminoglycans Long polysaccharide molecules made of pairs of sugar units, one of which is always an amino sugar. A major component of extracellular matrix.

glycosylation Covalent addition of sugars, usually to a protein or lipid molecule.

Golgi apparatus A membranous organelle in which proteins and lipids are modified and sorted for transport to different destinations. (Box 4.1)

gonadal mosaic—*see* germinal mosaic

GT–AG rule   The rule that almost all human introns begin with GT (GU in the RNA) and end in AG. A few follow an AT–AC rule and use a different spliceosome machine.

haploid   Describing a cell (typically a gamete) which has only a single copy of each chromosome (e.g. the 23 chromosomes in human sperm and eggs).

haploinsufficiency   A locus shows haploinsufficiency if producing a normal phenotype requires more gene product than the amount produced by a single functional allele. (Section 13.4)

haplotype   A series of alleles found at linked loci on a single chromosome.

Hardy–Weinberg distribution   The simple relationship between gene frequencies and genotype frequencies that is found in a population under certain conditions. (Section 3.5)

heat map   A form of data display used particularly for expression array data. A table of cells, with each row representing a gene, each column a sample, and the color of each cell representing the level of expression of that gene in that sample. (Box 8.7 Figure 2 and Figure 17.21)

helicase   A protein that acts to separate the two strands of double-stranded nucleic acid, as part of the machinery for replication, recombination and repair.

hematopoietic stem cells   Self-renewing bone marrow cell that gives rise to all the various types of blood cell. (Figure 4.17)

hemizygous   Having only one copy of a gene or DNA sequence in diploid cells. Males are hemizygous for most genes on the sex chromosomes. Deletions occurring on one autosome produce hemizygosity in males and in females.

heritability   The proportion of the causation of a character that is due to genetic causes. (Section 3.4)

heterochromatin   Chromatin that is highly condensed and shows little or no evidence of active gene expression. Facultative heterochromatin may reversibly decondense to form euchromatin, depending on the requirements of the cell. Constitutive heterochromatin remains condensed throughout the cell cycle. It is found at centromeres plus some other regions. *Cf.* euchromatin. (Figure 11.9)

heteroduplex   Double-stranded DNA in which there is some mismatch between the two strands. Important in mutation detection.

heteroplasmy   Mosaicism, usually within a single cell, for mitochondrial DNA variants. (Section 3.2)

heterozygote   An individual having two different alleles at a particular locus.

heterozygote advantage   The situation when somebody heterozygous for a mutation has a reproductive advantage over both homozygotes. Sometimes called over-dominance. Heterozygote advantage is one reason why severe recessive diseases may remain common. (Box 3.7)

histone code   The idea that the pattern of covalent modification of histones in nucleosomes determines the activity of the DNA in the vicinity. In fact, histone modification is only one of several factors that, between them, determine gene expression. (Section 11.2)

histones   A family of small basic proteins that complex with DNA to form nucleosomes. (Figures 2.8 and 11.8)

homeobox   A 180 bp module found in many genes that have functions in development. The products of homeobox genes regulate the expression of target genes through a 60 amino acid DNA-binding homeodomain.

homoduplex   Double-stranded DNA in which the two strands match perfectly. *Cf.* heteroduplex.

homologs (chromosomes)   The two copies of a chromosome in a diploid cell. Unlike sister chromatids, homologous chromosomes are not copies of each other; one was inherited from the father and the other from the mother.

homologs (genes)   Two or more genes whose sequences are significantly related because of a close evolutionary relationship, either between species (orthologs) or within a species (paralogs).

homoplasmy   Of a cell or organism, having all copies of the mitochondrial DNA identical. *Cf.* heteroplasmy.

homozygote   An individual having two identical alleles at a particular locus. For clinical purposes a person is often described as homozygous *AA* if they have two normally-functioning alleles, or homozygous *aa* if they have two pathogenic alleles at a locus, regardless of whether the alleles are in fact completely identical at the DNA sequence level. Homozygosity for alleles identical by descent is called autozygosity.

housekeeping gene   A gene that provides some basic aspect of cell function, common to most or all cells of an organism.

humanized antibodies   Monoclonal antibodies made in rodent systems but where some or most of the rodent-specific sequence has been replaced by human-specific sequence.

humanized mice   Mice that have been modified so that some chosen aspect of their genetics or physiology more closely resembles its human equivalent.

hybrid cells   Cells containing genetic material from more than one species. Human-rodent hybrid cells, containing a full rodent genome and one or more human chromosomes or fragments, were widely used in physical mapping of the human genome.

hybridization   Of nucleic acids, allowing complementary single strands to base-pair (anneal).

hybridization stringency   The degree to which the conditions (temperature, salt concentration) during a hybridization assay permit sequences with some mismatches to hybridize. High stringency conditions allow only perfect matches. (Figure 7.3)

hybridization, molecular   Hybridization of two complementary single-stranded nucleic acids.

hybridoma   A cell line made by fusing an antibody-producing B cell with a cell of a B lymphocyte tumor. The source of monoclonal antibodies.

hydrogen bond   A weak chemical bond that forms when a hydrogen atom lies in line between two oxygen, nitrogen or fluorine atoms. The basis of base-pairing in nucleic acids.

hydrophilic   Of a chemical group, having energetically favorable interactions with water and other polar molecules.

A property of charged or polar groups.

hydrophobic   Of a chemical group, repelled by water and other polar groups. Hydrophobic groups associate together in the interior of protein molecules, membranes etc.

hypoblast   The layer of cells in the pre-gastrulation embryo which gives rise to extraembryonic endoderm. (Figure 5.13)

hypomorph   An allele that produces a reduced amount or activity of product.

identity by descent (IBD)   Alleles in an individual or in two people that are known to be identical because they have both been inherited from a demonstrable common ancestor. (Figure 15.1)

identity by state (IBS)   Alleles that appear identical, but may or may not be identical by descent because there is no demonstrable common source. (Figure 15.1)

immunoblotting   Using an antibody to identify proteins that have been fractionated by size and charge by electrophoresis and then transferred to a nitrocellulose membrane. (Figure 8.21)

immunogen   Any molecule that elicits an immune response.

immunological memory   The ability of the adaptive immune system to mount a rapid and strong response to an antigen that it has previously encountered.

imprinting   In genetics, determination of the expression of a gene by its parental origin. (Table 11.5)

imprinting control center (IC)   A short sequence within an imprinted gene cluster where differential methylation controls the imprinting status of genes within the cluster. (Figure 11.22)

in situ hybridization   Hybridization of a labeled DNA or RNA probe to an immobilized nucleic acid target. The target may be denatured DNA within a chromosome preparation (chromosome *in situ* hybridization), RNA within the cells of a tissue section on a microscope slide (tissue *in situ* hybridization) or RNA within a whole embryo (whole mount *in situ* hybridization).

in vitro mutagenesis   Techniques to introduce a specific desired sequence change into the DNA of a cell through manipulation of vectors. (Figures 6.8 and 6.19)

inbreeding   Marrying a blood relative. The term is comparative, since ultimately everybody is related. The coefficient of inbreeding is the proportion of a person's genes that are identical by descent. (Box 3.6)

indels   Insertion / deletion variants.

induced pluripotent stem (iPS) cells   Somatic cells that have been treated with specific genes or gene products to reprogram them to resemble pluripotent stem cells. They can then be induced to differentiate into desired cell types. A great hope for regenerative medicine.

induction   In development, the process whereby one tissue changes the state or fate of an adjacent tissue.

innate immune system   System of nonspecific response to a pathogen using the natural defenses of the body. *Cf.* adaptive immune system.

inner cell mass (ICM)   A group of cells located internally within the blastocyst which will give rise to the embryo proper.

insertional mutagenesis   Mutation (usually abolition of function) of a gene by insertion of an unrelated DNA sequence into the gene.

insulators   DNA elements that act as barriers to the spread of chromatin changes or the influence of *cis*-acting elements.

interallelic complementation—*see* complementation

interference   In meiosis, the tendency of one crossover to inhibit further crossing over within the same region of the chromosomes. (Section 14.1)

interphase   All the time in the cell cycle when a cell is not dividing.

interphase FISH   Fluorescence *in situ* hybridization of a probe to interphase cell nuclei. Used to detect aneuploidies or other chromosomal abnormalities without the need to culture cells, or to examine the subnuclear localization of chromosomes in nondividing cells. (Figures 2.17 and 17.4)

intrabodies   Engineered nonsecreted intracellular antibodies that can be used to inactivate selected molecules inside a cell.

intracytoplasmic sperm injection (ICSI)   A human infertility treatment in which sperm heads are injected into unfertilized eggs. Sometimes used experimentally to make transgenic animals by first coating the sperm head with the desired transgene DNA.

intron   Segments of a transcript that are cut out during splicing (Figure 1.16). Introns appear to be largely nonfunctional, but they may contain elements that modify transcription or splicing of their host gene, and some introns are the source of small nucleolar RNAs (Figure 11.22), microRNAs (Figure 9.17), and other functional RNAs.

iron-response element (IRE)   A sequence element in certain mRNA species that changes the activity of the mRNA in response to excess or deficiency of Fe^{++}. (Figure 11.32)

isochromosome   An abnormal symmetrical chromosome, consisting of two identical arms, which are normally either the short arm or the long arm of a normal chromosome.

K_a/K_s ratio   An indicator of selection affecting a gene sequence: the ratio of non-synonymous to synonymous codon changes in a comparison of two organisms. (Box 10.5)

karyogram   A display of the chromosomes of a cell, sorted into pairs. (Figure 2.15)

karyotype   A summary of the chromosome constitution of a cell or person, such as 46,XY. Often used more loosely to mean an image showing the chromosomes of a cell sorted in order and arranged in pairs (strictly, a karyogram).

kinetochore   The structure at chromosomal centromeres to which the spindle fibers attach.

lagging strand   In DNA replication, the strand that is synthesized as Okazaki fragments. (Figure 1.11)

lateral inhibition   A process during embryogenesis in which cells that differentiate inhibit neighboring cells from doing the same. The result is to develop spaced sets of differentiated cells.

LCR—*see* locus control region

leader sequence   A sequence of a dozen or so amino acids at the N-terminal end of some proteins that serves as a signal defining the location to which the protein must be transported. Leader sequences are usually cleaved off once the sorting process is completed.

leading strand   In DNA replication, the strand that is synthesized continuously. (Figure 1.11)

ligand   Any molecule that binds specifically to a receptor or other molecule. (Figure 4.7)

ligase   DNA ligase is an enzyme that can seal single-strand nicks in double-stranded DNA or covalently join two oligonucleotides that are hybridized at adjacent positions on a DNA strand.

LINE (long interspersed nuclear element)   A class of repetitive DNA sequences that make up about 20% of the human genome. Some are active transposable elements. (Figures 9.20 and 9.21)

linkage disequilibrium (LD)   A statistical association between particular alleles at separate but linked loci, normally the result of a particular ancestral haplotype being common in the population studied. An important tool for high resolution mapping. (Figure 14.14, and Boxes 15.1 and 15.2)

linker (adapter) oligonucleotide   A double-stranded oligonucleotide that can be ligated to a DNA molecule of interest and which has been designed to contain some desirable characteristic, e.g. a favorable restriction site or a binding site for a PCR primer.

liposome   A synthetic lipid vesicle used to transport a molecule of interest into a cell. (Figure 21.11)

locus   A unique chromosomal location defining the position of an individual gene or DNA sequence.

locus control region (LCR)   A stretch of DNA containing regulatory elements which control the expression of genes in a gene cluster that may be located tens of kilobases away. (Figure 11.7)

locus heterogeneity   Determination of the same disease or phenotype by mutations at different loci. A common problem in mapping genetic diseases. (Section 3.2)

lod score (z)   A measure of the likelihood of genetic linkage between loci. The log (base 10) of the odds that the loci are linked (with recombination fraction θ) rather than unlinked. For Mendelian characters a lod score greater than +3 is evidence of linkage; one that is less than –2 is evidence against linkage. (Box 14.2 and Figure 14.10)

loop out   The way a long strand, e.g. of chromatin, can form a series of loops attached at their bases to a central scaffold.

loss-of-function mutations   Mutations that cause the gene product to lose its function, partially or totally. (Section 13.4)

loss of heterozygosity (LOH)   Homozygosity or hemizygosity in a tumor or other somatic cell when the constitutional genotype is heterozygous. Evidence of a somatic genetic change. (Figures 17.9 and 17.10)

lymphocytes   White blood cells involved in the immune response. The two main classes are T cells and B cells. (Table 4.7)

lyonization   X-chromosome inactivation, the process by which cells adapt to differing numbers of X chromosomes. (Section 3.2)

lysosomes   Membrane-bound organelles full of digestive enzymes where macromolecules are broken down to their basic subunits.

M13   A specialized cloning vector used to produce single-stranded DNA for sequencing. (Figure 6.9)

major groove   In a DNA double helix, the larger of the two spiral grooves that run the length of the molecule. Many DNA-binding proteins recognize sequence-specific features in the major groove.

major histocompatibility complex (MHC) proteins   Proteins encoded by the Class I and Class II regions of the MHC that function in antigen recognition by binding fragments of antigens and presenting them on the surface of T cells. (Box 4.4)

major pseudoautosomal region   A 2.6 Mb homologous region at the tips of Xp and Yp. These sequences pair in male meiosis and have an obligatory recombination.

manifesting heterozygote   A female carrier of an X-linked recessive condition who shows some clinical symptoms, presumably because of skewed X-inactivation. (Section 3.2)

mapping function   A mathematical equation describing the relation between recombination fraction and genetic distance. The mapping function depends on the extent to which interference prevents close double recombinants. (Section 14.1)

marker (molecular)   A chemical group or molecule that can be assayed in some way.

marker gene   In a cloning vector, a gene whose product assists cloning by enabling cells containing the vector to be selected, and/or cells containing recombinant vector to be recognized.

matrilineal inheritance   Transmission from just the mother, but to children of either sex; the pattern of mitochondrial inheritance. (Figure 3.10)

matrix-assisted laser desorption/ionization (MALDI)   A method for analyzing large nonvolatile molecules on a mass spectrometer. The molecules are mixed into a light-absorbing matrix which is vaporized by a laser pulse. (Box 8.8)

meiosis   The specialized reductive form of cell division used only to produce gametes. (Figures 2.6 and 2.7)

melting temperature (Tm)   In denaturing double-stranded DNA, the temperature at the mid-point of the transition from double to single strands. (Table 7.1)

Mendelian   A character whose pattern of inheritance suggests it is caused by variation at a single genetic locus; a monogenic character.

mesenchymal stem cells   The stem cells of connective tissues.

mesoderm   Embryonic tissue that is the precursor to muscle, connective tissue, the skeleton and many internal organs. (Figures 5.16 and 5.17)

messenger RNA (mRNA)   A processed gene transcript that carries protein-coding information to the ribosomes.

MHC (major histocompatibility complex) restriction The process that restricts recognition of foreign antigens by T cells to fragments that are associated with an MHC molecule on the surface of an antigen-presenting cell.

microcell-mediated chromosome transfer A technique for introducing a selected single chromosome into a mutant cell, to see if it can correct the mutant phenotype. (Figure 16.22)

microRNAs (miRNAs) Short (21–22 nt) RNA molecules encoded within normal genomes that have a role in regulation of gene expression and maybe also of chromatin structure. (Figures 9.16, 9.17, and 11.33)

microsatellite instability A type of genomic instability seen in some colon and other cancers where replication errors are not corrected; observed as the production of extra alleles at many polymorphic microsatellites. (Figure 17.19)

microsatellite Small run (usually less than 0.1 kb) of tandem repeats of a very simple DNA sequence, usually 1–4 bp, for example (CA)n. Often polymorphic, providing the primary tool for genetic mapping during the 1990s. Sometimes also described as STR (short tandem repeat) polymorphism. (Figure 13.2)

microtubules Long hollow cylinders constructed from tubulin polymers. Form the spindle fibers that move chromosomes in mitosis and meiosis, and contribute to the cytoskeleton.

minisatellite An array (typically 0.1–20 kb long) of tandemly repeated 10–50 bp DNA sequences. (Section 13.1)

minor groove In a DNA double helix, the smaller of the two spiral grooves that run along the length of the molecule.

missense changes Changes in a coding sequence that cause one amino acid in the gene product to be replaced by a different one. (Section 13.3)

mitogen A substance that stimulates cells to divide.

mitosis The normal process of cell division, which produces daughter cells genetically identical to the parent cell. (Figure 2.3)

MLPA—*see* multiplex ligation-dependent probe amplification

monoclonal antibody (mAb) A pure antibody with a single specificity, produced by hybridoma technology, as distinct from polyclonal antibodies that are raised by immunization. (Box 8.6)

morphogenesis The formation of structures during embryonic development. (Table 5.2)

morphogens Signaling molecules that can impose a pattern on a field of cells in response to a gradient of concentration of the morphogen.

Morpholino A stable chemically modified RNA analog used to inhibit expression of a gene under study. (Figure 20.14)

morula An early stage of embryonic development; a loosely packed ball of cells that will give rise to the blastocyst. (Figure 5.9)

mosaic An individual who has two or more genetically different cell lines derived from a single zygote. The differences may be point mutations, chromosomal changes, etc. (Figures 3.22 and 3.23)

motif A short sequence or structure (usually in a protein) that forms a recognizable signature of a structure or function.

mRNA—*see* messenger RNA

mtDNA (mitochondrial DNA) DNA of the 16,569 nt mitochondrial genome. (Figure 9.3)

multifactorial A character that is determined by some unspecified combination of genetic and environmental factors. *Cf.* polygenic.

multiplex ligation-dependent probe amplification (MLPA) A method for testing a DNA sample for specific copy-number changes, usually deletion or duplication of whole exons. (Figure 18.15)

multipoint mapping Genetic mapping based on considering the simultaneous segregation of more than two marker in pedigrees.

mutagenesis Creation of mutations, *in vitro* or *in vivo*.

mutation–selection hypothesis The hypothesis that genetic susceptibility to complex diseases is mainly the result of a heterogeneous collection of mutations that turn over relatively quickly because of natural selection (in distinction to the common disease-common variant hypothesis, *q.v.*). (Section 15.6)

necrosis Cell death as a result of irreparable external damage.

neural crest A group of migratory cells in an embryo that form along the lateral margin of the neural folds and give rise to many different tissues. Neural crest derivatives include part of the peripheral nervous system, melanocytes, some bone and muscle, the retina and other structures. (Box 5.4)

neural plate In an embryo, the precursor of the neural tube. (Figure 5.18)

neural tube In vertebrate embryos, a tube of ectoderm that will form the brain and spinal cord. (Figure 5.18)

node (1) In an early embryo, the anterior end of the primitive streak. (Figure 5.12) (2) In a phylogenetic tree, the point at which two lineages diverge. (Figure 10.24)

non-allelic homologous recombination (NAHR) Recombination between misaligned DNA repeats, either on the same chromosome, on sister chromatids or on homologous chromosomes. NAHR generates recurrent deletions, duplications or inversions. (Figure 13.20)

noncoding RNA (ncRNA) RNA that does not contain genetic code for a protein. Noncoding RNAs have many different functions in cells. (Table 9.9)

nondisjunction Failure of chromosomes (sister chromatids in mitosis or meiosis II; paired homologs in meiosis I) to separate (disjoin) at anaphase. The major cause of numerical chromosome abnormalities.

non-parametric In linkage analysis, a method such as affected sib pair analysis that does not depend on a specific genetic model.

non-parametric lod (NPL) score The statistical output of non-parametric linage analysis. (Table 15.6)

non-penetrance   The situation when somebody carrying an allele that normally causes a dominant phenotype does not show that phenotype. An effect of other genetic loci or of the environment. A pitfall in genetic counseling. (Figure 3.14)

non-recombinant   In a pedigree, two loci are non-recombinant in a gamete that contains the same combination of alleles as the person received from his or her parent. (Figure 14.1)

nonsense mutation   A mutation that replaces the codon for an amino acid with a premature termination codon. (Figure 13.12)

nonsense-mediated mRNA decay   A cellular mechanism that degrades mRNA molecules that contain a premature termination codon (>50 nt upstream of the last splice junction). (Figure 13.12)

northern blot   A membrane bearing RNA molecules that have been size-fractionated by gel electrophoresis, used as a target for a hybridization assay. Used to detect the presence and size of transcripts of a gene of interest. (Figure 7.11)

notochord   A flexible rod-like structure that in mammalian embryos induces formation of the central nervous system. (Figure 5.15)

nuclear reprogramming   Large-scale epigenetic changes to convert the pattern of gene expression in a cell to that typical of another cell type or state.

nucleic acid   DNA or RNA.

nucleolar organizer region (NOR)   The satellite stalks of human chromosomes 13, 14, 15, 21 and 22. NORs contain arrays of ribosomal RNA genes and can be selectively stained with silver. Each NOR forms a nucleolus in telophase of cell division; the nucleoli fuse in interphase.

nucleolus   The site within the nucleus where ribosomal RNA is transcribed and assembled into the ribosomal subunits.

nucleoside   A purine or pyrimidine base linked to a sugar (ribose or deoxyribose). (Table 1.1)

nucleosome   The basic structural unit of chromatin, comprising 147 bp of DNA wound round an octamer of histone molecules. (Figure 2.8)

nucleotide   A nucleoside phosphate. The basic building block of DNA and RNA. (Table 1.1)

odds ratio   In case-control studies, the relative odds of a person with or without a factor under study being a case. (Box 19.3)

Okazaki fragments   Short strands of DNA; the immediate product of lagging-strand replication.

oligogenic   A character that is determined by a small number of genes acting together.

oligosaccharide   A molecule consisting of a few linked sugar units.

oncogene   A gene involved in control of cell proliferation which, when overactive can help to transform a normal cell into a tumor cell (Table 17.1). Originally the word was used only for the activated forms of the gene, and the normal cellular gene was called a proto-oncogene, but this distinction is now widely ignored.

one gene–one enzyme hypothesis   The hypothesis advanced by Beadle and Tatum in 1941 that the primary action of each gene was to specify the structure of an enzyme. Historically very important, but now seen to be only part of the range of gene functions.

origin of replication   A site on DNA where replication can be initiated.

ortholog   Orthologous genes are genes present in different organisms that are related through descent from a common ancestral gene. (Figure 10.10)

P1 artificial chromosome (PAC)   A vector based on P1 bacteriophage that allows inserts of ≅100 kb to be cloned in *E. coli* cells.

paired-end mapping   Comparing the number of nucleotides separating two known sequences in a person's DNA with the number in a reference genome, as a way of identifying structural rearrangements. (Figure 13.6)

palindrome   A DNA sequence such as ATCGAT that reads the same when read in the 5′→3′ direction on each strand. DNA-protein recognition, for example by restriction enzymes, often relies on palindromic sequences.

paralog   One of a set of homologous genes within a single species. (Figure 10.10)

parametric   In linkage analysis, a method such as standard lod score analysis, that requires a tightly specified genetic model.

paramutation   An inherited phenotype that mimics the result of a DNA sequence change, but where any change is epigenetic rather than a change of the DNA sequence. (Figure 11.24)

passenger mutations   In cancer, mutations that arise incidentally during development of a tumor and do not play any causative role in the process. *Cf.* driver mutations.

PCR—*see* polymerase chain reaction

penetrance   The frequency with which a genotype manifests itself in a given phenotype.

peptide mass fingerprinting (PMF)   A way of identifying proteins in a mixture by digesting with trypsin, analyzing the mixture of peptides on a mass spectrometer, and comparing the resulting peaks against a database showing the patterns produced by known proteins. (Figure 8.27)

phage   A bacteriophage. A virus that replicates in bacterial cells.

phage display   An expression cloning method in which foreign genes are inserted into a phage vector and are expressed to give polypeptides that are displayed on the surface (protein coat) of the phage. (Figure 6.13)

phagemid vector   A vector containing sequences derived from phage and from plasmids, that allows inserts of several kilobases to be cloned in *E. coli* and then released as single-stranded DNA ready for sequencing. (Figure 6.10)

pharmacodynamics   The response of a target organ or cell to a drug.

pharmacogenetics   The study of the influence of individual genes or alleles on the metabolism or function of drugs.

pharmacogenomics The use of genome resources (genome sequences, expression profiles etc.) to identify new drug targets.

pharmacokinetics The absorption, activation, catabolism and elimination of a drug.

phase-known In a pedigree, a person in whom the phase of two or more loci (i.e. the combination of alleles inherited from each individual parent) is known. (Section 14.3)

phase-unknown—*see* phase-known

phenocopy A person (or organism) who has a phenotype normally caused by a certain genotype, but who does not have that genotype. Phenocopies may be the result of a different genetic variant, or of an environmental factor.

phenome The totality of phenotypes. *Cf.* genome.

phenotype The observable characteristics of a cell or organism, including the result of any test that is not a direct test of the genotype.

phylogeny Classification of organisms according to perceived evolutionary relatedness. (Figures 10.24 to 10.26)

physical map A map showing the locations of some physical entities on a chromosome or genome. The entities might be DNA sequences or features such as natural or radiation-induced breakpoints. *Cf.* genetic map.

pitch Of a spiral, the distance occupied by a single turn, which is 3.4 nm in the standard B-DNA double helix.

plasma membrane The membrane that surrounds a cell.

plasmid A small circular DNA molecule that can replicate independently in a cell. Modified plasmids are widely used as cloning vectors.

pleiotropy The common situation where variation in one gene affects several different aspects of the phenotype.

ploidy The number of complete sets of chromosomes in a cell. Cells can be haploid, diploid, triploid… polyploid.

polar body In female meiosis, the small product of the asymmetrical division of the cell mass during each division of meiosis. The polar bodies eventually degenerate.

poly(A) tail The string of 200 or so A residues that are attached to the 3′ end of a mRNA. The poly(A) tail is important for stabilizing mRNA. (Figure 1.21)

polyadenylation Addition of the poly(A) tail to the 3′ end of a mRNA. (Figure 1.21)

polyclonal antibodies Natural antibodies produced by the adaptive immune system in response to an antigen. Polyclonal antibodies are typically a mixture of species that respond to different epitopes of the stimulating antigen.

polygenic A character determined by the combined action of a number of genetic loci. Mathematical polygenic theory (Section 3.4) assumes there are very many loci, each with a small effect.

polylinker In a cloning vector, a short sequence containing recognition sites for several different restriction enzymes, as an aid to making recombinant molecules. (Figure 6.4)

polymerase chain reaction (PCR) The standard technique used to amplify short DNA sequences. (Figure 6.16 and Box 6.2)

polymorphism Strictly, the existence of two or more variants (alleles, phenotypes, sequence variants, chromosomal structure variants) at significant frequencies in the population. Looser usages among molecular geneticists include (1) any sequence variant present at a frequency >1% in a population (2) any nonpathogenic sequence variant, regardless of frequency. (Section 13.1)

polymorphism information content (PIC) A measure of how often genotyping a polymorphism can make a meiosis informative for linkage. For most purposes the average heterozygosity at the locus is used for this purpose, but the PIC is a more accurate metric.

polypeptide A string of amino acids linked by peptide bonds (Figure 1.3). Proteins may consist of one or more polypeptide chains.

population attributable risk (PAR) In epidemiology, the contribution that a particular factor or combination of factors makes to the overall incidence of a condition. Can be used to measure how much of the overall genetic susceptibility to a disease can be accounted for by the factors thus far identified. (Section 19.4)

position effect Complete or partial silencing of a gene when a chromosomal rearrangement moves it close to heterochromatin. (Section 16.2)

positional cloning Identifying a disease gene using knowledge of its chromosomal location. The way the great majority of Mendelian disease genes were identified. (Figure 16.2)

positional information Information supplied to or possessed by cells according to their position in a multicellular organism.

positive selection Selection in favor of a particular genotype.

potency Of a cell, its potential for dividing into different cell types. Cells can be totipotent, pluripotent or committed to one fate.

premutation alleles Among diseases caused by dynamic mutations (expanding repeats), a repeat expansion that is large enough to be unstable on transmission, but not large enough to cause disease. (Section 13.3)

primary structure Of a polypeptide or nucleic acid, the linear sequence of amino acids or nucleotides in the molecule. (Table 1.8)

primary transcript The RNA product of transcription of a gene by RNA polymerase, before splicing. The primary transcript of a gene contains all the exons and introns.

primer A short oligonucleotide, often 15–25 bases long, which base-pairs specifically to a target sequence to allow a polymerase to initiate synthesis of a complementary strand.

primitive streak In early embryos, a transient structure that defines the longitudinal axis. (Figure 5.12)

primordial germ cells (PGCs) Cells in the embryo and fetus which will ultimately give rise to germ-line cells.

proband (or propositus) The person through whom a family was ascertained.

probe A known DNA or RNA fragment (or a collection of such fragments) used in a hybridization assay to identify closely related DNA or RNA sequences within a complex, poorly understood mixture of nucleic acids. In standard hybridization assays, the probe is labeled but in reverse hybridization assays the target is labeled.

programmed cell death Programmed death of an animal cell, in which a 'suicide' program is activated in the cell. (Table 4.5 and Figure 4.16)

prokaryotes Single-celled microorganisms (bacteria or archaea) that lack a membrane-bound nucleus.

prometaphase In mitosis, late prophase, when chromosomes are well separated but not yet maximally contracted; the optimum stage for normal cytogenetic analysis. (Figure 2.15)

promoter A combination of short sequence elements, normally just upstream of a gene, to which RNA polymerase binds in order to initiate transcription of the gene. (Figure 11.1)

proofreading An enzymic mechanism by which DNA replication errors are identified and corrected.

propositus—*see* proband

protein A molecule consisting of one or more polypeptide chains folded into a specific 3-dimensional structure.

protein domain A structural (and often functional) subunit of a protein; a structural module that may be found in several different proteins.

proteome The totality of proteins in a cell or organism.

proteomics Global or large-scale studies of the proteins in a cell or organism. (Figure 12.6)

proto-oncogenes Normal cellular genes whose function is to promote cell proliferation, and in which tumors may carry activating mutations.

proximal Of a chromosomal location, comparatively close to the centromere.

pseudoautosomal regions (PAR) Regions at each tip of the X and Y chromosomes containing X-Y homologous genes (Figures 10.17 and 10.18). Because of X-Y recombination, alleles in these regions show an apparently autosomal mode of inheritance. (Figure 3.9)

pseudogene A DNA sequence that shows a high degree of sequence homology to a non-allelic functional gene, but which is itself nonfunctional. (Box 9.2)

purifying selection (negative selection) Selection against unfavorable genotypes.

purines Nitrogenous bases having a specific double-ring chemical structure. Adenine and guanine are purines. (Figure 1.2)

pyrimidines Nitrogenous bases having a specific single-ring chemical structure. Cytosine, thymine and uracil are pyrimidines. (Figure 1.2)

pyrosequencing A technique for sequencing a few nucleotides from a defined start point. (Figure 8.8)

qPCR—*see* quantitative PCR

quadrupole mass analyzer A method of separating analytes by mass/charge ratio for mass spectrometry. (Box 8.8)

quantitative character A character like height, which everybody has, but to differing degree – as compared with a dichotomous character like polydactyly, which some people have and others do not.

quantitative PCR (qPCR) PCR methods that allow accurate estimation of the amount of template present. Reliable qPCR methods are based on real-time techniques. (Box 8.5)

quantitative trait locus (QTL) A locus that contributes to determining the phenotype of a continuous character.

quaternary structure The overall structure of a multimeric protein. (Table 1.8)

reading frame During translation, the way the continuous sequence of the mRNA is read as a series of triplet codons. There are three possible reading frames for any mRNA, and the correct reading frame is set by correct recognition of the AUG initiation codon.

real-time PCR A PCR process in which the accumulation of product is followed in real time, which allows accurate quantitation of the amount of template present. (Box 8.5)

recessive A character is recessive if it is manifest only in the homozygote.

recombinant In linkage analysis, a gamete that contains a combination of alleles that is different from the combination which the parent inherited from their parent. (Figure 14.1)

recombinant DNA An artificially constructed hybrid DNA containing covalently linked sequences from two or more different sources. (Figure 6.3)

recombinant proteins Proteins produced in expression cloning systems. Although the vector is recombinant, the protein is not actually recombinant.

recombination fraction For a given pair of loci, the proportion of meioses in which they are separated by recombination. Usually signified as θ. θ, values vary between 0 and 0.5. (Section 14.1)

regression to the mean The phenomenon whereby parents with extreme values of quantitative characters have, on average, children with less extreme values. This is a purely statistical phenomenon, and has no bearing on whether or not a character is genetically determined. (Section 3.4)

relative risk In epidemiology, the relative risks of developing a condition in people with and without a susceptibility factor.

replicate Make an exact copy.

replication fork In DNA replication, the point along a DNA strand where the replication machinery is currently at work.

replicon Any nucleic acid that is capable of self-replication. Many cloning vectors use extrachromosomal replicons (as in the case of plasmids), while others use chromosomal replicons, either directly (as in the case of yeast artificial chromosome vectors), or indirectly, by allowing integration into chromosomal DNA.

reporter gene A gene used to test the ability of an upstream sequence joined on to it to cause its expression. Putative *cis*-acting regulatory sequences can be coupled to a reporter gene and transfected into suitable cells to study their function. Alternatively, transgenic animals (and other organisms) are often made with a promotorless reporter gene integrated at random into the chromosomes, so that expression of the reporter marks the presence of an efficient promoter. (Figure 20.9)

reporter molecule   A molecule whose presence is readily detected (for example, a fluorescent molecule) that is attached to a DNA sequence we wish to monitor. (Figures 7.7 and 7.8)

restriction endonucleases   A bacterial enzyme that cuts double-stranded DNA at a short (normally 4, 6 or 8 bp) recognition sequence. (Box 6.1 and Table 6.1)

restriction fragment length polymorphism (RFLP)   A DNA polymorphism that creates or abolishes a recognition sequence for a restriction endonuclease. When DNA is digested with the relevant enzyme, the sizes of the fragments will differ, depending on the presence or absence of the restriction site. (Figures 8.14 and 13.1C)

restriction site polymorphism (RSP)—*see* restriction fragment length polymorphism

retrogene   A functional gene that appears to be derived from a reverse-transcribed RNA. (Table 9.8)

retrovirus   An RNA virus with a reverse transcriptase function, enabling the RNA genome to be copied into cDNA prior to integration into the chromosomes of a host cell. (Figure 17.3)

reverse genetics   Inferring phenotypes from knowledge of genes (a reversal of the classical pathway in which genes are identified through the study of phenotypes)

reverse transcriptase   An enzyme, often of viral origin, the makes a DNA copy of an RNA template; a RNA-dependent DNA polymerase.

reverse transcriptase PCR (RT-PCR)   Indirect PCR amplification of RNA by first making a cDNA copy using reverse transcriptase.

ribonuclease   An enzyme that breaks down RNA.

ribonuclease protection assay   A method for quantitating one specific RNA transcript in a complex mixture. Uses a labeled antisense probe to protect the transcript of interest from degradation by ribonuclease.

ribosomal DNA   The DNA from which ribosomal RNA is transcribed. In human cells, located on the short arms of the acrocentric chromosomes (13, 14, 15, 21 and 22). (Figure 1.22)

ribosome   The large cytoplasmic protein-RNA complex where polypeptides are assembled using information in a messenger RNA.

ribozyme   A natural or synthetic catalytic RNA molecule.

risk ratio (λ)   In family studies, the relative risk of disease in a relative of an affected person, compared to a member of the general population. (Section 15.1)

RNA editing   Insertion, deletion or substitution of nucleotides in an mRNA after transcription. An unusual event in humans. (Figure 11.30)

RNA interference (RNAi)   The use of siRNAs to knock down (but rarely completely abolish) expression of specified genes. A powerful tool for studying gene function. (Box 9.6)

RNA polymerase   An enzyme that can add ribonucleotides to the 3′ end of an RNA chain. Most RNA polymerases use a DNA template, but some use an RNA template, and hence synthesize double-stranded RNA. (Table 1.5)

RNA processing   The processes required to convert a primary transcript into a mature messenger RNA – capping, splicing and polyadenylation.

RNA splicing—*see* splicing

RT-PCR—*see* reverse transcriptase PCR

SAGE (serial analysis of gene expression)   A method of expression profiling based on sequencing. (Box 11.1)

satellite (chromosome)   Stalked knobs variably seen on the ends of the short arms of the acrocentric chromosomes (13, 14, 15, 21, and 22)

satellite (DNA)   Originally described a DNA fraction that forms separate minor bands on density gradient centrifugation because of its unusual base composition. The DNA is composed of very long arrays of tandemly repeated DNA sequences.

second messenger   Intracellular signaling molecules that relay signals from cell surface receptors to downstream targets. (Table 4.4)

secondary structure   The path of the backbone of a folded polypeptide or single-stranded nucleic acid, determined by weak interactions between residues in different parts of the sequence. (Table 1.8)

second-degree relatives   Uncles, aunts, nephews, nieces, grandparents, grandchildren and half-sibs.

segmental aneuploidy syndrome   A syndrome caused by deletion or duplication of a segment of a chromosome. (Table 13.2)

segmental duplication   The existence of very highly related DNA sequence blocks on different chromosomes or at more than one location within a chromosome. (Figure 9.6)

segregation   (1) The distribution of allelic sequences between daughter cells at meiosis. Allelic sequences are said to segregate, non-allelic sequences to assort. (2) In pedigree analysis, the probability of a child inheriting a phenotype from a parent.

segregation analysis   The statistical methodology for inferring modes of inheritance. (Section 15.2)

segregation ratio   The proportion of offspring who inherit a given gene or character from a parent. (Section 15.2)

semi-conservative   Of DNA replication, each daughter double helix contains one parental and one newly synthesized strand. (Figure 1.10)

semi-discontinuous   Of DNA replication, where one newly synthesized strand has to be made in short pieces (Okazaki fragments) because DNA polymerase can only extend a chain in the 5′→3′ direction. (Figure 1.11)

sense strand   The DNA strand of a gene that is complementary in sequence to the template (antisense) strand, and identical to the transcribed RNA sequence (except that DNA contains T where RNA has U). Quoted gene sequences are always of the sense strand, in the 5′→3′ direction. (Figure 1.13)

sensitivity   Of a test, the proportion of all true positives that it is able to detect.

sequence identity   The degree to which two nucleic acid or protein sequences are identical.

sequence similarity   A looser measure of sequence identity that takes into account synonymous codon replacements in DNA and conservative amino acid replacements in proteins.

sequence-tagged site (STS)   Any unique piece of DNA for which a specific PCR assay has been designed, so that any DNA sample can be easily tested for its presence or absence. (Box 8.1)

sex chromatin   The Barr body (*q.v.*)

short interfering RNA (siRNA)   21–22 nt double-stranded RNA molecules that can dramatically shut down expression of genes (RNA interference). siRNAs are a major tool for studying gene function. (Box 9.6)

short tandem repeat (STR) polymorphism   A polymorphic microsatellite.

shotgun sequencing   Sequencing a genome by mass sequencing of random fragments, then using computers to assemble the mass of short sequence reads into a final overall sequence. Works well for simple genomes, but assembly is difficult for large genomes with much repetitive DNA, as with humans.

silencer   Combination of short DNA sequence elements which suppress transcription of a gene.

silent change   A nucleotide substitution in a coding sequence that does not alter the amino acid encoded. Silent mutations may nevertheless cause problems by interfering with splicing. (Figure 13.13C)

SINE (short interspersed nuclear element)   A class of moderate to highly repetitive DNA sequence families, of which the best known in humans is the Alu repeat family. (Figures 9.20 and 9.21)

single nucleotide polymorphism (SNP)   A position in the genome where two or occasionally three alternative nucleotides are common in the population. May be pathogenic or neutral. The dbSNP database lists human SNPs, but includes some rare pathogenic variants and some variants that involve two or more contiguous nucleotides.

single selection   Collecting a small subset of families in a population by collecting consecutive cases with the condition under study. Families with two affected cases will be twice as likely to feature in the collection as those with only one affected case, and families with three affected cases will be three times as likely. This introduces specific biases in the epidemiology of the sample compared to the general population, which need appropriate statistical treatment. (Box 3.4)

single-stranded binding protein   A protein that binds and stabilizes single-stranded DNA. Important in recombination and DNA repair.

sister chromatid   Two chromatids present within a single chromosome and joined by a centromere. Non-sister chromatids are present on different but homologous chromosomes.

small nucleolar RNA (snoRNA)   A large family of small RNA molecules present in the nucleolus that act as guides to modify specific bases in other RNA molecules, especially ribosomal RNAs.

SNP—*see* single nucleotide polymorphism

somatic cell   Any cell in the body that is not part of the germ line.

somatic cell nuclear transfer (SCNT)   An experimental manipulation in which the nucleus of an unfertilized egg is removed and replaced by the nucleus of a somatic cell from another animal. The technique used to produce Dolly the sheep. (Figure 20.4)

somatopleure   In an embryo, the somatic mesoderm. (Figure 5.16)

somites   Paired blocks of segmental mesoderm that will establish the segmental organization of the body by giving rise to most of the axial skeleton (including the vertebral column), the voluntary muscles and part of the dermis of the skin. (Figure 5.16)

Southern blot   Transfer of DNA fragments from an electrophoretic gel to a nylon or nitrocellulose membrane (filter), in preparation for a hybridization assay. (Figure 7.10)

specificity   In testing, a measure of the performance of a test. Specificity = (1 − false positive rate). (Box 19.2)

splanchnopleure   In an embryo, the visceral mesoderm. (Figure 5.16)

spliceosome   The large ribonucleoprotein complex that splices primary transcripts to remove introns.

splicing   Cutting out the introns from an RNA primary transcript and joining together the exons. (Figures 1.16 and 1.18)

SSCP (single strand conformation polymorphism)   A method for scanning a short piece of DNA (up to 200 bp) for sequence variants compared to a control sample. (Figure 18.2)

start codon   In mRNA, the AUG codon at which the ribosome initiates protein synthesis. Start codons are embedded in the Kozak consensus sequence, GCCRCC<u>AUG</u>G (R = purine).

stem cell   A cell that can act as a precursor to differentiated cells but which retains the capacity for self-renewal.

stem cell niches   The special locations of stem cells. (Figure 4.18)

sticky end   A short single-stranded protrusion at one end of a double-stranded nucleic acid. Molecules with complementary sticky ends can associate and then be covalently joined by DNA ligase. This is the key step in making recombinant DNA.

stop codon   In mRNA, a UAA, UAG or UGA triplet. When the ribosome encounters an in-frame stop codon it dissociates from the mRNA and releases the nascent polypeptide.

stratification   A population is stratified if it consists of several sub-populations that do not interbreed freely. Stratification is a source of error in association studies and risk estimation.

stringency (of hybridization)   The choice of conditions that will allow either imperfectly matched sequences or only perfectly matched sequences to hybridize.

subfunctionalization   The independent specialization of the two gene copies formed by gene duplication. (Figure 10.11)

subtraction cloning   A method of cloning DNA sequences that are present in one DNA sample and absent in a second, generally similar, sample. Has been used to clone a gene that is deleted in a patient with a disease by subtraction against normal DNA.

supercoiling   Coiling an already coiled strand.

susceptibility gene   A gene, variation in which influences susceptibility or resistance to a complex disease.

synapsis   A close functional association of two partners, e.g. homologous chromosomes in prophase I of meiosis (Figure 2.6) or neurons in the nervous system. (Figure 4.13)

synaptonemal complex   A proteinaceous substance that helps link paired homologous chromosomes during prophase I of meiosis.

synonymous   Of codons, two codons that specify the same amino acid.

synteny   Loci are syntenic if they are on the same chromosome. Syntenic loci are not necessarily linked: loci sufficiently far apart on the chromosome assort at random, with 50% recombinants.

synthetic biology   The attempt to produce wholly synthetic living organisms.

systems biology   The attempt to get a full understanding of how cells and organisms function by quantitative modeling of the network of interactions between genes, pathways and metabolism that link inputs and outputs.

T cells (T lymphocytes)   A heterogeneous set of lymphocytes including T helper cells and cytotoxic T lymphocytes that, between them, are responsible for adaptive cell-mediated immunity.

tag-SNPs   Single nucleotide polymorphisms selected because the combined genotypes of a small number of such tag-SNPs serve to identify haplotype blocks and make it unnecessary to genotype every SNP in the block. (Figure 15.7)

target   In hybridization assays, the sequence to which the probe is intended to hybridize.

TATA box   A short sequence, TATAAA or a close variant, that is part of the promoter of many genes that are transcribed by RNA polymerase II in a tissue-specific or stage-specific way.

taxon   A phylogenetic group. (Figure 10.22)

terminal differentiation   The state of a cell that has ceased dividing and has become irreversibly committed to some specialized function.

tertiary structure   The 3-dimensional structure of a polypeptide. (Table 1.8)

third-degree relatives   The parents or children of second-degree relatives of a person, most commonly the first cousins.

time-of-flight (TOF) mass analyzer   A method of analysis in mass spectrometry based on the time taken for ions to travel down a flight tube. (Box 8.8)

tissue   A set of contiguous functionally related cells.

tissue hybridization   Hybridization of a labeled probe to RNA molecules in a tissue section to show their distribution. (Figure 7.12)

Tm—*see* melting temperature

topoisomerase   An enzyme that can unwind DNA, relax coiling or even pass one DNA double helix through another by making temporary cuts and then rejoining the ends.

totipotent   A cell that is able to give rise to all the cell types in an organism.

trait   A character or phenotype.

***trans*-acting**   Of a regulatory factor, affecting expression of all copies of the target gene, irrespective of chromosomal location. Trans-acting regulatory factors are usually proteins which can diffuse to their target sites.

transchromosomic mice   Mice engineered to carry a whole exogenous (normally human) chromosome. Named by analogy to transgenic mice.

transcription factors   DNA-binding proteins that promote transcription of genes. (Section 11.1)

transcriptome   The total set of different RNA transcripts in a cell or tissue. *Cf.* genome, proteome.

transduction   (1) Relaying a signal from a cell surface receptor to a target within a cell. (2) Using recombinant viruses to introduce foreign DNA into a cell.

transfection   Direct introduction of an exogenous DNA molecule into a cell without using a vector.

transformation   Of a cell. (1) Uptake by a competent bacterial cell of naked high molecular weight DNA from the environment. (2) Alteration of the growth properties of a normal eukaryotic cell as a step towards evolving into a tumor cell.

transgene   An exogenous gene that has been transfected into cells of an animal or plant. It may be present in some tissues (as in human gene therapies) or in all tissues (as in germ-line engineering, e.g. in mouse). Introduced transgenes may be episomal and be transiently expressed, or can be integrated into host cell chromosomes.

transgenic animal   An animal in which artificially introduced foreign DNA (a transgene) becomes stably incorporated into the germ line. (Figures 20.2 and 20.3)

transit amplifying cells   The immediate progeny by which stem cells give rise to differentiated cells. Transit amplifying cells go through many cycles of division, but eventually differentiate.

translocation   Transfer of chromosomal regions between nonhomologous chromosomes. (Figure 2.23)

transmission disequilibrium test (TDT)   A statistical test of allelic association. (Box 15.3)

transposon   A mobile genetic element. (Figures 9.20 and 9.21)

triplet   Three consecutive nucleotides.

trophoblast (or **trophectoderm**)   Outer layer of polarized cells in the blastocyst which will go on to form the chorion, the embryonic component of the placenta.

tropism   The specificity of a virus for a particular cell type, determined in part by the interaction of viral surface structures with receptors present on the surface of the cell.

tumor suppressor gene (TSG)   A gene whose normal function is to inhibit or control cell division. TSG are typically inactivated in tumors. (Section 17.3)

ultraconserved sequence   Genomic sequences >200 bp long that are 100% conserved between human, rat and mouse genomes. They are likely to be important regulatory elements.

unclassified variant   A DNA sequence change seen in diagnostic testing, where the laboratory is unable to decide whether or not it is pathogenic.

uniparental diploidy   A 46,XX diploid conceptus where both genomes derive from the same parent. Such conceptuses never develop normally.

uniparental disomy   A cell or organism in which both copies of one particular chromosome pair are derived from one parent. Depending on the chromosome involved, this may or may not be pathogenic.

unrelated   Ultimately everybody is related; the word is used in this book to mean people who do not have an identified common ancestor in the last four or so generations.

untranslated region (5′UTR, 3′UTR)   Regions at the 5′ end of mRNA before the AUG translation start codon, or at the 3′ end after the UAG, UAA or UGA stop codon. (Figure 1.23)

variable expression   Variable extent or intensity of phenotypic signs among people with a given genotype. (Figure 3.17)

vector   A nucleic acid that is able to replicate and maintain itself within a host cell, and that can be used to confer similar properties on any sequence covalently linked to it. (Table 6.2)

Watson–Crick rules   Describe the normal base-pairing in double-stranded nucleic acid: A with T (or U); G with C.

western blotting   A procedure analogous to Southern blotting but involving proteins separated by charge and size on electrophoretic gels, blotted onto a membrane and detected using antibodies or stains. (Figure 8.21)

whole genome amplification   *In vitro* amplification of all the DNA of a genome as a way of making a precious small genomic DNA sample go further. Done using phi29 DNA polymerase rather than PCR. (Box 6.2)

wobble pairing   A special relaxed base-pairing that occurs between the 3′ nucleotide of a codon and a tRNA anticodon. (Table 1.6)

X-inactivation (lyonization)   The epigenetic inactivation of all but one of the X chromosomes in the cells of mammals that have more than one X. (Figures 11.19 and 11.20)

X-inactivation center (*Xic*)   The location on the proximal long arm of the human X chromosome from which the spreading X-inactivation is initiated. (Section 11.3)

XY body   In male meiosis, the X and Y chromosomes associate in a condensed body of heterochromatin.

YAC—*see* yeast artificial chromosome

yeast artificial chromosome (YAC)   A vector able to propagate inserts of up to 2 megabases in yeast cells. (Figure 6.7)

yeast two-hybrid system   A technique for identifying protein-protein interactions. Proteins that physically associate are identified by their ability to bring together two separated parts of a transcription factor, and so stimulate transcription of a reporter gene in yeast cells. (Figure 12.7)

zinc finger nucleases   Synthetic enzymes that combine an endonuclease module with a sequence-specific targeting module, so as to cleave DNA at a selected sequence. (Figure 20.13)

zona pellucida   A layer of glycoprotein that surrounds an unfertilized egg and acts as a barrier to fertilization.

zone of polarizing activity (ZPA)   During limb development, cells of the ZPA secrete a morphogen that controls the patterning of the digits. (Figure 5.6)

zoo blot   A Southern blot containing DNA samples from a range of different species, used to check for conserved sequences.

zygote   The fertilized egg cell.

Index

Notes: Please note that all genes mentioned in the text are listed in the index under 'genes (list)' using their official HUGO nomenclature where possible. Entries followed by 'f' refer to figures and those followed by 't' and 'b' refer to tables and boxed material respectively. Entries followed by 'ff' (or 'bb' or 'tt') refer to figures (boxes; tables) which span consecutive pages. Please note that all entries refer to humans unless otherwise stated